AIP Physics Desk Reference

Third Edition

Springer
New York
Berlin
Heidelberg
Hong Kong
London
Milan
Paris
Tokyo

E. Richard Cohen
David R. Lide
George L. Trigg
Editors

AIP Physics Desk Reference

Third Edition

With 125 Illustrations

E. Richard Cohen
17735 Corinthian Drive
Encino, CA 91316
USA
ercohen@aol.com

David R. Lide
13901 Riding Loop Drive
Gaithersburg, MD 20878
USA
drlide@post.harvard.edu

George L. Trigg
18 Duzine Road
New Paltz, NY 12561-1304
USA
georgeltrigg@alum.wustl.edu

Cover design: Joan Greenfield/Gooddesign Resource.

Library of Congress Cataloging-in-Publication Data
AIP Physics Desk Reference. — 3rd ed. / editors, E. Richard Cohen, David R. Lide, George L. Trigg.
p. cm.
Includes bibliographical references and index.
ISBN 0-387-98973-0 (alk. paper)
1. Physics—Handbooks, manuals, etc. 2. Astronomy—Handbooks, manuals, etc. I. Cohen, E. Richard, 1922– II. Lide, David R., 1928– III. Trigg, George L., 1925–.
QC61 .P49 2000
530—dc21 99-059693

ISBN 0-387-98973-0 Printed on acid-free paper.

AIP Press is an imprint of Springer-Verlag New York, Inc.

Printed in the United States of America.

9 8 7 6 5 4 3 2 1 SPIN 10751687

www.springer-ny.com

Springer-Verlag New York Berlin Heidelberg
A member of BertelsmannSpringer Science+Business Media GmbH

Preface to the Third Edition

The first edition of this book appeared in 1981 under the title *Physics Vade Mecum*. The idea for that work arose in planning the commemoration of the fiftieth anniversary of the American Institute of Physics, which had been founded in 1931. Several members of the planning committee, including the late Herbert L. Anderson, suggested the preparation of a concise volume of essential definitions, equations, and data from all fields of physics. As explained in his Preface to the First Edition, which is reprinted here, Anderson was influenced by his association with Enrico Fermi, who, in his later years, began to collect basic formulas and data for personal use in solving physics problems. The proposal was accepted by AIP, and Anderson agreed to become Editor.

The second edition, published in 1989 under the title *A Physicist's Desk Reference*, had the same structure as the first, with a modest degree of updating. This third edition has been completely revised and expanded. New chapters have been added, others have been reorganized or subdivided, and most of the authors are new. Topics such as chaos and environmental physics, which have assumed much more importance in the 20 years since the first edition appeared, are now covered. While the original concept of emphasizing the key equations and data in each sub-field of physics has been retained, most subjects are now covered in more depth than in the first two editions, with a consequent increase in the size of the book.

The editors and authors hope that this volume will prove helpful to students, teachers, and research workers in physics and related scientific fields.

E. Richard Cohen
David R. Lide
George L. Trigg

November 2002

Preface to the First Edition

Those of us who were fortunate enough to watch Enrico Fermi at work marveled at the speed and ease with which he could produce a solution to almost any problem in physics brought to his attention. When he had heard enough to know what the problem was, he proceeded to the blackboard and let the solution flow out of his chalk. He kept in trim by doing a lot of problems, either for the courses that he taught, the talks he gave, or the papers that he wrote. Most frequently, he worked out his own solutions to problems he heard about, in seminars, or in discussions with those who came to talk physics with him. Fermi's solutions were almost always simpler and easier to understand than the ones obtained by the person who raised the question in the first place.

As he grew older, Fermi became concerned about his speed in solving problems. He claimed that his memory was becoming less reliable and less certain and that he had to spend more and more time rederiving even the simple formulas he could previously write down without hesitation. To save time, he decided to organize an "artificial memory" kept close at hand and available for ready reference as needed. Unfortunately, he died before his artificial memory was well developed. However, the idea remained that a handy collection of useful physics formulas and data would help us all deal with the problems in physics that crossed our paths.

The occasion of the 50th anniversary of the American Institute of Physics seemed to be an opportunity to make a start in that direction. A committee was appointed under the chairmanship of David R. Lide, Jr. The members were Herbert L. Anderson, Hans Frauenfelder, Laurence W. Fredrick, John N. Howard, Robert H. Marks, and Murray Strasberg. The task was to assemble a carry-with-you compendium that would be useful to the wide spectrum of physicists associated with the AIP through its member societies. For our Vade Mecum we chose 22 subjects broadly representative of the fields physicists are supposed to know something about. Each subeditor was charged to compile within 10 pages the most useful information, formulas, numerical data, definitions, and references most physicists would like to have at hand. In the General Section there are collected the fundamental constants, the SI units and prefixes, conversion factors, magnitudes, basic mathematical and physics formulas, formulas useful in practical physics applications, and a list of physics data centers. Although the formulary is brief, it is intended to serve as a starting point.

There were a number of good examples of what could be done in special fields. The shirt-pocket-size booklet *Particle Properties* distributed by the Particle Data Group at LBL and CERN is used actively by many high-energy physicists. Plasma physicists like to keep with them the NRL *Plasma Formulary*, compiled by David L. Book of the Naval Research

Laboratory. Another useful compendium is *Astrophysical Quantities* by C. W. Allen, whose style and content influenced the present work. There is, of course, the AIP *Physics Handbook*, but this goes far beyond what we had in mind in size. We refer to it as a backup for more detail in the tables and other material we present. We refer also to the *Handbook of Optics* recently issued by the Optical Society of America. The section on medical physics was drawn from the recently published *Handbook of Medical Physics*. An excellent little compendium is W.H.J. Childs, *Physical Constants*. There are also "pocket" handbooks that come close to ours in general coverage but are larger in size. Among them are H. Ebert, *Physics Handbook*, and B. Yavorsky and A. Detlef, *Handbook of Physics*. We have excerpted some material from all of these.

The handbook is arranged for quick and easy location of the material being sought. The idea is to make each formula and each table usable with a minimum of reading. The explanatory material is placed nearby. Unlike a text book, it is not necessary to read the first page to understand what is written later on. As far as possible, each page is made complete in itself with the name of the principal contributor on its heading. The book should serve as an aid to quick recall of the information physicists most need to have readily available. In future editions it should be possible to remove pages and add new ones without disturbing the utility of the rest. In this way revisions can easily be made.

We wish to thank those who were kind enough to read the manuscripts in their specialty and make useful comments. In particular, David R. Lide, Jr., Emilio Segrè, D. Allan Bromley, Robert T. Beyer, Murray Strasberg, Stefan L. Wipf, Albert Petschek, Nicolas C. Metropolis, John R. Cunningham, and George W. Smith gave good advice, which we were happy to have.

We hope to make the Vade Mecum grow in value but not in size with each passing year.

Herbert L. Anderson
Editor in Chief, 1981

Contents

Contributors

Editors:
E. Richard Cohen, 17735 Corinthian Drive, Encino, CA 91316
E-mail: **ercohen@aol.com**

David R. Lide, 13901 Riding Loop Drive, Gaithersburg, MD 20878
E-mail: **drlide@post.harvard.edu**

George L. Trigg, 18 Duzine Road, New Paltz, NY 12561-1304
E-mail: **georgeltrigg@alum.wustl.edu**

Chapter 1:
E. Richard Cohen, 17735 Corinthian Drive, Encino, CA 91316
E-mail: **ercohen@aol.com**

Chapter 2:
E. Richard Cohen, 17735 Corinthian Drive, Encino, CA 91316
E-mail: **ercohen@aol.com**

Chapter 3:
Robert T. Beyer, Box 1843, Department of Physics, Brown University,
Providence, RI 02912
E-mail: **beyer@physics.brown.edu**

Chapter 4:
Jay M. Pasachoff, Hopkins Observatory, Williams College, Williamstown, MA 01267
E-mail: **jay.m.pasachoff@williams.edu**

Chapter 5:
Virginia Trimble, Department of Physics and Astronomy, University of California,
Irvine, CA 92697
and
Astronomy Department, University of Maryland, College Park, MD 20742
E-mail: **vtrimble@astro.umd.edu**

Chapter 6:
M. Raymond Flannery, School of Physics, Georgia Institute of Technology,
Atlanta, GA 30332
E-mail: **ray.flannery@physics.gatech.edu**

Chapter 7:
Wolfgang Wiese, A267 Physics Building, National Institute of Standards and Technology,
Gaithersburg, MD 20899
E-mail: **wolfgang.wiese@nist.gov**

Chapter 8:
Victor A. Bloomfield, Department of Biochemistry, University of Minnesota,
1479 Gortner Ave., St. Paul, MN 55108
E-mail: **victor@umn.edu**

Elias Greenbaum, Oak Ridge National Laboratory, P.O. Box 2008, 4500N,
Oak Ridge, TN 37831-6194
E-mail: **greenbaum@ornl.gov**

Chapter 9:
Vicky Lynn Karen, Ceramics Division (852), 100 Bureau Drive, Stop: 8520,
National Institute for Science and Technology, Gaithersburg, MD 20899
E-mail: **vicky.karen@nist.gov**

Chapter 10:
Ferris Webster, College of Marine Studies, University of Delaware, 700 Pilottown Road,
Lewes, DE 19958
E-mail: **ferris@udel.edu**

Chapter 11:
David Griffiths, Department of Physics, Reed College, 3203 SE Woodstock Boulevard,
Portland, OR 97202
E-mail: **griffith@reed.edu**

Chapter 12:
Heidi Schellman, Department of Physics and Astronomy, Northwestern University,
2145 Sheridan Road, Evanston, IL 60208
E-mail: **schellman@fnal.gov**

Chapter 13:
Stavros Tavoularis, Department of Mechanical Engineering, University of Ottawa,
Ottawa, ON K1N 6N5 Canada
E-mail: **tav@eng.uottawa.ca**

Chapter 14:
Florian Scheck, Theoretische Elementarteilchenphysik, Institut für Physik,
Johannes-Gutenberg Universität, D-55099 Mainz, Germany
E-mail: **scheck@zino.physik.uni-mainz.de**

Chapter 15:
Michael Yester, GSB 301F, University of Alabama, Birmingham, AL 35294
E-mail: **myester@uab.edu**

William Hendee, Medical College of Wisconsin, 8701 Watertown Plank Road,
Milwaukee, WI 53226

Chapter 16:
Peter F. Bernath, Department of Chemistry, University of Waterloo,
Waterloo, ON N2L 3G1 Canada
E-mail: **bernath@uwaterloo.ca**

Chapter 17:
Paul Manneville, Laboratoire d'Hydrodynamique, CNRS-UMR 7646,
École Polytechnique, F-91128 Palaiseau cedex, France
E-mail: **paul.manneville@ladhyx.polytechnique.fr**

Chapter 18:
Kenneth S. Krane, Department of Physics, 301 Weniger Hall, Oregon State University,
Corvallis, OR 97331
E-mail: **kranek@physics.orst.edu**

Chapter 19:
Joseph Reader, Atomic Physics Division, A167 Physics Building, National Institute of
Standards and Technology, Gaithersburg, MD 20899
E-mail: **jreader@nist.gov**

Chapter 20:
Donald A. Edwards, Fermilab MS 306, P.O. Box 500, Batavia, IL 60510-0500
E-mail: **edwards@fnal.gov**

Kai Desler, Fermilab MS 306 Batavia, IL 60510-0500
E-mail: **kai.desler@desy.de**

Chapter 21:
David L. Book, Physics Department, Naval Postgraduate School, Monterey, CA 93943
E-mail: **dlbook@nps.navy.mil**

Chapter 22:
Stephen Z.D. Cheng, Maurice Morton Institute and Department of Polymer Science,
The University of Akron, Akron, OH 44325-3909
E-mail: **cheng@polymer.uakron.edu**

Chapter 23:
Ronald L. Mallett, Department of Physics, University of Connecticut,
Box U-46, 2152 Hillside Road, Storrs, CT 06269
E-mail: **rlmallett@aol.com**

Mark P. Silverman, Department of Physics, Trinity College, Hartford, CT 06106
E-mail: **mark.silverman@trincoll.edu**

Chapter 24:
Constantinos M. Soukoulis, Ames Laboratory, Iowa State University, Ames, IA 50011
and
Foundation for Research and Technology—Hellas (FORTH), P.O. Box 1527,
71110 Heraklion, Crete, Greece
E-mail: **soukoulis@ameslab.gov**

E.N. Economou, Department of Physics, University of Crete, GR-71110 Iraklion,
Crete, Greece
E-mail: **economou@admin.forth.gr**

Chapter 25:
Bruce E. Koel, Department of Chemistry, SSC 606, University of Southern California,
920 W. 37th Street, University Park, CA 90089
E-mail: **koel@usc.edu**

Roland Resch
Infineon Technologies, VI PPE-D1, Siemensstrasse 2, A-9500, Villach, Austria

Chapter 26:
Martin Trusler, Department of Chemical Engineering Technology,
Imperial College of Science and Technology, London SW7 2BY United Kingdom
E-mail: **m.trusler@ic.ac.uk**

William A. Wakeham, University of Southampton, Highfield, Southampton S017 1BJ
United Kingdom
E-mail: **vice-chancellor@soton.ac.uk**

Chapter 27:
David R. Lide, 13901 Riding Loop Drive, Gaithersburg, MD 20878
E-mail: **drlide@post.harvard.edu**

Fundamental Physical Constants—Frequently used constants

Quantity	Symbol	Value	Unit	Relative std. uncert. u_r
speed of light in vacuum	c, c_0	299 792 458	$\mathrm{m\,s^{-1}}$	(exact)
magnetic constant	μ_0	$4\pi \times 10^7$	$\mathrm{N\,A^{-2}}$	
		$= 12.566\,370614\ldots \times 10^{-7}$	$\mathrm{N\,A^{-2}}$	(exact)
electric constant $1/\mu_0 c^2$	ε_0	$8.854\,187\,817\ldots \times 10^{-12}$	$\mathrm{F\,m^{-1}}$	(exact)
Newtonian constant of gravitation	G	$6.673(10) \times 10^{-11}$	$\mathrm{m^3\,kg^{-1}\,s^{-2}}$	1.5×10^{-3}
Planck constant	h	$6.626\,068\,76(52) \times 10^{-34}$	J s	7.8×10^{-8}
$h/2\pi$	$\hbar$	$1.054\,571\,596(82) \times 10^{-34}$	J s	7.8×10^{-8}
elementary charge	e	$1.602\,176\,462(63) \times 10^{-19}$	C	3.9×10^{-8}
magnetic flux quantum $h/2e$	Φ_0	$2.067\,833\,636(81) \times 10^{-15}$	Wb	3.9×10^{-8}
conductance quantum $2e^2/h$	G_0	$7.748\,091\,696(28) \times 10^{-5}$	S	3.7×10^{-9}
electron mass	m_e	$9.109\,381\,88(72) \times 10^{-31}$	kg	7.9×10^{-8}
proton mass	m_p	$1.672\,621\,58(13) \times 10^{-27}$	kg	7.9×10^{-8}
proton-electron mass ratio	m_p/m_e	1.836.152 6675(39)		2.1×10^{-9}
fine-structure constant $e^2/4\pi\epsilon_0\hbar c$	α	$7.297\,352\,533(27) \times 10^{-3}$		3.7×10^{-9}
inverse fine-structure constant	α^{-1}	137.035 999 76(50)		3.7×10^{-9}
Rydberg constant $\alpha^2 m_e c/2h$	R_∞	10 973 731.568 549(83)	$\mathrm{m^{-1}}$	7.6×10^{-12}
Avogadro constant	N_A, L	$6.022\,141\,99(47) \times 10^{-23}$	$\mathrm{mol^{-1}}$	7.9×10^{-8}
Faraday constant $N_a e$	F	96 485.3415(39)	$\mathrm{C\,mol^{-1}}$	4.0×10^{-8}
molar gas constant	R	8.314 472(15)	$\mathrm{J\,mol^{-1}\,K^{-1}}$	1.7×10^{-6}
Boltzmann constant R/N_A	k	$1.380\,6503(24) \times 10^{-23}$	$\mathrm{J\,K^{-1}}$	1.7×10^{-6}
Stefan–Boltzmann constant $(\pi^2/60)k^4/\hbar^3 c^2$	σ	$5.670\,400(40) \times 10^{-8}$	$\mathrm{W\,m^{-2}\,K^{-4}}$	7.0×10^{-6}
		Non-SI units accepted for use with the SI		
electron volt: (e/C) J	eV	$1.602\,176\,462(63) \times 10^{-19}$	J	3.9×10^{-8}
(unified) atomic mass unit $1\mathrm{u} = m_\mathrm{u} = \frac{1}{12}m(^{12}\mathrm{C}) = 10^{-3}\ \mathrm{kg\,mol^{-1}}/N_A$	U	$1.660\,538\,73(13) \times 10^{-27}$	kg	7.9×10^{-8}

Source: Peter J. Mohr and Barry N. Taylor, CODATA Recommended Values of the Fundamental Physical Constants: 1998. http://physics.nist.gov/constants.

1

Symbols, Units, and Nomenclature

E. Richard Cohen
California State University, Northridge

Contents

List of Tables

1.1. PHYSICAL QUANTITIES

A physical quantity is *an attribute of a phenomenon, body, or substance that may be distinguished qualitatively and determined quantitatively.*[1] The value of a physical quantity is expressed as the product of a numerical value and a unit of measurement:

$$Q = \text{physical quantity} = \text{number} \times \text{unit} = \{Q\}\,[Q], \tag{1.1}$$

where $\{Q\}$ denotes the numerical value of Q and $[Q]$ denotes the unit of Q. Neither the name nor the symbol for a physical quantity should imply any particular choice of unit; e.g., in spite of historical precedent, the term "voltage" should be avoided in favor of the terms "potential difference" or "electrical potential."[1]

If the unit is changed from $[Q]_1$ to $[Q]_2 = a[Q]_1$, the numerical value changes from $\{Q\}_1$ to $\{Q\}_2 = a^{-1}\{Q\}_1$.

$$Q = \{Q\}_1[Q]_1 = \{Q\}_2[Q]_2 = \{Q\}[Q] \tag{1.2}$$

Example $\lambda = 6.058\times10^{-7}$ m $= 605.8$ nm $= 23.8504$ μin.

1.1.1. Symbols for quantities

Symbols for physical quantities are usually single letters of the Latin, Greek, or other alphabet in *sloping* or *italic* type that may be modified by the use of subscripts and superscripts. In general, subscripts and superscripts referring to physical quantities are printed in sloping type, whereas descriptive or numerical subscripts and superscripts are printed in Roman type.

Complicated subscripts or superscripts may be avoided by using parentheses or brackets: instead of $\rho_{\mathrm{H_2SO_4,20°C}}$, write $\rho(\mathrm{H_2SO_4}, 20\,°\mathrm{C})$.

It is common to use standard weight type symbols for scalars and boldface type symbols for aggregates (vectors and tensors). A distinction is made between boldface type for the vector as an entity and standard weight type for the components of that vector: The vector $\boldsymbol{A}$ has components A_k. IUPAP recommends that vectors representing physical quantities be printed in slanted type, *a*, *A*, whereas the AIP Style Guide sets all vectors as upright boldface, **a**, **A**. Tensors can be printed in bold sans-serif type, T.

A vector may also be indicated by an arrow, $\vec{a}$, $\vec{B}$, and a tensor by a double arrow $\overleftarrow{\overrightarrow{S}}$, or by a double-headed arrow $\overleftrightarrow{S}$. For higher dimensional tensors, an index notation should be used uniformly for vectors and tensors:

$$A_i, \qquad S_{ij}, \qquad R^{ij}_{kl}, \qquad R^{i\,.\,.\,l}_{\,.\,jk\,.}$$

1.1.2. Symbols for units

The name of a unit is an common noun and is not capitalized, even if it is derived from a proper name. When the name of the unit is derived from a proper name, its symbol is composed of one or two letters, the first of which is capitalized; otherwise, the symbol is in lowercase. (Because of the possibility of confusing l with the number 1, an exception to this rule allows the symbol L in addition to l for liter.)

Examples meter m, mole mol, ampere A, weber Wb

Symbols for units do not contain a period, and they remain unaltered in the plural.

Examples 2 m 3.1416 m *and not* 2 m. *or* 2 ms

The multiplication of units is written, as recommended by the International Organization for Standardization (ISO)[2] and accepted by both IUPAP and IUPAC[3] as well as the AIP Style Guide,[5] in either of the following two forms:

$$\mathrm{N \cdot m} \qquad \mathrm{N\,m}$$

whereas the American National Standard ANSI/IEEE Std 268-1982[6] states that in U.S. practice, the first form (the centered dot) is to be preferred.

1.2. PHYSICAL UNITS

1.2.1. Base units

In a physical system consisting of a set of quantities and the relational equations connecting them, an arbitrary number of quantities are defined by convention to be dimensionally independent. These quantities form the set of *base quantities* for the system. All other physical quantities are *derived quantities*, defined in terms of the base quantities and expressed algebraically as products of powers of the base quantities.[3]

1.2.2. Dimension

Within a given measurement system, every physical quantity is assigned a *dimension*; the set of all units by which the physical quantity can be expressed. Thus, the dimension length L is the set [m, mm, km, in, ft, yd, Å, ly, . . .]; the dimension energy E is the set [erg, joule, kilowatt-hour, BTU, calorie, . . .]. A quantity that arises from dividing one physical quantity by another with the same dimension (e.g., relative density, refractive index), or, more generally, any quantity that has a dimension whose dimensional exponents are all equal to zero, $\mathsf{L}^0\mathsf{M}^0\mathsf{T}^0\ominus^0$, has a dimension symbolized by the number 1. Such a quantity, a quantity of dimension one, is often loosely called a *dimensionless quantity*.

In the field of mechanics, three base quantities (length L, mass M, and time T) are usually considered to be adequate to define a set of base dimensions, whereas heat and thermodynamics introduces temperature, dimension $\ominus$, as an additional independent base quantity. The dimension of any other quantity is then expressed in terms of the dimensional product of the base dimensions:

$$\dim \mathsf{Q} = \mathsf{L}^{\alpha}\mathsf{M}^{\beta}\mathsf{T}^{\gamma}\ominus^{\delta} . \tag{1.3}$$

The powers to which the various base quantities or base dimensions are raised are called the *dimensional exponents*.

Homogeneity

The mathematical relation among physical quantities may be a simple monomial or a sum of such terms, but each term of the sum must have the same dimension. Such a relation can always be expressed as $\phi(x_1, x_2, \dots) = 0$, in which the arguments x_j are variables of dimension 1. When a physical quantity Y is expressed as a function of other variables, the relation must then be expressible (although such a form is usually not explicitly shown) as $Y = AF(x_1, x_2, \dots)$, where F and its arguments x_j are quantities of dimension 1 and A has the same dimension as Y.

A set of conventionally defined samples of the base quantities are chosen as *base units* to form the foundation for a *system of units*. Although it is possible to define an independent unit for every quantity, it is much more convenient to relate the units of the derived quantities to the base units as products of powers, corresponding to the expressions in the system of quantities:

$$\mathrm{U}_Q = k_Q \mathrm{U}_\mathrm{L}^\alpha \mathrm{U}_\mathrm{M}^\beta \mathrm{U}_\mathrm{T}^\gamma \mathrm{U}_\Theta^\delta. \tag{1.4}$$

When all derived units are expressed in terms of the base units by relations with numerical factors $k_Q = 1$, the system is said to form a *coherent set* of units; in such a system, the numerical equations and the physical equations have the same form. Since there are no numerical quantities introduced in the definition of derived units, the equation expressing the value of a physical quantity has the same form as the equation expressing the numerical value of the quantities.

Derived units and their symbols are expressed algebraically in terms of base units by means of the mathematical signs for multiplication and division. Some derived units receive special names and symbols, which can then be used in forming names and symbols of other derived units. A derived unit can be expressed in different ways using the names of base units and the special names of derived units. A given physical quantity has a unique dimension in terms of base dimensions, even though its unit may be expressed in more than one way using appropriate combinations of base and derived units. A given unit or dimension, however, may correspond to more than one quantity; e.g., the quantities kinematic viscosity (η/ρ) and diffusion coefficient both have the dimension $\mathsf{L}^2\mathsf{T}^{-1}$ (length squared divided by time), whereas heat capacity and entropy both have the dimension $\mathsf{ML}^2\mathsf{T}^{-2}P\Theta^{-1}$ (energy per degree).

1.3. THE INTERNATIONAL SYSTEM OF UNITS (SYSTÈME INTERNATIONAL, SI)

1.3.1. Base units

The name Système International d'Unités (International System of Units) with the international abbreviation SI was adopted by the Conférence Générale des Poids et Mesures (CGPM) in 1960. It is a coherent system based on seven base units (CGPM 1960 and 1971):

meter: The meter is the length of the path traveled by light in vacuum during a time interval of 1/299 792 458 of a second.

TABLE 1.1. SI base units

		SI Unit	
Base quantity	Dimension	Name	Symbol
Length	L	Meter	m
Mass	M	Kilogram	kg
Time	T	Second	s
Electric current	I	Ampere	A
Thermodynamic temperature	Θ	Kelvin	K
Amount of substance	N	Mole	mol
Luminous intensity	J	Candela	cd

kilogram: The kilogram is the unit of mass; it is equal to the mass of the international prototype of the kilogram.

second: The second is the duration of 9 192 631 770 periods of the radiation corresponding to the transition between the two hyperfine levels of the ground state of the cesium-133 atom.

ampere: The ampere is that constant current which, if maintained in two straight parallel conductors of infinite length, of negligible circular cross section, and placed 1 meter apart in vacuum, would produce between these conductors a force equal to 2×10^{-7} newton per meter of length.

kelvin: The kelvin, the unit of thermodynamic temperature, is the fraction $1/273.16$ of the thermodynamic temperature of the triple point of water. The unit kelvin and its symbol K should be used to express both the thermodynamic temperature and an interval or a difference of temperature. In addition to the thermodynamic temperature (symbol T),

TABLE 1.2. Prefixes used in SI

These prefixes are used to indicate multiples or submultiples of the base unit, except that units for mass are formed by applying the prefix to the symbol g, i.e., Mg not kkg and mg not μkg. Only a single prefix is permitted. Use ns rather than mμs, pF rather than μμF, and GW rather than kMW. The first syllable of the prefix retains its stress in compounds.

Factor	Prefix	Symbol	Factor	Prefix	Symbol
10^{1}	deka	da	10^{-1}	deci	d
10^{2}	hecto	h	10^{-2}	centi	c
10^{3}	kilo	k	10^{-3}	milli	m
10^{6}	mega	M	10^{-6}	micro	μ
10^{9}	giga	G	10^{-9}	nano	n
10^{12}	tera	T	10^{-12}	pico	p
10^{15}	peta	P	10^{-15}	femto	f
10^{18}	exa	E	10^{-18}	atto	a
10^{12}	zetta	Z	10^{-21}	zepto	z
10^{24}	yotta	Y	10^{-24}	yocto	y

there is also the Celsius temperature (symbol t) defined by the equation

$$t = T - T_{\circ},$$

where $T_{\circ} = 273.15$ K. Celsius temperature is expressed in degree Celsius (symbol, °C). The unit "degree Celsius" is equal to the unit "kelvin," and a temperature interval or a difference of temperature may be expressed either in kelvin or in degrees Celsius.

mole: The mole is the amount of substance of a system which contains as many elementary entities as there are atoms in 0.012 kilogram of carbon 12. When the mole is used, the elementary entities must be specified and may be atoms, molecules, ions, electrons, other particles, or specified groups of such particles. (In this definition, it is understood that the carbon 12 atoms are unbound, at rest, and in their ground state.)

candela: The candela is the luminous intensity, in a given direction, of a source that emits monochromatic radiation of frequency 540×10^{12} Hz and that has a radiant intensity in that direction of $(1/683)$ watt per steradian.

1.3.2. Conversion factors to SI units

1. *Angle*

1 second (″)	$= 4.48481 \times 10^{-6}$ rad	1 degree (°)	$= 0.0174\,532$ rad
1 minute (′)	$= 2.9089 \times 10^{-4}$ rad	1 rad	$= 206\,264.806''$

2. *Area*

1 barn (b)	$\equiv 10^{-28}$ m^2	1 are	$\equiv 100$ m^2
1 circular mil	$= 5.0671 \times 10^{-10}$ m^2	1 acre	$= 4046.856$ m^2
1 in^2	$\equiv 6.4516 \times 10^{-4}$ m^2	1 hectare	$\equiv 10\,000$ m^2
1 ft^2	$\equiv 0.092\,903\,04$ m^2	1 mi^2	$= 2.5900 \times 10^{6}$ m^2
1 yd^2	$\equiv 0.836\,127\,36$ m^2		

3. *Concentration, Density*

1 grain/gal (US)	$= 0.017\,118$ kg/m^3	1 long ton/yd^3	$= 1328.9$ kg/m^3
1 lb/ft^3	$= 16.018$ kg/m^3	1 oz(avdp)/in^3	$= 1730.0$ kg/m^3
1 lb/gal (US)	$= 119.83$ kg/m^3	1 lb/in^3	$= 27\,680$ kg/m^3
1 short ton/yd^3	$= 1186.6$ kg/m^3		

4. *Energy*

th, thermochemical calorie; 15, 15 °C calorie; ST, International Steam Table calorie

1 erg	$\equiv 1 \times 10^{-7}$ J		
1 ton TNT (equivalent)	$\equiv 4.184 \times 10^{9}$ J		
1 ft·lbf	$= 1.3558$ J	1 therm(EC)	$\equiv 1.055\,06 \times 10^{8}$ J
1 cal$_{th}$	$\equiv 4.184$ J	1 therm(US)	$\equiv 1.054\,804 \times 10^{8}$ J
1 cal$_{15}$	$\equiv 4.1855$ J	1 Btu$_{th}$	$= 1054.350$ J
1 cal$_{ST}$	$\equiv 4.1868$ J	1 Btu$_{15}$	$= 1054.728$ J
1 watt second (W·s)	$\equiv 1$ J	1 Btu$_{ST}$	$= 1055.056$ J
1 watt hour (W·h)	$\equiv 3600$ J	1 ($\equiv 10^{15}$ Btu)	$\approx 10^{18}$ J $= 1$ EJ

5. *Force*

1 dyne	$\equiv 10^{-5}$ N	1 kilogram-force	$\equiv 9.80665$ N
1 ounce-force	$= 0.27801$ N	1 kip (1000 lbf)	$= 4448.2$ N
1 pound-force	$= 4.4482$ N	1 ton-force (2000 lbf)	$= 8896.4$ N

6. *Heat*

Heat capacity

1 Btu/(lb. °F) ≡ 4186.8 J/(kg·K)

Thermal conductivity:

1 Btu·ft/(h·ft^2 °F)	= 1.730 735 W·m^{-1} K^{-1}
1 Btu·in/(s ft^2·°F)	= 519.2204 W·m^{-1}·K^{-1}

Thermal resistance:

1°F·h·ft^2/Btu	= 0.176 11 K·m^2/W

Thermal resistivity:

1°F·h·ft^2/Btu·in	= 6.933 47 K·m/W

7. *Length*

1 fermi	≡ 10^{-15} m = 1 fm	1 statute foot	
1 angström (Å)	≡ 10^{-10} m	[(1200/3937) m]	= 0.304 8006 m
1 microinch	≡ 2.54×10^{-8} m	1 yard (yd)	≡ 0.9144 m
1 mil	≡ 2.54×10^{-5} m	1 fathom (2 yd)	≡ 1.8288 m
1 point (pt)		1 rod (5.5 yd)	≡ 5.0292 m
[0.013 837 in]	= 0.351 46 mm	1 chain (4 rod)	≡ 20.1168 m
1 pica (12 pt)	= 4.2175 mm	1 furlong (220 yd)	≡ 201.168 m
1 inch (in)	≡ 0.0254 m	1 mile (8 furlong)	≡ 1609.344 m
1 hand (4 in)	≡ 0.1016 m	1 statute mile	= 1609.34722 m
1 foot (12 in)	≡ 0.3048 m	1 nautical mile	≡ 1852 m

8. *Light*

1 foot-candle	= 10.764 lx	1 stilb	≡ 10 000 cd/m^2
1 phot	≡ 10 000 lx	1 lambert	= 3183.10 cd/m^2
1 cd/in^2	= 1550.003 cd/m^2	1 foot-lambert	= 3.426 26 cd/m^2

9. *Mass*

1 pound (avdp, 7000 gr)	≡ 0.453 592 37 kg	1 ounce (troy, 480 gr)	= 31.1035 g
1 pound (troy, 5760 gr)	≡ 0.373 241 7216 kg	1 carat (metric)	= 0.2 g
1 grain (gr)	≡ 64.79891 mg	1 stone 14 lb)	= 6.350 29 kg
1 scruple (20 gr)	= 1.2960 g	1 slug	= 14.5939 kg
1 pennyweight(24 gr)	= 1.5552 g	1 hundredweight (long)	= 50.8023 kg
1 dram(60 gr)	= 3.8879 g	1 ton (short)	= 907.185 kg
1 ounce (avdp, 437.5 gr)	= 28.3495 g	1 ton (long)	= 1016.047 kg

10. *Power*

1 Btu_{ST} = 1.000 669 Btu_{th}

1 cal_{ST}/s	≡ 4.1868 W	1 metric horsepower	
1 cal_{th}/s	≡ 4.184 W	(force de cheval)	= 735.50 W
1 erg/s	≡ 1×10^{-7} W	1 horsepower	
1 ft·lbf/h	= 3.7662×10^{-4} W	(550 ft lbf/s)	= 745.70 W
1 Btu_{th}/h	= 0.292 875 W	1 electric horsepower	≡ 746 W
1 Btu_{ST}/h	= 0.293 071 W		

11. *Pressure, Stress*

standard atmosphere ≡ 101 325 Pa

1 dyne/cm^2	≡ 0.1 Pa	1 torr [(101325/760) Pa]	= 133.3224 Pa
1 N/cm^2	≡ 10 000 Pa	1 cm of mercury (0 °C)	= 1333.224 Pa
1 bar	≡ 100 000 Pa	1 in of water (4 °C)	= 249.08 Pa
1 lbf/ft^2	= 47.880 Pa	1 in of mercury (4 °C)	= 3386.4 Pa
1 cm water (4°C)	= 98.0637 Pa	1 lbf/in^2 (psi)	= 6894.8 Pa
1 gmf/cm^2	≡ 98.0665 Pa	1 kgf/cm^2	≡ 98 066.5 Pa

12. *Torque*

1 dyne cm	≡ 10^{-7} N m	1 kgf m	≡ 9.80665 N m
1 ozf·in	= 0.007 0616 N m	1 lbf·in	= 0.112 985 N m
1 lbf·ft	= 1.35582 N m		

13. *Viscosity*

1 poise	≡ 0.1 Pa s	1 lb/ft s	= 1.4882 Pa s
1 lb/ft h	= 4.1338×10^{-4} Pa s	1 poundal·s/ft^2	= 1.4882 Pa s
1 rhe	≡ 10 Pa^{-1}s^{-1}	1 stokes	≡ 10^{-4} m^2/s
1 ft^2/s	= 0.092 903 m^2/s	1 slug/ft s	= 47.880 Pa s
1 lbf·s/ft^2	= 47.880 Pa s	1 lbf·s/in^2	= 6894.8 Pa s

14. *Volume*

1 stere	≡ 1 m^3	1 liter	≡ 0.001 m^3
1 ft^3	= 0.0283 168 m^3	1 in^3	= 1.6387×10^{-5} m^3
		1 acre·foot	= 1233.48 m^3
1 dram (US fluid)	= 3.6967×10^{-6} m^3	1 ounce(UK fluid)	= 2.8413×10^{-5} m^3
1 teaspoon(tsp)	= 4.9289×10^{-6} m^3	1 gill(UK)	= 1.4207×10^{-4} m^3
1 tablespoon (tbsp)	= 1.4787×10^{-5} m^3	1 pint(UK)	= 5.6826×10^{-4} m^3
1 ounce(US fluid)	= 2.9574×10^{-5} m^3	1 quart(UK)	= $1.136\,52\times10^{-3}$ m^3
1 gill(US)	= 1.1829×10^{-4} m^3	1 gallon(UK)	≡ $4.546\,09\times10^{-3}$ m^3
1 pint(US fluid)	= 4.7318×10^{-4} m^3		= 1.200 950 gal (US)
1 quart(US fluid)	= 9.4635×10^{-4} m^3	1 pint(US dry)	= 5.5061×10^{-4} m^3
1 gallon(US fluid) [231 in^3]	= 3.7854×10^{-3} m^3	1 quart(US dry)	= 1.1012×10^{-3} m^3
1 wine barrel [31.5 gal(US)]	= 0.119 240 m^3	1 gallon(US dry)	= 4.4049×10^{-3} m^3
1 barrel (petroleum) [42 gal(US)]	= 0.158 987 m^3	1 peck	= 8.8098×10^{-3} m^3
		1 bushel(US) [2150.42 in^3]	= 3.5239×10^{-2} m^3

1.4. RECOMMENDED SYMBOLS FOR PHYSICAL QUANTITIES

1.4.1. General physics

The list of symbols given here is not intended to be exhaustive, and the absence of a symbol should not of itself prohibit its use. Many of the symbols listed are general; they may be made more specific by adding superscripts, subscripts, or other decorations and by using both lowercase and uppercase forms. An expression given with the name of a symbol should be considered to be descriptive rather than definitive.

The dimension of a quantity is given in terms of the base quantities of SI (Table 1.1).

		Dimension	SI Unit
a	annihilation operator	1	
	relative chemical activity	1	
	specific activity	$\mathsf{M}^{-1}\mathsf{T}^{-1}$	Bq/kg
	thermal diffusivity: $\lambda/\rho c_p$	$\mathsf{L}^2\mathsf{T}{-1}$	$\mathrm{m^2/s}$
$\mathrm{a}^{\dagger}$	creation operator	1	
$\mathbf{a}$	acceleration	LT^{-2}	$\mathrm{m/s^2}$
b	breadth, impact parameter	L	m
	mobility ratio: $\mu_{\mathrm{n}}/\mu_{\mathrm{p}}$	1	
	phonon annihilation operator	1	
b_{B}	molality of solute B	$\mathsf{M}^{-1}\mathsf{N}$	mol/kg
$b^{\dagger}$	phonon creation operator	1	
$\mathbf{b}$	Burgers's vector	L	m
c	concentration: $c = n/V$	$\mathsf{L}^{-3}\mathsf{N}$	$\mathrm{mol/m^3}$
	specific heat capacity	$\mathsf{L}^2\mathsf{T}^{-2}\Theta^{-1}$	$\mathrm{J/(kg{\cdot}K)}$
	speed	LT^{-1}	m/s
	speed of light, speed of sound	LT^{-1}	m/s
c_{ijkl}	elasticity tensor: $\tau_{ij} = c_{ijkl}\epsilon_{lk}$	$\mathsf{L}^{-1}\mathsf{MT}^{-2}$	Pa, $\mathrm{N/m^2}$
$\bar{c}$	average speed	LT^{-1}	m/s
$\hat{c}$	most probable speed	LT^{-1}	m/s
$\langle c\rangle$	average speed	LT^{-1}	m/s
$\mathbf{c}$	velocity, average velocity	LT^{-1}	m/s
$\langle\mathbf{c}\rangle$	average velocity	LT^{-1}	m/s
d	relative density	1	
d	diameter, distance, thickness	L	m
	lattice plane spacing	L	m
e	linear strain	1	
	specific energy	$\mathsf{L}^2\mathsf{T}^{-2}$	J/kg
$\mathbf{e}$	polarization vector	1	
f	focal distance	L	m
	frequency	T^{-1}	Hz
f_{B}	activity coefficient of B (in a mixture)	1	
g	acceleration of free fall	LT^{-2}	$\mathrm{m/s^2}$
	g-factor: $\mu/I\mu_{\mathrm{N}}$	1	
	statistical weight (degeneracy)	1	
h	height	L	m
	heat transfer coefficient	$\mathsf{MT}^{-3}\Theta^{-1}$	$\mathrm{W/(m^2\ K)}$
h_1, h_2, h_3			
h, k, l	Miller indices	1	
j	electric current density	$\mathsf{L}^{-2}\mathsf{I}$	$\mathrm{A/m^2}$
j_i	total angular momentum		
	quantum number	1	

		Dimension	SI Unit
k	angular wavenumber	L^{-1}	m^{-1}
k_T	thermal diffusion ratio	1	
$\mathbf{k}$	angular wave vector, propagation vector	L^{-1}	m^{-1}
l	length, mean free path	L	m
l_i	orbital angular momentum quantum number	1	
l_e	mean free path of electrons	L	m
l_{ph}	mean free path of phonons	L	m
m	mass	M	kg
	molality of solution	$M^{-1}N$	mol/kg
m^*	effective mass	M	kg
m_a	atomic mass	M	kg
m_i	magnetic quantum number	1	
m_r	reduced mass: $m_1 m_2/(m_1+m_2)$	M	kg
m_u	atomic mass unit: $\frac{1}{12}m_a(^{12}C)$	M	kg
m_N	nuclear mass	M	kg
$\mathbf{m}$	magnetic dipole moment	L^2I	$A \cdot m^2$
n	amount of substance	N	mol
	electron density (conduction band)	L^{-3}	m^{-3}
	number density of particles	L^{-3}	m^{-3}
	order of reflection	1	
	principal quantum number	1	
	refractive index	1	
n_a	acceptor number density	L^{-3}	m^{-3}
n_d	donor number density	L^{-3}	m^{-3}
n_i	instrinsic number density: $(np)^{1/2}$	L^{-3}	m^{-3}
n_n	electron number density	L^{-3}	m^{-3}
n_p	hole number density	L^{-3}	m^{-3}
p	acoustic pressure, pressure	$L^{-1}MT^{-2}$	Pa, J/m^3
	hole density (conduction band)	L^{-3}	m^{-3}
$\mathbf{p}$	electric dipole moment	LTI	$C \cdot m$
	momentum: $m\mathbf{v}$	LMT^{-1}	$kg \cdot m/s$
$\mathbf{p}$, p_i	generalized momentum: $(\partial L/\partial q_i)$	varies	
q	electric charge	TI	$A \cdot s$
	flow rate	L^3T^{-1}	m^3/s
q_m	mass flow rate (specific flow rate)	MT^{-1}	kg/s
q_D	Debye angular wavenumber	L^{-1}	m^{-1}
$\mathbf{q}$	(phonon) propagation vector	L^{-1}	m^{-1}
$\mathbf{q}$, q_i	generalized coordinate	varies	1
r	distance, radius	L	m
	molar ratio of solute	1	
$\mathbf{r}$	position vector	L	m
s	path length	L	m
	long range order parameter	1	
	symmetry number	1	

Symbol	Quantity	Dimension	SI Unit
s_i	spin quantum number	1	
s_{klji}	compliance tensor: $\epsilon_{kl} = s_{klji}\tau_{ij}$	$\mathsf{M}^{-1}\mathsf{LT}^2$	$\mathrm{m^2/N}$
$\mathbf{s}$	position vector	L	m
t	time	T	s
	temperature	$\ominus$	K
u	average speed	LT^{-1}	m/s
	electromagnetic energy density	$\mathsf{L}^{-1}\mathsf{MT}^{-2}$	$\mathrm{J/m^3}$
$\mathbf{u}$	displacement vector	L	m
	velocity	LT^{-1}	m/s
v	specific volume	$\mathsf{L}^{-3}\mathsf{M}$	$\mathrm{m^3/kg}$
	speed: ds/dt	LT^{-1}	m/s
	vibrational quantum number	1	
	volume	L^3	$\mathrm{m^3}$
v_{dr}	drift velocity (speed)	LT^{-1}	m/s
$\bar{v}$	average speed	LT^{-1}	m/s
$\hat{v}$	most probable speed	LT^{-1}	m/s
$\langle v\rangle$	average speed	LT^{-1}	m/s
$\mathbf{v}$	velocity	LT^{-1}	m/s
$\mathbf{v}_\circ$	average velocity	LT^{-1}	m/s
$\langle\mathbf{v}\rangle$	average velocity	LT^{-1}	m/s
w	electromagnetic energy density	$\mathsf{L}^{-1}\mathsf{MT}^{-2}$	$\mathrm{J/m^3}$
	mass fraction	1	
x	molar fraction	1	
z	ionic charge number	1	
	reduced activity: $[2\pi mkT/h^2]^{3/2}\lambda$	1	
A	activity (radioactivity)	T^{-1}	$\mathrm{s^{-1}}$, Bq
	area	L^2	$\mathrm{m^2}$
	chemical affinity	$\mathsf{L}^2\mathsf{MT}^{-2}\mathrm{N}^{-1}$	J/mol
	Helmholz free energy	$\mathsf{L}^2\mathsf{MT}^{-2}$	J
	nucleon number, mass number	1	
	Richardson constant: $j = AT^2\exp(-\Phi/kT)$	$\mathrm{M}^{-2}\mathrm{I}\ominus^{-2}$	$\mathrm{A/(m^2\cdot K^2)}$
A_{H}	Hall coefficient	$\mathsf{M}^3\mathsf{T}^{-1}\mathsf{I}^{-1}$	$\mathrm{m^3C}$
A_{r}	relative atomic mass: $m_{\mathrm{a}}/m_{\mathrm{u}}$	1	
$\mathbf{A}$	magnetic vector potential	$\mathsf{LMT}^{-2}\mathsf{I}^{-1}$	Wb/m, N/A
B	susceptance	$\mathsf{L}^{-2}\mathsf{M}^{-1}\mathsf{T}^3\mathsf{I}^2$	S
$\mathbf{B}$	magnetic flux density	$\mathsf{MT}^{-2}\mathsf{I}^{-1}$	T, $\mathrm{N/(A\cdot m)}$
C	capacitance	$\mathsf{L}^{-2}\mathsf{M}^{-1}\mathsf{T}^4\mathsf{I}^2$	$\mathrm{A\cdot s/V}$
	heat capacity	$\mathsf{L}^2\mathsf{MT}^{-2}\ominus^{-1}$	J/K
D	Debye–Waller factor	1	
	diffusion coefficient	$\mathsf{L}^2\mathsf{T}^{-1}$	$\mathrm{m^2/s}$
D_{td}	thermal diffusion coefficient	$\mathsf{L}^2\mathsf{T}^{-1}$	$\mathrm{m^2/s}$
$\mathbf{D}$	electric displacement	$\mathsf{L}^{-2}\mathsf{TI}$	$\mathrm{C/m^2}$

		Dimension	SI Unit
E	electromotive force	$L^2MT^{-3}I^{-1}$	V, J/C
	energy	L^2MT^{-2}	J
	irradiance	MT^{-3}	J/m^2
	illuminance	$L^{-2}J$	lm/m^2, lx
	Young's modulus	$L^{-1}MT^{-2}$	Pa, N/m^2
E_a	acceptor ionization energy	L^2MT^{-2}	J, eV
E_{ab}	thermoelectromotive force	$L^2MT^{-3}I^{-1}$	V
E_d	donor ionization energy	L^2MT^{-2}	J, eV
E_g	energy gap	L^2MT^{-2}	J, eV
E_k	kinetic energy	L^2MT^{-2}	J
E_p	potential energy	L^2MT^{-2}	J
E_F	Fermi energy	L^2MT^{-2}	J, eV
$\mathbf{E}$	electric field	$LMT^{-3}I^{-1}$	N/C, V/m
$\mathcal{E}$	electromotive force	$L^2MT^{-3}I^{-1}$	V, J/C
F	hyperfine quantum number	1	
	Helmholtz free energy	L^2MT^{-2}	J
F_m	magnetomotive force	I	A
$\mathbf{F}$	force	LMT^{-2}	N, $kg{\cdot}m/s^2$
G	conductance	$L^{-2}M^{-1}T^3I^2$	S
	Gibbs free energy	L^2MT^{-2}	J
	shear modulus	$L^{-1}MT^{-2}$	Pa, N/m^2
$\mathbf{G}$	reciprocal lattice vector, $\mathbf{G}{\cdot}\mathbf{R} = 2\pi m$	L^{-1}	m^{-1}
	enthalpy	L^2MT^{-2}	J
H_c	superconductor critical field strength	$L^{-1}I$	A/m
$\mathbf{H}$	angular impulse: $\int \mathbf{M}\, dt$	L^2MT	$N{\cdot}m{\cdot}s$
	magnetic field strength	$L^{-1}I$	A/m
I	electric current	I	A
	luminous intensity	J	cd
	moment of inertia	L^2M	$kg{\cdot}m^2$
	nuclear spin quantum number (atomic physics)	1	
	radiant intensity	L^2MT^{-2}	W/sr
$\mathbf{I}$	impulse: $\int \mathbf{F}\, dt$	LMT^{-1}	$N{\cdot}s$
J	action integral: $\oint p\, dq$	L^2MT^{-1}	$J{\cdot}s$
	electric current density	$L^{-2}I$	A/m^2
	exchange integral	L^2MT^{-2}	J
	nuclear spin quantum number (nuclear physics)	1	
	rotational quantum number	1	
	total angular momentum quantum number	1	

		Dimension	SI Unit
K	bulk modulus	$\mathsf{L^{-1}MT^{-2}}$	Pa, N/m^2
	equilibrium constant	1	
	heat transfer coefficient	$\mathsf{MT^{-3}\Theta^{-1}}$	W/(m^2·K)
	kerma (kinetic energy released in matter)	$\mathsf{L^2T^{-2}}$	Gy, J/kg
	kinetic energy	$\mathsf{L^2MT^{-2}}$	J
	luminous efficacy	$\mathsf{L^{-2}M^{-1}T^3J}$	lm/W
	relative permittivity	1	
	rotational quantum number	1	
L	length	L	m
	Lorenz coefficient: $\lambda/\sigma T$	$\mathsf{L^4M^2T^{-6}I^{-2}\Theta^{-2}}$	V^2 K^2
	luminance	$\mathsf{M^{-2}J}$	cd/m^2
	orbital angular momentum quantum number	1	
	self-inductance	$\mathsf{L^2MT^{-2}I^{-2}}$	H, Wb/A
L_p	sound pressure level	1	bel, db
L_N	loudness level	1	bel, dB
L_W	sound power level	1	bel, dB
L_{12}	mutual inductance	$\mathsf{L^2MT^{-2}I^{-2}}$	H, Wb/A
$\mathbf{L}$	angular momentum: $\mathbf{r}\times\mathbf{p}$	$\mathsf{L^2MT^{-1}}$	kg·m^2/s, J·s
M	magnetic quantum number	1	
	molar mass	M	kg
	mutual inductance	$\mathsf{L^2MT^{-2}I^{-2}}$	H, Wb/A
	radiant exitance	$\mathsf{MT^{-3}}$	W/m^2
	torque, bending moment	$\mathsf{L^2MT^{-2}}$	N·m
M_r	relative molecular mass, relative molar mass	1	
$\mathbf{M}$	magnetization	$\mathsf{L^{-1}I}$	A/m
N	neutron number: A-Z	1	
	number of particles	1	
N_E	density of states: $dN(E)/dE$	$\mathsf{L^{-5}M^{-1}T^2}$	J^{-1}·m^{-3}
N_ω	(spectral) density of vibrational modes	$\mathsf{L^{-3}T}$	s/m^3
P	power	$\mathsf{L^2MT^{-3}}$	W, J/s
	pressure	$\mathsf{L^{-1}MT^{-2}}$	Pa, J/m^3
	probability density	$\mathsf{L^{-3}}$	m^{-3}
$\mathbf{P}$	electric polarization: $\mathbf{D}-\epsilon_\circ\mathbf{P}$	$\mathsf{L^{-2}TI}$	C/m^2
Q	quadrupole moment	$\mathsf{L^2TI}$	C·m^2
	quality factor	1	
	quantity of electricity, charge	TI	C
	quantity of heat	$\mathsf{L^2MT^{-2}}$	J
	quantity of light	TJ	lm·s
	reaction energy, disintegration energy	$\mathsf{L^2MT^{-2}}$	J, eV

		Dimension	SI Unit
R	radius, nuclear radius, range	L	m
	resistance, reluctance	$L^2MT^{-1}I^{-2}$	Ω
	thermal resistance	$L^{-2}m^{-1}T^3$	K/W
R_H	Hall coefficient	$M^3T^{-1}I^{-1}$	m^3/C
$\mathbf{R}$	lattice vector	L	m
S	area	L^2	m^2
	entropy	$L^2MT^{-2}\Theta^{-1}$	J/K
	spin quantum number	1	
	stopping power	LMT^{-2}	J/m, eV/m
S_a	atomic stopping power	L^4MT^{-2}	$J\cdot m^2$, $eV\cdot m^2$
S_{ab}	Seebeck coefficient	$L^2MT^{-3}I^{-1}\Theta^{-1}$	V/K
$\mathbf{S}$	Poynting vector	$L^{-1}MT^{-3}$	W/m^2
T	kinetic energy	L^2MT^{-2}	J
	period, periodic time	T	s
	torque, moment of a couple	L^2MT^{-2}	N·m
$T_{1/2}$	half life	T	s
T_c	superconductor critical transition temperature	Θ	K
T_C	Curie temperature	Θ	K
T_N	Néel temperature	Θ	K
U	(electrical) potential difference	$L^2MT^{-3}I^{-1}$	V
	potential energy, thermodynamic energy	L^2MT^{-2}	J
U_m	magnetic potential difference	I	A
V	electric potential difference	$L^2MT^{-3}I^{-1}$	V
	potential energy	L^2MT^{-2}	J
	volume	L^3	m^3
W	energy	L^2MT^{-2}	J
	weight	LMT^{-2}	N
	work: $\int \mathbf{F}\, ds$	L^2MT^{-2}	J
X	reactance	$L^2MT^{-1}I^{-2}$	Ω
	exposure (x- or γ-ray)	$M^{-1}TI$	C/kg
Y	admittance: $Y = 1/Z = G + jB$	$L^2MT^{-1}I^{-2}$	Ω
	Young's modulus	$L^{-1}MT^{-2}$	Pa, N/m^2
Z	atomic number	1	
	impedance: $R + jX$	$L^2MT^{-1}I^{-2}$	LΩ
α	absorption factor, absorbance	1	
	angular acceleration	T^{-2}	s^{-2}
	annihilation operator	1	
	attenuation factor	L^{-1}	m^{-1}
	cubic expansion coefficient	Θ^{-1}	K^{-1}
	internal conversion coefficient	1	
	Madelung constant	1	
	plane angle	1	rad
	(electric) polarizability	$M^{-1}T^4I^2$	$C\cdot m^2/V$
	recombination coefficient	L^3T^{-1}	m^3/s

		Dimension	SI Unit
$\alpha^\dagger$	creation operator	1	
α_T	thermal diffusion factor	1	
β	annihilation operator	1	
	plane angle	1	rad
$\beta^\dagger$	creation operator	1	
γ	conductivity: $1/\rho$	$L^{-3}M^{-1}T^3I^2$	S/m
	growth rate	T^{-1}	s^{-1}
	gyromagnetic ratio: ω/B	$M^{-1}TI$	s^{-1}/T, C/kg
	plane angle	1	rad
	shear strain	1	
	surface tension	MT^{-2}	J/m^2, N/m
γ, Γ	Grüneisen parameter: $\alpha/(\kappa_T c_V \rho)$	1	
δ	damping coefficient	T^{-1}	s^{-1}
	loss angle: arctan$(1/Q)$	1	rad
	thickness	L	m
ϵ	emissivity	1	
	linear strain	1	
	permittivity	$L^{-3}M^{-1}T^4I^2$	F/m
ϵ_{ab}	Seeback coefficient	$L^2MT^{-3}I^{-1}\Theta^{-1}$	V/K
ϵ_{ij}	strain tensor	1	
ϵ_r	relative permittivity	1	
ϵ_F	Fermi energy	L^2MT^{-2}	J, eV
η	viscosity	L^2MT^{-1}	Pa·s
θ	plane angle, scattering angle	1	
ϑ	Bragg angle, scattering angle	1	
	Celsius temperature	Θ	K, °C
κ	bulk modulus, compressibility	$LM^{-1}T^2$	M^2/N, Pa^{-1}
	electrolytic conductivity	$L^{-3}M^{-1}T^3I^2$	S/m
	Landau–Ginzburg parameter	1	
λ	absolute activity: $\exp(\mu/kT)$	1	
	damping coefficient, disintegration constant, decay constant	T^{-1}	s^{-1}
	mean free path	L	m
	thermal conductivity	$LMT^{-3}\Theta^{-1}$	W/(m·K)
	wavelength	L	m
λ_B	absolute activity of B	1	
λ_C	Compton wavelength: h/mc	L	m
λ_L	London penetration depth	L	m
μ	chemical potential	$L^2MT^{-2}N^{-1}$	J/mol
	linear attenuation coefficient	L^{-1}	m^{-1}
	electric dipole moment	LTI	C·m
	magnetic dipole moment	L^2I	A·m^2
	permeability	$LMT^{-2}I^{-2}$	N/A^2
	Poisson ratio	1	
	reduced mass: $m_1m_2/(m_1+m_2)$	M	kg
	shear modulus	$L^{-1}MT^{-2}$	Pa, N/m^2

		Dimension	SI Unit
μ_a	atomic attenuation coefficient	L^2	m^2
μ_m	mass attenuation coefficient	L^2M^{-1}	m^2/kg
μ_r	relative permeability: $\mu/\mu_\circ$	1	
ν	amount of substance	N	mol
	frequency	T^{-1}	s^{-1}
	kinematic viscosity: η/ρ	L^2T^{-1}	m^2/s
ν_B	stoichiometric number of substance B	1	
ξ	coherence length	L	m
	particle displacement	L	m
ρ	charge density	$L^{-3}TI$	C/m^3
	reflection coefficient	1	
	resistivity	$L^3MT^{-3}I^{-1}$	$\Omega\cdot m$
	(mass) density	$L^{-3}M$	kg/m^3
ρ_R	residual resistivity	$L^3MT^{-3}I^{-1}$	$\Omega\cdot m$
σ	conductivity: $1/\rho$	$L^{-3}M^{-1}T^3I$	S/m
	cross section	L^2	m^2
	normal stress	$L^{-1}MT^{-2}$	Pa, N/m^2
	short range order parameter	1	
	surface charge density	$L^{-2}TI$	C/m^2
σ	surface tension	$L^{-1}MT^{-2}$	N/m, J/m^3
	wave number	L^{-1}	m^{-1}
$\boldsymbol{\sigma}$	wave vector	L^{-1}	m^{-1}
τ	mean life, relaxation time	T	s
	shear stress	$L^{-1}MT^{-2}$	Pa, N/m^2
	transmission coefficient	1	
τ_{ij}	stress tensor	1	
τ_m	mean life	T	s
$\tau_{1/2}$	half life	T	s
ϕ	electric potential	$L^2MT^{-3}I^{-1}$	V
	osmotic coefficient	1	
	(particle) fluence rate, flux density	$L^{-2}T$	m^{-2}/s
	phase difference	1 rad	
	plane angle	1	
	volume fraction	1	
χ	(magnetic) susceptibility	1	
χ_e	electric susceptibility	1	
χ_m	magnetic susceptibility	1	
ψ	radiant energy fluence rate	MT^{-3}	W/m^2
ω	angular frequency: $2\pi f$	T^{-1}	s
	solid angle	1	sr
ω_D	Debye angular frequency	T^{-1}	s^{-1}
ω_L	Larmor circular frequency	T^{-1}	s^{-1}
Γ	level width	L^2MT^{-2}	J, eV

		Dimension	SI Unit
Δ	superconductor energy gap	L^2MT^{-2}	J, eV
Θ_{rot}	characteristic rotational temperature: $h^2/8\pi^2 kI$	Θ	K
Θ_{vib}	characteristic vibrational temperature: $h\nu/k$	Θ	K
Θ_D	Debye temperature: $h\nu_D/k$	Θ	K
Θ_E	Einstein temperature: $h\nu_E/k$	Θ	K
Θ_W	Weiss temperature	Θ	K
Λ	logarithmic decrement	1	Np
	mean free path of phonons	L	m
Π	osmotic pressure	$L^{-1}MT^{-2}$	Pa, N/m^2
Π_{ab}	Peltier coefficient	$L^2MT^{-3}I^{-1}$	V
Σ	macroscopic cross section: $n\sigma$	L^{-1}	m^{-1}
Φ	luminous flux	J	lm
	magnetic flux	$L^2MT^{-2}I^{-2}$	Wb, V·s
Ψ	electric flux	TI	C
	solid angle	1	sr

TABLE 1.3. Temperature scale conversions

Both the Kelvin scale and the Rankine scale measure the same physical quantity, *thermodynamic temperature*, but in different units. Strictly speaking, a temperature on the Celsius (or Fahrenheit) *scale* (particularly when referred to ITS-90) is a scale number; a temperature difference, however, measures a physical quantity.

Triple point of natural water:	$T_\circ = 273.16$ K
Absolute to Rankine:	$T_R = \frac{9}{5} T/\mathrm{K}$
Absolute to Celsius:	$t = T/\mathrm{K} - 273.15$
Absolute to Fahrenheit:	$t_F = \frac{9}{5} T/\mathrm{K} - 459.67$
Celsius to Fahrenheit:	$t_F = \frac{9}{5} t + 32$
	$5(t_F + 40) = 9(t + 40)$

1.4.2. Quantum mechanics

Matrix element and operator symbols

A_{ij}	matrix element:	$\int \phi_i^*(A\phi_j)\,d\tau$
$A^\dagger$	Hermitian conjugate of A:	$(A^\dagger)_{ij} = A_{ji}^*$
$\langle A \rangle$	expectation value of A:	$\mathrm{Tr}\,(A)$
$[A, B]$	commutator of A and B:	$AB - BA$
$[A, B]_-$	commutator of A and B:	$AB - BA$
$[A, B]_+$	anticommutator of A and B:	$AB + BA$

$\langle \ldots |$	Dirac bra vector
$| \ldots \rangle$	Dirac ket vector

Pauli matrices: σ, I

$$\sigma_x = \begin{pmatrix} 0 & 1 \\ 1 & 0 \end{pmatrix}, \quad \sigma_y = \begin{pmatrix} 0 & -\mathrm{i} \\ \mathrm{i} & 0 \end{pmatrix}, \quad \sigma_z = \begin{pmatrix} 1 & 0 \\ 0 & -1 \end{pmatrix}, \quad I = \begin{pmatrix} 1 & 0 \\ 0 & 1 \end{pmatrix}$$

Dirac matrices: α, β

$$\alpha_x = \begin{pmatrix} 0 & \sigma_x \\ \sigma_x & 0 \end{pmatrix}, \quad \alpha_y = \begin{pmatrix} 0 & \sigma_y \\ \sigma_y & 0 \end{pmatrix}, \quad \alpha_z = \begin{pmatrix} 0 & \sigma_z \\ \sigma_z & 0 \end{pmatrix}, \quad \beta = \begin{pmatrix} I & 0 \\ 0 & -I \end{pmatrix}$$

1.4.3. Crystallography

Fundamental translation vectors for the crystal lattice:

$$\mathbf{R} = n_1\mathbf{a}_1 + n_2\mathbf{a}_2 + n_3\mathbf{a}_3, \qquad (n_1, n_2, n_3, \text{ integers})$$

Fundamental translation vectors for the reciprocal lattice:

$$\mathbf{b}_1, \mathbf{b}_2, \mathbf{b}_3 \qquad \mathbf{a}^*, \mathbf{b}^*, \mathbf{c}^*$$

(In solid-state physics $\mathbf{a}_i \cdot \mathbf{b}_k = 2\pi\delta_{ik}$, but in crystallography $\mathbf{a}_i \cdot \mathbf{b}_k = \delta_{ik}$, where δ_{ik} is the Kronecker symbol.)

h, k, l (or h_1, h_2, h_3) are the Miller indices of Bragg reflection

(h, k, l) single plane or set of parallel planes in a lattice

$\{h, k, l\}$ set of all symmetry-equivalent lattice planes

$[u, v, w]$ indices of a lattice direction

$\langle u, v, w \rangle$ set of all symmetry-equivalent lattice directions

When the letter symbols are replaced by numbers it is customary to omit the commas and to represent negative values by a bar over the number; e.g., with $h = 1, k = 2, l = -2$, the planes (hkl) are denoted by the symbol $(12\bar{2})$, and the lattice spacing d_{hkl} is denoted by $d_{12\bar{2}}$.

TABLE 1.4. Symbols for nuclear particles

The common designations for particles used as projectiles or products in nuclear reactions are listed. In addition to the symbols given in the table, an accepted designation for a general heavy ion (where there is no chance of ambiguity) is HI.

Photon	γ	Nucleon	N
Neutrino	ν, ν_e, ν_μ, ν_τ	Proton ($^1H^+$)	p
Electron	e, β	Neutron	n
Muon	μ	Deuteron ($^2H^+$)	d
Tauon	τ	Triton ($^3H^+$)	t
Pion	π	Helion ($^3He^{2+}$)	h[a]
		Alpha particle ($^4He^{2+}$)	α

[a]The symbol τ has been used in older literature for the helion, but that symbol should be reserved for the tauon (heavy lepton).

1.4.4. Nuclear and fundamental particles

If no charge is indicated in connection with the symbols p and e, these symbols refer to the positive proton and the negative electron, respectively. The bar ¯ or the tilde ˜ above the symbol for a particle is used to indicate the corresponding anti-particle; the notation $\bar{p}$ is preferable to p^- for the anti-proton, but both $\bar{e}$ and e^+ (or $\bar{\beta}$ and β^+) are commonly used for the positron.

TABLE 1.5. Symbols for fundamental particles

Except for proton p, neutron n, and electron e, the symbols for the fundamental particles in this table follow the recommendations of the Particle Data Group.

Gauge bosons	γ, g, $W^\pm$, Z
Leptons	$e^\pm$, $\nu_e^\pm$, $\mu^\pm$, $\nu_\mu^\pm$, $\tau^\pm$, $\nu_\tau^\pm$
Quarks (q)	u, d, s, c, b, t
Mesons ($q\bar{q}$)	
unflavored ($S = C = B = 0$)	$\pi^\pm$, π^0, η, σ, ρ, ω, η', f, a, ϕ, h, b, f'
strange ($S = \pm 1$, $C = B = 0$)	$K^\pm$, K^0 (K_S, K_L)
charmed ($C = \pm 1$)	$D^\pm$, D^0
charmed strange ($C = S = \pm 1$)	$D_s^\pm$
bottom ($S = 0$, $B = \pm 1$)	$B^\pm$, B^0
$c\bar{c}$ mesons	η_c, J/ψ, χ_c
$b\bar{b}$ mesons	Υ, χ_b
Baryons (qqq)	
($S = 0$)	p, n, N^+, N^0, Δ
($S = -1$)	Λ^0, $\Sigma^\pm$, Σ^0
($S = -2$)	Ξ^0, Ξ^-
($S = -3$)	Ω^-
charmed baryons ($C = +1$)	Λ_c^+, Σ_c, Ξ_c^+, Ξ_c^0, Ω
bottom baryon ($B = -1$)	Λ_b^0

Although IUPAP recommends that the symbols for particles be printed in upright type (so that, for example, the symbol e (upright) for the electron will not be confused with the symbol *e* (slanted) for the elementary charge), the Particle Data Group[7] and the AIP Style Guide[5] use slanted type for particle symbols.

The names of many fundamental particles are simply the names for their symbols. The names "up," "down," "charm," "strange," "top" ("truth"), and "bottom" ("beauty") for quarks are to be considered only as mnemonics; the names of quarks are the symbols.

1.5. NOMENCLATURE CONVENTIONS IN NUCLEAR PHYSICS

1.5.1. Nuclide

A species of atoms with a given atomic number (proton number) and mass number (nucleon number) is a "nuclide"; different nuclides having the same atomic number (proton number) are *isotopic nuclides* or *isotopes*. Different nuclides having the same mass number are *isobaric nuclides* or *isobars*.

The symbolic expression representing a nuclear reaction follows the pattern:

$$\begin{matrix}\text{initial}\\ \text{nuclide}\end{matrix}\left(\begin{matrix}\text{incoming particle}\\ \text{or photon}\end{matrix},\ \begin{matrix}\text{outgoing particle}\\ \text{or photon}\end{matrix}\right)\begin{matrix}\text{final}\\ \text{nuclide}\end{matrix}$$

Examples:

$$^{14}\mathrm{N}(\alpha, \mathrm{p})^{17}\mathrm{O} \quad ^{59}\mathrm{Co}(\mathrm{n}, \gamma)^{60}\mathrm{Co} \quad ^{23}\mathrm{Na}(\gamma, 3\mathrm{n})^{20}\mathrm{Na} \quad ^{31}\mathrm{P}(\gamma, \mathrm{pn})^{29}\mathrm{Si}$$

1.5.2. Characterization of interactions

Multipolarity of a transition:

electric or magnetic monopole	E0 or M0	
electric or magnetic dipole	E1 or M1	
electric or magnetic quadrupole	E2 or M2	(1.5)
electric or magnetic octopole	E3 or M3	
electric or magnetic 2^n-pole	E*n* or M*n*	

Notation for covariant character of coupling:

S	Scalar	A	Axial vector
V	Vector	P	Pseudoscalar
T	Tensor		

Sign of polarization vector (Basel convention): In a nuclear interaction, the positive polarization direction for particles with spin $\frac{1}{2}$ is taken in the direction of the vector product:

$$\mathbf{k}_\mathrm{i} \times \mathbf{k}_\mathrm{o},$$

where $\mathbf{k}_\mathrm{i}$ and $\mathbf{k}_\mathrm{o}$ are the wave vectors of the incoming and outgoing particles, respectively.

Description of polarization effects (Madison convention): In the symbolic expression for a nuclear reaction A(b,c)D, an arrow placed over a symbol denotes a particle that is initially in a polarized state or whose state of polarization is measured.

Examples:

$A(\vec{b}, c)D$	polarized incident beam
$A(\vec{b}, \vec{c})D$	polarized incident beam; polarization of the outgoing particle c is measured (polarization transfer)
$A(b, \vec{c})D$	unpolarized incident beam; polarization of the outgoing particle c is measured
$\vec{A}(b, c)D$	unpolarized beam incident on a polarized target
$\vec{A}(b, \vec{c})D$	unpolarized beam incident on a polarized target; polarization of the outgoing particle c is measured
$A(\vec{b}, c)\vec{D}$	polarized incident beam; polarization of the residual nucleus is measured

1.6. REFERENCES

[1] *International Vocabulary of Basic and General Terms in Metrology*, Sponsored by ISO, IEC, BIPM, OIML, IUPAP, IUPAC, IFCC, Geneva, 1993.

[2] *ISO Standards Handbook 2*, 3rd ed. (International Organization for Standardization, Geneva, 1993) [This Handbook combines ISO Standard 1000 (Units) and the series on standard symbols in several fields of physics (ISO Standards 31.0–31.12).]

[3] The International System of Units, 7th ed., 1998. Bureau International des Poids et Mesures (BIPM).

[4] *Quantities, Units and Symbols in Physical Chemistry*, edited by E. Richard Cohen, T. Cvitas et al. (Royal Society of Chemistry, London, 2001).

[5] *AIP Style Guide*, 4th ed. (American Institute of Physics, New York, 1990).

[6] ANSI/IEEE, *American National Standard for Metric Practice* (Std 268-1992) (IEEE, New York, 1992).

[7] Particle Data Group, Eur. Phys. J. C **15**, 1–878 (2000).

[8] R. D. Cowan, *The Theory of Atomic Structure and Spectra* (Univ. of California Press, Berkeley, 1981).

[9] J. Chem. Phys. **23**, 1997 (1955).

2

Mathematical Basics[1],[2]

E. Richard Cohen
California State University, Northridge

Contents

List of Tables

2.1. FACTORIALS

$$0! = 1, \quad n! = n \cdot (n-1)!, \tag{2.1}$$

$$(2n)!! = 2 \cdot 4 \cdot 6 \cdots (2n) = 2^n \cdot n!, \tag{2.2}$$

$$(2n+1)!! = 1 \cdot 3 \cdot 5 \cdots (2n+1) = \frac{(2n+1)!}{(2n)!!} = \frac{(2n+1)!}{2^n \cdot n!}, \tag{2.3}$$

$$1! = 1, \quad 2! = 2, \quad 3! = 6, \quad 4! = 24, \quad 5! = 120, \ \ldots, 10! = 3\,628\,800, \ \ldots,$$

$$1!! = 1, \quad 2!! = 2, \quad 3!! = 3, \quad 4!! = 8, \quad 5!! = 15, \ \ldots, 9!! = 945, \quad 10!! = 3840, \ \ldots.$$

For additional relations, see Sec. 2.15.

2.2. PROGRESSION AND SERIES

Arithmetic progression: A sequence of numbers in which each term is the sum of the preceding term and a fixed constant; $a, a+d, a+2d, a+3d, \ldots$. The sums of n-successive terms of an arithmetic progression form an arithmetic series:

$$\begin{aligned} S_n &= a + (a+d) + (a+2d) + (a+3d) + \cdots + [a + (n-1)\,d] \\ &= na + \tfrac{1}{2}n(n-1)d = \frac{n}{2} \quad [\text{first term} + \text{last term}]. \end{aligned} \tag{2.4}$$

Geometric progression: A sequence of numbers in which each term is the product of the preceding term and a fixed constant; $a, ar, ar^2, ar^3, \ldots$. The sums of n successive terms of a geometric progression form a geometric series:

$$G_n = a + ar + ar^2 + ar^3 + \cdots + ar^{n-1} = \frac{a(1-r^n)}{1-r}. \tag{2.5}$$

If $r^2 > 1$ the series is divergent with increasing n; if $-1 \le r < 1$, the series converges to the limit $a/(1-r)$.

Harmonic progression: A sequence of numbers is a harmonic progression if the terms are the reciprocals of an arithmetic progression. Thus, the terms $1/a$, $1/(a+d)$, $1/(a+2d)$, $1/(a+3d), \ldots$ are in harmonic progression.

2.3. MEANS

For a set of n items, $a_1, a_2, \ldots a_n$

Arithmetic mean: $m_{\mathrm{a}} = (a_1 + a_2 + a_3 + \cdots + a_n)/n$.

Geometric mean: $m_{\mathrm{g}} = (a_1 a_2 a_3 \cdots a_n)^{1/n}$.

Harmonic mean: $m_\mathrm{h} = n/[1/a_1 + 1/a_2 + 1/a_3 + \cdots + 1/a_n]$.
If $a_j > 0$ for all j, $m_\mathrm{a} \ge m_\mathrm{g} \ge m_\mathrm{h}$. Equality is achieved, $m_\mathrm{a} = m_\mathrm{g} = m_\mathrm{h}$, if and only if $a_1 = a_2 = \cdots = a_n$.

2.4. SUMMATION FORMULAS

$$1 + 2 + 3 + \cdots + n = n(n+1)/2, \tag{2.6}$$

$$1^2 + 2^2 + 3^2 + \cdots + n^2 = n(n+1)(2n+1)/6, \tag{2.7}$$

$$1^3 + 2^3 + 3^3 + \cdots + n^3 = n^2(n+1)^2/4, \tag{2.8}$$

$$1^4 + 2^4 + 3^4 + \cdots + n^4 = n(n+1)(3n[n+1]-1)(2n+1)/30. \tag{2.9}$$

2.5. BINOMIAL THEOREM

$$(1+z)^n = 1 + nz + \frac{n(n-1)}{2!}z^2 + \frac{n(n-1)(n-2)}{3!}z^3 + \cdots + \frac{n!}{r!(n-r)!}z^r + \cdots.$$

The coefficient of z^r is the binomial coefficient and is denoted by $\binom{n}{r}$.

$$(1+z)^n = \sum_{r=0} \binom{n}{r} z^r = \sum_{r=0} \frac{n!}{(n-r)!r!} z^r. \tag{2.10}$$

If n is a positive integer, $\binom{n}{r} = 0$ for $r > n$ and the summation terminates with the rth term; otherwise, the series is infinite and converges only if $|z| < 1$.

$$(1+z)^{\frac{1}{2}} = 1 + \frac{1}{2}z - \frac{1\cdot 1}{2\cdot 4}z^2 + \frac{1\cdot 1\cdot 3}{2\cdot 4\cdot 6}z^3 - \frac{1\cdot 1\cdot 3\cdot 5}{2\cdot 4\cdot 6\cdot 8}z^4 + \cdots, \tag{2.11}$$

$$(1+z)^{-\frac{1}{2}} = 1 - \frac{1}{2}z + \frac{1\cdot 3}{2\cdot 4}z^2 - \frac{1\cdot 3\cdot 5}{2\cdot 4\cdot 6}z^3 + \frac{1\cdot 3\cdot 5\cdot 7}{2\cdot 4\cdot 6\cdot 8}z^4 - \cdots. \tag{2.12}$$

2.6. QUADRATIC EQUATION

If $a = 0$, the two roots of $ax^2 + bx + c = 0$ are

$$x = \frac{-b \pm (b^2 - 4ac)^{\frac{1}{2}}}{2a} = \frac{-2c}{b \pm (b^2 - 4ac)^{\frac{1}{2}}}. \tag{2.13}$$

There is one double root if $D = b^2 - 4ac = 0$. If the coefficients are real, the two roots are real for $D > 0$ and complex conjugates for $D < 0$.

The square roots of $z = x + iy$ (x and y real) are $\sqrt{z} = \pm(a + ib)$, with

$$a = \sqrt{\tfrac{1}{2}\left(\sqrt{x^2 + y^2} + x\right)}, \tag{2.14}$$

$$b = \sqrt{\tfrac{1}{2}\left(\sqrt{x^2+y^2}-x\right)}. \tag{2.15}$$

To ensure that the roots are in the correct quadrants of the complex plane and to avoid loss of numerical accuracy, compute a if $x > 0$ or b if $x < 0$, and obtain the other quantity from the condition $2ab = y$.

2.7. DIFFERENTIATION

a, c, n are constants; u, v, w are variables.

$$\frac{d}{dx}(au) = a\frac{du}{dx} \tag{2.16}$$

$$\frac{d}{dx}(u+v) = \frac{du}{dx} + \frac{dv}{dx} \tag{2.17}$$

$$\frac{d}{dx}U(w) = \frac{dU}{dw}\frac{dw}{dx} \tag{2.18}$$

$$\frac{d}{dx}(uv) = \frac{du}{dx}v + u\frac{dv}{dx} \tag{2.19}$$

$$\frac{d}{dx}u^n = nu^{n-1}\frac{du}{dx} \tag{2.20}$$

$$\frac{d}{dx}u^v = vu^{v-1}\frac{du}{dx} + u^v \ln u \frac{dv}{dx} \tag{2.21}$$

$$\frac{d^n}{dx^n}(uv) = \frac{d^n u}{dx^n}v + n\frac{d^{n-1}u}{dx^{n-1}}\frac{dv}{dx} + \cdots + \binom{n}{k}\frac{d^{n-k}u}{dx^{n-k}}\frac{d^k v}{dx^k} + \cdots \tag{2.22}$$

2.7.1. Taylor series

$$\begin{aligned} f(x+h) = f(x) + hf'(x) + \frac{h^2}{2!}f''(x) + \cdots \\ + \frac{h^{n-1}}{(n-1)!}f^{(n-1)}(x) + \frac{h^n}{n!}f^{(n)}(x+\theta h) \end{aligned} \tag{2.23}$$

for a suitable choice of θ, $0 < \theta < 1$.

$$\begin{aligned} f(x+h, y+k) = f(x,y) + \left(h\frac{\partial f(x,y)}{\partial x} + k\frac{\partial f(x,y)}{\partial y}\right) \\ + \frac{1}{2!}\left(h^2\frac{\partial^2 f(x,y)}{\partial x^2} + 2hk\frac{\partial^2 f(x,y)}{\partial x \partial y} + k^2\frac{\partial^2 f(x,y)}{\partial y^2}\right) + \end{aligned}$$

$$+\frac{1}{3!}\left(h^3\frac{\partial^3 f(x,y)}{\partial x^3}+3h^2k\frac{\partial^3 f(x,y)}{\partial x^2\partial y}\right.$$
$$\left.+3hk^2\frac{\partial^3 f(x,y)}{\partial x\partial y^2}+k^3\frac{\partial^3 f(x,y)}{\partial y^3}\right)+\cdots \tag{2.24}$$

2.8. INTEGRATION

$$\int f(u)dx=\int f(u)\frac{dx}{du}du \tag{2.25}$$

$$\int f(a+bx)dx=\frac{1}{b}\int f(y)dy \tag{2.26}$$

$$\int f(\sqrt{a+bx})dx=\frac{2}{b}\int f(y)y\,dy \tag{2.27}$$

Integration by parts:

$$\int U dv=Uv-\int v\,dU \tag{2.28}$$

$$\int UV dx=\left(\int U dx\right)V-\int\left(\int U dx\right)\frac{dV}{dx}dx \tag{2.29}$$

$$\int x^a dx=\begin{cases}\dfrac{x^{a+1}}{a+1}, & (a\neq -1)\\ \ln|x|, & (a=-1)\end{cases} \tag{2.30}$$

$$\int\frac{dx}{1+x^2}=\arctan x,\quad(-\pi/2<\arctan x<\pi/2) \tag{2.31}$$

$$\int\frac{dx}{a^2+b^2x^2}=\frac{1}{ab}\arctan\frac{bx}{a} \tag{2.32}$$

$$\int\frac{dx}{a^2-b^2x^2}=\frac{1}{2ab}\ln\frac{a+bx}{a-bx} \tag{2.33}$$

$$\int\frac{dx}{\sqrt{a^2+b^2x^2}}=\frac{1}{b}\operatorname{arsinh}\frac{bx}{a} \tag{2.34}$$

$$=\frac{1}{b}\ln\left(bx+\sqrt{a^2+b^2x^2}\right) \tag{2.35}$$

$$\int\frac{dx}{\sqrt{a^2-b^2x^2}}=\frac{1}{b}\arcsin\frac{bx}{a} \tag{2.36}$$

$$\int\frac{dx}{\sqrt{b^2x^2-a^2}}=\frac{1}{b}\ln\left(bx+\sqrt{b^2x^2-a^2}\right) \tag{2.37}$$

2.9. LOGARITHMIC FUNCTIONS

$$\ln x \equiv \log_e x, \quad \lg x \equiv \log_{10} x, \quad \ln 10 = 2.302\,585\,092\ldots$$

$$\log_a b \cdot \log_b c = \log_a c, \quad \log_b a = 1/\log_a b, \quad \log_a x = \frac{\log_b x}{\log_b a}$$

2.9.1. Series

$$\ln(1+x) = x - \frac{x^2}{2} + \frac{x^3}{3} - \frac{x^4}{4} + \cdots, \quad (-1 < x \le 1) \tag{2.38}$$

$$\ln x = \ln a + \frac{x-a}{a} - \frac{(x-a)^2}{2a^2} + \frac{(x-a)^3}{3a^3} + \cdots, \quad (0 < x \le 2a) \tag{2.39}$$

$$\ln \frac{1+x}{1-x} = 2\left[x + \frac{x^3}{3} + \frac{x^5}{5} + \frac{x^7}{7} + \cdots\right], \quad (-1 < x \le 1) \tag{2.40}$$

$$\ln x = \frac{2}{p}\left[\frac{x^p - 1}{x^p + 1} + \frac{(x^p-1)^3}{3(x^p+1)^3} + \frac{(x^p-1)^5}{5(x^p+1)^5} + \cdots\right] \tag{2.41}$$

Euler's constant:

$$\gamma = \lim_{m\to\infty}\left[1 + \frac{1}{2} + \frac{1}{3} + \cdots + \frac{1}{m} - \ln m\right] = 0.577\,215\,6649\ldots$$

$$\frac{d \ln g(x)}{dx} = \frac{1}{g(x)}\frac{dg(x)}{dx} \tag{2.42}$$

$$\int_1^x u^{a-1} \ln u \, du = \begin{cases} \dfrac{1}{a}\left(x^a \ln x + \dfrac{1-x^a}{a}\right), & (a \ne 0) \\ \frac{1}{2}(\ln x)^2, & (a = 0) \end{cases} \tag{2.43}$$

2.10. EXPONENTIAL FUNCTIONS

$$e^x = \exp x = 1 + \frac{x}{1!} + \frac{x^2}{2!} + \cdots + \frac{x^n}{n!} + \cdots \tag{2.44}$$

$$e^x = \lim_{n\to\infty}\left[1 + \frac{x}{n}\right]^{n+\frac{1}{2}x} \tag{2.45}$$

2.10.1. Derivatives

$$\frac{de^x}{dx} = e^x, \quad \frac{d}{dx}e^{u(x)} = e^{u(x)}\frac{du}{dx} \tag{2.46}$$

$$\frac{da^x}{dx} = a^x \ln a, \quad \frac{dx^x}{dx} = (1 + \ln x)x^x \tag{2.47}$$

$$\frac{du^v}{dx} = \left[\frac{dv}{dx} \ln u + \frac{v}{u}\frac{du}{dx}\right] \tag{2.48}$$

2.10.2. Integrals

$$\int e^{ax}\,dx = \frac{e^{ax}}{a} \tag{2.49}$$

$$\int xe^{ax}\,dx = \frac{e^{ax}}{a^2}(ax - 1) \tag{2.50}$$

$$\int x^2 e^{ax}\,dx = 2!\frac{e^{ax}}{a^3}\left(\frac{a^2x^2}{2} - ax + 1\right) \tag{2.51}$$

$$\int \frac{e^{ax}}{x^n} dx = -\frac{e^{ax}}{(n-1)x^{n-1}} + \frac{a}{n-1}\int \frac{e^{ax}}{x^{n-1}}\,dx \tag{2.52}$$

2.11. TRIGONOMETRIC FUNCTIONS

$$\sin^2 + \cos^2 A = 1 \tag{2.53}$$

$$\sin(A + B) = \sin A \cos B + \cos A \sin B \tag{2.54}$$

$$\cos(A + B) = \cos A \cos B - \sin A \sin B \tag{2.55}$$

$$2 \sin A \sin B = \cos(A - B) - \cos(A + B) \tag{2.56}$$

$$2 \sin A \cos B = \sin(A + B) + \sin(A - B) \tag{2.57}$$

$$2 \cos A \cos B = \cos(A + B) + \cos(A - B) \tag{2.58}$$

$$\sin 2A = 2 \sin A \cos A \tag{2.59}$$

$$\cos 2A = \cos^2 A - \sin^2 A = 1 - 2\sin^2 A = 2\cos^2 A - 1 \tag{2.60}$$

$$e^{ix} = \cos x + i \sin x \tag{2.61}$$

$$(\cos x + i \sin x)^\nu = \cos \nu x + i \sin \nu x \quad \text{(DeMoivre's theorem)} \tag{2.62}$$

$$\cos x = \tfrac{1}{2}(e^{ix} + e^{-ix}) \tag{2.63}$$

$$\sin x = \frac{1}{2i}(e^{ix} - e^{-ix}) \tag{2.64}$$

$$e^{i\pi} + 1 = 0 \quad \text{(Euler's formula)} \tag{2.65}$$

2.11.1. Series expansions

$$\sin z = z - \frac{z^3}{3!} + \frac{z^5}{5!} - \frac{z^7}{7!} + \frac{z^9}{9!} - \cdots$$

$$\cos z = 1 - \frac{z^2}{2!} + \frac{z^4}{4!} - \frac{z^6}{6!} + \frac{z^8}{8!} - \cdots$$

$$\tan z = z + \frac{z^3}{3} + \frac{2z^5}{15} + \frac{17z^7}{315} + \frac{62z^9}{2835} + \cdots, \qquad |z| < \tfrac{\pi}{2}$$

$$\cot z = \frac{1}{z} - \frac{z}{3} - \frac{z^3}{45} - \frac{2z^5}{945} - \frac{z^7}{4725} + \cdots, \qquad |z| < \pi$$

$$\sec z = 1 + \frac{z^2}{2} + \frac{5z^4}{24} + \frac{61z^6}{720} + \frac{277z^8}{8064} + \cdots, \qquad |z| < \tfrac{\pi}{2}$$

$$\csc z = \frac{1}{z} + \frac{z}{6} + \frac{7z^3}{360} + \frac{31z^5}{15120} + \frac{127z^7}{604800} + \cdots, \qquad |z| < \pi$$

2.11.2. Derivatives

$$\frac{d}{dx}\sin x = \cos x, \qquad \frac{d}{dx}\cot x = -\csc^2 x \tag{2.66}$$

$$\frac{d}{dx}\cos x = -\sin x, \qquad \frac{d}{dx}\sec x = \sec x \tan x \tag{2.67}$$

$$\frac{d}{dx}\tan x = \sec^2 x, \qquad \frac{d}{dx}\csc x = -\csc x \cot x \tag{2.68}$$

2.11.3. Integrals

$$\int \sin x dx = -\cos x \tag{2.69}$$

$$\int x \sin x dx = \sin x - x \cos x \tag{2.70}$$

$$\int \sin^2 x dx = \frac{x}{2} - \frac{\sin 2x}{4} = \frac{1}{2}(x - \sin x \cos x) \tag{2.71}$$

$$\int \cos x dx = \sin x \tag{2.72}$$

$$\int x \cos x dx = \cos x + x \sin x \tag{2.73}$$

$$\int \cos x \sin x dx = \sin^2 x \tag{2.74}$$

$$\int \cos^2 x dx = x - \int \sin^2 x dx \tag{2.75}$$

$$\int \tan x dx = \ln(\sec x) = -\ln \cos x \tag{2.76}$$

$$\int \tan^2 x dx = \tan x - x \tag{2.77}$$

$$\int \cot x dx = \ln \sin x \tag{2.78}$$

$$\int \sec x dx = \ln(\sec x + \tan x) = \ln\left[\tan\left(\frac{\pi}{4} + \frac{x}{2}\right)\right] \tag{2.79}$$

$$\int \csc x dx = \ln \tan(\tfrac{1}{2}x) = \ln(\csc x - \cot x) \tag{2.80}$$

2.11.4. Plane triangles

Let a, b, and c be the sides opposite the angles A, B, and C, respectively.

$$A + B = 180° = \pi \text{ rad} \tag{2.81}$$

$$\frac{\sin A}{a} = \frac{\sin B}{b} = \frac{\sin C}{c} \tag{2.82}$$

$$\frac{a+b}{a-b} = \frac{\tan \frac{1}{2}(A+B)}{\tan \frac{1}{2}(A-B)} \tag{2.83}$$

$$c = a\cos B + b\cos A, \qquad c^2 = a^2 - 2ab\cos C + b^2 \tag{2.84}$$

$$\text{Area} = S = \tfrac{1}{2}ab\sin C = \frac{a^2 \sin B \sin C}{2\sin A} \tag{2.85}$$

$$= \frac{1}{4}\sqrt{[(a+b)^2 - c^2][c^2 - (a-b)^2]} \tag{2.86}$$

$$= \sqrt{s(s-a)(s-b)(s-c)} \tag{2.87}$$

where $s = (a+b+c)/2$ is the semiperimeter.

2.11.5. Spherical triangles

On the unit sphere, let a, b, and c be the sides opposite the angles A, B, and C, respectively.

$$A + B + C > \pi, \qquad \text{Spherical excess, } E = A + B + C - \pi$$

$$\text{Area of a sphere: } S = 4\pi, \qquad \text{Area of a spherical triangle: } S = E$$

$$\frac{\sin A}{\sin a} = \frac{\sin B}{\sin b} = \frac{\sin C}{\sin c} \tag{2.88}$$

$$\cos c = \cos a \cos b + \sin a \sin b \cos C = \frac{\cos a \cos(b - \theta)}{\cos \theta} \tag{2.89}$$

where $\tan\theta = \tan a \cos C$, and $\cos C = -\cos A \cos B + \sin A \sin B \cos c$.

2.12. INVERSE TRIGONOMETRIC FUNCTIONS

2.12.1. Series expansions

$$\arcsin x = x + \frac{x^3}{2\cdot 3} + \frac{1\cdot 3\, x^5}{2\cdot 4\cdot 5} + \cdots + \frac{(2k-1)!!x^{2k+1}}{(2k)!!(2k+1)} + \cdots \tag{2.90}$$

$$\arctan x = \begin{cases} x - \dfrac{x^3}{3} + \dfrac{x^5}{5} - \dfrac{x^7}{7} + \cdots, & x^2 < 1 \\ \dfrac{\pi}{2} - \dfrac{1}{x} + \dfrac{1}{3x^3} - \dfrac{1}{5x^5} + \cdots, & x^2 > 1 \end{cases} \tag{2.91}$$

2.12.2. Derivatives

$$\frac{d}{dx} \arcsin \frac{x}{a} = \frac{1}{\sqrt{a^2 - x^2}} \tag{2.92}$$

$$\frac{d}{dx} \arccos \frac{x}{a} = \frac{-1}{\sqrt{a^2 - x^2}} \tag{2.93}$$

$$\frac{d}{dx} \arctan \frac{x}{a} = \frac{a}{a^2 + x^2} \tag{2.94}$$

$$\frac{d}{dx} \operatorname{arccot} \frac{x}{a} = \frac{-a}{a^2 + x^2} \tag{2.95}$$

$$\frac{d}{dx} \operatorname{arcsec} \frac{x}{a} = \frac{a}{x\sqrt{x^2 - a^2}} \tag{2.96}$$

$$\frac{d}{dx} \operatorname{arccsc} \frac{x}{a} = \frac{-a}{x\sqrt{x^2 - a^2}} \tag{2.97}$$

2.12.3. Integrals

$$\int \arcsin \frac{x}{a} dx = x \arcsin \frac{x}{a} + \sqrt{a^2 - x^2} \tag{2.98}$$

$$\int \arccos \frac{x}{a} dx = x \arccos \frac{x}{a} - \sqrt{a^2 - x^2} \tag{2.99}$$

$$\int \arctan \frac{x}{a} dx = x \arctan \frac{x}{a} - \frac{a}{2} \ln a^2 + x^2 \tag{2.100}$$

$$\int \operatorname{arccot} \frac{x}{a} dx = \operatorname{arccot} \frac{x}{a} + \frac{a}{2} \ln a^2 + x^2 \tag{2.101}$$

$$\int \operatorname{arcsec} \frac{x}{a} dx = x \operatorname{arcsec} \frac{x}{a} \pm a \ln \left[x + \sqrt{x^2 - a^2} \right]$$

$$\begin{cases} \pi/2 < \operatorname{arcsec}(x/a) < \pi \\ 0 < \operatorname{arcsec}(x/a) < \pi/2 \end{cases} \tag{2.102}$$

$$\int \operatorname{arccsc} \frac{x}{a} dx = \operatorname{arccsc} \frac{x}{a} + a \ln \left[x + \sqrt{x^2 - a^2} \right]$$

$$(0 < \operatorname{arccsc}(x/a) < \pi/2) \tag{2.103}$$

$$= \operatorname{arccsc} \frac{x}{a} - a \ln \left[x + \sqrt{x^2 - a^2} \right]$$

$$(-\pi/2 < \operatorname{arccsc}(x/a) < 0) \tag{2.104}$$

$$\int x \arcsin \frac{x}{a} dx = \tfrac{1}{4} \left[(2x^2 - a^2) \arcsin \frac{x}{a} + x\sqrt{a^2 - x^2} \right] \tag{2.105}$$

$$\int x \arctan \frac{x}{a} dx = \tfrac{1}{2} \left[(a^2 + x^2) \arctan \frac{x}{a} - ax \right] \tag{2.106}$$

2.13. HYPERBOLIC FUNCTIONS

Any trigonometric relation will yield a corresponding hyperbolic relation by the replacements $x \to ix$, $\cos x \to \cosh x$, and $\sin x \to i \sinh x$.

$$\sin ix = i \sinh x, \qquad \cos ix = \cosh x, \qquad \tan ix = i \tanh x$$

$$\csc ix = -i \operatorname{csch} x, \qquad \sec ix = \operatorname{sech} x, \qquad \cot ix = -i \coth x$$

$$\sin(x + iy) = \sin x \cosh y + i \cos x \sinh y \tag{2.107}$$

$$\cos(x + iy) = \cos x \cosh y - i \sin x \sinh y \tag{2.108}$$

$$\cosh^2 x + \sinh^2 x = 1 \tag{2.109}$$

$$\cosh x + \sinh x = e^x \tag{2.110}$$

$$\cosh x - \sinh x = e^{-x} \tag{2.111}$$

$$\cosh x = \tfrac{1}{2}(e^x + e^{-x}) \tag{2.112}$$

$$\sinh x = \tfrac{1}{2}(e^x - e^{-x}) \tag{2.113}$$

$$\sinh(A + B) = \sinh A \cosh B + \cosh A \sinh B \tag{2.114}$$

$$\cosh(A + B) = \cosh A \cosh B + \sinh A \sinh B \tag{2.115}$$

$$2 \sinh A \sinh B = \cosh(A + B) - \cosh(A - B) \tag{2.116}$$

$$2 \sinh A \cosh B = \sinh(A + B) + \sinh(A - B) \tag{2.117}$$

$$2 \cosh A \cosh B = \cosh(A + B) + \cosh(A + B) \tag{2.118}$$

2.13.1. Series expansions

$$\cosh x = 1 + \frac{x^2}{2!} + \frac{x^4}{4!} + \frac{x^6}{6!} + \frac{x^8}{8!} + \cdots \tag{2.119}$$

$$\sinh x = x + \frac{x^3}{3!} + \frac{x^5}{5!} + \frac{x^7}{7!} + \frac{x^9}{9!} + \cdots \tag{2.120}$$

$$\tanh x = x - \frac{x^3}{3} + \frac{2x^5}{15} - \frac{17x^7}{315} + \cdots \tag{2.121}$$

$$\coth x = \frac{1}{x} + \frac{x}{3} - \frac{x^3}{45} + \frac{2x^5}{945} + \cdots \tag{2.122}$$

$$\operatorname{sech} x = 1 - \frac{x^2}{2} + \frac{5x^4}{24} - \frac{61x^6}{720} + \cdots \tag{2.123}$$

$$\operatorname{csch} x = \frac{1}{x} - \frac{x}{6} + \frac{7x^3}{360} - \frac{31x^5}{1520} + \frac{127x^7}{423360} + \cdots \tag{2.124}$$

2.13.2. Derivatives

$$\frac{d}{dx} \sinh x = \cosh x, \qquad \frac{d}{dx} \coth x = -\operatorname{csch}^2 x \tag{2.125}$$

$$\frac{d}{dx} \cosh x = \sinh x, \qquad \frac{d}{dx} \operatorname{sech} x = -\operatorname{sech} x \tanh x \tag{2.126}$$

$$\frac{d}{dx} \tanh x = \operatorname{sech}^2 x, \qquad \frac{d}{dx} \operatorname{csch} x = -\operatorname{csch} x \coth x \tag{2.127}$$

2.13.3. Integrals

$$\int \sinh x dx = \cosh x \tag{2.128}$$

$$\int x \sinh x dx = x \cosh x - \sinh x \tag{2.129}$$

$$\int \cosh x dx = \sinh x \tag{2.130}$$

$$\int x \cosh x dx = x \sinh x - \cosh x \tag{2.131}$$

$$\int \tanh x dx = \ln(\cosh x) \tag{2.132}$$

$$\int \coth x\,dx = \ln(\sinh x) \tag{2.133}$$

$$\int \operatorname{sech} x\,dx = \arctan(\sinh x) \tag{2.134}$$

$$\int \operatorname{csch} x\,dx = \ln\left(\tanh \frac{x}{2}\right) \tag{2.135}$$

2.14. INVERSE HYPERBOLIC FUNCTIONS

$$\operatorname{arsinh} x = \int_0^x \frac{dt}{(t^2+1)^{1/2}} = \ln[x + (x^2+1)^{1/2}] \tag{2.136}$$

$$\operatorname{arcosh} x = \int_0^x \frac{dt}{(t^2-1)^{1/2}} = \ln[x + (x^2-1)^{1/2}] \tag{2.137}$$

$$\operatorname{artanh} x = \int_0^x \frac{dt}{1-t^2} = \frac{1}{2}\ln\frac{1+x}{1-x} \tag{2.138}$$

2.14.1. Series expansions

$$\operatorname{arsinh} x = \begin{cases} x - \dfrac{x^3}{2\cdot 3} + \dfrac{1\cdot 3x^5}{2\cdot 4\cdot 5} + \dfrac{1\cdot 3\cdot 5x^7}{2\cdot 4\cdot 6\cdot 7} + \cdots, & |x| < 1 \\ \ln 2x + \dfrac{1}{2\cdot 2x^2} - \dfrac{1\cdot 3}{2\cdot 4\cdot 4x^4} + \dfrac{1\cdot 3\cdot 5}{2\cdot 4\cdot 6\cdot 6x^6} + \cdots, & |x| > 1 \end{cases} \tag{2.139}$$

$$\operatorname{arcosh} x = \ln 2x - \frac{1}{2\cdot 2x^2} - \frac{1\cdot 3}{2\cdot 4\cdot 4x^4} - \frac{1\cdot 3\cdot 5}{2\cdot 4\cdot 6\cdot 6x^6} - \cdots, \quad (|x| > 1) \tag{2.140}$$

$$\operatorname{artanh} x = \operatorname{arcoth}\frac{1}{x} = x + \frac{x^3}{3} + \frac{x^5}{5} + \frac{x^7}{7} + \cdots, \quad (|x| < 1) \tag{2.141}$$

2.14.2. Derivatives

$$\frac{d}{dx}\operatorname{arsinh}\frac{x}{a} = (a^2+x^2)^{-1/2} \tag{2.142}$$

$$\frac{d}{dx}\operatorname{arcosh}\frac{x}{a} = (x^2-a^2)^{-1/2} \tag{2.143}$$

$$\frac{d}{dx}\arctan\frac{x}{a} = a\big/(a^2-x^2) \tag{2.144}$$

$$\frac{d}{dx}\operatorname{arcoth}\frac{x}{a} = -a\big/(x^2-a^2) \tag{2.145}$$

$$\frac{d}{dx}\operatorname{arsech}\frac{x}{a} = -\frac{a}{x(a^2-x^2)^{1/2}}, \quad x = 0 \tag{2.146}$$

$$\frac{d}{dx}\operatorname{arcsch}\frac{x}{a} = -\frac{a}{|x|(a^2+x^2)^{1/2}}, \quad x = 0 \tag{2.147}$$

2.14.3. Integrals

$$\int \operatorname{arsinh}\frac{x}{a}dx = x\operatorname{arsinh}\frac{x}{a} - \sqrt{a^2+x^2} \tag{2.148}$$

$$\int \operatorname{arcosh}\frac{x}{a}dx = x\operatorname{arcosh}\frac{x}{a} - \sqrt{x^2-a^2} \tag{2.149}$$

$$\int \operatorname{artanh}\frac{x}{a}dx = x\operatorname{artanh}\frac{x}{a} + \frac{a}{2}\ln(a^2-x^2) \tag{2.150}$$

$$\int \operatorname{arcoth}\frac{x}{a}dx = x\operatorname{arcoth}\frac{x}{a} + \frac{a}{2}\ln(x^2-a^2) \tag{2.151}$$

$$\int \operatorname{arsech}\frac{x}{a}dx = x\operatorname{arsech}\frac{x}{a} + a\operatorname{arsinh}\frac{x}{a} \tag{2.152}$$

$$\int \operatorname{arcsch}\frac{x}{a}dx = x\operatorname{arcsch}\frac{x}{a} + a\operatorname{arcsch}\frac{x}{a} \tag{2.153}$$

$$\int x\operatorname{arsinh}\frac{x}{a}dx = \tfrac{1}{4}\left[(2x^2+a^2)\operatorname{arsinh}\frac{x}{a} - x\sqrt{x^2+a^2}\right] \tag{2.154}$$

$$\int x\operatorname{artanh}\frac{x}{a}dx = \tfrac{1}{2}\left[(x^2-a^2)\arctan\frac{x}{a} + ax\right] \tag{2.155}$$

2.15. GAMMA FUNCTION

The gamma function is the analytic continuation into the complex plane of the factorial $n!$.

For n a positive integer, $\Gamma(n) = (n-1)!$

$$\Gamma(z+1) = z\Gamma(z), \quad \Gamma(1) = 1, \quad \Gamma\left(\tfrac{1}{2}\right) = \sqrt{\pi} \tag{2.156}$$

$$\Gamma(z) = \int_0^\infty t^{z-1}e^{-t}\,dt, \quad \Re(z) > 0 \tag{2.157}$$

$$\left(n-\frac{1}{2}\right)! = \Gamma\left(n+\frac{1}{2}\right) = \frac{(2n-1)!!}{2^n}\sqrt{\pi} = \frac{(2n)!}{2^{2n}n!}\sqrt{\pi} \tag{2.158}$$

$$\Gamma(z)\Gamma(1-z) = \frac{\pi}{\sin\pi z}, \quad \Gamma(z)\Gamma(-z) = -\frac{\pi}{z\sin\pi z}$$

$$\Gamma(2z) = \frac{2^{2z-1}\Gamma(z)\Gamma\left(z+\frac{1}{2}\right)}{\Gamma\left(\frac{1}{2}\right)}$$

$$\sqrt{2\pi}z^{z-1/2}e^{-z}\left[1+\frac{1}{12z}+\frac{1}{288z^2}-\frac{139}{51840z^3}-\frac{571}{2488320z^4}+\cdots\right] \tag{2.159}$$

$$\Gamma(z) \sim \sqrt{2\pi} z^{z-1/2} \exp\left[-z + \frac{1}{12z} - \frac{1}{360z^3} + \frac{1}{1260z^5} - \frac{1}{1680z^7} + \cdots \right] \tag{2.160}$$

Stirling's formula:

$$\lim_{n\to\infty} \frac{n!\,e^n}{\sqrt{n}\,n^n} = \sqrt{2\pi}$$

This gives approximate values of $n!$ for large n.

2.15.1. Definite integrals

$$\int_1^\infty \frac{dx}{x^m} = \frac{1}{m-1} \tag{2.161}$$

$$\int_0^a \frac{dx}{\sqrt{a^2 - x^2}} = \frac{\pi}{2} \tag{2.162}$$

$$\int_0^\infty \frac{x\,dy}{x^2 + y^2} = \begin{cases} -\frac{\pi}{2}, & x < 0 \\ 0, & x = 0 \\ \frac{\pi}{2}, & x > 0 \end{cases} \tag{2.163}$$

$$\frac{2}{\pi} \int_0^{\pi/2} \sin^{2n} x\,dx = \frac{2}{\pi} \int_0^{\pi/2} \cos^{2n} x\,dx = \frac{(2n-1)!!}{(2n)!!} \tag{2.164}$$

$$= \frac{(2n)!}{(2^n n!)^2} = \frac{1}{2^{2n}} \binom{2n}{n} = \frac{\Gamma(n+\frac{1}{2})}{\Gamma(\frac{1}{2})\Gamma(n+1)} \tag{2.165}$$

$$\frac{2}{\pi} \int_0^\pi \sin^{2n} x \cos^{2m}\,dx = \frac{(2n)!(2m)!}{2^{2n+2m} n! m! (n+m)!} \tag{2.166}$$

$$2\int_0^{\pi/2} \sin^{2z-1} x \cos^{2w-1} x\,dx = \int_0^1 t^{z-1}(1-t)^{w-1} dt = \frac{\Gamma(z)\Gamma(w)}{\Gamma(z+w)} \tag{2.167}$$

$$\int_0^x \frac{\sin t}{t}\,dt = \mathrm{Si}\,(x) = x - \frac{x^3}{3\cdot 3!} + \frac{x^5}{5\cdot 5!} - \frac{x^7}{7\cdot 7!} + \cdots \tag{2.168}$$

$$\int_0^x \frac{1 - \cos t}{t}\,dt = \mathrm{Cin}\,(x) = \frac{x^2}{2\cdot 2!} - \frac{x^4}{4\cdot 4!} + \frac{x^6}{6\cdot 6!} - \cdots \tag{2.169}$$

$$= \ln x + \gamma - \mathrm{Ci}\,(x) \tag{2.170}$$

$$\int_1^\infty \frac{e^{-xt}}{t^n}\,dt = E_n(x) \tag{2.171}$$

$$E_0(x) = \frac{e^{-x}}{x} \tag{2.172}$$

$$E_{n+1}(x) = \frac{1}{n}[e^{-x} - x E_n(x)] \tag{2.173}$$

$$E_1(x) = -\ln x - \gamma + x - \frac{x^2}{2\cdot 2!} + \frac{x^3}{3\cdot 3!} - \frac{x^4}{4\cdot 4!} + \cdots \tag{2.174}$$

$$\int_0^\infty e^{-a^2x^2}\,dx = \frac{\sqrt{\pi}}{2a} \tag{2.175}$$

$$\int_0^\infty xe^{-a^2x^2}\,dx = \frac{1}{2a^2} \tag{2.176}$$

$$\int_0^\infty x^2e^{-a^2x^2}\,dx = \frac{\sqrt{\pi}}{4a^3} \tag{2.177}$$

$$\int_0^\infty x^{2n}e^{-px^2}\,dx = \frac{(2n-1)!!}{2^{n+1}p^n}\sqrt{\frac{\pi}{p}}, \quad n \text{ integer} \tag{2.178}$$

$$\int_0^\infty x^{2n+1}e^{-px^2}\,dx = \frac{n!}{2p^{n+1}}, \quad n \text{ integer} \tag{2.179}$$

2.16. DELTA FUNCTION

The discrete delta function or Kronecker symbol is the quantity

$$\delta_{ij} = \begin{cases} 1 & \text{if } i = j \\ 0 & \text{if } i = j \end{cases} \tag{2.180}$$

The Dirac delta function $\delta(x)$ is an improper function such that for $a < b$,

$$x\delta(x) = 0 \quad \text{and} \quad \int_a^b \delta(x)\,dx = \begin{cases} 1, & \text{if } a < 0 < b \\ 0, & \text{otherwise} \end{cases} \tag{2.181}$$

For an arbitrary function $f(x)$, and $a < y < b$

$$\int_a^b f(x)\delta(x-y)\,dx = f(y) \tag{2.182}$$

$$\int_a^b f(x)\delta'(x-y)\,dx = -f'(y) \tag{2.183}$$

where the prime denotes differentiation with respect to the argument.

If the δ function has as an argument a function $g(x)$ of the independent variable x with simple zeros at x_i [$g(x_i) = 0$, $g'(x_i) = 0$], it can be transformed according to the rule:

$$\delta\left(g(x)\right) = \sum_i \frac{\delta(x - x_i)}{|g'(x_i)|} \tag{2.184}$$

2.17. VECTOR ALGEBRA

2.17.1. Notation

A, B, y, z, etc., are scalars; **A**, **B**, **y**, **z**, etc., are vectors. In particular, a vector and its scalar magnitude are denoted by the same symbol, with the vector in bold type: $v = |\mathbf{v}| = \sqrt{\mathbf{v} \cdot \mathbf{v}}$.

[Mathematical vectors are usually denoted by Roman boldface type, but *physical quantities* expressed as vectors (in accord with International Organization for Standardization recommendations[3]) should preferably be set in *italic* boldface type.]

2.17.2. Dot product (scalar product or inner product)

$$\mathbf{A} \cdot \mathbf{B} = (\mathbf{A}, \mathbf{B}) = AB \cos \phi$$

ϕ being the angle between **A** and **B**.

In an orthogonal coordinate system defined by unit vectors $\mathbf{e}_i$, $(i = 1, 2, 3)$, $\mathbf{e}_i \cdot \mathbf{e}_j = \delta_{ij}$,

$$\mathbf{A} = A_1 \mathbf{e}_1 + A_2 \mathbf{e}_2 + A_3 \mathbf{e}_3 \tag{2.185}$$

$$\mathbf{B} = B_1 \mathbf{e}_1 + B_2 \mathbf{e}_2 + B_3 \mathbf{e}_3 \tag{2.186}$$

the scalar product is

$$\mathbf{A} \cdot \mathbf{B} = A_1 B_1 + A_2 B_2 + A_3 B_3$$

2.17.3. Cross product (vector product or outer product)

In a three-dimensional space, $\mathbf{C} = \mathbf{A} \times \mathbf{B}$ is a vector perpendicular to the plane determined by **A** and **B** whose absolute value is $C = AB \sin \phi$ and whose direction is such that **A**, **B**, and **C** form a right-handed screw. In an orthogonal coordinate system, the cross product is expressed by the determinant

$$\mathbf{C} = \begin{vmatrix} \mathbf{e}_1 & \mathbf{e}_2 & \mathbf{e}_3 \\ A_1 & A_2 & A_3 \\ B_1 & B_2 & B_3 \end{vmatrix}$$

This can be expressed succinctly using the totally antisymmetric symbol ϵ_{ijk}: $\epsilon_{123} = 1$, and ϵ_{ijk} changes sign if any two indices are exchanged: $\epsilon_{ijk} = -\epsilon_{ikj}$, so that ϵ_{ijk} is zero if any two indices are the same, $+1$ if the indices are an even permutation of 123 and -1 if they are an odd permutation. Then,

$$C_i = \sum_{jk} \epsilon_{ijk} A_j B_k \tag{2.187}$$

2.17.4. Vector identities

$$\mathbf{A} \cdot \mathbf{B} = \mathbf{B} \cdot \mathbf{A}, \qquad \mathbf{A} \times \mathbf{B} = -\mathbf{B} \times \mathbf{A} \tag{2.188}$$

$$[\mathbf{ABC}] \equiv \mathbf{A}\cdot(\mathbf{B} \times \mathbf{C}) = (\mathbf{A} \times \mathbf{B})\cdot C = \begin{vmatrix} A_1 & A_2 & A_3 \\ B_1 & B_2 & B_3 \\ C_1 & C_2 & C_3 \end{vmatrix} \tag{2.189}$$

$$[\mathbf{ABC}] = [\mathbf{BCA}] = [\mathbf{CAB}] = -[\mathbf{CBA}] = -[\mathbf{ACB}] = -[\mathbf{BAC}] \tag{2.190}$$

$$\mathbf{A} \times (\mathbf{B} \times \mathbf{C}) + \mathbf{B} \times (\mathbf{C} \times \mathbf{A}) + \mathbf{C} \times (\mathbf{A} \times \mathbf{B}) = 0 \tag{2.191}$$

$[\mathbf{ABC}]$ is the volume of the parallelepiped defined by the vectors $\mathbf{A}$, $\mathbf{B}$, $\mathbf{C}$:

$$\mathbf{A} \times (\mathbf{B} \times \mathbf{C}) = (\mathbf{C} \times \mathbf{B}) \times \mathbf{A} = \mathbf{B}(\mathbf{A}\cdot\mathbf{C}) - \mathbf{C}(\mathbf{A}\cdot\mathbf{B}) \tag{2.192}$$

$$(\mathbf{A} \times \mathbf{B})\cdot(\mathbf{C} \times \mathbf{D}) = \mathbf{A}\cdot[\mathbf{B} \times (\mathbf{C} \times \mathbf{D})] = (\mathbf{A}\cdot\mathbf{C})(\mathbf{B}\cdot\mathbf{D}) - (\mathbf{A}\cdot\mathbf{D})(\mathbf{B}\cdot\mathbf{C}) \tag{2.193}$$

$$(\mathbf{A} \times \mathbf{B}) \times (\mathbf{C} \times \mathbf{D}) = \mathbf{C}[\mathbf{ABD}] + \mathbf{D}[\mathbf{CBA}] = \mathbf{A}[\mathbf{CBD}] + \mathbf{B}[\mathbf{ACD}] \tag{2.194}$$

2.17.5. Vector linear equations

Given $\mathbf{A}\cdot\mathbf{X} = p$, $\mathbf{B} \times \mathbf{X} = \mathbf{C}$:

$$\mathbf{X} = \frac{p\mathbf{B} + \mathbf{C} \times \mathbf{A}}{\mathbf{A}\cdot\mathbf{B}} \tag{2.195}$$

Given $\mathbf{A}\cdot\mathbf{X} = p$, $\mathbf{B}\cdot\mathbf{X} = q$, $\mathbf{C}\cdot\mathbf{X} = r$ with $\mathbf{A}\cdot\mathbf{B} \times \mathbf{C} = 0$:

$$\mathbf{X} = \frac{p\mathbf{B} \times \mathbf{C} + q\mathbf{C} \times \mathbf{A} + r\mathbf{A} \times \mathbf{B}}{[\mathbf{ABC}]} \tag{2.196}$$

2.17.6. Differential operators

The *gradient* of a scalar function $f(\mathbf{r}) = f(x, y, z)$ gives the absolute value and direction of the maximum change in f. The gradient defines a vector field perpendicular to the surfaces $f = \text{const}$.

$$\operatorname{grad} f = \nabla f = \lim_{V\to 0} \frac{1}{V} \int_S f d\mathbf{S} = \frac{\partial f}{\partial x} i + \frac{\partial f}{\partial y} j + \frac{\partial f}{\partial z} k \tag{2.197}$$

The *divergence* of a vector $\mathbf{A}(\mathbf{r}) = \mathbf{A}(x, y, z)$ is the net outflow of the vector flux through the surface of an infinitesimal volume element containing the point $\mathbf{r}$, divided by the volume:

$$\operatorname{div} \mathbf{A} = \nabla\cdot\mathbf{A} = \lim_{V\to 0} \frac{1}{V} \int_S \mathbf{A}\cdot d\mathbf{S} = \frac{\partial A_1}{\partial x_1}\mathbf{e}_1 + \frac{\partial A_2}{\partial x_2}\mathbf{e}_2 + \frac{\partial A_3}{\partial x_3}\mathbf{e}_3 \tag{2.198}$$

If C is a closed curve in the vector field $\mathbf{A}$, the circulation of $\mathbf{A}$ with respect to C is the contour integral $\oint \mathbf{A}\cdot d\mathbf{s}$. Assigning to the circulation a direction given by the thumb when the fingers of the right hand follow the contour in the positive direction, the *curl* of $\mathbf{A}$ is the vector representing the magnitude and the direction of the maximum circulation per unit area, $\mathbf{n}\oint \mathbf{A}\cdot d\mathbf{s}/S$, of a contour enclosing a small area S encircling the field point.

$$\operatorname{curl} \mathbf{A} = \lim_{S\to 0} \frac{\mathbf{n}}{S} \oint_C \mathbf{A}\cdot d\mathbf{s} \tag{2.199}$$

For a finite area, one then has

$$\oint_C \mathbf{A}\cdot d\mathbf{s} = \int_S \text{curl}\,\mathbf{A}\cdot\mathbf{n}\,dS = \int_S \nabla\times\mathbf{A}\cdot\mathbf{n}\,dS = \int_S \nabla\cdot\mathbf{A}\times\mathbf{n}\,dS \tag{2.200}$$

and

$$\text{curl}\,\mathbf{A} = \nabla\times\mathbf{A} = \lim_{V\to 0}\frac{1}{V}\int_S \mathbf{n}\times\mathbf{A}\,dS = \begin{vmatrix} \mathbf{e}_1 & \mathbf{e}_2 & \mathbf{e}_3 \\ \dfrac{\partial}{\partial x} & \dfrac{\partial}{\partial y} & \dfrac{\partial}{\partial z} \\ A_1 & A_2 & A_3 \end{vmatrix} \tag{2.201}$$

The vector differential operator may be represented by

$$\nabla\cdots = \lim_{V\to 0}\int_S \mathbf{n}\cdots dS \tag{2.202}$$

in which the dots $(\cdots)$ can be replaced by any valid expression: $f, \cdot\mathbf{A}, \times\mathbf{B}$, etc.:

$$\nabla(fg) = \nabla(gf) = f\nabla g + g\nabla f \tag{2.203}$$

$$\nabla\cdot(f\mathbf{A}) = f\nabla\cdot\mathbf{A} + \mathbf{A}\cdot\nabla f \tag{2.204}$$

$$\nabla\times(f\mathbf{A}) = f\nabla\times\mathbf{A} + (\nabla f)\times\mathbf{A} \tag{2.205}$$

$$\nabla\cdot(\mathbf{A}\times\mathbf{B}) = (\nabla\times\mathbf{A})\mathbf{B} - (\nabla\times\mathbf{B})\mathbf{A} \tag{2.206}$$

$$\nabla\times(\mathbf{A}\times\mathbf{B}) = \mathbf{A}(\nabla\cdot\mathbf{B}) + (\mathbf{B}\cdot\nabla)\mathbf{A} - \mathbf{B}(\nabla\cdot\mathbf{A}) - (\mathbf{A}\cdot\nabla)\mathbf{B} \tag{2.207}$$

$$\mathbf{A}\times(\nabla\times\mathbf{B}) = (\nabla\mathbf{B})\cdot\mathbf{A} - (\mathbf{A}\cdot\nabla)\mathbf{B} \tag{2.208}$$

$$\begin{aligned}\nabla(\mathbf{A}\cdot\mathbf{B}) = {} & \mathbf{A}\times(\nabla\times\mathbf{B}) + \mathbf{B}\times(\nabla\times\mathbf{A}) \\ & + (\mathbf{A}\cdot\nabla)\mathbf{B} + (\mathbf{B}\cdot\nabla)\mathbf{A}\end{aligned} \tag{2.209}$$

$$\nabla\cdot\mathbf{AB} = (\nabla\cdot\mathbf{A})\mathbf{B} + (\mathbf{A}\cdot\nabla)\mathbf{B} \tag{2.210}$$

For a tensor $\mathbf{T} = \sum_{i,j=1}^{3} T_{ij}\mathbf{e}_i\mathbf{e}_j$,

$$(\nabla\cdot\mathbf{T})_j = \sum_i \frac{\partial T_{ij}}{\partial x_i} \tag{2.211}$$

$$\nabla\cdot(g\mathbf{T}) = (\nabla g)\cdot\mathbf{T} + g\nabla\cdot\mathbf{T} \tag{2.212}$$

$$\nabla^2 f = \nabla\cdot\nabla f \tag{2.213}$$

$$\nabla^2\mathbf{A} = \nabla(\nabla\cdot\mathbf{A}) - \nabla\times(\nabla\times\mathbf{A}) \tag{2.214}$$

$$\nabla\times\nabla f = 0 \tag{2.215}$$

$$\nabla\cdot(\nabla\times\mathbf{A}) = 0 \tag{2.216}$$

If $\mathbf{r}$ is a position vector with magnitude $|\mathbf{r}| = r$,

$$\begin{aligned} \nabla \cdot \mathbf{r} &= 3, \qquad & \nabla \times \mathbf{r} &= 0 \\ \nabla r &= \frac{\mathbf{r}}{r}, & \nabla \frac{1}{r} &= -\frac{\mathbf{r}}{r^3} \\ \nabla^2 \frac{1}{r} &= -\nabla \cdot \left(\frac{\mathbf{r}}{r^3}\right) = 4\pi \delta(\mathbf{r}) \end{aligned} \tag{2.217}$$

If V is a volume enclosed by a surface S and $d\mathbf{S} = \mathbf{n} dS$, where $\mathbf{n}$ is the unit normal pointing outward from the volume V,

$$\int_V \nabla \cdots dV = \int_S \mathbf{n} \cdots dS \tag{2.218}$$

$$\int_V \nabla f \, dV = \int_S \mathbf{n} f \, dS \tag{2.219}$$

$$\int_V \nabla \cdot \mathbf{A} \, dV = \int_S \mathbf{n} \cdot \mathbf{A} \, dS \tag{2.220}$$

$$\int_V \nabla \cdot \mathsf{T} \, dV = \int_S \mathbf{n} \cdot \mathsf{T} \, dS \tag{2.221}$$

$$\int_V \nabla \times \mathbf{A} \, dV = \int_S \mathbf{n} \times \mathbf{A} \, dS \tag{2.222}$$

$$\int_V (f \nabla^2 g - g \nabla^2 f) \, dV = \int_S \mathbf{n} \cdot (f \nabla g - g \nabla f) \, dS \tag{2.223}$$

$$= \int_V [\mathbf{A} \cdot (\nabla \times (\nabla \times \mathbf{B})) - \mathbf{B} \cdot (\nabla \times (\nabla \times \mathbf{A}))] \, dV \tag{2.224}$$

$$= \int_S \mathbf{n} \cdot [\mathbf{B} \times (\nabla \times \mathbf{A}) - \mathbf{A} \times (\nabla \times \mathbf{B})] \, dS \tag{2.225}$$

If S is an open surface bounded by the contour C, of which the line element is $d\mathbf{s}$,

$$\int_S \mathbf{n} \times \nabla g \, dS = \oint_C g d\mathbf{s} \tag{2.226}$$

$$\int_S \mathbf{n} \cdot \nabla \times \mathbf{A} \, dS = \oint_C \mathbf{A} \cdot d\mathbf{s} \tag{2.227}$$

$$\int_S (\mathbf{n} \times \nabla) \times \mathbf{A} \, dS = -\oint_C \mathbf{A} \times d\mathbf{s} \tag{2.228}$$

$$\int_S \mathbf{n} \cdot (\nabla f \times \nabla g) dS = \oint_C f \nabla g \cdot d\mathbf{s} = -\oint_C g \nabla f \cdot d\mathbf{s} \tag{2.229}$$

2.18. ORTHOGONAL COORDINATE SYSTEMS

In a general orthogonal coordinate system, the three unit vectors are functions of position; therefore, when taking derivatives of vectors in such systems, the derivatives of the unit vectors as well as the derivatives of the components must be considered. The element of distance can be written as $d\mathbf{s} = h_1 dx_1 \mathbf{e}_1 + h_2 dx_2 \mathbf{e}_2 + h_3 dx_3 \mathbf{e}_3$.

rectangular: $x_1 = x, x_2 = y, x_3 = z, \quad h_1 = h_2 = h_3 = 1$

cylindrical: $x_1 = \rho, x_2 = \phi, x_3 = z, \quad h_1 = 1, h_2 = \rho, h_3 = 1$

spherical: $x_1 = r, x_2 = \theta, x_3 = \phi, \quad h_1 = 1, h_2 = r, h_3 = r \sin\theta$

2.18.1. Gradient of f, ∇f

$$\nabla f = \frac{\mathbf{e}_1}{h_1}\frac{\partial f}{\partial x_1} + \frac{\mathbf{e}_2}{h_2}\frac{\partial f}{\partial x_2} + \frac{\mathbf{e}_3}{h_3}\frac{\partial f}{\partial x_3}$$

rectangular: $$= \mathbf{e}_1\frac{\partial f}{\partial x} + \mathbf{e}_2\frac{\partial f}{\partial y} + \mathbf{e}_3\frac{\partial f}{\partial z} \tag{2.230}$$

cylindrical: $$= \mathbf{e}_1\frac{\partial f}{\partial \rho} + \frac{\mathbf{e}_2}{\rho}\frac{\partial f}{\partial \phi} + \mathbf{e}_3\frac{\partial f}{\partial z} \tag{2.231}$$

spherical: $$= \mathbf{e}_1\frac{\partial f}{\partial r} + \frac{\mathbf{e}_2}{r}\frac{\partial f}{\partial \theta} + \frac{\mathbf{e}_3}{r\sin\theta}\frac{\partial f}{\partial \phi} \tag{2.232}$$

2.18.2. Divergence of A, $\nabla \cdot \mathbf{A}$

$$\nabla\cdot\mathbf{A} = \sum_j \frac{1}{h_1h_2h_3}\left[\frac{\partial}{\partial x_1}(A_1h_2h_3) + \frac{\partial}{\partial x_2}(h_1A_2h_3) + \frac{\partial}{\partial x_3}(h_1h_2A_3)\right]$$

rectangular: $$= \frac{\partial A_1}{\partial x} + \frac{\partial A_2}{\partial y} + \frac{\partial A_3}{\partial z} \tag{2.233}$$

cylindrical: $$= \frac{1}{\rho}\frac{\partial(\rho A_1)}{\partial \rho} + \frac{1}{\rho}\frac{\partial A_2}{\partial \phi} + \frac{\partial A_3}{\partial z} \tag{2.234}$$

spherical: $$= \frac{1}{r^2}\frac{\partial}{\partial r}(r^2A_1) + \frac{1}{r\sin\theta}\left[\frac{\partial}{\partial\theta}(A_2\sin\theta) + \frac{\partial A_3}{\partial\phi}\right] \tag{2.235}$$

2.18.3. Curl of A, $\nabla \times \mathbf{A}$

$$\nabla\times\mathbf{A} = \frac{1}{h_1h_2h_3}\begin{vmatrix} h_1\mathbf{e}_1 & h_2\mathbf{e}_2 & h_3\mathbf{e}_3 \\ \dfrac{\partial}{\partial x} & \dfrac{\partial}{\partial y} & \dfrac{\partial}{\partial z} \\ h_1A_1 & h_2A_2 & h_3A_3 \end{vmatrix} \tag{2.236}$$

$$= \frac{\mathbf{e}_1}{h_2 h_3}\left[\frac{\partial h_3 A_3}{\partial x_2} - \frac{\partial h_2 A_2}{\partial x_3}\right] + \frac{\mathbf{e}_2}{h_3 h_1}\left[\frac{\partial h_1 A_1}{\partial x_3} - \frac{\partial h_3 A_3}{\partial x_1}\right] + \frac{\mathbf{e}_3}{h_1 h_2}\left[\frac{\partial h_2 A_2}{\partial x_3} - \frac{\partial h_1 A_1}{\partial x_2}\right] \tag{2.237}$$

$$\textit{rectangular}: = \mathbf{e}_1\left[\frac{\partial A_3}{\partial y} - \frac{\partial A_2}{\partial z}\right] + \mathbf{e}_2\left[\frac{\partial A_1}{\partial z} - \frac{\partial A_3}{\partial x}\right] + \mathbf{e}_3\left[\frac{\partial A_2}{\partial x} - \frac{\partial A_1}{\partial y}\right] \tag{2.238}$$

$$\textit{cylindrical}: = \mathbf{e}_1\left[\frac{1}{\rho}\frac{\partial A_3}{\partial \phi} - \frac{\partial A_2}{\partial z}\right] + \mathbf{e}_2\left[\frac{\partial A_1}{\partial z} - \frac{\partial A_3}{\partial \rho}\right] + \frac{\mathbf{e}_3}{\rho}\left[\frac{\partial(\rho A_2)}{\partial \rho} - \frac{\partial A_1}{\partial \phi}\right] \tag{2.239}$$

$$\textit{spherical}: = \frac{\mathbf{e}_1}{r\sin\theta}\left[\frac{\partial(A_3\sin\theta)}{\partial\theta} - \frac{\partial A_2}{\partial\phi}\right] + \frac{\mathbf{e}_2}{r\sin\theta}\left[\frac{\partial A_1}{\partial\phi} - \sin\theta\frac{\partial(rA_3)}{\partial r}\right] + \frac{\mathbf{e}_3}{r}\left[\frac{\partial(rA_2)}{\partial r} - \frac{\partial A_1}{\partial\theta}\right] \tag{2.240}$$

2.18.4. Projected derivative, components of A·∇B

$$\mathbf{A}\cdot\nabla\mathbf{B} = \sum_j \mathbf{e}_j\left[\mathbf{A}\cdot\nabla B_j + \sum_i\left(A_j\frac{\partial h_j}{\partial x_i} - A_i\frac{\partial h_i}{\partial x_j}\right)\frac{B_i}{h_i h_j}\right] \tag{2.241}$$

$$\textit{rectangular}: = \mathbf{e}_1\left[A_1\frac{\partial B_1}{\partial x} + A_2\frac{\partial B_1}{\partial y} + A_3\frac{\partial B_1}{\partial z}\right] + \mathbf{e}_2\left[A_1\frac{\partial B_2}{\partial x} + A_2\frac{\partial B_2}{\partial y} + A_3\frac{\partial B_2}{\partial z}\right] + \mathbf{e}_3\left[A_1\frac{\partial B_3}{\partial x} + A_2\frac{\partial B_3}{\partial y} + A_3\frac{\partial B_3}{\partial z}\right] \tag{2.242}$$

$$\textit{cylindrical}: = \mathbf{e}_1\left[A_1\frac{\partial B_1}{\partial\rho} + \frac{A_2}{\rho}\frac{\partial B_1}{\partial\phi} + A_3\frac{\partial B_1}{\partial z} - \frac{A_2 B_2}{\rho}\right] + \mathbf{e}_2\left[A_1\frac{\partial B_2}{\partial\rho} + \frac{A_2}{\rho}\frac{\partial B_2}{\partial\phi} + A_3\frac{\partial B_2}{\partial z} + \frac{A_2 B_1}{\rho}\right] + \mathbf{e}_3\left[A_1\frac{\partial B_3}{\partial\rho} + \frac{A_2}{\rho}\frac{\partial B_2}{\partial\phi} + A_3\frac{\partial B_3}{\partial z}\right] \tag{2.243}$$

$$\textit{spherical:} = \mathbf{e}_1 \left[A_1 \frac{\partial B_1}{\partial r} + \frac{A_2}{r} \frac{\partial B_1}{\partial \theta} + \frac{A_3}{r \sin\theta} \frac{\partial B_1}{\partial \phi} - \frac{A_2 B_2 + A_3 B_3}{r} \right]$$

$$+ \mathbf{e}_2 \left[A_1 \frac{\partial B_2}{\partial r} + \frac{A_2}{r} \frac{\partial B_2}{\partial \theta} + \frac{A_3}{r \sin\theta} \frac{\partial B_2}{\partial \phi} + \frac{A_2 B_1 - A_3 B_3 \cot\theta}{r} \right]$$

$$+ \mathbf{e}_3 \left[A_1 \frac{\partial B_3}{\partial r} + \frac{A_2}{r} \frac{\partial B_3}{\partial \theta} + \frac{A_3}{r \sin\theta} \frac{\partial B_3}{\partial \phi} + \frac{A_3}{r} (B_1 + B_2 \cot\theta) \right] \quad (2.244)$$

2.18.5. Divergence of a tensor $\nabla \mathbf{T}$

In an orthogonal coordinate system, a tensor $\mathbf{T}$ can be written in dyadic form $\mathbf{T} = \sum_{i,j} \mathbf{e}_i T_{ij} \mathbf{e}_j$.

$$\nabla \cdot \mathbf{T} = \sum_j \mathbf{e}_j \left[\nabla \cdot T_{\cdot j} + \sum_j \frac{1}{h_i h_j} \left(T_{ji} \frac{\partial h_j}{\partial x_i} - T_{ii} \frac{\partial h_i}{\partial x_j} \right) \right]$$

$$\textit{rectangular:} = \mathbf{e}_1 \left[\frac{\partial T_{11}}{\partial x} + \frac{\partial T_{21}}{\partial y} + \frac{\partial T_{31}}{\partial z} \right]$$

$$+ \mathbf{e}_2 \left[\frac{\partial T_{12}}{\partial x} + \frac{\partial T_{22}}{\partial y} + \frac{\partial T_{32}}{\partial z} \right]$$

$$+ \mathbf{e}_3 \left[\frac{\partial T_{13}}{\partial x} + \frac{\partial T_{23}}{\partial y} + \frac{\partial T_{33}}{\partial z} \right] \quad (2.245)$$

$$\textit{cylindrical:} = \mathbf{e}_1 \left[\frac{\partial T_{11}}{\partial \rho} + + \frac{1}{\rho} \frac{\partial T_{21}}{\partial \phi} + \frac{\partial T_{31}}{\partial z} + \frac{T_{11} - T_{22}}{\rho} \right]$$

$$+ \mathbf{e}_2 \left[\frac{\partial T_{12}}{\partial \rho} + \frac{1}{\rho} \frac{\partial T_{22}}{\partial \phi} + \frac{\partial T_{32}}{\partial z} + \frac{T_{12} + T_{21}}{\rho} \right]$$

$$+ \mathbf{e}_3 \left[\frac{\partial T_{13}}{\partial \rho} + \frac{T_{13}}{\rho} + \frac{1}{\rho} \frac{\partial T_{23}}{\partial \phi} + \frac{\partial T_{33}}{\partial z} \right] \quad (2.246)$$

$$\textit{spherical:} = \mathbf{e}_1 \left[\frac{\partial T_{11}}{\partial r} + \frac{1}{r \sin\theta} \left(\frac{\partial}{\partial \theta} (\sin\theta \, T_{21}) + \frac{\partial T_{31}}{\partial \phi} \right) \right.$$

$$\left. + \frac{2T_{11} - T_{22} - T_{33}}{r} \right]$$

$$+ \mathbf{e}_2 \left[\frac{\partial T_{12}}{\partial r} + \frac{1}{r \sin\theta} \left(\frac{\partial}{\partial \theta} (\sin\theta \, T_{22}) + \frac{\partial T_{32}}{\partial \phi} \right) \right.$$

$$\left. + \frac{2T_{12} + T_{21} - T_{33} \cot\theta}{r} \right]$$

$$+\,\mathbf{e}_3\left[\frac{\partial T_{13}}{\partial r}+\frac{1}{r\sin\theta}\left(\frac{\partial}{\partial\theta}(\sin\theta T_{23})+\frac{\partial T_{33}}{\partial\phi}\right)+\frac{2T_{13}+T_{31}+T_{32}\cot\theta}{r}\right] \tag{2.247}$$

2.18.6. Scalar Laplacian $\nabla^2 f$

$$\nabla^2 f = \operatorname{div}\operatorname{grad} f$$

$$= \frac{1}{h_1h_2h_3}\left[\frac{\partial}{\partial x_1}\left(\frac{h_2h_3}{h_1}\frac{\partial f}{\partial x_1}\right)+\frac{\partial}{\partial x_2}\left(\frac{h_3h_1}{h_2}\frac{\partial f}{\partial x_2}\right)+\frac{\partial}{\partial x_3}\left(\frac{h_1h_2}{h_3}\frac{\partial f}{\partial x_3}\right)\right] \tag{2.248}$$

$$\textit{rectangular:} = \frac{\partial^2 f}{\partial x^2}+\frac{\partial^2 f}{\partial y^2}+\frac{\partial^2 f}{\partial z^2} \tag{2.249}$$

$$\textit{cylindrical:} = \frac{1}{\rho}\frac{\partial}{\partial\rho}\left(\rho\frac{\partial f}{\partial\rho}\right)+\frac{1}{\rho^2}\frac{\partial^2 f}{\partial\phi^2}+\frac{\partial^2 f}{\partial z^2} \tag{2.250}$$

$$\textit{spherical:} = \frac{1}{r^2}\frac{\partial}{\partial r}\left(r^2\frac{\partial f}{\partial r}\right)+\frac{1}{r^2\sin^2\theta}\left[\sin\theta\frac{\partial}{\partial\theta}\left(\sin\theta\frac{\partial f}{\partial\theta}\right)+\frac{\partial^2 f}{\partial\phi^2}\right] \tag{2.251}$$

2.18.7. Vector Laplacian $\nabla^2\mathbf{A}$

$$\nabla^2\mathbf{A} = \nabla(\nabla\cdot\mathbf{A}) - \nabla\times(\nabla\times\mathbf{A})$$

$$\textit{rectangular:} = \mathbf{e}_1\nabla^2 A_1+\mathbf{e}_2\nabla^2 A_2+\mathbf{e}_3\nabla^2 A_3$$

$$= \mathbf{e}_1\left[\frac{\partial^2 A_1}{\partial x^2}+\frac{\partial^2 A_1}{\partial y^2}+\frac{\partial^2 A_1}{\partial z^2}\right] + \mathbf{e}_2\left[\frac{\partial^2 A_2}{\partial x^2}+\frac{\partial^2 A_2}{\partial y^2}+\frac{\partial A_2}{\partial z^2}\right]+\mathbf{e}_3\left[\frac{\partial^2 A_3}{\partial x^2}+\frac{\partial^2 A_3}{\partial y^2}+\frac{\partial^2 A_3}{\partial z^2}\right] \tag{2.252}$$

$$\textit{cylindrical:} = \mathbf{e}_1\left[\nabla^2 A_1-\frac{1}{\rho^2}\left(2\frac{\partial A_2}{\partial\phi}+A_1\right)\right] + \mathbf{e}_2\left[\nabla^2 A_2+\frac{1}{\rho^2}\left(2\frac{\partial A_1}{\partial\phi}-A_2\right)\right]+\mathbf{e}_3\left[\nabla^2 A_3\right] \tag{2.253}$$

$$\textit{spherical:} = \mathbf{e}_1\left[\nabla^2 A_1-\frac{2}{r^2}\left(A_1n+\frac{\partial A_2}{\partial\theta}+\cot\theta A_2+\csc\theta\frac{\partial A_3}{\partial\phi}\right)\right] + \mathbf{e}_2\left[\nabla^2 A_2+\frac{2}{r^2}\frac{\partial A_1}{\partial\theta}-\frac{1}{r^2\sin^2\theta}\left(2\cos\theta\frac{\partial A_3}{\partial\phi}+A_2\right)\right] + \mathbf{e}_3\left[\nabla^2 A_3+\frac{1}{r^2\sin^2\theta}\left(2\sin\theta\frac{\partial A_1}{\partial\phi}+2\cos\theta\frac{\partial A_2}{\partial\phi}-A_3\right)\right] \tag{2.254}$$

2.19. FOURIER SERIES AND FOURIER TRANSFORMS

A function $f(x)$ in the interval $-a/2 \leq x \leq a/2$ may be expanded in a trigonometric series

$$f(x) = \frac{1}{2}A_0 + \sum_{m=1}^{\infty}\left[A_m \cos\frac{2\pi mx}{a} + B_m \sin\frac{2\pi mx}{a}\right],$$

where

$$A_m = \frac{2}{a}\int_{-a/2}^{a/2} f(x)\cos\frac{2\pi mx}{a}\,dx, \tag{2.255}$$

$$B_m = \frac{2}{a}\int_{-a/2}^{a/2} f(x)\sin\frac{2\pi mx}{a}\,dx. \tag{2.256}$$

The functions

$$\sqrt{\frac{2}{a}}\sin\frac{2\pi mx}{a} \quad \text{and} \quad \sqrt{\frac{2}{a}}\cos\frac{2\pi mx}{a}$$

form an orthogonal set.

When the interval becomes infinite, the expansion becomes the Fourier integral:

$$f(x) = \frac{1}{\sqrt{2\pi}}\int_{-\infty}^{\infty} F(k)e^{-ikx}\,dk,$$

where

$$F(k) = \mathcal{F}[f(x); k] = \frac{1}{\sqrt{2\pi}}\int_{-\infty}^{\infty} f(x)e^{ikx}\,dx$$

is the Fourier transform of $f(x)$.

If $F(k)$ is the Fourier transform of $f(x)$, then $f^*(k)$, the complex conjugate of $f(k)$, is the Fourier transform of $F^*(x)$.

For arbitrary fixed constants a and b,

$$\mathcal{F}[f(ax); k] = aF(ak), \qquad \mathcal{F}[af(x) + bg(x); k] = aF(k) + bG(k),$$

$$\text{If } \lim_{|x|\to\infty} f'(x) = 0, \qquad \mathcal{F}[f'(x); k] = ikF(k),$$

$$\mathcal{F}[f(x+s); k] = e^{-iks}F(k). \tag{2.257}$$

2.19.1. Orthogonality condition

$$\frac{1}{2\pi}\int_{-\infty}^{\infty} e^{i(k-k')x}\,dx = \delta(k-k').$$

2.19.2. Completeness

$$\frac{1}{2\pi}\int_{-\infty}^{\infty} e^{ik(x-x')}dk = \delta(x-x').$$

2.19.3. Convolution

The convolution of two functions $f(x)$ and $g(x)$ is defined by

$$(f \circ g)(x) \equiv \int_{-\infty}^{\infty} f(y)g(x-y)\,dy,$$

with the properties

$$(f \circ g)(x) = (g \circ f)(x), \qquad ((f \circ g) \circ h)(x) = (f \circ (g \circ h))(x),$$

$$\mathcal{F}[(f \circ g)(x); k] = \sqrt{2\pi}\mathcal{F}[f(x); K]\mathcal{F}[g(x); K] = \sqrt{2\pi}F(x)G(x),$$

$$\mathcal{F}[f(x)g(x); K] = \frac{1}{\sqrt{2\pi}}\mathcal{F}[(F \circ G)(K); x].$$

2.20. LAPLACE TRANSFORMS

If $f(t)$ is a function of the real variable t, $t > 0$; the Laplace transform of $f(t)$ is

$$\mathcal{L}[f(t); s] = F(s) = \int_0^{\infty} f(t)e^{-st}\,dt, \tag{2.258}$$

where s is a complex variable. If the integral is convergent for some real value of s, $s = s_0$, then it converges for all s with $\Re(s) > s_0$, and $F(s)$ is a single-valued analytic function of s in the half-plane $\Re(s) > s_0$.

2.20.1. General properties

Inversion

$$f(t) = \frac{1}{2\pi i}\int_{c-i\infty}^{c+i\infty} e^{ts}F(s)\,ds \tag{2.259}$$

2.20.2. Linearity

$$\mathcal{L}[af(t) + bg(t); s] = aF(s) + bG(s) \tag{2.260}$$

2.20.3. Differentiation

$$\mathcal{L}[f'(t); s] = sF(s) - f(+0) \tag{2.261}$$

$$\mathcal{L}[f^{(n)}(t);s] = s^n F(s) - s^{n-1} f(+0) - s^{n-2} f'(+0) - \cdots - f^{(n-1)}(+0) \quad (2.262)$$

$$n \text{ integer}, \quad n > 0 \quad (2.263)$$

$$\mathcal{L}[t^n f(t);s] = (-)^n F^{(n)}(s) \quad (2.264)$$

Integration

$$\mathcal{L}\left[\int_0^t f(t')\,dt';s\right] = \frac{1}{s}F(s) \quad (2.265)$$

$$\mathcal{L}\left[\frac{1}{t}f(t);s\right] = \int_s^\infty F(x)\,dx \quad (2.266)$$

Convolution

$$\mathcal{L}[(f\circ g)(t);s] = \mathcal{L}\left[\int_0^\infty f(t-t')g(t')\,dt'\right] = F(s)G(s) \quad (2.267)$$

Translation and scaling

$$\mathcal{L}[e^{at}f(t);s] = F(s-a) \quad (2.268)$$

$$\mathcal{L}[f(t/c);s] = cF(cs), \qquad c > 0 \quad (2.269)$$

$$\mathcal{L}[e^{at}f(bt);s] = F\left((s-a)/b\right), \qquad b > 0 \quad (2.270)$$

$$\mathcal{L}[f(t-b)U(t-b);s] = e^{-bs}F(s), \qquad b > 0 \quad (2.271)$$

$$\text{where} \quad U(t) = \begin{cases} 0, & t < 0 \\ \frac{1}{2}, & t = 0 \\ 1, & t > 0 \end{cases} \quad (2.272)$$

Periodic functions

If $f(t)$ is periodic with period a, $f(t+a) = f(t)$,

$$\mathcal{L}[f(t);s] = (1-e^{-sa})^{-1}\int_0^a e^{-st}f(t)\,dt. \quad (2.273)$$

2.20.4. Bessel functions

Differential equation

The argument z and the order ν are complex; x is real and nonnegative; n is a positive integer or 0.

$$z^2\frac{d^2w}{dz^2} + z\frac{dw}{dz} + (z^2-\nu^2)w = z\frac{d}{dz}\left(z\frac{dw}{dz}\right) + (z^2-\nu^2)w = 0.$$

w is regular throughout the z-plane cut along the negative real axis with a branch point at the origin. Solutions of this equation are the Bessel functions of the first kind $J_\nu(z)$, the Weber functions (Bessel functions of the second kind) $Y_\nu(z)$, and the Hankel functions (Bessel functions of the third kind) $H_\nu^{(1)}(z)$, and $H_\nu^{(2)}(z)$. $J_\nu(z)$ and $J_{-\nu}(z)$ are linearly independent solutions when ν is not an integer; for $\nu = \pm n$, $J_{-n}(z) = (-)^n J_n(z)$, there is no branch point at $z = 0$ and $J_n(z)$ is an entire function of z.

$$Y_\nu(z) = N_\nu(z) = \frac{J_\nu(z)\cos(\nu z) - J_\nu(z)}{\sin(\nu z)} \tag{2.274}$$

$$Y_n(z) = \lim_{\epsilon\to 0} Y_{n+\epsilon}(z) \tag{2.275}$$

$$H_\nu^{(1)}(z) = J_\nu(z) + iY_\nu(z) = i\frac{e^{-i\pi\nu}J_\nu(z) - J_{-\nu}(z)}{\sin(\nu z)} \tag{2.276}$$

$$H_\nu^{(2)}(z) = J_\nu(z) - iY_\nu(z) = i\frac{J_{-\nu}(z) - e^{i\pi\nu}J_\nu(z)}{\sin(\nu z)} \tag{2.277}$$

$$H_{-\nu}^{(1)}(z) = e^{i\pi\nu}H_\nu^{(1)}(z) \qquad H_{-\nu}^{(2)}(z) = e^{-i\pi\nu}H_\nu^{(2)}(z) \tag{2.278}$$

The modified Bessel functions $I_{\pm\nu}(z)$ and $K_\nu(z)$ are solutions of the differential equation

$$z^2\frac{d^2v}{dz^2} + z\frac{dv}{dz} - (z^2 + \nu^2)v = z\frac{d}{dz}\left(z\frac{dv}{dz}\right) - (z^2 + \nu^2)v = 0$$

$$K_\nu(z) = \frac{\pi}{2}\frac{I_{-\nu}(z) - I_\nu(z)}{\sin(\nu\pi)} \tag{2.279}$$

$$I_{-n}(z) = I_n(z), \qquad K_{-\nu}(z) = K_\nu(z) \tag{2.280}$$

$$I_\nu(z) = \begin{cases} e^{-i\nu\pi/2}J_\nu(e^{i\pi/2}z), & -\pi < \arg z \le \pi/2 \\ e^{3i\nu\pi/2}J_\nu(e^{-3i\pi/2}z), & \pi/2 < \arg z \le \pi \end{cases} \tag{2.281}$$

$$K_\nu(z) = \begin{cases} \frac{i\pi}{2}e^{i\nu\pi/2}H_\nu^{(1)}(e^{i\pi/2}z), & -\pi < \arg z \le \pi/2 \\ -\frac{i\pi}{2}e^{-i\nu\pi/2}H_\nu^{(1)}(e^{-i\pi/2}z), & \pi/2 < \arg z \le \pi \end{cases} \tag{2.282}$$

For $z \to 0$,

$$J_\nu(z) \sim (\tfrac{1}{2}z)^\nu/\Gamma(\nu+1)$$

$$Y_\nu(z) \sim -iH_\nu^{(1)}(z) \sim iH_\nu^{(2)} \sim \Gamma(\nu)(\tfrac{1}{2}z)^{-\nu}/\pi, \quad \text{for} \quad \Re\,\nu > 0$$

$$Y_0(z) \sim -iH_0^{(1)}(z) \sim iH_0^{(2)} \sim (2/\pi)\ln z$$

For $x \to \infty$

$$J_\nu(x) \sim \sqrt{2/\pi x}\,\cos[x - (2\nu+1)\pi/4]$$

$$Y_\nu(x) \sim \sqrt{2/\pi x}\,\sin[x - (2\nu+1)\pi/4]$$

$$H_\nu^{(1)}(x) \sim \sqrt{2/\pi x}\ \exp i[x - (2\nu + 1)/4]$$

$$H_\nu^{(2)}(x) \sim \sqrt{2/\pi x}\ \exp -i[x - (2\nu + 1)/4]$$

Series expansion

$$J_\nu(z) = \sum_{k=0}^{\infty} (-)^k \frac{\left(\frac{1}{2}z\right)^{2k+\nu}}{\Gamma(\nu + k + 1)k!} \tag{2.283}$$

$$I_\nu(z) = \sum_{k=0}^{\infty} \frac{\left(\frac{1}{2}z\right)^{2k+\nu}}{\Gamma(\nu + k + 1)k!} \tag{2.284}$$

$$Y_n(z) = -\frac{1}{\pi}\left(\frac{1}{2}z\right)^{-n} \sum_{k=0}^{n-1} \frac{(n-k-1)!}{k!}\left(\frac{1}{2}z\right)^{2k} + \frac{2}{\pi}\ln\left(\frac{1}{2}z\right) J_n(z)$$

$$- \frac{2}{\pi} \sum_{k=1}^{\infty} (-)^k \left[\psi(k+1) + \frac{1}{2}\sum_{p=1}^{n} \frac{1}{k+p}\right] \frac{\left(\frac{1}{2}z\right)^{2k+n}}{(n+k)!k!} \tag{2.285}$$

with $\psi(1) = -\gamma$, and $\psi(k+1) = \sum_{p=1}^{k} \frac{1}{p} - \gamma$ for $k > 0$.

$$J_0(z) = 1 - (\tfrac{1}{2}z)^2 + \frac{(\frac{1}{2}z)^4}{2!^2} - \frac{(\frac{1}{2}z)^6}{3!^2} + \frac{(\frac{1}{2}z)^8}{4!^2} - \cdots$$

$$Y_0(z) = \frac{2}{\pi}\left[\ln(\tfrac{1}{2}z) + \gamma\right] J_0(z) + (\tfrac{1}{2}z)^2 - (1 + \tfrac{1}{2})\frac{(\frac{1}{2}z)^4}{2!^2}$$

$$+ (1 + \tfrac{1}{2} + \tfrac{1}{3})\frac{(\frac{1}{2}z)^6}{3!^2} + (1 + \tfrac{1}{2} + \tfrac{1}{3} + \tfrac{1}{4})\frac{(\frac{1}{2}z)^8}{4!^2} - \cdots \tag{2.286}$$

Integral representations

$$J_0(z) = \frac{2}{\pi}\int_0^{\pi/2} \cos(z\sin\theta)\, d\theta \tag{2.287}$$

$$Y_0(z) = \frac{4}{\pi^2}\int_0^{\pi/2} \cos(z\cos\theta)[\gamma + \ln(2z\sin^2\theta)]\, d\theta \tag{2.288}$$

$$I_0(z) = \frac{2}{\pi}\int_0^{\pi/2} \cosh(z\sin\theta)\, d\theta \tag{2.289}$$

$$K_0(z) = -\frac{2}{\pi}\int_0^{\pi/2} \cosh(z\cos\theta)[\gamma + \ln(2z\sin^2\theta)]\, d\theta \tag{2.290}$$

$$J_n(z) = \frac{z^n}{\pi(2n-1)!!}\int_0^{\pi/2}\cos(z\sin\theta)\cos^{2n}\theta\,d\theta \tag{2.291}$$

$$= \frac{z^n}{\pi(2n-1)!!}\int_0^{1}(1-t^2)^{n-1/2}\cos(zt)\,d\theta \tag{2.292}$$

$$= \frac{1}{\pi}\int_0^{\pi}\cos(z\sin\theta - n\theta)\,d\theta \qquad (|\arg z| < \tfrac{1}{2}\pi) \tag{2.293}$$

$$Y_n(z) = \frac{1}{\pi}\int_0^{\pi}\cos(z\sin\theta - n\theta)\,d\theta \qquad (|\arg z| < \tfrac{1}{2}\pi) \tag{2.294}$$

Recursion relations

Let $C_\nu(z)$ be any linear combination of J_ν, Y_ν, $H_\nu^{(1)}$, and $H_\nu^{(2)}$ with coefficients independent of z and ν.

$$C_{\nu-1}(z) + C_{\nu+1}(z) = \frac{2\nu}{z}C_\nu(z) \tag{2.295}$$

$$C_{\nu-1}(z) - C_{\nu+1}(z) = \frac{2\nu}{z}C'_\nu(z) \tag{2.296}$$

$$C_{\nu-1}(z) - \frac{\nu}{z}C_\nu(z) = C'_\nu(z) \tag{2.297}$$

$$\frac{\nu}{z}C_\nu(z) - C_{\nu+1}(z) = C'_\nu(z) \tag{2.298}$$

$$C'_0(z) = -C_1(z), \qquad J'_0(z) = -J_1(z) \tag{2.299}$$

Generating functions and series

$$e^{(t-1/t)z/2} = \sum_{k=-\infty}^{\infty} t^k J_k(z), \qquad t = 0 \tag{2.300}$$

$$\cos(z\sin\theta) = J_0(z) + 2\sum_{k=1}^{\infty} J_{2k}(z)\cos(2k\theta) \tag{2.301}$$

$$\sin(z\sin\theta) = 2\sum_{k=0}^{\infty} J_{2k+1}(z)\sin[(2k+1)\theta] \tag{2.302}$$

$$e^{z\cos\theta} = I_0(z) + 2\sum_{k=1}^{\infty} I_k(z)\cos(k\theta) \tag{2.303}$$

$$e^{z\sin\theta} = I_0(z) + 2\sum_{k=1}^{\infty}(-)^k\left[I_{2k}(z)\cos(2k\theta) - I_{2k-1}(z)\sin[(2k-1)\theta]\right] \tag{2.304}$$

Addition theorems

$$J_n(u \pm v) = \sum_{k=-\infty}^{\infty}(\pm 1)^k J_{n-k}(u)J_k(v) \tag{2.305}$$

For $w\cos\phi = u - v\cos\alpha$, $w\sin\phi = v\sin\alpha$, and $w^2 = u^2 - 2uv\cos\theta + v^2$,

$$J_n(w)e^{in\phi} = \sum_{k=-\infty}^{\infty} J_{n+k}(u)J_k(v)e^{ik\alpha}. \tag{2.306}$$

2.20.5. Spherical harmonics

Legendre polynomials $P_n(x)$

Definition:

$$\sum_{n=0}^{\infty} P_n(x)t^n = \frac{1}{(1-2xt+t^2)^{1/2}} = \frac{1}{2^n\, n!}\frac{d^n}{dx^n}(x^2-1)^n \tag{2.307}$$

$$P_n(x) = \frac{1}{n}[(2n-1)P_{n-1}(x) - (n-1)P_{n-2}(x)] \tag{2.308}$$

$$(1-x^2)P_n'' - 2xP_n' + n(n+1)P_n = 0 \tag{2.309}$$

$$P_0(x) = 1, \qquad P_3(x) = \tfrac{1}{2}(5x^3 - 3x)$$

$$P_1(x) = x, \qquad P_4(x) = \tfrac{1}{8}(35x^4 - 30x^2 + 3)$$

$$P_2(x) = \tfrac{1}{2}(3x^2 - 1), \qquad P_5(x) = \tfrac{1}{8}(63x^5 - 70x^3 + 15x)$$

$$P_n(1) = 1, \qquad P_n(-1) = (-1)^n$$

2.20.6. Potential at $\mathbf{r}_1$ due to a unit source at $\mathbf{r}_2$

$$\frac{1}{|\mathbf{r}_1 - \mathbf{r}_2|} = \frac{1}{r_>}\sum_{n=0}^{\infty}\left(\frac{r_<}{r_>}\right)^n P_n\left(\frac{\mathbf{r}_1\cdot\mathbf{r}_2}{r_1 r_2}\right),$$

where $r_<$ is the smaller and $r_>$ is the larger of the two distances r_1 and r_2.

Tesseral harmonics

Definition:

$$Y_{\ell,m} = N_{\ell,m}e^{im\phi}\sin^{|m|}\theta\,\frac{d^{|m|}P_\ell(\cos\theta)}{d(\cos\theta)^{|m|}},$$

where, for $m \geq 0$,

$$N_{\ell,m} = (-1)^m\sqrt{\frac{(2\ell+1)}{4\pi}\frac{(\ell-m)!}{(\ell+m)!}}, \qquad N_{\ell,-m} = (-1)^m N_{\ell,m}.$$

Normalization:

$$\iint Y^*_{\ell,m}(\theta,\phi)Y_{\ell',m'}(\theta,\phi)\,d\Omega = \delta_{\ell,\ell'}\delta_{m,m'},$$

where the asterisk denotes the complex conjugate.

Completeness relation:

$$\sum_{\ell=0}^{\infty} \sum_{m=-\ell}^{\ell} Y_{\ell,m}^*(\theta, \phi) Y_{\ell,m}(\theta', \phi') = \delta(\theta - \theta')\delta(\phi - \phi') \tag{2.310}$$

Specific expressions:

$$Y_{0,0} = \sqrt{\frac{1}{4\pi}} \tag{2.311}$$

$$Y_{1,0} = \sqrt{\frac{3}{4\pi}} \cos\theta \tag{2.312}$$

$$Y_{1,\pm 1} = \mp\sqrt{\frac{3}{8\pi}} \sin\theta e^{\pm i\phi} \tag{2.313}$$

$$Y_{2,0} = \frac{1}{2}\sqrt{\frac{5}{4\pi}} (3\cos^2\theta - 1) \tag{2.314}$$

$$Y_{2,\pm 1} = \mp\sqrt{\frac{15}{8\pi}} \sin\theta \cos\theta e^{\pm i\phi} \tag{2.315}$$

$$Y_{2,\pm 2} = \frac{1}{4}\sqrt{\frac{15}{2\pi}} \sin^2\theta e^{\pm 2i\phi} \tag{2.316}$$

$$Y_{3,0} = \frac{1}{2}\sqrt{\frac{7}{4\pi}} (5\cos^2\theta - 3)\cos\theta \tag{2.317}$$

$$Y_{3,\pm 1} = \mp\frac{1}{4}\sqrt{\frac{21}{4\pi}} \sin\theta (5\cos^2\theta - 1) e^{\pm i\phi} \tag{2.318}$$

$$Y_{3,\pm 2} = \frac{1}{4}\sqrt{\frac{105}{2\pi}} \sin^2\theta \cos\theta e^{\pm 2i\phi} \tag{2.319}$$

$$Y_{3,\pm 3} = \mp\frac{1}{4}\sqrt{\frac{35}{4\pi}} \sin^3\theta e^{\pm 3i\phi} \tag{2.320}$$

Addition theorem

Given two vectors $\mathbf{r}$ and $\mathbf{r}'$ with spherical coordinates (r, θ, ϕ) and (r', θ', ϕ'), respectively, with angle ψ between them,

$$\cos\psi = P_1(\cos\psi) = \cos\theta \cos\theta' + \sin\theta' \sin\theta' \cos(\phi - \phi') \tag{2.321}$$

and

$$P_\ell(\cos\psi) = \frac{4\pi}{2\ell + 1} \sum_{m=-\ell}^{\ell} Y_{\ell,m}^*(\theta', \phi') Y_{\ell,m}(\theta, \phi) \cos(\phi - \phi'). \tag{2.322}$$

2.20.7. Clebsch-Gordan (Wigner) coefficients

The spherical harmonics $Y_{\ell,m}(\theta,\phi)$ transform according to the irreducible representations $D^{(\ell)}$ of the rotation group. The eigenvectors of

$$\frac{1}{\sin\theta}\frac{\partial}{\partial\theta}\sin\theta\frac{\partial}{\partial\theta}+\frac{1}{\sin^2\theta}\frac{\partial^2}{\partial\phi^2}U=\ell(\ell+1)U$$

are represented in Dirac notation by $|\ell, m\rangle$. The coefficients of the expansion

$$|J,M\rangle=\sum_{j_1m_1,j_2m_2}(J,M\mid j_1,m_1;\,j_2,m_2)|j_1,m_1\rangle|j_2,m_2\rangle$$

are the Clebsch-Gordan coefficients (Wigner coefficients, vector addition coefficients).

The coefficients are constrained by the conditions that $2j_i$ and $j_i - m_i$ $(i = 1, 2)$ are nonnegative integers, $0 \le j_i - m_i \le 2j_i$, $M = m_1 + m_2$, and

$$J=|j_1-j_2|,|j_1-j_2|+1,\ldots,j_1+j_2-1,j_1+j_2.$$

The coefficients satisfy the symmetry relations

$$(j_3,M\mid j_1,m_1;\,j_2,m_2)=(j_3,-M\mid j_2,-m_2;\,j_1,-m_1) \tag{2.323}$$

$$=(-)^{j_1+j_2-j_3}(j_3,-M\mid j_1,-m_1;\,j_2,-m_2) \tag{2.324}$$

$$=(-)^{j_1-m_1}\sqrt{\frac{2j_3+1}{2j_1+1}}(j_2,-m_2\mid j_1,m_1;\,j_3,-M) \tag{2.325}$$

$$=(-)^{j_2+m_2}\sqrt{\frac{2J+1}{2j_1+1}}(j_1,-m_1\mid j_3,-M;\,j_2,m_2). \tag{2.326}$$

TABLE 2.1. Clebsch-Gordan coefficients: Transformation matrices for 1 × $\frac{1}{2}$.

	J, M					
$m_1; m_2$	$\frac{3}{2}, +\frac{3}{2}$	$\frac{3}{2}, +\frac{1}{2}$	$\frac{1}{2}, +\frac{1}{2}$	$\frac{3}{2}, -\frac{1}{2}$	$\frac{1}{2}, -\frac{1}{2}$	$\frac{3}{2}, -\frac{3}{2}$
$+1; +\frac{1}{2}$	1					
$+1; +\frac{1}{2}$		$\sqrt{\frac{1}{3}}$	$\sqrt{\frac{2}{3}}$			
$0; +\frac{1}{2}$		$\sqrt{\frac{2}{3}}$	$-\sqrt{\frac{1}{3}}$			
$0; -\frac{1}{2}$				$\sqrt{\frac{2}{3}}$	$\sqrt{\frac{1}{3}}$	
$-1; -\frac{1}{2}$				$\sqrt{\frac{1}{3}}$	$-\sqrt{\frac{2}{3}}$	
$-1; -\frac{1}{2}$						1

Since the wavefunctions are orthonormal, the inverse transformation is

$$|j_1, m_1\rangle |j_2, m_2\rangle = \sum_J (J, m_1 + m_2 \mid j_1, m_1; j_2, m_2) |J, m_1 + m_2\rangle.$$

Table 2.1, Clebsch-Gordon coefficients for $j_1 = 1$, $j_2 = \frac{1}{2}$, indicates the application of Table 2.2, which gives a condensed display of the transformations from $D^{(j_1)} \times D^{(j_2)}$ for $1 \leq j_1 + j_2 \leq 4$. Tables 2.3 and 2.4 give, respectively, the Clebsch-Gordon coefficients for $j \times \frac{1}{2}$ and $j \times 1$.

Thus, reading down the table, the total spin wavefunctions are expressed in terms of the individual wave functions:

$$\left|\tfrac{3}{2}, -\tfrac{1}{2}\right\rangle = \sqrt{\tfrac{1}{3}}\, |1, -1\rangle \left|\tfrac{1}{2}, \tfrac{1}{2}\right\rangle + \sqrt{\tfrac{2}{3}}\, |1, 0\rangle \left|\tfrac{1}{2}, -\tfrac{1}{2}\right\rangle,$$

and reading across the table, the individual wave functions are expressed in terms of the total spin wave functions:

$$|1, 0\rangle \left|\tfrac{1}{2}, -\tfrac{1}{2}\right\rangle = \sqrt{\tfrac{2}{3}} \left|\tfrac{3}{2}, -\tfrac{1}{2}\right\rangle + \sqrt{\tfrac{1}{3}} \left|\tfrac{1}{2}, -\tfrac{1}{2}\right\rangle.$$

2.21. REFERENCES

[1] *Handbook of Mathematical Functions*, edited by M. Abramowitz and I. Stegun (U.S. Government Printing Office, Washington, DC, 1964) NBS AMS-55.

[2] *Tables of Integrals, Series, and Products*, edited by I. S. Gradsteyn and I. M. Ryzhik, English translation by A. Jeffrey (Academic Press, New York, 1980).

[3] T. W. Körner, *Fourier Analysis* (Cambridge University Press, Cambridge, U.K., 1988).

[4] *ISO Standards Handbook 2* (ISO 31/11) (International Organization for Standardization, Geneva, 1993).

[5] E. R. Cohen, Thesis, Calif. Inst. Tech., 1949, unpublished; *Tables of the Clebsch-Gordon Coefficients* (North American Aviation Science Center, Thousand Oaks, CA, 1974).

[6] E. P. Wigner, *Group Theory* (Academic Press, New York, 1959).

[7] M. E. Rose, *Elementary Theory of Angular Momentum* (Wiley, New York, 1957).

TABLE 2.2. Clebsch-Gordon Coefficients. From Ref. [5].

The sign convention is that of E. P. Wigner,[6], also used by E. U. Condon and G. H. Shortly, *The Theory of Atomic Spectra* (Cambridge University Press, New York, 1953), M. E. Rose,[7], and E. R. Cohen[5].

A $\sqrt{\ }$ is to be understood in every entry; e.g., 3/5 should be read as $\sqrt{3/5}$, 4/7 should be read as $2/\sqrt{7}$, and $-3/7$ should be read as $-\sqrt{3/7}$.

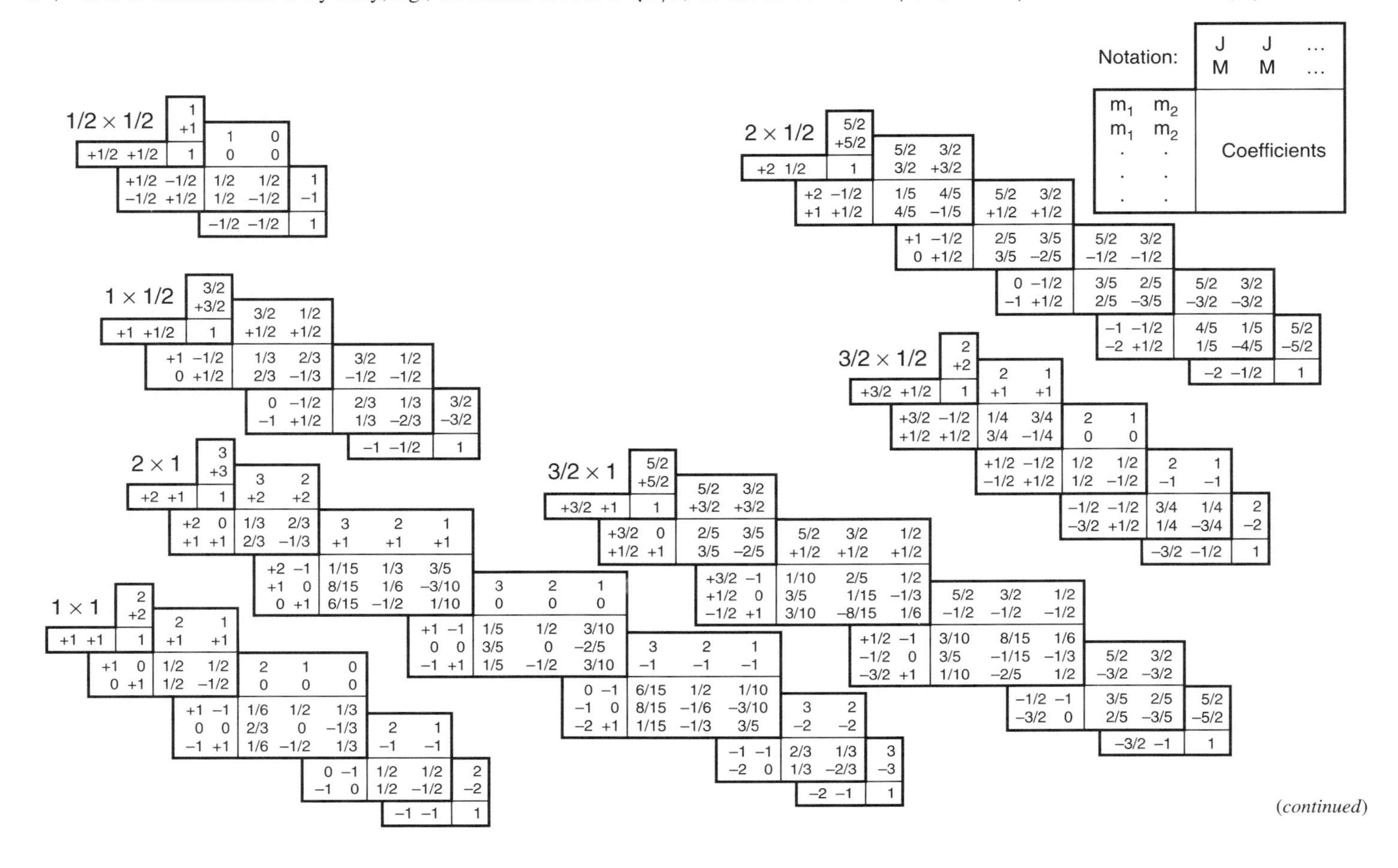

Notation:

m_1 m_2	J M	J M	...
m_1 m_2	Coefficients		
. .			

1/2 × 1/2

m_1 m_2	1, +1
+1/2 +1/2	1

m_1 m_2	1, 0	0, 0
+1/2 −1/2	1/2	1/2
−1/2 +1/2	1/2	−1/2

m_1 m_2	1, −1
−1/2 −1/2	1

1 × 1/2

m_1 m_2	3/2, +3/2
+1 +1/2	1

m_1 m_2	3/2, +1/2	1/2, +1/2
+1 −1/2	1/3	2/3
0 +1/2	2/3	−1/3

m_1 m_2	3/2, −1/2	1/2, −1/2
0 −1/2	2/3	1/3
−1 +1/2	1/3	−2/3

m_1 m_2	3/2, −3/2
−1 −1/2	1

2 × 1

m_1 m_2	3, +3
+2 +1	1

m_1 m_2	3, +2	2, +2
+2 0	1/3	2/3
+1 +1	2/3	−1/3

m_1 m_2	3, +1	2, +1	1, +1
+2 −1	1/15	1/3	3/5
+1 0	8/15	1/6	−3/10
0 +1	6/15	−1/2	1/10

m_1 m_2	3, 0	2, 0	1, 0
+1 −1	1/5	1/2	3/10
0 0	3/5	0	−2/5
−1 +1	1/5	−1/2	3/10

m_1 m_2	3, −1	2, −1	1, −1
0 −1	6/15	1/2	1/10
−1 0	8/15	−1/6	−3/10
−2 +1	1/15	−1/3	3/5

m_1 m_2	3, −2	2, −2
−1 −1	2/3	1/3
−2 0	1/3	−2/3

m_1 m_2	3, −3
−2 −1	1

1 × 1

m_1 m_2	2, +2
+1 +1	1

m_1 m_2	2, +1	1, +1
+1 0	1/2	1/2
0 +1	1/2	−1/2

m_1 m_2	2, 0	1, 0	0, 0
+1 −1	1/6	1/2	1/3
0 0	2/3	0	−1/3
−1 +1	1/6	−1/2	1/3

m_1 m_2	2, −1	1, −1
0 −1	1/2	1/2
−1 0	1/2	−1/2

m_1 m_2	2, −2
−1 −1	1

3/2 × 1

m_1 m_2	5/2, +5/2
+3/2 +1	1

m_1 m_2	5/2, +3/2	3/2, +3/2
+3/2 0	2/5	3/5
+1/2 +1	3/5	−2/5

m_1 m_2	5/2, +1/2	3/2, +1/2	1/2, +1/2
+3/2 −1	1/10	2/5	1/2
+1/2 0	3/5	1/15	−1/3
−1/2 +1	3/10	−8/15	1/6

m_1 m_2	5/2, −1/2	3/2, −1/2	1/2, −1/2
+1/2 −1	3/10	8/15	1/6
−1/2 0	3/5	−1/15	−1/3
−3/2 +1	1/10	−2/5	1/2

m_1 m_2	5/2, −3/2	3/2, −3/2
−1/2 −1	3/5	2/5
−3/2 0	2/5	−3/5

m_1 m_2	5/2, −5/2
−3/2 −1	1

2 × 1/2

m_1 m_2	5/2, +5/2
+2 1/2	1

m_1 m_2	5/2, 3/2	3/2, +3/2
+2 −1/2	1/5	4/5
+1 +1/2	4/5	−1/5

m_1 m_2	5/2, +1/2	3/2, +1/2
+1 −1/2	2/5	3/5
0 +1/2	3/5	−2/5

m_1 m_2	5/2, −1/2	3/2, −1/2
0 −1/2	3/5	2/5
−1 +1/2	2/5	−3/5

m_1 m_2	5/2, −3/2	3/2, −3/2
−1 −1/2	4/5	1/5
−2 +1/2	1/5	−4/5

m_1 m_2	5/2, −5/2
−2 −1/2	1

3/2 × 1/2

m_1 m_2	2, +2
+3/2 +1/2	1

m_1 m_2	2, +1	1, +1
+3/2 −1/2	1/4	3/4
+1/2 +1/2	3/4	−1/4

m_1 m_2	2, 0	1, 0
+1/2 −1/2	1/2	1/2
−1/2 +1/2	1/2	−1/2

m_1 m_2	2, −1	1, −1
−1/2 −1/2	3/4	1/4
−3/2 +1/2	1/4	−3/4

m_1 m_2	2, −2
−3/2 −1/2	1

(continued)

2 × 2

	4
	4
+2 +2	1

	4	3
	+3	+3
+2 +1	1/2	1/2
+1 +2	1/2	−1/2

	4	3	2
	+2	+2	+2
+2 0	3/14	1/2	2/7
+1 1	4/7	0	−3/7
0 2	3/14	−1/2	2/7

	4	3	2	1
	+1	+1	+1	+1
+2 −1	1/14	3/10	3/7	1/5
+1 0	3/7	1/5	−1/14	−3/10
0 1	3/7	−1/5	−1/14	3/10
−1 2	1/14	−3/10	3/7	−1/5

	4	3	2	1	0
	0	0	0	0	0
+2 −2	1/70	1/10	2/7	2/5	1/5
+1 −1	8/35	2/5	1/14	−1/10	−1/5
0 0	18/35	0	−2/7	0	1/5
−1 1	8/35	−2/5	1/14	1/10	−1/5
−2 2	1/70	−1/10	2/7	−2/5	1/5

	4	3	2	1
	−1	−1	−1	−1
+1 −2	1/14	3/10	3/7	1/5
0 −1	3/7	1/5	−1/14	−3/10
−1 0	3/7	−1/5	−1/14	3/10
−2 1	1/14	−3/10	3/7	−1/5

	4	3	2
	−2	−2	−2
0 −2	3/14	1/2	2/7
−1 −1	4/7	0	−3/7
−2 0	3/14	−1/2	2/7

	4	3
	−3	−3
−1 −2	1/2	1/2
−2 −1	1/2	−1/2

	4
	−4
−2 −2	1

2 × 3/2

	7/2
	+7/2
+2 +3/2	1

	7/2	5/2
	+5/2	+5/2
+2 +1/2	3/7	4/7
+1 +3/2	4/7	−3/7

	7/2	5/2	3/2
	+3/2	+3/2	+3/2
+2 −1/2	1/7	16/35	2/5
+1 1/2	4/7	1/35	−2/5
0 3/2	2/7	−18/35	1/5

	7/2	5/2	3/2	1/2
	+1/2	+1/2	+1/2	+1/2
+2 −3/2	1/35	6/35	2/5	2/5
+1 −1/2	12/35	5/14	0	−3/10
0 1/2	18/35	−3/35	−1/5	1/5
−1 3/2	4/35	−27/70	2/5	−1/10

	7/2	5/2	3/2	1/2
	−1/2	−1/2	−1/2	−1/2
+1 −3/2	4/35	27/70	2/5	1/10
0 −1/2	18/35	3/35	−1/5	−1/5
−1 1/2	12/35	−5/14	0	3/10
−2 3/2	1/35	−6/35	2/5	−2/5

	7/2	5/2	3/2
	−3/2	−3/2	−3/2
0 −3/2	2/7	18/35	1/5
−1 −1/2	4/7	−1/35	−2/5
−2 1/2	1/7	−16/35	2/5

	7/2	5/2
	−5/2	−5/2
−1 −3/2	4/7	3/7
−2 −1/2	3/7	−4/7

	7/2
	−7/2
−2 −3/2	1

3/2 × 3/2

	3
	+3
+3/2 +3/2	1

	3	2
	+2	+2
+3/2 +1/2	1/2	1/2
+1/2 +3/2	1/2	−1/2

	3	2	1
	+1	+1	+1
+3/2 −1/2	1/5	1/2	3/10
+1/2 +1/2	3/5	0	−2/5
−1/2 +3/2	1/5	−1/2	3/10

	3	2	1	0
	0	0	0	0
+3/2 −3/2	1/20	1/4	9/20	1/4
+1/2 −1/2	9/20	1/4	−1/20	−1/4
−1/2 +1/2	9/20	−1/4	−1/20	1/4
−3/2 +3/2	1/20	−1/4	9/20	−1/4

	3	2	1
	−1	−1	−1
+1/2 −3/2	1/5	1/2	3/10
−1/2 −1/2	3/5	0	−2/5
−3/2 +1/2	1/5	−1/2	3/10

	3	2
	−2	−2
−1/2 −3/2	1/2	1/2
−3/2 −1/2	1/2	−1/2

	3
	−3
−3/2 −3/2	1

TABLE 2.3. Clebsch-Gordan coefficients for $j \times \frac{1}{2}$: $(J, M \mid M - m_2; \frac{1}{2}, m_2)$

$J =$	$m_2 = +\frac{1}{2}$	$-\frac{1}{2}$
$j + \frac{1}{2}$	$\left[\frac{1}{2} + \frac{M}{2j+1}\right]^{\frac{1}{2}}$	$\left[\frac{1}{2} - \frac{M}{2j+1}\right]^{\frac{1}{2}}$
$j - \frac{1}{2}$	$-\left[\frac{1}{2} - \frac{M}{2j+1}\right]^{\frac{1}{2}}$	$\left[\frac{1}{2} + \frac{M}{2j+1}\right]^{\frac{1}{2}}$

TABLE 2.4. Clebsch-Gordan coefficients for $j \times 1$: $(J, M \mid M - m_2; 1, m_2)$

$J =$	$m_2 = +1$	0	-1
$j + 1$	$\left[\frac{(j+M)(j+M+1)}{2(j+1)(2j+1)}\right]^{\frac{1}{2}}$	$\left[\frac{(j-M+1)(j+M+1)}{(j+1)(2j+1)}\right]^{\frac{1}{2}}$	$\left[\frac{(j-M)(j-M+1)}{2(j+1)(2j+1)}\right]^{\frac{1}{2}}$
j	$-\left[\frac{(j+M)(j-M+1)}{2j(j+1)}\right]^{\frac{1}{2}}$	$\frac{M}{j(j+1)^{\frac{1}{2}}}$	$\left[\frac{(j-M)(j+M+1)}{2j(j+1)}\right]^{\frac{1}{2}}$
$j - 1$	$\left[\frac{(j-M)(j-M+1)}{2j(2j+1)}\right]^{\frac{1}{2}}$	$-\left[\frac{(j-M)(j+M)}{j(2j+1)}\right]^{\frac{1}{2}}$	$\left[\frac{(j+M)(j+M+1)}{2j(2j+1)}\right]^{\frac{1}{2}}$

3

Acoustics

Robert T. Beyer
Brown University

Contents

List of Tables

List of Figures

3.1. IMPORTANT ACOUSTICAL UNITS

Acoustic ohm. Unit of acoustic impedance. An acoustic impedance (including acoustic resistance and reactance) has a magnitude of 1 acoustic ohm (cgs) when a sound pressure of 1 μbar produces a volume velocity of 1 cm^3/s. The specific acoustic impedance (impedance for a unit area) of a plane wave is the product of the density of the medium and the velocity of sound. Typical values are air, 41.5 g $\cdot$ cm$^{-2}\cdot$ s^{-1}; water, 145,000 g $\cdot$ cm$^{-2}\cdot$ s^{-1}; and iron, 4×10^6 g $\cdot$ cm$^{-2}\cdot$ s^{-1}.

Cent. The interval between two musical sounds having as a basic frequency ratio the 1200th root of 2. The number of cents between frequencies f_1 and f_0 is

$$1200 \log_2(f_1/f_0) = 3986 \log_{10}(f_1/f_0).$$

Decibel. Unit expressing the magnitude of the ratio of two sound power or intensity levels. The number of decibels between powers P_1 and P_2 is $10 \log_{10}(P_1/P_2)$. For pressure, the number of decibels is $20 \log_{10}(P_1/P_2)$. Abbreviation: dB. Ordinary sound intensities in air: $10^{-16} - 10^{-4}$ W $\cdot$ cm^{-2}. The audible threshold is 10^{-16} W $\cdot$ cm^{-2}, $2 \times 10^{-4}\mu$bars.

Mechanical ohm. Unit of mechanical impedance. A mechanical impedance has a magnitude of 1 mechanical ohm (cgs) when an applied pressure of 1 dyn/cm^2 produces a velocity of 1 cm/s.

Mel. Unit of subjectively estimated pitch. The pitch of a 1000-Hz tone at 40 dB above threshold is taken to be 1000 mels. The pitch of any sound judged to be double that pitch is taken as 2000 mels, and so on.

Microbar. Unit of pressure commonly used in acoustics. 1 μbar = 1 dyn/cm^2. The SI unit is the micropascal, 1 μPa = 10^{-5} μbars.

Neper. Unit expressing the ratio between two amplitudes as a natural (Naperian) logarithm. 1 Np = 8.686 dB.

Noy. Unit of noisiness related to the perceived noise level through the relation

$$10 \log_2(N') = N'' - 40,$$

where N' noy is the perceived loudness (perceived noise level, PN) and N'' phon is defined as the sound-pressure level of a reference sound of the order of one octave wide centered at 1000 Hz, which is subjectively judged to be equally noisy as the sound being measured.

Phon. Unit of loudness level. The loudness level of a given sound is the sound-pressure level in dB of a pure tone of frequency 1000 Hz (relative to 2×10^{-4} μbars), which is assessed by normal observers as being equally as loud as the sound in question. Thus, phons are effectively expressed in dB. The value at the audible threshold is 0 phons, the threshold of feeling is about 140 phons. A loudness level of 74 phons corresponds to a sound pressure of 1 μbar of a 1000-Hz tone.

Rayl. Unit of specific acoustic impedance. A specific acoustic impedance has a magnitude of 1 rayl (cgs) when a sound pressure of 1 μbar produces a linear velocity of 1 cm/s. See *Acoustic ohm.*

Sabin. Unit of absorption of surface covering in room acoustics. In the English system the unit is ft^2; in SI the unit is m^2.

Sone. Unit of subjective loudness. If S is the loudness and P is the loudness level,

$$\log_{10}(S/\text{sone}) = 0.0301\,(P/\text{phon} - 40)$$

3.2. OSCILLATIONS OF A LINEAR SYSTEM

The equation of motion of a dissipative system with one degree of freedom subject to a linear restoring force, with equivalent localized mass m, resistance (damping factor) R, and stiffness k, subject to no external force, is

$$m\ddot{\xi} + R\dot{\xi} + k\xi = 0, \tag{3.1}$$

where ξ is the displacement from equilibrium. The general solution is

$$\xi = e^{-Rt/2m}\left(Ae^{+(R^2/4m^2 - k/m)^{1/2}t} + Be^{-(R^2/4m^2 - k/m)^{1/2}t}\right), \tag{3.2}$$

where A and B are arbitrary constants to be fixed by the initial conditions. If

$$R^2/4m^2 < k/m, \tag{3.3}$$

the motion is a damped oscillation with frequency

$$f = (1/2\pi)(k/m - R^2/4m^2)^{1/2} \tag{3.4}$$

and amplitude that varies with the time through the term $e^{-Rt/2m}$. The logarithmic decrement (the logarithm to the base e of the ratio of successive amplitudes) is

$$\delta = R/2mf. \tag{3.5}$$

If

$$R^2/4m^2 \geq k/m, \tag{3.6}$$

no oscillations take place. The case of equality is called "critical damping."

If the system is subject to a harmonic force $F_0 e^{i\omega t}$, with the angular frequency $\omega = 2\pi f$, the displacement in the steady state

$$\xi = \frac{F_0 e^{i\omega t}}{i\omega R + k - \omega^2 m}. \tag{3.7}$$

The velocity ($\dot{\xi}$) amplitude has its maximum for the angular frequency

$$\omega = (k/m)^{1/2}, \tag{3.8}$$

which is called the *resonance* frequency of the system.

An acoustic example of a lumped oscillating system: Helmholtz resonator, a hollow air-filled sphere of volume V, having an inlet opening for sound and an outlet placed in the external ear, is shown in Fig. 3.1.

FIGURE 3.1. Helmholtz resonator.

It is used as a resonant cavity in sound analysis. The acoustic elements of the oscillator are as follows: m is the equivalent mass of moving air in the opening,

$$m = \rho_0 S^2 / C_0, \tag{3.9}$$

R is the resistance due to radiation of sound from the opening,

$$R = 2\pi \rho_0 f^2 S^2, \tag{3.10}$$

and k is the equivalent stiffness,

$$k = \rho_0 c^2 / V, \tag{3.11}$$

with ρ_0 being the equilibrium density of the fluid in the resonator, V being the volume of the resonator cavity, c being the velocity of sound in the fluid medium, S being the area of the resonator opening, and C_0 being the acoustic conductivity. The approximate value of the resonance frequency is

$$f_r = (c/2\pi)(C_0/V)^{1/2}. \tag{3.12}$$

For an inlet consisting of a tube of length l and cross-sectional radius r, the acoustic conductivity becomes

$$C_0 = \frac{\pi r^2}{l + \pi r/2}.$$

For very small l compared with r, C_0 reduces to $2r$.

For resonance frequencies and normal modes of other oscillating systems, see Secs. 3.12 and 3.16.

3.2.1. Analogies

In the normal classical analogy between an oscillating electrical circuit of inductance L, resistance R, and capacitance C subject to an alternating voltage E_0, L is analogous to m, R to the mechanical resistance R, and C to the reciprocal of the stiffness k.

In the so-called mobility analogy, electric current corresponds to mechanical force, electromotive force to mechanical velocity, mechanical resistance to the reciprocal of electric resistance, and mechanical compliance (reciprocal of stiffness) to electric inductance.

Mechanical impedance is the ratio of the mechanical force to the velocity in the absence of dissipation.

3.3. GENERAL LINEAR ACOUSTICS; WAVE PROPAGATION IN FLUIDS

3.3.1. Plane waves

A plane wave of acoustic excess pressure p in the x-direction with velocity c, and in the absence of dissipation, is subject to the equation

$$\frac{\partial^2 p}{\partial x^2} = \frac{1}{c^2}\frac{\partial^2 p}{\partial t^2}, \tag{3.13}$$

with the general solution

$$p = \phi_1(x - ct) + \phi_2(x + ct), \tag{3.14}$$

where ϕ_1 and ϕ_2 are arbitrary continuous and differentiable functions. For a harmonic wave of frequency f in the positive x-direction,

$$\phi_1 = p_0 \cos[(2\pi f/c)(x - ct) + \theta_0], \tag{3.15}$$

where p_0 is the pressure amplitude and θ_0 is the initial phase at $x = 0$.

The acoustic impedance is defined as the complex ratio of the sound pressure on a given surface lying in a wave front to the volume velocity through the unit area in that surface. The real part of the impedance is the acoustic *resistance*, and the imaginary part is the acoustic *reactance*. The acoustic *stiffness* for a fluid medium traversed by a harmonic wave of frequency f is $2\pi f$ times the acoustic resistance. The acoustic *compliance* is the reciprocal of the acoustic stiffness. For a plane harmonic wave, the acoustic impedance is real and has the value

$$Z = \rho_0 c/S, \tag{3.16}$$

where S is the area of the wave front and ρ_0 is the equilibrium density of the fluid. The specific impedance, or the impedance for unit area of wave front, is

$$Z_S = \rho_0 c. \tag{3.17}$$

The acoustic intensity defined as the average rate at which wave energy is transmitted in a specified direction at a given point through unit area normal to the direction at this point is

$$I = p_0^2/2Z_S. \tag{3.18}$$

The coherent SI unit is Wm^{-2}. But see *Decibel* in Sec. 3.1.

The velocity of sound in a fluid medium of density ρ_0 and adiabatic bulk modulus B is

$$c = (B/\rho_0)^{1/2}. \tag{3.19}$$

For an ideal gas, this becomes

$$c = (\gamma p/\rho)^{1/2} = (\gamma R_m T)^{1/2} = c_0(T/T_0)^{1/2}, \tag{3.20}$$

where γ is the ratio of specific heat at constant pressure to specific heat at constant volume, T is the absolute temperature, R_m is the molar gas constant, and c_0 is the velocity of sound at temperature T_0. (See Sec. 3.5.)

Classic acoustic wave attenuation is shown in the equation

$$p = Ae^{-ax}\cos[(\omega/c)(x - ct) + \theta_0], \tag{3.21}$$

with the attenuation coefficient α given by

$$\alpha = \frac{\omega^2}{\rho_0 c^3}\left[\frac{2\eta}{3} + \left(\frac{\gamma - 1}{\gamma}\right)\frac{M\kappa}{2C_v}\right], \tag{3.22}$$

where η is the coefficient of viscosity, κ is the thermal conductivity, M is the molecular weight, C_v is the molar specific heat at constant volume, and γ is the ratio of specific heat at constant pressure to that at constant volume.

For relaxation attenuation, see Sec. 3.8.

3.3.2. Spherical waves

A spherical wave from a monopole source at the origin is governed by

$$\frac{\partial^2}{\partial r^2}(rp) = \frac{1}{c^2}\frac{\partial^2(rp)}{\partial t^2}. \tag{3.23}$$

The specific acoustic impedance is

$$Z_s = Z_{s_1} + iZ_{s_2}, \tag{3.24}$$

with Z_{s_1} as the specific acoustic resistance,

$$Z_{s_1} = \frac{\omega^2 \rho_0 r^2}{c(1 + \omega^2 r^2/c^2)} \tag{3.25}$$

and Z_{s_2} as the specific acoustic reactance,

$$Z_{s_2} = Z_{s_1}\left(\frac{c}{\omega r}\right). \tag{3.26}$$

3.3.3. Reflection and transmission

The reflection and transmission of a plane wave at a plane interface are given by

$$\frac{I_r}{I_i} = \frac{(Z_2 \cos\theta_i - Z_1 \cos\theta_t)^2}{(Z_2 \cos\theta_i + Z_1 \cos\theta_t)^2}, \tag{3.27}$$

where I_i is the intensity of the incident wave, I_r is the reflected intensity, θ_i is the angle of incidence, θ_t is the angle of refraction, Z_1 is the specific acoustic resistance of the incident wave, and Z_2 is the specific acoustic resistance of the transmitted wave.

Similarly,

$$\frac{I_t}{I_i} = \frac{4Z_2/Z_1}{(Z_2/Z_1 + \cos\theta_t/\cos\theta_i)^2}, \tag{3.28}$$

where I_t is the intensity of the transmitted wave. The law of refraction is

$$\frac{\sin\theta_i}{\sin\theta_t} = \frac{c_1}{c_2}, \tag{3.29}$$

where c_1 and c_2 are the sound velocities in the two media, respectively.

3.3.4. Velocity and attenuation of sound

The following tables present relevant acoustical data on the velocity and attenuation of sound.

TABLE 3.1. Velocity of sound in selected liquids. From Ref. [1].

Liquid	Velocity (m/s)
Mercury (20°C, 1 atm)	1451
Pentane (20°C, 1 atm)	1008
CS_2 (25°C, 1 atm, 2 mHz)	1140
CCl_4 (25°C, 1 atm, 4.85 mHz)	930
Ether (25°C, 1 atm)	976
Acetone (20°C, 1 atm)	1203
Ethyl alcohol (20°C, 1 atm)	1161.8
Methyl alcohol (20°C, 1 atm)	1121.2
Liquid helium (4 K, 15 mHz, 1 atm)	211

TABLE 3.2. Velocity of sound in pure water and in sea water (salinity 3.5%) at 1 atm. From Ref. [1].

	Velocity (m/s)	
Temperature (°C)	Pure water	Sea water
0	1402.3	1449.4
10	1447.2	1490.4
20	1482.3	1522.2
30	1509.0	1546.2
50	1542.5	
70	1554.7	
100	1543.0	

TABLE 3.3. Velocity of sound in selected gases at sonic frequencies. From Ref. [1].

Gas	Formula	Velocity at 0°C (m/s)
Air (dry)	—	331.45
Ammonia	NH_3	415
Argon	Ar	319
Carbon monoxide	CO	338
Carbon dioxide	CO_2	259.0
Carbon disulfide	CS_2	189
Chlorine	Cl_2	206
Ethylene	C_2H_4	317
Helium	He	965
Hydrogen	H_2	1284
Illuminating gas (coal)	—	453
Methane	CH_4	430
Neon	Ne	435
Nitric oxide (10°C)	NO	325
Nitrogen	N_2	334
Nitrous oxide	N_2O	263
Oxygen	O_2	316
Steam (134°C)	H_2O	494

TABLE 3.4. Elastic constants, sound velocities, and specific impedances of selected solids. Y_0 is Young's modulus, μ is the shearing modulus, and λ is the Lamé coefficient $B - \frac{2}{3}\mu$, where B is the bulk modulus; V_t is the longitudinal compressional wave velocity, V_s is the shear wave velocity, and V_{ext} is the velocity of a wave in a thin rod. From Ref. [1].

Materials	Y_0 (10^{10} N/m^2)	μ (10^{10} N/m^2)	λ (10^{10} N/m^2)	Poisson's ratio σ	$V_t = [(\lambda+2\mu)/\rho]^{1/2}$ (m/s)	$V_s = (\mu/\rho)^{1/2}$ (m/s)	$V_{ext} = (Y_0/\rho)^{1/2}$ (m/s)	$Z_t = [\rho(\lambda+2\mu)]^{1/2}$ (10^6 kg/m^{2}s)	$Z_s = (\rho\mu)^{1/2}$ (10^6 kg/m^{2}s)
Aluminum, rolled	6.8–7.1	2.4–2.6	6.1	0.355	6420	3040	5000	17.3	8.2
Beryllium	30.8	14.7	1.6	0.05	12890	8880	12870	24.1	16.6
Brass, yellow, 70 Cu, 30 Zn	10.4	3.8	11.3	0.374	4700	2110	3480	40.6	18.3
Constantan	16.1	6.09	11.4	0.327	5177	2625	4270	45.7	23.2
Copper, rolled	12.1–12.8	4.6	13.1	0.37	5010	2270	3750	44.6	20.2
Duralumin 17S	7.15	2.67	5.44	0.335	6320	3130	5150	17.1	8.5
Gold, hard-drawn	8.12	2.85	15.0	0.42	3240	1200	2030	62.5	23.2
Iron, cast	15.2	5.99	6.92	0.27	4994	2809	4480	37.8	21.35
Iron electrolytic	20.6	8.2	11.3	0.29	5950	3240	5120	46.4	25.3
Armco	21.2	8.24	11.35	0.29	5960	3240	5200	46.5	25.3
Lead, rolled	1.5–1.7	0.54	3.3	0.43	1960	690	1210	22.4	7.85
Magnesium, drawn, annealed	4.24	1.62	2.56	0.306	5770	3050	4940	10.0	5.3
Monel metal	16.5–18	6.18–6.86	12.4	0.327	5350	2720	4400	47.5	24.2
Nickel	21.4	8.0	16.4	0.336	6040	3000	4900	53.5	26.6
Nickel silver	10.7	3.92	11.2	0.37	4760	2160	3575	40.0	18.1
Platinum	16.7	6.4	9.9	0.303	3260	1730	2800	69.7	37.0
Silver	7.5	2.7	8.55	0.38	3650	1610	2680	38.0	16.7
Steel, K9	21.6	8.29	10.02	0.276	5941	3251	5250	46.5	25.4
347 stainless steel	19.6	7.57	11.3	0.30	5790	3100	5000	45.7	24.5
Tin, rolled	5.5	2.08	4.04	0.34	3320	1670	2730	24.6	11.8
Titanium	11.6	4.40	7.79	0.32	6070	3125	5090	27.3	14.1
Tungsten, drawn	36.2	13.4	31.3	0.35	5410	2640	4320	103	50.5
Tungsten carbide	53.4	21.95	17.1	0.22	6655	3984	6240	91.8	55.0
Zinc, rolled	10.5	4.2	4.2	0.25	4210	2440	3850	30	17.3
Fused silica	7.29	3.12	1.61	0.17	5968	3764	5760	13.1	8.29
Pyrex glass	6.2	2.5	2.3	0.24	5640	3280	5170	13.1	7.6
Heavy silicate flint	5.35	2.18	1.77	0.224	3980	2380	3720	15.4	9.22
Light borate crown	4.61	1.81	2.2	0.274	5100	2840	4540	11.4	6.35
Lucite	0.40	0.143	0.562	0.4	2680	1100	1840	3.16	1.3
Nylon 6-6	0.355	0.122	0.511	0.4	2620	1070	1800	2.86	1.18
Polyethylene	0.076	0.026	0.288	0.458	1950	540	920	1.75	0.48
Polystyrene	0.360	0.133	0.319	0.353	2350	1120	1840	2.49	1.19

TABLE 3.5. Attenuation of sound in air as a function of temperature, humidity, and frequency, in dB/100 m for an atmospheric pressure of 1.013 x 10^5 Pa (1 atm). From Ref. [2].

T	Relative humidity (%)	125	250	500	1000	2000	4000
		Frequency (Hz)					
30°C (86°F)	10	0.09	0.19	0.35	0.82	2.6	8.8
	20	0.06	0.18	0.37	0.64	1.4	4.4
	30	0.04	0.15	0.38	0.68	1.2	3.2
	50	0.03	0.10	0.33	0.75	1.3	2.5
	70	0.02	0.08	0.27	0.74	1.4	2.5
	90	0.02	0.06	0.24	0.70	1.5	2.6
20°C (68°F)	10	0.08	0.15	0.38	1.21	4.0	10.9
	20	0.07	0.15	0.27	0.62	1.9	6.7
	30	0.05	0.14	0.27	0.51	1.3	4.4
	50	0.04	0.12	0.28	0.50	1.0	2.8
	70	0.03	0.10	0.27	0.54	0.96	2.3
	90	0.02	0.08	0.26	0.56	0.99	2.1
10°C (50°F)	10	0.07	0.19	0.61	1.9	4.5	7.0
	20	0.06	0.11	0.29	0.94	3.2	9.0
	30	0.05	0.11	0.22	0.61	2.1	7.0
	50	0.04	0.11	0.20	0.41	1.2	4.2
	70	0.04	0.10	0.20	0.38	0.92	3.0
	90	0.03	0.10	0.21	0.38	0.81	2.5
0°C (32°F)	10	0.10	0.30	0.89	1.8	2.3	2.6
	20	0.05	0.15	0.50	1.6	3.7	5.7
	30	0.04	0.10	0.31	1.08	3.3	7.4
	50	0.04	0.08	0.19	0.60	2.1	6.7
	70	0.04	0.08	0.16	0.42	1.4	5.1
	90	0.03	0.08	0.15	0.36	1.1	4.1

TABLE 3.6. Classical attenuation α/f^2 for selected gases at 20 °C and 1 atm, where α is the attenuation coefficient in cm^{-1} and f the frequency. From Ref. [3].

	$10^{13}\alpha/f^2 (s^2/cm)$		
Gas	Due to heat conduction	Due to shear viscosity	Total
He	0.216	0.309	0.525
Ar	0.77	1.08	1.85
H_2	0.052	0.117	0.169
N_2	0.39	0.94	1.33
O_2	0.49	1.16	1.65
Air	0.38	0.99	1.37
CO_2	0.31	1.09	1.40

TABLE 3.7. Attenuation in selected gases due to vibrational relaxation in diatomic molecules [$(\alpha c/f)_{max}$]. f_r is the relaxation frequency; α is the attenuation coefficient in m^{-1}, c is the velocity of sound in the gas, and f is the frequency. From Ref. [3].

Gas	T (K)	f_r (kHz)	$(\alpha c/f)_{max}$
N_2	300	10^{-2}	0.03
NO	300	400	0.14
O_2	288	5×10^{-2}	0.56
F_2	301	84	4.3
Cl_2	293	39	8.3
Br_2	301	240	11.8

TABLE 3.8. Attenuation of sound in H_2O at 1 atm. α is the attenuation coefficient in m^{-1}, and f is the frequency. From Ref. [1].

Temperature (°C)	$10^{15}\alpha/f^2(s^2/m)$
0	56.9
10	36.1
20	25.3
50	12.0
80	7.9

TABLE 3.9. Attentuation of sound in selected liquids at 1 atm. α is the attenuation coefficient in m^{-1}, and f is the frequency.

Liquid	T(°C)	$10^{15}\alpha/f^2(s^2/m)$
Mercury[a]	25	5.7
Sodium[a]	100	12
Lead[a]	340	9.4
Tin[a]	240	5.6
Methyl alcohol[a]	30	30.2
Ethyl alcohol[a]	30	48.6
Benzene[b]	25	870
CCl_4[b]	25	540
CS_2[b]	25	5700

[a]Reference [1].
[b]Reference [3].

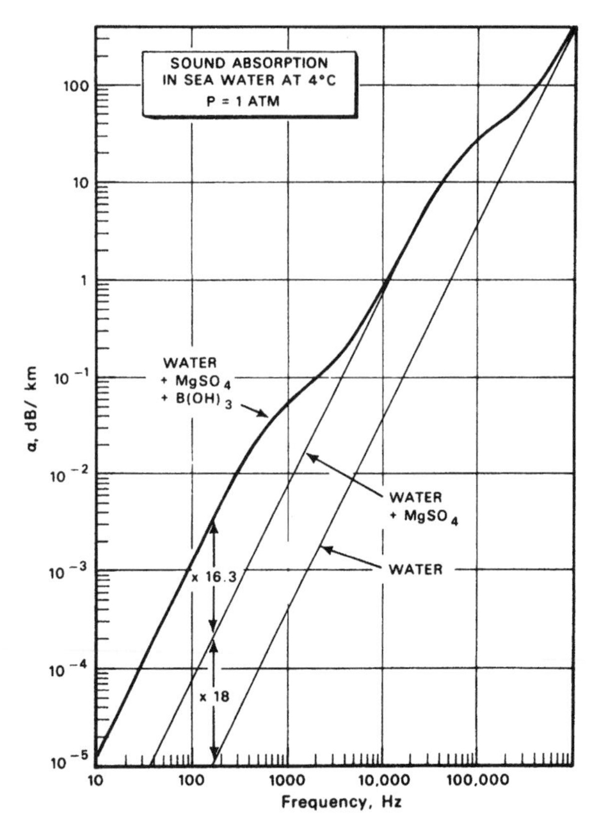

FIGURE 3.2. Attenuation of sound in sea water. These curves were calculated from laboratory acoustic measurements by V. P. Simmons from 6 to 350 kHz in a 200-liter spherical resonator using Lyman and Fleming sea water of salinity = 3.5% and *p*H = 8.0. This refers to artificial sea water, as defined by Lyman and Fleming, [4] which is sufficiently close to most actual sea water to be illustrative. From Ref. [5].

3.4. HIGH-INTENSITY SOUND; NONLINEAR ACOUSTICS

3.4.1. Lagrangian form of wave equation

Given below is the Lagrangian form of the wave equation for transmission in one direction in a fluid in which the small-amplitude sound velocity is $c_v = (\gamma p_0/\rho_0)^{1/2}$, where γ is the usual specific heat ratio and p_0 and ρ_0 are the equilibrium values of the fluid pressure and density, respectively. This scheme follows the fate of a particle of fluid that was at the particular point $x = a$ at time $t = 0$ and has reached any other point x at time t. Then, x is a function of a and t, which are the independent variables. Putting $x = a + \xi$, we have the wave equation

$$\frac{\partial^2 \xi}{\partial t^2} = \frac{c_0^2}{(1 + \partial\xi/\partial a)^{2+B/A}} \frac{\partial^2 \xi}{\partial a^2}, \tag{3.30}$$

where

$$A = \rho_0 \left(\frac{\partial p}{\partial \rho} \right)_{S,\rho=\rho_0} = \rho_0 c_0^2, \tag{3.31}$$

$$B = \rho_0^2 \left(\frac{\partial^2 p}{\partial \rho^2} \right)_{S,\rho=\rho_0}, \tag{3.32}$$

where the derivatives are taken at constant entropy. For a gas,

$$B/A = \gamma - 1. \tag{3.33}$$

The ratio B/A is known as the parameter of nonlinearity.

3.4.2. Solutions of Eulerian wave equation

Earnshaw's implicit solution of the Eulerian wave equation associated with Eq. (3.30) for harmonic waves with frequency f is

$$u(x, f) = u_0 \sin \left[\omega t - \frac{\omega x}{c_0} \left(1 + \frac{B}{2A} \frac{u}{c_0} \right)^{-2A/B-1} \right]. \tag{3.34}$$

From Eq. (3.34), the Fubini explicit solution can be obtained in the form

$$u = 2u_0 \sum_{n=1}^{\infty} \frac{J_n(nx/l)}{nx/l} \sin n \left(\omega t - \frac{\omega x}{c_0} \right). \tag{3.35}$$

Table 3.10 presents data relevant to nonlinear acoustics (B/A values).

TABLE 3.10. Nonlinear parameters for selected liquids. Values of B/A at 1 atm (see text). From Ref. [6].

Liquid	T(°C)	B/A
Water (distilled)	0	4.2
	20	5.0
Sea water (3.5% sal.)	0	4.9[a]
	20	5.25[a]
Methyl alcohol	20	9.6
Ethyl alcohol	20	10.5
Acetone	20	9.2
Benzene	20	9.0
Mercury	30	7.8
Sodium	110	2.7
Bismuth	318	7.1

[a] Reference [7].

3.4.3. Burgers' equation and other approximations

The one-dimensional wave equation with account of dissipation is approximated by Burger's equation[8]

$$c_0^3 u_x - bc_0 u u_\tau = \frac{1}{2}\frac{b\eta}{\rho_0} u_{\tau\tau} = \frac{c_0^3 \alpha}{\omega^2} u_{\tau\tau}. \tag{3.36}$$

Here, a single subscript denotes differentiation by that variable, α = absorption coefficient, $b = \frac{4}{3} + \frac{\eta'}{\eta}$ is the viscosity number, η' is the bulk viscosity, η is the shear viscosity, $\tau = t - x/c_0$.

A technique for handling bounded beams in a nondissipative medium was developed by Zabolotskaya and Khokhlov (1969). This procedure led to the equation

$$\frac{\alpha}{2}\frac{\partial^2 \rho}{\partial \tau^2} - \frac{\partial^2 \rho}{\partial \tau \partial x} + \frac{c_0}{2}\left(\frac{\partial^2 \rho}{\partial y^2} + \frac{\partial^2 \rho}{\partial z^2}\right) = 0. \tag{3.37}$$

Here, $\alpha = (\gamma + 1)/2\rho_0 c_0$, $\tau = t - x/c_0$, and ρ is the density of the medium.

The dissipative case was considered by Kuznetsov (1970), leading to the equation

$$\frac{\alpha}{2}\frac{\partial^2 \rho}{\partial \tau^2} - \frac{\partial^2 \rho}{\partial \tau \partial x} + \frac{c_0}{2}\left(\frac{\partial^2 \rho}{\partial y^2} + \frac{\partial^2 \rho}{\partial z^2}\right) + K\frac{\partial^3 \rho}{\partial \tau^3} = 0, \tag{3.38}$$

where $K = b'\eta/2\rho_0 c_0^3$ and b' is the combination of viscosity and thermal conductivity terms

$$b' = \frac{\eta'}{\eta} + \frac{4}{3} + \kappa\left(\frac{1}{c_v} - \frac{1}{c_p}\right). \tag{3.39}$$

Here, κ is the thermal conductivity, and c_v and c_p are the specific heats at constant volume and pressure, respectively. Equation 3.39 is referred to in the literature as the KZK equation.

As a high-intensity wave progresses through a dissipative medium, the wave front gradually assumes a sawtooth character, with the front part of the wave profile very steep. It then becomes a shock wave, characterized by a relatively large change in excess pressure across a very small region of space. The well-known sonic boom is a good example of an acoustic shock wave.

3.4.4. Radiation pressure

The Rayleigh radiation pressure p_R is the difference between the time average of the pressure at any point in a fluid traversed by a compressional wave and that which would have existed in a fluid of the same mean density at rest. For an ideal gas, we have

$$p_R = \tfrac{1}{4}(\gamma + 1)\bar{E}, \tag{3.40}$$

with γ as the usual specific heat ratio and $\bar{E}$ as the average energy density in the sound wave. For a liquid,

$$p_R = \tfrac{1}{2}(1 + B/2A)\bar{E}, \tag{3.41}$$

where B and A have the meanings in Eqs. 3.31 and 3.32.

The Langevin radiation pressure p_L is the difference between the average pressure at a reflecting or absorbing wall and that behind the wall in the same acoustic medium at rest:

$$p_L = \bar{E}. \tag{3.42}$$

3.5. ATMOSPHERIC ACOUSTICS

3.5.1. Velocity of sound in air

The velocity of sound in air is given by Eq. (3.20). For normal air, $\gamma \simeq 1.4$ and $R_m \simeq 0.288$ J/kg · K.

3.5.2. Refraction in a fluid medium

For a nonhomogeneous, stratified fluid medium like the atmosphere, the general ray equation for refraction is

$$c \sec\theta - c_0 \sec\theta_0 = u_{x0} - u_x \tag{3.43}$$

for a ray in the xz-plane, where the stratification is in the z-direction. The sound velocity in any particular layer for still air is c, with initial value c_0. The ray makes the angle θ with the x-axis (horizontal), θ_0 being the initial angle. The instantaneous large-scale velocity of the air in the x-direction is u_x.

3.5.3. Attenuation in atmosphere

For attenuation in the atmosphere, see Sec. 3.3.

3.5.4. Doppler effect

This phenomenon is the change in the observed frequency of a sound wave caused by a time rate of change in the effective length of the path of travel between the sound source and place of observation.

The fundamental formula is

$$f_o = \frac{1 + v_o/c}{1 - v_s/c} f_s, \tag{3.44}$$

where f_o is the observed frequency, f_s is the frequency at the source, v_o is the component of velocity (relative to the medium) of the observation point toward the source, v_s is the component of velocity (relative to the medium) of the source toward the observation point, and c is the velocity of sound in a stationary medium.

3.6. UNDERWATER SOUND

The *Lloyd mirror effect* is the interference between the direct sound in the water from source to receiver and that reflected on the way by the sea water surface. The fundamental equation is

$$I_r/I_d = 1 + R^2 - 2R\cos(\delta_1 - \delta_2), \tag{3.45}$$

where I_d is the intensity produced at the receiver by the direct radiation from the source located below the water surface; I_r is the resultant intensity at the receiver due to the combination of the direct sound and that reflected from the surface; R is the ratio of acoustic pressure due to the reflected radiation to that due to the direct radiation; $\delta_1 = 2\pi f l/c$, the phase difference between the disturbance at the receiver and that at the source, where l is the direct distance from source to receiver, c is the velocity of sound in the water, and f is the frequency; and δ_2 is the corresponding phase difference for the reflected sound.

Figure 3.3 shows underwater sound rays in a thermocline (sound velocity approximately uniform from the surface to C, decreases from C to C', and at lower depths is more or less constant, save for change of pressure and density with depth). S is the source of sound. The shaded region between rays 2 and 3 is an acoustic shadow zone.

The ray equations mentioned are most readily obtained from the solution of the eikonal equation. For a stratified ocean medium in which the index of refraction $n(z)$ (ratio of the sound velocity at a given depth to some standard sound velocity) is a function of the depth only (z axis), this has the form

$$\left(\frac{\partial\psi}{\partial x}\right)^2 + \left(\frac{\partial\psi}{\partial z}\right)^2 = [n(z)]^2. \tag{3.46}$$

ψ is the eikonal. When this equation is solved for $\psi(x, z)$, the differential equation for the rays in the xz-plane is

$$\frac{dz}{dx} = \frac{\partial\psi/\partial z}{\partial\psi/\partial x}. \tag{3.47}$$

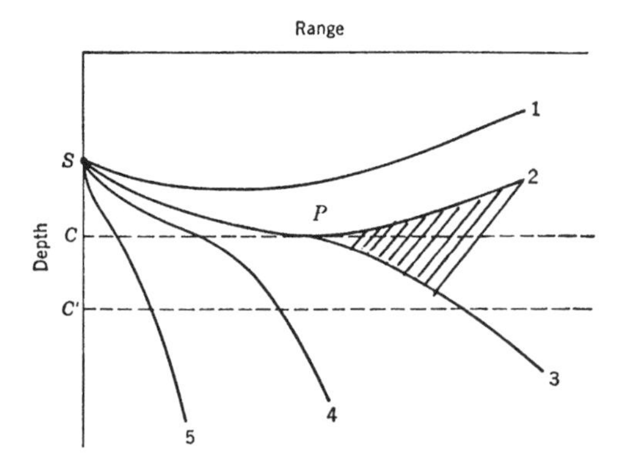

FIGURE 3.3. Underwater sound rays; shadow zone. From Ref. [9].

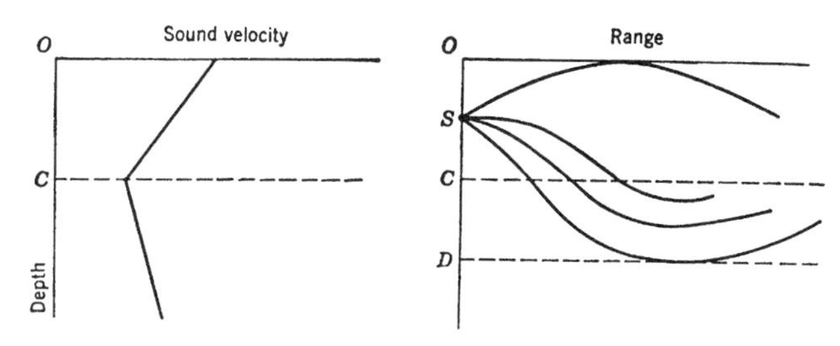

FIGURE 3.4. Underwater sound-channeling effect. From Ref. [9].

In Figure 3.4, the second figure shows the sound-channeling effect in the sea due to the velocity profile shown in the first figure.

3.7. ACOUSTIC TRANSMISSION IN SOLIDS

3.7.1. Velocity in an extended polycrystalline solid

The velocity of a compressional (sound) wave in an extended polycrystalline solid is given by

$$c = \left(\frac{B + \frac{4}{3}\mu}{\rho}\right)^{1/2} = \left(\frac{\lambda + 2\mu}{\rho}\right)^{1/2}, \tag{3.48}$$

where B is the bulk modulus, μ is the shear modulus, and λ is the dilatation modulus. (λ and μ are Lamé's elasticity constants.) For the special case of a thin solid rod,

$$c = (Y/\rho)^{1/2}, \tag{3.49}$$

where Y is Young's modulus. For shear waves in a solid,

$$c_s = (\mu/\rho)^{1/2}. \tag{3.50}$$

3.7.2. Configurational dispersion in a solid rod

The configurational dispersion in a solid rod is the change in compressional (sound) wave velocity with frequency due to the interaction between the longitudinal wave and the associated lateral vibrations. The ratio of the longitudinal wave velocity in such a rod to the shear wave velocity increases with frequency. The effect is most noted in the ultrasonic range.

3.7.3. Attenuation of sound in solids

The attenuation of sound in solids depends on many factors, including temperature, heat flow, magnetization, and dislocations. In general, the attenuation may be considered exponential [see Eq. (3.21)], with the attenuation coefficient varying directly either with the fre-

TABLE 3.11. Log decrement for selected solids (resonant frequencies in the range 10–50 kHz). From Ref. [10].

Solid	$10^3\delta$
Aluminum (annealed)	
200 K	0.03
275 K	0.1
Copper (unannealed)	
100 K	13.5
250 K	0.65
Lead	
225 K	3.15
250 K	9.5
Silver (annealed), 250 K	
(little change with temperature)	0.04
Polystyrene, room temperature	48

quency or its square. This is a very complicated subject, and the elaborate special literature should be consulted for details. For finite solids, the attenuation is usually best represented by the logarithmic decrement [see Eq. (3.5)]. Table 3.11 lists some typical values of the decrement.

3.8. MOLECULAR ACOUSTICS; RELAXATION PROCESSES

3.8.1. Propagation of sound

The propagation of sound is fundamentally a molecular process, and the interaction between elastic wave propagation and molecular behavior has a significant effect on sound dispersion and attenuation, particularly at ultrasonic frequencies.

If a sound wave in a fluid disturbs any particular equilibrium molecular aggregation (involving, for example, transfer of translational energy into internal energy modes of the polyatomic molecules), it takes a certain time τ, called the relaxation time, for the original state to be restored after the passage of the crest of the wave. The process is usually called thermal relaxation.

For a polyatomic gas, it is found that the quantity $\alpha/\pi f$, where α is the attenuation coefficient [see Eq. (3.21)] and f is the frequency, when plotted as a function of frequency has a maximum at

$$f = 1/2\pi\tau \quad \text{or} \quad \omega\tau = 1 \tag{3.51}$$

and that the course of the attenuation as a function of frequency is given by

$$\frac{\alpha}{\pi f} = \left(\frac{\alpha}{\pi f}\right)_{\max} \frac{2\omega\tau}{1 + \omega^2\tau^2}. \tag{3.52}$$

The value of τ is usually evaluated in practice experimentally from Eq. (3.51). For rotational relaxation in hydrogen at 20 °C, the value of τ is of the order of 10^{-8} s. For oxygen, vibrational relaxation prevails, with τ of the order 10^{-3} s. The values of α discussed here are, of course, in excess of those due to viscosity and heat conduction (Sec. 3.3).

3.8.2. Excess sound attenuation in liquids

Excess sound attenuation in liquids is attributable to a variety of relaxation processes connected with changes in the structural aggregation of molecules and with chemical reactions and dissociations. Chemical dissociation (as affected by sound propagation) connected with dissolved magnesium sulfate in sea water has been invoked to explain the abnormally high sound attenuation in this medium. The literature must be consulted for details.

3.8.3. Ultrasonic propagation at very low temperatures

There are numerous anomalies in ultrasonic propagation at very low temperatures. This is particularly evident in superconductors, as Figure 3.5 shows.

In superfluid liquid helium, four different varieties of sound and sound velocity in addition to the normal kind have been detected.

Normal sound in liquid helium II is called first or zero sound. Second sound in this liquid is a temperature wave propagating at a speed of about 20 m/s between 1 and 2 K, decreasing to zero at the λ-transition point. Third sound is propagation of thickness variation in a thin film of liquid helium II. The velocity depends on the film thickness, ranging from about 10 m/s for 15 atomic layers to 60 m/s for 5 atomic layers. Fourth sound is a compressional wave in liquid helium II in narrow pores in which the normal component is locked in and only the superfluid component participates in the wave propagation. The velocity varies with the temperature, ranging from about 220 m/s at 1.2 K to 100 m/s at 2.1 K.

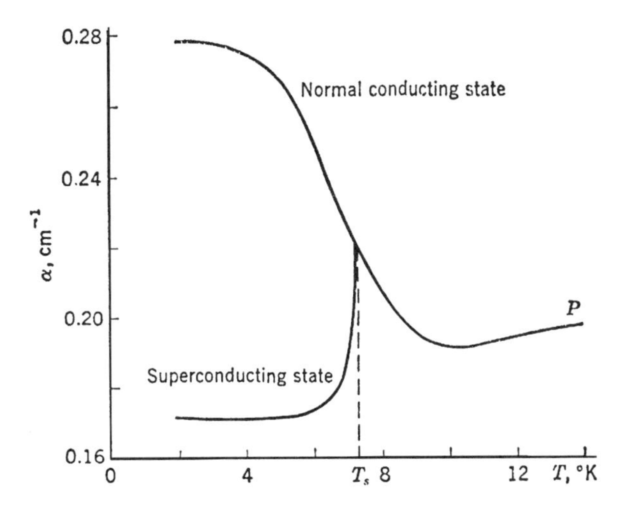

FIGURE 3.5. Variation of acoustic attenuation with temperature for a superconductor. From Ref. [9].

Fifth sound results if in experiments leading to second sound, one immobilizes the normal component and applies a pressure-release boundary condition. The velocity is a function of temperature, reaching a maximum of 10 m/s at 2 K and vanishing at the λ point. See Ref. [11].

3.9. BUBBLES, CAVITATION, SONOLUMINESCENCE

3.9.1. Cavitation

Cavitation refers to the formation of holes in liquids, and it is a matter of indifference whether the holes are produced by local heating in a kettle of water, by the slashing of a propeller blade through the water, or by the oscillation of liquid particles under the action of a sound beam. Acoustic cavitation can itself be divided into two types. The first of these (gaseous cavitation) results from the growth of gaseous bubbles in the liquid during the passage of a sound wave, whereas the second (vapor cavitation) describes the appearance and growth of holes in a degassed liquid, in which the only gaseous content is the vapor of the host liquid.

The dynamics of cavitation bubble growth and collapse has been widely studied,[8] and several equations have been developed to describe their behavior. In the Noltingk-Neppiras (1950, 1951) equation, the liquid is assumed to be incompressible, so that the analysis does not hold in the stage of bubble collapse. The governing equation here is

$$\frac{dU}{dt} + \frac{3}{2}\frac{U^2}{R} = \frac{1}{\rho_{l0} R}\left[\left(p_0 + \frac{2\sigma}{R_0}\right)\left(\frac{R_0}{R}\right)^{3\Gamma} + p_v - \frac{2\sigma}{R} - p_{l0} + p_m \sin \omega t\right]. \quad (3.53)$$

In the Herring-Flynn (1941, 1965) equation, compressibility of the liquid is taken roughly into account, but the results are not accurate if the rate of growth of the bubble $(U - dR/dt)$ is comparable with or larger than the speed of sound. The Herring-Flynn equation is

$$\left(1 - \frac{2U}{c_0}\right)\frac{dU}{dt} + \frac{3}{2}\left(1 - \frac{4}{3}\frac{U}{c_0}\right)\frac{U^2}{R} = \frac{1}{\rho_{l0} R}\left[p_l(t) - p_\infty(t) + \frac{R}{c_0}\left(1 - \frac{1}{c_0}U\right)\frac{dp_l}{dt}\right]. \quad (3.54)$$

Finally, there is the Kirkwood-Bethe equation (1942), developed for underwater explosions. This takes the form

$$\left(1 - \frac{u}{c}\right)\frac{dU}{dt} + \frac{3}{2}\left(1 - \frac{U}{3c}\right)\frac{U^2}{R} = \left(1 + \frac{U}{c}\right)H + \frac{U}{c}\left(1 - \frac{U}{c}\right)\frac{dH}{dR} = 0. \quad (3.55)$$

Here,

$$H = \int_{p_\infty}^{p(R)} \frac{dp}{\rho}.$$

In the above equations, Γ is the effective specific-heat ratio, $p_\infty = p_{l0} - p_m \sin \omega t$, R is the radius of the bubble, R_0 is the initial radius of the bubble, p_{l0} and p_l are the pressure in the liquid at $R = R_0$ and $R = R_1$, p_v is the vapor pressure, $U = dR/dt$, σ is the surface tension, and p_m is the pressure amplitude of the sound wave of angular frequency, ω.

3.9.2. Sonoluminescence

Sonoluminescence refers to the luminescence that appears when an ultrasonic signal is applied to water in which air or oxygen is dissolved. Recent studies have divided into the investigation of a single bubble and of multiple bubbles.[12] The duration of the light flashes is very short, less than 50 ps,[13] suggesting that very high temperatures are created. The precise origin of the light is not yet known, although it has been attributed to the dynamic Casimir effect.[14]

3.10. NONDESTRUCTIVE TESTING, ULTRASONIC IMAGING

3.10.1. Nondestructive testing

The large difference in acoustic impedance between solids and fluids and the creation of ultrasonic beams made it possible to detect flaws in solids (nondestructive testing). An ultrasonic pulse of short duration, passing through a solid and encountering a hole, a crack, or any other discontinuity, will be reflected back to the source. The one-dimensional picture of the pulse and its echo on an oscilloscope locates the flaw. The technique is known as an A scan.[15] Instead of measuring the intensity of the echo by the height of the pulse on the oscilloscope screen, one can also measure it by the brightness of the resultant spot. If a series of such scans of a sample is taken from different positions in the same plane, a two-dimensional view of the flaw is then obtained.[16] This is known as a B scan.

3.10.2. Ultrasonic imaging

The creation of two-dimensional pictures of the cross section of samples led to numerous application in medicine.[17] Ultrasonic pictures of the human fetus (ultrasonograms) became commonplace, and echocardiography, brain and kidney scans, and other developments abounded.[18] The focusing of shock waves within the human body has made possible the destruction of kidney and gall stones (lithotripsy).[19]

3.11. NOISE AND ITS CONTROL

Noise is unwanted sound. Ambient noise is the composite of all unwanted sound in a given environment. Common sources of noise are machinery operating inside and outside of buildings, transportation vehicles (in particular, aircraft), and human beings and animals. Intense noise can produce temporary and even permanent hearing loss, particularly if the exposure is long continued.

It can be seen from Fig. 3.7 that the human ear is far less sensitive to sound frequencies below 200 Hz than it is over higher frequencies. A weighting scheme that takes this into account is known as *A*-weighting, and the decibel readings with such a weighting are called dBa. This weighting, incorporated into the sound level meter, matches its sensitivity as far as possible with that of the human ear.

The following tables and figure relate to noise and its effects, as well as to recommended levels for safety and comfort.

3.11.1. Active suppression of noise

Active suppression of noise is the creation, by means of microphones and digitizing equipment, of a signal that mirrors that of the noise but is opposite in phase, so that the playback by a loudspeaker will cancel the original sound.[21] The same technique can be used for active vibration suppression.[22]

TABLE 3.12. Typical noise levels due to various sources (approximate values at the source). Mainly from Ref. [2].

Sound power (dB) relative to 10^{-12} W	Source
200	Large rocket engine
160	Aircraft turbo jet engine
140	Light airplane, cruising
115	Crawler tractor, 150 hp
105	100-hp electric motor at 2600 rpm
100	Pneumatic drill
90	Subway with train passing
85	Vacuum cleaner
75	Busy traffic
65	Conversational speech
40	Whispered speech

TABLE 3.13. Recommended acceptable average noise levels in unoccupied rooms. The levels are "weighted," i.e., measured with a standard sound-level meter incorporating an A (40-dB) frequency-weighting network. From Ref. [20].

Rooms	Noise level (dB)
Radio, recording, and television studios	25–30
Music rooms	30–35
Legitimate theaters	30–35
Hospitals	35–40
Motion-picture theaters, auditoriums	35–40
Churches	35–40
Apartments, hotels, homes	35–45
Classrooms, lecture rooms	35–40
Conference rooms, small offices	40–45
Courtrooms	40–45
Private offices	40–45
Libraries	40–45
Large public offices, banks, stores, etc.	45–55
Restaurants	50–55

TABLE 3.14. NIPTS (noise-induced permanent threshold shift) at 4000 Hz for population median after approximately 10 years of daily exposure to noises having A-weighted sound levels given along the horizontal axis. Open circles indicate more recent data. From Ref. [2].

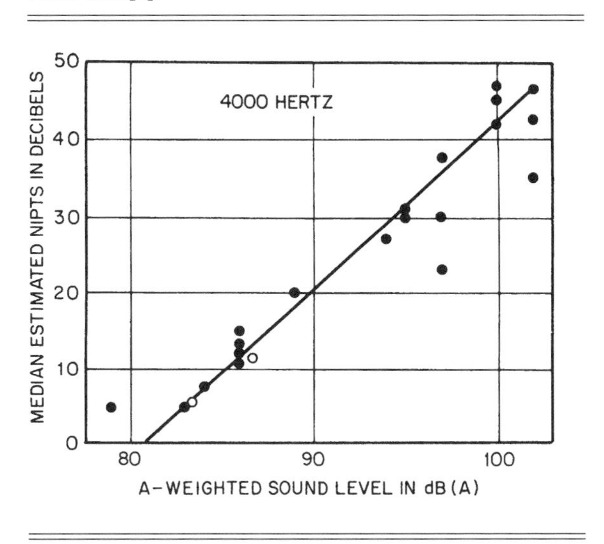

3.12. ROOM AND ARCHITECTURAL ACOUSTICS

3.12.1. Sabine's formula

In room acoustics, the reverberation time is the time after a source of given frequency has been stopped for the sound intensity level in the room to decrease by 60 dB. Sabine's formula for the reverberation time is

$$T = c_R V/A, \tag{3.56}$$

where T is the reverberation time, V is the volume of the room, A is the absorption, and $c_R = 0.049$ s/ft $= 0.161$ s/m.

Sabine's reverberation time formula has been modified by C. F. Eyring to the form

$$T = \frac{kV}{-S\ln(1-\bar{\alpha}) + 4mV}, \tag{3.57}$$

where k is a constant, V is the volume of the room, S is the total surface area of the room, and $\bar{\alpha}$ is the average absorption coefficient of the room surfaces. The correction term $4mV$ is due to the absorption of the air in the room. The Eyring formula has been considered particularly applicable to "dead" rooms.

3.12.2. Practical architectural acoustics

The following table and figure present material relating to practical architectural acoustics.

TABLE 3.15. Absorption coefficients (in sabins) for building materials. From Ref. [23].

	Frequency (Hz)					
Materials	125	250	500	1000	2000	4000
Brick, unglazed	0.03	0.03	0.03	0.04	0.05	0.07
Painted	0.01	0.01	0.02	0.02	0.02	0.03
Carpet, heavy, on concrete	0.02	0.06	0.14	0.37	0.60	0.65
On 40-oz hairfelt or foam rubber	0.08	0.24	0.57	0.69	0.71	0.73
With impermeable latex backing on 40-oz hairfelt or foam rubber	0.08	0.27	0.39	0.34	0.48	0.63
Concrete block						
Coarse	0.36	0.44	0.31	0.29	0.39	0.25
Painted	0.10	0.05	0.06	0.07	0.09	0.08
Fabrics						
Light velour, 10 oz/yd^2 hung straight, in contact with wall	0.03	0.04	0.11	0.17	0.24	0.35
Medium velour, 14 oz/yd^2, draped to half area	0.07	0.31	0.49	0.75	0.70	0.60
Heavy velour, 18 oz/yd^2, draped to half area	0.14	0.35	0.55	0.72	0.70	0.65
Floors						
Concrete or terrazzo	0.01	0.01	0.015	0.02	0.02	0.02
Linoleum, asphalt, rubber, or cork tile on concrete	0.02	0.03	0.03	0.03	0.03	0.02
Wood	0.15	0.11	0.10	0.07	0.06	0.07
Wood parquet in asphalt on concrete	0.04	0.04	0.07	0.06	0.06	0.07
Glass						
Large panes of heavy plate glass	0.18	0.06	0.04	0.03	0.02	0.02
Ordinary window glass	0.35	0.25	0.18	0.12	0.07	0.04
Gypsum board, $\frac{1}{2}$ in. nailed to 2 × 4's 16 inches o.c.	0.29	0.10	0.05	0.04	0.07	0.09
Marble or glazed tile	0.01	0.01	0.01	0.01	0.02	0.02
Openings						
Stage, depending on furnishing			0.25–0.75			
Deep balcony, upholstered seats			0.50–1.00			
Grills, ventilating			0.15–0.50			
Plaster, gypsum, or lime						
Smooth finish on tile or brick	0.013	0.015	0.02	0.03	0.04	0.05
Rough finish on lath	0.02	0.03	0.04	0.05	0.04	0.03
Smooth finish on lath	0.02	0.02	0.03	0.04	0.04	0.03
Plywood paneling, $\frac{3}{8}$ inch thick	0.28	0.22	0.17	0.09	0.10	0.11
Water surface, as in a swimming pool	0.008	0.008	0.013	0.015	0.020	0.025

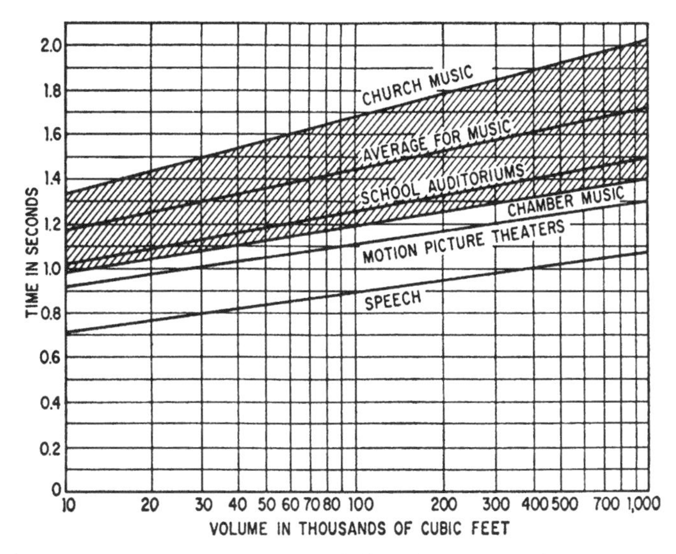

FIGURE 3.6. Optimum reverberation time at 500 Hz for different types of rooms as a function of room volume. From Ref. [1] (after Ref. [20]).

3.13. PHYSIOLOGICAL AND PSYCHOLOGICAL ACOUSTICS

Physiological acoustics is the study of the physiological response in animals, including man, to acoustic stimuli. It includes reference to the anatomy of the outer, middle, and inner ear, with the physiology of the cochlea and the auditory central nervous system. Physics enters essentially in the mechanics of the middle ear and cochlea and in the electrical transmissions of nerve impulses from the cochlea to the brain.

Psychological acoustics is the study of the psychological response of animals, including man, to acoustic stimuli. It includes reference to the following topics.

3.13.1. Loudness and loudness level

Loudness is the subjective impression of the intensity of sound. The unit is the sone (see Sec. 3.1). Loudness level is the sound-pressure level in decibels of a pure tone of frequency 1000 Hz (relative to 2×10^{-4} dyn/cm^2), which is assessed by normal observers as being equally loud as the sound being measured. The unit is the phon (see Sec. 3.1).

The loudness S of a sound of 1000 Hz measured in sones is given by

$$\log_{10} S = 0.0301(P - 40), \tag{3.58}$$

where P is the loudness level in phons.

3.13.2. Auditory sensation area

The following figure plots the average minimum audible field threshold for young adults as a function of frequency.

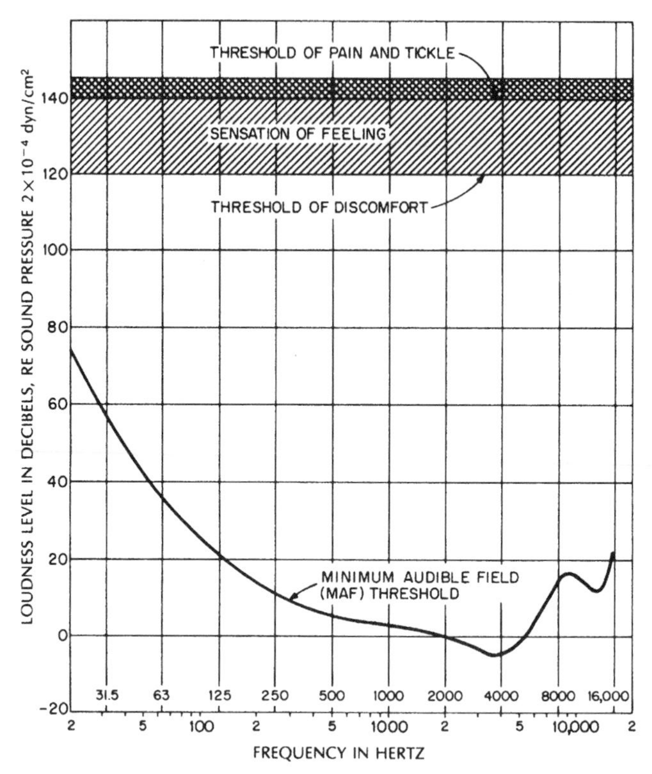

FIGURE 3.7. Average minimum audible field threshold as a function of frequency. Upper limits are set by the thresholds of discomfort, feeling, pain, and tickle. Lower limit is minimum audible field (MAF) threshold for young adults. From Ref. [2].

3.13.3. Masking

Masking is the process by which the threshold of hearing of one sound is raised due to the presence of another sound. It is primarily in general due to noise and is measured in decibels. An important quantity in this connection is the *signal-to-noise* ratio, the ratio of the signal intensity (peak or root-mean-square) to that of the noise, expressed in decibels.

3.13.4. Temporary threshold shift (TTS)

The TTS is the temporary increase in the hearing threshold induced by noise. In such a case, hearing ultimately returns to normal. The following figure is illustrative. For permanent threshold shift, see Sec. 3.11.

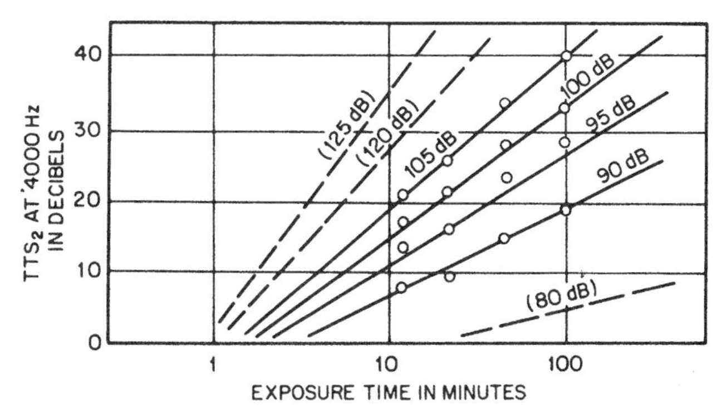

FIGURE 3.8. Temporary threshold shift at 4000 Hz measured 2 minutes after exposure to an octave band of noise centered at 1700 Hz at the sound-pressure levels and durations indicated. From Ref. [2].

3.13.5. Pitch

Pitch is the subjective estimate of a tone as higher or lower on a scale. It is measured in mels (see Sec. 3.1). Its relation to sound frequency is indicated in the following table.

TABLE 3.16. Pitch of a pure tone (in mels) as a function of frequency. From Ref. [1].

Frequency	Mels	Frequency	Mels	Frequency	Mels
20	0	350	460	1750	1428
30	24	400	508	2000	1545
40	46	500	602	2500	1771
60	87	600	690	3000	1962
80	126	700	775	3500	2116
100	161	800	854	4000	2250
150	237	900	929	5000	2478
200	301	1000	1000	6000	2657
250	358	1250	1154	7000	2800
300	409	1500	1296	10000	3075

3.13.6. Binaural hearing

Binaural hearing is hearing by use of the two ears, through which the direction of the source of sound may be determined. It also assists in the extraction of the signal from disturbing noise.

3.13.7. Audiogram

An audiogram is a graph showing hearing loss as a function of frequency. It is produced by the use of an instrument called an audiometer. Hearing loss due to aging is called presbycusis. It mainly affects tones of higher frequency.

3.14. SPEECH COMMUNICATION

Speech is the result of the alteration in frequency and intensity of voiced and unvoiced sounds produced by the lungs and larynx. Modification of the cavities of the mouth and nose serve to introduce spectral content into speech. The normal range of speech frequencies is approximately 400–6000 Hz.

Total, average radiated conversational speech power has been estimated at about $30 \mu W$.

Among important acoustical aspects of speech are spectral analysis of speech, speech synthesis (artificial speech), perception of speech by individuals and machines, speech transmission systems, and phonetics (speech and linguistics). Some important quantities in speech communication are as follows:

Articulation. The percentage of speech units spoken by a speaker correctly identified by a listener.

Formant of a speech sound. The frequency range of the spectrum of the sound within which the partials (harmonics) have relatively large amplitudes. The central frequency within the format is called the formant frequency.

Phoneme. The smallest phonetic unit of a speech sound.

3.15. BIOACOUSTICS

Bioacoustics is the study of the acoustical characteristics of biological media. It also includes the effects of sound on living systems, the generation and detection of sound by animals, as well as medical diagnosis and therapy with acoustic radiation. The role of ultrasonic radiation is particularly stressed. The following table is illustrative.

TABLE 3.17. Ultrasonic propagation properties of some tissues at 1 MHz, listed in order of increasing attenuation coefficient. Except for bone, the attenuation coefficient follows the relation $\alpha = \alpha_1 (F/\text{MHz})^{1.1}$, where α_1 is the attenuation coefficient at 1 MHz (from table). Condensed from Ref. [24].

Tissue	Attenuation coefficient (Np/cm)	Velocity (m/s)	Density (g/cm^3)	Impedance (10^5 rayl)	Trends
Blood	0.014	1566	1.04	1.63	Increasing structural protein content ↓; Decreasing water content
Fat	0.07	1478	0.92	1.36	
Nerve	0.10				
Muscle	0.14	1552	1.04	1.62	
Blood vessel	0.20	1530	1.08	1.65	
Skin	0.31	1519		1.58	
Tendon	0.56	1750			
Cartilage	0.58	1665			
Bone	1.61	3445	1.82	6.27	

3.16. MUSICAL ACOUSTICS

3.16.1. Resonance frequencies for an organ pipe

Resonance frequencies of an open organ pipe of length l (open at both ends) are given by

$$f = \frac{nc}{2(l + x_1 + x_2)}, \tag{3.59}$$

where c is the velocity of sound in the medium; x_1 = end correction = $0.3d$ for the unimpeded end, where d is the diameter of the pipe; $x_2 = 1.4d$ for the mouth of the pipe; and n is an integer giving the particular harmonic of the pipe.

For an organ pipe closed at one end,

$$f = \frac{nc}{4(l + x)}, \tag{3.60}$$

where $x = 1.4d$.

3.16.2. Resonance frequencies for a rectangular membrane

Resonance frequencies for a rectangular membrane with sides a_1 and a_2 are given by

$$f = \frac{c}{2}\left[\left(\frac{n_1}{a_2}\right)^2 + \left(\frac{n_2}{a_2}\right)^2\right]^{1/2}, \tag{3.61}$$

where n_1 and n_2 are any integers, and c is the velocity of transverse waves in the membrane, with

$$c = (T/\sigma)^{1/2}, \tag{3.62}$$

where T is the surface tension and σ is the surface density.

3.16.3. Resonance frequencies for a circular membrane

Resonance frequencies for a circular membrane of radius a, clamped at the periphery, are given by

$$J_n(2\pi f a/c) = 0, \tag{3.63}$$

where c is given by Eq. (3.62). J_n is the Bessel function of order n. The lowest resonance frequencies of the clamped circular membrane are

$$0.7655c/2a, \quad 1.2197c/2a, \quad 1.6347c/2a. \tag{3.64}$$

3.17. ACOUSTICAL MEASUREMENTS AND INSTRUMENTS

All acoustical measurements demand the availability of sound sources capable of producing both continuous and pulsed radiation with controllable frequency and intensity. Equally suitable sound receivers (microphones) are also necessary. These are all referred to as acoustic transducers. The most common transducers are based on the use of the piezoelectric and ferroelectric effects, electrostatic and electrodynamic effects (loudspeakers), and magnetostriction. Sources of this kind are usually coupled to plane diaphragms (so-called piston sources), radiating sound that varies in intensity with the distance from the source as well as the direction. For a given piston, a directivity factor can be calculated. This is he ratio of the intensity at a given distance from the source on a presented axis from the transducer to the intensity that would be produced at the same point by a spherical source centered at the transducer and radiating the same total acoustic power into a free field. By the use of appropriate baffles, a piston source may be made to produce approximate plane radiation at a sufficiently large distance from the source.

Most transducers can be adapted to the production of short pulses of radiation, which are useful in many measurements.

3.17.1. Absolute measurement of sound intensity

This can be carried out by means of a Rayleigh disk, which is a solid circular disk suspended in a tube so that it can rotate about a diametral axis normal to the axis of the tube. The torque M on such a disk of radius a with angle θ between the normal to the disk and the direction of propagation of the sound wave is given by

$$M = \tfrac{4}{3}\rho a^3 u^2 \sin 2\theta, \tag{3.65}$$

where ρ is the density of the medium and u is the rms particle velocity in the sound wave, which can be measured by the disk with the use of this equation, thus, providing a direct measurement of the intensity of the sound wave.

3.17.2. Calibration of microphones

This is now usually carried out by a method based on the acoustical reciprocity principle. In its simplest form, this says that in a region containing a simple source, such a source at point A produces the same sound pressure at another point B in the medium as would have been produced at A had the source been located at B. The specific literature should be consulted for details.

3.17.3. Frequency measurement

Acoustical frequency standardization is based on a comparison with standard radio frequencies accurate to one part in 10^7.

The precision, electrically driven tuning fork is also in use as a secondary frequency standard. The cathode ray oscilloscope is used for the comparison of two frequencies by means of Lissajous figures, which are the resultant patterns on an oscilloscope screen when sound is detected at two separate microphones connected separately to the x- and y-plates of the oscilloscope. Phase differences between the sounds at the microphones give rise to different geometric figures. This provides a means of measuring phase differences (and, hence, sound velocity) or of comparing the frequencies of two sounds.

3.17.4. Acoustic filter

An acoustical structure with periodic variations in acoustical properties or mode of confinement of the medium constitutes an acoustic filter in the sense that (disregarding irreversible dissipation) sound radiation of certain frequencies is transmitted through the structure (passbands), whereas for other frequencies, there is complete attenuation (attenuation bands). As an example, consider a fluid-filled tube of circular cross-sectional area S with side branches spaced a distance l-apart and with branch impedance equal to Z_b for each branch. The transmission through such a structure (assumed infinite in extent) is governed by

$$\cos W = \cos \frac{4\pi f l}{c} + \frac{iZ}{2Z_b} \sin \frac{4\pi f l}{c}, \tag{3.66}$$

where f is the frequency, c is the velocity of sound in the fluid, and $Z = \rho_0 c/S$, the acoustic resistance of a plane wave progressing along the tube. For $|\cos W| \leq 1$, there is complete transmission; for $|\cos W| > 1$, there is complete attenuation. Thus, the plot of $|\cos W|$ as a function of frequency gives the pass and attenuation bands.

3.18. REFERENCES

[1] *American Institute of Physics Handbook*, 3rd ed., edited by Dwight E. Gray (McGraw-Hill, New York, 1972).

[2] *Handbook of Noise Control*, 2nd ed., edited by Cyril M. Harris (McGraw-Hill, New York, 1979).

[3] R. T. Beyer and S. V. Letcher, *Physical Ultrasonics* (Academic, New York, 1969).

[4] J. Lyman and R. H. Fleming, J. Mar. Res. **3**, 134 (1940).

[5] F. H. Fisher and V. P. Simmons, J. Acoust. Soc. Am. **62**, 558 (1977).

[6] R. T. Beyer, *Nonlinear Acoustics* (Naval Sea Systems Command, Washington, DC, 1974). [Reprinted 1997 by American Institute of Physics for Acoustical Society of America.]

[7] A. B. Coppens *et al.*, J. Acoust. Soc. Am. **38**, 797 (1965).

[8] R. T. Beyer, *Encyclopedia of Applied Physics* (VCH Pub., New York, 1991), Vol. I.

[9] R. B. Lindsay, *Mechanical Radiation* (McGraw-Hill, New York, 1960).

[10] P. G. Bordoni, J. Acoust. Soc. Am. **26**, 495 (1954).

[11] I. Rudnick, J. Acoust. Soc. Am. **68**, 36 (1980).

[12] L. A.Crum and R. A. Roy, Science **266**, 233 (1994).

[13] B. P. Barber and S. J. Putterman, Nature **352**, 318 (1991).

[14] J. Schwinger, Proc. Nat. Acad. Sci. **89**, 4091 (1992).

[15] F. A. Firestone, Metal Prog. **48**, 505 (1945).

[16] E. Papadakis (private communication).

[17] J. J. Wild and J. M. Reid, Am. J. Pathol. **28**, 839 (1952); Cancer Res. **14**, 277 (1954).

[18] H. Feigenbaum, *Echocardiography*, 3rd. ed. (Lee and Febiger, Philadelphia, PA, 1981).

[19] L. A. Crum, C. C. Church, and D. T. Blackstock, J. Acoust. Soc. Am. **85**, 2225 (1989).

[20] V. O. Knudsen and C. M. Harris, *Acoustic Designing in Architecture* (Wiley, New York, 1950). [Reprinted 1976 by American Institute of Physics for Acoustical Society of America.]

[21] P. A. Nelson and S. J. Elliott, J. Acoust. Soc. Am. **91**, 1195 (P) (1992).

[22] C. R. Fuller, S. J. Elliott and P. A. Nelson, *Active Control of Vibration* (Academic, London, 1996).

[23] Acoust. Mater. Assoc. Bull. **29** (1969).

[24] S. A. Goss, R. L. Johnston, and F. Dunn, J. Acoust. Soc. Am. **64**, 423 (1978).

4

Astronomy

Jay M. Pasachoff
Hopkins Observatory, Williams College, Williamstown, MA 01267 and Harvard-Smithsonian Center for Astrophysics, Cambridge, MA 02138

Contents

List of Tables

List of Figures

4.1. BASIC DATA

TABLE 4.1. Astronomical constants

Day	$= 86\,400$ s
Julian century	$= 36\,525$ days
	$= 3.155\,76 \times 10^9$ s
Tropical (ephemeris) year (J2000)	$= 31\,556\,925.1875$ s
Sidereal year (J2000)	$= 31\,558\,149.5$ s
Astronomical Unit	$1\ \text{AU} = 1.495\,97870 \times 10^{11}$ m
$\tau_A c = 1$ AU	$\tau_A = 499.047\,82$ s
Solar parallax	$\pi_\odot = 8.794\,148$ arcsec
Light year (Julian year)	$1\ \text{ly} = 9.460\,7305 \times 10^{15}$ m
Parsec	$1\ \text{pc} = 206\,264.8062$ A.U.
	$= 3.0856776 \times 10^{16}$ m
	$= 3.2616$ ly
Mass of the Sun	$M_\odot = 1.9891 \times 10^{30}$ kg
Radius of the Sun	$R_\odot = 696\,000$ km
Luminosity of the Sun	$L_\odot = 3.827 \times 10^{26}$ J/s
Mass of the Earth	$M_E = 5.9743 \times 10^{24}$ kg
Equatorial radius of the Earth	$R_E = 6378.140$ km
Center of Earth to center of Moon (mean)	$= 384\,403$ km
Mass of the Moon	$M_M = 7.35 \times 10^{22}$ kg
Radius of the Moon	$R_M = 1738$ km
Solar constant	$S = 1.368$ W/m^2
Direction of the galactic center (J2000)	$\alpha = 17^h\,45.0383^m \pm 0.0007^m$
	$\delta = -29°\,00'\,28.069'' \pm 0.014''$

TABLE 4.2. Precession

For epoch J2000.0 (JD2451545.0):	
Inclination of the ecliptic, ϵ_o	$23°\,26'\,21.448''$
Rate of precession, Ω	$50.290\,966''/\text{year} = 3.352\,731^s/\text{year}$
Change in δ (dec.)	$\Omega t \sin\epsilon_o \cos\alpha$
Change in α (R.A.)	$\Omega t[\cos\epsilon_o + \sin\epsilon_o \tan\delta]$

4.2. SOLAR SYSTEMS

TABLE 4.3. Intrinsic and rotational properties[a]

Name	Equatorial radius km	Equatorial radius ÷ Earth's	Mass ÷ Earth's	Mean density (kg/m^3)	Oblateness	Surface gravity ($g_E = 1$)	Sidereal rotation period[b]	Inclination of equator to orbit	Apparent magnitude during 1997
Mercury	2439	0.3824	0.0553	5430	0	0.378	$58^d15^h30^m$	0.0°	−2.2 to +4.8
Venus	6052	0.9489	0.8150	5240	0	0.894	$243^d00^h14^m$ R	177.3°	−4.7 to −3.9
Earth	6378.140	1	1	5515	0.0034	1	$23^h56^m04.1^s$	23.45°	—
Mars	3397	0.5326	0.1074	3940	0.005	0.379	$24^h37^m22.662^s$	25.19°	−1.3 to +1.2
Jupiter	71492	11.194	317.896	1330	0.064	2.54	9^h50^m to > 9^h55^m	3.12°	−2.8 to −1.9
Saturn	60268	9.41	95.18	700	0.108	1.07	$10^h39.9^m$	26.73°	+0.2 to +1.0
Uranus	25559	4.0	14.537	1300	0.03	0.8	17^h14^m R	97.86°	+5.7 to +5.9
Neptune	24764	3.9	17.151	1760	0.017	1.2	16^h07^m	29.56°	+1.9 to +8.0
Pluto	1151	0.2	0.0025	2030	?	0.01	$6^d09^h17^m$ R	120°	+13.7 to +13.8

[a]The masses and radii for Mercury, Venus, Earth, and Mars are the values recommended by the International Astronomical Union in 1976. The radii are from *The Astronomical Almanac 2002* (U.S. Government Printing Office, 2001). Surface gravities were calculated from these values.
The length of the Martian day is from G. de Vaucouleurs (1979). Most densities, oblatenesses, inclinations, and magnitudes are from *The Astronomical Almanac 1997*. Pluto values are from David J. Tholen. Neptune data are from *Science*, December 15, 1989, and August 9, 1991. Values for the masses of the giant planets are based on Voyager data for the mass of the Sun divided by the mass of the planet [E. Myles Standish, Jr., *Astron. J.*, **105**, 2000, (1993)]: Jupiter: 1047.3486; Saturn: 3497.898; Uranus: 22902.94; Neptune: 19412.24.
[b]R signifies retrograde rotation.

TABLE 4.4. Orbital properties*

Name	Semimajor axis AU	Semimajor axis 10^6 km	Sidereal period Years	Sidereal period Days	Synodic period (Days)	Eccentricity	Inclination to ecliptic
Mercury	0.387099	57.909	0.24084	87.96	115.9	0.20563	7.00487°
Venus	0.723332	108.209	0.61518	224.68	583.9	0.00677	3.39471°
Earth	1	149.598	0.99998	365.25	—	0.01671	0.00005°
Mars	1.523662	227.939	1.8807	686.95	779.9	0.09341	1.85061°
Jupiter	5.203363	778.298	11.857	4337	398.9	0.04839	1.30530°
Saturn	9.537070	1429.394	29.424	10760	378.1	0.05415	2.48446°
Uranus	19.191264	2875.039	83.75	30700	369.7	0.047168	0.76986°
Neptune	30.068963	4504.450	163.72	60200	367.5	0.00859	0.76917°
Pluto	39.481687	5915.799	248.02	90780	366.7	0.24881	17.14175°

*Mean elements of planetary orbits for 2000, referred to the mean ecliptic and equinox of J2000. (E.M. Standish, X.X. Newhall, J.C. Williams, and D.K. Yeomans, *Explanatory Supplement to the Astronomical Almanac*, edited by P. K. Seidelmann, University Science Books, Mill Valley, CA, 1992). Periods are calculated from them.

TABLE 4.5. Planetary satellites[a]

Satellite		Orbital period[b]	Maximum elongation at mean opposition	Semi-major axis	Orbital eccentricity	Inclination of orbit to planet equator	Mass	Radius	Sidereal period of rotation[f]	Mean magnitude at opposition
		d	° ′ ″	$(a/10^3\text{km})$		°	m/m_{Planet}	km[h]	d	$V^{h,i}$
Earth										
	Moon	27.321661		384.400	0.054900	18.28 −28.58	0.01230002	1737.4	S	0.12
Mars										
I	Phobos	0.31891023	0 00 25	9.378	0.015	1.0	1.65×10^{-8}	13.4 × 11.2 × 9.2	S	0.07
II	Deimos	1.2624407	0 01 02	23.459	0.0005	0.9–2.7	3.71×10^{-9}	7.5 × 6.1 × 5.2	S	0.08
Jupiter										
I	Io	1.769137786	0 02 18	422	0.004	0.04	4.70×10^{-5}	1830 × 1819 × 1815	S	0.63
II	Europa	3.551181041	0 03 40	671	0.009	0.47	2.53×10^{-5}	1565	S	0.67
III	Ganymede	7.15455296	0 05 51	1070	0.002	0.21	7.80×10^{-5}	2634	S	0.44
IV	Callisto	16.6890184	0 10 18	1883	0.007	0.51	5.67×10^{-5}	2403	S	0.20
V	Amaltbea	0.49817905	0 00 59	181	0.003	0.40	3.8×10^{-9}	131 × 73 × 67	S	0.07
VI	Himalia	250.5662	1 02 46	11490	0.15798	27.63	5.0×10^{-9}	85	0.40	0.03
VII	Elara	259.6528	1 04 10	11737	0.20719	24.77	0.4×10^{-9}	40		0.03
VIII	Pasiphae	735 R	2 08 46	23500	0.378	145	0.1×10^{-9}	18		0.10
IX	Sinope	758 R	2 09 31	23700	0.275	153	0.4×10^{-10}	14	0.548	0.05
X	Lysithea	259.22	1 04 04	11720	0.107	29.02	0.4×10^{-10}	12	0.533	0.06
XI	Carme	692 R	2 03 31	22600	0.20678	164	0.5×10^{-10}	15	0.433	0.06
XII	Ananke	631 R	1 55 52	21200	0.16870	147	0.2×10^{-10}	10	0.35	0.06
XIII	Leda	238.72	1 00 39	11094	0.14762	26.07	0.3×10^{-11}	5		0.07
XIV	Thebe	0.6745	0 01 13	222	0.015	0.8	0.4×10^{-9}	55 × 45	S	0.04
XV	Adrastea	0.29826	0 00 42	129			0.1×10^{-10}	13 × 10 × 8	S	0.05
XVI	Metis	0.294780	0 00 42	128			0.5×10^{-10}	20	S	0.05
Saturn										
I	Mimas	0.942421813	0 00 30	195.52	0.0202	1.53	6.60×10^{-8}	209 × 196 × 191	S	0.5
II	Enceladus	1.370217955	0 00 38	238.02	0.00452	0.0	0.1×10^{-6}	256 × 247 × 245	S	1.0
III	Tethys	1.887902160	0 00 48	294.66	0.00000	1.86	1.10×10^{-6}	536 × 528 × 526	S	0.9
IV	Dione	2.736914742	0 01 01	377.40	0.002230	0.02	1.93×10^{-6}	560	S	0.7
V	Rhea	4.517500436	0 01 25	527.04	0.00100	0.35	4.06×10^{-6}	764	S	0.7
VI	Titan	15.94542069	0 03 17	1221.83	0.029192	0.33	2.37×10^{-4}	2575	S	0.22
VII	Hyperion	21.2766088	0 03 59	1481.1	0.104	0.43	0.4×10^{-7}	180 × 140 × 113		0.3
VIII	Iapetus	79.3301825	0 09 35	3561.3	0.02829	14.72	2.8×10^{-6}	718	S	0.2[g]
IX	Phoebe	550.48 R	0 34 51	12952	0.16326	177[c]	0.7×10^{-9}	110	0.4	0.06:
X	Janus	0.6945	0 00 24	151.472	0.007	0.14	3.38×10^{-9}	97 × 95 × 77	S	0.9:
XI	Epimetheus	0.6942	0 00 24	151.422	0.009	0.34	9.5×10^{-10}	69 × 55 × 55	S	0.8:
XII	Helene	2.7369	0 01 01	377.40	0.005	0.0		18 × 16 × 15		0.7:
XIII	Telesto	1.8878	0 00 48	294.66				15 × 125 × 7.5		1.0:
XIV	Calypso	1.8878	0 00 48	294.66				15 × 8 × 8		1.0:
XV	Atlas	0.6019	0 00 22	137.670	0.0	0.3		18.5 × 17.2 × 13.5		0.8:
XVI	Prometheus	0.6130	0 00 23	139.353	0.003	0.0		74 × 50 × 34		0.5:
XVII	Pandora	0.6285	0 00 23	141.700	0.004	0.0		55 × 44 × 31		0.7:
XVIII	Pan	0.5750	0 00 21	133.583				10		0.5:

(*continued*)

TABLE 4.5. Planetary satellites (continued)

Satellite	Orbital period[b]	Maximum elongation at mean opposition	Semi-major axis	Orbital eccentricity	Inclination of orbit to planet equator	Mass	Radius	Sidereal period of rotation[f]	Mean magnitude at opposition
	d	° ′ ″	$(a/10^3\text{km})$		°	m/m_{Planet}	km[h]	d	$V^{h,i}$
Uranus									
I Ariel	2.52037935	0 00 14	191.02	0.0034	0.3	1.55×10^{-5}	581 × 578 × 578	S	0.35
II Umbriel	4.1441772	0 00 20	266.30	0.0050	0.36	1.35×10^{-5}	585	S	0.19
III Titania	8.7058717	0 00 33	435.91	0.0022	0.14	4.06×10^{-5}	789	S	0.28
IV Oberon	13.4632389	0 00 44	583.52	0.0008	0.10	3.47×10^{-5}	761	S	0.25
V Miranda	1.41347925	0 00 10	129.39	0.0027	4.2	0.08×10^{-5}	240 × 234 × 233	S	0.27
VI Cordelia	0.3350338	0 00 04	49.77	0.00026	0.08		13		0.07:
VII Ophelia	0.376400	0 00 04	53.79	0.0099	0.10		15		0.07:
VIII Bianca	0.43457899	0 00 04	59.17	0.0009	0.19		21		0.07:
IX Cressida	0.46356960	0 00 05	61.78	0.0004	0.01		31		0.07:
X Desdemona	0.47364960	0 00 05	62.68	0.00013	0.11		27		0.07:
XI Juliet	0.49306549	0 00 05	64.35	0.00066	0.07		42		0.07:
XII Portia	0.51319592	0 00 05	66.09	0.0000	0.06		54		0.07:
XIII Rosalind	0.55845953	0 00 05	69.94	0.0001	0.28		27		0.07:
XIV Belinda	0.62352747	0 00 06	75.26	0.00007	0.03		33		0.07:
XV Puck	0.76183287	0 00 07	86.01	0.00012	0.32		77		0.075
XVI Caliban	579 R	0 08 56	7169	0.082	139.7[c]		30		0.07:
XVII Sycorax	1289 R	0 15 26	12214	0.509	152.7[c]		60		0.07:
XVIII S1986 U10[j]	0.638		76.4				20		
Neptune									
I Triton	5.8768541 R	0 00 17	354.76	0.000016	157.345	2.09×10^{-4}	1353	S	0.77
II Nereid	360.13619	0 04 21	5513.4	0.7512	27.6[d]	0.2×10^{-6}	170		0.4
III Naiad	0.294396	0 00 02	48.23	0.000	4.74		29:		0.06:
IV Thalassa	0.311485	0 00 02	50.07	0.000	0.21		40:		0.06:
V Despina	0.334655	0 00 02	52.53	0.000	0.07		74		0.06
VI Galatea	0.428745	0 00 03	61.95	0.000	0.05		79		0.06
VII Larissa	0.554654	0 00 03	73.55	0.00139	0.20		104 × 89		0.06
VIII Proteus	1.122315	0 00 06	117.65	0.0004	0.55		218 × 208 × 201	S	0.06
Rings									
Galle			42.0						
Leverier			53.3						
Lassell			53.2–57.5						
Arago			57.4						
Adams			53.2						
Ring Arcs									
Courage			62.9						
Liberté			62.9						
Egalité			62.9						
Fraternité			62.9						
Pluto									
I Charon	6.38725	< 1	19.6	< 0.001	99[d]	0.125	593	S	0.5

Source: Adapted from *The Astronomical Almanac for the Year 2002* (U.S. Government Printing Office, 2001), and J. M. Pasachoff, *Field Guide to the Stars and Planets, 2000* (Houghton Mifflin, Boston).

[a] The asteroids 253 Ida, 45 Eugenia, 762 Pulcova, 90 Antiope, 617 Patroclus, 22 Kalliope, 87 Sylvia, $1999TC_{36}$, $1998WW_{31}$, $1998DP_{107}$, and asteroid 26308 have one satellite each. Ida's satellite is named Dactyl and Eugenia's is Petit-Prince. See http://cfa-www.harvard.edu/iau/Headlines.html.

[b] Sidereal periods, except tropical periods for the satellites of Saturn. R = retrograde.

[c] Relative to the ecliptic plane.

[d] Referred to the equator of 1950.0.

[e] On the ecliptic plane.

[f] S = Synchronous, rotation and orbital periods are the same.

[g] Bright side, 0.5; faint side, 0.05.

[h] A colon following an entry indicates an uncertain or nominal value.

[i] $V_{\odot} = -26.75$.

[j] Photographed by Voyager 2, January 1986; recognized as a satellite, May 1999; unconfirmed as of 2002.

See http://ssd.jpl.nasa.gov/sat_props.html and http://ssd.jpl.nasa.gov/sat_elem.html.

TABLE 4.6. (a) Confirmed planets around main sequence stars. Global statistics: 67 planetary systems, 75 planets, and 7 multiple-planet systems

Star	$\frac{M \sin i}{M J}$	Semi-major axis (AU)	Period (days)	Eccentricity
HD 83443	0.35	0.038	2.9861	0.08
	0.16	0.174	29.83	0.42
HD 16141	0.215	0.35	75.82	0.28
HD 168746	0.24	0.066	6.409	0.
HD 46375	0.249	0.041	3.024	0.
HD 108147	0.34	0.098	10.881	0.558
HD 75289	0.42	0.046	3.51	0.054
51 Peg	0.47	0.05	4.2293	0.0
BD −10 3166	0.48	0.046	3.487	0.
HD 6434	0.48	0.15	22.09	0.30
HD 187123	0.52	0.042	3.097	0.03
HD 209458	0.69	0.045	3.524738	0.0
υ And	0.71	0.059	4.6170	0.034
	2.11	0.83	241.2	0.18
	4.61	2.50	1266.6	0.41
HD 192263	0.76	0.15	23.87	0.03
ϵ Eridani	0.86	3.3	2502.1	0.608
HD 38529	0.81	0.1293	14.41	0.280
HD 4208	0.81	1.69	829.0	0.04
HD 179949	0.84	0.045	3.093	0.05
55 Cnc	0.84	0.11	14.648	0.051
	> 5 ?	> 4	> 2920 ?	—
HD 82943	0.88	0.73	221.6	0.54
	1.63	1.16	444.6	0.41
HD 121504	0.89	0.32	64.6	0.13
HD 114783	0.9	1.20	501.0	0.10
HD 37124	1.04	0.585	155	0.19
HD 130322	1.08	0.088	10.724	0.048
ρ CrB	1.1	0.23	39.645	0.028
HD 52265	1.13	0.49	118.96	0.29
HD 177830	1.28	1.00	391	0.43
HD 217107	1.28	0.07	7.11	0.14
HD 210277	1.28	1.097	437.	0.45
HD 142	1.36	0.980	338.0	0.37
HD 27442	1.43	1.18	423	0.02
16 Cyg B	1.5	1.70	804	0.67
HD 74156	1.56	0.276	51.61	0.649
	> 7.5	4.47	2300.0	0.395
HD 134987	1.58	0.78	260	0.25
HD 4203	1.64	1.09	406.0	0.53
HD 68988	1.90	0.071	6.276	0.14
HD 160691	1.97	1.65	743	0.62
HD 19994	2.0	1.3	454	0.2
Gliese 876	1.98	0.21	61.02	0.27
	0.56	0.13	30.1	0.12

(continued)

TABLE 4.6. Confirmed planets around main sequence stars (continued)

Star	$\frac{M \sin i}{MJ}$	Semi-major axis (AU)	Period (days)	Eccentricity
HD 8574	2.23	0.76	228.8	0.40
HR810	2.26	0.925	320.1	0.161
47 UMa	2.41	2.10	1095	0.096
	0.76	3.73	2594	< 0.1
HD 23079	2.54	1.48	627.3	0.06
HD 12661	2.83	0.789	264.5	0.33
HD 169830	2.96	0.823	230.4	0.34
14 Her	3.3	2.5	1619	0.3537
GJ 3021	3.31	0.49	133.82	0.505
HD 80606	3.41	0.439	111.78	0.927
HD 195019	3.43	0.14	18.3	0.05
HD 92788	3.8	0.94	340	0.36
GI 86	4	0.11	15.78	0.046
HD 213240	4.5	2.03	951	0.45
τ Boo	3.87	0.0462	3.3128	0.018
HD 50554	4.9	2.38	1279.0	0.42
HD 190228	4.99	2.31	1127	0.43
HD 168443	7.2	0.29	57.9	0.54
	17.1	2.87	2135	0.2
HD 222582	5.4	1.35	576	0.71
HD 28185	5.6	1.0	385	0.06
HD 178911	6.292	0.32	71.487	0.1243
HD 10697	6.59	2.0	1083	0.12
70 Vir	6.6	0.43	116.6	0.4
HD 106252	6.81	2.61	1500.0	0.54
HD 89744	7.2	0.88	256	0.7
HD 33636	7.71	2.62	1553.0	0.39
HIP 75458	8.64	1.34	550.651	0.71
HD 141937	9.7	1.49	658.8	0.40
HD 39091	10.37	3.34	2083	0.62
HD 114762	11.	0.3	84.03	0.334
HD 136118	11.9	2.335	1209.6	0.366
HD 162020	13.75	0.072	8.428198	0.277

(b) Confirmed pulsar planets

Star	$M \sin i$ Jup. mass: (J) Earth mass: (E)	Semi-major axis (AU)	Period years (y) days (d)	Eccentricity
PSR 1257 + 12	0.015 (E)	0.19	25.34 (d)	0.0
	3.4 (E)	0.36	66.54 (d)	0.0182
	2.8 (E)	0.47	98.22 (d)	0.0264
	$\sim$100 (E)	$\sim$40	$\sim$170 (y)	—
PSR B1620 − 26	$1.2 < M \sin i < 6.7$ (J)	10 − 64	61.8 − 389 (y)	0. − 0.5

From the Extrasolar Planets Encyclopaedia, as of March 2002. http://www.obspm.fr/planets. Maintained by Jean Schneider, Observatoire de Paris.

4.3. STARS AND THE MILKY WAY

TABLE 4.7. Nearest stars*

	Name	R.A. (2000.0) h	m	Dec. °	′	Parallax π ″/yr	Distance $(1/\pi)$ pc	Proper motion μ ″/yr	θ °	Radial velocity km/s	Spectral type	V	B–V	M_V	Luminosity ($L_\odot = 1$)
1	Sun										G2 V	−26.75	0.65	4.82	1.
2	Proxima Cen	14	29.7	−62	41	0.772	1.29	3.86	282	−22	M5.5 e	11.05	1.90	15.49	0.00005
	α Cen A	14	39.6	−60	05	.742	1.35	3.71	278	−22	G2 V	.02	0.65	4.37	1.51
	α Cen B							3.69	281	−18	K0 V	1.36	0.85	5.71	0.44
3	Barnard's star	17	57.8	+04	42	.549	1.82	10.37	356	−111	M4 V	9.54	1.74	13.24	0.0004
4	Wolf 359 (CN Leo)	10	56.5	+07	01	.419	2.39	4.69	235	−13	M6 V	3.45	2.0	16.56	0.00002
5	BD +36°2147 HD95735	11	03.4	+35	58	.392	2.55	4.81	187	−85	M2 V	7.49	1.51	10.46	0.006
	(Lalande 21185)														
6	Sirius A	6	45.1	−16	43	.379	2.64	1.34	204	−8	AI V	−1.45	0.00	1.44	22.49
	Sirius B							1.34	204		DA2	8.44	−0.03	11.33	0.0025
7	L 726–8, BL Cet = A	1	39.0	−17	57	.374	2.68	3.37	81	+29	M5.5 V	12.41	1.87	15.27	0.00007
	UV Cet = B							3.37	81	+28	M6 V	13.25		16.11	0.00003
8	Ross 154 (V1216 Sgr)	18	49.8	−23	50	.337	2.97	0.67	107	−12	M3.5 V	10.45	1.76	13.08	0.0005
9	Ross 248 (HH And)	23	41.9	+44	10	.316	3.16	1.62	177	−78	M5.5 V	12.29	1.91	14.79	0.0001
10	ϵ Eri	3	32.9	−09	27	.311	3.22	0.98	271	+16	K2V	3.72	0.88	6.18	0.286
11	CD −36°15693 HD 217987	23	05.9	−35	51	.304	3.29	6.90	79	+10	M2 V	7.35	1.49	9.76	0.01
	(Lacaille 9352)														
12	Ross 128 (FI Vir)	11	47.7	+00	48	.300	3.34	1.36	154	−31	M4 V	11.12	1.76	13.50	0.00034
13	L 789–6 (EZ Aqr) = A	22	38.5	−15	18	.290	3.45	3.25	47	−60	M5 V	12.69	1.99	15.00	0.00008
	= B							3.25	47			13.6		15.9	0.00004
14	61 Cyg A	21	0.69	−38	45	.286	3.49	5.28	52	−65	K5 V	5.22	1.17	7.51	0.084
	61 Cyg B							5.17	53	−64	K7 V	6.04	1.36	8.32	0.0398
15	Procyon A	7	39.3	+05	14	.286	3.50	1.26	215	−4	F5 IV-V	0.36	0.42	2.64	7.45
	Procyon B							1.26	215		DA	10.75		13.03	0.0005
16	BD +43° 44 A(GX And)	0	18.4	+44	01	.280	3.57	2.92	82	+12	M1.5 V	8.08	1.56	10.32	0.006
	BD +43° 44 B (GQ And)	0	18.4	+44	02	.280	3.57	2.92	82	+11	M3.5 V	11.05	1.81	13.29	0.0004
17	BD +59 1915 A	18	42.7	+59	38	.280	3.57	2.24	324	−1	M3 V	8.90	1.53	11.14	0.00
	BD +59 1915 B	18	42.8	+59	38	.280	3.57	2.27	323	+2	M3.5 V	9.69	1.59	11.93	0.0014
	Struve 2398AB = ADS11632AB														
18	ϵ Ind	22	03.4	−56	47	.276	3.63	4.71	123	−40	K5 Ve	4.69	1.05	6.89	0.148
19	G 51 −15 (DX Cnc)	8	29.8	+26	47	.276	3.63	1.29	243	+25	M6.5 V	14.79	2.07	16.99	0.00001
20	τ Cen	1	44.1	−15	56	.274	3.65	1.92	296	−17	G8 Vp	3.49	0.72	5.68	0.45
21	L 372−58 = LHS 1565	3	36.0	−44	31	.270	3.70	0.84	119	−20	M5.5 V	13.01	1.90	15.17	0.00007
22	L 725−32 (YZ Cet)	1	12.5	−17	00	.269	3.72	1.37	62	+28	M4.5 V	12.05	1.83	14.20	0.0002
23	BD +5° 1668	7	27.4	+05	14	.263	3.80	3.74	171	+18	M3.5 V	9.85	1.57	11.95	0.0014
24	Kapteyn's star	5	11.7	−45	01	.255	3.92	8.66	131	+246	M1 p V	8.85	1.56	10.89	0.004
25	CD−39° 14192 (AX Mic)	21	17.3	−38	52	.253	3.95	3.45	251	+26	M0.5 V	6.68	1.41	8.70	0.028
	(Lacaille 8760)														
26	BD +56°2783 A	22	28.0	+57	42	.250	4.01	0.99	242	−33	M3 V	9.79	1.65	11.78	0.002
	BD +56°2783 B (DO Cep)							0.99	242	−32	M4 V	11.46	1.8	13.45	0.0004
	Krüger 60 AB														
27	Ross 614 A	6	29.4	−02	49	.243	4.12	0.93	132	+17	M4.5 V	11.14	1.72	13.07	0.0005
	Ross 614 B (V577 Mon)							0.93	132			14.47		16.40	0.00002
28	BD −12°4523	16	30.3	−12	40	.235	4.26	1.19	185	−22	M3 V	10.08	1.59	11.93	0.0014
29	van Maanen's star	0	49.2	+05	23	.233	4.30	2.98	156	+43	DZ7	12.39	0.55	14.22	0.0002
30	CD −37°15492	0	05.4	−3	21	.229	4.36	6.10	113	+23	M3 V	8.55	1.45	10.35	0.006
31	Wolf 424 A	12	33.3	+09	01	.228	4.39	1.81	278	−2	M5.5 V	13.04	1.84	14.83	0.0001
	Wolf 424 B (FL Vir)							1.81	278		M7	13.3		15.1	0.00008
32	L 1159−16 (TZ Ari)	2	00.2	+13	03	.225	4.45	2.10	148	−22	M4.5 V	12.27	1.81	14.03	0.0002
33	L143−23 = LHS 288	10	44.5	−61	12	.223	4.49	1.66	348		M5.5	13.87	1.83	15.61	0.00005
34	BD +68°946	17	36.4	+68	20	.221	4.53	1.31	194	−27	M3 V	9.17	1.49	10.89	0.0037

(*continued*)

TABLE 4.7. Nearest stars (continued)

		R.A.		Dec.		Parallax	Distance	Proper motion		Radial					
		(2000.0)				π	$(1/\pi)$	μ	θ	velocity	Spectral				Luminosity
	Name	h	m	°	′	″/yr	(pc)	″/yr	°	km/s	type	V	B–V	M_V	$(L_\odot = 1)$
35	CD −46°11540	17	28.7	−45	54	.220	4.54	1.05	147	−16	M3 V	9.38	1.55	11.10	0.0031
36	LP 731−58 = LHS 292	10	48.2	−11	20	.220	4.54	1.64	159	−2	M16.5 V	15.60	2.10	17.32	0.00001
37	G 208−44 = A	19	53.9	+44	25	.220	4.54	0.74	143	+42	M5.5 V	13.48	1.92	15 9	0.00007
	G 208−45 = B							0.74	143	+73	M6 V	14.01	1.97	15 2	0.00004
	G 208−44 = C							0.74	143			16.66		18.07	0.000004
38	G41−14	8	58.9	+08	28	.219	4.57	0.50	130	−25	M3.5 V	10.90	1.67	12.6	0.0008
39	L 145−141	11	45.7	−64	50	.216	4.62	2.69	97		DQ6	11.50	0.19	13.18	0.00045
40	G 158−27	0	06.7	−07	32	.213	4.69	2.04	204	−28	M5.5 V	13.76	1.97	15.40	0.00006
41	BD−15°6290	22	53.3	−14	16	.213	4.70	1.17	125	−2	M3.5 V	10.16	1.59	11.80	0.0016
42	BD +44°2051 = A	11	05.5	+43	32	.207	4.83	4.51	282	+69	M1 V	8.76	1.55	10.34	0.006
	BD +44°2051 = B (WX Uma)	11	05.5	+43	31	.207	4.83	4.51	282	+68	M5.5 V	14.42	2.02	16.00	0.00003
43	BD +50°1725	10	11.4	+49	27	.205	4.87	1.45	250	−26	K7 V	6.59	1.36	8.15	0.0466
44	BD +20°2465 (AD Leo)	10	19.6	+19	52	.205	4.89	0.50	265	+12	M3 V	9.41	1.54	10.96	0.0035
45	CD −49°13515 HD204961	21	33.6	−49	01	.203	4.94	0.82	183	+4	M1 V	8.67	1.48	10.20	0.007
46	LP 944−20	3	39.6	−35	26	.201	4.97	0.44	49		≥ M9 V				
47	CD −44°11909	17	37.0	−44	19	.198	5.04	1.18	217	−41	M4.5 V	10.95	1.66	12.44	0.0009
	40 Eri A	4	15.3	−07	39	.198	5.05	4.09	213	−43	KI Ve	4.42	0.82	5.91	0.366
	40 Eri B	4	15.4	−07	39	.198	5.05	4.09	213	−17	DA4	9.51	0.04	11.00	0.0034
	40 Eri C (DY Eri)							4.09	213	−46	M4.5 V	11.21	1.64	12.70	0.0007
49	BD +43°4305 (EV Lac)	22	46.8	+44	20	.198	5.05	0.84	237	+1	M3.5 V	10.23	1.61	11.71	0.0018
50	BD +02°3482 = A	18	05.5	+02	30	.197	5.09	0.97	173	−07	K0 Ve	4.24	0.78	5.71	0.44
	BD +02°3482 = B							0.97	173	−10	K5 Ve	6.01		7.48	0.086

*Parallaxes and distances are the new ones from the Hipparcos satellite (1997), courtesy of Hartmut Jahreiss. From J. M. Pasachoff, *Astronomy: From the Earth to the Universe, 6th ed.* (Brooks/Cole Publishers, Pacific Grove, CA, 2002). See www.solarcorona.net.

TABLE 4.8. Brightest stars

Star	Name	Position (2000.0) R.A.	Decl.	Apparent magnitude (V)	Spectral type	Absolute magnitude (M_V)	Distance D (ly)	Proper Motion μ ″/yr	θ °	Radial vel. (km/s)
1. α CMa A	Sirius	06 45 09	−16 42 58	−1.46	A1 V	+1.5	9	1.324	204	−8
2. α Car	Canopus	06 23 57	−52 41 44	−0.72	A9 Ib	−5.4	313	0.034	50	+21
3. α Boo	Arcturus	14 15 31	+19 10 57	−0.04	K2 IIIp	−0.6	37	2.281	209	−5
4. α Cen A	Rigil Kentaurus	14 39 37	−60 50 02	0.00	G2 V	+4.2	4	3.678	28	−25
5. α Lyr	Vega	18 36 56	+38 47 01	0.03	A0 V	+0.6	25	0.348	35	−14
6. α Aur	Capella	05 16 41	+45 59 53	0.08	G6 + G2	−0.8	42	0.430	169	+30
7. β Ori A	Rigel	05 14 32	−08 12 06	0.12	B8 Ia	−6.6	773	0.004	236	+21
8. α CMi A	Procyon	07 39 18	+05 13 30	0.38	F2 IV-V	+2.8	11	1.248	214	−3
9. α Eri	Achernar	01 37 43	−57 14 12	0.46	B3 V	−2.9	144	0.108	105	+19
10. α Ori	Betelgeuse	05 55 10	+07 24 36	0.50	M2 Iab	−5.0	522	0.028	68	+21
11. β Cen AB	Hadar	14 03 49	−60 22 22	0.61	B1 III	−5.5	526	0.030	221	−12
12. α Aql	Altair	19 50 47	+08 52 06	0.77	A7 IV-V	+2.1	17	0.662	54	−26
13. α Tau A	Aldebaran	04 35 55	+16 30 33	0.85	K5 III	−0.8	65	0.200	161	+54
14. α Sco A	Antares	16 29 24	−26 25 55	0.96	M1.5 Iab	−5.8	604	0.024	197	−3
15. α Vir	Spica	13 25 12	−11 09 41	0.98	B1 V	−3.6	262	0.054	232	+1
16. β Gem	Pollux	07 45 19	+28 01 34	1.14	K0 IIIb	+1.1	34	0.629	265	+3
17. α Ps A	Formalhaut	22 57 39	−29 37 20	1.16	A3 V	+1.6	25	0.373	116	+7
18. α Cyg	Deneb	20 41 26	+45 16 49	1.25	A2 Ia	−7.5	1467	0.005	11	−5
19. β Cru		12 47 43	−59 41 19	1.25	B0.5 III	−4.0	352	0.042	246	+20
20. α Leo A	Regulus	10 08 22	+11 58 02	1.35	B7 V	−0.6	77	0.264	271	+4
21. α Cru A		12 26 35	−63 05 56	1.41	B0.5 IV	−4.0	321	0.030	236	−11
22. ϵ CMa A	Adara	06 58 38	−28 59 20	1.50	B2 II	−4.1	431	0.002	27	+27
23. λ Sco	Shaula	17 33 36	−37 06 14	1.63	B1.5 IV	−3.6	359	0.029	178	0
24. γ Ori	Bellatrix	05 25 08	+06 20 59	1.64	B2 III	−2.8	243	0.018	221	+18
25. β Tau	Alnath	05 26 18	+28 36 27	1.65	B7 III	−1.3	131	0.178	172	+8

TABLE 4.9. Short-period variable stars[a]

Name	Type[b]	Position (2000.0) R.A. h m	Dec. ° ′	Magnitude range	Period (Days)
Algol (β Persei)	E	03 08.2	+40 57	2.1–3.4	2.87
λ Tauri	E	04 00.7	+12 29	3.4–3.9	3.95
RT Aurigae	C	06 28.6	+30 30	5.1–5.8	3.73
ζ Geminorum	C	07 04.1	+20 34	3.7–4.2	10.15
δ Librae	E	15 01.0	−08 31	4.9–5.9	2.33
u Herculis	E	17 17.3	+33 06	4.8–5.4	2.05
W Sagittarii	C	18 05.0	−29 35	4.3–5.1	7.59
β Lyrae	E	18 50.1	+33 22	3.3–4.3	12.94
RR Lyrae	RR	19 25.5	+42 47	7.0–8.1	0.57
η Aquilae	C	19 52.5	+01 00	3.5–4.5	5.18
δ Cephei	C	22 29.2	+58 25	3.5–4.4	5.37

[a]From J. M. Pasachoff, *Field Guide to the Stars and Planets, 4th ed.* (Houghton Mifflin, Boston, 2000). Drawn from *Sky Catalog 2000.0*, Volume 2, by Roger W. Sinnott and Alan Hirshfeld. Copyright ©1985 by Sky Publishing Corporation. Reprinted with permission of the publisher.
[b]C = Cepheid variable; E = eclipsing binary; RR = RR Lyrae variable (cluster variable).

TABLE 4.10. Long-period variable stars

		Position (2000.0)			
Name	Type[a]	R.A. h m	Dec. ° ′	Magnitude range[b]	Period (Days)
T Cas	M	00 23.2	+55 48	[6-9–11-9]	445
o Cet	M	02 19.3	−02 59	[3.4–9.2]	332
R Tri	M	02 37.0	+34 16	[6.2–11.7]	266
R Hor	M	02 53.9	−49 53	[6.0–13-0]	404
X Cam	M	04 45.7	+75 06	[8.1–12.6]	144
R Pic	SR	04 46.2	−49 15	6.7–10.0	164
L^2 Pup	SR	07 13.5	−44 39	2.6–6.2	140
S CMi	M	07 32.7	+08 19	[7.5–12.6]	333
U Gem	UG	07 55.1	+22 00	8.2–14.9	103
R Car	M	09 32.2	−62 47	[4.6–9.6]	309
ZZ Car	C	09 45.2	−62 30	3.3–4.2	36
R Leo	M	09 47.6	+11 26	[5.8–10.0]	312
S Car	M	10 09.4	−61 33	[5.7–8.5]	150
R UMa	M	10 44.6	+68 47	[7.5–13.0]	302
T UMa	M	12 36.4	+59 29	[7.7–12.9]	257
S UMa	M	12 43.9	+61 06	[7.8–11.7]	226
T Cen	SR	13 41.8	−33 36	5.5–9.0	90
R CVn	M	13 49.0	+39 33	[7.7–11.9]	329
R Cen	M	14 16.6	−59 55	5.3–11.8	546
S Boo	M	14 22.9	+53 49	[8.4–13.3]	271
V Boo	SR	14 29.8	+38 52	7.0–12.0	258
R CrB	RCB	15 48.6	+28 09	5.7–14.8	[c]
R Ser	M	15 50.7	+15 08	[6.9–13.41	356
U Her	M	16 25.8	+18 54	[7.5–12.5]	406
R Dra	M	16 32.7	+66 45	[7.6–12.4]	245
R Oph	M	17 07.8	−16 06	[7.6–13.3]	303
T Her	M	18 09.1	+31 01	[8.0–12.8]	165
R Sct	RV	18 47.5	−05 42	4.5–8.2	140
R Cyg	M	19 36.8	+50 12	[7.5–13.9]	426
R Vul	M	21 04.4	+23 49	[9.1–12.6]	136
T Cep	M	21 09.5	+68 29	[6.0–10.3]	388
SS Cyg	M	21 42.7	+43 35	8.2–12.4	50
R Peg	M	23 06.6	+10 33	[7.8–13.2]	378
V Cas	M	23 11.7	+59 42	6.9–13.4	229

From J. M. Pasachoff, *Field Guide to the Stars and Planets, 4th ed.* (Houghton Mifflin, Boston, 2000). Drawn from *Sky Catalog 2000.0*, Volume 2, by Roger W. Sinnott and Alan Hirshfeld. Copyright ©1985 by Sky Publishing Corporation. Reprinted with permission of the publisher.

[a]C = Cepheid variable; M = Mira variable; RCB = R Coronae Borealis variable; RV = RV Tauri variable; SR = semiregular variable; UG = U Geminorum variable.

[b]Brackets indicate an average range of magnitudes.

[c]Irregular variation, no fixed period.

TABLE 4.11. Messier catalogue*

M	NGC	(2000.0) α h m	δ ° ′	m_V	Description
1	1952	5 34.5	+22 01	8.4	Crab Nebula (Tau)
2	7089	21 33.5	−00 49	6.5	Globular cluster (Aqr)
3	5272	13 42.2	+28 23	6.4	Glob. cluster (CVn)
4	6121	16 21.6	−26 32	5.9	Glob. cluster (Sco)
5	5904	15 18.6	+02 05	5.8	Glob. cluster (Ser)
6	6405	17 40.1	−32 13	4.2	Open cluster (Sco)
7	6475	17 53.9	−34 49	3.3	Open cluster (Sco)
8	6523	18 03.8	−24 23	5.8	Lagoon Nebula (Sgr)
9	6333	17 19.2	−18 31	7.9	Glob. cluster (Oph)
10	6254	16 57.1	−04 06	6.6	Glob. cluster (Oph)
11	6705	18 51.1	−05 16	5.8	Open cluster (Scu)
12	6218	16 47.2	−01 57	6.6	Glob. cluster (Oph)
13	6205	16 41.7	+36 28	5.9	Glob. cluster (Her)
14	6402	17 37.6	−03 15	7.6	Glob. cluster (Oph)
15	7078	21 30.0	+12 10	6.4	Glob. cluster (Peg)
16	6611	18 18.8	−13 47	6.0	Open cl. & nebula (Ser)
17	6618	18 20.8	−16 11	7	Omega nebula (Sgr)
18	6613	18 19.9	−17 08	6.9	Open cluster (Sgr)
19	6273	17 02.6	−26 16	7.2	Clob. cluster (Oph)
20	6514	18 02.6	−23 02	8.5	Trifid Nebula (Sgr)
21	6531	18 04.6	−22 30	5.9	Open cluster (Sgr)
22	6656	18 36.4	−23 54	5.1	Glob. cluster (Sgr)
23	6494	17 56.8	−19 01	5.5	Open cluster (Sgr)
24	6603	18 16.9	−18 29	4.5	Open cluster (Sgr)
25	IC4725	18 31.6	−19 15	4.6	Open cluster (Sgr)
26	6694	18 45.2	−09 24	8.0	Open cluster (Scu)
27	6853	19 59.6	+22 43	8.1	Dumbbell N., PN (Val)
28	6626	18 24.3	−24 52	6.9	Glob. cluster (Sgr)
29	6913	20 23.9	+38 32	6.6	Open cluster (Cyg)
30	7099	21 40.4	−23 11	7.5	Glob. cluster (Cap)
31	224	0 42.7	+41 16	3.4	Andromeda galaxy (Sb)
32	221	0 42.7	+40 52	8.2	Elliptical galaxy (And)
33	598	1 33.9	+30 39	5.7	Spiral galaxy (Sc)(Tri)
34	1039	2 42.0	+42 47	5.2	Open cluster (Per)
35	2168	6 08.9	+24 20	5.1	Open cluster (Gem)
36	1960	5 36.1	+34 08	6.0	Open cluster (Aur)
37	2099	5 52.4	+32 33	5.6	Open cluster (Aur)
38	1912	5 28.7	+35 50	6.4	Open cluster (Aur)
39	7092	21 32.2	+48 26	4.6	Open cluster (Cyg)
40		12 22.4	+58 05	8	Double star (UMa)
41	2287	6 46.0	−20 44	4.5	Open cluster (CMa)
42	1976	3 35.4	− 5 27	4	Orion Nebula (Ori)
43	1982	5 35.6	− 5 16	9	Orion Nebula; smaller (Ori)
44	2632	8 40.1	+19 46	3.1	Praesepe; open cl. (Can)
45		3 47.0	+24 07	1.2	Pleiades; open cl. (Tau)
46	2437	7 41.8	−14 49	6.1	Open cluster (Pup)
47	2422	7 36.6	−14 30	4.4	Open cluster (Pup)
48	2548	8 13.8	− 5 48	5.8	Open cluster (Hyd)
49	4472	12 29.8	+ 8 00	8.4	Elliptical galaxy (Vir)
50	2323	7 02.8	− 8 20	5.9	Open cluster (Mon)
51	5194	13 29.9	+47 12	8.1	Whirlpool Galaxy (Sc)(CVn)
52	7654	23 24.2	+61 35	6.9	Open cluster (Cas)
53	5024	13 12.9	+18 10	7.7	Glob. cluster (Com)
54	6715	18 53.1	−30 29	7.7	Glob. cluster (Sgr)
55	6809	19 40.0	−30 58	7.0	Glob. cluster (Sgr)
56	6779	19 16.6	+30 11	8.2	Glob. cluster (Lyr)
57	6720	18 53.6	+33 02	9.0	Ring N; planetary (Lyr)
58	4579	12 37.7	+11 49	9.8	Spiral galaxy (SBb)(Vir)
59	4621	12 42.0	+11 39	9.8	Elliptical galaxy (Vir)
60	4649	12 43.7	+11 33	8.8	Elliptical galaxy (Vir)
61	4303	12 21.9	+4 28	9.7	Spiral galaxy (Sc)(Vir)
62	6266	17 01.2	−30 07	6.6	Glob. cluster (Sco)
63	5055	13 15.8	+42 02	8.6	Spiral galaxy (Sb)(CVn)
64	4826	12 56.7	+21 41	8.5	Sprial galaxy (Sb)(Com)
65	3623	11 18.9	+13 05	9.3	Spiral galaxy (Sa)(Leo)
66	3627	11 20.2	+12 59	9.0	Spiral galaxy (Sb)(Leo)
67	2682	8 51.4	+11 49	6.9	Open cluster (Can)
68	4590	12 39.5	−26 45	8.2	Glob. cluster (Hyd)
69	6637	18 31.4	−32 21	7.7	Glob. cluster (Sgr)
70	6681	18 43.2	−32 18	8.1	Glob. cluster (Sgr)
71	6838	19 53.8	+18 47	8.3	Glob. cluster (Sge)
72	6981	20 53.5	−12 32	9.4	Clob. cluster (Aqr)
73	6994	20 58.9	−12 38		Glob. cluster (Aqr)
74	628	1 36.7	+15 47	9.2	Spiral galaxy (Sc) in Pisces
75	6864	20 06.1	−21 55	8.6	Glob. cluster (Sgr)
76	650-1	1 42.4	+51 34	11.5	Planetary nebula (Per)
77	1068	2 42.7	−0 01	8.8	Spiral galaxy (Sb)(Cer)
78	2068	5 46.7	+0 03	8	Small emission nebula (Ori)
79	1904	5 24.5	−24 33	8.0	Glob. cluster (Lep)
80	6093	16 17.0	−22 59	7.2	Glob. cluster (Sco)
81	3031	9 55.6	+69 04	6.8	Spiral galaxy (Sb)(UMa)
82	3034	9 55.8	+69 41	8.4	Irregular galaxy (UMa)
83	5236	13 37.0	−29 52	7.6	Spiral galaxy (Sc)(Hyd)
84	4374	12 25.1	+12 53	9.3	Elliptical galaxy (Vir)
85	4382	12 25.4	+18 11	9.2	S0 galaxy (Com)
86	4406	12 26.2	+12 57	9.2	Elliptical galaxy (Vir)
87	4486	12 30.8	+12 24	8.6	Elliptical galaxy (Ep)(Vir)
88	4501	12 32.0	+14 25	9.5	Spiral galaxy (Sb)(Com)
89	4552	12 35.7	+12 33	9.8	Elliptical galaxy (Vir)
90	4569	12 36.8	+13 10	9.5	Spiral galaxy (SBb)(Vir)
91	4548	12 35.4	+14 30	10.2	M58? (Vir)
92	6341	17 17.1	+43 08	6.5	Glob. cluster (Her)
93	2447	7 44.6	−23 52	6.2	Open cluster in (Pup)
94	4736	12 50.9	+41 07	8.1	Spiral galaxy (Sb)(CVa)
95	3351	10 44.0	+11 42	9.7	Barred spiral g. (SBb)(Leo)
96	3368	10 46.8	+11 49	9.2	Spiral galaxy (Sa)(Leo)
97	3587	11 14.8	+55 01	11.2	Owl Nebula; planetary (UMa)
98	4192	12 13.8	+14 54	10.1	Spiral galaxy (Sb)(Com)
99	4254	12 18.8	+14 25	9.8	Spiral galaxy (Sc)(Com)
100	4321	12 22.9	+15 49	9.4	Spiral galaxy (Sc)(Com)
101	5457	14 03.2	+54 21	7.7	Spiral galaxy (Sc)(UMa)
102					M101; duplication (UMa)
103	581	1 33.2	+60 42	7.4	Open cluster (Cas)
104	4594	12 40.0	−11 37	8.3	Sombrero N.; spiral (Sa) (Vir)
105	3379	10 47.8	+12 35	9.3	Elliptical galaxy (Leo)
106	4258	12 19.0	+47 18	8.3	Spiral galaxy (Sb)(CVn)
107	6171	16 32.5	−13 03	8.1	Glob. cluster (Oph)
108	3556	11 11.5	+55 40	10.0	Spiral galaxy (Sb)(UMa)
109	3992	11 57.6	+53 23	9.8	Barred spiral g. (SBc)(UMa)
110	205	0 40.4	+41 41	8.0	Elliptical galaxy (And)

[a]From *Sky Catalog 2000.0*, Volume 2, by Roger W. Sinnott and Alan Hirshfeld. Copyright ©1985 by Sky Publishing Corporation. Reprinted with permission of the publisher.

TABLE 4.12. Milky Way globular clusters*

Keys to columns:
(1) Cluster identification number
(2) Other commonly used cluster name
(3,4) Right ascension and declination (epoch 2000)
(5) Integrated V magnitude of the cluster
(6) Absolute visual magnitude (cluster luminosity)
(7) Distance from Sun (kiloparsecs)
(8) Distance from Galactic center (kpc), assuming $R_0 = 8.0$ kpc

ID	Name	RA (2000)	Dec	V	M_V	R_{Sun}	R_{gc}
NGC 104	47 Tuc	00 24 05.2	−72 04 51	3.95	−9.42	4.5	7.4
NGC 288		00 52 47.5	−26 35 24	8.09	−6.60	8.3	11.6
NGC 362		01 03 14.3	−70 50 54	6.40	−8.40	8.5	9.3
NGC 1261		03 12 15.3	−55 13 01	8.29	−7.81	16.4	18.2
Pal 1		03 33 23.0	+79 34 50	13.18	−2.47	10.9	17.0
AM 1	ESO1	03 55 02.7	−49 36 52	15.72	−4.71	121.9	123.2
Eridanus		04 24 44.5	−21 11 13	14.70	−5.14	90.2	95.2
Pal 2		04 46 05.9	+31 22 51	13.04	−8.01	27.6	35.4
NGC 1851		05 14 06.3	−40 02 50	7.14	−8.33	12.1	16.7
NGC 1904	M79	05 24 10.6	−24 31 27	7.73	−7.86	12.9	18.8
NGC 2298		06 48 59.2	−36 00 19	9.29	−6.30	10.7	15.7
NGC 2419		07 38 08.5	+38 52 55	10.39	−9.58	84.2	91.5
Pyxis		09 07 57.8	−37 13 17	12.90	−5.73	39.4	41.4
NGC 2808		09 12 02.6	−64 51 47	6.20	−9.36	9.3	11.0
E 3		09 20 59.3	−77 16 57	11.35	−2.77	4.3	7.6
Pal 3		10 05 31.4	+00 04 17	14.26	−5.70	92.7	95.9
NGC 3201		10 17 36.8	−46 24 40	6.75	−7.49	5.2	9.0
Pal 4		11 29 16.8	+28 58 25	14.20	−6.02	109.2	111.8
NGC 4147		12 10 06.2	+18 32 31	10.32	−6.16	19.3	21.3
NGC 4372		12 25 45.4	−72 39 33	7.24	−7.77	5.8	7.1
Rup 106		12 38 40.2	−51 09 01	10.90	−6.35	21.2	18.5
NGC 4590	M68	12 39 28.0	−26 44 34	7.84	−7.35	10.2	10.1
NGC 4833		12 59 35.0	−70 52 29	6.91	−8.01	6.0	6.9
NGC 5024	M53	13 12 55.3	+18 10 09	7.61	−8.77	18.3	18.8
NGC 5053		13 16 27.0	+17 41 53	9.47	−6.72	16.4	16.9
NGC 5139	ω Cen	13 26 45.9	−47 28 37	3.68	−10.29	5.3	6.4
NGC 5272	M3	13 42 11.2	+28 22 32	6.19	−8.93	10.4	12.2
NGC 5286		13 46 26.5	−51 22 24	7.34	−8.61	11.0	8.4
AM 4		13 56 21.2	−27 10 04	15.90	−1.60	29.9	25.5
NGC 5466		14 05 27.3	+28 32 04	9.04	−7.11	17.0	17.2
NGC 5634		14 29 37.3	−05 58 35	9.47	−7.75	25.9	21.9
NGC 5694		14 39 36.5	−26 32 18	10.17	−7.81	34.7	29.1
IC 4499		15 00 18.5	−82 12 49	9.76	−7.33	18.9	15.7
NGC 5824		15 03 58.5	−33 04 04	9.09	−8.84	32.0	25.8
Pal 5		15 16 05.3	−00 06 41	11.75	−5.17	23.2	18.6
NGC 5897		15 17 24.5	−21 00 37	8.53	−7.29	12.8	7.7
NGC 5904	M5	15 18 33.8	+02 04 58	5.65	−8.81	7.5	6.2

(continued)

TABLE 4.12. Milky Way globular clusters (continued)

ID	Name	RA (2000)	Dec	V	M_V	R_{Sun}	R_{gc}
NGC 5927		15 28 00.5	−50 40 22	8.01	−7.80	7.6	4.5
NGC 5946		15 35 28.5	−50 39 34	9.61	−7.60	12.8	7.4
BH 176		15 39 07.3	−50 03 02	14.00	−4.20	14.5	8.8
NGC 5986		15 46 03.5	−37 47 10	7.52	−8.42	10.5	4.8
Lynga 7		16 11 03.0	−55 18 52			7.2	4.2
Pal 14	AvdB	16 11 04.9	+14 57 29	14.74	−4.73	73.9	69.0
NGC 6093	M80	16 17 02.5	−22 58 30	7.33	−8.23	10.0	3.8
NGC 6121	M4	16 23 35.5	−26 31 31	5.63	−7.20	2.2	5.9
NGC 6101		16 25 48.6	−72 12 06	9.16	−6.91	15.3	11.1
NGC 6144		16 27 14.1	−26 01 29	9.01	−7.05	10.3	3.6
NGC 6139		16 27 40.4	−38 50 56	8.99	−8.36	10.1	3.6
Terzan 3		16 28 40.1	−35 21 13	12.00	−4.61	7.5	2.4
NGC 6171	M107	16 32 31.9	−13 03 13	7.93	−7.13	6.4	3.3
1636-283	ESO452-SC11	16 39 25.5	−28 23 52	12.00	−3.97	7.8	2.0
NGC 6205	M13	16 41 41.5	+36 27 37	5.78	−8.70	7.7	8.7
NGC 6229		16 46 58.9	+47 31 40	9.39	−8.07	30.7	30.0
NGC 6218	M12	16 47 14.5	−01 56 52	6.70	−7.32	4.9	4.5
NGC 6235		16 53 25.4	−22 10 38	9.97	−6.14	10.0	2.9
NGC 6254	M10	16 57 08.9	−04 05 58	6.60	−7.48	4.4	4.6
NGC 6256		16 59 32.6	−37 07 17	11.29	−6.02	6.6	2.1
Pal 15		17 00 02.4	−00 32 31	14.00	−5.49	44.6	37.9
NGC 6266	M62	17 01 12.6	−30 06 44	6.45	−9.19	6.9	1.7
NGC 6273	M19	17 02 37.7	−26 16 05	6.77	−9.08	8.7	1.6
NGC 6284		17 04 28.8	−24 45 53	8.83	−7.87	14.7	6.9
NGC 6287		17 05 09.4	−22 42 29	9.35	−7.16	8.5	1.7
NGC 6293		17 10 10.4	−26 34 54	8.22	−7.77	8.8	1.4
NGC 6304		17 14 32.5	−29 27 44	8.22	−7.32	6.1	2.1
NGC 6316		17 16 37.4	−28 08 24	8.43	−8.35	11.0	3.2
NGC 6341	M92	17 17 07.3	+43 08 11	6.44	−8.20	8.2	9.6
NGC 6325		17 17 59.2	−23 45 57	10.33	−7.35	9.6	2.0
NGC 6333	M9	17 19 11.8	−18 30 59	7.72	−8.04	8.2	1.7
NGC 6342		17 21 10.2	−19 35 14	9.66	−6.44	8.6	1.7
NGC 6356		17 23 35.0	−17 48 47	8.25	−8.52	15.2	7.6
NGC 6355		17 23 58.6	−26 21 13	9.14	−7.48	7.2	1.0
NGC 6352		17 25 29.2	−48 25 22	7.96	−6.48	5.7	3.3
IC 1257		17 27 08.5	−07 05 35	13.10	−6.15	25.0	17.9
Terzan 2	HP 3	17 27 33.4	−30 48 08	14.29	−5.27	8.7	0.9
NGC 6366		17 27 44.3	−05 04 36	9.20	−5.77	3.6	5.0
Terzan 4	HP 4	17 30 38.9	−31 35 44	16.00	−6.09	9.1	1.3
HP 1	BH 229	17 31 05.2	−29 58 54	11.59	−6.43	7.4	0.8
NGC 6362		17 31 54.8	−67 02 53	7.73	−7.06	8.1	5.3
Liller 1		17 33 24.5	−33 23 20	16.77	−7.63	10.5	2.6
NGC 6380	Ton 1	17 34 28.0	−39 04 09	11.31	−7.46	10.7	3.2
Terzan 1	HP 2	17 35 47.8	−30 28 11	15.90	−3.30	6.2	1.8

(continued)

TABLE 4.12. Milky Way globular clusters (continued)

ID	Name	RA (2000)	Dec	V	M_V	R_{Sun}	R_{gc}
Ton 2	Pismis 26	17 36 10.5	−38 33 12	12.24	−6.14	8.1	1.4
NGC 6388		17 36 17.0	−44 44 06	6.72	−9.82	115	4.4
NGC 6402	M 14	17 37 36.1	−03 14 45	7.59	−9.02	8.9	3.9
NGC 6401		17 38 36.9	−23 54 32	9.45	−7.62	7.7	0.8
NGC 6397		17 40 41.3	−53 40 25	5.73	−6.63	2.3	6.0
Pal 6		17 43 42.2	−26 13 21	11.55	−7.37	7.3	0.8
NGC 6426		17 44 54.7	+03 10 13	11.01	−6.65	20.4	14.2
Djorgovski 1		17 47 28.3	−33 03 56	13.60	−6.40	8.8	1.0
Terzan 5	Terzan 11	17 48 04.9	−24 48 45	13.85	−7.91	7.6	0.7
NGC 6440		17 48 52.6	−20 21 34	9.20	−8.75	8.4	1.3
NGC 6441		17 50 12.9	−37 03 04	7.15	−9.47	11.2	3.5
Terzan 6	HP 5	17 50 46.4	−31 16 31	13.85	−7.67	9.5	1.6
NGC 6453		17 50 51.8	−34 35 55	10.08	−7.05	11.2	3.3
UKS 1		17 54 27.2	−24 08 43	17.29	−6.88	8.3	0.8
NGC 6496		17 59 02.0	−44 15 54	8.54	−7.23	11.5	4.3
Terzan 9		18 01 38.8	−26 50 23	16.00	−3.93	7.7	0.6
Djorgovski 2	E456-SC38	18 01 49.1	−27 49 33	9.90	−6.98	6.7	1.4
NGC 6517		18 01 50.6	−08 57 32	10.23	−8.28	10.8	4.3
Terzan 10		18 02 57.4	−26 04 00	14.90	−6.31	5.7	2.4
NGC 6522		18 03 34.1	−30 02 02	8.27	−7.67	7.8	0.6
NGC 6535		18 03 50.7	−00 17 49	10.47	−4.73	6.7	3.9
NGC 6528		18 04 49.6	−30 03 21	9.60	−6.93	9.1	1.3
NGC 6539		18 04 49.8	−07 35 09	9.33	−8.30	8.4	3.1
NGC 6540	Djorgovski 3	18 06 08.6	−27 45 55	9.30	−5.38	3.7	4.4
NGC 6544		18 07 20.6	−24 59 51	7.77	−6.56	2.6	5.4
NGC 6541		18 08 02.2	−43 42 20	6.30	−8.37	7.0	2.2
NGC 6553		18 09 15.6	−25 54 28	8.06	−7.99	5.6	2.5
NGC 6558		18 10 18.4	−31 45 49	9.26	−6.46	7.4	1.0
IC 1276	Pal 7	18 10 44.2	−07 12 27	10.34	−6.67	5.4	3.7
Terzan 12		18 12 15.8	−22 44 31	15.63	−4.14	4.8	3.4
NGC 6569		18 13 38.9	−31 49 35	8.55	−7.88	8.7	1.2
NGC 6584		18 18 37.7	−52 12 54	8.27	−7.68	13.4	7.0
NGC 6624		18 23 40.5	−30 21 40	7.87	−7.50	8.0	1.2
NGC 6626	M28	18 24 32.9	−24 52 12	6.79	−8.33	5.7	2.6
NGC 6638		18 30 56.2	−25 29 47	9.02	−6.83	8.4	1.6
NGC 6637	M69	18 31 23.2	−32 20 53	7.64	−7.52	8.6	1.6
NGC 6642		18 31 54.3	−23 28 35	9.13	−6.57	7.7	1.6
NGC 6652		18 35 45.7	−32 59 25	8.62	−6.57	9.6	2.4
NGC 6656	M22	18 36 24.2	−23 54 12	5.10	−8.50	3.2	4.9
Pal 8		18 41 29.9	−19 49 33	11.02	−5.52	12.9	5.6
NGC 6681	M70	18 43 12.7	−32 17 31	7.87	−7.11	9.0	2.1
NGC 6712		18 53 04.3	−08 42 22	8.10	−7.50	6.9	3.5
NGC 6715	M54	18 55 03.3	−30 28 42	7.60	−10.01	27.2	19.6
NGC 6717	Pal 9	18 55 06.2	−22 42 03	9.28	−5.67	7.4	2.3

(continued)

TABLE 4.12. Milky Way globular clusters (continued)

ID	Name	RA (2000)	Dec	V	M_V	R_{Sun}	R_{gc}
NGC 6723		18 59 33.2	−36 37 54	7.01	−7.86	8.8	2.6
NGC 6749	Be42	19 05 15.3	+01 54 03	12.44	−6.70	7.9	5.0
NGC 6752		19 10 51.8	−59 58 55	5.40	−7.73	4.0	5.2
NGC 6760		19 11 12.1	+01 01 50	8.88	−7.86	7.4	4.8
NGC 6779	M56	19 16 35.5	+30 11 05	8.27	−7.38	10.1	9.7
Terzan 7		19 17 43.7	−34 39 27	12.00	−5.05	23.2	16.0
Pal 10		19 18 02.1	+18 34 18	13.22	−5.79	5.9	6.4
Arp 2		19 28 44.1	−30 21 14	12.30	−5.29	28.6	21.4
NGC 6809	M55	19 39 59.4	−30 57 44	6.32	−7.55	5.4	3.8
Terzan 8		19 41 45.0	−34 00 01	12.40	−5.05	26.0	19.1
Pal 11		19 45 14.4	−08 00 26	9.80	−6.81	12.9	7.8
NGC 6838	M71	19 53 46.1	+18 46 42	8.19	−5.56	3.9	6.7
NGC 6864	M75	20 06 04.8	−21 55 17	8.52	−8.35	18.8	12.8
NGC 6934		20 34 11.6	+07 24 15	8.83	−7.65	17.4	14.3
NGC 6981	M72	20 53 27.9	−12 32 13	9.27	−7.04	17.0	12.9
NGC 7006		21 01 29.5	+16 11 15	10.56	−7.68	41.5	38.8
NGC 7078	M15	21 29 58.3	+12 10 01	6.20	−9.17	10.3	10.4
NGC 7089	M2	21 33 29.3	−00 49 23	6.47	−9.02	11.5	10.4
NGC 7099	M30	21 40 22.0	−23 10 45	7.19	−7.43	8.0	7.1
Pal 12		21 46 38.8	−21 15 03	11.99	−4.48	19.1	15.9
Pal 13		23 06 44.4	+12 46 19	13.80	−3.51	26.9	27.8
NGC 7492		23 08 26.7	−15 36 41	11.29	−5.77	25.8	24.9

*Extracted from the catalogue compiled by William E. Harris, McMaster University. W. E. Harris, *Astron. J.* **112**, 1487 (1996). University of Chicago Press.

TABLE 4.13. Properties of the principal spectral types*

Spectral type	Apparent color	Color index (B-V)	Surface temperature (K)	Primary absorption lines in spectrum	Examples
O	blue	< −0.2	25,000—40,000	Strong lines of ionized helium and highly ionized metals; hydrogen lines weak	ζ Orionis (O 9.5)
B	blue	0.2 — 0.0	11,000—25,000	Lines of neutral helium prominent; hydrogen lines stronger than in type O	Spica (B1) Rigel (B8)
A	blue to white	0.0 — 0.3	7,500 —11,000	Strong lines of hydrogen, ionized calcium, and other ionized metals; weak helium lines	Vega (A0) Sirius (Al)
F	white	0.3 — 0.6	6,000 — 7,500	Hydrogen lines weaker than in type A; ionized calcium strong; lines of neutral metals becoming prominent	Canopus (F0) Procyon (F5) Polaris (F8)
G	white to yellow	0.6 — 1.1	5,000 — 6,000	Numerous strong lines of ionized calcium and other ionized and neutral metals; hydrogen lines weaker than in type F	Sun (G2) Capella (G5)
K	orange to red	1.1 — 1.5	3,500 — 5,000	Numerous strong lines of neutral metals	Arcturus (K2) Aldebaran (K5)
M	red	> 1.5	3,000 — 3,500	Numerous strong lines of neutral metals; strong molecular bands (primarily titanium oxide)	Antares (M1) Betelgeuse (M2)
L	infrared		1,300 — 2,900	Metallic hydrides and neutral alkali metals	
T	infrared		< 1,300	Methane	

*The number after the letter in each spectral type (see last column, above) indicates a further subdivision within each type; e.g., Sirius (type Al) is hotter than Deneb (type A2). From *Sky Catalog 2000.0*, Volume 2, by Roger W. Sinnott and Alan Hirshfeld.

TABLE 4.14. The constellations

Latin name	Genitive	Abbreviation	Translation	Latin name	Genitive	Abbreviation	Translation
Andromeda	Andromedae	And	Andromeda[a]	Circinus	Circini	Cir	Compass
Antlia	Antliae	Ant	Pump	Columba	Columbae	Col	Dove
Apus	Apodis	Aps	Bird of Paradise	Coma Berenices	Comae Berenices	Com	Berenice's Hair[a]
Aquarius	Aquarii	Aqu	Water Bearer	Corona Australis	Coronae Australis	CrA	Southern Crown
Aquila	Aquilae	Aql	Eagle	Corona Borealis	Coronae Borealis	CrB	Northern Crown
Ara	Arae	Ara	Altar	Corvus	Corvi	Crv	Crow
Aries	Arietis	Ari	Ram	Crater	Crateris	Crt	Cup
Auriga	Aurigae	Aur	Charioteer	Crux	Crucis	Cru	Southern Cross
Boötes	Boötis	Boo	Herdsman	Cygnus	Cygni	Cyg	Swan
Caelum	Caeli	Cae	Chisel	Delphinus	Delphini	Del	Dolphin
Camelopardalis	Camelopardalis	Cam	Giraffe	Dorado	Dorados	Dor	Swordfish
Cancer	Cancri	Cnc	Crab	Draco	Draconis	Dra	Dragon
Canes Venatici	Canum Venaticorum	CVn	Hunting Dogs	Equuleus	Equulei	Equ	Little Horse
Canis Major	Canis Majoris	CMa	Big Dog	Eridanus	Eridani	Eri	River Eridanus[a]
Canis Minor	Canis Minoris	CMi	Little Dog	Fornax	Fornacis	For	Furnace
Capricornus	Capricorni	Cap	Goat	Gemini	Geminorum	Gen	Twins
Carina	Carinae	Car	Ship's Keel[b]	Grus	Gruis	Gru	Crane
Cassiopeia	Cassiopeiae	Cas	Cassiopeia[a]	Hercules	Herculis	Her	Hercules[a]
Centaurus	Centauri	Cen	Centaur[a]	Horologium	Horologii	Hor	Clock
Cepheus	Cephei	Cep	Cepheus[a]	Hydra	Hydrae	Hya	Hydra[a] (water monster)
Cetus	Ceti	Cet	Whale	Hydrus	Hydri	Hyd	Sea serpent
Chamaeleon	Chamaeleonis	Cha	Chameleon	Indus	Indi	Ind	Indus[a] (Indian River)

(continued)

TABLE 4.14. The constellations (continued)

Latin name	Genitive	Abbreviation	Translation	Latin name	Genitive	Abbreviation	Translation
Lacerta	Lacertae	Lac	Lizard	Piscis Austrinus	Piscis Austrini	PsA	Southern Fish
Leo	Leonis	Leo	Lion	Puppis	Puppis	Pup	Ship's Stern[b]
Leo Minor	Leonis Minoris	LMi	Little Lion	Pyxis	Pyxidis	Pyx	Ship's Compass[b]
Lepus	Leporis	Lep	Hare	Reticulum	Reticuli	Ret	Net
Libra	Librae	Lib	Scales	Sagitta	Sagittae	Sgi	Arrow
Lupus	Lupi	Lup	Wolf	Sagittarius	Sagittarii	Sgt	Archer
Lynx	Lyncis	Lyn	Lynx	Scorpius	Scorpii	Sco	Scorpion
Lyra	Lyrae	Lyr	Harp	Sculptor	Sculptoris	Scl	Sculptor
Mensa	Mensae	Men	Table (mountain)	Scutum	Scuti	Sct	Shield
Microscopium	Microscopii	Mic	Microscope	Serpens	Serpentis	Ser	Serpent
Monoceros	Monocerotis	Mon	Unicorn	Sextans	Sextantis	Sex	Sextant
Musca	Muscae	Mus	Fly	Taurus	Tauri	Tau	Bull
Norma	Normae	Nor	Level (square)	Telescopium	Telescopii	Tel	Telescope
Octans	Octantis	Oct	Octant	Triangulum	Trianguli	Tri	Triangle
Ophiuchus	Ophiuchi	Oph	Ophiuchus[a] (serpent bearer)	Triangulum Australe	Trianguli Australia	TrA	Southern Triangle
Orion	Orionis	Ori	Orion[a]	Tucana	Tucanae	Tuc	Toucan
Pavo	Pavonis	Pav	Peacock	Ursa Major	Ursae Majoris	UMa	Big Bear
Pegasus	Pegasi	Peg	Pegasus[a] (winged horse)	Ursa Minor	Ursae Minoris	UMi	Little Bear
Perseus	Persei	Per	Perseus[a]	Vela	Velorum	Vel	Ship's Sails[b]
Phoenix	Phoenicia	Phe	Phoenix	Virgo	Virginia	Vir	Virgin
Pictor	Pictoris	Pic	Easel	Volans	Volantis	Vol	Flying Fish
Pisces	Piscium	Psc	Fish	Vulpecula	Vulpeculae	Vul	Little Fox

[a]Proper names.
[b]Formerly formed the constellation Argo Navis, the Argonauts' ship.

4.4. TIME AND PLANETARY POSITIONS

TABLE 4.15. Julian day and planetary longitudes

Year	Date		Julian Day 2450000+	Sun	Mercury	Venus	Mars	Jupiter	Saturn	Uranus	Neptune	Pluto
2000	Jan	1	1545	279	271	240	327	25	40	314	303	251
2000	Jan	11	1555	290	286	253	335	25	40	315	303	251
2000	Jan	21	1565	300	303	265	343	26	40	315	303	252
2000	Jan	31	1575	310	320	277	350	27	40	316	304	252
2000	Feb	10	1585	320	337	289	358	29	41	317	304	252
2000	Feb	20	1595	330	346	302	6	30	41	317	305	252
2000	Mar	1	1605	340	341	314	13	32	42	318	305	252
2000	Mar	11	1615	350	333	326	21	34	43	318	305	252
2000	Mar	21	1625	0	334	339	28	36	44	319	305	252
2000	Mar	31	1635	10	342	351	35	38	45	319	306	252
2000	Apr	10	1645	20	355	3	43	41	46	320	306	252
2000	Apr	20	1655	30	11	16	50	43	47	320	306	252
2000	Apr	30	1665	39	29	28	57	45	49	320	306	252
2000	May	10	1675	49	50	40	64	48	50	320	306	252
2000	May	20	1685	59	71	53	71	50	51	320	306	251
2000	May	30	1695	68	89	65	78	53	52	320	306	251
2000	Jun	9	1705	78	102	77	84	55	54	320	306	251
2000	Jun	19	1715	88	109	90	91	57	55	320	306	251
2000	Jun	29	1725	97	108	102	98	59	56	320	305	250
2000	Jul	9	1735	107	103	114	104	61	57	320	305	250
2000	Jul	19	1745	116	100	126	111	63	58	319	305	250
2000	Jul	29	1755	126	106	139	118	65	59	319	305	250
2000	Aug	8	1765	135	121	151	124	67	59	318	304	250
2000	Aug	18	1775	145	141	163	130	68	60	318	304	250
2000	Aug	28	1785	154	160	176	137	69	60	318	304	250
2000	Sep	7	1795	164	178	188	143	70	60	317	304	250
2000	Sep	17	1805	174	194	200	149	70	60	317	304	250
2000	Sep	27	1815	184	208	212	156	71	60	317	303	250
2000	Oct	7	1825	194	219	225	162	71	60	317	303	250
2000	Oct	17	1835	203	225	237	168	70	59	316	303	251
2000	Oct	27	1845	213	220	249	175	69	59	316	303	251
2000	Nov	6	1855	223	210	261	181	68	58	316	303	251
2000	Nov	16	1865	233	214	273	187	67	57	317	304	252
2000	Nov	26	1875	244	228	285	193	66	56	317	304	252
2000	Dec	6	1885	254	243	297	199	65	56	317	304	252
2000	Dec	16	1895	264	258	308	205	63	55	317	304	253
2000	Dec	26	1905	274	274	320	211	62	54	318	305	253
2001	Jan	5	1915	284	290	331	217	61	54	318	305	253
2001	Jan	15	1925	294	307	341	223	61	54	319	305	254
2001	Jan	25	1935	305	322	351	228	61	54	319	306	254
2001	Feb	4	1945	315	330	1	234	61	54	320	306	254
2001	Feb	14	1955	325	323	8	239	61	54	321	306	254
2001	Feb	24	1965	335	315	14	244	62	54	321	307	255
2001	Mar	6	1975	345	318	17	249	63	55	322	307	255
2001	Mar	16	1985	355	328	16	254	65	56	322	307	255
2001	Mar	26	1995	5	341	12	258	66	57	323	308	255
2001	Apr	5	2005	15	357	6	262	68	58	323	308	255

(continued)

TABLE 4.15. Julian day and planetary longitudes (continued)

Year	Date		Julian Day 2450000+	Sun	Mercury	Venus	Mars	Jupiter	Saturn	Uranus	Neptune	Pluto
2001	Apr	15	2015	25	16	2	265	70	59	324	308	255
2001	Apr	25	2025	34	36	1	267	72	60	324	308	254
2001	May	5	2035	44	57	5	268	74	61	324	308	254
2001	May	15	2045	54	74	11	268	76	63	324	308	254
2001	May	25	2055	63	85	18	267	78	64	324	308	254
2001	Jun	4	2065	73	89	27	265	81	65	324	308	253
2001	Jun	14	2075	83	86	37	262	83	66	324	308	253
2001	Jun	24	2085	92	81	47	259	85	68	324	308	253
2001	Jul	4	2095	102	82	58	256	87	69	324	308	253
2001	Jul	14	2105	111	91	69	255	90	70	324	307	252
2001	Jul	24	2115	121	107	80	255	92	71	323	307	252
2001	Aug	3	2125	130	127	91	256	94	72	323	307	252
2001	Aug	13	2135	140	147	103	258	96	73	322	307	252
2001	Aug	23	2145	149	165	115	262	98	73	322	306	252
2001	Sep	2	2155	159	181	126	266	100	74	322	306	252
2001	Sep	12	2165	169	194	138	271	101	74	321	306	252
2001	Sep	22	2175	179	205	151	277	102	74	321	306	252
2001	Oct	2	2185	188	209	163	283	104	74	321	306	252
2001	Oct	12	2195	198	203	175	289	104	74	321	306	253
2001	Oct	22	2205	208	194	188	296	105	74	320	305	253
2001	Nov	1	2215	218	200	200	302	105	73	320	306	253
2001	Nov	11	2225	228	214	213	309	105	73	320	306	254
2001	Nov	21	2235	238	230	225	317	105	72	321	306	254
2001	Dec	1	2245	248	246	238	324	104	71	321	306	254
2001	Dec	11	2255	259	262	250	331	103	70	321	306	255
2001	Dec	21	2265	269	278	263	338	102	70	321	307	255
2001	Dec	31	2275	279	294	275	346	100	69	322	307	255
2002	Jan	10	2285	289	308	288	353	99	68	322	307	256
2002	Jan	20	2295	299	314	301	0	98	68	323	308	256
2002	Jan	30	2305	309	304	313	8	97	68	323	308	256
2002	Feb	9	2315	320	298	326	15	96	68	324	308	257
2002	Feb	19	2325	330	303	338	22	95	68	325	309	257
2002	Mar	1	2335	340	314	351	29	95	68	325	309	257
2002	Mar	11	2345	350	328	3	36	95	68	326	309	257
2002	Mar	21	2355	0	344	16	43	96	69	326	310	257
2002	Mar	31	2365	10	2	28	50	96	70	327	310	257
2002	Apr	10	2375	19	22	40	57	98	71	327	310	257
2002	Apr	20	2385	29	43	52	64	99	72	328	310	257
2002	Apr	30	2395	39	59	65	71	100	73	328	310	257
2002	May	10	2405	49	68	77	77	102	74	328	310	256
2002	May	20	2415	58	69	89	84	104	75	328	310	256
2002	May	30	2425	68	64	101	90	106	77	328	310	256
2002	Jun	9	2435	78	61	113	97	108	78	328	310	256
2002	Jun	19	2445	87	65	124	104	110	79	328	310	255
2002	Jun	29	2455	97	76	136	110	112	80	328	310	255
2002	Jul	9	2465	106	92	147	117	114	82	328	310	255

(continued)

TABLE 4.15. Julian day and planetary longitudes (continued)

Year	Date		Julian Day 2450000+	Sun	Mercury	Venus	Mars	Jupiter	Saturn	Uranus	Neptune	Pluto
2002	Jul	19	2475	116	113	159	123	116	83	328	309	255
2002	Jul	29	2485	125	134	170	129	119	84	327	309	255
2002	Aug	8	2495	135	152	180	136	121	85	327	309	254
2002	Aug	18	2505	144	168	190	142	123	86	326	309	254
2002	Aug	28	2515	154	181	200	148	125	87	326	308	254
2002	Sep	7	2525	164	190	209	155	127	88	326	308	254
2002	Sep	17	2535	173	192	216	161	129	88	325	308	255
2002	Sep	27	2545	183	185	222	168	131	88	325	308	255
2002	Oct	7	2555	193	178	225	174	133	89	325	308	255
2002	Oct	17	2565	203	186	224	180	134	89	325	308	255
2002	Oct	27	2575	213	201	220	187	135	88	324	308	255
2002	Nov	6	2585	223	218	214	193	136	88	324	308	256
2002	Nov	16	2595	233	234	210	199	137	87	324	308	256
2002	Nov	26	2605	243	250	210	206	137	87	325	308	256
2002	Dec	6	2615	253	265	214	212	138	86	325	308	257
2002	Dec	16	2625	263	280	220	219	137	85	325	309	257
2002	Dec	26	2635	274	293	228	225	137	84	326	309	258
2003	Jan	5	2645	284	297	237	232	136	84	326	309	258
2003	Jan	15	2655	294	287	247	238	135	83	326	310	258
2003	Jan	25	2665	304	282	258	245	134	82	327	310	259
2003	Feb	4	2675	314	289	269	251	132	82	328	310	259
2003	Feb	14	2685	324	301	280	257	131	82	328	311	259
2003	Feb	24	2695	334	315	292	264	130	82	329	311	259
2003	Mar	6	2705	344	331	304	270	129	82	329	311	259
2003	Mar	16	2715	354	349	315	277	128	82	330	312	259
2003	Mar	26	2725	4	9	327	283	128	83	330	312	259
2003	Apr	5	2735	14	28	339	289	128	83	331	312	259
2003	Apr	15	2745	24	44	351	295	128	84	331	312	259
2003	Apr	25	2755	34	50	4	301	128	85	332	313	259
2003	May	5	2765	44	47	16	307	129	86	332	313	259
2003	May	15	2775	53	42	28	313	130	87	332	313	259
2003	May	25	2785	63	41	40	318	131	88	332	313	259
2003	Jun	4	2795	72	48	52	323	133	89	332	313	258
2003	Jun	14	2805	82	61	64	328	134	91	332	312	258
2003	Jun	24	2815	92	78	75	332	136	92	332	312	258
2003	Jul	4	2825	101	99	89	336	138	93	332	312	257
2003	Jul	14	2835	111	120	101	338	140	95	332	312	257
2003	Jul	24	2845	120	139	113	339	142	96	332	312	257
2003	Aug	3	2855	130	154	125	339	144	97	331	311	257
2003	Aug	13	2865	139	167	138	338	146	98	331	311	257
2003	Aug	23	2875	149	174	150	336	149	99	330	311	257
2003	Sep	2	2885	159	175	163	333	151	100	330	311	257
2003	Sep	12	2895	168	167	175	331	153	101	330	310	257
2003	Sep	22	2905	178	162	187	330	155	102	329	310	257
2003	Oct	2	2915	188	171	200	330	157	102	329	310	257
2003	Oct	12	2925	198	188	212	331	159	103	329	310	257
2003	Oct	22	2935	208	205	225	333	161	103	329	310	257

(continued)

TABLE 4.15. Julian day and planetary longitudes (continued)

Year	Date		Julian Day 2450000+	Sun	Mercury	Venus	Mars	Jupiter	Saturn	Uranus	Neptune	Pluto
2003	Nov	1	2945	218	222	237	337	163	103	328	310	258
2003	Nov	11	2955	228	237	250	341	164	103	328	310	258
2003	Nov	21	2965	238	253	262	346	165	102	328	310	258
2003	Dec	1	2975	248	267	274	351	167	102	329	310	259
2003	Dec	11	2985	258	279	287	356	167	101	329	311	259
2003	Dec	21	2995	268	281	299	2	168	100	329	311	260
2003	Dec	31	3005	278	269	311	8	168	99	330	311	260
2004	Jan	10	3015	289	267	324	14	168	99	330	311	26
2004	Jan	20	3025	299	275	336	20	168	98	330	312	261
2004	Jan	30	3035	309	288	348	27	167	97	331	312	261
2004	Feb	9	3045	319	302	0	33	166	96	332	313	261
2004	Feb	19	3055	329	318	12	39	165	96	332	313	261
2004	Feb	29	3065	339	336	23	46	164	96	333	313	262
2004	Mar	10	3075	349	355	34	52	163	96	333	314	262
2004	Mar	20	3085	359	14	45	59	161	96	334	314	262
2004	Mar	30	3095	9	28	55	65	160	96	334	314	262
2004	Apr	9	3105	19	31	64	71	159	97	335	314	262
2004	Apr	19	3115	29	25	73	78	159	97	335	315	262
2004	Apr	29	3125	39	21	80	84	158	98	336	315	261
2004	May	9	3135	48	23	84	91	158	99	336	315	261
2004	May	19	3145	58	32	86	97	159	100	336	315	261
2004	May	29	3155	67	46	83	103	159	101	336	315	261
2004	Jun	8	3165	77	64	78	109	160	102	336	315	260
2004	Jun	18	3175	87	85	72	116	161	104	336	315	260
2004	Jun	28	3185	96	107	69	122	162	105	336	314	260
2004	Jul	8	3195	106	125	70	128	164	106	336	314	260
2004	Jul	18	3205	115	141	75	135	166	108	336	314	260
2004	Jul	28	3215	125	152	81	141	167	109	335	314	259
2004	Aug	7	3225	134	158	89	147	169	110	335	313	259
2004	Aug	17	3235	144	156	98	154	171	111	335	313	259
2004	Aug	27	3245	154	148	108	160	173	112	334	313	259
2004	Sep	6	3255	163	146	119	166	175	113	334	313	259
2004	Sep	16	3265	173	157	129	173	178	114	334	312	259
2004	Sep	26	3275	183	175	141	179	180	115	333	312	259
2004	Oct	6	3285	193	193	152	186	182	116	333	312	259
2004	Oct	16	3295	202	210	164	192	184	116	333	312	260
2004	Oct	26	3305	212	225	176	199	186	117	332	312	260
2004	Nov	5	3315	222	240	188	205	188	117	332	312	260
2004	Nov	15	3325	232	254	200	212	190	117	332	312	260
2004	Nov	25	3335	243	264	213	219	192	117	332	312	261
2004	Dec	5	3345	253	264	225	225	193	116	333	313	261
2004	Dec	15	3355	263	252	237	232	195	116	333	313	262
2004	Dec	25	3365	273	252	250	239	196	115	333	313	262
2005	Jan	4	3375	283	261	262	246	197	114	334	313	262
2005	Jan	14	3385	293	275	275	253	198	113	334	314	263
2005	Jan	24	3395	304	290	287	260	198	113	334	314	263

(continued)

TABLE 4.15. Julian day and planetary longitudes (continued)

Year	Date		Julian Day 2450000+	Sun	Mercury	Venus	Mars	Jupiter	Saturn	Uranus	Neptune	Pluto
2005	Feb	3	3405	314	306	300	267	198	112	335	315	263
2005	Feb	13	3415	324	323	312	274	198	111	336	315	264
2005	Feb	23	3425	334	341	325	281	198	111	336	315	264
2005	Mar	5	3435	344	359	337	288	197	110	337	316	264
2005	Mar	15	3445	354	12	350	295	196	110	337	316	264
2005	Mar	25	3455	4	12	2	303	195	110	338	316	264
2005	Apr	4	3465	14	4	15	310	193	110	338	317	264
2005	Apr	14	3475	24	1	27	317	192	110	339	317	264
2005	Apr	24	3485	33	7	40	324	191	111	339	317	264
2005	May	4	3495	43	17	52	332	190	112	340	317	264
2005	May	14	3505	53	32	64	339	189	112	340	317	263
2005	May	24	3515	62	50	76	346	189	113	340	317	263
2005	Jun	3	3525	72	72	89	353	188	114	340	317	263
2005	Jun	13	3535	82	93	101	0	189	115	340	317	263
2005	Jun	23	3545	91	111	113	7	189	117	340	317	262
2005	Jul	3	3555	101	126	125	14	190	118	340	317	262
2005	Jul	13	3565	110	136	137	20	190	119	340	316	262
2005	Jul	23	3575	120	140	149	26	192	120	340	316	262
2005	Aug	2	3585	129	136	161	32	193	122	339	316	262
2005	Aug	12	3595	139	129	173	38	194	123	339	316	261
2005	Aug	22	3605	149	130	185	42	196	124	339	315	261
2005	Sep	1	3615	158	143	197	47	198	125	338	315	261
2005	Sep	11	3625	168	161	209	50	200	126	338	315	261
2005	Sep	21	3635	178	180	220	52	202	127	338	315	261
2005	Oct	1	3645	187	197	232	53	204	128	337	314	262
2005	Oct	11	3655	197	213	243	52	206	129	337	314	262
2005	Oct	21	3665	207	227	254	50	208	130	337	314	262
2005	Oct	31	3675	217	240	264	47	211	130	336	314	262
2005	Nov	10	3685	227	249	274	44	213	131	336	314	263
2005	Nov	20	3695	237	248	283	40	215	131	336	314	263
2005	Nov	30	3705	247	236	291	38	217	131	336	315	263
2005	Dec	10	3715	258	237	297	38	219	131	337	315	264
2005	Dec	20	3725	268	248	301	38	221	130	337	315	264
2005	Dec	30	3735	278	262	300	40	222	130	337	315	264
2006	Jan	9	3745	288	277	296	43	224	129	338	316	265
2006	Jan	19	3755	298	293	290	46	225	128	338	316	265
2006	Jan	29	3765	308	310	286	50	227	127	338	316	265

TABLE 4.16. Local sidereal time at 00:00 local standard time*

Date	Jan	Feb	Mar	Apr	May	Jun	Jul	Aug	Sep	Oct	Nov	Dec
1	6:42	8:45	10:35	12:37	14:35	16:38	18:36	20:38	22:40	0:39	2:41	4:39
2	6:46	8:49	10:39	12:41	14:39	16:42	18:40	20:42	22:44	0:43	2:45	4:43
3	6:50	8:52	10:43	12:45	14:43	16:46	18:44	20:46	22:48	0:47	2:49	4:47
4	6:54	8:56	10:47	12:49	14:47	16:50	18:48	20:50	22:52	0:51	2:53	4:51
5	6:58	9:00	10:51	12:53	14:51	16:53	18:52	20:54	22:56	0:54	2:57	4:55
6	7:02	9:04	10:55	12:57	14:55	16:57	18:56	20:58	23:00	0:58	3:01	4:59
7	7:06	9:08	10:59	13:01	14:59	17:01	19:00	21:02	23:04	1:02	3:05	5:03
8	7:10	9:12	11:03	13:05	15:03	17:05	19:04	21:06	23:08	1:06	3:09	5:07
9	7:14	9:16	11:07	13:09	15:07	17:09	19:08	21:10	23:12	1:10	3:12	5:11
10	7:18	9:20	11:10	13:13	15:11	17:13	19:11	21:14	23:16	1:14	3:16	5:15
11	7:22	9:24	11:14	13:17	15:15	17:17	19:15	21:18	23:20	1:18	3:20	5:19
12	7:26	9:28	11:18	13:21	15:19	17:21	19:19	21:22	23:24	1:22	3:24	5:23
13	7:30	9:32	11:22	13:25	15:23	17:25	19:23	21:26	23:28	1:26	3:28	5:27
14	7:34	9:36	11:26	13:28	15:27	17:29	19:27	21:29	23:32	1:30	3:32	5:30
15	7:38	9:40	11:30	13:32	15:31	17:33	19:31	21:33	23:36	1:34	3:36	5:34
16	7:42	9:44	11:34	13:36	15:35	17:37	19:35	21:37	23:40	1:38	3:40	5:38
17	7:45	9:48	11:38	13:40	15:39	17:41	19:39	21:41	23:43	1:42	3:44	5:42
18	7:49	9:52	11:42	13:44	15:43	17:45	19:43	21:45	23:47	1:46	3:48	5:46
19	7:53	9:56	11:46	13:48	15:46	17:49	19:47	21:49	23:51	1:50	3:52	5:50
20	7:57	9:59	11:50	13:52	15:50	17:53	19:51	21:53	23:55	1:54	3:56	5:54
21	8:01	10:03	11:54	13:56	15:54	17:57	19:55	21:57	23:59	1:58	4:00	5:58
22	8:05	10:07	11:58	14:00	15:58	18:00	19:59	22:01	0:03	2:01	4:04	6:02
23	8:09	10:11	12:02	14:04	16:02	18:04	20:03	22:05	0:07	2:05	4:08	6:06
24	8:13	10:15	12:06	14:08	16:06	18:08	20:07	22:09	0:11	2:09	4:12	6:10
25	8:17	10:19	12:10	14:12	16:10	18:12	20:11	22:13	0:15	2:13	4:16	6:14
26	8:21	10:23	12:14	14:16	16:14	18:16	20:15	22:17	0:19	2:17	4:19	6:18
27	8:25	10:27	12:17	14:20	16:18	18:20	20:18	22:21	0:23	2:21	4:23	6:22
28	8:29	10:31	12:21	14:24	16:22	18:24	20:22	22:25	0:27	2:25	4:27	6:26
29	8:33		12:25	14:28	16:26	18:28	20:26	22:29	0:31	2:29	4:31	6:30
30	8:37		12:29	14:32	16:30	18:32	20:30	22:33	0:35	2:33	4:35	6:34
31	8:41		12:33		16:34		20:34	22:36		2:37		6:37

*Compared with the sidereal times listed here:
Subtract 3 minutes from January 1 to February 28, 2000, 2004, 2008, ...
Add 1 minute from February 29 to December 31, 2000, 2004, 2008, ...
Subtract 1 minute for years 2002, 2006, 2010, ...
Subtract 2 minutes for years 2003, 2007, 2011, ...

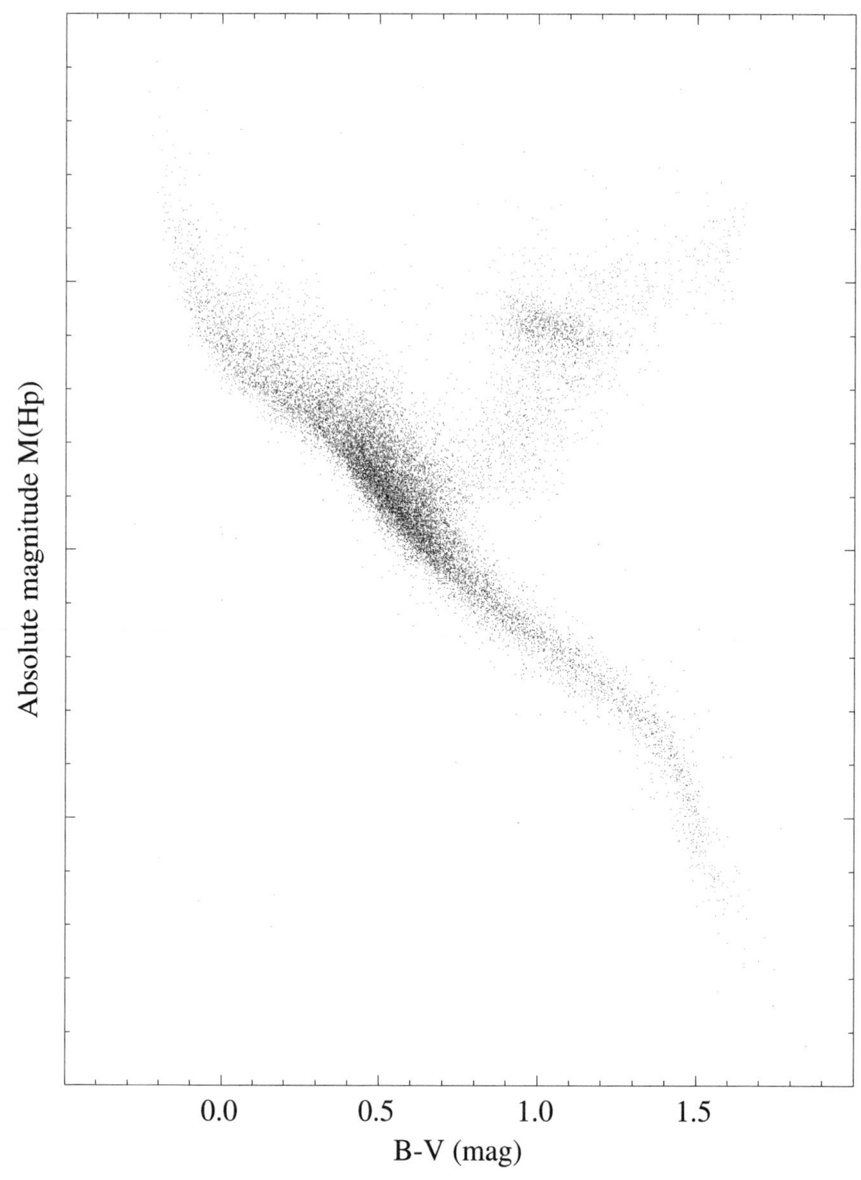

FIGURE 4.1. The observational Hertzsprung-Russell diagram, Hipparcos absolute magnitude versus color, from the Hipparcos Catalogue for the 20,853 stars with parallaxes known to 10% and uncertainties in color less than 0.25 magnitude. From M. A. C. Perryman et al., "The Hipparcos Catalogue," 1997, *Astronomy and Astrophysics*, vol. 323, pp. L49–L52. Used with permission.

5

Astrophysics and Cosmology

Virginia Trimble
University of California, Irvine, CA, and University of Maryland, College Park, MD

Contents

List of Tables

List of Figures

5.1. STELLAR ASTRONOMY

5.1.1. Stellar structure and evolution

The equations of stellar structure

The internal structure of a star can be described by four nonlinear differential equations representing the dependence of mass M, luminosity L, pressure P, and temperature T on radius r (or the other four in terms of M):

$$\frac{dM}{dr} = 4\pi r^2 \rho, \tag{5.1}$$

$$\frac{dL}{dr} = 4\pi r^2 \rho \epsilon + \text{additional terms if not in hydrostatic equilibrium}, \tag{5.2}$$

$$\frac{dP}{dr} = -G\rho \frac{M}{r^2}, \tag{5.3}$$

$$\frac{dT}{dr} = -\frac{3L\kappa\rho}{16\pi r^2 acT^3} \qquad \text{(if energy transport is radiative)}, \tag{5.4}$$

$$\frac{dT}{dr} = \left(1 - \frac{1}{\gamma}\right)\frac{T}{P}\frac{dP}{dr} \qquad \text{(if energy transport is convective and convection is completely efficient)}. \tag{5.5}$$

Radiative transport gives way to convective when dT/dr from Eq. (5.4) is equal to the value in Eq. (5.5).

The auxiliary quantities ϵ, κ, and $P(\rho, T)$ represent, respectively, energy generation, opacity to radiation, and the equation of state. Each is a function of temperature, density, and chemical composition. They are often stored as tables.

Methods of solution

The appropriate boundary conditions are not known for all of the variables at any one radius r. $M(r{=}0){=}0$ and $L(r{=}0){=}0$, but T and P at $r{=}R$ must come from a model atmosphere. Thus, integrations must start from $r = 0$ and $r = R$ and proceed to a matching point in between, with the missing boundary values guessed. The derivatives in mismatch as the guesses are changed can be used to improve the boundary values until convergence is regarded as satisfactory (Schwarzschild method). After an initial model has been computed, it can be used as a first guess at subsequent ones (changing as nuclear reactions change the composition) and improved to converence (Henyey method.)

Evolutionary tracks and isochrones

A time-sequence of models (typically plotted in an HR diagram) constitutes an evolutionary track. A set of models of fixed age, but for stars of different masses (and identical initial composition), constitutes an isochrone and can be compared with the observations of star clusters, also usually in the HR diagram. Figure 5.1 shows typical results.

The Eddington relations

The Eqs. (5.1)–(5.4) can be used to estimate the central pressure and temperature of the Sun and stars and the approximate relationships among M, L, T, and R (without knowledge of the energy source) by regarding each as a difference equation, where $dr \to r/2$, $dM \to M/2$, $dP \to P_c$, and so forth. This is left as an exercise for the student, and the results are

$$P_c \approx 2\bar{\rho}\frac{GM}{R} \qquad (\approx 6 \times 10^{14} \text{ for the Sun}),$$

$$T_c \approx P_c\frac{\overline{m}}{k\overline{\rho}} \approx 10^7\ \text{K for the Sun} \quad (\text{where } \overline{m} \text{ is the mean molecular weight}),$$

$$L \propto M^3 \quad (\text{hence, lifetime} \propto \text{M}^{-2}),$$

$$T_c \propto M^{1/2} \quad \text{and} \quad r \propto M^{1/2}.$$

5.1.2. Stellar atmospheres

In principle, the basic equations are the same as for stellar interior structure, except that there is no nuclear energy generation; thus, L is changed only by expansion or contraction of the gas:

$$\frac{dL}{dM} = -\frac{dU}{dt} - P\frac{d}{dt}(1/\rho),$$

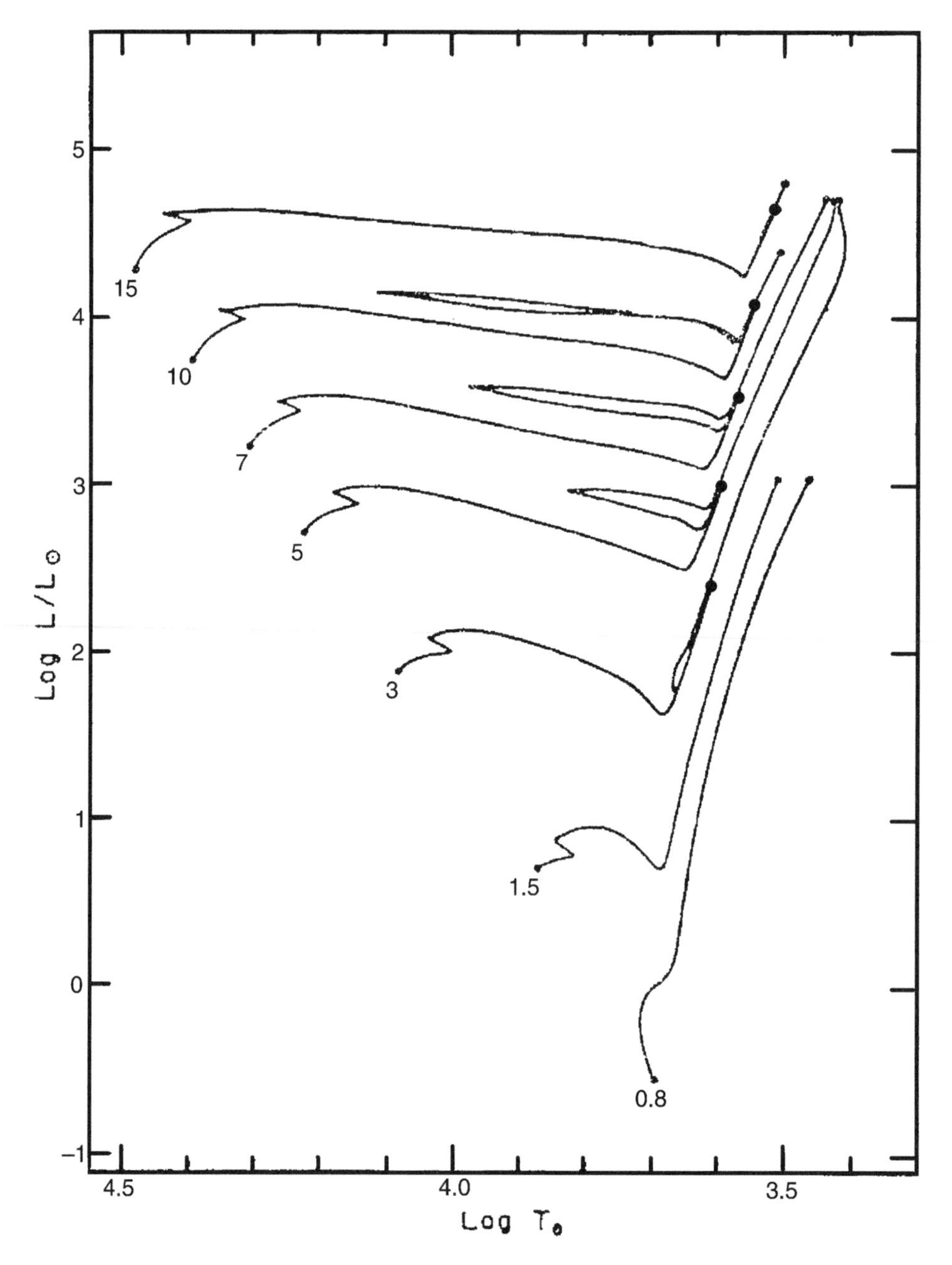

FIGURE 5.1. Evolutionary tracks on the HR diagram. Population I stars ($X = 0.7, Z = 0.03$). The numbers at the beginning of each track give stellar mass in units of solar mass. Large dots indicate the position of the homogeneous main sequence models and the position of the models at the times of helium and carbon ignition in their cores. (Reprinted with permission from Bohdan Paczyński, *Acta Astronomica*, **20**, 47.)

and, more important, the gas is no longer in complete thermal equilibrium (otherwise, there could be no absorption lines in the spectrum, and so forth). Thus, in the most general case, one must define an optical depth $d\tau_\nu = k_\nu \rho\, dr$ at each frequency ν and follow the transfer of radiation outward from a deep, equilibrium layer to $\tau \leq 1$, via an equation of radiative transfer,

$$\frac{dI_\nu}{d\tau_\nu} = -I_\nu + S_\nu,$$

where S_ν is the ratio of the emission coefficient to the absorption coefficient for a volume of gas. Because k_ν and S_ν will have contributions from bound-bound and bound-free transitions of many elements in several states of ionization and excitation, a complete model atmosphere keeps track of these level populations as a function of r or of optical depth in the mean, defined by $d\tau = k\rho\, dr$, where

$$\frac{1}{\kappa} = \frac{\pi}{acT^3} \int_0^\infty \frac{1}{\kappa_\nu} \frac{\partial B_\nu}{\partial T}\, d\nu \tag{5.6}$$

defines the Rosseland mean opacity, weighted by a blackbody spectrum at the local temperature T. Ionization and excitation in a given layer can have significant contributions from photons coming from some distance away and not described by the local T (non-local-thermodynamic-equilibrium, non-LTE).

No existing atmospheric code can do this in full generality, and the lore of stellar atmosphere modeling is largely the art of approximating as many items as possible. A grid of model atmospheres constitutes the outer boundary conditions for calculations of stellar interiors.

5.1.3. Nuclear reactions, energy generation, and nucleosynthesis

Hydrogen burning

Fusion of 4 H to ^{4}He can be accomplished in two ways. It releases about 6.85×10^{18} erg/gm [685 TJ/kg] (approximately corrected for neutrino losses), and the rate (governed primarily by barrier penetration) can balance losses by the star for central temperature of $(10\text{–}20) \times 10^6$ K. The first set has a lower barrier and therefore dominates in low-mass stars (the Sun, and stars up to $1.5 M_\odot$), whereas the second set (which is the primary source of N in the universe) dominates in more massive stars.

Proton-proton chain:

$$\mathrm{p(p,\, e^+\nu)d\ (p, \gamma\,)^3He} \qquad \text{(twice)},$$
$$\mathrm{^3He\ (^3He, 2p)^4He},$$

or

$$\mathrm{^3He\,(\alpha, \gamma)\,^7Be} \rightarrow \begin{cases} \mathrm{^7Be(e^-, \nu)\,^7Li(p, \alpha)\,^4He}, \\ \mathrm{^7Be\,(p, \gamma)\,^8B(\ ,\ e^+\nu)\,^8Be\,(\ , \alpha)\,^4He}. \end{cases} \tag{5.7}$$

Virtually all of the energy generation comes from the first part of the chain, but the minor, ^{7}Be and ^{8}B, branches produce high-energy neutrinos that have been historically important.

CNO Tricycle:

$^{12}\mathrm{C}\,(p,\,\gamma)^{13}\mathrm{N}\,(\,,\,\mathrm{e}^+\nu)^{13}\mathrm{C}\,(p,\gamma)^{14}\mathrm{N}$ (which has a small capture cross section, so material hangs up there),

$^{14}\mathrm{N}\,(\mathrm{p},\gamma)^{15}\mathrm{O}(\,,\,\mathrm{e}^+\nu)\,^{15}\mathrm{N}\,(\mathrm{p},\alpha)^{12}\mathrm{C}$ (completing the simplest cycle),

or

$^{15}\mathrm{N}\,(\mathrm{p},\gamma)^{16}\mathrm{O}\,(\mathrm{p},\gamma)^{17}\mathrm{F}(\,,\,\mathrm{e}^+\nu)^{17}\mathrm{O}\,(\mathrm{p},\alpha)^{14}\mathrm{N}$

or: $^{17}\mathrm{O}\,(\mathrm{p},\gamma)^{18}\mathrm{F}(\,,\,\mathrm{e}^+\nu)^{18}\mathrm{O}\,(\mathrm{p},\alpha)^{15}\mathrm{N}$ (with the potential to extend upward to F, Ne, and Na at high temperatures). (5.8)

Helium burning (triple-alpha reaction)

Hydrogen fusion releases nearly 90% of the total energy available from fusion. The rest is released about half in helium burning and half on the rest of the way up to ^{56}Ni or ^{56}Fe. Because neither ^{5}Li nor ^{8}B is stable, fusion beyond helium can continue only at high density, where three-body encounters occur. Typical stellar conditions are $T \approx 10^8$ K and $\rho = (10^6\text{–}10^9)\,\mathrm{kg/m^3}$. At the upper end of this range, the helium is partially degenerate, resulting in a "helium flash" at ignition.

The two reactions: $^4\mathrm{He}(\alpha,\gamma)\,^8\mathrm{B}(\alpha,\gamma)\,^{12}\mathrm{C}$ and $^{12}\mathrm{C}(\alpha,\gamma)\,^{16}\mathrm{O}$ occur under rather similar conditions (owing to a fortunately placed excited level of ^{12}C) so that stellar processes produce comparable amounts of C and O.

Heavy element burning

Alpha captures do not continue beyond ^{16}O owing to the absence of low-lying levels of ^{20}Ne (etc.) with appropriate spin and parity. Instead, fuels burn in the sequence carbon, neon, oxygen, silicon, at gradually increasing temperatures near 10^9 K, generally not by the most obvious reaction. Some typical reactions include

$$^{12}\mathrm{C} + {}^{12}\mathrm{C} \rightarrow \begin{cases} ^{20}\mathrm{Ne} + {}^4\mathrm{He}, & \\ ^{23}\mathrm{Mg} + \mathrm{n}, & \text{with decay to } {}^{23}\mathrm{Na}, \\ ^{23}\mathrm{Na} + \mathrm{p}, & \end{cases}$$

$$^{16}\mathrm{O} + {}^{16}\mathrm{O} \rightarrow {}^{28}\mathrm{Si} + {}^4\mathrm{He}.$$

Notice that there are loose p's, n's, and α-particles available for additional captures, so that the full range of common and rare isotopes is produced, ending with

^{28}Si (photodissociated to 7 ^{4}He),

$^{28}\mathrm{Si} + 7\,^4\mathrm{He} \rightarrow {}^{56}\mathrm{Ni}$ which β-decays to ^{56}Fe during and after a type II supernova explosion.

Beyond the iron peak

All elements heavier than $A \approx 60$ are produced by neutron captures, primarily on iron-peak seeds. There are two potential neutron sources:

$$^{13}\mathrm{C}(\alpha,\ \mathrm{n})\,^{16}\mathrm{O},$$

$$^{22}\mathrm{Ne}(\alpha,\ \mathrm{n})\,^{25}\mathrm{Mg},$$

and two contexts where neutron capture occurs:

1. In the double-shell burning phase of intermediate mass stars, on a time scale slower than beta-decays (the s-process).
2. In core-collapse supernovae of massive stars, on a time scale more rapid than beta-decays (the r-process).

Some fine-tuning results from the addition of protons or (more probably) photoremoval of neutrons from a few nuclides to produce and n-poor, rare isotopes (the p-process).

Nucleosynthesis: Summary

1. H, ^{2}H, ^{3}He, ^{4}He, and about 10% of the ^{7}Li are left from the Big Bang.
2. Additional helium and elements up to the iron peak are made by hydrostatic reactions in stable stars, especially massive ones.
3. Neutron captures produce the elements beyond iron (only the r-process reaches Th and U).
4. ^{6}Li, more ^{7}Li, ^{9}Be, and ^{10,11}B come from spallation; cosmic rays breaking up CNO atoms in the interstellar medium (this process also contributes to some rare odd nuclides).
5. With a little good will, these can all be made to add to the totals in Table 5.1.

TABLE 5.1. Standard solar composition: Elemental abundances in the solar photosphere and in meteorites ($\log_{10} A$, normalized to $\log_{10} A(\mathrm{H}) = 12.00$). Reprinted, with permission, from Edward Anders, 1998.

El	Photosphere[a]	Meteorites	Ph − Met	El	Photosphere[a]	Meteorites	Ph − Met
01 H	12.00	—	—	42 Mo	1.92 ± 0.05	1.97 ± 0.02	−0.05
02 He	[10.930 ± 0.004]	—	—	44 Ru	1.84 ± 0.07	1.83 ± 0.04	+0.01
03 Li	1.10 ± 0.10	3.31 ± 0.04	−2.21	45 Rh	1.12 ± 0.12	1.10 ± 0.04	+0.02
04 Be	1.40 ± 0.09	1.42 ± 0.04	−0.02	46 Pd	1.69 ± 0.04	1.70 ± 0.04	−0.01
05 B	(2.55 ± 0.30)	2.79 ± 0.05	(−0.24)	47 Ag	(0.94 ± 0.25)	1.24 ± 0.04	(−0.30)
06 C	8.52 ± 0.06	—	—	48 Cd	1.77 ± 0.11	1.76 ± 0.04	+0.01
07 N	7.92 ± 0.06	—	—	49 In	(1.66 ± 0.15)	0.82 ± 0.04	(+0.84)
08 0	8.83 ± 0.06	—	—	50 Sn	2.0 ± (0.3)	2.14 ± 0.04	−0.14
09 F	[4.56 ± 0.3]	4.48 ± 0.06	+0.08	51 Sb	1.0 ± (0.3)	1.03 ± 0.07	−0.03
10 Ne	[8.08 ± 0.06]	—	—	52 Te	—	2.24 ± 0.04	—
11 Na	6.33 ± 0.03	6.32 ± 0.02	+0.01	53 1	—	1.51 ± 0.08	—
12 Mg	7.58 ± 0.05	7.58 ± 0.01	0.00	54 Xe	—	2.17 ± 0.08	—
13 Al	6.47 ± 0.07	6.49 ± 0.01	−0.02	55 Cs	—	1.13 ± 0.02	—
14 Si	7.55 ± 0.05	7.56 ± 0.01	−0.01	56 Ba	2.13 ± 0.05	2.22 ± 0.02	−0.09
15 P	5.45 ± (0.04)	5.56 ± 0.06	−0.11	57 La	1.17 ± 0.07	1.22 ± 0.02	−0.05
16 S	7.33 ± 0.11	7.20 ± 0.06	+0.13	58 Ce	1.58 ± 0.09	1.63 ± 0.02	−0.05
17 Cl	[5.5 ± 0.30]	5.28 ± 0.06	+0.22	59 Pr	0.71 ± 0.08	0.80 ± 0.02	−0.09
18 Ar	[6.40 ± 0.06]	—	—	60 Nd	1.50 ± 0.06	1.49 ± 0.02	+0.01
19 K	5.12 ± 0.13	5.13 ± 0.02	−0.01	62 Sm	1.01 ± 0.06	0.98 ± 0.02	+0.03
20 Ca	6.36 ± 0.02	6.35 ± 0.01	+0.01	63 En	0.51 ± 0.08	0.55 ± 0.02	−0.04
21 Sc	3.17 ± 0.10	3.10 ± 0.01	+0.07	64 Gd	1.12 ± 0.04	1.09 ± 0.02	+0.03
22 Ti	5.02 ± 0.06	4.94 ± 0.02	+0.08	65 Tb	(−0.1 ± 0.3)	0.35 ± 0.02	(−0.45)
23 V	4.00 ± 0.02	4.02 ± 0.02	−0.02	66 Dy	1.14 ± 0.08	1.17 ± 0.02	−0.03
24 Cr	5.67 ± 0.03	5.69 ± 0.01	−0.02	67 Ho	(0.26 ± 0.16)	0.51 ± 0.02	(−0.25)
25 Mn	5.39 ± 0.03	5.53 ± 0.01	−0.14	68 Er	0.93 ± 0.06	0.97 ± 0.02	−0.04
26 Fe	7.50 ± 0.05	7.50 ± 0.01	0.00	69 Tm	(0.00 ± 0.15)	0.15 ± 0.02	(−0.15)
27 Co	4.92 ± 0.04	4.91 ± 0.01	+0.01	70 Yb	1.08 ± (0.15)	0.96 ± 0.02	+0.12
28 Ni	6.25 ± 0.04	6.25 ± 0.01	0.00	71 Lu	0.06 ± 0.10	0.13 ± 0.02	−0.07
29 Cu	4.21 ± 0.04	4.29 ± 0.04	−0.08	72 Hf	0.88 ± (0.08)	0.75 ± 0.02	+0.13
30 Zn	4.60 ± 0.08	4.67 ± 0.04	−0.07	73 Ta	—	0.13 ± 0.02	—
31 Ga	2.88 ± (0.10)	3.13 ± 0.02	−0.25	74 W	(1.11 ± 0.15)	0.69 ± 0.03	(+0.42)
32 Ge	3.41 ± 0.14	3.63 ± 0.04	−0.22	75 Re	—	0.28 ± 0.03	—
33 As	—	2.37 ± 0.02	—	76 Os	1.45 ± 0.10	1.39 ± 0.02	+0.06
34 Se	—	3.41 ± 0.03	—	77 Ir	1.35 ± (0.10)	1.37 ± 0.02	−0.02
35 Br	—	2.63 ± 0.04	—	78 Pt	1.8 ± 0.3	1.69 ± 0.04	+0.11
36 Kr	—	3.31 ± 0.08	—	79 Au	(1.01 ± 0.15)	0.85 ± 0.04	(+0.16)
37 Rb	2.60 ± (0.15)	2.41 ± 0.02	+0.19	80 Hg	1.13 ± 0.08	—	
38 Sr	2.97 ± 0.07	2.92 ± 0.02	+0.05	81 TI	(0.9 ± 0.2)	0.83 ± 0.04	(+0.07)
39 Y	2.24 ± 0.03	2.23 ± 0.02	+0.01	82 Pb	1.95 ± 0.08	2.06 ± 0.04	−0.11
40 Zr	2.60 ± 0.02	2.61 ± 0.02	−0.01	83 Bi	—	0.71 ± 0.04	—
41 Nb	1.42 ± 0.06	1.40 ± 0.02	+0.02	90 Th	—	0.09 ± 0.02	—
				92 U	(< −0.47)	−0.50 ± 0.04	—

[a] Values between square brackets are not derived from the photosphere, but from sunspots, solar corona and solar wind particles; values between parentheses are less accurate results.

TABLE 5.2. Summary of phases of stellar evolution

Phase	Time Scale	Energy Source	Mass Range	Comments
Giant molecular cloud	10^8 yr	Turbulence	all	Lasts much longer than free-fall time
Collapse and fragmentation	10^{-3} MS time	Gravitational potential	all	Magnetic field, turbulence, rotation important
Main Sequence	$10^{10}(M/M_\odot)(L/L_\odot)^{-1}$ yr $\approx 10^{10}(M/M_\odot)^{-2}$ yr	Hydrogen fusion	all	p-p cycle at $M \leq 1.5\,M_\odot$ CNO cycle $M \geq 1.5 M_\odot$
Core collapse	$\leq 10^{-3}$ MS time	Gravitational potential	all	Hertzsprung gap in HR diagrams of open clusters
Red giant or red supergiant	0.1 MS time	Shell hydrogen fusion	$M \gtrsim 0.3\,M_\odot$	Always CNO cycle fusion. Lower masses fully convective, and no core developes
Helium white dwarf	unlimited	Kinetic energy of nucleons	$M \lesssim 0.4 M_\odot$	Currently found only in binary products
He core burning	10^{-2} MS time	C fusion to C and O; H shell burning	$M \gtrsim 0.4\,M_\odot$	Onset violent for $M < 1.5\,M_\odot$; horizontal branch phase in globular clusters; clump phase in old open clusters
Double shell burning	10^{-3} MS time	He and H fusion in thin shells; flashes at lower masses	$M \gtrsim 0.4\,M_\odot$	*s*-process occurs; nuclear reaction products mixed to the surface; asymptotic giant branch phase in globular clusters
Superwind	10^{-4} MS time	Continued fusion	$M \gtrsim 0.4\,M_\odot$	Leads to extended circumstellar material
Planetary nebula	10^4 yr	Thermal	0.4–8 $M_\odot$	CO core seen as source of ionization
CO white dwarf	unlimited	Kinetic energy of nuclei	0.4–8 $M_\odot$	Eventual fate of the Sun
Heavy element burning	$\leq 10^3$ yr	Fusion of C, Ne, 0, Si	$M \geq 8 M_\odot$	
Core collapse	seconds	Gravitational potential	$M \geq 8 M_\odot$	Type II Supernova; SN remnant and pulsar or black hole left

MS = main sequence; WD = white dwarf; NS = neutron star.

5.2. BINARY STARS

5.2.1. Significance

Half or more of all stars are in gravitationally bound pairs, with separations ranging from the sum of the stellar radii to about 0.1 parsec. They are detected via eclipses (eclipsing binaries), changes in radial velocity around the center of mass (spectroscopic binaries), detection of double images (visual binaries), or periodic changes in location of the light centroid (astrometric binaries).

All measured stellar masses come from either visual or spectroscopic eclipsing binaries, interpreted using Newtonian gravitation and Kepler's third law,

$$G(M_1 + M_2)P^2 = 4\pi^2 a^3, \tag{5.9}$$

where P is the orbit period and a is the semimajor axis.

The measured masses (which range from about 0.1 $M_\odot$ to 50 $M_\odot$) are used to calibrate calculations of stellar structure and evolution. A number of interesting astronomical phenomena occur only in binary systems, including Algol-type variability, nova explosions, and (probably) Type Ia supernovae and gamma ray bursts.

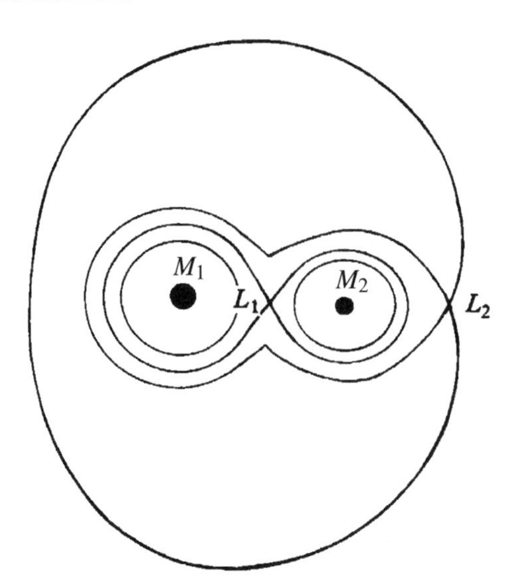

FIGURE 5.2. The Roche Geometry.

5.2.2. Evolution of binary stars

Binary evolution differs from that of single stars primarily because material can be transferred from one star to the other, particularly when one fills its Roche lobe or inner Lagrangian surface (the equipotential that envelopes both stars). In a coordinate system that rotates with the orbiting stars, material that reaches L_1 is free to flow down onto either star; material that reaches L_2 is free to leave the system, typically carrying away more than its proportionate share of angular momentum.

The first phase of mass transfer occurs when the initially more massive star begins to expand as a red giant. Beta Lyrae, Algol, and other kinds of variability occur in such systems. After M_1 completes its evolution as a white dwarf (WD), neutron star (NS), or black hole (BH), a second phase of mass transfer gives rise to novae and other cataclysmic variables (WD remnant), x-ray binaries (NS or BH remnant), Type Ia supernova explosion (merger of two WD remnants), or gamma ray burst (merger of two NS or NS + BH remnants), the latter two occuring after M_2 has also completed its evolution. Intermediate stages, including binary WDs and binary NS (pulsars) are observed.

Collapse of a single, massive star producing magnetically collimated jets is another possible source of gamma ray bursts.

5.3. STAR CLUSTERS, INTERSTELLAR MEDIUM, AND THE MILKY WAY

5.3.1. Star clusters

Probably all star formation (at least in galaxies like the Milky Way) occurs in clusters fragmenting out of dense clouds of molecular gas. We currently observe two kinds of clusters, as follows.

TABLE 5.3. Stellar populations

Component	Radial scale length	Radial extent	Vertical scale height or c/a	ρ near Sun	M_{tot}	$L_{\mathrm{B}}^{\mathrm{tot}}$	M/L within $2R_{\circ}$	[Fe/H]	V_{rotn} at $R_{\odot}$	Contents
	kpc	kpc		$M_{\odot}/\mathrm{pc}^3$	$M_{\odot}$	$L_{\odot}$	$M_{\odot}/L_{\odot}$		km/s	
Bulge	0.2	1	0.4		10^{10}	2×10^9	5	$+0.3 \pm 0.2$		Old stars (not Pop II) Molecular gas
Spheroid	2.9	≥ 40	0.6–1	2.6×10^{-4}	$(2\text{–}10)\times 10^9$	$(1\text{–}2)\times 10^9$	1–5	$-1.5^{+0.5}_{-3.0}$	0–50	Metal-poor globular clusters; Old stars (Pop II)
Disk		20–25		0.15	$(3\text{–}6) \times 10^{10}$	1.9×10^{10}	4.6	-0.3 ± 0.3		
thick disk	4–5	20–25?	1.3 kpc	0.026	$(2\text{–}4) \times 10^9$	2×10^8	10?	-0.6 ± 0.3	180	Metal-rich globular clusters; Old stars; Hot gas
thin disk	3.5–5	20–25	325 pc	0.124	$(3\text{–}6) \times 10^{10}$	1.7×10^{10}	4.5	-0.1 ± 0.2	220	Pop I stars; Atomic and molecular gas
Dark halo	2–3	≥ 100	1?	0.009	$(6\text{–}10) \times 10^{11}$		≥ 650		0?	Unknown

TABLE 5.4. Phases of the interstellar medium

Phase	Temp. (K)	Particle density (cm^{-3})	Scale height (pc)	Fraction of mass	Gas bulk speeds (km/s)	Fraction of volume	Observed phenomena	Heating processes	Cooling processes
Dust	5–40	10^{-12}	40	1%	2–4	Small	Reflection, extinction, polarization, reddening of starlight	Collisions, cosmic rays, photoabsorption	Thermal IR emission
Molecular clouds	10–30	10^{2-5}	60	$\frac{1}{3}-\frac{1}{2}$	2–4	Small	Molecular emission lines	Stellar radiation; collisions	Molecular lines, esp. CO
Neutral gas									
Cool clouds[a]	80	30	100	$\frac{1}{4}-\frac{1}{3}$	7	35% 5	21 cm absorption & emission	Photoelectric heating by dust	Fine-structure lines of C, C^{+}, O, etc.
Warm intercloud[a]	8000	0.1–0.3	200	$\frac{1}{4}-\frac{1}{3}$	9	30%	H II emission lines	Photoelectric heating by dust	Atomic emission lines, esp, H
Ionized gas									
Warm ionized gas[a]	8000	0.1–0.3	≥ 1000	10%	10–20	65% 15	Diffuse Hα emission	Stellar photo-ionization	Line emission of C, N, O, Ne
coronal gas[a]	10^{5-6}	$10^{3\pm1}$	≥ 2000	Small	Up to 1000	50%	Soft x rays, coronal lines	Supernovae	Line emission and continua of H, C, O, He
H II regions	8000	> 1–100	60	Small	10–30	Small	H recombination & ion emission lines	Individual stars	Emission lines

[a]Phases in approximate pressure equilibrium.

Globular clusters: 10^4–10^6 stars, characteristic core sizes 1–10 parsecs, ages 10–18 Gyr, distributed through the spheroidal galactic halo, though concentrated toward the center, composition deficient in heavy elements by factors 3–300 relative to the Sun.

Open or galactic clusters and associations: 30–3000 stars, less centrally condensed than globular clusters (sizes also parsecs), ages 1 Myr to 6 Gyr, located in the galactic disk and including sites of on-going star formation.

Other galaxies seem to have young globular clusters forming in regions of very dense or shocked gas.

5.3.2. The galactic center

The center of our galaxy is located about 8.5 kpc from us in the direction of the constellation Sagitarius. The radio source Sgr A* coincides with the dynamical center and is the site of a condensed mass of (2–3) $\times\ 10^6\ M_\odot$, presumably a black hole. Our rotation speed around the galactic center is about 220 km/s.

5.3.3. The Milky Way: General properties

Age	12–17 Gyr
Radius of luminous disk	15 kpc
Radius of dark halo	$\geq$50 kpc
Luminosity in visible light	$2.3 \times 10^{10} L_\odot$
Bolometric luminosity	
(of which 1/3 is in infrared reemitted by dust)	$3.6 \times 10^{10}\ L_\odot$
Total mass to $R = 8.5$ kpc (based on rotation)	$9.5 \times 10^{10} M_\odot,\ M/L_\mathrm{B} = 4.2 M_\odot/L_{\mathrm{B}\odot}$
Total mass to $R = 230$ kpc	
(based on dynamics of the Local Group)	$1.3 \times 10^{12} M_\odot,\ M/L_\mathrm{B} = 56 M_\odot/L_{\mathrm{B}\odot}$

Conversion from equatorial coordinates (right ascension α, declination δ) to galactic coordinates, centered on the galactic center and with the $\ell = 0°$-plane equal to the plane of the galactic disk.

b and ℓ (for Epoch 2000):

$$\sin b = \sin\delta \cos 62.^\circ 87 - \cos\delta \sin(\alpha - 282.^\circ 86) \sin 62.^\circ 87,$$

$$\cos b \cos(\ell - 32.^\circ 93) = \cos\delta \cos(\alpha - 282.^\circ 86).$$

5.3.4. Backgrounds

Energy densities near the Sun in magnetic field, galactic cosmic rays, gas cloud turbulence, starlight, and microwave background radiation: each $(5 - 20)$ keV/m^3.

Cosmic rays:	Flux of particles with energies greater than E	
	for $E = (10 - 10^7)$ GeV/nucleon:	$10^{20}(E/\text{eV})^{-1.74}$ particles/$(\text{m}^2 \cdot \text{s} \cdot \text{sr})$
	Spectrum continues to at least 10^{12} GeV with steeper fall-off.	
	Confinement time:	3×10^7 yr
	Mass density traversed:	(30–60) kg/m^2
	Composition:	roughly cosmic, dominated by protons at lower energies. Highest energy composition not understood.
Visible light:	In the galactic plane:	$2.2 \times 10^{-10}\,\text{J}/(\text{m}^2 \cdot \text{s} \cdot \text{sr} \cdot \text{Å})$
	Toward galactic poles:	$0.55 \times 10^{-10}\,\text{J}/(\text{m}^2 \cdot \text{s} \cdot \text{sr} \cdot \text{Å})$
	Absorption in the Milky Way:	$A_v =$
	Toward poles	0.11 magnitude
	In the galactic plane:	1 magnitude/kpc
	Toward galactic center:	up to 40 magnitudes or more
Soft x-rays:	local background:	$(4 - 7) \times 10^5\text{keV/m}^2 \cdot \text{s} \cdot \text{sr} \cdot \text{keV}$

Integrated ultraviolet—optical—IR background, 100 nm to 100 μm:

$$\frac{dI}{d \ln \nu} = (10^{-6} - 10^{-7})\,\text{W/m}^2 \cdot \text{sr}$$

smaller by a factor of 10 toward the galactic poles

Total galactic Luminosities:

$L(\geq 100\,\text{MeV})$:	$(1 - 2) \times 10^{32}$ W
$L(2 - 10\,\text{keV})$:	10^{33} W
$L(4 - 100\,\mu\text{m})$:	$1.2 \times 10^{10} L_\odot = (4 - 8) \times 10^{36}$ W

5.4. GALAXIES

5.4.1. Types and their properties

E Elliptical (spheroidal component dominates) from E0 (round) to E7 (most flattened). $b/a = 1 - n/10$. (i.e., $b/a = 0.3$ is E7). True shapes probably triaxial. Residual gas hot; little or no current star formation. Luminous masses (10^7–10^{12}) $M_\odot$ (dwarf elliptical to giant elliptical, dE, gE) Have recognizable nuclei. Luminous diameter 10–50 kpc.

S0 Lenticular (spheroid and disk but no recent star formation). Spiral arms not well marked. Relatively little cold (atomic or molecular) gas.

S Spiral or disk galaxies (old spheroid of stars and disk with arms marked by gas and young stars) from S0a (least dominant arms) through Sa, Sb, Sc, Sd (most dominant arms) to Sm (Magellanic faint spirals). SA = unbarred; SB = central bars. Luminous masses (10^9–10^{11}) $M_\odot$. There are few or no true dwarf spirals. Luminous diameter: 10–30 kpc.

Irr — Irregular (spheroid not easily recognized; gas and young stars not in coherent arms). Magellanic type (intrinsically irregular) Irr, Im, IBm (barred) with masses $(10^7\text{–}10^{9.5})\ M_\odot$. Merger products (disturbed and active galaxies) = Irr II or I0. dIrr = dwarf irregulars at low end of mass range. Up to 50% gas. Sizes a few kpc.

dSph — Dwarf spheroidals (masses $(10^5\text{–}10^7)\ M_\odot$; much less centrally condensed than globular clusters; many are companions to E or S galaxies). Probably the commonest kind of galaxy in a volume-limited sample. Sizes a few kpc.

Special Types:

cD — (supergiant diffuse) galaxies at centers of rich clusters (probably consist of merger products and stars torn from member galaxies).

BCD — (blue compact dwarf) galaxies = dIrr, with very active star formation.

Star burst galaxies can occur in any type with significant cold gas; often triggered by interactions or mergers.

TABLE 5.5. Brighter members of the local group

Galaxy	RA (Epoch = 2000) h m	Decl (Epoch = 2000) ° ′	Diam arcmin	b/a	B_T	$(B-V)_\circ$	V_r km/s	Type	Dist. kpc	M_V	V_{rot} km/s	L_X (W)	[Fe/H]
Milky Way								Sbc I-II		−20.6	220	3.0×10^{32}	
M31 = NGC224	00 43	+41 16	190.5	0.32	4.36	0.68	−121	Sb I-II	725	−21.1	254	3.6×10^{32}	
M32 = NGC221	00 43	+40 52	8.7	0.74	9.03	0.88	−28	E2	725	−16.4		5.4×10^{30}	
M110 = NGC205	00 40	+41 41	21.9	0.5	8.92	0.82	−60	Sph/E5p	725	−16.3		$< 9 \times 10^{29}$	−0.85
SMC	00 53	−72 49	316.2	0.6	2.70	0.36	+34	Im IV-V	58	−16.2	50:	6.1×10^{30}	−0.75
NGC185	00 39	+48 20	11.7	0.85	10.10	0.73	−64	dSph/dE3p	620	−15.3			
NGC147	00 33	+48 30	13.2	0.6	10.47	0.78	+28	dSph/dE5	645	−15.1			−1.20
And I	00 46	+38 02	1.4	1.0	13.50	0.75		dSph/E5p	725	−11.8			−1.40
Sculptor	01 00	−33 42	39.8	0.8	10.5		+115	dSph	78	−10.7			−1.80
And III	00 35	+36 31	0.75	0.4	13.5			dSph	725	−10.3			
IC 1613 = DDO 8	01 05	+02 08	16.2	0.9	9.88	0.67	−152	IBm/Irr V	715	−14.9			−0.90
Psc = LGS3	01 04	+21 51					−149	dIrr	725				
M33 = NGC598	01 34	+30 34	70.8	0.6	6.27	0.47	−46	Sc, cdII-III	795	−18.9	101	1.1×10^{32}	
And II	01 16	+33 26	1.6	0.7	13.5			dSph	725	−11.8			
Phoenix	01 51	−44 26	4.9	0.8				dIm/dSph	417	−9.9			−2.0
DDO 221 = WLM	00 02	−15 26	11.5	0.35	11.03	0.31	−61	IBm IV-V	940	−14.1			
Fornax	02 40	−34 27	63.1	0.7	9.04		−41	dSph	131	−13.7			−1.40
LMC	05 24	−69 45	645.7	0.85	0.91	0.43	+119	SBm/Ir III-IV	55	−18.1	100:	6.6×10^{31}	−0.40
Carina	06 42	−50 58	23.4	0.66			+13	dSph	107	−7.6			−1.52
Leo I	10 08	+12 18	9.8	0.8	11.18	0.63	+60	dSph	229	−11.7			−1.30
Sextans	10 13	−01 36	90	0.6				dSph	87	−10.0			−1.5?
Leo II	11 14	+22 09	12.0	0.9	12.6	0.59	+36	dSph/E0p	234	−9.4			−1.90
Ursa Minor	15 09	+67 12	30.2	0.6	11.9		−47	dSph	69	−8.9		$< 3 \times 10^{28}$	−2.20
Draco	17 20	+57 55	35.5	0.7	10.8		−87	dSph	76	−8.6			−2.10
NGC 6822	19 46	−64 25	15.5	0.9	9.31		+44	IBm IV-V	495	−16.4		1×10^{30}	−0.6
DDO 210	20 47	−12 50	2.2	0.5	14.0	0.01	−23	IBm V	725	−11.5			
Tucana	22 49	−64 25	5.5	0.55	14.0, R			dSph/dE5	890	−9.5			

TABLE 5.6. Nearby bright galaxies, $B_T < 10.0$

Galaxy	RA (Epoch = 2000.0) h m	Decl ° ′	Diam	b/a	B_T	$(B-V)_o$	Type	V_r km/s	Dist. Mpc	M_{BT}	Vrot km/s	log M_{Tot}
NGC 55	00 15	−39 13	32.4	0.17	8.42	0.54	SBc/m III	+94	1.3	−18.13	83	9.87
NGC 247	00 47	−20 46	21.4	0.32	9.67	0.54	SABc/d III-IV	+176	2.1	17.98	92	10.02
NGC 253	00 48	−25 17	27.5	0.25	8.04		SABC II	+251	3.0	−20.02	197	10.87
NGC 300	00 54	−37 41	21.9	0.71	8.72	0.58	Sc/d 11-IV	+98	1.2	−16.88	100	9.89
NGC 628 (M74)	01 37	+15 47	10.5	0.91	9.95	0.51	SAc II	+753	9.7	−20.32	face on	
NGC 1068 (M77)	02 43	−00 01	7.1	0.85	9.61	0.70	SAb II	+1144	14.4	−21.39	face on	
NGC 1291	03 17	−41 06	9.8	0.83	9.39	0.91	SBO/a	+712	8.6	−20.26	face on	
NGC 1313	03 17	−66 30	9.1	0.76	9.20	0.48	SBc/d III-IV	+292	3.7	−18.60	136	10.31
NGC 1316 (Fornax A)	03 23	−37 12	12.0	0.71	9.42	0.87	SABO/a pec	+1674	16.9	−21.47		
NGC 2403	07 37	+65 36	21.9	0.56	8.93	0.39	Scd III	+226	4.2	−19.68	124	10.67
NGC 2903	09 32	+21 30	12.6	9.48	9.68	0.55	SABbc 1-11	+476	6.3	−19.85	196	10.91
NGC 3031 (M81)	09 56	+69 04	26.9	0.52	7.39	0.82	SAab 1-II	+69	1.4	−18.29	236	10.73
NGC 3034 (M82)	09 56	+69 41	11.2	0.38	9.30	0.79	IO/Amorphous	+323	5.2	−19.42		
NGC 3115	10 05	−07 43	7.2	0.84	9.87	0.94	S0	+492	6.7	−19.18		
NGC 3521	11 06	−00 02	11.0	0.47	9.83	0.68	SABbc II	+673	7.2	−19.88	295	10.99
NGC 3627 (M66)	11 20	+13 00	9.1	0.46	9.65	0.60	SABb II	+643	6.6	−19.66	188	10.74
NGC 4258 (M106)	12 19	+47 18	14.8	0.39	9.10	0.55	SABbc II-III	+510	6.8	−20.59	213	11.17
NGC 4449	12 28	+46 06	6.2	0.71	9.99	0.41	IB/Sm IV	+255	3.0	−17.66		
NGC 4472 (M49)	12 30	+08 00	10.2	0.81	9.37	0.95	El-2/SO	+846	16.8	−21.82		
NGC 4486 (M87)	12 31	+12 23	8.3	0.79	9.59	0.93	cD, EOp	+1229	16.8	−21.64		
NGC 4594 (M104)	12 40	−11 37	8.7	0.41	8.98	0.45	Sa/ab	+969	20.0	−22.98	369	11.38
NGC 4631	12 42	+32 32	15.5	0.17	9.75	0.55	SBc/d III	+629	6.9	−20.12	142	10.70
NGC 4649 (M60)	12 44	+11 33	7.4	0.81	9.81	0.95	SO/E2	+970	16.8	−21.36		
NGC 4736 (M94)	12 51	+41 07	11.2	0.81	8.99	0.72	SAab II	+360	4.3	−19.37	186	10.78
NGC 4826 (M64)	12 57	+21 41	10.0	0.54	9.36	0.71	Sab II	+403	4.1	−19.15		
NGC 4945	13 05	−49 28	20.0	0.19	9.30		SBcd IV	+383	5.2	−20.65		
NGC 5055 (M63)	13 16	+42 02	12.6	0.58	9.31	0.64	SAbc II-III	+571	7.2	−20.42	224	11.15
NGC 5128 (Cen A)	13 25	−43 01	25.7	0.79	7.84	0.88	SOp	+398	4.9	−20.97		
NGC 5194 (M55)	13 30	+47 12	11.2	0.62	8.96	0.53	SAbc I-IIp	+551	7.7	−20.75		
NGC 5236 (M83)	13 37	−29 52	12.9	0.89	8.20	0.61	SBC II	+384	4.7	−20.31	face on	
NGC 5457 (M101)	14 03	+54 21	28.8	0.93	8.31	0.44	SABcd I	+360	5.4	−20.45	face on	
NGC 6744	19 10	−63 51	20.0	0.65	9.14		Sbc II	+746	10.4	−21.39	185	11.37
NGC 6946	20 35	+60 09	11.5	0.85	9.61	0.40	Scd II	+277	5.5	−20.78	153	10.81
NGC 7793	23 58	−32 35	9.3	0.68	9.63		SAd IV	+228	2.8	−17.69	104	9.95

5.4.2. Dark matter

TABLE 5.7. The evidence for dark matter: Dynamical masses of galaxies on various scales

Scale	M/L $(M_\odot/L_\odot = 1)$	Contribution to Ω	Kinds of evidence
Visible stars and clusters parsecs to kiloparsecs	1–3	0.001	Stellar velocities and scale heights
Visible parts of galaxies (10–30) kpc	10	0.01	Stellar velocity dispersions, rotation of spirals, orbits of companion galaxies and globular clusters; x-ray temperatures of E's
Binary galaxies and small groups (0.1–1) Mpc	10–100	0.01–0.1	Velocity dispersion of galaxies, gravitational lensing, x-ray emission
Rich clusters and superclusters (1–100) Mpc	100–300	0.3 ± 0.1	Velocity dispersions of galaxies, gravitational lensing, x-ray emission
Closure density	1000	~ 1.0	Assumption that the universe is closed by some form of matter

TABLE 5.8. Dark matter candidates

Candidate/Particle[c]	Mass range	Predicted by	Astrophysical effects / comments
Baryonic (hot intergalactic gas, brown dwarfs)	GeV	Standard Model	Limited to 10% of closure by Big Bang nucleosynthesis considerations
$G(R)$	—	Non-Newtonian gravity	Mimics DM on large scales; must also pass tests for general relativity
Λ, (Cosmological constant)[d]	—	General relativity	Permits flat space without DM
Gravitational radiation	—	General relativity	Very low limits on some, but not all, wavelength scales
Quintessence	$\geq$ GeV	Various	Critical property is $P = -\omega\rho$, $0 < \omega < 1$
Axion	10^{-5} eV	quantum electrodynamics, Peccei-Quinn symmetry breaking	Cold DM (because particles are formed at rest)
Neutrinos with nonzero rest-mass	$\leq$ 15 eV	GUT[a]	Hot DM (promotes structure on large scales). There is some independent evidence for nonzero neutrino rest mass.
Lowest-mass supersymmetric partner = WIMP = ino[b] (gravitino, Higgsino, etc.)	$\geq$ MeV to GeV	supersymmetry	Cold DM (Promotes structure on small scales)
Primordial (mini) balack holes	$(10^{12}$–$10^{27})$ kg	general relativity	Cold DM. Evade limits from nucleosynthesis by forming earlier
Topological singularities:			
0d Monopoles	10^{16} GeV	GUTs	
1d Cosmic Strings	10^{19} GeV	GUTs	Limits set by large-scale magnetic fields.
2d Domain walls	large	GUTs	Not a major DM contributor, but can
3d Textures	large	GUTs	seed galaxy information

[a]GUT = Grand Unified Theory.
[b]WIMP = Weakly Interacting Massive Particle.
[c]A large number of other candidates have been suggested at various times, but are currently out of favor.
[d]Or dark energy.

5.4.3. Formation and evolution of galaxies

Galaxies must arise from fluctuations in the density of matter (etc.) in the early universe whose amplitude is constrained to be $(1 - 2) \times 10^{-5}$ by the isotropy of the 3 K microwave background radiation. The dominant process is presumably gravitational clustering, but modeling is currently at the many-parameter stage, in which the modeler chooses an underlying cosmological model, some kind or kinds of dark matter, and some spectrum of primordial fluctuations. None of the simulations currently reproduces the full range of clustering and other galaxy properties very well.

Models of the subsequent dynamical and chemical evolution of galaxies are considerably more advanced. Interactions and mergers of young galaxies are clearly an important contributor. Both Type Ia and Type II supernovae, with different time histories, contribute to the gradual enrichment of galaxies in heavy elements. These models can be tested by comparison with data on galaxies at large redshift and with the chemical histories revealed by old stars.

5.4.4. Collective properties, clustering, and large-scale structure

Galaxy counts

The number of galaxies per square degree brighter than m_B (for $12 \leq m_B \leq 15$) in the south galactic polar cap is

$$\log N(m) = 0.62m_B - 9.7. \qquad (5.10)$$

The number in the north polar cap is larger by $\Delta \log N(m) = 0.4$ at $m_B = +12$, tapering to $\Delta \log N(m) = 0.0$ at $m_B = +14$. Fainter than $b_J = +20$, most surveys find differential counts per square degree per unit magnitude

$$\log N(b_J) = 0.45b_J + C, \text{ with normalization } C = 7.7 \text{ to } 8.2. \qquad (5.11)$$

The luminosity distribution of galaxies

In most cases, number density versus luminosity can be fit by

$$\phi(L)L = \phi^*(L/L^*)^\alpha \exp(-L/L^*)\,(L/L^*), \qquad (5.12)$$

where

$$\phi^* = (0.015 \pm 0.001)\,h^3 \text{ galaxies/Mpc}^3, \alpha = -1.1 \pm 0.1 (\alpha = -1.25 \text{ in rich clusters}), \qquad (5.13)$$

and L^* is the luminosity corresponding to $M_B^* = -19.5 \pm 0.2$ on the scale where $M_{B\odot} = +5.48$.

In reality, the luminosity distribution varies a great deal with the type of galaxy (in particular, the dwarf numbers continue to rise faintward of $M_B = -14$, though not sharply enough for faint galaxies to dominate the cosmic luminosity density).

The luminosity density of the universe that comes from integrating the above Schechter luminosity function is

$$(1.4\text{--}2.0) \times 10^8 h\, L_\odot\, \text{Mpc}^{-3}. \qquad (5.14)$$

Clustering

Truly isolated galaxies are rare or nonexistent. Structures range from small groups of a few to a few dozen galaxies (like the Local Group), whose brightest members are spirals, with velocity disperions of 100–300 km/sec and masses of $(1\text{–}3) \times 10^{12} M_\odot$ to rich clusters with thousands of members (like the Abell clusters), whose brightest members (especially at the center) are ellipticals, with velocity dispersions of 1000–2000 km/s and total masses of $10^{15}\, M_\odot$ or more. The latter are nearly always sources of bright x-rays emitted by gas at 10^7 K–10^8 K, whereas the former are sometimes fainter x-ray sources. Clusters are typically not relaxed systems, and the rich ones often show substructure suggestive of very recent assembly from smaller clusters. Our Local Group is on the outskirts of a supercluster whose dominant member is the Virgo cluster of galaxies at a distance of about 18 Mpc.

Superclustering, very large-scale structure, and streaming

The clusters in turn define larger scale structures typically filamentary or sheet-like, surrounding voids where galaxies are underdense by factors of five or more. The largest scale superclusters and voids range up to at least $150h^{-1}$ Mpc. There are corresponding deviations from smooth Hubble expansion flow of galaxy motions, amounting to at least 600 km/sec on similar scales. Our Local Group is moving at about this speed relative to the cosmic microwave background (seen as a Doppler effect in the cosmic microwave background), and mass concentrations out to 100 Mpc or more have contributed to our peculiar velocity. The basic topology of the clusters and voids is apparently that of a sponge, as opposed to either swiss cheese or meatball topology; that is, both the regions of clusters and the regions of voids are connected.

5.5. HIGH-ENERGY ASTROPHYSICS

5.5.1. Basic physical mechanisms

The space-time around a nonrotating, spherical or point mass is described by the Schwarzschild metric

$$ds^2 = dt^2\left(1 - \frac{2GM}{rc^2}\right) - dr^2\left(1 - \frac{2GM}{rc^2}\right)^{-1} - r^2(\theta^2 + \sin^2\theta\phi^2), \tag{5.15}$$

where M is the central mass. The Schwarzschild radius, $R = 2GM/c^2$ is the distance at which the redshift becomes infinite. It is a coordinate singularity only; a real, physical singularity occurs at $r = 0$. General relativity characteristically introduces corrections to Newtonian calculations that are of order $(1 - 2GM/Rc^2)^{\frac{1}{2}}$.

Radiation from compact astrophysical objects and radiation at x-ray and gamma ray wavelengths is typically bremsstrahlung or nonthermal (synchrotron or inverse Compton) radiation.

Thermal bremsstrahlung (recognizable from its very flat spectrum with exponential cutoff at high frequency) produces

$$6.8 \times 10^{-38} n_e n_i Z^2 (T/K)^{-\frac{1}{2}} \exp(-h\nu/kT)\bar{g}_{ff} \text{ erg}/(\text{cm}^3 \cdot \text{s}), \tag{5.16}$$

where Z is the charge on the dominant ion, n_e and n_i, z_i are electron and ion densities, and $\bar{g}_{ff}$ is a velocity-averaged Gaunt factor, whose value is never very different from unity.

Synchrotron radiation, where one electron in a magnetic field B radiates

$$P = \frac{4}{3}\sigma_T c\beta^2\gamma^2\left(\frac{B^2}{8\pi}\right) \quad \text{(unrationalized cgs)}, \tag{5.17}$$

where $\beta = v/c$, $\gamma = (1 - v^2/c^2)^{-\frac{1}{2}}$, and $\sigma_T = (8\pi/3)(e^2/m_e c^2)^2$ is the Thomson cross section.

Astrophysical synchrotron radiation is usually electron synchrotron, and the particles characteristically have a power-law distribution,

$$N(\gamma) = N_\circ \gamma^{-x}, \tag{5.18}$$

where x is in the range 1–3. The total spectrum can be obtained by convolving the power per electron with the electron energy distribution. The resulting power-law spectrum and polarization are signatures of synchrotron radiation. Incoherent synchrotron effects cannot produce brightness temperatures in excess of 10^{12} K unless there is relativistic beaming of the radiating electrons toward the observer, because the electron energy will be rapidly lost to inverse Compton radiation.

Inverse Compton radiation is the analog of ordinary Compton scattering, but with energy transfered from a relativistic electron to less energtic photons, which may belong to the 3 K cosmic microwave background, to ambient visible or infrared light, or to synchrotron radiation being emitted locally (in which case, the process is called synchro-Compton radiation).

One electron in a photon bath of energy density U_{light} will radiate

$$P = \frac{4}{3}\sigma_{\text{T}}\beta^2\gamma^2 U_{\text{light}}, \tag{5.19}$$

and the emission from a distribution of electrons is found as a convolution of the radiation per electron and electron distribution. The emitted radiation is peaked at $\nu = \gamma^2\nu_{\text{inc}}$.

5.5.2. Neutron stars and black holes as endpoints of stellar evolution

Neutron stars and stellar-mass black holes can form either directly from the collapse of the cores of massive stars (associated with Type II supernovae) or from the accretion-induced collapse of less compact objects (WD to NS; NS to BH) in close binary systems.

Neutron stars

Measured masses of neutron stars in binary systems with a second NS or WD are all close to $1.4\,M_\odot$. Radii (based on theory and observed radiation) are close to 10 km. Magnetic fields at birth can range from 10^{12} G to 10^{14} G [10^7T to 10^9T] and initial rotation periods are apparently about 0.01 s (break-up instability is closer to 0.001 s). Magnetic fields of 10^8 G to 10^{10} G and rotation periods of milliseconds are seen in pulsars that have been "recycled" by accretion from close companions.

Black holes

Only black holes in binary systems can be detected. Measured masses for 10 examples range from $3\,M_\odot$ to $10\,M_\odot$. Indirect evidence suggests that some may have large angular momenta, and quasi-periodic phenomena indicate that gas disks extend down very close to the Schwarzschild radius.

5.5.3. Pulsars and x-ray binaries

Pulsars

A neutron star of radius R, with magnetic dipole field B at the poles, making an angle α with the rotation axis and rotating with angular frequency ω will radiate a power

$$L = \frac{B^2 R^6 \omega^6 \sin^2\alpha}{6c^3}. \tag{5.20}$$

The power comes at the expense of the kinetic energy of rotation of the star, and if the initial rotation period was very short, the actual age is $t = P/2\dot{P}$ and $B = 3.3 \times 10^{19}\sqrt{(P\dot{P})}$ G.

For the prototypical pulsar in the Crab Nebula, period $P = 0.033$ s, stored energy $E =$ (2–3) $\times$ 10^{49} erg, total luminosity $L =$ (2–6) $\times$ 10^{38} erg/sec (most of which is eventually seen as x-rays from the nebula around the pulsar and as kinetic energy input to the nebula). Most pulsars fade below delectability by the time P reaches a few seconds. It is uncertain whether the magnetic field always decays with time or only in those cases in which material has accreted onto the surface of the neutron star. The oldest strong-field pulsars have ages of about 10^7 years.

Short-period, weak-field, millisecond pulsars have ages up to 10^{10} years and, because they are much fainter, are greatly underrepresented in most surveys. The total number of pulsars or former pulsars in the Milky Way is probably about 10^8.

X-ray binaries

Accretion onto an object of mass M and radius R at rate $\dot{M}$ releases energy at the rate (Newtonian approximation)

$$L = GM\dot{M}/R. \tag{5.21}$$

For accretion onto a black hole, the efficiency of liberation of available energy in radiation is probably about 10% (or, sometimes, much lower, if the accreting gas remains hot and takes its energy with it, called advection).

The maximum luminosity (Eddington luminosity) available from accretion occurs when the outward force of radiation equals the inward force of gravity:

$$L_{\text{Edd}} = 4\pi GMm_{\text{p}}/\sigma_{\text{T}} = 1.3 \times 10^{31}\ \text{W}(\text{M}/\text{M}_{\odot})\ \text{W}. \tag{5.22}$$

X-ray binaries (XRBs) with both neutron star and black hole primaries occur (the latter are less common by a factor of 10 or more). Both occur with young, massive donor stars (high-mass XRBs) and with old, low-mass donors (LMXRBs). The main signature of a black hole is its mass. Many neutron star binaries show periodicities associated with rotation and x-ray bursts caused by explosive burning of helium on the surface after accreted hydrogen has steadily burned to helium. Because many XRBs are extremely variable in luminosity, the several hundred known galactic systems are very much a lower limit.

LMXRBs and binary pulsars occur much more often in globular clusters than would be expected by chance. It is believed that two- or three-body captures and star exchanges are responsible for this excess of roughly a factor 100.

5.5.4. Supernovae

Supernova explosions occur at a rate of a few per century in the Milky Way and in other large, bright galaxies. Most galactic events are hidden in dust, and the local rate is derived from the birthrates of pulsars and supernova remnants. The energy released is at least 10^{42} J in electromagnetic radiation and 10^{44} J in kinetic energy of ejecta from typical events.

Type II supernovae are defined by the presence of hydrogen features in their spectra. All or most result from the collapse of the iron core of a star of more than about 8 $M_{\odot}$

down to NS densities (10^{18} kg/m^3). The total energy liberated is roughly 10^{46} J (GM^2/R for a neutron star), most of which comes out in neutrinos, as confirmed by the detection of a neutrino burst from the event SN 1987A in the nearby Large Magellanic Cloud. The details of how 1% of that energy is transferred to the ejecta are not well understood.

Type I supernovae are defined by the absence of hydrogen features in their spectra. Types Ib and Ic are probably core collapse events in stripped stars. The classic Type Ia events are nuclear explosions of about 1 $M_{\odot}$ of carbon and oxygen into iron peak elements, completely disrupting the original star. The progenitors must be WDs, which accrete appreciable material (without hydrogen) from binary companions. A pair of WDs, spiraling together because of the loss of angular momentum in gravitational radiation, fits the observations well, but no systems of this type are actually known. The explosion is most easily understood if the white dwarf or merged pair is very close to the Chandrasekhar maximum mass for degenerate matter supported by electron pressure, about 1.4 $M_{\odot}$. White dwarfs close to this limit have been seen, but they are single stars, or found only in wide binary pairs. Known close pairs have much smaller masses. The class of cataclysmic binary called the recurrent novae could also give rise to SNe Ia.

Type II supernovae are the main source of oxygen in the universe, and Type Ia events are the main source of iron. Contributions from both can be traced in the chemical evolution of the Milky Way and other galaxies.

5.5.5. Quasars and other active galaxies

Observations increasingly indicate that almost all galaxies have at their nuclei compact objects (black holes) of mass 10^6 $M_{\odot}$ (as in the Milky Way) up to at least 10^{10} $M_{\odot}$ (as in the radio-emitting elliptical galaxy M87). These are quiescent most of the time, but when fueled with gas from a captured companion, tidally disrupted stars, or other sources can radiate at the Eddington luminosity, up to $10^{14} L_{\odot}$.

The kinds of radiation, variability, and other phenomena will depend on the following:

1. The mass of the black hole.
2. The accretion rate.
3. Whether the accretion goes through a thin disk, thick disk, or spherically.
4. The extent to which the disk is magnetized and collimates relativistic jets perpendicular to itself.
5. The type of the host galaxy and, therefore, the extent to which the jets interact with ambient material.
6. The angle from which we observe the source; straight along a jet gives rise to the most extreme variability and to the highest energy photons.

The main types of active galaxies, active galactic nuclei, or AGNs, include the following:

Seyfert galaxies: Spirals with nuclear luminosity comparable with that of the rest of the galaxy, broad emission lines, and x-ray sources. The types Seyfert 1 and 2 are thought to be those observed along the jets and along the disks, respectively. These make up a few percent of spiral galaxies.

Radio galaxies: Ellipticals with radio emission coming from a region much larger than the visible galaxy. Narrow jets "feeding" the radiolobes are often seen on one (or occasionally both) sides. The radio luminosity is comparable with the optical. X-ray radiation is generally weak. The large radio extent and detectable jets tell us that we must be seeing these more or less perpendicular to the jet direction.

Quasi-stellars: Quasars with strong radio emission and quasi-stellar objects (much more common) without it. The nuclear luminosity greatly outshines that of the rest of the galaxy, so that the hosts were only detected with great difficulty some time after the class was discovered in 1963. The radio-emitters are often ellipticals, but the correlation is not perfect. Most quasi-stellars are variable (especially at x-ray wavelengths) on time scales from decades down to hours.

Blazars: The subset where we seem to be looking almost directly along the jet. They are the most variable at all wavelengths and are often gamma ray sources extending to MeV or even GeV energies. Actual jet speeds are large, with $\Gamma = (1 - v^2/c^2)^{-\frac{1}{2}}$ up to 10 or more. Projection effects can, therefore, give the impression of structures changing faster than light could travel across the object. Such sources are called "superluminal" but without any intention of suggesting any violation of special relativity.

When first observed, QSOs with redshifts near 4 were the most distant known objects. By the summer of 2001, QSOs and bright galaxies were competing for the record, with redshifts of 5 to 6.

5.5.6. Gamma ray bursters

Gamma ray bursters (GRBs) were an accidental discovery of the Vela (U.S.) and Kosmos (USSR) satellite projects intended to monitor for atmospheric tests of nuclear weapons, and the most powerful ones deposit about 10^{-6} J/m^2 at the top of the earth's atmosphere.

The Compton Gamma Ray Observatory found GRBs at the rate of about 1 per day (down to 10^{-10} J/m^2 with a power-law distribution of number versus flux received). Time scales range from a few milliseconds up to a few minutes. Many have energy peaks near 100 keV, but there is an enormous range. Spectra frequently soften through the burst. A number of correlations of flux, time scale, spectrum, and so forth are seen, but none applies to all events. Detected events are isotropic on the sky.

Serious modeling of the GRBs was greatly hampered by the absence of either transient or steady counterparts at any other wavelength. Starting in 1997, a new generation of x-ray satellites, with rapid response times, began finding x-ray tails to the bursts and providing accurate enough positions to permit optical and radio identification. By 2001, many events with these counterparts had been found in external galaxies, with redshifts ranging from 0.01 to 3.4, the closest of these apparently being associated with an atypical supernova. Modeling has focused on processes involving neutron stars and black holes, including mergers of binary pairs, such as are known to exist among the binary pulsars in the Milky Way, and on collapses of rapidly rotating cores of very massive stars, which produce black holes and collimated jets. One event per million years per galaxy will suffice to maintain the rate seen, and this is plausible based on the galactic inventory of binary neutron stars.

Soft gamma repeaters, of which four are known in the Milky Way and one in the Large Magellanic Cloud, are actually young neutron stars with magnetic fields of 10^{14} G or more, rotation periods of 5–10 seconds, and ages (based on spin-down rate and kinematics of surrounding supernova remants) of 1000–3000 years.

5.6. COSMOLOGY

5.6.1. Evidence that the universe is expanding and experienced a Big Bang

1. **Theoretical:** Within the framework of general relativity, a homogeneous, isotropic universe must expand or contract and cannot be static. Conservation of mass-energy guarantees that temperature and density will have been larger in the past.
2. **Observational:**
   (a) The linear velocity-distance relation (Hubble's law) is expected in an expanding universe. Obvious alternatives can be ruled out: tired light by the discovery of time dilation in light curves of distant supernovae, and expansion into existing space by the absence of edge effects and by the details of apparent brightnesses of galaxies as a function of redshift.
   (b) The microwave background and primoridal composition of roughly 1/4 helium are expected from a hot, dense phase. The alternative of producing the helium and radiation in galaxies, quasars, or other sources would result in galaxies having to be about ten times as bright as they are, and in space having to be opaque at the wavelengths of the microwave-submm background to produce the extreme isotropy and thermalization that we see.

5.6.2. The Friedmann-Robertson-Walker metric

The most general solution to the Einstein field equations for a homogeneous, isotropic universe may be written as

$$ds^2 = dt^2 - R^2(t)\left[\frac{dr^2}{1-kr^2} + r^2\left(\theta^2 + \sin^2\theta\,\phi^2\right)\right], \tag{5.23}$$

where r is a dimensionless radial variable, k is the spatial curvature index [$k > 0$, closed (spherical); $k = 0$, asymptotically flat (parabolic); $k < 0$, open (hyperbolic)]; and $R(t)$ is a scale parameter containing the actual time history of the universe.

$$\ddot{R}/R = \frac{1}{3}[\Lambda - 4\pi(\rho + 3p/c^2)G], \tag{5.24}$$

$$\left(\dot{R}/R\right)^2 = \frac{8\pi}{3}G\rho - \frac{kc^2}{R^2} + \frac{\Lambda}{3}, \tag{5.25}$$

where Λ is Einstein's infamous cosmological constant; $\dot{R}/R$ is the Hubble constant, the logarithmic slope of $R(t)$; and $q = -\ddot{R}R/\dot{R}^2$ is the deceleration parameter. $H_\circ$ is often used for the current value of the Hubble constant, and because the actual value remains uncertain, the parametrization $h = H_\circ/(100(\text{km/s})\text{Mpc}^{-1}) = \text{H}_\circ/(3.24 \times 10^{18}\text{s}^{-1})$ is common. The cosmological constant can be reduced to corresponding units as $\lambda = \Lambda/3H^2$.

The critical density, $\rho_c = 3H^2/8\pi G$, is the density that for $\Lambda = 0$ will asymptotically (in infinite time) stop the expansion.

A number of morphologically different forms of $R(t)$ are possible; some do not include a singular (hot, dense) state in the past and so are ruled out. Table 5.9 shows the four classes of "interesting" solutions and some of their properties. Numerical values of H (or h), $\Omega = \rho/\rho_c$, and Λ (or λ) suffice to select a unique model.

TABLE 5.9. Possible model universes

	R cycloid (vs. t) "Closed"	R parabola (vs. t) "Critical"	R hyperbola (vs. t) "Open"	R (vs. t): parabolic, coasting phase, exponential expansion — Nonzero Λ
k	$+1$	0	-1	All values possible, $k = 0$ interesting
ρ	$> \rho_c$	$= \rho_c$	$> \rho_c$	All values possible, $\rho < \rho_c$ interesting
Age	$< \frac{2}{3}\frac{1}{H}$	$\equiv \frac{1}{H}$	$\left(\frac{2}{3} \text{ to } 1\right)\frac{1}{H}$	$> \frac{1}{H}$ possible
Cosmological constant	0	0	0	$\Lambda > 0$ interesting
Geometry, $\Sigma\angle$'s of a Δ	$> 180°$	$= 180°$	$< 180°$	All possible, $= 180°$ interesting

5.6.3. Big-Bang nucleosynthesis

Knowing the present temperature ($T_\circ = 2.7$ K) enables one to extrapolate back in redshift (the fundamental time and space coordinate for cosmology) to $T(z) = T_\circ(1 + z)$, $z = R(\text{now})/R(t$ when the photons were emitted). The density in ordinary matter scales as $(1 + z)^3$, and the energy density in photons scales as $(1 + z)^4$. Thus, nuclear reactions in the early universe can be calculated as a function of the fraction of Ω present in baryonic material, called Ω_b, and is closely related to the ratio of baryons to photons, $\eta = 2.68 \times 10^{-8}\Omega h^2$.

Figure 5.3 shows results of such a calculation and the measured values of the abundances of helium (by mass) and of deuterium and lithium-7 (by number, relative to hydrogen) that must have come out of the big bang to account for what is now seen. A reasonable fit occurs for $\eta = (1.5\text{–}3.5) \times 10^{-10}$ [or $\Omega_b = (0.006\text{–}0.013)h^{-2}$]. This is larger than the luminous matter density, implying the existence of dark baryons (in diffuse gas or brown dwarfs presumably), but smaller than the dynamical density, $\Omega_M = 0.3 \pm 0.1$, implying also the existence of dark non-baryons.

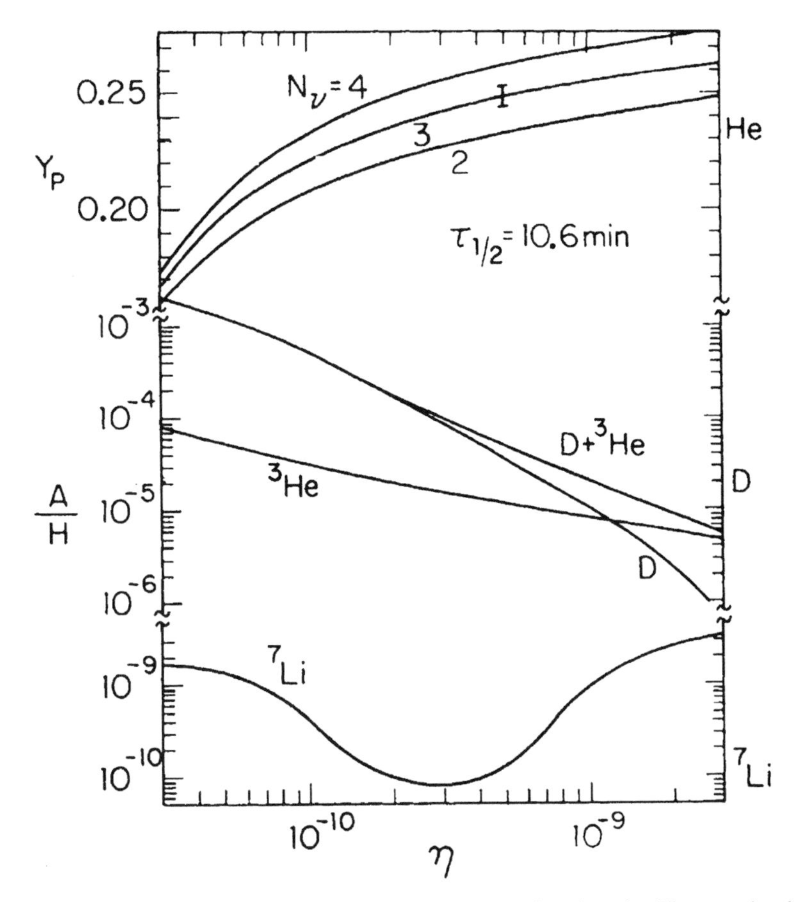

FIGURE 5.3. Primordial abundances of the light elements as a function of η. The error bar indicates the change in Y_p for $\Delta\tau_{1/2} = \pm 0.2$ min.

5.6.4. "Best values" of the parameters for a relativistic universe

All of the quantities H, Ω_M, Ω_b, q, Ω_Λ, and k are, in principle, measurable, but none of them very easily or directly. There is still nearly a factor of 2 uncertainly in distance scales outside of the Local Group; the observed correlation of apparent brightness with redshift is sensitive to $\Omega_M - \Omega_\Lambda$, not to either separately, and fluctuations in the microwave background near multipoles $\ell = 100$–200 are sensitive to $\Omega_M + \Omega_\Lambda$. Many (but not all) published values and error bars are consistent with the following:

$H = (50$–$75)$ km/sec/Mpc Hubble time, $1/H = (13$–$20)$ Gyr,

Hubble radius, $c/H = (4000$–$6000)$ Mpc,

$h^2\Omega_b = 0.02 \pm 0.01$ actual age, $t = 14 \pm 4$ Gyr,

$\Omega_M = 0.35 \pm 0.1$ (perhaps divided as hot DM $= 0.1$, cold DM $= 0.9$),

$q = 0.0 \pm 0.5$

$k = 0$ (flat space),

$\Omega_\Lambda = 0.65 \pm 0.10$ (nonzero cosmological constant, dark energy, or quintessence).

5.6.5. Connections with particle physics

Astronomical observations, interpreted within the framework of the standard hot Big Bang leave open a number of questions, as follows:

1. What is the nature of the dark matter (or kinds of dark matter)?
2. What is the source of the fluctuations in density that gave rise to large-scale structure?
3. Why do the several parameters have the values we observe?
4. Why is the universe so very nearly homogeneous and isotropic on large scales, when those scales have never been in causal connection?
5. Why is the universe so very close to being flat (given that an Ω in the range 0.1 to 1.0 now implies a value exceedingly close to 1 in the past)?
6. If the forces are unified at very high energy and become discrete as the universe expands and cools, where are (and what are) the topological singularities that should result from phase transitions in (very small) causally connected regions?
7. If the cosmological constant in particular is nonzero, why is it so very small?

None of these questions currently has a definitive answer within either astronomy or particle physics, but some of the implications of grand unified theories, supersymmetry, supergravity, and so forth, include an assortment of particles (weakly interacting massive particles, axions, non-zero-rest-mass neutrinos) that may contribute to dark matter, and a phase of exponential expansion near $t = 10^{-30}$ s that takes a small, causally connected region and expands it to a radius much larger than the radius of the observable universe, R_H.

5.7. REFERENCES

Additional and more technical information on stars, galaxies, and the universe can be found in

Allen's Astrophysical Quantities, 4th ed., edited by A. N. Cox (Springer-Verlag, New York, 2000).

K. R. Lang, *Astrophysical Formulae*, 3rd ed. (Springer-Verlag, New York, 1999), two volumes; *Astrophysical Data* (Springer-Verlag, New York, 1992).

V. Trimble, "Astrophysics" in American Inst. of Physics Encyclopedia of Applied Physics, vol. 2, 117–143 (New York, VCH Publishing, 1991); "Dark Matter in the Universe: Where and What and Why," Contemporary Physics, **29**, 373–392, 1988.

6

Atomic and Molecular Collision Processes

M. R. Flannery
Georgia Institute of Technology, Atlanta, Georgia

Contents

*The research is supported by the Air Force Office of Scientific Research and by the National Science Foundation.

List of Tables

List of Figures

6.1. INTRODUCTION

The emphasis here is on providing the basic principles, definitions, collision properties, key formulae, and theory useful in the study of atomic and molecular collision processes. Elastic and inelastic collisions between structured particles, photon scattering by particles, and recombination processes involving radiative stabilization are treated. Equilibrium distributions and long-range interactions are also included. Illustrative examples and an appropriate list of general references are provided.

6.2. COLLISIONS

6.2.1. Differential and integral cross sections

A uniform monoenergetic beam of test or projectile particles A with number density N_A and velocity v_A is incident on a single field or target particle B of velocity v_B. The direction of the relative velocity $\mathbf{v} = \mathbf{v}_A - \mathbf{v}_B$ is along the Z-axis of a Cartesian XYZ-frame of reference. The incident current (or intensity) is then $\mathbf{j}_i = N_A\mathbf{v}$, which is the number of test particles crossing unit area normal to the beam in unit time. The differential cross section for scattering of the test particles into unit solid angle $d\Omega = d(\cos\psi)\,d\phi$ about the direction $\hat{\mathbf{v}}'(\psi, \phi)$ of the final relative motion is

$$\frac{d\sigma(v;\psi,\phi)}{d\Omega} = \frac{\begin{array}{c}\text{number of test particles scattered per unit time}\\ \text{by one field particle into unit solid angle}\end{array}}{\text{current } j_i \text{ of incident beam}}.$$

The number of particles scattered per unit time by the field particle and detected within solid angle $d\Omega$ is then

$$\frac{d\mathcal{N}_d}{dt} = j_i \frac{d\sigma}{d\Omega} d\Omega = N_A\, v \frac{d\sigma}{d\Omega}\, d\Omega = N_A\, v \frac{d\sigma}{d\Omega} \frac{dA}{r^2},$$

where the detector, located along the scattered direction $\hat{\mathbf{v}}'(\psi, \phi)$, subtends an angle $d\Omega = dA/r^2$ at the scattering center and projects an area $dA = r^2 d(\cos\theta)\,d\phi$ normal to the scattered beam. Thus, $[d\sigma/d\Omega]\,d\Omega$ is the cross sectional area of the beam that is intercepted by one target particle and scattered into the solid angle dA/r^2 of a cone with axis along $\hat{\mathbf{v}}'(\psi, \phi)$ and vertical angle $d\psi$. In classical terms (Figure 6.1), the number of particles detected per second about direction (ψ, ϕ) is the number $N_A v(b\,db\,d\phi)$ of incident particles crossing the initial areal element $b db\, d\phi$ per second. Hence,

$$\frac{d\sigma}{d\Omega} = \frac{b\,db}{d(\cos\psi)}.$$

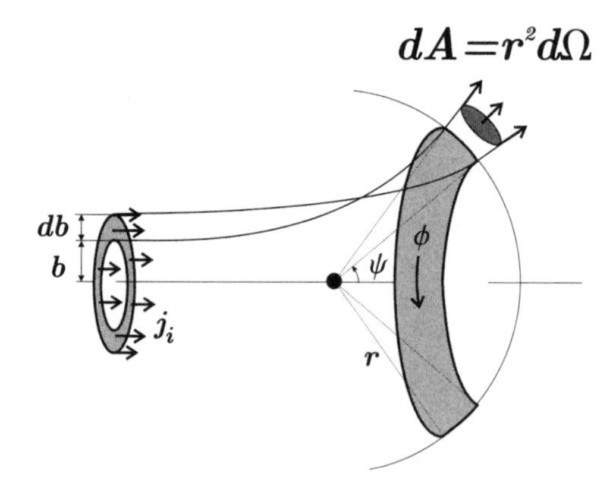

FIGURE 6.1. Scattering of a beam with current $j_i = N_A v$ particles per unit area incident between two cylinders of radii b and $b + db$ by one particle at rest in the Laboratory.

For an incident current flowing between two cylinders of radii b and $b+db$, $j_i\,2\pi[d\sigma/d\Omega]$ $d(\cos\psi)$ is the number of particles scattered per second between the two cones of semivertical angles ψ, $\psi+d\psi$ (Figure 6.1).

The integral cross section for scattering over all directions is

$$\sigma(v) = \int_{-1}^{+1} d(\cos\psi) \int_{0}^{2\pi} \left[\frac{d\sigma}{d\Omega}(v;\psi,\phi)\right] d\phi.$$

The integral cross section is, therefore, the effective area presented by each field particle B for scattering of the test particles A into all directions. The probability that the test particles are scattered into a given direction $\hat{\mathbf{v}}'(\psi,\phi)$ is the ratio

$$\mathcal{P}(v;\psi,\phi) = \frac{d\sigma(v;\psi,\phi)}{d\Omega}/\sigma(v)$$

of the differential to integral cross sections.

6.2.2. Collision rates, collision frequency, and path length

An electron or atomic beam of (projectile or test) particles A with density N_A of particles per cm^3 travels with speed v and energy E through an infinitesimal thickness dx of (target or field) gas particles B at rest with density N_B particles per cm^3. The particles are scattered out of the beam by $(A-B)$-collisions with integral cross section $\sigma(E)$ at a rate (cm^{-3} sec^{-1}) given by the total number of collisions between A and B particles

$$\begin{aligned}\frac{dN_A(E)}{dt} &= -\left[v\sigma(E)N_B\right]N_A(E)\\ &= -\left[k(E)N_B\right]N_A(E)\\ &= -\nu_B(E)N_A(E)\end{aligned}$$

in unit time and unit volume. The microscopic rate coefficient (cm^3 s^{-1}) for the scattering of 1 test particle by 1 field particle is $k(E)=v\sigma(E)$. The frequency (s^{-1}) of collision between one test particle and N_B field particles (cm^{-3}) is $\nu_B = k(E)N_B$. Since $v = dx/dt$, the variation with x of intensity j_i of the attenuated beam is governed by

$$\frac{dj_i}{dx} = -[N_B\sigma(E)]j_i(x).$$

For constant density N_B and speed v, the solution is

$$\begin{aligned}j_i(E,x) &= j_i(E,0)\exp\left[-N_B\sigma(E)x\right]\\ &= j_i(E,0)\exp(-x/\lambda),\end{aligned}$$

where $\lambda \equiv 1/N_B\sigma(E) = v/\nu_B$ is the path length between collisions. Since $j_i = N_A v$, the density $N_A(E,x)$ obeys a similar equation. These equations describe the attenuation of a particle beam A traveling through a target gas B. For target gas particles with a distribution

$f_B(\mathbf{v}_B)\,d\mathbf{v}_B$ in velocities $\mathbf{v}_B$, the above microscopic rate then becomes

$$k(E) = \int |\mathbf{v}_A - \mathbf{v}_B|\sigma(|\mathbf{v}_A - \mathbf{v}_B|) f_B(\mathbf{v}_B)\,d\mathbf{v}_B,$$

where $E = \frac{1}{2}M_A v_A^2$ is the kinetic energy of the projectile beam. For an isothermal beam with an energy distribution $f_A(E)\,dE$ at temperature T, the macroscopic rate coefficient ($\mathrm{cm}^3\,\mathrm{s}^{-1}$) or thermal rate constant is

$$k(T) = \int_0^\infty k(E) f_A(E)\,dE.$$

6.2.3. Energy and angular momentum: Center of mass and relative

The velocity of the center of mass of the projectile and target particles of respective masses M_A and M_B is

$$\mathbf{V} = (M_A\mathbf{v}_A + M_B\mathbf{v}_B)/(M_A + M_B).$$

The relative velocity is

$$\mathbf{v} = \mathbf{v}_A - \mathbf{v}_B.$$

The velocities of A and B in terms of $\mathbf{V}$ and $\mathbf{v}$ are

$$\mathbf{v}_A = \mathbf{V} + \frac{M_B}{M_A + M_B}\mathbf{v},$$

$$\mathbf{v}_B = \mathbf{V} - \frac{M_A}{M_A + M_B}\mathbf{v}.$$

The total kinetic energy then decomposes into the sum

$$\begin{aligned} E &= \tfrac{1}{2}M_A v_A^2 + \tfrac{1}{2}M_B v_B^2 = \tfrac{1}{2}MV^2 + \tfrac{1}{2}M_{AB}v^2 \\ &= E_{\mathrm{cm}}(A + B) + E_{\mathrm{rel}}(AB) \end{aligned}$$

of the energy $E_{\mathrm{cm}} = \frac{1}{2}MV^2$ of the CM, with mass $M = (M_A + M_B)$, and the energy $E_{\mathrm{rel}} = \frac{1}{2}M_{AB}v^2$ of relative motion, where the reduced mass M_{AB} is $M_A M_B/(M_A + M_B)$. Let $\mathbf{R}$ be the position of the CM relative to a fixed origin O and $\mathbf{r}$ be the inter-particle separation. The total angular momentum about O similarly decomposes into the sum

$$\begin{aligned} \mathbf{L} &= \mathbf{R} \times M\mathbf{V} + \mathbf{r} \times M_{AB}\mathbf{v} \\ &= \mathbf{L}_{\mathrm{cm}}(A + B) + \mathbf{L}_{\mathrm{rel}}(AB) \end{aligned}$$

of angular momenta of the CM and of relative motion. For any collision in the absence of any external field, the energy E_{cm} and angular momentum $\mathbf{L}_{\mathrm{cm}}$ of the CM are always conserved for all types of collision. The two species A and B may be electrons, ions, atoms, or molecules, with or without any internal structure, and may therefore possess internal energy and angular momentum, which must be taken into account. For structured particles, E_{rel} and $\mathbf{L}_{\mathrm{rel}}$ can change in a collision.

6.2.4. Elastic scattering

$$A(\alpha) + B(\beta) \rightarrow A(\alpha) + B(\beta).$$

Elastic scattering involves no permanent changes in the internal structures (states α and β) of A and B. Both the energy E_{rel} and angular momentum $\mathbf{L}_{\mathrm{rel}}(AB)$ of relative motion are therefore all conserved.

6.2.5. Inelastic scattering

$$A(\alpha) + B(\beta) \rightarrow A(\alpha') + B(\beta').$$

Inelastic scattering produces a permanent change in the internal energy and angular momentum state of one or both structured collision partners A and B, which retain their original identity after the collision. For inelastic $i \equiv (\alpha, \beta) \rightarrow f \equiv (\alpha', \beta')$ collisional transitions, the energy $E_{i,f} = \frac{1}{2}M_{AB}v_{i,f}^2$ of relative motion, before (i) and after (f) the collision, satisfies the energy conservation condition

$$E_i + \epsilon_\alpha(A) + \epsilon_\beta(B) = E_f + \epsilon_{\alpha'}(A) + \epsilon_{\beta'}(B),$$

where $\epsilon_{\alpha,\beta}$ are the internal energies of A and B in states α and β, respectively. The maximum amount of kinetic energy that can be transferred to internal energy is limited to the initial kinetic energy of relative motion, $E_{\mathrm{rel}}(AB) = \frac{1}{2}M_{AB}v_i^2$. Excitation implies $\epsilon_i \equiv \epsilon_\alpha(A) + \epsilon_\beta(B) < \epsilon_{\alpha'}(A) + \epsilon_{\beta'}(B) \equiv \epsilon_f$, deexcitation (or superelastic) implies $\epsilon_i > \epsilon_f$, and energy resonance or excitation transfer implies $\epsilon_i = \epsilon_f$. Changes in angular momentum are limited by the conservation requirement that

$$\mathbf{L}_{\mathrm{rel}}(i) + \mathbf{L}_\alpha(A) + \mathbf{L}_\beta(B) = \mathbf{L}_{\mathrm{rel}}(f) + \mathbf{L}_{\alpha'}(A) + \mathbf{L}_{\beta'}(B),$$

where $\mathbf{L}_{\alpha,\beta}$ denotes the internal angular momentum of each isolated species. Collisions, in which only angular momentum is transferred without any energy change, are called quasi-elastic collisions.

6.2.6. Reactive scattering

$$A + B \rightarrow C + D$$

Reactive scattering or a chemical reaction is characterized by a rearrangment of the component particles within the collision system, thereby resulting in a change of physical and chemical identity of the original collision reactants $A + B$ into different collision products $C + D$. Total mass is conserved. The reaction is exothermic when $E_{\mathrm{rel}}(CD) > E_{\mathrm{rel}}(AB)$ and is endothermic when $E_{\mathrm{rel}}(CD) < E_{\mathrm{rel}}(AB)$. A threshold energy is required for the endothermic reaction.

6.2.7. Center-of-mass to laboratory cross-section conversion

Theorists calculate cross sections in the center-of-mass frame, whereas experimentalists usually measure cross sections in the laboratory frame of reference. The laboratory (Lab) system is the coordinate frame in which the target particle B is at rest before the collision, i.e., $\mathbf{v}_B = 0$. The center of mass (CM) zystem (or Barycentric System) is the coordinate frame in which the CM is at rest, i.e., $\mathbf{V} = 0$. Since each scattering of projectile A into (ψ, ϕ) is accompanied by a recoil of target B into $(\pi - \psi, \phi + \pi)$ in the CM frame, the cross sections for scattering of A and B are related by

$$\left\{\frac{d\sigma(\psi,\phi)}{d\Omega}\right\}_{\text{CM}} \equiv \left\{\frac{d\sigma_A(\psi,\phi)}{d\Omega}\right\}_{\text{CM}} = \left\{\frac{d\sigma_B(\pi-\psi,\phi+\pi)}{d\Omega}\right\}_{\text{CM}}$$

$$\left\{\frac{d\sigma_B(\psi,\phi)}{d\Omega}\right\}_{\text{CM}} = \left\{\frac{d\sigma(\pi-\psi,\phi-\pi)}{d\Omega}\right\}_{\text{CM}}.$$

In the Lab frame, the projectile is scattered by θ_A and the target, originally at rest, recoils through angle θ_B. The number of particles scattered into each solid angle in each frame remains the same, the relative speed v is now v_A and $j_i = N_A v$ in each frame. Hence,

$$\left\{\frac{d\sigma_A(\theta_A,\phi)}{d\Omega_A}\right\}_{\text{Lab}} = \left\{\frac{d\Omega}{d\Omega_A}\right\}\left[\frac{d\sigma(\psi,\phi)}{d\Omega}\right]_{\text{CM}},$$

$$\left\{\frac{d\sigma_B(\theta_B,\phi)}{d\Omega_B}\right\}_{\text{Lab}} = \left\{\frac{d\Omega}{d\Omega_B}\right\}\left[\frac{d\sigma(\psi,\phi)}{d\Omega}\right]_{\text{CM}}.$$

Two-body elastic scattering

$$A(\alpha) + B(\beta) \rightarrow A(\alpha) + B(\beta).$$

The scattering and recoil angles θ_A and θ_B in the Lab frame are related to the CM scattering angle ψ by

$$\tan\theta_A = \frac{\sin\psi}{1+\gamma\cos\psi}, \qquad \gamma = M_A/M_B$$

$$\theta_B = \tfrac{1}{2}(\pi-\psi) \quad 0 \le \theta_B \le \tfrac{1}{2}\pi.$$

The elastic cross sections for scattering and recoil in the Lab-frame are related to the cross section in the CM frame by

$$\left\{\frac{d\sigma_A(\theta_A,\phi)}{d\Omega_A}\right\}_{\text{Lab}} = \frac{(1+\gamma^2+2\gamma\cos\psi)^{3/2}}{|1+\gamma\cos\psi|}\left[\frac{d\sigma(\psi,\phi)}{d\Omega}\right]_{\text{CM}},$$

$$\left\{\frac{d\sigma_B(\theta_B,\phi)}{d\Omega_B}\right\}_{\text{Lab}} = |4\sin\tfrac{1}{2}\psi|\left[\frac{d\sigma(\psi,\phi)}{d\Omega}\right]_{\text{CM}}.$$

Two-body inelastic or reactive scattering process $A + B \rightarrow C + D$

The energies E_i and E_f of relative motion of A and B and of C and D, respectively, satisfy $E_f/E_i = 1 - \epsilon_{fi}/E_i$, where $\epsilon_{fi} = \epsilon_f - \epsilon_i$ is the increase in internal energy. The scattering and recoil angles are

$$\tan\theta_C = \frac{\sin\psi}{(\gamma_C + \cos\psi)}, \qquad \gamma_C = \left[\frac{M_A M_C}{M_B M_D}\right]^{1/2} \left(\frac{E_i}{E_f}\right)^{1/2},$$

$$\tan\theta_D = \frac{\sin\psi}{(|\,\gamma_D\,| - \cos\psi)}, \qquad \gamma_D = -\left[\frac{M_A M_D}{M_B M_C}\right]^{1/2} \left(\frac{E_i}{E_f}\right)^{1/2}.$$

The Lab and CM cross sections are then related by

$$\left\{\frac{d\sigma_j(\theta_j, \phi)}{d\Omega_j}\right\}_{\text{Lab}} = \frac{\left[1 + 2\gamma_j \cos\psi + \gamma_j^2\right]^{3/2}}{|\,1 + \gamma_j \cos\psi\,|} \left[\frac{d\sigma(\psi, \phi)}{d\Omega}\right]_{\text{CM}},$$

where j denotes C or D. The scattering of a beam from a stationary target are governed by the above equations. A crossed beam experiment in which two beams intersect at an angle is not in the Lab frame. In this case the measured quantities can be similarly transformed [1] to CM for comparison with theoretical calculations.

6.3. GENERAL COLLISION PROPERTIES

6.3.1. Momentum transfer

Let A and B be point particles 1 and 2. The momentum transferred from 1 to 2 is therefore

$$\mathbf{q} = M_1(\mathbf{v}_1 - \mathbf{v}_1') = M_2(\mathbf{v}_2{}' - \mathbf{v}_2) = M_{12}(\mathbf{v} - \mathbf{v}'),$$

where primes denote corresponding velocities after the collision. The momentum transferred has magnitude $q = M_{12}(v^2 + v'^2 - 2vv'\cos\psi)^{1/2}$, which for elastic scattering reduces to $2M_{12}v\sin\frac{1}{2}\psi$. The vector momentum transferred from test particle 1 to field particle 2 by elastic or inelastic scattering into direction (ψ, ϕ) varies with ψ and ϕ according to

$$\mathbf{q}(\mathbf{v}_1, \mathbf{v}_2; \psi, \phi) = M_{12}[(v - v'\cos\psi)\hat{\mathbf{k}} - v'\sin\psi(\cos\phi\hat{\mathbf{i}} + \sin\phi\hat{\mathbf{j}})],$$

where $\hat{\mathbf{i}}$, $\hat{\mathbf{j}}$, and $\hat{\mathbf{k}}$ are the unit vectors along the fixed *XYZ*- Cartesian set of coordinate axes.

Collision-averaged momentum transfer

The momentum transferred due to scattering into all directions is

$$\mathbf{q}(\mathbf{v}_1, \mathbf{v}_2) = \int_{-1}^{+1} d(\cos\psi) \int_0^{2\pi} \mathbf{q}(v; \psi, \phi)\mathcal{P}(v; \psi, \phi)\, d\phi,$$

where $\mathcal{P}(v; \psi, \phi)$ is the probability $[(d\sigma(v; \psi, \phi)/d\Omega)]/\sigma(v)$ for scattering along direction $\hat{\mathbf{v}}'(\psi, \phi)$.

6.3.2. Momentum transfer cross section

The relative momentum lost by the incident beam only along the Z-direction is $\mathbf{q}\cdot\hat{\mathbf{k}} = M_{12}(v - v'\cos\psi)$. When averaged over all scattering angles, the fraction of momentum lost by the beam in the Z-direction is

$$\mathcal{P}_M^Z(v) = \frac{\mathbf{q}(\mathbf{v}_1, \mathbf{v}_2) \cdot \hat{\mathbf{k}}}{M_{12} v} = \sigma_D(v)/\sigma(v),$$

where the *momentum-transfer* or *diffusion cross section* is defined to be

$$\sigma_D(v) = 2\pi \int_{-1}^{+1} \left(1 - \frac{v'}{v} \cos\psi\right) \frac{d\sigma}{d\Omega}(v; \psi)\, d(\cos\psi).$$

Scattering for most cases is independent of the azimuthal angle ϕ; i.e., the scattering is axially symmetric, and $\mathcal{P}$ is independent of ϕ. The probability of momentum loss or the fraction of the initial momentum lost by particle A via (axially-symmetric) elastic or inelastic scattering into all directions reduces also to

$$\mathcal{P}_M(v) = \mathbf{q}(\mathbf{v}_1, \mathbf{v}_2)/M_{12} v = \sigma_D(v)/\sigma(v),$$

the ratio of the diffusion cross section to the integral cross section. The momentum transfer in the X- and Y-directions (Sec. 6.3.1) when integrated over ϕ, each yield zero average for axially symmetric scattering.

6.3.3. Energy transfer

The kinetic energy transferred $\mathcal{E} = \frac{1}{2} M_2 (v_2'^2 - v_2^2) = \frac{1}{2} M_1 (v_1^2 - v_1'^2)$ from 1 to 2 in an elastic collision is

$$\mathcal{E} = \mathbf{q} \cdot \left(\mathbf{v}_2 + \frac{\mathbf{q}}{2m_2}\right) = \mathbf{q} \cdot \left(\mathbf{v}_1 - \frac{\mathbf{q}}{2m_1}\right) = \mathbf{q} \cdot \mathbf{V}.$$

When the initial relative velocity $\mathbf{v} = \mathbf{v}_1 - \mathbf{v}_2$ is directed along the Z-axis and the final relative velocity $\mathbf{v}'(v, \psi, \phi) = \mathbf{v}_1' - \mathbf{v}_2'$ is directed into (ψ, ϕ), the energy transferred varies with ψ and ϕ according to

$$\mathcal{E}(\mathbf{v}_1, \mathbf{v}_2; \psi, \phi) = \Delta\mathcal{E}_1 (1 - \cos\psi) - \Delta\mathcal{E}_2 \sin\psi \cos\phi,$$

where $\Delta\mathcal{E}_1 = M_{12} \mathbf{V} \cdot \mathbf{v}$ and $\Delta\mathcal{E}_2 = M_{12} |\, \mathbf{V} \times \mathbf{v}\,|$. These depend only on the precollision speeds v_1, v_2 and the relative speed v and reduce to

$$\begin{aligned}\Delta\mathcal{E}_1(\mathbf{v}_1, \mathbf{v}_2) &= \frac{2M_1 M_2}{(M_1 + M_2)^2} \left[E_1 - E_2 - \tfrac{1}{2}(M_1 - M_2)\mathbf{v}_1 \cdot \mathbf{v}_2\right] \\ &= \tfrac{1}{2} M_{12} \left[v_1^2 - v_2^2 + \frac{M_1 - M_2}{M_1 + M_2} v^2\right],\end{aligned}$$

where $E_j = \frac{1}{2} M_j v_j^2$ are the initial kinetic energies of each particle $j = 1, 2$, and to

$$\begin{aligned}\Delta\mathcal{E}_2(\mathbf{v}_1, \mathbf{v}_2) &= M_{12} \left[v_1^2 v_2^2 - (\mathbf{v}_1 \cdot \mathbf{v}_2)^2\right]^{1/2} \\ &= \tfrac{1}{2} M_{12} \left[v_+^2 - v^2\right]^{1/2} \left[v^2 - v_-^2\right]^{1/2},\end{aligned}$$

where $v_\pm = |\, v_1 \pm v_2 \,|$ are the maximum and minimum relative speeds, corresponding to the precollision velocities $\mathbf{v}_i$ having parallel or anti-parallel directions. The energy $\Delta\mathcal{E}_1$ has a physical significance in that it is the (ψ, ϕ)-averaged value

$$\langle \mathcal{E}(\mathbf{v}_i, \mathbf{v}_j; \psi, \phi) \rangle = \Delta\mathcal{E}_1$$

of the overall energy transfer. The quantity

$$\begin{aligned}\gamma^2 &= \Delta\mathcal{E}_1^2 + \Delta\mathcal{E}_2^2 \\ &= \frac{4M_1M_2}{(M_1+M_2)^2}\,[E_1 + E_2 - E_{\rm rel}]\,E_{\rm rel} \\ &= M_{12}^2 v^2 V^2\end{aligned}$$

is conserved in the elastic collision.

Special Case: The fraction of energy transferred to field particles 2 initially at rest ($\mathbf{v}_2 = 0$) is

$$\frac{\mathcal{E}(\mathbf{v}_1)}{E_1} = \frac{4M_1M_2}{(M_1+M_2)^2}\sin^2\frac{\psi}{2}.$$

Generalization: The energy transferred from 1 to the internal relative motion of a composite structure (2, 3) by an (1–3) elastic collision that leaves the velocity of the (spectator) particle 3 of mass M_3 unaffected is

$$\begin{aligned}\mathcal{E} &= \tfrac{1}{2}\mathcal{M}_1[v_1^2 - v_1'^2] = \tfrac{1}{2}\mathcal{M}_2[v_2'^2 - v_2^2] \\ &= \tfrac{1}{2}M_{23}[v_{23}'^2 - v_{23}^2],\end{aligned}$$

where the velocities $\mathbf{v}_1$ and $\mathbf{v}_2$ are taken with respect to the (2, 3) initial center of mass and $\mathbf{v}_{23}$ is the velocity of the (2, 3) relative motion. The effective two-body masses are

$$\begin{aligned}\mathcal{M}_1 &= \frac{M_1(M_2+M_3)}{M_1(M_1+M_2+M_3)}, \\ \mathcal{M}_2 &= M_2\left(1 + \frac{M_2}{M_3}\right).\end{aligned}$$

The energy transferred is as above but with $M_{1,2}$ replaced by $\mathcal{M}_{1,2}$ and $E_i = \frac{1}{2}\mathcal{M}_i v_i^2$. Note that $\mathcal{M}_{12} = M_{12}$. Also, γ^2 is replaced by $M_{12}^2 v^2 \tilde{V}^2$, where

$$\tilde{V}^2 = \left[a v_1^2 + v_2^2 - \frac{a}{1+a}g^2\right]/(1+a),$$

with $a = \mathcal{M}_1/\mathcal{M}_2$.

The momentum transferred from 1 to the (2, 3) relative motion is

$$\mathbf{q}_{23} = \frac{M_3}{M_2+M_3}\mathbf{q},$$

where

$$\mathbf{q} = \mathcal{M}_1(\mathbf{v}_1 - \mathbf{v}_1') = \mathcal{M}_2(\mathbf{v}_2' - \mathbf{v}_2)$$

is the momentum transferred from 1 to 2.

Collision-averaged energy transfer

Since $\mathcal{P}(\psi, \phi)$ is the probability for elastic scattering into direction $\hat{\mathbf{v}}(\psi, \phi)$, the amount of energy transferred to 2 in all elastic collisions is

$$\mathcal{E}(\mathbf{v}_1, \mathbf{v}_2) = \int_{-1}^{+1} d(\cos\psi) \int_0^{2\pi} \mathcal{E}(\mathbf{v}_1, \mathbf{v}_2; \psi, \phi)\mathcal{P}(\psi, \phi)\, d\phi,$$

the ϕ, ψ-average of that in Sec. 6.3.3. For azimuthally symmetric scattering, the scattering probability $\mathcal{P} = \mathcal{P}(\psi)$ is ϕ-independent. The ϕ-averaged energy transfer is then

$$\mathcal{E}(\mathbf{v}_1, \mathbf{v}_2; \psi) = \Delta\mathcal{E}_1(\mathbf{v}_1, \mathbf{v}_2)[2\pi(1 - \cos\psi)] \left\{ \frac{d\sigma}{d\Omega}(v)/\sigma(v) \right\}.$$

The collision (ψ, ϕ)-averaged energy transfer is

$$\begin{aligned} \mathcal{E}(\mathbf{v}_1, \mathbf{v}_2) &= \Delta\mathcal{E}_1(\mathbf{v}_1, \mathbf{v}_2)\, \{\sigma_D(v)/\sigma(v)\} \\ &= \frac{2M_1M_2}{(M_1 + M_2)^2} \left[E_1 - E_2 - \frac{1}{2}(M_1 - M_2)\mathbf{v}_1 \cdot \mathbf{v}_2 \right] \mathcal{P}_M(v), \end{aligned}$$

where $\mathcal{P}_M(v) = \sigma_D(v)/\sigma(v)$ is the probability (Sec. 6.3.2) for momentum-loss. The probability of $1 \to 2$ energy transfer, or fractional energy lost by 1, via scattering by 2 into all directions, is then

$$\mathcal{P}_E(\mathbf{v}_1, \mathbf{v}_2) = \frac{\Delta\mathcal{E}_1(\mathbf{v}_1, \mathbf{v}_2)}{E_1} \mathcal{P}_M(v).$$

Special cases

1. The averaged energy transferred between equal-mass species is

$$\mathcal{E}(\mathbf{v}_1, \mathbf{v}_2) = \tfrac{1}{2}[E_1 - E_2]\mathcal{P}_M(v).$$

2. The probability for $1 \to 2$ energy transfer to test particles 2 initially at rest ($\mathbf{v}_2 = 0$ is

$$\mathcal{P}_E(v_1) = \mathcal{E}/E_1 = \frac{2M_1M_2}{(M_1 + M_2)^2} \mathcal{P}_M(v_1),.$$

3. For isotropic or hard sphere collisions, $\mathcal{P}_M = 1$. For isotropic collisions between heavy (h) and light (l) particles $\mathcal{P}_E(v_1) \to (2m_l/M_h)$, irrespective of which particle is initially at rest.

6.3.4. Energy transfer cross sections

Cross section for energy transfer in elastic Coulomb collisions

The differential cross section for elastic scattering of particles with charges Z_1e and Z_2e by the Coulomb interaction

$$V(r) = \frac{Z_1 Z_2 e^2}{r}$$

may also be expressed in the form

$$\frac{d\sigma}{d\Omega}(v, \psi) = \frac{(b_0/2)^2}{\sin^4 \psi/2},$$

where $b_0 = (Z_1 Z_2 e^2 / M_{12} v^2)$. In terms of the classical impact parameter b, the scattering angle (in CM frame) is determined by

$$\tan \frac{\psi}{2} = \frac{b_0}{b}$$

so that particles with impact parameters $b \leq b_0$ are scattered through angles $\geq 90°$. The energy transferred to field particles 2 initially at rest is therefore

$$\mathcal{E}(b) = \left[\frac{4M_1 M_2}{(M_1 + M_2)^2} E_1 \right] \left(\frac{b_0^2}{b^2 + b_0^2} \right),$$

which decreases monotonically with b. The cross section for energy transfer $\geq \mathcal{E}$ is therefore

$$\pi b^2(\mathcal{E}) = \pi b_0^2 \left[\frac{4M_1 M_2}{(M_1 + M_2)^2} \frac{E_1}{\mathcal{E}} - 1 \right].$$

The cross section for energy transfers in the range between $\mathcal{E}$ and $\mathcal{E}'$ is

$$\sigma(\mathcal{E}, \mathcal{E}') = 2\pi \int_b^{b'} b\, db = \pi b_0^2 \left[\frac{4M_1 M_2}{(M_1 + M_2)^2} E_1 \right] \left(\frac{1}{\mathcal{E}} - \frac{1}{\mathcal{E}'} \right)$$

$$= \frac{2\pi (Z_1 Z_2 e^2)^2}{M_2 v_1^2} \left(\frac{1}{\mathcal{E}} - \frac{1}{\mathcal{E}'} \right).$$

On setting $\mathcal{E}' = \mathcal{E} + d\mathcal{E}$, the differential (per unit $d\mathcal{E}$) cross section for energy transfer in a Coulomb collision is

$$\frac{d\sigma(\mathcal{E})}{d\mathcal{E}} = \pi b_0^2 \left[\frac{4M_1 M_2}{(M_1 + M_2)^2} \right] \frac{E_1}{\mathcal{E}^2}.$$

Energy-transfer cross sections for general interactions $V(r)$ have also been derived [2] in terms of the corresponding differential CM cross section $\sigma(v, \psi)$ for 1–2 elastic and inelastic scattering.

Thomson cross section for ionization

Thomson assumed that ionization of an atom by a fast incident beam of electrons of mass m_e and speed v_e occurs when an elastic binary encounter between 1 and each of the N_e atomic electrons (assumed at rest) transfers energy $\mathcal{E}$ in the range $I_n < \mathcal{E} < E_e = \frac{1}{2} m_e v_e^2$. Since $b_0 = (\varepsilon_0 / E_e) a_0$, the cross section is then

$$\sigma(E_e) = \pi a_0^2 \frac{\varepsilon_0}{E_e} \left(\frac{\varepsilon_0}{I_n} - \frac{\varepsilon_0}{E_e} \right) N_e,$$

TABLE 6.1. Atomic units (a.u) and related quantities

Quantity	Symbol			Value
Electron mass	m_e			$9.109\,389\,7 \times 10^{-31}$ kg
Proton mass	m_p			$1.672\,623\,1 \times 10^{-27}$ kg
Neutron mass	m_n			$1.674\,928\,6 \times 10^{-27}$ kg
Atomic mass unit (AMU)	$m_u \equiv u$	$\frac{1}{12}M(^{12}\text{C})$		$1.660\,540\,2 \times 10^{-27}$ kg
Elementary charge (SI)	e			$1.602\,177\,33 \times 10^{-19}$ C
Elementary charge (CGS)	e			$4.806\,531\,99 \times 10^{-10}$ Stat-C
Bohr radius (a.u)	a_0	$\hbar^2/m_e e^2$	$\hbar/m_e v_0$	$0.529\,177\,249 \times 10^{-10}$ m
Bohr velocity (a.u)	v_0	$e^2/\hbar$	αc	$2.187\,691\,42 \times 10^6$ m s^{-1}
Energy (a.u)	ε_0	e^2/a_0	$m_e v_0^2$	$4.359\,748\,21 \times 10^6$ J
Time (a.u)	τ_0	a_0/v_0	$\hbar/\varepsilon_0$	$2.418\,884\,326\,555 \times 10^{-17}$ s
Wavenumber	k_0	$m_e v_0/\hbar$	a_0^{-1}	$1.889\,726 \times 10^{10}$/m
Planck's constant	h	$2\pi\hbar$		$6.626\,075\,5 \times 10^{-34}$ J s
Angular momentum (action)	$\hbar$	$h/2\pi$	$m_e a_0 v_0$	$1.054\,572\,66 \times 10^{-34}$ J s
Cross section	a_0^2			$0.280\,028\,560\,9 \times 10^{-20}$ m^2
Rate	$a_0^2 v_0$	$a_0^2 e^2/\hbar$		$6.126\,160\,8 \times 10^{-15}$ m^3 s^{-1}
Angular frequency (a. u.)	ω_0	v_0/a_0	$\varepsilon_0/\hbar$	$4.134\,137\,33 \times 10^{16}$ Hertz or s^{-1}
Rydberg constant R_∞	$\frac{1}{2}\varepsilon_0/(hc)$	$\alpha/(4\pi a_0)$	$\alpha^2 m_e c/(2h)$	$1.097\,373\,156\,834 \times 10^7$/m
Line angular frequency	$\omega(a.u.)$	$hc/\varepsilon_0\lambda$	$1/(2R_\infty\lambda)$	$45.563\,352\,6/\lambda$ (nm)
Speed of light (SI)	c			$2.997\,924\,58 \times 10^8$ m s^{-1}
Speed of light (a.u)	c/v_0	α^{-1}		137.035 989 5
Relativistic energy unit	$m_e c^2$			510.990 6 keV
Atomic mass energy unit	$m_u c^2$			931.494 32 MeV
Fine structure constant	α	v_0/c	$e^2/\hbar c$	$7.297\,353\,08 \times 10^{-3}$
Emission rate, A_0(s^{-1})	α^3/τ_0	$\alpha^3\omega_0$	$e^4/m_e^2 c^3 a_0^3$	$1.606\,501 \times 10^{10}$ s^{-1}
Compton wavelength, $\bar{\lambda}_C$	$\lambda_C/2\pi$	αa_0	$\hbar/m_e c$	$3.861\,593\,23 \times 10^{-13}$ m
Electron radius	r_e	$e^2/m_e c^2$	$\alpha^2 a_0$	$2.817\,940\,92 \times 10^{-15}$ m
Thomson cross section	σ_T	$8\pi r_e^2/3$	$8\alpha^4\pi a_0^2/3$	$0.665\,246\,16 \times 10^{-28}$ m^2
Boltzmann constant (SI)	k_B			$1.380\,658 \times 10^{-23}$ J K^{-1}
Boltzmann constant (CGS)	k_B			$1.380\,658 \times 10^{-16}$ ergs K^{-1}
Boltzmann constant (eV K^{-1})	k_B			$8.617\,385 \times 10^{-5}$ eV K^{-1}
Boltzmann constant (ε_0 K^{-1})	k_B			$3.166\,830 \times 10^{-6}\varepsilon_0$ K^{-1}
Mean electron-speed (cm/s)	$\langle v_e \rangle$	$(8k_B T/\pi m_e)^{1/2}$		$1.076\,042 \times 10^7 (T/300)^{1/2}$ cm s^{-1}
Mean atom-speed (cm/s)	$\langle v_A \rangle$	$(8k_B T/\pi M_A)^{1/2}$		$2.520\,279 \times 10^5 (T/300M_u)^{1/2}$ cm s^{-1}
Natural radius, R_e	$e^2/k_B T$	557.8 Å $(300/T)$	14.42 Å$/T_{\text{eV}}$	$1.673\,314\,47 \times 10^{-5}/T$ m
Natural cross section	πR_e^2			$0.997\,377\,9 \times 10^{-10}(300/T)^2$ cm^2
Natural rate	$\pi R_e^2 v_A$			$2.463\,265\,613 \times 10^{-5}(300/TM_u)^{3/2}$ cm^3 s^{-1}

$m_e/m_p = 5.446\,170\,13 \times 10^{-4}$; $m_p/m_e = 1836.152\,701$; $m_p/m_u = 1.007\,276\,47$.

where N_e is the number of equivalent electrons in the level n with ionization energy I_n and a_0 and ε_0 are the atomic units of length and energy.

6.3.5. Atomic units

Atomic and molecular calculations based on bound and continuum wave function solutions of the Schrödinger equation and the collision matrix elements are facilitated by the use of atomic units (a.u). The final result is then converted to the correct SI-units, as listed in Table 6.1. In atomic units

$$\hbar = m_e = e = 4\pi\epsilon_0 = 1,$$

where ϵ_0 is the permittivity in vacuum. The atomic units and related quantities are listed in Table 6.1. These are based on the 1986 CODATA recommendations [3]. For the lowest $1s$ state of hydrogen (with infinite nuclear mass), a_0 is the Bohr radius, v_0 is the Bohr velocity, $2\pi\tau_0$ is the time taken to complete a Bohr orbit, and ε_0 or one Hartree is twice the ionization energy, which is one Rydberg unit of energy.

6.3.6. Energy conversion factors

These are listed in Table 6.2. The table shows, for example, that 1 eV is equivalent to a temperature of 11, 604.45 K and that 1 a.u ≡ 627.509 kcal/mol. A mole of particles each with atomic or molecular mass $M(AMU)$ has a mass of M gms. The number of particles in a mole is Avogadro's number $N_A = 6.0221367 \times 10^{23}$ mol^{-1}. The Gas Constant $R = N_A k_B = 8.314511$ J mol^{-1} K^{-1} is the energy of one mole of particles at 1 K.

TABLE 6.2. Some energy conversion factors

Quantity	1 eV	1 a.u	1 cm^{-1}
Frequency ($\nu = E/h$)	$2.417\,988\,36 \times 10^{14}$ Hz	$6.579\,683\,93 \times 10^{15}$ Hz	$2.997\,924\,58 \times 10^{10}$ Hz
Wavelength ($\lambda = hc/E$)	$1.239\,844\,24 \times 10^{-6}$ m	$4.556\,341\,89 \times 10^{-8}$ m	10^{-2} m
Wavenumber ($\tilde{\nu} = E/hc$)	80,6554.1 0 m^{-1}	$2.194\,746\,32 \times 10^{7}$ m^{-1}	100 m^{-1}
Temperature ($T_K = E/k_B$)	11,604.45 K	$3.157\,732\,87 \times 10^{5}$ K	1.438 768 95 K
Atomic unit ($E_{au} = E/\varepsilon_0$)	$3.674\,930\,9 \times 10^{-2}$a.u	1 a.u.	$4.556\,335\,28 \times 10^{-6}$a.u
eV ($E_{eV} = E/e$)	1eV	27.211 396 2 eV	$1.239\,842\,45 \times 10^{-4}$ eV
Joules	$1.602\,177\,33 \times 10^{-19}$ J	$4.35\,748\,21 \times 10^{-18}$J	$1.986\,447\,44 \times 10^{-23}$J
kcal/mol	23.060 54 kcal/mol	627.509 5 kcal/mol	$2.859\,143\,6 \times 10^{-3}$ kcal/mol
kJ/mol	96.485 309 kJ/mol	2625.5 kJ/mol	$1.196\,265\,8 \times 10^{-2}$ kJ/mol
AMU ($= E/m_u c^2$)	$1.073\,5 \times 10^{-9}u$	$2.921\,14 \times 10^{-8}u$	$1.330\,970\,86 \times 10^{-13}u$

1 cal = 4.184 J; 1 eV/particle ≡ 23.06032 kcal/mol.

TABLE 6.3. General *n*-dependence and characteristic properties of Rydberg states

Property	n-Dependence	$n=10$	$n=50$	$n=100$
Mean radius (cm) $r_n = [\langle r^{-1}\rangle]^{-1}$	$(n^2/Z)a_0$	5.3×10^{-7}	1.3×10^{-5}	5.3×10^{-5}
Area (cm^2) πr_n^2	$(n^4/Z^2)\pi a_0^2$	8.8×10^{-13}	5.5×10^{-10}	8.8×10^{-9}
Velocity (cm/s) $v_n = \sqrt{\langle v_e^2\rangle}$	$(Z/n)v_0$	2.18×10^{7}	4.4×10^{6}	2.18×10^{6}
Ionization energy (eV) $I_n = \mid E_n \mid = \frac{1}{2}m_e v_n^2$	$(Z^2/2n^2)\varepsilon_0$	1.36×10^{-1}	63.2 (K)	15.79 (K)
Transition energy (eV) $\Delta E_{n',n} = \mid I_{n\pm1} - I_n \mid$	$(Z^2/n^3)\varepsilon_0$	316 (K)	2.526 (K)	0.316 (K)
Classical period (s) $\tau_n = 2\pi r_n/v_n = 2\pi/\omega_n$	$(2\pi n^3/Z^2)\tau_0$	1.5×10^{-13}	1.9×10^{-11}	1.5×10^{-10}
Transition frequency (s^{-1}) $(\omega_{n,n\pm1} = \omega_n)^*$	$(Z^2/n^3)\omega_0$	4.1×10^{13}	3.3×10^{11}	4.1×10^{10}
Wavelength (cm) $\lambda_{n,n\pm1} = \tau_n c$	$(2\pi n^3/Z^2)(a_0/\alpha)$	4.6×10^{-3}	0.570	4560.9
Radiative lifetime (s)	$\frac{n^5}{Z^4}[\ln(2n-1)-0.365]^{-1} B_0^{\dagger}$	1.81×10^{-6}	3.45×10^{-3}	9.5×10^{-2}

$^{\dagger}B_0 = 3/4A_0 = 4.66853\times10^{-11}$s [ref. [4]].

6.3.7. Rydberg properties

A Rydberg state refers to any state of hydrogenic species or to a one-electron highly excited state of any atom or molecule for which the energy of the Rydberg electron is

$$E_{n\ell} = -\frac{Z^2\varepsilon_0}{2(n-\mu_\ell)^2},$$

where μ_ℓ is the quantum defect that originates from the non-Coulomb part of the Rydberg electron-core interaction and is a measure of the penetration of the more eccentric (lower ℓ) Bohr orbits with the atomic core. Characteristic properties of Rydberg levels n are listed in Table 6.3. The wavelength λ, frequency ν, and energy $E_{ul} = E_u - E_l$ of radiation emitted in a $n_u \to n_l$ transition in hydrogenic systems are given by

$$\lambda^{-1} = R_\infty Z^2\left\{n_l^{-2} - n_u^{-2}\right\} \qquad R_\infty = \varepsilon_0/2hc,$$

$$\nu = R_\nu Z^2\left\{n_l^{-2} - n_u^{-2}\right\}, \qquad R_\nu = R_\infty c,$$

$$E_{ul} = R_H Z^2\left\{n_l^{-2} - n_u^{-2}\right\}, \qquad R_H = R_\infty hc.$$

The wavelengths of the discrete transitions to the ground level $n_l = 1$ converge to the *Rydberg constant* R_∞ for the $n_u = \infty \to 1)$ transition. The dipole moment $\langle nlm|\, e\mathbf{r}\,|nlm\rangle$ varies as n^2. Level nl decays radiatively to level $n'(l' = l \pm 1)$ at the rate (Sec. 6.14.2)

$$A_{nl,n'l'} = \frac{4}{3}A_0(E_{nn'}/\varepsilon_0)^3(2l+1)^{-1}\sum_{m,m'} \mid \langle n'l'm'|\,\mathbf{r}\,|nlm\rangle\mid^2,$$

which varies as n^{-5}. The basic unit of transition rate $A_0 = \alpha^3\omega_0$ is given in Table 6.1. Spontaneous transitions between Rydberg states have therefore very little probability, mainly due to the external ν^3-factor. Level nl decays radiatively to all lower levels $n', l' = l \pm 1$ at rate (s^{-1})

$$\nu_r(nl) = \sum_{nl>n'l'} A_{nl,n'l'} \approx \frac{2}{3}A_0\left[\frac{Z^4}{n^3(l+\frac{1}{2})^2}\right] = 1.071\,10^{10}\frac{Z^4}{n^3(l+\frac{1}{2})^2}\ \mathrm{s}^{-1}$$

which varies as n^{-5} for $l = n-1$ and as n^{-3} for $l = 0$. The circular ($l = n-1$) states are therefore very long lived and decay at frequency

$$\nu_r(n, l = n-1) \approx \frac{10^{10}}{n^5} \ll \nu_r(np \to 1s) \approx \frac{4}{9}\frac{10^{10}}{n^3}\ (s^{-1}),$$

very much less than that for the ($np \to 1s$) transition. Radiative lifetimes $\tau_{nl} = \nu_r^{-1} = 9.337\,10^{-11}n^3(l+\frac{1}{2})^2/Z^4$ s therefore increase dramatically from $1.68\,10^{-9}$ s for $n = 2$ to as long as 1 s for $n > 166$ circular states (Table 6.3). The main transitions are those to the lowest $n = 1, 2$, with the largest energy jumps, and to the neighboring $n-1$ level, with the largest oscillator strength. Additional Rydberg properties are that the polarizability and the fine structure interval vary as n^7 and n^{-3}, respectively.

Virial theorem

The expectation values of $T = \frac{1}{2}m_e v_e^2$ and of $V = -Ze^2/r$ are related via the Virial theorem by

$$E_n = -\frac{Z^2e^2}{n^2a_0} = \langle T\rangle + \langle V\rangle = \tfrac{1}{2}\langle V\rangle = -\langle T\rangle = \frac{1}{2}m_e v_n^2.$$

Quantal expectation values of r^p for $H(n\ell)$

$$\langle r\rangle = \tfrac{1}{2}[3n^2 - \ell(\ell+1)],$$

$$\langle r^2\rangle = \tfrac{1}{2}n^2[5n^2 + 1 - 3\ell(\ell+1)],$$

$$\langle r^3\rangle = \tfrac{1}{8}n^2[35n^2(n^2-1) - 30n^2(\ell+2)(\ell-1) + 3(\ell+2)(\ell+1)\ell(\ell-1)],$$

$$\langle r^4\rangle = \tfrac{1}{8}n^4[63n^4 - 35n^2(2\ell^2+2\ell-3) + 5\ell(\ell+1)(3\ell^2+3\ell-10) + 12],$$

$$\langle r^{-1}\rangle = \frac{1}{n^2},$$

$$\langle r^{-2}\rangle = \frac{2}{n^3(2\ell+1)},$$

$$\langle r^{-3}\rangle = \frac{2}{n^3\ell(\ell+1)(2\ell+1)},$$

$$\langle r^{-4}\rangle = \frac{4[3n^2 - \ell(\ell+1)]}{n^5\ell(\ell+1)(2\ell+3)(2\ell+2)(2\ell-1)}.$$

These expectation values (in units of a_0^p) also apply to $\langle (Zr)^p \rangle$ for hydrogenic systems of nuclear charge Z.

Bohr and Heisenberg correspondence principles

Rydberg atoms satisfy the Bohr correspondence principle, which states that the angular frequency $\omega_{n,n\pm s}$ for neighboring transitions $n \to (n \pm s)$ equals s-times the frequency ω_n of electron motion in Rydberg level n; i.e., $\mid E_{n\pm s} - E_n \mid = 2\pi/s\omega_n$. The general energy difference between any two Rydberg levels with separation $s = n' - n$ is

$$E_{n+s} - E_n = s\hbar\omega_n[1 - \tfrac{3}{2}(s/n) + 4(s/n)^2 + \cdots].$$

The levels therefore become equally spaced and the Bohr correspondence is satisfied for $s \ll n$. The one-dimensional quantum matrix element

$$F^q_{n'n}(\lambda) = \int_0^\infty \psi^*_{n'}(r) F(r,\lambda)\psi_n(r)\, dr$$

connecting two Rydberg states can be evaluated by the classical approximation

$$F^c_{n'n}(\lambda) = \frac{\omega_n}{2\pi}\int_0^{2\pi/\omega_n} F(r(t),\lambda)\exp(is\omega_n t)\, dt,$$

where $r(t)$ is the classical trajectory and $s = n - n'$. This equivalence is the Heisenberg correspondence principle, which is useful for the calculation of classical expectation values ($s = 0$), oscillator strengths, interaction matrix elements, and transition probabilities. It can be regarded as the usual classical correspondence $\langle n|\, F(r)\, |n\rangle \approx F^c_{nn} = \tau_n^{-1} \oint F(t)\, dt$ for diagonal elements generalized to nondiagonal matrix elements.

Classical expectation values of r^p for H($n\ell$)

$$\langle (r/r_n)^p \rangle = \frac{1}{2\pi}\oint (1 - \epsilon\cos\psi)^{p+1}\, d\psi,$$

$$\langle (r_n/r)^p \rangle = \frac{1}{2\pi}(n/\ell_c)^{2p-3}\oint (1 + \epsilon\cos\theta)^{p-2}\, d\theta,$$

where $r_n = n^2 a_0$ and ϵ, the eccentricity of the classical ellipse, is determined from $\epsilon^2 = 1 - \ell_c^2/n^2$, where the classical angular momentum is $L_c = \ell_c\hbar$. The classical elliptical orbit is

$$r(\theta) = \frac{(\ell_c^2/n^2) r_n}{1 + \epsilon\cos\theta}$$

in terms of the polar angle θ (true anomaly). It is

$$r(\psi) = r_n(1 - \epsilon \cos \psi)$$

in terms of the eccentric anomaly ψ. The classical expectation values are related by

$$\langle (r/r_n)^p \rangle = (\ell_c/n)^{2p+3} \langle (r_n/r)^p \rangle$$

and compare favorably with the quantal values, particularly when ℓ_c is replaced by its semiclassical counterpart $(\ell + \frac{1}{2})$.

6.4. EQUILIBRIUM DISTRIBUTIONS

Collisions (particle-particle, photon-particle, and photon-photon) promote the attainment of thermodynamic equilibrium. Equilibrium distributions provide detailed microscopic balance between the forward and reverse rates of collisional mechanisms that are direct inverses of one another.

Examples

1. Maxwell: $e^-(v_1) + e^-(v_2) \rightleftharpoons e^-(v_1') + e^-(v_2')$, where the kinetic energies of the particles are redistributed by free electron-free electron collisions.
2. Boltzmann: $e^- + H(n, \ell) \rightleftharpoons e^- + H(n', \ell')$ between collisional $(n, \ell \rightarrow n', \ell')$ excitation and collisional $(n', \ell' \rightarrow n, \ell)$ deexcitation by electron impact.
3. Saha: $e^- + H(n\ell) \rightleftharpoons e^- + H^+ + e^-$ between direct electron-impact ionization from and direct electron-ion recombination into a given level $n\ell$.
4. Planck: $e^- + H^+ + h\nu \rightleftharpoons H(n\ell) + 2h\nu$, which involves the interaction between radiation and atoms in photoionization/(stimulated) recombination to a given level $n\ell$.

6.4.1. Maxwell velocity distribution for free particles

Particles A of mass m, moving freely, attain equilibrium by collision with a gas bath of similar particles at temperature T when they develop the velocity distribution

$$N_A(v)\, dv = N_A f_M(v)\, dv,$$

where N_A is the number density (cm^{-3}) of particles with all speeds. For a nonrelativistic nondegenerate gas, the one-particle Maxwell-velocity distribution (normalized to unity) is isotropic and equal to,

$$f_M(v)\, dv = \frac{4}{\sqrt{\pi}} u^2 \exp(-u^2)\, du,$$

where $u = v/v_p$. The most probable speed, defined by $df_M/dv = 0$, is $v_p = \sqrt{2k_B T/m}$. The mean speed is

$$\langle v \rangle \equiv \bar{v} = \int_0^\infty v f_{MB}(v)\, dv = \sqrt{\frac{8k_B T}{\pi m}},$$

and the mean square speed is

$$\langle v^2 \rangle = \int_0^\infty v^2 f_{MB}(v)\, dv = \frac{3k_B T}{m}.$$

These speeds are in the ratio, $v_p/\langle v \rangle/\langle v^2 \rangle^{1/2} = 1/1.1284/1.2248$. The mean speeds for electrons of mass m_e and ions of mass m_i are

$$\langle v_e \rangle = 1.076\,042 \times 10^7 \left[\frac{T_e(K)}{300}\right]^{1/2} \text{cm/s}$$

$$= 6.692\,38 \times 10^7\, T_{\text{eV}}^{1/2} \text{ cm/s},$$

$$\langle v_i \rangle = 2.511\,16 \times 10^5 \left[\frac{T_i(K)}{300}\right]^{1/2} \left(m_{\text{p}}/m_i\right)^{1/2} \text{ cm/s},$$

where m_{p} is the proton mass and $(m_{\text{p}}/m_e)^{1/2} = 42.850\,352$. The Maxwell distribution in kinetic energy $E = \frac{1}{2}mv^2$ is

$$N_A(E)\, dE = N_A f_M(E)\, dE,$$

where the one-particle distribution is

$$f_M(E)\, dE = \frac{2}{\pi^{1/2}} \sqrt{\epsilon} \exp(-\epsilon),\, d\epsilon,$$

with $\epsilon = E/k_B T = v^2/v_p^2$. The averaged energy is

$$\langle E \rangle \equiv \bar{E} \equiv \int_0^\infty E f_M(E)\, dE = \frac{3}{2} k_B T.$$

The fraction of particles whose energies exceed D is

$$\int_D^\infty f_M(E)\, dE = 1 - \left[\Phi(\sqrt{d} - \frac{2}{\pi^{1/2}} \sqrt{d} \exp(-d)\right],$$

where $d = D/k_B T$ and $\Phi(x) = (2/\sqrt{\pi}) \int_0^x \exp(-y^2)\, dy$ is the probability or error integral. The Maxwell-Boltzmann distribution for an isothermal bath of particles in an external field of potential $V(R)$ is

$$G_{MB}(E, R)\, dE = \frac{2}{\pi^{1/2}} \left[\frac{E - V(R)}{k_B T}\right]^{1/2} \exp\left(-\frac{E}{k_B T}\right) dE.$$

6.4.2. Two temperature Maxwell distributions

For Maxwellian distributions of ensembles of A and B at different temperatures T_A and T_B, respectively, the product distribution

$$f_M(\mathbf{v}_A)\, d\mathbf{v}_A f_M(\mathbf{v}_B)\, d\mathbf{v}_B = f_M(\tilde{\epsilon}_{\text{cm}})\, d\tilde{\epsilon}_{\text{cm}} f_M(\tilde{\epsilon}_{\text{rel}})\, d\tilde{\epsilon}_{\text{rel}}$$

separates naturally into the corresponding product of distributions associated with the energies

$$\tilde{\epsilon}_{\rm cm} = \frac{1}{2}\left(\frac{M_A T_B + M_B T_A}{k_B T_A T_B}\right)\tilde{V}^2_{\rm CM},$$

with

$$\tilde{V}_{\rm CM} = (M_A T_B \mathbf{v}_A + M_B T_A \mathbf{v}_B)/(M_A T_B + M_B T_A)$$

and

$$\tilde{\epsilon}_{\rm rel} = \frac{1}{2}\frac{M_A M_B v^2}{k_B(M_A T_B + M_B T_A)}.$$

6.4.3. Boltzmann distribution

Collisions among a bath of atoms at temperature T also promote equilibrium among their bound states. Under equilibrium conditions, the populations of atoms, (or ions of the same species and stage of ionization), occupying level j, with energy E_j and electronic statistical weight g_j, and level i, with energy E_i and statistical weight g_i, are in the ratio

$$\frac{N_j}{N_i} = \frac{g_j \exp(-E_j/k_B T)}{g_i \exp(-E_i/k_B T)}.$$

The numerator and denominator are called Boltzmann factors associated with the energy levels E_j and E_i, respectively. The relative population is the Boltzmann equilibrium distribution of atoms over bound energy levels. The probability that the atom has energy E_j is the distribution

$$\mathcal{P}_j = \frac{N_j}{N(\Sigma)} = \frac{g_j}{\mathcal{Z}_{el}(T)}\exp(-E_{j0}/k_B T),$$

where $E_{j0} = E_j - E_0 > 0$ is the excitation energy of level i relative to the ground energy level E_0. The electronic partition function is the sum

$$\mathcal{Z}_{el}(T) = \sum_j g_j \exp(-E_{j0}/k_B T)$$

of Boltzmann factors for all states. It may be interpreted as the maximum number of quantum states available to the atom/molecule at temperature T. The distribution $\mathcal{P}_j$ is then normalized to unity. The electronic partition function $\mathcal{Z}^+_{el}$ of the ion is related to $\mathcal{Z}_{el}$ for the atom with ionization energy I by

$$\mathcal{Z}^+_{el} = \mathcal{Z}_{el}\exp(-I/k_B T).$$

The electronic degeneracy factor or statistical weight g_n is

$$g_n = \begin{cases} 2 & \text{for a free electron, spin up and down,} \\ 2(2\ell+1) & \text{for a bound electron in state } n, \ell, \\ 2n^2 & \text{for a bound electron in } n\text{-shell.} \end{cases}$$

The internal partition function for the electronic and rovibrational states of a diatomic AB is similarly

$$\mathcal{Z}_{AB} = \frac{\omega_{AB}}{\sigma} \sum_{v,J} (2J+1) \exp(-\epsilon_{vJ}/k_B T),$$

where ω_{AB} is now reserved solely for the electronic statistical weight and ϵ_{vJ} is the energy of rovibrational level vJ above the $v = 0$, $J = 0$ level. The symmetry factor σ is the number of equivalent arrangements or transpositions of identical atoms plus one. For example,

$$\sigma = \begin{cases} 2 & \text{for homonuclear diatomics } A_2, \\ 1 & \text{for heteronuclear diatomics } AB. \end{cases}$$

Only half of the levels are filled for a linear molecule that is symmetrical to reflection in a plane perpendicular to the molecular axis passing through the center of gravity.

AB oscillator-rotor partition function

The energy of an oscillator-rotor for state (v, J) separates as

$$\epsilon_{vJ} = (v + \tfrac{1}{2})h\nu + BJ(J+1).$$

The natural frequency of vibration is ν, and the rotational constant is $B = \hbar^2/2I$, where I is the moment of inertia of AB. The rovibrational partition function is

$$\begin{aligned} \mathcal{Z}_{vr} &= \frac{1}{\sigma} \sum_{v=0}^{\infty} \sum_{J=0}^{\infty} (2J+1) \exp(-\epsilon_{vJ}/k_B T)\, dJ \\ &= \frac{1}{\sigma} \left(\frac{2Ik_B T}{\hbar^2} \right) [1 - \exp -(h\nu/k_B T)]^{-1}, \end{aligned}$$

which naturally separates into the product of rotational and vibrational partition functions $\mathcal{Z}_{\rm rot} = 2Ik_B T/\hbar^2\sigma$ and $\mathcal{Z}_{\rm vib} = [1 - \exp -(h\nu/k_B T)]^{-1}$, respectively. On taking the zero of energy at the bottom of the potential well, $\mathcal{Z}_{\rm vib} = \exp -(h\nu/2k_B T)[1 - \exp -(h\nu/k_B T)]^{-1}$. For vibration under a general interaction $V(R)$, this decomposition is not normally possible. The partition function for a molecule can be separated into partition functions appropriate to each independent degree of freedom,

$$\mathcal{Z} = \mathcal{Z}_t \mathcal{Z}_{el} \mathcal{Z}_{vr}.$$

The vibrational partition function for a molecule considered as a collection of harmonic oscillators with natural frequency ν_i is

$$\mathcal{Z}_{\rm vib} = \prod_i \left[1 - \exp\left(-\frac{h\nu_i}{k_B T} \right) \right]^{-1}.$$

The rotational partition function for a nonlinear polyatomic molecule with principal moments of inertia $I_a I_b I_c$, is

$$\mathcal{Z}_{\text{rot}} = \frac{(8\pi I_a I_b I_c)^{1/2}}{\sigma \hbar^3} (k_B T)^{3/2}.$$

6.4.4. Classical statistical weights

Free particles

Free particles of general mass M with momenta $\mathbf{p}$ move in volume $\mathcal{V}$. Since each state occupies a volume $(2\pi\hbar)^3$ of phase space, the number of translational states in the momentum interval $[\mathbf{p}, \mathbf{p} + d\mathbf{p}]$ and volume $\mathcal{V}$ is $d\mathbf{p}\mathcal{V}/(2\pi\hbar)^3$, which reduces to the number

$$g(E) \equiv \rho(E)\, dE = \frac{2}{\pi^{1/2}} \mathcal{V} \frac{(2\pi M)^{3/2}}{h^3} E^{1/2}\, dE$$

of states with kinetic energy $E = p^2/2M$ in the interval $E, E + dE$. The number of states per unit interval dE is the energy density of states $\rho(E)$. The translational partition function is

$$\mathcal{Z}_t = \mathcal{V} \int \frac{d\mathbf{p}}{(2\pi\hbar)^3} \exp\left(-\frac{p^2}{2Mk_B T}\right)$$

$$= \int_0^\infty \rho(E)\, dE = \mathcal{V} \frac{(2\pi M k_B T)^{3/2}}{h^3}.$$

The normalized energy distribution is then

$$f_M(E)\, dE = \rho(E) \exp -(E/k_B T)\, dE/\mathcal{Z}_t,$$

which is the Maxwell distribution

$$f_M(E)\, dE = \frac{2}{\pi^{1/2}} \epsilon^{1/2} \exp(-\epsilon)\, d\epsilon,$$

where $\epsilon = E/k_B T$.

deBroglie wavelength

The deBroglie wavelength

$$\lambda_{dB} = \frac{h}{\langle p \rangle} = \frac{h}{(2\pi M k_B T)^{1/2}}.$$

The numerical value for electrons of mass m_e at temperature $T_e(K)$ is

$$\lambda_e = \frac{7.453818 \times 10^{-6}}{T_e^{1/2}(K)} \text{ cm} = 43.035 \left[\frac{300}{T_e(K)}\right]^{1/2} \text{Å} = \frac{6.9194}{T_e^{1/2}(eV)} \text{Å}.$$

The translational partition in one and three dimensions can be expressed as

$$\mathcal{Z}_t^{(1)} = L/\lambda_{dB} = \frac{(2\pi M k_B T)^{1/2} L}{h}$$

and

$$\mathcal{Z}_t^{(3)} = \mathcal{V}/\lambda_{dB}^3 = \frac{(2\pi M k_B T)^{3/2} \mathcal{V}}{h^3},$$

respectively. For example, the maximum electron density required to fill all of the translational states in a T_e (eV)-plasma is

$$n_e^{\max} = \frac{\mathcal{Z}_t}{\mathcal{V}} = \lambda_e^{-3} = 3.02\, T_e^{3/2} \times 10^{27}\ \mathrm{m}^3.$$

Bound particles: Energy densities

The number (weights) of particle-pairs AB with interseparation $\mathbf{r}$ and relative momentum $\mathbf{p}$ in the volume element $\mathbf{r} + d\mathbf{r}, \mathbf{p} + d\mathbf{p}$ of phase space is $d\mathbf{r}\, d\mathbf{p}/(2\pi\hbar)^3$. The weight $g_{AB}(E, \mathbf{R})$ of these states with specific internal energy in the interval dE about E and interseparation in the interval $d\mathbf{R}$ about $\mathbf{R}$ is

$$\rho(E, \mathbf{R})\, dE\, d\mathbf{R} = \left\{ \int \delta(H - E)\, \delta(\mathbf{r} - \mathbf{R}) \frac{d\mathbf{r}\, d\mathbf{p}}{(2\pi\hbar)^3} \right\} dE\, d\mathbf{R}.$$

The Hamiltonian $H(r, p)$ is $p^2/2M_{AB} + V(r)$, the interaction between the particles is $V(r)$, and M_{AB} is the reduced mass of particles A and B. The $(E, \mathbf{R})$-density reduces to

$$\rho_{AB}(E, \mathbf{R}) = \frac{4\pi M_{AB}\, p}{(2\pi\hbar)^3} = \frac{2}{\pi^{1/2}} \frac{(2\pi M_{AB})^{3/2}}{(2\pi\hbar)^3} [E - V(R)]^{1/2}.$$

The weight $g_{AB}(E)$ of bound levels with energy E in the interval dE about E is $\rho_{AB}(E)\, dE$, where the (energy) density of states is then

$$\rho_{AB}(E) = \frac{4\pi M_{AB}}{(2\pi\hbar)^3} \int_0^{R_E} \{2M_{AB}[E - V(R)]\}^{1/2} 4\pi R^2\, dR,$$

where R_E is the largest apocenter defined by $E = V(R_E)$. For free particles with no interaction in volume $\mathcal{V}$, the weight is

$$g_{AB}(E) = \rho_{AB}^{(f)}(E)\, dE = \frac{2}{\pi^{1/2}} \frac{(2\pi M_{AB})^{3/2} \mathcal{V}}{h^3} E^{1/2}\, dE.$$

For particles interacting via the Coulomb field $V = -Z_A Z_B e^2/R$, and $R_E = Z_A Z_B e^2/|E|$,

$$\rho_{AB}^{(c)}(E)\, dE = \left\{ \frac{(Z_A Z_B e^2)^2 M_{AB}}{\hbar^2} \right\}^{3/2} \frac{dE}{|2E|^{5/2}}.$$

The energy spacing between highly excited levels n is given by the Bohr correspondence $dE = \hbar\omega(E) = h/\tau(E)$, where $\tau(E)$ is the orbital period. The classical statistical weight g_H for $H(n)$ simply reduces to n^2, in agreement with the quantal result. The total number of bound levels is

$$g_{AB} = \int_{-D}^{0} \rho(E) \exp\left(-\frac{E}{k_B T}\right) dE$$

$$= \frac{(2\pi M_{AB} k_B T)^{3/2}}{(2\pi\hbar)^3} \int_{-D}^{0} dE \int_{0}^{R_E} 4\pi R^2 G_{MB}(E, R)\, dR$$

$$= \frac{(2\pi M_{AB} k_B T)^{3/2}}{(2\pi\hbar)^3} \int_{R_0}^{0} 4\pi R^2\, dR \int_{V_0}^{0} G_{MB}(E, R)\, dE,$$

where D is the dissociation (ionization) energy of the lowest state and the Maxwell-Boltzmann distribution is

$$G_{MB}(E, R)\, dE = \frac{2}{\pi^{1/2}} \left[\frac{E - V(R)}{k_B T}\right]^{1/2} \exp\left(-\frac{E}{k_B T}\right) dE.$$

It is sometimes convenient to reverse the order of E, R-integration by taking the limits to be determined from $V_0 = \min[V(R), 0]$ and $V(R_0) = 0$.

Bound particles: Energy and angular momentum weights

The fraction of the number of states in the band $(E, E + dE; L, L + dL; L_z, L_z + dL_z)$ to the total number of states in the phase volume $d\mathbf{R}\, d\mathbf{p}$ of the system is

$$f_{E,L,L_z}(\mathbf{R}, \mathbf{p}) = \{\delta(H - E)\, dE\, \delta(|\mathbf{L}| - L)\, dL\, \delta(\mathbf{L} \cdot \hat{z} - L_z)\, dL_z\},$$

where the Hamiltonian H, angular momentum $\mathbf{L} = \mathbf{R} \times \mathbf{p}$, and projection of the angular momentum on z-axis $\mathbf{L} \cdot \hat{z}$ are conserved quantities and specify the state (E, L, L_z) of the system. The weight of states within interval $dR\, dE\, dL\, dL_z$ is

$$g_{AB}(R, E, L, L_z) = \frac{R^2\, dR}{(2\pi\hbar)^3} \int f_{E,L,L_z}(\mathbf{R}, \mathbf{p})\, d\hat{\mathbf{R}}\, d\mathbf{p},$$

which reduces upon integration to

$$g_{AB}(R; E, L, L_z) = (2\, dR/\dot{R})(dE/h)(dL\, dL_z/\hbar^2).$$

Integration over R provides the weight of states in the interval $dE\, dL\, dL_z$ to be

$$g_{AB}(E, L, L_z) = \frac{dE}{\hbar\omega(E, L)} \frac{dL\, dL_z}{\hbar^2}.$$

The classical period $2\pi/\omega$ for radial motion for a round trip between the pericenter R_- and apocenter R_+ is

$$\tau(E, L) = \oint \frac{dR}{\dot{R}} = (2M)^{1/2} \int_{R_-(L)}^{R_+(L)} [E - V(R) - L^2/2MR^2]^{-1/2}\, dR.$$

For bound levels, $dE = \hbar\omega(E, L) = h/\tau(E, L)$, $dL = \hbar$, and $dL_z = \hbar$ are the spacings among neighboring levels, according to the Bohr-Sommerfeld correspondence, and g_{AB} is then unity. For a state in the energy continuum $R_+(L) \to \infty$. The weight of states in the $dR\, dE\, dL$ interval is

$$g_{AB}(R; E, L) = (2\,dR/\dot{R})(dE/h)(2L\,dL/\hbar^2).$$

Upon R-integration, the weight of the $dE\, dL$ interval is

$$g_{AB}(E, L) = \frac{dE}{\hbar\omega(E, L)} \frac{2L\,dL}{\hbar^2}.$$

The maximum L for a specified E and R is determined from $L_0^2 = 2M_{AB}[E - V(R)]R^2$. Upon L-integration, the weight of the $dR\, dE$ interval is

$$g_{AB}(R; E) = \frac{(2\pi M_{AB} k_B T)^{3/2}}{(2\pi\hbar)^3} \times 4\pi R^2 \left\{ \frac{2}{\pi^{1/2}} \left[\frac{E - V(R)}{k_B T} \right]^{1/2} \right\} dR\, dE,$$

as before. The density $\rho(\Gamma)$ of states $\Gamma \equiv (R, E, L, L_z)$ is defined as the statistical weight $g_{AB}(\Gamma) = \rho(\Gamma)\, d\Gamma$ per unit interval $d\Gamma$. When A, B and AB are in equilibrium at temperature T, the ratio of the number densities $n_{AB} = g_{AB}/\mathcal{V}$ of bound pairs AB with parameters in the interval $\Gamma, \Gamma + d\Gamma$ to the number densities $n_A = g_A/\mathcal{V}$ and $n_B = g_B/\mathcal{V}$ of free atoms A and B is

$$\left[\frac{n_{AB}(\Gamma) d\Gamma}{n_A n_B} \right]_{\text{equil}} = \frac{\omega_{AB}}{\omega_A \omega_B} \left\{ \frac{\mathcal{Z}_t(A + B)\mathcal{V}}{\mathcal{Z}_t(A)\mathcal{Z}_t(B)\sigma} \right\} \rho(\Gamma) \exp(-E/k_B T)\; d\Gamma$$

$$= \frac{\omega_{AB}}{\omega_A \omega_B} \left\{ \frac{(2\pi\hbar)^3}{(2\pi M_{AB} k_B T)^{3/2} \sigma} \right\} \rho(\Gamma) \exp(-E/k_B T)\; d\Gamma,$$

where M_{AB} is the reduced mass. For example,

$$\left[\frac{n_{AB}(R, E) dR dE}{n_A n_B} \right]_{\text{equil}} = \left[\frac{\omega_{AB}}{\omega_A \omega_B \sigma} \right] 4\pi R^2 G_{MB}(E, R)\, dR\, dE$$

For e-ion pairs, $M_{AB} = m_e$ and $\sigma = 1$. For $A - A$ pairs, $\sigma = 2$.

Classical AB-partition function

The dissociation energy D is $\epsilon_{vJ} - E$, where E is measured with respect to zero at the dissociation limit and the rovibrational energy ϵ_{vJ} is with respect to the $v = 0, J = 0$ ground-state energy. The classical internal partition function for the rovibrational states of a diatomic molecule AB is then

$$\mathcal{Z}_{vr}(AB) = \sigma^{-1} \int \rho(R, E, L) \exp(-\epsilon_{vJ}/k_B T)\, dR\, dE\, dL,$$

where $L = J\hbar$. On integrating over L,

$$\mathcal{Z}_{vr}(AB) \equiv \sigma^{-1} \exp(-D/k_BT) \int_{-D}^{0} \rho(E) \exp\left(-\frac{E}{k_BT}\right) dE$$

$$= \sigma^{-1} \exp(-D/k_BT) \int_{R_0}^{\infty} dR \int_{-V_0}^{0} \rho(R, E) \exp(-E/k_BT)\, dE$$

$$= \frac{(2\pi M_{AB} k_B T)^{3/2}}{(2\pi\hbar)^3 \sigma} \exp(-D/k_BT) \int_{R_0}^{\infty} 4\pi R^2\, dR \int_{V_0}^{0} G_{MB}(R, E)\, dE,$$

where the limits are determined from $V_0 = \min[V(R), 0]$ and $V(R_0) = 0$. The number of bound states AB with internuclear separation R is the integral

$$\int_{-V_0}^{0} G_{MB}(R, E)\, dE = \left[\Phi(X^{\frac{1}{2}}) - \frac{2}{\pi^{1/2}} X^{\frac{1}{2}} \exp(-X)\right]$$

of the Maxwell-Boltzmann distribution, where $X = V/k_BT$ and $\Phi(x) = (2/\sqrt{\pi}) \int_0^x \exp(-y^2)\, dy$ is the probability integral. See Refs. [5] and [6] for classical statistical weights and partition functions in atomic and molecular physics.

6.4.5. Association/dissociation equation

On including the translational partition function $\mathcal{Z}_t(A + B)$ for the motion of the center of mass, and the electronic weight $\omega(AB)$, the total partition function of AB is

$$\mathcal{Z}(AB) = \mathcal{Z}_t(A + B)\omega_{AB}\mathcal{Z}_{vr}(AB) = n_{AB}\mathcal{V},$$

where $n_{(AB)}$ is the number density of molecules in volume $\mathcal{V}$. In terms of the various partition functions, the ratio of number densities of dissociated to bound states is

$$\left(\frac{n_A n_B}{n_{AB}}\right)_{\text{equil}} = \frac{\omega_A \omega_B}{\omega_{AB}} \left\{\frac{\mathcal{Z}_t(A)\mathcal{Z}_t(B)}{\mathcal{Z}_t(A + B)\mathcal{V}\mathcal{Z}_{vr}(AB)}\right\} \exp\left(-D/k_BT\right),$$

where D is the (positive) energy for dissociation of the molecule AB. Using the classical $\mathcal{Z}_{vr}(AB)$,

$$\left(\frac{n_{AB}}{n_A n_B}\right)_{\text{equil}} = \frac{\omega_{AB}}{\omega_A \omega_B \sigma} \int_{R_0}^{\infty} 4\pi R^2\, dR \int_{V_0}^{0} G_{MB}(R, E)\, dE.$$

6.4.6. Saha's ionization equations

The Saha equations are particular examples of the general association-dissociation equation above.

Formula 1

The ionization process

$$A^{z+}(i) \rightleftharpoons A^{(z+1)+}(1) + e^-$$

is maintained in thermodynamic equilibrium by electron-ion, electron-electron, and ion-ion collisions. The equilibrium ratio of the RHS number density $N_1^{(z+1)}$ of ions $A^{(z+1)+}(1)$ with net positive charge $(z+1)+$ in their ground level 1 to the LHS number density N_i^z of (recombined) ions $A^{z+}(i)$ with net charge $z^+ = 0, 1, 2, \ldots$ in electronic level i is provided by the Saha equilibrium distribution,

$$\left[\frac{n_e N_1^{(z+1)}}{N_i^z}\right]_{\text{equil}} = \frac{2\omega_1^{(z+1)}}{\omega_i^z}\frac{(2\pi m_e k_B T)^{3/2}}{h^3}\exp[-I_i/k_B T],$$

appropriate to each stage $z+1$ of the ionization process. Here, n_e is the number density of free electrons, I_i is the (positive) ionization energy of the lower energy recombined ion $A^{z+}(i)$, and ω_j^z denotes the statistical weight of an ion of net charge z in level j.

Example: Since $\omega_n^z = 2n^2$ for a recombined hydrogenic ion with electronic level n, and $\omega_1^{(z+1)} = 1$ for the bare ion of charge $z+1$, the equilibrium ratio of the densities is

$$\left[\frac{N_n^z}{n_e N^{(z+1)}}\right]_{\text{equil}} = n^2 \frac{h^3}{(2\pi m_e k_B T)^{3/2}} \exp(I_n/k_B T)$$

$$= 4.2 \times 10^{-16} T^{-3/2} n^2 \exp[157,890\,(z+1)^2/n^2\, T(K)],$$

where $\varepsilon_0 = e^2/a_0 = 27.211\,3961(81)$ eV is the atomic unit of energy and T is in degrees Kelvin. The distribution displays a minimum where $(z+1)^2\varepsilon_0/(2n_*^2) = k_B T$. The level n_* therefore acts as a bottleneck to an electron-ion recombination process.

Formula 2

Generalization to the equilibrium process

$$A^{z+}(i) \rightleftharpoons A^{(z+1)}(j) + e^-$$

involving the number density $N_j^{(z+1)}$ of ions $A^{(z+1)+}(j)$ in specified electronic level j is then obtained via the use of the appropriate Boltzmann factors to yield

$$\frac{n_e N_j^{(z+1)}}{N_i^z} = \frac{2\omega_j^{(z+1)}}{\omega_i^z}\frac{(2\pi m_e k_B T)^{3/2}}{h^3}\exp-[(I_i + E_{j1}^{z+1})/k_B T],$$

where $E_{j1}^z = E_j^z - E_1^z$ is the (positive) excitation energy measured from the ground state $k = 1$ of the ion $A^{(z+1)+}$.

Formula 3

It is also useful to relate the total densities N^z and $N^{(z+1)}$ of the same species in two subsequent stages z and $(z+1)$ of ionization by

$$\left[\frac{n_e N^{(z+1)}}{N^z}\right]_{\text{equil}} = \frac{2\mathcal{Z}_{el}^{(z+1)}(T)}{\mathcal{Z}_{el}^z(T)} \frac{(2\pi m_e k_B T)^{3/2}}{h^3} \exp[-I_1^z/k_B T],$$

where $\mathcal{Z}_{el}^z(T) = \sum_i \omega_i^z \exp(-E_{i0}^z/kT)$ is the electronic partition function of the ion A^{z+} and I_1^z is the ionization energy of the ground state of the recombined ion A^{z+}. This is the Saha distribution of atoms over successive stages of ionization, without regard to electron configuration.

It is worth noting that the Saha distribution differs from the general association/dissociation distribution in that most of the recombined species (atoms) are in the ground state, in contrast to the many rovibrational states populated at T and described by the presence of the rovibrational partition function $\mathcal{Z}_{vr}$.

6.4.7. Macroscopic detailed balance

Recombination/ionization

The ionization and recombination rate coefficients, α_I and α_R, respectively, for the forward and reverse processes of the collisional reaction

$$e^- + A^{(z+1)+}(j) + M \underset{k_I(i)}{\overset{\alpha_R(j)}{\rightleftharpoons}} A^{z+}(i) + M,$$

where M is any third body, are defined via the equation

$$\frac{dn_i(t)}{dt} = \alpha_R(j) n_e(t) N_j^{(z+1)+}(t) - k_I(i) N_i^{z+}(t)$$

for rate of growth of the density $n_i(t)$ of the recombined species. The density of ions in stage z and level i is N_i^{z+}. Under equilibrium conditions, $dn_i/dt = 0$ and the Saha equation can then be used to provide

$$\frac{\alpha_R(j)}{k_I(i)} = \left[\frac{\omega_i^z}{2\omega_j^{(z+1)}} \frac{h^3}{(2\pi m_e k_B T)^{3/2}} \exp[(I_i + E_{j1}^{z+1})/k_B T]\right],$$

the detailed balance relation between the recombination rate $\alpha_R(j)$ and the ionization rate $k_I(i)$. For ground state $A^{(z+1)+}(j=1)$, E_{j1}^{z+1} is zero.

Association/dissociation

The rate coefficients k_{A} and k_{D} for association and dissociation in the nonequilibrium process

$$A + B \underset{k_{\mathrm{D}}}{\overset{e^{k_{\mathrm{A}}}}{\rightleftharpoons}} AB$$

and defined by the rate equation

$$\frac{dn_{AB}(t)}{dt} = k_A n_A(t) n_B(t) - k_D n_{AB}(t)$$

are interconnected by

$$\left(\frac{n_A n_B}{n_{AB}}\right)_{\rm equil} = \frac{k_D(T)}{k_A(T)} \equiv K_{\rm equil},$$

where the equilibrium constant $K_{\rm equil}$, in terms of the various partition functions, is

$$\left(\frac{n_A n_B}{n_{AB}}\right)_{\rm equil} = \frac{\omega_A \omega_B}{\omega_{AB}} \left\{\frac{(2\pi M_{AB} k_B T)^{3/2}}{(2\pi\hbar)^3 \mathcal{Z}_{vr}}\right\} \exp\left(-D/k_B T\right),$$

where D is the (positive) energy for dissociation of the molecule AB. When the densities have their equilibrium ratio, $dn_{AB}(t)/dt$ vanishes.

6.4.8. Planck's equilibrium distribution

The energy absorbed within the frequency band $[\nu, \nu + d\nu]$ and radiated by unit volume of a perfect absorber/emitter, i.e., a blackbody, maintained at temperature T, is the energy density

$$\tilde{\rho}_E(\nu)\, d\nu = \left(\frac{8\pi h\nu^3}{c^3}\right) \left[\exp h\nu/k_B T - 1\right]^{-1} d\nu.$$

This is Planck's law for a radiation field in equilibrium at temperature T. In the high limit $h\nu \gg k_B T$ of photon energy, $\tilde{\rho}_E(\nu) \to (8\pi h\nu^3/c^3) \exp -h\nu/k_B T$, which is Wien's law. In the opposite limit $h\nu \ll k_B T$ for low energy photons, $\tilde{\rho}_E(\nu) \to (8\pi \nu^2 k_B T/c^3)$, which is Rayleigh-Jean's law. The above spectral energy distribution can be decomposed as

$$\tilde{\rho}_E(\nu)\, d\nu = (h\nu)\, \tilde{n}_\nu\, \rho(\nu)\, d\nu.$$

The number of photons in unit volume and in the frequency band $(\nu, \nu + d\nu)$ with propagation vectors $\mathbf{k}_\nu$ pointing in all directions is $\tilde{n}_\nu \rho(\nu)\, d\nu$. The average number of photons in this frequency mode is

$$\tilde{n}_\nu = \left[\exp h\nu/k_B T - 1\right]^{-1},$$

the photon occupation number for radiation in thermal equibrium. The number of modes per unit volume in the frequency band $d\nu$ about ν is

$$\rho(\nu)\, d\nu = \frac{8\pi\nu^2}{c^3}\, d\nu,$$

which always holds, irrespective of equilibrium. The total amount of electromagnetic energy of all frequencies is

$$U(T) = \int_0^\infty \tilde{\rho}_E(\nu)\, d\nu = \frac{\pi}{15} \frac{(k_B T)^4}{(\hbar c)^3}$$

under equilibrium conditions per unit volume. This relation provides the Stefan-Boltzmann law $I = \sigma T^4$ for the energy intensity I or emitted power $(c/4)U(T)$ radiated per unit area.

6.4.9. Boltzmann equation

The one-particle distribution function $f(\mathbf{r}, \mathbf{p}, t)$ is defined so that $f\, d\mathbf{r}\, d\mathbf{p}$ is the probability that the particle is located at time t within the interval $[\mathbf{r}, \mathbf{r} + \mathbf{dr}]$ and with momentum within the interval $[\mathbf{p}, \mathbf{p} + \mathbf{dp}]$. The distribution $f(\mathbf{r}, \mathbf{p}, t)$ in phase-space evolves in time according to Boltzmann's equation,

$$\frac{\partial f}{\partial t} + \frac{\mathbf{p}}{m} \cdot \nabla_{\mathbf{r}} f + \mathbf{F} \cdot \nabla_{\mathbf{p}} f = \left(\frac{\partial f}{\partial t}\right)_{\text{collisions}}.$$

Streaming term

The LHS describes how f changes by virtue of the independent (collisionless) motion of the ensemble of test particles, which move from phase-space element $(\mathbf{r}, \mathbf{p})$ at time t to element $(\mathbf{r} + \mathbf{p}t/m, \mathbf{p} + \mathbf{F}t)$ at time $t + dt$. This is accomplished by changes that occur with t at fixed $\mathbf{r}$ and $\mathbf{p}$ (first term), with free motion of test particles where some move into $\mathbf{dr}$ and some move out of $\mathbf{dr}$ at fixed momentum $\mathbf{p}$ (second term), and with the action of an external force $\mathbf{F}$ at fixed $\mathbf{r}$ that alters $\mathbf{p}$ (third term). For conservative forces, $\mathbf{F} = -\nabla V_{\mathbf{r}}(\mathbf{r})$.

Collision term

The RHS is the change in f due to collisions and is

$$\left(\frac{\partial f(\mathbf{r}, \mathbf{p}, t)}{\partial t}\right)_{\text{collisions}} = \sum_j \int d\mathbf{P}_j \int \left[f(\mathbf{r}, \mathbf{p}', t) f_j(\mathbf{r}, \mathbf{P}'_j, t) - f(\mathbf{r}, \mathbf{p}, t) f_j(\mathbf{r}, \mathbf{P}_j, t)\right]$$

$$v_j \left\{\frac{d\sigma(v_j, \psi, \phi)}{d\Omega}\right\} d\Omega.$$

The input term is the frequency at which test particles enter element $\mathbf{dp}\,\mathbf{dr}$ from element $\mathbf{dp}'\,\mathbf{dr}$ by collision at relative speed v_j with species j with momentum $\mathbf{P}'_j$. The momentum distribution of species j is $f_j(\mathbf{r}, \mathbf{P}_j, t)\, d\mathbf{P}_j$. The output term is the collisional frequency for the reversed transition. For conservative forces $\mathbf{F} = -\nabla_{\mathbf{r}} V(\mathbf{r})$, both sides of Boltzmann's equation separately vanish for the Maxwell-Boltzmann equilibrium distribution

$$f_{MB}(\mathbf{r}, \mathbf{p}) = (2\pi M_{AB} k_B T)^{-3/2} \exp -(p^2/2M_{AB} k_B T) \exp\left\{-\frac{V(r)}{k_B T}\right\}.$$

6.5. MACROSCOPIC RATE COEFFICIENTS

6.5.1. Scattering rate

A distribution $f_A(\mathbf{v}_A)$ of $N_A(t)$ test particles (cm^{-3}) of species A in a beam collisionally interacts with a distribution $f_B(\mathbf{v}_B)$ of $N_B(t)$ field particles of species B. Collisions with

B will scatter A out of the beam at the loss rate (cm^{-3} s^{-1})

$$\frac{dN_A}{dt} = -kN_A(t)N_B(t) = -\nu_B N_A(t).$$

The macroscopic rate coefficient k(cm^3 s^{-1}) for elastic collisions between the ensembles A and B is

$$k(\text{cm}^3\ \text{s}^{-1}) = \int f_A(\mathbf{v}_A)\, d\mathbf{v}_A \int [v\,\sigma(v)]\, f_B(\mathbf{v}_B)\, d\mathbf{v}_B$$

in terms of the integral cross section $\sigma(v)$ for $A - B$ elastic scattering at relative speed $v = |\mathbf{v}_A - \mathbf{v}_B|$. The microscopic rate coefficient is $v\sigma(v)$. The frequency ν_B (s^{-1}) of collision between one test particle A with N_B field particles is kN_B.

The rate coefficient for elastic scattering between two species with nonisothermal Maxwellian distributions is then

$$k(\text{cm}^3\ \text{s}^{-1}) = \tilde{v}_{AB} \int_0^\infty \sigma(\tilde{\epsilon}_{\text{rel}})\, \tilde{\epsilon}_{\text{rel}}\, \exp(-\tilde{\epsilon}_{\text{rel}})\, d\tilde{\epsilon}_{\text{rel}},$$

where

$$\tilde{v}_{AB} = \left[\frac{8k_B}{\pi}\left(\frac{T_A}{M_A} + \frac{T_B}{M_B}\right)\right]^{1/2}$$

and

$$\tilde{\epsilon}_{\text{rel}} = \frac{1}{2}\frac{M_A M_B v^2}{k_B(M_A T_B + M_B T_A)}.$$

For isothermal distributions $T_A = T_B = T$, the rate is

$$k(T) = \langle v_{AB}\rangle \int_0^\infty \sigma(\epsilon)\, \epsilon\, \exp(-\epsilon)\, d\epsilon\ (\text{cm}^3\ \text{s}^{-1}),$$

where $\epsilon = \frac{1}{2}M_{AB}v^2/k_B T$ and $\langle v_{AB}\rangle = (8k_B T/\pi M_{AB})^{1/2}$. The rate of collisions of electrons A at temperature T_e with a gas of heavy particles B at temperature T_B is

$$k(T_e) = \langle v_e\rangle \int_0^\infty \sigma(\epsilon_e)\, \epsilon_e\, \exp(-\epsilon_e)\, d\epsilon_e\ (\text{cm}^3\ \text{s}^{-1}),$$

where $\epsilon_e = \frac{1}{2}m_e v^2/k_B T_e$ and $\langle v_e\rangle = (8k_B T_e/\pi m_e)^{1/2}$.

6.5.2. Energy transfer rate

Each of the species A transfers energy $\mathcal{E}_{AB}$ to each species B. The amount of energy transferred in unit time from ensemble A to unit volume of ensemble B is

$$\frac{d}{dt}[N_A\langle\mathcal{E}_{AB}\rangle] = -k_E N_A(t)N_B(t) = -\nu_{EB}N_A(t),$$

where the macroscopic rate coefficient k_E (energy cm^3 s^{-1}) for the averaged energy loss $\langle \mathcal{E}_{AB} \rangle$ is

$$k_E = \int f_A(\mathbf{v}_A)\, d\mathbf{v}_A \int f_B(\mathbf{v}_B)\, d\mathbf{v}_B \int \mathcal{E}_{AB}(\mathbf{v}_A, \mathbf{v}_B; \psi, \phi) v \left(\frac{d\sigma}{d\Omega}\right) d\Omega.$$

The amount of energy lost in unit time, the energy-loss frequency, is $\nu_{EB} = k_E N_B(t)$. The energy-loss rate coefficient for two-temperature Maxwellian distributions is

$$k_E(T_A, T_B) = \frac{2M_A M_B}{(M_A + M_B)^2} k_B (T_A - T_B) \tilde{v}_{AB} \int_0^\infty \sigma_D(\tilde{\epsilon}_{\rm rel})(\tilde{\epsilon}_{\rm rel})^2 \exp(-\tilde{\epsilon}_{\rm rel})\, d\tilde{\epsilon}_{\rm rel},$$

where $\sigma_D(\tilde{\epsilon}_{\rm rel})$ is the momentum transfer cross section at reduced energy $\tilde{\epsilon}_{\rm rel}$. For isothermal distributions, $T_A = T_B$ and the energy rate coefficient k_E of course then vanishes.

6.5.3. Transport cross sections and collision integrals

Transport cross sections are defined for integral $n \geq 1$ as

$$\sigma^{(n)}(E) = 2\pi \left[1 - \frac{1 + (-1)^n}{2(n+1)}\right]^{-1} \int_{-1}^{+1} \left[1 - \cos^n \theta\right] \frac{d\sigma}{d\Omega}\, d(\cos\theta).$$

The diffusion and viscosity cross sections are given by the transport cross sections $\sigma^{(1)}$ and $\frac{2}{3}\sigma^{(2)}$, respectively.

Collision integrals for integral s are defined as

$$\Omega^{(n,s)}(T) = \left[(s+1)!(k_B T)^{s+2}\right]^{-1} \int_0^\infty \sigma^{(n)}(E)\, E^{s+1} \exp(-E/k_B T)\, dE$$

$$= [(s+1)!]^{-1} \int_0^\infty \sigma^{(n)}(\epsilon)\, \epsilon^{s+1} \exp -\epsilon\, d\epsilon,$$

where $\epsilon = \frac{1}{2} M_{AB} v^2 / k_B T$. The external factors are chosen so that the above expressions for $\sigma^{(n)}$ and $\Omega^{(n,s)}$ reduce to πd^2 for classical rigid spheres of diameter d. The rate coefficient k(cm^3 s^{-1}) for scattering can then be expressed in terms of the collision integral, as $= \tilde{v}_{AB}\Omega^{(0,0)}$. The amount of energy lost per cm^3 per second by collision can be expressed in terms of $\Omega^{(1,1)}$. Tables of transport cross sections and collision integrals for (n, 6, 4) ion -neutral interactions are available. [7, 8]

Chapman-Enskog mobility formula

When ions move under equilibrium conditions in a gas and an external electric field, the energy gained from the electric field $\mathbf{E}$ between collisions is lost to the gas upon collision so that the ions move with a constant drift speed $\mathbf{v}_d = K\mathbf{E}$. The mobility K of ions of charge e in a gas of density N is given in terms of the collision integral by the Chapman-Enskog formula [7]

$$K = \frac{3e}{16N} \left(\frac{\pi}{2Mk_B T}\right)^{1/2} \left[\Omega^{(1,1)}(T)\right]^{-1}.$$

6.6. QUANTAL TRANSITION RATES AND CROSS SECTIONS

6.6.1. Microscopic rate of transitions

In the general elastic/inelastic collision process

$$A(\alpha) + B(\beta) \rightarrow A(\alpha') + B(\beta'),$$

the external scattering or deflection of a beam of projectile particles A (electrons, ions, atoms) by target particles B (atoms, molecules) is accompanied by transitions (electronic, vibrational, rotational) within the internal structure of either or both collision partners. For a beam with incident momentum $\mathbf{p}_i = \hbar\mathbf{k}_i$ in the range $(\mathbf{p}_i, \mathbf{p}_i + d\mathbf{p}_i)$ or directed energy $\mathbf{E}_i \equiv (E_i, \hat{\mathbf{p}}_i)$ in the range $(\mathbf{E}_i, \mathbf{E}_i + d\mathbf{E}_i)$, the translational states representing the $A - B$ relative or external motion undergo free-free transitions $(\mathbf{E}_i, \mathbf{E}_i + d\mathbf{E}_i) \rightarrow (\mathbf{E}_f, \mathbf{E}_f + d\mathbf{E}_f)$ within the translational continuum, whereas the structured particles undergo bound-bound (excitation, deexcitation, excitation transfer) or bound-free (ionization, dissociation) transitions $i \equiv (\alpha, \beta) \rightarrow f \equiv (\alpha', \beta')$ in their internal electronic, vibrational, or rotational structure. The transition frequency (s^{-1}) for the above collision is

$$\frac{dW_{if}}{d\hat{\mathbf{p}}_f}(s^{-1}) = \frac{2\pi}{\hbar}\frac{1}{g_i}\sum_{i,f} |\, V_{fi}\,|^2 \rho_f(E_f),$$

which is an average over the g_i initial degenerate internal states i and a sum over all g_f final degenerate internal states f of the isolated systems A and B. It is therefore the probability per unit time for scattering from a specified $\mathbf{E}_i$-(external) continuum state into unit solid angle $d\hat{\mathbf{p}}_f$ accompanied by a transition from any one of the g_i-initial states (α, β) to all final internal states (α', β') of degeneracy g_f and to all final translational states $\rho_f(E_f)\,dE_f$ of relative motion consistent with energy conservation. The double summation $\sum_{i,f}$ is over the g_i initial and g_f final internal states of A and B with total energy ϵ_i and ϵ_f, respectively.

Check: The dimension of $[|\, V_{ij}\,|^2\rho]$ is E, $[\hbar] = Et$ so that $dW_{if}/d\hat{\mathbf{p}}_f$ indeed has the correct dimension of t^{-1}.

Interaction matrix element

The matrix element

$$V_{fi} = \langle N_f\Phi_f|\, V(\mathbf{r}_A, \mathbf{r}_B, \mathbf{R})\, |N_i\Psi_i^+\rangle_{\mathbf{r},\mathbf{R}} = V_{if}^*$$

is an integration over the internal coordinates $\mathbf{r} \equiv \mathbf{r}_A, \mathbf{r}_B$ for the particles of the structures A and B and over the channel vector $\mathbf{R}$ for $A-B$ relative motion. The matrix element of the mutual electrostatic interaction $V(\mathbf{r}_A, \mathbf{r}_B, \mathbf{R})$ couples the eigenfunction $N_i\Psi_i^+(\mathbf{R}, \mathbf{r}_A, \mathbf{r}_B)$ of $[\hat{H}_{\rm rel} + \hat{H}_{\rm int} + V]$ for the complete collision system for all $\mathbf{R}$ to the final $R \rightarrow \infty$ asymptotic state $N_f\Phi_f(\mathbf{R}, \mathbf{r}_A, \mathbf{r}_B)$, which is an eigenfunction only of the unperturbed Hamiltonian $[\hat{H}_{\rm rel} + \hat{H}_{\rm int}]$. The wave function

$$\Psi_i^+(\mathbf{R}, \mathbf{r}_A, \mathbf{r}_B) = \sum_j \Phi_j^+(\mathbf{R})\psi_\alpha(\mathbf{r}_A)\phi_\beta(\mathbf{r}_B) \equiv \sum_j \Phi_j^+(\mathbf{R})\psi_j^{\rm int}(\mathbf{r})$$

for the full collision system with Hamiltonian $\hat{H}_{\text{rel}} + \hat{H}_{\text{int}} + V$ tends at asymptotic R to

$$\Psi_i^+ \sim \sum_j \left[e^{i\mathbf{k}_j \cdot \mathbf{R}} \delta_{ij} + f_{ij}(\theta, \phi) \frac{e^{ik_j R}}{R} \right] \psi_j^{\text{int}}(\mathbf{r}),$$

which represents an incoming plane wave of unit amplitude in the incident elastic channel i and an outgoing spherical waves of amplitude f_{ij} in all channels j, including i. The Kroneker symbol means $\delta_{ij} = 1, i = j$ and $\delta_{ij} = 0, i \neq j$. The final state at infinite separation R is

$$\Phi_f(\mathbf{R}, \mathbf{r}_A, \mathbf{r}_B) = e^{i\mathbf{k}_f \cdot \mathbf{R}} \psi_{\alpha'}(\mathbf{r}_A) \phi_{\beta'}(\mathbf{r}_B) \equiv e^{i\mathbf{k}_f \cdot \mathbf{R}} \psi_f^{\text{int}}(\mathbf{r}),$$

which is an eigenfunction only of $\hat{H}_{\text{rel}} + \hat{H}_{\text{int}}$. The plane wave of unit amplitude describes the external relative motion with Hamiltonian $\hat{H}_{\text{rel}}$, and $\phi_{\alpha'}(\mathbf{r}_A)\psi_{\beta'}(\mathbf{r}_B)$ describes the isolated internal normalized eigenstates of A and B with internal Hamiltonian $\hat{H}_{\text{int}}$. The factors $N_{i,f}$ provide the possibility of having translational (scattering) states with arbitrary amplitudes that are not necessarily unity.

Transition operator

The interaction matrix element can also be written as

$$V_{fi} = \langle N_f \Phi_f | \hat{T} | N_i \Phi_i \rangle,$$

where the transition operator, $\hat{T}$ is defined by $\hat{T}\Phi = V\Psi$. The transition operator $\hat{T}$ therefore couples states that are eigenfunctions of the same unperturbed Hamiltonian $\hat{H}_{\text{rel}} + \hat{H}_{\text{int}}$, in contrast to V, which couples states Ψ_i^+ and Φ_f belonging to different Hamiltonians.

6.6.2. Detailed balance between rates

The frequency (number per second) of $i \to f$ transitions from all g_i degenerate initial internal states and from the $\rho_i \, d\mathbf{E}_i$ initial external translational states is equal to the reverse frequency from the g_f degenerate final internal states and the $\rho_f \, d\mathbf{E}_f$ final external translational states. The detailed balance relation between the forward and reverse frequencies is therefore

$$[g_i \rho_i \, dE_i \, d\hat{\mathbf{p}}_i] \left(\frac{dW_{if}}{d\hat{\mathbf{p}}_f} \right) d\hat{\mathbf{p}}_f = [g_f \rho_f \, dE_f \, d\hat{\mathbf{p}}_f] \left(\frac{dW_{fi}}{d\hat{\mathbf{p}}_i} \right) d\hat{\mathbf{p}}_i,$$

since $V_{if} = V_{fi}^*$. From energy conservation $\epsilon_i + E_i = \epsilon_f + E_f$, $dE_i = dE_f$. The differential frequencies

$$\frac{dR_{if}}{d\hat{\mathbf{p}}_f} \equiv g_i \rho_i \frac{dW_{if}}{d\hat{\mathbf{p}}_f} = g_f \rho_f \frac{dW_{fi}}{d\hat{\mathbf{p}}_i} \equiv \frac{dR_{fi}}{d\hat{\mathbf{p}}_i}$$

for the forward and reverse transitions, $i \rightleftharpoons f$, are therefore equal.

6.6.3. Energy density of continuum states

The continuum wave functions $\phi_{\mathbf{p}}(\mathbf{R})$ for the states of the $A - B$ relative motion satisfy the orthonormality condition

$$\int \rho(\mathbf{E})\, d\mathbf{E} \int \phi_{\mathbf{p}}(\mathbf{R})\phi^*_{\mathbf{p}'}(\mathbf{R})\, d\mathbf{R} = 1.$$

The number of translational states per unit volume $d\mathbf{R}$ with directed energies $\mathbf{E} \equiv (E, \hat{\mathbf{p}})$ in the range $[\mathbf{E}, \mathbf{E}+d\mathbf{E}]$ is $\rho(\mathbf{E})\, d\mathbf{E}$. This orthonormality condition for continuum states is analogous to the condition $\sum_j \mid \langle \phi_j|\phi_i\rangle \mid^2 = 1$ for bound states. For plane waves, $\phi_{\mathbf{p}}(\mathbf{R}) = N \exp(i\mathbf{p}\cdot\mathbf{R}/\hbar)$ so that

$$\begin{aligned} \langle \phi_{\mathbf{p}'} \mid \phi_{\mathbf{p}} \rangle &= \mid N \mid^2 (2\pi\hbar)^3 \delta(\mathbf{p} - \mathbf{p}') = \mid N \mid^2 (2\pi)^3 \delta(\mathbf{k} - \mathbf{k}') \\ &= \mid N \mid^2 \frac{(2\pi\hbar)^3}{mp} \delta(\mathbf{E} - \mathbf{E}') = \frac{1}{\rho(\mathbf{E})} \delta(\mathbf{E} - \mathbf{E}'). \end{aligned}$$

Note, irrespective of the method chosen to normalize the wave functions, that

$$\mid N \mid^2 \rho(\mathbf{E}) = \frac{mp}{(2\pi\hbar)^3}$$

always. The amplitude $\mid N \mid$ does however depend on the various methods adopted for normalization of continuum wave functions.

1. For momentum normalized states, $\langle \phi_{\mathbf{p}'} \mid \phi_{\mathbf{p}} \rangle = \delta(\mathbf{p} - \mathbf{p}')$, $\mid N \mid = (2\pi\hbar)^{-3/2}$ and the density of states $\rho(\mathbf{E}) = mp$.
2. For wave-vector normalized states, $\langle \phi_{\mathbf{p}'} \mid \phi_{\mathbf{p}} \rangle = \delta(\mathbf{k} - \mathbf{k}')$, $\mid N \mid = (2\pi)^{-3/2}$ and the density of states $\rho(\mathbf{E}) = (mp/\hbar^3)$.
3. For energy-normalized states, $\langle \phi_{\mathbf{p}'} \mid \phi_{\mathbf{p}} \rangle = \delta(\mathbf{E} - \mathbf{E}')$, $\rho(\mathbf{E}) = 1$, and $\mid N \mid = (mp/h^3)^{1/2}$.
4. For waves with unit amplitude, $\mid N \mid = 1$ and $\rho(\mathbf{E}) = (mp/h^3)$.

Note that $\langle \phi_{\mathbf{p}'} \mid \phi_{\mathbf{p}} \rangle \rho(\mathbf{E})\, d\mathbf{E}$ is dimensionless for all cases and yields unity for a single particle when integrated over all $\mathbf{E}$. The number of states in the phase-space element $d\mathbf{E}\, d\mathbf{R}$ is

$$dn = \mid N \mid^2 \rho(\mathbf{E})\, d\mathbf{E}\, d\mathbf{R} = d\mathbf{p}\, d\mathbf{R}/(2\pi\hbar)^3;$$

i.e., each translational state occupies a cell of phase volume $(2\pi\hbar)^3$. The density of states in the interval $[E, E+dE]$ is $\rho(E) = 4\pi\rho(\mathbf{E})$. The number of translational states per unit volume with energy in the scalar range $[E, E+dE]$ is

$$\mid N \mid^2 \rho(E)\, dE = \frac{2}{\pi^{1/2}} \frac{(2\pi m)^{3/2}}{h^3} E^{1/2}\, dE.$$

Check: The number of free particles with all momenta $\mathbf{p}$ in equilibrium with a gas bath of volume $\mathcal{V}$ at temperature T is the translational partition function $\mathcal{Z}_t$. Since the fraction of particles with energy E is $\exp(-E/k_BT)/\mathcal{Z}_t$, the Maxwell distribution

$$f_M(E)dE = \frac{|N|^2\mathcal{V}\rho(E)\,dE}{\mathcal{Z}_t}\exp-(E/k_BT)$$
$$= \frac{2}{\pi^{1/2}}(E/k_BT)^{1/2}\exp(-E/k_BT)\,d(E/k_BT)$$

is then recovered.

Current

Current is the number of particles crossing unit area in unit time. The current in a beam with directed energy $\mathbf{E}$ within the range $(\mathbf{E}, \mathbf{E}+d\mathbf{E})$ is

$$j\,d\mathbf{E} = v|N|^2\rho(\mathbf{E})\,d\mathbf{E} = (p^2/h^3)\,d\mathbf{E}.$$

The current per unit $d\mathbf{E}$ is the current density $j = (p^2/h^3)$. The quantal expression for current

$$\mathbf{J} \equiv \frac{\hbar}{2mi}[\phi_{\mathbf{p}}^*\nabla\phi_{\mathbf{p}} - \phi_{\mathbf{p}}\nabla\phi_{\mathbf{p}}^*],$$

when applied to the plane wave $\phi_{\mathbf{p}} = N\exp(i\mathbf{p}\cdot\mathbf{R}/\hbar)$, gives $\mathbf{J} = |N|^2\mathbf{v}$. The current in a $(\mathbf{E}, \mathbf{E}+d\mathbf{E})$-beam of plane waves is then $\mathbf{J}[\rho(\mathbf{E})\,d\mathbf{E}]$ so that the current density is $j(\mathbf{E}) = J\rho(\mathbf{E}) = |N|^2v\rho(\mathbf{E})$, as above.

6.6.4. Inelastic cross sections

The differential cross section $d\sigma_{if}/d\hat{\mathbf{p}}_f$ for $i \to f$ transitions from any one of the g_i initial states is defined as $[dR_{if}/d\hat{\mathbf{p}}_f]/g_i j_i$, the transition rate per unit incident current. Since current is the number of particles crossing unit area in unit time, the cross section is therefore the effective area presented by the target toward $i \to f$ internal transitions in the internal structures of the collision partners, which are scattered into unit solid angle $d\hat{\mathbf{p}}_f$ about direction $\hat{\mathbf{p}}_f$ in the CM frame.

Basic expression for cross section

The differential cross section for

$$A + B \to C + D$$

collisions is therefore defined as

$$\frac{d\sigma_{if}}{d\hat{\mathbf{p}}_f} = \frac{1}{g_i j_i}\frac{dR_{if}}{d\hat{\mathbf{p}}_f} = \frac{2\pi}{\hbar}\left(\frac{\rho_i\rho_f}{j_i}\right)\frac{1}{g_i}\sum_{i,f}|\langle N_f\Phi_f \mid V(\mathbf{r}_A, \mathbf{r}_B, \mathbf{R})|N_i\Psi_i^+\rangle|^2,$$

which is an average over the g_i initial internal degenerate states and a sum over the g_f final degenerate states.

Since $j_i = |N_i|^2 v_i \rho_i = p_i^2/h^3$, an alternative form [9] for the cross section is

$$\frac{d\sigma_{if}}{d\hat{\mathbf{p}}_f} = \frac{2\pi}{\hbar v_i}\left(\frac{\rho_f}{g_i}\right)\sum_{i,f} |\langle N_f \Phi_f \mid V(\mathbf{r}_A, \mathbf{r}_B, \mathbf{R})|\Psi_i^+\rangle|^2.$$

6.6.5. Detailed balance between cross sections

When cast in terms of cross sections, the detailed balance relation in Sec. 6.6.2 is

$$g_i j_i(\mathbf{E}_i)\frac{d\sigma_{if}}{d\hat{\mathbf{p}}_f} = g_f j_f(\mathbf{E}_f)\frac{d\sigma_{fi}}{d\hat{\mathbf{p}}_i}.$$

The basic relationship satisfied by the differential cross sections for the forward and reverse $i \rightleftharpoons f$ transitions is

$$g_i p_i^2 \frac{d\sigma_{if}(E_i)}{d\hat{\mathbf{p}}_f} = g_f p_f^2 \frac{d\sigma_{fi}(E_f)}{d\hat{\mathbf{p}}_i}.$$

Collision strengths

Collision strengths Ω_{if} exploit this detailed balance relation by being defined as

$$\Omega_{if} = g_i p_i^2 \sigma_{if}(E_i) = g_f p_f^2 \sigma_{fi}(E_f) = \Omega_{fi}.$$

They are therefore symmetrical in i and f.

Reactive processes

For any reactive process

$$A + B \rightleftharpoons C + D,$$

the detailed balance relations involving differential/integral cross sections are

$$g_A g_B p_{AB}^2 \left[\frac{d\sigma_{if}(E_{AB})}{d\hat{\mathbf{p}}_{CD}}\right] = g_C g_D p_{CD}^2 \left[\frac{d\sigma_{fi}(E_{CD})}{d\hat{\mathbf{p}}_{AB}}\right],$$

$$g_A g_B p_{AB}^2 \sigma_{if}(E_{AB}) = g_C g_D p_{CD}^2 \sigma_{fi}(E_{CD}),$$

where $p_{JK}^2 = 2M_{JK}E_{JK}$, in terms of the reduced mass M_{JK} and relative energy E_{JK} of species J and K.

6.6.6. Examples of detailed balance

Excitation—deexcitation

$$e^-(E_i) + A_i \rightleftharpoons e^-(E_f) + A_f,$$

$$\sigma_{if}(E_i) = \left(\frac{g_f}{g_i}\right)\left(\frac{E_f}{E_i}\right)\sigma_{fi}(E_f).$$

With energy conservation $E_i = E_f + (\epsilon_f - \epsilon_i) \equiv E_f + \epsilon_{fi}$, the cross section for superelastic collisions $(E_f > E_i)$ can be obtained from σ_{if} at energy (E_i) by

$$\sigma_{fi}(E_i - \epsilon_{fi}) = \left(1 - \frac{\epsilon_{fi}}{E_i}\right)^{-1} \left(\frac{g_i}{g_f}\right) \sigma_{if}(E_i).$$

Dissociative recombination/associative ionization

Dissociative recombination and associative ionization are represented by the forward and backward directions of

$$e^- + AB^+ \rightleftharpoons A + B^*.$$

The respective cross sections σ_{DR} and σ_{AI} are related by

$$\sigma_{DR}(E_e) = \left(\frac{g_A g_B}{g_e g_{AB^+}}\right) \left(\frac{p_{AB}^2}{p_e^2}\right) \sigma_{AI}(E_{AB}),$$

where the statistical weight of each species j involved is denoted by g_j.

Radiative recombination/photoionization

Similarly, the cross sections for radiative recombination and for photoionization, the forward and reverse directions of

$$A^+ + e^- \rightleftharpoons h\nu + A(n\ell),$$

are related by

$$\sigma_{RR}(E_e) = \left(\frac{g_A g_\nu}{g_e g_{A^+}}\right) \left(\frac{p_\nu^2}{p_e^2}\right) \sigma_{PI}(h\nu).$$

The photon statistical weight is $g_\nu = 2$, corresponding to the two directions of polarization of the photon. The photon energy E is related to its momentum p_ν and wavenumber k_ν and to the ionization energy $I_{n\ell}$ of the atom $A(n\ell)$ by

$$E = h\nu = p_\nu c = \hbar k_\nu c = I_{n\ell} + E_e,$$

where c is the speed of light. The ratio above is

$$\frac{p_\nu^2}{p_e^2} = \frac{(h\nu)^2}{(2E_e m_e c^2)} = \frac{\alpha^2 (h\nu)^2}{2E_e \varepsilon_0}.$$

6.6.7. Four useful expressions for the cross section

Final expressions to be used for calculation of cross sections depend on the particular choice of normalization of the continuum wave function for relative motion. Since it is often a vexing problem and is a continued source of confusion and error in the literature, these final expressions are worked out below. The external relative-motion part of the system wave function $N\Psi^+ \equiv \Psi_{\mathbf{p}}(\mathbf{R}, \mathbf{r}_A, \mathbf{r}_B)$ is $N\Phi_{\mathbf{p}}(\mathbf{R})$. Since $|N|^2 \rho = mp/h^3 = j/v$,

TABLE 6.4. Continuum wavefunction normalization, density of states, and cross section factors

Type	$\langle \Phi_{\mathbf{p}'} \mid \Phi_{\mathbf{p}} \rangle$	N	N_ℓ	$\rho(E)$	γ_{if}
Unit ampl.	$(2\pi\hbar)^3\delta(\mathbf{p}-\mathbf{p}')$	1	$\dfrac{4\pi}{k}$	mp/h^3	$\dfrac{v_f}{v_i}\left(\dfrac{1}{4\pi}\right)^2\left(\dfrac{2m_f}{\hbar^2}\right)^2$
Wavenumber	$\delta(\mathbf{k}-\mathbf{k}')$	$(2\pi)^{-3/2}$	$\left(\dfrac{2}{\pi}\right)^{1/2}\dfrac{1}{k}$	$mp/\hbar^3$	$\dfrac{v_f}{v_i}(2\pi^2)^2\left(\dfrac{2m_f}{\hbar^2}\right)^2$
Momentum	$\delta(\mathbf{p}-\mathbf{p}')$	$(2\pi\hbar)^{-3/2}$	$\left(\dfrac{2}{\pi\hbar}\right)^{1/2}\dfrac{1}{k}$	mp	$\dfrac{v_f}{v_i}(2\pi^2\hbar^3)^2\left(\dfrac{2m_f}{\hbar^2}\right)^2$
Directed energy	$\delta(\mathbf{E}-\mathbf{E}')$	$(mp/h^3)^{1/2}$	$\left(\dfrac{2m}{\hbar^2}\dfrac{1}{\pi k}\right)^{1/2}$	1	$\dfrac{(2\pi)^4}{k_i^2}$

the density $\rho(\mathbf{E})$ of continuum states therefore depends on the choice of normalization factor N adopted for the continuum wave. For future reference, the amplitude N and the energy densities $\rho(\mathbf{E})$ associated with four common methods adopted for normalization of continuum waves are provided in Table 6.4. Also included are the amplitude N_ℓ of the corresponding radial partial wave

$$R_{\epsilon\ell}(r) \sim \frac{N_\ell}{r}\sin(kr - \frac{1}{2}\ell\pi + \eta_\ell)$$

of Sec. 6.8.1. The external multiplicative factors $\gamma_{if} = (2\pi/\hbar)\left(\rho_i\rho_j/j_i\right)$ in the basic formula in Sec. 6.6.4 for the cross section are also provided in Table 6.4 for the various normalization schemes. The reduced masses before and after the collision are $m_i = M_A M_B/(M_A + M_B)$ and $m_f = M_C M_D/(M_C + M_D)$, respectively.

Energy normalized initial and final states

The wave functions

$$\chi_{\mathbf{p}} = \rho^{1/2} N\Psi = (mp/2\pi\hbar^3)^{1/2}\Psi_{\mathbf{p}}(\mathbf{R}, \mathbf{r}_A, \mathbf{r}_B)$$

are energy-normalized according to

$$\langle \chi_{\mathbf{p}'} \mid \chi_{\mathbf{p}} \rangle = \delta(\mathbf{E} - \mathbf{E}').$$

The basic formula in Sec. 6.6.4 with $j_i = p_i^2/(2\pi\hbar)^3$ yields

$$\frac{d\sigma_{if}}{d\hat{\mathbf{p}}_f} = \frac{\pi}{k_i^2}\frac{1}{g_i} \mid T_{if}\mid^2 = \left(\frac{h^2}{8\pi m_i E_i}\right)\frac{1}{g_i}\mid T_{if}\mid^2.$$

The transition probability is

$$P_{if} = |T_{if}|^2 = \sum_{i,f}\int |2\pi\langle\tilde{\chi}_f|V|\chi_i^+\rangle|^2 \, d\hat{\mathbf{k}}_i,$$

the magnitude squared of the element T_{if} of the transition matrix T between χ_i^+ and $\tilde{\chi}_f$, the two energy-normalized eigenfunctions of $\hat{H}_{\rm rel} + \hat{H}_{\rm int} + V$ and $\hat{H}_{\rm rel} + \hat{H}_{\rm int}$, respectively. The detailed balance relation in this case is simply

$$|T_{if}| = |T_{fi}|^2,$$

thereby verifying that $|T_{if}|^2$ is indeed the $i \to f$ transition probability for transitions between all g_i initial and g_f final states. This type of normalization is convenient for rearrangement collisions such as dissociative, radiative, and dielectronic recombination.

Unit amplitude initial and final states

Here, the initial and final wave functions with unit amplitude are Ψ_i^+ and Φ_f. They are each normalized according to

$$\langle \Psi_{\mathbf{p}'} \mid \Psi_{\mathbf{p}} \rangle = (2\pi\hbar)^3 \delta(\mathbf{p} - \mathbf{p}').$$

The basic expression in Sec. 6.6.4 with $j_i = |N_i|^2 v_i \rho_i$ and $|N_f|^2 \rho_f = m_f p_f / h^3$ reduces to

$$\frac{d\sigma_{if}}{d\hat{p}_f} = \frac{v_f}{v_i} |f_{if}(\theta, \varphi)|^2,$$

where the scattering amplitude is

$$f_{if} = -\frac{1}{4\pi}(2m_f/\hbar^2)\langle \Phi_f | V | \Psi_i^+ \rangle,$$

which couples scattering states Ψ_i^+ and Φ_f of unit amplitude. This expression is also applicable for rearrangement collisions $A + B \to C + D$ by including the reduced mass $m_f = M_C M_D/(M_C + M_D)$ of the reacted species after the collision. The integral cross section consistent with the above scattering amplitude is

$$\sigma_{if}(E) = \frac{v_f}{v_i} \int_0^{\pi} d(\cos\theta) \int_0^{2\pi} |f_{if}(\theta, \varphi)|^2 \, d\varphi.$$

At relative energy, $E = k_i^2 \hbar^2 / 2M_{AB}$. The scattering amplitude consistent with the common use of

$$\sigma_{if}(E) = \frac{k_f}{k_i} \int_0^{\pi} d(\cos\theta) \int_0^{2\pi} |\tilde{f}_{if}(\theta, \varphi)|^2 \, d\varphi$$

for rearrangement collisions is

$$\tilde{f}_{if} = -\frac{1}{4\pi}(2\sqrt{m_i m_f}/\hbar^2)\langle \Phi_f | V | \Psi_i^+ \rangle.$$

Both conventions are identical only for direct collisions $A(\alpha) + B(\beta) \to A(\alpha') + B(\beta')$. This normalization is customary [10] for elastic and inelastic scattering processes.

For symmetric potentials $V(r)$, scattering is confined to a plane and f_{if} depends only on scattering angle $\theta = \hat{\mathbf{k}}_i \cdot \hat{\mathbf{k}}_f$.

Momentum normalized initial and final states

Here, the initial and final wave functions $\xi_{\mathbf{p}} = (2\pi\hbar)^{-3/2}\Psi_i^+$ and $\tilde{\xi}_{\mathbf{p}'} = (2\pi\hbar)^{-3/2}\Phi_f$ are normalized according to

$$\langle \xi_{\mathbf{k}'} \mid \xi_{\mathbf{k}} \rangle = \delta(\mathbf{p} - \mathbf{p}').$$

The cross section in Sec. 6.6.4 is then

$$\frac{d\sigma_{if}}{d\hat{p}_f} = \frac{v_f}{v_i} \mid f_{if}(\theta, \varphi) \mid^2,$$

where the scattering amplitude [11]– [13] is now

$$f_{if} = -(2\pi^2\hbar^3)(2m_f/\hbar^2) \mid \langle \tilde{\xi}_f \,|V|\, \xi_i^+ \rangle \mid.$$

Energy normalized final and unit amplitude initial states

Here, the basic formula in Sec. 6.6.4 yields

$$\frac{d\sigma_{if}}{d\hat{\mathbf{p}}_f} = \frac{2\pi}{\hbar v_i} \mid \langle \tilde{\chi}_f \,|\, V \,|\Psi_i^+ \rangle \mid^2,$$

which couples the initial scattering state Ψ_i^+ of unit amplitude with the energy-normalized final state $\tilde{\chi}_f = \rho_f^{1/2} N_f \Phi_f$. This normalization is customary for photoionization problems.

6.7. BORN CROSS SECTIONS

Here, an (undistorted) plane wave of unit amplitude is adopted for the channel wave function $\Phi_j(\mathbf{R})$ in the wave function $\Psi_i^+ = \Sigma_j \Phi_j(\mathbf{R}) \psi_j^{\text{int}}(\mathbf{r})$ for the complete system. The differential cross section for elastic ($i = f$) or inelastic scattering ($i \neq f$) into $\hat{k}_f(\theta, \phi)$ is then

$$\frac{d\sigma_{if}}{d\Omega} = \frac{v_f}{v_i} \mid f_{if}(\theta) \mid^2.$$

The Born scattering amplitude for $A - B$ collisions is

$$f_{if}^{(B)}(\mathbf{K}) = -\frac{1}{4\pi}\frac{2M_{AB}}{\hbar^2} \int V_{fi}(\mathbf{R}) \exp(i\mathbf{K} \cdot \mathbf{R})\, d\mathbf{R},$$

which is the Fourier transform of the interaction potential

$$V_{fi}(\mathbf{R}) = \langle \psi_f^{\text{int}}(\mathbf{r}) |\, V(\mathbf{r}, \mathbf{R}) \,| \psi_i^{\text{int}}(\mathbf{r}) \rangle,$$

which couples the initial and final isolated states $\psi_j^{\text{int}}(\mathbf{r}) = \phi_j(\mathbf{r}_A)\psi_j(\mathbf{r}_B)$ of the atoms. The diagonal potential $V_{ii}(\mathbf{R})$ is the static interaction for elastic scattering. The Born scattering amplitude is a pure function only of the collisional momentum change

$$\mathbf{q} = \hbar\mathbf{K} = M_{AB}(\mathbf{v}_i - \mathbf{v}_f) = \hbar(\mathbf{k}_i - \mathbf{k}_f),$$

where $\mathbf{v}$ is the $A-B$ relative velocity. Since $K^2 = k_i^2 + k_f^2 - 2k_i k_f \cos\theta$, the Born integral cross section is

$$\sigma_{if}^B(k_i) = \frac{2\pi}{M_{AB}^2 v_i^2} \int_{q_-}^{q_+} | f_{if}^{(B)}(q) |^2 q\, dq$$

$$= \frac{2\pi}{(k_i a_0)^2} \int_{K_- a_0}^{K_+ a_0} | f_{if}^{(B)}(K) |^2 (K a_0)\, d(K a_0),$$

where $q_\pm = \hbar K_\pm = \hbar | k_i \pm k_f |$ are the maximum and minimum momentum changes consistent with energy conservation. For symmetric interactions $V_{fi}(R)$,

$$f_{if}^B(K) = -\frac{2M_{AB}}{\hbar^2} \int V_{fi}(R) \frac{\sin KR}{KR} R^2\, dR.$$

6.7.1. Fermi golden rules

Rule A. The transition rate (probability per unit time) for a transition from state Φ_i of a quantum system to a number $\rho(\mathbf{E})\, d\mathbf{E}$ of continuum states $\Phi_{\mathbf{E}}$ by an external perturbation V is

$$w_{if} = \frac{2\pi}{\hbar} | \langle \Phi_{\mathbf{E}} | V | \Phi_i \rangle |^2 \rho_f(\mathbf{E}) \equiv \frac{2\pi}{\hbar} | V_{i\epsilon} |^2 \rho_f(\mathbf{E})$$

to first order in V. Since $| V_{i\epsilon} |^2 \rho_f$ has the dimension of energy, w_{if} has the dimension t^{-1}.

Rule B. When the direct coupling $V_{i\epsilon}$ from only the initial state to the continuum vanishes but the coupling $V_{n\epsilon} \neq 0$ for $n \neq i$, the transition can then occur via the intermediate states n at the rate

$$w_{if} = \frac{2\pi}{\hbar} \sum_n \left| \frac{V_{in} V_{n\epsilon}}{E - E_n} \right|^2 \rho_f(\mathbf{E}).$$

These rules A and B (which are not exact) are useful for both scattering and radiative processes and are often referenced as Fermi's rules 2 and 1, respectively.

Scattering example

The cross section for inelastic scattering of beam of particles by potential $V(\mathbf{r}, \mathbf{R})$ is

$$\frac{d\sigma_{if}}{d\hat{\mathbf{p}}_f} = \frac{w_{if}}{J_i}.$$

A plane-wave monoergic beam, $\Phi = N \exp(i\mathbf{p} \cdot \mathbf{R}/\hbar)\psi(\mathbf{r})$, has current $J_i = | N_i |^2 v_i$ and density determined from $| N_f |^2 \rho_f(\mathbf{E}) = M_{AB} p_f / (2\pi\hbar)^3$. Hence,

$$\frac{d\sigma}{d\hat{\mathbf{p}}_f} = \frac{v_f}{v_i}\left|\frac{1}{4\pi}\frac{2M_{AB}}{\hbar^2}\int V_{fi}(\mathbf{r})e^{i(\mathbf{p}_i-\mathbf{p}_f)\cdot\mathbf{r}/\hbar}\,d\mathbf{r}\right|^2.$$

Since this agrees with the first Born differential cross section for (in)elastic scattering, Fermi's Rule 2 is therefore valid to first order in the interaction V.

6.7.2. Ion (electron)-atom collisions

The electrostatic interaction between a structureless projectile ion P of charge $Z_p e$ and an atom A with nuclear charge $Z_A e$ is

$$V(\mathbf{r},\mathbf{R}) = \frac{Z_A Z_p e^2}{R} - \sum_{j=1}^{N_A}\frac{Z_p e^2}{|\mathbf{R}-\mathbf{r}_j|}.$$

With the use of Bethe's integral

$$\int \frac{e^{i\mathbf{K}\cdot\mathbf{R}}}{|\mathbf{R}-\mathbf{r}_j|}\,d\mathbf{R} = \frac{4\pi}{K^2}e^{i\mathbf{K}\cdot\mathbf{r}_j},$$

the Born scattering amplitude (6.7) reduces to

$$|f_{if}^{(B)}(q)|^2 = \frac{4M_{PA}^2 Z_p^2 e^4}{q^4}|Z_A\delta_{if} - F_{if}^A(q)|^2,$$

which is a function only of momentum transfer $q = \hbar K$. The dimensionless inelastic form factor for $i \to f$ inelastic transitions between states $\phi_{i,f}$ of atom A with Z_A electrons is defined as

$$F_{fi}^A(\mathbf{q}) = \left\langle \phi_f(\mathbf{r})\left|\sum_{j=1}^{Z_A} e^{i\mathbf{q}\cdot\mathbf{r}_j/\hbar}\right|\phi_i(\mathbf{r})\right\rangle,$$

where the integration is over all electron positions denoted collectively by $\mathbf{r} \equiv \{\mathbf{r}_j\}$. The integrated cross section is

$$\sigma_{if}(v_{PA}) = \frac{8\pi Z_p^2 e^4}{v_{PA}^2}\int_{q_-}^{q_+}|Z_A\delta_{if} - F_{if}^A(q)|^2\frac{dq}{q^3}$$

$$= \frac{8\pi Z_p^2 a_0^2}{(v_{PA}/v_0)^2}\int_{K_-a_0}^{K_+a_0}|Z_A\delta_{if} - F_{if}^A(K)|^2\frac{d(Ka_0)}{(Ka_0)^3},$$

where $v_0 = e^2/\hbar$ is the a.u. of velocity. The dimensionless momentum-change $q/m_e v_0$ is Ka_0. In the heavy particle or high energy limit, $q_+ \to \infty$ and

$$q_- \approx \frac{|\Delta E_{fi}|}{v_{PA}}\left[1 + \frac{\Delta E_{fi}}{2M_{PA}v_{PA}^2}\right],$$

where $\Delta E_{fi} = E_f - E_i$ is the energy lost by the projectile. Since

$$f_{if}^B(q) = f_C^{Z_P Z_A}(q)\delta_{if} + f_C^{eZ_P}(q)F_{if}(q)$$

can be expressed in terms of the individual two-body amplitudes f_C for Coulomb elastic scattering, the Born cross section for inelastic collisions can be written in the useful form [14]–[16]

$$\sigma_{if}^B(v_i) = \frac{2\pi}{M_{eP}^2 v_i^2}\int_{q_-}^{q_+} P_{fi}(q)\left(\frac{d\sigma}{d\Omega}\right)_{el} q\,dq,$$

where $P_{fi}(q) = |\,F_{fi}^A(\mathbf{q})\,|^2$ is the transition probability for the impulsive transfer of momentum q to atom A and

$$\left(\frac{d\sigma}{d\Omega}\right)_{el} = \frac{4M_{eP}^2 Z_P^2 e^4}{q^4}$$

is the differential cross section for elastic (Coulomb) scattering with momentum q transferred from the projectile of charge $Z_P e$ to one electron of atom A.

6.7.3. Atom-atom collisions

The Born integral cross section for specific $(\alpha\beta) \to (\alpha'\beta')$ transitions in the collision

$$A(\alpha) + B(\beta) \to A(\alpha') + B(\beta')$$

in terms of the atomic Form Factors is

$$\sigma_{\alpha\alpha'}^{\beta\beta'}(v_i) = \frac{8\pi a_0^2}{(v_i/v_0)^2}\int_{K_-a_0}^{K_+a_0} |\,Z_A\delta_{\alpha\alpha'} - F_{\alpha\alpha'}^A(K)\,|^2 \times |\,Z_B\delta_{\beta\beta'} - F_{\beta\beta'}^B(K)\,|^2 \frac{d(Ka_0)}{(Ka_0)^3}.$$

6.7.4. Quantal and classical impulse cross sections

In the impulse approximation [11, 14], the integral cross section for $i \equiv (\alpha, \beta) \to f \equiv (\alpha', \beta')$ transitions is

$$\sigma_{if}(v_i) = \frac{2\pi}{M_{eB}^2 v_i^2}\int_{q_-}^{q_+} |\,f_{eB}^{\beta\beta'}(q)\,|^2|\,F_{\alpha,\alpha'}^A(q)\,|^2 q\,dq$$

where $f_{eB}^{\beta\beta'}$ is the scattering amplitude for elastic $\beta = \beta'$ or inelastic $\beta \neq \beta'$ collisions between projectile B and an orbital electron of A. For structureless ions B, the Coulomb $f_{eB}^{\beta\beta}(q)$ for elastic electron-ion collisions reproduces the Born approximation for B-A collisions. When Born amplitudes $f_{eB}^{\beta\beta'}(q)$ are used for fast atom $B - e$ collisions, then the

Born approximation for atom-atom collisions is also for general scattering amplitudes recovered. For slow atoms B, f_{eB} is dominated by s-wave elastic scattering so that $f_{eB} = -a$ and $\sigma_{eB} = 4\pi a^2$ where a is the scattering length. Then

$$\sigma_{if}(v_i) = \frac{2\pi a^2}{(v_i/v_0)^2} \int_{K_- a_0}^{K_+ a_0} |\, F_{if}^A(K)\,|^2 (Ka_0)\, d(Ka_0)$$

which is a good approximation for collisional transitions $nl \to n'l'$ in Rydberg atoms A. The full quantal impulse cross section [11, 14] for general $f_{eB}^{\beta\beta'}$ has recently been presented in a valuable new form [15] which is the appropriate representation for direct classical correspondence. The classical impulse cross section was then defined [15] to yield the first general expression for the classical impulse cross section for $n\ell - n'\ell'$ and $n\ell - \epsilon\ell'$ electronic transitions. The cross section satisfies the optical theorem and detailed balance. Direct connection with the classical binary encounter approximation (BEA) was established and the derived $n\ell - n'$ and $n\ell - \epsilon$ cross sections reproduce the standard BEA cross sections.

6.7.5. Atomic form factor and generalized oscillator strength

In terms of the form factor $F_{if}(K)$, the generalized oscillator strength is defined as

$$f_{if}(K) = \left(\frac{2m_e E_{fi}}{q^2}\right) |\, F_{if}(q)\,|^2 = \frac{2\, E_{fi}^{a.u.}}{(Ka_0)^2} |\, F_{if}(K)\,|^2,$$

which tends to the dipole oscillator strength in the $K \to 0$ limit.

Sum rules

$$\sum\!\!\!\!\!\!\!\int_f f_{if}(K) = \sum_f f_{if}(K) + \int_0^\infty \frac{df_{iE}}{dE}\, dE = N,$$

$$\sum\!\!\!\!\!\!\!\int_f |F_{if}(K)|^2 = \sum_f F_{if}(K) + \int_0^\infty \frac{dF_{iE}}{dE}\, dE$$

$$= N + \sum_{j<k}^{N} |\, \langle \Psi_i \,|\exp(i\vec{K}\cdot(\vec{r}_j - \vec{r}_k))|\, \Psi_i\rangle\,|^2,$$

where N is the number of electrons. The summation $\sum\!\!\!\!\!\!\!\int$ extends over all discrete and continuum states.

Energy change moments

The energy change moments are defined as

$$S(\alpha, K) = \sum_{f\neq i} (2\Delta E_{fi}^{a.u.})^\alpha f_{if}(K) = \sum_{f\neq i} (2\Delta E_{fi}^{a.u.})^{\alpha+1} |\, F_{if}(K)\,|^2 (Ka_0)^{-2}.$$

The exact energy change moments for $H(1s)$ are

$$S(-1,K) = \{1 - [1 + \tfrac{1}{4}(Ka_0)^2]^{-4}\}(Ka_0)^{-2},$$
$$S(0,K) = 1,$$
$$S(1,K) = (Ka_0)^2 + 4/3,$$
$$S(2,K) = (Ka_0)^4 + 4(Ka_0)^2 + 16/3.$$

6.7.6. Form factors for atomic hydrogen

The probability of a transition $i \to f$ resulting from any external perturbation that impulsively transfers momentum $\mathbf{q}$ to the internal momenta of the electrons of the target system is

$$P_{if}(\mathbf{q}) = |\, F_{fi}(\mathbf{q}) \,|^2.$$

The impulse can be due to a sudden collision with particles or to exposure to a pulse of electromagnetic radiation. The physical significance of the form factor is that P_{if} is the impulsive transition probability for any atom. For $nlm \to n'l'm'$ subshell transitions, the amplitude

$$F_{nlm,n'l'm'}(\mathbf{q}) = \langle \Psi_{nlm}(\mathbf{r}) |\, e^{i\mathbf{q}\cdot\mathbf{r}/\hbar} \,| \Psi_{n'l'm'}(\mathbf{r}) \rangle$$

decomposes, under $\Psi(\mathbf{r}) = R_{nl}(r) Y_{lm}(\hat{r})$, to

$$F_{nlm,n'l'm'}(\mathbf{q}) = 4\pi \sum_{L=|l-l'|}^{l+l'} \imath^L w^{(L)}_{lm'l'm'}\, f^{(L)}_{nl,n'l'}(q) Y_{L,M}(\hat{\mathbf{q}}),$$

where $M = m - m'$ and the radial integral in terms of the modified Bessel function j_L is

$$f^{(L)}_{nl,n'l'}(q) = \int_0^\infty R_{nl}(r) R_{n'l'}(r) j_L(qr) r^2\, dr.$$

The coefficients, in terms of the Wigner's $3j$-symbol, $\{\ldots\}$, are

$$w^{(L)}_{lm'\,l'm'} = \left[\frac{(2l+1)(2l'+1)(2L+1)}{4\pi}\right]^{\frac{1}{2}} \times \begin{Bmatrix} L & l & l' \\ M & -m & m' \end{Bmatrix} \begin{Bmatrix} L & l & l' \\ 0 & 0 & 0 \end{Bmatrix}.$$

For $nl \to n'l'$ transitions in atomic hydrogen,

$$P_{nl,n'l'}(\mathbf{q}) = \sum_{m,m'} |\, \langle \Psi_{nlm}(\mathbf{r}) \,|\, e^{i\mathbf{q}\cdot\mathbf{r}/\hbar} \,|\, \Psi_{n'l'm'}(\mathbf{r}) \rangle \,|^2,$$

and then decomposes as

$$P_{nl,n'l'}(\mathbf{q}) = (2l+1)(2l'+1) \sum_{L=|l-l'|}^{l+l'} (2L+1) \times \begin{Bmatrix} L & l & l' \\ 0 & 0 & 0 \end{Bmatrix}^2 \left[f^{(L)}_{nl,n'l'}(q)\right]^2.$$

Exact algebraic expressions for the probability

$$P_{n,n'}(q) = \sum_{l'l'} \sum_{m,m'} | \langle n'l'm' \mid e^{i\mathbf{q}\cdot\mathbf{r}/\hbar} \mid nlm \rangle |^2$$

of general $n \to n'$ transitions in atomic hydrogen have been recently derived [16] as analytical functions of n and n'.

6.7.7. Rotational excitation

For ion-point dipole D interactions, only $\Delta J = \pm 1$ transitions are allowed. For ion-point quadrupole Q interactions, only $\Delta J = 0, \pm 2$ transitions are allowed. The Born differential cross sections for $J \to J'$ transitions are

$$\frac{d\sigma^{(d)}}{d\hat{\mathbf{k}}_f}(J \to J+1) = \frac{4}{3}\frac{k_f}{k_i}\left(\frac{J+1}{2J+1}\right)\frac{D^2}{K^2},$$

$$\frac{d\sigma^{(q)}}{d\hat{\mathbf{k}}_f}(J \to J) = \frac{4}{45}\frac{J(J+1)}{(2J-1)(2J+3)}Q^2,$$

$$\frac{d\sigma^{(q)}}{d\hat{\mathbf{k}}_f}(J \to J+2) = \frac{2}{15}\frac{k_f}{k_i}\frac{(J+1)(J+2)}{(2J+1)(2J+3)}Q^2,$$

which are all spherical symmetrical. The sum

$$\sum \frac{d\sigma^{(q)}}{d\hat{\mathbf{k}}_f}(J \to J, J \pm 2) = \frac{4}{45}Q^2$$

is independent of the initial value of J. The integral cross sections

$$\sigma^{(d)}(J \to J+1) = \frac{8\pi}{3k_i^2}\left(\frac{J+1}{2J+1}\right)\ln\left(\frac{k_i+k_f}{k_i-k_f}\right)D^2,$$

$$\sigma^{(q)}(J \to J, J+2) = \frac{8\pi}{15}\frac{k_f}{k_i}\frac{(J+1)(J+2)}{(2J+1)(2J+3)}Q^2$$

all satisfy the detailed balance relation

$$k_i^2(2J_i+1)\sigma(J_i \to J_f) = k_f^2(2J_f+1)\sigma(J_f \to J_i).$$

The summed diffusion cross sections are

$$\sigma^{(d)} = \int \left[\sum_{J\pm 1} \frac{d\sigma}{d\hat{\mathbf{k}}_f}(J \to J')\right](1-\cos\theta)\, d\hat{\mathbf{k}}_f = \left(\frac{8\pi}{3k_i^2}\right)D^2,$$

$$\sigma^{(q)} = (16\pi/45)Q^2.$$

6.7.8. List of Born cross sections for model potentials

$$k^2 = (2M_{AB}/\hbar^2)E, \qquad K = 2k\sin\tfrac{1}{2}\theta,$$
$$U = (2M_{AB}/\hbar^2)V, \qquad U/k^2 = V/E.$$

The scattering amplitude and integral cross section for elastic scattering by a symmetric potential reduce to

$$f_{\rm B}(K) = -\int U(R)\frac{\sin KR}{KR}R^2\,dR,$$

$$\sigma_{\rm B}(E) = \frac{2\pi}{k^2}\int_0^{2k} |\, f_{\rm B}(K)\,|^2 K\,dK.$$

The cross section is independent of the sign of the interaction.

Exponential:

$$V(R) = V_0\exp(-\alpha R),$$
$$f_{\rm B}(K) = -\frac{2\alpha U_0}{(\alpha^2+K^2)^2},$$
$$\sigma_{\rm B}(E) = \frac{16}{3}\pi U_0^2\left[\frac{3\alpha^4+12\alpha^2k^2+16k^4}{\alpha^4(\alpha^2+4k^2)^3}\right] \overset{E\to\infty}{\to} \frac{4}{3}\pi\left(\frac{V_0}{E}\right)\left(\frac{U_0}{\alpha^4}\right).$$

Gaussian:

$$V(R) = V_0\exp(-\alpha^2R^2),$$
$$f_{\rm B}(K) = -\left(\frac{\pi^{1/2}U_0}{4\alpha^2}\right)\exp(-K^2/4\alpha^2),$$
$$\sigma_{\rm B}(E) = \left(\frac{\pi^2U_0}{8\alpha^4}\right)\left(\frac{V_0}{E}\right)\left[1-\exp(-2k^2/\alpha^2)\right].$$

1. *Spherical well/barrier:* $V(R) = V_0$ for $R < a$, $V(R) = 0$ for $R > a$, $U_0 = (2M_{AB}/\hbar^2)V_0$

$$f_{\rm B}(K) = -\frac{U_0}{K^3}\left[\sin Ka - Ka\cos Ka\right],$$
$$\sigma_{\rm B}(E) = \frac{\pi}{2}\frac{V_0}{E}(U_0a^4)\left[1-(ka)^{-2}+(ka)^{-3}\sin 2ka-(ka)^{-4}\sin^2 2ka\right].$$

At low energies, $f_{\rm B} \to (2M_{AB}/\hbar^2)V_0a^3/3$ and the scattering is isotropic. At high energies, $\sigma_{\rm B}(E) \sim E^{-1}$.

2. *Screened Coulomb interaction:* $V(R) = V_0 \exp(-\alpha R)/R$.

$$f_B(K) = -\frac{U_0}{\alpha^2 + K^2},$$

$$\sigma_B(E) = \frac{4\pi U_0^2}{\alpha^2(\alpha^2 + 4k^2)},$$

where $U_0 = 2Z/a_0$. At low energies, $f_B = -U_0/\alpha^2$ is isotropic. At high energies, $\sigma_B \to \pi\left(\frac{V_0}{E}\right)\left(\frac{U_0}{\alpha^2}\right)$.

3. *Electron-atom model static interaction:*

$$V(R) = -N(e^2/a_0)\,[Z + a_0/R]\exp(-2ZR/a_0),$$

$$f_B(\theta) = \frac{2N}{a_0}\left[\frac{2\alpha^2 + K^2}{(\alpha^2 + K^2)^2}\right], \quad \alpha = 2Z/a_0,$$

$$\sigma_B(E) = \frac{\pi a_0^2 N^2\left[12Z^4 + 18Z^2k^2a_0^2 + 7k^4a_0^2\right]}{3Z^2(Z^2 + k^2a_0^2)^3}.$$

For atomic H(1s), $N = 1$ and $Z = 1$. For He($1s^2$), approximate parameters are $N = 2$ and $Z = 27/16$.

Polarization potential:

$$V(R) = V_0/(R^2 + R_0^2)^2,$$

$$f_B(K) = -\frac{1}{4}\pi\left(\frac{U_0}{R_0}\right)\exp(-KR_0),$$

$$\sigma_B(E) = \left(\frac{\pi^3 U_0}{32R_0^4}\right)\left(\frac{V_0}{E}\right)\left[1 - (1 + 4kR_0)\exp(-4kR_0)\right].$$

6.8. QUANTAL POTENTIAL SCATTERING

The Schrödinger equation

$$\left(-\frac{\hbar^2}{2M_{AB}}\nabla_{\mathbf{r}}^2 + V(\mathbf{r})\right)\Psi_{\mathbf{k}}^+(\mathbf{r}) = E\Psi_{\mathbf{k}}^+(\mathbf{r})$$

solved subject to the asymptotic condition

$$\Psi_{\mathbf{k}}^+(\mathbf{r}) \sim \exp(i\mathbf{k}\cdot\mathbf{r}) + \frac{1}{r}f(\theta,\phi)\exp(ikr)$$

for outgoing spherical waves is equivalent to the solution of the Lippman-Schwinger integral equation

$$\Psi_{\mathbf{k}}^+(\mathbf{r}) = \Phi_{\mathbf{k}}^+(\mathbf{r}) + \int G(\mathbf{r},\mathbf{r}')U(\mathbf{r}')\Psi_{\mathbf{k}}^+(\mathbf{r}')\,d\mathbf{r}',$$

where the outgoing Green's function for a free particle is

$$G(\mathbf{r},\mathbf{r}') = -\frac{1}{4\pi}\frac{\exp\{ik|\mathbf{r}-\mathbf{r}'|\}}{|\mathbf{r}-\mathbf{r}'|}.$$

Solution of the scattering amplitude may then be determined from the asymptotic form of $\Psi_{\mathbf{k}}^{+}(\mathbf{r})$ directly or from the integral representation

$$f(\theta,\phi) = -\frac{1}{4\pi}(2M_{AB}/\hbar^2)\langle\exp(i\mathbf{k}_f\cdot\mathbf{r})|\,V(\mathbf{r})\,|\Psi_i^{+}\rangle.$$

The differential cross section for elastic scattering is

$$\frac{d\sigma}{d\Omega} = |\,f(\theta,\phi)\,|^2.$$

6.8.1. Partial wave expansion

A plane wave of unit amplitude can be decomposed according to

$$\Phi_{\mathbf{k}}(\mathbf{r}) = \exp\imath\mathbf{k}\cdot\mathbf{r} = 4\pi\sum_{\ell,m}\imath^{\ell}j_{\ell}(kr)Y_{\ell m}^{*}(\hat{\mathbf{k}})Y_{\ell m}(\hat{\mathbf{r}}),$$

where j_ℓ is the spherical Bessel function, which varies asymptotically as

$$j_{\ell}(kr) \sim \frac{1}{kr}\sin\left(kr - \frac{1}{2}\ell\pi\right).$$

The addition theorem for the spherical harmonics is

$$4\pi\sum_{m}Y_{\ell m}^{*}(\hat{\mathbf{k}})Y_{\ell m}(\hat{\mathbf{r}}) = (2\ell+1)P_{\ell}(\hat{\mathbf{k}}\cdot\hat{\mathbf{r}}).$$

Another useful identity is

$$\frac{\sin Kr}{Kr} = 4\pi\sum_{\ell,m}j_{\ell}(k_i r)j_{\ell}(k_f r)Y_{\ell m}(\hat{\mathbf{k}}_i)Y_{\ell m}^{*}(\hat{\mathbf{k}}_f),$$

where $K^2 = k_i^2 + k_f^2 - 2k_ik_f\cos\theta$. The system wave function $\Psi_{\mathbf{k}}^{+}(\mathbf{r}) \sim N\Phi_{\mathbf{k}}$ with amplitude N is expanded according to

$$\Psi_{\mathbf{k}}^{+}(\mathbf{r}) = \sum_{\ell,m}\imath^{\ell}e^{\imath\eta_\ell}R_{\epsilon\ell}(r)Y_{\ell m}^{*}(\hat{\mathbf{k}})Y_{\ell m}(\hat{\mathbf{r}})$$

$$= \frac{4\pi N}{kr}\sum_{\ell,m}\imath^{\ell}F_{\epsilon\ell}(r)Y_{\ell m}^{*}(\hat{\mathbf{k}})Y_{\ell m}(\hat{\mathbf{r}}).$$

The radial wave F_ℓ is the solution of the radial Schrödinger equation

$$\frac{d^2F_\ell}{dr^2} + \left[k^2 - \left\{U(r) + \frac{\ell(\ell+1)}{r^2}\right\}\right]F_{\ell}(r) = 0.$$

The reduced potential and energy are $U(r) = (2M_{AB}/\hbar^2)V(R)$ and $k^2 = (2M_{AB}/\hbar^2)E$, respectively. They both have dimensions of $[a_0^{-2}]$. Also, $(ka_0)^2 = (2E/\varepsilon_0)(M_{AB}/m_e)$. Each ℓ-partial wave is separately scattered since the angular momentum of relative motion is conserved for central forces. The radial waves $R_{\epsilon\ell}(r)$ and $F_{\epsilon\ell}(r)$ vary asymptotically as

$$R_{\epsilon\ell}(r) \sim \frac{N_\ell}{r}\sin(kr - \tfrac{1}{2}\ell\pi + \eta_\ell),$$

$$F_{\epsilon\ell}(r) \sim e^{\iota\eta_\ell}\sin(kr - \tfrac{1}{2}\ell\pi + \eta_\ell).$$

The amplitude of the partial radial wave wave $R_{\epsilon\ell}(r)$ is $N_\ell = 4\pi N/k$. In Table 6.1 the amplitudes N and N_ℓ appropriate to various choices for normalization of the continuum wave functions $\Psi_{\mathbf{k}}(\mathbf{r})$ are displayed.

6.8.2. Scattering amplitudes

For symmetric interactions $V = V(r)$, the wave functions Ψ_i^+ and $\exp(i\mathbf{k}_f \cdot \mathbf{r})$ are decomposed into partial waves. From their asymptotic forms, the following partial wave expansions for the scattering amplitude:

$$f(\theta) = \frac{1}{2ik}\sum_{\ell=0}^{\infty}(2\ell+1)\left[\exp(2i\eta_\ell) - 1\right]P_\ell(\cos\theta),$$

$$f(\theta) = \frac{1}{2ik}\sum_{\ell=0}^{\infty}(2\ell+1)\left[S_\ell(k) - 1\right]P_\ell(\cos\theta),$$

$$f(\theta) = \frac{1}{2ik}\sum_{\ell=0}^{\infty}(2\ell+1)T_\ell(k)P_\ell(\cos\theta),$$

can be deduced. The scattering, transition, and reactance matrix elements are defined, in terms of the phase shift η_ℓ suffered by each partial wave, as

$$S_\ell(k) = \exp(2i\eta_\ell),$$

$$T_\ell(k) = 2i\sin\eta_\ell\exp(i\eta_\ell),$$

$$K_\ell(k) = \tan\eta_\ell.$$

The asymptotic ($kr \to \infty$) form of F_ℓ may then be written in terms of the following linear combinations:

$$F_\ell(kr) \sim \sin(kr - \ell\pi/2) + \left(\frac{T_\ell}{2i}\right)e^{i(kr-\ell\pi/2)},$$

$$= -\frac{1}{2i}[e^{-i(kr-\ell\pi/2)} - S_\ell e^{i(kr-\ell\pi/2)}],$$

$$= e^{i\eta_\ell}\cos\eta_\ell[\sin(kr - \ell\pi/2) + K_\ell\cos(kr - \ell\pi/2)],$$

expressed as a combinations of standing waves (trigonometric functions), of incoming $(-)$ and outgoing $(+)$ spherical waves (exponential functions) and of a standing wave and an outgoing spherical wave. The physical significance of the admixture coefficients S_ℓ, T_ℓ, and K_ℓ is then transparent. The elements are connected by

$$S_\ell = 1 + T_\ell = (1 + iK_\ell)/(1 - iK_\ell).$$

K_ℓ is real, whereas both S_ℓ and T_ℓ are complex. In term of the full solutions F_ℓ of the radial Schrödinger equation, the T-matrix element for elastic scattering is

$$T_\ell = -\frac{2i}{k}\int_0^\infty F_\ell^{(0)}(r)U(r)F_\ell(r)\,dr,$$

where $F_\ell^{(0)} = (kr)j_\ell(kr)$ is the radial component of the final plane wave. The Born approximation to T_ℓ is obtained on the substitution $F_\ell^{(0)} = F_\ell$.

6.8.3. Integral cross sections

$$\begin{aligned}\sigma(E) &= \frac{4\pi}{k^2}\sum_{\ell=0}^{\infty}(2\ell+1)\sin^2\eta_\ell \\ &= \frac{\pi}{k^2}\sum_{\ell=0}^{\infty}(2\ell+1)|\,T_\ell\,|^2 \\ &= \frac{2\pi}{k}\sum_{\ell=0}^{\infty}(2\ell+1)[1-\mathrm{Re}S_\ell].\end{aligned}$$

The semiclassical version is obtained by the substitution $mvb = (\ell + \frac{1}{2})\hbar$ so that $k^2b^2 = (\ell + \frac{1}{2})^2$ in terms of the impact parameter b. Regarding ℓ as a continuous variable,

$$\sigma(E) = \frac{\pi}{k^2}\int_0^\infty (2\ell+1)|\,T_\ell\,|^2\,d\ell = 2\pi\int_0^\infty |\,T(b)\,|^2 b\,db.$$

The transition matrix $|\,T(b)\,|^2$ is therefore the probability of scattering of particles with impact parameter b.

6.8.4. Differential cross sections

The differential cross section for elastic scattering is

$$\frac{d\sigma}{d\Omega} = |\,f(\theta)\,|^2 = A(\theta)^2 + B(\theta)^2,$$

where the real and imaginary parts of $f(\theta)$ are, respectively,

$$A(\theta) = \frac{1}{2k}\sum_{\ell=0}^{\infty}(2\ell+1)\sin 2\eta_\ell P_\ell(\cos\theta),$$

$$B(\theta) = \frac{1}{2k}\sum_{\ell=0}^{\infty}(2\ell+1)\,[1-\cos 2\eta_\ell]\,P_\ell(\cos\theta).$$

Their individual contributions to the integral cross sections are

$$\int A(\theta)^2\,d\Omega = \frac{4\pi}{k^2}\sum_{\ell=0}^{\infty}(2\ell+1)\sin^2\eta_\ell\cos^2\eta_\ell,$$

$$\int B(\theta)^2\,d\Omega = \frac{4\pi}{k^2}\sum_{\ell=0}^{\infty}(2\ell+1)\sin^4\eta_\ell.$$

Expansion in Legendre polynomials

When expanded as a series of Legendre polynomials $P_L(\cos\theta)$, the differential cross section has the following form:

$$\frac{d\sigma(E,\theta)}{d\Omega} = \frac{1}{k^2}\sum_{L=0}^{\infty}a_L(E)P_L(\cos\theta),$$

where the coefficients

$$a_L = \sum_{\ell=0}^{\infty}\sum_{\ell'=|\ell-L|}^{\ell+L}(2\ell+1)(2\ell'+1)(\ell\ell'00\mid\ell\ell'L0)^2\sin\eta_\ell\sin\eta_{\ell'}\cos(\eta_\ell-\eta_{\ell'})$$

are determined by the phase shifts η_ℓ and the Clebsch-Gordon coefficients

$$(\ell\ell'mm'\mid\ell\ell'LM).$$

Example: Three-term expansion in $\cos\theta$

The differential cross section can be expanded as

$$\frac{d\sigma(E,\theta)}{d\Omega} = \frac{1}{k^2}\left[(a_0-\tfrac{1}{2}a_2)+a_1\cos\theta+\tfrac{3}{2}a_2\cos^2\theta\right].$$

The coefficients are

$$a_0 = \sum_{\ell=0}^{\infty}(2\ell+1)\sin^2\eta_\ell,$$

$$a_1 = 6\sum_{\ell=0}^{\infty}(\ell+1)\sin\eta_\ell\sin\eta_{\ell+1}\cos(\eta_\ell\ell+1-\eta_\ell)$$

$$a_2 = 5\sum_{\ell=0}^{\infty}\left[b_\ell\sin^2\eta_\ell\ell + c_\ell\sin\eta_\ell\ell\sin^2\eta_\ell\ell + 2\cos(\eta_{\ell+2}-\eta_\ell)\right],$$

where

$$b_\ell = \frac{\ell(\ell+1)(2\ell+1)}{(2\ell+1)(2\ell_3)},$$

$$c_\ell = \frac{3(\ell+1)(\ell+2)}{(2\ell+3)}.$$

Example: S- and P-wave contributions

The combined S, P-wave ($\ell = 0, 1$) contributions to the differential and integral cross sections are

$$\frac{d\sigma}{d\Omega} = \frac{1}{k^2}\left[\sin^2\eta_0 + [6\sin\eta_0\sin\eta_1\cos(\eta_1-\eta_0)]\cos\theta + 9\sin^2\eta_1\cos^2\right],$$

$$\sigma(E) = \frac{4\pi}{k^2}\left[\sin^2\eta_0 + 3\sin^2\eta_1\right].$$

For pure S-wave scattering, the differential cross section (DCS) is isotropic. For pure P-wave scattering, the DCS is symmetric about $\theta = \pi/2$, where it vanishes; the DCS rises to equal maxima at $\theta = 0, \pi$. For combined S- and P-wave scattering, the DCS is asymmetric with forward-backward asymmetry.

6.8.5. Optical theorem

The optical theorem relates the integral cross section to the imaginary part of the forward scattering amplitude by

$$\sigma(E) = (4\pi/k)\,\mathrm{Im} f(0).$$

This relation is a direct consequence of the conservation of flux. The target casts a shadow in the forward direction, where the intensity of the incident beam becomes reduced by just that amount that appears in the scattered wave. This decrease in intensity or shadow results from interference between the incident wave and the scattered wave in the forward

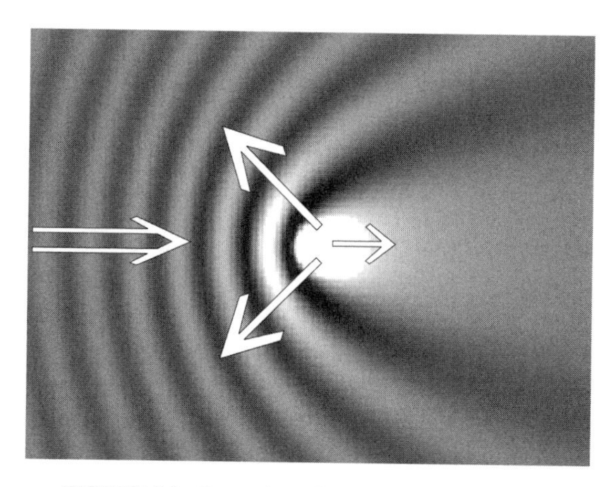

FIGURE 6.2. Scattering of an incident plane wave.

direction. Figure 6.2 for the density $|\Psi_{\mathbf{k}}^{+}(\mathbf{r})|$ of Sec. 6.8 illustrates how this interference tends to illuminate the shadow region at the RHS-side of the target. Flux conservation also implies that the phase shifts η_ℓ are always real. Thus,

$$|S_\ell| = 1, \quad |T_\ell|^2 = \mathrm{Im} T_\ell.$$

6.8.6. Levinson's theorem

For a local potential $V(r)$ that supports n_ℓ bound states of angular momentum ℓ and energy $E_n < 0$, the phase shift $\lim_{k\to 0} \eta_\ell(k))$ tends in the limit of zero collision energy to $n_\ell \pi$. When the well becomes deep enough to introduce an additional bound level $E_{n+1} = 0$ at zero energy, $\lim_{k\to 0} \eta_0(k) = (n_0 + \frac{1}{2})\pi$.

6.8.7. Partial wave expansion for transport cross sections

The transport cross sections

$$\sigma^{(n)}(E) = 2\pi \left[1 - \frac{1+(-1)^n}{2(n+1)}\right]^{-1} \int_{-1}^{+1} \left[1 - \cos^n\theta\right] \frac{d\sigma}{d\Omega}\, d(\cos\theta)$$

for $n = 1 - 4$ have the following phase shift expansions:

$$\sigma^{(1)}(E) = \frac{4\pi}{k^2} \sum_{\ell=0}^{\infty} (\ell+1) \sin^2(\eta_\ell - \eta_{\ell+1}),$$

$$\sigma^{(2)}(E) = \frac{4\pi}{k^2} \left(\frac{3}{2}\right) \sum_{\ell=0}^{\infty} \frac{(\ell+1)(\ell+2)}{(2\ell+3)} \sin^2(\eta_\ell - \eta_{\ell+2}),$$

$$\sigma^{(3)}(E) = \frac{4\pi}{k^2} \sum_{\ell=0}^{\infty} \frac{(\ell+1)}{(2\ell+5)} \left[\frac{(\ell+2)(\ell+3)}{(2\ell+3)} \sin^2(\eta_\ell - \eta_{\ell+3}) + \frac{3(\ell^2+2\ell-1)}{(2\ell-1)} \sin^2(\eta_\ell - \eta_{\ell+1})\right],$$

$$\sigma^{(4)}(E) = \frac{4\pi}{k^2} \left(\frac{5}{4}\right) \sum_{\ell=0}^{\infty} \frac{(\ell+1)(\ell+2)}{(2\ell+3)(2\ell+7)} \times \left[\frac{(\ell+3)(\ell+4)}{(2\ell+5)} \sin^2(\eta_\ell - \eta_{\ell+4}) + \frac{2(2\ell^2+6\ell-3)}{(2\ell-1)} \sin^2(\eta_\ell - \eta_{\ell+2})\right].$$

The momentum-transfer or diffusion cross section is $\sigma^{(1)}$, and the viscosity cross section is $\frac{2}{3}\sigma^{(2)}$.

6.8.8. Born phase shifts

For a symmetric interaction, the Born amplitude is

$$f_{\mathrm{B}}(K) = -\int U(R)\frac{\sin KR}{KR}R^2\,dR,$$

where $U(r) = (2M_{AB}/\hbar^2)V(R)$. Comparison with the partial wave expansion for $f_{\mathrm{B}}(K)$ and

$$\frac{\sin KR}{KR} = \sum_{\ell=0}^{\infty}(2\ell+1)\,[j_\ell(kR)]^2\,P_\ell(\cos\theta)$$

provides the Born phase shift

$$\tan\eta_\ell^B(k) = -k\int_0^\infty U(R)\,[j_\ell(kR)]^2\,R^2\,dR.$$

Examples of the Born S-wave phase shift

$$\tan\eta_0^{\mathrm{B}}(k) = -\frac{1}{k}\int_0^\infty U(R)\sin^2(kR)\,dR.$$

For the potential $U = U_0\dfrac{e^{-\alpha R}}{R}$,

$$\tan\eta_0^{\mathrm{B}} = -\frac{U_0}{4k}\ln\left[1 + 4k^2/\alpha^2\right].$$

For the potential $U = \dfrac{U_0}{(R^2+R_0^2)^2}$,

$$\tan\eta_0^{\mathrm{B}} = -\frac{\pi U_0}{4kR_0^3}\left[1 - (1+2kR_0)e^{-2kR_0}\right].$$

Born phase shifts (large ℓ)

For $\ell \gg ka$,

$$\tan\eta_\ell^{\mathrm{B}} = -\frac{k^{2\ell+1}}{[(2\ell+1)!!]^2}\int_0^\infty U(R)R^{2\ell+2}\,dR,$$

valid only for finite range interactions $U(R > a) = 0$.

Example

For $U = -U_0$, $R \le a$ and $U = 0$, $R > a$, then

$$\tan\eta_\ell^{\mathrm{B}}(\ell \gg ka) = U_0 a^2\frac{(ka)^{2\ell+1}}{[(2\ell+1)!!]^2\,(2\ell+3)}.$$

The ratio $\eta_{\ell+1}/\eta_\ell \sim (ka/2\ell)^2$.

6.8.9. Coulomb scattering

For elastic scattering by the interaction $V(r) = Z_A Z_B e^2/r$, the Coulomb wave can be decomposed as

$$\Psi_{\mathbf{k}}^{(C)}(\mathbf{r}) = \frac{4\pi}{kr} \sum_{\ell,m} \imath^{\ell} e^{\imath \eta_\ell} F_{\epsilon\ell}(r) Y_{\ell m}^*(\hat{\mathbf{k}}) Y_{\ell m}(\hat{\mathbf{r}}),$$

where the radial wave varies asymptotically as

$$F_\ell \sim \sin(kR - \tfrac{1}{2}\ell\pi + \eta_\ell^{(C)} - \beta \ln 2kR),$$

where the parameter β is $Z_A Z_B e^2/\hbar v = Z_A Z_B (v_0/v)$. The Coulomb phase shift is

$$\eta_\ell^{(C)} = arg\Gamma(\ell + 1 + i\beta) = \operatorname{Im} \ln \Gamma(\ell + 1 + i\beta)$$

to give the Coulomb S-matrix element

$$S_\ell^{(C)} = \exp\left[2i\eta_\ell^{(C)}\right] = \frac{\Gamma(\ell + 1 + i\beta)}{\Gamma(\ell + 1 - i\beta)}.$$

The Coulomb scattering amplitude is

$$f_C(\theta) = -\frac{\beta \exp\left[2i\eta_\ell^{(C)} - i\alpha \ln\left(\sin^2 \tfrac{1}{2}\theta\right)\right]}{2k \sin^2 \tfrac{1}{2}\theta}.$$

The Coulomb differential cross section $| f_C |^2$ is

$$\left(\frac{d\sigma}{d\Omega}\right)_{\text{Coul}} = \frac{Z_A^2 Z_B^2 e^4}{16E^2} \csc^4 \frac{1}{2}\theta.$$

This is the Rutherford scattering cross section. It is interesting to note that Born and classical theory also reproduce this cross section. Moreover,

$$\left(\frac{d\sigma}{d\Omega}\right)_{\text{Coul}} (q) = \frac{4M_{AB}^2 Z_A^2 Z_B^2}{q^4} = 4a_0^2 (M_{AB}/m_e)^2 \left[\frac{Z_A^2 Z_B^2}{(Ka_0)^4}\right]$$

is a function only of the momentum transferred $q = \hbar K = 2\hbar k \sin \frac{1}{2}\theta$ in the collision. Note that $q^2 = 8M_{AB} E \sin^2 \frac{1}{2}\theta$.

6.9. COLLISIONS BETWEEN IDENTICAL PARTICLES

The identical colliding particles, each with spin s, are in a resolved state with total spin S_t in the range $(0 \to 2s)$. The spatial wave function, with respect to particle interchange, satisfies $\Psi(\mathbf{R}) = (-1)^{S_t} \Psi(-\mathbf{R})$. Wave functions for identical particles with even or odd total spin S_t are therefore symmetric (S) or antisymmetric (A) with respect to particle interchange. The appropriate combinations are $\Psi_{S,A}(\mathbf{R}) = \Psi(\mathbf{R}) \pm \Psi(-\mathbf{R})$, where the positive sign (symmetric wave function S) and the negative sign (antisymmetric wave function A) are associated with even and odd values of the total spin S_t, respectively. The scattering wave function for a pair of identical particles in spatially symmetric (+) or antisymmetric (−) states behaves asymptotically as

$$\Psi_{S,A}(\mathbf{R}) \to \left[\exp(i\mathbf{k}\cdot\mathbf{R}) \pm \exp(-i\mathbf{k}\cdot\mathbf{R})\right] + [f(\theta,\phi) \pm f(\pi-\theta,\phi+\pi)]\,\frac{\exp(ikR)}{R}.$$

The differential cross section for scattering of both the projectile and target particles into direction θ is

$$\left(\frac{d\sigma}{d\Omega}\right)_{S,A} = |\, f(\theta,\phi) \pm f(\pi-\theta,\phi+\pi)\,|^2$$

in the CM frame, where scattering of the projectile into polar direction$(\pi - \theta, \phi + \pi)$ is accompanied by scattering of the identical target particle into direction (θ, ϕ). This is related to the probability that both identical particles are scattered into θ. In the classical limit, where the particles are distinguishable, the classical cross section is

$$\left(\frac{d\sigma}{d\Omega}\right)_{C} = |\, f(\theta,\phi)\,|^2 + |\, f(\pi-\theta,\phi+\pi)\,|^2$$

the sum of the cross sections for observation of the projectile and target particles in the direction (θ, ϕ). Since $P_\ell[\cos(\pi-\theta)] = (-1)^\ell P_\ell(\cos\theta)$, the differential cross section for ϕ-independent amplitudes f is then

$$\left(\frac{d\sigma}{d\Omega}\right)_{S,A} = \frac{1}{4k^2}\left|\sum_{\ell=0}^{\infty}\omega_\ell(2\ell+1)\left[\exp 2i\eta_\ell - 1\right]P_\ell(\cos\theta)\right|^2.$$

For scattering in the symmetric (S) channel where S_t is even, $\omega_\ell = 2$ for ℓ even and $\omega_\ell = 0$ for ℓ odd. For scattering in the antisymmetric channel where S_t is odd, $\omega_\ell = 0$ for ℓ even and $\omega_\ell = 2$ for ℓ odd. The integral cross section is

$$\sigma_{S,A}(E) = \frac{8\pi}{k^2}\sum_{\ell=0}^{\infty}\omega_\ell(2\ell+1)\sin^2\eta_\ell.$$

Let g_A and g_S be the fractions of states with odd and even total spins $S_t = 0, 1, 2, \ldots, 2s$. When the $2s+1$ spin-states S_t are unresolved, the appropriate combination of symmetric and antisymmetric cross sections is the weighted mean

$$\frac{d\sigma}{d\Omega} = g_S\left(\frac{d\sigma}{d\Omega}\right)_S + g_A\left(\frac{d\sigma}{d\Omega}\right)_A,$$

$$\sigma(E) = g_S\sigma_S(E) + g_A\sigma_A(E).$$

6.9.1. Fermion and Boson scattering

1. Fermions: For fermions with half integral spin s, the statistical weights are $g_S = s/(2s+1)$ and $g_A = (s+1)/(2s+1)$. The differential cross section for fermion-

fermion scattering is then

$$\frac{d\sigma_F}{d\Omega} = |\, f(\theta)\,|^2 + |\, f(\pi-\theta)\,|^2 - \left(\frac{2}{2s+1}\right)\mathrm{Re}\left[f(\theta) f^*(\pi-\theta)\right].$$

The integral cross section fermion-fermion collisions is

$$\sigma_F = \tfrac{1}{2}\left[\sigma_S + \sigma_A\right] - \tfrac{1}{2}\left[\sigma_S - \sigma_A\right]/(2s+1),$$

which reduces, for fermions with spin $1/2$, to

$$\sigma_F(E) = \frac{2\pi}{k^2}\left[\sum_{\ell=\text{even}}^{\infty}(2\ell+1)\sin^2\eta_\ell + 3\sum_{\ell=\text{odd}}^{\infty}(2\ell+1)\sin^2\eta_\ell\right].$$

2. Bosons: The statistical weights for bosons with integral spin s are $g_S = (s+1)/(2s+1)$ and $g_A = s/(2s+1)$. The differential cross section for boson-boson scattering is

$$\frac{d\sigma_B}{d\Omega} = |\, f(\theta)\,| + |\, f(\pi-\theta)\,| + \left(\frac{2}{2s+1}\right)\mathrm{Re}\left[f(\theta) f^*(\pi-\theta)\right].$$

The integral cross section boson-boson collisions is

$$\sigma_B = \tfrac{1}{2}\left[\sigma_S + \sigma_A\right] + \tfrac{1}{2}\left[\sigma_S - \sigma_A\right]/(2s+1),$$

which reduces, for bosons with zero spin, to

$$\sigma_B(E) = \frac{8\pi}{k^2}\left[\sum_{\ell=\text{even}}^{\infty}(2\ell+1)\sin^2\eta_\ell\right].$$

Symmetry oscillations therefore appear in the differential cross sections for fermion-fermion and boson-boson scattering. They originate from the interference between unscattered incident particles in the forward ($\theta = 0$) direction and backward scattered particles ($\theta = \pi, \ell = 0$). A general differential cross section for scattering of spin s particles is

$$\frac{d\sigma}{d\Omega} = |\, f(\theta)\,|^2 + |\, f(\pi-\theta)\,|^2 + \frac{(-1)^{2s}}{2s+1} 2\,\mathrm{Re}\left[f(\theta) f^*(\pi-\theta)\right].$$

6.9.2. Coulomb scattering of two identical particles

1. Two spin-zero bosons (e.g., ^{4}He – ^{4}He):

$$\frac{d\sigma}{d\Omega} = \frac{\beta^2}{4k^2}\left[\csc^4\tfrac{1}{2}\theta + \sec^4\tfrac{1}{2}\theta + 2\csc^2\tfrac{1}{2}\theta\sec^2\tfrac{1}{2}\theta\cos\gamma\right].$$

2. Two spin-$\frac{1}{2}$ fermions (e.g., $H^+ - H^+$, $e^\pm - e^\pm$):

$$\frac{d\sigma}{d\Omega} = \frac{\beta^2}{4k^2}\left[\csc^4\tfrac{1}{2}\theta + \sec^4\tfrac{1}{2}\theta - \csc^2\tfrac{1}{2}\theta\sec^2\tfrac{1}{2}\theta\cos\gamma\right].$$

3. Two spin-1 bosons (e.g., deuteron-deuteron):

$$\frac{d\sigma}{d\Omega} = \frac{\beta^2}{4k^2}\left[\csc^4 \tfrac{1}{2}\theta + \sec^4 \tfrac{1}{2}\theta + \tfrac{2}{3}\csc^2 \tfrac{1}{2}\theta \sec^2 \tfrac{1}{2}\theta \cos\gamma\right].$$

These are the Mott Formulae with $\beta = (Ze)^2/\hbar v$ and $\gamma = 2\beta \ln(\tan \frac{1}{2}\theta)$.

6.9.3. Scattering of identical atoms

Two ground-state hydrogen atoms, for example, interact via the $X^1\Sigma_g^+$ and $b^3\Sigma_u^+$ electronic states of H_2. The nuclei are interchanged by rotating the atom pair by π, then by reflecting the electrons first through the midpoint of **R**, and then through a plane perpendicular to the original axis of rotation. The midpoint reflection changes the sign only of the ungerade state wave function and both Σ^+ states are symmetric with respect to the plane reflection.

The cross section for scattering by the gerade potential is then the combination

$$\left(\frac{d\sigma}{d\Omega}\right)_g = \tfrac{1}{4}\left(\frac{d\sigma}{d\Omega}\right)_S + \tfrac{3}{4}\left(\frac{d\sigma}{d\Omega}\right)_A$$

of S and A cross sections, which involve the phase shifts η_ℓ^S calculated under the singlet interaction. For scattering by the ungerade triplet interaction,

$$\left(\frac{d\sigma}{d\Omega}\right)_u = \tfrac{1}{4}\left(\frac{d\sigma}{d\Omega}\right)_A + \tfrac{3}{4}\left(\frac{d\sigma}{d\Omega}\right)_S,$$

where the S and A cross sections involve the phase shifts η_ℓ^T calculated under the triplet interaction. Since the electrons have statistical weights 1/4 and 3/4 for the ${}^1\Sigma_g^+$ and ${}^3\Sigma_u^+$ states, the differential cross section for $H(1s) - H(1s)$ scattering by both potentials is

$$\frac{d\sigma}{d\Omega} = \tfrac{1}{4}\left(\frac{d\sigma}{d\Omega}\right)_g + \tfrac{3}{4}\left(\frac{d\sigma}{d\Omega}\right)_u.$$

The above combinations also hold for the integral cross sections.

Scattering of incident beam alone

Since the current of incident particles $j_i = 2v$, the cross sections presented by the target (i.e., the number of incident particles removed from the beam in unit time per unit incident current) are $\frac{1}{2}$ of all those above. For example,

$$\left(\frac{d\sigma}{d\Omega}\right)_{S,A}^I = \tfrac{1}{2}|\, f(\theta) \pm f(\pi - \theta)\,| = \tfrac{1}{2}\left(\frac{d\sigma}{d\Omega}\right)_{S,A}$$

and $\sigma_{S,A}^I(E) = \frac{1}{2}\sigma_{S,A}(E)$.

6.10. CLASSICAL POTENTIAL SCATTERING

A particle of mass M with *impact parameter* b and energy $E = \frac{1}{2}Mv^2$ is scattered by a spherically symmetric potential $V(R)$. Both the energy

$$E = \frac{1}{2}Mv^2 = \frac{1}{2}M\dot{R}^2 + V(R) + \frac{L^2}{2MR^2}$$

and angular momentum

$$L = Mvb = \left[MR^2(t)\dot{\psi}(t)\right]$$

are constants of the motion. The direction of the vector angular momentum $\mathbf{L} = \mathbf{R} \times M\mathbf{v}$ is also constant so that the orbit $\mathbf{R}(R, \psi)$ is confined to a plane normal to $\mathbf{L}$. The turning points R_i, where the radial velocity $\dot{R}$ vanishes, are the roots of $E = V_{\text{eff}}(R)$. The effective potential is

$$V_{\text{eff}}(R) = V(R) + \frac{L^2}{2MR^2} = V(R) + \frac{b^2}{R^2}E.$$

The smallest R_i provides the distance of closest approach $R_c(E, b)$, which is the pericenter of the planar orbit. The apse line joins the scattering origin O to the pericenter. The time taken to proceed from R_c to R is

$$t(R) = \frac{2}{v}\int_{R_c}^{R}\left[1 - \frac{V(R)}{E} - \frac{b^2}{R^2}\right]^{-1/2} dR,$$

which may be "turned-inside-out" to provide the full trajectory $R = R(t; E, b)$. The collision delay time is

$$\tau(E, b) = \frac{2}{v}\int_{R_c}^{R}\left[1 - \frac{V(R)}{E} - \frac{b^2}{R^2}\right]^{-1/2} dR - \frac{2}{v}\int_{R_c}^{R}\left[1 - \frac{b^2}{R^2}\right]^{-1/2} dR.$$

The time dependence of the angle $\psi(t)$ between $\mathbf{R}$ and the apse line is determined by

$$\psi(t; E, b) = vb\int_0^t R^{-2}(t)\, dt$$

when the trajectory $R = R(t)$ is known. The variation with R is given by

$$\psi(R; E, b) = b\int_{R_c}^{R} \frac{dR/R^2}{[1 - V(R)/E - b^2/R^2]^{1/2}}.$$

6.10.1. Deflection functions

For scattering by potential $V(R)$, the deflection function is

$$\chi(E, b) = \pi - 2b \int_{R_c}^{\infty} \frac{dR/R^2}{[1 - V(R)/E - b^2/R^2]^{1/2}},$$
$$= \pi - 2 \int_0^1 \left[\left\{ \frac{1 - V(R_c/x)/E}{1 - V(R_c)/E} \right\} - x^2 \right]^{-1/2} dx.$$

An expression that avoids spurious divergences is

$$\chi(E, b) = \pi + 2\frac{\partial}{\partial b} \int_{R_c}^{\infty} \left[1 - V(R)/E - b^2/R^2\right]^{1/2} dR.$$

Since $-\infty \leq \chi < \pi$, the scattered particle may wind or spiral many times around ($\chi \to -\infty$) the scattering center. The experimentally observed quantity is the scattering angle θ ($0 \leq \theta \leq \pi$), which is associated with the various deflections

$$\chi_i = +\theta, -\theta, -2\pi \pm \theta, -4\pi \pm \theta, \ldots (i = 1, 2, \ldots n)$$

resulting from n-different impact parameters b_i.

6.10.2. Classical cross sections

The differential cross section for is

$$\frac{d\sigma(\theta, E)}{d\Omega} = \sum_{i=1}^{n} \left| \frac{b_i\, db_i}{d(\cos\chi_i)} \right|.$$

The integral cross section for scattering by angles $\theta \geq \theta_0$ is

$$\sigma_0(E) = 2\pi \int_0^{\theta_0} b\, db,$$

where θ_0 results from one impact parameter $b_0 = b(\theta_0)$. When three impact parameters b_1, b_2, b_3, for example, produce the same scattering angle θ ($0 \leq \theta \leq \pi$), the cross section is

$$\sigma_0(E) = 2\pi \int_0^{b_1} b\, db + 2\pi \int_{b_2}^{b_3} b\, db = \pi \left[b_1^2 + b_3^2 - b_2^2\right].$$

The momentum-transfer (diffusion) cross section is

$$\sigma_{\rm d}(E) = 2\pi \int_0^{\infty} [1 - \cos\theta(E, b)]\, b\, db.$$

The viscosity cross section is

$$\sigma_{\rm v}(E) = 2\pi \int_0^{\infty} \left[1 - \cos^2\theta(E, b)\right] b\, db.$$

Examples

Hard Sphere:

$$V(R) = \infty,\ R \leq a, 0,\ R > a.$$

The scattering angle is independent of energy and varies with impact parameter b as

$$\theta(b) = \chi = \begin{cases} \pi - 2\sin^{-1} b/a, & b \leq a, \\ 0, & b > a. \end{cases}$$

Since $b(\theta) = a\cos\frac{1}{2}\theta$, the differential and integral cross sections $\frac{d\sigma}{d\Omega} = 1/4a^2$ and $\sigma = \pi a^2$ are geometric and independent of collision energy.

Potential Barrier:

$$V(R) = V_0,\ R \leq a0,\ R > a.$$

Classical scattering at energies $E < V_0$, is the same as for hard sphere reflection. For energies $E > V_0$,

$$\theta(b) = \begin{Bmatrix} 2\left[\sin^{-1}(b/na) - \sin^{-1}(b/a)\right], & 0 \leq b \leq b_0, \\ \pi - 2\sin^{-1}(b/a), & b_0 \leq b \leq a, \end{Bmatrix}$$

where $n^2 = 1 - V_0/E$ and $b_0 = na$. The two impact parameters that contribute to a given θ are

$$b_1(\theta) = \frac{an\sin\frac{1}{2}\theta}{[1 - 2n\cos\frac{1}{2}\theta + n^2]^{1/2}}, \qquad 0 < b_1 \leq b_0,$$

$$b_2(\theta) = a\cos\tfrac{1}{2}\theta, \qquad b_0 < b_2 \leq a.$$

There is no scattering through angles $\theta \geq \theta_{\max} = 2\cos^{-1} n$. The differential cross section for scattering is

$$\frac{d\sigma}{d\Omega} = \frac{1}{4}a^2 + \frac{a^2n^2(n\cos\frac{1}{2}\theta - 1)(n - \cos\frac{1}{2}\theta)}{4\cos\frac{1}{2}\theta\left[1 + n^2 - 2\cos\frac{1}{2}\theta\right]}$$

for $\theta \leq \theta_{\max}$ and zero otherwise. The integral cross section is then

$$\sigma = \int_{\theta=0}^{\theta_{\max}} \left(\frac{d\sigma}{d\Omega}\right) d\Omega = \pi a^2,$$

the geometric cross section.

Potential Well:

$$V(R) = -V_0,\ \ R \leq a,\ \ 0,\ R > a.$$

Results are similar to the potential barrier case above, except that there is only a single scattering trajectory with $\theta = -\chi$, and $n = (1 + V_0/E)^{1/2}$ is the effective index of refraction for the equivalent problem in geometrical optics. Refraction occurs on entering and exiting the well. The collision properties are

$$\theta(b) = -2\left[\sin^{-1}(b/na) - \sin^{-1}(b/a)\right],$$

$$\theta(b = a) = \theta_{\max} = 2\cos^{-1}(1/n),$$

$$b(\theta) = \frac{-an\sin\frac{1}{2}\theta}{[1 - 2n\cos\frac{1}{2}\theta + n^2]^{1/2}},$$

$$\frac{d\sigma}{d\Omega} = \frac{a^2n^2\left[n\cos\frac{1}{2}\theta - 1\right]\left[n - \cos\frac{1}{2}\theta\right]}{4\cos\frac{1}{2}\theta\left[n^2 + 1 - 2n\cos\frac{1}{2}\theta\right]^2},$$

$$\sigma = \pi a^2.$$

Coulomb Attraction or Repulsion:

$$V(R) = \pm k/R,$$

$$\theta(b, E) = |\chi| = 2\csc^{-1}\left[1 + (2bE/k)^2\right]^{1/2},$$

$$b(\theta, E) = (k/2E)\cot\tfrac{1}{2}\theta,$$

$$\sigma(\theta) = \frac{d\sigma}{d\Omega} = \left(\frac{k}{4E}\right)^2 \csc^4\tfrac{1}{2}\theta.$$

Finite Range Coulomb:

$$V(R) = -k/R + k/R_s, \quad R \leq R_s = 0, R > R_s,$$

$$R_0(E) = \frac{k}{2E}, \quad \alpha(E) = R_0(E)/R_s,$$

$$\frac{d\sigma}{d\Omega} = \frac{R_0^2}{4}\left[\frac{1+\alpha}{\alpha^2 + (1+2\alpha)\sin^2\frac{1}{2}\theta}\right]^2.$$

6.10.3. Orbiting cross sections

Attractive interactions $V(R) = -C/R^n$ ($n \geq 2$) can support quasibound states with positive energy within the angular momentum barrier. Particles with $b < b_0$ spiral toward the scattering center. Those with $b = b_0$ are in unstable circular orbits of radius R_0. The radius R_0 is determined from the two conditions

$$\left(\frac{dV_{\text{eff}}}{dR}\right)_{R_0} = 0, \quad E = V_{\text{eff}}(R_0),$$

which, when combined, yields

$$E = V_{\text{eff}}(R_0) = V(R_0) + \tfrac{1}{2}R_0\left(\frac{dV}{dR}\right).$$

The angular momentum L_0 of the circular orbit is

$$L_0^2 = (2ME)b_0^2 = 2MR_0^2\,[E - V(R_0)].$$

The orbiting and spiraling cross section is then

$$\sigma_{\text{orb}}(E) = \pi b_0^2 = \pi R_0^2 F,$$

where the focusing factor is

$$F = 1 - \frac{V(R_0)}{E} = \frac{1}{2}\left(\frac{R_0}{E}\right)\left(\frac{dV}{dR}\right)_{R_0}.$$

Attractive power law potentials

$$V(R) = -C/R^n,\ (n > 2),$$

$$V_{\text{eff}}(R_0) = (1 - \tfrac{1}{2}n)V(R_0), \quad n > 2,$$

$$R_0(E) = \left[\frac{(n-2)C}{2E}\right]^{1/n}, \quad F = \left[\frac{n}{(n-2)}\right],$$

$$\sigma_{\text{orb}}(E) = \pi\left[\frac{n}{(n-2)}\right]\left[\frac{(n-2)C}{2E}\right]^{2/n}.$$

Langevin cross section

The Langevin cross section for spiraling encounters in ion-neutral atom (molecule) collisions due to the long range attraction $V(R) = -\alpha_{\text{d}}e^2/2R^4$, where α_{d} is the polarizability of the neutral, is

$$\sigma_{\text{L}}(E) = 2\pi R_0^2 = 2\pi\left(\frac{\alpha_{\text{d}}e^2}{2E}\right)^{1/2},$$

which varies as v^{-1}. The Langevin rate

$$k_{\text{L}} = v\sigma_{\text{L}}(E) = 2\pi(\alpha_{\text{d}}e^2/M)^{1/2}$$

is therefore independent of E. The orbiting cross section for van der Waals attraction $V(R) = -C/R^6$ is

$$\sigma_{\text{orb}}(E) = \tfrac{3}{2}\pi R_0^2 = \tfrac{3}{2}\pi\,(2C/E)^{1/3}.$$

6.11. QUANTAL INELASTIC HEAVY-PARTICLE COLLISIONS

The wave function for the complete $A - B$ collision system satisfies the Schrödinger equation

$$\mathcal{H}(\mathbf{r},\mathbf{R})\Psi(\mathbf{r},\mathbf{R}) = \left[\hat{H}_{\text{int}}(\mathbf{r}) - \frac{\hbar^2}{2M_{AB}}\nabla_{\mathbf{R}}^2 + V(\mathbf{r},\mathbf{R})\right]\Psi(\mathbf{r},\mathbf{R}) = E\Psi(\mathbf{r},\mathbf{R}),$$

where the internal Hamiltonian is the sum $\hat{H}_{\text{int}}(\mathbf{r}) = H_A(\mathbf{r}_A) + H_B(\mathbf{r}_B)$ of individual Hamiltonians $H_{A,B}$ for each isolated atomic or molecular species. The total energy (internal plus relative)

$$E = \frac{\hbar^2 k_i^2}{2M_{AB}} + \epsilon_i = \frac{\hbar^2 k_f^2}{2M_{AB}} + \epsilon_f$$

remains constant for all channels f throughout the collision. The combined internal energy ϵ_j of A and B at infinite separation R is $\epsilon_j(A) + \epsilon_j(B)$, which are the eigenvalues of the internal Hamiltonian $\hat{H}_{\text{int}}$ corresponding to the combined eigenstates $\Phi_A(\mathbf{r}_A)\Phi_B(\mathbf{r}_B)$. There are two limiting formulations (diabatic and adiabatic) for describing the relative motion. These depend on whether the mutual electrostatic interaction $V(\mathbf{r},\mathbf{R})$ between A and B at nuclear separation R, or the variation in the kinetic energy of relative motion, is considered to be a perturbation to the system, i.e., on whether the incident speed v_i is fast or slow in comparison with the internal motions, e.g., with the electronic speeds of the electrons bound to A and B.

6.11.1. Adiabatic formulation (kinetic coupling scheme)

When relaxation of the internal motion during the collision is fast compared with the slow collision speed v_i, or when the relaxation time is short compared with the collision time, the kinetic energy operator $(\hbar^2/2M_{AB})\nabla_R^2$ is then considered as a small perturbation to the quasi-molecular $A - B$ system at fixed $\mathbf{R}$. The system wave function $\Psi(\mathbf{r},\mathbf{R}) = \sum_n F_n(\mathbf{R})\Phi_n(\mathbf{r},\mathbf{R})$ is therefore expanded in terms of the known "adiabatic" molecular wave functions $\Phi_n(\mathbf{r},\mathbf{R})$ for the quasi-molecule AB at fixed nuclear separation $\mathbf{R}$. This set of orthornormal eigenfunctions satisfies

$$[\hat{H}_{\text{int}}(\mathbf{r}) + V(\mathbf{r},\mathbf{R})]\Phi_n(\mathbf{r},\mathbf{R}) = E_n(R)\Phi_n(\mathbf{r},\mathbf{R}).$$

As $R \to \infty$, both $\Phi_n(\mathbf{r},\mathbf{R})$ and the eigenenergies $E_n(R)$ tend in the limit of infinite nuclear separation $\mathbf{R}$ to the (diabatic) eigenfunctions $\Phi_n(\mathbf{r}_A,\mathbf{r}_B) = \psi_i(\mathbf{r}_A)\phi_j(\mathbf{r}_B)$ and eigenenergies ϵ_n, of $\hat{H}_{\text{int}}$, respectively. The substitution $\Psi(\mathbf{r},\mathbf{R}) = \sum_n F_n(\mathbf{R})\Phi_n(\mathbf{r},\mathbf{R})$ into the Schrödinger equation results in the following set of coupled equations for the relative motion functions F_n:

$$[\nabla_R^2 + \mathcal{K}_n^2(\mathbf{R})]F_n(\mathbf{R}) = \sum_j [X_{nj}\cdot\nabla_{\mathbf{R}} + T_{nj}(\mathbf{R})]F_j(\mathbf{R}).$$

The local wavenumber $\mathcal{K}_n$ is determined from $\mathcal{K}_n^2 = 2M_{AB}[E - E_n(R)]/\hbar^2$ and the coupling matrix elements are

$$X_{nj}(\mathbf{R}) = -2\langle\Phi_n(\mathbf{r}, \mathbf{R})|\, \nabla_\mathbf{R}\, |\Phi_j(\mathbf{r}, \mathbf{R})\rangle_\mathbf{r}$$

and

$$T_{nj}(\mathbf{R}) = -\langle\Phi_n(\mathbf{r}, \mathbf{R})|\, \nabla^2_\mathbf{R}\, |\Phi_j(\mathbf{r}, \mathbf{R})\rangle_\mathbf{r}.$$

Solution of the above set for $F_n(\mathbf{R})$ represents the adiabatic close coupling method. The adiabatic states are normally determined (via standard computational techniques of quantum chemistry) relative to a set of axes (X', Y', Z') with Z'-axis directed along nuclear separation $\mathbf{R}$. On transforming to this set, which rotates during the collision, $\Psi(\mathbf{r}', \mathbf{R}')$, for the diatomic $A - B$ case, satisfies

$$\left[\hat{H}_0(\mathbf{r}') + V(\mathbf{r}', \mathbf{R}') - \frac{\hbar^2}{2M_{AB}R'^2}\hat{K}\right]\Psi(\mathbf{r}', \mathbf{R}') = E\Psi(\mathbf{r}', \mathbf{R}'),$$

where the perturbation operator to the molecular wave functions in the rotating frame is

$$\hat{K} = \frac{\partial}{\partial R'}\left(R'^2\frac{\partial}{\partial R'}\right) - (\hat{L}_{X'} - \hat{J}_{X'})^2 - (\hat{L}_{Y'} - \hat{J}_{Y'})^2$$

in terms of the operators $\hat{L}$ and $\hat{J}$ for the total and internal angular momentum $\mathbf{L}$ and $\mathbf{J}$, respectively, of the collision system. Note $L_{Z'} = J_{Z'}$, for diatoms. An advantage of using this rotating system in the adiabatic treatment is that radial perturbations, which cause vibrational $v \to v'$ and electronic $nl \to n'l$ transitions, originate from the first term (radial) of $\hat{K}$, whereas angular perturbations (torques) that cause rotational $J \to J'$ and electronic $nl \to nl'$ transitions originate from the angular momentum operator products $[\hat{L}_{X'}\hat{J}_{X'} + \hat{L}_{Y'}\hat{J}_{X'}]$. The use of a rotating frame causes some complication, however, to the direct use of the asymptotic boundary condition for $\Psi(\mathbf{r}', \mathbf{R}')$.

6.11.2. Diabatic formulation (potential coupling scheme)

When relaxation of the internal motion is slow compared with the fast relative speed v_i, Ψ is expanded in terms of the known unperturbed (diabatic) orthonormal eigenstates $\Phi_j(\mathbf{r}_A, \mathbf{r}_B) = \psi_i(\mathbf{r}_A)\phi_k(\mathbf{r}_B)$ of $\hat{H}_{\text{int}}$ according to

$$\Psi(\mathbf{r}, \mathbf{R}) = \sum_j F_j(\mathbf{R})\Phi_j(\mathbf{r}).$$

Substituting into the Schrödinger equation, multiplication by $\Phi_n^*(\mathbf{r})$ and integration over $\mathbf{r}$, the unknown functions $F_n(\mathbf{R})$ for the relative motion in channel n satisfy the infinite set of coupled equations

$$[\nabla^2_R + \mathcal{K}^2_n(\mathbf{R})]F_n(\mathbf{R}) = \sum_{j\neq i} U_{nj}(\mathbf{R})F_j(\mathbf{R}).$$

The reduced potential matrix elements that couple the internal states n and j are

$$U_{nj}(\mathbf{R}) = \frac{2M_{AB}}{\hbar^2}V_{nj}(\mathbf{R}) = U^*_{jn}(\mathbf{R}),$$

where the electrostatic interaction averaged over states n and j is

$$V_{nj}(\mathbf{R}) = \int \Phi_n^*(\mathbf{r}) V(\mathbf{r}, \mathbf{R}) \Phi_j(\mathbf{r})\, d\mathbf{r}.$$

The local wavenumber $\mathcal{K}_n$ of relative motion under the static interaction V_{nn} is given by

$$\mathcal{K}_n^2(\mathbf{R}) = k_n^2 - U_{nn}(\mathbf{R}).$$

The diagonal elements U_{nn} are the distortion matrix elements that distort the relative motion from plane waves in elastic scattering, whereas the off-diagonal matrix elements, U_{if} U_{ij} and U_{jf} that couple states i and f either directly or via intermediate channels j cause inelastic scattering and polarizations contributions to elastic scattering. In contrast to the adiabatic formulation, radial and angular transitions originate in the diabatic formulation from the radial and angular components to the potential coupling elements $V_{nj}(\mathbf{R})$. The set of coupled are solved subject to the usual asymptotic ($R \to \infty$) requirement that

$$F_j(\mathbf{R}) \sim \exp(ik_i Z)\delta_{ij} + f_{ij} \exp(ik_j R)/R$$

for the elastic $i = j$ and inelastic $i \neq j$ scattered waves. In terms of the amplitude f_{ij} for scattering into direction (θ, ϕ), the differential and integral cross sections for $i \to j$ transitions are

$$\frac{d\sigma_{ij}}{d\Omega} = \frac{v_j}{v_i} | f_{ij}(\theta, \phi) |^2$$

and

$$\sigma_{ij} = \frac{v_j}{v_i} \int_0^{\pi} d(\cos\theta) \int_0^{2\pi} | f_{ij}(\theta, \phi) |^2 \, d\phi.$$

As well as obtaining the scattering amplitude from the above asymptotic boundary conditions, f_{ij} can also be obtained from the integral representation, for the scattering amplitude is

$$f_{ij}(\theta) = \langle \Phi_j(\mathbf{r}) \exp(i\mathbf{k}_j \cdot \mathbf{R}) | \, V(\mathbf{r}, \mathbf{R}) \, | \Psi(\mathbf{r}, \mathbf{R}) \rangle_{\mathbf{r},\mathbf{R}}.$$

6.11.3. Inelastic scattering by a central field

When the atom-atom or atom-molecule interaction is spherically symmetric in the channel vector $\mathbf{R}$, i.e., $V(\mathbf{r}, \mathbf{R}) = V(\mathbf{r}, R)$, the orbital $\mathbf{l}$ and rotational $\mathbf{j}$ angular momenta are each conserved throughout the collision so that a ℓ-partial wave decomposition of the translational wave functions for each value of j is possible. The translational wave is decomposed according to

$$F_j(\mathbf{R}) = \frac{4\pi N}{k_i R} \sum_{\ell,m} \imath^{\ell} F_{j\ell}(R) Y_{\ell m}^*(\hat{\mathbf{k}}) Y_{\ell m}(\hat{\mathbf{R}})$$

and inserted into the diabatic set of coupled equations (of Sec. 6.11.2). The radial wave function $F_{j\ell}$ is then the solution of

$$\frac{d^2F_{j\ell}}{dR^2} + \left[k_i^2 - \left\{U_{ii}(R) + \frac{\ell(\ell+1)}{R^2}\right\}\right]F_{j\ell}(R) = \sum_{j\neq i} U_{ij}(R)F_{j\ell}(R),$$

which is the direct generalization of the quantal radial equation for potential scattering to directly include other channels $j \neq i$. The coupled equations are now solved subject to the requirements that

$$F_{i\ell}(k_iR) \sim \sin(k_iR - \ell\pi/2) + \left\{\frac{T_{ii}^{\ell}}{2i}\right\} e^{i(k_iR-\ell\pi/2)}$$

for the elastic scattered wave and

$$F_{j\ell}(k_jR) \sim \left(\frac{k_i}{k_j}\right)^{\frac{1}{2}} \left\{\frac{T_{ij}^{\ell}}{2i}\right\} e^{i(k_jR-\ell\pi/2)}$$

for the inelastic wave. The transition-matrix elements for elastic and inelastic scattering are

$$T_{ij}^{\ell} = -\frac{2i}{(k_ik_j)^{\frac{1}{2}}}\int_0^{\infty} F_{j\ell}^{(0)}(r)U_{ji}(r)F_{i\ell}(r)\,dr,$$

where $F_{j\ell}^{(0)} = (k_fr)j_\ell(k_fr)$ and $F_{i\ell}(r)$ are the solutions of the above-coupled radial equations. The differential cross section for inelastic scattering is

$$\frac{d\sigma_{ij}}{d\Omega} = (1/4k_i^2)\left|\sum_{\ell=0}^{\infty}(2\ell+1)T_{ij}^{\ell}P_\ell(\cos\theta)\right|^2.$$

The integral inelastic cross section is

$$\sigma_{ij}(E) = \frac{\pi}{k_i^2}\sum_{\ell=0}^{\infty}(2\ell+1)|\,T_{ij}^{\ell}\,|.$$

The transition matrix $\mathbf{T}^{\ell} = \{T_{ij}^{\ell}\}$ is symmetrical, $T_{ij}^{\ell} = T_{ji}^{\ell}$, and the cross sections satisfy detailed balance. Each transition matrix element $|\,T_{ij}^{\ell}\,|^2$ is the probability of an $i \to f$ transition in the target for each value ℓ of the (orbital) angular momentum of relative motion.

6.11.4. Two-state treatment

Here, all couplings are ignored except the direct couplings between the initial and final states, as in a two-level atom. The coupled equations to be solved are

$$[\nabla^2 + k_i^2 - U_{ii}(\mathbf{R})]\psi_i(\mathbf{R}) = U_{if}(\mathbf{R})\psi_f(\mathbf{R}),$$
$$[\nabla^2 + k_f^2 - U_{ff}(\mathbf{R})]\psi_f(\mathbf{R}) = U_{fi}(\mathbf{R})\psi_i(\mathbf{R}).$$

Distorted-wave approximation

Here, all matrix elements in the two-level equations (Sec. 6.11.4) are included, except the back coupling $V_{if}\Psi_f$ term, which provides the influence of the inelastic channel on the elastic channel and is required to conserve probability. Distortion of the elastic and outgoing inelastic waves by the averaged (static) interactions V_{ii} and V_{ff}, respectively, is therefore included. The two state equations can then be decoupled and effectively reduced to one-channel problems. An analogous static-exchange distortion approximation, where exchange between the incident and one of the target particles, also follows from the two-level treatment.

Born approximation

Here, the distortion (diagonal) and back coupling matrix elements in the two-level equations (Sec. 6.11.4) are ignored so that $\psi_i(R) = \exp(i\mathbf{k}_i \cdot \mathbf{R})$ remains an undistorted plane wave. The asymptotic solution for ψ_f when compared with the asymptotic boundary condition then provides the Born elastic ($i = f$) or inelastic scattering amplitudes

$$f_{if}^B(\theta, \phi) = -\frac{1}{4\pi}\frac{2M_{AB}}{\hbar^2}\int V_{fi}(\mathbf{R}) \exp i\mathbf{K}\cdot\mathbf{R}\, d\mathbf{R}.$$

The momentum change resulting from the collision is $\mathbf{Q} = \hbar\mathbf{K}$, where $\mathbf{K} = \mathbf{k}_i - \mathbf{k}_f$. The Born amplitude also follows by inserting $\Psi(\mathbf{r}, \mathbf{R}) = \Phi_i(\mathbf{r}) \exp i(\mathbf{k}_i \cdot \mathbf{R})$ in the integral representation. Comparison with potential scattering shows that the elastic scattering of structured particles occurs in the Born approximation via the averaged electrostatic interaction $V_{ii}(\mathbf{R})$.

For electron-ion or ion-ion collisions, the plane waves $\exp(i\mathbf{k}_{i,f}\cdot\mathbf{R})$ are simply replaced by Coulomb waves to provide the Coulomb-Born approximation.

6.11.5. Exact resonance

The two-state equations of Sec. 6.11.4 cannot in general be solved analytically except for the specific case of exact resonance when $k_i = k_f = k$ and $U_{ii} = U_{ff} = U$, $U_{if} = U_{fi}$. Then, the equations can be decoupled by introducing the linear combinations $\psi^{\pm}(\mathbf{R}) = \frac{1}{\sqrt{2}}[\psi_i(\mathbf{R}) \pm \psi_f(\mathbf{R})]$, so the two-state set can be converted to two one-channel decoupled equations

$$[\nabla^2 + k^2 - (U \pm U_{if})]\psi^{\pm}(\mathbf{R}) = 0.$$

The problem has therefore been reduced to potential scattering by the interactions $U_{\pm} = (U \pm U_{if})$ associated with elastic scattering amplitudes $f^{\pm}$. Hence, the elastic ($i = f$) and "inelastic" ($i \neq f$) amplitudes are

$$f_{ii} = (f^+ + f^-)/2 \quad f_{if} = (f^+ - f^-)/2.$$

In terms of the phase shifts $\eta_l^{\pm}$ associated with potential scattering by $U_{\pm}$, the amplitudes for elastic and inelastic scattering are then

$$f_{in}(\theta) = \frac{1}{2ik}\sum_{l=0}^{\infty}(2l+1)[(e^{2i\eta_l^+} + e^{2i\eta_l^-})/2 - 1]P_l(\cos\theta)$$

and

$$f_{if}(\theta) = \frac{1}{2ik}\sum_{l=0}^{\infty}(2l+1)[(e^{2i\eta_l^+} - e^{2i\eta_l^-})/2 - 1]P_l(\cos\theta).$$

The corresponding differential cross sections $| f_{if} |^2$ will therefore exhibit interference oscillations. The integral cross sections are

$$\sigma_{ii} = \frac{4\pi}{k^2}\sum_{l=0}^{\infty}(2l+1)\left[\frac{1}{2}(\sin^2\eta_l^+ + \sin^2\eta_l^-) - \frac{1}{4}\sin^2(\eta_l^+ - \eta_l^-)\right]$$

and

$$\sigma_{if} = \frac{\pi}{k^2}\sum_{l=0}^{\infty}(2l+1)\sin^2(\eta_l^+ - \eta_l^-),$$

respectively.

Examples: atomic collisions with identical nuclei

Important cases of exact resonance are the symmetrical resonance charge transfer collision

$$\mathrm{He}_f^+(1s) + \mathrm{He}_s(1s^2) \rightarrow \mathrm{He}_f(1s^2) + \mathrm{He}_s^+(1s),$$

which converts a fast ion beam f to a fast neutral beam via electron transfer and the excitation transfer collision

$$\mathrm{He}(1s2s\ ^3S) + \mathrm{He}(1s^2\ ^1S) \rightarrow \mathrm{He}(1s^2\ ^1S) + \mathrm{He}(1s2s\ ^3S),$$

which transfers the internal excitation in the projectile beam fully to the target atom. The electronic molecular wave functions divide into even (gerade) or odd (ungerade) classes on reflection about the midpoint of the internuclear line ($\mathbf{R} \rightarrow -\mathbf{R}$). In the separated atom limit, $\Psi_{g,u} \sim \phi(\mathbf{r}_A) \pm \phi(\mathbf{r}_B)$.The potentials $U_\pm$ in the former case are the gerade and ungerade interactions $V_{g,u}$. The phase shifts for elastic scattering by the resulting gerade (g) and ungerade (u) molecular potentials of A_2^+ are, respectively, η_ℓ^g and η_ℓ^u. The charge transfer (X) and transport cross sections are then

$$\sigma_X(E) = \frac{\pi}{k^2}\sum_{\ell=0}^{\infty}(2\ell+1)\sin^2(\eta_\ell^g - \eta_\ell^u),$$

$$\sigma_{u,g}^{(1)}(E) = \frac{4\pi}{k^2}\sum_{\ell=0}^{\infty}(\ell+1)\sin^2(\beta_\ell - \beta_{\ell+1}),$$

$$\sigma_{u,g}^{(2)}(E) = \frac{4\pi}{k^2}\left(\frac{3}{2}\right)\sum_{\ell=0}^{\infty}\frac{(\ell+1)(\ell+2)}{(2\ell+3)}\sin^2(\beta_\ell - \beta_{\ell+2}).$$

For ungerade potentials, $\beta_\ell = \eta_\ell^g$ for ℓ even and η_ℓ^u for ℓ odd. For gerade potentials, $\beta_\ell = \eta_\ell^u$ for ℓ even and η_ℓ^g for ℓ odd. The diffusion cross section $\sigma_{u,g}^{(1)}$ contains (g/u) interference. The viscosity cross section $\sigma_{u,g}^{(2)}$ does not. For charge transfer between the heavier rare gas ions Rg^+ with their parent atoms Rg, the degenerate states at large internuclear separations are not s-states but p-states. The states are then $\Sigma_{g,u}$, which originate from the p-state with $m = 0$ and $\Pi_{g,u}$, which originates from $m = \pm 1$ with space quantization along the molecular axis. Since there is no coupling between molecular states of different electronic angular momentum, the scattering by the $^2\Sigma_{g,u}$ pair and the $^2\Pi_{g,u}$ pair of Ne_2^+ potentials (for example) is independent. The cross section is therefore the combination

$$\sigma_{el,X}(E) = \tfrac{1}{3}\sigma_\Sigma(E) + \tfrac{2}{3}\sigma_\Pi(E)$$

of cross sections σ_Σ and σ_Π for the individual contributions originating from the isolated $^2\Sigma_{g,u}$ and $^2\Pi_{g,u}$ states to elastic el or charge-transfer X scattering. See Refs. [27, 28] for further details on excitation transfer and charge transfer collisions.

Singlet-triplet spin flip cross section

This cross section is

$$\sigma_{\mathrm{ST}}(E) = \frac{\pi}{k^2}\sum_{\ell=0}^{\infty}(2\ell+1)\sin^2(\eta_\ell^{\mathrm{s}} - \eta_\ell^{\mathrm{t}}),$$

where $\eta_\ell^{\mathrm{s,t}}$ are the phase shifts for individual potential scattering by the singlet and triplet potentials, respectively.

6.11.6. Partial wave analysis

In order to reduce the three-dimensional diabatic or adiabatic set of coupled equations for atom-atom and atom-molecule scattering to a corresponding working set of coupled radial equations, analogous to those in Sec. 6.11.3, the orbital angular momentum $\mathbf{l}$ of relative motion must be distinguished from the combined internal angular momentum $\mathbf{j}$ associated with the internal (rotational and electronic) degrees of freedom of the partners A and B at rest at infinite separation R. Both the orbital angular momentum $\mathbf{l}$ of relative motion and the internal angular momentum $\mathbf{j}$ of the atomic electrons or of molecular rotation are in general coupled. The total angular momentum $\mathbf{J} = \mathbf{l} + \mathbf{j}$ and its component J_z along some fixed direction (of incidence) are each conserved. Angular momentum may therefore be exchanged between the internal (rotational) and translational (orbital) degrees of freedom via the couplings $V_{nm}(\mathbf{R})$ or $\hat{K}$. Partial wave analysis is an exercise in angular momentum coupling and is well established (e.g., Ref. [29]) for both the diabatic and adiabatic treatments of heavy-particle collisions.

6.11.7. Close coupling equations for electron-atom (ion) collisions

A partial wave decomposition provides the full close-coupling quantal method for treating $A - B$ collisions, electron-atom, electron-ion, or atom-molecule collisions. The method [30] is provided here for the inelastic processes

$$e^- + A_i \to e^- + A_f$$

at collision speeds less than or comparable with those target electrons actively involved in the transition. It is based on an expansion of the total wave function Ψ for the $(e^- - A)$ multielectron system in terms of a sum of products of the known atomic target state wave functions $|\Phi_i>$ and the unknown functions $F_i(r)$ for the relative motion. Here,

$$\Psi^\Gamma(\mathbf{r}_1, \ldots, \mathbf{r}_N; \mathbf{r}) = \mathcal{A} \sum\!\!\!\!\!\!\int_i \Phi_i^\Gamma(\mathbf{r}_1, \mathbf{r}_2, \ldots \mathbf{r}_N; \hat{\mathbf{r}}) \frac{1}{r} F_i^\Gamma(r)$$

involves a sum over all discrete and an integral over the continuum states of the target. The operator $\mathcal{A}$ antisymmetrizes the summation with respect to exchange of all pairs of electrons in accordance with the Pauli exclusion principle. The angular and spin momenta of the projectile electron are denoted collectively by $\hat{\mathbf{r}}$ and have been coupled with the orbital and spin angular momenta of the target states $|\Phi_i>$ to produce the "channel functions" $\Phi_i^{LS\pi}(\mathbf{r}_1, \mathbf{r}_2, \ldots, \mathbf{r}_i; \hat{\mathbf{r}})$, which are eigenstates of the total orbital $\mathbf{L}$, total spin $\mathbf{S}$, angular momentum, their Z - components M_L, M_S, and parity π. The set $\Gamma \equiv LSM_LM_S\pi$ of quantum numbers are therefore conserved throughout the collision. By substituting the expansion for Ψ^Γ into the Schrödinger equation,

$$H_{N+1}\Psi^\Gamma = \left[\sum_{i=1}^{N+1}\left(-\frac{1}{2}\nabla_i^2 - \frac{Z}{r_i}\right) + \sum_{i>j=1}^{N+1} \frac{1}{r_{ij}}\right]\Psi^\Gamma,$$

expressed in atomic units, the radial functions for the motion of the scattered electron satisfy the infinite set of coupled integro-differential equations

$$\left[\frac{d^2}{dr^2} + k_i^2 - \frac{\ell_i(\ell_i+1)}{r^2} + \frac{2(Z-N)}{r}\right] F_i^\Gamma(r) = 2\sum_j \left[V_{ij}^\Gamma(r) + W_{ij}^\Gamma(r)\right] F_j^\Gamma(r).$$

The direct potential couplings are represented by

$$V_{ij}^\Gamma(r) = Z_p \left[\frac{Z_t}{r}\delta_{ij} + \sum_{k=1}^{N}\left\langle \Phi_i^\Gamma \left| \frac{1}{|\mathbf{r}_k - \mathbf{r}|} \right| \Phi_j^\Gamma \right\rangle\right].$$

The nonlocal exchange couplings are represented by

$$W_{ij} F_j^\Gamma(r) = \sum_{k=1}^{N}\left\langle \Phi_i^\Gamma \left| \frac{1}{|\mathbf{r}_k - \mathbf{r}|} \right| (\mathcal{A} - 1)\Phi_j^\Gamma F_j^\Gamma \right\rangle.$$

The direct potential gives rise to the long-range polarization attraction that is very important for low-energy scattering. The exchange potentials are short range and are extremely complicated. Additional nonlocal potentials that originate from various correlations (which cannot be included directly but which can be constructed from pseudostates) can also be added to the RHS of the equations.

Numerical solution of the above set of close coupled equations is feasible only for a limited number of close target states. For each N, several sets of independent solutions F_{ij} of the resulting close coupled equations are determined subject to $F_{ij} = 0$ at $r = 0$ and to the reactance **K**-matrix asymptotic boundary conditions,

$$F_{ij}^{\Gamma} \sim \sin\theta_i \delta_{ij} + K_{ij}^{\Gamma} \cos\theta_i$$

for n-open channels characterized by $k_i^2 = 2(E - E_i) > 0$. The argument is

$$\theta_i = k_i r - \frac{1}{2}\ell_i \pi + \frac{Z-N}{k_i} \ln(2k_i r) + \sigma_i,$$

where ℓ_i is the orbital angular momentum of the scattered elecron and where $\sigma_i = \arg\Gamma[\ell_i + 1 - i(Z-N)/k_i]$ is the Couloumb phase. For closed channels, $k_i^2 < 0$ and $F_{ij} \sim C_{ij} \exp(-|k_i|r)$ as $r \to \infty$. The scattering amplitude can then be expressed in terms of the elements T_{ij} of the $(n \times n)$ **T**-matrix, which is related to the **K** and **S** matrices by

$$\mathbf{T}^{\Gamma} = \frac{2i\mathbf{k}^{\Gamma}}{\mathbf{I} - i\mathbf{K}^{\Gamma}} = \mathbf{S}^{\Gamma} - \mathbf{I}.$$

The integral cross section for the transition $i \equiv \alpha_a L_i S_i \to f \equiv \alpha_f L_f S_f$ in the target atom, where α denotes the additional quantum numbers required to completely specify the state, is then

$$\sigma_{if}(k_i^2) = \frac{\pi}{k_i^2} \sum_{LS\pi,\ell_i,\ell_j} \frac{(2L+1)(2S+1)}{2(2L_i+1)(2S_i+1)} |T_{ij}^{\Gamma}|^2.$$

According to detailed balance, the collision strength

$$\Omega_{if} = k_i^2 (2L_i+1)(2S_i+1)\sigma_{if}(k_i^2)$$

is therefore dimensionless and is symmetric with respect to the $i \to f$ interchange. Further extensions, simplifications, and calculational schemes of the basic close coupling and related methods are found in Refs. [30]–[32].

6.12. SEMICLASSICAL INELASTIC SCATTERING

The term semiclassical is used in scattering theory to denote many different situations.

1. The use of some time-dependent classical path $\mathbf{R}(t)$ within a time-dependent quantal treatment of the response of the internal degrees of freedom of A and B to the time-varying field $V(\mathbf{R}(t))$ created by the approach of A toward B along the classical trajectory $\mathbf{R}(t)$. This procedure generalizes classical theory for potential scattering to structured collision partners and inelastic transitions.
2. The use of the three-dimensional eikonal phase $S(\mathbf{R})$, which is the solution of the Hamilton-Jacobi Equation, for the channel wave function $\Psi^+(\mathbf{R})$, within the full quantal expression for the cross section.
3. The use of JWKB-approximate solutions of the radial Schrödinger equation for the radial wave function $R_{\epsilon,\ell}$ for $A-B$ relative motion within the full quantum treatment of the $A-B$ collision.

6.12.1. Classical path theory

The basic assumption here is the existence over the inelastic scattering region of a common classical trajectory $\mathbf{R}(t)$ for the relative motion under an appropriately averaged central potential $\bar{V}[R(t)]$. The interaction $V[\mathbf{r}, \mathbf{R}(t)]$ between A and B may then be considered as time dependent. The system wave function therefore satisfies

$$i\hbar \frac{\partial \Psi(\mathbf{r}, t)}{\partial t} = [\hat{H}_{\text{int}}(\mathbf{r}) + V(\mathbf{r}, \mathbf{R}(t))]\Psi(\mathbf{r}, t)$$

and can be expanded in terms of the eigenfunctions Φ_n of $\hat{H}_{\text{int}}$ as

$$\Psi(\mathbf{r}, t) = \sum_n A_n(t)\Phi_n(\mathbf{r}) \exp(-iE_n t).$$

The transition amplitudes A_n then satisfy the set

$$i\hbar \frac{\partial A_n(b, t)}{\partial t} = \sum_n A_j(b, t) V_{nj}(\mathbf{R}(t) \exp(i\omega_{nj} t)$$

of first-order equations coupled by the matrix elements $V_{nj}(\mathbf{R})$ between states n and j with energy separation $\hbar\omega_{nj} = E_n - E_j$. Once the classical trajectory $\mathbf{R} \equiv (R(t), \theta(t), \phi = \text{const.})$ is established from the classical equations

$$\frac{dR}{dt} = \pm v[1 - b^2/R^2 - \bar{V}(R)/E]^{\frac{1}{2}}, \qquad \frac{db}{dt} = \frac{vb}{R^2}$$

of motion for impact parameter b and kinetic energy $E = \frac{1}{2}M_{AB}v^2$, the coupled equations are solved subject to the requirement $A_n(b, t \to -\infty) = \delta_{ni}$. Since the probability for an $i \to f$ transition is $P_{if} = |A_f(b, t \to \infty|^2$, the differential cross section for inelastic scattering is

$$\frac{d\sigma_{if}}{d\Omega} = \sum_n P_{if}(b_n, \phi) \left\{ \frac{d\sigma_{el}}{d\Omega} \right\},$$

where $d\sigma_{el}/d\Omega$ is the differential cross section $|b db/d(\cos\theta)|$ for elastic scattering by $\bar{V}(R)$ and the summation is over all trajectories b_n that pass through (θ, ϕ). The integral cross section is

$$\sigma_{if} = 2\pi \int_0^\infty |A_f(b, \infty)|^2 \, b \, db.$$

Impact parameter method

This normally refers to the use of the straight line trajectory $R(t) = (b^2 + v^2t^2)^{1/2}$, $\theta(t) = \arctan(b/vt)$ within the classical path treatment. See Bates [17, 18] for examples and further discussion.

6.12.2. Landau-Zener cross section

The Landau-Zener transition probability is derived from an approximation to the full two-state impact-parameter treatment of the collision. The single passage probability for a transition between the diabatic surfaces $H_{11}(\mathbf{R})$ and $H_{22}(\mathbf{R})$, which cross at R_X is the Landau-Zener transition probability

$$P_{12}(R_X; b) = 1 - \exp(-2\pi |H_{12}(R_X)|^2/\hbar v_X |H'_{11} - H'_{22}|),$$

where H_{12} is the interaction coupling states 1 and 2. The diabatic curves are assumed to have linear shapes in the vicinity of the crossing at R_X, i.e., $(H'_{11} - H'_{22} = \Delta F)$ and H_{12} is assumed constant. The adiabatic surfaces

$$W^{\pm} = \tfrac{1}{2}(H_{11} + H_{22}) \pm \tfrac{1}{2}[(H_{11} - H_{22})^2 + 4H_{12}]$$

do not cross (avoided crossing). They are separated at R_X by $W^+ - W^- = 2H_{12}(R_X)$. The probability for remaining on the adiabatic surface is $P_{12}(R_X)$. The probability for remaining on the diabatic surface or for pseudocrossing between the adiabatic curves is $1 - P_{12}(R_X)$. The overall transition probability for both the incoming and outgoing legs of the trajectory $\mathbf{R}(t)$ is then

$$\mathcal{P}_{12} = 2P_{12}(1 - P_{12}).$$

The Landau-Zener cross section is

$$\sigma_{12} = 4\pi \int_0^{R_X} P_{12}(1 - P_{12}) b\, db,$$

where the variation of P_{12} on impact parameter b originates from the speed $v_X \approx v_0[1 - V(R_X)/E - b^2/R_X^2]^{1/2}$ at the crossing point R_X. For rectilinear trajectories $\mathbf{R} = \mathbf{b} + \mathbf{v}_0 t$,

$$\sigma_{12}(v_0) = 4\pi R_X^2 [E_3(\alpha) - E_3(2\alpha)],$$

where $E_n(\alpha) = \int_1^\infty y^{-n} \exp(-\alpha y)\, dy$ is the exponential integral with argument $\alpha = 2\pi |H_{12}|^2/\hbar v_0 \Delta F$. See Nikitin [19, 20] for more elaborate models that include interference effects originating from the phases or eikonals associated with the incoming and outgoing legs of the trajectory.

6.12.3. Eikonal theories

Here, the relative motion wave function $F_n(\mathbf{R})$ is decomposed as [21]

$$F_n(\mathbf{R}) = A_n(\mathbf{R}) \exp i S_n(\mathbf{R}) \exp(-\chi_n(\mathbf{R})).$$

The classical action, or solution of the Hamilton-Jacobi Equation $\nabla S_n(\mathbf{R}) = \kappa_n(\mathbf{R})$, for relative motion under the channel interaction $V_{nn}(\mathbf{R})$, is

$$S_n(\mathbf{R}) = \mathbf{k}_n \cdot \mathbf{R} + \int_{\mathbf{R}_0}^{\mathbf{R}} (\kappa_n - \mathbf{k}_n) \cdot \mathbf{dR}_n,$$

where $\mathbf{R}_0$ is the initial point on the associated trajectory $\mathbf{R}_n(t)$, where $\kappa_n = k_n$. The current $\mathbf{J}_n$ in channel n, assumed elastic, satisfies the conservation condition $\nabla \mathbf{J}_n = 0$, so that χ_n is the solution of

$$\nabla^2_{\mathbf{R}} S_n - 2(\nabla_{\mathbf{R}} S_n) \cdot (\nabla_{\mathbf{R}} \chi_n) = 0.$$

Flux in channel n is therefore lost only via transition to another state f with probability controlled solely by A_f. When many wavelengths of relative motion can be accomodated within the range of V_{nn}, as at the higher energies favored by the diabatic scheme, the fast $\mathbf{R}$-variation of F_n is mainly controlled by S_n, and the original diabatic set of coupled equations then reduce to the simpler set

$$i\hbar \frac{\partial A_f(t)}{\partial t} = \sum_{n \neq f} A_n(t) V_{fn}[\mathbf{R}_f(t)] \exp i(S_n - S_f) \exp -(\chi_n - \chi_f)$$

of first-order coupled equations.When a common trajectory $\mathbf{R}_n(t) = \mathbf{R}(t)$ under some averaged interaction $\bar{V}(R)$ can be assumed for all channels n,

$$S_n(\mathbf{R}) - S_f(\mathbf{R}) = \omega_{fn} t + \hbar^{-1} \int_{t_0}^{t} [V_{ff}(\mathbf{R}(t)) - V_{nn}(\mathbf{R}(t))]\, dt$$

and the classical path equations are recovered. [21]

Averaged potential

The orbit common to all channels is found by choosing the potential governing the relative motion as the average [21]

$$\mathcal{V}(\mathbf{R}) = \langle \Psi(\mathbf{r}, \mathbf{R}) |\, \hat{H}_{\text{int}}(\mathbf{r}) + V(\mathbf{r}, \mathbf{R})\, | \Psi(\mathbf{r}, \mathbf{R}) \rangle_{\mathbf{r}}.$$

Hamilton's equations of motion for this interaction

$$\mathcal{V}(\mathbf{R}) = \sum_n \left[| A_n |^2 + \sum_f A_f^* A_n V_{fn} \exp i(S_n - S_f) \right]$$

are therefore coupled to the set of first-order equations for the transition amplitudes $A_f(\mathbf{R})$. An essential feature is that total energy is always conserved, being continually redistributed between the relative motion and the internal degrees of freedom, as motion along the trajectory proceeds. In terms of the solutions $A_f(b_j, t)$ and the differential cross section $d\sigma_{el}^{(j)}/d\Omega$ for elastic scattering of particles with impact parameter $b_j(\theta)$ through θ by $\bar{V}(R)$, the semiclassical scattering amplitude is

$$f_{if}^{(j)}(\theta) = A_f[b_j(\theta), \infty] \exp i S_f^{'}(b_j(\theta), \infty) \left\{ d\sigma_{el}^{(j)}/d\Omega \right\}^{1/2}.$$

The accumulated classical action for orbit $b_j(\theta)$ is

$$S_f^{'}(b_j, t) = \int_{\mathbf{R}_0}^{\mathbf{R}(t)} [\kappa_f(\mathbf{R}) - k_n(\mathbf{R})] \cdot \mathbf{dR}.$$

When the same scattering angle originates from more than one impact parameter b_j, interference effects originate from the different actions associated with the different orbits $b_j(\theta)$. The contributions arising from N-orbits, which are well separated, combine according to

$$f_{if}(\theta) = -i \sum_{j=1}^{N} \alpha_j \beta_j \, f_{if}^{(j)}(\theta).$$

The coefficients $\alpha_j = \exp \pm\pi/4$ depend on whether the scattered particle emerges on the same side (+) of the axis as it entered, as in a collision overall repulsive, or on the opposite side (−), as in an overall attractive collision. The coefficient β_j is $\exp \pm\pi/4$ according to whether the sign of $db/d\theta$ is (+) or (−). The differential cross section will therefore exhibit characteristic oscillations, directly attributable to interference between the action phases $S_f^{'}(b_j)$ associated with each contributing classical path $b_j(\theta)$. The analysis can be extended, as in the uniform Airy function approximation to cover orbits that are not widely separated, as for the case of rainbow scattering or of caustics, in general, where the density of paths become infinite. The theory above provides the basis of the multistate orbital treatment [21] successful for rotational and vibrational excitation in atom-molecule and ion-molecule collisions at higher energies $E_{AB} \geq 1$1eV. Other semiclassical treatments based on the JWKB-approximation to the corresponding set of coupled equations for the radial wave function for relative motion can be found in Refs. [22, 23, 24].

In Figure 6.3 cross sections for the quenching process are displayed

$$He + H_2(v = 1, J = 0) \rightarrow He + H_2(v^{'} = 0, J = 0)$$

for collision energies E ranging from the ultracold to 1 Kev. The full quantal results [33] are shown together with those calculated [34] from the semiclassical multistate orbital

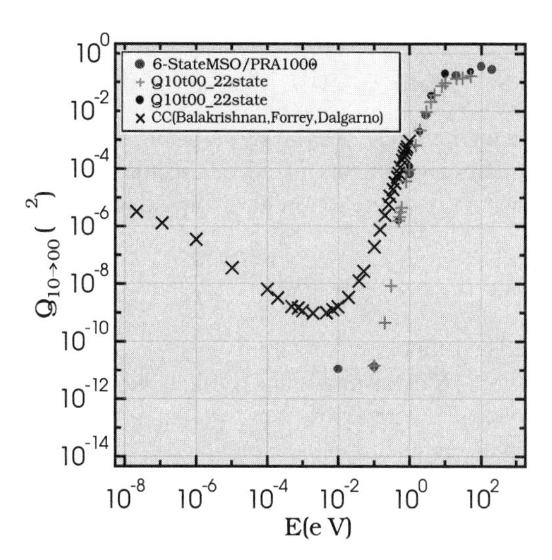

FIGURE 6.3. Vibrational relaxation cross sections (quantal and semiclassical) as a function of collision energy E.

method [21]. It is seen that results from both methods complement and connect with each other very well, in that the quantal treatment is calculationally feasible up to $E \sim 1Ev$, whereas semiclassical procedures are feasible at the higher collision energies.

Multichannel Eikonal method

For electronic transitions in electron-atom and heavy particle collisions at high impact energies, the major contribution to inelastic cross sections originates from scattering in the forward direction.The trajectories implicit in the action phases and the set of coupled equations can be taken as rectilinear. The integral representation

$$f_{if}(\theta) = \langle \phi_f(\mathbf{r}) \exp(i\mathbf{k}_f \cdot \mathbf{R})| \, V(\mathbf{r}, \mathbf{R}) \, |\Psi(\mathbf{r}, \mathbf{R})\rangle_{\mathbf{r},\mathbf{R}}$$

for the scattering amplitude, where

$$\Psi(\mathbf{r}, \mathbf{R}) = \sum_n A_n(\mathbf{R})\phi_n(\mathbf{r}) \exp i S_n(\mathbf{R})$$

then provides the basis of the multichannel eikonal treatment [15] valuable, in particular, for heavy particle collisions and for electron (ion)-excited atom collisions, where due to the large effect of atomic polarization (charge-induced dipole), the collision is dominated by scattering in the forward direction.

6.13. LONG-RANGE INTERACTIONS

6.13.1. Polarization, electrostatic, and dispersion interactions

The long-range interaction $V(\mathbf{R})$ between two atomic/molecular species can be decomposed into

$$V_{\text{polarization}}(\mathbf{R}) + V_{\text{electrostatic}}(\mathbf{R}) + V_{\text{dispersion}}(\mathbf{R}).$$

Polarization interactions: These originate from the interaction between the ion of charge Ze and the multipole moments it induces in the atom or molecule AB. The dominant polarization interaction is the ion-induced dipole interaction

$$V_{\text{pol}}(Ze; \text{ind}\, \mathbf{D}) = -\frac{\alpha_d Z e^2}{2R^4}[1 + (\alpha'_d/\alpha_d)P_2(\hat{\mathbf{s}} \cdot \hat{\mathbf{R}})],$$

where the averaged dipole polarizability is $\alpha_d = (\alpha_\parallel + 2\alpha_\perp)/3$ and $\alpha_\parallel$ and $\alpha_\perp$ are the polarizabilities of AB in the directions parallel and perpendicular to the molecular axis $\hat{s}$ of AB. The anisotropic polarizability is $\alpha'_d = 2(\alpha_\parallel - \alpha_\perp)/3$. The next polarization interaction is the charge-induced quadrupole interaction, averaged over all molecular orientations

$$V_{\text{pol}}(Ze; \text{ind}\, \mathbf{Q}) = -\bar{\alpha}_q Z e^2/2R^6,$$

where $\bar{\alpha}_q$ is the averaged quadrupole polarizability. Additional polarization terms originate from permanent multipole moments of one partner and the dipole (or multipole) it induces in the other, averaged over all directions. The leading term is

$$V_{\text{pol}}(\mathbf{D}; \text{ind } \mathbf{D}) = -\frac{1}{R^6}(D_n^2 \bar{\alpha}_{di} + D_i^2 \bar{\alpha}_{dn}),$$

where the subscripts i and n label the permanent dipole moments D and the dipole-polarizabilities α_d of the ion and neutral, respectively. The R^{-6}-variation is similar to that for the charge-induced quadrupole interaction.

Electrostatic interactions result from the interaction of the ion with the permanent multipole moments of the neutral. For cylindrically symmetric neutrals or linear molecules, the ion-neutral multipole interaction is

$$V_{el}(Ze; \mathbf{D}, \mathbf{Q}) = -\frac{(Ze)D_n}{R^2} P_1(\hat{\mathbf{s}} \cdot \hat{\mathbf{R}}) + \frac{(Ze)Q_n}{R^3} P_2(\hat{\mathbf{s}} \cdot \hat{\mathbf{R}}),$$

where $D_n = -e \int \mathbf{r}\rho(\mathbf{r})\, d\mathbf{r}$ and $Q_n = -e \int (3z^2 - r^2)\rho(\mathbf{r})\, d\mathbf{r}$ are the permanent dipole and quadrupole moments of the neutral. The ion dipole-neutral dipole interaction is

$$\begin{aligned} V_{el}(\mathbf{D}_i; \mathbf{D}_n) &= [\mathbf{D}_i \cdot \mathbf{D}_n - 3(\mathbf{D}_i \cdot \hat{\mathbf{R}})(\mathbf{D}_n \cdot \hat{\mathbf{R}})]/R^3 \\ &= -\frac{D_i D_n}{R^3}[2\cos\theta_i \cos\theta_n - \sin\theta_i \sin\theta_n \cos(\phi_i - \phi_n)], \end{aligned}$$

where θ_i and θ_n are the angles made by the ionic and molecular dipoles $\mathbf{D}_i$ and $\mathbf{D}_n$ with the line $\mathbf{R}$ of centers and ϕ_i and ϕ_n are the azimuthal angles of rotation about the line of centers. The dipole-molecular quadrupole interaction is

$$V_{el}(\mathbf{D}_i, \mathbf{Q}_n) = \frac{3D_i Q_n}{4R^4}[(3\cos^2\theta_n - 1)\cos\theta_i - 2\sin\theta_i \sin\theta_n \cos\theta_n \cos(\phi_i - \phi_n)].$$

The dispersion interaction originates between the fluctuating multipoles and the moments they induce and can occur even between spherically symmetric ions and neutrals. Thus,

$$V_{\text{dispersion}} \sim -\frac{C_6}{R^6} - \frac{C_8}{R^8} - \frac{C_{10}}{R^{10}} \cdots$$

represents the interaction of the fluctuating dipole interacting with the induced dipole C_6-term and quadrupole C_8-term, respectively. The leading R^{-6}-term represents the Van der Waal's attraction.

6.14. RADIATIVE PROCESSES

6.14.1. Photon scattering by free and bound electrons

Compton scattering refers to scattering of photons by free electrons. Thomson scattering is Compton scattering of low-energy photons whose frequency is unaffected by the scattering. Photon scattering by an atom is regarded as a two-photon process in which the incident photon of frequency ν_i is absorbed and a second photon of frequency ν_f is emitted with simultaneous transition of the atomic electron from initial state ψ_i to state ψ_f. Rayleigh scattering denotes elastic $i = f$ photon scattering from bound atomic electrons. Raman scattering is the inelastic $i \neq f$ process. Stokes scattering and anti-Stokes scattering occur when $\nu_f < \nu_i$ and when $\nu_f > \nu_i$, respectively.

Thomson scattering

The differential cross section for the (elastic) scattering

$$h\nu(\hat{\epsilon}_i) + e^-(E_e) \to h\nu'(\hat{\epsilon}_f) + e^-(E'_e)$$

of low-frequency photons with propagation vector $\mathbf{k}_\nu$ along the Z-direction into direction $\hat{\mathbf{k}}_\nu(\theta, \phi)$ by a free electron is

$$\frac{d\sigma}{d\mathbf{k}'_\nu} = r_e^2(\hat{\epsilon}_i \cdot \hat{\epsilon}_f)^2,$$

where $r_e = e^2/m_e c^2 = \alpha_0^2 = 2.818 \times 10^{-15}$ m is the classical electron radius. The scattering plane contains $\mathbf{k}_\nu$ and $\mathbf{k}'_\nu$. The two independent directions of the initial polarization vector $\hat{\epsilon}_i$ of the photon are $\hat{\mathbf{i}}$ and $\hat{\mathbf{j}}$. The two independent directions of the final photon-polarization vector $\hat{\epsilon}_f$ are $(\cos\theta\cos\phi, \cos\theta\sin\phi, -\sin\theta)$, parallel ($\|$) to the scattering plane, and $(-\sin\phi, \cos\phi, 0)$, perpendicular ($\perp$) to the plane. The individual and combined differential cross sections are

$$\frac{d\sigma^{\|}}{d\mathbf{k}'_\nu} = r_e^2(\cos^2\theta\cos^2\phi),$$

$$\frac{d\sigma^{\perp}}{d\mathbf{k}'_\nu} = r_e^2\sin^2\phi,$$

$$\frac{d\sigma}{d\mathbf{k}'_\nu} = r_e^2(1 - \sin^2\theta\ \cos^2\phi).$$

When the incident radiation is unpolarized,

$$\frac{d\sigma}{d\mathbf{k}'_\nu} = \tfrac{1}{2} r_e^2(1 + \cos^2\theta).$$

The integral cross section for low-energy photon scattering into all directions $\hat{\mathbf{k}}'_\nu$ is the Thomson cross section,

$$\sigma_T = \frac{8\pi}{3} r_e^2 = 6.65245 \times 10^{-25}\,\text{cm}^2.$$

Compton scattering

A photon with energy $h\nu_i$ upon collision with an electron at rest in the Lab frame is scattered through an angle θ and has its energy reduced to

$$h\nu_f = h\nu_i\left[1 + \frac{h\nu_i}{m_e c^2}(1 - \cos\theta)\right]^{-1}, \quad \gamma \ll 1.$$

The kinetic energy of the recoiling electron is

$$E_e = h\nu \left[\frac{\gamma(1-\cos\theta)}{1+\gamma(1-\cos\theta)} \right],$$

where $\gamma = h\nu_i/mc^2$. The photon energy $h\nu_i$ becomes equal to the electron mass energy mc^2 at the Compton frequency and wavelength

$$\nu_C = \frac{mc^2}{h} \equiv 1.23 \times 10^{20}\,\text{Hz},$$

$$\lambda_C = \frac{h}{mc} \equiv 2.426 \times 10^{-12}\,\text{m}.$$

The associated change in its wavelength is

$$\lambda_f - \lambda_i = \lambda_C(1-\cos\theta).$$

Compton cross section

The differential cross section for Compton scattering by free electrons, averaged over the initial and summed over the final photon and electron polarizations, is

$$\frac{d\sigma^{KN}}{d\mathbf{k}'_\nu} = \tfrac{1}{2} r_e^2 \left(\frac{\nu'}{\nu}\right)^2 \left(\frac{\nu}{\nu'} + \frac{\nu'}{\nu} - \sin^2\theta\right)$$

for unpolarized low-frequency ($\nu \ll \nu_C$) incident radiation. The Thomson cross section is recovered in the ($\nu' \to \nu$)-limit. The Klein-Nishina relativistic formulation for the integral cross section yields

$$\sigma^{KN} = \sigma_T \left[1 - \frac{2h\nu}{mc^2} + \frac{26}{5}\left(\frac{h\nu}{mc^2}\right) + \cdots\right], \quad \nu \ll \nu_C$$

for low-frequency photons, and

$$\sigma^{KN} = \frac{3}{8}\sigma_T \left(\frac{mc^2}{h\nu}\right)\left[\ln\left(\frac{2h\nu}{mc^2}\right) + \frac{1}{2}\right], \quad \nu \gg \nu_C$$

for high-frequency photons.

Rayleigh scattering

The scattering of plane wave radiation of angular frequency $\omega = 2\pi\nu$ by atoms or molecules with characteristic frequency ω_0 is given by the classical Rayleigh cross section

$$\sigma_R = \tfrac{8}{3}\pi r_e^2 \frac{\omega^4}{(\omega^2-\omega_0^2)^2 + \omega^2\Gamma^2}.$$

The damping (decay) constant Γ of the internal motion is the reciprocal of the lifetime τ of the bound state with energy $\hbar\omega_0$. In the ($\omega_0 \to 0$) and $\Gamma = 0$ limits, the Thomson cross section is recovered. At low frequencies $\omega \ll \omega_0$, the Rayleigh scattering law,

$$\sigma_R = \tfrac{8}{3}\pi r_e^2 \left(\frac{\lambda_0}{\lambda}\right)^4 = 6.65 \times 10^{-25} \left(\frac{\lambda_0}{\lambda}\right)^4 \text{ cm}^2$$

explains the scattering of visible sunlight of wavelength λ by the earth's atmosphere—blue skies and red sunsets.

Quantal cross section

A first-order quantal differential cross section for Rayleigh scattering yields

$$\frac{d\sigma}{d\mathbf{k}'_\nu} = \tfrac{1}{2} r_e^2 (1 + \cos^2\theta) |\, F(Q)\,|^2,$$

where F is the elastic Form Factor so that $|\, F(Q)\,|^2$ is the probability (Sec. 6.7.6) for impulsive transfer of momentum $\hbar Q = (2\hbar\omega/c)\sin(\theta/2)$. The quantal Kramers cross section for the scattering of incident radiation of frequency ν or wavelength λ is

$$\sigma_K(i) = \tfrac{8}{3}\pi r_e^2 \left[\sum_j f_{ij} \left\{ \frac{\nu_{ij}^2}{\nu^2} - 1 \right\}^{-1} \right]^2 \approx \frac{128\pi^5 \alpha_d^2}{3\lambda^4},$$

where f_{ij} is the oscillator strength for the $i \to j$ transition in the atom or molecule, with dipole polarizability α_d.

6.14.2. Radiative emission rate

The number of transitions per second or probability frequency (s^{-1}) of transitions from one of the upper substates of level u with g_u-degenerate states i to all of the g_ℓ-degenerate states f of lower level ℓ is

$$w_{u\ell}(s^{-1}) = \frac{2\pi}{\hbar} \frac{1}{g_u} \sum_{i,f} |\, V_{fi}\,|^2 \rho(E_\nu)\, d\hat{\mathbf{k}}_\nu.$$

This is a sum over all final internal states $|\phi_f\rangle$ and an average over all initial internal states $|\phi_i\rangle$. The states are coupled within the matrix element

$$V_{fi} = \langle \phi_f |\, V(\mathbf{r})\, | \phi_i \rangle$$

by the electric dipole photon-atom interaction

$$V(\mathbf{r}) = \left[\frac{2\pi h\nu}{\mathcal{V}}\right]^{1/2} \hat{\epsilon} \cdot e\mathbf{r}$$

in a volume $\mathcal{V}$. The direction of polarization is along the electric field direction $\hat{\epsilon}$, which is perpendicular to the direction $\hat{k}_\nu$ of photon propagation.

The probability frequencies for emission of one photon and absorption of one photon are related by

$$g_i w_{u\ell} = g_f w_{\ell u}.$$

The number of photons in volume $\mathcal{V}$ and propagation vectors $\mathbf{k}_\nu$ within the range $[\mathbf{k}_\nu, \mathbf{k}_\nu + d\mathbf{k}_\nu]$ is

$$dN(\mathbf{k}_\nu) = g_\nu n_\nu \frac{d\mathbf{p}_\nu \mathcal{V}}{(2\pi\hbar)^3} = 2n_\nu \mathcal{V}\frac{d\mathbf{k}_\nu}{(2\pi)^3}$$

where there are two states ($g_\nu = 2$) of polarization, where $n_\nu(\mathbf{k}_\nu)$ is the number of photons in each mode (momentum state) and $d\mathbf{p}_\nu/(2\pi\hbar)^3$ is the number of momentum states per unit volume. Since $E_\nu = p_\nu c = h\nu = \hbar\omega$ and $\mathbf{p}_\nu = \hbar\mathbf{k}_\nu$, the number of photons in volume $\mathcal{V}$ and with propagation vectors $\mathbf{k}_\nu$ in all directions

$$dN_\nu = n_\nu \rho(E_\nu) dE_\nu = n_\nu \rho(\nu) d\nu = n_\nu \rho(\omega) d\omega,$$

where the photons in volume $\mathcal{V}$ have their frequencies distributed according to

$$\rho(\nu)\, d\nu = \left(\frac{8\pi\nu^2}{c^3}\right)\mathcal{V}\, d\nu$$

$$\rho(\omega)\, d\nu, = \left(\frac{\omega^2}{\pi^2 c^3}\right)\mathcal{V}\, d\omega.$$

The rate (s^{-1}) for a transition from one of the upper states of level u to all of the final g_f-states of level ℓ accompanied by emission of one photon ($n_\nu = 1$), with specified polarization $\hat{\epsilon}$, into all directions $\hat{\mathbf{k}}_\nu$ is

$$w_{u\ell}(\hat{\epsilon}) = \frac{2\pi}{\hbar}\left(\frac{2\omega^3}{\pi c^3}\right)\frac{1}{g_u}\sum_{i,f} |\, \mathbf{d}_{fi} \cdot \hat{\epsilon}\,|^2.$$

The dipole matrix element is

$$\mathbf{d}_{fi} = \langle \phi_f |\, e\mathbf{r}\, | \phi_i \rangle.$$

Integration over all polarization directions $\hat{\epsilon}$ yields

$$\langle |\, \mathbf{d}_{fi} \cdot \hat{\epsilon}\, | \rangle = \tfrac{1}{3} |\, d_{fi}\, |^2.$$

Einstein A- and B-coefficient

The rate $w_{u\ell}(s^{-1})$ for spontaneous photon emission of a photon of energy $h\nu = E_{u\ell} = \epsilon_u - \epsilon_\ell$ in a $u \to \ell$ transition is the Einstein A-coefficient

$$A_{u\ell} \equiv w_{u\ell}(s^{-1}) = \frac{4e^2}{3\hbar}\left(\frac{\omega}{c}\right)^3 \frac{1}{g_u} S_{u\ell} a_0^2$$

$$\equiv \frac{4}{3}\left(\frac{\alpha^3}{\tau_0}\right)\left(\frac{E_{u\ell}}{\varepsilon_0}\right)^3 \frac{1}{g_u} S_{u\ell}.$$

The dimensionless line strength

$$S_{u\ell} = \sum_i \sum_f |\langle \phi_f | \mathbf{r}/a_0 | \phi_i \rangle|^2 = S_{\ell u} \equiv S$$

is a sum over all upper and lower degenerate states and is therefore symmetrical in u and ℓ. The ω^3–factor in $A_{u\ell}$ ensures that the transitions with highest $h\nu$ will dominate to the radiative decay rate and that the dominant decay is to the lowest state possible. The atomic units (Sec. 6.1) of time and energy are τ_0 and ε_0, respectively.

Absorption and emission oscillator strength

The absorption oscillator strength, in terms of the line strength is,

$$f_{\ell u} = \frac{2m}{3\hbar^2}(\epsilon_u - \epsilon_\ell)\frac{1}{g_\ell}S = \frac{2}{3}\left[\frac{\epsilon_u - \epsilon_\ell}{\varepsilon_0}\right]\frac{1}{g_u}S$$

and is positive since $\epsilon_u > \epsilon_\ell$. The emission oscillator strength

$$f_{u\ell} = -\left(\frac{g_\ell}{g_u}\right) f_{\ell u}$$

is always negative. The oscillator strength is dimensionless and satisfies the sum rule

$$\sum_f f_{if} = N,$$

where N is the number of atomic electrons. The emission A-coefficient, written in terms of the absorption oscillator strength, is

$$A_{u\ell} = \left(\frac{2e^2\omega^2}{mc^3}\right)\left(\frac{g_\ell}{g_u}\right) f_{\ell u} = 2\left(\frac{\alpha^3}{\tau_0}\right)\left(\frac{E_{u\ell}}{\varepsilon_0}\right)^2\left(\frac{g_\ell}{g_u}\right) f_{\ell u}.$$

The external factor $A_0 = \alpha^3/\tau_0 = 1.607 \times 10^{10}\ \mathrm{s}^{-1}$. The transition rate for $n_u \to n_\ell$ transitions in hydrogenic systems with nuclear charge Z is

$$A_{u,\ell} = \tfrac{1}{2}A_0 Z^4\left[n_u^{-2} - n_\ell^{-2}\right]\left(\frac{g_\ell}{g_u}\right) f_{\ell u}.$$

When the wavelength $\lambda = c/\nu = 2\pi c/\omega$ is expressed in Angströms Å and S in atomic units (a.u),

$$A_{u\ell}(\mathrm{s}^{-1}) = \frac{6.6702 \times 10^{15}}{\lambda^2(\text{Å})}\left(\frac{g_\ell}{g_u}\right) f_{\ell u} = \frac{2.0261 \times 10^{18}}{\lambda^3(\text{Å})}(S_{a.u}/g_u).$$

Lifetimes

The averaged rates of radiative emission are

$$A_{nl,n'l'} = \frac{1}{(2l+1)}\sum_{m,m'} A_{nlm,n'l'm'},$$

$$A_{n,n'} = \frac{1}{n^2} \sum_{l,l'} (2l+1) A_{nl,n'l'}.$$

The radiative lifetime τ_{nl} of an excited nl-sublevel is then

$$\frac{1}{\tau_{nl}} = \sum_{n'<n} \left[A_{nl,n'l+1} + A_{nl,n'l-1} \right].$$

The lifetime τ_n of level n is defined by

$$\frac{1}{\tau_n} = \frac{1}{n^2} \sum_{l} (2l+1) \sum_{n'<n} \left[A_{nl,n'l+1} + A_{nl,n'l-1} \right] = \sum_{n'<n} A_{n,n'} = \frac{1}{n^2} \sum_{l} \frac{(2l+1)}{\tau_{nl}}.$$

Expressions for lifetimes of excited atoms A toward collision with atom B with number density N_B are similarly constructed by replacing A_{ij} by the collisional frequency $N_B \langle v\sigma_{ij} \rangle$ and summing over all levels $n'l' \neq nl$.

Photon current, emitted energy, and energy intensity

The photon current (number of photons per cm^2/sec), entering volume $\mathcal{V}$ is

$$J_\nu \, d\nu \equiv c \, n_\nu \, \rho(\nu) d\nu \, / \mathcal{V} = \left(\frac{8\pi \nu^2}{c^2} \right) n_\nu \, d\nu.$$

The energy $(h\nu) dN_\nu / \mathcal{V}$ of n_ν photons in unit volume is

$$\rho_E(\nu) \equiv (n_\nu h\nu) \, \rho(\nu) d\nu / \, \mathcal{V} = n_\nu \left(\frac{8\pi h\nu^3}{c^3} \right) d\nu,$$

where the subscript E is used to distinguish the spectral energy density $\rho_E d\nu$ (energy per unit volume) from the frequency distribution $\rho(\nu) \, d\nu$. The energy intensity (energy per cm^2/sec) supplied by the current is

$$I_\nu \, d\nu \equiv (h\nu) J_\nu \, d\nu = c \, \rho_E(\nu) \, d\nu = \left(\frac{8\pi h\nu^3}{c^2} \right) n_\nu \, d\nu.$$

Einstein B-coefficient

In the presence of radiation, an atom can be stimulated to emit or absorb a photon at the rate of $B_{u\ell} \rho_E(\nu)$ or $B_{\ell u} \rho_E(\nu)$ per unit time, where the B's are the Einstein B-coefficients. In the process

$$X(u) \rightleftharpoons X(\ell) + h\nu,$$

photon emission from N_u atoms X proceeds at a rate of

$$R_{em}(s^{-1}) = N_u \left[A_{u\ell} + B_{u\ell} \rho_E(\nu) \right]$$

and photons are absorbed by N_ℓ atoms at a rate of

$$R_{abs}(\mathrm{s}^{-1}) = N_\ell B_{\ell u} \rho_E(\nu),$$

where the Einstein B-coefficients for emission and absorption are defined in terms of the energy density ρ_E. Under equilibrium conditions at temperature T, the atomic densities satisfy the Boltzmann distribution

$$\frac{\tilde{N}_u}{\tilde{N}_\ell} = \left(\frac{g_u}{g_\ell}\right) \exp -h\nu/k_B T$$

since $h\nu = E_u - E_\ell$ and the photon energy satisfies the Planck distribution

$$\tilde{\rho}_E(\nu) = \left(\frac{8\pi h\nu^3}{c^3}\right) \left[\exp h\nu/k_B T - 1\right]^{-1}.$$

The equilibrium photon number or occupation number is then

$$\tilde{n}_\nu = \left[\exp h\nu/k_B T - 1\right]^{-1}.$$

The Einstein A and B coefficients are then related by

$$A_{u\ell} = \left(\frac{8\pi h\nu^3}{c^3}\right) B_{u\ell}$$

and by

$$g_u B_{u\ell} = g_\ell B_{\ell u}.$$

Since $B_{u\ell}\rho_E = A_{u\ell} n_\nu$, the rate of emission from N_u atoms can be conveniently expressed as

$$R_{em} = N_u A_{ul}(n_\nu + 1).$$

The rate of absorbtion from N_ℓ atoms is similarly

$$R_{abs} = N_\ell A_{\ell u} n_\nu,$$

where $A_{\ell u} = (g_u/g_\ell) A_{u\ell}$.

Other conventions for B-coefficient

Two other conventions are sometimes used to define the Einstein B-coefficient for stimulated emission in terms of the number n_ν of photons and energy intensity I_ν, respectively, according to

$$B_{u\ell}\rho_E(\nu) = B'_{u\ell} n_\nu = B''_{u\ell} I_\nu.$$

The various B-coefficients are then related by

$$B'_{u\ell} = \left(\frac{8\pi h\nu^3}{c^3}\right) B_{u\ell} = A_{u\ell},$$

$$B''_{u\ell} = cB_{u\ell}.$$

6.14.3. Cross section for radiative recombination

Radiative recombination is the capture of an electron by an ion, with the spontaneous emission of a photon, i.e.,

$$e^-(\mathbf{E}_e) + A^+(i) \rightarrow A(f) + h\nu.$$

The differential rate per unit density $\rho(\mathbf{E}_e)\, d\mathbf{E}_e$ of initial translational states in the range $[\mathbf{E}_e, \mathbf{E}_e + d\mathbf{E}_e]$ is

$$A_f(E_e, \hat{\mathbf{k}}_e) = \frac{2\pi}{\hbar} \mid D_{fi}(\mathbf{k}_e) \mid^2,$$

where the atom-radiation dipole interaction coupling in this free-bound $(E_e, \hat{\mathbf{k}}_e) \rightarrow f$ electronic transition is

$$\mid D_{fi}(\mathbf{k}_e) \mid^2 = \frac{2e^2\omega^3}{3\pi c^3} \sum_j \mid \langle \Psi_j \mid \mathbf{r} \mid \Psi_{\mathbf{k}} \rangle \mid^2$$

$$= \frac{2}{3\pi} \frac{(\alpha h\nu)^3}{\varepsilon_0^2} \sum_j \mid \langle \Psi_j \mid \mathbf{r}/a_0 \mid \Psi_{\mathbf{k}} \rangle \mid^2,$$

which has dimensions of energy. The summation is over g_f-degenerate final states j. The cross section for radiative recombination is provided by the frequency relation

$$\left[j_i\, dE_e\, d\hat{\mathbf{k}}_e\right] \sigma_R^f(\mathbf{k}_e) = \left[\rho(\mathbf{E}_e)\, dE_e d\hat{\mathbf{k}}_e\right] A_f(\mathbf{k}_e).$$

The current density j_i in terms of energy distribution $\rho(\mathbf{E}_e)$ of continuum electrons with directed energy $\mathbf{E}_e \equiv (E_e, \hat{\mathbf{k}}_e)$ is $v_e\rho(\mathbf{E}_e) = p_e^2/h^3$. When averaged over all incident directions $\hat{\mathbf{k}}_e$, the cross section

$$\sigma_R^f(k_e) = \frac{1}{4\pi} \int \sigma_f(\mathbf{k}_e)\, d\hat{\mathbf{k}}_e = \left(\frac{h^3}{8\pi m E_e}\right) \rho(\mathbf{E}_e) \int A_f(\mathbf{k}_e)\, d\hat{\mathbf{k}}_e$$

can be expressed in terms of the transition probability

$$\mid T_f(k_e) \mid^2 = 4\pi^2 \rho(\mathbf{E}_e) \int \mid D_{fi}(\mathbf{k}_e) \mid^2 d\hat{\mathbf{k}}_e$$

as

$$\sigma_R^f(k_e) = \frac{\pi a_0^2}{(k_e a_0)^2} | T_f(k_e) |^2.$$

The radiative recombination cross section into level $n\ell$ is

$$\sigma_R^{n\ell}(k_e) = \frac{8\pi^2}{3k_e^2} \frac{(\alpha h\nu)^3}{\varepsilon_0^2} \rho(\mathbf{E}) | \mathbf{r}_{\epsilon,n\ell} |^2,$$

where the dipole matrix element

$$| \mathbf{r}_{\epsilon,n\ell} |^2 = \int d\hat{\mathbf{k}}_e \left[\sum_m | \langle \psi_{n\ell m} | \mathbf{r}/a_0 | \Psi(\mathbf{k}_e) \rangle |^2 \right]$$

has dimensions of $\langle \Psi_{\mathbf{k}_e} \mid \Psi_{\mathbf{k}_e'} \rangle$. This cross section is so written as to conveniently incorporate various choices of normalization for Ψ. For unit amplitude plane-wave normalization, $\langle \Psi_{\mathbf{k}_e} \mid \Psi_{\mathbf{k}_e'} \rangle = (2\pi)^3 \delta(\mathbf{k}_e - \mathbf{k}_e')$ has dimension a_0^3 and $\rho(\mathbf{E}_e) = mp_e/h^3$. For energy normalization, $\langle \Psi_{\mathbf{k}_e} \mid \Psi_{\mathbf{k}_e'} \rangle = \delta(E_e - E_e')\, \delta(\hat{\mathbf{k}}_e - \hat{\mathbf{k}}_e')$, has dimension ε_0^{-1} and $\rho(\mathbf{E}_e) = 1$.

Semiclassical recombination cross section

The semiclassical (*Kramers*) cross section for radiative recombination of electrons with energy E_e into level n of hydrogenic systems with statistical weight $g_n = 2n^2$, charge Z and binding energy I_n is

$$_{\mathrm{K}}\sigma_{\mathrm{R}}^n(E_e) = \sigma_{\mathrm{R0}}(E_e) \left(\frac{2}{n} \right) \left(\frac{I_n}{I_n + E_e} \right),$$

where

$$\sigma_{\mathrm{R0}}(E_e) = \left(\frac{8\pi a_0^2 \alpha^3}{3\sqrt{3}} \right) \frac{(Z^2 e^2/a_0)}{E_e}.$$

The numerical value of the cross section is

$$_{\mathrm{K}}\sigma_{\mathrm{R}}^n(E_e) = 3.897 \times 10^{-20} \left[n\varepsilon(13.606 + n^2\varepsilon^2) \right]^{-1} \mathrm{cm}^2,$$

where ε is in units of eV and is given by

$$\varepsilon = E_e(\mathrm{eV})/Z^2 \equiv (2.585 \times 10^{-2}/Z^2)\,(T_e(K)/300).$$

The expression illustrates that radiative recombination into low n at low electron energies E_e is favored.

6.14.4. Radiative recombination rate

The recombination rate ($\mathrm{cm}^3\ \mathrm{s}^{-1}$) is

$$\alpha_{RR}(T_e) = \langle v_e \rangle \int_0^\infty \sigma_{RR}(\epsilon_e)\epsilon_e \exp(-\epsilon_e)\, d\epsilon_e$$

for a Maxwellian distribution of electrons when

$$\epsilon_e = \tfrac{1}{2} m_e v^2 / k_B T_e \quad \text{and} \quad \langle v_e \rangle = (8 k_B T_e / \pi m_e)^{1/2}.$$

It may be calculated directly from the recombination cross section σ_{RR} or indirectly from the cross section σ_{PI} for the reverse process of photoionization via use of the detailed balance relation (Sec. 6.6.6)

$$\sigma_{RR}(E_e) = \left(\frac{g_A g_\nu}{g_e g_{A^+}} \right) \frac{(h\nu)^2}{(2 E_e m_e c^2)} \sigma_{PI}(h\nu),$$

where the photon statistical weight $g_\nu = 2$. This provides

$$\alpha_{RR}(T_e) = \left[\frac{h^3}{(2\pi m_e k_B T)^{3/2}} \frac{g_A}{g_e g_{A^+}} \exp(I_n / k_B T) \right] \times \left\{ \int_{\nu_n}^{\infty} c \sigma_{PI}(\nu) \left(\frac{8\pi \nu^2}{c^3} \right) \exp(-h\nu / k_B T)\, d\nu \right\}.$$

The ionization energy of level n is $I_n = h\nu_n$. By comparison with the detailed balance rates in Sec. 6.4.7, the frequency (s^{-1}) of photoionization is

$$k_I = \int_{\nu_n}^{\infty} c\, \sigma_{PI}(\nu) \left(\frac{8\pi \nu^2}{c^3} \right) \exp(-h\nu / k_B T)\, d\nu.$$

Milne detailed balance relation

When expressed in reduced units $\tilde{\omega} = h\nu / I_n$, $\tilde{T} = k_B T_e / I_n$, the recombination rate can be written, in terms of the frequency-averaged cross section, as

$$\langle \sigma_{PI}^{n\ell}(T_e) \rangle = \frac{e^{1/\tilde{T}}}{\tilde{T}} \int_1^\infty \tilde{\omega}^2 \sigma_{PI}^{n\ell}(\tilde{\omega}) \exp(-\tilde{\omega}/\tilde{T})\, d\tilde{\omega}$$

for photoionization of level $n\ell$, as

$$\alpha_{\mathrm{R}}^{n\ell}(T_e) = \bar{v}_e \left(\frac{g_A}{2 g_A^+} \right) \left(\frac{k_B T_e}{mc^2} \right) \left(\frac{I_n}{k_B T_e} \right)^2 \langle \sigma_{PI}^{n\ell}(T_e) \rangle.$$

This form is known as the Milne relation. When $\sigma_I^{n\ell}\omega$ is expressed in Mb (10^{-18} cm^2),

$$\alpha_{\mathrm{R}}^{n\ell}(T_e) = 6.032 \times 10^{-13} \left(\frac{300}{T_e} \right)^{1/2} \left(\frac{I_n}{\varepsilon} \right)^2 \left(\frac{g_A}{2 g_A^+} \right) \langle \sigma_{PI}^{n\ell}(T_e) \rangle \quad \mathrm{cm^3\, s^{-1}}.$$

Energy-loss rate and radiated power

The amount of energy lost in unit time by a Maxwellian distribution of electrons is

$$\left\langle \frac{dE}{dt} \right\rangle = n_e \langle v_e \rangle (k_B T_e) \int_0^\infty \varepsilon^2 \sigma_{RR}^{n\ell}(\varepsilon) e^{-\varepsilon}\, d\epsilon,$$

where n_e is the number density of electrons and $\epsilon = E_e/k_B T_e$. The power radiated by the photons for recombination into level $n\ell$ is

$$\left\langle \frac{d(h\nu)}{dt} \right\rangle = n_e \langle v_e \rangle \int_0^\infty \varepsilon h\nu \sigma_{RR}^{n\ell}(\varepsilon)\, e^{-\varepsilon}\, d\epsilon.$$

6.14.5. Dielectronic recombination cross section

The dielectronic recombination process is represented by the two-stage sequence

$$\mathrm{e}^- + A^{(Z+1)+}(i) \rightleftharpoons \left[A^{(Z+1)+}(k) - \mathrm{e}^-\right]_{n\ell} \rightarrow A^{Z+}_{n'\ell'}(f) + h\nu$$

of resonance capture of the electron into Rydberg level $n\ell$ of the ion, which is simultaneously excited from electronic state i to k. The doubly excited bound state $(k, n\ell)$ of the ion A^{Z+} is stabilized against autoionization by radiative decay. The cross section for the two-step (capture-radiative decay) process is

$$\sigma_{\mathrm{DLR}}^{n\ell} E_e = \frac{\pi a_0^2}{(k_e a_0)^2} |\, T_{\mathrm{DLR}}(E)\,|^2 \rho(E_e),$$

where the transition matrix or transition probability is

$$|\, T_{\mathrm{DLR}}(E_e)\,|^2 = 4\pi^2 \int d\hat{k}_e \sum_j \left| \frac{\langle \Psi_f \rangle |\hat{\mathbf{D}}| \Psi_j \rangle \langle \Psi_j | V | \Psi_i(\mathbf{k}_e) \rangle}{(E_e - \varepsilon_j + i\Gamma_j/2)} \right|.$$

This construction is similar to the Fermi Golden Rule 1 of Sec. 6.7.1. It includes the effect of intermediate doubly excited autoionizing states $|\Psi_j\rangle$ with energy ε_j in resonance to within width Γ_j of the initial continuum state Ψ_i of energy E_e. The electrostatic interaction $V = e^2 \sum_{i=1}^{N} (\mathbf{r}_i - \mathbf{r}_{N+1})^{-1}$ between the continuum electron at $\mathbf{r}_{N+1}$ and the bound electrons at $\mathbf{r}_i$ initially produces dielectronic capture by coupling the initial state i with the doubly excited resonant states $j = n\ell$, which become stabilized by coupling via the dipole radiation field interaction $\mathbf{D} = (2\omega^3/3\pi c^3)^{1/2} \sum_{i=1}^{N+1} (e\mathbf{r}_i)$ to the final stabilized state f. This cross section is valid for isolated, nonoverlapping resonances.

6.14.6. Bremsstrahlung

Bremsstrahlung is the radiation in general resulting from the scattering of any free charged particle accelerating in any external field. In electron-ion collisions,

$$e^-(\mathbf{E}) + A^+ \rightarrow e^-(\mathbf{E}') + A^+ + h\nu$$

the energy of the Bremsstrahlung radiation is

$$h\nu = \frac{1}{2m}(p_e^2 - p_e'^2) = E_e - E_e'.$$

6.14.7. Bremsstrahlung cross section

The cross section can be obtained from the rate $A(\mathbf{k}_e, \mathbf{k}'_e)$ of free-free transitions via the frequency relation

$$\left[j(\mathbf{E}_e)dE_e\,d\hat{\mathbf{k}}_e\right]\frac{d^2\sigma_{if}}{dE'_e\,d\hat{\mathbf{k}}'_e}\,dE'_e\,d\mathbf{k}'_e = \left[\rho(\mathbf{E}_e)\,dE_e\,d\hat{\mathbf{k}}_e\right]A(\mathbf{k}_e,\mathbf{k}'_e)\left[\rho(\mathbf{E}'_e)\,dE'_e\,d\hat{\mathbf{k}}'_e\right].$$

The rate of free-free transitions per unit densities $\rho(\mathbf{E}_e)\,dE_e$ of initial and final translational states is

$$A(\mathbf{k}_e,\mathbf{k}'_e) = \frac{4}{3}\frac{e^2}{\hbar}\left(\frac{\omega}{c}\right)^3|\langle\Psi(\mathbf{k}'_e)|\mathbf{r}|\Psi(\mathbf{k}_e)\rangle|^2 = \frac{4\alpha^3}{3\hbar}\frac{(h\nu)^3}{\varepsilon_0^2}|\langle\Psi(\mathbf{k}'_e)|\mathbf{r}/a_0|\Psi(\mathbf{k}_e)\rangle|^2.$$

When averaged over all incident directions $\hat{\mathbf{k}}_e$,

$$\frac{d^2\sigma_{if}}{dE'_e d\hat{\mathbf{k}}'_e} = \left(\frac{h^3}{4\pi p_e^2}\right)\rho(\mathbf{E}_e)\rho(\mathbf{E}'_e)\int A(\mathbf{k}_e,\mathbf{k}'_e)\,d\hat{\mathbf{k}}_e.$$

When integrated over all final directions $\hat{\mathbf{k}}'_e$, the Bremsstrahlung energy-differential cross section (per unit range dE'_e) is

$$\frac{d\sigma_{BR}}{dE'_e} = \frac{8\pi^2\alpha^3 a_0^2}{3(k_e a_0)^2}\frac{(h\nu)^3}{\varepsilon_0^2}\rho(\mathbf{E}_e)\rho(\mathbf{E}'_e)|\,\mathbf{r}_{\epsilon\epsilon'}\,|^2,$$

where the dipole matrix element

$$|\,\mathbf{r}_{\epsilon\epsilon'}\,|^2 = \sum_{\ell=0}^{\infty}\int d\hat{\mathbf{k}}_e\,d\hat{\mathbf{k}}'_e\sum_m\left|\left\langle\Psi(\mathbf{k}'_e)\left|\frac{\mathbf{r}}{a_0}\right|\Psi(\mathbf{k}_e)\right\rangle\right|^2$$

has dimension $|\,\langle\Psi_{\mathbf{k}_e}\mid\Psi_{\mathbf{k}'_e}\rangle\,|^2$ and can be calculated in terms of the radial integrals involving the continuum radial wave functions $R_{\epsilon\ell}$ and $R_{\epsilon'\ell'}$. When both continuum wave functions are energy normalized, $\rho(\mathbf{E}) = 1$ and

$$\frac{d\sigma_{BR}}{dE'_e} = \frac{8\pi^2\alpha^3 a_0^2}{3(k_e a_0)^2}\frac{(h\nu)^3}{\varepsilon_0^4}(|\,\mathbf{r}_{\epsilon\epsilon'}\,|^2\varepsilon_0^2).$$

For unit amplitude plane-wave normalization, the Bremsstrahlung cross section is

$$\frac{d\sigma_{BR}}{dE'_e} = \frac{\alpha^3 a_0^2}{24}\frac{k'_e}{k_e}\left\{\frac{(h\nu)^3}{\varepsilon_0^4}\right\}\left(\frac{|\,\mathbf{r}_{\epsilon\epsilon'}\,|^2}{a_0^6}\right),$$

since $|\langle\Psi_{\mathbf{k}_e}\mid\Psi_{\mathbf{k}'_e}\rangle|^2$ has dimension a_0^6 and $\rho(\mathbf{E}_e) = m_e p_e/h^3$.

6.14.8. Dipole transition matrix elements

The dipole matrix elements that appear in the rates and cross sections for the above bound-bound, free-bound and free-free transitions in central field systems are

$$|\mathbf{r}_{n\ell,n'\ell'}|^2 = \sum_{mm'} |\langle\phi_{n\ell m}|\mathbf{r}|\phi_{n'\ell'm'}\rangle|^2,$$

$$|\mathbf{r}_{n\ell,\epsilon'\ell'}|^2 = \int d\hat{\mathbf{k}} \sum_{mm'} |\langle\phi_{n\ell m}|\mathbf{r}|\psi_{\mathbf{k}}\rangle|^2,$$

$$|\mathbf{r}_{\epsilon\ell,\epsilon'\ell'}| = \int d\hat{\mathbf{k}} \int d\hat{\mathbf{k}}' \sum_{mm'} |\langle\psi_{\mathbf{k}}|\mathbf{r}|\psi_{\mathbf{k}'}\rangle|^2,$$

where $\ell' = \ell \pm 1$. Under the substitutions

$$\phi_{n\ell m}(\mathbf{r}) = R_{n\ell}(r)\, Y_{\ell m}(\hat{\mathbf{r}}),$$

$$\psi_{\mathbf{k}}(\mathbf{r}) = \sum_{\ell,m} R_{\epsilon\ell}(r) Y^*_{\ell m}(\hat{\mathbf{k}}) Y_{\ell m}(\hat{\mathbf{r}})$$

for the bound and continuum wave functions, the angular integrations can be accomplished to eventually yield

$$|\mathbf{r}_{n\ell,n'\ell'}|^2 = \ell \left[R_{n\ell}^{n',\ell-1}\right]^2 + (\ell+1)\left[R_{n\ell}^{n',\ell+1}\right]^2,$$

$$|\mathbf{r}_{n\ell,\epsilon'}|^2 = \ell \left[R_{n\ell}^{\epsilon,\ell-1}\right]^2 + (\ell+1)\left[R_{n\ell}^{\epsilon,\ell+1}\right]^2,$$

$$|\mathbf{r}_{\epsilon,\epsilon'}|^2 = \sum_{\ell=0}^{\infty} (\ell+1) \left\{\left[R_{\epsilon,\ell+1}^{\epsilon',\ell}\right]^2 + \left[R_{\epsilon,\ell}^{\epsilon',\ell+1}\right]^2\right\}.$$

The various radial integrals are defined as

$$R_{n\ell}^{n'\ell'} = \int_0^\infty (R_{n\ell}\, r\, R_{n'\ell'})\, r^2\, dr,$$

$$R_{n\ell}^{\epsilon\ell} = \int_0^\infty (R_{n\ell}\, r\, R_{\epsilon\ell})\, r^2\, dr,$$

$$R_{\epsilon\ell}^{\epsilon'\ell'} = \int_0^\infty (R_{\epsilon\ell}\, r\, R_{\epsilon'\ell'})\, r^2\, dr.$$

6.15. ATOMIC AND MOLECULAR DATABASES

World-Wide Web Addresses:

1. **Physics reference data:**
   http://physics.nist.gov/PhysicsRefData
2. **Physical constants:**
   /PhysicsRefData/contents-constants.html
3. **Atomic spectroscopic data:** Ground levels and ionization energies for the neutral atoms; atomic transition probabilities, atomic spectral line broadening:
   /PhysicsRefData/contents-atomic.html

4. **Molecular spectroscopic data:**
   /PhysicsRefData/contents-mol.html
5. **Electron impact total ionization cross sections for polyatomic molecules:**
   /PhysicsRefData/Ionization/Xsection.html
6. **Photon cross sections, x-ray form factors:**
   /PhysicsRefData/xCom/Text/XCOM.html
7. **For list of other databases**, check out:
   http://plasma-gate.weizmann.ac.il/DBfAPP.html

See also ref. 35 for fundamental constants and further Atomic and Molecular databases.

6.16. GENERAL REFERENCES

For further reading and reference, the following edited books and volumes are recommended, in addition to those cited in the list of references.

Atomic, Molecular and Optical Physics Handbook, edited by Gordon W. Drake (American Institute of Physics Press, 1996).

Guide to Bibliographies, Books, Reviews and Compendia of Data on Atomic Collisions, by E. W. McDaniel and E. J. Mansky in *Advances in Atomic and Molecular Physics*, edited by B. Bederson and H. Walther Vol. **33**, 389 (1994).

Advances in Atomic and Molecular Physics, Vols. **1**, 1965–**1**, 1973, edited by D.R. Bates and I. Estermann; Vols. **10**, 1974–**25**, 1988, edited by D.R. Bates and B. Bederson.

Advances in Atomic Molecular and Optical Physics, Vols. **26**, 1989-**31**, 1993, edited by D.R. Bates and B. Bederson. Vol. **32**, 1994, edited by B. Bederson and A. Dalgarno. Vols. **33**, 1994–**38**, 1998, edited by B. Bederson and H. Walther.

Atomic and Molecular Beam Methods, edited by G. Scoles (Oxford University Press, New York, 1988).

Theory of Chemical Reaction Dynamics, Vols. 1–4, edited by M. Baer (CRC Press, Boca Raton, Florida, 1985).

Atom-Molecule Collision Theory: A Guide for the Experimentalist, edited by R. B. Bernstein (Plenum Press, New York, 1979).

Dynamics of Molecular Collisions, Parts A and B, edited by W.H. Miller (Plenum Press, New York, 1976).

Case Studies in Atomic Collision Physics, edited by E.W. McDaniel and M.R.C. McDowell, (North-Holland, Amsterdam, Vol. 1 (1969), Vol. 2, (1972))

Ion-Molecule Reactions, edited by E. W. McDaniel, V. Čermák, A. Dalgarno, E. E. Ferguson, and L. Friedman (Wiley, New York, 1970).

Atomic and Molecular Processes, edited by D. R. Bates (Academic Press, New York, 1962).

Quantum Theory I. Elements, edited by D. R. Bates (Academic Press, New York, 1961).

Electronic and Ionic Impact Phenomena, H. S. W. Massey, E. H. S. Burhop, and H. B. Gilbody, editors (Clarendon Press, Oxford, 1969–1974), Vols. 1–5.

Atomic Collisions: Electron and Photon Projectiles, E. W. McDaniel, (Wiley, New York, 1989).

Atomic Collisions: Heavy Particle Projectiles, E. W. McDaniel, J. B. A. Mitchell, and M. E. Rudd, (Wiley, New York, 1993).

Rydberg Atoms, T. F. Gallagher, (Cambridge University Press, Cambridge, 1994).

Atomic Spectra and Radiative Transitions, I. I. Sobelman, (Springer Series on Atoms and Plasmas, New York, 1996).

Radiative Processes in Atomic Physics, V. P. Krainov, H. Reiss and B. M. Smirnov, (Wiley, New York, 1997).

Physics of Highly Excited Atoms and Ions, V. S. Lebedev and I. L. Beigman, (Springer Series on Atoms and Plasmas, New York, 1998).

Polarization, Alignment and Orientation in Atomic Collisions, N. Andersen and K. Bartschat, (Springer Series on Atomic, Optical and Plasma Physics, New York, 2000).

6.17. REFERENCES

[1] G. L. Catchen, J. Husain, and R. N. Zare, J. Chem. Phys. **69** 1737 (1978).

[2] M. R. Flannery, Ann. Phys. (NY) **79**, 480 (1973).

[3] E. R. Cohen and B. N. Taylor, Rev. Mod. Phys. **59**, 1121 (1987).

[4] E. S. Chang, Phys. Rev. A **31**, 495 (1985).

[5] D. R. Bates and C. S. Mc Kibbin, J. Phys. B **6**, 2458 (1973); Proc. Roy. Soc. Lond. A **339**, 13, (1974).

[6] M. R. Flannery, J. Chem. Phys. **95**, 8205 (1991).

[7] E. A. Mason and E. W. McDaniel, *Transport Properties of Ions in Gases* (Wiley, New York, 1988).

[8] L. A. Viehland, E. A. Mason, W. F. Morrison, and M. R. Flannery, Atom. Data Nucl. Data Tables **16**, 495 (1975).

[9] L. S. Rodberg and R. M. Thaler, editors, *Introduction to the Quantum Theory of Scattering* (Academic Press, New York, 1967), p. 226.

[10] N. F. Mott and H. S. W. Massey, editors, *The Theory of Atomic Collisions* (Clarendon Press, Oxford, 1965), 3rd edition.

[11] M. L. Goldberger and K. M. Watson, editors, *Collision Theory* (Wiley, New York, 1964).

[12] R. G. Newton, editor, *Scattering Theory of Waves and Particles* (McGraw-Hill, New York, 1966).

[13] C. J. Joachain, editor, *Quantum Collision Theory* (North-Holland, Amsterdam, 1975) p. 383.

[14] M. R. Flannery, Phys. Rev. A **22**, 2408 (1980).

[15] M. R. Flannery and D. Vrinceanu, Phys. Rev. Letts. **85** (2000).

[16] D. Vrinceanu and M. R. Flannery, Phys. Rev. A **6**, 1053 (1999).

[17] D. R. Bates, in *Quantum Theory I. Elements*, edited by D. R. Bates (Academic Press, New York, 1961), Ch. 8.

[18] D. R. Bates, in *Atomic and Molecular Processes*, edited by D. R. Bates (Academic Press, New York, 1962), Ch. 14.

[19] E. E. Nikitin, in *Atomic, Molecular and Optical Physics Handbook*, edited by Gordon W. Drake (American Institute of Physics Press, New York, 1996), Ch. 47.

[20] E. E. Nikitin and S. Ya. Umanskiĭ, *Theory of Slow Atomic Collisions*, (Springer-Verlag, Berlin, 1984).

[21] K. J. McCann and M. R. Flannery, J. Chem. Phys. **12**, 5275 (1978); **63**, 4695 (1975).

[22] D. R. Bates and D. S. F. Crothers, Proc. Roy. Soc. A **315**, 465 (1970).

[23] D. S. F. Crothers, Adv. Phys. **20**, 405 (1971).

[24] M. S. Child, *Semiclassical Mechanics with Molecular Applications* (Clarendon Press, Oxford, 1991).

[25] M. R. Flannery and K. J. McCann, Phys. Rev. **9**, 1947 (1974).

[26] H. S. W. Massey, E. H. S. Burhop, and H. B. Gilbody, editors, *Electronic and Ionic Impact Phenomena* (Clarendon Press, Oxford, 1969–1974), Vols. 1–5.

[27] B. H. Bransden and M. R. C. McDowell, *Charge Exchange and the Theory of Ion-Atom Collisions* (Clarendon Press, Oxford, 1992).

[28] B. H. Bransden, *Atomic Collision Theory*, (Benjamin-Cummings, Menlo Park, CA, 1983), 2nd edition.

[29] M. S. Child, *Molecular Collision Theory* (Dover Publications, Inc., New York, 1996).

[30] P. G. Burke, in *Atomic, Molecular and Optical Physics Handbook*, edited by Gordon W. Drake (American Institute of Physics Press, New York, 1996), Ch. 45.

[31] *Computational Atomic Physics*, edited by Klaus Bartschat (Springer, New York, 1996).

[32] I. E. McCarthy and E. Weigold, *Electron-Atom Collisions* (Cambridge University Press, Cambridge, 1995).

[33] N. Balakrishnan, R. C. Forrey and A. Dalgarno, Phys. Rev. Letts. **80**, 3224 (1998).

[34] M. R. Flannery and K. J. McCann (unpublished).

[35] Atomic and Molecular Data and their applications, Edited Volume, AIP Conference Proceedings, Volume 543, K. A. Berrington and K. L. Bell (Editors) (AIP Press, New York, 2000).

7

Atomic Spectroscopy

Wolfgang L. Wiese
National Institute of Standards and Technology, Gaithersburg, Maryland

Contents

List of Tables

7.1. INTRODUCTION

This chapter outlines the main concepts of atomic structure and spectroscopy. In particular, the wavelengths, intensities, transition probabilities, and shapes and shifts of spectral lines are discussed, and a few remarks on continuous spectra are included. As key spectroscopy data, the ionization energies and ground state configurations for all neutral atoms and the transition data for persistent lines of selected atoms, including transition probabilities, are presented in two tables.

A large number of treatises on atomic spectroscopy have been written, and Refs. [1] and [2] are examples of recent ones.

7.2. PHOTON ENERGIES, FREQUENCIES, AND WAVELENGTHS

7.2.1. Photon energy

The photon energy due to an electron transition between an upper atomic level k (of energy E_k) and a lower level i is $\Delta E = E_k - E_i = h\nu = hc\sigma = hc/\lambda_{\text{vac}}$, where ν is the frequency,

σ the wave number in vacuum, and λ_{vac} is the wavelength in vacuum. E's are usually listed in atomic energy level tables in wavenumber units (see below), with the ground state energy E_0 being the reference point ($E_0 = 0$). [3]–[6]

7.2.2. Frequency, wavelength, wavenumber

The frequency ν, wavelength λ, and wavenumber σ are the most accurately determined spectroscopic quantities. If any of these three is precisely known for an atomic transition, the other two quantities are equally well known, since the speed of light c is an exactly defined number. The unit of frequency is the Hertz (1 Hz $= 1\text{s}^{-1}$), and transition frequencies rather than wavelengths are often determined in low-frequency and high-precision spectroscopy, with MHz, GHz, and THz as the customary units. The common wavelength units are the nanometer (nm), the Angström (1 Å $= 10^{-1}$ nm) and the micrometer (μm).

7.2.3. Spectral wavelength ranges

The main spectral wavelength ranges have been named as follows:

2 μm to 20 μm midinfrared (ir),
750 nm to 2000 nm near ir,
400 nm to 750 nm visible spectrum,
200 nm to 400 nm near (or air) ultraviolet (uv),
100 nm to 200 nm vacuum ultraviolet (also far uv or deep uv),
10 nm to 100 nm extreme uv (euv or xuv),
1 nm to 10 nm soft x-ray,
<1 nm x-ray.

However, the limits of these wavelength ranges should not be taken as exact. Considerable variations with respect to the definitions of their extent are found in the literature.

7.2.4. Wavelengths in air

Wavelengths in standard air are normally used in visible-spectrum spectroscopy and are tabulated in most wavelength tables. These wavelengths can be converted to vacuum wavelengths with available conversion tables or conversion formulas. [7]

7.2.5. Wavelength standards

In 1992, the Comité International des Poids et Mesures recommended values for six frequencies for the absorbing molecule $^{127}I_2$ (515 nm to 640 nm), one frequency for the ^{40}Ca atom (657 nm), and one for methane (3.39 μm) as standards for suitably stabilized lasers. [8] These frequencies range from 88 376 181.600 18 MHz (3.392 231 397 327 μm, fractional uncertainty 3×10^{-12}) for a particular hyperfine-structure (hfs) component in methane to 582 490 603 .37 MHz (514.673 466 4 nm, fractional uncertainty 2.5×10^{-10}) for a particular hfs component in $^{127}I_2$.

A 1974 compilation gives reference wavelengths for some 5400 lines of 38 elements covering the range from 1.5 nm to 2.5 μm, with most uncertainties between 10^{-5} nm and 2×10^{-4} nm. [9] The wavelengths for some 1000 Fe lines selected from the Fe/Ne hollow-cathode spectrum have been recommended for reference standards over the range 183 nm to 4.2 μm, with wavenumber uncertainties 0.001 cm^{-1} to 0.002 cm^{-1}. [10] Wavelengths for about 3000 vuv and uv lines of Pt I and Pt II (100 nm to 400 nm) from a Pt/Ne hollow-cathode lamp have been determined with uncertainties of 0.0002 nm or less. [11]

7.2.6. Energy conversion factors

Energy equivalents of 1eV:

$$\begin{aligned}
1\ \text{eV} &= 1.602\,176\,462\,(63) \times 10^{-19}\ \text{J} && \text{(energy in joules)},\\
&= 8065.544\,77\,(32)\ \text{cm}^{-1} \times hc && \text{(wavenumber)},\\
&= 2.417\,989\,491\,(95) \times 10^{14}\ \text{Hz} \times h && \text{(frequency)},\\
&= 1.160\,4506\,(20) \times 10^{4}\ \text{K} \times k_\text{B} && \text{(temperature)},\\
&= 3.674\,932\,60\,(14) \times 10^{-2}\ E_\text{h} && \text{(atomic unit, Hartree)}.
\end{aligned}$$

(k_B is the Boltzmann constant, E_h the Hartree energy.)

7.3. ATOMIC STATES, ATOMIC SHELL STRUCTURE

7.3.1. Quantum numbers

Electronic states in an atom are described and fully defined by a set of four quantum numbers nlm_lm_s:

n = principal quantum number, in positive integers ($n = 1, 2, 3, \ldots, \infty$),

l = orbital angular momentum quantum number (the customary designations for l are

$l = s, p, d, f, g, \ldots$ corresponding to values of $l = 0, 1, 2, 3, 4, \ldots$,

with $l_\text{max} = n - 1$),

m_l = projection of the angular momentum (quantized), i.e.,

$m_l = -l, -(l+1), \ldots, +l$,

m_s = quantized projection of the electron spin angular momentum vectors

$s, m_\text{s} = \pm 1/2$.

Alternatively, one may also use the total angular momentum j, which is the vector sum of the orbital and spin angular momenta, $j = l + s$ (with $j = l \pm \frac{1}{2} \geq \frac{1}{2}$), in a set $nljm_j$, with $m_j = -j, -(j+1), \ldots + j$.

7.3.2. Pauli exclusion principle, atomic shells

According to the Pauli exclusion principle, for a given set of four specific quantum numbers, only one electron in an atom or atomic ion may occupy this state. This restriction yields the pronounced order of atomic shell structure: A group of electrons with the same principal quantum number n belongs to a shell for that n. Electrons with the same n and l belong into the same *subshell* and are called equivalent electrons. The maximum number of equivalent electrons according to the Pauli exclusion principle is $2(2l + 1)$; then, the subshell is called full or complete.

The general notation is nl^m, where m is the number of equivalent electrons. For example, $3d^{10}$ is the notation for a complete subshell of ten 3d-electrons, whereas $2p^4$ would be an example of an unfilled or open subshell.

The parity of a configuration is termed even or odd according to whether $\Sigma_i l_i$ is even or odd, the sum being taken over all electrons. Odd configurations are denoted with a superscript "o."

7.4. THE HYDROGEN SPECTRUM

This one-electron spectrum consists of many regular series of lines, with their wavelengths λ given by:

$$\text{Lyman Series: } \lambda^{-1} = R_{\text{H}}\left(1 - \frac{1}{n^2}\right), \text{ with } n = 2, 3, 4, \ldots,$$

where R_{H} is the Rydberg constant for the hydrogen atom, ($R_{\text{H}}(Z = 1) = 109\,677.6\,\text{cm}^{-1}$). The lines are specifically denoted as $\text{L}_\alpha(n = 1 \to 2)$, $\text{L}_\beta(n = 1 \to 3)$, $\text{L}_\gamma(n = 1 \to 4)$, and so forth.

$$\text{Balmer Series: } \lambda^{-1} = R_{\text{H}}\left(\frac{1}{2^2} - \frac{1}{n^2}\right), \text{ with } n = 3, 4, 5, \ldots.$$

The Balmer Lines are denoted as $\text{H}_\alpha(n = 2 \to 3)$; $\text{H}_\beta(n = 2 \to 4)$; $\text{H}_\gamma(n = 2 \to 5)$; and so forth. The next spectral series, in ascending order, is known as the Paschen and Brackett series. Some key spectroscopic data for leading lines of the Lyman, Balmer, and Paschen series are given in Table 7.1. Hydrogen-like ions and their scaling relationships are discussed in Sec. 7.10.

Nuclear spin and relativistic and quantum electrodynamic effects cause small shifts and splittings of the levels. [1, 2] For example, the wavenumber of the transition between the hyperfine components of the ground state of hydrogen is $\sigma = 0.047380\ \text{cm}^{-1}$ ($\equiv$ 1420.406 MHz). The hyperfine splitting is due to the interaction of the nuclear magnetic moment with the magnetic field produced by the electron and gives rise to a famous transition—the 21 cm line radiated by interstellar hydrogen.

7.5. ALKALI SPECTRA

The electronic structure of the alkali atoms (Li, Na, K, ...) and alkali-like ions, such as Be^+ and Mg^+, is similar to hydrogen, with only one optically active electron outside an atomic core with filled shells. Again, distinct spectral series—named principal series (ns-n'p) sharp series (np-n's), diffuse series (np-n'd), and so forth—are observed (n is the

TABLE 7.1. Spectral series of hydrogen.

Transition	Customary notation[a]	λ (Å)	Statistical weight g_i[b]	g_k	A_{ki} (10^8 s^{-1})
1–2	L_α	1215.67	2	8	4.699
1–3	L_β	1025.73	2	18	5.575(−1)[c]
1–4	L_γ	972.537	2	32	1.278(−1)
1–5	L_δ	949.743	2	50	4.125(−2)
1–6	L_ε	937.803	2	72	1.644(−2)
2–3	H_α	6562.80	8	18	4.410(−1)
2–4	H_β	4861.32	8	32	8.419(−2)
2–5	H_γ	4340.46	8	50	2.530(−2)
2–6	H_δ	4101.73	8	72	9.732(−3)
2–7	H_ε	3970.07	8	98	4.389(−3)
3–4	P_α	18751.0	18	32	8.986(−2)
3–5	P_β	12818.1	18	50	2.201(−2)
3–6	P_γ	10938.1	18	72	7.783(−3)
3–7	P_δ	10049.4	18	98	3.358(−3)
3–8	P_ε	9545.97	18	128	1.651(−3)

[a] L_α is often called Lyman α, H_α = Balmer α, P_α = Paschen α, and so forth.
[b] For transitions in hydrogen, $g_{i(k)} = 2(n_{i(k)})^2$, where $n_{i(k)}$ is the principal quantum number of the lower (upper) shell.
[c] The number in parentheses indicates the power of 10 by which the value has to be multiplied.

(fixed)) principal quantum number of the lower state, n' is the running number of the upper state ($n' = n, n+1, n+2$, etc.). However, because of the electrostatic interaction of the valence electron with the core electrons, there is a strong l-dependence of the spectral series, as seen from the explicit appearance of l in the above listed spectral series notations.

7.6. ATOMIC STATES AND SPECTRA FOR MANY-ELECTRON ATOMS

7.6.1. Typical features, general quantum designations

A *typical feature* is the appearance of multiplets, which is especially characteristic for lighter atoms and ions, in which muliplets form small groups of closely spaced lines.

Two general *quantum designations* characteristic of atomic states (in the absence of external fields) regardless of the electron coupling scheme are

1. The *parity*, determined by $\sum_i l_i$ (see 7.3.2).
2. The *total angular momentum quantum number J*.

7.6.2. Russell-Saunders or *LS*-coupling

Coupling between outer (valence) electrons occurs in several ways. For the prominent lower atomic states of many elements, the *Russell-Saunders* or *LS-coupling* scheme is a good approximation. According to this most important coupling scheme:

1. The orbital angular momenta of the valence electrons couple among themselves vectorially: $\mathbf{L} = \sum_i \mathbf{l}_i$. $\mathbf{L}$ must be a positive integer.
2. The spin momenta couple among themselves: $\mathbf{S} = \sum_i \mathbf{s}_i \leq 1/2m_o$, where m_o is the number of electrons in unfilled subshells. $\mathbf{S}$ may be an integer or half-integer.
3. Electrostatic interactions among the electrons are strong, and spin-orbit interactions (magnetic interactions) are weak (ideally: negligible).
4. The possible values of the orbital angular momentum $\mathbf{L}$ and the spin $\mathbf{S}$ are bounded by the total angular momenta: $\mathbf{J} = \mathbf{L} + \mathbf{S}$, with $|L - S| \leq J \leq |L + S|$.

7.6.3. Customary notation, sample case

The *customary designations* for L are letters (same as for l above), such that if $L \rightarrow 0, 1, 2, 3, 4, 5, 6, 7, 8, \ldots$, the designation for L is $L \rightarrow$ S, P, D, F, G, H, I, K, L, and so on, in alphabetical order.

A *sample* spectroscopic designation of an atomic energy level in *LS*-coupling is

$$(1s^2 2s^2 2p^6 3s^2)3p^5\,{}^2P^o_{3/2} \quad \text{(ground state of chlorine).}$$

Interpretation: Closed electron subshells (in the sample: $1s^2 2s^2 2p^6 3s^2$) are usually not given and are therefore listed here in parentheses ($1s^2$ = 2 electrons in filled 1s subshell, etc.). $3p^5$ indicates that 5 equivalent electrons (i.e., possessing the same values of n and l) with principal quantum number $n = 3$ and orbital angular momentum quantum number $l = 1$ (or p) are in the unfilled 3p subshell; these form by vector addition a resulting state of $L = 1$ (or P); the superscript 2 indicates the multiplicity $2S + 1$ (in this case, $S = |\sum \mathbf{s}| = 1/2$); subscript 3/2 indicates the total angular momentum quantum number J (either 1/2 or 3/2 is possible for the given L and S). Finally, the superscript o (for "odd") indicates a level of odd parity ($\Sigma l = 5$). (No symbol appears when the parity is even.)

Note that the Pauli exclusion principle restricts the possible values of L and S in configurations involving two or more equivalent electrons (see 7.3.2 and Refs. [1] and [2]).

7.6.4. Other coupling schemes

For heavier atoms, highly stripped atomic ions, and higher atomic levels (larger n and especially larger l), relativistic effects, such as spin-orbit interactions between the l and s of the same electron, can become large compared with interactions between the $\mathbf{l}_j$ (or $\mathbf{s}_j$) of different electrons. In this case, the l and s of each electron are coupled:

$$\mathbf{l}_i + \mathbf{s}_i = \mathbf{j}_i (j_i = l_i \pm 1/2 \geq 1/2),$$

and the j_i couple to a resultant $J = |\sum \mathbf{j}_i|$ (jj coupling). Often, the coupling within a spectrum changes from *LS* to jj as n and l increase, and there are regions of *intermediate coupling* where electrostatic and magnetic interactions are of the same magnitude.

Several other types of coupling are possible. One—the $J_l l$ (pair) coupling scheme—is a very good approximation for higher nl-states of the *noble gases*. [For the noble gases, the (historical) Paschen notation is still sometimes used for the more prominent lines, with the levels characterized by an integer (related to the principal quantum number n), the orbital quantum number l, and a running number (subscript).]

Full descriptions of these and other coupling schemes and their notation are given in Refs. [1], [2], and [12]. The ground levels of most neutral atoms may be well described in terms of LS-coupling notation. These important states are given in Table 7.2, together with the ionization energies. Only for Pa, U, and Np are the ground levels more appropriately given in terms of $J_1 j$-coupling.

TABLE 7.2. Ground state configurations, ground levels, and ionization energies of neutral atoms.

Z	Element	Ground state configuration[a]	Ground level	Ionization energy (eV)	Z	Element	Ground state configuration[a]	Ground level	Ionization energy (eV)
1	H	$1s$	$^2S_{1/2}$	13.5984	30	Zn	$[Ar]3d^{10}4s^2$	$^1S_0$	7.3942
2	He	$1s^2$	$^1S_0$	24.5874	31	Ga	$[Ar]3d^{10}4s^24p$	$^2P^\circ_{1/2}$	5.9993
3	Li	$1s^22s$	$^2S_{1/2}$	5.3917	32	Ge	$[Ar]3d^{10}4s^24p^2$	$^3P_0$	7.8994
4	Be	$1s^22s^2$	$^1S_0$	9.3227	33	As	$[Ar]3d^{10}4s^24p^3$	$^4S^\circ_{3/2}$	9.7886
5	B	$1s^22s^22p$	$^2P^\circ_{1/2}$	8.2980	34	Se	$[Ar]3d^{10}4s^24p^4$	$^3P_2$	9.7524
6	C	$1s^22s^22p^2$	$^3P_0$	11.2603	35	Br	$[Ar]3d^{10}4s^24p^5$	$^2P^\circ_{3/2}$	11.8138
7	N	$1s^22s^22p^3$	$^4S^\circ_{3/2}$	14.5341	36	Kr	$[Ar]3d^{10}4s^24p^6$	$^1S_0$	13.9996
8	O	$1s^22s^22p^4$	$^3P_2$	13.6181	37	Rb	$[Kr]5s$	$^2S_{1/2}$	4.1771
9	F	$1s^22s^22p^5$	$^2P^\circ_{3/2}$	17.4228	38	Sr	$[Kr]5s^2$	$^1S_0$	5.6949
10	Ne	$1s^22s^22p^6$	$^1S_0$	21.5646	39	Y	$[Kr]4d5s^2$	$^2D_{3/2}$	6.2171
11	Na	$[Ne]3s$	$^2S_{1/2}$	5.1391	40	Zr	$[Kr]4d^25s^2$	$^3F_2$	6.6339
12	Mg	$[Ne]3s^2$	$^1S_0$	7.6462	41	Nb	$[Kr]4d^45s$	$^6D_{1/2}$	6.7589
13	Al	$[Ne]3s^23p$	$^2P^\circ_{1/2}$	5.9858	42	Mo	$[Kr]4d^55s$	$^7S_3$	7.0924
14	Si	$[Ne]3s^23p^2$	$^3P_0$	8.1517	43	Tc	$[Kr]4d^55s^2$	$^6S_{5/2}$	7.28
15	P	$[Ne]3s^23p^3$	$^4S^\circ_{3/2}$	10.4867	44	Ru	$[Kr]4d^75s$	$^5F_5$	7.3605
16	S	$[Ne]3s^23p^4$	$^3P_2$	10.3600	45	Rh	$[Kr]4d^85s$	$^4F_{9/2}$	7.4589
17	Cl	$[Ne]3s^23p^5$	$^2P^\circ_{3/2}$	12.9676	46	Pd	$[Kr]4d^{10}$	$^1S_0$	8.3369
18	Ar	$[Ne]3s^23p^6$	$^1S_0$	15.7596	47	Ag	$[Kr]4d^{10}5s$	$^2S_{1/2}$	7.5762
19	K	$[Ar]4s$	$^2S_{1/2}$	4.3407	48	Cd	$[Kr]4d^{10}5s^2$	$^1S_0$	8.9938
20	Ca	$[Ar]4s^2$	$^1S_0$	6.1132	49	In	$[Kr]4d^{10}5s^25p$	$^2P^\circ_{1/2}$	5.7864
21	Sc	$[Ar]3d4s^2$	$^2D_{3/2}$	6.5615	50	Sn	$[Kr]4d^{10}5s^25p^2$	$^3P_0$	7.3439
22	Ti	$[Ar]3d^24s^2$	$^3F_2$	6.8281	51	Sb	$[Kr]4d^{10}5s^25p^3$	$^4S^\circ_{3/2}$	8.6084
23	V	$[Ar]3d^34s^2$	$^4F_{3/2}$	6.7462	52	Te	$[Kr]4d^{10}5s^25p^4$	$^3P_2$	9.0096
24	Cr	$[Ar]3d^54s$	$^7S_3$	6.7665	53	I	$[Kr]4d^{10}5s^25p^5$	$^2P^\circ_{3/2}$	10.4513
25	Mn	$[Ar]3d^54s^2$	$^6S_{5/2}$	7.4340	54	Xe	$[Kr]4d^{10}5s^25p^6$	$^1S_0$	12.1298
26	Fe	$[Ar]3d^64s^2$	$^5D_4$	7.9024	55	Cs	$[Xe]6s$	$^2S_{1/2}$	3.8939
27	Co	$[Ar]3d^74s^2$	$^4F_{9/2}$	7.8810	56	Ba	$[Xe]6s^2$	$^1S_0$	5.2117
28	Ni	$[Ar]3d^84s^2$	$^3F_4$	7.6398	57	La	$[Xe]5d6s^2$	$^2D_{3/2}$	5.5769
29	Cu	$[Ar]3d^{10}4s$	$^2S_{1/2}$	7.7264	58	Ce	$[Xe]4f5d6s^2$	$^1G^\circ_4$	5.5387

(continued)

TABLE 7.2. Continued

Z	Element	Ground state configuration[a]	Ground level	Ionization energy (eV)	Z	Element	Ground state configuration[a]	Ground level	Ionization energy (eV)
59	Pr	[Xe]$4f^3 6s^2$	$^4I^\circ_{9/2}$	5.473	82	Pb	[Xe]$4f^{14}5d^{10}6s^2 6p^2$	$^3P_0$	7.4167
60	Nd	[Xe]$4f^4 6s^2$	$^5I_4$	5.5250	83	Bi	[Xe]$4f^{14}5d^{10}6s^2 6p^3$	$^4S^\circ_{3/2}$	7.2856
61	Pm	[Xe]$4f^5 6s^2$	$^6H^\circ_{5/2}$	5.582	84	Po	[Xe]$4f^{14}5d^{10}6s^2 6p^4$	$^3P_2$	8.417?
62	Sm	[Xe]$4f^6 6s^2$	$^7F_0$	5.6436	85	At	[Xe]$4f^{14}5d^{10}6s^2 6p^5$	$^2P^\circ_{3/2}$	
63	Eu	[Xe]$4f^7 6s^2$	$^8S^\circ_{7/2}$	5.6704	86	Rn	[Xe]$4f^{14}5d^{10}6s^2 6p^6$	$^1S_0$	10.7485
64	Gd	[Xe]$4f^7 5d6s^2$	$^9D^\circ_2$	6.1501	87	Fr	[Rn]$7s$	$^2S_{1/2}$	4.0727
65	Tb	[Xe]$4f^9 6s^2$	$^6H^\circ_{15/2}$	5.8638	88	Ra	[Rn]$7s^2$	$^1S_0$	5.2784
66	Dy	[Xe]$4f^{10}6s^2$	$^5I_8$	5.9389	89	Ac	[Rn]$6d7s^2$	$^2D_{3/2}$	5.17
67	Ho	[Xe]$4f^{11}6s^2$	$^4I^\circ_{15/2}$	6.0215	90	Th	[Rn]$6d^2 7s^2$	$^3F_2$	6.3067
68	Er	[Xe]$4f^{12}6s^2$	$^3H_6$	6.1077	91	Pa	[Rn]$5f^2(^3H_4)6d7s^2$	$(4, 3/2)_{11}$	5.89
69	Tm	[Xe]$4f^{13}6s^2$	$^2F^\circ_{7/2}$	6.1843	92	U	[Rn]$5f^3(^4I^\circ_{9/2})6d7s^2$	$(9/2, 3/2)$	6.1941
70	Yb	[Xe]$4f^{14}6s^2$	$^1S_0$	6.2542	93	Np	[Rn]$5f^4(^5I_4)6d7s^2$	$(4, 3/2)_{11}$	6.2657
71	Lu	[Xe]$4f^{14}5d6s^2$	$^2D_{3/2}$	5.4259	94	Pu	[Rn]$5f^6 7s^2$	$^7F_0$	6.0262
72	Hf	[Xe]$4f^{14}5d^2 6s^2$	$^3F_2$	6.8251	95	Am	[Rn]$5f^7 7s^2$	$^8S^\circ_{7/2}$	5.9738
73	Ta	[Xe]$4f^{14}5d^3 6s^2$	$^4F_{3/2}$	7.5496	96	Cm	[Rn]$5f^7 6d7s^2$	$^9D^\circ_2$	5.9915
74	W	[Xe]$4f^{14}5d^4 6s^2$	$^5D_0$	7.8640	97	Bk	[Rn]$5f^9 7s^2$	$^6H^\circ_{15/2}$	6.1979
75	Re	[Xe]$4f^{14}5d^5 6s^2$	$^6S_{5/2}$	7.8335	98	Cf	[Rn]$5f^{10}7s^2$	$^5I_8$	6.2817
76	Os	[Xe]$4f^{14}5d^6 6s^2$	$^5D_4$	8.4382	99	Es	[Rn]$5f^{11}7s^2$	$^4I^\circ_{15/2}$	6.42
77	Ir	[Xe]$4f^{14}5d^7 6s^2$	$^4F_{9/2}$	8.9670	100	Fm	[Rn]$5f^{12}7s^2$	$^3H_6$	6.50
78	Pt	[Xe]$4f^{14}5d^9 6s$	$^3D_3$	8.9587	101	Md	[Rn]$5f^{13}7s^2$	$^2F^\circ_{7/2}$	6.58
79	Au	[Xe]$4f^{14}5d^{10}6s$	$^2S_{1/2}$	9.2255	102	No	[Rn]$5f^{14}7s^2$	$^1S_0$	6.65
80	Hg	[Xe]$4f^{14}5d^{10}6s^2$	$^1S_0$	0.4375	103	Lr	[Rn]$5f^{14}7s^2 7p$?	$^2P^\circ_{1/2}$	4.9?
81	Tl	[Xe]$4f^{15}5d^{10}6s^2 6p$	$^2P^\circ_{1/2}$	6.1082	104	Rf	[Rn]$5f^{14}6d^2 7s^2$?	$^3F_2$?	6.0?

[a] Ground state configurations of elements will filled electron subshells—indicated by their respective element symbols in square brackets—are used for abbreviation. For example, [Mg] = [Ne]$3s^2$ = $1s^2 2s^2 2p^6 3s^2$.

7.7. ATOMIC STRUCTURE HIERARCHIES, SELECTION RULES FOR DISCRETE TRANSITIONS

7.7.1. Atomic structure hierarchies

Groups of atomic states and transitions, specified by the following sets of quantum numbers, have the designations listed in Table 7.3. They constitute hierarchies starting with the largest, all-inclusive entity at the left.

TABLE 7.3. Designations for atomic states and transitions.

Atomic entity →	Configuration	Polyad	Term	Level[a]	State[b]
Required quantum number specifications[c]	n, l	n, l, S	n, l, L, S	n, l, L, S, J	n, l, L, S, J, M
Corresponding transition	Transition array	Supermultiplet	Multiplet	Spectral line	Component of line

[a] Splittings of terms into levels (and, thus, of muliplets into lines) constitute fine structure.
[b] Splittings of levels into states (and, thus, of lines into components) occur only in magnetic fields; these are identified by magnetic quantum numbers $M(|M| \leq J)$.
[c] Designations are usually confined to n and l of the electrons in the unfilled subshell(s) (valence electrons).

7.7.2. Selection rules for discrete transitions

The selection rules for discrete transitions are given in Table 7.4, with the rigorous rules listed first. For allowed lines (E1), the parity changes (see 7.6.1) between lower and upper level, whereas for the most common forbidden lines (M1, E2), no parity change occurs.

TABLE 7.4. Selection rules for discrete transitions.

	Electric dipole (E1) ("allowed" lines)	Magnetic dipole (M1) ("forbidden lines")	Electric quadrupole (E2) ("forbidden lines")
Rigorous rules	1. $\Delta J = 0, \pm 1$ (except $J = 0 \leftrightarrow 0$)	$\Delta J = 0, \pm 1$ (except $0 \leftrightarrow 0$)	$\Delta J = 0, \pm 1, \pm 2$ (except $0 \leftrightarrow 0$, $1/2 \leftrightarrow 1/2, 0 \leftrightarrow 1$)
	2. $\Delta M = 0, \pm 1$ (except $0 \leftrightarrow 0$ when $\Delta J = 0$)	$\Delta M = 0, \pm 1$ (except $0 \leftrightarrow 0$ when $\Delta J = 0$)	$\Delta M = 0, \pm 1, \pm 2$
	3. Parity change	No parity change	No parity change
With negligible configuration interaction	4. One electron jumping, with $\Delta l = \pm 1$, Δn arbitrary	No change in electron configuration; i.e., for all electrons, $\Delta l = 0$, $\Delta n = 0$	No change in electron configuration; or one electron jumping, with $\Delta l = 0, \pm 2$, Δn arbitrary
For *LS*-coupling only	5. $\Delta S = 0$	$\Delta S = 0$	$\Delta S = 0$
	6. $\Delta L = 0, \pm 1$ (except $L = 0 \leftrightarrow 0$)	$\Delta L = 0$ $\Delta J = \pm 1$	$\Delta L = 0, \pm 1, \pm 2$ (except $0 \leftrightarrow 0, 0 \leftrightarrow 1$)

7.8. SPECTRAL LINE INTENSITIES, ATOMIC TRANSITION PROBABILITIES, *f*-VALUES, AND LINE STRENGTHS

7.8.1. Emission intensities

The total power ε radiated in a spectral line of frequency ν per unit source volume and per unit solid angle is

$$\varepsilon_{\text{line}} = (4\pi)^{-1} h\nu A_{ki} N_k,$$

where A_{ki} is the atomic transition probability and N_k is the number of excited atoms or ions per unit volume (number density) in the upper (initial) level k. For a homogeneous light source of length l and for the optically thin case, in which all radiation escapes without getting reabsorbed, the total emitted line intensity (SI quantity: radiance) is

$$I_{\text{line}} = \varepsilon_{\text{line}} l = \int_0^{+\infty} I(\lambda)\, d\lambda = (4\pi)^{-1}(hc/\lambda_0) A_{ki} N_k l,$$

where $I(\lambda)$ is the specific intensity at wavelength λ and λ_0 is the wavelength at line center.

7.8.2. Absorption intensities

For measurements in absorption, the reduced absorption

$$W(\lambda) = [I(\lambda) - I'(\lambda)]/I(\lambda)$$

is used, where $I(\lambda)$ is the incident intensity at wavelength λ, e.g., from a source providing a continuum background, and $I'(\lambda)$ is the intensity after passage through the absorbing medium. The reduced line intensity after passage through a homogeneous and optically thin absorbing medium of length l follows as:

$$W_{ik} = \int_0^{+\infty} W(\lambda)\, d\lambda = \frac{e^2}{4\varepsilon_0 m_e c^2} \lambda_0^2 N_i f_{ik} l,$$

where f_{ik} is the atomic (absorption) oscillator strength (dimensionless).

7.8.3. Line strengths

A_{ki} and f_{ik} are the principal atomic quantities related to line intensities. In theoretical work, the *line strength* S is computed (not to be confused with the spin angular momentum quantum number):

$$S = S(i, k) = S(k, i) = |\, R_{ik}\,|^2, \; R_{ik} = \langle \psi_k |\, P \,| \psi_i \rangle,$$

where ψ_i and ψ_k are the initial- and final-state wave functions and R_{ik} is the *transition matrix element* of the appropriate multipole operator P (R_{ik} involves an integration over spatial and spin coordinates of all N electrons of the atom or ion). This operator for the allowed (electric dipole) transition is

$$P = -|\, e\,| \sum_{m=1}^{N} \mathbf{r}_m,$$

where the $\mathbf{r}_m$ are the position vectors of the electrons. S is normally given in atomic units (for electric dipole transitions: $a_0^2e^2 = 7.1883 \times 10^{-59}\ \mathrm{m^2C^2}$).

7.8.4. Relationships among *A*, *f*, and *S*

The relationships among A, f, and S for electric dipole (allowed) transitions in SI units (A in s^{-1}, λ in m, S in $\mathrm{m^2C^2}$) are

$$A_{ki} = \frac{2\pi e^2}{m_e c\varepsilon_0\lambda^2}\frac{g_i}{g_k}f_{ik} = \frac{16\pi^3}{3h\varepsilon_0\lambda^3 g_k}S.$$

Numerically, in customary units (A in s^{-1}, λ in Å, S in atomic units),

$$A_{ki} = \frac{6.6702 \times 10^{15}}{\lambda^2}\frac{g_i}{g_k}f_{ik} = \frac{2.0261 \times 10^{18}}{\lambda^3 g_k}S,$$

and for S and ΔE in atomic units,

$$f_{ik} = \frac{2}{3}(\Delta E/g_i)S.$$

g_i and g_k are the statistical weights, which are obtained from the appropriate angular momentum quantum numbers. Thus, for the lower (upper) level of a spectral line,

$$g_{i(k)} = 2J_{i(k)} + 1,$$

and for the lower (upper) term of a muliplet,

$$\bar{g}_{i(k)} = \sum_{J_{i(k)}}(2J_{i(k)} + 1) = (2L_{i(k)} + 1)(2S_{i(k)} + 1).$$

For forbidden transitions, i.e., electric quadrupole, magnetic dipole lines, and so forth, f is not in use. Conversions between S and A_{ki} for the principal forbidden transitions are given in Table 7.5.

TABLE 7.5. Conversions between *S* and A_{ki} for forbidden transitions.

	SI Units[a]	Numerically, in customary units[b]
Electric quadrupole	$A_{ki} = \dfrac{16\pi^5}{15h\varepsilon_0\lambda^5 g_k}S$	$A_{ki} = \dfrac{1.120 \times 10^{18}}{g_k\lambda^5}S$
Magnetic dipole	$A_{ki} = \dfrac{16\pi^3\mu_0}{3h\lambda^3 g_k}S$	$A_{ki} = \dfrac{2.697 \times 10^{13}}{g_k\lambda^3}S$

[a] A in s^{-1}, λ in m. Electric quadrupole: S in $\mathrm{m^4C^2}$. Magnetic dipole: S in $\mathrm{J^2T^{-2}}$.
[b] A in s^{-1}, λ in Å. S in atomic units: $a_0^4e^2 = 2.013 \times 10^{-79}\ \mathrm{m^4C^2}$ (electric quadrupole), $e^2h^2/16\pi^2m_e^2 = \mu_B^2 = 8.601 \times 10^{-47}\ \mathrm{J^2T^{-2}}$ (magnetic dipole). μ_B is the Bohr magneton.

7.8.5. Relationships between spectral line and multiplet values

The relations between total line strengths and f-values of multiplets (M) and the corresponding quantities for spectral lines within multiplets of allowed transitions are:

1. $S_M = \sum S_{\text{line}}$,
2. $f_M = (\bar{\lambda}\bar{g}_i)^{-1} \sum_{J_k, J_i} g_i \lambda(J_i, J_k) f(J_i, J_k)$,
3. $A_M = (\bar{\lambda}\bar{g}_k)^{-3} \sum_{J_k, J_i} g_k \lambda(J_i, J_k)^3 A(J_i, J_k)$,

where $\bar{g}_i$, $\bar{g}_k$ are the statistical weights for the lower and upper terms of the multiplet (see Sec.7.8.4).

$\bar{\lambda}$ is the weighted ("multiplet") wavelength:

$$\bar{\lambda} = hc/\bar{\Delta}\bar{E},$$

where

$$\bar{E} = \bar{E}_k - \bar{E}_i = (\bar{g}_k)^{-1} \sum_{J_k} g_k E_k - (\bar{g}_i)^{-1} \sum_{J_i} g_i E_i.$$

Often, the wavelength differences for the lines within a multiplet are small, so that the wavelength factors may be neglected.

7.8.6. Tabulations

A_{ki}-values for persistent lines of selected elements are given in Table 7.6. For extensive numerical tables of A, f, and S, including forbidden lines, see Refs. [13]–[18].

TABLE 7.6. Transition data for persistent spectral lines of neutral atoms. Wavelengths λ, upper energy levels E_k, statistical weights g_i and g_k of lower and upper levels, and transition probabilities A_{ki} are presented. Many tabulated lines are resonance lines (marked "g"), in which the lower energy level belongs to the ground term.

Spectrum	λ^a Å	E_k (cm^{-1})	g_i	g_k	A_{ki} 10^8 s^{-1}	Accuracy[b]	Spectrum	λ^a Å	E_k (cm^{-1})	g_i	g_k	A_{ki} 10^8 s^{-1}	Accuracy[b]
Ag	3280.7g	30,473	2	4	1.4	B	Ar	7635.1	106,238	5	5	0.245	C
	3382.9g	29,552	2	2	1.3	B		7948.2	107,132	1	3	0.186	C
	5209.1	48,744	2	4	0.75	D		8115.3	105,463	5	7	0.331	C
	5465.5	48,764	4	6	0.86	D	As	1890.4g	52,898	4	6	2.0	D
Al	3082.2g	32,435	2	4	0.63	C		1937.6g	51,610	4	4	2.0	D
	3092.7g	32,437	4	6	0.74	C		2288.1	54,605	6	4	2.8	D
	3944.0g	25,348	2	2	0.493	C		2349.8	53,136	4	2	3.1	D
	3961.5g	25,348	4	2	0.98	C	Au	2428.0g	41,174	2	4	1.99	B+
Ar	1048.2g	95,400	1	3	5.32	B		2676.0g	37,359	2	2	1.64	B+
	4158.6	117,184	5	5	0.014	B	B	1825.9g	54,767	2	4	1.76	B
	4259.4	118,871	3	1	0.039	B		1826.4g	54,767	4	6	2.11	B

(continued)

TABLE 7.6. Continued

Spectrum	λ^a Å	E_k (cm^{-1})	g_i	g_k	A_{ki} 10^8 s^{-1}	Accuracy[b]	Spectrum	λ^a Å	E_k (cm^{-1})	g_i	g_k	A_{ki} 10^8 s^{-1}	Accuracy[b]
B	2496.8g	40,040	2	2	0.864	C	Cs	3876.1g	25,792	2	4	0.0038	C
	2497.7g	40,040	4	2	1.73	C		4555.3g	21,946	2	4	0.0188	C
Ba	5535.5g	18,060	1	3	1.19	B		4593.2g	21,765	2	2	0.0080	C
	6498.8	24,980	7	7	0.86	D		8521.1g	11,732	2	4	0.3276	AA
	7059.9	23,757	7	9	0.71	D		8943.5g	11,178	2	2	0.287	A
	7280.3	22,947	5	7	0.53	D	Cu	2178.9g	45,879	2	4	0.913	B
Be	2348.6g	42,565	1	3	5.547	AA		3247.5g	30,784	2	4	1.39	B
	2650.6[c]	59,696	9	9	4.24	AA		3274.0g	30,535	2	2	1.37	B
Bi	2228.3g	44,865	4	4	0.89	D		5218.2	49,942	4	6	0.75	C
	2898.0	45,916	4	2	1.53	C	F	954.83g	104,731	4	4	5.77	C
	2989.0	44,865	4	4	0.55	C		6856.0	116,987	6	8	0.494	C
	3067.7g	32,588	4	2	2.07	C		7398.7	115,918	6	6	0.285	C+
Br	1488.5g	67,184	4	4	1.2	D		7754.7	117,623	4	6	0.382	C+
	1540.7g	64,907	4	4	1.4	D	Fe	3581.2	34,844	11	13	1.02	B+
	7348.5	78,512	4	6	0.12	D		3719.9g	26,875	9	11	0.162	B+
C	1561.4g	64,087	5	7	1.18	A		3734.9	33,695	11	11	0.902	B+
	1657.0g	60,393	5	5	2.52	A		3745.6g	27,395	5	7	0.115	B+
	1930.9	61,982	5	3	3.51	B+		3859.9g	25,900	9	9	0.0970	B+
	2478.6	61,982	1	3	0.340	B+		4045.8	36,686	9	9	0.863	B+
Ca	4226.7g	23,652	1	3	2.18	B+	Ga	2874.2g	34,782	2	4	1.2	C
	4302.5	38,552	5	5	1.36	C+		2943.6g	34,788	4	6	1.4	C
	5588.8	38,259	7	7	0.49	D		4033.0g	24,789	2	2	0.49	C
	6162.2	31,539	5	3	0.354	C		4172.0g	24,789	4	2	0.92	C
	6439.1	35,897	7	9	0.53	D	Ge	2651.6g	37,702	1	3	0.85	C
Cd	2288.0g	43,692	1	3	5.3	C		2709.6g	37,452	3	1	2.8	C
	3466.2	59,498	3	5	1.2	D		2754.6g	37,702	5	3	1.1	C
	3610.5	59,516	5	7	1.3	D		3039.1	40,021	5	3	2.8	C
	5085.8	51,484	5	3	0.56	C	He	537.03g	186,209	1	3	5.66	AA
Cl	1347.2g	74,226	4	4	4.19	C		584.33g	171,135	1	3	17.99	AA
	1351.7g	74,866	2	2	3.23	C		3888.7[c]	185,565	3	9	0.09475	AA
	4526.2	96,313	4	4	0.051	C		4026.2[c]	193,917	9	15	0.1160	AA
	7256.6	85,735	6	4	0.15	C		4471.5[c]	191,445	9	15	0.2458	AA
Co	3405.1	32,842	10	10	1.0	C+		5875.7[c]	186,102	9	15	0.7070	AA
	3453.5	32,431	10	12	1.1	C+	Hg	2536.5g	39,412	1	3	0.0800	B
	3502.3	32,028	10	8	0.80	C+		3125.7	71,396	3	5	0.656	B
	3569.4	35,451	8	8	1.6	C		4358.3	62,350	3	3	0.577	B
Cr	3578.7g	27,935	7	9	1.48	B		5460.7	62,350	5	3	0.487	B
	3593.5g	27,820	7	7	1.50	B	I	1782.8g	56,093	4	4	2.71	C
	3605.3g	27,729	7	5	1.62	B		1830.4g	54,633	4	6	0.16	D
	4254.3g	23,499	7	9	0.315	B	In	3039.4g	32,892	2	6	1.3	D
	4274.8g	23,386	7	7	0.307	B		3256.1g	32,915	4	4	1.3	D
	5208.4	26,788	5	7	0.506	B		4101.8g	24,373	2	2	0.56	C

(continued)

TABLE 7.6. Continued

Spectrum	λ[a] Å	E_k (cm^{-1})	g_i	g_k	A_{ki} 10^8 s^{-1}	Accuracy[b]
In	4511.3g	24,373	4	2	1.02	C
K	4044.1g	24,720	2	4	0.0124	C
	4047.2g	24,701	2	2	0.0124	C
	7664.9g	13,043	2	4	0.387	B+
	7699.0g	12,985	2	2	0.382	B+
Kr	5570.3	97,919	5	3	0.021	D
	5870.9	97,945	3	5	0.018	D
	7601.5	93,123	5	5	0.31	D
	8112.9	92,294	5	7	0.36	D
Li	3232.7g[c]	30,925	2	6	0.0117	B
	4602.9[c]	36,623	6	10	0.223	B
	6103.6[c]	31,283	6	10	0.6860	AA
	6707.8g[c]	14,904	2	6	0.3691	AA
Mg	2025.8g	49,347	1	3	0.84	D
	2852.1g	35,051	1	3	4.95	B
	4703.0	56,308	3	5	0.255	C
	5183.6	41,197	5	3	0.575	B
Mn	2794.8g	35,770	6	8	3.7	C
	2798.3g	35,726	6	6	3.6	C
	2801.1g	35,690	6	4	3.7	C
	4030.8g	24,820	6	8	0.17	C+
	4033.1g	24,788	6	6	0.165	C+
	4034.4g	24,779	6	4	0.158	C+
N	1199.6g	83,365	4	6	4.01	B+
	1492.6	86,221	6	4	3.13	B+
	4935.1	106,478	4	2	0.0176	B
	7468.3	96,751	6	4	0.193	B+
	8216.3	95,532	6	6	0.223	B
Na	5890.0g	16,973	2	4	0.611	AA
	5895.9g	16,956	2	2	0.610	AA
	5682.6	34,549	2	4	0.103	C
	8183.3	29,173	2	4	0.453	C
Ne	735.90g	135,889	1	3	6.11	B
	743.72g	134,459	1	3	0.486	B
	5852.5	152,971	3	1	0.682	B
	6402.2	149,657	5	7	0.514	B
Ni	3101.6	33,112	5	7	0.63	C+
	3134.1	33,611	3	5	0.73	C+
	3369.6g	29,669	9	7	0.18	C
	3414.8	29,481	7	9	0.55	C
	3524.5	28,569	7	5	1.0	C
	3619.4	31,031	5	7	0.66	C
O	1302.2g	76,795	5	3	3.14	A
	4368.2[c]	99,681	3	9	0.0072	B
	5436.9	105,019	7	5	0.0180	C+
	7156.7	116,631	5	5	0.505	B
	7771.9	86,631	5	7	0.369	A
P	1775.0g	56,340	4	6	2.17	C
	1782.9g	56,090	4	4	2.14	C
	2136.2	58,174	6	4	2.83	C
	2535.6	58,174	4	4	0.95	C
Pb	2802.0g	46,329	5	7	1.6	D
	2833.1g	35,287	1	3	0.58	D
	3683.5g	34,960	3	1	1.5	D
	4057.8g	35,287	5	3	0.89	D
Rb	4201.8g	23,793	2	4	0.018	C
	4215.5g	23,715	2	2	0.015	C
	7800.3g	12,817	2	4	0.370	B
	7947.6g	12,579	2	2	0.340	B
S	1474.0g	67,843	5	7	1.6	D
	1666.7	69,238	5	5	6.3	C
	1807.3g	55,331	5	3	3.8	C
	4694.1	73,921	5	7	0.0067	D
Sc	3907.5g	25,585	4	6	1.28	C+
	3911.8g	25,725	6	8	1.37	C+
	4020.4g	24,866	4	4	1.65	C+
	4023.7g	25,014	6	6	1.44	C+
Si	2506.9g	39,955	3	5	0.466	C
	2516.1g	39,955	5	5	1.21	C
	2881.6	40,992	5	3	1.89	C
	5006.1	60,962	3	5	0.028	D
	5948.65	57,798	3	5	0.022	D
Sn	2840.0g	38,629	5	5	1.7	D
	3034.1g	34,641	3	1	2.0	D
	3175.1g	34,914	5	3	1.0	D
	3262.3	39,257	5	3	2.7	D
Sr	2428.1g	41,172	1	3	0.17	C
	4607.3g	21,698	1	3	2.01	B
Ti	3642.7g	27,615	7	9	0.774	B
	3653.5g	27,750	9	11	0.754	C+
	3998.6g	25,388	9	9	0.408	B
	4981.7	26,911	11	13	0.660	C+
	5120.4g	19,574	9	9	0.0357	C+

(continued)

TABLE 7.6. Continued

Spectrum	λ^a Å	E_k (cm^{-1})	g_i	g_k	A_{ki} 10^8 s^{-1}	Accuracy[b]	Spectrum	λ^a Å	E_k (cm^{-1})	g_i	g_k	A_{ki} 10^8 s^{-1}	Accuracy[b]
Tl	2767.9g	36,118	2	4	1.26	C	V	4384.7	25,112	8	10	1.1	C
	3519.2g	36,200	4	6	1.24	C	Xe	1192.0g	83,890	1	3	6.2	C
	3775.7g	26,478	2	2	0.625	B		1295.6g	77,186	1	3	2.5	C
	5350.5g	26,478	4	2	0.705	B		1469.6g	68,046	1	3	2.8	B
U	3566.6g	28,650	11	11	0.24	B		4671.2	88,470	5	7	0.010	D
	3571.6	38,338	17	15	0.13	C		7119.6	92,445	7	9	0.066	D
	3584.9g	27,887	13	15	0.18	B	Zn	2138.6g	46,745	1	3	7.09	B
V	3183.4g	31,541	6	8	2.4	C+		3302.6	62,772	3	5	1.2	B
	4111.8	26,738	10	10	1.01	B		3345.0	62,777	5	7	1.7	B
	4379.2	25,254	10	12	1.1	C		6362.3	62,459	3	5	0.474	C

[a] A "g" following the wavelength indicates that the lower level of the transition belongs to the ground term; i.e., the line is a resonance line. Wavelengths below 2000 Å are in vacuum, and those above 2000 Å are in air.

[b] Accuracy estimates pertain to A_{ki} values: AA, uncertainty within 1%; A, within 3%; B, within 10%; C, within 25%; D, within 50%.

[c] The superscript "c" following the wavelength indicates that A_{ki}, λ, E_k, g_i, and g_k are multiplet values, since the wavelengths of the component lines almost overlap.

If multiplet strengths S_M are known, and individual line strengths are needed, these may be obtained for *LS*-coupling from general tables, given, for example, in Refs. [1] and [19].

7.9. ATOMIC (RADIATIVE) LIFETIMES

The radiative lifetime τ_k of an atomic level k is related to the sum of transition probabilities to all levels i lower in energy than k:

$$\tau_k = \left(\sum_i A_{ki}\right)^{-1}.$$

The branching fraction of a particular transition, say to state i', is defined as

$$A_{ki'} \Big/ \sum_i A_{ki} = A_{ki'} \tau_k.$$

The relation between τ_k and the (absorption) oscillator strengths of transitions to levels lower in energy than k is

$$\tau_k = \frac{m_e c \varepsilon_0 g_k}{2\pi e^2} \left(\sum_{i, E(i)<E(k)} \frac{g_i f_{ik}}{[\lambda(J_i, J_k)]^2} \right)^{-1}.$$

If only one branch (i') exists or is *dominant*, one obtains $A_{ki'}\tau_k \simeq 1$ and $\tau_k = 1/A_{ki'}$. This is typically the case for resonance lines.

7.10. SCALING, SYSTEMATIC TRENDS (REGULARITIES), AND IMPORTANT CHARACTERISTICS OF SPECTRA

7.10.1. Hydrogenic (one-electron) species

1. *Energy levels:* The nonrelativistic *energy levels* with respect to the ground level ($1s\,^2S_{1/2}$) of a hydrogenic ion (including the neutral H-atom) of nuclear charge Z ($Z = 1$ for H, 2 for He^+, ...) and nuclear mass M_Z are

$$E_{Z,n} = \frac{m_e e^4 Z^2 (1 - 1/n^2)}{8h^2\varepsilon_0^2(1 + m_e/M_Z)}.$$

The factor $1 + m_e/M_Z$ takes into account the reduced mass of the electron-nucleus system.

In terms of the Rydberg constant $R_\infty (R_\infty = m_e e^4/8h^3 c\varepsilon_0^2)$, one obtains

$$E_{Z,n} = \frac{R_\infty hcZ^2(1 - 1/n^2)}{1 + m_e/M_Z}.$$

Thus, if $R_Z \equiv R_\infty/(1 + m_e/M_Z)$,

$$E_{Z,n} = R_Z hcZ^2(1 - 1/n^2).$$

$E_{Z,n}$ is obtained in wavenumber units by omitting the factor hc, and with $R_\infty = 109\,737.32\ \mathrm{cm^{-1}}$.

2. *Transitions:* The nonrelativistic *energy* of a hydrogenic transition (see also Sec. 7.4):

$$(\Delta E)_Z = (E_k - E_i)_Z = R_Z hcZ^2(1/n_i^2 - 1/n_k^2).$$

(The corresponding wavenumber is obtained by omitting the factor hc.) Isotope shifts arise from differences in the masses M_Z of the various isotopes of a given ion of nuclear charge Z.

3. *Hydrogenic Z-scaling:* The spectroscopic quantities for a hydrogenic ion of nuclear charge Z relate to the equivalent quantities in hydrogen ($Z = 1$) as follows (neglecting small differences in the values of R_Z):

$$(\Delta E)_Z = Z^2(\Delta E)_H, \quad (\lambda_{vac})_Z = Z^{-2}(\lambda_{vac})_H,$$
$$S_Z = Z^{-2}S_H, \quad f_Z = f_H, \quad A_Z = Z^4 A_H.$$

For large values of Z, roughly $Z > 20$, relativistic corrections become noticeable and must be taken into account. [20]

4. *f-value trends:* f-values for high series members (large n'-values) of hydrogenic ions decrease according to

$$f(n, l \to n', l \pm 1) \propto (n')^{-3}.$$

7.10.2. Atoms and ions with two or more electrons

1. *Isoelectronic sequences:* These comprise all atomic species with the same number of electrons but different Z (e.g., Li sequence = Li, Be^+, B^{2+}, C^{3+}, N^{4+}, ...). Atomic quantities for a given state or transition may be expressed as power series expansions in Z^{-1}:

$$Z^{-2}E = \varepsilon_0 + \varepsilon_1 Z^{-1} + \varepsilon_2 Z^{-2} + \cdots,$$
$$Z^2 S = S_0 + S_1 Z^{-1} + S_2 Z^{-2} + \cdots,$$
$$f = f_0 + f_1 Z^{-1} + f_2 Z^{-2} + \cdots,$$

where E_0, f_0, and S_0 are hydrogenic quantities. [21] For transitions in which n does not change ($n_i = n_k$), $f_0 = 0$, since states i and k are degenerate for hydrogenic species.

2. *Spectral series:* Appearing in simple atomic structures, mainly the alkalis, they are groups of spectral lines from a given lower state (n, l) to all upper states with increasing $n' = n, n+1, n+2, \ldots$ and constant orbital angular momentum quantum number (either $l+1$ or $l-1$). The line positions and f-values will generally show a regular pattern with increasing n', since the orbital quantum number of the electron undergoing the transition as well as the angular structure ${}^{2S'+1}L'_{j'}$ are the same for all n'.

3. *Homologous atoms:* They have similar valence structures but different numbers of closed shells of electrons. For example, atoms with a single valence electron outside of closed shells, i.e., in the spectroscopic sense, the "alkalis," are

$$\begin{aligned} &\text{Li:} && 1s^2 && + \text{ valence electron,}\\ &\text{Na:} && 1s^2 2s^2 2p^6 && + \text{ valence electron,}\\ &\text{Cu:} && 1s^2 2s^2 2p^6 3s^2 3p^6 3d^{10} && + \text{ valence electron.}\\ &\vdots \end{aligned}$$

For equivalent transitions, f-values vary gradually within a group of homologous atoms. Transitions to be compared in the case of the "alkalis" are

$$\begin{aligned} (nl - n'l')_{\text{Li}} &\to [(n+1)l - (n'+1)l']_{\text{Na}}\\ &\to [(n+2)l - (n'+2)l']_{\text{Cu}}\\ &\to \ldots. \end{aligned}$$

In complex atomic structures, as well as in cases of appreciable cancellation in the integrand of the transition integral (see below), this regular behavior is not observed.

7.10.3. Important characteristics of complex spectra

1. *Quantum defect:* For atoms or ions of more than one electron, the effective principal quantum number n^* corresponding to a particular energy level E is often used:

$$n^* = (Z - N + 1)\sqrt{\frac{R_Z hc}{I - E}}.$$

N is the total number of electrons, and I is the appropriate ionization limit. The quantum defect μ is *defined as* $\mu = n - n^*$ (n is the principal quantum number of the excited electron). Within a spectral series, transition energies are $\Delta E \propto 1/(n'^*)^2 - 1/(n'^*)^2$.

Quantum defect theory (QDT) exploits regularities of quantum defects within spectral series to predict the positions and relative intensities of transitions to high-lying series members (large values of n). [22] Multichannel QDT accounts for the fact that different Rydberg series converge to different limits, and that the appropriate coupling scheme changes for higher members of each series.

2. *Configuration interaction:* Because of the mutual interactions among the electrons of an atom or ion, electronic wave functions usually cannot be accurately described by a single configuration. Since both the energy levels and especially the transition probabilities may be drastically affected by the mixing of many configurations, recent atomic structure calculations have been carried out with extensive multiconfiguration approaches, sometimes containing hundreds of configurations, such as the "multiconfiguration self-consistent field" (MCSCF), the "superposition of configurations" (SOC) methods, and other configuration-interaction programs. [23]–[25] For example, the ground term of a Be-like ion ($2s^2\ ^1S$) is better described as

$$a_1 2s^2\ ^1S + a_2 2p^2\ ^1S + a_3 2s3s\ ^1S$$
$$+ \text{ other configurations of even parity that form a } ^1S\text{-term.}$$

The a_i's are mixing coefficients ($\sum_i |a_i|^2 = 1$). The squares of the principal mixing coefficients, $|a_i|^2$, multiplied by 100, are included as "leading percentages" in recent energy level tables. [4, 6]

3. *Cancellation effects:* In calculations of transition matrix elements, numerous cases occur in which the positive and negative contributions to the transition integrand are comparable, so that the integral becomes a fairly small number and is very sensitive to the numerical integration method used. This may make calculated results rather uncertain.

7.11. SPECTRAL LINE SHAPES, WIDTHS, AND SHIFTS

The principal causes of spectral line broadening in emission sources, i.e., discharges or plasmas, are as follows.

7.11.1. Doppler broadening

Doppler broadening is due to the thermal motion of emitters. For a Maxwellian velocity distribution, the line shape is *Gaussian*; the full-width at half maximum intensity (FWHM) is (in Å)

$$\Delta\lambda_{1/2}^{D} = 7.16 \times 10^{-7} \lambda (T/M)^{1/2}.$$

T is the temperature of the emitters, λ is the wavelength in Å, and M is the atomic weight in atomic mass units.

7.11.2. Pressure broadening

Pressure broadening is due to collisions between emitters and the surrounding particles. Shapes are often approximately Lorentzian, i.e., $I(\lambda) \propto [1 + (\lambda - \lambda_0/\Delta\lambda_{1/2})^2]^{-1}$. In the following formulas, all FWHM's and wavelengths are expressed in Å and the particle densities N in cm^{-3}.

1. *Resonance broadening:* Self-broadening occurs only between identical species and is essentially confined to lines with upper or lower states combining by a dipole transition (i.e., the resonance line) with the ground state. The FWHM may be estimated as

$$\Delta\lambda_{1/2}^{\mathrm{R}} \simeq 8.6 \times 10^{-30}(g_i/g_k)^{1/2}\lambda^2\lambda_r f_r N_i.$$

   λ is the wavelength of the observed line, f_r and λ_r are the oscillator strength and wavelength of the resonance line, g_k and g_i are the statistical weights of its upper and lower state, and N_i is the ground state number density.

2. *Van der Waals broadening:* This arises from the dipole interaction of an excited atom with the induced dipole of a ground state atom. (In the case of foreign gas broadening, both the perturber and the radiator may be in their respective ground states.) An approximate formula for the FWHM, strictly applicable to hydrogen and similar atomic systems only, is

$$\Delta\lambda_{1/2}^{\mathrm{W}} \simeq 3.0 \times 10^{-16}\lambda^2 C_6^{2/5}(T/\mu)^{3/10}N,$$

   where μ is the atom-perturber reduced mass in amu, N is the perturber density, and C_6 is the interaction constant. C_6 may be roughly estimated as follows: $C_6 = C_k - C_i$, with $C_{i(k)} = 9.8 \times 10^{-10}\alpha_P R_{i(k)}^2$ (α_P in cm^3, R^2 in a_0^2). The mean atomic polarizability is $\alpha_P \approx 6.7 \times 10^{-25}(3I_{\mathrm{H}}/4E^*)^2$ cm^3, where I_{H} is the ionization energy of hydrogen and E^* is the energy of the first excited level of the perturber atom. $R_{i(k)}^2 \approx 2.5[I_{\mathrm{H}}/(I - E_{i(k)})]^2$, where I is the ionization energy of the radiator. Van der Waals broadened lines are red shifted by about two-thirds the size of the FWHM.

3. *Stark broadening:* This is due to charged perturbers, i.e., ions and electrons, that usually dominate resonance and van der Waals broadening in discharges and plasmas. The FWHM for hydrogen lines is

$$\Delta\lambda_{1/2}^{\mathrm{S}} = 2.50 \times 10^{-9}\alpha_{1/2}N_e^{2/3},$$

   where N_e is the electron density. The half-width parameter $\alpha_{1/2}$ for the widely used H_β-line at 4861 Å is tabulated in Table 7.7 for some typical temperatures and electron densities. [26] This reference also contains $\alpha_{1/2}$-parameters for other hydrogen lines, as well as Stark width and shift data for numerous lines of many other elements, i.e., neutral atoms and singly charged ions (for the latter, Stark widths and

TABLE 7.7. Values of the Stark-broadening parameter $\alpha_{1/2}$ of the H_β-line of hydrogen at 4861 Å for various temperatures and electron densities.

	$N_e(\text{cm}^{-3})$			
T(K)	10^{15}	10^{16}	10^{17}	10^{18}
5 000	0.0787	0.0808	0.0765	—
10 000	0.0803	0.0840	0.0851	0.0781
20 000	0.0815	0.0860	0.0902	0.0896
30 000	0.0814	0.0860	0.0919	0.0946

shifts depend linearly on N_e). Other tabulations of complete hydrogen Stark profiles exist. [27]

7.12. SPECTRAL CONTINUUM RADIATION

Continuum radiation principally arises from two processes: the recombination of free electrons with ions or atoms, which produces free-bound or recombination continua (including negative ion continua), and radiative transitions of free electrons, where the electrons lose energy, but remain free, giving rise to free-free or bremsstrahlung continua. The inverse processes in absorption are photoionization and continuous absorption.

7.12.1. Hydrogenic species

Precise quantum-mechanical calculations exist only for hydrogenic species. The total power $\varepsilon_{\text{cont}}$ radiated (per unit source volume and per unit solid angle, and expressed in SI units) in the wavelength interval $\Delta\lambda$ is the sum of radiation due to the recombination of a free electron with a bare ion (free-bound transitions) and bremsstrahlung (free-free transitions):

$$\varepsilon_{\text{cont}} = \frac{e^6}{2\pi\varepsilon_0^3(6\pi m_e)^{3/2}} N_e N_Z Z^2 \times \frac{1}{(kT)^{1/2}} \exp\left(-\frac{hc}{\lambda kT}\right)\frac{\Delta\lambda}{\lambda^2}$$
$$\times \left\{ \frac{2Z^2 I_{\text{H}}}{kT} \sum_{n\geq(Z^2 I_{\text{H}}/hc)^{1/2}}^{n'} \frac{\gamma_{fb}}{n^3} \exp\left(\frac{Z^2 I_{\text{H}}}{n^2 kT}\right) + \gamma_{fb}\left[\exp\left(\frac{Z^2 I_{\text{H}}}{(n'+1)^2 kT}\right) - 1\right] + \gamma_{ff} \right\},$$

where N_e is the electron density, N_Z is the number density of hydrogenic (bare) ions of nuclear charge Z, I_{H} is the ionization energy of hydrogen, and n' is the principal quantum number of the lowest level for which spacings to adjacent levels are so close that they approach a continuum and the summation over n may be replaced by an integral. (The choice of n' is rather arbitrary—n' as low as 6 is found in the literature.) γ_{fb} and γ_{ff} are the Gaunt factors, which are generally close to unity. [28] (For the higher free-bound continua, starting with $n' + 1$, an average Gaunt factor $\bar{\gamma}_{fb}$ is used.) For moderate or low temperature neutral hydrogen plasmas, the recombination continuum forming H^- also becomes important. [29]

7.12.2. Many-electron systems

For many-electron systems, only approximate calculational treatments for the continuum radiation exist, based, for example, on the quantum-defect method [30] and the Biberman-Norman factors. [31, 32] Results of calculations for the noble gases and C, N, O are given in Refs. [32] and [33], and experimental work is centered on the noble gases. [34]

Near the ionization limit, the f-values for bound-bound transitions of a spectral series ($n' \to \infty$) make a smooth connection to the differential oscillator strength distribution in the continuum. [1, 2]

7.13. SOURCES OF SPECTROSCOPIC DATA

Major compilations of critically evaluated spectroscopy data are Refs. [3, 4, 5, 6] and [9] for energy levels and wavelengths, and Refs. [13]–[18] for atomic transition probabilities.

An extensive and unified spectroscopic database—the Atomic Spectra Database (ASD), containing mostly material from the above-mentioned references—is maintained by the NIST Physics Laboratory on the Internet at the following address:

http://physics.nist.gov/asd

This address also contains links to extensive bibliographic databases on atomic transition probabilities and spectral line shape parameters.

7.14. REFERENCES

[1] R. D. Cowan, *The Theory of Atomic Structure and Spectra* (University of California Press, Berkeley, 1981).

[2] I. I. Sobelman, *Atomic Spectra and Radiative Transitions*, 2nd ed. (Springer, Berlin, 1992).

[3] C. E. Moore, *Atomic Energy Levels*, Natl. Stand. Ref. Data Ser., Natl. Bur. Stand. 35 (U.S. Govt. Print. Off., Washington, D.C., 1971).

[4] W. C. Martin, R. Zalubas, and L. Hagan, *Atomic Energy Levels—The Rare-Earth Elements*, Natl. Stand. Ref. Data Ser., Natl. Bur. Stand. 60 (U.S. Govt. Print. Off., Washington, D.C., 1978).

[5] C. E. Moore, *Tables of Spectra of Hydrogen, Carbon, Nitrogen, and Oxygen Atoms and Ions*, edited by J. W. Gallagher (CRC Press, Boca Raton, FL, 1993).

[6] J. Sugar and C. Corliss, J. Phys. Chem. Ref. Data **14**, (Suppl. 2) (1985).

[7] E. R. Peak and K. Reeder, J. Opt. Soc. Am. **62**, 958 (1972).

[8] T. J. Quinn, Metrologia **30**, 523 (1994).

[9] V. Kaufman and B. Edlén, J. Phys. Chem. Ref. Data **3**, 825 (1974).

[10] G. Nave *et al.*, Astrophys. J. Suppl. Ser. **94**, 221 (1994).

[11] J. E. Sansonetti, J. Reader, C. J. Sansonetti, and N. Acquista, J. Res. Natl. Inst. Stds. Tech. **97**, 1 (1992).

[12] W. C. Martin and W. L. Wiese, in *Atomic, Molecular and Optical Physics Handbook*, edited by G. W. F. Drake (AIP Press, Woodbury, NY, 1996), pp. 135–153.

[13] W. L. Wiese, M. W. Smith, and B. M. Glennon, *Atomic Transition Probabilties—Vol. I, Hydrogen through Neon*, Natl. Stand. Ref. Data Ser., Natl. Bur. Stand. 4 (U.S. Govt. Print. Off., Washington, D.C., 1966).

[14] W. L. Wiese, M. W. Smith, and B. M. Miles, *Atomic Transition Probabilties—Vol. II, Sodium through Calcium*, Natl. Stand. Ref. Data Ser., Natl. Bur. Stand. 22 (U.S. Govt. Print. Off., Washington, D.C., 1969).

[15] G. A. Martin, J. R. Fuhr, and W. L. Wiese, J. Phys. Chem. Ref. Data **17**, (Suppl. 3) (1988).

[16] J. R. Fuhr, G. A. Martin, and W. L. Wiese, J. Phys. Chem. Ref Data **17**, (Suppl. 4) (1988).

[17] W. L. Wiese, J. R. Fuhr, and T. M. Deters, J. Phys. Chem. Ref. Data, Monograph 7 (1996).

[18] J. R. Fuhr and W. L. Wiese, in *Handbook of Chemistry and Physics*, 79th ed., edited by D. R. Lide (CRC Press, Boca Raton, FL, 1998).

[19] C. R. Cowley, W. L. Wiese, J. R. Fuhr, and L. A. Kuznetsova, Chapter 4 "Spectra," in *Allen's Astrophysical Quantities*, 4th ed., edited by A. Cox (AIP Press, New York, 2000).

[20] S. M. Younger and A. W. Weiss, J. Res. Natl. Bur. Stand., Sect. A **79**, 629 (1975).

[21] W. L. Wiese and A. W. Weiss, Phys. Rev. **175**, 50 (1968).

[22] M. J. Seaton, Rep. Prog. Phys. **46**, 167 (1983).

[23] C. F. Fischer, in *Atomic, Molecular and Optical Physics Handbook*, edited by G. W. F. Drake (AIP Press, Woodbury, NY, 1996), pp. 243–257.

[24] A. W. Weiss, Adv. At. Mol. Phys. **9**, 1 (1973).

[25] A. Hibbert, Comput. Phys. Commun. **9**, 141 (1975).

[26] H. R. Griem, *Spectral Line Broadening by Plasmas* (Academic, New York, 1974).

[27] R. Vidal, J. Cooper, and E. W. Smith, Astrophys. J. Suppl. Ser. **25**, 37 (1973).

[28] W. J. Karzas and R. Latter, Astrophys. J. Suppl. Ser. **6**, 167 (1961).

[29] J. R. Roberts and P. A. Voigt, J. Res. Natl. Bur. Stand. Sect. A **75**, 291 (1971).

[30] M. J. Seaton, Mon. Not. R. Astron. Soc. **118**, 504 (1958).

[31] L. M. Biberman and G. E. Norman, Opt. Spectrosc. **8**, 230 (1960).

[32] D. Schlueter, Z. Phys. **210**, 80 (1968).

[33] D. Hofsaess, At. Data Nucl. Data Tables **24**, 285 (1979).

[34] A. T. M. Wilbers, G. M. W. Kroesen, C. J. Timmermans, and D. C. Schram, J. Quant. Spectrosc. Radiat. Transfer. **45**, 1 (1991).

8

Biological Physics

Elias Greenbaum

Oak Ridge National Laboratory, Oak Ridge, Tennessee, and The University of Tennessee, Knoxville, Tennessee

Victor Bloomfield

University of Minnesota, Minneapolis, Minnesota

Contents

List of Tables

List of Figures

8.1. INTRODUCTION

Modern biological physics encompasses the study of biology at the molecular and cellular levels using the concepts and methods of molecular physics and physical chemistry, imaging, and computational science. Research on a broad range of systems, including proteins, nucleic acids, membranes, and their biologically functional complexes, such as channels, receptors, contractile systems, and nucleoproteins, is unified when approached with this common set of intellectual tools.

8.2. INTERMOLECULAR FORCES

The structures and dynamics of biological macromolecules are ultimately determined by the covalent and noncovalent forces involving biomolecules and (usually aqueous) solvent.

8.2.1. Elementary electrostatic and dispersion interactions

The easiest interactions to calculate from first principles are energies between ions or dipoles of valence Z, dipole moment μ, and polarizability α, at a distance r and orientation function $f(\theta)$ in a medium of dielectric constant ϵ (Table 8.1). However, because

TABLE 8.1. Noncovalent interactions between molecules (adapted from Ref [18]).

Type of Interaction	Potential Energy V	Order of Magnitude (kJ/mol)
Ion-ion	$\frac{Z_1 Z_2 e^2}{\varepsilon r}$	60
Ion-dipole	$\frac{Z_1 e \mu_1 f(\theta)}{D r^2}$	−8 to +8
Dipole-dipole	$\frac{\mu_1 \mu_2 f'(\theta)}{\varepsilon r^2} - 3\frac{[\mu_1 r f''(\theta)][\mu_2 r f'''(\theta)]}{\varepsilon r^5}$	−2 to +2
Ion-induced dipole	$\frac{Z_1 e^2 \alpha_2}{2\varepsilon^2 r^4}$	0.2
Dispersion	$\frac{3h\nu_0\alpha^2}{4r^6}$	0 to 40

the spatial dependence of ϵ is complex and poorly known, these simple equations are only semiquantitative.

8.2.2. Force fields

Computer simulations of biopolymer structure and dynamics (e.g., molecular mechanics, molecular dynamics, Brownian dynamics, Monte Carlo) use an analytical "force field" to represent intramolecular and intermolecular interactions. The force field used in Eq. (8.1) contains potential energy terms for bond stretching, bond bending, torsional rotation, and out-of-plane distortion [top line of Eq. (8.1)]. Nonbonded energies include exchange repulsion, dispersion, and electrostatic terms, and an additional "H-bond term" exists that is not included in many analytical models in the literature, but it can be used to simulate "charge transfer" and other nonbonded interaction energies [bottom line of Eq. (8.1)]:

$$V_{\text{total}} = \sum_{\text{bonds}} K_r (r - r_{\text{eq}})^2 + \sum_{\text{angles}} K_\theta (\theta - \theta_{\text{eq}})^2 + \sum_{\text{dihedrals}} \frac{V_n}{2}[1 + \cos(n\phi - \gamma)]$$
$$+ \sum_{i<j} \left[\frac{A_{ij}}{R_{ij}^{12}} - \frac{B_{ij}}{R_{ij}^{6}} + \frac{q_i q_j}{\varepsilon R_{ij}} \right] + \sum_{\text{H-bonds}} \sum_{i<j} \left[\frac{C_{ij}}{R_{ij}^{12}} - \frac{D_{ij}}{R_{ij}^{10}} \right]. \tag{8.1}$$

The coefficients are optimized to reproduce experimental data on model systems.

8.2.3. Interactions in water

Some of the most important noncovalent interactions originate from, or are strongly modified by, the complex properties of water. The standard free energy of a peptide hydrogen bond is about 20 kJ/mol *in vacuo*, but only 2–6 kJ/mol *in aqua* because of competition with H-bonding from the water. Large positive and negative changes in heat capacity originate from exposure of nonpolar and polar surface areas (ΔA_{np} and ΔA_{p}, respectively [39])

$$\Delta C_p = (1.34 \pm 0.17)\,\Delta A_{\rm np} - (0.59 \pm 0.17)\,\Delta A_{\rm p}\ \text{kJ/mol-K}, \tag{8.2}$$

where ΔA is in Å^2. The entropy change in the hydrophobic (HΦ) effect because of exposure of nonpolar surfaces in water disappears at 386 K: $\Delta S^0_{\rm H\Phi} = 0.32\Delta A_{\rm np} \ln(T/386)$ kJ/mol-K. In the approach of polar surfaces, the polar groups structure the water, leading to a hydration force [33], which has an exponential decay length

$$\lambda = \frac{1}{2}\left[\lambda_w^{-2} + \left(\frac{2\pi}{\alpha}\right)^2\right]^{-\frac{1}{2}}, \tag{8.3}$$

where λ_w is the water correlation length (4–5 Å) and α is the periodicity of the surface lattice.

8.2.4. Ionic solutions and polyelectrolytes

The Bjerrum length $l_B = e^2/\epsilon kT$ is the distance at which the Coulomb energy between two univalent charges equals the thermal energy. In water at 25°C, $l_B = 7.14$ Å. Ionic solutions are characterized by the Debye-Hückel inverse length $\kappa = (8\pi l_B I)^{1/2}$, where I is the ionic strength. Numerically, for water at 25°C, if I is expressed in molar units, $\kappa = 3.3 \times 10^7 I^{1/2}$ cm^{-1}. Coulombic interaction between point charges is reduced by the factor $\exp(-\kappa r)$. Linear polyelectrolytes can be characterized by a charge spacing b along their backbone. If the ratio $\xi \equiv l_B/b > 1$, counterions condense along the backbone to reduce the effective linear charge density parameter $\xi_{\rm eff}$ to 1. [35] If the counterions have valence Z, a fraction $1 - 1/Z\xi$ of the polyion charge is neutralized by counterion condensation. For B-DNA, $b = 1.7$ Å; so $\xi = 4.2$. Thus, 76%, 88%, 92%, and 94% of the DNA charge is neutralized by counterions of valence 1, 2, 3, and 4.

8.3. NUCLEIC ACIDS

8.3.1. Structures of nucleic acid bases and nucleotides

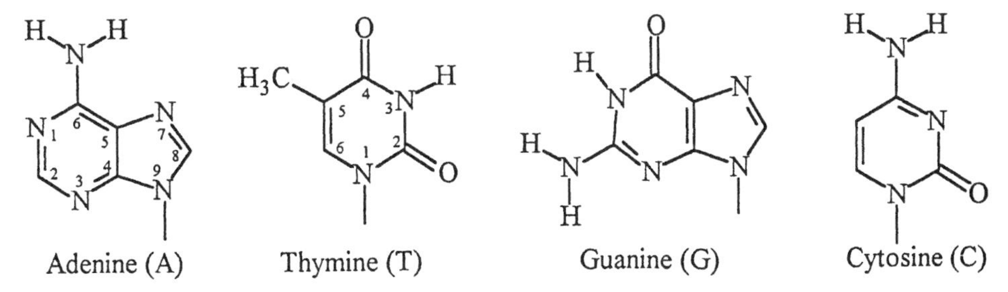

FIGURE 8.1. Structures of the purine (A, G) and pyrimidine (T, C) bases in DNA, showing heterocyclic ring numbering convention. In RNA, uracil (U) appears instead of T, with the C5–CH_3 replaced by –H.

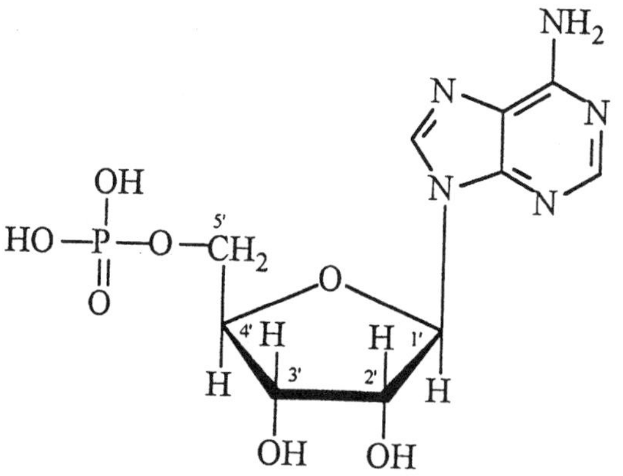

FIGURE 8.2. Structure of the nucleotide adenoside 5′-monophosphate (AMP), showing sugar numbering convention and *anti* conformation of glycosidic bond. The deoxynucleotide dAMP would have –H rather than –OH at C2′.

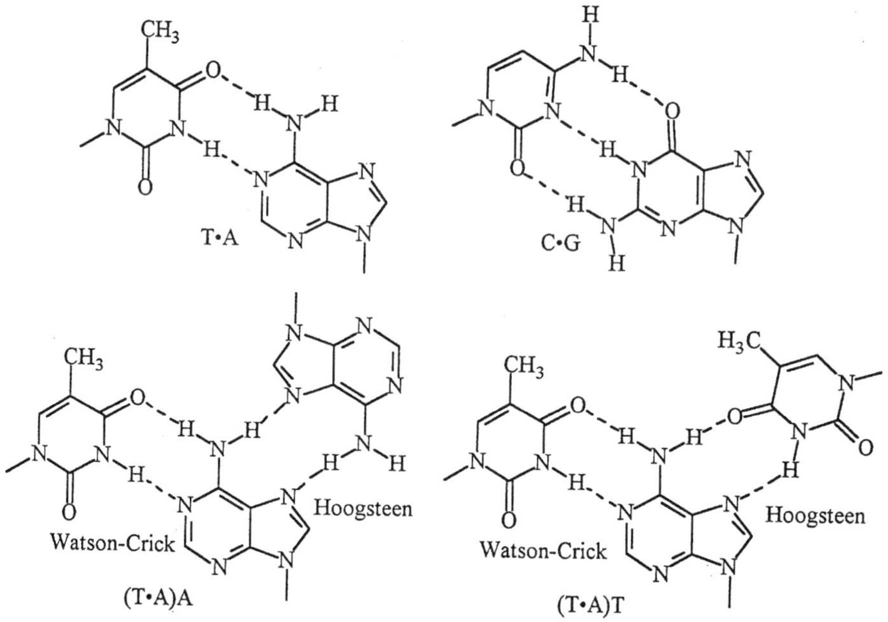

FIGURE 8.3. Watson-Crick and Hoogsteen base pairing. The dashed lines represent hydrogen bonds that result in base pairing of A with T and G with C. Top row: Watson-Crick base pairs are those commonly found in double-helical DNA. CG base pairs have three H-bonds; hence, they are more stable than TA base pairs, which have only two. Bottom row: Watson-Crick and the alternative arrangement of Hoogsteen base pairs found in triplex DNA.

TABLE 8.2. Average structural parameters for various helical forms of DNA. [2]

	A-DNA	B-DNA	Z-DNA
Helix handedness	Right	Right	Left
bp/repeating unit	1	1	2
bp/turn	11	10	12
Helix twist, °	32.7	36.0	-10^a, -50^b
Rise/bp, Å	2.9	3.4	-3.9^a, -3.5^b
Helix pitch, Å	32	34	45
Base pair inclination, °	12	2.4	−6.2
P distance from helix axis, Å	9.5	9.4	6.2^a, 7.7^b
X displacement from bp to helix axis, Å	−4.1	0.8	3.0
Glycosidic bond orientation	anti	anti	anti[c], syn[d]
Sugar conformation	C3′-endo	C2′-endo[e]	C2′-endo[c] C3′-endo[d]
Major groove depth	13.5	8.5	convex
width, Å	2.7	11.7	
Minor groove depth	2.8	7.5	9
width, Å	11.0	5.7	4

[a] CpG step.
[b] GpC step.
[c] Cytosine.
[d] Guanine.
[e] Range of conformations.

8.3.2. Energetics of bending and twisting

The free energy required to bend a length L of DNA through radius of curvature R_c or angle $\theta = L/R_c$ and to twist it through ϕ radians is

$$\Delta G_{\text{bend}} = \tfrac{1}{2}\kappa_p L/R_c^2 = \tfrac{1}{2}\kappa_p \theta^2 L, \tag{8.4}$$

$$\Delta G_{\text{twist}} = \tfrac{1}{2}C(\phi/L)^2 L. \tag{8.5}$$

The bending rigidity κ_p is generally about 2×10^{-19} erg-cm, and the twist rigidity C is about 2.5×10^{-19} erg-cm. The former value corresponds to a persistence length of 50 nm.

8.3.3. Supercoiled DNA

Naturally occurring DNA is often found in a topologically constrained, supercoiled form, in which the ends of double helical DNA are brought together and covalently joined or fixed relative to one another. Supercoiled DNA can be described by three quantities: linking number *Lk*, the number of times one strand passes around the other; twist *Tw*, the number of times one strand passes about the other if the DNA is confined to a plane; and writhe Wr, the tortuosity of the duplex axis in space, calculated by counting crossings in a plane projection. Lk is a topological constant if the strands are not cut, but Tw and Wr are geometrical variables. They are related by $Lk = Tw + Wr$.

If the DNA contains N base pairs (bp), and the average number of bp per turn in solution for B-DNA is 10.5, the linking number for relaxed DNA is $Lk_0 = N/10.5$, and one defines $\Delta Lk = Lk - Lk_0$. The specific linking difference (superhelix density), $\sigma = Lk/Lk_0 - 1$, is typically about -0.05. If the DNA is not too short, the free energy difference per bp $\Delta g_{\Delta Lk}$ is approximately constant for fixed ΔLk per bp:

$$\Delta g_{\Delta Lk} = \Delta G_{\Delta Lk}/N = NK(\Delta Lk/N)^2. \tag{8.6}$$

$NK \approx 1200RT$, or about 700 kcal at 20°C. With $\sigma \approx -0.05$, $\Delta g_{\Delta Lk}$ is about 67 J/bp, corresponding to 0.71 kJ/turn.

8.4. PROTEINS AND AMINO ACIDS

Proteins are characterized by four levels of structure: primary (sequence of amino acids), secondary (helical or beta sheet structure of polypeptide chains), tertiary (folding of secondary structure elements into compact globular domains), and quaternary (noncovalent association of polypeptide chains).

TABLE 8.3. Structures and properties of amino acids. The first three columns give the name, three-letter code, and one-letter code. Molecular weight M_{res} is of un-ionized amino acid minus that of water. Van der Waals volumes V of residues are from atomic volumes [F. M. Richards, J. Mol. Biol. 82, 1 (1974)]. Values of pK_a from Ch. 8 of Ref [6]; pK_a of α-COOH $\approx 2.1 \pm 0.3$, of α-$NH_3^+ \approx 9.6 \pm 0.5$, depending strongly on adjacent residue. Normalized frequencies P_α, P_β, P_t of finding residue in α-helix, β-sheet, or reversed turn from Table 6.5 of Ref. [5]. Hydrophobicities HΦ from solubilities in water relative to ethanol or dioxane [Y. Nozaki and C. Tanford, J. Biol. Chem. 246, 2211 (1971); M. Levitt, J. Mol. Biol. 104, 59 (1976)].

			Side Chain R	M_{res}	V Å^3	pK$_a$	P_α	P_β	P_t	HΦ kcal/mol
Alanine	Ala	A	$-CH_3$	71.09	67		1.41	0.72	0.82	−0.5
Arginine	Arg	R	$-CH_2CH_2CH_2-N-C(NH_2^+)(NH_2)$	156.19	148	12.5	1.21	0.84	0.90	3.0
Asparagine	Asn	N	$-CH_2-CONH_2$	114.11	96		0.76	0.48	1.34	0.2
Aspartate	Asp	D	$-CH_2-COO^-$	115.09	91	3.9	0.99	0.39	1.24	2.5
Cysteine	Cys	C	$-CH_2-SH$	103.15	86	8.3	0.66	1.40	0.54	−1.0
Glutamine	Gln	Q	$-CH_2-CH_2-CONH_2$	128.14	114		1.27	0.98	0.84	0.2
Glutamate	Glu	E	$-CH_2-CH_2-COO^-$	129.12	109	4.3	1.59	0.52	1.01	2.5
Glycine	Gly	G	$-H$	57.05	48		0.43	0.58	1.77	0.0
Histidine	His	H	$-CH_2$ C(imidazole ring: H–N, HC, CH, C—NH+)	137.14	118	6.0	1.05	0.80	0.81	−0.5

(continued)

TABLE 8.3. Continued

			Side Chain R	M_{res}	V Å^3	pKa	P_a	P_β	P_t	HΦ kcal/mol
Isoleucine	Ile	I	$-CH(CH_3)-CH_2-CH_3$	113.16	124		1.09	1.67	0.47	−1.8
Leucine	Leu	L	$-CH_2-CH(CH_3)_2$	113.16	124		1.34	1.22	0.57	−1.8
Lysine	Lys	K	$-CH_2CH_2CH_2CH_2NH_3^+$	128.17	135	10.8	1.23	0.69	1.07	3.0
Methionine	Met	M	$-CH_2CH_2-S-CH_3$	131.19	124		1.30	1.14	0.52	−1.3
Phenylalanine	Phe	F	$-CH_2-\phi$	147.18	135		1.16	1.33	0.59	−2.5
Proline	Pro	P	$^{-}OOC-CH-CH_2-CH_2-CH_2-NH_2^+$ (ring)	97.12	90		0.34	0.31	1.32	−1.4
Serine	Ser	S	$-CH_2OH$	87.08	73		0.57	0.96	1.22	0.3
Threonine	Thr	T	$-CH(OH)CH_3$	101.11	93		0.76	1.17	0.90	−0.4
Trytophan	Trp	W	$-CH_2-C=CH-NH-$ (indole)	186.21	163		1.02	1.35	0.65	−3.4
Tyrosine	Tyr	Y	$-CH_2-\phi-OH$	163.18	141	10.9	0.74	1.45	0.76	−2.3
Valine	Val	V	$-CH(CH_3)_2$	99.14	105		0.90	1.87	0.41	−1.5

8.4.1. Peptide bond and polypeptide conformations

TABLE 8.4. Geometric parameters for polypeptide helices. All of these helices have a *trans* peptide bond, except polyproline I, in which it is *cis*. (Adapted from Ref. [5], Table 5.2.)

	ϕ, deg	ψ, deg	Residues per turn	Translation per residue, Å
Antiparallel β-sheet	−139	135	2.0	3.4
Parallel β-sheet	−119	113	2.0	3.2
Right-handed α-helix	−57	−47	3.6	1.50
3_{10}-helix	−49	−26	3.0	2.00
π-helix	−57	−70	4.4	1.15
Polyproline I	−83	158	3.33	1.9
Polyproline II	−78	149	3.00	3.12
Polygylcine II	−80	150	3.0	3.1

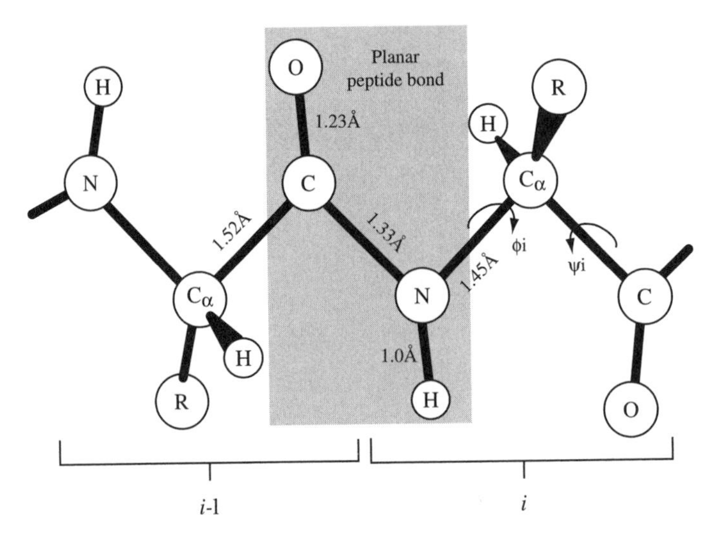

FIGURE 8.4. Schematic diagram of dipeptide.

8.4.2. Helix-coil transition and protein folding

A physically insightful model of the helix–random coil transition in polypeptide assumes an all-or-none transition in which each hydrogen bond contributes enthalpy change Δh and each amino acid residue immobilized contributes entropy change Δs. In an α-helix, the $C = O$ of residue I-4 forms an H bond with the N–H of residue I. If the polypeptide chain contains N amino acids, $N - 4$ H-bonds can be formed, but $N - 2$ residues are immobilized. Thus, the Gibbs energy change is $\Delta G = (N-4)\,\Delta h - T(N-2)\,\Delta s$. At the midpoint of the transition, $\Delta G = 0$; so the melting temperature is $T_m = (N-4)\,\Delta h/(N-2)\,\Delta s$. This theory correctly predicts that T_m is lower, and the transition is broader, for small N. However, it is oversimplified in assuming an all-or-none transition. More rigorous statistical thermodynamic treatments, based on the Ising model, are discussed in Ch. 20 of Ref. [3].

In fact, protein folding is probably driven more by the clustering of hydrophobic residues in the protein core than by helix formation. Theoretical understanding of protein folding is extremely difficult, because of its great sensitivity to poorly known potential energy functions and many-body interactions in aqueous solution. However, it is being actively pursued because of the desire to predict the protein structure from amino acid sequence determined from DNA sequence in the human genome project. One such approach is the energy landscape theory of protein folding, illustrated in Figure 8.5.

8.5. BINDING THERMODYNAMICS

Most biologically important events are regulated by noncovalent interactions between biological macromolecules and ligands, which may be small molecules, such as drugs, or other macromolecules. If the macromolecule M has n binding sites, and the molar con-

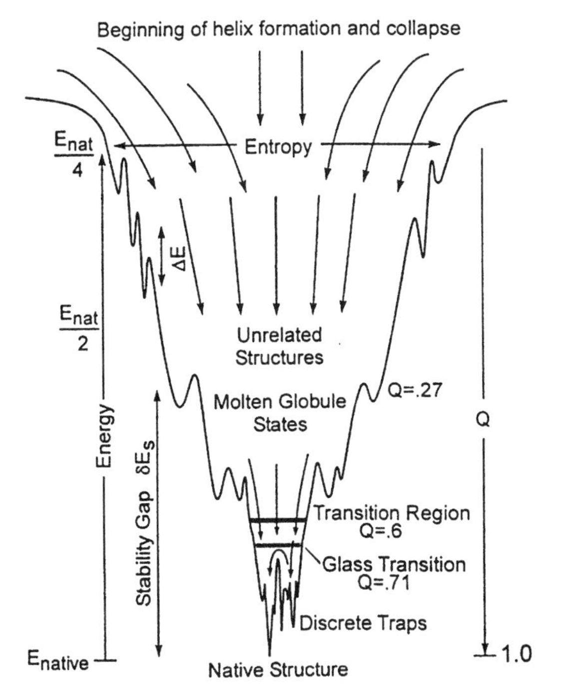

FIGURE 8.5. A one-dimensional cross section through the energy landscape of a protein. [From P. G. Wolynes and Z. Luthey-Schulten, in *Physics of Biological Systems*, edited by H. Flyvberg *et al.* (Springer-Verlag, 1997, p. 68, reprinted by permission.]

centration of species with i ligands bound is $[M_i]$, the average number of ligands bound is

$$\bar{\nu} = \frac{\sum_{i=1}^{n} i[M_i]}{\sum_{i=0}^{n} [M_i]} \tag{8.7}$$

and the fractional occupancy is $\theta = \bar{\nu}/n$. In general, the binding constants, $K_i = [M_i]/[M_{i-1}][L]$ $(i = 1 \ldots n)$, where $[L]$ is the free-ligand concentration, are independent of each other, but often are assumed to obey simple relations in some widely used binding models. Some of these constants are listed in Table 8.5.

8.6. NUCLEAR MAGNETIC RESONANCE

The first nuclear magnetic resonance (NMR) experiments were performed in 1946 by Bloch, Hansen, and Packard and by Purcell, Torrey, and Pound. A nucleus with nonzero spin angular momentum, **I**, measured in units of $\hbar$ (where $\hbar = h/2\pi$), possesses a mag-

TABLE 8.5. Some common binding isotherms. (See Refs. [3] and [17].) The first model, with $n = 1$, can be rearranged to give the Henderson-Hasselbalch equation $\text{pH} = \text{p}K_a + \log([\text{base}]/[\text{acid}])$, where $\text{p}K_a = -\log(K_a)$, and $K_a = 1/K$ is the acid dissociation constant.

Model	Binding equation
n identical, independent sites with association constant K	$\bar{\nu} = \dfrac{nK[L]}{1 + K[L]}$
j classes of identical, independent sites $\{n_j, K_j\}$	$\bar{\nu} = \sum_{i=1}^{j} \dfrac{n_i K_i[L]}{1 + K_i[L]}$
n identical sites, uncharged when unoccupied and charge z when occupied, with intrinsic association constant K_0 at net charge $Z = 0$, and electrostatic interaction between the sites at $\mid Z \mid = \bar{\nu} \mid z \mid > 0$[a].	$\bar{\nu} = \dfrac{nK_0 e^{-2wzZ}[L]}{1 + K_0 e^{-2wzZ}[L]}$
For spherical macromolecule of a radius R, counterion distance of closest approach a, Debye length $1/\kappa$	$w = \dfrac{q^2}{\varepsilon kTR}\left(1 - \dfrac{\kappa R}{1 + \kappa a}\right)$
n total sites, ligands occupy l contiguous sites[b]	$\bar{\nu} = nK[L](1 - l\bar{\nu}/n)\left(\dfrac{1 - l\bar{\nu}/n}{1 - (l-1)\bar{\nu}/n}\right)^{l-1}$
A protein with n subunits (e.g., $n = 4$ for hemoglobin), each of which can be in T or R state[c]	$\bar{\nu} = n\dfrac{\alpha(1+\alpha)^{n-1} + \lambda c\alpha(1 - c\alpha)^{n-1}}{(1+\alpha)^n + \lambda(1 + c\alpha)^n}$
	$\lambda = [\text{T}_0]/[\text{R}_0]$, $c = K_{\text{T}}/K_{\text{R}}$, $\alpha = [L]K_{\text{R}}$

[a] Ref. [17], Ch. 8, 537–544.
[b] J. D. McGhee and P. H. von Hippel, *J. Mol. Biol.* **86**, 469 (1974).
[c] J. Monod, J. Wyman, and J.-P. Changeux, *J. Mol. Biol.* **12**, 88 (1965).

netic dipole moment $\boldsymbol{\mu} = \gamma\hbar\mathbf{I}$, where γ is the nuclear gyromagnetic ratio. If the nucleus is placed in a magnetic field in the z-direction, its energy of interaction is

$$E = -\boldsymbol{\mu} \cdot \mathbf{H}_0 = -\gamma\hbar\mathbf{I} \cdot \mathbf{H}_0 \tag{8.8}$$

and the spacing between the $2I + 1$ energy levels is

$$\Delta E = \gamma\hbar H_0. \tag{8.9}$$

The resonance condition for absorption of a photon is $\nu_0 = \gamma H_0/2\pi$. The actual magnetic field at the site of the nucleus is shifted by the local diamagnetic properties of its chemical environment, the "chemical shift," and is responsible for the significance of high-resolution NMR in structural studies. Many interesting variations on the NMR theme have been developed, such as two-dimensional (2D) NMR spectroscopy, which is especially useful for the resolution of dense spectra, and magnetic resonance imaging (MRI). The introduction of multidimensional spectra was a major advance in NMR spectroscopy along with FT-NMR and MRI; it simplifies spectra and provides additional information. Two-dimensional experiments follow logically from one-dimensional (1D) methodology. Two-dimensional NMR consists of four steps: sample preparation, evolution, mixing time, and detection. Sample preparation and detection are similar to that of 1D NMR. After sample preparation, the spins are allowed to precess freely for a given time t_1, during which magnetization is labeled with the chemical shift of a first nucleus. During mixing time, magnetization is transferred from a first nucleus to a second. At the end of the experiment, magnetization is labeled with the chemical shift of the second nucleus.

Figure 8.6 is an illustration of the energy level diagram of a spin-1/2 nucleus in an external magnetic field. Table 8.6 is a summary of the properties of the most important nuclei for NMR studies in biological physics. The orientation of nuclear spins in an external magnetic field leads to a macroscopic, measurable magnetization $M = \chi H_0$, where χ is the susceptibility. The magnitude of M can be obtained from the distribution of spins in the two states, α and β, which is given by the Boltzmann distribution

$$\frac{n_\alpha}{n_\beta} = e^{\frac{\Delta E}{kT}} = e^{\frac{\gamma\hbar H_0}{kT}} \approx 1 + \frac{\gamma\hbar H_0}{kT}, \tag{8.10}$$

where n_α and n_β are the numbers of spin states α and β, respectively, k is the Boltzmann constant, and T is the absolute temperature. The differential equations describing relaxation processes are

$$\frac{dM_x}{dT} = -\frac{M_x}{T_2} + \omega_0 M_y, \tag{8.11}$$

$$\frac{dM_y}{dT} = -\frac{M_y}{T_2} + \omega_0 M_x, \tag{8.12}$$

$$\frac{dM_z}{dT} = -\frac{M_0 - M_z}{T_1}. \tag{8.13}$$

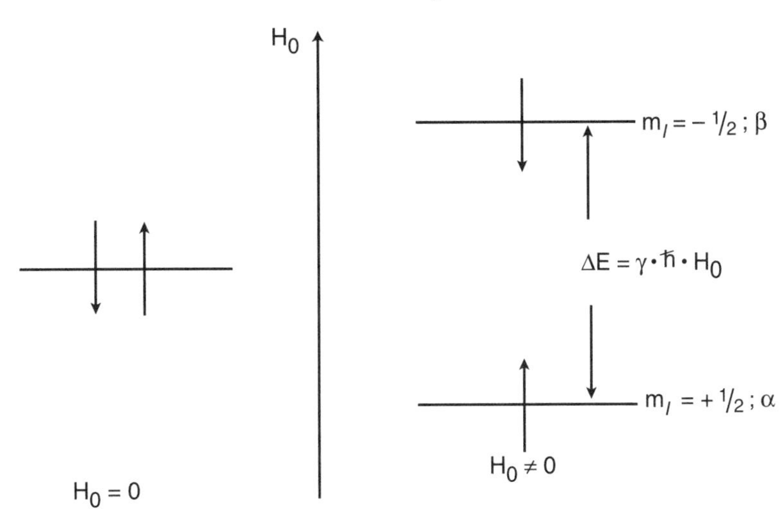

FIGURE 8.6. Energy diagram of a nucleus with nuclear spin quantum number $I = 1/2$ in an external magnetic field. [From H.-H. Paul *et al.* in *Biophysics*, edited by W. Hoppe *et al.* (Springer-Verlag, 1983, p. 190, reprinted by permission.]

TABLE 8.6. Properties of the most important nuclei for nuclear magnetic resonance in biophysics. [From H.-H. Paul *et al.*, in *Biophysics*, edited by W. Hoppe *et al.* (Springer-Verlag, 1983), p. 190, by permission.]

Nuclide	Spin quantum number I	Electric quadrupole moment (10^{-24} cm^2)	NMR frequency (MHz) ($H_0 = 2.35$ T)	Rel. nat. abundance (%)	Sensitity rel. to [^{1}H] (at constant field)	Usual standard
^{1}H	1/2	—	100	99.98	1	TMS
^{2}H	1	$2.73 \cdot 10^{-3}$	15.351	0.015	$9.65 \cdot 10^{-3}$	—
^{13}C	1/2	—	25.145	1.108	$1.59 \cdot 10^{-2}$	TMS
^{14}N	1	$7.1 \cdot 10^{-2}$	7.224	99.63	$1.01 \cdot 10^{-3}$	—
^{15}N	1/2	—	10.137	0.37	$1.04 \cdot 10^{-3}$	CH_3NO_2 or NO_3^-
^{17}O	5/2	$-2.6 \cdot 10^{-2}$	13.557	0.037	$2.91 \cdot 10^{-2}$	—
^{19}F	1/2	—	94.094	100	0.834	CCl_3F
^{23}Na	3/2	0.12	26.451	100	$9.25 \cdot 10^{-2}$	Na^+aq.
^{31}P	1/2	—	40.48	100	$6.63 \cdot 10^{-2}$	85% H_3PO_4

T_1 is the longitudinal or spin relaxation time, T_2 is the transverse or spin-spin relaxation time, and M_0 is the equilibrium magnetization in the direction of the external field (along with its associated components). Additional information is presented in Sec. 8.14.

8.7. ELECTRON PARAMAGNETIC RESONANCE

Many biological systems are paramagnetic. For example, they may contain transition-metal complexes or free-radical intermediates caused by thermally activated electron transfer, light, or ionizing radiation. Electron paramagmetic resonance (EPR) spectroscopy is based on the fact that electrons possess an intrinsic spin angular momentum with a corresponding magnetic dipole moment, $\boldsymbol{\mu}_e = g\beta\mathbf{S}$. Here, $\mathbf{S}$ is the spin vector of the electron in units of $h/2\pi$, β is the Bohr magneton, and g is the Landé g-factor, a dimensionless number comprising the contributions of spin and orbital angular momenta to the total angular momentum of the electron. For a free electron, $g_e = 2.0023$. The potential energy of an electron in an external magnetic field $\mathbf{H}$ is $E = \boldsymbol{\mu}_e \cdot \mathbf{H}$. For the simple case of a free electron, a spin-$1/2$ particle, only two energy levels exist, $\pm(1/2)g\beta H$.

For general applications of EPR spectroscopy, all possible interactions must be considered. This process requires writing a Hamiltonian that is a sum of operators:

$$\mathcal{H} = \mathcal{H}_{\text{EL}} + \mathcal{H}_{\text{CF}} + \mathcal{H}_{\text{LS}} + \mathcal{H}_{\text{ZE}} + \mathcal{H}_{\text{SS}} + \mathcal{H}_{\text{SI}} + \mathcal{H}_{\text{ZI}} + \mathcal{H}_{\text{Q}}.$$

The terms of this Hamiltonian do not contribute equally. Table 8.7 contains a definition of each term and associated magnitude of energy. The components of an EPR spectrometer are illustrated schematically in Figure 8.7.

TABLE 8.7. Definitions and orders of magnitude of the energies associated with the EPR Hamiltonian.* Unit = cm^{-1}.

$\mathcal{H}_{EL}$	electronic energy of the free ion ($10^4 - 10^5$)
$\mathcal{H}_{CF}$	crystal field interaction ($10^3 - 10^4$)
$\mathcal{H}_{LS}$	spin-orbit interaction ($10^1 - 10^3$)
$\mathcal{H}_{ZE}$	electronic Zeeman interaction ($0 - 1$)
$\mathcal{H}_{SS}$	spin-spin interaction ($0 - 1$)
$\mathcal{H}_{SI}$	hyperfine interaction ($0 - 10^{-2}$)
$\mathcal{H}_{ZI}$	nuclear Zeeman interaction ($0 - 10^{-3}$)
$\mathcal{H}_{O}$	nuclear electric quadrupole interaction ($0 - 10^{-2}$)

*From H. Neubacher and W. Lohmann, in *Biophysics*, edited by W. Hoppe *et al.* (Springer-Verlag, 1983), p. 179.

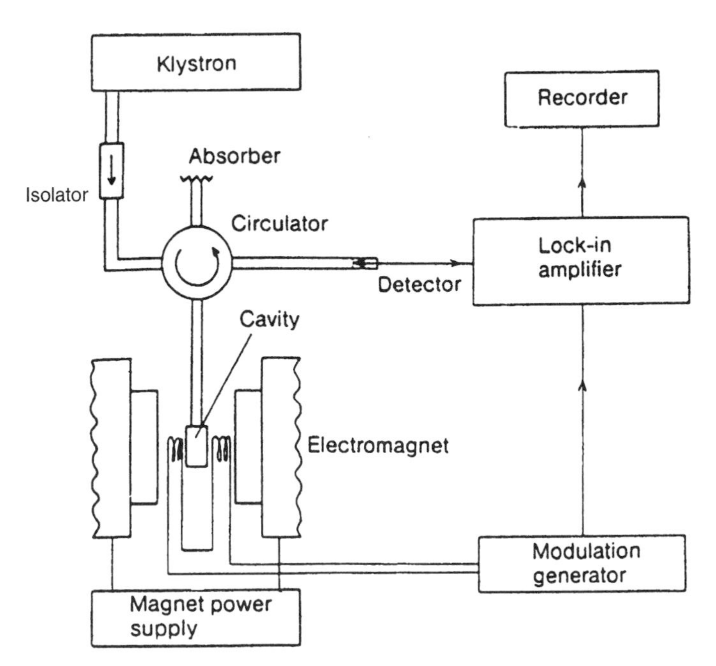

FIGURE 8.7. Schematic illustration of an EPR spectrometer. Three major components are illustrated: the microwave system for generation, absorption, and detection of microwaves; the electronics used for measurement and recording; and the electromagnet and stable power supply for linear sweep of the magnetic field. [From H. Naubacher and W. Lohmann, in *Biophysics*, edited by W. Hoppe *et al.* (Springer-Verlag, 1983), p. 185, reprinted by permission.]

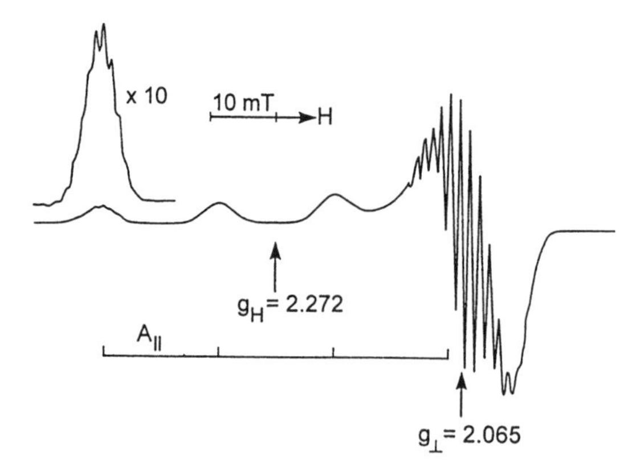

FIGURE 8.8. EPR spectrum of Cu(II) imidazole complex dissolved in ethanol at 77 K. [From H. Neubacher and W. Lohmann, in *Biophysics*, edited by W. Hoppe *et al.* (Springer-Verlag, 1983), p. 187, reprinted by permission.]

Transition metals occur in a variety of forms in important biomolecules, such as copper or iron proteins. EPR spectroscopy has been used effectively to study the magnitude and symmetry of the electric and magnetic environment of these transitions metals, which are usually at the catalytically active center of the protein. Figure 8.8 is an illustration of the EPR spectrum of a Cu(II) imidazole complex.

Multiple-resonance techniques have been developed that offer enhanced experimental capability. One such technique is ENDOR, electron-nuclear double resonance. In ENDOR spectroscopy, changes produced in EPR signals can be used to detect NMR radio-frequency transitions. In principle, the sensitivity of EPR is combined with the resolution of NMR. An interesting application of ENDOR spectroscopy to an important biological problem is illustrated in Figure 8.9, in which it is shown that the EPR signal for bateriochlorophyll (BChl) *in vitro* is 40% larger than in chromatophore membranes (*in vivo*). These data were used by Norris *et al.* [37] to suggest convincingly that the primary donor in bacterial photosynthesis is a dimer of BChl. The narrowing (by a factor of $1/\sqrt{2}$) of the EPR signal *in vivo* is a purely quantum mechanical phenomenon associated with the sharing of the positive charge ("hole") in $BChl^+$ between two BChl molecules.

8.8. THERMODYNAMICS, MITOCHRONDRIA, AND CHLOROPLASTS

8.8.1. Free-energy change of a chemical reaction

For the reaction

$$n_A A + n_B B \rightarrow n_C C + n_D D, \tag{8.14}$$

the free energy change, ΔG, is

$$\Delta G = \Delta G^0 + RT \ln \frac{(C_C)^{n_C}(C_D)^{n_D}}{(C_A)^{n_A}(C_B)^{n_B}}. \tag{8.15}$$

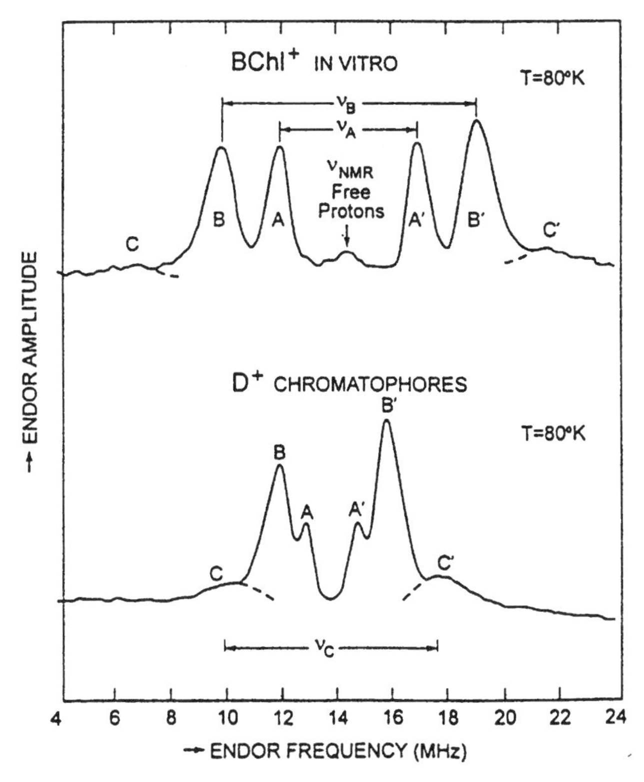

FIGURE 8.9. Identification of the primary donor in bacterial photosynthesis. A comparison of ENDOR spectra from $BChl^{+}$ *in vitro* (top) and the oxidized primary donor D^{+} (bottom). [From Ref. [26].]

At equilibrium $\Delta G = 0$, therefore,

$$\Delta G^0 = -RT \ln \left\{ \frac{(C_C)^{n_C}(C_D)^{n_D}}{(C_A)^{n_A}(C_B)^{n_B}} \right\}_{\text{equil}}. \tag{8.16}$$

The quantity within the braces is defined as the equilibrium constant, K_{eq}:

$$\Delta G^0 = -RT \ln K_{\text{eq}} = -2.3RT \log_{10} K_{\text{eq}}. \tag{8.17}$$

$\Delta G^0 < 0$ when $K_{\text{eq}} > 1$, and $\Delta G^0 > 0$ when $K_{\text{eq}} < 1$.

8.8.2. Electrical and chemical work

$$\Delta G_{(\text{elec})} = n_i z F \Delta\psi \tag{8.18}$$

for the movement of n_i moles of charge z through a potential difference of $\Delta\psi$. F, the Faraday constant, equals 96 487 C/mol. The chemical potential μ_i of compound I is the change in free energy with respect to n_i:

$$\left[\frac{\partial G}{\partial n_i}\right]_{T,p,n_j\neq n_i} \equiv \mu_i. \tag{8.19}$$

The expression for the electrochemical potential is

$$\tilde{\mu} = \mu^0 + RT\ln(c/c_0) + zF\psi. \tag{8.20}$$

8.8.3. Ion gradients, active transport, and ATP synthesis

The proton motive force, Δp, is

$$\Delta p = \frac{\Delta\tilde{\mu}_{H^+}}{F} = \Delta\psi - \frac{2.3RT}{F}\cdot\Delta\text{pH}. \tag{8.21}$$

Symport, uniport, and antiport reactions are illustrated in Figure 8.10. According to the chemiosmotic hypothesis, the source of free energy for the synthesis of ATP is the proton electrochemical gradient across the membrane from the "p" side to the "n" side.

$$\text{ADP} + P_i + nH_p^+ \rightarrow \text{ATP} + \text{H}_2\text{O} + nH_n^+, \tag{8.22}$$

$$\log_{10}\frac{(\text{ATP})}{(\text{ADP})(P_i)} = -\frac{1}{2.3RT}(\Delta G^0 + n\cdot\Delta\tilde{\mu}_{H^+}). \tag{8.23}$$

Table 8.8 presents the standard free energy of hydrolysis of selected phosphate anhydride and ester compounds.

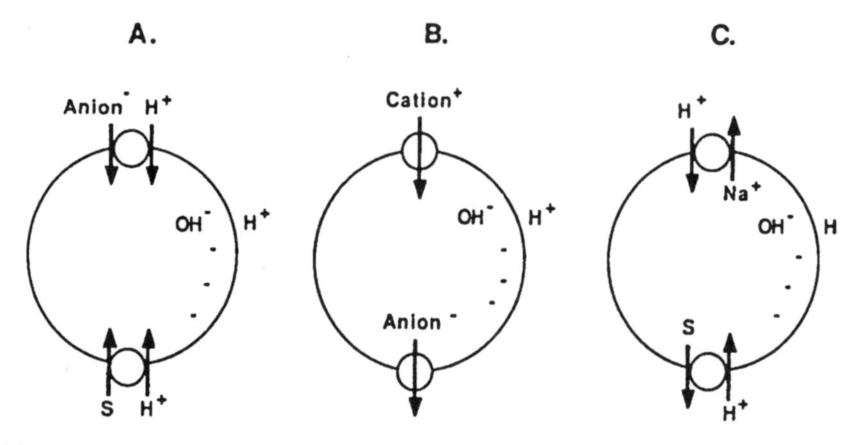

FIGURE 8.10. Diagrammatic representation of (A) symport, (B) uniport, and (C) antiport reactions. S represents solute. [Based on Rosen and Kashket, 1978. After W. A. Cramer and D. B. Knaff, *Energy Transduction in Biological Membranes* (Springer-Verlag, New York, 1991), p. 20, reprinted by permission.]

TABLE 8.8. Standard free energy of hydrolysis of some phosphate anhydride and ester compounds.

Compound	ΔG° (kcal/mol)[a]
Phosphoenolpyruvate	−14.8
1.3-Diphosphoglyceric acid	−11.8
Creatine phosphate	−10.0
Acetyl phosphate	−10.0
Phosphoarginine (pH 8.0)	−8.0
ATP $\rightarrow$ ADP + $P_i(+Mg^{++})$	−7.7
ATP $\rightarrow$ ADP (pH 8.0)	−8.4
ATP $\rightarrow$ ADP (pH 9.5)	−10.4
Glucose-1-phosphate	−5.0
Pyrophosphate	−4.0
Fructose-6-phosphate	−3.8
Glucose-6-phosphate	−3.3
Glycerol-1-phosphate	−2.2

[a] pH = 7.0, except where noted; temperature, 25°C. From Jencks, W. P. (1976) in *Handbook of Biochemistry and Molecular Biology* (G. D. Fasman, ed.), *Vol. I*, p. 302, CRC Press, Cleveland; Bridger, W. A. and J. F. Henderson (1983) *Cell ATP*, John Wiley, New York.

8.8.4. Mitochondria

Tables 8.9 and 8.10 summarize selected properties of mitochondria. Figure 8.11 illustrates the technique of patch-clamping of mitochondrial membranes. Patch-clamp recording, introduced in 1976, was a major advance in membrane biophysics. In this technique, micropipettes are sealed against enzyme-treated, connective, tissue-free cell membranes,

TABLE 8.9. Comparison of respiratory rates of mitochondria from various tissues and cells. [From W. A. Cramer and D. B. Knaff, *Energy Transduction in Biological Membranes* (Springer-Verlag, New York, 1991), p. 82, reprinted by permission.]

	Respiratory rates[a]		
	Substrate		
Organism	Succinate	α-Ketoglutarate	NADH
Guinea pig liver[b]	90	30	—
Beef heart	120	90	—
Potato tubers	510	260	330
Mung bean hypocotyls	450	—	510
Spinach leaves	240	—	320
Neurospora crassa	260	140	450
Saccharomyces cerevisiae	200	140	530

[a] Units: nmol O_2/min/mg protein; multiply by 4 to obtain electron transfer rate.
[b] Liver cells contain 1,000–2,000 mitochondria, ~ 1/5 of the cell volume. From Douce, R. (1985), *Mitochondria in Higher Plants: Structure, Function, Biogenesis*. Academic Press, Orlando.

TABLE 8.10. Proteins and protein complexes of the bovine heart mitochondrial membrane. [From W. A. Cramer and D. B. Knaff, *Energy Transduction in Biological Membranes* (Springer-Verlag, New York, 1991), p. 87, reprinted by permission.]

Component	Concentration (nmol/mg protein)	Activities of complexes (μ mol e^-/min/mg)	Redox groups	E_m (mV)
Complex I[a]	0.1	NADH→ CoQ,[b] 150	FMN, [2Fe-2S], [4Fe-4S]	−370 to −20 (see Chap. 4)
Complex II	0.2	Succinate→ Q_2,[b] 50	FAD, [2Fe-2S], [4Fe-4S], [3Fe-XS]	−270 to +140 (see Chap. 4)
Complex III	0.25–0.5	$Q_2H_2 \rightarrow$ cyt $c(Fe^{3+})$, 300–600	b_{566}, b_{562}, c_1 [2Fe-2S]	−90, +50 +220, +290
Complex IV	0.6–1	Cyt $c(Fe^{2+}) \rightarrow$ oxidase, 25–50	a, a_3, 2Cu	+300–400
Cytochrome c	0.8–1.0	—	c heme	+260
ATP synthase	0.5	—	—	—
ADP/ATP translocase	4.0	—	—	—
NADH-NADP$^+$ transhydrogenase	0.05	—	—	—
Ubiquinone	6–8	—	—	—
Phospholipid	500	—	—	—

[a] Approximate stoichiometries in mammalian mitochrondria: Complex I: Complex III: Cytochrome c: Complex IV: UQ = 1:3:8:8:64. Complex IV: ATPase: ADP-ATP translocase = 1:1:5. (cf. Wainio, W. W. (1970), *The Mammalian Mitochondrial Respiratory Chain*, Academic Press, New York.)
[b] CoQ is ubiquinone-10, Q_2 is ubiquinone-2. From Capaldi, R. (1982) Biochim. Biophys. Acta 654: 291–706.

which results in a high-resistance seal, (2-3) $\times$ $10^7\,\Omega$, that permitted Neher and Salmon to observe the opening and closing of single channels. Around 1980, Neher's group developed a method to make even higher resistance seals, $10^{10} - 10^{11}\,\Omega$, that reduced noise even further and allowed patches of membrane to be torn from the cell while adhering to the pipette. See Sec. 8.10 for additional information on membrane biophysics and electrophysiology.

8.8.5. Chloroplasts

Tables 8.11 and 8.12 summarize some of the physicochemical properties of chloroplasts.

8.9. SIGNALING AND TRANSPORT ACROSS CELL MEMBRANES

To communicate with their environment and with each other, cells and intracellular organelles must send signals across their membranes. These signals may either be conformational changes in proteins that span the membrane or transport of molecules or ions across the membrane.

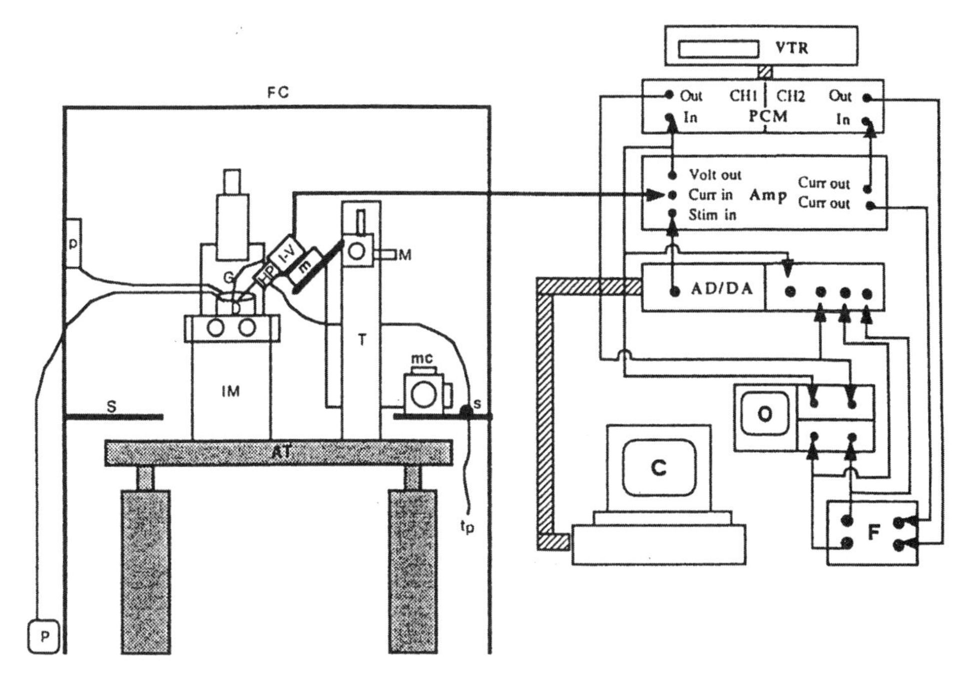

FIGURE 8.11. The patch-clamp setup. Mechanical setup: AT, antivibrational table; S, support isolated from AT; FC, Faraday cage; IM, inverted microscope; D, experimental dish; T, tower; M, macromanipulator; m, micromanipulator; mc, micromanipulator remote control; I-V, headstage (I-V converter); HP, holder and pipette; G, ground electrode; P, peristaltic pump; p, syringe with the perfusing solution; tp, tube for pressure control; s, switch to hold the pressure. Electronic setup: VTR, video tape recorder; PCM, pulse code modulator; Amp, amplifier; O, oscilloscope; AD/DA, analog-to-digital/digital-to-analog converter; F, filter; C, personal computer. [From C. Ballarin and M. Catia Soragato, in *Bioenergetics: A Practical Approach*, edited by G. C. Brown and C. E. Cooper (Oxford University Press, 1995), p. 14, reprinted by permission.]

TABLE 8.11. Properties of isolated chloroplast electron transport complexes. [From W. A. Cramer and D.B. Knaff, *Energy Transduction in Biological Membranes* (Springer-Verlag, New York, 1991), p. 88, reprinted by permission.]

Complex	Activity	No. polypeptides
Photosystem II	$2H_2O \rightarrow O_2 + 4e^- + 4H^+$ (600μmol O_2/mg Chl-hr)[a]	4–9
Cytochrome $b_6 - f$	$QH_2 \rightarrow$ plastocynanin ($100\ s^{-1}$, with isolated complex)	4–5
Photosystem I	plastocyanin (red) $\rightarrow NADP^+$ (200μmol e^-/mg Chl-hr)	~ 7

[a] pH optimum shifted from 8 to 5 in isolated PSII; multiply by 4 to obtain electron transfer activity.

TABLE 8.12. Properties of chloroplast electron transfer chain components. [From D. O. Hall and K. K. Rao, *Photosynthesis*, 6th ed. (Cambridge University Press, UK, 1999)]

	Component and symbol	Molecular mass (daltons)	E_m; mid-point redox potential (volts)[a]	Detection, probable function, etc.
PSII complex; predominantly in the grana (appressed membranes)	'Water-oxidizing' complex 'M' (hypothetical)		+0.82	EPR, EXAFS, reconstitution studies, cyanobacterial mutants. Probably contains 4 Mn ions and C^{2+} and Cl^-. Oxidizes H_2O. Probable ligands are EP33 and D_1-D_2 proteins donating electrons to Yz and releasing protons into the lumen of the thylakoid.
	Yz, primary e^- donor to PSII reaction centre		+1.0	EPR, site-directed mutagenesis; tyrosine 161 bound to the D_1 protein of the PSII reaction centre (RC); $1e^-$ mediator between 'M' and P680. Probably in its neutral state Yz, abstracts H^+ from Mn-bound water in M. TYM 161 of the D_2 protein
	P680	1 or 2 X 892	+1.0 - 1.0 (excited state)	Flash absorption spectroscopy and EPR; special Chl *a* monomer or dimer bound to the D_1-D_2 protein or to 47 kDa RC protein; energy trap of PSII.
	Pheophytin (Pheo) 1	868	-0.45	EPR; metastable intermediate e^- carrier from P*680 to Q_A.
	Plastoquinones			UV absorption or fluorescence; first PSII stable e^- acceptor; $1e^-$ mediator between pheo and Q_B; may be bound to the D_2 protein.
	Q_A	740 (PQ A)	-0.15	
	Q_B			Flash absorption; two e^- gate between Q_A and PQ pool; reduced to semiquinone and then to quinol; semiquinone form binds to D_1.
Cyt b + f complex in grana and stroma lamellae	Cytochrome *b*559	9000 & 4000 Two subunits	+0.05 low potential +0.35 high potential	Absorption spectroscopy; may be involved in O_2 evolution, cyclic e^- transport around PSII and photoprotection
	Cytochrome *b*563 (or b_6)	23440	-0.05 high potential -0.17 low potential	Absorption spectroscopy; PQ→ PQH_2 redox (Q) cycle, i.e. energy transduction; cyclic e^- transport; may be in protein phosphorylation.
	Cytochrome *f*	34000	+~0.35	Absorption spectroscopy; c-type cytochrome; $1e^-$ donor to PC (plastocyanin).
	Rieske iron-sulphur protein $[Fe\text{-}S]_R$	20000	+0.29	EPR; $1e^-$ acceptor from PQ pool; Q cycle. e^- donor to cyt *f*.

(continued)

TABLE 8.12. Continued

	Component and symbol	Molecular mass (daltons)	E_m; mid-point redox potential (volts)[a]	Detection, probable function, etc.
PSI complex; predominantly in the stroma lamellae (non-appressed membranes)	P700	2 X 892	+0.48	Optical absorption, EPR; Chl *a* dimer bound to PSI A-B proteins; energy trap for PSI.
	A_0 (Chl a)	892	-1.0	Flash EPR, metastable primary e^- acceptor from P700*; Chl *a* monomer.
	A_1 (Phylloquinone)	450	-0.80	Flash EPR, reconstruction studies; transient $1e^-$ mediator between A_0 and F_X.
	F_X	Not known	~-0.73	EPR, redox titrations; $1e^-$ mediator from A_1 to F_A and F_B; $[FE\text{-}S]_4$ centre bound to PSI-A and PSI-B proteins.
	Bound [Fe-S] centers, F_A	Not known	-0.55	EPR, redox titrations; $1e^-$ mediators between F_X and Fd; two [4FE-4S] centres bound to PSI-C protein.
	F_B	Not known	-0.59	
	Ferredoxin-NADP reductase (FNR)	40000	-0.38	Isolation from outer lamellae; 1 FAD/mole; e^- mediator between Fd and NADP; can act as diaphorase and transhydrogenase.
Mobile e- carriers	Plastoquinones (PQ)	Variable	~-0.10	Optical absorption; abundant in chloroplasts; shuttles e^- and H^+ between PSII and cyt *bf* complex; energy transduction and cyclic e^- transport from reduced Fd.
	Plastocyanin (PC)	10500 per Cu	+0.38	Isolation from leaves; 4 or 2 Cu per mole; lumen side of thylakoids; $1e^-$ mediator between cyt *f* and P700.
	Ferredoxin (Fd)	11000	-0.42	EPR, optical spectrum; easily isolated from leaves and algae; soluble non-haem iron protein with [2Fe-2S] centre; cyclic and non-cyclic e^- transport; $1e^-$ donor to various chloroplast constituents.

8.9.1. Receptors [13]

A cell-surface receptor is a protein, located in the plasma membrane with its ligand-binding site exposed to the external medium, that binds a hormone, neurotransmitter, drug, or intracellular messenger (such as cyclic AMP, or cyclic GMP) to initiate a change in cell function. An important example is the class of receptor tyrosine kinases. Binding of insulin and growth factors to their extracellular domain causes a conformational change that turns on the kinase activity of their cystosolic domain. This process causes phosphorylation of tyrosine residues in target proteins, activating the targets and starting cascades of enzymatic activity that induce cell growth and differentiation. Mutants with persistent kinase activity cause cancer.

Binding of ligands to receptors can usually be characterized by one of the binding isotherms in Table 8.5; cooperative binding behavior is common. Receptors can diffuse laterally in the cell membrane, with mean square displacement $\langle r^2 \rangle = 4Dt$; this process can lead to clustering of receptors and more complex binding behavior.

8.9.2. Transporters

A transporter is a membrane protein that catalyzes the passage of molecules from one face of a membrane to the other. Transport across a membrane may occur by passive diffusion of lipid-soluble molecules, but it more commonly occurs by facilitated diffusion or active transport. Facilitated diffusion occurs through interaction with molecularly specific transporters, which enable passage of molecules to which the membrane would otherwise by impermeable. The equilibrium distribution reached is the same as that achieved by passive diffusion. Active transport is any energy-dependent process by which molecules or ions are transported across membranes against a chemical potential gradient. The transport reaction may be symport, uniport, or antiport (Figure 8.10).

8.9.3. Channels and pumps [10, 34]

Channels and pumps that enable the passage of ions across membranes are particularly important types of transporters. Channels allow ions to move toward lower electrochemical potential, whereas pumps drive ions against an electrochemical gradient by coupling to a source of free energy, such as ATP hydrolysis or light (Sec. 8.8.3).

The most important channels regulate the flow of Na^+, K^+, and Ca^+. The structure of the Na^+ channel of the eel electric organ is typical: a single polypeptide chain of about 2000 amino acids containing four repeating domains (I–IV), with each repeat probably folding into six transmembrane helices connected by extensive cytosolic and extracellular loops. The packing of the helices forms passages through which ions flow. The channels are highly selective for particular ions, largely (though not entirely) on the basis of ion size and ΔG of hydration (Table 8.13). For example, the K^+ channel has a diameter of 3 Å, so that only dehydrated ions can pass through. Even though Na^+ is smaller than K^+, its free energy of dehydration cost is larger, whereas the ΔG of dehydration of K^+ is largely gained back by closer interactions with the oxygen atoms lining the channel.

TABLE 8.13. Ion-channel permeabilities. Relative permeabilities of cations through sodium and potassium channels, related to ionic radius and hydration free energy.

			Relative permeability	
Ion	Ionic radius, Å	ΔG of hydration kJ/mol	Na^+ channel	K^+ channel
Li^+	0.60	−410	0.93	< 0.01
Na^+	0.95	−301	1.00	< 0.01
K^+	1.33	−230	0.09	1.00
Rb^+	1.48	−213	< 0.01	0.91
Cs^+	1.69	−197	< 0.01	< 0.08

Channels may be either voltage-gated or ligand-gated to control the flow of ions. Na^+ and K^+ channels in nerve axon membranes are voltage-gated: The transmembrane potential interacts with the dipoles of the transmembrane helices to cause allosteric conformational changes that open or close the channels. The acetylcholine receptor channel is ligand-gated, open to both Na^+ and K^+ when acetylcholine is bound, and closing via an allosteric switch when acetylcholine is removed.

8.10. ELECTROPHYSIOLOGY

8.10.1. Impulses in nerve and muscle cells

A nerve cell conducts an electrochemical impulse because of changes that take place in the cell membrane. These charges allow movement of ions through the membrane, setting up currents that flow through the membrane and along the cell. Similar impulses travel along muscle cells before they contract. Nerve cells are composed of dendrites (input), cell body, axon (long conducting portion), and an output end.

8.10.2. Properties of nerve and muscle cells

As illustrated in Figure 8.12, membrane potential can be measured with the aid of fine-tipped glass capillaries with a tip diameter of $< 1\mu$m inserted through the cell membrane into the intracellular fluid. In resting cells, the intracellular fluid is always electrically negative with respect to the extracellular fluid, and varies between -10 mV (for red blood cells) and -90 mV (for skeletal muscle cells). Figure 8.13 illustrates the magnitude and time course of a typical nerve impulse. Figure 8.14 illustrates the ion concentrations in a typical nerve and in the extracellular fluid surrounding the nerve.

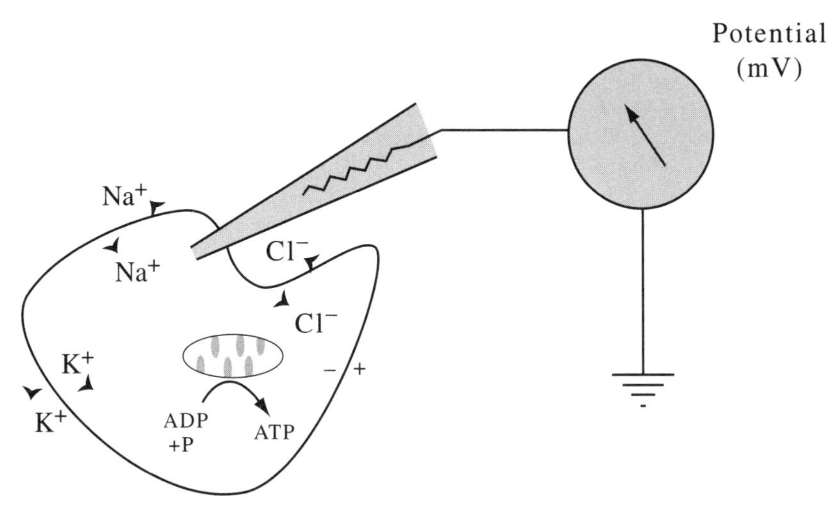

FIGURE 8.12. Schematic illustration of a cell impaled by a microelectrode. [© 1998 O. S. Andersen, reprinted by permission.]

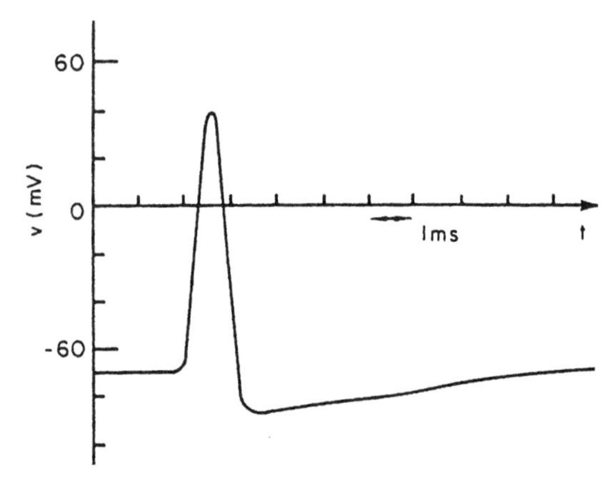

FIGURE 8.13. A typical nerve impulse or action potential plotted as a function of time. [From R. K. Hobbie, *Intermediate Physics for Medicine and Biology*, 3rd ed. (Springer/AIP, 1997), p. 137, reprinted by permission of publisher.]

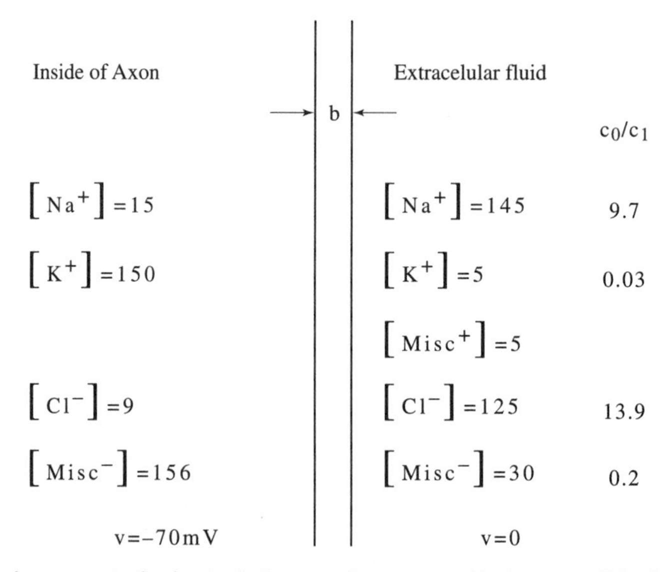

FIGURE 8.14. Ion concentration in a typical mammalian nerve and in the extracellular fluid surrounding the nerve. Concentrations are in mmol L^{-1}; c_0/c_i is the concentration ratio. [From R. K. Hobbie, *Intermediate Physics for Medicine and Biology*, 3rd ed. (Springer/AIP, 1997), p. 138, reprinted by permission of publisher.]

TABLE 8.14. Properties of unmyelinated and myelinated axons of the same radius.

Quantity	Unmyelinated	Myelinated
Axon inner radius, a	5×10^{-6} m	5×10^{-6} m
Membrane thickness, b'	6×10^{-9} m	
Myelin thickness, $b \approx 0.4a$		2×10^{-6} m
$\kappa\varepsilon_0$	$6.20 \times 10^{-11}\ \mathrm{s}^{-1}\,\Omega^{-1}\,\mathrm{m}^{-1}$	$6.20 \times 10^{-11}\ \mathrm{s}^{-1}\,\Omega^{-1}\,\mathrm{m}^{-1}$
Axoplasm resistivity, ρ_i	1.1Ω m	1.1Ω m
Membrane (resting) or myelin resistivity, ρ_m	$10^7\Omega$ m	$10^7\Omega$ m
Time constant $\tau = \kappa\varepsilon_0\rho_m$	6.2×10^{-4} s	6.2×10^{-4} s
Space constant, λ	$\lambda = \sqrt{\dfrac{ab\rho_m}{2\rho_i}} = 0.165\sqrt{a} = 370 \times 10^{-6}m$	$\lambda = \sqrt{\dfrac{ab\rho_m}{2\rho_i}} = \sqrt{\dfrac{0.4a^2\rho_m}{2\rho_i}} = a\sqrt{\dfrac{0.4\rho_m}{2\rho_i}} = 1350a = 6.8 \times 10^{-3}$ m
Node spacing, D		$D = 280a = 1.4 \times 10^{-3}$ m

8.10.3. Axons: The cable model

Applying Kirchoff's laws to a segment of axon produces a cable model that assumes that the potential inside the axon has no radial or azimuthal dependence:

$$c_m \frac{\partial v}{\partial t} = -j_m + \frac{1}{2\pi a r_i} \frac{\partial^2 v}{\partial x^2}. \tag{8.24}$$

In this equation, c_m is the membrane capacitance per unit area, v is the potential difference, t is the time, j_m is the membrane current per unit area, a is the axon inner radius, x is the distance, and r_i is the resistance per unit length along the inside of the axon. Table 8.14 presents properties of typical unmyelinated and myelinated axons.

8.10.4. Models for membrane current density

The *electrotonus* model assumes that j_m obeys Ohm's law, which leads to the telegrapher's equation, $\lambda^2(\partial^2 v/\partial x^2) - v - \tau(\partial v/\partial t) = -v_{\text{resting}}$, where the constants are related to the membrane parameters by $\lambda^2 = 1/(2\pi a r_i g_m) = ab\rho_m/2\rho_i, \tau = c_m/g_m = \kappa\epsilon_0\rho_m$. This model may be useful for small perturbations of the transmembrane potential or for myelinated fibers in the region between the non-Ohmic Nodes of Ranvier.

Conduction of an action potential requires a nonlinear model for the membrane. The Hodgkin-Huxley model assumes that the current is

$$j_m = g_{\text{Na}}(\nu - \nu_{\text{Na}}) + g_K(\nu - \nu_{\text{K}}) + j_L, \tag{8.25}$$

where ν_{Na} and ν_K are the sodium and potassium Nernst potentials, e.g.,

$$\nu_{\text{Na}} = \nu_i - \nu_0 = \frac{k_B T}{e} \ln\left(\frac{[\text{Na}_o]}{[\text{Na}_i]}\right). \tag{8.26}$$

Their conductances per unit area are

$$g_{\text{K}}(\nu, t) = g_{\text{K}\infty} n^4(\nu, t), \tag{8.27}$$

$$g_{\text{Na}}(\nu, t) = g_{\text{Na}\infty} m^3 h. \tag{8.28}$$

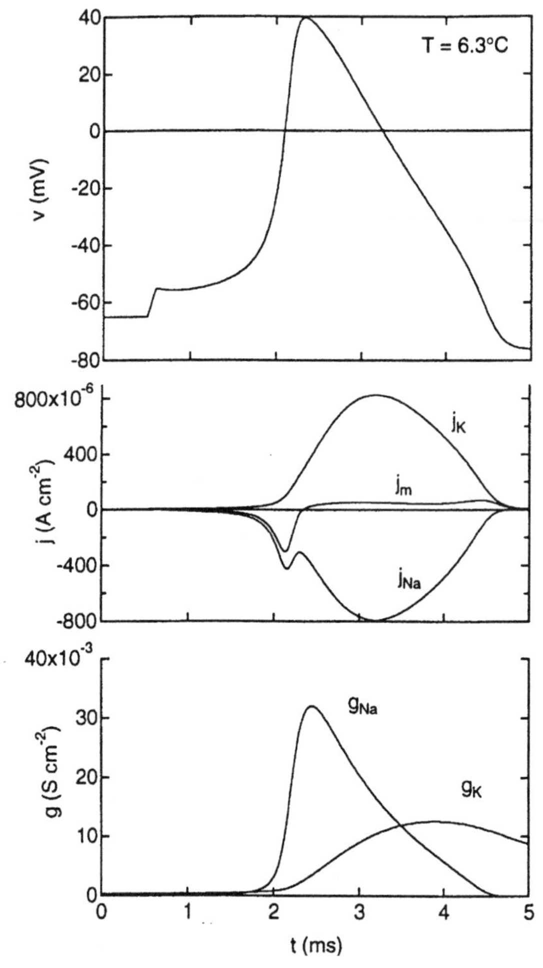

FIGURE 8.15. Computed response to a pulse in a space-clamped squid axon at $T = 6.3°\text{C}$. The axon was stimulated at $t = 0.5$ ms for 0.1 ms. [From R. K. Hobbie, *Intermediate Physics for Medicine and Biology*, 3rd ed. (Springer/AIP, 1997), p. 165, reprinted by permission of publisher.]

Parameters n, m, and h evolve with time according to three first-order rate equations of the form

$$\frac{dm}{dt} = \alpha_m(1-m) - \beta_m m, \tag{8.29}$$

where the six parameters $\alpha_{n,m,h}$ and $\beta_{n,m,h}$ are experimentally determined functions of the voltage. [29] Figure 8.15 shows computed parameters as a function of time during an action potential. The middle panel of Figure 8.15 illustrates the current densities per unit area of the cell membrane. The lower panel shows the conductances per unit area. $j_K = g_K(\nu - \nu_K)$, where ν_K is the potassium Nernst potential (similarly for ν_{Na}). These parameters correlate with the $\nu(t)$ profiles in the upper panel of Figure 8.15 and Figure 8.13.

8.11. PHOTOBIOPHYSICS

Photobiophysics comprises molecular studies of photon interactions with condensed matter. Like biological physics, it is a broad subdiscipline. Figure 8.16 summarizes fundamental photobiophysical phenomena as well as research areas of photobiophysics.

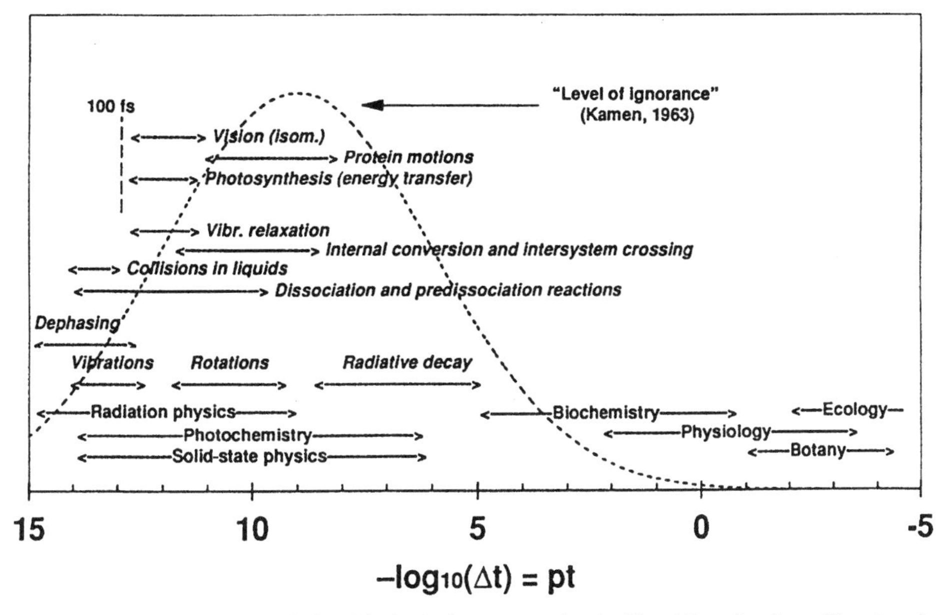

FIGURE 8.16. Classification of photobiophysical processes by significant time durations. The pt scale and curve centered at pt = 9 were suggested by Kamen (1963), and adapted by Feher (1998). Entries in italic type are those of Zewail (1996), with the addition of dephasing. The upper group of three are associated with biology, the next four with chemistry and physics, and the lower four (left) "fundamental." The three rows at the bottom (right) of the chart contain Kamen's classifications of disciplines. His 1963 "level of ignorance" has clearly shifted considerably to the left in modern times. [Provided by R. S. Knox (1999).]

Fundamental photobiophysical processes play an important role in clinical medicine in the area of photodynamic therapy. [See Section 8.14.]

The application of x-rays to problems in biology is, arguably, the most important application of physics to the study of modern biology. A major advance in photobiophysics occurred with the publications of Deisenhofer *et al.* [25] on the structure of the protein subunits in the photosynthetic reaction center of *Rhodopseudomonas viridis* at 3 Å resolution. [24, 25] Recently, Pebay-Peyroula *et al.* have determined the structure of bacteriorhodopsin at 2.5 Å resolution. Bacteriorhodopsin, like the *viridis* reaction center, is a membrane-bound protein. Unlike chlorophyll-based photosynthesis, which couples proton translocation to electron transport, bacteriorhodopsin converts light energy into chemical energy by directly pumping protons. Figure 8.17 is a ribbon model of the bacteriorhodopsin monomer. The current model of the proton pump mechanism of bateriorhodopsin is illustrated in Figure 8.18.

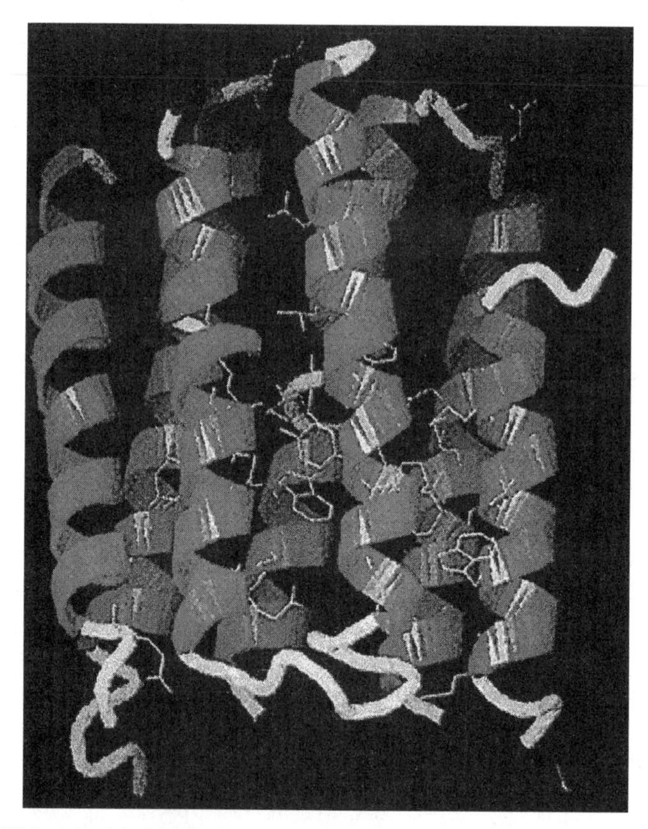

FIGURE 8.17. Ribbon model of the bacteriorhodopsin monomer. The bundle made of the seven transmembrane helices is perpendicular to the plane of the cell membrane. [From F. Hucho (1998). Figure provided by E. Pebay-Peyroula and J. Rosenbusch.]

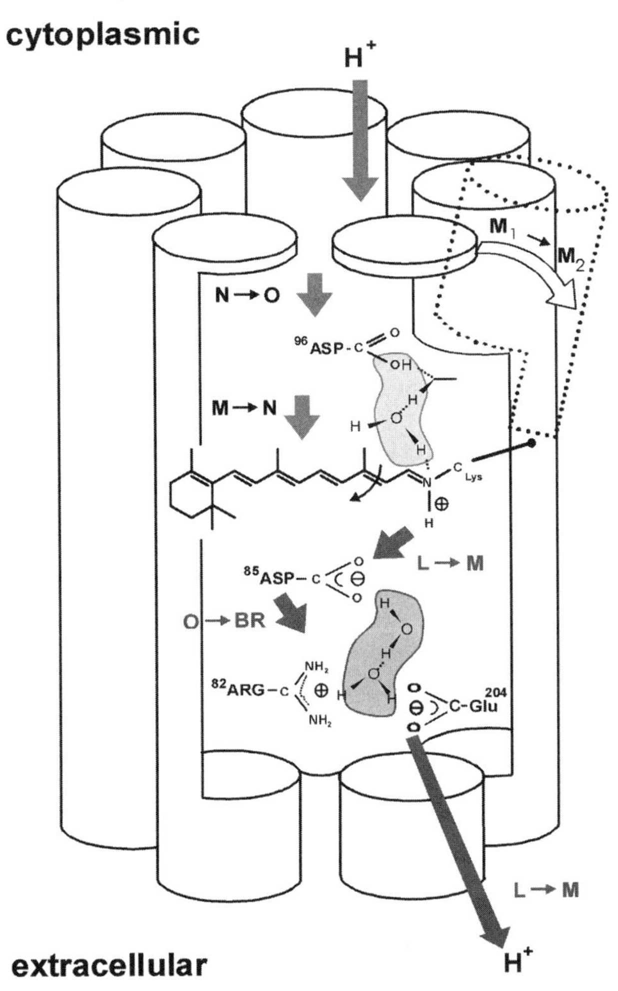

FIGURE 8.18. The current model of the proton pump mechanism of bacteriorhodopsin. It is the collective work of many research groups. The primary event is photon absorption that triggers the all-*trans* to 13-*cis* isomerization of the retinal chromophore, whose molecular structure is illustrated at the center of the model. The absorption of light sets in motion a series of concerted protonation/deprotonation/reprotonation reactions that results in the transport of a proton from the cytoplasmic side of the membrane to the extracellular side. [Figure provided by K. Gerwert.]

8.12. MUSCLE AND CONTRACTILITY

Coordinated movement is driven by molecular motors. Three major types of molecular motors in eukaryotes, all of which depend on ATP hydrolysis, are (1) muscle contraction, via sliding of interdigitating myosin and actin filaments; (2) beating of cilia and flagella, from interaction of dynein and tubulin; and (3) movement of vesicles on microtubules, mediated by kinesin and tubulin. Binding of ATP to the motor proteins myosin, dynein, and kinesin causes conformational changes that are reversed by hydrolysis of ATP and ADP and inorganic phosphate. The repeating sequence of conformational changes advances the motor protein on its complementary scaffold of actin or tubulin.

Bacteria are propelled by the rotation of flagellar motors, located in their cytoplasmic membrane. The sliding of motor proteins over an array of complementary proteins is similar to that in eukaryotes, but the process is driven by proton-motive force rather than by ATP hydrolysis.

Vertebrates contain three kinds of muscle: striated (under voluntary control), cardiac, and smooth (responsible for contraction of stomach, bladder, intestines, uterus, and artery walls). Striated muscle has the most readily visualized structure and has been most extensively studied. Muscle cells are surrounded by an electrically excitable plasma membrane, which regulates the flow of Ca^{2+} in and out, and are composed of parallel organelles called myofibrils, about 1μm in diameter, surrounded by cytoplasm. The functional unit of the myofibril is the sarcomere, which appears in micrographs to have a banded structure, diagrammed in Figure 8.19. In cross section, the thick and thin filaments are arrayed in interdigitating hexagonal lattices.

The thick filaments contain primarily myosin; the thin filaments contain actin, tropomyosin, and the troponin complex. A variety of other proteins, only some of which are listed in Tables 8.15 and 8.16, serve a variety of regulatory and anchoring functions.

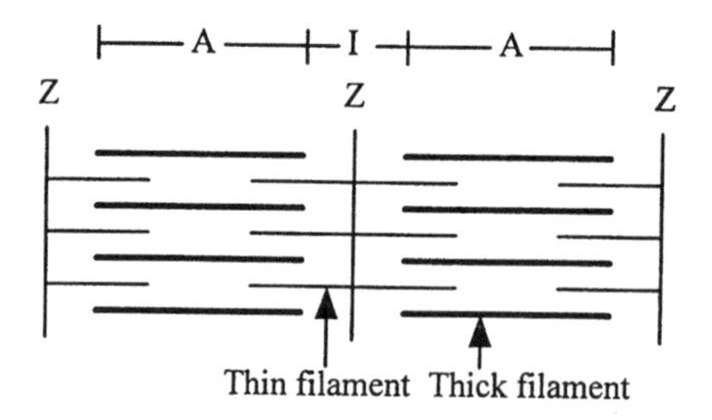

FIGURE 8.19. Band structure of vertebrate striated muscle sarcomere.

TABLE 8.15. Some structural proteins of vertebrate skeletal muscle (Ref. [1], Table A.2).

Protein	Mol. Wt.	Wt. % of muscle	Location	Function
Myosin	520,000	43	A-band	Motor protein
Actin	42,000	22	I-band	Slides over myosin
Tropomyosin	$2 \times 33,000$	5	I-band	Regulates actin
Troponin C	18,000	1	I-band	Binds Ca^{2+}
Troponin I	21,000	1	I-band	Inhibits myosin interaction
Troponin T	31,000	2	I-band	Binds to tropomyosin
α-Actinin	$2 \times 95,000$	2	Z-line	Cross-links actin filaments
β-Actinin	37,000	< 1	End I-band	Caps actin filaments
Titin	2,500,000	10	A + I bands	Links myosin to Z-line
Nebulin	800,000	5	I-band	Regulates actin filament length?

TABLE 8.16. Values for vertebrate skeletal muscle properties (frog sartorius muscle at 0°C). Rates and velocities are 5–10 times higher for rabbit psoas muscle at 20°C (Ref. [1], Table A.1).

Sarcomere length	2.0–3.0 μm (2.5-μm rest length)
Thick filament length	1.6 μm (about 300 myosin molecules)
Thick filament shaft diameter	15 nm
Thin filament length	1.0 μm (about 380 actin molecules)
Thin filament diameter	9 μm
Bare zone length	0.15 μm
Thick filament spacing (center-center)	42 nm (at rest length)
Thick-thin filament spacing (center-center)	22–30 nm (26 nm at rest length)
Thick-thin filament ratio	1:2 (in cross section); 1:4 (in total sarcomere)
Total actin content	600 nmol g^{-1} muscle (about 600 μM)
Total myosin content	120 nmol g^{-1} muscle (about 240 μM heads)
ATPase rate (isometric contraction)	0.5 μmol ATP/g muscle/s = 2.1 ATP/head/s
(isotonic contraction)	1.5 μmol ATP/g muscle/s = 6.3 ATP/head/s
(relaxed)	< 0.5 nmol ATP/g muscle/s $< 2 \times 10^{-3}$ ATP/head/s
ΔG for ATP hydrolysis	60 kJ $mol^{-1} = 10^{-19}$ J per molecule
Isometric tension	20 N $cm^{-2} \approx 1.6$ pN per myosin head
Elastic modulus	4 kN cm^{-2} (isometric); 0.025 kN cm^{-2} (relaxed)
Maximum shortening velocity, V_0	2 muscle lengths/s = 2.5 μm/s/half-sarcomere
Maximum power	44 mW g^{-1} muscle (at $V_0/3$)
Maximum thermodynamic efficiency	50% (at $V_0/3$)
Heat production (isometric)	13 mW g^{-1} muscle
Half-time to peak tension	50 ms
Half-time for relaxation	400 ms
Myosin subfragment 1 ATPase	0.01 s^{-1} (in solution)
Actin-activated subfragment 1 ATPase	4.5 s^{-1} (in solution)

8.13. CHARACTERIZING BIOPOLYMERS IN SOLUTION

8.13.1. Sedimentation and diffusion

Spinning a solution of molecules of molecular weight M and partial specific volume $\bar{\nu}$ in a solvent of density ρ at angular velocity ω at distance r from the axis of rotation produces a velocity dr/dt. The proportionality constant between velocity and angular acceleration is the sedimentation coefficient S:

$$S = \frac{dr/dt}{\omega^2 r} = \frac{M(1 - \bar{\nu}\rho)}{N_A f}. \tag{8.30}$$

The friction coefficient f is also related to the translational diffusion coefficient D by the Einstein equation $D = kT/f$.

8.13.2. Sedimentation equilibrium

At equilibrium, the fluxes caused by sedimentation and those caused by diffusion are equal and opposite, leading to a concentration distribution

$$c(r) = c(a)\exp\left[\frac{M(1-\bar{v}\rho)\omega^2(r^2-a^2)}{2RT}\right], \tag{8.31}$$

where a is a reference distance and R is the gas constant. For banding in a density gradient $\partial\rho/\partial r$ around r_0, this distribution becomes $c(r) = c(r_0)\exp[-(r-r_0)^2/2\sigma^2]$, where $\sigma^2 = RT\omega^2 r_0/M\bar{v}(\partial\rho/\partial r)$.

8.13.3. Rotational motion

Molecular orientation can be produced by flow, electric or magnetic fields, or photoselection, and measured by birefringence, dichroism, or singlet or triplet anistropy. Rotational diffusion and friction coefficients D_r and f_r about an axis are related by the Einstein equation. Rotation is also characterized by rotational relaxation times $\tau^{(1)} = 1/2\,D_r$ or $\tau^{(2)} = 1/6\,D_r$, depending on whether the orientational measure is proportional to $\cos(\theta)$ or $\cos^2(\theta) - 1$.

8.13.4. Frictional coefficients

Translational and rotational friction coefficients for a sphere of radius R are $f = 6\pi\eta R$, $f_r = 8\pi\eta R^3$, where η is the solvent viscosity. Table 8.17 gives values of $F = f/f_{\text{sphere}}$ for regular nonspherical shapes of axial ratio $p = a/b$ and $q = 1/p$.

8.13.5. Electrophoresis and gel electrophoresis

For an isolated particle of charge Q in a field E, the electrophoretic mobility $\mu = QE/f$. Backflow of counterions and solvent prevents accurate determination of Q for most polyelectrolyte solutions. Gel electrophoresis in polyacrylamide or agarose is commonly used, with calibration, to determine molecular weights of proteins and nucleic acids. Linear plots are frequently obtained from $\log M$ versus μ_r or M versus $1/\mu_r$, where μ_r is the relative mobility.

8.13.6. Scattering

The intensity of light $I(\theta)$ scattered at angle θ from unit volume of a polymer solution of concentration (g/ml) c and molecular weight M irradiated with polarized light of intensity I_0 is

$$I(\theta) = I_0\frac{4\pi^2 n^2(\partial n/\partial c)^2\sin^2(\theta)P(q)}{N_A\lambda^4 r^2(1/M + 2Bc + 3Cc^2 + \cdots)}, \tag{8.32}$$

TABLE 8.17. Frictional coefficients of ellipsoids and rods. Hydrodynamic properties of ellipsoids of semiaxes *a*, *b*, and rods of length *L* relative to spheres of the same volume.

	Prolate Ellipsoid[a]	Oblate Ellipsoid[a]	Cylindrical Rod[b]
R_e	$(ab^2)^{1/3}$	$(ab^2)^{1/3}$	$(3/2p^2)^{1/3}(L/2)$
F_t	$\dfrac{(1-q^2)^{1/2}}{q^{2/3}\ln\left\{\dfrac{[1+(1-q^2)^{1/2}]}{q}\right\}}$	$\dfrac{(q^2-1)^{1/2}}{q^{2/3}\arctan(q^2-1)^{1/2}}$	$\dfrac{(2p^2/3)^{1/3}}{\ln p+\gamma}$ $\gamma = 0.312 + \dfrac{0.565}{p} + \dfrac{0.100}{p^2}$
$F_{r,a}$	$\dfrac{4(1-q^2)}{3(2-2q^{4/3}/F_t)}$	$\dfrac{4(1-q^2)}{3(2-2q^{4/3}/F_t)}$	$0.64\left(1+\dfrac{0.677}{p}-\dfrac{0.183}{p^2}\right)$
$F_{r,b}$	$\dfrac{4(1-q^4)}{3q^2[2q^{-2/3}(2-q^2)/F_t-2]}$	$\dfrac{4(1-q^4)}{3q^2[2q^{-2/3}(2-q^2)/F_t-2]}$	$\dfrac{2p^2}{9(\ln p+\delta_a)}$ $\delta_a = -0.662 + \dfrac{0.917}{p} - \dfrac{0.505}{p^2}$

[a]Perrin, F. (1934) J. Phys. Rad. 5:497–511; Perrin, F. (1936) J. Phys. Rad. 7:1–11; Koenig, S. (1975) Biopolymers 14:2421–2423.
[b]Tirado, M. M. & J. Garcia de la Torre (1979) J. Chem. Phys. 71:2581–2587; Tirado, M. M. & J. Garcia de la Torre (1980) J. Chem. Phys. 73:1986–1993; Garcia de la Torre, J. & V. A. Bloomfield (1981) Q. Rev. Biophys. 14:81–139.

where n is the solvent refractive index, $\partial n/\partial c$ is the refractive index increment, $\lambda = \lambda_0/n$ is the wavelength in the solvent, r is the distance from the detector, B and C are second and third virial coefficient, and $P(q)$ is the particle scattering form factor, a function of the magnitude of the scattering vector $q = (4\pi/\lambda)\sin(\theta/2)$. In a polydisperse solution, M is the weight-average molecular weight. For spherical macromolecules of specific volume ν, $B = 4\nu/M$; for rods of length L and diameter d, $B = L\nu/dM$.

In the limit of small qR_G, $P(q) = 1 - q^2R_G^2/3 \simeq \exp(-q^2R_G^2/3)$, regardless of particle shape, where R_G is the particle rms radius (radius of gyration). It equals $\sqrt{(3/5)}\,R$ for spheres of radius R and $L/\sqrt{12}$ for thin rods of length L. Over a wider range of qR_G, $P(q)$ has the following analytical values for model shapes:

$$\text{spheres,} \qquad \left\{\frac{3}{(qR)^3}[\sin(qR) - qR\cos(qR)]\right\}^2; \tag{8.33}$$

$$\text{rods,} \qquad \frac{2}{qL}Si(qL) - \left[\frac{\sin(qL/2)}{qL/2}\right]^2; \tag{8.34}$$

$$\text{random coils,} \qquad \frac{2}{w^2}(e^{-w} + w - 1), \quad w = q^2R_G^2. \tag{8.35}$$

These equations are valid only when qR_G is ≤ 1; for larger particles, Mie scattering theory must be used.

Equation (8.32) is also valid for x-ray and neutron scattering, except that the contrast function $\partial n/\partial c$ is replaced by a function of the difference in electron density, or neutron scattering length. Since the wavelengths are lower than that of light, smaller size scales may be probed, but length scales from hundreds to thousands of angstroms may be measured using small-angle x-ray or neutron scattering (SAXS or SANS).

Thermal neutron scattering provides a particularly powerful tool, complementary in many ways to x-ray diffraction and scattering. Neutron diffraction can be used for high-precision structural determinations regardless of the atomic number of constituent atoms, in contrast to x-ray methods that are more sensitive to heavy atoms. For example, neutron diffraction can yield the location of protons and water molecules. The isotopic dependence of neutron cross sections enables the use of contrast matching in scattering from aqueous solutions. In this method, one or more constituents are selectively deuterated to enhance or suppress the signal. Some future improvements of neutron diffraction for biologically relevant materials may involve measurements at low temperatures and high magnetic fields to reduce the intrinsic incoherent scattering background originating from protons in the system. Finally, inelastic neutron scattering and specialized related techniques can be used in principle to study collective atomic motions and microscopic diffusion. The Spallation Neutron Source, an international user facility under construction at the Oak Ridge National Laboratory, will greatly enhance the application of neutron science to problems in biology.

8.13.7. Dynamic light scattering

Dynamic laser-light scattering (quasielastic light scattering, photon correlation spectroscopy) is the most common method for determining translational diffusion coefficients of macromolecules in solution. The homodyne autocorrelation function decays exponentially with time:

$$g^{(2)}(\tau) = 1 + \beta e^{-2Dq^2\tau}, \tag{8.36}$$

where β is an instrumental coherence factor. If the solution is polydisperse, cumulant analysis yields the z-average diffusion coefficient:

$$D_z = \frac{\sum_i D_i c_i M_i P_i(q)}{\sum_i c_i M_i P_i(q)}. \tag{8.37}$$

8.14. BIOPHYSICS, THE HEALTH SCIENCES, AND EMERGING TECHNOLOGY

The physical sciences have made substantial contributions to our fundamental understanding of biology. We note in this section several examples of how basic research in biophysics has contributed to medicine and economic well-being and how it may continue to do so in the future.

Many interesting variations on the NMR theme have been developed, such as 2D NMR spectroscopy, which is especially useful for the resolution of dense spectra and MRI. Unlike conventional NMR, which uses a uniform external magnetic field, MRI uses an external magnetic field gradient such that each volume element of the sample has a unique resonance frequency. MRI has become an important medical diagnostic imaging tool. Figure 8.20 is an illustration of the MRI technique.

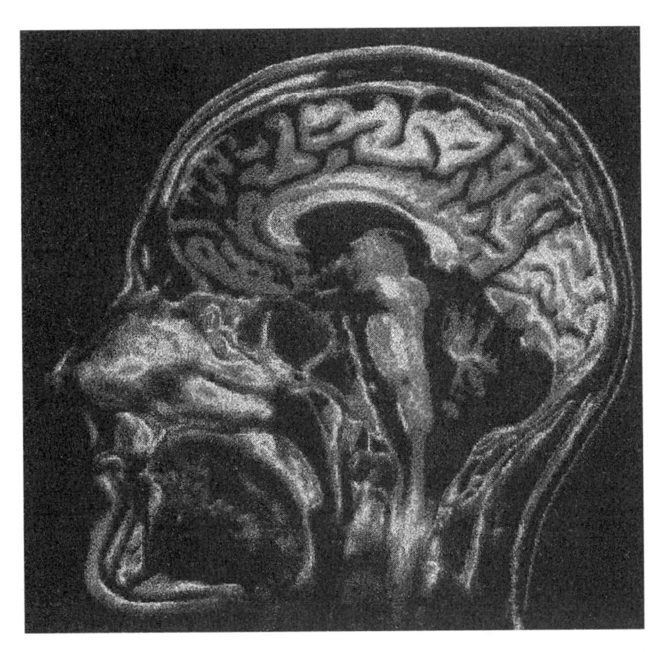

FIGURE 8.20. Magnetic resonance imaging scan of healthy human brain in sagittal section.

Photodynamic therapy (PDT) is an important application of photobiophysics to medicine. The molecular physics of PDT is summarized in the energy level (Jabłońsky) diagram of Figure 8.21. S_0 and S_1 are the photosensitizer singlet ground and singlet first electronic excited states, respectively. S_1 is populated through absorption of light. T_1 is the lowest-lying triplet state of the photosensitizer, which is populated by intersystem crossing from S_1. Yields of triplet formation for current PDT photosensitizers are high, typically, 0.5 or better. Ground-state oxygen ($^3O_2$) present in the biological system quenches these triplets efficiently. The oxygen-triplet quenching rate constant is approximately 1×10^9 M^{-1}s^{-1}. This interaction results in energy transfer from the sensitizer triplet to oxygen and in the formation of singlet oxygen, $^1O_2$, which is the lowest-lying electronic excited state of oxygen and understood to be the main species responsible for cell damage and death. The "A" in the diagram refers to unspecified targets of singlet oxygen reaction.

Although photosynthesis is generally regarded as the conversion of light energy into stored chemical energy, modern knowledge of the molecular architecture of photosynthetic membranes allows an appreciation of photosynthesis from a biomolecular optoelectric point of view. As its names implies, the photosynthetic reaction center is a molecular photovoltaic device. However, it has recently been discovered that in addition to its optoelectronic properties, the Photosystem I reaction center of photosynthesis is a nanometer-scale molecular diode. [32] The I-V characteristics of a single Photosystem I reaction center is illustrated in Figure 8.22. Also illustrated is the construction of a NAND logic gate from elementary diode elements. Any combinatorial logic function can be performed with NAND logic gates.

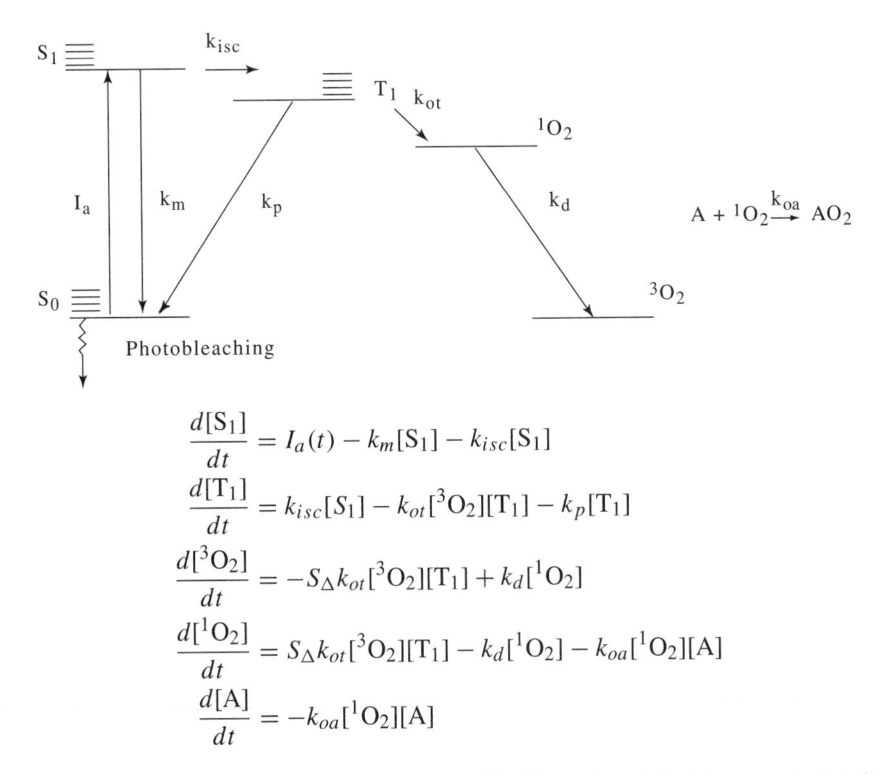

FIGURE 8.21. Energy-level diagram for the photosensitized formation of singlet oxygen. In the singlet oxygen model of PDT, the concentrations of excited states of the photosensitizer, ground-state oxygen, singlet oxygen, and the singlet-oxygen acceptor are described by the coupled set of differential equations indicated in the figure. k_m is the sum of radiative and nonradiative decay rates from S_1 to S_0 (s^{-1}). I_a, rate of formation of sensitizer excited states ($M \cdot s^{-1}$); A, singlet oxygen accepter within a target cell (M); k_{isc}, intersystem crossing rate from S_1 to T_1 (s^{-1}); k_{ot}, rate of sensitizer triplet quenching by $^3O_2$ ($M^{-1} \cdot s^{-1}$); k_p, sum of radiative and nonradiative decay rates from T_1 to S_0 (s^{-1}); S_Δ, the fraction of triplet quenching collisions with $^3O_2$ that result in $^1O_2$; k_d, $^1O_2$ to $^3O_2$ decay rate (s^{-1}); k_{oa}, the rate of reaction of $^1O_2$ with acceptor A ($M^{-1} \cdot s^{-1}$). [Provided by T. H. Foster.]

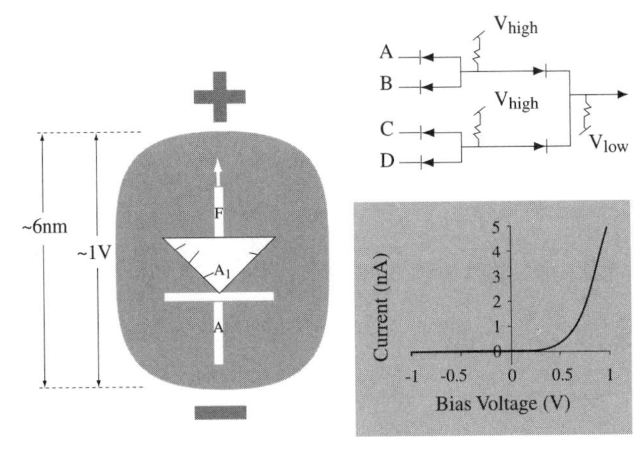

FIGURE 8.22. Single, isolated Photosystem I reaction centers are molecular diodes. [From Ref. [32].]

TABLE 8.18. Selected equations of biological physics.

Equation	Description/Comment
$\mathrm{Mb} + \mathrm{O_2} \leftrightarrow \mathrm{Mb\,O_2}, \mathrm{Mb} + \mathrm{CO} \leftrightarrow \mathrm{Mb\,CO}$	Binding/dissociation of O_2 and CO to myoglobin
$k(H, T) = A\exp(-H/RT)$	Arrhenius law, rate of unimolecular reactions in terms of barrier height and temperature
$N(t) = N(0)\exp(-[k(T)t])^{\beta}$	Nonexponential relaxation in complex systems
$k(H^*, T) \approx A\exp[-(H^*/RT)^2]$	Non-Arrhenius behavior, the Ferry relation
$N(t) = N(0)\int g(H)\exp[-H/RT]\,dH$	Nonexponential binding of O_2 or CO at low temperatures, distribution of barriers. (Austin, 1975)
$f_{DW} = \exp[-16\pi^2\langle x^2\rangle \sin^2\theta/\lambda^2]$	Debye-Waller factor
$k_t \approx A\exp[-\pi^2 d(2ME)^{1/2}/h]$	Approximate tunneling rate for particle mass M through a barrier height H and width d. h is Planck's constant
$\kappa_M = \dfrac{E\,d^3}{12(1-\nu^2)}$	Bending modulus of a membrane with Young's modulus, E, thickness d, and Poisson's ratio ν
$H_{\text{bend}} = \frac{1}{2}\kappa_M \int_{\text{surface}} dA \left(\dfrac{\partial^2 u}{\partial x^2} + \dfrac{\partial^2 u}{\partial y^2} - C_0\right)^2$	The bending energy stored in a membrane where u is the magnitude of the membrane normal vector and C_0 is the spontaneous curvature of the membrane due to assymmetric sides (Gruner, 1994)
$\kappa_p = E\int x^2\,dx\,dy$	Mechanical rigidity, κ_p, of a polymer is a function of the modulus of elasticity, E, and cross-sectional shape
$P = \kappa_p/k_B T$	Persistence length of a polymer (DNA), the length scale on which the directionality of polymer is maintained.

[Adapted from H. Frauenfelder, P. G. Wolynes, and R. H. Austin, *Rev. Mod. Phys.* **71**, S419–S430 (1999).]

8.15. REFERENCES

8.15.1. General references

[1] C. R. Bagshaw, *Muscle Contraction*, 2nd ed. (Chapman & Hall, London, 1993).

[2] V. A. Bloomfield, D. M. Crothers, and I. Tinoco, Jr., *Nucleic Acids: Structures, Properties, and Functions* (University Science Books, Mill Valley, CA, 2000).

[3] C. R. Cantor and P. R. Schimmel, *Biophysical Chemistry*, 3 vols. (Freeman, New York, 1980).

[4] W. A. Cramer and D. B. Knaff, *Energy Transduction in Biological Membranes* (Springer-Verlag, New York, 1991).

[5] T. E. Creighton, *Proteins: Structures and Molecular Properties*, 2nd ed. (Freeman, New York, 1994).

[6] J. T. Edsall and J. Wyman, *Biophysical Chemistry* (Academic, New York, 1958).

[7] J. J. Eggermont and A. J. Hoff, *Handbook of Biological Physics*, Multivolume Series (Elsevier Science, New York).

[8] H. Frauenfelder and M. C. Marden, in *A Physicist's Desk Reference*, 2nd ed., edited by H. L. Anderson (American Institute of Physics, 1989).

[9] H. Frauenfelder, P. G. Wolynes, and R. H. Austin, Rev. Mod. Phys. **71**, S419–S430 (1999).

[10] B. Hille, *Ionic Channels of Excitable Membranes*, 2nd ed. (Sinauer, 1992).

[11] R. Hobbie, *Intermediate Physics for Medicine and Biology*, 3rd ed. (Springer/AIP Press, 1997).

[12] W. Hoppe, W. Lohmann, H. Markl, and H. Ziegler, *Biophysics* (Springer-Verlag, 1983).

[13] D. A. Lauffenburger and J. J. Linderman, *Receptors: Models for Binding, Trafficking, and Signaling* (Oxford University Press, 1996).

[14] P. Läuger, *Electrogenic Ion Pumps* (Sinauer, 1991).

[15] R. R. Sinden, *DNA Structure and Function* (Academic, San Diego, 1994).

[16] L. Stryer, *Biochemistry*, 4th ed. (Freeman, New York, 1995).

[17] C. Tanford, *Physical Chemistry of Macromolecules* (Wiley, New York, 1961).

[18] K. W. van Holde, W. C. Johnson, and P. S. Ho, *Principles of Physical Biochemistry* (Prentice-Hall, Upper Saddle River, NJ, 1998).

8.15.2. On-line resources

[19] American Physical Society, Division of Biological Physics, <http://www.aps.org/DBP/>

[20] Biophysical Society, <http://www.biophysics.org/biophys/society/biohome.htm>, especially the multivolume online biophysics textbook

[21] Protein Data Bank, <http://www.pdb.bnl.gov/>

[22] Nucleic Acid Database, <http://ndbserver.rutgers.edu:80/NDB/ndb.html>

8.15.3. Specific references

[23] R. H. Austin *et al.*, Biochemistry **13**, 5355–5373 (1975).

[24] J. Deisenhofer and H. Michel, EMBO J. **8**, 2149–2169 (1989).

[25] J. Deisenhofer *et al.*, Nature (London) **318**, 618–624 (1985).

[26] G. Feher, A. J. Hoff, R. A. Isaacson, and L. C. Ackerson, Ann. NY Acad. Sci. **244**, 239–259 (1975).

[27] T. H. Foster *et al.*, Radiation Res. **126**, 296–303 (1991).

[28] S. M. Gruner, in *ACS Adv. Chem. Ser. No. 235, Biomem. Electrochem.*, edited by M. Blank and I. Vodanoy (ACS Books, 1994).

[29] A. L. Hodgkin and A. F. Huxley, *J. Physiol.* **117**, 500–544 (1952).

[30] H. Flyvbjerg *et al.*, editors, *Physics of Biological Systems* (Springer-Verlag, 1997).

[31] B. Katz, *Nerve, Muscle and Synapse* (McGraw-Hill, New York, 1966).

[32] I. Lee, J. W. Lee, and E. Greenbaum, Phys. Rev. Lett. **79**, 3294–3297 (1997).

[33] S. Leikin, V. A. Parsegian, D. C. Rau, and R. P. Rand, Annu. Rev. Phys. Chem. **44**, 369–395 (1993).

[34] C. Luo and Y. Rudy, Circ. Res. **74**, 1091–1113 (1994).

[35] G. S. Manning, Quart. Revs. Biophys. **11**, 179–246 (1978).

[36] O. G. Mouritsen and O. S. Andersen, editors, in *Search of New Membrane Model* (Royal Danish Society, 1998).

[37] J. R. Norris, R. A. Uphaus, H. L. Crespi, and J. J. Katz, Proc. Natl. Acad. Sci. USA **68**, 625–628 (1971).

[38] B. J. Roth and J. P. Wikswo, Jr., Proc. IEEE **84**, 379–391 (1996).

[39] R. S. Spolar, J. R. Livingston, and M. T. Record, Jr., Biochemistry **31**, 3947–3955 (1992).

[40] S. J. Weiner *et al.*, J. Amer. Chem. Soc. **106**, 765–784 (1984).

[41] J. M. van Egerat and J. P. Wikswo, Jr., Biophys. J. **64**, 1287–1928 (1993).

9

Crystallography

George A. Jeffrey, Vicky Lynn Karen

Contents

List of Tables

List of Figures

9.1. HISTORICAL SKETCH

Prior to the discovery of x-ray diffraction by crystals by Friedrich, Knipping, and Laue [1] crystallography was concerned with the external morphology of crystals. Groth's five-volume *Chemische Krystallographie* published between 1906 and 1919 contains the results of very accurate measurements of interfacial angles of crystals of 7350 chemical compounds. Lack of knowledge about the atomic structure of crystals did not prevent the classical crystallographers from obtaining significant insight into the symmetry properties of crystals. The concepts of unit cell (Haüy, 1800), lattices (Bravais, 1840), and space groups (Barlow, 1883; Federov, 1890; Schoenflies, 1891) were well developed, albeit hypothetical, until W. H. and W. L. Bragg used Laue's discovery of x-ray diffraction by crystals for the experimental determination of crystal structures. [2] With the Braggs, the frontier of crystallography moved out of mineralogy into physics. There it remained, generally, with some excursions into physical metallurgy and material sciences, until the early 1950s, when the availability of the general-purpose digital computer made crystal structure determination an important component of chemistry. This development into chemistry was pioneered primarily by Linus Pauling and his colleagues at the California Institute of

Technology and by Monteath Robertson at the Royal Institution in London. An excellent example is Lipscomb's work on the boron hydrides, [3] following the first crystal structure of these compounds by Kasper, Lucht, and Harker. [4] By 1960, the work of Watson and Crick, [5] Hodgkin, [6] Perutz, [7] and Kendrew [8] had sufficient impact on the biological sciences that crystallography in the 1970s began to play the same crucial role in the science of molecular biology as it had for chemistry some twenty years earlier. Crystallography continues to be a valuable discipline for studying the atomic and electronic structures of minerals and for both crystal structural and texture studies in metallurgy and the material sciences.

The development of the "direct methods" for solving the crystal structure phase-problem, recognized by the Nobel Prize in 1986 to Karle and Hauptman made crystal structure analysis an indispensable tool for determining the configuration and conformation of molecules with molecular weights of less than 500. X-ray crystal structure analysis is the only source of the complete structure determination of macromolecules, such as proteins, nucleic acids, and polysaccharides. Charge density studies have begun to provide information concerning the electronic distribution that relates directly to the chemical reactivity of molecules.

The study of crystals by x-rays has been the major activity of crystallographers since the Laue experiment in the summer of 1912. The use of electrons [9] and neutrons [10] has developed complementary diffraction methodologies for studies in which they provide particular advantages. In the case of electron diffraction, these advantages are in the study of very small crystals and of crystal surfaces and for revealing the systematics of solid-state defect structures. In the case of neutrons, the advantages are for locating light atoms, especially hydrogen, with an accuracy comparable to that of heavier atoms, and for studying magnetic spin structures.

New tools for crystallographers include the synchrotron x-ray sources and spallation neutrons. It will be interesting to see which of the many facets of crystallography benefit most from these powerful pulsed-beam sources. The interplay between crystallography and the other disciplines is shown by the "citation map" in Figure 9.1. [11]

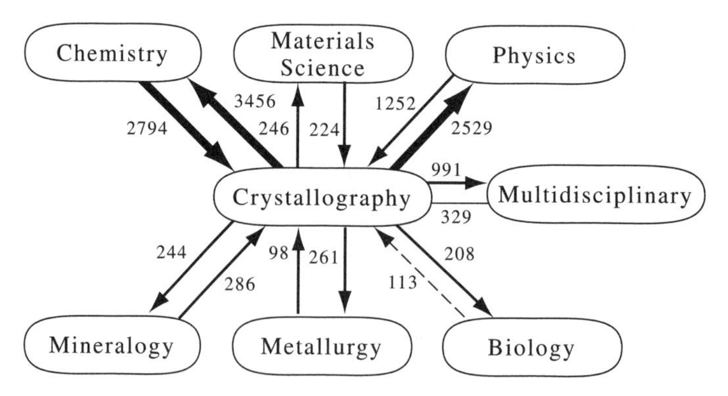

FIGURE 9.1. Citation map for crystallography and other disciplines with the number of citations between those shown.

Scientists in many fields now use crystallographic models to visualize, explain, and predict the behavior of chemicals, materials, and biological compounds. The data required for such models can be obtained from the databases described in the Appendix to this chapter.

9.2. CRYSTAL DATA AND SYMMETRY

9.2.1. Crystal system, space group, lattice constants, and structure type

The seven *crystal systems* are based on the principal axial symmetry of the 32 *point-group symmetries* of the crystal lattice, as shown in Table 9.1. The lattice divides the crystal structure into *unit cells*, which are defined by three concurrent cell edges, **a**, **b**, **c**, which, by convention, form a right-handed system. The corresponding scalar *lattice constants* are a, b, c, α, β, γ. Also commonly used in crystallography is the reciprocal lattice [12] $\mathbf{a}^*$, $\mathbf{b}^*$, $\mathbf{c}^*$, where

TABLE 9.1. Point-group symmetries (class), Laue symmetries, and lattice constants for the seven crystal systems.

System	Point-group symmetry, crystal class	Laue symmetry	Symmetry directions	Lattice constraints	Lattice constants
Triclinic (Tric)	$1, \bar{1}$	$\bar{1}$	None	None	a, b, c α, β, γ
Monoclinic (M)	$2, \bar{2}, 1/m$	$2/m$	b	$\alpha = \gamma = 90°$	a, b, c, β
Orthorhombic (O)	222 $mm2$ mmm	mmm	a, b, c	$\alpha = \beta = \gamma = 90°$	a, b, c
Tetragonal (Tetr)	$4, \bar{4}, 4/m$ $422, 4mm$ $\bar{4}2m, 4/mmm$	$4/m$ $4/mmm$	c, a, b $[110]$	$a = b$ $\alpha = \beta = \gamma = 90°$	a, c
Hexagonal (H)	$6, \bar{6}, 6/m$ $622, 6mm,$ $\bar{6}m2, 6/mmm$	$6/m$ $6/mmm$	c, a, b $[110]$	$a = b$ $\alpha = \beta = 90°$ $\gamma = 120°$	a, c
Rhombohedral (R) or Trigonal (Trig)	$3, \bar{3}$ $32, 3m$ $\bar{3}m$	$\bar{3}$ $\bar{3}m$	c $[110]$	$a = b = c$ $\alpha = \beta = \gamma$ $a = b \neq c$ $\alpha = \beta = 90°,$ $\gamma = 120°$	a, γ a, c
Cubic (C)	$23, m3$ $432, \bar{4}3m$ $m3m$	$m3$ $m3m$	a, b, c $[110], [111]$	$a = b = c$ $\alpha = \beta = \gamma = 90°$	a

$$\mathbf{a}\cdot\mathbf{a}^* = \mathbf{b}\cdot\mathbf{b}^* = \mathbf{c}\cdot\mathbf{c}^* = 1,$$

$$\mathbf{a}\cdot\mathbf{b}^* = \mathbf{a}\cdot\mathbf{c}^* = \mathbf{b}\cdot\mathbf{a}^* = \mathbf{b}\cdot\mathbf{c}^* = \mathbf{c}\cdot\mathbf{a}^* = \mathbf{c}\cdot\mathbf{b}^* = 0.$$

The corresponding scalar reciprocal-lattice constants are a^*, b^*, c^*, α^*, β^*, γ^*. The point-group symmetry of the lattice entails the specialization of the lattice constants shown in Table 9.1. The lattices are defined by a, b, c, α, β, γ for triclinic crystals; a, b, c, β for monoclinic; a, b, c for orthorhombic; a, c for tetragonal and hexagonal; a, α for rhombohedral; and a for cubic crystals. In the monoclinic system, the unique symmetry axis, by convention, is **b**. In the tetragonal and hexagonal systems, it is **c**. The *Laue symmetry*, of which there are 11 types, describes the point-group symmetry of the diffraction spectra.

Table 9.2 gives crystallographic data for selected elements, intermetallic phases, and inorganic and organic compounds. More extensive compilations of crystal data in print form are available elsewhere. [13–19]

The *Hermann-Mauguin (HM) point- and space-group symbols* are used by crystallographers. The space-group symbol starts with a lattice descriptor: P is primitive; A, B, C are (100), (010), (001) face centered; I is body centered; F is all faces centered. R is a rhombohedral lattice. It is followed by the symmetry operator in the direction of the principal symmetry axis, followed by the other symmetry-independent axes. The space-group symbol, given in Table 9.3, is that used in the *International Tables for X-ray Crystallography*, Vol. I. [20] It is the *short symbol*, containing only those symmetry elements necessary to generate the full symmetry of the space group. The order of precedence in use of symbols is glide and mirror planes, screw axes, and axes. In a few examples, a redundant symmetry element is introduced to indicate a distinction in the orientation of a symmetry element, as in $P\bar{6}m2$ and $P\bar{6}2m$. Further details are given in Int. Tables, Vol. I, p. 29. The position of the symbol denotes the direction of the symmetry element with respect to the axes **a**, **b**, **c**. A list of the diffraction symmetry for each space group is given in Int. Tables, Vol. I, pp. 349–352.

n-fold rotation axes are denoted by 1, 2, 3, 4, and 6, n-fold rotation-inversion axes are denoted by $\bar{1}$, $\bar{2}$, $\bar{3}$, $\bar{4}$, and $\bar{6}$. 1 implies absence of symmetry and $\bar{1}$ a center of symmetry; $\bar{2}$ is a mirror plane, also denoted by the symbol m. The symbol n/m denotes a mirror plane perpendicular to an n-fold rotation axis. Screw axes are denoted by 2_1, 3_1, 3_2, 4_1, 4_2, 4_3, 6_1, 6_2, 6_3, 6_4, and 6_5, where the subscript denotes the fractional translation in the direction of the screw axis. Glide planes are denoted by a, b, c, n, and d, indicating the translation direction.

There are 230 space groups, of which 50 can be uniquely identified from their diffraction symmetry, as can nine enantiomorphous pairs and two special pairs, leaving 158 space groups that cannot be determined by inspection of the diffraction patterns.

Most space groups are sparsely occupied. [21] Sixty-five percent of the organic compounds occur in six space groups, [22] $P\bar{1}$, $P2_1$, $P2_1/c$, $C2/c$, $P2_12_12_1$, and $Pbca$, of which only $P2_1$ and $P2_12_12_1$ occur in natural products.

The older *Schoenflies point-group notation* is still used by spectroscopists, inorganic and theoretical chemists, and some solid-state physicists. It is based on rotation axes and combinations of rotation axes and mirror planes, known as rotation-reflexion, or alternating

TABLE 9.2. Crystallographic data for selected crystalline materials.

Formula or name	Crystal system	HM space group	Z	Lattice constants a, b, c(Å), α, β, γ (deg)	Structure type and comments
				Elements	
As, α	R	$R\bar{3}m$	2	4.1320(2), 54.12(1)°	$A7$ There are 5 other allotropes
Au	C	$Fm3m$	4	4.0782(2)	$A1$ Face-centered cubic, cubic close packed (ccp)
B, rhomb-12	R	$R\bar{3}m$	12	5.057, 58.06°	Structure contains nearly regular icosahedra; 16 other allotropes have been reported
C, diamond	C	$Fd3m$	8	3.56688(15)	$A4$
C, hexag. diam	H	$P6_3/mmc$	4	2.52, 4.12	Found in meteorites, and made synthetically
C, graphite	H	$P6_3mc$	4	2.4612(2), 6.7090(12)	$A9$
Fe, α	C	$Im3m$	2	2.8664(2)	$A2$ Body-centered cubic
γ	C	$Fm3m$	4	3.6467(2)	$A1$
δ	C	$Im3m$	2	2.9315(2)	$A2$
Hg, α	R	$R\bar{3}m$	1	3.005, 70.53°	$A10$ There are 2 other allotropes
Mg	H	$P6_3/mmc$	2	3.209, 5.221	$A3$ Hexagonal close packed (hcp)
Mn, α	C	$I\bar{4}3m$	58	8.9129(6)	$A12$ There are 3 other allotropes
S_6	R	$R\bar{3}$	1	6.373, 115.2°	More than 50 allotropes of S have been described; many are doubtful
S_8	O	$Fddd$	16	10.46(1), 12.879(2), 24.478(5)	
Se, α	H	$P3_121$ $P3_221$	3	4.3655(10), 4.9576(24)	$A8$ There are 5 other allotropes
Sn, α	C	$Fd3m$	8	6.4892	$A4$
β	Tetr	$I4_1/amd$	4	5.8316(2), 3.1815(2)	$A5$
U, α	O	$Cmcm$	4	2.838(4), 5.868(2), 4.956(1)	$A20$
β	Tetr	$P4_2/mnm$	30	10.759, 5.656	
γ	C	$Im3M$	2	3.524	$A2$
				Intermetallic phases	
βCuZn	C	$Im3m$	1	2.9907 (871°C)	$A2$ Hume Rothery phases
β'CuZn	C	$Pm3m$	1	2.9539(47% Zn, 20°C)	$B2$ Hume Rothery phases
γCuZn	C	$I\bar{4}3m$	26	8.852 (63% Zn, 20°C)	$D8_2$ Hume Rothery phases
$MgCu_2$	C	$Fd3m$	8	7.034	$C15$ Laves phases
$MgNi_2$	H	$P6_3/mmc$	8	4.805, 15.77	$C36$ Laves phases
$MgZn_2$	H	$P6_3/mmc$	4	5.17, 8.50	$C14$ Laves phases
Mg_2Cu	O	$Fddd$	16	9.05, 18.21, 5.273	
AuCu	Tetr	$P4/mmm$	2	3.9512, 3.6798 (350°C)	$L1_o$
$AuCu_3$	C	$Pm3m$	1	3.7432 (20°C)	$L1_2$ Zintl phase
$AuMg_3$	H	$P6_3/mmc$	2	4.63, 8.44	DO_{18}
$CaCu_5$	H	$P6/mmm$	1	5.082, 4.078	$D2_d$
$AlFe_3$	C	$Fm3m$	4	5.780	DO_3
$AlCu_2Mn$	C	$Fm3m$	4	5.937	$L2_1$ Heusler alloys
Cu_5Zn_8	C	$I\bar{4}3m$	4	8.854 (65% Zn)	$D8_2$
Cr_3Ge	C	$Pm3n$	2	4.614	$A15$
Cr_5Si_3	Tetr	$I4/mcm$	4	9.170, 4.636	$D8_m$
$NaZn_{13}$	C	$Fm3c$	8	12.2836	$D2_3$
$Al_{12}W$	C	$Im3$	2	7.5803	
$Cr_{23}C_6$	C	$Fm3m$	4	10.638	$D8_4$
Fe_3C	O	$Pnma$	4	5.0787, 6.7297, 4.5144	DO_{11} Cementite

(continued)

TABLE 9.2. Continued

Formula or name	Crystal system	HM space group	Z	Lattice constants a, b, c(Å), α, β, γ (deg)	Structure type and comments
				Solid-state physics	
Fe_3O_4	C	$Fd3m$	8	8.3940	$H1_1$ Ferrimagnet, Magnetite
	O	$Imcm$	4	5.945, 8.388, 5.912 (−195°C)	
MnO	C	$Fm3m$	4	4.445	$B1$ Antiferromagnet, Magnanosite
MnF_2	Tetr	$P4_2/mnm$	2	4.8734, 3.3099	$C4$ Rutile structures antiferromagnet
$CsNiCl_3$	H	$P6_3/mmc$	2	7.17, 5.94	$BaNiO_3$ structure
$CrBr_3$	R	$R\bar{3}$	2	7.05, 52°36′	Ferromagnet
$MnAu_2$	Tetr	$I4/mmm$	2	3.197, 7.871	$C11_b$ Helimagnet
αFe_2O_3	R	$R\bar{3}c$	2	5.4271, 55.26°	$D5$ Weak ferromagnet
$KMnF_3$	C	$Pm3m$	1	4.190	Weak ferromagnet
$YFeO_3$	C	$Pm3m$	1	3.785	Weak ferromagnet
$CuCl_2 \cdot 2H_2O$	O	$Pbmn$	2	7.38, 8.04, 3.72	$E2_1$ 4-sublattice antiferromagnet
$\alpha CoSO_4$	O	$Cmcm$	4	5.198, 7.871, 6.522	4-sublattice antiferromagnet
$BaTiO_3$	C	$Pm3m$	1	4.0118 (201°C)	
	Tetr	$P4mm$	1	3.9947, 4.0336 (25°C)	Ferroelectric
	O	$Amm2$	2	3.990, 5.669, 5.682 (5°C)	
$SrTiO_3$	C	$Pm3m$	1	3.9051	Ferroelectric
KH_2PO_4	Tetr	$I\bar{4}2d$	4	7.453, 6.959	Ferroelectric
$PbNb_2O_6$	Tetr	$P4/*b*$	5	12.46, 3.907 (600°C)	Paraelectric
$BaMnO_4$	O	$Pbnm$	4	7.304, 9.065, 5.472	Barite-type structure
Cr_2O_3	R	$R\bar{3}C$	2	5.35, 55.0°	$D5_1$ Magnetoelectric
$NaKC_4H_4O_6 \cdot 4H_2O$	O	$P2_12_12$	4	11.93, 14.30, 6.17	Rochelle salt, anomalous dielectric
$YBa_2Cu_3O_{6.9}$	O	$Pmmm$	1	3.8218, 3.8913, 11.677	High T_c superconductors
Y_2BaCuO_5	O	$Pnma$	4	12.176, 5.655, 7.130	"
$YBa_2Cu_3O_{7-x}$	T	$P4/mmm$	1	3.8683, 11.708	"
$Y_2BaCu_3O_{6-x}$	T	$P\bar{4}m2$	1	3.859, 11.71	"
$La_{1.85}Ba_{0.15}CuO_4$	T	$I4/mmm$	1	3.7817, 13.2487	"
				Inorganics (*see also* **Minerals**)	
CsCl	C	$Pm3m$	1	4.123	$B2$
H_2O I	H	$P6_3/mmc$	4	4.48, 7.31	Normal ice
Ic	C	$Fd3m$	8	6.35	Low-temperature phase
II	R	$R\bar{3}$	12	7.78, 113.1°	Ices II–VII are high-pressure phases
III	Tetr	$P4_{1,3}2,2$	12	6.73, 6.83	
IV					
V	M	$A2/a$	28	9.22, 7.54, 10.35, 109.2°	
VI	Tetr	$P4_2/nmc$	10	6.27, 5.79	Self-clathrate
VII	C	$Pn3m$	2	3.41	Self-clathrate
VIII	C		32	9.70 (−30°)	
H_2O_2	Tetr	$P4_12_12$	4	4.06, 8.00	Below −0.9°C
N_2O_2	M	$P2_1/n$	2	5.811, 3.96, 6.55, 114.90°	At −175°C

(continued)

TABLE 9.2. Continued

Formula or name	Crystal system	HM space group	Z	Lattice constants a, b, c(Å), α, β, γ (deg)	Structure type and comments
6X, 2Y, $46H_2O$	C	$Pm3n$	1	~ 12.0	Type I, gas hydrate, X is a molecule not larger than CH_3Cl
8X, 16Y, $136H_2O$	C	$Fd3m$	1	~ 17.3	Type II, gas hydrate, Y is a molecule not larger than CCl_4
Fe_2Ti_4O	C	$Fd3m$	16	11.275	$E9_3$
$Na_2O \cdot 11Al_2O3$	H	$P6_3/mmc$	2	5.595, 22.49	β-Alumina
$Na_2SO_4 \cdot 10H_2O$	M	$P2_1/c$	4	11.51, 10.38, 12.83, 107°45′	Glaubers salt
$KAl(SO_4)_2 \cdot 12H_2O$	C	$Pa3$	4	12.158	Typical of alums
				Minerals	
NaCl	C	$Fm3m$	4	5.64056 (26°C)	$B1$ Halite
ZnS	H	$P6_3mc$	2	3.835, 6.268	$B3$ Wurtzite
ZnS	C	$F\bar{4}3m$	4	5.4093 (26°C)	$B4$ Sphalerite
NiAs	H	$P6_3/mmc$	2	3.602, 5.009	$B8_1$ Niccolite
CaF_2	C	$Fm3m$	4	5.463	$C1$ Fluorite
FeS_2	C	$Pa3$	4	5.40667 (26°C)	$C2$ Pyrite
FeS_2	O	$Pnnm$	2	4.436, 5.414, 3.381	$C18$ Marcasite
SiO_2	H	$P3_121$	3	4.9027, 5.3934	α Quartz
	H	$P6_322$	3	5.45, 4.99	β Quartz
	H	$P6_3/mmc$	4	5.03, 8.22	β Tridymite
	Tetr	$P4_12_12$	4	4.97, 6.92	α Cristobalite
	C	$Fd3m$	8	7.05	β Cristobalite
TiO_2	Tetr	$P4_2/mnm$	2	4.5937, 2.9581	$C4$ Rutile
TiO_2	Tetr	$I4/amd$	4	3.785, 9.514	Anatase
TiO_2	O	$Pbca$	8	9.184, 5.447, 5.145	Brookite
Cu_2O	O	$Pn3m$	2	4.2696 (26°C)	$C3$ Cuprite
Al_2O_3	R	$R\bar{3}c$	2	5.128, 55.28°	$D5_1$ Corundum
$BaSO_4$	O	$Pnma$	4	8.8701, 5.4534, 7.1507	Barite
$CaCO_3$	R	$R\bar{3}c$	2	6.75, 46°5′	Calcite
$CaCO_3$	O	$Pnam$	4	5.72, 7.94, 4.94	Aragonite
$CuFeS_2$	Tetr	$I\bar{4}2d$	4	5.24, 10.30	$E1_1$ Chalcopyrite
Al_2MgO_4	C	$Fd3m$	8	8.0800 (26°C)	$H1_1$ Spinel
Fe_2SiO_4	O	$Pbnm$	4	4.820, 10.485, 6.093	Olivine
$MgAl_2O_4$	C	$Fd3m$	8	9.083	Spinel
$ZrSiO_4$	Tetr	$I4_1/amd$	4	6.607, 5.982	Zircon
$CaSO_4$	O	$Amma$	4	6.991, 6.996, 6.238	Anhydrite
$CaSO_4 \cdot 2H_2O$	M	$C2/c$	4	6.284, 15.208, 5.678, 114.09°	Gypsum
$CaWO_4$	Tetr	$I4_1/a$	4	5.243, 11.376	Scheelite
$FeWO_4$	M	$P2/c$	2	4.75, 5.72, 4.97, 90.17°	Wolframite
$Ca_6F(PO_4)_3$	H	$P6_3m$	2	9.3684, 6.8841	Fluorapatite
$Al_2Mg_3(SiO_4)_3$	C	$Ia3d$	8	11.459	A synthetic garnet
$Be_3Al_2(SiO_3)_6$	H	$P6/mcc$	2	9.212, 9.187	Beryl
$[Al(F, OH)]_2SiO_4$	O	$Pbnm$	4	4.6499, 8.7968, 8.3909	Topaz
				Organics	
C_2H_6	H	$P6/mmc$	2	8.19, 4.46	Ethane
C_2H_4	O	$Pnnm$	2	4.87, 6.46, 4.15	Ethylene
C_8H_8	R	$R\bar{3}$	1	5.340, 74°26′	Cubane
CH_3OH	M	$P2_1/m$	2	4.53, 4.91, 4.69, 90.0°	α Methanol
CH_3OH	O	$Cmcm$	4	4.67, 7.24, 6.43	β Methanol
C_6H_6	O	$Pbca$	4	7.034, 7.460, 9.666	Benzene −3°C

(continued)

TABLE 9.2. Continued

Formula or name	Crystal system	HM space group	Z	Lattice constants a, b, c(Å), α, β, γ (deg)	Structure type and comments
$C_{10}H_8$	M	$P2_1/c$	2	7.832, 5.940, 8.108, 114.32°	Naphthalene
$C_{14}H_{10}$	M	$P2_1/c$	2	9.379, 6.003, 8.496, 103.07°	Anthracene
$C_{10}H_{16}$	C	$F\bar{4}3m$ (or $Fm3m$)	4	9.426	Adamantine
	Tetr	$P\bar{4}2_1c$	2	6.60, 8.81 (−65°)	
CH_4N_2O	Tetr	$P\bar{4}2_1m$	2	5.670, 4.726	Urea
C_2H_5NO	R	$R3c$	6	7.914, 93.34°	Acetamide
$C_5H_{12}O_4$	Tetr	$I\bar{4}$	2	6.10, 8.73	Pentaerythritol
$C_6H_{16}N_2$	O	$Pbca$	4	6.94, 5.77, 19.22	Hexamethylenediamine
$C_2H_2O_4$	M	$P2_1/c$	2	5.30, 6.09, 5.51, 115°30′	Oxalic acid
$C_2H_2O_4$	M	$P2_1/n$	2	6.119, 3.604, 12.051, 106°12′	Oxalic acid dihydrate
$C_4H_6O_6$	M	$P2_1$	2	7.72, 6.00, 6.20, 100°10′	*D*-Tartaric acid
$C_2H_5NO_2$	M	$P2_1/n$	4	5.0835, 11.820, 5.458, 111.95°	α Glycine
$C_{10}H_{14}$	M	$P2_1/a$	2	11.57, 5.77, 7.03, 112.3°	1,2,4,5-Tetramethylbenzene
$C_{12}H_{18}$	Tric	$P\bar{1}$	1	8.92, 8.86, 5.30, 44°27′, 116°43′, 119°34′	Hexamethylbenzene
$C_{25}H_{20}$	Tetr	$P\bar{4}2_1c$	2	10.87, 7.23	Tetraphenylmethane
$C_{24}H_{12}$	M	$P2_1/a$	2	16.10, 4.695, 10.15, 110°8′	Coronene
$C_6H_{12}O_6$	O	$P2_12_12_1$	4	10.36, 14.84, 4.93	α Glucose
$C_6H_8O_6$	M	$P2_1$	4	16.95, 6.32, 6.38, 102.5°	Ascorbic acid
$C_{16}H_{17}N_2O_4SNa$	M	$P2_1$	2	8.48, 6.33, 15.63, 94.2°	Penicillin
				Organic macromolecules	
Cellulose I	M	$P2_1$	8	16.34, 15.72, 10.38, 97°	
Cellulose II	M	$P2_1$	2	8.01, 9.04, 10.36, 117.1°	Fortisan
α-Chitin	O	$P2_12_12_1$	4	4.74, 18.86, 10.32	Lobster tendon
β-Chitin	M	$P2_1$	2	4.85, 9.26, 10.38, 97.5°	
DNA, A-form	M	$C2$	4	22.1, 40.4, 28.1, 97.1°	SNa, salt, fiber axis c
Transfer RNA, yeast phenylalanine	O	$P2_122_1$	4	330, 560, 161	Molec. wt. 23 788
Myoglobin	M	$P2_1$	2	64.6, 31.3, 34.8, 105.5°	Sperm whale
Haemoglobin	M	$C2_1$	2	108.95, 63.51, 54.92, 110°53′	Horse
Satellite tobacco necrosis virus	M	$C2$	4	318.4, 305.0, 185.3, 94°37′	Molec. wt. 1.7×10^6
DNA, β-form	O		4	22.7, 31.2, 33.7	Salt
Polyoma virus	C	$I23$		490	Molec. wt. 3.6×10^6
Southern bean mosaic virus	R	$R32$	32	757.5, 63.6°	Molec. wt. asymmetric unit, 9×10^6

TABLE 9.3. Conversion from Schoenflies to Hermann-Mauguin point-group notations. In crystals, $n = 1, 2, 3, 4, 6$.

Schoenflies	HM	Schoenflies	HM
C_n	n	$S_2 \equiv C_i$	$\bar{1}$
$C_s \equiv C_1^h (\equiv C_1^v \equiv S_1)$	m	S_4	$\bar{4}$
C_n^h	n/m	$S_4 \equiv C_3^i$	$\bar{3}$
C_2^v	mm	D_2	222
C_3^v	$3m$	D_3	32
C_4^v	$4mm$	D_4	42
C_6^v	$6mm$	D_6	62
D_2^d	$\bar{4}2m$	D_2^h	mmm
D_3^d	$\bar{3}m$	D_3^h	$\bar{6}m (\equiv 3/mm)$
T	23	D_4^h	$4/mmm$
T_h	$m3$	D_6^h	$6/mmm$
T_d	$\bar{4}3m$	O	43
		O_h	$m3m$

axes. Table 9.3 gives the conversion from Schoenflies to Hermann-Mauguin notation. The Schoenflies space-group notation is obsolete.

The lattice constants, in Å, define the unit cell with volume V in Å [3]. This is related to the calculated crystal densities, by

$$Vd_c = ZM/N_A,$$

where d_c is the calculated density of the crystal in $\mathrm{Mg/m^3}$; Z is the number of asymmetric formula units in the unit cell, usually an integral; M is the molecular weight of the formula unit; and N_A is Avogadro's constant, 6.02214×10^{23}.

Experimental crystal densities d_m are generally measured by flotation in mixtures of suitable liquids of known densities. Suggested liquids and their densities at 25°C are given in Int. Tables, Vol. III, p. 19.

9.2.2. Reduced cells

The reduced cell is a unique, primitive cell that is based on the three shortest noncoplanar vectors of the lattice and satisfies a specified set of mathematical conditions. The reduced cell is often chosen for triclinic crystals. For a cell to be reduced, the cell must be in normal representation and both the main and special conditions for reduction must be satisfied.

For a unit cell represented by

$$\mathbf{P} = \begin{pmatrix} \mathbf{a}\cdot\mathbf{a} & \mathbf{b}\cdot\mathbf{b} & \mathbf{c}\cdot\mathbf{c} \\ \mathbf{b}\cdot\mathbf{c} & \mathbf{a}\cdot\mathbf{c} & \mathbf{a}\cdot\mathbf{b} \end{pmatrix},$$

the *positive reduced cell* (type I) has all angles $< 90°$:

$$\mathbf{a}\cdot\mathbf{a} \le \mathbf{b}\cdot\mathbf{b} \le \mathbf{c}\cdot\mathbf{c},$$

$$\mathbf{b}\cdot\mathbf{c} \le \tfrac{1}{2}\mathbf{b}\cdot\mathbf{b}, \quad \mathbf{a}\cdot\mathbf{c} \le \tfrac{1}{2}\mathbf{a}\cdot\mathbf{a}, \quad \mathbf{a}\cdot\mathbf{b} \le \tfrac{1}{2}\mathbf{a}\cdot\mathbf{a}.$$

The *negative reduced cell* (Type II) has all angles $\ge 90°$:

$$\mathbf{a}\cdot\mathbf{a} \le \mathbf{b}\cdot\mathbf{b} \le \mathbf{c}\cdot\mathbf{c}, \quad |\mathbf{b}\cdot\mathbf{c}| \le \tfrac{1}{2}\mathbf{b}\cdot\mathbf{b}, \quad |\mathbf{a}\cdot\mathbf{c}| \le \tfrac{1}{2}\mathbf{a}\cdot\mathbf{a},$$

$$|\mathbf{a}\cdot\mathbf{b}| \le \tfrac{1}{2}\mathbf{a}\cdot\mathbf{a}, \quad |\mathbf{b}\cdot\mathbf{c}| + |\mathbf{a}\cdot\mathbf{c}| + |\mathbf{a}\cdot\mathbf{b}| \le \tfrac{1}{2}(\mathbf{a}\cdot\mathbf{a} + \mathbf{b}\cdot\mathbf{b}).$$

Other special conditions and the transformations from an unreduced to a reduced cell are given in Int. Tables (1969 edition), Vol. I, pp. 530–535.

9.2.3. Physical properties of crystals

The properties of the crystal classes according to the type of piezoelectric moment and optical activity are given in Int. Tables, Vol. I, p. 42. Of the 32 crystal classes, 21 are noncentrosymmetrical. Optical activity occurs in 15 classes. Pyroelectricity can theoretically occur only when there is a unique polar axis in the crystal, but owing to the accompanying piezoelectricity of the measurement, it can occur in all noncentrosymmetrical crystals. Piezoelectricity can occur in all noncentrosymmetrical crystals, except those in class 432, where the moduli are all zero owing to the high symmetry.

The relations between electrical, magnetic, thermal, optical, and elastic stress and strain variables and their scalar and tensorial properties are described in detail in Sec. 9a-2 of the *American Institute of Physics Handbook*, 3rd ed. (McGraw-Hill, New York, 1972).

9.3. CRYSTAL DIFFRACTION

9.3.1. Conditions for diffraction

Diffraction occurs if the wavelength λ of the incident radiation on a crystal is comparable with the periodicity of the atomic structure expressed by the lattice constants. Thus, crystals diffract x-rays, electrons, and neutrons with wavelengths from 0.1 to 10 Å. Crystallography uses the *angstrom*, Å, as the unit of length. The conversion factor to the *kx unit*, 10^{-10} m, is kx $= 1.002\,077\,6(54)$. X-ray wavelengths calculated from excitation potential from $\lambda = 12.398\ 10$ keV are given in Int. Tables, Vol. IV, pp. 5–43.

The condition for the occurrence of a diffraction spectrum, hkl, is given by

$$\mathbf{d}^*_{hkl} = \frac{1}{d_{hkl}} = \frac{2\sin\theta}{\lambda} = \mathbf{h}a^* + \mathbf{k}b^* + \mathbf{l}c^*,$$

where $\mathbf{d}^*_{hkl}$ is the reciprocal-lattice vector, or *scattering vector*, of the spectrum hkl (also referred to as ρ_{hkl} and σ_{hkl}). [27, 28] d_{hkl} is the spacing of the hkl crystal planes, θ is the Bragg reflection angle, and 2θ is the angle between incident beam $\mathbf{s}_0$ and diffracted beam $\mathbf{s}$. The direction of the diffracted beam $\mathbf{s}$ relative to that of the incident beam $\mathbf{s}_0$ and the reciprocal-lattice vector $\mathbf{d}^*_{hkl}$ is given by $\mathbf{s} - \mathbf{s}_0 = \mathbf{d}^*_{hkl}$. This is conveniently represented graphically by the Ewald construction [23] shown in Figure 9.2.

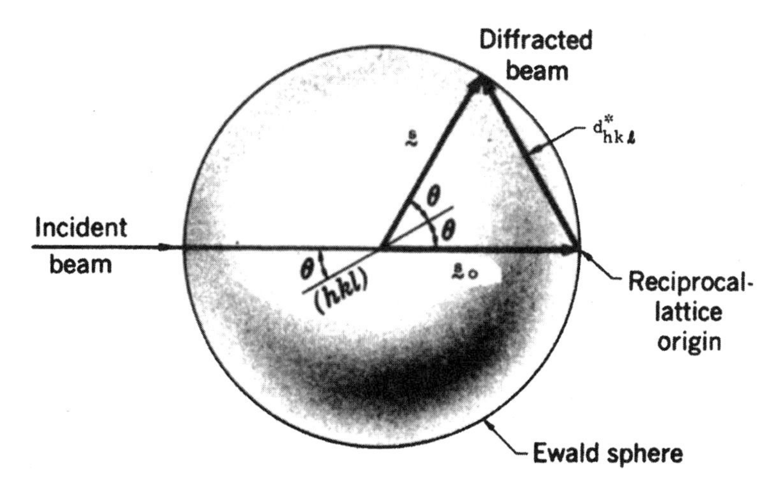

FIGURE 9.2. Ewald reciprocal-lattice condition for x-ray diffraction with monochromatic radiation.

For a particular wavelength λ, the number of diffraction spectra that can be observed is limited to those lying within the *sphere of diffraction* of radius $2\lambda^{-1}$. The number of diffraction spectra is therefore $32\pi V_p/3\lambda^3$, where V_p is the volume of the primitive unit cell.

Diffraction will occur for a stationary crystal, when the value of λ varies. This is the *Laue method.* A more common procedure is to use *monochromatic radiation*, with fixed λ, and oscillate, rotate, or precess the crystal. In these moving-crystal methods, diffraction occurs when the reciprocal-lattice point d^*_{hkl} passes through the Ewald sphere. Both the reciprocal-lattice point and the surface of the sphere are finite, the former because of the mosaic structure and size of the diffracting crystal, and the latter because of the finite width of monochromatic K wavelengths used ($\Delta\lambda/\lambda \approx 10^{-4}$); the substructure of the radiation, e.g., $\lambda_{\mathrm{Cu}K\alpha_1}$, $\lambda_{\mathrm{Cu}K\alpha_2}$; and the finite size of the focal spot of the x-ray tube or monochromator.

The most useful theory for relating the intensities of the x-ray diffracted spectra to the atomic coordinates and thermal parameters of a crystal structure is the *kinematical theory*. This classical theory is based on a model in which the radiation is represented as a wave that transverses the whole crystal. The amplitude phase differences between the radiation scattered at different points in the crystal depend only on the difference in length of the paths of the incident and diffracted waves to and from those points. This theory applies only to perfectly *mosaic crystals*. Departures from ideality due to the crystal being too perfect, or insufficiently mosaic, are treated as corrections. For diffraction by *perfect crystals*, the *dynamical theory* [24] is more appropriate. Perfection is as hard to come by in crystals as in humans, [25] and the dynamical theory is applicable only to a limited class of crystals.

In the kinematical theory, the intensity of the diffracted spectra depends on the wavelength of the incident radiation, the crystal structure and the nature of the diffracting specimen, the direction of diffraction, and the experimental conditions of the measurement. Intensity measurements are generally made with monochromatic radiation produced by means of appropriate filters, balanced filters, or crystal monochromators. The properties of filters and of various monochromating crystals are given in Int. Tables, Vol. III, pp. 73–87.

For a small crystal, volume V, completely bathed in the incident x-ray beam,

$$P_{hkl} = Q_{hkl}V = I_{hkl}\omega/I_0,$$

where P_{hkl} is the integrating diffracting power of the crystal, with the dimensions of area, Q_{hkl} is the integrated diffracting power per unit volume, I_0 and I_{hkl} are the energy per unit area per unit time of the incident and diffracted beams, and ω is the constant angular velocity with which the reciprocal-lattice point passes through the *Ewald sphere*.

For extended faces of crystals or powder specimens that are larger than the incident beam, a more useful expression is

$$I'_{hkl} = \int_{\theta-\epsilon}^{\theta+\epsilon} R_{hkl}(\theta)\, d\theta,$$

where R_{hkl} is the ratio of the power (energy per second) of the diffracted beam to that of the incident beam as the crystal moves through the diffracting position, starting at $\theta - \epsilon$ and finishing at $\theta + \epsilon$, beyond which no diffraction occurs. I' is dimensionless.

For x-ray diffraction by a small crystal, volume V, entirely in the x-ray incident beam,

$$I_{hkl} = \frac{I_0}{\omega} = \frac{1+\cos^2 2\theta}{2\sin 2\theta}\left(\frac{e^2}{4\pi\epsilon_0 mc^2}\right)^2 N^2\lambda^3 F_{hkl}^2 V,$$

where N is the number of unit cells per unit volume of the crystal. The θ-dependent factors are the Lorentz factor $1/\sin 2\theta$ and the polarization factor $(1 + \cos^2 2\theta)/2$. F_{hkl} is the structure factor for the diffracted spectrum hkl. The classical radius of an electron is $e^2/4\pi\epsilon_0 mc^2$. For a particular wavelength and crystal, the quantity

$$(4\pi\epsilon_0)^{-1}e^4m^{-2}c^{-4}\lambda^3N^2V$$

is a constant. It is generally not calculated, since for a given crystal and radiation, the absolute values of I_{hkl} are rarely measured experimentally. Note, however, that the diffracting power is proportional to λ^3; therefore, Cu$K\alpha$ radiation diffracts ten times more efficiently than does Mo$K\alpha$. Very small crystals with very large unit cells, e.g., proteins, diffract more poorly than do large crystals with small unit cells, because of N^2. Because of m^{-2}, the scattering of x-rays by the protons is negligible.

Neutron diffraction uses wavelengths similar to those of x-rays. The scattering is by the atomic nuclei, and the constant term in the intensity expression is $m^2\lambda^3h^{-2}N^2$. The kinematic theory can be used in the same way as for x-ray diffraction, except that the departures from the ideal theory are greater.

In *electron diffraction*, the wavelengths provided in Int. Tables, Vol. IV, p. 174, depend on the accelerating voltage. They are an order of magnitude shorter. The scattering involves both atomic nuclei and electrons.

For qualitative (rough) crystal structure determination, the kinematic theory can be used, or a compromise between the kinematical and two-beam dynamical theory, wherein $I_{hkl} \propto | F_{hkl} |^{\alpha}$, where $1 < \alpha < 2$. [26] For quantitative crystal structure analyses, n-beam dynamical theory must be used. [27] Because of the dynamical scattering effects, these methods are applicable only to very thin crystals of light-atom compounds with carefully controlled morphologies. [28]

The principal formulas for x-ray diffraction experiments are given in Int. Tables, Vol. II, p. 314.

The *polarization factor* $(1 + \cos^2 2\theta)/2$ is for an unattenuated or attenuated, i.e., filtered, x-ray beam. If a crystal monochromator is used, this factor becomes, for an ideally imperfect monochromating crystal,

$$\frac{\cos^2 2\theta_m + \cos^2 2\theta}{1 + \cos^2 2\theta_m}.$$

For a perfect monochromating crystal, it is

$$\frac{\cos 2\theta_m + \cos^2 2\theta}{1 + \cos^2 2\theta_m}$$

where θ_m is the Bragg angle at the monochromator. These differences are not significant for Mo$K\alpha$ radiation ($< 5\%$ for $0 \leq 2\theta \leq 90°$), but they may be significant for very accurate measurements with Cu$K\alpha$ radiation.

The *Lorentz factor* $1/\sin 2\theta$ depends on the angular velocity with which the reciprocal-lattice point hkl passes through the Ewald reflecting sphere. Its form varies with the geometry of the instrument, which moves the crystal and records the diffraction spectra. The Lorentz factors for different methods of recording diffraction spectra are given in Int. Tables, Vol. III, pp. 266–267.

Polarization, Lorentz, and other angle factors are usually combined in the computer programs that reduce intensities to structure amplitudes. Tabulated values are given in Int. Tables, Vol. II, pp. 268–273.

9.3.2. Single-crystal diffractometer [29, 30]

The standard x-ray or neutron diffractometer used in the crystallography is a four-angle instrument. The angles θ, Φ, χ, and ω are defined in Figure 9.3.

The *diffraction plane* is defined by the source (focal spot), the crystal, and the detector. The angle subtended by the source and the detector at the crystal is $180° - 2\theta$, and the bisector of this angle is the *diffraction vector*. Rotation about the diffraction vector is denoted by ψ.

The *orientation matrix* **U** is such that

$$\mathbf{A}^* = \mathbf{U}\mathbf{A}_G,$$

when

$$A^* = \begin{pmatrix} \mathbf{a}^* \\ \mathbf{b}^* \\ \mathbf{c}^* \end{pmatrix},$$

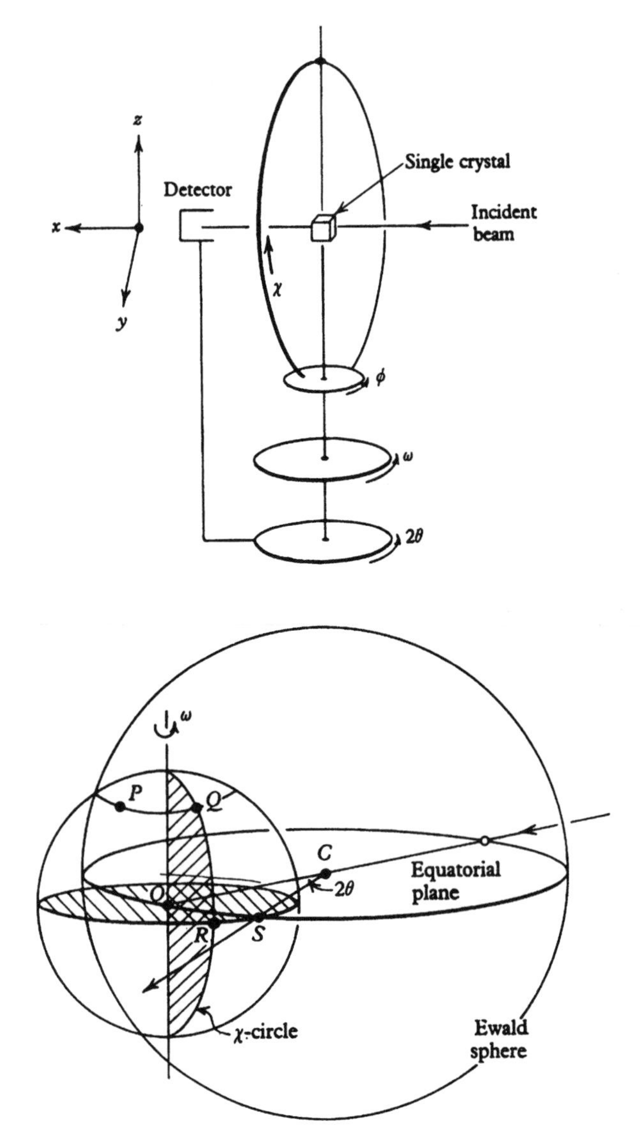

FIGURE 9.3. Geometry of four-circle diffractometer. Reciprocal-lattice point P can be moved to reflecting position S on the Ewald sphere as follows: $P \overset{\omega}{\to} Q \overset{\chi}{\to} R \overset{\omega}{\to} S$.

with metric

$$G^{-1} = \begin{pmatrix} a^*a^* & a^*b^* & a^*c^* \\ a^*b^* & b^*b^* & b^*c^* \\ a^*c^* & b^*c^* & c^*c^* \end{pmatrix};$$

$$A_g = A_D$$

when (Φ, χ, ω) are zero, where A_g refers to the orientation of the crystal on its goniometer head and A_D is the orientation of the diffractometer axes.

The search for crystal diffraction spectra, determination of the orientation matrix, and indexing and determination of the lattice parameters is a computer software component supplied with or developed for the particular computer-controlled diffractometer. The formulas for calculating setting angles and determining the orientation matrix are given in Int. Tables, Vol. IV, pp. 278–884.

9.3.3. Absorption

The diffracted intensity I_{hkl} is reduced by a transmission factor A_{hkl} that is less than unity (or absorption factor A^*_{hkl}):

$$A_{hkl} = \frac{1}{V}\int \exp[-\mu(r_\alpha - r_\beta)]\,dV,$$

where μ is the linear absorption coefficient in cm^{-1}, and r_α and r_β are the path lengths in the crystal transversed by the incident and diffracted beams to and from the element of volume dV. The *linear absorption coefficient* is independent of the physical state of the material and is calculated from the *mass absorption coefficient* μ/ρ, where ρ is the density. To a good approximation, mass absorption coefficients are additive properties given by

$$\mu/\rho = \sum_i g_i(\mu/\rho)_i,$$

where g_i is the mass fraction contributed by the element i whose mass absorption (attenuation) coefficient is $(\mu/\rho)_i$. For a material of known formula weight, the most convenient formula is

$$\mu = \frac{Z}{V}\sum_i(\mu/\rho)_i,$$

where the summation is over one formula weight in the unit cell. Values of the x-ray mass atomic absorption coefficient (in $cm^2\,g^{-1}$), for the most commonly used radiations, Cu$K\alpha$ and Mo$K\alpha$, are given in Table 9.4. Those for other wavelengths are given in Int. Tables, Vol. IV, pp. 61–66.

9.3.4. X-ray absorption corrections

X-ray absorption corrections are very important in accurate x-ray crystal structure analysis. The quantity A_{hkl} is calculated by analytical procedures, [31, 32] requiring as input μ, (hkl), and Δ_{hkl}, where Δ_{hkl} is the perpendicular distance of the hkl face to an origin *within* the crystal; hkl need not be integral, so that noncrystallographic faces can be defined. The determination of Δ_{hkl} requires careful measurements of the crystal using an optical goniometer or the optics of a diffractometer. If the faces can be indexed, the calculated face normals can be refined by least-squares fit to the known axial ratios.

TABLE 9.4. Mass attenuation coefficients (μ/σ in $cm^2\ g^{-1}$) of the atoms for $\left.\begin{matrix}CuK\alpha\\MoK\alpha\end{matrix}\right\}$ radiation

IA																	VIIIA
H																	He
0.391																	0.284
0.173	IIA											IIIA	IVA	VA	VIA	VIIA	0.202
Li	Be											B	C	N	O	F	Ne
0.477	1.001											2.142	4.219	7.142	11.01	15.95	22.13
0.197	0.245											0.345	0.515	0.790	1.147	1.584	2.209
Na	Mg											Al	Si	P	S	Cl	Ar
30.30	40.88											50.23	65.32	77.28	92.53	109.2	119.5
2.939	3.979	IIIB	IVB	VB	VIB	VIIB		VIII		IB	IIB	5.041	6.23	7.87	9.63	11.66	12.62
K	Ca	Sc	Ti	V	Cr	Mn	Fe	Co	Ni	Cu	Zn	Ga	Ge	As	Se	Br	Kr
148.4	171.6	186.0	202.4	222.6	252.1	272.5	104.4	338.6	48.8	51.5	59.5	62.1	67.9	75.7	82.9	90.3	97.0
16.20	19.00	21.04	23.25	25.24	29.25	11.86	17.74	41.0	47.2	49.3	55.5	56.9	60.5	66.0	68.8	74.7	79.1
Rb	Sr	Y	Zr	Nb	Mo	Tc	Ru	Rh	Pd	Ag	Cd	In	Sn	Sb	Te	I	Xe
106.3	115.3	127.1	136.8	168.8	158.3	167.7	180.8	194.1	205.0	218.1	229.3	242.1	253.3	266.5	273.4	291.7	309.8
83.0	88.0	97.6	16.1	16.97	18.44	19.78	21.11	23.05	24.4	26.38	27.73	29.11	31.18	33.01	33.92	36.33	38.31
Cs	Ba	La	Hf	Ta	W	Re	Os	Ir	Pt	Au	Hg	Tl	Pb	Bi	Po	At	Rn
325.6	336.1	353.5	157.7	161.5	170.5	178.1	181.8	192.9	198.2	207.8	216.2	222.2	232.1	242.9			263.7
29.5	31.0	45.4	59.7	89.5	95.8	98.7	100.2	101.4	108.6	111.3	114.7	119.4	122.8	125.9			117.2
Fr	Ra	Ac															
				Ce	Pr	Nd	Pm	Sm	Eu	Gd	Tb	Dy	Ho	Er	Tm	Yb	Lu
				378.8	402.2	417.9	441.1	453.5	417.9	426.7	321.9	336.6	128.4	134.1	140.2	144.7	152.0
				48.6	50.8	53.3	55.5	58.0	61.2	62.8	66.8	68.9	72.1	75.4	79.0	80.2	84.2
				Th	Pa	U	Np	Pu	Am	Cm	Bk	Cf	Es	Fm	Md	No	Lr
				306.8		305.7		352.9									
				99.5		96.7		48.8									

9.3.5. Extinction

The intensity formulas given above are based on the kinematic theory of x-ray diffraction by a crystal. This assumes that the crystal is ideally mosaic, i.e., that the crystal lattice is not coherent over large regions, and consequently, the diffracted amplitudes within each mosaic block are small enough that interaction between the incident and diffracted waves can be neglected. If this assumption does not hold, the dynamical theory of crystal diffraction may be more appropriate.

In practice, crystals that are perfect enough for application of the dynamical theory are rare. Since the dynamical theory gives smaller values for I_{hkl} than does the kinematic theory ($I_{hkl} \approx | F_{hkl} |$), deviations from the *ideally mosaic model* are therefore treated as *extinction corrections* to the kinematic theory. [33]

If the mosaic blocks are too large, interference occurs between the diffracted and incident beams within each block. This is known as *primary extinction*. If the mosaic blocks are small, but so well aligned that several blocks are simultaneously in the reflecting position for the same incident x-ray beam, then lower blocks receive incident beams of lower intensity. This is referred to as *secondary extinction*.

Electrons with the wavelengths used in diffraction experiments are very strongly absorbed, and diffraction by transmission is only possible with thin crystals (less than ≈ 500 Å). The small penetration of low-energy electrons is used to study the surfaces of crystals by the LEED (low-energy electron diffraction) technique. [34, 35]

Neutron absorption coefficients [10] are generally negligible (i.e., $\mu/\rho < 0.5$), with the exception of elements such as lithium (3.5), boron (24), cadmium (14), samarium (47), europium (6), and gadolinium (73), where the neutron capture resonance energies are in the region of the wavelengths used for crystal diffraction.

Extinction is generally of lesser importance than is absorption in x-ray diffraction, but is very important in neutron and electron diffraction, so much so in electron diffraction that the dynamical theory generally has to be applied.

Satisfactory treatments of extinction in x-ray and neutron diffraction are available. [36–39] These use the transmission factors A_{hkl} calculated for the absorption corrections, and permit the determination of anisotropic extinction corrections by including an extinction parameter tensor $\mathbf{g}_{ij}$ in the least-squares refinement of the atomic and thermal parameters.

9.3.6. Multiple reflections

When a crystal is oriented with respect to the beam so that several reciprocal-lattice points lie on the Ewald sphere simultaneously,

$$\mathbf{d}_i^* = \frac{\mathbf{s}_i - \mathbf{s}_0}{\lambda} \quad \text{and} \quad \mathbf{d}_i^* - \mathbf{d}_j^* = \frac{\mathbf{s}_i - \mathbf{s}_j}{\lambda}.$$

Thus, the diffracted beam corresponding to $\mathbf{d}_j^*$ can act as an incident beam for $\mathbf{d}_i^*$. This is known as the *Renninger effect*. [40, 41]

The intensity of the multiple-diffracted beam is small compared with the single-diffracted beam. However, when one of the intensities is large, this can introduce significant errors. [42, 43] It can also lead to space-group ambiguities in the identification of systematically absent reflections. The occurrence of multiple reflections is increased with shorter values of λ (i.e., larger Ewald spheres) and larger unit cells (closer density of reciprocal-lattice points). Since the occurrence is greater when $\mathbf{s}_0$ is in the same plane as a well-populated layer of reciprocal-lattice points, it is customary to offset the crystal axis from the rotation axis by a few degrees to reduce this condition. The importance of this effect for a particular reflection is examined by recording the diffracted intensity while rotating the crystal about the diffraction vector $\mathbf{d}^*_{hkl}$.

9.3.7. Diffraction by perfect crystals

Since the diffracting power depends on the *crystal perfection*, it can be used to investigate faults in crystals, especially for investigating imperfections in nearly perfect crystals. This is the basis of the Berg-Barrett method. [44–46] The image of the face of a crystal can be obtained by illuminating it with an x-ray beam placed about 30 cm away, and recording on a film parallel to the face and as close as possible. The resolving power, of the order of microns, can reveal dislocations in the crystal surface. Thin crystal plates can also be examined by transmission in this way. This method is most sensitive for nearly perfect crystals, and it ceases to be useful for the mosaic crystals, where the kinematic theory of diffraction is a good approximation.

9.3.8. "Borrmann" or "anomalous transmission" effect [47]

Transmitted and diffracted x-ray beam intensities are observed under circumstances where the absorption is such that no penetration of the x-ray beam would be expected. This occurs in x-ray transmission through a thick and highly perfect crystal, which is in the symmetrical Laue reflection orientation. In dynamical theory, if the nodes of the standing waves generated in the crystal correspond to the main absorption centers, i.e., the heavier atoms, energy is transmitted unabsorbed. This phenomenon is rarely observed in other than large, very perfect crystals, of semiconductors, for example. The transmitted beams are highly monochromatic, parallel, and polarized, but with insufficient intensity to be used as monochromated sources, except perhaps with synchrotron radiation. They are used in *x-ray topography*. If the transmitted and reflected beams are recorded with a fine-grain emulsion, they provide images of the dislocations and other departures from lattice perfection in the crystal.

9.3.9. Kossel and Kikuchi lines

When a perfect crystal is the target of the x-ray tube, the absorption of the characteristic diffraction spectra from the source within the crystal produces deficiency lines on the scattering pattern, which are known as Kossel lines. [48] Similar patterns can be produced by divergent beam diffraction when a divergent monochromatic x-ray beam is transmitted through a thin slice of perfect crystal. Measurement of the point of intersection of these lines, which are conic sections, can provide very accurate lattice parameters. [49, 50] These patterns are also characteristic of the crystal perfection and have been used to distinguish between type-I and type-II diamonds. [51] A quantum-mechanical interpretation has been provided for this phenomenon. [52]

The same phenomenon, when it occurs in electron diffraction, is referred to as Kikuchi lines. [53] Owing to the shorter waves, the lines appear straight rather than as curved conic sections.

X-ray diffraction from thin, wedge-shaped, perfect crystals produces interference patterns known as *pendellösung fringes*. [54] This phenomenon has been explored experimentally [55] and is used in x-ray topography. The dynamical theory has been developed by Kato. [56] It can be used for very accurate measurements of structure amplitudes in special cases. [57]

9.3.10. Powder diffractometry

The diffraction spectra from a crystalline powder lie on the surface of cones with semivertical angles of $2\theta_{hkl}$ with respect to the direct beam. When the powder consists of suitably small crystals, the diffraction spectra form uniform *powder lines*, which are the intersection of the diffraction cones with the recording device. The *powder diffraction pattern* is a one-dimensional record of I_{hkl} versus 2θ, for all reciprocal-lattice points within the sphere of diffraction. It takes the form of a record of 2θ or d, and I/I_0 for each symmetry-independent hkl diffraction spectra. It is characteristic of the crystal, and it can be used as a *fingerprint* for identification purposes. In favorable cases, the presence of impurities of 1–2% can be detected. A *powder diffraction* file [58] is available for identification purposes (see 9.10.3).

When very small quantities of material are available, film methods using a 57.3-mm-radius powder camera are commonly used. Larger quantities permit the use of powder diffractometers with slabs of powder, making use of the *Bragg-Brentano parafocusing* instrumentation. A very effective film instrument for high resolution of powder lines is the Guinier focusing powder camera. Back-reflection powder patterns with large-radius evacuated (or helium-filled) cameras are used for high-precision lattice parameter measurements of high-symmetry crystals. For cubic crystals, systematic errors can be reduced [59] by extrapolation of a to $\theta = 90°$ by plotting the lattice parameters versus

$$\frac{1}{2}\left(\frac{\cos^2\theta}{\theta} + \frac{\cos^2\theta}{\sin\theta}\right).$$

This function is tabulated in Int. Tables, Vol. II, pp. 228–229. For noncubic crystals, an analytical least-squares refinement of the lattice parameters is preferable. [60] If the crystals are too small, there is line broadening, which is additional to that characteristic of the collimation and specimen. The increase in width of the diffraction line $\Delta 2\theta \approx \lambda/t\cos\theta$, where t is the linear dimension of the crystal.

When the crystalline powder is sufficiently coarse, discrete diffraction spots appear on the powder line. For metals and alloys, this provides a method of studying *preferred orientation* in the polycrystalline *grain texture*. Preferred orientation is also common in compressed powders. This can seriously interfere with qualitative and quantitative analysis by means of powder diffractometry. Under special circumstances, information concerning the structure of the grain boundaries can be obtained from "extra" weak or diffuse spectra observed on x-ray or electron diffraction photographs. [61]

9.3.11. Powder diffraction profile refinement: Rietveld method

The whole-pattern-fitting method of interpreting powder diffraction was first introduced by Rietveld [62, 63] for neutron powder diffraction patterns, for which the theory is mathematically simpler. Many crystal structure analyses have been successfully determined from neutron powder data. [64] Later, the theory was adapted to x-ray patterns, [65, 66] and a number of structures have been determined. The method can also be applied to mixtures of several phases. [67] The method has been combined with a well-known single-crystal atomic parameter refinement method. [68]

The powder pattern intensity is measured stepwise across the whole pattern. The scan is usually $\approx 0.05°$ in 2θ. A set of data will therefore contain about 2000 separate I_i values. The method aims to minimize by least squares the residual

$$R = \sum_i \omega_i \left(I_i(\text{obs}) - \frac{1}{k} I_i(\text{calc}) \right)^2,$$

where ω_i is a weight assigned to the ith datum point, usually derived from the counting statistics, and k is a scale factor parameter. The success of the method depends on how well $I_i(\text{calc})$ can be derived from an expression such as

$$I_i(\text{calc}) = I_{ib} + \sum_{\mathbf{s}} G(\theta_i - \theta_{\mathbf{s}}) m_{\mathbf{s}} T_{\mathbf{s}} (\text{Lp})_i |\, F(\mathbf{s}) \,|,$$

where I_{ib} is the background intensity; $G(\theta_i - \theta_{\mathbf{s}})$ is a convolution, in analytical form, of the instrumental profile function and the intrinsic diffraction profile, both of which are dependent on θ; $m_{\mathbf{s}}$ is the multiplicity of the diffraction spectrum; $T_{\mathbf{s}}$ is a preferred orientation function; $(\text{LP})_i$ is the Lorentz-polarization factor at θ_i; and $F(\mathbf{s})$ is the structure factor corresponding to diffraction vector, **s**.

9.4. STRUCTURE FACTOR

The structure factor $F(\mathbf{s})$, F_{hkl}, or F_{h} is a dimensionless quantity. It depends on the scattering properties of the matter in the crystal. For Daltonian convenience, it is separated into the scattering power of the atoms and their geometrical arrangement with respect to the crystal lattice. The former are referred to as *atomic scattering factors*, and the latter are referred to as *geometrical structure factors*.

9.4.1. Atomic scattering factors $f_i(\mathbf{s})$

For x-rays,

$$f_i^x(\mathbf{s}) = \int_{\text{atom}} \psi_f^*(\mathbf{r}) \exp(i\mathbf{s} \cdot \mathbf{r}) \psi_i(\mathbf{r})\, dr,$$

where ψ_i^* and ψ_i refer to the wave functions for the initial and final states. For coherent scattering, these states are the same. $|\mathbf{s}| = 4\pi\lambda^{-1}\sin\theta$. Except in special problems, it is assumed that atoms are spherical; then,

$$f(\mathbf{s}) = \int \frac{r^2 P(r)^2 \sin(sr)}{sr}\, dr,$$

where $P(r)$ is the radial component of $\psi(r)$. Numerical values for the x-ray scattering factors for free atoms and chemically significant ions as a function of $(\sin\theta)/\lambda$ are given in Int. Tables, Vol. IV, pp. 71–98. An analytical expression convenient for computer generation of atomic scattering factors is

$$f\left(\frac{\sin\theta}{\lambda}\right) = \sum_{i=1}^{4} a_i \exp\left(-b_i \frac{\sin^2\theta}{\lambda}\right) + c,$$

the coefficients a_i, b_i, and c of which are given in Int. Tables, Vol. IV, pp. 99–102. These expressions reproduce those from relativistic Hartree-Fock or relativistic Dirac-Slater calculations within a mean error generally less than 0.010 and seldom greater than 0.020. Scattering factors for spherically bonded hydrogen atoms are given in Int. Tables, Vol. IV, p. 102.

For electrons,

$$f_i^e(\mathbf{s}) = 2\frac{m^2}{h^2}\frac{Z_i - f_i^x(\mathbf{s})}{s^2},$$

where Z_i is the nuclear charge. If θ is small,

$$f_i^e(\mathbf{s}) = 0.023\,93[Z_i - f_i^x(\mathbf{s})]\lambda^2/\theta^2.$$

Because of the denominator, electron scattering falls off much more rapidly with scattering angle than does x-ray scattering. Tables of electron scattering factors, in Å, for neutral atoms are given in Int. Tables, Vol. IV, pp. 155–174.

For neutrons, [69] the wavelengths used for diffraction, ≈ 1 Å, are much larger than nuclear dimensions. As a result, nuclei act as point scatterers:

$$f_i^n(\mathbf{s}) = \frac{m}{2\pi\hbar^2} a \left(\frac{4}{3}\pi r_0^3\right),$$

where a is the Fermi pseudopotential, which is zero outside the radius $r_0 (\approx 10^{-3}$ Å). $f_n(\mathbf{s})$ is independent of scattering angles and is denoted by b in 10^{12} cm, the nuclear scattering length, which is a constant for each isotropic species of the elements.

The nuclear neutron scattering factors [70] are shown in Table 9.5. Nuclei that have magnetic moments give additional neutron scattering. The neutron magnetic scattering factor b_m is given by

$$b_m = (e^2\gamma/mc^2)Sf,$$

where γ is the neutron magnetic moment in nuclear magnetons, S is the electronic-spin quantum number, and f is the atomic scattering factor of the unpaired nuclear electron

TABLE 9.5. Neutron scattering factors (cross sections) of the atoms, in 10^{-13} cm, and in [ ], mass absorption coefficients (μ/ρ cm^2 gm^{-1}) for $\lambda = 1.08$ Å.

IA	IIA	IIIB	IVB	VB	VIB	VIIB	VIII			IB	IIB	IIIA	IVA	VA	VIA	VIIA	VIIIA
H −3.7409 [25.3]	D 6.674																He 3.26
Li −2.03 [3.5]	Be 7.79 [0.0003]											B 5.35 [24]	C 6.6484 [0.0001]	N 9.3 [0.048]	O 5.805 [0.0000]	F 5.665 [0.0002]	Ne 4.55 [0.006]
Na 3.63 [0.007]	Mg 5.375 [0.001]											Al 3.449 [0.003]	Si 4.149 [0.002]	P 5.13 [0.002]	S 2.847 [0.0055]	Cl 9.579 [0.33]	Ar 1.884 [0.006]
K 3.67 [0.018]	Ca 4.90 [0.0037]	Sc 12.3 [0.25]	Ti −3.3438 [0.044]	V −0.380 [0.033]	Cr 3.635 [0.021]	Mn −3.73 [0.083]	Fe 9.54 [0.015]	Co 2.53 [0.21]	Ni 10.3 [0.028]	Cu 7.772 [0.021]	Zn 5.680 [0.0055]	Ca 7.29 [0.015]	Ge 8.193 [0.011]	As 6.58 [0.020]	Se 7.97 [0.056]	Br 6.79 [0.029]	Kr 7.85 [0.13]
Rb 7.22 [0.003]	Sr 7.02 [0.005]	Y 7.75 [0.006]	Zr 7.16 [0.0006]	Nb 7.054 [0.004]	Mo 6.95 [0.009]	Tc 6.8	Ru 7.21 [0.009]	Rh 5.93 [0.53]	Pd 5.91 [0.023]	Ag [0.20]	Cd 0.50 [14]	In 4.06 [0.6]	Sn 6.228 [0.002]	Sb 5.64 [0.016]	Te 5.80 [0.013]	I 5.28 [0.018]	Xe 4.89 [0.083]
Cs 5.42 [0.077]	Ba 5.25 [0.0027]	La 8.27 [0.023]	Hf 7.7 [0.20]	Ta 6.91 [0.044]	W 4.77 [0.036]	Re 9.2 [0.16]	Os 10.7 [0.028]	Ir 10.6 [0.80]	Pt 9.5	Au 7.63 [0.17]	Hg 12.66 [0.63]	Tl 8.79 [0.006]	Pb 9.401 [0.0003]	Bi 8.5233 [0.0000]	Po	At [0.79]	Rn
				Ce 4.84 [0.0021]	Pr 4.45 [0.029]	Nd 7.69 [0.11]	Pm 12.6	Sm −5.0 [47]	Eu 6.0 [6]	Gd 9.5 [73]	Tb 7.38 [0.09]	Dy 16.9 [2.0]	Ho 8.08 [0.15]	Er 8.03 [0.36]	Tm 7.05 [0.25]	Yb 12.4 [0.076]	Lu 7.3 [0.22]
				Th 9.84 [0.01]	Pa 9.1	U 8.42 [0.005]	Np 10.6	Pu 7.7	Am 8.3								

(dependent on $\sin\theta/\lambda$, $f = 1$ for $\theta = 0°$). Magnetic scattering is of the same order as nuclear scattering.

These atomic scattering factors for x-rays, electrons, and neutrons are for isolated atoms *at rest*. The effects of thermal motion on the scattering factors are discussed separately. The relative dependence of the scattering factors with scattering angle for the types of radiation used in crystallography is illustrated in Figure 9.4 for an iron atom.

9.4.2. Dispersion corrections for x-ray atomic scattering factors

The x-ray atomic scattering factors take account of the spatial distribution of electrons in the atom, but they are calculated on the assumption that the electronic binding energy is so small compared with the energy of the x-ray photon that the scattering power of each electron is like that of a free electron. When this is true, $f_{hkl} = f_{\overline{hkl}}$, $F(\mathbf{s}) = F(-\mathbf{s})$, and $F_{hkl} = F_{\overline{hkl}}$. This is known as *Friedel's law*. When the incident x-ray wavelength approaches a characteristic resonance frequency of an electronic transition in the atom, the scattering power of a bound electron will be greater than or less than that of a free electron, and the phase of the scattered wave will be different. These effects are taken into account by representing the atomic scattering factor f as a complex number:

$$f = f_0 + \Delta f' + i\Delta f'',$$

where $\Delta f'$ and $\Delta f''$ are the real and imaginary dispersion corrections.

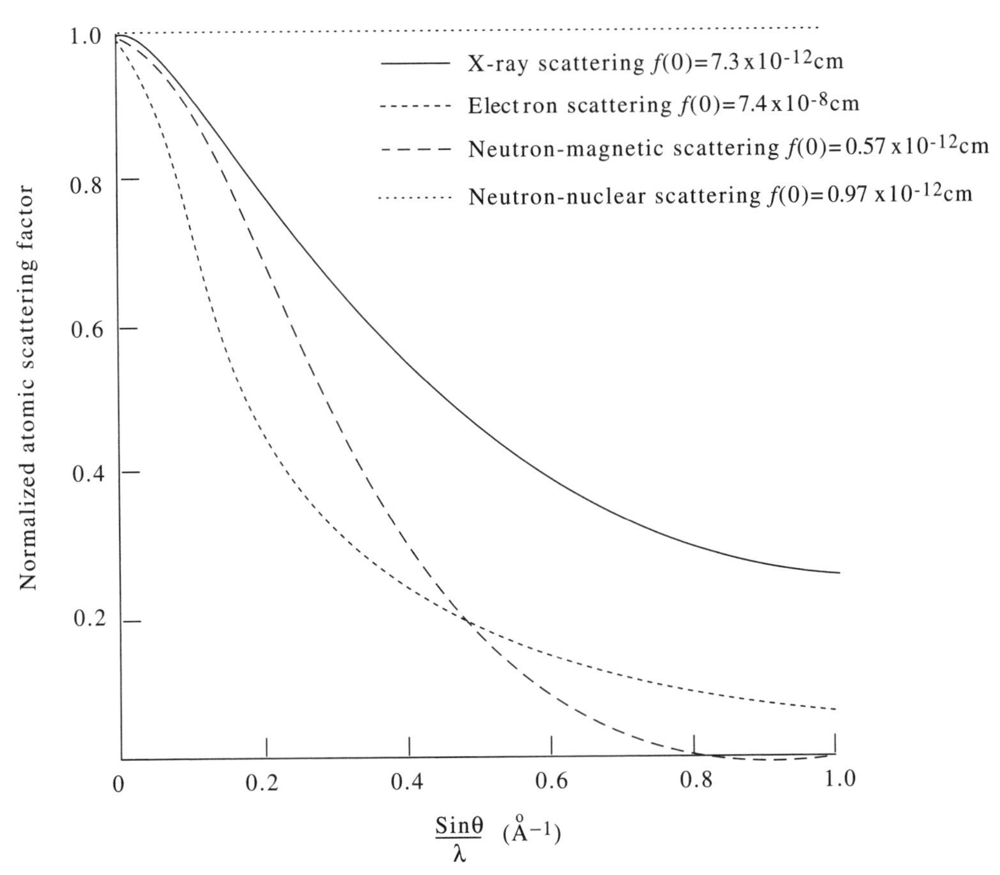

FIGURE 9.4. Relative dependence of scattering factors with scattering angle for the types of radiation used in crystallography for an iron atom.

The dispersion corrections depend on the x-ray wavelength λ and the diffraction angle θ. They are less sensitive functions of θ than is f_0, because the tightly bound electrons responsible for these effects are concentrated in a small volume near the atomic nucleus. Values of f' generally increase, and those of f'' decrease, by about 10% in going from $(\sin\theta)/\lambda = 0$ to 1.0 (see Int. Tables, Vol. III, pp. 214–216).

Values for $\Delta f'$ and $\Delta f''$ for Cu$K\alpha$ and Mo$K\alpha$ are given in Table 9.6. The values for other radiations are given in Int. Tables, Vol. IV, pp. 149–150. It is difficult to assess the accuracy of these values owing to approximations in the theory and the absence of any accurate experimental measurements. The use of tunable synchrotron x-radiation has made possible some experimental measurements of $\Delta f'$ and $\Delta f''$ close to the L-shell absorption edges of particular elements. [71]

TABLE 9.6. Real and imaginary dispersion corrections for x-ray atomic scattering factors for Cu$K\alpha$ and Mo$K\alpha$ radiation.

IA	IIA	IIB	IVB	VB	VIB	VIIB		VIII		IB	IIB	IIIA	IVA	VA	VIA	VIIA	VIIIA
H																	He
Li 0.001 0.000 0.000 0.000	Be 0.003 0.000 0.001 0.000											B 0.008 0.000 0.004 0.001	C 0.017 0.002 0.009 0.002	N 0.029 0.004 0.018 0.003	O 0.407 0.008 0.032 0.006	F 0.069 0.014 0.053 0.010	Ne 0.097 0.021 0.083 0.106
Na 0.129 0.030 0.124 0.025	Mg 0.165 0.042 0.177 0.036											Al 0.204 0.056 0.246 0.052	Si 0.244 0.072 0.330 0.071	P 0.283 0.090 0.434 0.095	S 0.319 0.110 0.557 0.124	Cl 0.348 0.132 0.702 0.159	Ar 0.366 0.155 0.872 0.201
K 0.365 0.179 1.066 0.250	Ca 0.341 0.203 1.286 0.306	Sc 0.285 0.226 1.533 0.372	Ti 0.189 0.248 1.807 0.466	V 0.035 0.267 2.110 0.530	Cr −0.198 0.284 2.443 0.624	Mn −0.568 0.295 2.808 0.729	Fe −1.179 0.301 3.204 0.845	Co −2.464 0.299 3.608 0.973	Ni −2.956 0.285 0.506 1.113	Cu −2.019 0.263 0.589 1.266	Zn −1.612 0.222 0.678 1.431	Ga −1.354 0.163 0.777 1.609	Ge −1.163 0.081 0.886 1.801	As −1.011 −0.030 1.006 2.007	Se −0.879 −0.178 1.139 2.223	Br −0.767 −0.374 1.283 2.456	Kr −0.665 −0.652 1.439 2.713
Rb −0.574 1.608	Sr −0.465 1.820	Y −0.386 2.025	Zr −0.314 2.245	Nb −0.248 2.482	Mo −0.191 2.735	Tc −0.145 3.005	Ru −0.105 3.296	Rh −0.077 3.605	Pd −0.059 3.934	Ag −0.060 4.282	Cd −0.079 4.653	In −0.126 5.045	Sn −0.194 5.459	Sb −0.287 5.894	Te −0.418 6.352	I −0.579 6.835	Xe −0.783 7.348
Cs −1.022 7.904	Ba −1.334 8.460	La −1.716 9.036	Hf −6.715 4.977	Ta −6.351 5.271	W −6.048 5.577	Re −5.790 5.891	Os −5.581 6.221	Ir −5.391 6.566	Pt −5.233 6.925	Au −5.096 7.297	Hg −4.990 7.686	Tl −4.883 8.089	Pb −4.818 8.505	Bi −4.776 8.930	Po −4.756 9.383	At −4.772 9.842	Rn −4.787 10.317
Fr −4.833 10.803	Ra −4.898 11.296	Ac −4.994 11.799															
				Ce −2.170 9.648	Pr −2.939 10.535	Nd −3.431 10.933	Pm −4.357 11.614	Sm −5.696 12.320	Eu −7.718 11.276	Gd −9.242 11.946	Tb −9.498 9.242	Dy −10.423 9.748	Ho −12.255 3.704	Er −9.733 3.937	Tm −8.488 4.181	Yb −7.701 4.432	Lu −7.133 4.693
				Th −5.091 12.330	Pa −5.216 12.868	U −5.359 13.409	Np −5.529 13.969	Pu −5.712 14.536	Am −5.930 15.087	Cm −6.176 15.634	Bk −6.498 16.317	Cf −6.798 16.930	Es	Fm	Md	No	Lw

The left column for each element refers to Cu$K\alpha$ and the right column (if present) to Mo$K\alpha$. The top and bottom numbers are $\Delta f'$ and $\Delta f''$, respectively.

9.4.3. Geometrical structure factor

The structure factor

$$F(\mathbf{s}) = \sum_{i=1}^{n} f_i \exp(2\pi i \mathbf{r}_j \cdot \mathbf{s}),$$

or

$$F_{hkl} = \sum_{i=1}^{n} f_i A + \sum_{i=1}^{n} f_i B,$$

where

$$A = \cos 2\pi \mathbf{r}_j \cdot \mathbf{s} = \cos 2\pi (hx_i + ky_i + lz_i),$$
$$B = i \sin 2\pi \mathbf{r}_j \cdot \mathbf{s} = i \sin 2\pi (hx_i + ky_i + lz_i),$$

where x, y, and z are the *fractional atomic coordinates* x_i/a, y_i/b, and z_i/c.

The geometrical structure factors A and B can be simplified by summing over the symmetry-equivalent atomic positions characteristic of the space group. The trigonometric expressions for A, B, and α appropriate for each space group are given in Int. Tables, Vol. I, pp. 353–525. These simplified expressions are sometimes advantageous for computer-programming structure factor calculations for highly symmetrical space groups.

The *phase angle* α_{hkl} is given by

$$\alpha_{hkl} = \tan^{-1}(B/A).$$

Then,

$$F_{hkl} = \sum_{i=1}^{n} f_i \cos 2\pi (hx_i + ky_i + lz_i - \alpha_{hkl}).$$

The phase angle plays a key role in crystal structure determination, because the experimentally measured quantity $I_{hkl} \approx F_{hkl}^2$, and therefore, α_{hkl} cannot be measured experimentally for a particular diffracted spectrum, except under the special conditions of anomalous scattering. The quantity that can be measured experimentally by normal diffraction methods, $| F_{hkl} |$, is referred to as the *structure amplitude*.

The structure factor expression, including the dispersion effect, is [72]

$$F(\mathbf{s}) = \sum_{i=1}^{n} f_i A + \sum_{i'=1}^{n} (\Delta f_i' A' - \Delta f_{i'}'' B') + \sum_{i=1}^{n} f_i B + \sum_{i=1}^{n'} (\Delta f_{i'}' B' + \Delta f_i'' B'),$$

$$F(\bar{\mathbf{s}}) = \sum_{i=1}^{n} f_i A + \sum_{i'=1}^{n'} (\Delta f_{i'}' A' + \Delta f_{i'}'' B') - \sum_{i=1}^{n} f_i B - \sum_{i'=1}^{n'} (\Delta f_i' A' - \Delta f_{i'}'' B'),$$

where A and B are summed over *all* atoms, and A' and B' over only those atoms with significant values of the dispersion corrections $\Delta f'$ and $\Delta f''$.

9.4.4. Unitary and normalized structure factors

Unitary and normalized structure factors are used particularly in phase determination. The *unitary structure factor* [73]

$$U_{hkl} = F_{hkl} / \left(\sum_i^n f_i \right)^{-1}.$$

The *normalized structure factor (E value)*

$$E_{hkl} = F_{hkl} / \left(\epsilon \sum_i^n f_i^2 \right)^{-1/2}.$$

ϵ is an integral multiple to account for the multiplicity of certain classes of hkl. It is the integral multiple that occurs in the reduced geometrical structure for the class of *hkl* (see Int. Tables, Vol. I, pp. 353–525).

9.5. THERMAL MOTION [74, 75]

The thermal motion of the atoms in a crystal reduces the *Bragg intensities* of the x-ray, neutron, or electron diffracted spectra relative to that of an atomic structure *at rest*. This is accounted for by means of atomic temperature factors, which are included with the atomic scattering factors in the calculation of the structure factors. These atomic thermal parameters are included with the atomic coordinates in the variables that are refined by least-squares methods and are an integral part of the results of a modern crystal structure analysis. (An alternative approach [76] is to determine the molecular rigid-body motion parameters directly by constrained least-squares refinement of the diffraction data without determining individual atomic vibration parameters.) The structure factor expression that includes thermal motion in matrix-vector notation is then

$$F(\mathbf{h}) = \sum_j f_i(\mathbf{h}) \exp(2\pi i \mathbf{h}^t \mathbf{x}_j - 2\pi^2 h^t \mathbf{Q}^t \mathbf{U} \mathbf{Q} \mathbf{h}),$$

where $\mathbf{h}$ is the Miller index triplet vector, h^t its transpose, and $\mathbf{x}_j$ is the position parameter vector,

$$\begin{aligned} \mathbf{Q}_{ij} &= \mathbf{a}_i^* \qquad \text{for} \quad i = j \\ &= 0 \qquad \text{otherwise,} \end{aligned}$$

with $\mathbf{Q}^t$ its transpose, where $\mathbf{a}_i^*$, $i = 1, 2, 3$, are the reciprocal-lattice lengths and $\mathbf{U}$ is the symmetric 3×3 matrix describing the thermal motion.

In addition to the dynamic time-dependent motion of an atom due to vibrational displacements relating to molecular vibrations or lattice mode vibrations, there may be static lattice-dependent effects due to disordering of the atoms. Unless these are specifically separated by appropriate experimental or interpretative methods, the atomic parameters, i.e., coordinates and thermal parameters, describe the crystal structure "averaged" over the *space* of the crystal lattice and the *time* of the measurement of the diffracted spectra.

The commonly used method for including the effect of thermal motion on the atomic scattering factor is to assume that the atom vibrates anisotropically in a harmonic potential field and can therefore be represented by a Gaussian smearing function. [77] These vibrations can then be described by a symmetric tensor **U** with six independent components such that the mean-square amplitude of vibration $\langle u^{-2}\rangle$ in the direction of the unit vector with components I_i is

$$\langle u^2\rangle \equiv \sum_{i=1}^{3}\sum_{j=1}^{3} U_{ij} I_i I_j .$$

The smearing function to be applied to each atomic scattering factor f_i^{hkl} is then

$$\exp[-2\pi^2(h^2a^{*2}U_{11} + k^2b^{*2}U_{22} + I^2c^{*2}U_{33} + 2hka^*b^*U_{12} + 2hla^*c^*U_{13} + 2klb^*c^*U_{23})]$$

This is also expressed as

$$\exp[-(\beta_{11}h^2 + \beta_{22}k^2 + \beta_{33}l^2 + 2\beta_{12}hk + 2\beta_{13}hl + 2\beta_{23}kl)].$$

The reporting of U_{ij} rather than the dimensionless β_{ij} values has the advantage that it reveals the anisotropy of the atomic thermal motion more directly. The diagonal elements are equal in magnitude to the mean-square displacements in Å^2 along the reciprocal axes. The exponential factor in the equation for $F(h)$ is the characteristic function or Fourier transform of a Gaussian probability density function. The density function is

$$P(u) = \frac{\det(P_j)^{1/2}}{(2\pi)^{3/2}} \exp[-\tfrac{1}{2}(u - x_j)^t P_j (u - x_j)],$$

where P is the matrix inverse of $Q^t U Q$. Properties of this density function are given in Int. Tables, Vol. IV, Sec. 5.

Graphical representations of the atomic parameters, which include the thermal motion, are obtained by plotting these probability functions using the program ORTEP. [78] In these representations, an example of which is shown in Figure 9.5, the atoms appear as ellipsoids. The distances to the ellipsoidal surface along the the principal axes of the ellipsoids are equal to the root-mean-square displacements in those directions. The ellipsoid may contain a chosen fraction of the total probability function. For room-temperature measurements, a commonly used level is 50%, corresponding to a scale of 1.5. For low-temperature work, 74% may be used, corresponding to a scale of 2.0.

If the motion is isotropic, as in cubic crystals, the isotropic temperature factor $B = 8\pi^2\langle u_{\text{iso}}^2\rangle$, where $\langle u_{\text{iso}}^2\rangle$ is the mean-square displacement of the atom from its equilibrium position. For cubic crystals,

$$\langle u_{\text{iso}}^2\rangle = \frac{3\hbar^2}{4\pi^2 m\Theta_D}\left(\frac{\phi(\Theta_D/T)}{\Theta_D/T} + \frac{1}{4}\right),$$

where Θ_D is the *Debye characteristic temperature*, defined by

$$\hbar\nu_m = k_B\Theta_D,$$

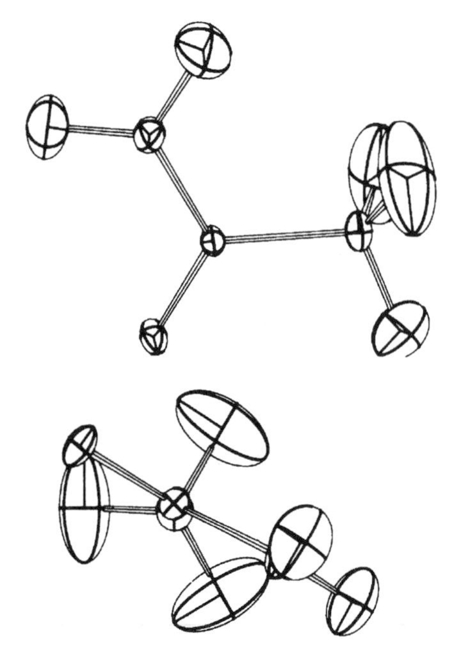

FIGURE 9.5. ORTEP representation, at 75% probability, of the nuclear thermal motion in acetamide,

$$H_2N—\underset{\underset{O}{\|}}{C}—CH_3$$

at 23 K, viewed in the plane of the molecule and down the C–C bond, from a neutron diffraction analysis of the rhombohedral crystal form: G. A. Jeffrey, J. R. Ruble, R. K. McMullan, D.J. DeFrees, J. S. Binkley, and J. A. Pople, Acta Crystallogr. Sect. B 36. 2292 (1980).

where ν_m is the maximum (Debye cutoff) frequency of the elastic vibrations of the crystal, k_B is Boltzmann's constant, and T is the absolute temperature. The Debye integral function $\Phi(x)$ is defined by

$$\Phi(x) = \frac{1}{s}\int_0^x \frac{y}{e^y - 1}\,dy.$$

At high temperatures, $T > \Theta_D$ or $x < 1$; then,

$$\langle u_{\text{iso}}^2 \rangle = 3\hbar^2 T/4\pi m k_B \Theta_D^2,$$

so that $\langle U^2 \rangle$ is proportional to the absolute temperature T. At low temperatures, $T < \Theta_D$ or $1/x < 1$; zero-point motion is dominant. $\Phi(x) = 0$, and

$$\langle u_{\text{iso}}^2 \rangle = 3\hbar^2/4\pi m k_B \Theta_D,$$

which is independent of T. Values of the Debye characteristic temperatures and vibration amplitudes for some cubic elemental crystal structures are given in Int. Tables, Vol. III, pp. 234–244.

Even when an atom is undergoing anisotropic thermal motion, it is useful to have a single parameter to represent the degree of thermal motion. Such a parameter is the *equiv-*

alent isotropic temperature B_{eq}. In general,

$$B_{\text{eq}} = \tfrac{8}{3}\pi^2\hbar(Q^t U Q),$$

expressed in a Cartesian axis system. For an orthogonal axis system,

$$B_{\text{eq}} = \tfrac{8}{3}\pi^2(U_{11} + U_{22} + U_{33}).$$

For cases in which the Gaussian probability density is an inadequate model, more elaborate models are needed. Gram-Charlier series expansion, cumulant expansion of the Gaussian model, [79] and models based on curvilinear density functions are described in Int. Tables, Vol. IV, Sec. 5.

9.6. DIFFRACTING DENSITY FUNCTION

For x-rays, the diffracting density function is the electron density function in the unit cell; for electrons, it is the nuclear and electron potential; and for neutrons, it is the nuclear scattering distribution. In all cases,

$$\rho(\mathbf{r}) = \int_v F(\mathbf{s}) \exp(-2\pi i \mathbf{s} \cdot \mathbf{r})\, dr.$$

This is the Fourier transform of the structure factor expression

$$F(\mathbf{s}) = V \int_v \rho(r) \exp 2\pi i \mathbf{r} \cdot \mathbf{s}\, ds.$$

In a crystal, the electron, or scattering density, distribution ρ_{xyz} is a Fourier synthesis of waves, the amplitudes and phases of which are given by the structure factors F_{hkl}: [80]

$$\rho_{xyz} = \frac{1}{V} \int_{-\infty}^{\infty} \int_{-\infty}^{\infty} \int_{-\infty}^{\infty} F_{hkl} \exp[-2\pi i(hx + ky + lz)].$$

Separating out the measurable structure amplitude $|\, F_{hkl}\,|$ from the unknown phase angle α_{hkl},

$$\rho_{xyz} = \frac{1}{V} \sum_{-\infty}^{\infty} \sum_{-\infty}^{\infty} \sum_{-\infty}^{\infty} |\, F_{hkl}\,| \cos[2\pi(hx + ky + lz - \alpha_{hkl})].$$

In some cases, it is desirable to remove from the electron density map a known part of the crystal structure. A *difference Fourier synthesis* is then calculated:

$$\rho'_{xyz} = \frac{1}{V} \sum_{-\infty}^{\infty} \sum_{-\infty}^{\infty} \sum_{-\infty}^{\infty} |\Delta\, F_{hkl}\,| \cos[2\pi(hx + ky + lz - \alpha'_{hkl})],$$

where $|\Delta\, F_{hkl}\,|$ is the difference between the observed structure amplitudes and those calculated for the known atomic positions, and α'_{hkl} is calculated from the known atomic positions. This method is commonly used for completing the analysis of a structure containing heavy atoms, or for determining hydrogen atom positions when the nonhydrogen atoms have been located. It is often used in protein crystallography.

Normalized structure factors E_{hkl} are frequently used instead of F_{hkl} in Fourier syntheses, since these give sharper and better-resolved atomic peaks. The three-dimensional plot of these Fourier syntheses is referred to as an E map.

9.7. PHASE PROBLEM

Since $I_{hkl} = kF_{hkl}^2$, only the structure amplitudes $|F_{hkl}|$ can be measured experimentally. The determination of the phases α_{hkl} is therefore the central problem in crystal structure analysis.

9.7.1. Phase-solving methods

There are two basic approaches. One makes use of the *Patterson function*; the other, called the *direct method*, uses the statistical properties of the structure factors that arise from the fact that the electron density distribution in a crystal is always positive. (Although the neutron scattering density in a crystal can be negative, owing to the negative scattering cross section of deuterium and other nuclei, the direct method has been applied successfully to neutron crystal structure analysis.)

Phase-solving methods based on these alternative approaches are given in Table 9.7.

TABLE 9.7. Phase-solving methods used in crystal structure analysis.

Based on use of Patterson synthesis	Based on use of structure factor statistics
Heavy-atom method [82]	Inequalities method [92]
Deconvolution methods by superposition, image seeking, or vector search [83–85]	Sayre equation [93]
	Symbolic addition [94–96]
Rotation function [86–88]	Tangent formula [97–100]
Multiple isomorphous replacement [89–91]	Multisolution methods [101]
Anomalous scattering	Structure invariants and semi-invariants [102–104]
	Magic integers [105]
	Anomalous scattering [106]

9.7.2. Patterson synthesis

The Patterson Fourier synthesis [81] $P(UVW)$ can be calculated from experimental data alone, since

$$P(\mathbf{U}) = P(UVW) = \frac{1}{V} \sum_{\infty}^{\infty} \sum_{\infty}^{\infty} \sum_{\infty}^{\infty} F_{hkl}^2 \cos[2\pi(hU + kV + lW)].$$

Since

$$F_{\mathbf{s}}^2 = \sum_{i=1}^{N} \sum_{j=1}^{N} f_i f_j \cos[2\pi(\mathbf{r}_i - \mathbf{r}_j) \cdot \mathbf{s}],$$

this synthesis gives a vectorial pattern of the interatomic distances in the unit cell. Interpretation of the "Patterson" is complex except for very simple structures, since n atoms in the cell will give rise to $n(n-1)$ peaks on the Patterson Fourier synthesis. Nevertheless, this synthesis is the basis of several very important methods of phase determination.

Since the overlap of the interatomic vector peaks is a serious obstacle to interpretation of the Patterson synthesis, a function commonly used is the *sharpened Patterson synthesis, with origin peak removed.* This is obtained by using normalized structure amplitudes $E_{hkl}^2 - 1$ as the Fourier series coefficients.

9.7.3. Direct methods

The "direct methods" are now the most powerful methods of phase determination. Methods such as MULTAN, SHELX, and MITHRIL are used for the majority of crystal structure analyses. Only in exceptional cases and for structures having more than 1000 atomic positional parameters do they fail. Programs of this type are incorporated within the x-ray diffractometer computer software.

The direct methods of phase determination depend on the nonnegative electron density inequality [107]

$$\begin{vmatrix} F_0 & F_{\bar{\mathbf{h}}_1} & F_{\bar{\mathbf{h}}_{n-1}} \\ F_{\mathbf{h}_1} & F_0 & F_{\mathbf{h}_1\bar{\mathbf{h}}_{n-1}} \\ F_{\mathbf{h}_{n-1}} & F_{\mathbf{h}_{n-1}\bar{\mathbf{h}}_1} & F_0 \end{vmatrix} \geq 0,$$

where $\mathbf{h}_n$ refer to different *hkl*s.

9.8. CRYSTAL STRUCTURE REFINEMENT: METHOD OF LEAST SQUARES

In crystal structure analysis, the number m of observations $|F_{hkl}|$ generally exceeds the number n of unknown parameters (approximately nine times the number of atoms in the asymmetric unit). The method of least squares is therefore commonly used for parameter refinement. [108] The functions most frequently minimized are

$$R_1 = \sum \omega_{hkl}(| F_{hkl}^{\text{obs}} | - k| F_{hkl}^{\text{calc}} |)^2,$$

$$R_2 = \sum \omega'_{hkl}(| F_{hkl}^{\text{obs}} | - k| F_{hkl}^{\text{calc}} |)^2,$$

where the sums are over the sets of crystallographically independent hkl reflections. k is a scaling factor between $| F_{hkl}^{\text{obs}} |$ and $| F_{hkl}^{\text{calc}} |$. The weighting factors

$$\omega_{hkl} = 1/\sigma_{hkl}^2,$$

where σ is the standard deviation of $| F_{hkl}^{\text{obs}} |$ or $| F_{hkl}^{\text{calc}} |^2$. The n simultaneous linear equations in m unknowns are then

$$\sum_{i=1}^{n} \Delta u_i \left(\sum \omega_{hkl} \frac{\delta F_{hkl}^{\text{calc}}}{\delta u_j} \frac{\delta F_{hkl}^{\text{calc}}}{\delta u_i} \right) = -\sum \omega_{hkl} \Delta_{hkl} \frac{\delta F_{hkl}^{\text{calc}}}{\delta u_j} \qquad (j = 1, 2, \ldots, m),$$

where $\Delta_{hkl} = | F_{hkl}^{\text{obs}} | - | F_{hkl}^{\text{calc}} |$.

For R_1,

$$\frac{\delta \Delta_{hkl}}{\delta u_j} = -\frac{\delta | F_{hkl}^{\text{calc}} |}{\delta u_j}.$$

For R_2,

$$\frac{\delta \Delta_{hkl}}{\delta u_j} = -2| F_{hkl}^{\text{obs}} | \frac{\delta | F_{hkl}^{\text{calc}} |}{\delta u_j}.$$

In matrix form, $\mathbf{Nx} = \mathbf{e}$, where $\mathbf{x}$ and $\mathbf{e}$ are column vectors of order n and m, respectively; $\mathbf{N}$ is a matrix of order mn. The elements of $\mathbf{N}$ are

$$N_{ji} = \sum_{hkl} \omega_{hkl} \frac{\delta | F_{hkl}^{\text{calc}} |}{\delta u_j} \frac{\delta | F_{hkl}^{\text{calc}} |}{\delta u_i}.$$

The elements of $\mathbf{e}$ are

$$E_j = \sum \omega_{hkl} \frac{\delta | F_{hkl}^{\text{calc}} |}{\delta u_j} \Delta_{hkl},$$

where

$$\Delta_{hkl} = | F_{hkl}^{\text{obs}} | - | F_{hkl}^{\text{calc}} |, \quad \mathbf{x} = \Delta u_i.$$

The normal set of equations is

$$\tilde{\mathbf{N}}\mathbf{Nx} = \tilde{\mathbf{N}}\mathbf{e},$$

where $\tilde{\mathbf{N}}$ is the transpose of $\mathbf{N}$.

The solution vector for the x's is

$$x = (\tilde{\mathbf{N}}\mathbf{N})^{-1}\tilde{\mathbf{N}}\mathbf{e}.$$

The variance of the derived parameters u_i is given by

$$\sigma^2(u_i) = \frac{\mathbf{M}_{ii}^{-1}\Sigma\omega\Delta u_i^2}{m-n},$$

where $\mathbf{M}_{ii}^{-1}$ is the inverse of the normal equation matrix.

The covariance between derived parameters u_i and u_j is given by

$$\sigma(u_i u_j) = \frac{\mathbf{M}_{ij}^{-1}\Sigma\omega\Delta u_i^2}{m-n}.$$

For a parameter x_i such as an atomic coordinate or thermal parameter, its variance

$$\sigma(x_i)^2 = \mathbf{M}_{ii}^{-1}\frac{\Sigma\omega(|F_0| - |F_c|)^2}{m-n}.$$

The goodness or error of fit,

$$S = \frac{\Sigma\omega(|F_0| - |F_c|)^2}{m-n},$$

is a more meaningful measure. It should be unity if the model is complete and correct and the weights are properly assigned.

The choice of weights ω is important. Since the individual values of ω_{hkl} are not determined experimentally (it would require repeated measurement of each intensity), a weighting scheme is used. The simplest, and least useful, are unit weights. If the weights are inversely proportional to the overall atomic scattering factors, including the thermal motion, the atomic positions obtained by the least-squares minimization are the same as those given by the zero gradients of an unweighted difference Fourier synthesis. [109]

With the advent of automatic diffractometers, it is usual to use the σ_c from counting statistics, combined with some additional terms that correct for systematic trends noted in the values of ΔF_h. Such expressions generally take the form suggested by Cruickshank, [110]

$$\omega^{-1/2} = (a + b|F_h| + c|F_h|^2)^{-1},$$

where $a = \sigma_c^2$ from counting statistics, and b and c are empirically determined constants. Values of $F_h < 2\sigma$ or 3σ are considered to be unobserved. However, the rejection of low-intensity data leads to an underestimate of the scale and thermal parameters. [111] The information content of the unobserved reflections and how best to weight them has been discussed. [112] The consistency of the weights as a function of 2θ, I_{hnk}, or time, should be examined for systematic trends, as described in Int. Tables, Vol. IV, pp. 293–294.

9.9. REFERENCES

[1] W. Friedrich, P. Knipping, and M. Laue, Sitzungsber. Bayer. Akad. Wiss. **1912**, 303 [reprinted in Naturawissenschaften **39**, 361 (1952)].

[2] W. L. Bragg, Proc. R. Soc. London **A89**, 248 (1913); **A89**, 468 (1913).

[3] W. N. Lipscomb, *Boron Hydrides* (Benjamin, New York, 1963).

[4] J. S. Kasper, C. M. Lucht, and D. Harker, Acta Crystallogr: **3**, 436 (1950).

[5] J. D. Watson and F. H. C. Crick, Nature **171**, 737 (1953); **171**, 964 (1953).

[6] D. C. Hodgkin, J. Kamper, J. Lindsey, M. MacKay, J. Pickworth, J. H. Robertson, C. B. Shoemaker, J. G. White, R. J. Prosen, and K. N. Trueblood, Proc. R. Soc. London **A242**, 228 (1957).

[7] M. F. Perutz, M. G. Rossman, A. F. Cullis, H. Muirhead, G. Will, and A. C. T. North, Nature **185**, 416 (1960).

[8] J. C. Kendrew, H. C. Watson, B. E. Strandberg, R. E. Dickerson, D. C. Phillips, and V. C. Shore, Nature **190**, 666 (1961).

[9] J. M. Cowley, *Diffraction Physics* (North-Holland, Amsterdam, 1975).

[10] G. E. Bacon, *Neutron Diffraction*, 3rd ed. (Clarendon, Oxford, 1975).

[11] D. T. Hawkins, Acta Crystallogr. Sect. A **36**, 475 (1980).

[12] P. P. Ewald, Z. Krist. **56**, 129 (1921).

[13] J. Donohue, *The Structure of the Elements* (Wiley, New York, 1974).

[14] W. B. Pearson, *A Handbook of Lattice Spacings and Structures of Metals and Alloys* (Pergamon, New York, 1958), Vols. 1 and 2.

[15] J. D. H. Donnay and Helen M. Ondik, *Crystal Data Determinative Tables, 3rd Ed.* (International Centre for Diffraction Data, Newtown Square, PA), Vol. 1 (1972), Vol. 2 (1973), Vol. 3 (1978), Vol. 4 (1978).

[16] *Structure Berichte* (Johnson, New York, 1966), Vols. 1–7.

[17] R. W. G. Wykoff, *Crystal Structures* (Interscience, New York, 1963–1971), Vols. 1–6.

[18] *Structure Reports* (Oosthoek, Utrecht, 1940–1975), Vols. 8–41.

[19] *Molecular Structures and Dimensions, Guide to the Literature, 1935–1976, Organic and Organo-metallic Crystal Structures*, and *Molecular Structures and Dimensions*, Vols. 1–11 (1935–1979), edited by O. Kennard *et al.* (Reidel, Hingham, MA).

[20] The definitive reference sources for this chapter are Vols. I–IV of the *International Tables for X-ray Crystallography*, published for the International Union of Crystallography (Kynock, Birmingham, England, 1959–1974). These are abbreviated herein as Int. Tables, Vol. xx, pp. yy. New editions of the first three volumes of the *International Tables for Crystallography* have been published. These are *Volume A, Space-Group Symmetry*, 4th ed., edited by T. Hahn (1995); *Volume B, Reciprocal Space*, edited by U. Shmueli (1992); and *Volume C, Mathematical, Physical, and Chemical Tables*, edited by A. J. C. Wilson (1992); all published for the International Union of Crystallography by Kluwer Academic Publishers, Dordrecht.

[21] A. L. Mackay, Acta Crystallogr. **22**, 329 (1967).

[22] W. Nowacki, T. Matsumoto, and A. Edenharter, Acta Crystallogr. **22**, 935 (1967).

[23] P. P. Ewald, Phys. Z. **14**, 465 (1913).

[24] L. V. Azaroff, *et al.*, *X-ray Diffraction* (McGraw-Hill, New York, 1974).

[25] Extinction effects in x-ray crystal structure analysis are recognized by $|F_{\text{obs}}^{hkl}| \ll |F_{\text{calc}}^{hkl}|$ for small values of h, k, l. These data are omitted from the least-squares refinement calculations.

[26] B. K. Vainshtein, *Structure Analysis by Electron Diffraction* (Pergamon, Oxford, 1964).

[27] P. S.Turner and J. M. Cowley, Acta Crystallogr. Sect. A **25**, 475 (1969).

[28] P. Goodman, Acta Crystallogr. Sect. A **32**, 793 (1976).

[29] U. W. Arndt and B. T. M. Willis, *Single Crystal Diffractometry* (Cambridge University, London, 1966).

[30] W. R. Busing and H. A. Levy, Acta Crystallogr. **22**, 457 (1967).
[31] W. R. Busing and H. A. Levy, Acta Crystallogr. **10**, 180 (1957).
[32] J. de Meulenaer and H. Tompa, Acta Crystallogr. **19**, 1014 (1965).
[33] W. H. Zachariasen, Acta Crystallogr. **23**, 558 (1967).
[34] R. M. Stern, Trans. Am. Cryst. Assoc. **4**, 14 (1968).
[35] G. A. Somorjal and L. L. Kesmodel, Trans. Am. Cryst. Assoc. **13**, 67 (1977).
[36] W. C. Hamilton, Acta Crystallogr. Sect. A **25**, 194 (1969).
[37] P. Coppens and W. C. Hamilton, Acta Crystallogr. Sect. A **26**, 71 (1970).
[38] P. J. Becker and P. Coppens, Acta Crystallogr. Sect. A **30**, 129 (1974); **31**, 417 (1975).
[39] F. R. Thornley and R. J. Nelmes, Acta Crystallogr. Sect. A **30**, 748 (1974).
[40] O. Berg. Veroeff. Siemens Konzern **5**, 89 (1926).
[41] M. Renninger, Z. Krist. **97**, 107 (1937).
[42] W. H. Zachariasen, Acta Crystallogr. **18**, 705 (1965).
[43] R. D. Burbank, Acta Crystallogr. **19**, 957 (1965).
[44] C. S. Barrett, *Structure of Metals* (McGraw-Hill, New York, 1952).
[45] A. P. L. Turner, T. Vreeland Jr., and D. P. Pope, Acta Crystallogr. Sect. A **24**, 452 (1968).
[46] C. S. Barrett and T. B. Massalski, *Structure of Metals* (McGraw-Hill, New York, 1966).
[47] B. Borie, Acta Crystallogr. **21**, 470 (1966).
[48] W. Kossel, V. Loeck, and H. Voges, Z. Phys. **94**, 139 (1935).
[49] W. Kossel and H. Voges, Ann. Phys. (Paris) **23**, 677 (1935).
[50] B. J. Isherwood and C. A. Wallace, Acta Crystallogr. Sect. A **27**, 119 (1971).
[51] K. Lonsdale, Nature **151**, 52 (1943); **153**, 22 (1944).
[52] M. Kohler, Berl. Sitzunsber., 1935, 334.
[53] S. Kiluchi, Proc. Jpn. Acad. Sci. **4**, 271 (1928); **4**, 275 (1928); **4**, 354 (1928); **4**, 475 (1928).
[54] P. P. Ewald, Ann. Phys. (Paris) **54**, 519 (1917).
[55] N. Kato and A. R. Lang, Acta Crystallogr. **12**, 787 (1959).
[56] N. Kato, Acta Crystallogr. **14**, 526 (1961); J. Appl. Phys. **39**, 2225 (1968); **39**, 2231 (1968).
[57] M. Hart and A. D. Milne, Acta Crystallogr. Sect. A **26**, 223 (1970).
[58] International Centre for Diffraction Data, 12 Campus Boulevard, Newtown Square, PA 19073-3273; former name was Joint Committee on Powder Diffraction Standards (JCPDS).
[59] J. B. Nelson and D. P. Riley, Proc. Phys. Soc. London **57**, 160 (1954).
[60] M. U. Cohen, Rev. Sci. Instrum. **6**, 68 (1935).
[61] S. L. Sass, J. Appl. Crystallogr. **13**, 109 (1980).
[62] H. M. Reitveld, Acta Crystallogr. **22**, 151 (1967).
[63] H. M. Reitveld, J. Appl. Cryst. **2**, 65 (1969).
[64] A. K. Cheetham and J. C. Taylor, J. Solid State Chem. **21**, 253 (1977).
[65] W. Parrish, T. C. Huang, and G. L. Ayers, Trans. Am. Cryst. Assoc. **12**, 55 (1976).
[66] R. A. Young, P. E. Mackie, and R. D. Van Dreche, J. Appl. Phys. **10**, 262 (1977).
[67] P. E. Werner, S. Salome, G. Malmros, and J. O. Thomas, J. Appl. Crystallogr. **12**, 107 (1979).
[68] G. S. Pawley, J. Appl. Crystallogr. **13**, 630 (1980).
[69] C. G. Shull, Trans. Am. Cryst. Assoc. **3**, 1 (1967).

[70] L. Koester, *Neutron Physics*, Vol. 80 of *Springer Tracts in Modern Physics* (Springer, New York, 1977).

[71] J. C. Phillips, D. H. Templeton, L. K. Templeton, and K. O. Hodgson, Science **201**, 257 (1978).

[72] C. H. Dauben and D. H. Templeton, Acta Crystallogr. **8**, 841 (1955).

[73] A. L. Patterson, Phys. Rev. **46**, 372 (1934).

[74] B. T. M. Willis and A. W. Pryor, *Thermal Vibrations in Crystallography* (Cambridge University, London, 1975).

[75] Int. Tables, Vol. IV, pp. 314–319.

[76] G. S. Pawley, Acta Crystallogr. **20**, 631 (1966).

[77] D. W. J. Cruickshank, Acta Crystallogr. **9**, 747 (1956).

[78] C. K. Johnson, ORTEP II (Oak Ridge National Laboratory, Oak Ridge, TN, 1976), Report ORNL-5138.

[79] C. K. Johnson, Acta Crystallogr. Sect. A **25**, 187 (1969).

[80] W. H. Bragg, Philos. Trans. R. Soc., London **A216**, 254 (1915).

[81] A. L. Patterson, Z. Krist. **90**, 517 (1935).

[82] J. M. Robertson, J. Chem. Soc. 1936, 1195.

[83] D. M. Wrinch, Philos. Mag. J. Sci. **27**, 98 (1939).

[84] M. J. Buerger, *Vector Space, and Its Application in Crystal Structure Investigation* (Wiley, New York, 1959).

[85] C. E. Nordman, in *Computing in Crystallography*, edited by R. Diamond, S. Rameseshan, and K. Venkatesan (Indian Academy of Science, Bangalore, 1980).

[86] M. G. Rossman and D. M. Blow, Acta Crystallogr. **15**, 24 (1962).

[87] M. G. Rossman, *The Molecular Replacement Method*, Vol. 13 of *International Science Review Series* (Gordon and Breach, New York, 1972).

[88] P. Tollin and W. Cochran, Acta Crystallogr. **17**, 1322 (1964).

[89] D. Harker, Acta Crystallogr. **9**, 1 (1956).

[90] C. Bokhoven, J. C. Schoone, and J. M. Bijvoet, Acta Crystallogr. **4**, 245 (1951).

[91] Y. Okaya, Y. Saito, and R. Pepinsky, Phys. Rev. **98**, 1857 (1955).

[92] D. Harker and J. S. Kasper, Acta Crystallogr. **1**, 70 (1948).

[93] D. Sayre, Acta Crystallogr. **5**, 60 (1952).

[94] H. A. Hauptman and J. Karle, ACA Mongr. **3** (1953).

[95] W. H. Zachariasen, Acta Crystallogr. **5**, 68 (1952).

[96] I. L. Karle and J. Karle, Acta Crystallogr. **16**, 969 (1963); **17**, 835 (1964).

[97] J. Karle and H. Hauptman, Acta Crystallogr. **9**, 635 (1956).

[98] J. Karle and I. L. Karle, Acta Crystallogr. **21**, 849 (1966).

[99] G. Germain, P. Main, and M. M. Wolfson, Acta Crystallogr. Sect. A **27**, 368 (1971); P. Main, I. Lessinger, M. M. Woolfson, G. Germain, and J. P. Declercq, MULTAN-77 (University of York, UK, and Louvain, Belgium, 1977).

[100] G. M. Sheldrick, SHELX, Program for Crystal Structure Determination (Cambridge University, UK).

[101] P. Main, in *Computing in Crystallography*, edited by R. Diamond, S. Rameseshan, and K. Venkatesan (Indian Academy of Science, Bangalore, 1980).

[102] J. Karle and H. Hauptman, Acta Crystallogr. **14**, 217 (1961).

[103] H. A. Hauptman, *Crystal Structure Determination; the Role of the Cosine Semiinvariants* (Plenum, New York, 1972).

[104] H. Schenk, in *Computing in Crystallography*, edited by R. Diamond, S. Rameseshan, and K. Venkatesan (Indian Academy of Science, Bangalore, 1980).

[105] P. S. White and M. M. Woolfson, Acta Crystallogr. Sect. A **31**, 53 (1975).

[106] J. Karle, Acta Crystallogr. Sect. A **41**, 387 (1985); Sect A. **42**, 246 (1986).

[107] J. Karle and H. Hauptman, Acta Crystallogr. **3**, 181 (1950).

[108] E. W. Hughes, J. Am. Chem. Soc. **63**, 1737 (1941).

[109] W. Cochran, Acta Crystallogr. **1**, 138 (1948).

[110] D. W. J. Cruickshank, in *Computing Methods and the Phase Problem*, edited by R. Pepinsky, J. M. Robertson, and J. C. Speakman (Pergamon, New York, 1961).

[111] F. L. Hirshfeld and D. Rabinovich, Acta Crystallogr. Sect. A **29**, 510 (1973).

[112] L. Arnberg, S. Hovmoller, and S. Westman, Acta Crystallogr. Sect. A **35**, 497 (1979).

9.10. APPENDIX: CRYSTALLOGRAPHIC DATA SOURCES

9.10.1. Introduction

Crystallography has a long and successful history of self-organization and was one of the first areas to create numerical scientific databases. Virtually all structure determinations have been archived in databases that allow ready access and complete coverage. Crystallographic databases and computational archives support research on a daily basis for thousands of scientists worldwide. Although the earliest uses of these databases typically focused on one entry at a time for the purposes of identifying or finding related compounds, gradually this began to shift toward using larger subsets or even the entire database as a basis for research. Today, scientists use crystallographic data models to visualize, explain, and predict behavior of chemicals, materials, or biological compounds. With increasing use comes an increasing range of computational techniques to analyze and correlate data, and to help researchers concentrate experimental work in directions that optimize the discovery process. Selective cross-linking of the crystallographic data with other database systems is one important step in realizing the potential envisioned by today's molecular and materials designers. Interoperability with other data sources and software tools appears to be one of the emerging driving forces for innovation today.

9.10.2. Categories, quality, and description

There are two major categories of crystallographic databases: the *full structural* and the *identification* databases. The full structural databases are the primary repositories and are essentially nonoverlapping data sources. These databases form a comprehensive set covering inorganics, minerals, metals, intermetallics, small molecule organics and organometallics, nucleic acids, and proteins. The crystallographic databases used for identification are, to a large extent, currently derived from the major structural databases. The focus of the full structural databases is on the three-dimensional nature of the structure, whereas the focus of the identification databases is on the use of powder and single-crystal diffraction for compound identification.

The importance of quality throughout all stages of the data flow, including collection, publication, and incorporation into a database, has been widely recognized and promoted throughout the crystallographic community. Each database provider mentioned herein has well-established procedures in place to evaluate the specific category of data archived. Several types of evaluation are performed by experts in the specific discipline and by specialized computer programs. The first type examines an individual data item, such as a density, and checks to see that it falls within a reasonable range of values. The second type of evaluation looks for consistency within a complete entry. For example, interatomic distances are calculated on the basis of atomic coordinates and then compared with distances estimated from the ionic radii, and a chemical formula calculated from the coordinates, site occupancy, and multiplicity can be compared with the formula reported by the author. The third type examines the relationship of an individual entry to the entire database. This can be as simple as comparing a new space group with a table of space group statistics, or as comparing literature references and author lists.

The theory and algorithms developed to evaluate and standardize crystallographic information have had wide impact on data collection experiments, crystallographic publications, and database producers and users. One practical development is the adoption by the International Union of Crystallography (IUCr) of a standard file format for crystallographic information interchange (CIF format). In addition to promoting cooperation and facilitating standardization in the field of crystallography, the IUCr serves as a point of contact between crystallography and other sciences. Complete references and documentation for the CIF format, links to national crystallographic associations, as well as links to data resources for specialized applications and relevant software can be obtained from the IUCr Web site.

International Union of Crystallography
2 Abbey Square
Chester
CH1 2HU
England
e-mail: execsec@iucr.org
Web site: http://www.iucr.org

9.10.3. Major sources of crystallographic data

This section aims to identify the principal sources of crystallographic data, describe the scope of coverage, and provide a point of contact for further information.

Structural databases

Inorganic Crystal Structure Database The Inorganic Crystal Structure Database (ICSD) contains information on all structures that have no C–C and/or C–H bonds, and that include at least one of the nonmetallic elements H, He, B, C, N, O, F, Ne, Si, P, S, Cl, Ar, As, Se, Br, Kr, Te, I, Xe, At, and Rn. The ICSD contains full structural and bibliographic information for all structures from 1915 through the present. There are more than 45,000 entries in the current version, and approximately 2000 new entries are added per year. Data items include bibliographic information, such as the article title, authors' names, and literature citation; compound designation, such as chemical name, chemical formula, and mineral name; and crystallographic parameters, such as unit cell, space group, element symbol with numbering, oxidation state, multiplicity for Wyckoff position, x, y, z-coordinates, site occupation, thermal parameters, and reliability index R, among others. The ICSD is marketed through a variety of computerized media.

The ICSD is produced cooperatively by the Fachinformationszentrum (FIZ) Karlsruhe and the National Institute of Standards and Technology (NIST).

Fachinformationszentrum Karlsruhe
Gesellschaft für wissenshaftlich-technische Information mbH
D-76344 Eggenstein-Leopoldshafen
Germany
Web site: http://www.fiz-karlsruhe.de/

National Institute of Standards and Technology
Standard Reference Data Program
100 Bureau Drive
Gaithersburg, MD 20899 USA
Web site: http://www.nist.gov/srd/

Metals Crystallographic Data File The Metals Crystallographic Data File (CRYSTMET) is a computerized database containing crystallographic, chemical, and bibliographic data for metals, including alloys, intermetallics, and minerals. Entries included in the database are composed of elements to the left of the Zintl line in the periodic table, and phases formed between these elements and the elements immediately to the right. The crystallographic definition requires only that the space group, unit cell, and composition be clearly defined, resulting in about one-half of the entries having atomic coordinates. CRYSTMET contains approximately 50,000 entries, with about 1500 new entries added each year. CRYSTMET is available on CD-ROM; Web-searchable versions are planned.

Originally developed by the National Research Council Canada, CRYSTMET is now supported and maintained by Toth Information Systems, Inc., 2045 Quincy Avenue, Gloucester, ON, Canada K1J 6B2; Web site: http://www.tothcanada.com/

Cambridge Structural Database Experimentally determined crystal structures for organic and organometallic compounds are searchable through the Cambridge Structural Database (CSD). The CSD contains data for more than 200,000 entries starting from 1935, and it has an annual growth rate of approximately 15,000 new entries. Graphical search, retrieval, and data visualization software are distributed as part of the CSD system, allowing the user to access comprehensive bibliographic, two-dimensional chemical, and three-dimensional crystallographic information. The combination of file structure and accompanying software support complex searches, such as combining a two-dimensional substructure or chemical fragment search with three-dimensional constraints describing the geometry of the fragment. In addition, the Cambridge Crystallographic Data Centre develops and distributes an extensive list of software products that make use of the data in the CSD.

Cambridge Crystallographic Data Centre
12 Union Road
Cambridge CB2 1EZ
United Kingdom
Web site: http://www.ccdc.cam.ac.uk/

Protein Data Bank The Protein Data Bank (PDB) is a computerized repository of biological macromolecules whose three-dimensional structures have been determined experimentally by crystallography or NMR techniques. Information for more than 9500 PDB entries includes coordinates, NMR restraint files, and structure factors. The PDB accepts and distributes information in the mmCIF and PDB formats. The mmCIF data dictionary is composed of more than 1600 definitions. The PDB can be browsed using a query and reporting interface; searches and reports can be generated for single or multiple studies. The PDB is distributed via the Web, ftp, and CD-ROM.

The PDB is managed by the Research Collaboratory for Structural Bioinformatics (RCSB), a nonprofit consortium among Rutgers, the State University of New Jersey; the San Diego Supercomputer Center of the University of California, San Diego; and the National Institute of Standards and Technology. The PDB is managed by an RCSB Project Team; information about the PDB can be obtained from the Web (http://www.rcsb.org/), by e-mail to info@rcsb.org, or from the PDB Newsletter (c/o University of California, San Diego; San Diego Supercomputer Center, MC 0537; 9500 Gilman Drive, La Jolla, CA 92093-0537).

Nucleic Acid Database The focus of the Nucleic Acid Database (NDB) is on crystal structures that contain nucleic acids. Currently, there are over 900 entries in the NDB. The types of information archived include coordinate information, data collection methods, refinement statistics, crystallization conditions, as well as derived information, such as distances, torsion angles and base morphology. The NDB can be searched, and reports generated, through a variety of interfaces; search results link to the NDB Atlas and the NDB Archives. The NDB Atlas highlights important features of each structure and provides a graphical representation, whereas the NDB Archives maintain coordinate files, nucleic acid dictionaries, and prepared reports. The NDB is distributed via the Web (http://ndbserver.rutgers.edu), and it is located in the Department of Chemistry, Rutgers, the State University; Wright-Reiman Laboratories, 610 Taylor Road, Piscataway, NJ 08854-8087. Additional information may be obtained from the NDB Newsletter (ndbnews@ndbserver.rutgers.edu), or by contacting ndbadmin@ndbserver.rutgers.edu.

The NDB is a member of the Research Collaboratory for Structural Bioinformatics (RCSB; http://www.rcsb.org/).

Identification databases

Powder Diffraction File The Powder Diffraction File (PDF) is a collection of evaluated, single-phase, x-ray powder diffraction patterns used for the identification of crystalline materials. The distribution of patterns in the PDF is organized in a series of subfiles, such as inorganics, organics, forensics, common phases, and many others. Each entry must contain a list of the *d*-spacing and intensity pairs, chemical formula, and reference to the original source of the data, and it may contain supplemental data, such as the unit cell, physical constants, experimental details, preparative techniques, plus additional comments.

The PDF is the main database of the International Centre for Diffraction Data (ICDD), covering more than 100,000 materials. Recently, the ICDD has entered collaborative agreements with the Fachinformationszentrum Karlsruhe (Germany) and the Cambridge Crystallographic Data Centre (England) to allow the ICDD to calculate powder patterns from defined portions of the full structural databases (the ICSD and the CSD, respectively). Information regarding the addition of calculated patterns, distributed products, and available software may be obtained from the ICDD.

International Centre for Diffraction Data
12 Campus Boulevard
Newtown Square, PA 19073-3273 USA
Web site: http://icdd.com

NIST Crystal Data NIST Crystal Data is a comprehensive database containing chemical, physical, and crystallographic information on all types of well-characterized crystalline compounds. With approximately 240,000 entries, NIST Crystal Data covers all categories of materials, including inorganics, organics, organometallics, metals, intermetallics, and minerals. Among the data items archived are author's cell, Crystal Data cell, reduced cell, space group, Pearson's symbol, chemical name and formula, empirical formula, density, literature references, cross-references to other databases, plus others.

One important use of NIST Crystal Data is for the identification of unknown materials via lattice-matching. The lattice, as defined by a unit cell, is highly characteristic of a compound and, like a fingerprint, may be used for identification. When coupled with limited chemical information, this method typically yields unique results. This identification procedure has been incorporated directly into single-crystal diffractometer software.

NIST Crystal Data is compiled by the NIST Crystallographic Data Center, located within the Materials Science and Engineering Laboratory, and is distributed through the Standard Reference Data Program at NIST.

National Institute of Standards and Technology
Standard Reference Data Program
100 Bureau Drive
Gaithersburg, MD 20899 USA
Web site: http://www.nist.gov/srd/

NIST/Sandia/ICDD Electron Diffraction Database The Electron Diffraction Database is produced cooperatively with the National Institute of Standards and Technology, the International Centre for Diffraction Data, and Sandia National Laboratory. Designed for phase identification in electron diffraction, this database contains crystallographic and chemical information for over 80,000 inorganic materials, including calculated *d*-spacings

("R-spacings"), space group, unit cell, chemical formula, chemical name, and literature reference.

International Centre for Diffraction Data
12 Campus Boulevard
Newtown Square, PA 19073-3273 USA
Web site: http://icdd.com

National Institute of Standards and Technology
Standard Reference Data Program
100 Bureau Drive
Gaithersburg, MD 20899 USA
Web site: http://www.nist.gov/srd/

10

Earth, Ocean, and Atmosphere Physics

Ferris Webster
University of Delaware, Lewes, Delaware

Contents

List of Tables

List of Figures

10.1. INTRODUCTION

The earth sciences draw heavily on several fields of physics. In this chapter, we present a number of properties of planet Earth, summarize equations and definitions that are important in the physics of the ocean and atmosphere, and discuss some of the factors influencing global climate.

Even though the ocean and atmosphere have many physical and dynamic similarities, the equations and notation used often differ. To minimize those differences, we have chosen to use the following conventions in this chapter:

φ = latitude: $+\,90°$ at North pole, $-\,90°$ at South pole.

x = horizontal coordinate measured eastward.

y = horizontal coordinate measured northward.

z = vertical coordinate measured upward, in meters.

= 0 at sea level.

u, v, w = velocity components directed along the x-, y-, and z-axes.

10.2. PROPERTIES OF PLANET EARTH

10.2.1. Planetary dimensions and constants

Polar radius: 6357 km.

Equatorial radius: 6378 km.

Mean radius (a sphere having the same volume as Earth): 6371 km.

Mass of Earth: 5.977×10^{24} kg.

Mass of ocean: 1.4×10^{21} kg.

Mass of atmosphere: 5.3×10^{18} kg.

Mass of water in lakes and rivers: 5×10^{17} kg.

Mass of water in sediments and rocks: 2×10^{20} kg.

Mass of ice on Earth: 2.2×10^{19} kg.

Mass of water vapor in atmosphere: 1.3×10^{16} kg.

Dry mass of the atmosphere: 5.13×10^{18} kg.

(Continued)

Area of Earth: 5.10×10^{14} m^2.

Area of ocean: 3.62×10^{14} m^2.

Area of land: 1.49×10^{14} m^2.

Area of ice sheets and glaciers: 1.62×10^{13} m^2.

Area of sea ice: 1.75×10^{13} m^2 in March, 2.84×10^{13} m^2 in September.

Mean depth of ocean: 3.73 km [for comparison, mean elevation of land, 0.84 km].

Deepest ocean depth: 11.02 km, in the Marianas Trench [for comparison, Mt. Everest, 8.84 km].

Earth's angular velocity: $\Omega = 7.292115 \times 10^{-5}$ s^{-1}.

Coriolis parameter:

$$f = 2\Omega \sin \varphi = 1.03 \times 10^{-4} \text{ s}^{-1} \text{ at } 45^\circ \text{ N latitude.}$$

The tangential velocity of rotation at any latitude is

$$V = a\Omega \cos \varphi,$$

where a = mean radius of Earth, 6371 km.

At the equator, V becomes

$$464.58 \text{ m s}^{-1}.$$

Acceleration due to gravity:

Mean surface value of g:

$$g = 9.80665 \text{ m s}^{-1}$$

Inertial period (coriolis period; half-pendulum day):

$$T = \frac{2\pi}{f} = \frac{12 \text{ hr}}{\sin \varphi}$$

10.2.2. Ocean areas, volumes, and depths

In Table 10.1, the mediterranean seas are defined as follows:

Arctic Mediterranean: North Polar Sea, Norwegian Sea, Baffin Bay, and Canadian Archipelago.

Asiatic Mediterranean: The eastern boundary is a line extending from the coast of China to Taiwan, the Philippine Islands, Moluccas, New Guinea, to Cape York, Australia. The western boundary follows Malaysia and the Indonesian archipelago.

American Mediterranean: Caribbean Sea and Gulf of Mexico.

TABLE 10.1. Ocean areas, volumes, and depths. From Ref. [1].

Sea	Area	Volume	Mean depth	Max depth
	10^6 km^2	10^6 km^3	km	km
Oceans, without adjacent seas				
Pacific	166.24	696.19	4.188	11.022
Atlantic	84.11	322.98	3.844	9.219
Indian	73.43	284.34	3.872	7.455
Total	323.78	1303.51	4.026	
Mediterranean seas				
Arctic Mediterranean	12.26	13.70	1.117	5.449
Asiatic Mediterranean	9.08	11.37	1.252	7.440
American Mediterranean	4.36	9.43	2.164	7.680
European Mediterranean & Black Sea	3.02	4.38	1.450	5.092
Total	28.72	38.88	1.354	
Intracontinental mediterranean seas				
Hudson Bay	1.23	0.16	0.128	0.218
Red Sea	0.45	0.24	0.538	2.604
Baltic Sea	0.39	0.02	0.055	0.459
Persian Gulf	0.24	0.01	0.025	0.170
Total	2.31	0.43	0.184	
Marginal sea				
Bering Sea	2.26	3.37	1.491	4.096
Sea of Okhotsk	1.39	1.35	0.971	3.372
Yellow & East China Seas	1.20	0.33	0.275	2.719
Sea of Japan	1.01	1.69	1.673	4.225
Gulf of California	0.15	0.11	0.733	3.127
North Sea	0.58	0.05	0.093	0.725
Gulf of St. Lawrence	0.24	0.03	0.125	0.549
Irish Sea	0.10	0.01	0.060	0.272
Remaining seas	0.30	0.15	0.470	
Total	7.23	7.09	0.979	
Oceans including adjacent seas				
Pacific	181.34	714.41	3.940	11.022
Atlantic	106.57	350.91	3.293	9.219
Indian	74.12	284.61	3.840	7.455
Total	362.03	1349.93	3.729	11.022

10.3. OCEAN

10.3.1. Seawater properties

Units

Most quantities in oceanographic usage today are expressed in SI units. However, for practical use, a few special units are widely used, as described in Table 10.2.

TABLE 10.2. Some units used in oceanography.

Quantity	Symbol	Units
Salinity	S	see below
Temperature	t	°C
Temperature	T	K
Pressure	p	Pa
Sea pressure	P	bar
Density	ρ	kg m^{-3}
Specific volume	$\nu = 1/\rho$	m^3 kg^{-1}

Sverdrup (ocean current volume transport): 1 Sverdrup (Sv) $= 10^6$ m^3 s^{-1}.

Nautical mile: The nautical mile was originally defined as equal to one minute of arc on the Earth's surface at the equator and, thus, approximately as one minute of longitude. By convention, the International Nautical Mile is defined as exactly

$$1 \text{ International Nautical Mile} = 1.852 \text{ km}.$$

Pressure

Atmospheric pressure, $p = 10^5$ Pa $= 1$ bar $= 0.987$ atm:

$$1 \text{ decibar (dbar)} = 10^4 \text{ Pa} = 0.1 \text{ bar} = 0.0987 \text{ atm}.$$

Pressure in the ocean increases approximately 1 dbar/m of depth. At depths shallower than 4000 m, the difference between hydrostatic pressure (in decibars) and geometric depth (in meters) is within 2%. Consequently, oceanographers frequently express depth in the ocean in decibars.

In oceanographic practice, the quantity P refers to the sea pressure, which is the pressure due only to the column of water above a point in the sea. Thus, at the ocean surface, at atmospheric pressure, 1 bar, oceanographic usage gives the sea pressure as $P = 0$. In accordance with traditional oceanographic usage, throughout the ocean part of this chapter, the term P will be used to mean the incremental pressure due to the column of water.

Salinity

Absolute Salinity (symbol S_A) is defined as the ratio of mass of dissolved material in sea water to the mass of sea water. In practice, absolute salinity cannot be measured directly with accuracy. A complete chemical analysis of seawater must be made to determine the true salinity of seawater. Since this is too time-consuming, in the past, chlorinity was traditionally used to determine salinity. Over the years, a number of relationships between chlorinity and salinity have been used, the most recent of which is

$$S(‰) = 1.80655\ \mathrm{Cl}(‰) \quad \text{(see ref. [2])}.$$

Because of the difficulties involved in defining salinity in terms of the chlorinity-salinity relationship, a *Practical Salinity*, based on a salinity-conductivity ratio was defined in 1978 for reporting oceanographic observations.

The **Practical Salinity** (symbol S) of a sample of sea water is defined in terms of the ratio K_{15} of the electrical conductivity of the seawater sample at the temperature of 15°C and the pressure of one standard atmosphere, to that of a potassium chloride (KCl) solution, in which the mass fraction of KCl is 32.4356×10^{-3}, at the same temperature and pressure. The K_{15} value exactly equal to 1 corresponds, by definition, to a Practical Salinity exactly equal to 35 on the Practical Salinity Scale. The Practical Salinity is defined in terms of the ratio K_{15} by the following equation:

$$S = a_0 + a_1 K_{15}^{1/2} + a_2 K_{15} + a_3 K_{15}^{3/2} + a_4 K_{15}^2 + a_5 K_{14}^{5/2},$$

where

$$\begin{aligned} a_0 &= 0.0080, \\ a_1 &= -0.1692, \\ a_2 &= 25.3851, \\ a_3 &= 14.0941, \\ a_4 &= -7.0261, \\ a_5 &= 2.7081, \\ \sum a_i &= 35.0000, \end{aligned}$$

and

$$2 \le S \le 42.$$

This equation is valid for a Practical Salinity, S, from 2 to 42. [3]

Equation of state for seawater

The 1980 equation of state for seawater (EOS-80) was adopted by the UNESCO Joint Panel on Oceanographic Tables and Standards (JPOTS) in 1981: [4]

The specific volume is

$$\nu(S, t, P) = \nu(S, t, 0)\left[1 - \frac{P}{K(S, t, P)}\right],$$

with pressure expressed in bars.

The density at surface pressure is

$$\rho(S,t,0) = \frac{1}{\nu(S,t,0)} = A + BS + CS^{3/2} + DS^2.$$

The secant bulk modulus is

$$K(S,t,P) = E + FS + GS^{3/2} + (H + IS + JS^{3/2})P + (M + NS)P^2.$$

TABLE 10.3. Coefficients in equation of state.

	A	B	C
t^0	+999.842594	+8.24493E-1	−5.72466E-3
t^1	+6.793952E-2	−4.0899E-3	+1.0227E-4
t^2	−9.095290E-3	+7.6438E-5	−1.6546E-6
t^3	+1.001685E-4	−8.2467E-7	
t^4	−1.120083E-6	+5.3875E-9	
t^5	+6.536332E-9		
	D	E	F
t^0	+4.8314E-4	+19652.21	+54.6746
t^1		+148.4206	−0.603459
t^2		−2.327105	+1.09987E-2
t^3		+1.360477E-2	−6.1570E-5
t^4		−5.155288E-5	
	G	H	I
t^0	+7.944E-2	+3.239908	+2.2838E-3
t^1	+1.6483E-2	+1.43713E-3	−1.0981E-5
t^2	−5.3009E-4	+1.16092E-4	−1.6078E-6
t^3		−5.77905E-7	
	J	M	N
t^0	+1.91075E-4	+8.50935E-5	−9.9348E-7
t^1		−6.12293E-6	+2.0816E-7
t^2		+5.2787E-8	+9.1697E-10

Density properties [2]

Density anomaly: The maximum variation of density in magnitude over the oceanic range of salinity, temperature, and pressure is only 7%. Numerical resolution can be improved by using anomalies of specific volume and density. The density anomaly γ[kg m^{-3}] is defined by

$$\gamma(S,t,P) = \rho(S,t,P) - 1000.0 \quad [\text{kg m}^{-3}].$$

An older equivalent of the density anomaly is the specific gravity anomaly

$$\sigma_t = 10^3(\rho/\rho_m - 1),$$

where the maximum density of pure water was taken as $\rho_m = 1 \text{ g cm}^{-3}$.

The older values of specific gravity anomaly may be converted to density anomaly using

$$\gamma = 10^{-3}\rho_m\sigma_t + (\rho_m - 1000) \quad [\text{kg m}^{-3}].$$

With the recently accepted value of maximum density of standard mean ocean water, $\rho_m = 999.975 \text{ kg m}^{-3}$, the conversion becomes

$$\gamma = 0.000075\sigma_t - 0.025.$$

Specific volume (steric) anomaly: The specific volume anomaly, $\delta[\text{m}^3 \text{ kg}^{-1}]$, is defined as

$$\delta(S, t, P) = \nu(S, t, P) - \nu(35, 0, P),$$

where

$$\nu(35, 0, P) = \nu(35, 0, 0)\left[1 - \frac{P}{K(35, 0, P)}\right],$$

$$K(35, 0, P) = E_{35} + H_{35}P + M_{35}P^2.$$

Pressure in bars:

$$E_{35} = +21582.27,$$

$$H_{35} = +3.359406,$$

$$M_{35} = +5.03217\text{E-}5,$$

$$\nu(35, 0, 0) = 972.662039 \times 10^{-6} \text{ m}^3 \text{ kg}^{-1}.$$

Potential temperature

The temperature that a water parcel would attain if raised adiabatically to the sea surface without gain or loss of heat to the surroundings is called the *potential temperature* and is designated θ.

For $30 < S < 40$, $1 < t < 30$°C, and $0 < P < 1000$ bars (see Ref. [5]):

$$\begin{aligned}\theta(S, t, P) = t - P(&3.6504 \times 10^{-4} + 8.3198 \times 10^{-5}t \\ &- 5.4065 \times 10^{-7}t^2 + 4.0274 \times 10^{-9}t^3) \\ &- P(S - 35)(1.7439 \times 10^{-5} - 2.9778 \times 10^{-7}t) \\ &- P^2(8.9309 \times 10^{-7} - 3.1628 \times 10^{-8}t + 2.1987 \times 10^{-10}t^2) \\ &+ P^2(S - 35)(4.1057 \times 10^{-9}) \\ &- P^3(-1.6056 \times 10^{-10} + 5.0484 \times 10^{-12}t).\end{aligned}$$

TABLE 10.4. Major constituents in surface seawater. From Ref. [6].

Constituent	g/kg of seawater, $S = 35$	Percentage by mass
Cl^-	19.353	55.03
Na^+	10.781	30.65
SO_4^{--}	2.712	7.71
Mg^{++}	1.284	3.65
Ca^{++}	0.4119	1.17
K^+	0.399	1.13
HCO_3^-	0.126	0.36
Br^-	0.0673	0.19
Sr^{++}	0.008	0.02
$B(OH)_3$	0.0257	0.07
F^-	0.00130	0.01
Total	35.169	99.99
All other constituents	~ 0.03	
Totals	35.199	
Water	~ 964.80	

Specific heat of seawater

The specific heat of seawater c_p is the heat in joules required to raise the temperature of 1 kg of seawater 1 °C at constant pressure. Specific heat increases with t and decreases with S and P. A table of values is given in Ref. [7].

$$c_p = 3993.9 \text{ J kg}^{-1}\,{}^\circ\text{C}^{-1}, \qquad \text{at atmospheric pressure, } t = 20\,{}^\circ\text{C},\ S = 35.$$

For fresh water:

$$c_p(0, t, 0) = 4217.4 - 3.720283t + 0.1412855t^2 - 2.654387(10^{-3}t^3 + 2.093236 \times 10^{-5}t^4.$$

For seawater:

$$c_p(S, t, 0) = c_p(0, t, 0) + S(-7.643575 + 0.1072763t - 1.38385 \times 10^{-3}t^2)$$
$$+ S^{3/2}(0.1770383 - 4.07718 \times 10^{-3}t + 5.148 \times 10^{-5}t^2) \quad \text{(see Ref. [4])}.$$

Speed of sound in seawater [8]

$$c(S, t, z) = c_0 + \alpha_0(t - 10) + \beta_0(t - 10)^2 + \gamma_0(t - 18)^2 + \delta_0(S - 35)$$
$$+ \varepsilon_0(t - 18)(S - 35) + \zeta_0\,|z|\,, \quad [\text{m s}^{-1}]$$

where

$$c_0 = 1493.0 \text{ m s}^{-1},$$

$$\alpha_0 = 3.0,$$

$$\beta_0 = -0.006,$$

$$\gamma_0 = -0.04,$$

$$\delta_0 = 1.2,$$

$$\varepsilon_0 = -0.01,$$

$$\zeta_0 = 0.164.$$

At depth 1000 m, $t = 10°\text{C}$, $S = 35$, the speed of sound in seawater is 1506.3 m s^{-1}.

Freezing point

The freezing point temperature is [9]

$$t_f(S, P) = -0.0575S + 1.710523 \times 10^{-3} S^{3/2} - 2.154996 \times 10^{-4} S^2 - 7.53 \times 10^{-2} P,$$

where P is the pressure in bars.

$$t_f = -1.33°\text{C, at } S = 24.7.$$

Molecular diffusion

Molecular diffusion of sea salt:

$$F_s = -\rho\kappa_s \frac{\partial S}{\partial n} \quad [\text{kg m}^{-2}\,\text{s}^{-1}],$$

where n is defined as normal to the surface of constant S. The negative sign denotes diffusion down the salt gradient, thus, tending to reduce it. For seawater, the coefficient of molecular diffusivity of salt is

$$\kappa_S \cong 1.5 \times 10^{-9}\ \text{m}^2\,\text{s}^{-1}.$$

Molecular diffusion of heat:

$$Q = -\rho\kappa_Q c_p \frac{\partial t}{\partial n} \quad [\text{J m}^{-2}\,\text{s}^{-1}],$$

where n is defined as normal to the surface of constant t. The negative sign denotes diffusion down the temperature gradient, thus, tending to reduce it. For seawater, the coefficient of molecular diffusivity of heat is

$$\kappa_Q \cong 1.5 \times 10^{-7}\ \text{m}^2\,\text{s}^{-1}.$$

Molecular diffusion of momentum: Momentum diffusion is equivalent to the force per unit area on a plane parallel to flow, or the stress per unit surface area. It is

$$\tau = -\mu \frac{dv}{dn} \quad [\text{N m}^{-2}],$$

where n is defined as the direction of the velocity gradient, normal to the surface of constant v. The negative sign denotes diffusion down the velocity gradient, thus, tending to reduce it.

For seawater, the dynamic viscosity is

$$\mu \cong 1 \times 10^3 \text{ kg m}^{-1} \text{ s}^{-1}.$$

Turbulent diffusion

Diffusion through turbulent processes can be described by replacing the molecular coefficients with eddy coefficients. Oceanic eddy coefficients cover a wide range of values, since turbulent eddies in the ocean can have scales that vary from a few tens of centimeters to tens of kilometers. Furthermore, because stratification inhibits vertical motions, horizontal eddy coefficients are generally several orders of magnitude greater than vertical eddy coefficients. Though the molecular diffusion coefficients for salt and heat differ by two orders of magnitude, under conditions of turbulent control of diffusion, the same eddy coefficients are generally used for both salt and heat.

For vertical diffusion:

$$A_z \sim 0.1 \rightarrow 10 \text{ kg m}^{-1} \text{ s}^{-1}.$$

For horizontal diffusion:

$$A_h \sim 10^2 \rightarrow 10^6 \text{ kg m}^{-1} \text{ s}^{-1}.$$

10.3.2. Air-sea interaction

Evaporation bulk formula

The rate of evaporation E is also known as the latent heat flux across the air-sea interface:

$$E = \rho_a c_E (q_s - q_a) W,$$

where

$$\begin{aligned}
E &= \text{evaporation rate} \quad [\text{W m}^{-2}], \\
c_E &\sim 1.5 \times 10^3, \\
q_a &= \text{specific humidity of air at standard level}, \\
q_s &= \text{specific humidity at sea surface}, \\
\rho_a &= \text{air density} \quad [\text{kg m}^{-3}], \\
W &= \text{wind speed} \quad [\text{m s}^{-1}].
\end{aligned}$$

Sensible heat bulk formula

The turbulent flux of heat across the air-sea interface Q_h is proportional to the difference between the sea temperature t_s and the air temperature t_a:

$$\begin{aligned}
Q_h &= \rho_a c_p c_H (t_s - t_a) W, \\
Q_h &= \text{upward heat flux} \quad [\text{W m}^{-2}], \\
\rho_a c_p &= \text{heat capacity per unit volume of air} \quad [\text{J m}^{-3}{}^\circ\text{C}^{-1}],
\end{aligned}$$

$$W = \text{wind speed} \quad [\text{m s}^{-1}],$$
$$t_s = \text{sea-surface temperature} \quad [^\circ\text{C}],$$
$$t_a = \text{air temperature at standard level} \quad [^\circ\text{C}],$$
$$c_H \sim 0.83 \times 10^{-3} \text{ for stable conditions,}$$
$$\sim 1.10 \times 10^{-3} \text{ for unstable conditions.}$$

Wind stress bulk formula

The stress of the wind on the sea surface τ varies quadratically with wind speed:

$$\tau = c_d \rho_a |\mathbf{u}_w| \mathbf{u}_w \quad [\text{N m}^{-2}],$$

where

$$\rho_a = 1.225 \text{ kg m}^{-3} \quad [\text{density of air, kg m}^{-3}],$$
$$\mathbf{u}_w = \text{wind vector at 10 m height,} \quad [\text{m s}^{-1}].$$

The aerodynamic drag coefficient c_d increases with wind speed. At low wind speeds, $c_d \sim 0.95 \times 10^{-3}$. At higher wind speeds,

$$c_d = \left(\frac{\kappa}{\ln(\rho_a g z / a\tau)} \right)^2 .$$
$$\kappa = 0.4, a = 0.0185 \quad \text{(see Ref. [10])}.$$

z is the anemometer height [m].

10.3.3. Tides

TABLE 10.5. The most important constituents of the tide-generating force. From Ref. [11], p. 267.

	Symbol	Period (solar hours)	Amplitude ($M_2 = 100$)	Constituent description
Semidiurnal tides	M_2	12.42	100.0	main lunar semidiurnal
	S_2	12.00	46.6	main solar semidiurnal
	N_2	12.66	19.1	larger lunar elliptic
	K_2	11.97	12.7	luni-solar semidiurnal
Diurnal tides	K_1	23.93	58.4	luni-solar diurnal
	O_1	25.82	41.5	main lunar diurnal
	P_1	24.07	19.3	main solar diurnal
Long-period tides	M_f	327.86	17.2	lunar fortnightly
	M_m	661.30	9.1	lunar monthly
	S_{sa}	2191.43	8.0	solar semiannual

10.3.4. Waves

Small-amplitude surface gravity waves

TABLE 10.6. Characteristics of small-amplitude surface gravity waves

Quantity	Symbol	Short waves, deep water wavelength controls speed $\frac{h}{\Lambda} > \frac{1}{2}$	Long waves, shallow water water depth controls speed $\frac{h}{\Lambda} < \frac{1}{20}$
Phase velocity	C	$C_s = \sqrt{\frac{g}{k}}$	$C_l = \sqrt{gh}$
Wavelength (crest-to-crest)	Λ	$\Lambda_s = \frac{gT^2}{2\pi}$	$\Lambda_l = \sqrt{gh}\,T$
Surface displacement	η	$A\cos(kx - \omega t)$	$A\cos(kx - \omega t)$
Particle velocity	u	$A\omega\,e^{kz}\cos(kx - \omega t)$	$\frac{A\omega}{kh}\cos(kx - \omega t)$
components	w	$-A\omega\,e^{kz}\sin(kx - \omega t)$	$-A\omega\left(1 + \frac{z}{h}\right)\sin(kx - \omega t)$
Differential pressure	Δ_p	$\rho g A\,e^{kz}\cos(kx - \omega t)$	$\rho g A\cos(kx - \omega t)$
Semimajor & semiminor axes of ellipse	$\mathbf{A}$ / $\mathbf{B}$	$\mathbf{A} = \mathbf{B} = A\,e^{kz}$	$\mathbf{A} = \frac{A}{kh}$; $\mathbf{B} = A\left(\frac{h+z}{h}\right)$
Group velocity	V	$V_s = \frac{1}{2}\sqrt{\frac{g}{k}} = \frac{1}{2}C_s$	$V_l = \sqrt{gh} = C_l$

$k = \frac{2\pi}{\Lambda}$ = wavenumber $\qquad \omega = \frac{2\pi}{T}$ = frequency

A = wave amplitude $\qquad T$ = wave period

h = water depth $\qquad t$ = time

g = acceleration due to gravity $\qquad \rho$ = water density

x = horizontal dimension $\qquad z$ = vertical dimension, measured upward

Capillary waves

The dispersion relation for capillary waves is

$$c^2 = \frac{g}{k} + \frac{k}{\rho}\Upsilon,$$

where

$$k = \text{wavenumber} \quad [\text{m}^{-1}],$$
$$\rho = \text{water density} \quad [\text{kg m}^{-3}],$$
$$\Upsilon = \text{surface tension} \quad [\text{N m}^{-1}].$$

To account for the effect of salinity and temperature on surface tension, we may use an empirical formula:

$$\Upsilon = 0.0756 - 1.44 \times 10^{-4} t + 2.12 \times 10^{-5} S \quad [\text{N m}^{-1}].$$

However, even small amounts of organic matter, as commonly occurs on the sea surface, can lower surface tension dramatically. A typical value for the sea surface could thus be $\Upsilon \sim 0.05\ \text{N m}^{-1}$.

10.3.5. Geophysical fluid dynamics

Geostrophic balance

For horizontal frictionless unaccelerated flow in the ocean or atmosphere, there is a balance between Coriolis acceleration and pressure gradient force:

$$fv = \frac{1}{\rho}\frac{\partial p}{\partial x},$$
$$fu = -\frac{1}{\rho}\frac{\partial p}{\partial y},$$

where p is pressure, ρ is density, and f is the Coriolis parameter.

Thermal wind equation

The vertical gradient of velocity can be determined from the horizontal gradient of density for flow that is frictionless and unaccelerated:

$$\frac{\partial(\rho f v)}{\partial z} = -g\frac{\partial \rho}{\partial x},$$
$$\frac{\partial(\rho f u)}{\partial z} = g\frac{\partial \rho}{\partial y}.$$

Ekman flow

Wind stress acting on the sea surface can generate a net transport at right angles to the direction of the stress in a frictionless, unaccelerated ocean:

$$\tau_x = -M_y f,$$
$$\tau_y = M_z f,$$

τ_x, τ_y are the components of horizontal wind stress acting in the x- and y-directions.

$$\left.\begin{aligned} M_x &= \int_{-z}^{0} \rho u \, dz \\ M_y &= \int_{-z}^{0} \rho v \, dz \end{aligned}\right\}$$ are the *Ekman transports* per unit distance in the x-and y-directions.

10.4. ATMOSPHERE

10.4.1. Principal atmospheric constituents

Table 10.7 gives values for dry air. For gases increasing in concentration, values are estimated for the year 1990. [6]

TABLE 10.7. Average atmospheric composition.

Gaseous constituent	Molecular form	Parts per million by volume	Notes
Nitrogen	N_2	780,840	
Oxygen	O_2	209,460	
Argon	Ar	9,340	
Carbon dioxide	CO_2	354	Increasing $\sim$ 0.4%/yr
Helium	He	5.22	
Krypton	Kr	1.14	
Xenon	Xe	0.087	
Methane	CH_4	1.6	Increasing $\sim$ 1%/yr
Nitrous oxide	N_2O	0.5	Increasing

10.4.2. Properties of moist air

See Ref. [12], Appendix 4.

Latent heats of water

Latent heat of vaporization:

$$\begin{aligned} L_v(t) &= (2.5008 - 0.00239t) \times 10^6 \quad \text{J kg}^{-1} \\ &= 2.453 \times 10^6 \text{ J kg}^{-1} \text{ at } t = 20°\text{C} \\ &= 586 \text{ cal g}^{-1}. \end{aligned}$$

Latent heat of fusion:

$$L_f = 0.335 \times 10^6 \text{ J kg}^{-1} \quad \text{at } t = 0°\text{C}$$

Moisture content of air

Molar mass of dry air: $m_a = 28.966 \quad \text{g mol}^{-1}$.
Molar mass of water: $m_w = 18.016 \quad \text{g mol}^{-1}$.
Molecular weight ratio: $e = m_w/m_a = 0.62197$.

Vapor concentration or absolute humidity The vapor concentration ρ_v is the mass of water vapor per unit volume of moist air.

Specific humidity, q, mass of vapor per unit mass of moist air:

$$q = \frac{\rho_v}{\rho}.$$

Mixing ratio, r, ratio of the mass of vapor to the mass of dry air:

$$r = \frac{q}{1 - q}.$$

Vapor pressure, e', of water vapor in moist air is a function of pressure, p, and specific humidity, q:

$$\begin{aligned} \frac{e'}{p} &= \frac{q}{e + (1 - e)q} \\ &= \frac{r}{e + r} \\ &= \frac{r}{0.62197 + r}. \end{aligned}$$

Relative humidity, U, is the ratio of the mixing ratio, r, to the saturation mixing ratio, r_w, of moist air relative to a plane water surface:

$$U = \frac{r}{r_w} = \frac{q(1 - q_w)}{q_w(1 - q)}.$$

10.4.3. Properties of dry air

(at temperature of 0°C, pressure of one standard atmosphere)

Specific heat at constant volume: $c_v = 719.6\ \mathrm{J\,kg^{-1}\,K^{-1}}$.
Specific heat at constant pressure: $c_p = 1004.2\ \mathrm{J\,kg^{-1}\,K^{-1}}$.
Dry adiabatic lapse rate: $\Gamma = (g/c_p)$ (to within 0.3%).
Density: $\rho = 1.293\ \mathrm{kg\,m^{-3}}$.
Speed of sound in air: $331.4\ \mathrm{m\,s^{-1}}$.

Specific heat of water vapor

$$c_p = 2076.6\ \mathrm{J\,kg^{-1}\,K^{-1}} \quad @100°\mathrm{C},\ 1\ \mathrm{bar}.$$

Specific heat of liquid water

$$c_p = 4188.5\ \mathrm{J\,kg^{-1}\,K^{-1}} \quad @15°\mathrm{C},\ 1\ \mathrm{bar}.$$

10.4.4. U.S. Standard Atmosphere (1976)

A Standard Atmosphere is a hypothetical vertical distribution of atmospheric temperature, pressure, and density that is roughly representative of year-round, midlatitude conditions. Typical uses are to serve as a basis for pressure altimeter calibrations, aircraft performance calculations, aircraft and rocket design, ballistic tables, meteorological diagrams, and various types of atmospheric modeling, The air is assumed to be dry and to obey the perfect gas law and the hydrostatic equation that, taken together, relate temperature, pressure, and density with vertical position. The atmosphere is considered to rotate with the earth and to be an average over the diurnal cycle, the semiannual variation, and the range from active to quiet geomagnetic and sunspot conditions.

The U.S. Standard Atmosphere (1976) [13, 14] is an idealized, steady-state representation of mean annual conditions of the Earth's atmosphere from the surface to 1000 km at latitude 45° N, as it is assumed to exist during a period with moderate solar activity. The defining meteorological elements are sea-level temperature and pressure and a temperature-height profile to 1000 km. The 1976 Standard Atmosphere uses the following sea-level values, which have been standard for many decades:

Temperature:	288.15 K (15°C).
Pressure:	101.325 kPa (1013.25 mbar, 760 mm of Hg, or 29.92 in. of Hg).
Density:	1.225 kg m^{-3} (1.225 g/L).
Mean molar mass:	28.964 g/mol.

The parameters included in this condensed version of the U.S. Standard Atmosphere are as follows:

Z = Height (geometric) above mean sea level in meters.

T = Temperature [K].

P = Pressure [Pa].

ρ = Density [kg m^{-3}].

n = Number density [molecules m^{-3}].

ν = Mean collision frequency [s^{-1}].

l = Mean free path [m].

η = Absolute viscosity [Pa s].

k = Thermal conductivity [J m^{-1} s^{-1} kg^{-1} or W m^{-1} K^{-1}].

v_s = Speed of sound [m s^{-1}].

g = Acceleration of gravity [m s^{-2}].

For the following table, the sea-level composition (percent by volume) is taken to be:

N_2	78.084	Ne	0.001818	Xe	0.0000087
O_2	20.9476	He	0.000524	CH_4	0.0002
Ar	0.934	Kr	0.000114	H_2	0.00005
CO_2	0.0314				

TABLE 10.8. Values for U.S. Standard Atmosphere (1976)

Z m	T K	P Pa	ρ kg m^{-3}	n m^{-3}	ν s^{-1}	l m	η Pa s	k J m^{-1}s^{-1}K^{-1}	v_S m s^{-1}	g m s^{-2}
−5000	320.68	1.778E + 05	1.931	4.015E + 25	1.151E + 10	4.208E − 08	1.942E − 05	0.02788	359.0	9.822
−4500	317.42	1.685E + 05	1.849	3.845E + 25	1.096E + 10	4.395E − 08	1.927E − 05	0.02763	357.2	9.830
−4000	314.17	1.596E + 05	1.770	3.680E + 25	1.044E + 10	4.592E − 08	1.912E − 05	0.02738	355.3	9.819
−3500	310.91	1.511E + 05	1.693	3.520E + 25	9.933E + 09	4.800E − 08	1.897E − 05	0.02713	353.5	9.818
−3000	307.66	1.430E + 05	1.619	3.366E + 25	9.448E + 09	5.019E − 08	1.882E − 05	0.02688	351.6	9.816
−2500	304.41	1.352E + 08	1.547	3.217E + 25	8.982E + 09	5.252E − 08	1.867E − 05	0.02663	349.8	9.814
−2000	301.15	1.278E + 05	1.478	3.102E + 25	8.623E + 09	5.447E − 08	1.852E − 05	0.02638	347.9	9.813
−1500	297.90	1.207E + 05	1.411	2.935E + 25	8.106E + 09	5.757E − 08	1.836E − 05	0.02613	346.0	9.811
−1000	294.65	1.139E + 05	1.347	2.801E + 25	7.693E + 09	6.032E − 08	1.821E − 05	0.02587	344.1	9.810
−500	291.40	1.075E + 05	1.285	2.672E + 25	7.298E + 09	6.324E − 08	1.805E − 05	0.02562	342.2	9.808
0	288.15	1.013E + 05	1.225	2.547E + 25	6.919E + 09	6.633E − 08	1.789E − 05	0.02533	340.3	9.807
500	284.90	9.546E + 04	1.167	2.427E + 25	6.556E + 09	6.961E − 08	1.774E − 05	0.02511	338.4	9.805
1000	281.65	8.988E + 04	1.112	2.311E + 25	6.208E + 09	7.310E − 08	1.758E − 05	0.02485	336.4	9.804
1500	278.40	8.456E + 04	1.058	2.200E + 25	5.874E + 09	7.680E − 08	1.742E − 05	0.02459	334.5	9.802
2000	275.15	7.950E + 04	1.007	2.093E + 25	5.555E + 09	8.073E − 08	1.726E − 05	0.02433	332.5	9.801
2500	271.91	7.469E + 04	0.957	1.990E + 25	5.250E + 09	8.491E − 08	1.710E − 05	0.02407	330.6	9.799
3000	268.66	7.012E + 04	0.909	1.891E + 25	4.959E + 09	8.937E − 08	1.694E − 05	0.02381	328.6	9.797
3500	265.41	6.579E + 04	0.863	1.795E + 25	4.680E + 09	9.411E − 08	1.678E − 05	0.02355	326.6	9.796
4000	262.17	6.166E + 04	0.819	1.704E + 25	4.414E + 09	9.917E − 08	1.661E − 05	0.02329	324.6	9.794
4500	258.92	5.775E + 04	0.777	1.616E + 25	4.160E + 09	1.046E − 07	1.645E − 05	0.02303	322.6	9.793
5000	255.68	5.405E + 04	0.736	1.531E + 25	3.918E + 09	1.103E − 07	1.628E − 05	0.02277	320.6	9.791
5500	252.43	5.054E + 04	0.697	1.450E + 25	3.687E + 09	1.165E − 07	1.612E − 05	0.02250	318.5	9.790
6000	249.19	4.722E + 04	0.660	1.373E + 25	3.467E + 09	1.231E − 07	1.595E − 05	0.02224	316.5	9.788
6500	245.94	4.408E + 04	0.664	1.299E + 25	3.258E + 09	1.302E − 07	1.578E − 05	0.02197	314.4	9.787
7000	242.70	4.111E + 04	0.590	1.227E + 25	3.058E + 09	1.377E − 07	1.561E − 05	0.02170	312.3	9.785
7500	239.46	3.830E + 04	0.557	1.159E + 25	2.869E + 09	1.458E − 07	1.544E − 05	0.02144	310.2	9.784
8000	236.22	3.565E + 04	0.526	1.093E + 25	2.689E + 09	1.545E − 07	1.527E − 05	0.02117	308.1	9.782
8500	232.97	3.315E + 04	0.496	1.031E + 25	2.518E + 09	1.639E − 07	1.510E − 05	0.02090	306.0	9.781
9000	229.73	3.080E + 04	0.467	9.711E + 24	2.356E + 09	1.740E − 07	1.493E − 05	0.02063	303.9	9.779
9500	226.49	2.858E + 04	0.440	9.141E + 24	2.202E + 09	1.848E − 04	1.475E − 05	0.02036	301.7	9.777
10000	223.25	2.650E + 04	0.414	8.598E + 24	2.056E + 09	1.965E − 04	1.458E − 05	0.02009	299.5	9.776

(continued)

TABLE 10.8. Continued

Z m	T K	P Pa	ρ kg m^{-3}	n m^{-3}	ν s^{-1}	l m	η Pa s	k J m^{-1}s^{-1}K^{-1}	v_s m s^{-1}	g m s^{-2}
10500	220.01	2.454E + 04	0.389	8.079E + 24	1.918E + 09	2.091E − 07	1.440E − 05	0.01982	297.4	9.774
11000	216.77	2.270E + 04	0.365	7.585E + 24	1.787E + 09	2.227E − 07	1.422E − 05	0.01954	295.2	9.773
11500	216.65	2.098E + 04	0.337	7.016E + 24	1.653E + 09	2.408E − 07	1.422E − 05	0.01953	295.1	9.771
12000	216.65	1.940E + 04	0.312	6.486E + 24	1.528E + 09	2.605E − 07	1.422E − 05	0.01953	295.1	9.770
12500	216.65	1.793E + 04	0.288	5.996E + 24	1.412E + 09	2.818E − 07	1.422E − 05	0.01953	295.1	9.768
13000	216.65	1.658E + 04	0.267	5.543E + 24	1.306E + 09	3.048E − 07	1.422E − 05	0.01953	295.1	9.767
13500	216.65	1.533E + 04	0.246	5.124E + 24	1.207E + 09	3.297E − 07	1.422E − 05	0.01953	295.1	9.765
14000	216.65	1.417E + 04	0.228	4.738E + 24	1.116E + 09	3.566E − 07	1.422E − 05	0.01953	295.1	9.764
14500	216.65	1.310E + 04	0.211	4.380E + 24	1.032E + 12	3.857E − 07	1.422E − 05	0.01953	295.1	9.762
15000	216.65	1.211E + 04	0.195	4.049E + 24	9.538E + 08	4.172E − 07	1.422E − 05	0.01953	295.1	9.761
16000	216.65	1.035E + 04	0.166	3.461E + 24	8.153E + 08	4.881E − 07	1.422E − 05	0.01953	295.1	9.758
17000	216.65	8.850E + 03	0.142	2.959E + 24	6.969E + 08	5.710E − 07	1.422E − 05	0.01953	295.1	9.754
18000	216.65	7.565E + 03	0.122	2.529E + 24	5.958E + 08	6.680E − 07	1.422E − 05	0.01953	295.1	9.751
19000	216.65	6.467E + 03	0.104	2.162E + 24	5.093E + 08	7.814E − 07	1.422E − 05	0.01953	295.1	9.748
20000	216.65	5.529E + 03	8.891E − 02	1.849E + 24	4.354E + 08	9.139E − 04	1.422E − 05	0.01953	295.1	9.745
21000	217.58	4.729E + 03	7.572E − 02	1.574E + 24	3.716E + 08	1.073E − 03	1.427E − 05	0.01961	295.1	9.742
22000	218.57	4.048E + 03	6.451E − 02	1.341E + 24	3.173E + 08	1.260E − 06	1.432E − 05	0.01970	296.4	9.739
23000	219.57	3.467E + 03	5.501E − 02	1.144E + 24	2.712E + 08	1.477E − 06	1.438E − 05	0.01978	297.1	9.736
24000	220.56	2.972E + 03	4.694E − 02	9.759E + 23	2.319E + 08	1.731E − 06	1.443E − 05	0.01986	297.7	9.733
25000	221.55	2.549E + 03	4.008E − 02	8.334E + 23	1.985E + 08	2.027E − 06	1.448E − 05	0.01995	298.4	9.730
26000	222.54	2.188E + 03	3.426E − 02	7.123E + 23	1.700E + 08	2.372E − 06	1.454E − 05	0.02003	299.1	9.727
27000	223.54	1.880E + 03	2.930E − 02	6.092E + 23	1.458E + 08	2.773E − 06	1.459E − 05	0.02011	299.7	9.724
29000	224.53	1.610E + 03	2.508E − 02	5.214E + 23	1.250E + 08	3.240E − 06	1.465E − 05	0.02020	300.4	9.721
29000	225.52	1.390E + 03	2.148E − 02	4.466E + 23	1.073E + 08	3.783E − 06	1.470E − 05	0.02028	301.1	9.718
30000	226.51	1.197E + 03	1.841E − 02	3.828E + 23	9.219E + 07	4.414E − 06	1.475E − 05	0.02036	301.7	9.715
31000	227.50	1.031E + 03	1.579E − 02	3.283E + 23	7.925E + 07	5.146E − 06	1.481E − 05	0.02044	302.4	9.712
32000	228.49	8.891E + 02	1.356E − 02	2.813E + 23	6.818E + 07	5.995E − 06	1.486E − 05	0.02053	303.0	9.709
33000	230.97	7.673E + 02	1.157E − 02	2.406E + 23	5.852E + 07	7.021E − 06	1.499E − 05	0.02073	304.7	9.706
34000	233.74	6.634E + 02	9.887E − 03	2.056E + 23	5.030E + 07	8.218E − 06	1.514E − 05	0.02096	306.5	9.703
35000	236.51	5.746E + 02	8.463E − 03	1.760E + 23	4.331E + 07	9.601E − 06	1.529E − 05	0.02119	308.3	9.700
36000	239.28	4.985E + 02	7.258E − 03	1.509E + 23	3.736E + 07	1.120E − 05	1.543E − 05	0.02142	310.1	9.697

(continued)

TABLE 10.8. Continued

Z m	T K	P Pa	ρ kg m^{-3}	n m^{-3}	ν s^{-1}	l m	η Pa s	k J m^{-1}s^{-1}K^{-1}	ν_s m s^{-1}	g m s^{-2}
38000	244.82	3.771E + 02	5.367E – 03	1.116E + 23	2.794E + 07	1.514E – 05	1.572E – 05	0.02188	313.7	9.690
40000	250.35	2.871E + 02	3.996E – 03	8.308E + 22	2.104E + 07	2.034E – 05	1.601E – 05	0.02233	317.2	9.684
42000	255.88	2.200E + 02	2.995E – 03	6.227E + 22	1.594E + 07	2.713E – 05	1.629E – 05	0.02278	320.7	9.678
44000	261.40	1.695E + 02	2.259E – 03	4.697E + 22	1.215E + 07	3.597E – 05	1.657E – 05	0.02323	324.1	9.672
46000	266.93	1.313E + 02	1.714E – 03	3.564E + 22	9.318E + 06	4.740E – 05	1.685E – 05	0.02376	327.5	9.666
48000	270.65	1.023E + 02	1.317E + 00	2.738E + 22	7.208E + 06	6.171E – 05	1.704E – 05	0.02397	329.8	9.660
50000	270.65	7.978E + 01	1.027E + 00	2.135E + 22	5.620E + 06	7.913E – 05	1.703E – 05	0.02397	329.8	9.654
52000	269.03	6.221E + 01	8.056E – 04	1.675E + 22	4.397E + 06	1.009E – 04	1.696E – 05	0.02384	328.8	9.648
54000	263.52	4.834E + 01	6.390E – 04	1.329E + 22	3.452E + 06	1.272E – 04	1.660E – 05	0.02340	325.4	9.642
56000	258.02	3.736E + 01	5.045E – 04	1.049E + 22	2.696E + 06	1.611E – 04	1.640E – 05	0.02296	322.0	9.636
58000	252.52	2.872E + 01	3.963E – 04	8.239E + 21	2.095E + 06	2.051E – 04	1.612E – 05	0.02251	318.6	9.632
60000	247.02	2.196E + 01	3.097E – 04	6.439E + 21	1.620E + 06	2.624E – 04	1.584E – 05	0.02206	315.1	9.624
65000	233.29	1.093E + 01	1.632E – 04	3.393E + 21	8.294E + 05	4.979E – 04	1.512E – 05	0.02093	306.2	9.609
0000	219.59	5.221	8.283E – 05	1.722E + 21	4.084E + 05	9.810E – 04	1.438E – 05	0.01978	297.1	9.594
75000	208.40	2.388	3.992E – 05	8.300E + 20	1.918E + 05	2.035E – 03	1.376E – 05	0.01883	289.4	9.579
80000	198.64	1.052	1.846E – 05	3.838E + 20	8.656E + 04	4.402E – 03	1.321E – 05	0.01800	282.5	9.564
85000	188.89	4.457E – 01	8.220E – 06	1.709E + 20	3.766E + 04	9.886E – 03	1.265E – 05	0.01716	275.5	9.550
90000	186.87	1.836E – 01	3.416E – 06	7.116E + 19	1.560E + 04	2.370E – 02				9.535
95000	188.42	7.597E – 02	1.393E – 06	2.920E + 19	6.440E + 03	5.790E – 02				9.520
100000	195.08	3.201E – 02	5.604E – 07	1.189E + 19	2.680E + 03	1.420E – 01				9.505
110000	240.00	7.104E – 03	9.708E – 08	2.144E + 18	5.480E + 02	7.880E – 01				9.476
120000	360.00	2.538E – 03	2.222E – 08	5.107E + 17	1.630E + 02	3.310				9.447
130000	469.27	1.251E – 03	8.152E – 09	1.930E + 17	7.100E + 01	8.800				9.418
140000	559.63	7.203E – 04	3.831E – 09	9.322E + 16	3.800E + 01	1.800E + 01				9.389
150000	634.39	4.542E – 04	2.076E – 09	5.186E + 16	2.300E + 01	3.300E + 01				9.360
160000	696.29	3.040E – 04	1.233E – 09	3.000E + 00	1.500E + 01	5.300E + 01				9.331
170000	747.57	2.121E – 04	7.815E – 10	2.055E + 16	1.000E + 01	8.200E + 01				9.302
180000	790.07	1.527E – 04	5.194E – 10	1.400E + 16	7.200	1.200E + 02				9.274
190000	825.16	1.127E – 04	3.581E – 10	9.887E + 15	5.200	1.700E + 02				9.246
200000	854.56	8.474E – 05	2.541E – 10	7.182E + 15	3.900	2.400E + 02				9.218
220000	899.01	5.015E – 05	1.367E – 10	4.040E + 15	2.300	4.200E + 02				9.162

(continued)

TABLE 10.8. Continued

Z m	T K	P Pa	ρ kg m^{-3}	n m^{-3}	ν s^{-1}	l m	η Pa s	k J m^{-1}s^{-1}K^{-1}	v_s m s^{-1}	g m s^{-2}
240000	929.73	3.106E − 05	7.858E − 11	2.420E + 15	1.400	7.000E + 02				9.106
260000	950.99	1.989E − 05	4.742E − 11	1.515E + 15	9.300E − 01	1.100E + 03				9.051
280000	965.75	1.308E − 05	2.971E − 11	9.807E + 14	6.100E − 01	1.700E + 03				8.997
300000	976.01	8.770E − 06	1.916E − 11	6.509E + 14	4.200E − 01	2.600E + 03				8.943
320000	983.16	5.980E − 06	1.264E − 11	4.405E + 14	2.900E − 01	3.800E + 03				8.889
340000	988.15	4.132E − 06	8.503E − 12	3.029E + 14	2.000E − 01	5.600E + 03				8.836
360000	991.65	2.888E − 06	5.805E − 12	2.109E + 14	1.400E − 01	8.000E + 03				8.784
380000	994.10	2.038E − 06	4.013E − 12	1.485E + 14	1.000E − 01	1.100E + 04				8.732
400000	995.83	1.452E − 06	2.803E − 12	1.056E + 14	7.200E − 02	1.600E + 04				8.680
450000	998.22	6.447E − 07	1.184E − 12	4.678E + 13	3.300E − 02	3.600E + 04				8.553
500000	999.24	3.024E − 07	5.215E − 13	2.192E + 13	1.600E − 02	7.700E + 04				8.429
550000	999.67	1.514E − 07	2.384E − 13	1.097E + 13	8.400E − 03	1.500E + 05				8.307
600000	999.85	8.213E − 08	1.137E − 13	5.950E + 12	4.800E − 03	2.800E + 05				8.188
650000	999.93	4.887E − 08	5.712E − 14	3.540E + 12	3.100E − 03	4.800E + 05				8.072
700000	999.97	3.191E − 08	3.070E − 14	2.311E + 12	2.200E − 03	7.300E + 05				7.958
750000	999.98	2.260E − 08	1.788E − 14	1.637E + 12	1.700E − 03	1.000E + 06				7.846
800000	999.99	1.704E − 08	1.136E − 14	1.234E + 12	1.400E − 03	1.400E + 06				7.737
850000	1000.00	1.342E − 08	7.824E − 15	9.717E + 11	1.200E − 03	1.700E + 06				7.630
900000	1000.00	1.087E − 08	5.759E − 15	7.876E + 11	1.000E − 03	2.100E + 06				7.525
950000	1000.00	8.982E − 09	4.453E − 15	6.505E + 11	8.700E − 04	2.600E + 06				7.422
1000000	1000.00	7.514E − 09	3.561E − 15	5.442E + 11	7.500E − 04	3.100E + 06				

The T and P columns for the troposphere and lower stratosphere were generated from the following formulas:

Height [m]	T/K	P/Pa
$H \leq 11{,}000$	$288.15 - 0.0065\,H$	$101325(288.15/T)^{-5.25577}$
$11{,}000\ \mathrm{m} < H \leq 20{,}000$	216.65	$22632\mathrm{e}^{-0.00015768832(H-11{,}000)}$
$20{,}000\ \mathrm{m} < H \leq 32{,}000$	$216.65 + 0.0010(H - 20{,}000)$	$5474.87(216.65/T)^{34.16319}$

where $H = rZ/(r + Z)$ is the geopotential height in meters and r is the mean earth radius at 45° N, taken as 6356.766 km. For altitudes up to 32 km, $\rho = 0.003483677(P/T)$. Formulas for the other quantities may be found in Refs. [13] and [14].

10.5. GLOBAL CLIMATE

10.5.1. Earth's radiation balance

Stefan-Boltzmann law

The radiative energy Q of a blackbody is

$$Q = cT^4 \quad [\mathrm{W\,m^{-2}}],$$
$$c = 5.6705 \times 10^{-8}\ \mathrm{W\,m^{-2}\,K^{-4}}.$$

Wien's law

The frequency λ_{max} of maximum emission of a blackbody is

$$\lambda_{\mathrm{max}} = \frac{c'}{T} \quad [\mu\mathrm{m}],$$
$$c' = 2897.8\ \mu\mathrm{m\,K}.$$

Mean solar constant

The intensity of solar radiance at the top of the atmosphere is

$$1376\ \mathrm{W\,m^{-2}}.$$

This value of solar radiance falls on a disk of area πR_{E}^2, where R_{E} is the Earth's radius. As the Earth rotates, the solar radiance is distributed over the entire spherical surface area of the planet, of $4\pi R_{\mathrm{E}}^2$, thus, giving an average incoming solar radiation at the top of the atmosphere of 342 W m^{-2}.

Greenhouse effect; Earth's radiation and energy budget

Figure 10.1 shows the annual global-average budget of energy. All values are in W m^{-2}.

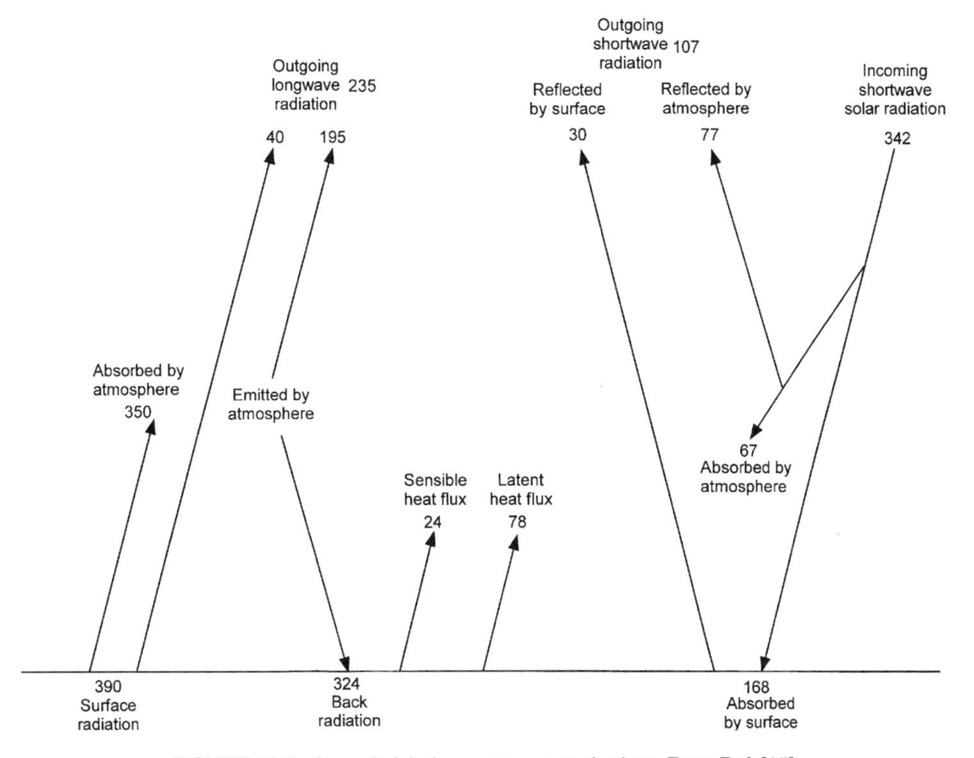

FIGURE 10.1. Annual global-average energy budget. From Ref. [15].

Corresponding to a solar temperature of 5780 K, from Wien's law, the wavelength of maximum radiance is 0.52 μm [shortwave radiation].

Corresponding to a mean terrestrial temperature of about 285 K, the wavelength of maximum radiance is about 10.1 μm [longwave radiation].

The absorption and reemission of radiant energy by the Earth's atmosphere is known as the *greenhouse effect.*

10.5.2. Global temperature trends

These plots of mean global temperature trends are based on data by provided by P. D. Jones,[1] D. E. Parker,[2] T. J. Osborn, and K. R. Briffa1.

The data series are kept up to date by and may be obtained from the Carbon Dioxide Information Analysis Center at http://cdiac.esd.ornl.gov/trends/trends.htm.

[1] Climatic Research Unit, School of Environmental Sciences, University of East Anglia, Norwich NR4 7TJ, United Kingdom

[2] Hadley Centre for Climate Prediction and Research, Meteorological Office, Bracknell, Berkshire, United Kingdom

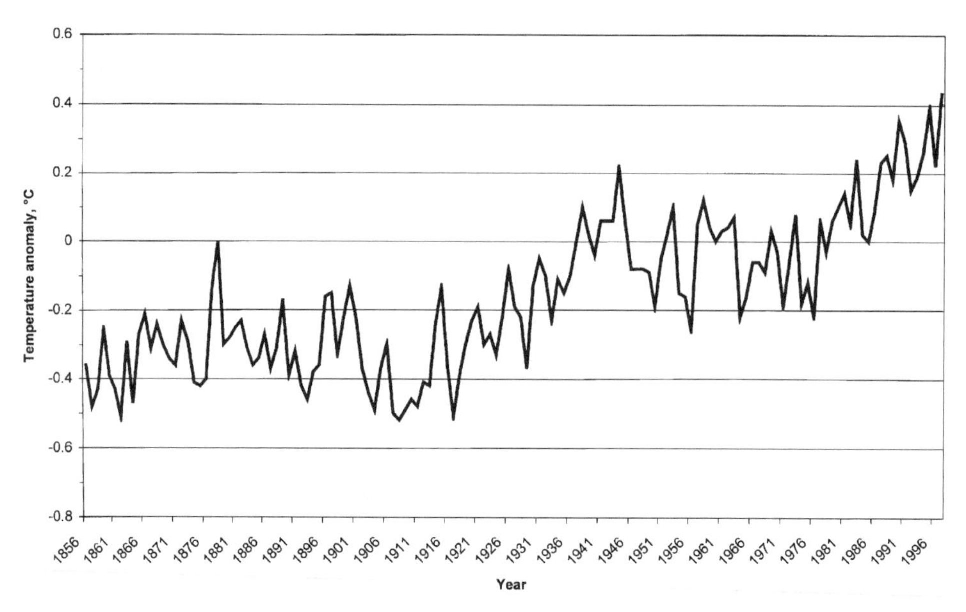

FIGURE 10.2. Global annual temperature anomaly.

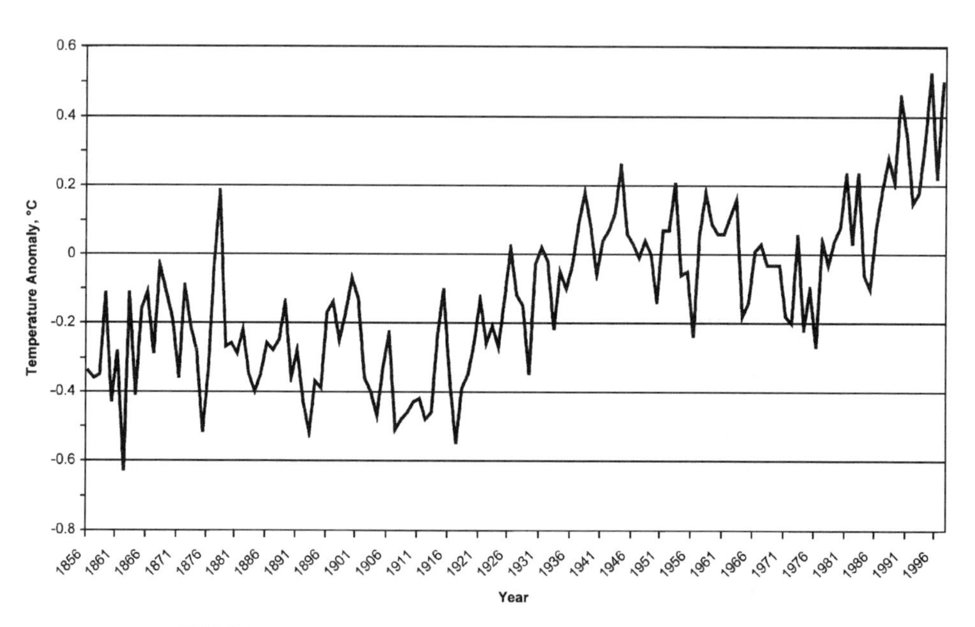

FIGURE 10.3. Northern Hemisphere annual temperature anomaly.

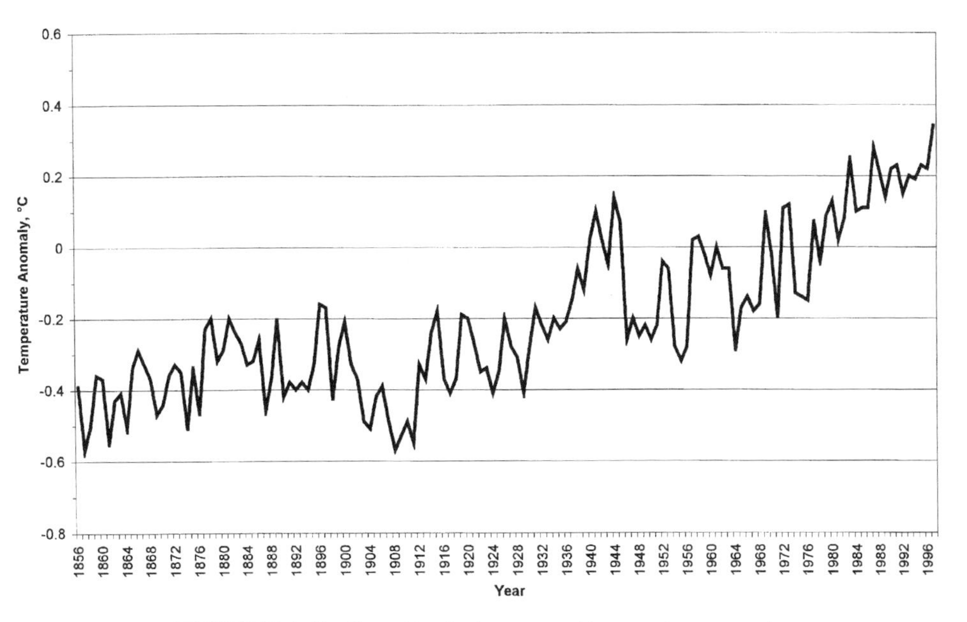

FIGURE 10.4. Southern Hemisphere annual temperature anomaly.

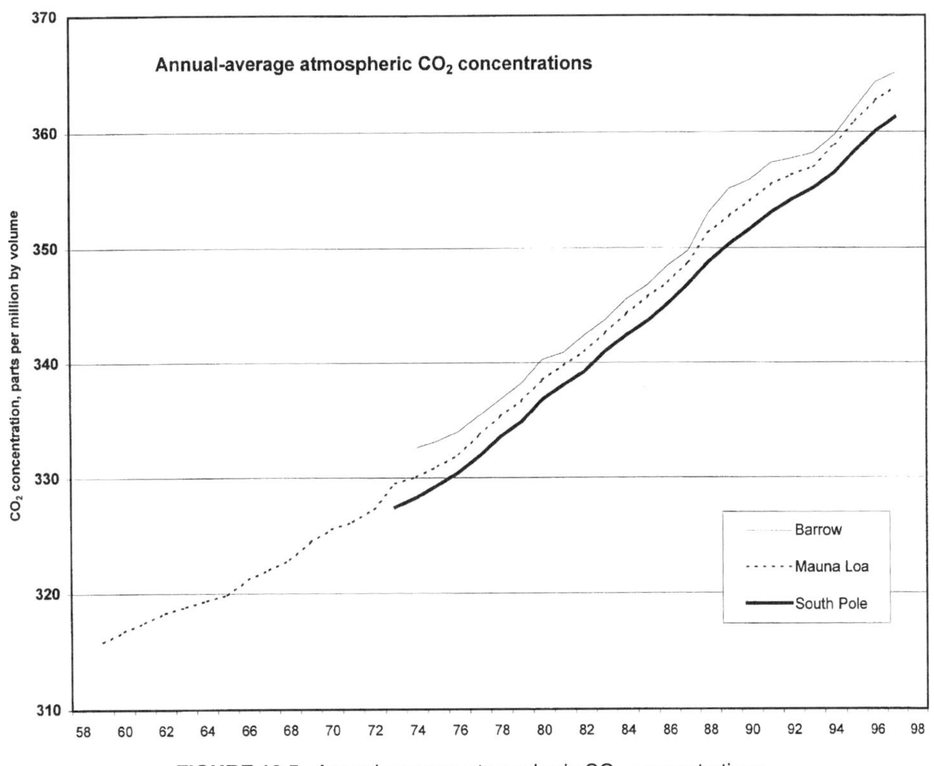

FIGURE 10.5. Annual-average atmospheric CO_2 concentrations.

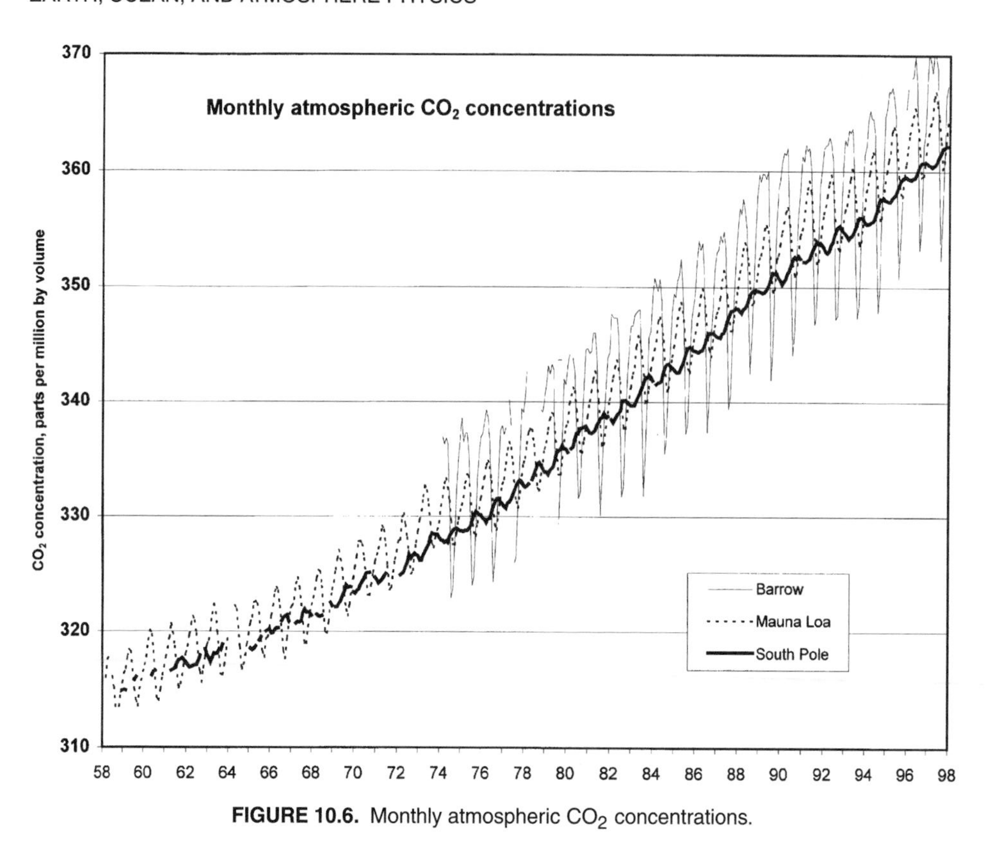

FIGURE 10.6. Monthly atmospheric CO_2 concentrations.

10.5.3. Atmospheric CO_2 concentrations

Figures 10.5 and 10.6 show concentrations of atmospheric CO_2 in parts per million by volume for the following three stations:

1. Barrow, Alaska, U.S.A., on the coast of the Arctic Ocean 71°19′ N, 156°36′ W, 11 m above mean sea level
2. Mauna Loa, Hawaii, U.S.A., a barren lava field of an active volcano 19°32′ N, 155°35′ W, 3397 m above mean sea level
3. South Pole, Antarctica, an ice- and snow-covered plateau 89°59′ S, 24°48′ W, 2810 m above mean sea level

The annual series shows the steady increase in CO_2 concentrations due to the burning of fossil fuels. The monthly series shows the influence of the annual cycle of vegetation, with winter increases due to respiration and summer decreases due to photosynthesis. Note the highest values of annual variation on the North American continent (Barrow). The lowest annual variation occurs at the South Pole—a region without vegetation. Note also the phase inversion between the Northern and Southern hemispheres.

These data were produced by C.D. Keeling and T.P. Whorf, Scripps Institution of Oceanography, University of California, La Jolla, California 92093-0220 USA.

The data are kept up to date and may be obtained from the Carbon Dioxide Information Analysis Center at http://cdiac.esd.ornl.gov/trends/trends.htm.

10.6. REFERENCES

[1] G. Dietrich, K. Kalle, W. Krauss, and G. Siedler, *General Oceanography*, 2nd ed. (Wiley, New York, 1980).

[2] JPOTS Editorial Panel, *Processing of Oceanographic Station Data* (Unesco, Paris, 1991).

[3] IAPSO Working Group on Symbols, Units, and Nomenclature in Physical Oceanography (SUN), *The International System of Units (SI) in Oceanography*, UNESCO Technical Papers in Marine Science No. 45, Paris, 1985.

[4] UNESCO, *Background Papers and Supporting Data on the International Equation of State of Seawater 1980*, UNESCO Technical Papers in Marine Science No. 38, 1981.

[5] H. Bryden, Deep-Sea Res. **20**, 401–408 (1973).

[6] M.E.Q. Pilson, *An Introduction to the Chemistry of the Sea* (Prentice-Hall, New York, 1998).

[7] UNESCO, *International oceanographic tables*, Vol. 4, UNESCO Technical Papers in Marine Science No. 40, 1987.

[8] W. D. Wilson, J. Acoust. Soc. Am. **32**, 641–644 (1960).

[9] F. Millero, in UNESCO Technical Papers in Marine Science No. 28, 1978.

[10] J. Wu, J. Phys. Oceanogr. **10**, 727–740 (1980).

[11] A. Defant, *Physical Oceanography* (Macmillan, New York, 1961), Vol. II.

[12] A. Gill, *Atmosphere-Ocean Dynamics* (Academic Press, New York, 1982).

[13] COESA, *U.S. Standard Atmosphere, 1976* (U.S. Government Printing Office, Washington, DC, 1976).

[14] *Handbook of Geophysics and the Space Environment*, edited by A. S. Jursa (Air Force Geophysics Laboratory, 1985).

[15] J. T. Kiehl and K. E. Trenberth, Bull. Am. Meteorol. Soc., **78**, 2, 197–208 (1997).

11

Electricity and Magnetism

David J. Griffiths
Department of Physics, Reed College, Portland, Oregon

Contents

List of Tables

List of Figures

11.1. INTRODUCTION

All electric and magnetic phenomena are due to the interaction of charged particles. This interaction is mediated by the electric field $\mathbf{E}$ and the magnetic field (or "magnetic induction") $\mathbf{B}$; in SI units (Sec. 11.10), the force $\mathbf{F}$ on a test charge q moving with velocity $\mathbf{v}$ is

$$\mathbf{F} = q(\mathbf{E} + \mathbf{v} \times \mathbf{B}) \qquad \text{(Lorentz force law).} \tag{11.1}$$

A continuous distribution of charge is described by the charge density ρ; a flow of charge is represented by the current density $\mathbf{J}$, whose magnitude is the charge per unit time, per unit area, passing through an infinitesimal surface whose normal is oriented parallel to the flow. The total current through an arbitrary surface $\mathcal{S}$ is the flux of $\mathbf{J}$:

$$I = \int_{\mathcal{S}} \mathbf{J} \cdot d\mathbf{a}. \tag{11.2}$$

In terms of the density and velocity of the moving charges,

$$\mathbf{J} = \rho_{\mathrm{m}} \mathbf{v}_{\mathrm{m}}. \tag{11.3}$$

Thus, the force on a charge and current distribution is

$$\mathbf{F} = \int (\rho \mathbf{E} + \mathbf{J} \times \mathbf{B})\, d^3 r. \tag{11.4}$$

In the special case of a neutral wire carrying a steady current I,

$$\mathbf{F} = I \int d\mathbf{l} \times \mathbf{B}. \tag{11.5}$$

The fields are produced by source charges and currents, as detailed below. They obey the superposition principle: The total field is the (vector) sum of the fields due to all of the source charges considered separately.

11.2. ELECTROSTATICS

In electrostatics, the source charge density is independent of time. The simplest example is a point charge q_1 at the origin, for which

$$\mathbf{E}(\mathbf{r}) = \frac{1}{4\pi\epsilon_0} \frac{q_1}{r^2} \hat{\mathbf{r}}; \tag{11.6}$$

the force on a second charge (q_2) is

$$\mathbf{F} = \frac{1}{4\pi\epsilon_0} \frac{q_1 q_2}{r^2} \hat{\mathbf{r}} \qquad \text{(Coulomb's law).} \tag{11.7}$$

For a continuous distribution,

$$\mathbf{E}(\mathbf{r}) = \frac{1}{4\pi\epsilon_0} \int \frac{\rho(\mathbf{r}')}{\mathscr{r}^2} \hat{\boldsymbol{\mathscr{r}}} d^3 r', \tag{11.8}$$

where

$$\boldsymbol{\mathscr{r}} \equiv \mathbf{r} - \mathbf{r}' \tag{11.9}$$

is the separation vector from the source point $\mathbf{r}'$ to the field point $\mathbf{r}$. It follows[1] that

$$\nabla \cdot \mathbf{E} = \rho/\epsilon_0 \qquad \text{(Gauss's law)} \tag{11.10}$$

and

$$\nabla \times \mathbf{E} = 0. \tag{11.11}$$

Using the divergence theorem, Gauss's law can be converted to integral form:

$$\oint_{\mathcal{S}} \mathbf{E} \cdot d\mathbf{a} = q_{\text{enc}}/\epsilon_0, \tag{11.12}$$

where $\mathcal{S}$ is any closed "gaussian surface," and q_{enc} is the total charge enclosed; this affords the most efficient means for calculating $\mathbf{E}$ in cases of spherical, cylindrical, and planar symmetry.

Equation (11.11) licenses the introduction of a scalar potential,

$$V(\mathbf{r}) \equiv -\int_{\mathcal{O}}^{\mathbf{r}} \mathbf{E} \cdot d\mathbf{l}, \tag{11.13}$$

where $\mathcal{O}$ is an arbitrary reference point (usually placed at infinity). Physically, $qV(\mathbf{r})$ is the work required to transport a charge q from $\mathcal{O}$ to $\mathbf{r}$. It follows that

$$\nabla V = -\mathbf{E}, \tag{11.14}$$

and, hence, from Eq. (11.10) that

$$\nabla^2 V = -\rho/\epsilon_0 \qquad \text{(Poisson's equation).} \tag{11.15}$$

For localized charge distributions, the solution is

$$V(\mathbf{r}) = \frac{1}{4\pi\epsilon_0} \int \frac{\rho(\mathbf{r}')}{\mathscr{r}} \, d^3r'. \tag{11.16}$$

In charge-free regions, Eq. (11.15) reduces to

$$\nabla^2 V = 0 \qquad \text{(Laplace's equation),} \tag{11.17}$$

which is most readily handled by separation of variables; solutions are known as harmonic functions.

The simplest static configuration (beyond the point charge, or "monopole") is the dipole: equal and opposite charges ($\pm q$) separated by a displacement $\mathbf{d}$ (pointing from $-q$ to $+q$). The electric dipole moment is

$$\mathbf{p} \equiv q\mathbf{d}. \tag{11.18}$$

[1] For vector calculus, see Chapter 2.

In a uniform electric field, the net force on a dipole is zero, but it does experience a torque,

$$\mathbf{N} = \mathbf{p} \times \mathbf{E}, \tag{11.19}$$

tending to align **p** parallel to **E**. The energy of the dipole is

$$U = -\mathbf{p} \cdot \mathbf{E}. \tag{11.20}$$

In a nonuniform field, the force is

$$\mathbf{F} = (\mathbf{p} \cdot \nabla)\mathbf{E}. \tag{11.21}$$

The potential of an ideal (point) dipole at the origin is

$$V_{\text{dip}}(\mathbf{r}) = \frac{1}{4\pi\epsilon_0} \frac{\mathbf{p} \cdot \hat{\mathbf{r}}}{r^2}, \tag{11.22}$$

and the field is

$$\mathbf{E}_{\text{dip}}(\mathbf{r}) = \frac{1}{4\pi\epsilon_0} \frac{1}{r^3} \left[3(\mathbf{p} \cdot \hat{\mathbf{r}})\,\hat{\mathbf{r}} - \mathbf{p}\right] - \frac{1}{3\epsilon_0} \mathbf{p}\, \delta^3(\mathbf{r}). \tag{11.23}$$

When a dielectric (insulating) material is placed in an external field, it becomes polarized: Polar molecules (such as water), which have permanent electric dipole moments, rotate, whereas atoms and nonpolar molecules stretch, developing induced dipole moments that are typically proportional to the field,

$$\mathbf{p} = \alpha \mathbf{E} \tag{11.24}$$

(the proportionality factor α is called the polarizability; representative values are listed in Table 11.1.) In either case, the result is a lot of atomic-scale dipoles lined up parallel to the field:

$$\mathbf{P} \equiv \text{electric dipole moment per unit volume} \tag{11.25}$$

is called the polarization of the material. Polarization results in the accumulation of bound charge,

$$\rho_{\text{b}} = -\nabla \cdot \mathbf{P} \tag{11.26}$$

(in particular, at the edges, there is a surface bound charge, $\sigma_{\text{b}} = \mathbf{P} \cdot \hat{\mathbf{n}}$). Introducing the electric displacement

$$\mathbf{D} \equiv \epsilon_0 \mathbf{E} + \mathbf{P}, \tag{11.27}$$

Gauss's law can be expressed in terms of the free charge only:

$$\nabla \cdot \mathbf{D} = \rho_{\text{f}}. \tag{11.28}$$

TABLE 11.1. Atomic polarizabilities ($\alpha/4\pi\epsilon_0$, in units of 10^{-30} m^3). [*Handbook of Chemistry and Physics*, 78th ed., edited by David R. Lide, Jr. (CRC Press, Boca Raton, FL, 1997).]

H	He	Li	Be	C	Ne	Na	Ar	K	Cs
0.667	0.205	24.3	5.60	1.76	0.396	24.1	1.64	43.4	59.6

In linear media, the polarization is proportional to the field:

$$\mathbf{P} = \epsilon_0 \chi_e \mathbf{E}, \tag{11.29}$$

and hence so too is the displacement:

$$\mathbf{D} = \epsilon \mathbf{E}; \tag{11.30}$$

χ_e is called the electric susceptibility, and $\epsilon \equiv \epsilon_0(1 + \chi_e)$ is the permittivity; $\epsilon_r \equiv \epsilon/\epsilon_0 = 1 + \chi_e$ is the relative permittivity, or dielectric constant, of the material—representative values are given in Table 11.2. The capacitance of a dielectric-filled capacitor is increased (as compared with vacuum-filled) by a factor of ϵ_r.

TABLE 11.2. Dielectric constants (unless otherwise specified, values given are for 1 atm, 20°C). [*Handbook of Chemistry and Physics*, 78th ed., edited by David R. Lide, Jr. (CRC Press, Boca Raton, FL, 1997).]

Material	Dielectric Constant	Material	Dielectric Constant
Vacuum	1	Benzene	2.28
Helium	1.000065	Diamond	5.7
Neon	1.00013	Salt	5.9
Hydrogen	1.00025	Silicon	11.8
Argon	1.00052	Methanol	33.0
Air (dry)	1.00054	Water	80.1
Nitrogen	1.00055	Ice (−30°C)	99
Water vapor (100°C)	1.00587	$KTaNbO_3$ (0°C)	34,000

11.3. MAGNETOSTATICS

In magnetostatics, ρ and $\mathbf{J}$ are independent of time, and the magnetic field is

$$\mathbf{B}(\mathbf{r}) = \frac{\mu_0}{4\pi} \int \frac{\mathbf{J}(\mathbf{r}') \times \hat{\boldsymbol{\imath}}}{\imath^2}\, d^3r' \qquad \text{(Biot-Savart law)}, \tag{11.31}$$

or, in the case of a current-carrying wire,

$$\mathbf{B}(\mathbf{r}) = \frac{\mu_0 I}{4\pi} \int \frac{d\boldsymbol{\ell} \times \hat{\boldsymbol{\imath}}}{\imath^2}. \tag{11.32}$$

It follows from Eq. (11.31) that

$$\nabla \cdot \mathbf{B} = 0 \tag{11.33}$$

and

$$\nabla \times \mathbf{B} = \mu_0 \mathbf{J} \qquad \text{(Ampére's law)}. \tag{11.34}$$

Using Stokes's theorem, Ampére's law can be converted to integral form:

$$\oint_C \mathbf{B} \cdot d\boldsymbol{\ell} = \mu_0 I_{\text{enc}}, \tag{11.35}$$

where $\mathcal{P}$ is any closed "amperian loop," and I_{enc} is the total current passing through it; this affords the most efficient means for calculating $\mathbf{B}$ in cases of cylindrical, planar, solenoidal, or toroidal symmetry.

Equation (11.33) licenses the introduction of a vector potential $\mathbf{A}$:

$$\mathbf{B} = \nabla \times \mathbf{A}. \tag{11.36}$$

It is convenient to choose $\nabla \cdot \mathbf{A} = 0$, so that

$$\nabla^2 \mathbf{A} = -\mu_0 \mathbf{J}. \tag{11.37}$$

This again is Poisson's equation, and the solution (for localized current distributions) is

$$\mathbf{A}(\mathbf{r}) = \frac{\mu_0}{4\pi} \int \frac{\mathbf{J}(\mathbf{r}')}{\mathscr{r}}\, d^3 r'. \tag{11.38}$$

Physically, Eq. (11.33) says that there are no magnetic monopoles; the simplest magnetostatic configuration is a current loop, or magnetic dipole. Its dipole moment is

$$\mathbf{m} \equiv I \mathbf{a}, \tag{11.39}$$

where I is the current and $\mathbf{a} \equiv \int d\mathbf{a}$ is the vector area of the loop. In a uniform magnetic field, the net force on a dipole is zero, but it does experience a torque,

$$\mathbf{N} = \mathbf{m} \times \mathbf{B}, \tag{11.40}$$

tending to align $\mathbf{m}$ parallel to $\mathbf{B}$. The energy of the dipole is

$$U = -\mathbf{m} \cdot \mathbf{B}. \tag{11.41}$$

In a nonuniform field, the force is

$$\mathbf{F} = \nabla(\mathbf{m} \cdot \mathbf{B}). \tag{11.42}$$

The vector potential of an ideal (point) dipole at the origin is

$$\mathbf{A}_{\text{dip}}(\mathbf{r}) = \frac{\mu_0}{4\pi} \frac{\mathbf{m} \times \hat{\mathbf{r}}}{r^2}, \tag{11.43}$$

and the field is

$$\mathbf{B}_{\text{dip}}(\mathbf{r}) = \frac{\mu_0}{4\pi} \frac{1}{r^3} \left[3(\mathbf{m} \cdot \hat{\mathbf{r}})\, \hat{\mathbf{r}} - \mathbf{m} \right] + \frac{2\mu_0}{3} \mathbf{m}\, \delta^3(\mathbf{r}). \tag{11.44}$$

In an external field, matter becomes magnetized, with atomic-scale magnetic dipoles lined up parallel (paramagnetism) or antiparallel (diamagnetism) to $\mathbf{B}$. In ferromagnetic materials, magnetization occurs spontaneously even in the absence of an external field. Paramagnetism is due to the torque, Eq. (11.40), acting on the dipole moment associated with electron spin; diamagnetism results from the distortion of electron orbital motion. Ferromagnetism is due to exchange forces between electrons in neighboring atoms. Whatever the cause, we describe the phenomenon by the magnetization vector,

$$\mathbf{M} \equiv \text{magnetic dipole moment per unit volume.} \tag{11.45}$$

Magnetization produces a bound current,

$$\mathbf{J}_\mathrm{b} = \nabla \times \mathbf{M} \tag{11.46}$$

(in particular, at the edges, there is a surface bound current, $\mathbf{K}_\mathrm{b} = \mathbf{M} \times \hat{\mathbf{n}}$). Introducing the auxiliary field $\mathbf{H}$ (sometimes called the "magnetic field")

$$\mathbf{H} \equiv \frac{1}{\mu_0}\mathbf{B} - \mathbf{M}, \tag{11.47}$$

Ampére's law can be expressed in terms of the free current only:

$$\nabla \times \mathbf{H} = \mathbf{J}_\mathrm{f}. \tag{11.48}$$

In linear media, the magnetization is proportional to $\mathbf{H}$:

$$\mathbf{M} = \chi_\mathrm{m}\mathbf{H}, \tag{11.49}$$

and hence so too is $\mathbf{B}$:

$$\mathbf{B} = \mu\mathbf{H}; \tag{11.50}$$

χ_m is called the magnetic susceptibility, $\mu \equiv \mu_0(1 + \chi_\mathrm{m})$ is the permeability, and $\mu_\mathrm{r} \equiv \mu/\mu_0 = 1+\chi_\mathrm{m}$ is the relative permeability, of the material—representative values are given in Table 11.3. Ferromagnetic materials are emphatically nonlinear, and their magnetization depends not just on the field, but also on their entire magnetic history (hysteresis).

TABLE 11.3. Magnetic susceptibilities (unless otherwise specified, values are for 1 atm, 20°C). [*Handbook of Chemistry and Physics*, 67th ed., edited by David R. Lide, Jr. (CRC Press, Boca Raton, FL, 1986).]

Material	Susceptibility	Material	Susceptibility
Diamagnetic:		*Paramagnetic:*	
Bismuth	-1.6×10^{-4}	Oxygen	1.9×10^{-6}
Gold	-3.4×10^{-5}	Sodium	8.5×10^{-6}
Silver	-2.4×10^{-5}	Aluminum	2.1×10^{-5}
Copper	-9.7×10^{-6}	Tungsten	7.8×10^{-5}
Water	-9.0×10^{-6}	Platinum	2.8×10^{-4}
Carbon dioxide	-1.2×10^{-8}	Liquid oxygen (−200°C)	3.9×10^{-3}
Hydrogen	-2.2×10^{-9}	Gadolinium	4.8×10^{-1}

11.4. ELECTRODYNAMICS

Nonconstant sources produce time-dependent fields. Equation (11.11) generalizes to

$$\nabla \times \mathbf{E} = -\frac{\partial \mathbf{B}}{\partial t} \qquad \text{(Faraday's law)}, \tag{11.51}$$

and Ampére's law, Eq. (11.34), picks up an extra term:

$$\nabla \times \mathbf{B} = \mu_0 \mathbf{J} + \mu_0 \epsilon_0 \frac{\partial \mathbf{E}}{\partial t} \qquad \text{(Ampére-Maxwell law)}. \tag{11.52}$$

We say that a changing magnetic field *induces* an electric field, and a changing electric field induces a magnetic field, though ρ and $\mathbf{J}$ are the ultimate sources of all fields. Maxwell called the quantity

$$\mathbf{J}_\mathrm{d} \equiv \epsilon_0 \frac{\partial \mathbf{E}}{\partial t} \tag{11.53}$$

"displacement current," but the term is misleading, since $\mathbf{J}_\mathrm{d}$ does not represent a flow of charge.

The complete set of Maxwell's equations is

$$\nabla \cdot \mathbf{E} = \rho/\epsilon_0 \qquad \text{(Gauss's law)}, \tag{11.54a}$$

$$\nabla \cdot \mathbf{B} = 0 \qquad \text{(no magnetic monopoles)}, \tag{11.54b}$$

$$\nabla \times \mathbf{E} = -\frac{\partial \mathbf{B}}{\partial t} \qquad \text{(Faraday's law)}, \tag{11.54c}$$

$$\nabla \times \mathbf{B} = \mu_0 \mathbf{J} + \mu_0 \epsilon_0 \frac{\partial \mathbf{E}}{\partial t} \qquad \text{(Ampère-Maxwell law)}. \tag{11.54d}$$

Together with the Lorentz force law, Maxwell's equations summarize the entire theoretical foundation of classical electrodynamics. If magnetic monopoles are included, they generalize to

$$\nabla \cdot \mathbf{E} = \rho_\mathrm{e}/\epsilon_0, \tag{11.55a}$$

$$\nabla \cdot \mathbf{B} = \mu_0 \rho_\mathrm{m}, \tag{11.55b}$$

$$\nabla \times \mathbf{E} = -\mu_0 \mathbf{J}_\mathrm{m} - \frac{\partial \mathbf{B}}{\partial t}, \tag{11.55c}$$

$$\nabla \times \mathbf{B} = \mu_0 \mathbf{J}_\mathrm{e} + \mu_0 \epsilon_0 \frac{\partial \mathbf{E}}{\partial t}, \tag{11.55d}$$

$$\mathbf{F} = q_\mathrm{e}(\mathbf{E} + \mathbf{v} \times \mathbf{B}) + q_\mathrm{m}(\mathbf{B} - \mathbf{v} \times \mathbf{E}/c^2), \tag{11.55e}$$

where q_e is the electric charge, q_m is the magnetic charge, ρ_e and ρ_m are the corresponding densities, and $\mathbf{J}_\mathrm{e}$, $\mathbf{J}_\mathrm{m}$ are the currents.

In the presence of polarizable and magnetizable materials, it is convenient (as before) to separate the charge and current into free and bound components,

$$\rho = \rho_\mathrm{f} - \nabla \cdot \mathbf{P}, \quad \mathbf{J} = \mathbf{J}_\mathrm{f} + \nabla \times \mathbf{M} + \frac{\partial \mathbf{P}}{\partial t}. \tag{11.56}$$

Maxwell's equations (11.54) become

$$\nabla \cdot \mathbf{D} = \rho_\mathrm{f}, \tag{11.57a}$$

$$\nabla \cdot \mathbf{B} = 0, \tag{11.57b}$$

$$\nabla \times \mathbf{E} = -\frac{\partial \mathbf{B}}{\partial t}, \tag{11.57c}$$

$$\nabla \times \mathbf{H} = \mathbf{J}_{\mathrm{f}} + \frac{\partial \mathbf{D}}{\partial t}. \tag{11.57d}$$

The advantage of this formulation is that it makes explicit reference to free quantities only; however, since it involves **D** (Eq. 11.27) and **H** (Eq. 11.47), as well as **E** and **B**, the system is underdetermined, and must be supplemented by appropriate "constitutive relations," specifying **D** and **H** in terms of **E** and **B**. The constitutive relations depend on the nature of the material; for linear media, they are Eqs. (11.30) and (11.50).

The homogeneous Maxwell equations (11.54b and c) license the introduction of scalar and vector potentials:

$$\mathbf{E} = -\nabla V - \frac{\partial \mathbf{A}}{\partial t}, \qquad \mathbf{B} = \nabla \times \mathbf{A}. \tag{11.58}$$

But these equations do not uniquely determine V and **A**; specifically, a gauge transformation

$$\mathbf{A} \to \mathbf{A} + \nabla \lambda, \qquad V \to V - \frac{\partial \lambda}{\partial t} \tag{11.59}$$

(for any scalar function λ) leaves the fields unchanged. This gauge freedom can be exploited to simplify the inhomogeneous Maxwell equations (11.54a and d):

$$\left.\begin{aligned} \nabla^2 V - \frac{1}{c^2}\frac{\partial^2 V}{\partial t^2} + \frac{\partial}{\partial t}\left(\nabla \cdot \mathbf{A} + \frac{1}{c^2}\frac{\partial V}{\partial t}\right) &= -\rho/\epsilon_0, \\ \nabla^2 \mathbf{A} - \frac{1}{c^2}\frac{\partial^2 \mathbf{A}}{\partial t^2} - \nabla\left(\nabla \cdot \mathbf{A} + \frac{1}{c^2}\frac{\partial V}{\partial t}\right) &= -\mu_0 \mathbf{J}, \end{aligned}\right\} \tag{11.60}$$

where

$$c \equiv \frac{1}{\sqrt{\epsilon_0 \mu_0}} \qquad \text{(the speed of light).} \tag{11.61}$$

In the Coulomb gauge, $\nabla \cdot \mathbf{A} = 0$; in the Lorentz gauge (ordinarily the most convenient choice)

$$\nabla \cdot \mathbf{A} = -\frac{1}{c^2}\frac{\partial V}{\partial t}, \tag{11.62}$$

and the potentials satisfy the inhomogeneous wave equation:

$$\left.\begin{aligned} \Box^2 V &= -\rho/\epsilon_0, \\ \Box^2 \mathbf{A} &= -\mu_0 \mathbf{J}, \end{aligned}\right\} \tag{11.63}$$

where

$$\Box^2 \equiv \nabla^2 - \frac{1}{c^2}\frac{\partial^2}{\partial t^2} \tag{11.64}$$

is the d'Alembertian.

For localized sources, the solutions to Eq. (11.63) are

$$\left.\begin{aligned} V(\mathbf{r}, t) &= \frac{1}{4\pi\epsilon_0}\int \frac{\rho(\mathbf{r}', t_r)}{\mathscr{r}}\, d^3r', \\ \mathbf{A}(\mathbf{r}, t) &= \frac{\mu_0}{4\pi}\int \frac{\mathbf{J}(\mathbf{r}', t_r)}{\mathscr{r}}\, d^3r', \end{aligned}\right\} \quad \text{(retarded potentials)} \tag{11.65}$$

where

$$t_r \equiv t - \frac{\mathscr{r}}{c} \tag{11.66}$$

is the retarded time (physically, the time a "message" left the source point such that—traveling at the speed of light—it reaches the field point at time t). Equations (11.63) also admit advanced solutions (in which t_r is replaced by the advanced time $t_a \equiv t + \mathscr{r}/c$), but these are ordinarily rejected on the ground that they violate causality.

Differentiation of the retarded potentials yields

$$\mathbf{E}(\mathbf{r}, t) = \frac{1}{4\pi\epsilon_0}\int \left(\frac{\rho(\mathbf{r}', t_r)}{\mathscr{r}^2}\hat{\boldsymbol{\mathscr{r}}} + \frac{\dot{\rho}(\mathbf{r}', t_r)}{\mathscr{r}c}\hat{\boldsymbol{\mathscr{r}}} - \frac{\dot{\mathbf{J}}(\mathbf{r}', t_r)}{\mathscr{r}c^2}\right) d^3r', \tag{11.67}$$

$$\mathbf{B}(\mathbf{r}, t) = \frac{\mu_0}{4\pi}\int \left(\frac{\mathbf{J}(\mathbf{r}', t_r)\times\hat{\boldsymbol{\mathscr{r}}}}{\mathscr{r}^2} + \frac{\dot{\mathbf{J}}(\mathbf{r}', t_r)\times\hat{\boldsymbol{\mathscr{r}}}}{\mathscr{r}c}\right) d^3r', \tag{11.68}$$

where a dot denotes the derivative with respect to time. These are the (retarded) solutions to Maxwell's equations—the generalizations of Coulomb's law and the Biot-Savart law for time-dependent sources; they are sometimes called Jefimenko's equations.

For a *point* charge q,

$$\left.\begin{aligned} V(\mathbf{r}, t) &= \frac{1}{4\pi\epsilon_0}\frac{q}{(\mathscr{r} - \boldsymbol{\mathscr{r}}\cdot\mathbf{v}/c)}, \\ \mathbf{A}(\mathbf{r}, t) &= \frac{\mu_0}{4\pi}\frac{q\mathbf{v}}{(\mathscr{r} - \boldsymbol{\mathscr{r}}\cdot\mathbf{v}/c)}, \end{aligned}\right\} \quad \text{(Liénard-Wiechert potentials)} \tag{11.69}$$

where $\boldsymbol{\mathscr{r}}$ and $\mathbf{v}$ (the velocity) are evaluated at the retarded time, and

$$\begin{aligned} \mathbf{E}(\mathbf{r}, t) &= \frac{1}{4\pi\epsilon_0}\frac{q}{(\mathscr{r} - \boldsymbol{\mathscr{r}}\cdot\mathbf{v}/c)^3}\left\{(1 - v^2/c^2)(\boldsymbol{\mathscr{r}} - \mathscr{r}\mathbf{v}/c) + \frac{\boldsymbol{\mathscr{r}}\times[(\boldsymbol{\mathscr{r}} - \mathscr{r}\mathbf{v}/c)\times\mathbf{a}]}{c^2}\right\}, \\ \mathbf{B}(\mathbf{r}, t) &= \frac{1}{c}(\hat{\boldsymbol{\mathscr{r}}}\times\mathbf{E}), \end{aligned} \tag{11.70}$$

where $\mathbf{a}$ is the acceleration at the retarded time. The first term in $\mathbf{E}$ is called the velocity (or generalized Coulomb) field; the second is the acceleration (or radiation) field. In the special case of motion at *constant* velocity,

$$\mathbf{E}(\mathbf{r}, t) = \frac{1}{4\pi\epsilon_0}\frac{q(1 - v^2/c^2)}{R^2[1 - (v\sin\theta/c)^2]^{3/2}}\frac{\hat{\mathbf{R}}}{R^2}; \quad \mathbf{B}(\mathbf{r}, t) = \frac{1}{c^2}\mathbf{v}\times\mathbf{E}(\mathbf{r}, t), \tag{11.71}$$

where $\mathbf{R}$ is the vector from the *present* (time t) location of the charge to the field point $\mathbf{r}$, and θ is the angle between $\mathbf{R}$ and $\mathbf{v}$.

11.5. CONSERVATION LAWS

Maxwell's equations (11.54) enforce a relation between ρ and $\mathbf{J}$:

$$\nabla \cdot \mathbf{J} = -\frac{\partial \rho}{\partial t} \qquad \text{(continuity equation)}, \tag{11.72}$$

which is the local statement of conservation of charge. In integral form,

$$-\frac{d}{dt}\int_{\mathcal{V}} \rho \, d^3r = \oint_{\mathcal{S}} \mathbf{J} \cdot d\mathbf{a}, \tag{11.73}$$

it says that the rate of decrease in the charge contained within a volume $\mathcal{V}$ is equal to the total current flowing out through the surface $\mathcal{S}$.

An analogous equation can be derived from Maxwell's equations and the Lorentz force law, expressing conservation of energy:

$$-\frac{d}{dt}\int_{\mathcal{V}} u \, d^3r = \oint_{\mathcal{S}} \mathbf{S} \cdot d\mathbf{a} + \frac{dW}{dt} \qquad \text{(Poynting's theorem)}. \tag{11.74}$$

Here

$$u \equiv \frac{1}{2}\left(\epsilon_0 E^2 + \frac{1}{\mu_0} B^2\right) \tag{11.75}$$

is the energy density in the fields,

$$\mathbf{S} \equiv \frac{1}{\mu_0}(\mathbf{E} \times \mathbf{B}) \qquad \text{(Poynting vector)} \tag{11.76}$$

is the energy flux (energy per unit area, per unit time, carried by the fields), and dW/dt is the rate at which work is done by the fields on the charged particles in $\mathcal{V}$ (W is their *mechanical* energy).

The corresponding equation for conservation of momentum can be formulated in terms of the Maxwell stress tensor,

$$T_{ij} \equiv \epsilon_0 \left(E_i E_j - \frac{1}{2}\delta_{ij} E^2\right) + \frac{1}{\mu_0}\left(B_i B_j - \frac{1}{2}\delta_{ij} B^2\right). \tag{11.77}$$

Physically, T_{ij} is the stress (force per unit area) in the i direction acting on a patch of surface oriented in the j direction; the "diagonal" elements (T_{xx}, T_{yy}, T_{zz}) represent pressures, and the "off-diagonal" elements (T_{xy}, T_{yz}, etc.) are shears. From Maxwell's equations and the Lorentz force law, it follows that

$$-\frac{d}{dt}\int_{\mathcal{V}} \wp \, d^3r = -\oint_{\mathcal{S}} \mathsf{T} \cdot d\mathbf{a} + \mathbf{F}, \tag{11.78}$$

where

$$\wp \equiv \epsilon_0(\mathbf{E} \times \mathbf{B}) \tag{11.79}$$

is the momentum density in the fields, $-\mathsf{T}$ is the momentum flux, and $\mathbf{F}$ is the total electromagnetic force on all the charged particles in $\mathcal{V}$ (and, hence, the rate of change of their collective mechanical momentum).

Similarly,

$$-\frac{d}{dt}\int_{\mathcal{V}} \boldsymbol{\ell}\, d^3r = -\oint_{\mathcal{S}} (\mathbf{r} \times \mathsf{T}) \cdot d\mathbf{a} + \mathbf{N}, \tag{11.80}$$

where

$$\boldsymbol{\ell} \equiv \epsilon_0 \mathbf{r} \times (\mathbf{E} \times \mathbf{B}) \tag{11.81}$$

is the angular momentum density in the fields, $-\mathbf{r} \times \mathsf{T}$ is the angular momentum flux, and $\mathbf{N}$ is the total electromagnetic torque on all of the charged particles in $\mathcal{V}$ (the rate of change of their collective mechanical angular momentum).

For electromagnetic fields in matter, one is usually concerned only with the *free* charges and currents. In linear media, the resulting formulas for u, $\mathbf{S}$, T, $\wp$, and $\boldsymbol{\ell}$ can be obtained by the substitution $\epsilon_0 \to \epsilon$, $\mu_0 \to \mu$; they are customarily written in terms of $\mathbf{E}$ and $\mathbf{H}$, as follows:

$$u = \tfrac{1}{2}(\epsilon E^2 + \mu H^2), \tag{11.82a}$$

$$\mathbf{S} = \mathbf{E} \times \mathbf{H}, \tag{11.82b}$$

$$\wp = \epsilon\mu(\mathbf{E} \times \mathbf{H}), \tag{11.82c}$$

$$\boldsymbol{\ell} = \epsilon\mu\left[\mathbf{r} \times (\mathbf{E} \times \mathbf{H})\right], \tag{11.82d}$$

$$\mathsf{T} = \epsilon(E_i E_j - \tfrac{1}{2}\delta_{ij}E^2) + \mu(H_i H_j - \tfrac{1}{2}\delta_{ij}H^2). \tag{11.82e}$$

11.6. ELECTROMAGNETIC WAVES

In vacuum ($\rho = 0$, $\mathbf{J} = 0$), $\mathbf{E}$ and $\mathbf{B}$ satisfy the homogeneous wave equation:

$$\nabla^2\mathbf{E} - \mu_0\epsilon_0\frac{\partial \mathbf{E}}{\partial t^2} = 0, \qquad \nabla^2\mathbf{B} - \mu_0\epsilon_0\frac{\partial \mathbf{B}}{\partial t^2} = 0, \tag{11.83}$$

indicating that empty space supports the propagation of electromagnetic waves, at the speed

$$c = 1/\sqrt{\epsilon_0\mu_0} = 2.99792458 \times 10^8 \text{m/s}. \tag{11.84}$$

In a linear medium with no *free* charge or current ($\rho_{\text{f}} = 0$, $\mathbf{J}_{\text{f}} = 0$), the fields satisfy the homogeneous wave equation with propagation speed

$$v = \frac{1}{\sqrt{\epsilon\mu}} = \frac{c}{n}, \tag{11.85}$$

where

$$n \equiv \sqrt{\epsilon_{\text{r}}\mu_{\text{r}}} \tag{11.86}$$

is the index of refraction of the medium.

Monochromatic plane waves with angular frequency ω, propagation vector $\mathbf{k}$, and polarization $\hat{\mathbf{n}}$ can be expressed as the real part of

$$\mathbf{E}(\mathbf{r}, t) = E_0 e^{i(\mathbf{k}\cdot\mathbf{r}-\omega t)}\,\hat{\mathbf{n}}; \quad \mathbf{B}(\mathbf{r}, t) = \frac{E_0}{v} e^{i(\mathbf{k}\cdot\mathbf{r}-\omega t)}\,(\hat{\mathbf{k}} \times \hat{\mathbf{n}}). \tag{11.87}$$

The waves are transverse:

$$\hat{\mathbf{n}} \cdot \hat{\mathbf{k}} = 0, \tag{11.88}$$

and they satisfy the linear dispersion relation

$$\omega = v|\,\mathbf{k}\,|. \tag{11.89}$$

The electromagnetic spectrum is summarized in Table 11.4.

TABLE 11.4. The electromagnetic spectrum.

Frequency (Hz)	Type	Wavelength (m)
10^{22}		10^{-13}
10^{21}	gamma rays	10^{-12}
10^{20}		10^{-11}
10^{19}		10^{-10}
10^{18}	X rays	10^{-9}
10^{17}		10^{-8}
10^{16}	ultraviolet	10^{-7}
10^{15}	visible	10^{-6}
10^{14}	infrared	10^{-5}
10^{13}		10^{-4}
10^{12}		10^{-3}
10^{11}		10^{-2}
10^{10}	microwave	10^{-1}
10^{9}		1
10^{8}	TV, FM	10
10^{7}		10^{2}
10^{6}	AM	10^{3}
10^{5}		10^{4}
10^{4}	RF	10^{5}
10^{3}		10^{6}

The visible range

Frequency (Hz)	Color	Wavelength (m)
1.0×10^{15}	near ultraviolet	3.0×10^{-7}
7.5×10^{14}	shortest visible blue	4.0×10^{-7}
6.5×10^{14}	blue	4.6×10^{-7}
5.6×10^{14}	green	5.4×10^{-7}
5.1×10^{14}	yellow	5.9×10^{-7}
4.9×10^{14}	orange	6.1×10^{-7}
3.9×10^{14}	longest visible red	7.6×10^{-7}
3.0×10^{14}	near infrared	1.0×10^{-6}

In linear media, the electric and magnetic contributions to the energy density are equal. Averaged over a complete cycle,

$$\langle u \rangle = \tfrac{1}{2}\epsilon E_0^2, \tag{11.90a}$$

$$\langle \mathbf{S} \rangle = \langle u \rangle v\,\hat{\mathbf{k}}, \tag{11.90b}$$

$$\langle \boldsymbol{\wp} \rangle = (\langle u \rangle / v)\,\hat{\mathbf{k}}. \tag{11.90c}$$

In particular, the intensity (power per unit area) is

$$\langle I \rangle \equiv \frac{1}{2}\sqrt{\frac{\epsilon}{\mu}}\,E_0^2, \tag{11.91}$$

and the pressure exerted on a perfectly absorbing surface by a plane wave at normal incidence is

$$\langle \mathcal{P} \rangle = \tfrac{1}{2}\epsilon E_0^2 \tag{11.92}$$

(on a perfectly reflecting surface, the pressure is twice as great).

At an interface between two linear media, Maxwell's equations prescribe the following boundary conditions:

$$\left.\begin{array}{llll} \text{(i)} & \epsilon_1 E_1^{\perp} = \epsilon_2 E_2^{\perp}, & \text{(iii)} & \mathbf{E}_1^{\parallel} = \mathbf{E}_2^{\parallel}, \\ \text{(ii)} & B_1^{\perp} = B_2^{\perp}, & \text{(iv)} & \dfrac{1}{\mu_1}\mathbf{B}_1^{\parallel} = \dfrac{1}{\mu_2}\mathbf{B}_2^{\parallel}, \end{array}\right\} \tag{11.93}$$

where $\perp$ denotes the component perpendicular to the surface and $\parallel$ refers to the parallel components. From these boundary conditions, all of the laws of geometrical optics follow, including Fresnel's equations for the reflection and transmission coefficients, which give the fraction of the incident power reflected and transmitted:

For polarization parallel to the plane of incidence:

$$R = \left(\frac{\alpha - \beta}{\alpha + \beta}\right)^2, \qquad T = \alpha\beta\left(\frac{2}{\alpha + \beta}\right)^2. \tag{11.94a}$$

For polarization perpendicular to the plane of incidence:

$$R = \left(\frac{1 - \alpha\beta}{1 + \alpha\beta}\right)^2, \qquad T = \alpha\beta\left(\frac{2}{1 + \alpha\beta}\right)^2. \tag{11.94b}$$

Here

$$\alpha \equiv \frac{\cos\theta_2}{\cos\theta_1}, \qquad \beta \equiv \sqrt{\frac{\epsilon_2\mu_1}{\epsilon_1\mu_2}}, \tag{11.95}$$

θ_1 is the angle of incidence, and θ_2 is the angle of refraction.

Electromagnetic waves confined to the interior of a perfectly conducting tube ("wave-guide") are subject to the boundary conditions

$$\mathbf{E}^{\parallel} = 0, \qquad B^{\perp} = 0 \tag{11.96}$$

at the surface. For TE ("transverse electric") waves, the longitudinal component of **E** is zero; for TM ("transverse magnetic") waves, the longitudinal component of **B** is zero. A hollow waveguide cannot support TEM waves (both fields transverse), but a coaxial cable, for example, with a conductor running down the center, can. The dispersion relation for a rectangular waveguide (with dimensions $a > b$) is

$$k = \frac{1}{c}\sqrt{\omega^2 - \omega_{mn}^2}, \tag{11.97}$$

with the "cutoff" frequency

$$\omega_{mn} \equiv \pi c \sqrt{(m/a)^2 + (n/b)^2}. \tag{11.98}$$

The mode indices (m and n) are nonnegative integers; they cannot *both* be zero for TE waves, and *neither* can be zero for TM waves. For frequencies below ω_{mn}, k is imaginary, and the waves are rapidly attenuated. Thus, to propagate a particular mode, the driving frequency must exceed the associated cutoff. The lowest cutoff frequency is $\omega_{10} = c\pi/a$; below this, no waves can propagate at all. The (phase) velocity is

$$v_p = \frac{\omega}{k} = \frac{c}{\sqrt{1 - (\omega_{mn}/\omega)^2}} > c; \tag{11.99}$$

however, energy is transported at the *group* velocity,

$$v_g = \frac{d\omega}{dk} = c\sqrt{1 - (\omega_{mn}/\omega)^2} < c. \tag{11.100}$$

11.7. RADIATION

When the source charges accelerate, the dominant term in $\mathbf{E}(\mathbf{r}, t)$ [and in $\mathbf{B}(\mathbf{r}, t)$] goes like $1/r$ at large distances, transporting energy off to infinity. We call this electromagnetic radiation. The power radiated is

$$P = \lim_{r \to \infty} \oint \mathbf{S} \cdot d\mathbf{a}, \tag{11.101}$$

where the integration is over a sphere of radius r.

In the case of an oscillating electric dipole at the origin,

$$\mathbf{p}(t) = \mathbf{p}_0 \cos(\omega t), \tag{11.102}$$

the fields in the radiation zone ($r \gg \omega/c$) are

$$\mathbf{B}(\mathbf{r}, t) = \left(\frac{\mu_0 \omega^2}{4\pi c}\right) \frac{\hat{\mathbf{r}} \times \mathbf{p}(t_r)}{r}, \quad \mathbf{E}(\mathbf{r}, t) = -c\,\hat{\mathbf{r}} \times \mathbf{B}(\mathbf{r}, t) \tag{11.103}$$

(where $t_r = t - r/c$), and the intensity (averaged over a full cycle) is

$$\langle I(\mathbf{r})\rangle = \left(\frac{\mu_0 \omega^4 p_0^2}{32\pi^2 c}\right) \frac{\sin^2\theta}{r^2} \tag{11.104}$$

(where θ is the angle between $\mathbf{p}_0$ and $\hat{\mathbf{r}}$). The intensity profile looks like a donut, with no radiation along the direction of $\mathbf{p}$ and the maximum in the "equatorial" plane. The (average) power radiated is

$$\langle P\rangle = \frac{\mu_0 p_0^2 \omega^4}{12\pi c}. \tag{11.105}$$

For an oscillating *magnetic* dipole,

$$\mathbf{m}(t) = \mathbf{m}_0 \cos(\omega t), \tag{11.106}$$

the corresponding formulas are

$$\mathbf{E}(\mathbf{r}, t) = -\left(\frac{\mu_0 \omega^2}{4\pi c}\right) \frac{\hat{\mathbf{r}} \times \mathbf{m}(t_r)}{r}, \quad \mathbf{B}(\mathbf{r}, t) = \frac{1}{c}\,\hat{\mathbf{r}} \times \mathbf{E}(\mathbf{r}, t), \tag{11.107}$$

and

$$\langle I(\mathbf{r})\rangle = \left(\frac{\mu_0 \omega^4 m_0^2}{32\pi^2 c^3}\right) \frac{\sin^2\theta}{r^2}, \quad \langle P\rangle = \frac{\mu_0 m_0^2 \omega^4}{12\pi c^3} \tag{11.108}$$

(this time θ is the angle between $\mathbf{m}_0$ and $\hat{\mathbf{r}}$). The intensity profile is the same as that of an electric dipole, but for comparable dimensions, the radiation is typically much weaker.

For a point charge q with nonrelativistic velocity $\mathbf{v}$ and acceleration $\mathbf{a}$, the (instantaneous) intensity is

$$I(\mathbf{r}) = \left(\frac{\mu_0 q^2}{16\pi^2 c}\right) \frac{a^2 \sin^2\theta}{r^2}, \tag{11.109}$$

where θ is now the angle between $\mathbf{a}$ and $\mathbf{r}$. Again, the intensity profile is a donut, with the maximum radiation in the plane perpendicular to $\mathbf{a}$. The total power radiated is given by the Larmor formula

$$P = \left(\frac{\mu_0}{6\pi c}\right) q^2 a^2. \tag{11.110}$$

In general (for arbitrary velocities),

$$I(\mathbf{r}) = \left(\frac{\mu_0 q^2}{16\pi^2 c}\right) \frac{\left|\hat{\mathbf{r}} \times \left[(\hat{\mathbf{r}} - \mathbf{v}/c) \times \mathbf{a}\right]\right|^2}{(1 - \hat{\mathbf{r}} \cdot \mathbf{v}/c)^5}, \tag{11.111}$$

and the total power radiated is given by the Liénard forumla,

$$P = \left(\frac{\mu_0}{6\pi c}\right) q^2 \gamma^6 \left[a^2 - \left(\frac{\mathbf{v} \times \mathbf{a}}{c}\right)^2\right], \tag{11.112}$$

where

$$\gamma \equiv \frac{1}{\sqrt{1 - v^2/c^2}}. \tag{11.113}$$

When $\mathbf{v}$ and $\mathbf{a}$ are collinear (as, for example, in straight line motion)

$$I(\mathbf{r}) = \left(\frac{\mu_0 q^2}{16\pi^2 c}\right) \frac{a^2 \sin^2\theta}{r^2(1 - v\cos\theta/c)^5} \tag{11.114}$$

(at ultrarelativistic velocities, the "donut" is squeezed forward into a narrow cone about the direction of $\mathbf{v}$—whether it is slowing down *or* speeding up), and

$$P = \left(\frac{\mu_0}{6\pi c}\right) \gamma^6 q^2 a^2. \tag{11.115}$$

(These formulas apply in particular to "braking radiation," or bremsstrahlung.) On the other hand, if $\mathbf{v}$ and $\mathbf{a}$ are *perpendicular* (as in circular motion),

$$P = \left(\frac{\mu_0}{6\pi c}\right) \gamma^4 q^2 a^2. \tag{11.116}$$

(This applies, in particular, to synchrotron radiation.)

Because of the energy lost to radiation, a charged particle might be expected to accelerate *less* (when subjected to a given force) than a neutral particle of the same mass. Physically, the fields exert a recoil force on the particle, known as the radiation reaction (or, in the case of oscillatory motion, radiation damping). For nonrelativistic speeds, it is given by the Abraham-Lorentz formula,

$$\mathbf{F}_{\text{rad}} = \left(\frac{\mu_0}{6\pi c}\right) q^2 \dot{\mathbf{a}}, \tag{11.117}$$

which has the notoriously embarrassing feature that it admits runaway solutions (exponentially increasing velocity in the absence of any applied force). These can be suppressed by the imposition of suitable initial conditions, but the cure is as bad as the disease, for it introduces acausal preacceleration (the response begins *before* a force is applied). These pathologies (which persist in the relativistic régime) are not fully understood; they seem to be artifacts of the point-charge limit, for they disappear in spherical models larger than the "classical" radius ($q^2/4\pi\epsilon_0 mc^2$). Perhaps, as some have argued, they are harbingers of quantum mechanics.

11.8. RELATIVISTIC FORMULATION

Classical electrodynamics is already consistent with special relativity—provided, of course, that it is applied in the context of relativistic mechanics. But it is useful and illuminating to reformulate the theory in manifestly covariant notation.[2] Charge itself is invariant, but an electrical phenomenon in one inertial frame can be magnetic in another. Specifically, if $\mathcal{S}'$ is an inertial system traveling at speed v in the x direction with respect to inertial system $\mathcal{S}$,

$$\begin{aligned} E'_x &= E_x, & E'_y &= \gamma(E_y - vB_z), & E'_z &= \gamma(E_z + vB_y), \\ B'_x &= B_x, & B'_y &= \gamma\left(B_y + \frac{v}{c^2}E_z\right), & B'_z &= \gamma\left(B_z - \frac{v}{c^2}E_y\right). \end{aligned} \tag{11.118}$$

Two cases are of particular interest:

$$\text{if } \mathbf{E} = 0, \quad \text{then} \quad \mathbf{E}' = (\mathbf{v} \times \mathbf{B}'); \tag{11.119}$$

$$\text{if } \mathbf{B} = 0, \quad \text{then} \quad \mathbf{B}' = -\frac{1}{c^2}(\mathbf{v} \times \mathbf{E}'). \tag{11.120}$$

Equations (11.118) mean that $\mathbf{E}$ and $\mathbf{B}$ transform as the off-diagonal elements in an antisymmetric field tensor:

$$F^{\mu\nu} = \begin{Bmatrix} 0 & E_x/c & E_y/c & E_z/c \\ -E_x/c & 0 & B_z & -B_y \\ -E_y/c & -B_z & 0 & B_x \\ -E_z/c & B_y & -B_x & 0 \end{Bmatrix}. \tag{11.121}$$

Associated with $F^{\mu\nu}$ is the dual tensor

$$G^{\mu\nu} = -\frac{1}{2}\epsilon^{\mu\nu\lambda\sigma}F_{\lambda\sigma} = \begin{Bmatrix} 0 & B_x & B_y & B_z \\ -B_x & 0 & -E_z/c & E_y/c \\ -B_y & E_z/c & 0 & -E_x/c \\ -B_z & -E_y/c & E_x/c & 0 \end{Bmatrix}. \tag{11.122}$$

In terms of $F^{\mu\nu}$, the inhomogeneous Maxwell equations (11.54a and d) read

$$\partial_\nu F^{\mu\nu} = \mu_0 J^\mu, \tag{11.123}$$

where

$$J^\mu = (c\rho, \mathbf{J}) \tag{11.124}$$

is the current density 4-vector. The continuity equation (11.72) says that J^μ is divergenceless:

$$\partial_\mu J^\mu = 0. \tag{11.125}$$

[2]Relativity is discussed in Chapter 14.

The homogeneous Maxwell equations (11.54b and c) can be written in terms of $G^{\mu\nu}$:

$$\partial_\nu G^{\mu\nu} = 0, \tag{11.126}$$

or, equivalently,

$$\partial_\lambda F_{\mu\nu} + \partial_\mu F_{\nu\lambda} + \partial_\nu F_{\lambda\mu} = 0. \tag{11.127}$$

There are two field invariants:

$$F^{\mu\nu} F_{\mu\nu} = -G^{\mu\nu} G_{\mu\nu} = 2(B^2 - E^2/c^2), \quad F^{\mu\nu} G_{\mu\nu} = -\frac{4}{c}(\mathbf{E} \cdot \mathbf{B}). \tag{11.128}$$

The Lorentz force law (11.1) takes the form

$$K^\mu = q\eta_\nu F^{\mu\nu}, \tag{11.129}$$

where η^μ is the particle's proper velocity and K^μ is the Minkowski force.

The fields can be expressed in terms of the 4-vector potential

$$A^\mu = (V/c, \mathbf{A}) \tag{11.130}$$

as follows:

$$F^{\mu\nu} = \partial^\mu A^\nu - \partial^\nu A^\mu. \tag{11.131}$$

In the Lorentz gauge (11.62)

$$\partial_\mu A^\mu = 0, \tag{11.132}$$

the potential satisfies the inhomogeneous wave equation (11.63):

$$\partial^\nu \partial_\nu A^\mu = -\mu_0 J^\mu. \tag{11.133}$$

The electromagnetic stress-energy tensor

$$S^{\mu\nu} = \frac{1}{\mu_0}\left[F^{\mu\lambda} F_\lambda^{\ \nu} + \frac{1}{4} g^{\mu\nu} F^{\lambda\sigma} F_{\lambda\sigma}\right] \tag{11.134}$$

contains the energy density (11.75):

$$S^{00} = \frac{1}{2}\left(\epsilon_0 E^2 + \frac{1}{\mu_0} B^2\right) = u, \tag{11.135}$$

the momentum density (11.79):

$$S^{0i} = \frac{1}{\mu_0 c}(\mathbf{E} \times \mathbf{B})_i = c\wp_i, \tag{11.136}$$

and the Maxwell stress tensor (11.77):

$$S^{ij} = -T_{ij}. \tag{11.137}$$

It is symmetric and traceless,

$$S^{\nu\mu} = S^{\mu\nu}, \qquad S^\mu_{\ \mu} = 0, \tag{11.138}$$

but not divergenceless:

$$\partial_\nu S^{\mu\nu} = J_\nu F^{\mu\nu}. \tag{11.139}$$

(A famously confusing consequence of the latter is that the "field energy"

$$\int_{\mathcal{V}} S^{00}\, d^3r \tag{11.140}$$

and the "field momentum"

$$\frac{1}{c}\int_{\mathcal{V}} S^{0i}\, d^3r \tag{11.141}$$

do not constitute a 4-vector, if the volume $\mathcal{V}$ contains charge or current. But the total stress-energy tensor—including nonelectromagnetic contributions—is always divergenceless, and the total energy and momentum do constitute a 4-vector.)

The Liénard formula (11.112), giving the power radiated by a point charge, takes the form

$$P = \frac{\mu_0}{6\pi c} q^2 \dot{\eta}^\mu \dot{\eta}_\mu \tag{11.142}$$

(where the dots now denote differentiation with respect to *proper* time), and the relativistic generalization of the Abraham-Lorentz formula (11.117) for the radiation reaction is

$$K^\mu = \left(\frac{\mu_0}{6\pi c}\right) q^2 \left(\ddot{\eta}^\mu + \frac{1}{c^2}\eta^\mu \ddot{\eta}^\nu \eta_\nu\right). \tag{11.143}$$

11.9. CIRCUITS

In many materials, the current density is proportional to the applied force per unit charge, **f**:

$$\mathbf{J} = \sigma \mathbf{f}; \tag{11.144}$$

σ is the conductivity of the medium (Table 11.5 lists some representative resistivities, $\rho \equiv 1/\sigma$). If the force is electromagnetic,

$$\mathbf{J} = \sigma(\mathbf{E} + \mathbf{v} \times \mathbf{B}). \tag{11.145}$$

In electric circuits (but not, for example, in plasmas), the magnetic contribution is usually negligible, and

$$\mathbf{J} = \sigma \mathbf{E}. \tag{11.146}$$

In particular, for a resistor,

$$V = IR, \tag{11.147}$$

where I is the current that flows when the potential difference between the terminals is V, and R is the resistance. For a cylindrical resistor of length l and cross-sectional area A,

$$R = \frac{l}{\sigma A}. \tag{11.148}$$

TABLE 11.5. Resistivities, in ohm-meters (all values are for 1 atm, 20°C). [*Handbook of Chemistry and Physics*, 78th ed., edited by David R. Lide, Jr. (CRC Press, Boca Raton, FL, 1997).]

Material	Resistivity	Material	Resistivity
Conductors:		*Semiconductors:*	
Silver	1.59×10^{-8}	Salt water (saturated)	4.4×10^{-2}
Copper	1.68×10^{-8}	Germanium	4.6×10^{-1}
Gold	2.21×10^{-8}	Diamond	2.7
Aluminum	2.65×10^{-8}	Silicon	2.5×10^{3}
Iron	9.61×10^{-8}	*Insulators:*	
Mercury	9.58×10^{-7}	Water (pure)	2.5×10^{5}
Nichrome	1.00×10^{-6}	Wood	$10^{8} - 10^{11}$
Manganese	1.44×10^{-6}	Glass	$10^{10} - 10^{14}$
Graphite	1.4×10^{-5}	Quartz (fused)	$\sim 10^{16}$

Equation (11.147) is the familiar form of Ohm's law, though the underlying physical principle is contained in Eq. (11.144).

The voltage across a capacitor is proportional to the charge Q:

$$V = \frac{Q}{C}, \tag{11.149}$$

where C is the capacitance. For an air-filled parallel-plate capacitor with area A and separation d,

$$C = \frac{\epsilon_0 A}{d}. \tag{11.150}$$

The voltage across an inductor is proportional to the rate of change of the current:

$$V = -L\frac{dI}{dt}, \tag{11.151}$$

where L is the inductance. For an air-filled solenoid of cross-sectional area A, length l, and N turns,

$$L = \mu_0 N^2 \frac{A}{l}. \tag{11.152}$$

In alternating current circuits, Ohm's law can be generalized to accommodate capacitors and inductors, as well as resistors:

$$\tilde{V} = \tilde{I} Z, \tag{11.153}$$

where a tilde denotes the *complex* quantity (of which the *physical* value is the real part), and Z is the impedance:

$$Z_R = R, \quad Z_C = \frac{1}{i\omega C}, \quad Z_L = i\omega L, \tag{11.154}$$

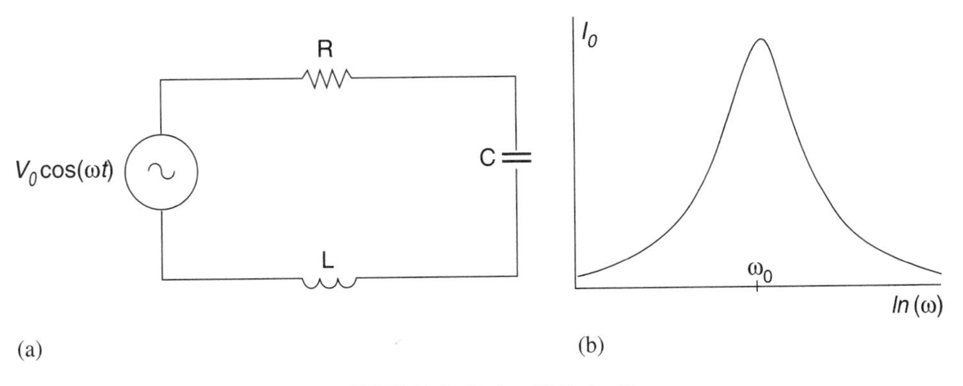

FIGURE 11.1. Series *RLC* circuit.

for angular frequency ω. In series, impedances add; in parallel, their *reciprocals* (admittances) add.

For example, in the series *RLC* circuit (Fig. 11.1a)

$$I(t) = \frac{V_0}{\sqrt{R^2 + (\omega L - 1/\omega C)^2}} \cos(\omega t - \delta), \tag{11.155}$$

where

$$\tan\delta = \frac{1}{R}\left(\omega L - \frac{1}{\omega C}\right). \tag{11.156}$$

The amplitude (I_0) is a maximum when the circuit is driven at the resonant frequency (Fig. 11.1b):

$$\omega_0 = \frac{1}{\sqrt{LC}}. \tag{11.157}$$

For the low-pass filter in Fig. 11.2a, the gain (ratio of the amplitudes of V_{out} to V_{in}) is

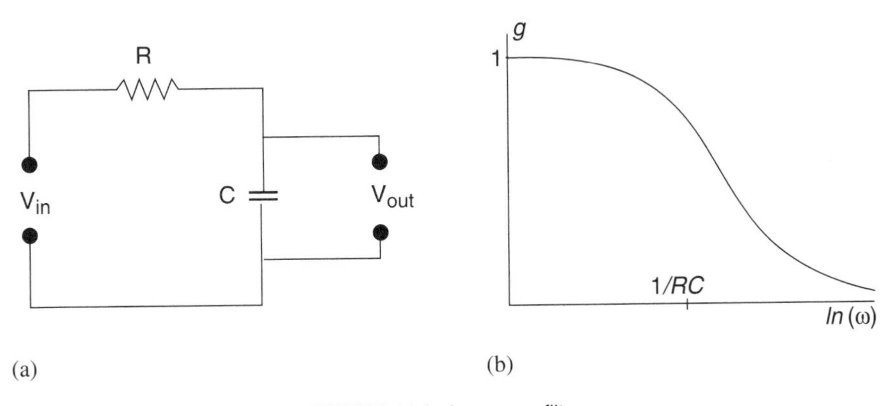

FIGURE 11.2. Low-pass filter.

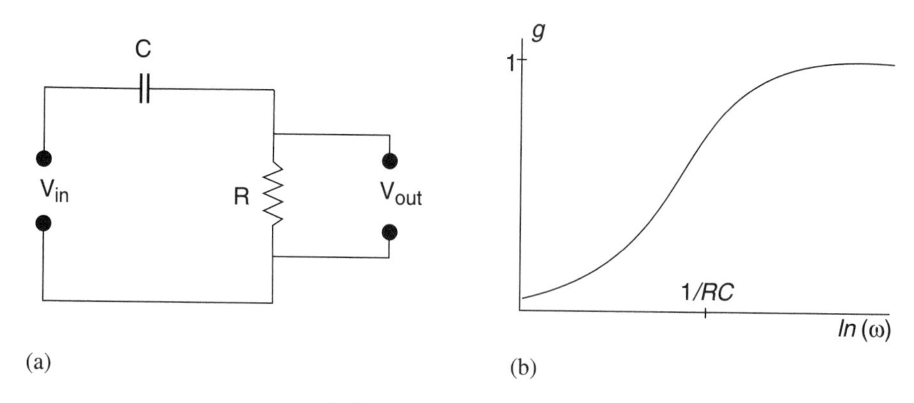

FIGURE 11.3. High-pass filter.

$$g = \frac{1}{\sqrt{1 + (\omega RC)^2}}; \tag{11.158}$$

it blocks out frequencies *above* about $1/RC$ (Fig. 11.2b). Reversing the roles of the resistor and capacitor produces a high-pass filter (Fig. 11.3a); its gain is

$$g = \frac{1}{\sqrt{1 + (1/\omega RC)^2}}, \tag{11.159}$$

and it blocks out frequencies *below* about $1/RC$ (Fig. 11.3b).

11.10. UNITS

The equations in this chapter have been expressed in SI *(Système International)* units, since these are used by most physicists and engineers. But atomic physics is usually done in gaussian units, particle physicists prefer Heaviside-Lorentz units, and some older books use electrostatic units (esu) or electromagnetic units (emu). Thus, Coulomb's law may appear variously as

$$\mathbf{F} = \frac{1}{4\pi\epsilon_0}\frac{q_1 q_2}{r^2}\hat{\mathbf{r}} \qquad \text{(SI)}, \tag{11.160a}$$

$$= \frac{q_1 q_2}{r^2}\hat{\mathbf{r}} \qquad \text{(gaussian and esu)}, \tag{11.160b}$$

$$= \frac{1}{4\pi}\frac{q_1 q_2}{r^2}\hat{\mathbf{r}} \qquad \text{(Heaviside-Lorentz)} \tag{11.160c}$$

$$= c^2\frac{q_1 q_2}{r^2}\hat{\mathbf{r}}. \qquad \text{(emu)}. \tag{11.160d}$$

Maxwell's equations and the Lorentz force law can be written in the universal form

$$\nabla \cdot \mathbf{E} = \alpha\rho, \tag{11.161a}$$

$$\nabla \cdot \mathbf{B} = 0, \tag{11.161b}$$

$$\nabla \times \mathbf{E} = -\frac{1}{\beta c^2}\frac{\partial \mathbf{B}}{\partial t}, \tag{11.161c}$$

$$\nabla \times \mathbf{B} = \alpha\beta\mathbf{J} + \beta\frac{\partial \mathbf{E}}{\partial t}, \tag{11.161d}$$

$$\mathbf{F} = q\left[\mathbf{E} + \frac{1}{\beta c^2}(\mathbf{v} \times \mathbf{B})\right], \tag{11.161e}$$

with the constants α and β given in Table 11.6. The SI system uses mks (meter-kilogram-second) units for mechanical quantities; the others all use cgs (centimeter-gram-second). Conversion factors relating gaussian and SI units are given in Table 11.7. A comparison

TABLE 11.6. The constants α and β (Eq. 11.161) for various systems of units.

System	α	β
SI	$1/\epsilon_0$	$\mu_0\epsilon_0(= 1/c^2)$
Gaussian	4π	$1/c$
Heaviside-Lorentz	1	$1/c$
electrostatic	4π	$1/c^2$
electromagnetic	$4\pi c^2$	$1/c^2$

TABLE 11.7. Conversion factors. [*Note:* Except in exponents, every "3" is short for $\kappa \equiv 2.99792458$ (the numerical value of the speed of light), "9" means κ^2, and "12" is 4κ.]

Quantity	SI	Factor	Gaussian
Length	meter (m)	10^2	centimeter
Mass	kilogram (kg)	10^3	gram
Time	second (s)	1	second
Force	newton (N)	10^5	dyne
Energy	joule (J)	10^7	erg
Power	watt (W)	10^7	erg/second
Charge	coulomb (C)	3×10^9	esu (statcoulomb)
Current	ampere (A)	3×10^9	esu/second (statampere)
Electric field	volt/meter	$(1/3) \times 10^{-4}$	statvolt/centimeter
Potential	volt (V)	$1/300$	statvolt
Displacement	coulomb/meter^2	$12\pi \times 10^5$	statcoulomb/centimeter^2
Resistance	ohm (Ω)	$(1/9) \times 10^{-11}$	second/centimeter
Capacitance	farad (F)	9×10^{11}	centimeter
Magnetic field	tesla (T)	10^4	gauss
Magnetic flux	weber (Wb)	10^8	maxwell
H	ampere/meter	$4\pi \times 10^{-3}$	oersted
Inductance	henry (H)	$(1/9) \times 10^{-11}$	second^2/centimeter

TABLE 11.8. Fundamental equations in SI and Gaussian units.

	SI	Gaussian
Maxwell's equations		
In general:	$\begin{cases} \nabla\cdot\mathbf{E} = \rho/\epsilon_0 \\ \nabla\times\mathbf{E} = -\partial\mathbf{B}/\partial t \\ \nabla\cdot\mathbf{B} = 0 \\ \nabla\times\mathbf{B} = \mu_0\mathbf{J} + \mu_0\epsilon_0\partial\mathbf{E}/\partial t \end{cases}$	$\nabla\cdot\mathbf{E} = 4\pi\rho$ $\nabla\times\mathbf{E} = -(1/c)\partial\mathbf{B}/\partial t$ $\nabla\cdot\mathbf{B} = 0$ $\nabla\times\mathbf{B} = (4\pi/c)\mathbf{J} + (1/c)\partial\mathbf{E}/\partial t$
In matter:	$\begin{cases} \nabla\cdot\mathbf{D} = \rho_f \\ \nabla\times\mathbf{E} = -\partial\mathbf{B}/\partial t \\ \nabla\cdot\mathbf{B} = 0 \\ \nabla\times\mathbf{H} = \mathbf{J}_f + \partial\mathbf{D}/\partial t \end{cases}$	$\nabla\cdot\mathbf{D} = 4\pi\rho_\mathrm{f}$ $\nabla\times\mathbf{E} = -(1/c)\partial\mathbf{B}/\partial t$ $\nabla\cdot\mathbf{B} = 0$ $\nabla\times\mathbf{H} = (4\pi/c)\mathbf{J}_\mathrm{f} + (1/c)\partial\mathbf{E}/\partial t$
D and H		
Definitions:	$\begin{cases} \mathbf{D} = \epsilon_0\mathbf{E} + \mathbf{P} \\ \mathbf{H} = (1/\mu_0)\mathbf{B} - \mathbf{M} \end{cases}$	$\mathbf{D} = \mathbf{E} + 4\pi\mathbf{P}$ $\mathbf{H} = \mathbf{B} - 4\pi\mathbf{M}$
Linear media:	$\begin{cases} \mathbf{P} = \epsilon_0\chi_e\mathbf{E}, \quad \mathbf{D} = \epsilon\mathbf{E} \\ \mathbf{M} = \chi_m\mathbf{H}, \quad \mathbf{H} = (1/\mu)\mathbf{B} \end{cases}$	$\mathbf{P} = \chi_e\mathbf{E}, \quad \mathbf{D} = \epsilon\mathbf{E}$ $\mathbf{M} = \chi_m\mathbf{H}, \quad \mathbf{H} = (1/\mu)\mathbf{B}$
Lorentz force law	$\mathbf{F} = q(\mathbf{E} + \mathbf{v}\times\mathbf{B})$	$\mathbf{F} = q\,[\mathbf{E} + (1/c)\,(\mathbf{v}\times\mathbf{B})]$
Energy and power		
Energy density:	$u = \frac{1}{2}[\epsilon_0 E^2 + (1/\mu_0)B^2]$	$u = (1/8\pi)(E^2 + B^2)$
Poynting vector:	$\mathbf{S} = (1/\mu_0)(\mathbf{E}\times\mathbf{B})$	$\mathbf{S} = (c/4\pi)\,(\mathbf{E}\times\mathbf{B})$
Larmor formula:	$P = q^2a^2/6\pi\epsilon_0c^3$	$P = 2q^2a^2/3c^3$

of the fundamental equations of electrodynamics in these two systems is given in Table 11.8.[3]

11.11. REFERENCES

[1] R. P. Feynman, R. B. Leighton, and M. Sands, *The Feynman Lectures on Physics* (Addison-Wesley, Reading, MA, 1964).

[2] E. M. Purcell, *Electricity and Magnetism*, 2nd ed. (McGraw-Hill, New York, 1985).

[3] D. J. Griffiths, *Introduction to Electrodynamics*, 3rd ed. (Prentice-Hall, Upper Saddle River, NJ, 1999).

[4] P. Lorrain, D. R. Corson, and F. Lorrain, *Electromagnetic Fields and Waves*, 3rd ed. (Freeman, New York, 1988).

[5] J. R. Reitz, F. J. Milford, and R. W. Christy, *Foundations of Electromagnetic Theory*, 3rd ed. (Addison-Wesley, Reading, MA, 1979).

[6] J. D. Jackson, *Classical Electrodynamics*, 3rd ed. (Wiley, New York, 1999).

[3]For further details, see the Appendix in J. D. Jackson, *Classical Electrodynamics*, 3rd ed. (Wiley, New York, 1999).

[7] L. D. Landau and E. M. Lifshitz, *Electrodynamics of Continuous Media*, 2nd ed. (Addison-Wesley, Reading, MA, 1984).

[8] W. K. H. Panofsky and M. Phillips, *Classical Electricity and Magnetism,* 2nd ed. (Addison-Wesley, Reading, MA, 1962).

[9] W. R. Smythe, *Static and Dynamic Electricity,* 3rd ed. (Hemisphere, New York, 1989).

[10] A. Sommerfeld, *Electrodynamics* (Academic, New York, 1952).

[11] J. A. Stratton, *Electromagnetic Theory* (McGraw-Hill, New York, 1941).

12

Elementary Particles

H. Schellman
Northwestern University, Evanston, Illinois

Contents

List of Tables

List of Figures

Unless otherwise stated, the material in this section is based on the *2000 Review of Particle Properties* [1] and the previous edition of *A Physicist's Desk Reference*. [2] Numerical values and formulae are appropriate for quick calculations, but the full Review and original references should be consulted for full precision.

12.1. THE STANDARD MODEL

The Standard Model of Elementary Particles describes the weak, electromagnetic, and strong forces in terms of gauge interactions between fundamental fermions and gauge bosons.

12.1.1. Fundamental fermions

The fermions consist of leptons (no strong interactions) and hadrons (strongly interacting). There are three known "generations" of both leptons and hadrons.

Leptons:

$$\psi = \begin{pmatrix} \nu_i \\ \ell_i \end{pmatrix} = \begin{pmatrix} \nu_e \\ e^- \end{pmatrix}, \begin{pmatrix} \nu_\mu \\ \mu^- \end{pmatrix}, \begin{pmatrix} \nu_\tau \\ \tau^- \end{pmatrix}.$$

Quarks:

$$\psi = \begin{pmatrix} \mathrm{u}_i \\ \mathrm{d}_i \end{pmatrix} = \begin{pmatrix} \mathrm{u} \\ \mathrm{d} \end{pmatrix}, \begin{pmatrix} \mathrm{c} \\ \mathrm{s} \end{pmatrix}, \begin{pmatrix} \mathrm{t} \\ \mathrm{b} \end{pmatrix}. \tag{12.1}$$

TABLE 12.1. Weak and electromagnetic charges of fundamental fermions.

	ψ_i	Electric charge	T^3	g_V	g_A
Left-handed	ν_i	0	$+\frac{1}{2}$	$+\frac{1}{2}$	$+\frac{1}{2}$
	ℓ_i	-1	$-\frac{1}{2}$	$-\frac{1}{2}+2\sin^2\theta_W$	$-\frac{1}{2}$
Right-handed	$\overline{\ell}_i$	$+1$	0	$-2\sin^2\theta_W$	0
Left-handed	u, c, t	$+\frac{2}{3}$	$+\frac{1}{2}$	$+\frac{1}{2}-\frac{4}{3}\sin^2\theta_W$	$+\frac{1}{2}$
	d, s, b	$-\frac{1}{3}$	$-\frac{1}{2}$	$-\frac{1}{2}+\frac{2}{3}\sin^2\theta_W$	$-\frac{1}{2}$
Right-handed	$\bar{\text{d}}, \bar{\text{s}}, \bar{\text{b}}$	$+\frac{1}{3}$	0	$\frac{2}{3}\sin^2\theta_W$	0

12.1.2. Electroweak couplings

The electro-weak gauge couplings for these fermions are described by the group $SU(2)_L \times U(1)_R$. Table 12.1 tabulates the weak and electromagnetic couplings for the fundamental fermions.

12.1.3. Electroweak Lagrangian

$$\mathcal{L}_W = \sum_i \bar{\psi}_i' \left[\left(i\,\partial\!\!\!/ - m_i - g\frac{m_i H}{2M_W} \right) - \frac{g}{2\sqrt{2}}\gamma^\mu(1-\gamma_5)(T^+ W_\mu^+ + T^- W_\mu^-) \right.$$
$$\left. -e q_i \gamma^\mu A_\mu - \frac{g}{2\cos\theta_W}\gamma^\mu(g_V^i - g_A^i\gamma_5)Z_\mu \right] \psi_i'$$
$$-\frac{1}{4}\sum_{\text{bosons}} F_{\mu\nu}F^{\mu\nu} + \frac{g}{2M_W}[2M_W^2 H W_\mu^+ W^{-\mu} + M_Z^2 H Z_\mu Z^\mu], \tag{12.2}$$

where i runs over fermion flavors, the prime on the fermion wave functions indicates that the weak eigenstates are meant, m_i is the fermion mass, W, Z, and A are the gauge boson fields, $F_{B(a)}^{\mu\nu} = \partial^\mu B_{(a)}^\nu - \partial^\nu B_{(a)}^\mu + g f_{(abc)} B_{(b)}^\mu B_{(c)}^\nu$ is the field tensor for boson field B (the third term introduces boson-boson couplings), H is the Higgs field, the γ are the Dirac matrices, the T are isospin changing operators, e is the positron charge, g is the weak coupling ($= e/\sin\theta_W$), and θ_W is the Weinberg angle. The q_i are the electromagnetic charges, and g_V and g_A are the vector and axial weak couplings.

$$G_F = \frac{g^2}{4\sqrt{2}M_W^2}, \tag{12.3}$$

$$g_V = T_i^3 - 2q_i \sin^2\theta_W, \tag{12.4}$$

$$g_A = T_i^3, \tag{12.5}$$

$$\cos\theta_W = \frac{M_W}{M_Z}. \tag{12.6}$$

This Lagrangian is consistent with the known behavior of the electroweak interactions to better than 1%. However, understanding of the details (and even validity) of the Higgs component of the model await direct experimental observation of the Higgs and its interactions.

12.1.4. Cabibbo-Kobayashi-Maskawa mixing matrix

The mass and electroweak eigenstates of quarks are not the same. The weak interactions can lead to mixings among the mass states; hence the primed notation for the weak doublets. In the quark sector, the mixing is described by the Kobayashi-Maskawa matrix, which relates the weak eigenstates d' to the mass eigenstates via

$$\begin{pmatrix} d' \\ s' \\ b' \end{pmatrix} = \begin{pmatrix} V_{ud} & V_{us} & V_{ub} \\ V_{cd} & V_{cs} & V_{cb} \\ V_{td} & V_{ts} & V_{tb} \end{pmatrix} \begin{pmatrix} d \\ s \\ b \end{pmatrix}. \tag{12.7}$$

In the Wolfenstein [4] parameterization,

$$\mathbf{V} = \begin{pmatrix} 1-\frac{1}{2}\lambda^2 & \lambda & A\lambda^3(\rho - i\eta) \\ -\lambda & 1-\frac{1}{2}\lambda^2 & A\lambda^2 \\ A\lambda^3(1-\rho-i\eta) & -A\lambda^2 & 1 \end{pmatrix}, \tag{12.8}$$

where A, ρ, and η are of order unity and $\lambda = V_{us} \approx 0.22$ is the Cabibbo angle.

If the CKM matrix is in fact unitary,

$$V_{ud}V_{ub}^* + V_{cd}V_{cb}^* + V_{td}V_{tb}^* = 0, \tag{12.9}$$

which can be illustrated graphically as a triangle in the complex plane (Figure 12.1), where $V_{cd}V_{cb}^*$ is assumed to be real and $A = V_{ud}V_{ub}^*$ is the apex of the triangle. The angles α, β, and γ should sum to π if the unitarity condition holds. If sides are rescaled by a factor of $V_{cd}V_{cb}^*$, the base of the triangle has length 1 and the apex A is located at (ρ, η).

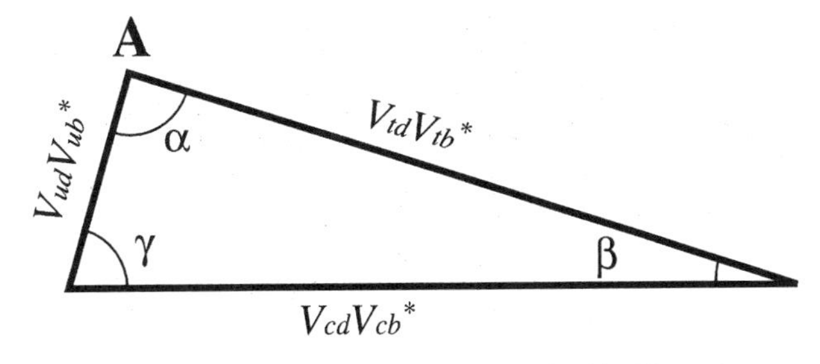

FIGURE 12.1. The unitarity triangle with $V_{cd}V_{cb}^*$ real.

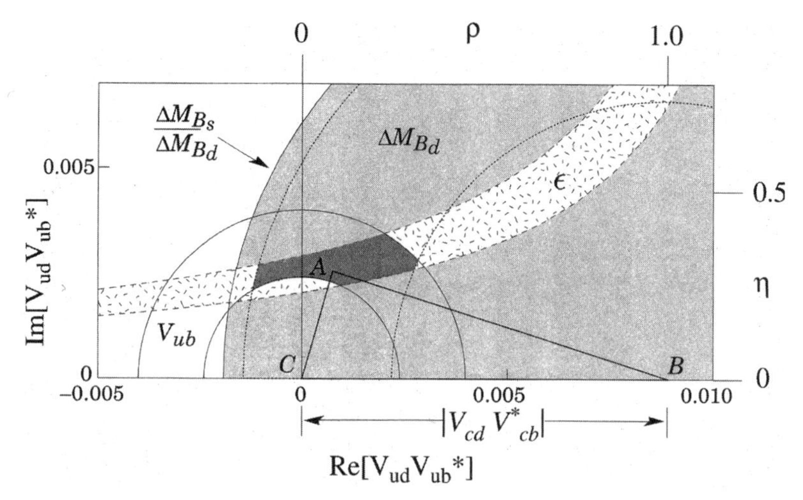

FIGURE 12.2. Constraints on the unitarity triangle as of 2000. The axes are labeled for both the elements VV^* and the (ρ, η) convention.

The $i\eta$ terms introduce *CP* violation, which can be eliminated by a phase rotation in processes involving only two generations, but which remains in processes involving all three. *CP* violation has been observed in neutral kaon decays and is expected to appear in B meson decays.

If the CKM matrix is unitary, the 90% confidence limits on the magnitudes of the matrix elements are [1]

$$\mathbf{V} = \begin{pmatrix} 0.9742\text{–}0.9757 & 0.219\text{–}0.226 & 0.002\text{–}0.005 \\ 0.219\text{–}0.225 & 0.9734\text{–}0.9749 & 0.037\text{–}0.043 \\ 0.004\text{–}0.014 & 0.035\text{–}0.043 & 0.990\text{–}0.9993 \end{pmatrix}. \tag{12.10}$$

Figure 12.2 shows limits on the complex phase as of 2000. Experimental measurements of *CP* violation in *B* meson decays, such as $B \to \psi K_S$, should provide tight constraints on the *CP*-violating parameters.

12.1.5. *CP* violation in the kaon system

The weak eigenstates of the neutral kaon are the $K^0(\bar{s}d)$ and $\bar{K}^0(\bar{d}s)$. The mass eigenstates are

$$| K_S >= p \mid K^0 > +q \mid \bar{K}^0 >, \qquad | K_L >= p \mid K^0 > -q \mid \bar{K}^0 >, \tag{12.11}$$

where K_S and K_L are short- and long-lived states and

$$\frac{p}{q} = \frac{(1+\tilde{\epsilon})}{(1-\tilde{\epsilon})}, \tag{12.12}$$

$$\tilde{\epsilon} = \epsilon + i Im[A_0]/Re[A_0], \tag{12.13}$$

where $\epsilon \approx 2.3 \times 10^{-3}$ and A_0 is the isospin 0 decay amplitude. The different magnitudes of p and q are a signature of CP violation that can come either via a $\Delta S = 2$ transition in $K^0 - \bar{K}^0$ mixing or via a direct $\Delta S = 1$ transition in the decay. The direct $\Delta S = 1$ contribution is parametrized by a factor $\frac{\epsilon'}{\epsilon}$ that is believed to be $\sim (2-3) \times 10^{-3}$ [5, 6].

12.1.6. Neutrino masses and mixing

If neutrinos have mass, their mass eigenstates may not correspond to flavor eigenstates and one can expect mixing of neutrino species. If two neutrino flavors mix, one can expect to see neutrino oscillations in which a beam of neutrino type a changes to type b with probability

$$\begin{aligned} P(\nu_a \to \nu_b) &= \sin^2 2\theta_{ab} \sin^2[\Delta M_{ab}^2 L/4E] \\ &= \sin^2 2\theta_{ab} \sin^2[1.27 \Delta M_{ab}^2 (\mathrm{eV})^2 L(\mathrm{km})/E(\mathrm{GeV})], \end{aligned} \tag{12.14}$$

where the first equation uses natural ($\hbar = c = 1$) units and the second shows the scales useful for experiments.

The electron neutrino mass is constrained to be less than 10–15 eV by the endpoint of tritium decay. However, solar and cosmic ray neutrino studies and accelerator-based neutrino production experiments indicate that some mixing between neutrino species does occur and imply a nonzero mass for at least one neutrino type.

By analogy with the quark sector, a three-flavor scheme has neutrino mass eigenstates ν_1, ν_2, ν_3 and weak eigenstates $(\nu_e, \nu_\mu, \nu_\tau)$. The probability of observing a neutrino oscillation from weak eigenstate α to state β can be written as

$$P(\nu_\alpha \to \nu_\beta) = \sum_{i,j} U_{\alpha i} U_{\beta i}^* U_{\alpha j}^* U_{\beta j} e^{-i\Delta m_{ij}^2 L/2E}, \tag{12.15}$$

where L is the distance from production to detection, E is the neutrino energy, and $U_{\alpha i}$ is the leptonic mixing matrix analogous to the CKM mixing matrix. The mixing matrix is parametrized [8] in terms of four angles, θ_{12}, θ_{23}, θ_{13}, and δ, where δ indicates the CP-violating phase.

$$\begin{pmatrix} c_{13}c_{12} & c_{13}s_{12} & s_{13}e^{-i\delta} \\ -c_{23}s_{12} - s_{13}s_{23}c_{12}e^{i\delta} & c_{23}c_{12} - s_{13}s_{23}s_{12}e^{i\delta} & c_{13}s_{23} \\ s_{23}s_{12} - s_{13}c_{23}c_{12}e^{i\delta} & -s_{23}c_{12} - s_{13}c_{23}s_{12}e^{i\delta} & c_{13}c_{23} \end{pmatrix} \tag{12.16}$$

Current measurements indicate that

$$\theta_{23} \simeq \frac{\pi}{4} \qquad \text{Atmospheric neutrinos, [9]} \tag{12.17}$$

$$\theta_{13} < 0.08 \qquad \text{Reactor experiments, [10]} \tag{12.18}$$

$$\theta_{12} \simeq 0 \quad \text{or} \quad \simeq \frac{\pi}{4} \qquad \text{Solar neutrinos. [11]} \tag{12.19}$$

12.1.7. Strong interactions

The strong interactions are believed to be described by quantum chromodynamics, a gauge theory in which quarks carry color charges and forces are mediated by eight colored gluons. There are three colors, and gauge group is SU(3).

The Langrangian is

$$\mathcal{L}_{\mathrm{QCD}} = -\frac{1}{4} F^{(a)}_{\mu\nu} F^{(a)\mu\nu} + i \sum_i \bar{\psi}^{(r)}_i \gamma^\mu D_{\mu(rs)} \psi^{(s)}_i - \sum_i m_i \bar{\psi}^{(r)}_i \psi_{i(r)}, \quad (12.20)$$

$$F^{(a)}_{\mu\nu} = \partial_\mu A^{(a)}_\nu - \partial_\nu A^{(a)}_\mu + g_S f_{(abc)} A^{(b)}_\mu A^{(c)}_\nu, \quad (12.21)$$

$$D_{\mu(rs)} = \delta_{rs}\partial_\mu - i g_S \sum \frac{1}{2} \lambda^{(a)}_{rs} A^{(a)}_\mu, \quad (12.22)$$

where r and s are the three quark color indices, a, b, c are the eight gluon indices, $\lambda_{(a)}$ are the generators of SU(3) transformations, and the f_{abc} is the SU(3) structure constant. g_S is a coupling constant.

The effective QCD coupling constant

$$\alpha_S(\mu) = \frac{g_S^2}{4\pi} = \frac{4\pi}{\beta_0 \ln(\mu^2/\Lambda^2)} \left[1 - 2\frac{\beta_1}{\beta_0^2} \frac{\ln(\ln(\mu^2/\Lambda^2))}{\ln(\mu^2/\Lambda^2)} + \cdots \right],$$

$$\beta_0 = 11 - \frac{2}{3}\eta_f, \quad \beta_1 = 51 - \frac{19}{3}\eta_f, \quad (12.23)$$

where μ is the energy scale, η_f is the number of quark flavors involved at scale μ, and Λ is of order 200–300 MeV for $\eta_f = 4$. The current standard is to quote α_S at the mass of the Z_0. The most recent particle data group value (2000)[1] is $\alpha_S(M_{Z^0}) = 0.118 \pm 0.002$. At a scale of 2GeV, α_S rises to 0.35.

12.2. SELECTED PARTICLE PROPERTIES

Tables 12.2, 12.3, and 12.4 describe some of the more useful particle properties derived from the *Particle Data Group Booklet 2000 edition.* [1] Most of the spin-0 mesons, but only those vector mesons and baryons that are easily produced or detected, are included.

TABLE 12.2. Gauge bosons $-J = 1$.

	Mass, GeV	Width, GeV	decays
G - gluon	0		
γ	0	stable	
W^+	80.4	2.0	$\bar{\ell}\nu$, $\mathrm{u}_i\bar{\mathrm{d}}_i$
Z_0	91.19	0.083	$\nu\bar{\nu}$, $\ell\bar{\ell}$, $\mathrm{q}\bar{\mathrm{q}}$

TABLE 12.3. Leptons and quarks.

	Leptons $-J^P = \frac{1}{2}^+$			
	Mass	Charge	Lifetime	Decay modes
e^-	0.51100 MeV	$-e$	stable	
μ^-	0.10566 GeV	$-e$	2.1970×10^{-6} sec	$e^- \bar{\nu}_e \nu_\mu$
τ^-	1.777 GeV	$-e$	2.90×10^{-13} sec	$\sim$17% $\nu_{\bar{\ell}} \bar{\nu} \ell \nu_\tau$, rest hadronic
ν_e	<10–15 eV	0	stable	
ν_μ	<0.19 MeV	0	stable	
ν_τ	<18.2 MeV	0	stable	

	Quarks $-J^P = \frac{1}{2}^+$	
	Current mass[a]	Charge
u	$\sim$3 MeV	$\frac{2}{3}e$
d	$\sim$6 MeV	$-\frac{1}{3}e$
s	$\sim$100 MeV	$-\frac{1}{3}e$
c	$\sim$1.3 GeV	$\frac{2}{3}e$
b	$\sim$4.2 GeV	$-\frac{1}{3}e$
t	$\sim$175 GeV	$\frac{2}{3}e$

[a] These are the masses relevant to the QCD Lagrangian discussed in Section 12.2.7

TABLE 12.4. Mesons and baryons.

	Pseudoscalar mesons				
	Quark content	$I(J^{PC})$	Mass GeV	Lifetime/width in sec or eV	Decay modes[a]
π^+	$u\bar{d}$	$1(0^-)$	0.1396	2.60×10^{-8} sec	$\mu^+\nu_\mu$ ($\sim$ 100%)
π^0	$u\bar{u} - d\bar{d}$	$1(0^{-+})$	0.1350	8.4×10^{-17} sec	2γ
K^+	$u\bar{s}$	$\frac{1}{2}(0^-)$	0.4937	1.239×10^{-8} sec	$\mu^+\nu_\mu$ (63%), 2π (21%), 3π (7%), $\pi^0\ell^+\nu_\ell$ ($\sim$ 4%)
$K^0/\bar{K}^0$	$d\bar{s}/\bar{d}s$	$\frac{1}{2}(0^-)$	0.4977	K_S, 8.93×10^{-11} sec K_L, 5.17×10^{-8} sec	$\pi^+\pi^-$ (69%), $2\pi^0$ (31%) $\pi e\nu$ (39%), $\pi\mu\nu$ (27%), 3π (34%)
η^0	$u\bar{u}$, $d\bar{d}$, $s\bar{s}$	$0(0^{-+})$	0.5473	1.18 KeV	2γ (39%), 3π (55%)
η'	$u\bar{u}$, $d\bar{d}$, $s\bar{s}$	$0(0^{-+})$	0.9578	203 KeV	$2\pi\eta$ (65%), $\rho^0\gamma$ (30%)
D^+	$c\bar{d}$	$\frac{1}{2}(0^-)$	1.869	1.06×10^{-12} sec	$\ell^+ + X$ (17%), $K^- + X$ ($\sim$ 25%), $K^0/\bar{K}^0 + X$ ($\sim$ 60%)
D^0	$c\bar{u}$	$\frac{1}{2}(0^-)$	1.865	4.15×10^{-13} sec	ℓ^+X (7%), $K^- + X$ ($\sim$ 50%), $K^+ + X$ ($\sim$ 3%), $K^0/\bar{K}^0 + X$ ($\sim$ 40%)
D_s^+	$c\bar{s}$	$\frac{1}{2}(0^-)$	1.969	4.7×10^{-13} sec	ℓ^+X ($\sim$ 8%)
η_c	$c\bar{c}$	$0^+(0^{-+})$	2.980	13 MeV	hadronic
B^+	$u\bar{b}$	$\frac{1}{2}(0^-)$	5.279	1.65×10^{-12} sec	ℓ^+X (10%)
B^0	$d\bar{b}$	$\frac{1}{2}(0^-)$	5.279	1.56×10^{-12} sec	ℓ^+X (10%)
B_s	$s\bar{b}$	$\frac{1}{2}(0^-)$	5.369	1.54×10^{-12} sec	$D_S + X$ ($\sim$ 90%)
B_c	$c\bar{b}$	$\frac{1}{2}(0^-)$	6.4	5×10^{-13} sec	$\psi\ell^+\nu$[b]
η_B	$b\bar{b}$	$0^+(0^{-+})$	?	?	not observed[c]

(continued)

TABLE 12.4. Continued

	Quark content	$I(J^{PC})$	Mass GeV	Lifetime/width in sec or eV	Decay modes[a]
Selected vector mesons					
$\rho(770)$	$u\bar{d} - d\bar{d}, d\bar{u}$	$1(1^{--})$	0.770	151 MeV	2π (100%)
ω	$u\bar{u}, d\bar{d}, s\bar{s}$	$0(1^{--})$	0.782	8.4 MeV	3π (89%) $\pi^0\gamma$ (8%)
ϕ	$s\bar{s}$	$0(1^{--})$	1.019	4.4 MeV	$2K$ (83%), 3π (15%)
K^{*+}	$u\bar{s}$	$\frac{1}{2}(1^-)$	0.892	51 MeV	$K\pi$ ($\sim$ 100%)
K^{*0}	$d\bar{s}$	$\frac{1}{2}(1^-)$	0.896	51 MeV	$K\pi$ ($\sim$ 100%)
D^{*0}	$c\bar{u}$	$\frac{1}{2}(1^-)$	2.007	< 2 MeV	$D^0\pi$ (62%), $D^0\gamma$ (38%)
D^{*+}	$c\bar{d}$	$\frac{1}{2}(1^-)$	2.010	< 0.131 MeV	$D\pi$ (99%), $D\gamma$ (1%)
J/ψ	$c\bar{c}$	$0^-(1^{--})$	3.097	87 KeV	$\ell^+\ell^-$ (6%), 88% hadronic
$\psi(2S)$	"	"	3.686	280 KeV	$\psi + X$ (55%), χ states ($\sim$ 30%)
$\psi(3770)$	"	"	3.770	24 MeV	DD ($\sim$ 100%)
$\Upsilon(1S)$	$b\bar{b}$	$0^-(1^{--})$	9.460	53 KeV	$\ell^+\ell^-$ (7.5%)
$\Upsilon(4S)$	"	"	10.580	10 MeV	$B\bar{B}$ ($\sim$ 100%)
Selected baryon states					
p	uud	$\frac{1}{2}\left(\frac{1}{2}^+\right)$	0.93827	$> 10^{31}$ years	
n	udd	$\frac{1}{2}\left(\frac{1}{2}^+\right)$	0.93957	887 sec	$pe^-\bar{\nu}_e$
Λ^0	uds	$0\left(\frac{1}{2}^+\right)$	1.116	2.63×10^{-10} sec	$p\pi^0$ (64%), $n\pi^+$ (36%)
Σ^+/Σ^-	uus/dds	$1\left(\frac{1}{2}^+\right)$	1.189/1.197	$0.80/1.48 \times 10^{-10}$ sec	$p\pi^0, n\pi^+/\eta_\pi^0$ ($\sim$ 100%)
Σ^0	uds	$1\left(\frac{1}{2}^+\right)$	1.193	7.4×10^{-20} sec	$\Lambda\gamma$ ($\sim$ 100%)
Ξ^0/Ξ^-	uss/dss	$\frac{1}{2}\left(\frac{1}{2}^+\right)$	1.315/1.321	$2.9/1.64 \times 10^{-10}$ sec	$\Lambda\pi$ ($\sim$ 100%)
Λ_c	udc	$0(\left(\frac{1}{2}^+\right)$	2.285	2.0×10^{-13} sec	$\ell + X$ (5%)
Λ_b	udb	$0\left(\frac{1}{2}^+\right)$	5.62	1.2×10^{-12} sec	$\ell + X$ (9%)
Δ	uuu . . . ddd	$\frac{3}{2}\left(\frac{3}{2}^+\right)$	1.23	120 MeV	$p\pi \ldots n\pi$ (100%)
Ω^-	sss	$0\left(\frac{3}{2}^+\right)$	1.672	8.2×10^{-11} sec	ΛK (68%), $\Xi\pi$ (32%)

[a]The use of ℓ implies that the μ and e branching fractions are similar. The rate stated is the rate to e or μ, not the total.
[b]The B_c parameters are from the original publications. [3]
[c]The η_β is the sole remaining unobserved pseudoscalar meson because top quarks do not form bound states.

12.3. KINEMATICS

12.3.1. Relativistic kinematics of reactions and decays

In this section, $\mathbf{p}$ indicates four-vectors, $\vec{p}$ indicates three-vectors, and p indicates the length of a three-vector.

$$\vec{\beta} = \frac{\vec{v}}{c}, \tag{12.24}$$

$$\gamma = \frac{1}{\sqrt{1-\beta^2}}, \tag{12.25}$$

where $\vec{v}$ is the velocity and c is the speed of light.

12.3.2. Boost of a four-vector

The boost of a four-vector $\mathbf{P} = (E, \vec{p}c)$ by velocity βc in the z direction is

$$\begin{pmatrix} E' \\ p'_z c \end{pmatrix} = \begin{pmatrix} \gamma & \gamma\beta \\ \gamma\beta & \gamma \end{pmatrix} \begin{pmatrix} E \\ p_z c \end{pmatrix}, \qquad (p'_x, p'_y) = (p_x, p_y). \tag{12.26}$$

12.3.3. Decay length distribution

The probability for a particle with rest lifetime τ_0 to decay with path length l is

$$P(l) = \frac{1}{\gamma\beta c\tau_0} \exp \frac{-l}{\gamma\beta c\tau_0}.$$

12.3.4. Two-body decays

For a particle of mass M at rest decaying to two particles of masses m_1 and m_2, the energy and momentum of the decay products are given by

$$E_1 = c^2 \frac{(M^2 + m_1^2 - m_2^2)}{2M}, \tag{12.27}$$

$$p_1 = c \frac{\sqrt{\lambda(M^2, m_1^2, m_2^2)}}{2M}, \tag{12.28}$$

where p_1 is the momentum of decay product 1 and

$$\lambda(M^2, m_1^2, m_2^2) = [M^2 - (m_1 + m_2)^2][M^2 - (m_1 - m_2)^2]. \tag{12.29}$$

12.3.5. Three-body decays

In three-body decays, the two-body invariant masses are related by

$$m_{12}^2 + m_{23}^2 + m_{31}^2 = M^2 + m_1^2 + m_2^2 + m_3^2, \tag{12.30}$$

where

$$m_{ij}^2 c^4 = (\mathbf{p}_i + \mathbf{p}_j)^2. \tag{12.31}$$

In the m_{12} rest frame

$$E_1 = c^2(m_{12}^2 + m_1^2 - m_2^2)/2m_{12}, \tag{12.32}$$

$$E_3 = c^2(M^2 - m_{12}^2 - m_3^2)/2m_{12}, \tag{12.33}$$

setting $E_1 = m_1$ or $E_3 = m_3$ determines the kinematic limits on m_{12}.

12.3.6. Two-body reactions

Figure 12.3 illustrates the two-body reaction $\mathbf{p}_1 + \mathbf{p}_2 \rightarrow \mathbf{p}_3 + \mathbf{p}_4$.

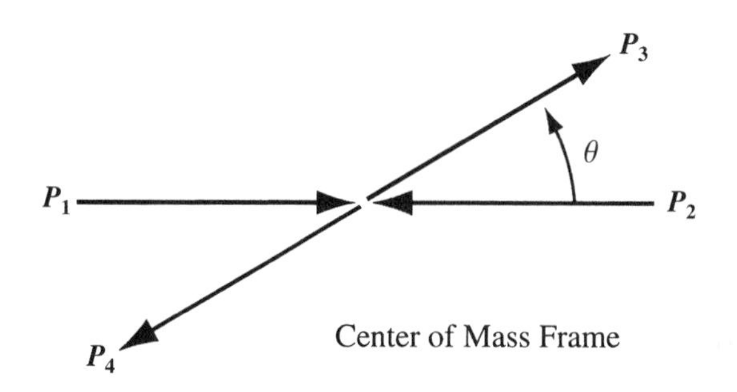

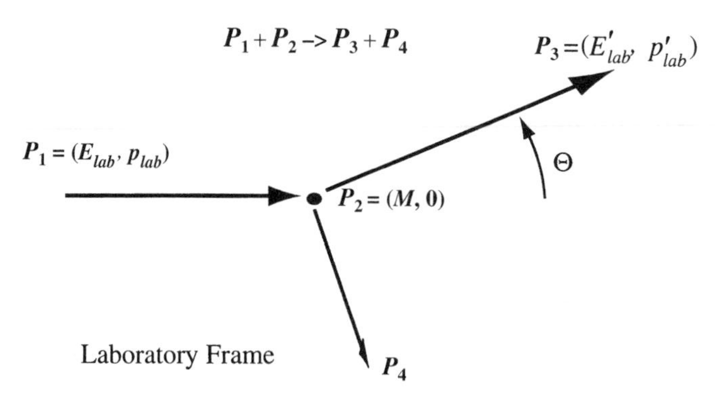

FIGURE 12.3. Two-body reaction kinematics.

In the center-of-mass frame, $\mathbf{p}_i = (E_i, \vec{p}_i c)$ and $\cos\theta$ is the cosine of the angle between $\vec{p}_1$ and $\vec{p}_3$. In the laboratory frame, particle 1 is the beam with energy E_{lab} and 2 is the target at rest with mass M.

12.3.7. Mandelstam variables

In all frames,

$$s = (\mathbf{p}_1 + \mathbf{p}_2)^2 = (\mathbf{p}_3 + \mathbf{p}_4)^2, \tag{12.34}$$

$$t = (\mathbf{p}_1 - \mathbf{p}_3)^2 = (\mathbf{p}_2 - \mathbf{p}_4), \tag{12.35}$$

$$u = (\mathbf{p}_1 - \mathbf{p}_4)^2 = (\mathbf{p}_2 - \mathbf{p}_3)^2, \tag{12.36}$$

$$s + t + u = c^2(m_1^2 + m_2^2 + m_3^2 + m_4^2). \tag{12.37}$$

s is the center-of-mass energy squared.

12.3.8. Transformations between the laboratory and the center of mass frames

$$\gamma = (E_{\text{lab}} + c^2 M)/\sqrt{s}, \tag{12.38}$$

$$\beta = \frac{p_{\text{lab}} c}{E_{\text{lab}} + c^2 M}, \tag{12.39}$$

$$\beta_3 = \frac{p_3 c}{E_3}, \tag{12.40}$$

$$\tan \Theta_{\text{lab}} = \frac{\sin \theta_{CM}}{\gamma(\cos \theta_{CM} + \beta/\beta_3)}. \tag{12.41}$$

12.3.9. In specific frames

$$s = m_1^2 c^4 + m_2^2 c^4 + 2Mc^2 E_{\text{lab}}, \tag{12.42}$$

$$t = m_2^2 c^4 + m_3^2 c^4 - 2E_1 E_3 + 2c^2 p_1 p_3 \cos \theta, \tag{12.43}$$

$$u = m_1^2 c^4 + m_4^2 c^4 - 2E_1 E_4 + 2E_1 E_4 + 2c^2 p_1 p_4 \cos \theta, \tag{12.44}$$

$$p_{\text{cm}_1} = p_{\text{lab}}(c^2 M)/\sqrt{s}. \tag{12.45}$$

12.3.10. Lepton scattering

In lepton scattering, the beams are often approximated by massless particles ($m_1, m_3 \approx 0$). If E'_{lab} denotes the laboratory frame energy of the scattered lepton $\mathbf{p}_3$,

$$\mathbf{q} = \mathbf{p}_1 - \mathbf{p}_3, \qquad Q^2 = -\mathbf{q}^2, \tag{12.46}$$

$$\nu = (\mathbf{p}_2 \mathbf{q})/M \approx E_{\text{lab}} - E'_{\text{lab}}, \tag{12.47}$$

$$x = Q^2/2M\nu, \tag{12.48}$$

$$y = M\nu/(\mathbf{p}_1 \mathbf{p}_2) = (1 + \cos \theta_{CM})/2 \approx \nu/E_{\text{lab}}, \tag{12.49}$$

$$W^2 = m_4^2 c^4 = 2Mc^2 \nu + M^2 c^4 - Q^2, \tag{12.50}$$

where $\mathbf{q}/c$ is the four-momentum transfer to the target; ν is the energy transfer; x is the Bjorken x variable, which can be interpreted as the fraction of the target momentum carried by a massless component; y is the scaled energy transfer; and W^2 is the invariant mass of the final state hadronic system squared.

The element of invariant phase space is

$$E' \frac{d\sigma}{d^3 \vec{p}'} = \frac{1}{2\pi M \nu} \frac{d\sigma}{dx\, dy} = \frac{E_{\text{lab}}}{\pi} \frac{d\sigma}{dQ^2\, d\nu} = \frac{x}{\pi y} \frac{d\sigma}{dx\, dQ^2}. \tag{12.51}$$

12.3.11. Inclusive particle production

In a scatter in which a single particle or jet is detected in the final state, $1 + 2 \to h + X$, define $p_\perp$ and $p_\parallel$ as the cm components of $\mathbf{p}_h$ relative to $\mathbf{p}_l$ and E_h as the energy.

$$\text{the rapidity } y = \frac{1}{2}\log\left(\frac{E_h + p_\parallel}{E_h - p_\parallel}\right), \tag{12.52}$$

$$\text{if } m_h \approx 0, \text{ the pseudo-rapidity } \eta = -\log\tan\theta/2 \approx y. \tag{12.53}$$

The rapidity y is not the same as the scaled energy transfer y in Sec. 12.3.10. The maximum possible rapidity for a particle of mass m is

$$y = \frac{1}{2}\log\frac{s}{m^2}. \tag{12.54}$$

The Lorentz-invariant phase space element is

$$E_h \frac{d\sigma}{d^3\vec{p}_h} = \frac{1}{\pi}\frac{d\sigma}{dy\,dp_\perp^2}. \tag{12.55}$$

12.4. DECAYS AND CROSS SECTIONS

The element of n-body invariant phase space is

$$d\Phi_{1\ldots n} = \left(\frac{1}{2\pi}\right)^{3n} \delta^4\left(\mathbf{P}_{\text{initial}} - \sum \mathbf{p}_n\right) \prod_{i=1}^{n} \frac{d^3\vec{p}_i}{2E_i}, \tag{12.56}$$

where n varies over final state particles.

The decay rate for a particle mass M is

$$d\sigma(M \to p_1 \ldots p_n) = \frac{(2\pi)^4}{2M} |\mathcal{M}|^2 \,\mathrm{d}\Phi_\mathrm{n}. \tag{12.57}$$

The cross section for two-body interactions $p_1 + p_2 \to p_3 \ldots p_n$ is

$$d\sigma(p_1 + p_2 \to p_3 \ldots p_n) = \frac{(2\pi)^4}{4(\mathbf{p}_1\mathbf{p}_2)^{\frac{1}{2}}} |\mathcal{M}|^2 \,\mathrm{d}\Phi_{3\ldots n}, \tag{12.58}$$

where $\mathcal{M}$ is the matrix element for the process.

12.4.1. Lepton scattering

The kinematic variables are defined in Sec. 33.10.

Lepton scattering from unpolarized targets with $Q \ll M_Z$:

$$\frac{d\sigma}{dx\,dQ^2} = \frac{C}{x}\frac{1}{(Q^2+M_E^2)^2}\left[y^2xF_1(x,Q^2) + \left(1-y-\frac{M^2xy}{s-M^2}\right)F_2(x,Q^2) \pm\left(y-\frac{y^2}{2}\right)xF_3(x,Q^2)\right], \tag{12.59}$$

$$C = 4\pi\alpha^2 \text{ and } M_E = M_\gamma \text{ for photon exchange,} \tag{12.60}$$

$$C = \frac{M_W^4 G_F^2}{2\pi} \text{ and } M_E = M_W \text{ for } W \text{ exchange,} \tag{12.61}$$

where M is the mass of the target and the F_i are structure functions.

In the quark parton model, quarks have probability $q_i(x)$ of carrying momentum fraction x of the proton momentum. The function $q_i(x)$ is called a parton distribution function or PDF. Typical PDFs are shown in Figure 12.4.

For photon exchange,

$$2xF_1^{CC}(x,Q^2) \simeq F_2^{CC}(x,Q^2) = \sum_i xq_i^2[q_i(x)+\bar{q}_i(x)], \tag{12.62}$$

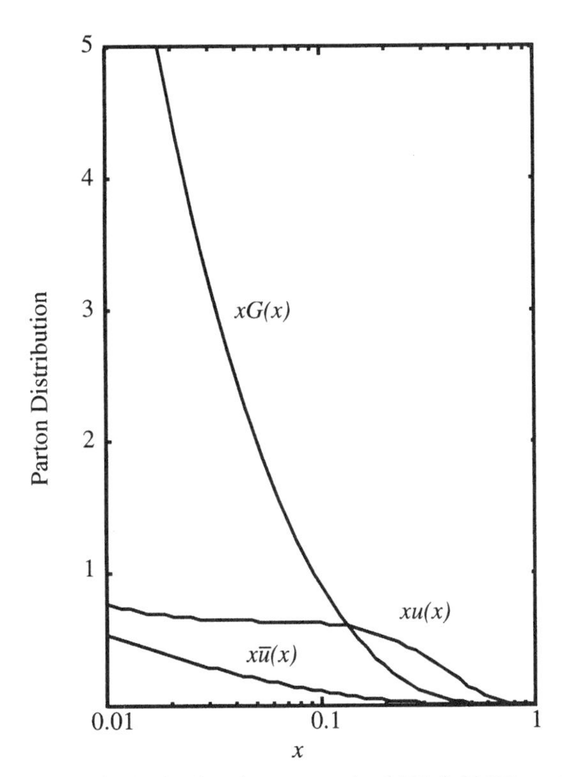

FIGURE 12.4. Typical parton distribution functions at a scale of 100 GeV. At low x the gluon distribution dominates.

where q_i represents all light (u, d, s) quark species. For W exchange,

$$2xF_1^{CC}(x, Q^2) \simeq F_2^{CC}(x, Q^2) = 2\sum_i x(u_i(x) + \bar{d}_i(x)), \tag{12.63}$$

$$xF_3^{CC}(x, Q^2) = 2\sum_i^i x(d_i(x) - u_i(x)) \quad \text{for } \nu p \to \ell X \text{ and } \bar{\ell} \to \bar{\nu} X \tag{12.64}$$

$$= 2\sum_i^i x(\bar{u}_i(x) - d_i(x)) \quad \text{for } \bar{\nu} p \to \bar{\ell} X \text{ and } \ell p \to \nu X, \tag{12.65}$$

where u_i and d_i represent the sum over light quark species of that type, normally, u, d, and s only.

In neutrino scattering, the structure functions quoted are generally an average over equal numbers of protons and neutrons in the target. Via isospin symmetry they sample both u- and d-type quarks where the weak interactions on protons would only be sensitive to d directly. The quoted structure functions are $F_2^\nu \simeq x\sum[q_i(x)+\bar{q}_i(x)]$ and $F_3^\nu \simeq \sum[q_i(x)-\bar{q}_i(x)]$, where the quark distributions are those for the photon. Neutrino data and charged lepton scattering on deuterium can be related via

$$2xF_2^{\gamma d}(x, Q^2) \simeq F_2^{\gamma d}(x, Q^2) = x\sum_i q_i^2[q_i(x) + \bar{q}_i(x)]$$
$$\approx \frac{5}{18}F_2^\nu(x, Q^2). \tag{12.66}$$

12.4.2. e^+e^- scattering

The unpolarized e^+e^- scattering cross section for $s \ll M_Z^2$ is

$$\frac{d\sigma}{d\Omega}(e^+e^- \to f_i\bar{f}_i) = C_i\frac{\alpha^2}{4s}\beta[1 + \cos^2\theta + \sin^2\theta(1 - \beta^2)], \tag{12.67}$$

where f_i is a final state fermion and $C_i = 1$ for μ and τ and $C_I = 3q_i^2$ for quarks. The process $e^+e^- \to e^+e^-$ includes additional t-channel processes.

12.4.3. Resonance production

The cross section for production of a resonance R of spin J is approximately

$$\sigma(e^+e^- \to R \to X) \approx \frac{(2J+1)\pi}{s}B(R \to e^+e^-)B(R \to X)\frac{\Gamma^2}{(E - M(R)c^2)^2 + \frac{\Gamma^2}{4}}, \tag{12.68}$$

where the B are the branching fractions for R to decay to e^+e^- and the observed final state.

12.4.4. Hadron scattering

The cross section for hadron-hadron scattering at high momentum transfer is a convolution of parton scattering cross sections with PDFs defined in Sec. 12.4.1. In hadron scattering, chargeless objects such as gluons also have PDFs.

$$\sigma(h_1 h_2 \to h+X) = \sum_{ij} \int p_i(x_i, Q^2) p_j(x_2, Q^2)\, dx_i\, dx_2 \hat{\sigma}(p_1+p_2 \to h+X), \quad (12.69)$$

where the p_i are PDFs, x_1, x_2 are the fractional momenta carried by the partons in the initial h_1 and h_2, Q^2 is the momentum transfer squared, and $\hat{\sigma}$ is the parton level cross section.

12.4.5. Fragmentation

Final state partons q hadronize to form jets of detectable particles. The fractional momenta

$$z = \frac{E^h + p_\parallel^h}{E^q + p_\parallel^q} \approx \frac{E^h}{E^q} \quad (12.70)$$

of individual final state particles h are distributed via a fragmentation function $D_h^{(q)}(z, Q^2)$, normalized such that the integral over z and sum over hadron types is 1.

For light quarks, $D_h^q(z)$ is peaked toward low z, whereas for heavy quarks, the heavy meson or baryon that carries the original heavy quark retains most of the momentum and follows the Peterson [12] form:

$$\frac{dN}{dz} = \frac{1}{z\left(1 - \frac{1}{z} - \frac{\epsilon}{1-z}\right)}. \quad (12.71)$$

The parameter ϵ ranges from 0.25 for charmed baryons to 0.04–0.14 for charmed mesons to 0.005 for b mesons.

12.4.6. Typical interaction cross sections

Hardron and photon interactions

Above laboratory energies of ~10 GeV, total hadronic cross sections vary as s^ϵ, where $\epsilon < 0.1$. At 50 GeV, the total hadronic cross sections are

$$\sigma(pp, p\bar{p}, pn) = 44 \text{ mb}, \quad (12.72)$$

$$\sigma(\pi^\pm p) = 24 \text{ mb}, \quad (12.73)$$

$$\sigma(K^\pm p) = 20 \text{ mb}, \quad (12.74)$$

$$\sigma(pd) = 82 \text{ mb}, \quad (12.75)$$

$$\sigma(\pi^\pm) = 46 \text{ mb}, \quad (12.76)$$

$$\sigma(K^{\pm}) = 39 \text{ mb}, \tag{12.77}$$

$$\sigma(\gamma p) = 0.1 \text{ mb}. \tag{12.78}$$

Neutrino interactions

The total neutrino cross section for $M_p^2 \ll s \ll M_W^2$ scales as

$$\sigma(\nu p \to \ell^- + X) \simeq 0.67 \times 10^{-38}(E,\ \text{GeV})\ \text{cm}^{-2}, \tag{12.79}$$

$$\sigma(\bar{\nu} p \to \ell^+ + X) \simeq 0.34 \times 10^{-38}(E,\ \text{GeV})\ \text{cm}^{-2}. \tag{12.80}$$

12.5. PARTICLE DETECTORS

12.5.1. Cherenkov radiation

If a charged particle traverses a material with velocity larger than the velocity of light in that material ($\beta > \frac{1}{n}$, where n is the index of refraction), it will emit Cherenkov radiation with a characteristic angle

$$\cos\theta_C = \frac{1}{n\beta}. \tag{12.81}$$

The number of photons emitted per unit photon energy and path length is

$$\frac{dN}{dE\,dx} = \frac{\alpha Z_1^2}{\hbar c}\sin^2\theta_C \sim 370 Z_1^2 \sin^2\theta_C/(\text{eV-cm}). \tag{12.82}$$

12.5.2. Ionization energy loss

The mean energy loss per unit length due to ionization of a particle (1) of charge Z_1 incident on a material (2) with atomic charge Z_2 and atomic number A_2 is [13, 1]

$$\frac{dE}{dx} = \frac{2\pi Z_1^2 e^4}{(4\pi\epsilon_0)^2 m_e c^2 \beta_1^2}\,\frac{Z_2 N_A \rho_2}{A_2} \times \log\left(\frac{2 m_e c^2 \beta_1^2 \gamma_1^2 T_{\max}}{I^2} - \beta_1^2 - \frac{\delta}{2}\right). \tag{12.83}$$

Here, N_A is Avagadro's number, ρ_2 is the mass density of the material, m_e is the mass of the electron, I is the minimum possible scattering energy (approximately the ionization potential -22 eV for hydrogen, $10 \times Z_2$ eV for $Z_2 > 20$), and δ is a density correction, important at high γ_1.

$$T_{\max} = \frac{2 m_e c^2 \beta_1^2 \gamma_1^2}{1 + 2\gamma_1 m_e/M_1 + \left(\frac{m_e}{M_1}\right)^2} \tag{12.84}$$

is the maximum energy transferable in a single collision.

There is a broad minimum in this distribution for $\gamma_1\beta_1 \sim 3$. Such minimum ionizing particles typically deposit 1.5–2 MeV/(gram/cm^2) of material traversed. This relation holds to within a factor for $\gamma_1 \beta_1 \sim 1 - 1000$. Losses in liquid hydrogen are a factor of two higher than for heavier materials.

At very high energies, other processes, such as bremsstrahlung and e^+e^- pair production, can begin to dominate the energy loss. Such processes produce rare, but very large, energy losses and cannot be described by a simple mean.

12.5.3. Multiple scattering through small angles

Scattering from atomic electrons also changes the direction of the incident particle. If we define the angle θ_0 to be the mean scattering angle in a single plane,

$$\theta_0 = \frac{13.6\,\text{MeV}}{\beta c p_1} Z_1 \sqrt{\frac{t}{X_0}} \left[1 + 0.038 \log \frac{t}{X_0}\right], \tag{12.85}$$

where t is the thickness of the scatterer, p_1 is the momentum of the particle in MeV/c, and X_0 is the radiation length tabulated in Table 12.5. [1]

12.5.4. Charged particle trajectories

A charged particle moving in a uniform magnetic field follows a helical path about the field axis. The trajectory maintains a uniform angle $\frac{\pi}{2} - \lambda$ with respect to the field axis. The radius of curvature R is related to the momentum p in GeV/c via:

$$p \cos \lambda = 0.29979 Z H R, \tag{12.86}$$

where H is the field in teslas, Z is the charge in units of e, and R is in meters.

The error on the curvature $\kappa = 1/R$ is usually Gaussian distributed with contributions from measurement error and from multiple scattering:

$$\delta\kappa_{\text{meas}} \simeq \sqrt{\frac{720}{N+5}} \frac{\delta x}{(L \cos \lambda)^2}, \tag{12.87}$$

$$\delta\kappa_{\text{scatt}} \simeq \frac{0.016 Z}{(L \cos \lambda)^2 p \beta} \sqrt{\frac{L}{X_0}}, \tag{12.88}$$

$$\delta\kappa = \sqrt{\delta\kappa_{\text{meas}}^2 + \delta\kappa_{\text{scatt}}^2}, \tag{12.89}$$

where N is the number measurements (uniformly spaced), δx is the measurement resolution, L is the total path length, and X_0 is the radiation length of the material.

The fractional momentum resolution scales with p when measurement limited and is approximately constant when scattering limited. Typical fractional resolutions on p range from $10^{-5} p$ (GeV) for small δx, large L, and large H in gaseous detectors to $\simeq 10\%$ for muons traversing meters of magnetized iron at 1.5 T.

TABLE 12.5. Atomic and nuclear properties of materials from the Review of Particle Properties [1]. Gases are evaluated at 20°C and 1 atm (in parentheses) or at STP [square brackets]. Densities and refractive indices without parentheses or brackets are for solids or liquids, or are for cryogenic liquids at the indicated boiling point (BP) at 1 atm. Refractive indices are evaluated at the sodium D line.

Material	Z	A	$\langle Z/A \rangle$	Nuclear[a] collision length λ_T {g/cm^2}	Nuclear[a] interaction length λ_I {g/cm^2}	$dE/dx\|^b_{\rm min}$ $\left\{\frac{MeV}{g/cm^2}\right\}$	Radiation length[c] X_0 {g/cm^2}	Radiation length[c] X_0 {cm}	Density {g/cm^2} ({g/ℓ} for gas)	Liquid boiling point at 1 atm (K)	Refractive index n ($(n-1) \times 10^6$ for gas)
H_2 gas	1	1.00794	0.99212	43.3	50.8	(4.103)	61.28[d]	(731000)	(0.0838)[0.0877]		[139.2]
H_2 liquid	1	1.00794	0.99212	43.3	50.8	4.034	61.28[d]	866	0.0708	20.39	1.112
D_2	1	2.0140	0.49652	45.7	54.7	(2.052)	122.4	724	0.169[0.179]	23.65	1.128[138]
He	2	4.002602	0.49968	49.9	65.1	(1.937)	94.32	756	0.1249[0.1786]	4.224	1.024[34.9]
Li	3	6.941	0.43221	54.6	73.4	1.639	82.76	155	0.534		
Be	4	9.012182	0.44384	55.8	75.2	1.594	65.19	35.28	1.848		
C	6	12.011	0.49954	60.2	86.3	1.745	42.70	18.8	2.265[e]		
N_2	7	14.00674	0.49976	61.4	87.8	(1.825)	37.99	47.1	0.8073[1.250]	77.36	1.205[298]
F_2	9	18.9984032	0.47372	65.5	95.3	(1.675)	32.93	21.85	1.507[1.696]	85.24	[195]
Ne	10	20.1797	0.49555	66.1	96.6	(1.724)	28.94	24.0	1.204[0.9005]	27.09	1.092[67.1]
Al	13	26.981539	0.48181	70.6	106.4	1.615	24.01	8.9	2.70		
Si	14	28.0855	0.49848	70.6	106.0	1.664	21.82	9.36	2.33		3.95
Ar	18	39.948	0.45059	76.4	117.2	(1.519)	19.55	14.0	1.396[1.782]	87.28	1.233[283]
Ti	22	47.867	0.45948	79.9	124.9	1.476	16.17	3.56	4.54		
Fe	26	55.845	0.46556	82.8	131.9	1.451	13.84	1.76	7.87		
Cu	29	63.546	0.45636	85.6	134.9	1.403	12.86	1.43	8.96		
Ge	32	72.61	0.44071	88.3	140.5	1.371	12.25	2.30	5.323		
Sn	50	118.710	0.42120	100.2	163	1.264	8.82	1.21	7.31		
Xe	54	131.29	0.41130	102.8	169	(1.255)	8.48	2.87	2.953[5.858]	165.1	[701]
W	74	183.84	0.40250	110.3	185	1.145	6.76	0.35	19.3		
Pt	78	195.08	0.39984	113.3	189.7	1.129	6.54	0.305	21.45		
Pb	82	207.2	0.39575	116.2	194	1.123	6.37	0.56	11.35		
U	92	238.0289	0.38651	117.0	199	1.082	6.00	≈0.32	≈18.95		

(continued)

TABLE 12.5. Continued

Air, (20°C, 1 atm), [STP]	0.49919	62.0	90.0	(1.815)	36.66	[30420]	(1.205)[1.2931]	78.8	(273)[293]
H_2O	0.55509	60.1	8.36	1.991	36.1	1.00	373.15	1.33	
CO_2 gas	0.49989	62.4	89.7	(1.819)	36.2	[18310]	[1.977]		[410]
CO_2 solid (dry ice)	0.49989	62.4	89.7	1.787	36.2	23.2	1.563	sublimes	
Shielding concrete f	0.50274	67.4	99.9	1.711	26.7	10.7	2.5		
SiO_2 (fused quartz)	0.49926	66.5	97.4	1.699	27.05	12.3	2.20^g		1.458
Dimethyl ether, $(CH_3)_2O$	0.54778	59.4	82.9		38.89			248.7	
Methane, CH_4	0.62333	54.8	73.4	(2.417)	46.22	[64850]	0.4224[0.717]	111.7	[444]
Ethane, C_2H_6	0.59861	55.8	75.7	(2.304)	45.47	[34035]	$0.509(1.356)^h$	184.5	$(1.038)^h$
Propane, C_3H_8	0.58962	56.2	76.5	(2.262)	45.20		(1.879)	231.1	
Isobutane, $(CH_3)_2CHCH_3$	0.58496	56.4	77.0	(2.239)	45.07	[16930]	[2.67]	261.42	[1900]
Octane, liquid, $CH_3(CH_2)_6CH_3$	0.5778	56.7	77.7	2.123	44.86	63.8	0.703	298.8	1.397
Paraffin wax, $CH_3(CH_2)_{n\approx 23}CH_3$	0.57275	56.9	78.2	2.087	44.71	48.1	0.93		
Nylon, type 6[i]	0.54790	58.5	81.5	1.974	41.84	36.7	1.14		
Polycarbonate (Lexan)[j]	0.52697	59.5	83.9	1.886	41.46	34.6	1.20		
Polyethylene terephthlate (Mylar)[k]	0.52037	60.2	85.7	1.848	39.95	28.7	1.39		
Polyethylene[l]	0.57034	57.0	78.4	2.076	44.64	≈47.9	0.92–0.95		
Polyimide film (Kapton)[m]	0.51264	60.3	85.8	1.820	40.56	28.6	1.42		
Lucite, Plexiglas[n]	0.53937	59.3	83.0	1.929	40.49	≈34.4	1.16–1.20		≈1.49
Polystyreme, scintillator[o]	0.53768	58.5	81.9	1.936	43.72	42.4	1.032		1.581
Polytetrafluoroethylene (Teflon)[p]	0.47992	64.2	93.0	1.671	34.84	15.8	2.20		
Polyvinyltoluene, scintillator[q]	0.54155	58.3	81.5	1.956	43.83	42.5	1.032		

(continued)

TABLE 12.5. Continued

Material	Z	A	⟨Z/A⟩	Nuclear[a] collision length λ_T {g/cm^2}	Nuclear[a] interaction length λ_I {g/cm^2}	$dE/dx\|^b_{min}$ $\left\{\frac{MeV}{g/cm^2}\right\}$	Radiation length[c] X_0 {g/cm^2}	Radiation length[c] X_0 {cm}	Density {g/cm^2} ({g/ℓ} for gas)	Liquid boiling point at 1 atm (K)	Refractive index n (($n-1$) × 10^6 for gas)
Aluminum oxide (Al_2O_3)			0.49038	67.0	98.9	1.647	19.27	4.85	3.97	1.761	
Barium fluoride (BaF_2)			0.42207	92.0	145	1.303	9.91	2.05	4.89	1.56	
Bismuth germanate (BGO)[r]			0.42065	98.2	157	1.251	7.97	1.12	7.1		2.15
Cesium iodide (Csl)			0.41569	102	167	1.243	8.39	1.85	4.53		1.80
Lithium fluoride (LiF)			0.46262	62.2	88.2	1.614	39.25	14.91	2.632		1.392
Sodium fluoride (NaF)			0.47632	66.9	98.3	1.69	29.87	11.68	2.558		1.336
Sodium iodide (Nal)			0.42697	94.6	151	1.305	9.49	2.59	3.67		1.775
Silica Aerogel[s]			0.52019	64	92	1.83	29.83	≈150	0.1–0.3		1.0 + 0.25ρ
NEMA G10 plate[t]				62.6	90.2	1.87	33.0	19.4	1.7		
H_2	(253.9)										
He	(64)										
Li			56	0.86	8.55(0°)	0.17					
Be		37	12.4	0.436	5.885(0^{c}irc)	0.38					
C		0.7	0.6–4.3	0.165	1375(0^{c}irc)	0.057					
N_2	(548.5)										
O_2	(495)										
Ne	(127)										
Al		10	23.9	0.215	2.65(20°)	0.53					
Si	11.9	16	2.8–7.3	0.162		0.20					
Ar	(517)										
Ti		16.8	8.5	0.126	50(0°)						
Fe		28.5	11.7	0.11	9.71(20°)	0.18					
Cu		16	16.5	0.092	1.67(20°)	0.94					
Ge	16.0		5.75	0.073		0.14					

(continued)

TABLE 12.5. Continued

Sn	6	20	0.052	11.5(20°)	0.16
Xe					
W	50	4.4	0.032	5.5(20°)	0.48
Pt	21	8.9	0.032	9.83(0°)	0.17
Pb	2.6	29.3	0.038	20.65(20°)	0.083
U		36.1	0.028	29(20°)	0.064

1. R. M. Sternheimer, M. J. Berger, and S. M. Seltzer, Atom. Data Nucl. Data Tables **30**, 261–271 (1984).
2. S. M. Seltzer and M. Berger, Int. J. Appl. Radiat. **33**, 1189–1218 (1982).
3. D. E. Groom, N. V. Mokhov, and S. I. Striganov, Atom. Data. Nucl. Data Tables (to be published). 4. S. M. Seltzer and M J. Berger, Int. J. Appl. Radiat. **35**, 665 (1984) and **http://physics.nist.gov/PhysRefData/Star/Text/contents.html.**

a. σ_T, λ_T, and λ_I are energy dependent. Values quoted apply to high-energy range, where energy dependence is weak. Mean free path between collision (λ_T) or inelastic interactions (λ_I), calculated from $\lambda^{-1} = N_A \sum w_j \sigma_j / A_j$, where N is Avogadro's number and w_j is the weight fraction of the jth element in the element, compound, or mixture. σ_{total} at 80–240 GeV for neutrons ($\approx \sigma$ for protons) from Murthy *et al.*, Nucl. Phys. **B92**, 269 (1975). This scales approximately as $A^{0.77}$. $\sigma_{\text{elastic}} = \sigma_{\text{total}} - \sigma_{\text{elastic}} - \sigma_{\text{quasielastic}}$; for neutrons at 60–375 GeV from Roberts *et al.*, Nucl. Phys. **B159**, 56 (1979). For protons and other particles, see Carroll *et al.*, Phys. Lett. **80B**, 319 (1979); note that $\sigma_I(p) \approx \sigma_I(n)$. σ_I scales approximately as $A^{0.71}$.

b. For minimum-ionizing muons (results are very slightly differnt for other particles). Minimum dE/dx from Ref. [3] of this table, using density effect correction coefficients from Ref. [1] above. For electrons and positrons, see Ref. [4] above. Ionization energy loss is discussed in Sec. 23 of the Review of Particle Properties [1].

c. From Y. S. Tsai, Rev. Mod. Phys. **46**, 815 (1974); X_0 data for all elements up to uranium are given. Corrections for molecular binding applied for H_2 and D_2. For atomic H, $X_0 = 63.05$ g/cm^2.

d. For molecular hydrogen (deuterium). For atomic H. $X_0 = 63.047$ g cm^{-2}.

e. For pure graphite; industrial graphite density may vary 2.1–2.3 g/cm^3.

f. Standard shielding blocks, typical composition O_2 52%, Si 32.5%, Ca 6%, Na 1.5%, Fe 2%, Al 4%, plus reinforcing iron bars. The attenuation length, $\ell = 115 \pm 5$ g/cm^2, is also valid for earth (typical $\rho = 2.15$), from CERN-LRL-RHEL Shielding exp., UCRL-17841 (1968).

g. For typical fused quartz. The specific gravity of crystalline quartz is 2.64.

h. solid ethane density at −60°C; gaseous refractive index at 0°C, 546 mm pressure.

i. Nylon, Type 6, $(NH(CH_2)_5CO)_n$. *j*. Polycarbonate (Lexane), $(C_{16}H_{14}O_3)_n$.

k. Polyethylene terephthlate, monomer, $C_5H_4O_2$. *l*. Polyethylene, monomer $CH_2 = CH_2$.

m. Polymide film (Kapton), $(C_{22}H_{10}N_2O_5)_n$. *n*. Polymethylmerthacrlate, monomer $CH_2 = C(CH_3)CO_2CH_3$.

o. Polystyrene, monomer $C_6H_5CH = CH_2$. *p*. Teflon, monomer $CF_2 = CF_2$.

q. Polyvinyltolulene, monomer 2–$CH_3C_6H_4CH = CH_2$. *r*. Bismuth germanate (BGO), $(Bi_2O_3)_2(Ge)_2)_3$. *s*. $n(SiO_2) + 2n(H_2O)$ used in Cerenkov counters, $\rho =$ density in g/cm^3. From M. Cantin *et al.*, Nucl. Instrum. Methods **118**, 177 (1974).

t. G10-plate, typically 60% SiO_2 and 40% epoxy.

12.5.5. Calorimetry

The total energy of interacting particles can be measured in electromagnetic and hadronic calorimeters. The radiation length X_0 and hadronic interaction length λ_I tabulated in Table 12.5 determine the shower dimensions.

Fabjan and Ludlam [14] provide the following estimates for shower development in electromagnetic sampling calorimeters. Energies are in GeV, and t is the thickness of active material in a sampling layer:

$$L_{\max} \simeq (\log[1.8Z(E,\ \mathrm{GeV})] - C_{\mathrm{em}}), \tag{12.90}$$

$$L_{95\%} \simeq L_{\max} + X_0\,(0.6\log(E,\ \mathrm{GeV}) + 0.08Z), \tag{12.91}$$

$$\sigma(E)/E \simeq (0.70/Z)\sqrt{\frac{t/X_0}{(E,\ \mathrm{GeV})}}, \tag{12.92}$$

where $L_{\max}$ is the depth of maximum energy deposition and $L_{95\%}$ is the depth for 95% containment. C_{em} is approximately 1 for electrons and 0.5 for photons. Typical sampling electromagnetic calorimeters have fractional resolutions in the range 5–15%$\sqrt{E}$. In uniform active materials, such as sodium and cesium iodide, resolutions of order 1% are achievable.

Response to hadrons depends on the interaction length and is improved in "compensating" calorimeters, where the response to electrons and hadrons is similar. Modern compensating calorimeters can achieve hadronic resolutions of $\sigma(E)/E \sim$ 30–50%$/\sqrt{E,\ \mathrm{GeV}}$, but require thicknesses of (7–8)λ_I to contain hadrons with energies above 100 GeV.

12.6. REFERENCES

[1] D. Groom *et.al.*, Euro. Phys. J. **C15**, 1 (2000).

[2] T. G. Trippe, in *A Physicist's Desk Reference*, 2nd ed. (AIP Press, New York, 1989).

[3] F. Abe *et. al*, Phys. Rev. Lett. **81**, 2432 (1998).

[4] L. Wolfenstein, Phys. Rev. Lett. **51**, 1945 (1983).

[5] V. Fanti *et al.* [NA48 Collaboration], Phys. Lett. **B465**, 335 (1999).

[6] A. Alavi-Harati *et al.* [KTeV Collaboration], Phys. Rev. Lett. **83**, 22 (1999).

[7] C. Cason *et al.*, Euro. Phys. J. **C3**, 1 (1998).

[8] Z. Maki, M. Nakagawa, and S. Sakata, Prog. Theor. Phys. **28**, 870 (1962).

[9] Y. Fukuda *et al.*, Phys. Rev. Lett. **82**, 1810 (1999).

[10] M. Apollonio *et al.*, Phys. Lett. **B420**, 397 (1998); Phys. Lett. **B466**, 415 (1999).

[11] Fits and references to the Homstake, Kamiokande, GALLEX, SAGE, and Super Kamiokande data include N. Hata and P. Langacker, Phys. Rev. **D56**, 6107 (1997); J. Bahcall, P. Krastev, and A. Smirnov, Phys. Rev. **D58**, 096016 (1998).

[12] C. Peterson *et al.*, Phys. Rev. **D27**, 105 (1983).

[13] R. Fernow, *Introduction to Experimental Particle Physics* (Cambridge University Press, Cambridge, 1986).

[14] C. W. Fabjan and T. Ludlam, Annu. Rev. Nucl. Sci. **32**, 335 (1982).

13

Fluid Dynamics

Stavros Tavoularis
University of Ottawa

Contents

List of Tables

List of Figures

13.1. INTRODUCTION

Due to its extensive range of applications in nature and technology, fluid dynamics has been a favorite subject of engineers of all disciplines [3, 18, 24, 41] and atmospheric scientists [25, 30, 40, 11]. Nevertheless, this subject is also of interest to mainstream physicists, [17] not only for being a branch of classical mechanics, but also because the large volume of information that has been carefully and over a long time compiled by fluid dynamicists becomes a handy design tool for the physics, as well as for the chemistry, biology, and geology, laboratories. The fluid dynamics and physics laboratories share several important instruments: the laser, the shock tube, digital image processing, pressure, temperature and velocity transducers, spectrum analyzers and computational algorithms, among so many others. Beyond the utilitarian function of fluid dynamics, however, the recent study of nonlinear phenomena in fluid flows has elevated this field into the center of activity of nonequilibrium physics. As a contemporary and exciting research subject, nonlinear fluid dynamics unfolds great complexity. When increasingly large stresses are applied to a fluid, its motion may undergo changes, including instabilities, bifurcations, temporal chaos, pattern formation, phase modulations, defects, growth of localized structures, interactions among dissimilar length scales and time scales, universal and anomalous scaling, intermittency, anomalous transport, and the like. [29] The mathematical and experimental approaches to the study of such phenomena have considerable overlap with current problems of nonequilibrium physics, and the exchange of information between fluid dynamicists and physicists can only benefit both. As modern physicists reconsider their place in science and society in the new millennium, they will likely find increasing use for fluid dynamics, both as part of a more applied physics curriculum and as part of a complementary science.

13.2. PROPERTIES OF COMMON FLUIDS

Properties of common fluids can be found in detail in standard handbooks.[1, 6] In the following, we reproduce some of the most frequently needed properties of air (Tables 13.1–13.4), water (Tables 13.5–13.6) and some common gases (Table 13.7).

To avoid refraction and optical distortion, one must match (quite precisely) the refractive index of transparent apparatus walls with that of the contained liquid. The refractive index of materials depends on their composition, manufacturing process, temperature, and wavelength of light. Quartz, common glass, and acrylic materials (perspex, plexiglass, lucite) all have refractive indices in the range 1.47 to 1.50, which is substantially higher than that of water. Various liquids and solutions are available for matching these values; for example, the refractive index of a saturated aqueous solution of ammonium thiocyanate is about 1.50. [8]

To generate low Reynolds number flows, highly viscous liquids are required. Depending on its concentration and temperature, the viscosity of an aqueous solution of glycerol at room temperature varies between that of water and a value over 1000 times higher. Note

TABLE 13.1. Properties of dry air at sea level and STP (standard temperature of $T = 287$ K and absolute pressure of $p = 101.325$ kPa). [18]

Density:	$\rho = 1.225\ \text{kg/m}^3$	Ratio of specific heats:	$\gamma = 1.40$
Viscosity:	$\mu = 1.789 \times 10^{-5}\ \text{Ns/m}^2$	Gas constant:	$R = 286.9\ \text{J/kg K}$
Kinematic viscosity:	$\nu = 1.46 \times 10^{-5}\ \text{m}^2/\text{s}$	Speed of sound:	$c = 339.5\ \text{m/s}$
Thermal conductivity:	$k = 0.0252\ \text{W/mK}$	Refractive index:	$\eta = 1.00$

TABLE 13.2. Average composition of dry air at sea level and essentially for altitudes up to 15,000 m. Other minor constituents include dust, pollen, bacteria, spores, smoke particles, SO_2, H_2S, hydrocarbons, and larger amounts of CO_2 and ozone, depending on various factors. The above composition does not include water vapor, which is an important constituent in all normal atmospheres. [6]

Gas	Molecular weight	Percentage by volume	Percentage by weight
N_2	28.016	78.09	75.55
O_2	32.000	20.95	23.13
Ar	39.944	0.93	1.27
CO_2	44.010	0.03	0.05
Ne	20.183	18×10^{-4}	12.9×10^{-4}
He	4.003	5.2×10^{-4}	0.74×10^{-4}
CH_4	16.04	2.2×10^{-4}	1.3×10^{-4}
Kr	83.8	1.0×10^{-4}	3.0×10^{-4}
N_2O	44.01	$1. \times 10^{-4}$	1.6×10^{-4}
H_2	2.0160	0.5×10^{-4}	0.03×10^{-4}
Xe	131.3	0.08×10^{-4}	0.37×10^{-4}
O_3	48.000	0.01×10^{-4}	0.02×10^{-4}
Rn	222	0.06×10^{-10}	—

TABLE 13.3. Viscosity and kinematic viscosity of air. [15]

T (°C)	μ ($\times 10^{-5}$ N s/m^2)	ν ($\times 10^{-5}$ m^2/s)	T (°C)	μ ($\times 10^{-5}$ N s/m^2)	ν ($\times 10^{-5}$ m^2/s)
0	1.709	1.32	260	2.806	4.24
20	1.808	1.50	280	2.877	4.51
40	1.904	1.69	300	2.946	4.81
60	1.997	1.88	320	3.014	5.07
80	2.088	2.09	340	3.080	5.35
100	2.175	3.30	360	3.146	5.65
120	2.260	2.52	380	3.212	5.95
140	2.344	2.74	400	3.277	6.25
160	2.425	2.98	420	3.340	6.56
180	2.505	3.22	440	3.402	6.88
200	2.582	3.46	460	3.463	7.20
220	2.658	3.71	480	3.523	7.52
240	2.733	3.97	500	3.583	7.85

TABLE 13.4. U.S. standard atmosphere. [6]

H (m)	T (K)	p (kPa)	ρ (kg/m^3)	c (m/s)	μ ($\times 10^{-5}$ N s/m^2)
0	288.15	101.33	1.225	340.3	1.789
200	286.85	98.95	1.202	339.5	1.783
600	284.25	94.32	1.156	338.0	1.771
1000	281.65	89.88	1.112	336.4	1.758
2000	275.15	79.50	1.007	332.5	1.726
3000	268.66	70.12	0.909	328.6	1.694
4000	262.17	61.66	0.819	324.6	1.661
5000	255.68	54.05	0.736	320.5	1.628
6000	249.19	47.22	0.660	316.5	1.595
8000	236.22	35.65	0.526	308.1	1.527
10000	223.25	26.50	0.414	299.5	1.458
15000	216.65	12.11	0.195	295.1	1.422
20000	216.65	5.53	0.0889	295.1	1.422
30000	226.51	1.20	0.0184	301.7	1.475
50000	270.65	0.0798	0.0010	329.8	1.704
100000	210.02	3×10^{-5}	5×10^{-7}	—	—

TABLE 13.5. Properties of water at 293 K. [18]

Density:	$\rho = 998$ kg/m^3	Vapor pressure:	$p_V = 2.34$ kPa
Viscosity:	$\mu = 1.00 \times 10^{-3}$ Ns/m^2	Modulus of elasticity:	$E_V = 2.24 \times 10^6$ N/m^2
Kinematic viscosity:	$\nu = 1.00 \times 10^{-6}$ m^2/s	Speed of sound:	$c = 473.8$ m/s
Surface tension in air:	$\sigma = 0.0727$ N/m	Refractive index:	$\eta = 1.333$

TABLE 13.6. Viscosity and kinematic viscosity of water. [15]

T (°C)	μ ($\times 10^{-3}$ Ns/m^2)	ν ($\times 10^{-6}$ m^2/s)	T (°C)	μ ($\times 10^{-3}$ N s/m^2)	ν ($\times 10^{-6}$ m^2/s)
0	1.792	1.792	40	0.656	0.661
5	1.519	1.519	45	0.599	0.605
10	1.308	1.308	50	0.549	0.556
15	1.140	1.141	60	0.469	0.477
20	1.005	1.007	70	0.406	0.415
25	0.894	0.897	80	0.357	0.367
30	0.801	0.804	90	0.317	0.328
35	0.723	0.727	100	0.284	0.296

TABLE 13.7. Thermodynamic properties of common gases at STP (standard temperature of 287 K and absolute pressure of 101.325 kPa). [18]

Gas	R (J/kg K)	γ —
CO_2	188.9	1.29
CO	296.8	1.40
He	2077	1.66
H_2	4124	1.41
CH_4	518.3	1.31
N_2	296.8	1.40
O_2	259.8	1.40
Steam	461.4	~1.30

that highly concetrated glycerol is hygroscopic and that its viscosity is extremely sensitive to temperature. Silicone oil and castor oil, pure or in mixtures with other oils, also have a viscosity that is several hundred times that of water.

13.3. MATHEMATICAL DESCRIPTION

In the present context, fluids are treated as *continua*, such that all of their properties are defined as continuously distributed in space.[3] A notable exception to this postulate is low-density (rarefied) gases.

13.3.1. Equations of motion

Equations for a fixed control volume:

In contrast to rigid body mechanics, the integral equations of motion for fluids are written for a control volume, rather than a closed system of bodies. [18]

- Continuity (mass conservation) equation:

$$\frac{\partial}{\partial t}\int_{\mathcal{V}} \rho \, d\mathcal{V} + \int_{\mathcal{A}} \rho \vec{V} \cdot d\vec{A} = 0.$$

ρ is the fluid density, $\vec{V}$ is the velocity vector, $\mathcal{V}$ is the control volume, and $\mathcal{A}$ is the area of its surface.

- Momentum equation (Newton's second law) for an inertial coordinate system:

$$\vec{F} = \frac{\partial}{\partial t}\int_{\mathcal{V}} \vec{V} \rho \, d\mathcal{V} + \int_{\mathcal{A}} \vec{V} \rho \vec{V} \cdot d\vec{A}.$$

$\vec{F}$ is the net external force acting on the control volume.

- Energy equation (first law of thermodynamics):

$$\dot{Q} - \dot{W} = \frac{\partial}{\partial t}\int_{\mathcal{V}} \left(u + \frac{1}{2}V^2 + gz\right) \rho \, d\mathcal{V} + \int_{\mathcal{A}} \left(u + \frac{p}{\rho} + \frac{1}{2}V^2 + gz\right) \rho \vec{V} \cdot d\vec{A}$$

$\dot{Q}$ is the rate of heat transfer from the surroundings to the control volume; $\dot{W}$ is the mechanical power produced by moving solid components, shear stresses acting on the boundary or electromagnetic forces, but not normal stresses acting on the control volume (the latter have been included on the right hand side); u is the specific internal energy; p is the pressure; z is an upward vertical axis; and g is the gravitational acceleration.

Differential equations of motion

Flow of an incompressible (i.e., constant-density), Newtonian (i.e., with a linear stress-strain relationship) fluid with constant properties is described by the following equations: [3]

- Continuity equation:

$$\operatorname{div} \vec{V} = 0.$$

- Momentum (Navier-Stokes) equation:

$$\frac{\partial \vec{V}}{\partial t} + \vec{V} \cdot \nabla \vec{V} = \vec{g} - \frac{1}{\rho}\nabla p + \nu \, \Delta \vec{V},$$

where ν is the kinematic viscosity.

Bernoulli's equation

For steady, incompressible, inviscid flow, along a streamline: [18, 3]

$$p + \tfrac{1}{2}\rho V^2 + \rho g z = \text{const.}$$

13.3.2. Dimensionless parameters

Among the many dimensionless parameters used in fluid mechanics and associated fields, most common are the following: [24]

- Reynolds number:

$$\mathrm{Re} = \frac{\rho V d}{\mu}.$$

  V is a characteristic velocity, d is a characteristic length, and μ is the viscosity. It represents the ratio of the "inertia forces" and the viscous forces.

- Mach number:

$$M = V/c.$$

  c is the speed of sound. It represents the ratio of the inertia and elastic forces (compressibility).

- Pressure coefficient (or Euler number):

$$C_P (\equiv Eu) = \frac{p}{\frac{1}{2}\rho V^2}.$$

  It represents the ratio of pressure and inertia forces. Some sources define *Eu* as one-half of the above.

- Drag coefficient for an immersed object:

$$C_D = \frac{F_D}{\frac{1}{2}\rho A V^2}.$$

  F_D is the drag force, and A is the frontal area for bluff objects or the planform area for streamlined objects.

- Lift coefficient:

$$C_L = \frac{F_L}{\frac{1}{2}\rho A V^2}.$$

  F_L is the lift force, and A is the planform area.

- Prandtl number:

$$Pr = \nu/\gamma.$$

  γ is the thermal diffusivity.

- Schmidt number:

$$Sc = \nu/\gamma_c.$$

  γ_c is the molecular diffusivity of a species c in a fluid.

- Froude number (for liquids):

$$Fr = \frac{V}{\sqrt{gL}}.$$

L is a characteristic length. Its square represents the ratio of inertia and gravitational forces. It applies mainly to liquid flows with a free surface.

- Weber number (for liquids):

$$We = \frac{\rho V^2 L}{\sigma}.$$

σ is the surface tension. It represents the ratio of the inertia and surface tension forces.

- Capillary number (for two-phase flows):

$$Ca = \mu V/\sigma.$$

- Cavitation number (for liquids):

$$\sigma_c = \frac{p - p_V}{\frac{1}{2}\rho V^2}.$$

p_V is the vapor pressure.

- Nusselt number:

$$Nu = \frac{hL}{k}.$$

h is the convective heat transfer coefficient, and k is the thermal conductivity.

- Grashof number:

$$Gr = \frac{\alpha g \, \Delta T L^3}{\nu^2}.$$

α is the thermal expansion coefficient, and ΔT is a temperature difference.

- Péclet number:

$$Pe = VL/\kappa.$$

- Rayleigh number:

$$Ra = \frac{\alpha g \, \Delta T L^3}{\nu\kappa} (\equiv GrPr).$$

- Richardson number (for density-stratified flows):

$$Ri = -\frac{g}{\rho}\frac{d\rho}{dz}\left(\frac{dV}{dz}\right)^{-2}.$$

- Taylor number:

$$Ta = \Omega^2 L^4/\nu^2.$$

Ω is the rotation rate.

- Rossby number:

$$Ro = \frac{V}{\Omega L}.$$

It represents the ratio of inertia and Coriolis forces.

- Strouhal number:

$$S = fL/V.$$

f is the frequency of vortex shedding.

- Knudsen number for gases:

$$Kn = \lambda/L.$$

λ is the mean free path.

13.3.3. Some laminar solutions

Some useful exact solutions for laminar flows are reproduced below:[3, 13, 41]

- Fully developed (Poiseuille) flow in a circular tube:

$$V(r) = -\frac{R^2}{4\mu}\frac{\partial p}{\partial x}\left(1 - r^2/R^2\right).$$

R is the tube's radius, x is the axial direction, and r is the radial direction.

- Flow between a still plane wall and a wall moving with speed V_h and with a streamwise pressure gradient:

$$V(r) = \frac{V_h y}{h} + \frac{h^2}{2\mu}\frac{\partial p}{\partial x}\frac{y}{h}(y/h - 1).$$

h is the gap height, and y is the distance from the still wall. The case with $\partial p/\partial x = 0$ is called Couette flow.

- Flow between two concentric cylinders with radii R_1 (inner) and R_2 (outer) and rotating with angular velocities Ω_1 (inner) and Ω_2 (outer):

circumferential velocity: $$V_\theta = \Omega_1 \frac{1 - (\Omega_2/\Omega_1)\left(R_2^2/R_1^2\right)}{1 - R_2^2/R_1^2} r + R_1^2\Omega_1 \frac{1 - \Omega_2/\Omega_1}{1 - R_1^2/R_2^2}\frac{1}{r}.$$

- Stokes (Re $\to$ 0) flow past a sphere (in spherical coordinates):

radial velocity:	$V_r = V_\infty \cos\theta(1 + R^3/2r^3 - 3R/2r)$,
circumferential velocity:	$V_\theta = V_\infty \sin\theta(-1 + R^3/4r^3 + 3R/4r)$,
pressure:	$p = p_\infty - \frac{3}{2}\mu V_\infty R \cos\theta/r^2$,
drag coefficient:	$C_D = 24/\text{Re}$.

V_∞ and p_∞ are the far-field velocity and pressure, and R is the radius of the sphere.

- Potential (inviscid, irrotational) flow past a sphere (in spherical coordinates):

radial velocity:	$V_r = V_\infty \cos\theta(1 - R^3/r^3)$,
circumferential velocity:	$V_\theta = -V_\infty \sin\theta(1 + R^3/2r^3)$.

- Potential flow past a circular cylinder (in cylindrical coordinates):

radial velocity:	$V_r = V_\infty \cos\theta(1 - R^2/r^2)$,
circumferential velocity:	$V_\theta = -V_\infty \sin\theta(1 + R^2/r^2)$.

- Periodic flow over a plate oscillating on its plane with velocity $V_o \cos\omega t$:

$$V = V_o e^{-y\sqrt{\omega/2\nu}} \cos(\omega t - y\sqrt{\omega/2\nu}).$$

13.4. INSTABILITY, TRANSITION, AND TURBULENCE

13.4.1. Hydrodynamic stability

When a characteristic parameter expressing the relative magnitude of stresses in a fluid, such as the Reynolds number, the Taylor number, or the Richardson number, exceeds a critical value, the fluid motion becomes unstable and reverts to a more stable motion. Well-studied problems of hydrodynamic stability are the Rayleigh-Benard convection, in which convection cells form in a fluid confined between two plates and heated from below, and the Taylor-Couette flow, in which toroidal vortices form in the fluid filling a thin gap between two coaxial cylinders, at least one of which rotates. [13]

13.4.2. Transition

Increasing the Reynolds number of a laminar boundary layer results in the formation of turbulent spots and, eventually, through the process known as transition, to turbulence. Typical Re-value for transition over smooth plane walls is up to 5,000,000 for extremely "quiet" flows, but, for practical flows, it is about 500,000 (based on streamwise distance from the edge); in circular pipes, under usual industrial or laboratory conditions it is 2,300 (based on diameter). This means that, for air at STP and a speed of 10 m/s, transition over a thin plate will occur at 0.75 m, and in a smooth pipe when its diameter exceeds 3.5 mm; the same values apply to water at a speed of 0.67 m/s. Transition is affected by wall roughness, streamwise pressure gradient, free stream disturbances, and compressibility. [18, 33]

13.4.3. Turbulence

Turbulence is a random, vortical motion with spatial and temporal irregularity. It results in increased forces, frictional losses, convective heat transfer, and rates of mixing and chemical reaction. Self-similarity types of analyses have provided useful predictions of the spatial growth rates of free shear flows; axisymmetric and two-dimensional turbulent jets, two-dimensional mixing layers, and axisymmetric and two-dimensional plumes, all grow linearly, whereas axisymmetric wakes grow as $x^{1/3}$ and two-dimensional wakes as $x^{1/2}$. [39, 38]

The most celebrated turbulence theory is Kolmogoroff's theory of universal equilibrium, which provides scaling laws for the fine structure of high Reynolds number turbulence. [19, 35] This theory predicts that the turbulent kinetic energy spectrum in the inertial subrange (namely, a range of equilibrium between energy gained from larger eddies and lost to smaller eddies through the process known as energy cascade) decreases as $\kappa^{-5/3}$, where κ is the wavenumber. Furthermore, it postulates that the dissipation of the turbulent kinetic energy to heat occurs mainly at scales of the order of magnitude of $(\nu^3/\epsilon)^{1/4}$, where ϵ is the dissipation rate.

Of interest has also been the formation and properties of large-scale coherent structures that occur in shear flows and influence the growth of turbulence and turbulent transport. [9, 22, 23] Special techniques, known as conditional sampling and phase averaging, have been developed to extract the typical features of these structures. The structure of turbulence has been studied experimentally and computationally, particularly with the use of direct numerical simulation (DNS). [28]

Analytical theories and models of turbulence are usually tested in simplified configurations before being applied to more complex flows. The simplest case is homogeneous and isotropic turbulence, whose statistical properties are invariant under translation, rotation, or reflection of the coordinate axes. This has been approximated in the wind tunnel by grid-generated turbulence; such flows have no energy production mechanism to counteract its dissipation, so that their energy decays with distance from the grid, following a power law with a typical exponent of -1.2. [10] Simplified turbulence that contains production and grows, rather than decays, is the homogeneous shear flow, with a constant transverse gradient of the mean velocity. Its experimental counterpart is uniformly sheared turbulence, whose energy grows exponentially with downstream distance. [37] The DNS realizations of homogeneous shear flow grow exponentially in time.

13.5. FRICTION AND DRAG

Some useful experimental results have been presented in Figures 13.1 to 13.3.

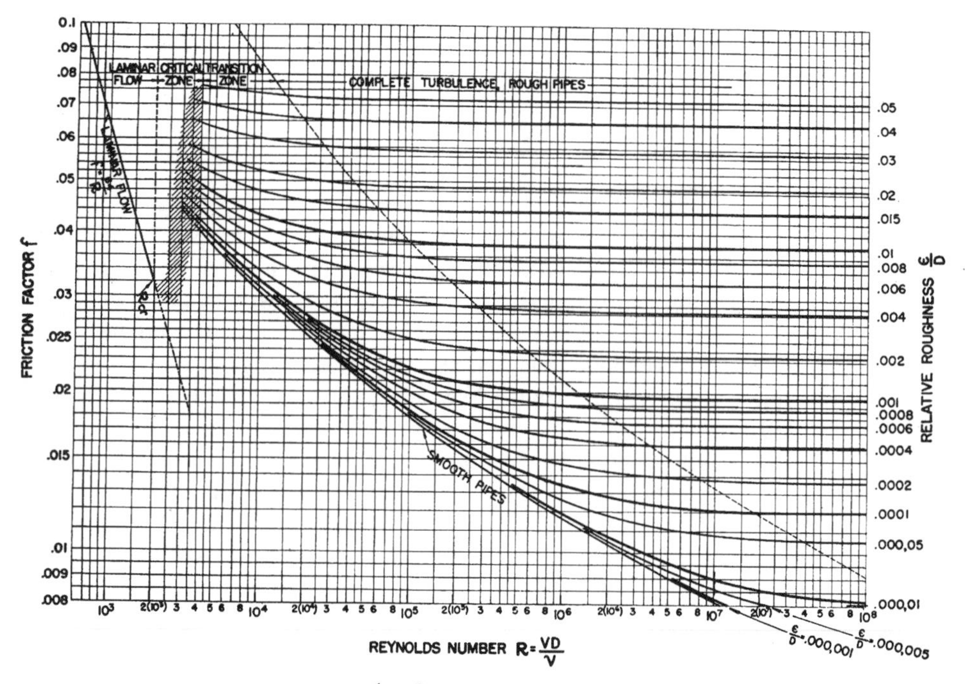

FIGURE 13.1. Friction factor, $f = \Delta p/(\frac{1}{2}\rho V^2 L/D)$, for fully developed, incompressible flow in long circular tubes. [5] Frictional losses in non-circular and curved pipes and ducts as well as in an assortment of accessories can be found in Chapter 6 of Ref. [5], which also contains losses for compressible and two-phase flows. (Reproduced with permission.)

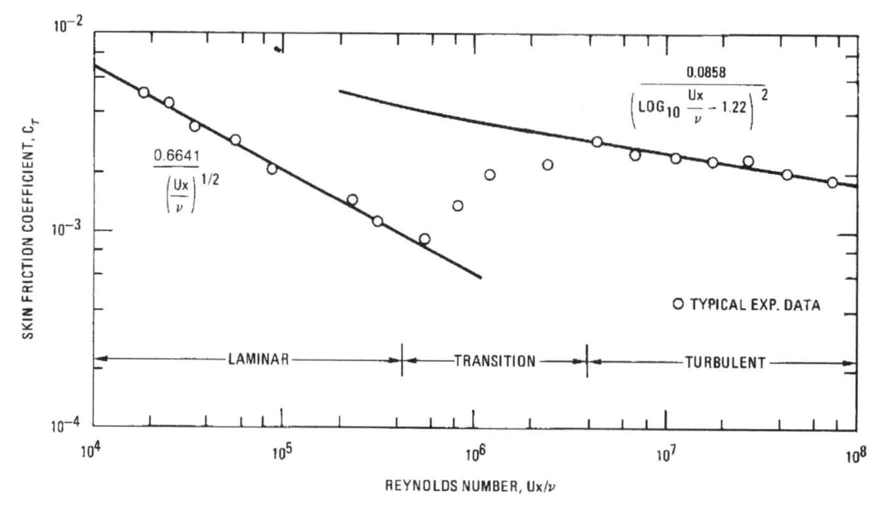

FIGURE 13.2. Skin friction coefficient for incompressible flow over a flat plate. [5] (Reproduced with permission.)

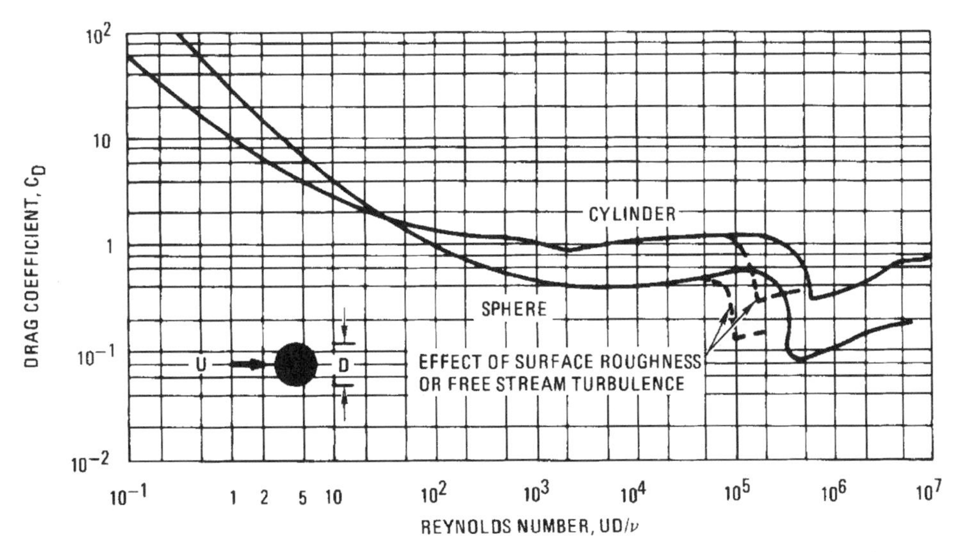

FIGURE 13.3. Drag coefficients for incompressible flow past spheres and circular cylinders in cross-flow. Drag coefficients for objects of different shapes have been tabulated by Blevins. [5] (Reproduced with permission.)

13.6. GAS DYNAMICS

13.6.1. Wave propagation in fluids

Liquid flows are normally treated as incompressible, except in situations involving wave propagation or cavitation. Gas flows are compressible, except when their speed is small compared to the speed of sound, in practice for $M < 0.2$, which, for air at STP, corresponds to $V < 60$ m/s. Small pressure disturbances in a fluid travel with the *speed of sound*, c. For homogeneous liquids, $c = \sqrt{E_V/\rho}$, where E_V is the bulk modulus of elasticity, while, for perfect gases, $c = \sqrt{\gamma RT}$, where R is the gas constant and T is the absolute temperature.

When $M < 1$, a flow is called *subsonic*, when $M = 1$, *sonic* and, when $M > 1$, *supersonic*. Flows in the range $0.8 < M < 1.2$ are referred to as *transonic*, while the term *hypersonic* usually refers to flows with $M > 5$. At supersonic speeds, the influence of a small disturbance is confined within the space surrounded by the surface of the *Mach cone*, having its apex at the source of disturbance and a generating angle of $\sin^{-1}(1/M)$ (*Mach angle*).

13.6.2. One-dimensional, isentropic, compressible flow

In *isentropic* (namely, reversible and adiabatic) flow, [18] all properties can be expressed in terms of the corresponding *stagnation* properties, achieved by the fluid if it would reach stagnation isentropically, and denoted by the subscript "o":

$$\frac{p_o}{p} = \left(1 + \frac{\gamma - 1}{2}M^2\right)^{\gamma/(\gamma-1)}, \quad \frac{T_o}{T} = 1 + \frac{\gamma - 1}{2}M^2, \quad \frac{\rho_o}{\rho} = \left(1 + \frac{\gamma - 1}{2}M^2\right)^{1/(\gamma-1)}.$$

When a gas flows isentropically through a duct with a variable cross-sectional area A, its speed will change as

$$\frac{dV}{V} = -\frac{1}{1-M^2}\frac{dA}{A}.$$

It follows that subsonic flow would accelerate through a converging duct (subsonic contraction) and decelerate through a diverging duct (subsonic diffuser), whereas supersonic flow would decelerate through a converging duct (supersonic diffuser) and accelerate through a diverging duct (supersonic contraction). Therefore, a rocket nozzle, designed to produce gas with the highest possible supersonic exit speed, has a converging-diverging shape. The above expression is singular at sonic conditions. The only physically possible situation in which isentropic flow in a duct may become sonic (*choked* flow) is when it goes through a minimum area, $A^\star$, called the *critical area*:

$$\frac{A}{A^\star} = \frac{1}{M}\left(\frac{1+\frac{\gamma-1}{2}M^2}{1+\frac{\gamma-1}{2}}\right)^{\frac{\gamma+1}{2(\gamma-1)}}.$$

The maximum possible mass flow rate through a duct is

$$\dot{m} = \frac{p_o A^\star}{\sqrt{RT_o}}\sqrt{\gamma}\left(\frac{2}{\gamma+1}\right)^{\frac{\gamma+1}{2(\gamma-1)}}.$$

13.6.3. Shock waves

Shock waves occur in supersonic flows. They are thin regions, across which there are dramatic increases of pressure, density, and temperature and a sudden reduction of velocity. In one-dimensional flow, the shocks are *normal*, and the properties before and after the shock are related as[18]

$$M_2 = \sqrt{\frac{(\gamma-1)M_1{}^2+2}{2\gamma M_1{}^2-(\gamma-1)}}, \quad \frac{p_2}{p_1} = \frac{1+\gamma M_1{}^2}{1+\gamma M_2{}^2}, \quad \frac{T_2}{T_1} = \frac{1+\frac{\gamma-1}{2}M_1{}^2}{1+\frac{\gamma-1}{2}M_2{}^2}.$$

Shock waves are irreversible processes, causing a decrease in total pressure and an increase in entropy, whereas the total enthalpy and the total temperature across them remain essentially unchanged.

Standing normal shocks can be created in converging-diverging nozzles, whereas traveling normal shocks can be created in *shock tubes* by bursting a diaphragm separating fluids under different pressures. When supersonic flow encounters an immersed object with a sharp nose, such as a cone or a wedge, it goes through an *oblique* shock, which is attached to the nose. The flow direction and property changes across an oblique shock may be computed by using the normal shock expressions for the velocity component normal to the shock while preserving the tangential component. When supersonic flow encounters an object with a blunt nose, it goes through a detached, curved ("*bow*") shock. Shock waves near body surfaces interact with the boundary layers and may cause transition to turbulence or flow separation from the surface.

13.7. MEASUREMENT IN FLUIDS

13.7.1. Bulk flow measurement

Among the most common, commercially available, flowmeters [2] are obstruction flow meters, including orifice plates, flow nozzles, Venturi tubes and averaging Pitot tubes, rotameters, turbine and paddlewheel flowmeters, positive displacement flowmeters, vortex shedding flowmeters, ultrasonic flowmeters, magnetic flowmeters, and Coriolis flowmeters.

13.7.2. Flow visualization

The observation of flow patterns is very useful in the preliminary stages of an experiment for a qualitative assessment of the flow, for example, to detect the formation or map the boundaries of a free turbulent shear flow, a separated flow region, or a second phase. In combination with various image recording and processing methods, flow visualization may also be used for measurement, for example, the determination of fractal properties of a turbulent/nonturbulent interface.

Flow visualization methods [27, 16] may be classified into two general classes: a) *optical methods*, namely, methods that utilize variations of the refractive index in the fluid; they include the shadowgraph, the Schlieren method, interferometry, holography, stream birefringence and radiation emission techniques, and b) *marker methods*, namely, methods that use the visibility of foreign objects, introduced into the flow; among the most common are tufts, surface markers (e.g., oil streaks, temperature and pressure sensitive paints, and liquid crystals), dyes, hydrogen bubbles, smoke and oil mists, and various solid powders and liquid droplets; when using flow markers, one must ascertain that they do not interfere with the flow and that they follow, sufficiently closely, the local fluid motion.

13.7.3. Pressure measurement

The precise measurement of absolute pressure, particularly of vacuum, requires sophisticated instrumentation and procedures. However, the usual requirements of pressure measurement in fluid mechanics are satisfied by many available, practical, fast-response, and relatively inexpensive, pressure transducers. [21, 4] The mean static pressure is commonly monitored by liquid manometers and mechanical pressure gages, whereas pressure fluctuations can be measured with piezoelectric or capacitor-type microphones. Care must be taken for a faithful transmission of the flow pressure to the measuring instrument. Wall pressure is sensed by small pressure taps, whereas in-flow pressure is sensed by static tubes. The measurement of in-flow static pressure fluctuations remains a difficult problem, due to the unavoidable interference of the transducer with the pressure field; a successful method for gases has used a bleed-type probe, which contains a frequency-compensated hot wire or hot film anemometer.

In recent years, the measurement of surface pressure distribution with fairly high spatial resolution has been achieved with the use of pressure-sensitive paints (PSP). [26] These consist of a photoluminescent medium suspended in a polymer binder. The luminescence depends on the oxygen concentration (oxygen interacts with the excited molecules), which, in turn, depends on the local static pressure. The PSP method is being used mainly for high-speed ($M > 0.3$) aerodynamic studies.

13.7.4. Velocity measurement

The local mean velocity in laboratory flows is usually measured with the use of Pitot tubes, [4, 36] which actually measure the time-averaged local total pressure, in combination with static pressure tubes or taps. Time-dependent and turbulent velocities can be measured with one of the following methods.

1. *Thermal anemometry*: [7, 36] This includes hot-wires (HWA) for gas flows and hot-films, mainly for liquid flows. Thermal anemometers are usually operated in the constant-temperature mode and provide a voltage output, E, which is related to the flow velocity, V, by the semiempirical King's law

$$\frac{E^2}{T_w - T_f} = A + BV^n,$$

   where T_w is the sensor temperature, T_f is the flow temperature, and A, B, and n are calibration constants ($n \approx 0.45$). In supersonic flows, the hot-wire is sensitive to the "mass flow rate," ρV, rather than to the flow velocity itself.
2. *Laser-Doppler velocimetry (LDV)*: [20, 12] This method is based on the Doppler phenomenon, namely, that the frequency of light emitted by a moving particle is shifted, with respect to the incident light frequency, by an amount that is proportional to the particle speed. LDV is nonintrusive and requires no calibration but, in general, has a lower temporal and spatial resolution, compared with HWA, and it usually requires flow seeding with small (typically, 1 μm) particles.
3. *Particle-image velocimetry (PIV)*: [32] In this method, pairs of successive images of small particles introduced into the flow are recorded and the flow velocity is computed from the distances between the corresponding image pairs. The best way to produce the images is to use a dual-pulsed YAG laser and synchronized CCD cameras. Compared with HWA and LDV, PIV has the advantage of providing the velocity distribution over an area rather than at a single point.

13.7.5. Temperature measurement

Common temperature transducers include: [34, 4]

1. liquid-in-glass thermometers;
2. mechanical (bimetallic) thermometers;
3. thermocouples, namely, junctions of two metallic wires, which produce a voltage depending on its temperature; thermocouples have a typical resolution of about 1 K and a low-level voltage output but, also, a wide range, extending over hundreds of degrees Kelvin; the J- (iron-constantan), K- (chromel-alumel), and T- (copper-constantan) types are common;

4. RTDs (resistive temperature devices), most commonly using platinum as the sensing element, in the form of a coil or film; RTDs are stable and accurate, but they require a current source and a high-gain amplifier for their operation;
5. thermistors, which are metal oxide semiconductors, whose resistance depends on temperature; thermistors are the most sensitive among these sensors, but they are highly nonlinear and their useful operation does not extend far outside of the usual ambient temperature range.

A variety of optical techniques have also been developed for the measurement of temperature in specific environments. Pyrometry [34, 4] is used for high-temperature materials; the most common pyrometer is the "disappearing filament optical pyrometer." Surface temperature can be mapped by coating a surface with liquid crystals [16] or, for high gas speeds, with temperature-sensitive paints (TSP). [26] Spectroscopic methods, [14] including Rayleigh scattering and Raman scattering, also provide the temperature of gases, although the main purpose of these methods is the measurement of concentration.

13.7.6. The fluid mechanics laboratory

A fluid mechanics laboratory contains flow generation apparatus and flow measurement instrumentation. The former may include wind tunnels, water tunnels, towing tanks, shock tubes, and other general purpose or specialized flow channels and loops. A typical wind tunnel contains a contraction to improve flow uniformity and turbulence management devices, including honeycombs, woven screens, or perforated plates. [31] Common instrumentation includes manometers, pressure transducers, thermometers, Pitot-static tubes, flow visualization components, and general purpose instruments such as a multimeter, an oscilloscope, and a digital data acquisition system connected to a personal computer. A turbulence laboratory has usually invested in hot wire anemometers, a laser-Doppler velocimeter, and, if afforded, in a particle image velocimeter. Much of the cost of these instruments may be defrayed by the construction of homemade substitutes, based on available designs.

13.8. REFERENCES

[1] *CRC Handbook of Chemistry and Physics (Latest Edition)*. The Chemical Rubber Co., Cleveland.

[2] *The Flow and Level Handbook*. Omega Engineering Inc., Stamford, Connecticut, Latest Edition.

[3] G.K. Batchelor. *An Introduction to Fluid Dynamics*, Cambridge University Press, Cambridge, UK, 1970.

[4] R.P. Benedict. *Fundamentals of Temperature, Pressure and Flow Measurements, Second Edition*, Wiley Interscience, New York, 1977.

[5] R.D. Blevins, *Applied Fluid Dynamics Handbook*. Van Nostrand Reinhold Company, New York, 1984.

[6] R.E. Bolz and G.L. Tuve. *Handbook of Tables for Applied Engineering Science (Second Edition)*. CRC Press, Boca Raton, Florida, 1986.

[7] H.H. Bruun, *Hot-Wire Anemometry*. Oxford University Press, Oxford, UK, 1995.

[8] R. Budwig. Refractive index matching methods for liquid flow investigations. *Experiments in Fluids*, 17:350–355, 1994.

[9] B.J. Cantwell. Organized motion in turbulent flow. *Annual Review of Fluid Mechanics*, 13:457–515, 1981.

[10] G. Comte-Bellot and S. Corrsin. The use of a contraction to improve the isotropy of grid-generated turbulence. *Journal of Fluid Mechanics*, 25:657–682, 1966.

[11] G.T. Csanady. *Turbulent Diffusion in the Environment*. D. Reidel Publishing Company, Boston, 1973.

[12] F. Durst, A. Melling and J.H. Whitelaw. *Principles and Practice of Laser Doppler Anemometry*. Academic Press, New York, 1976.

[13] L. Rosenhead (Editor). *Laminar Boundary Layers*. Oxford University Press, Oxford, UK, 1963.

[14] R.J. Emrich (Editor). Fluid dynamics. IN L. Marton and C. Marton (Editors in Chief), editors. *Methods of Experimental Physics, Vol. 18A and B,*, Academic Press, New York, 1981.

[15] S. Goldstein (Editor). *Modern Developments in Fluid Dynamics*. Dover Publications Inc., New York, 1965.

[16] W.-J. Yang (Editor). *Handbook of Flow Visualization*. Taylor and Francis, Washington, DC, 1989.

[17] T.E. Faber. *Fluid Dynamics for Physicists*. Cambridge University press, Cambridge, UK, 1995.

[18] R.W. Fox and A.T. McDonald. *Introduction to Fluid Mechanics (Fifth Edition)*. 1998, John Wiley and Sons, New York, 1998.

[19] U. Frisch. *Turbulence*. Cambridge University Press, Cambridge, UK, 1995.

[20] R.J. Goldstein. *Fluid Mechanics Measurements* (second edition). Taylor & Francis, Washington, DC, 1996.

[21] P.W. Harland. *Pressure Gauge Handbook*. Marcel Dekker, New York, 1985.

[22] P. Holmes, J.L. Lumley and G. Berkooz. *Turbulence, Coherent Structures, Dynamical Systems and Symmetry*. Cambridge University Press, Cambridge, UK, 1996.

[23] A.K.M.F. Hussain. Coherent structures—reality and myth. *Physics of Fluids*, 26:2816–2850, 1983.

[24] S.P. Parker (Editor in Chief). *Fluid Mechanics Source Book*. McGraw-Hill, New York, 1987.

[25] J. Lighthill. *Waves in Fluids*. Cambridge University Press, Cambridge, UK, 1979.

[26] T. Liu, T. Campbell, S. Burns, and J. Sullivan. Temperature and pressure sensitive luminescent paints in aerodynamics. *Applied Mechanics Reviews*, 50:227–246, 1997.

[27] W. Merzkirch. *Flow Visualization (Second Edition)*. Academic Press, New York, 1987.

[28] P. Moin and K. Mahesh. Direct numerical simulation: A tool in turbulence research. *Annual Review of Fluid Mechanics*, 30:539–578, 1998.

[29] Committee on Condensed-Matter and Materials Physics. Nonequilibrium physics. In *Condensed-Matter and Materials Physics*, chapter 4, pages 168–175. National Academy Press, Washington, D.C., 1999.

[30] O.M. Phillips. *The Dynamics of the Upper Ocean (Second Edition)*, Cambridge University Press, Cambridge, UK, 1980.

[31] W.H. Rae and A. Pope. *Low-Speed Wind Tunnel Testing*. John Wiley and Sons, New York, 1984.

[32] M. Raffel, C. Willard, and J. Kompenhans. *Particle Image Velocimetry—A Practical Guide*. Springer, Berlin, 1998.

[33] J.A. Schetz. *Boundary Layer Analysis*. Prentice Hall, Englewood Cliffs, New Jersey, 1993.

[34] J.F. Schooley. *Thermometry*. CRC Press, Boca Raton, 1986.

[35] K.R. Sreenivasan. Fluid turbulence. *Reviews of Modern Physics*, 71:S383–S395, 1999.

[36] S. Tavoularis. Techniques for turbulence measurement. In N.P. Cheremisinoff, editor, *Encyclopedia of Fluid Mechanics, Vol. 1*, chapter 36, pages 1207–1255, Gulf Publishing Co., Houston, 1986.

[37] S. Tavoularis and U. Karnik. Further experiments on the evolution of turbulent stresses and scales in uniformly sheared turbulence. *Journal of Fluid Mechanics*, 204:457–478, 1989.

[38] H. Tennekes and J.L. Lumley. *A First Course in Turbulence*. The MIT Press, Cambridge, Massachusetts, 1972.

[39] A.A. Townsend. *The Structure of Turbulent Shear Flow (Second Edition)*. Cambridge University Press, Cambridge, UK, 1976.

[40] J.S. Turner. *Buoyancy Effects in Fluids*. Cambridge University Press, Cambridge, UK, 1979.

[41] F.M. White. *Viscous Fluid Flow* (second edition). McGraw-Hill, New York, 1991.

14

Mechanics

Florian Scheck
Johannes Gutenberg-Universität, Mainz, Germany

Contents

List of Figures

14.1. INTRODUCTION

Mechanics ranges from Newtonian mechanics of nonrelativistic few-body problems to qualitative dynamics concerned with long-term stability and deterministically chaotic behavior. It develops formal frameworks, such as Lagrangian mechanics and Hamilton-Jacobian mechanics, which, when suitably generalized, provide a comprehensive framework for the formulation of field theories, and it prepares the ground for their extension to quantum theory. Whenever velocities are no longer small compared with the speed of light, the Galilei group is replaced by the Poincaré group in its role of covariance group of dynamics. Also, if the back reaction of mechanical objects on the underlying space-time manifold cannot be neglected, particle trajectories must be computed from the geodesic equation on space time equipped with the differential structure of a semi-Riemannian manifold.

14.2. NEWTONIAN MECHANICS

Newtonian mechanics deals with the kinematics and dynamics of a finite number N of particles subject to internal and external forces, but whose motions are not otherwise constrained by conditions on their coordinates [1, 2]. In most applications, velocities are small compared with c, the velocity of light, and nonrelativistic kinematics applies. The underlying spacetime is a *flat* Euclidean space $(\mathbb{R})_{\text{time}} \times (\mathbb{R}^3)_{\text{space}} \cong \mathbb{R}^4$ endowed with the Galilei group as the covariance group of the equations of motion. When dealing with the classic motion of elementary particles (e.g., motion of electrons in classic, external fields), however, relativistic effects can be important and the Galilei group is replaced by the Poincaré group. No back reaction of matter particles on the supporting space time is taken into account. Such effects become important only for large amounts of matter and energy placed in a given spatial volume. Classic mechanics is then replaced by general relativity. Time loses its absolute nature, and force-free or "free-fall" motion along straight lines in $\mathbb{R}^3$ becomes geodesic motion in curved space time.

14.2.1. Newton's laws and inertial frames

1. *Every body continues in its state of rest or of uniform rectilinear motion, except if it is compelled by forces acting on it to change that state.*

2. *The change of motion is proportional to the applied force and takes place in the direction of the straight line along which that force acts.*
3. *To every action, there is always an equal and contrary reaction; the mutual actions of any two bodies are always equal and oppositely directed along the same straight line.*

Law (1) singles out the class of *inertial frames*, i.e., the frames of reference with respect to which force-free motion is uniform (constant velocity) and rectilinear (geodesic in flat space). If **K** is an inertial frame, all other inertial frames are related to it by Galilei transformations, viz.

$$g : \begin{pmatrix} \boldsymbol{x} \\ t \end{pmatrix} \longmapsto \begin{pmatrix} \boldsymbol{x}' = \mathbf{R} \cdot \boldsymbol{x} + \boldsymbol{v}t + \boldsymbol{a} \\ t' = \lambda t + s \end{pmatrix}, \tag{14.1}$$

where $\mathbf{R} \in O(3)$ is a rotation if $\det \mathbf{R} = 1$, a rotation times space reflection if $\det \mathbf{R} = -1$; $\boldsymbol{v}$ is a constant velocity (relative velocity of two inertial frames); and $\boldsymbol{a}$, s are translations in space and time, respectively, with $\lambda = +1$ (-1) for orthochronous (time-reflecting) transformations. The four possibilities of choosing the signs of $\det \mathbf{R}$ and of λ indicate that the Galilei group has four disjoint branches, only one of which, $G_+^\uparrow$, the set of proper ($\lambda = 1$, "$\uparrow$") orthochronous ($\det \mathbf{R} = 1$, "$+$") Galilei transformations, is a subgroup. As is obvious from Eq. (14.1), the Galilei group depends on ten real parameters: three that parametrize the rotation, three for the velocity $\boldsymbol{v}$, and four for the space and time translations.

Law (2) relates the time derivative of the linear momentum of a particle to the force acting on it, $\dot{\boldsymbol{p}} = \boldsymbol{F}$. If the mass is invariant and independent of velocity, the left-hand side is the acceleration multiplied by m. With τ denoting the *proper* time (time of a comoving clock) law (2) reads

$$\frac{\mathrm{d}^2}{\mathrm{d}\tau^2} t(\tau) = 0 \quad \text{(a)}, \qquad \frac{\mathrm{d}^2}{\mathrm{d}\tau^2} \boldsymbol{x}(\tau) - \frac{1}{m} \left(\frac{\mathrm{d}t(\tau)}{\mathrm{d}\tau} \right)^2 \boldsymbol{F}(\boldsymbol{x}, \tau) = 0 \quad \text{(b)}. \tag{14.2}$$

Eq. [14.2(a)] expresses the absolute nature of time in Newtonian mechanics: Except for possibly different choices of unit and of origin, the time of any inertial observer is the same as the proper time. With the choice $\tau = t$, Eq. [14.2(b)] is the familiar second-order differential equation $m\ddot{\boldsymbol{x}} = \boldsymbol{F}$. For vanishing force, Eq. (14.2) is the geodesic equation on flat, four-dimensional Euclidean space $\mathbb{R} \times \mathbb{R}^3$.

Fundamental forces of nature are the gravitational force between two matter particles with masses m_i, m_k,

$$\boldsymbol{F}_{ki}^{\text{grav}} = (-G m_i m_k) \frac{\boldsymbol{x}_i - \boldsymbol{x}_k}{|\boldsymbol{x}_i - \boldsymbol{x}_k|^3} = -\nabla_i \left(\frac{-G m_i m_k}{|\boldsymbol{x}_i - \boldsymbol{x}_k|} \right), \tag{14.3}$$

and the Coulomb force acting between particles with charges e_i and e_k,

$$\boldsymbol{F}_{ki}^{C} = (\kappa_C e_i e_k) \frac{\boldsymbol{x}_i - \boldsymbol{x}_k}{|\boldsymbol{x}_i - \boldsymbol{x}_k|^3} = -\nabla_i \left(\frac{\kappa_C e_i e_k}{|\boldsymbol{x}_i - \boldsymbol{x}_k|} \right). \tag{14.4}$$

Here, G is Newton's constant $G = (6.67259 \pm 0.00085) \times 10^{-11}$ m^3 kg^{-1} s^{-2}; κ_C is a constant whose value depends on the choice of unit for electric charge. In the Gaussian system of electrodynamics $\kappa_C = 1$, in SI units, it is $\kappa_C = 1/(4\pi\varepsilon_0) = c^2 \times 10^{-7}$. In the realm of elementary particle physics, the Yukawa force

$$\boldsymbol{F}^Y_{ki} = -(g_i g_k)\nabla_i \left(\frac{1}{|\boldsymbol{x}_i - \boldsymbol{x}_k|} e^{-|\boldsymbol{x}_i - \boldsymbol{x}_k|/\lambda} \right) \tag{14.5}$$

may also be termed fundamental because it describes the interaction due to the exchange of a massive particle whose Compton wavelength is $\lambda_C = 2\pi\lambda$ with $\lambda = \hbar/(Mc)$, with M its mass and $\hbar$ Planck's constant divided by 2π. The constants g_i and g_k denote the strengths with which the particles i and k couple to the exchanged carrier of the force.

Being gradient fields of potentials that depend only on the distance $|\boldsymbol{x}_i - \boldsymbol{x}_k|$, all three fundamental forces are *central* forces.

Law (3), finally, expresses the well-known principle *actio* = *reactio*, the force of particle k acting on particle i equals minus the force of i on k, $\boldsymbol{F}_{ki} = -\boldsymbol{F}_{ik}$.

An important application of Newtonian mechanics is provided by the closed N-body system with internal, central forces. This system possesses 10 integrals of the motion that, by Noether's theorem, are related to the 10 parameters of the Galilei group. Invariance under translations $\boldsymbol{a}$ in space implies conservation of total linear momentum $\boldsymbol{P}$; invariance under time translations s implies conservation of total energy E; invariance under rotations $\mathbf{R}$ implies conservation of total angular momentum $\boldsymbol{L}$; invariance under special Galilei transformations $\boldsymbol{x}' = \boldsymbol{x} + \boldsymbol{v}t$, $t' = t$, corresponds to the center of mass principle $\boldsymbol{x}_S(t) - (\boldsymbol{P}/M)t = \boldsymbol{x}_S(0)$ (M is the total mass, $\boldsymbol{x}_S(t)$ denotes the trajectory of the center of mass). In the case of the *two*-body system, the 10 integrals suffice to render the problem *integrable*; this is not so, in general, for three or more bodies [3].

14.2.2. Kepler's laws for planetary motion

The Kepler problem is the classic example of a closed and integrable two-body system under the action of the gravitational force (14.3). The center of mass $\boldsymbol{r}_S = (m_1\boldsymbol{x}_1 + m_2\boldsymbol{x}_2)/M$ obeys $\ddot{\boldsymbol{r}}_S = 0$ and, hence, moves uniformly along a straight line. The motion in the relative coordinate $\boldsymbol{r} = \boldsymbol{x}_2 - \boldsymbol{x}_1$ is one of a particle with reduced mass $\mu = m_1m_2/(m_1 + m_2)$ under the action of (14.3). The momentum of the center of mass reads $\boldsymbol{\pi}_S = \boldsymbol{p}_1 + \boldsymbol{p}_2$ and is constant in time; the momentum of relative motion reads $\boldsymbol{\pi} = (m_1\boldsymbol{p}_2 - m_2\boldsymbol{p}_1)/M$.

Kepler's *first law* states that planetary orbits are ellipses. This result is exceptional and specific to all finite orbits in the attractive $1/r$ potential: The center of mass S, the pericenter P, and the apocenter A lie on a straight line, with S between A and P. In general, finite orbits have a rosette-like shape that can be constructed from the branch connecting an arbitrary pericenter with the successive apocenter, by the symmetry of the orbit with respect to the straight lines SP and SA; cf. Figure 14.1. In general, they do not close. A precise statement is provided by *Bertrand's theorem* (1873): *The central potentials $U(r) = \alpha/r$ with $\alpha < 0$, and $U(r) = br^2$ with $b > 0$, are the only ones for which* all *finite orbits close.* (A proof is found in Ref. [4].)

Kepler's *second law* states that the radius vector of the relative coordinate sweeps out equal areas in equal times. This is the law of conservation of the angular momentum $\boldsymbol{l}_{\text{rel}} = \mu\boldsymbol{r} \times \dot{\boldsymbol{r}}$, and it holds true in every central potential.

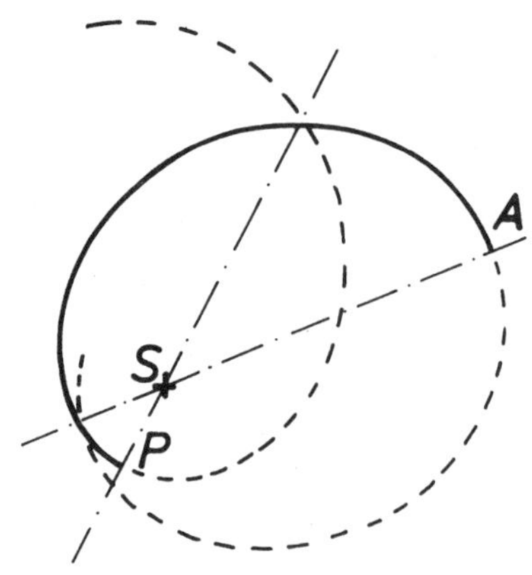

FIGURE 14.1. Branch of finite orbit from a pericenter P to the successive apocenter A, from which the entire rosette orbit can be obtained by reflection symmetry with respect to the straight lines SP and SA. (Figure taken from Ref. [1].)

Kepler's *third law*, like the first, is specific to the $1/r$ potential. It relates the semimajor axis a of a planetary ellipse to the period T of revolution:

$$\frac{a^3}{T^2} = \frac{G(m_1 + m_2)}{(2\pi)^2}. \tag{14.6}$$

In the case of the N-body problem, the above formulae for coordinates and momenta generalize to the following: Let $M_k = m_1 + m_2 + \cdots + m_k$ be the total mass of the first $k \leq N$ particles. Then,

$$\boldsymbol{r}_k = \boldsymbol{x}_{k+1} - \frac{1}{M_k}\sum_{i=1}^{k} m_i \boldsymbol{x}_i \quad (k = 1, \ldots, N-1); \ \boldsymbol{r}_N = \frac{1}{M_N}\sum_{i=1}^{N} m_i \boldsymbol{x}_i. \tag{14.7}$$

The coordinates (14.7) are called *Jacobian coordinates*. The corresponding momenta are

$$\boldsymbol{\pi}_k = \frac{1}{M_{k+1}}\left(M_k \boldsymbol{p}_{k+1} - m_{k+1}\sum_{i=1}^{k} \boldsymbol{p}_i\right), \ (k = 1, \ldots, N-1); \ \boldsymbol{\pi}_N = \sum_{i=1}^{N} \boldsymbol{p}_i. \tag{14.8}$$

The $(6N)$ variables (14.7) and (14.8), like the original ones $(\boldsymbol{x}_1, \ldots, \boldsymbol{x}_N; \boldsymbol{p}_1, \ldots, \boldsymbol{p}_N)$, are canonically conjugate in the sense of Hamilton-Jacobi mechanics (Sect. 14.3).

14.2.3. Phase space and determinism

Newtonian mechanics, like canonical mechanics, is best described in phase space $\mathbb{P}$ spanned by the coordinates and the conjugate momenta. In geometric terms it is the cotangent bundle $T^*\mathbb{Q}$ of the manifold $\mathbb{Q}$ of coordinates. The theorem about existence and uniqueness of solutions of ordinary differential equations [5]

$$\dot{x} = \mathcal{F}(x, t) \tag{14.9}$$

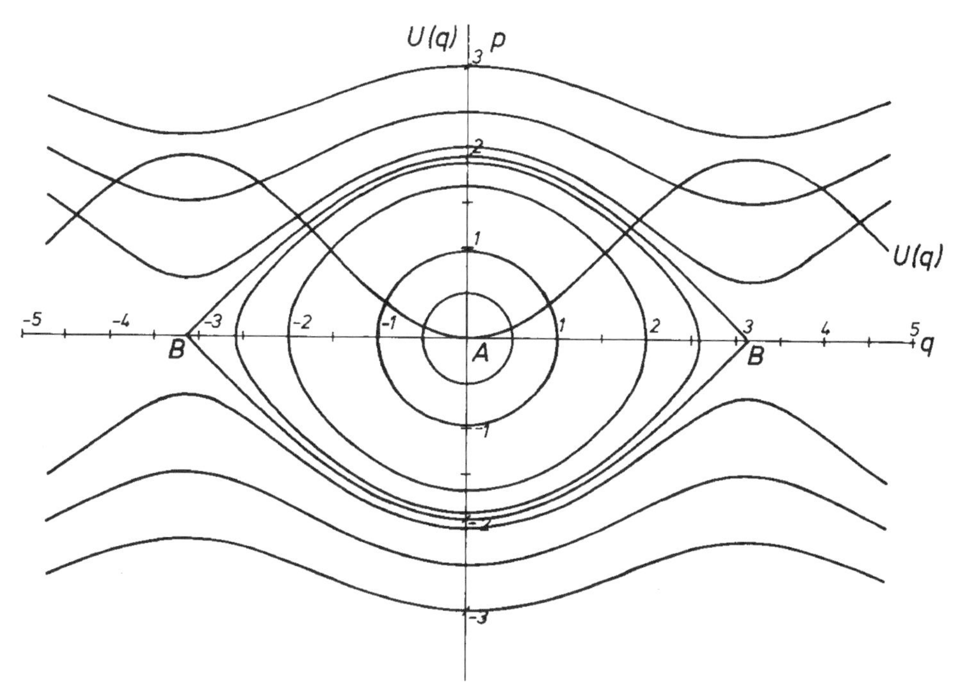

FIGURE 14.2. Planar mathematical pendulum in reduced, dimensionless coordinates (two figures): Potential $U(q) = 1 - \cos q$ as a function of q; phase portraits in phase space whose intersection with the ordinate gives the value of $\varepsilon := E/(mgl)$. (Figure taken from Ref. [1].)

guarantees that a point of phase space $x_0 \in \mathbb{P}$ (the initial condition) defines a unique solution of the equation of motion. The theorem states that if $\mathcal{F}(x, t)$ is continuous in x and t, as well continuously differentiable in x, for any x and any t, there is a neighborhood U of x and a time interval I around t such that for all initial conditions $(x_0 \in U, t_0 \in I)$ there is precisely one solution $\phi(t, t_0, x_0)$ of (14.9) that at time $t = t_0$ passes through the point x_0, $\phi(t_0, t_0, x_0) = x_0$, and that is continuously differentiable in t as well as in the initial conditions x_0 and t_0. Figure 14.2 shows the example of the mathematical pendulum in a plane.

A mechanical system which obeys Eq. (14.9) is generically called a *dynamical system*. Geometrically speaking, it relates the tangent vector field of physical orbits to a given dynamical vector field $\mathcal{F}$. In the case of Newtonian mechanics, x is a point in phase space $\mathbb{P}$ whose dimension is $2f$, with f as the number of degrees of freedom. In the N-body problem (where $f = 3N$), the first f equations are $\dot{\boldsymbol{x}}_i = \boldsymbol{p}_i/m_i$, and the second group are $\dot{\boldsymbol{p}}_i = \boldsymbol{F}_i$.

Thus, mechanical systems are *finite dimensional*, *differentiable*, and *deterministic* in the precise sense of the existence and uniqueness theorem: Under the assumptions of the theorem, the solution pertaining to the initial condition $(x_0 \in \mathbb{P}, t_0)$ can be followed in a unique manner into the past and into the future until it hits a singularity of the dynamical vector field.

14.3. CANONICAL MECHANICS

Canonical mechanics deals with constrained systems, such as the N-body system subject to a number of independent *holonomic* constraints,

$$f_\lambda(\boldsymbol{x}_1, \boldsymbol{x}_2, \ldots, \boldsymbol{x}_N; t) = 0, \quad \lambda = 1, 2, \ldots, \Lambda. \tag{14.10}$$

The constraints are said to be *rheonomic* when there is genuine time dependence; otherwise, they are called *scleronomic*. As the original coordinates are subject to constraints, they are replaced by independent, generalized coordinates $\boldsymbol{q} = (q^1, \ldots, q^f)$. There is no general prescription for how to find a suitable set of independent, generalized coordinates, but often the choice of the original coordinates can be adapted to the constraints so that all coordinates that are constrained can be eliminated [6].

There are two frameworks in which canonical mechanics is treated: *Lagrangian mechanics*, which is formulated on the tangent bundle $T\mathbb{Q}$, or velocity space, and makes use of the Lagrangian function $L : T\mathbb{Q} \times \mathbb{R}_t \to \mathbb{R}$; and *Hamiltonian mechanics*, which is formulated on the cotangent bundle $T^*\mathbb{Q}$, or phase space, and makes use of the Hamiltonian function $H : T^*\mathbb{Q} \times \mathbb{R}_t \to \mathbb{R}$, (the factor $\mathbb{R}_t$ appears whenever L or H has an explicit time dependence). The two are related by *Legendre transformation*, the mapping being bijective if L, and hence H, is sufficiently regular. Lagrangian mechanics is best adapted for the discussion of symmetries (Noether's theorem), and Hamiltonian mechanics is well suited for the treatment of canonical transformations, flows in phase space (Liouville's theorem), perturbation theory, and quantization of classical systems with a finite number of degrees of freedom.

14.3.1. Lagrangian functions and Euler-Lagrange equations

A *Lagrangian system* is a mechanical system with f degrees of freedom to which one associates a smooth function $L(\boldsymbol{q}, \dot{\boldsymbol{q}}, t)$, depending on $2f + 1$ variables $\boldsymbol{q} = (q^1, \ldots, q^f)$, $\dot{\boldsymbol{q}} = (\dot{q}^1, \ldots, \dot{q}^f)$, and t, such that the *action integral*

$$I[\boldsymbol{q}] := \int_{t_1}^{t_2} \mathrm{d}t \; L(\boldsymbol{q}(t), \dot{\boldsymbol{q}}(t), t) \tag{14.11}$$

is extremal for any solution of the equations of motion that assumes given boundary conditions $\boldsymbol{q}(t_i) = \boldsymbol{q}_i^{(0)}$ at the fixed initial and final times t_1 and t_2. The extremality of the functional (14.11) for physical orbits is postulated by *Hamilton's variational principle.* The equations of motion (*Euler-Lagrange equations*) that follow from it read

$$\frac{\mathrm{d}}{\mathrm{d}t}\left(\frac{\partial L}{\partial \dot{q}^k}\right) - \frac{\partial L}{\partial q^k} = 0, \quad k = 1, \ldots, f. \tag{14.12}$$

As these are f ordinary differential equations of second order in time, a specific orbit requires $2f$ data (initial conditions). In many applications, L can be taken to be the difference between kinetic energy T and potential energy U, $L = T - U$, the so-called *natural form* of L. Nevertheless, L is not unique and, hence, is not an observable. For instance, any

Lagrangian function L can be modified by a *gauge transformation* $L' = L + \mathrm{d}M(\boldsymbol{q}, t)/\mathrm{d}t$, with M a smooth function of the generalized coordinates and of time. Adding a total time derivative changes the Lagrangian function but does not change the equations of motion (14.12).

The Lagrangian framework is well adapted to study (or to introduce) continuous symmetries *(theorem by E. Noether)*: If the Lagrangian function of a time-independent (autonomous) mechanical system is invariant under a one-parameter group of transformations $\boldsymbol{q} \mapsto h(s, \boldsymbol{q})$, with $s \in \mathbb{R}$, such that $h(s = 0, \boldsymbol{q}) = \boldsymbol{q}$ is the identity, an integral of the motion exists, given by

$$I(\boldsymbol{q}, \dot{\boldsymbol{q}}) = \sum_{i=1}^{f} \frac{\partial L}{\partial \dot{q}^i} \left. \frac{\mathrm{d}h(s, q^i)}{\mathrm{d}s} \right|_{s=0} . \tag{14.13}$$

Important examples are as follows: If L is invariant under space translations along the direction $\hat{\boldsymbol{n}}$ in $\mathbb{R}^3$, Eq. (14.13) is the projection $P_n = \mathbf{P} \cdot \hat{\boldsymbol{n}}$ of total linear momentum onto that direction; if L is invariant under rotations about a given axis $\hat{\boldsymbol{n}}$, Eq. (14.13) is the projection of total angular momentum $\boldsymbol{L}$ onto that direction.

D'Alembert's principle of virtual displacements yields equations of motion which bear some similarity with Eq. (14.12):

$$\frac{\mathrm{d}}{\mathrm{d}t}\left(\frac{\partial T}{\partial \dot{q}^k}\right) - \frac{\partial T}{\partial q^k} = Q_k, \quad k = 1, \ldots, f. \tag{14.14}$$

Here, $T = \sum m_i \dot{x}_i^2/2$ is the kinetic energy, transformed to the variables $\boldsymbol{q}$ and $\dot{\boldsymbol{q}}$, and Q_k are the generalized forces given by $Q_k := \sum_{i=1}^{N} \boldsymbol{F}_i \cdot \partial \boldsymbol{x}_i / \partial q^k$ in terms of the true forces $\boldsymbol{F}_i$. If these are potential forces, $\boldsymbol{F}_i = -\nabla_i U$, T and U can be combined to a Lagrangian function in natural form $L = T - U$ and the equations of motion (14.14) are the same as (14.12). D'Alembert's equations (14.14) are more general, however, because they are also valid if the constraints are given in a differential form $\delta \boldsymbol{x}_i = \sum_{k=1}^{f} \left(\partial \boldsymbol{x}_i / \partial q^k\right) \delta q^k$ that cannot be integrated to holonomic equations such as (14.10).

14.3.2. Hamiltonian systems

A Hamiltonian or *canonical system* is a mechanical system that admits a Hamiltonian function $H(\boldsymbol{q}, \boldsymbol{p}, t)$ such that the equations of motion take the form

$$\dot{q}^k = \frac{\partial H}{\partial p_k}, \quad \dot{p}_k = -\frac{\partial H}{\partial q^k}, \quad k = 1, \ldots, f, \tag{14.15}$$

where p_i is the momentum canonically conjugate to q^i defined as

$$p_i := \partial L / \partial \dot{q}^i . \tag{14.16}$$

Unlike Eq. (14.12), the $2f$ equations of motion (14.15) are differential equations of *first* order in time; the number of initial data required to uniquely fix a solution is the same.

The Hamiltonian function is obtained from the Lagrangian function by Legendre transformation with respect to the variables $\dot{\boldsymbol{q}}$,

$$H(\boldsymbol{q},\boldsymbol{p},t) = \mathcal{L}L(\boldsymbol{q},\boldsymbol{p},t) := \sum_{k=1}^{f} p_k \dot{q}^k(\boldsymbol{q},\boldsymbol{p},t) - L(\boldsymbol{q},\dot{\boldsymbol{q}}(\boldsymbol{q},\boldsymbol{p},t),t).$$

Solving Eq. (14.16) for $\dot{\boldsymbol{q}}$, by the global implicit function theorem, requires the condition

$$\det\left(\frac{\partial^2 L}{\partial \dot{q}^i \partial \dot{q}^k}\right) \neq 0. \tag{14.17}$$

If this condition is fulfilled, the Legendre transformation is a bijective mapping and the determinant of the second derivatives of H with respect to p_i is the inverse of Eq. (14.17). (More precisely, if the condition (14.17) holds globally, the mapping from the Lagrangian formulation to the Hamiltonian formulation is a diffeomorphism.) Any Lagrangian system that fulfills the condition (14.17) is canonical. Conversely, any canonical system with $\det(\partial^2 H/(\partial p_k \partial p_l)) \neq 0$ obeys the Euler-Lagrange equations with L as obtained from H by Legendre transformation.

One should note that Lagrangian mechanics and Hamiltonian mechanics are defined on $T\mathbb{Q}$ and $T^*\mathbb{Q}$, respectively, i.e., on different spaces that, in general, cannot be identified. Therefore, moving from one formulation to the other is more than a mere change of variables [1, 4, 7, 8].

There are physical situations, e.g., in relativistic mechanics, in which L is not sufficiently regular so that the determinant (14.17) vanishes. In these cases, the Hamiltonian version must be treated by making use of *Dirac's method of constraints* [9].

14.3.3. Canonical transformations and Hamilton-Jacobi equation

Let $x := \{q^1, \ldots, q^f; p_1, \ldots, p_f\}$ be one choice of coordinates in phase space, and let $y := \{Q^1, \ldots, Q^f; P_1, \ldots, P_f\}$ be another such choice. The transformation $x \mapsto y$ is said to be a *canonical transformation*, generated by a function M, depending on old *and* new coordinates, possibly also on time, if it is a diffeomorphism and if it preserves the structure of the canonical equations (14.15). If $\{\boldsymbol{q}, \boldsymbol{p}, H(\boldsymbol{q}, \boldsymbol{p}, t)\}$ is canonical, then $\{\boldsymbol{Q}, \boldsymbol{P}, \widetilde{H}(\boldsymbol{Q}, \boldsymbol{P}, t)\}$ with $\widetilde{H} = H + \partial M/\partial t$ is also canonical.

A case of special importance is the one in which $M \equiv \Phi(\boldsymbol{q}, \boldsymbol{Q}, t)$ depends on old and new coordinates only. In this case,

$$p_i = \frac{\partial \Phi}{\partial q^i}, \quad P_k = -\frac{\partial \Phi}{\partial Q^k}, \quad \widetilde{H} = H + \frac{\partial \Phi}{\partial t}. \tag{14.18}$$

Another choice is obtained from this one by Legendre transformation with respect to Q such that $M = S(\boldsymbol{q}, \boldsymbol{P}, t) - \sum Q^k(\boldsymbol{q}, \boldsymbol{P}, t) P_k$. The equations analogous to Eq. (14.18) read

$$p_i = \frac{\partial S}{\partial q^i}, \quad Q^k = \frac{\partial S}{\partial P_k}, \quad \widetilde{H} = H + \frac{\partial S}{\partial t}. \tag{14.19}$$

Assume first the phase space to be a Euclidean space, $\mathbb{P} = \mathbb{R}^{2f}$. The matrix $\mathbf{M} = \{M_{\alpha\beta} = \partial x_\alpha / \partial y_\beta\}$ of partial derivatives of the old by the new coordinates is an element of the *real symplectic group* $\mathrm{Sp}_{2f}(\mathbb{R})$ over phase space $\mathbb{P}$; i.e., it satisfies

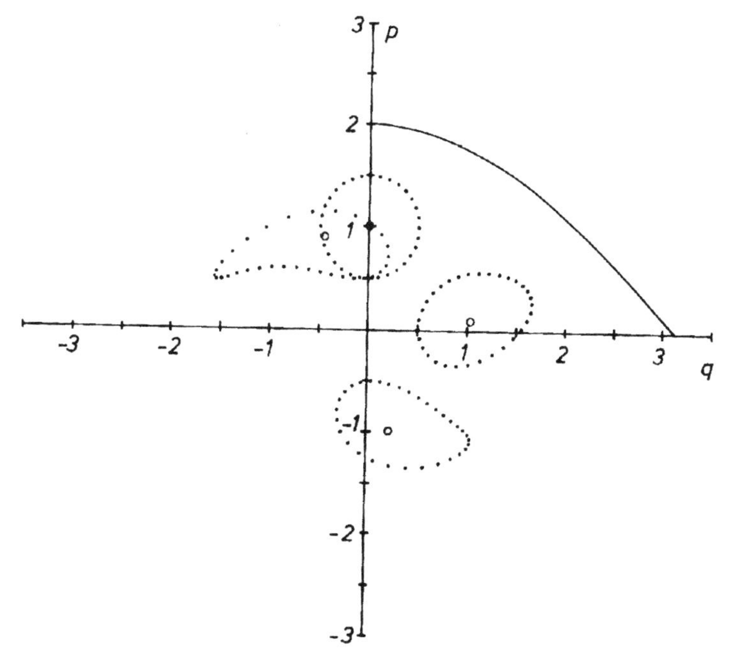

FIGURE 14.3. The flow of a circle of initial conditions (circle with radius 1 and center ($q = 0, p = 1$) at $\tau = 0$) with τ_0, the period of the harmonic oscillator obtained for *small* values of the amplitude. In a clockwise direction, later positions are shown at times $\tau = \tau_0/4$, $\tau = \tau_0/2$, and $\tau = \tau_0$. (Figure taken from Ref. [1].)

$$\mathbf{M}^T\mathbf{J}\mathbf{M} = \mathbf{J}, \ \text{with}\ \mathbf{J} = \begin{pmatrix} 0 & \mathbb{1}_{f\times f} \\ -\mathbb{1}_{f\times f} & 0 \end{pmatrix}, \tag{14.20}$$

and hence has determinant 1. Its inverse is $\mathbf{M}^{-1} = \mathbf{J}^{-1}\mathbf{M}^T\mathbf{J} = -\mathbf{J}\mathbf{M}^T\mathbf{J}$.

The dynamical system (14.15) is written equivalently as $\dot{x} = \mathbf{J}H_{,x}$, with $H_{,x}$ the gradient field of H. If $x = \varphi(t, t_0, y)$ denotes its flow, with initial condition y at time t_0, the matrix $(\mathbf{D}\varphi)$ of its partial derivatives is symplectic, $(\mathbf{D}\varphi)^T\mathbf{J}(\mathbf{D}\varphi) = \mathbf{J}$, and hence, its determinant is $\det(\mathbf{D}\varphi) = 1$. This is known as *Liouville's theorem*: The initial condition y flows to x; and the Jacobian $\partial x/\partial y$ has determinant 1, which means that *the flow of a canonical system preserves volume and orientation.* Figure 14.3 shows the example of a circular set of initial conditions, for the planar pendulum, flowing through phase space.

In many cases, $\mathbb{P}$ is not a flat space $\mathbb{R}^{2f}$ but is a more general smooth manifold with even dimension $2f$. Phase space is then a *symplectic manifold*, $(\mathbb{P}, \omega)$, i.e., a pair consisting of a differentiable manifold $\mathbb{P}$, with $\dim \mathbb{P} = (2f)$, and a nondegenerate, closed, skew-symmetric two-form ω. By Darboux's theorem, this two-form can be represented locally by

$$\omega = \sum_{i=1}^{f} \mathrm{d}x^i \wedge \mathrm{d}x^{i+f} = \sum_{i=1}^{f} \mathrm{d}q^i \wedge \mathrm{d}p_i,$$

whose matrix representation is just **J**, Eq. (14.20). Thus, Eq. (14.20) is the local form of the more general case. Darboux's theorem shows that, away from equilibrium points, symplectic manifolds with the same dimension locally all look the same. Genuine differences occur only globally [10].

Poisson brackets are defined by the scalar product of the gradient fields of two functions on phase space with the metric **J**,

$$\{f(\boldsymbol{q},\boldsymbol{p}), g(\boldsymbol{q},\boldsymbol{p})\} \equiv \{f(x), g(x)\} := -\left(\nabla_x f(x)\right)^T \mathbf{J}\left(\nabla_x g(x)\right). \tag{14.21}$$

They are invariant under canonical transformations. They are useful in identifying constants of the motion $I(x)$ that fulfill $\{H, I\} = 0$.

The class (14.19) of canonical transformations contains $S_{id} := \sum q^i P_i$, the identical mapping, and is useful in defining *infinitesimal canonical transformations*, generated by smooth functions $\sigma(x)$:

$$S(\boldsymbol{q}, \boldsymbol{Q}) \approx S_{id} + \varepsilon\, \sigma(\boldsymbol{q}, \boldsymbol{p}), \tag{14.22}$$

with $\varepsilon \ll 1$. The resulting changes of the coordinates and momenta are given by

$$\delta q^i = \{\sigma(\boldsymbol{q},\boldsymbol{p}), q^i\}\, \varepsilon, \quad \delta p_i = \{\sigma(\boldsymbol{q},\boldsymbol{p}), p_i\}\, \varepsilon.$$

If σ is taken to be the Hamiltonian function, these equations become the canonical equations (14.15): The Hamiltonian is the generating function that boosts the system in an infinitesimal time interval $\varepsilon = \delta t$. Infinitesimal canonical transformations are also useful in discussing constants of the motion and in making contact with Noether's theorem (14.13).

The class (14.19) is relevant when one aims at rendering all coordinates cyclic ones by requiring that $\widetilde{H}$ vanish. This leads to the *partial differential equation of Hamilton and Jacobi*,

$$H\left(\boldsymbol{q}, \boldsymbol{p} = \nabla_q S^*, t\right) + \frac{\partial S^*}{\partial t} = 0. \tag{14.23}$$

$S^*(\boldsymbol{q}, \boldsymbol{P} = \boldsymbol{\alpha}, t)$ is the *action function* and depends on the $f + 1$ variables $\boldsymbol{q}$ and t, and on the constant momenta $P_i = \alpha_i$. If the Hamiltonian function does not depend on time, one studies the *reduced action* $S(\boldsymbol{q}, \boldsymbol{\alpha}) = S^*(\boldsymbol{q}, \boldsymbol{\alpha}, t) - Et$ that obeys the *characteristic equation* of Hamilton and Jacobi,

$$H(\boldsymbol{q}, \nabla_q S) = E. \tag{14.24}$$

14.3.4. Action-angle variables, manifolds of motion

A central *theorem on integrable systems* (Liouville) states the following: Let $H(\boldsymbol{q},\boldsymbol{p})$ describe an autonomous canonical system, and let $\{H \equiv g_1(x), g_2(x), \ldots, g_f(x)\}$ be independent observables on phase space $\mathbb{P}$, all of which are in involution; i.e., $\{g_i(x), g_k(x)\} = 0$ for all $i, k = 1, \ldots, f$. Then, $S_c = \{x \in \mathbb{P} \mid g_i(x) = c_i,\ i = 1, \ldots, f\}$ is a smooth hypersurface invariant under the flow caused by H. If S_c is compact and connected, it is mapped diffeomorphically onto the torus T^f, which in turn is described by angles $\theta^i \in [0, 2\pi)$. The canonical equations are solved by quadratures; in particular, the equations of motion for the new coordinates become $\dot{\theta}^i = \omega^{(i)}$.

For an integrable system with f integrals in involution whose values are $c_1, \ldots, c_f$, the angular coordinates θ_i parametrizing the torus T^f are called *angle variables*. The *action variables*, canonically conjugate to the angle variables, are

$$I_k(c_1, \ldots, c_f) = \frac{1}{2\pi} \oint_{\Gamma_k} \sum p_i \mathrm{d}q^i,$$

where Γ_k is the image in $\mathbb{P}$ of $(\theta^k \in S^1 | \text{with } \theta^i = \text{const. for all } i \neq k)$. Hence, the manifold of motion of the integrable system is $\Delta_1 \times \ldots \times \Delta_f \times T^f$ with Δ_i intervals on the real axis.

Action-angle variables are of general use in perturbation theory and, more specifically, in applications of the *averaging principle*. In particular, perturbation theory in the neighborhood of integrable systems relies on action-angle variables. To witness, the *theorem of Kolmogorov, Arnol'd, and Moser* asserts that if the frequencies of an integrable system described by a Hamiltonian H_0 are rationally independent and if they are sufficiently irrational, then for small values of μ, the perturbed system described by the Hamiltonian $H = H_0 + \mu H_1$ has predominantly quasiperiodic solutions that differ only slightly from those of H_0; the perturbed system possesses nonresonant tori on which the orbits are dense. (The earlier literature and present state of the art may be traced back from Refs. [11, 12].)

14.4. RIGID BODIES

Barring degeneracies (such as in a rod whose mass is arranged along a straight line), the rigid body in $\mathbb{R}^3$ has $f = 6$ degrees of freedom. In describing the motion of rigid bodies in space, it is convenient to introduce three frames of reference: an *inertial* frame $\mathbf{K}$, a *body-fixed* frame $\overline{\mathbf{K}}$ centered in the center of mass S, and an auxiliary frame $\overline{\mathbf{K}}_0$, which is also attached to the center of mass but whose axes are parallel, at all times, to the axes of the inertial system; cf. Figure 14.4. Although the center of mass S moves like a point particle of mass M, rotational motion is described by $\mathbf{R}(t) \equiv \mathbf{R}\,(\phi(t), \theta(t), \psi(t))$, the rotation relating $\overline{\mathbf{K}}$ to $\overline{\mathbf{K}}_0$, and parametrized, for instance, in terms of Eulerian angles (ϕ, θ, ψ). Thus, the manifold of motion of the rigid body is $\mathbb{R}^3 \times SO(3)$: The center of mass coordinate $\boldsymbol{x}_S$ moves through $\mathbb{R}^3$, and the rotational movement sweeps out the rotation group $SO(3)$.

14.4.1. The inertia tensor

The inertia tensor is defined with respect to the body-fixed frame $\overline{\mathbf{K}}$ by

$$\mathbf{J} := \int \mathrm{d}^3x \; \rho(\boldsymbol{x}) \left[\boldsymbol{x}^2 \mathbb{1} - |\boldsymbol{x}\rangle\langle \boldsymbol{x}|\right] \quad \text{or } J_{ik} = \int \mathrm{d}^3x \; \rho(\boldsymbol{x}) \left[\boldsymbol{x}^2 \delta_{ik} - x_i x_k\right], \tag{14.25}$$

where $\rho(\boldsymbol{x})$ is the mass density. It is symmetric and positive definite (barring degeneracy), its eigenvalues, called *principal moments of inertia*, obey the inequalities

$$I_i > 0, (i = 1, 2, 3), \;\; I_1 + I_2 \geq I_3, \;\; I_2 + I_3 \geq I_1, \;\; I_3 + I_1 \geq I_2.$$

If the inertia tensor is to be evaluated with respect to another body-fixed system $\overline{\mathbf{K}}'$, which is shifted and rotated compared with $\overline{\mathbf{K}}$ by a translation $\boldsymbol{a}$ and a constant rotation $\mathbf{R}$,

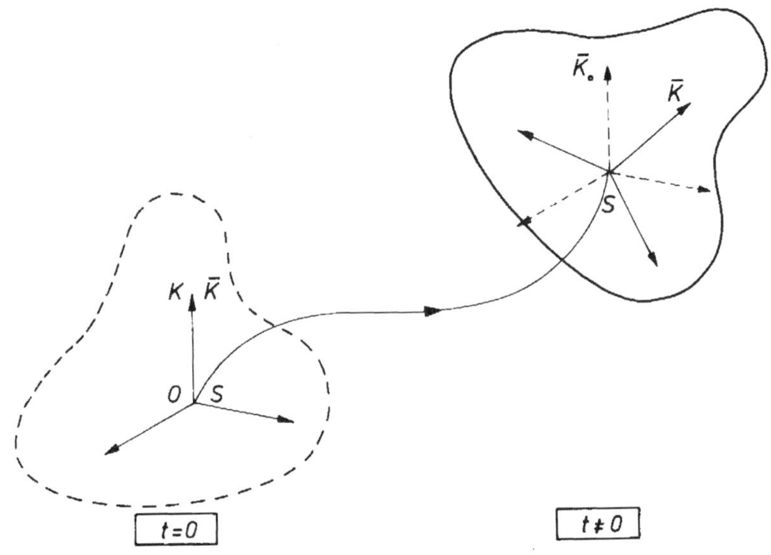

FIGURE 14.4. Motion of a rigid body that starts at $t = 0$ with its center of mass at the origin of the inertial frame **K**. (Figure taken from Ref. [1].)

respectively, it is given by (*Steiner's theorem*)

$$\mathbf{J}' = \mathbf{R}\left(\mathbf{J} + M\left[a^2 \mathbb{1} - |\boldsymbol{a}\rangle\langle\boldsymbol{a}|\right]\right)\mathbf{R}^T. \tag{14.26}$$

The inertia tensor is needed to express the angular momentum $\boldsymbol{L}$ in terms of the angular velocity $\boldsymbol{\omega}$, and the rotational kinetic energy T_{rot} in terms of $\boldsymbol{\omega}$ and of $\boldsymbol{L}$. If all quantities are expressed with respect to, say, the frame of reference $\overline{\mathbf{K}}_0$, these relations read

$$\boldsymbol{L} = \mathbf{J}^{(0)}\,\boldsymbol{\omega}, \tag{14.27}$$

$$T_{\text{rot}} = \tfrac{1}{2}\boldsymbol{\omega}\cdot\mathbf{J}^{(0)}\,\boldsymbol{\omega} = \tfrac{1}{2}\boldsymbol{\omega}\cdot\boldsymbol{L}. \tag{14.28}$$

Note that $\mathbf{J}^{(0)}$, being evaluated in $\overline{\mathbf{K}}_0$, is a function of time. It is related to $\mathbf{J}$ (evaluated in a body-fixed system, hence, constant in time) by Steiner's theorem with $\boldsymbol{a} = 0$, $\mathbf{R}(t)\mathbf{J}^{(0)}(t)\mathbf{R}^T(t) = \mathbf{J}$. When expressed with respect to the body-fixed system $\overline{\mathbf{K}}$, Eqs. (14.27) and (14.28) become $\overline{\boldsymbol{\omega}} = \mathbf{R}(t)\boldsymbol{\omega}$, $\overline{\boldsymbol{L}} = \mathbf{R}(t)\boldsymbol{L}$, $T_{\text{rot}} = \overline{\boldsymbol{\omega}}\cdot\mathbf{J}\,\overline{\boldsymbol{\omega}}/2 = \overline{\boldsymbol{\omega}}\cdot\overline{\boldsymbol{L}}/2$.

These formulae simplify further if $\overline{\mathbf{K}}$ is chosen to be a system of *principal axes*, i.e., such that $\mathbf{J} = \text{diag}\,(I_1, I_2, I_3)$. With this choice, the kinetic energy of rotation becomes $2T_{\text{rot}} = \sum_{k=1}^{3} I_k\overline{\omega}_k^2 = \sum_{k=1}^{3}\overline{L}_k^2/I_k$. The latter relation is useful for visualizing the motion of a force-free, asymmetric top: In the absence of forces, T_{rot} and $\boldsymbol{L}^2 = \sum\overline{L}_k^2$ are conserved. With a convention for labeling the axes such that $I_1 < I_2 < I_3$, one has

$$2T_{\text{rot}}I_1 \le \boldsymbol{L}^2 (= \overline{\boldsymbol{L}}^2) \le 2T_{\text{rot}}I_3. \tag{14.29}$$

For given values of the kinetic energy and of the square of angular momentum, the movement is such that the vector $\overline{L}$ moves along a curve of intersection of a sphere and an ellipsoid. Motion with $\overline{L}^2$ close to the lower or to the upper limit of (14.29) is stable against perturbations, and motion in the neighborhood of $\overline{L}^2 = 2T_{\text{rot}} I_2$ is unstable [1].

14.4.2. Euler's equations

Euler's equations describe the motion of a rigid body under the influence of external forces. They are written with respect to the body-fixed system $\overline{\mathbf{K}}$. With $\overline{D}$ the external torque, all other symbols as before,

$$\mathbf{J}\dot{\overline{\boldsymbol{\omega}}} + \overline{\boldsymbol{\omega}} \times \overline{L} = \overline{D}. \tag{14.30}$$

In the absence of external forces and choosing $\overline{\mathbf{K}}$ to be a principal-axes system, they reduce to three-coupled, nonlinear equations,

$$I_1 \dot{\overline{\omega}}_1 = (I_2 - I_3)\, \overline{\omega}_2 \overline{\omega}_3, \tag{14.31}$$

with cyclic permutation of $(1, 2, 3)$. In this case the system has six integrals of the motion in involution: the energy $H = T_{\text{rot}}$, Eq. (14.28), the momentum $\boldsymbol{P}$ of the center of mass, and the square and one component of angular momentum, $\boldsymbol{L}^2$ and, say, L_3. As $f = 6$, by the theorem on integrable systems, it is solved by quadratures.

If the external forces are potential forces, one has a Lagrangian function and a Hamiltonian function. Written in terms of Eulerian angles, they are

$$L = \frac{1}{2}\Big\{I_1(\dot{\theta}\cos\psi + \dot{\phi}\sin\theta\sin\psi)^2 + I_2(-\dot{\theta}\sin\psi + \dot{\phi}\sin\theta\cos\psi)^2 + I_3(\dot{\psi} + \dot{\phi}\cos\theta)^2\Big\} - U, \tag{14.32}$$

$$H = \frac{1}{2}\left\{\frac{(p_\phi - p_\psi\cos\theta)^2}{\sin^2\theta}\left(\frac{\sin^2\psi}{I_1} + \frac{\cos^2\psi}{I_2}\right) + p_\theta^2\left(\frac{\cos^2\psi}{I_1} + \frac{\sin^2\psi}{I_2}\right) + \frac{\sin\psi\cos\psi}{2\sin\theta} p_\theta(p_\phi - p_\psi\cos\theta)\left(\frac{1}{I_1} - \frac{1}{I_2}\right) + \frac{1}{2I_3}p_\psi^2\right\} + U. \tag{14.33}$$

Here, p_ϕ, p_θ, p_ψ are the momenta canonically conjugate to the Euler angles ϕ, θ, and ψ, respectively. The momentum $p_\phi = L_3$ is the projection of the angular momentum onto the space-fixed 3-axis of $\overline{\mathbf{K}}_0$; $p_\psi = \overline{L}_3$ is its projection onto the body-fixed 3-axis of $\overline{\mathbf{K}}$, and p_θ is its projection onto the intermediate axis in the $(1, 2)$-plane that defines the direction about which the second Euler rotation takes place. Eulerian angles and their conjugate momenta are used in quantizing symmetric or triaxial tops, thus obtaining the rotational spectra known from diatomic molecules and the nuclear collective model.

Apart from center of mass motion, the force-free top moves along geodesics of $SO(3)$. Indeed, for a system with Lagrangian

$$L = T_{\text{rot}} = \frac{1}{2} \sum_{i,k=1}^{f} g_{ik}(\boldsymbol{q}) \dot{q}^i \dot{q}^k,$$

an orbit γ is a solution of the Euler-Lagrange equations (14.12) if and only if it is geodesic on the manifold $(\mathbb{Q}, g)$. The geodesic equation

$$\ddot{q}^i + \sum_{jk} \Gamma^i_{jk}(\gamma) \dot{q}^j \dot{q}^k = 0 \tag{14.34}$$

is equivalent to the Euler-Lagrange equations. In the case of the top, the metric is determined by the inertia tensor $\mathbf{J}$. Considering the geodesic starting at the identity, $\mathbf{R}(t = 0) = \mathbb{1}$, the kinetic energy of rotation is [13, 1]

$$T_{\text{rot}} = \tfrac{1}{2} \overline{\boldsymbol{\omega}} \cdot \mathbf{J} \overline{\boldsymbol{\omega}} = \tfrac{1}{2} \dot{\boldsymbol{\varphi}} \mathbf{J} \dot{\boldsymbol{\varphi}} - \tfrac{1}{2} \dot{\boldsymbol{\varphi}} \cdot \mathbf{J}(\boldsymbol{\varphi} \times \dot{\boldsymbol{\varphi}}) + \mathcal{O}(\varphi^2).$$

Here, $\varphi = |\boldsymbol{\varphi}|$ is the angle of rotation, and $\boldsymbol{\varphi}/\varphi$ is the direction about which the rotation takes place. The geodesic equation, taken at $\varphi = 0$, is equivalent to (14.30) for $\overline{\boldsymbol{D}} = 0$.

14.4.3. Spinning tops

The equations of motion for symmetric tops in the homogeneous gravitational field of Earth, under certain circumstances, are integrable. The classic example is the cone-shaped *children's top* that rotates rapidly on its tip such that the rotational friction on the plane of support can be neglected. The canonical momenta $p_\phi = L_3$ and $p_\psi = \overline{L}_3$ as well as the energy are conserved. The equation of motion is reduced to a one-dimensional equation in the Euler angle $\theta(t)$, under the influence of an effective potential.

Inclusion of frictional forces leads to more complicated equations of motion that require numerical methods. A notable exception is the so-called *tippe top* for which an explicit analysis is known. This top has spherical shape, and its mass distribution is axially symmetric but not spherically symmetric. If it spins rapidly, supported by a horizontal plane, and if its moments of inertia $I_1 = I_2$ and I_3 fulfill the inequalities $(1 - \alpha)I_3 < I_1 < (1 + \alpha)I_3$ (up to terms in mgI_3^2), with α the ratio (distance from center of the sphere to center of mass) over (radius of the sphere), sliding friction with the plane of support drives it into an upside-down, inverted position. The center of mass is lifted (by the action of friction!), and the sense of rotation is the same before and after inversion. The motion manifold in the 10-dimensional configuration space is constrained by the conservation law $L_3 - \alpha \overline{L}_3 = \text{const.}$ (Jelett's integral). In the analysis of its long-term behavior, the total energy is a suitable Liapunov function: Under the above condition on the moments of inertia, any initial configuration is driven to the rotating state with complete inversion [14].

14.5. RELATIVISTIC KINEMATICS

Whenever velocities of mechanical objects are not small compared with the speed of light, $c = 2.99792458 \times 10^8$ m s^{-1}, the group of Galilei transformations (14.1) is replaced by the *Poincaré group* of transformations; this is the covariance group of any physical theory formulated on a flat space time $\mathbb{R}^4$ obeying special relativity theory (SRT). Although the *kinematics* can be dealt with in full generality, the *dynamics* depends very much on the theory one considers. If the back reaction of mass and energy onto the underlying space time cannot be neglected, SRT must be replaced by general relativity theory (GRT) on a four-dimensional curved, semi-Riemannian space time M^4 of type $(1, 3)$. The covariance group is then the full group of diffeomorphisms of M^4, with SRT applying *locally* in normal (or geodesic) coordinates.

14.5.1. Lorentz transformations and decomposition theorem

With relativistic kinematics, inertial frames are related by Poincaré transformations (Λ, a), acting on space-time variables $x^\mu = (x^0 = ct, \boldsymbol{x})$

$$(\Lambda, a): \quad x^\mu \longmapsto x'^\mu = \Lambda^\mu{}_\sigma x^\sigma + a^\mu, \tag{14.35}$$

(summation over repeated contragredient indices is implied.) Its translational part is the same as in (14.1); rotations $\mathbf{R}$ in 3-space are also the same. However, the special transformations (or *boosts*) that connect two frames moving at constant velocity relative to each other are different. The homogeneous part of (14.35) is the *Lorentz group* and is characterized by the requirement that it leave invariant the scalar product

$$(x, y) = x^0 y^0 - \boldsymbol{x} \cdot \boldsymbol{y} = x^\mu g_{\mu\nu} y^\nu, \tag{14.36}$$

with $g \equiv \{g_{\mu\nu}\} = \text{diag}\,(1, -1, -1, -1)$. This is equivalent to the condition

$$\Lambda^T g \Lambda = g. \tag{14.37}$$

The four possible choices $\det \Lambda = \pm 1$ and $\Lambda^0{}_0 \geq 0$ or $\Lambda^0{}_0 \leq 0$ indicate that the Lorentz group has four disjoint branches, $L_+^\uparrow$, $L_-^\uparrow$, $L_-^\downarrow$, $L_+^\downarrow$, among which only the *proper, orthochronous Lorentz group* $L_+^\uparrow$ is a subgroup. $L_-^\uparrow$ contains the space reflection $\mathbf{P}$, $L_-^\downarrow$ contains time reversal $\mathbf{T}$, and $L_+^\downarrow$ contains the product $\mathbf{PT}$. The latter three branches are obtained from the first by multiplying the elements $\Lambda \in L_+^\uparrow$ by $\mathbf{P}$, $\mathbf{T}$, and $\mathbf{PT}$, respectively.

Rotations in 3-space have the form

$$\mathcal{R} = \begin{pmatrix} 1 & 0 \\ 0 & \mathbf{R} \end{pmatrix} \quad \text{with } \mathbf{R} \in SO(3) \tag{14.38}$$

and form a subgroup of $L_+^\uparrow$. Denoting a vector in $\mathbb{R}^3$ by $|\boldsymbol{v}\rangle$ if it appears in a column, and by $\langle \boldsymbol{v}|$ its dual when it appears in a row (so that $\langle \boldsymbol{a}|\boldsymbol{b}\rangle = \boldsymbol{a} \cdot \boldsymbol{b}$ is the scalar product), special Lorentz transformations (or boosts) along a given 3-velocity $\boldsymbol{v}$ are given by

$$\mathbf{L}(\boldsymbol{v}) = \begin{pmatrix} \gamma & (\gamma/c)\langle \boldsymbol{v}| \\ (\gamma/c)|\boldsymbol{v}\rangle & \mathbb{1}_{3\times 3} + \gamma^2/((\gamma+1)c^2)|\,\boldsymbol{v}\rangle\langle \boldsymbol{v}\,| \end{pmatrix}, \gamma = \frac{1}{\sqrt{1 - \boldsymbol{v}^2/c^2}}. \tag{14.39}$$

Composition of two boosts $\mathbf{L}(v_1 \hat{\boldsymbol{n}})$ and $\mathbf{L}(v_2 \hat{\boldsymbol{n}})$ along the same direction yields another boost along the same direction with velocity $v_{12} = (v_1 + v_2)/(1 + v_1 v_2/c^2)$ (addition theorem for parallel relativistic velocities). If the two boosts act along different directions, their composition yields a rotation followed by a boost whose $(0, 0)$-component is $\gamma_{12} = \gamma_1 \gamma_2 (1 + \boldsymbol{v}_1 \cdot \boldsymbol{v}_2/c^2)$. As long as $\gamma_i \geq 1$, one concludes that also $\gamma \geq 1$; the speed of light is never exceeded.

The *decomposition theorem* asserts that every transformation $\Lambda \in L_+^\uparrow$ can be written uniquely as the product of a rotation, followed by a special Lorentz transformation,

$$\Lambda = \mathbf{L}(\boldsymbol{v})\mathcal{R}, \tag{14.40}$$

with

$$\frac{v^i}{c} = \frac{\Lambda^i{}_0}{\Lambda^0{}_0}, \quad R^{ik} = \Lambda^i{}_k - \frac{\Lambda^i{}_0\Lambda^0{}_k}{1+\Lambda^0{}_0}. \tag{14.41}$$

If the order in the decomposition (14.40) is chosen differently, i.e., $\Lambda = \mathcal{R}'\mathbf{L}(\boldsymbol{v}')$, the rotation is the same as in (14.40), $\mathcal{R}' = \mathcal{R} = \text{diag}\,(1, \mathbf{R})$, whereas $\boldsymbol{v}$ and $\boldsymbol{v}'$ are related by $\boldsymbol{v} = \mathbf{R}\boldsymbol{v}'$.

14.5.2. Causal orbits, energy, and momentum

Every physical trajectory of a massive particle in space-time has a line element

$$\mathrm{d}s^2 = c^2\mathrm{d}\tau^2 = c^2(\mathrm{d}t)^2 - (\mathrm{d}\boldsymbol{x})^2, \tag{14.42}$$

which is everywhere *timelike*, i.e., positive with our choice of metric, in accordance with $|\boldsymbol{v}| < c$. Here, τ is the proper time, and t and $\boldsymbol{x}$ are time and space coordinates with respect to a frame **K**. This is an expression of *causality* that implies that no physical signal can propagate at speeds exceeding the speed of light. At every point of its trajectory, the particle's future must be in the interior of the local forward light cone, and its past must be in the interior of the backward light cone. With m the (invariant) rest mass and $x^\mu(\tau)$ the orbit in spacetime, as a function of proper time, the four-momentum is

$$p^\mu = m\frac{\mathrm{d}}{\mathrm{d}\tau}x^\mu(\tau) \equiv m\dot{x}^\mu(\tau). \tag{14.43}$$

The invariant square of the four-velocity is normalized to $\dot{x}^2 = c^2$; cf. (14.42). When evaluated in a frame of reference **K** (not necessarily an inertial frame), p^μ decomposes in conventional energy and three-momentum variables:

$$p^0 = mc\gamma = E/c, \quad \boldsymbol{p} = m\gamma\boldsymbol{v}, \tag{14.44}$$

which are related by the (invariant) *mass shell condition* (cf. Figure 14.5),

$$p^2 = (p^0)^2 - \boldsymbol{p}^2 = (mc)^2 \quad \text{or} \quad E^2 - c^2\boldsymbol{p}^2 = (mc^2)^2. \tag{14.45}$$

The quantity mc^2 is the *rest energy*, and the kinetic energy is $T^{(\text{rel})} = E - mc^2$. For moderately relativistic motion, it is $T^{(\text{rel})} \approx (\boldsymbol{p}^2/2m)\left(1 - (\boldsymbol{p}^2/4m^2c^2)\right)$.

For *massless* particles such as photons and neutrinos, the four-momentum

$$p = (E/c = |\boldsymbol{p}|, \boldsymbol{p}) \tag{14.46}$$

is always lightlike, $p^2 = 0$, the modulus of the velocity $\boldsymbol{v} = \boldsymbol{p}c^2/E$ equals c, and the kinetic energy equals the total energy E.

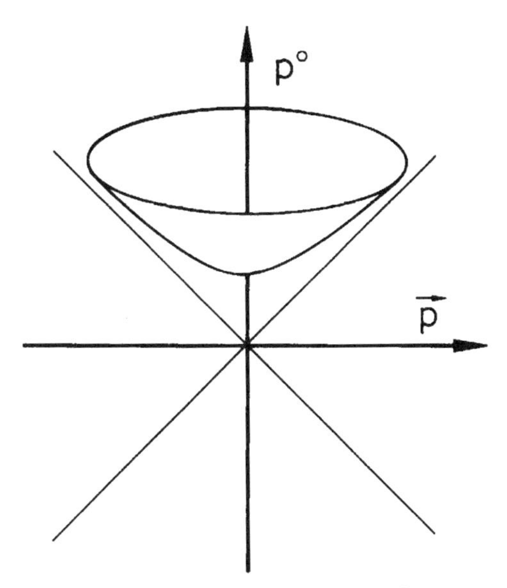

FIGURE 14.5. The space components $\boldsymbol{p}$ and the time component p^0 of the four-momentum of a particle with mass m lie on the hyperboloid (14.45). As the energy must be positive, only the upper half of the hyperboloid is physically relevant. (Figure taken from Ref. [1].)

Kinematics in reactions in which not all velocities are small compared with c is best evaluated using Lorentz-invariant quantities. In a two-body reaction $A+B \to C+X$, where A, B, and C are single particles, while X may be another single particle or a multiparticle state, useful invariant variables are

$$s := c^2(p^{(A)} + p^{(B)})^2, \quad t := c^2(p^{(A)} - p^{(C)})^2. \tag{14.47}$$

In the center of mass frame, s is the square of the total energy, and t is the invariant momentum transfer squared and is proportional to $1 - \cos\theta^*$ with θ^* the scattering angle. In a frame in which B is at rest (laboratory frame), $s = (m_A^2 + m_B^2)c^4 + 2m_B c^2 E^{(A)}$.

14.5.3. Time dilatation and scale contraction

A time interval $\Delta\tau$ in proper time (as measured on a comoving clock) is registered by an observer who moves at relative velocity $\boldsymbol{v}$, in the space-time interval $(\Delta t, \Delta\boldsymbol{x} = \boldsymbol{v}\Delta t)$ such that $\sqrt{(\Delta t)^2 - (\Delta\boldsymbol{x})^2/c^2} = \Delta t\sqrt{1 - (\boldsymbol{v}/c)^2} = \Delta\tau$. The apparent *time dilatation* observed,

$$\Delta t = \gamma\Delta\tau, \tag{14.48}$$

is proportional to $\gamma = 1/\sqrt{1 - (\boldsymbol{v}/c)^2}$. The effect is observable, for instance, by in-flight decays of unstable elementary particles. The lifetime, as observed in the measuring apparatus, is scaled by the factor $\gamma \in [1, \infty)$ compared with the true lifetime as measured in the particle's rest frame.

Scale contraction, although intimately related to time dilatation, is more difficult to visualize and to verify empirically. The phenomenon is this: Let two points $x = (ct, \boldsymbol{x})$ and $y = (ct, \boldsymbol{y})$ in space time have vanishing relative velocity, choose a frame of reference in which both are at rest, and let l_0 be their distance in that frame. A Galileian observer traveling with constant velocity $\boldsymbol{v}$ from x to y arrives there after (coordinate) time Δt and concludes that $l_0 = |\boldsymbol{v}|\Delta t$. A Lorentzian observer, on the other hand, reaches y after the interval of his proper time $\Delta\tau = \Delta t/\gamma$ and concludes that the distance is

$$l = |\boldsymbol{v}|\Delta\tau = l_0/\gamma, \quad \text{with } \gamma = \frac{1}{\sqrt{1-(v/c)^2}}. \tag{14.49}$$

To work out the apparent quenching of a macroscopic geometric object as seen from a moving observer is more difficult because one must take into account the finite propagation time of every signal traveling from the object to the observer. Qualitatively speaking, the object appears contracted in the direction of the relative velocity but it also appears rotated [15, 16].

14.5.4. Motion of free particles in SRT and GRT

Force-free, relativistic motion of a point particle with rest mass m in *flat* space is described by the action

$$I = -mc\int_x^y \mathrm{d}s = -mc\int_{\tau(x)}^{\tau(y)} \mathrm{d}\tau\,\sqrt{\frac{\mathrm{d}x^\mu}{\mathrm{d}\tau}\frac{\mathrm{d}x_\mu}{\mathrm{d}\tau}}, \tag{14.50}$$

with $\mathrm{d}s$ the line element and τ the proper time. Up to gauge transformations, the invariant Lagrangian function is

$$L_{\text{inv}} = -mc\sqrt{\frac{\mathrm{d}x^\mu}{\mathrm{d}\tau}\frac{\mathrm{d}x_\mu}{\mathrm{d}\tau}} = -mc\sqrt{\dot{x}^2}, \tag{14.51}$$

from which follows the conjugate momentum $p^\mu = \partial L/\partial\dot{x}_\mu = -mc\dot{x}^\mu/\sqrt{\dot{x}^2}$. The Euler-Lagrange equations (14.12) yield the expected equation of motion

$$mc\frac{\mathrm{d}}{\mathrm{d}\tau}\frac{\dot{x}^\mu}{\sqrt{\dot{x}^2}} = 0 \quad \text{or} \quad \frac{\mathrm{d}}{\mathrm{d}\tau}\,\dot{p}^\mu = 0.$$

The canonical momentum p fulfills the constraint $\Psi(p) := p^2 - m^2c^2 = 0$.

In passing to the Hamiltonian formulation by means of Legendre transformation with respect to $\dot{x}^\mu$, the condition (14.17) is found to be violated. The Jacobi matrix of second derivatives with respect to $\dot{x}$ vanishes. As a consequence, the Hamiltonian function $H = \dot{x}^\mu p_\mu - L_{\text{inv}}$ is zero. This is an example of a canonical system whose coordinates are not independent. It is solved by introducing the constraint via a Lagrange multiplier λ,

$$H' = H + \lambda\Psi(p),$$

from which $\dot{x}^\mu = \{H', x^\mu\} = 2\lambda p^\mu$, $\dot{p}^\mu = \{H', p^\mu\} = 0$, and $\lambda = -\sqrt{\dot{x}^2}/(2mc)$.

In *general relativity*, single-particle motion obeys the geodesic equation [17, 18]

$$\ddot{x}^{\lambda}(\tau) + \sum_{\mu\nu} \Gamma^{\lambda}_{\mu\nu} \dot{x}^{\mu}(\tau) \dot{x}^{\nu}(\tau) = 0 \tag{14.52}$$

on a semi-Riemannian manifold with dimension 4 and index 1 [19]. The Christoffel symbols are given in terms of the metric $g_{\mu\nu}$ and its inverse by

$$\Gamma^{\lambda}_{\mu\nu} = \frac{1}{2} \sum_{\sigma} g^{\lambda\sigma} \left\{ \frac{\partial g_{\nu\sigma}}{\partial x^{\mu}} + \frac{\partial g_{\mu\sigma}}{\partial x^{\nu}} - \frac{\partial g_{\mu\nu}}{\partial x^{\sigma}} \right\}. \tag{14.53}$$

Ray-tracing programs in four dimensions and visualization methods are found, e.g., in Ref. [16]. Important, analytically solvable cases are the perihelion advance of planets and light deflection in the exterior gravitational field of a point-like mass M (classic tests of GRT). The metric (*Schwarzschild metric*) is, in spherical coordinates,

$$g = \left(1 - \frac{2\mu}{r}\right) \mathrm{d}t \otimes \mathrm{d}t - \frac{\mathrm{d}r \otimes \mathrm{d}r}{1 - 2\mu/r} - r^2 \left(\mathrm{d}\theta \otimes \mathrm{d}\theta + \sin^2\theta \, \mathrm{d}\phi \otimes \mathrm{d}\phi\right), \tag{14.54}$$

with $\mu = GM/c^2$ (as an example the Sun has $\mu = 1.5$ km). For a massive particle moving through this static field, a Lagrangian is

$$\mathcal{L} = \frac{1}{2} \sum_{\mu\nu} g_{\mu\nu} \dot{x}^{\mu} \dot{x}^{\nu} = \tfrac{1}{2} c^2, \tag{14.55}$$

where the second equality holds when evaluated along solutions. Like in the case of $\mathbb{R}^3$, Sec. 14.4.2, the Euler-Lagrange equations are identical with the geodesic equations (14.52). With $\sigma(\phi) := 1/r(\phi)$, taking $\theta = \pi/2$, $\dot{\theta} = 0$, and with $L = r^2\dot{\phi}$, the conserved angular momentum per unit mass, the equation of motion is $\mathrm{d}\sigma/\mathrm{d}\phi = 0$ (for circular motion), or

$$\frac{\mathrm{d}^2\sigma}{\mathrm{d}\phi^2} + \sigma = \frac{GM}{L^2} + 3\mu\sigma^2 \tag{14.56}$$

(for noncircular motion). The equivalent Newtonian potential is $\Phi(r) = -GM/r - \mu L^2/r^3$. The resulting perihelion advance per revolution is

$$\tan\delta \approx \delta \approx \frac{6\pi\mu}{a(1-\varepsilon^2)} \tag{14.57}$$

with a the semimajor axis and ε the eccentricity of the Kepler ellipse. In the case of Mercury, $a = 5.8 \times 10^7$ km, $\varepsilon = 0.206$, giving an advance of $\sim 43''$ per century.

For massless particles, the null geodesics follow by putting $\mathcal{L} = 0$. The equation of motion is $\mathrm{d}\sigma/\mathrm{d}\phi = 0$ (circular orbits), or

$$\frac{\mathrm{d}^2\sigma}{\mathrm{d}\phi^2} + \sigma = 3\mu\sigma^2 \tag{14.58}$$

(other orbits). For a grazing trajectory on a spherically symmetric mass distribution with radius b, the total deflection is approximately $\Delta \approx 4\mu/b$ (example: Sun: $b = 7 \times 10^5$ km, $\Delta \approx 1.7''$).

14.6. HAMILTONIAN DYNAMICAL SYSTEMS

Qualitative dynamics deals with the long-term behavior of dynamical systems and the global properties of their flows. It investigates (classic) chaotic behavior, identifies routes to chaos, and defines quantitative measures of chaoticity. As these matters are treated in Chapter 17, we restrict this section to the case of Hamiltonian systems.

14.6.1. Long-term behavior of mechanical systems

A point x_0 is an equilibrium, or *critical point*, of the autonomous dynamical system $\dot{x} = \mathcal{F}(x)$ if $\mathcal{F}(x_0) = 0$. Its nature is determined by linearizing Eq. (14.9) in $y = x - x_0$, giving

$$\dot{y} = (\mathbf{D}\mathcal{F})|_{x=x_0}\, y \tag{14.59}$$

and by determining the eigenvalues of the matrix $(\mathbf{D}\mathcal{F}) =: \mathbf{A}$ (*characteristic exponents*). (1) If all eigenvalues α_i of $\mathbf{A}$ fulfill $\operatorname{Re}\alpha_i < -c < 0$, orbits in a neighborhood of x_0 converge to x_0 uniformly and at an exponential rate. (2) If x_0 is a stable equilibrium position, none of the eigenvalues of $\mathbf{A}$ has a positive real part. (3) Hamiltonian systems $\dot{x} = \mathbf{J}H_{,x}$ have linearizations (14.59) with

$$\mathbf{A} = (\mathbf{D}\mathcal{F}) = \mathbf{JB}, \quad \text{where } B_{ik} = \left.\frac{\partial^2 H}{\partial x^i \partial x^k}\right|_{x_0};$$

the matrix $\mathbf{A}$ fulfills $\mathbf{A}^T\mathbf{J} + \mathbf{JA} = 0$. This is the local version of Eq. (14.20), with $\mathbf{M} \simeq \mathbb{1} + \mathbf{A}$; hence, $\mathbf{A}$ is called *infinitesimally symplectic*. Then, if α is an eigenvalue of $\mathbf{A}$ with multiplicity k, so is $-\alpha$, with the same multiplicity. If an eigenvalue is zero, its multiplicity is even. Canonical systems cannot have asymptotically stable equilibria.

The stability of a critical point x_0 of a dynamical system (14.9) is decided by means of a *Liapunov function* that, by definition, vanishes in x_0 and is positive everywhere in an open neighborhood of x_0. If the orbital derivative of the Liapunov function is negative or zero, x_0 is stable. A more refined arsenal of stability criteria exists for dynamical systems having both equilibria and periodic orbits as *critical elements* [1, 7, 20].

An instructive Hamiltonian example is provided by the stability analysis of Euler's equations (14.31) for a triaxial top with $I_3 > I_2 > I_1$ close to the periodic orbits (A): $\overline{\boldsymbol{\omega}}_0^{(1)} = (\omega_1, 0, 0)$, (B): $\overline{\boldsymbol{\omega}}_0^{(2)} = (0, \omega_2, 0)$, (C): $\overline{\boldsymbol{\omega}}_0^{(3)} = (0, 0, \omega_3)$, where the top spins about one of the principal axes [1]. Linearizing near $\overline{\boldsymbol{\omega}}_0^{(2)}$, one characteristic exponent is zero, $\alpha_1^{(2)} = 0$, whereas the other two are real, $\alpha_2^{(2)} = -\alpha_3^{(2)} = \omega_2\sqrt{(I_3 - I_2)(I_2 - I_1)/I_1 I_3}$. As one of these has positive real part, the periodic orbit (B) is unstable. The characteristic exponents for the orbits (A) and (C) are either zero or pure imaginary. For example, at $\overline{\boldsymbol{\omega}}_0^{(1)}$ $\alpha_1^{(1)} = 0$, $\alpha_2^{(1)} = -\alpha_3^{(1)} = i\omega_1\sqrt{(I_2 - I_1)(I_3 - I_1)/I_2 I_3}$. Defining suitable Liapunov functions [1] in the vicinity of $\overline{\boldsymbol{\omega}}_0^{(1)}$ and in the vicinity of $\overline{\boldsymbol{\omega}}_0^{(3)}$, respectively, one finds that their orbital derivatives vanish, thus proving that the periodic orbits (A) (smallest moment of inertia) and (C) (largest moment of inertia) are stable.

14.6.2. Deterministic chaos in Hamiltonian systems

A common feature of chaotic regimes of classical systems is the *sensitive dependence on initial conditions* [21]. Close-by trajectories move apart exponentially, long-term behavior becomes unpredictable, and motion appears irregular. Although Hamiltonian systems are constrained by the conservation of phase volume (Liouville's theorem, Sec. 14.3.3), deterministic chaos can be present already in low dimensions. Quantitative measures of chaos are provided by the *Liapunov characteristic exponents* for which algorithms and methods of computation exist. If the leading characteristic exponent λ_1 is positive, there is at least one direction along which neighboring trajectories, on average, move apart at a rate $\exp(\lambda_1 t)$. There is sensitive dependence on initial conditions and, hence, chaotic behavior.

Celestial mechanics of the solar system provides striking examples for chaotic motion in Hamiltonian systems:

1. The spin dynamics of asymmetric ($I_1 \neq I_2 \neq I_3$) satellites on elliptic Kepler orbits around their mother planet is strongly chaotic [1, 24]. A prominent example is Hyperion, a satellite of Saturn, whose erratic tumbling was predicted and was observed by direct observation from Earth [25].
2. Extensive numerical studies indicate that the Kirkwood gaps in the asteroid belt between Mars and Jupiter are due to chaotic motion. The eccentricity of elliptic orbits close to rational ratios of periods (asteroid versus Jupiter), such as 3:1 or 2:1, varies in an irregular way as a function of time. The asteroids make long excursions (over hundreds of thousands of years) to orbits with large eccentricity that come close to Mars, or even to Earth, and are scattered out of their orbits. This interpretation of the Kirwood gaps is corroborated by the result that for other ratios such as 3:2, where there is little chaoticity, no gap is found.
3. Numerical studies of the long-term stability of the solar system [26] suggest that the motion of Pluto is chaotic. This planet differs from the other outer planets by the large eccentricity and high inclination of its orbit. The time scale for exponential divergence of neighboring orbits is found to be remarkably short, of the order of 20 million years.
4. The present state of long-term numerical studies of the solar system is reviewed in Ref. [27]. It is shown, in particular, that the obliquity of the Earth, like those of other planets, is intrinsically subject to erratic variations with, potentially, dramatic consequences for terrestial climates. These chaotic variations are found to be stabilized by the presence of the Moon [28].

14.7. REFERENCES

[1] F. Scheck, *Mechanics—From Newton's Laws to Deterministic Chaos*, 3d edition (Springer, New York, 1999).

[2] H. Goldstein, *Classical Mechanics* (Addison-Wesley, Reading, MA, 1980).

[3] W. Thirring, *A Course in Mathematical Physics, Vol. 1: Classical Dynamical Systems* (Springer, Heidelberg, 1992).

[4] V. I. Arnol'd, *Mathematical Methods of Classical Mechanics* (Springer, New York, 1989).

[5] V. I. Arnol'd, *Ordinary Differential Equations* (Springer, New York, 1992).

[6] L. D. Landau and E. M. Lifshitz, *Mechanics* (Pergamon, Oxford, and Addison-Wesley, Reading, MA, 1992).

[7] R. Abraham and J. E. Marsden, *Foundations of Mechanics* (Benjamin Cummings, Reading, MA, 1981).

[8] J. E. Marsden and T. S. Ratiu, *Introduction to Mechanics and Symmetry* (Springer, New York, 1994).

[9] P. A. M. Dirac, *Lectures on Quantum Mechanics* (Belfer Graduate School of Science, New York, 1964); C. A. Hurst, in *Proceedings of the XXVI Int. Universitätswochen für Kernphysik, Schladming 1987*, edited by H. Mitter and L. Pittner (Springer, Heidelberg, 1987), p. 18; L. D. Faddeev and A. A. Slavnov, *Gauge Fields: An Introduction to Quantum Theory* (Addison-Wesley, Redwood City, CA, 1991).

[10] H. Hofer and E. Zehnder, *Symplectic Invariants and Hamiltonian Dynamics* (Birkhäuser Advanced Texts, Basel, 1994).

[11] H. Rüssmann, *Invariant Tori in the Perturbation Theory of Weakly Non-Degenerate Integrable Hamiltonian Systems* (Department of Mathematics, Mainz University, 1998).

[12] L. H. Eliasson, Math. Phys. Electron. J. **2** (1996).

[13] V. I. Arnol'd, Ann. Inst. Four. **16**, 319 (1966).

[14] St. Ebenfeld and F. Scheck, Ann. Phys. (N.Y.) **243**, 195 (1995).

[15] G. F. R. Ellis and R. M. Williams, *Flat and Curved Space-Times* (Clarendon Press, Oxford, 1994).

[16] H.-P. Nollert, U. Kraus, and H. Ruder, in *Relativity and Scientific Computing*, edited by F. W. Hehl, R. A. Puntigam, and H. Ruder (Springer, Heidelberg, 1996).

[17] S. Weinberg, *Gravitation and Cosmology* (John Wiley and Sons, New York, 1972).

[18] Ch. W. Misner, K. S. Thorne, and J. A. Wheeler, *Gravitation* (W. H. Freeman and Company, San Francisco, 1973).

[19] B. O'Neill, *Semi-Riemannian Geometry, With Applications to Relativity* (Academic Press, New York, 1983).

[20] J. Guckenheimer and P. Holmes, *Nonlinear Oscillations, Dynamical Systems, and Bifurcations of Vector Fields* (Springer, Berlin and Heidelberg, 1990).

[21] S. Newhouse, D. Ruelle, and F. Takens, Comm. Math. Phys. **64**, 35 (1978); D. Ruelle and F. Takens, Comm. Math. Phys. **20**, 167 (1971) and **23**, 343 (1971).

[22] R. L. Devaney, *An Introduction to Chaotic Dynamical Systems* (Addison-Wesley, Reading, MA, 1989).

[23] P. Bergé, Y. Pomeau, and Ch. Vidal, *Order within Chaos; Towards a Deterministic Approach to Turbulence* (John Wiley and Sons, New York, 1987).

[24] J. Wisdom, Nucl. Phys. **B2** (Proc. Suppl.), 391 (1987).

[25] J. Klavetter, Science **246**, 998 (1989); Astron. J. **98**, 1855 (1989).

[26] G. J. Sussman and J. Wisdom, Science **241**, 433 (1988).

[27] J. Laskar, in *Proceedings of the XI-th International Conference on Mathematical Physics* (International Press, Paris, 1995); see also J. Laskar, Celestial Mech. **64**, 115 (1995).

[28] J. Laskar, F. Joutel, and P. Robutel, Nature **361**, 615 (1993).

15

Medical Physics

William R. Hendee
Medical College of Wisconsin, Milwaukee, Wisconsin

Michael Yester
University of Alabama at Birmingham, Birmingham, Alabama

Contents

List of Tables

List of Figures

15.1. INTRODUCTION

Medical physics is largely concerned with the use of various forms of ionizing and non-ionizing radiation to detect, diagnose, and treat human illness and injury. In these applications, radiation interacts with biological tissues according to fundamental characteristics of the tissues. Some of these characteristics are depicted in Table 15.1, where the elemental composition in grams is described in shorthand scientific notation (e.g., $3.5 + 03 = 3.5 \times 10^3 = 3500$).

TABLE 15.1. Characteristics of biological tissues.

	Organ							
	Adipose	Blood	Brain	Heart	Kidney	Liver	Muscle	Pancreas
Mass (g)	15,055	5394	1400	330	310	1800	28,000	100
Mass density (g/cm^3)	0.92	1.06	1.03	1.03	1.05	1.05	1.04	1.05
Electron density (electrons/g)	0.513	0.584	0.567	0.568	0.567	0.583	0.573	0.581
	Elemental Composition							
Calcium (20)	3.4 − 01	3.1 − 01	1.2 − 01	1.2 − 02	2.9 − 02	9.0 − 02	8.7 − 01	9.1 − 03
Carbon (6)	9.6 + 03	5.4 + 02	1.7 + 02	5.4 + 01	4.0 + 01	2.6 + 02	3.0 + 03	1.3 + 01
Chlorine (17)	1.8 + 01	1.5 + 01	3.2 + 00	5.4 − 01	7.4 − 01	3.6 + 00	2.2 + 01	1.6 − 01
Hydrogen (1)	1.8 + 03	5.5 + 02	1.5 + 02	3.4 + 01	3.2 + 01	1.8 + 02	2.8 + 03	9.7 + 00
Iron (26)	3.6 − 01	2.5 + 00	7.4 − 02	1.5 − 02	2.3 − 02	3.2 − 01	1.1 + 00	3.9 − 03
Magnesium (12)	3.0 − 01	2.1 − 01	2.1 − 01	5.4 − 02	4.0 − 02	3.1 − 01	5.3 + 00	1.6 − 02
Nitrogen (7)	1.2 + 02	1.6 + 02	1.8 + 01	8.8 + 00	8.5 + 00	5.1 + 01	7.7 + 02	2.1 + 00
Oxygen (8)	3.5 + 03	4.1 + 03	1.0 + 03	2.3 + 02	2.3 + 02	1.2 + 03	2.1 + 04	6.7 + 01
Phosphorus (15)	2.2 + 00	1.9 + 00	4.8 + 00	4.8 − 01	5.0 + 01	4.7 + 00	5.0 + 01	2.3 − 01
Potassium (19)	4.8 + 00	8.8 + 00	4.2 + 00	7.2 − 01	5.9 − 01	4.5 + 00	8.4 + 01	2.3 − 01
Sodium (11)	7.6 + 00	1.0 + 01	2.5 + 00	4.0 + 00	6.2 − 01	1.8 + 00	2.1 + 01	1.4 − 01
Sulfur (16)	1.1 + 00	5.5 + 00	2.4 + 00	5.4 − 01	0.0 + 00	5.2 + 00	6.7 + 01	
Zinc (30)	2.7 − 02	3.4 − 02	1.7 − 02	8.4 − 03	1.5 − 02	8.5 − 02	1.5 + 00	2.5 − 03

15.1.1. Imaging

Medical images depict certain properties such as opacity or reflectivity that are influenced directly by fundamental tissue characteristics, such as those described in Table 15.2. Images may be formed by radiation from a variety of sources. Image properties, tissue characteristics, and radiation sources available for medical imaging are listed in Table 15.2.

TABLE 15.2. Thumbnail sketch of medical imaging.

Image properties	Image influences	Image sources
Transmissivity	Mass density	X-rays
Opacity	Electron density	γ-rays
Emissivity	Proton density	Visible light
Reflectivity	Atomic number	Ultraviolet light
Conductivity	Velocity	Annihilation radiation
Magnetizability	Pharmaceutical location	Electric fields
	Current flow	Magnetic fields
	Relaxation	Infrared
		Ultrasound
		Applied voltage

15.2. IONIZING RADIATION: X- AND γ-RAYS

X- and γ-rays are major forms of medical radiation. Various units are used to characterize such radiation. Some of the more common ones are defined in Table 15.3.

15.2.1. Interaction of x- and γ-rays with tissue

Three principal mechanisms—photoelectric absorption, Compton scattering, and pair production—account for interactions in tissue over the x-ray energy range used for diagnosis and therapy. The prevalence of each of these interactions as a function of x-ray energy is shown in Figure 15.1. Figure 15.2 shows which interaction is prevalent in water (or muscle) at a given energy.

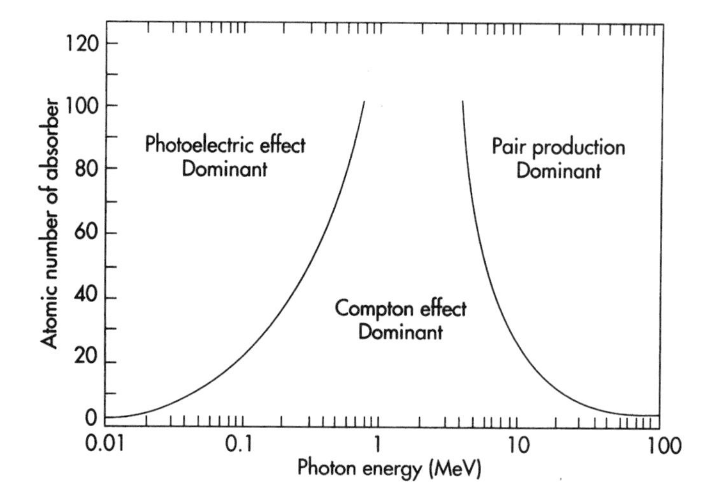

FIGURE 15.1. Relative importance of the three major types of γ-ray interaction. The lines show the values of Z and $h\nu$ for which two neighboring effects are equal.

TABLE 15.3. Radiation units (with permission from Ref. [1]).

Quantity	Symbol	Meaning	SI unit abbreviation	Fundamental units or conversions
Exposure	X	charge liberated per unit mass	C/kg	Q/m 1 Roentgen = 2.58×10^{-4} C/kg
Energy imparted	ε	net energy absorbed in matter (see text)	*	J
Absorbed dose	D	energy imparted per unit mass	gray/Gy	J/kg 1 Gy = 100 rad 1 Gy = 10^4 erg/g (kerma has same definition)
Absorbed dose rate	$\dot{D}$	D/t	*	J/kg.s
Quality factor	Q	weighting factor for varying biological effectiveness of radiation	*	
Dose equivalent	H	QD	sievert/Sv	1 Sv = 100 rem
Dose equivalent rate	$\dot{H}$			
Activity	A	disintigrations/second	becquerel/Bq	s^{-1} 1 Ci = 3.7×10^{10} Bq
Fluence	Φ	particles/unit area	*	m^{-2}
Fluence rate or flux density	ϕ	fluence/unit time	*	$m^{-2} \cdot s^{-1}$
Energy fluence	Ψ	energy/unit area	*	J/m^2
Intensity or energy flux	ψ	energy fluence/unit time	*	J/m^2s
Linear energy transfer	LET	energy lost by particle per unit pathlength	*	J/m
Lineal energy	y	energy imparted per mean chord length through volume	*	J/m

*No SI unit applies.

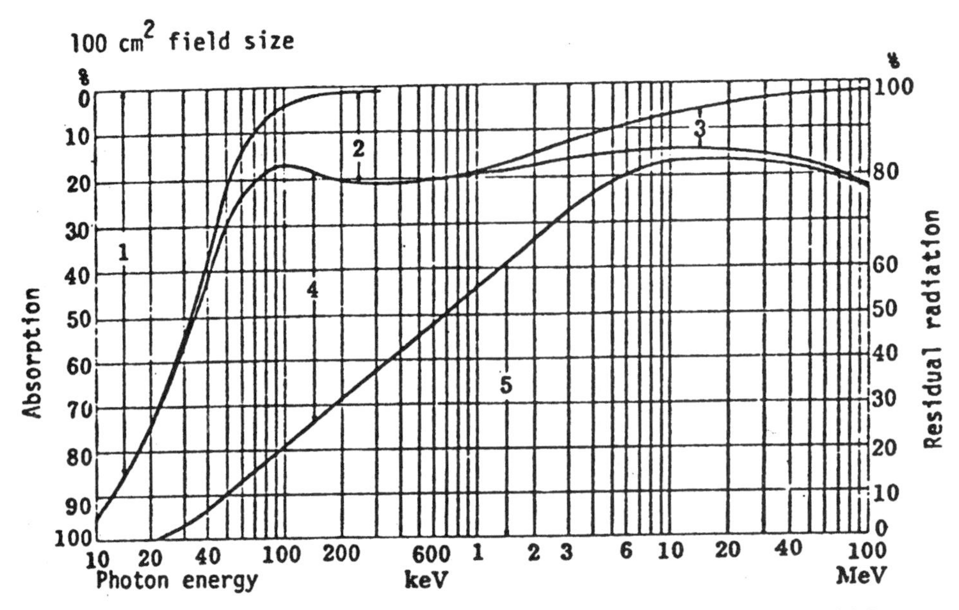

FIGURE 15.2. Attenuation of x-rays in a 10-cm layer of water: (1) photoelectric absorption, (2) Compton absorption, (3) pair production, (4) scattering, and (5) transmitted primary radiation. For example, if a 100-keV x-ray beam penetrates a 100-cm slab of water, 20% of the photons will be transmitted, 63% will be scattered, 12% will undergo Compton interaction, and 5% will undergo photoelectric absorption. (Used with permission from Ref. [2].)

X- and γ-ray beams are attenuated by absorption and scattering as they traverse matter. The attenuation is expressed as the mass attenuation coefficient μ/ρ in units of cm^2/g. This coefficient for water (which closely simulates muscle) is shown in Figure 15.3 as a function of x- or γ-ray energy in MeV. The mass energy absorption coefficient μ_{en}/ρ, also given in cm^2/g, is the fraction of x- or γ-ray energy absorbed per unit length (g/cm^2) of absorber. This coefficient for selected materials, and tissues as a function of x- or γ-ray energy is in Table 15.4.

The f factor is the energy absorbed per unit mass of material for a given exposure of the material to x- or γ-rays. This factor is $0.869[(\mu_{en}/\rho)_{med}/(\mu_{en}/\rho)_{air}]$, where the subscripts "med" and "air" designate that the mass energy absorption coefficients are for the material and for the air, respectively. The f factors for water, muscle, and bone as functions of x-ray energy are shown in Figure 15.4.

The half-value layer (HVL) is the amount of material in an x-ray beam required to reduce the exposure rate of the beam to one-half. HVLs for various materials in a heavily filtered diagnostic x-ray beam at various values of kVp are shown in Figure 15.5.

15.2.2. X-ray dosage

The dose rate, or rate of absorption of energy, from an x-ray beam increases with peak voltage (kVp) applied to the x-ray tube and decreases with the amount of filtration in the

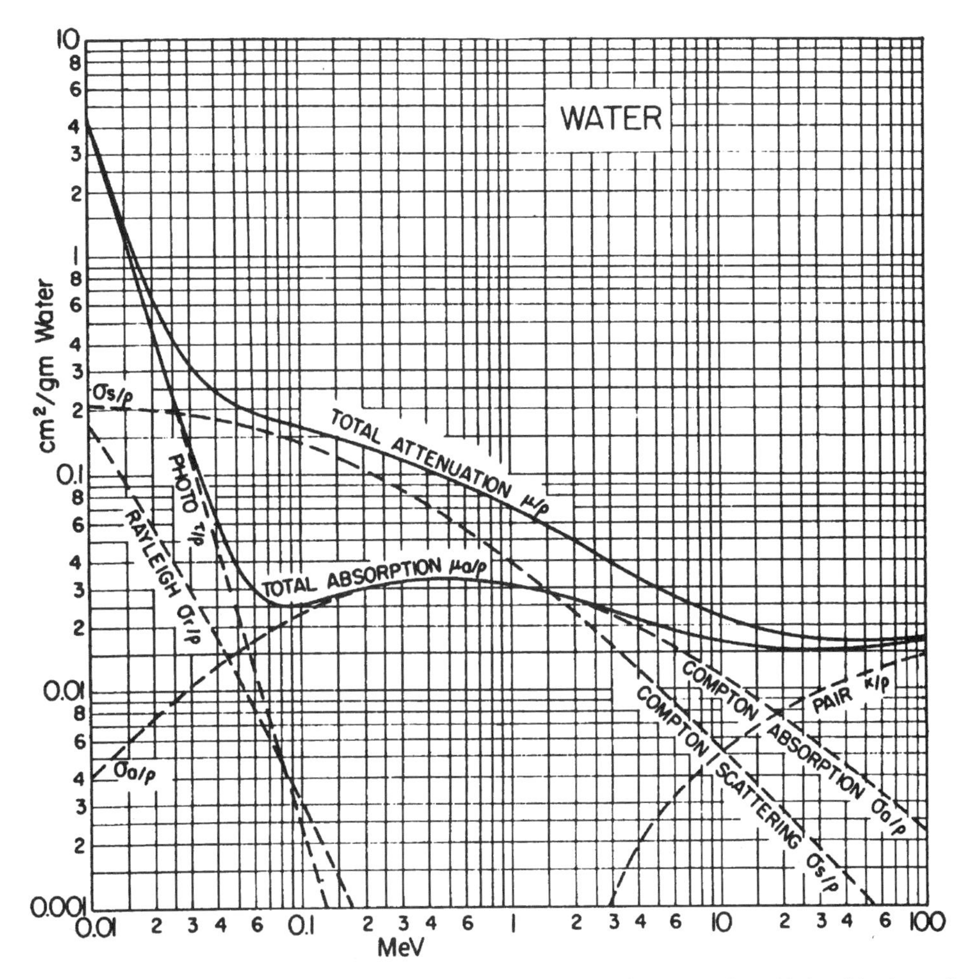

FIGURE 15.3. Mass attenuation coefficients for x- and γ-rays. (Courtesy of the National Institute of Standards and Technology, Technology Administration, U. S. Department of Commerce.

x-ray beam. Figure 15.6 shows the dose rate [in units of mGy/(mA · min)] at 100 cm from the target of the x-ray tube, for various values of kVp. The voltage across the x-ray tube is single-phase, full-wave rectified.

Attenuation results in a decrease of dose rate with increasing depth below the surface. This decrease, expressed as a percentage of the maximum dose rate, is shown in Figure 15.7. The tissue-maximum ratio (TMR) is the x- or γ-ray dose at depth divided by the maximum dose at the same location under a build-up thickness of tissue. TMRs for 6-MV x-rays as a function of field size are given in Table 15.5.

At higher energies, x- and γ-ray beams exhibit a dose build-up region over several millimeters depth below the surface. The build-up region, illustrated in Figure 15.8, creates a "skin-sparing" effect in high-energy x- and γ-ray beams used in the treatment of cancer.

TABLE 15.4. Mass energy absorption coefficients and f factors for selected materials. Mass energy absorption coefficient is defined as the fraction of photon energy absorbed per unit length (kg/m^2) of absorber. The signed two-digit number following each value is its power-of-ten multiplier (with permission from Ref. [3]).

Photon energy	Mass energy absorption coefficient μ_{en}/ρ (m^2/kg) (multiply by 10 for cm^2/g)								$f = 0.869\,(\mu_{en}/\rho)_{med}/(\mu_{en}/\rho)_{air}$		
	Air	Water (H_2O)	Polystyrene (C_8H_8)	Lucite ($C_5H_8O_2$)	Polyeth (CH_2)	Bakelite ($C_{43}H_{38}O_7$)	Compact bone	Muscle	Water	Compact bone	Muscle
10 keV	4.648 − 01	4.839 − 01	1.849 − 01	2.943 − 01	1.717 − 01	2.467 − 01	1.900 + 00	0.496 + 00	0.912	3.54	0.925
15	1.304 − 01	1.340 − 01	5.014 − 02	8.081 − 02	4.662 − 02	6.741 − 02	0.589 + 00	0.136 + 00	0.889	3.97	0.916
20	5.266 − 02	5.364 − 02	2.002 − 02	3.231 − 02	1.868 − 02	2.692 − 02	0.251 + 00	0.544 − 01	0.881	4.23	0.916
30	1.504 − 02	1.519 − 02	6.056 − 03	9.385 − 03	5.754 − 03	7.904 − 03	0.743 − 01	0.154 − 01	0.869	4.39	0.910
40	6.706 − 03	6.800 − 03	3.190 − 03	4.498 − 03	3.128 − 03	3.989 − 03	0.305 − 01	0.677 − 02	0.878	4.14	0.919
50	4.038 − 03	4.153 − 03	2.387 − 03	3.019 − 03	2.410 − 03	2.711 − 03	0.158 − 01	0.409 − 02	0.892	3.58	0.926
60	3.008 − 03	3.151 − 03	2.153 − 03	2.505 − 03	2.218 − 03	2.316 − 03	0.979 − 02	0.312 − 02	0.905	2.91	0.929
80	2.394 − 03	2.582 − 03	2.152 − 03	2.292 − 03	2.258 − 03	2.191 − 03	0.520 − 02	0.255 − 02	0.932	1.91	0.930
100	2.319 − 03	2.539 − 03	2.292 − 03	2.363 − 03	2.419 − 03	2.288 − 03	0.386 − 02	0.252 − 02	0.948	1.45	0.918
150	2.494 − 03	2.762 − 03	2.631 − 03	2.656 − 03	2.788 − 03	2.593 − 03	0.304 − 02	0.276 − 02	0.962	1.05	0.956
200	2.672 − 03	2.967 − 03	2.856 − 03	2.872 − 03	3.029 − 03	2.806 − 03	0.302 − 02	0.297 − 02	0.973	0.979	0.983
300	2.872 − 03	3.192 − 03	3.088 − 03	3.099 − 03	3.275 − 03	3.032 − 03	0.311 − 02	0.317 − 02	0.966	0.938	0.957
400	2.949 − 03	3.279 − 03	3.174 − 03	3.185 − 03	3.367 − 03	3.117 − 03	0.316 − 02	0.325 − 02	0.966	0.928	0.954
500	2.966 − 03	3.298 − 03	3.195 − 03	3.205 − 03	3.389 − 03	3.137 − 03	0.316 − 02	0.327 − 02	0.966	0.925	0.957
600	2.952 − 03	3.284 − 03	3.181 − 03	3.191 − 03	3.375 − 03	3.123 − 03	0.315 − 02	0.326 − 05	0.966	0.925	0.957
800	2.882 − 03	3.205 − 03	3.106 − 03	3.115 − 03	3.295 − 03	3.049 − 03	0.306 − 02	0.318 − 02	0.965	0.920	0.956
1.0 MeV	2.787 − 03	3.100 − 03	3.005 − 03	3.014 − 03	3.188 − 03	2.950 − 03	0.297 − 02	0.308 − 02	0.965	0.922	0.956
1.5	2.545 − 03	2.831 − 03	2.744 − 03	2.752 − 03	2.911 − 03	2.693 − 03	0.270 − 02	0.281 − 02	0.964	0.920	0.958
2.0	2.342 − 03	2.604 − 03	2.522 − 03	2.530 − 03	2.675 − 03	2.476 − 03	0.248 − 02	0.257 − 02	0.968	0.921	0.954
3.0	2.055 − 03	2.279 − 03	2.196 − 03	2.208 − 03	2.325 − 03	2.160 − 03	0.219 − 02	0.225 − 02	0.962	0.928	0.954
4.0	1.868 − 03	2.064 − 03	1.978 − 03	1.993 − 03	2.089 − 03	1.950 − 03	0.199 − 02	0.203 − 02	0.958	0.930	0.948
5.0	1.739 − 03	1.914 − 03	1.822 − 03	1.842 − 03	1.919 − 03	1.801 − 03	0.186 − 02	0.188 − 02	0.954	0.934	0.944
6.0	1.646 − 03	1.805 − 03	1.707 − 03	1.730 − 03	1.793 − 03	1.691 − 03	0.178 − 02	0.178 − 02	0.960	0.940	0.949
8.0	1.522 − 03	1.658 − 03	1.548 − 03	1.578 − 03	1.618 − 03	1.541 − 03	0.165 − 02	0.163 − 02	0.958	0.950	0.944
10	1.445 − 03	1.565 − 03	1.445 − 03	1.480 − 03	1.503 − 03	1.445 − 03	0.159 − 02	0.154 − 02	0.935	0.960	0.929
15	1.347 − 03	1.440 − 03	1.302 − 03	1.346 − 03	1.341 − 03	1.313 − 03					
20	1.306 − 03	1.384 − 03	1.233 − 03	1.284 − 03	1.206 − 03	1.251 − 03					

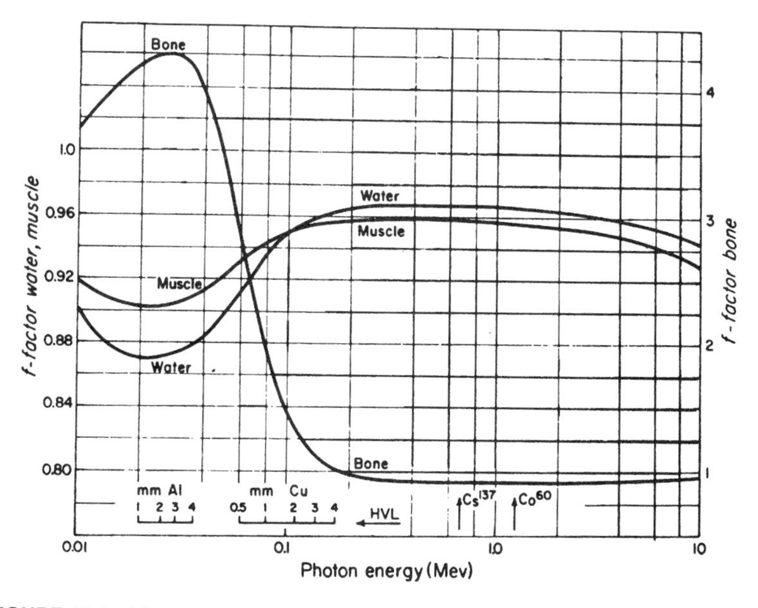

FIGURE 15.4. *f* factor as a function of x- or γ-ray energy for water, muscle, and bone.

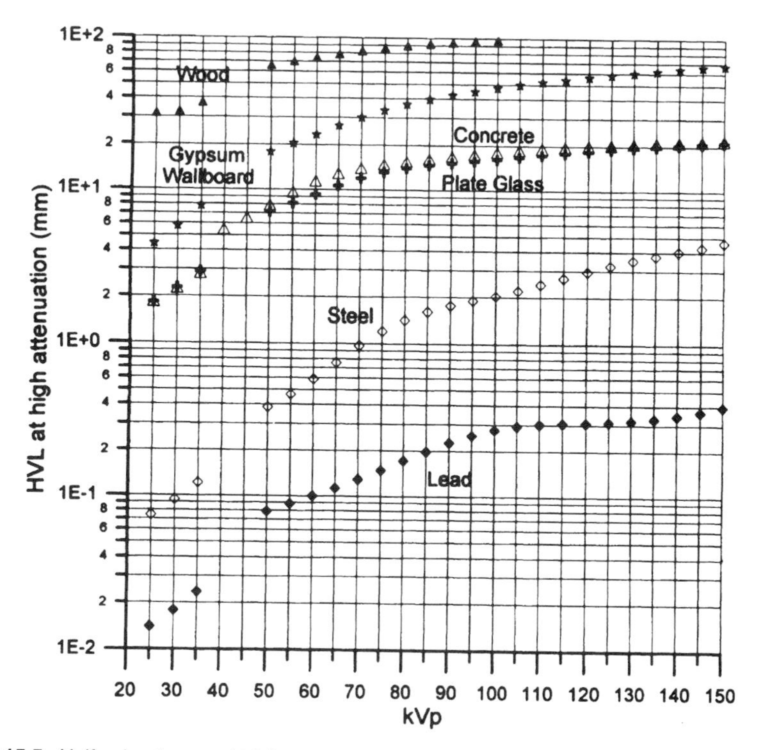

FIGURE 15.5. Half-value layers at high attenuation for diagnostic x-rays. (Used with permission from Ref. [4].)

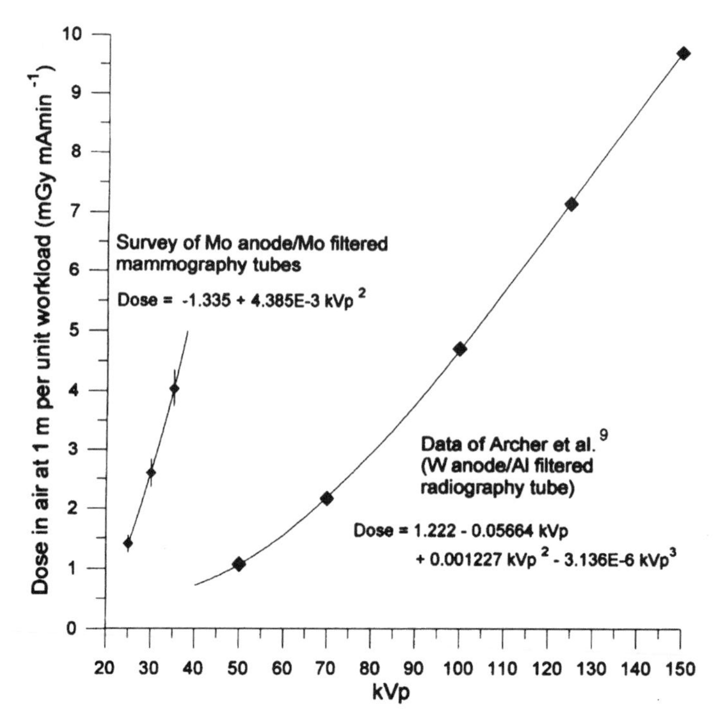

FIGURE 15.6. The primary x-ray dose in air per unit workload, measured at 1 m. Below 40 kVp Mo anode, Mo-filtered mammography beam is assumed. The graphed data are from a survey of modern mammography units. Above 40 kVp, the W anode, Al-filtered radiographic data of Simpkin *et al.* are shown. (Used with permission from Ref. [4].)

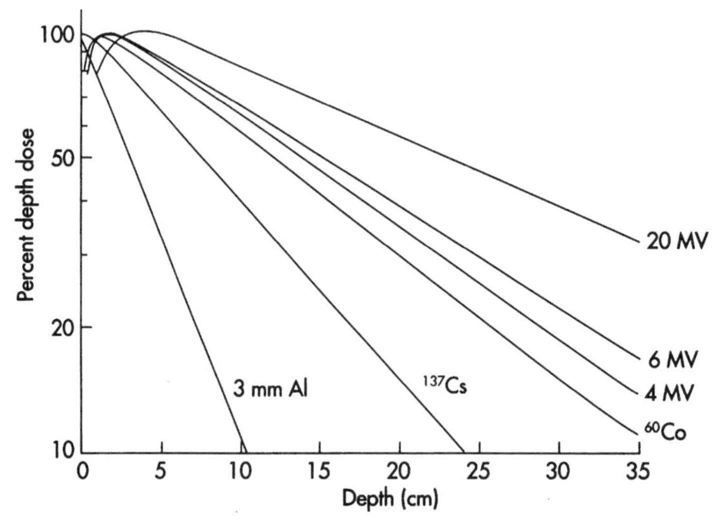

FIGURE 15.7. Percentage depth dose for 100-cm^2 area and x- and γ-ray beams of different energies as functions of depth in water. The SSD is 100 cm for all beams except the 30-mm Al (SSD = 15 cm) x-ray beam and the ^{137}Cs beam (SSD = 35 cm).

TABLE 15.5. Tissue maximum ratios for 6-MV x-rays (Varian Clinac 6-100, 100 CM SAD) (with permission from Ref. [6]).

	Field size (cm × cm)									
Depth (cm)	0 × 0	4 × 4	6 × 6	8 × 8	10 × 10	12 × 12	16 × 15	20 × 20	30 × 30	40 × 40
1.0	0.950	0.966	0.968	0.969	0.970	0.972	0.974	0.977	0.983	0.990
1.5	1.000	1.000	1.000	1.000	1.000	1.000	1.000	1.000	1.000	1.000
2.0	0.990	0.991	0.992	0.992	0.992	0.993	0.994	0.995	0.996	0.998
3.0	0.950	0.957	0.964	0.969	0.972	0.974	0.976	0.977	0.981	0.984
4.0	0.893	0.924	0.934	0.944	0.949	0.953	0.958	0.959	0.961	0.965
5.0	0.851	0.891	0.904	0.914	0.921	0.926	0.933	0.936	0.942	0.949
6.0	0.812	0.858	0.874	0.886	0.894	0.901	0.910	0.915	0.924	0.930
7.0	0.776	0.822	0.841	0.856	0.866	0.874	0.885	0.892	0.901	0.909
8.0	0.735	0.787	0.806	0.822	0.834	0.842	0.858	0.868	0.880	0.887
9.0	0.701	0.749	0.769	0.786	0.802	0.814	0.832	0.842	0.856	0.865
10.0	0.678	0.720	0.740	0.758	0.774	0.788	0.808	0.821	0.834	0.846
12.0	0.595	0.649	0.672	0.694	0.712	0.728	0.748	0.762	0.781	0.797
14.0	0.537	0.590	0.614	0.636	0.656	0.672	0.698	0.714	0.736	0.754
15.0	0.502	0.560	0.584	0.608	0.628	0.646	0.675	0.692	0.714	0.731
16.0	0.477	0.535	0.559	0.581	0.601	0.620	0.548	0.667	0.693	0.712
18.0	0.428	0.479	0.504	0.526	0.547	0.566	0.596	0.617	0.647	0.668
20.0	0.391	0.439	0.464	0.486	0.507	0.525	0.554	0.574	0.603	0.625
22.0	0.348	0.399	0.422	0.444	0.464	0.482	0.511	0.531	0.564	0.586
24.0	0.301	0.359	0.384	0.406	0.426	0.444	0.472	0.492	0.530	0.551
26.0	0.265	0.328	0.350	0.371	0.390	0.407	0.434	0.454	0.488	0.514
28.0	0.240	0.297	0.319	0.338	0.356	0.373	0.398	0.420	0.454	0.480
30.0	0.216	0.270	0.290	0.308	0.326	0.342	0.366	0.386	0.423	0.446

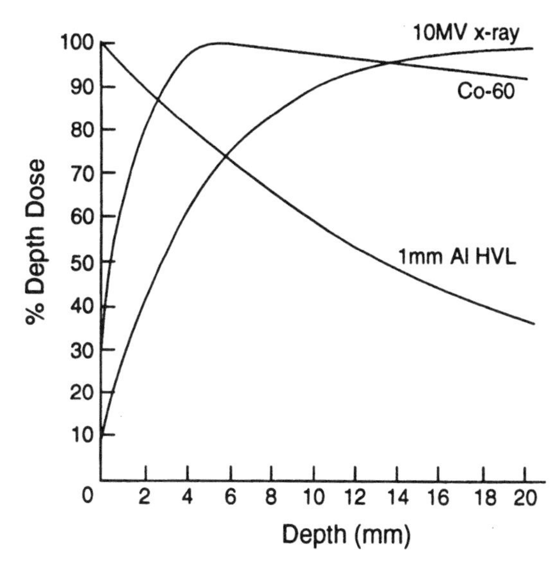

FIGURE 15.8. Percentage depth dose in the build-up region for various photon beams. (Used with permission from Ref. [7].)

TABLE 15.6. Contrast arising from a 1-cm block of tissue compared with a 1-cm block of muscle.

Tissue	μ (cm^{-1})	N/N_0 ($x = 1$ cm)	C w.r.t. muscle (%)
Muscle	0.180	0.835	0
Air	0	1	20
Blood	0.178	0.837	0.2
Bone	0.480	0.619	−26

15.2.3. X-ray image contrast

Tissue contrast, described as the difference in x-ray transmission between adjacent tissues, is depicted in Table 15.6.

15.3. IONIZING RADIATION: ELECTRONS

Electron beams can also be characterized by percentage dose versus depth in tissue (water). Central-axis depth dose curves for several electron energies are given in Figure 15.9.

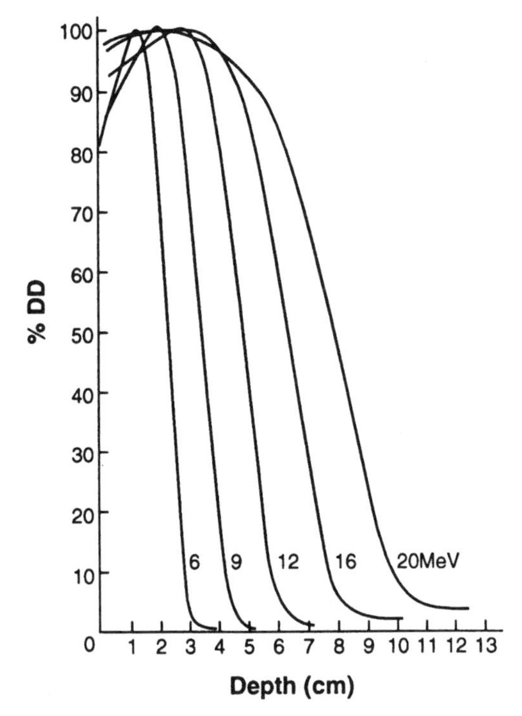

FIGURE 15.9. Central-axis depth dose curves of 6-, 9-, 12-, and 16-MeV electron beams. The depth dose falls off sharply, particularly at the lower energies, and the surface dose increases with increasing energy (used with permission from ref. [7]).

TABLE 15.7. Summary of risks from low-LET, low-dose rate ionizing radiation (with permission from Ref. [8]).

Source of estimate	Dose-response model	Quantitative estimate
Cancer: Excess lifetime cancer fatality (Normal incidence: 200,000 deaths per million)		
Source of estimate	Dose-response model	Quantitative estimate (Excess cancer deaths per 10,000 person-Sv)
BEIR V (Assume DREF of 2.0 for solid tumors)	Linear-quadratic (Leukemia) Linear (solid tumors)	460
Genetic: Mendelian and chromosomal disease and congenital abnormalities (Normal incidence: 47,300 serious genetic defects per million live born)		
Source of estimate	Dose-response model	Quantitative estimate (Excess genetic defects per 0.01 Sv per million live born)
BEIR V	Linear	15–35 first generation 110–200 at equilibrium
Genetic: Multifactorial disorders of complex etiology (Normal incidence: 650,000 to 1,200,000 per million live born)		
Source of estimate	Dose-response model	Quantitative estimate (Excess genetic defects per 0.01 Sv per million live born
BEIR V	Not estimated	Not estimated
ICRP	Linear	120 at equilibrium
Fetal: Malformations (Normal incidence: 27 serious malformations per thousand live born; 60–100 malformations of any type per thousand live born)		
Source of estimate	Dose-response model	Quantitative estimate (Excess malformations per 1.0 Sv per thousand live born)
BEIR V	Linear with probable thresholds	430 severe mental retardation cases at 1.0 Sv, threshold may be 0.2 to 0.4 Sv
Brent	Linear with threshold	Negligible malformations for fetal dose 0.05 Sv Signifcantly increased risk above 0.2 to 0.5 Sv
Fetal: Childhood carcinogenesis (all sites) (Normal incidence: 1400 per million incidence, 500 deaths per million in first ten years of life)		
Source of estimate	Dose-response model	Quantitative estimate (Excess childhood cancer per 10,000 fetus-Sv
Stovall et. al. (Oxford)	None determined	640 (Incidence), 217 (Death)
Stovall et. al. (A-bomb)	Linear quadratic	55 (Incidence)

TABLE 15.8. Radiation dose limits (with permission from Ref. [9]).

Radiation dose limits for occupationally exposed individuals:

- Lifetime total effective dose in tens of mSv should not exceed the individual's age in years, excluding exposure to medical and natural background radiation.
- Occupational exposure to radiation should not be permitted for persons below 18 years of age.
- Annual occupational total effective dose should be limited to 0.05 Sv, provided the limit on lifetime total effective dose is not exceeded.
- Exposures should always be maintained at levels as low as reasonably achievable (ALARA).
- All dose limits apply to the sum of external and internal exposures, with external exposures assessed as the effective dose and the internal exposures assessed as the committed effective dose.
- New facilities and new practices should be designed to limit annual effective doses to workers to a fraction of the 10 mSv year^{-1} implied by the lifetime limit in total effective data.

Additional dose limits for regions where deterministic radiation effects are the controlling influence:

- 150 mSv to the crystalline lens of the eye.
- 500 mSv to all other tissues and organs, including localized areas of the skin, hands, and feet.

15.4. HEALTH RISKS

Health risks of exposure to ionizing radiation can be classified as cancer, genetic, and multifactorial disorders of complex etiology. These risks are summarized in Table 15.7. The risks are controlled by establishing limits for exposure of humans to radiation. Current limits are summarized in Table 15.8.

15.5. ULTRASOUND

The fractional reflection of ultrasound energy at an interface between two tissues is a measure of the difference in acoustic impedance between the tissues, in which the acoustic impedance is the product of the mass density of the tissue by the velocity of ultrasound in the tissue. Acoustic impedances for various tissues are given in Table 15.9.

Ultrasound attenuation in tissue increases with the frequency of the ultrasound beam. In Table 15.10 the attenuation per unit frequency per unit path length in various tissues is given, expressed in units of dB/cm · MHz. The decibel (dB) is a unit of attenuation; the attenuation in dB is $10\log(I/I_0)$, where I_0 and I are the intensities before and after attenuation, respectively.

15.6. MAGNETIC RESONANCE

Magnetic resonance exploits the magnetic properties of biological constituents to yield well-differentiated images of subtle differences in tissue composition. Magnetic properties of biologically important elements are given in Table 15.11.

TABLE 15.9. Acoustic impedancies for different tissues (with permission from Ref. [10])

Tissue	Acoustic impedance (in Rayls[a]) $\times 10^6$
Liver	1.64
Muscle	1.70
Fat	1.38
Kidney	1.62
Bone	7.80
Lung	0.26

[a] 1 Rayl is 1 kg/m^2s.

TABLE 15.10. Ultrasound attenuation coefficients for different tissues (with permission from Ref. [10]).

Tissue	α (dB/cm MHz)
Fat	0.6
Muscle (transverse)	3.5
Muscle (longitudinal)	1.2
Liver (normal)	0.9
Liver (cirrhotic)	1.2
Amniotic fluid	0.005

TABLE 15.11. Magnetic characteristics of biologically relevant elements (with permission from Ref. [11]).

Nucleus	Quantum spin I	Physiological concentration (M)	Magnetic moment (μ)	Gyromagnetic ratio (γ) (Mhz/T)
^{1}H	1/2	100	2.79	42.58
^{13}C	1/2	—	0.69	10.71
^{23}Na	3/2	0.080	2.22	11.26
^{13}P	1/2	0.075	1.13	17.24

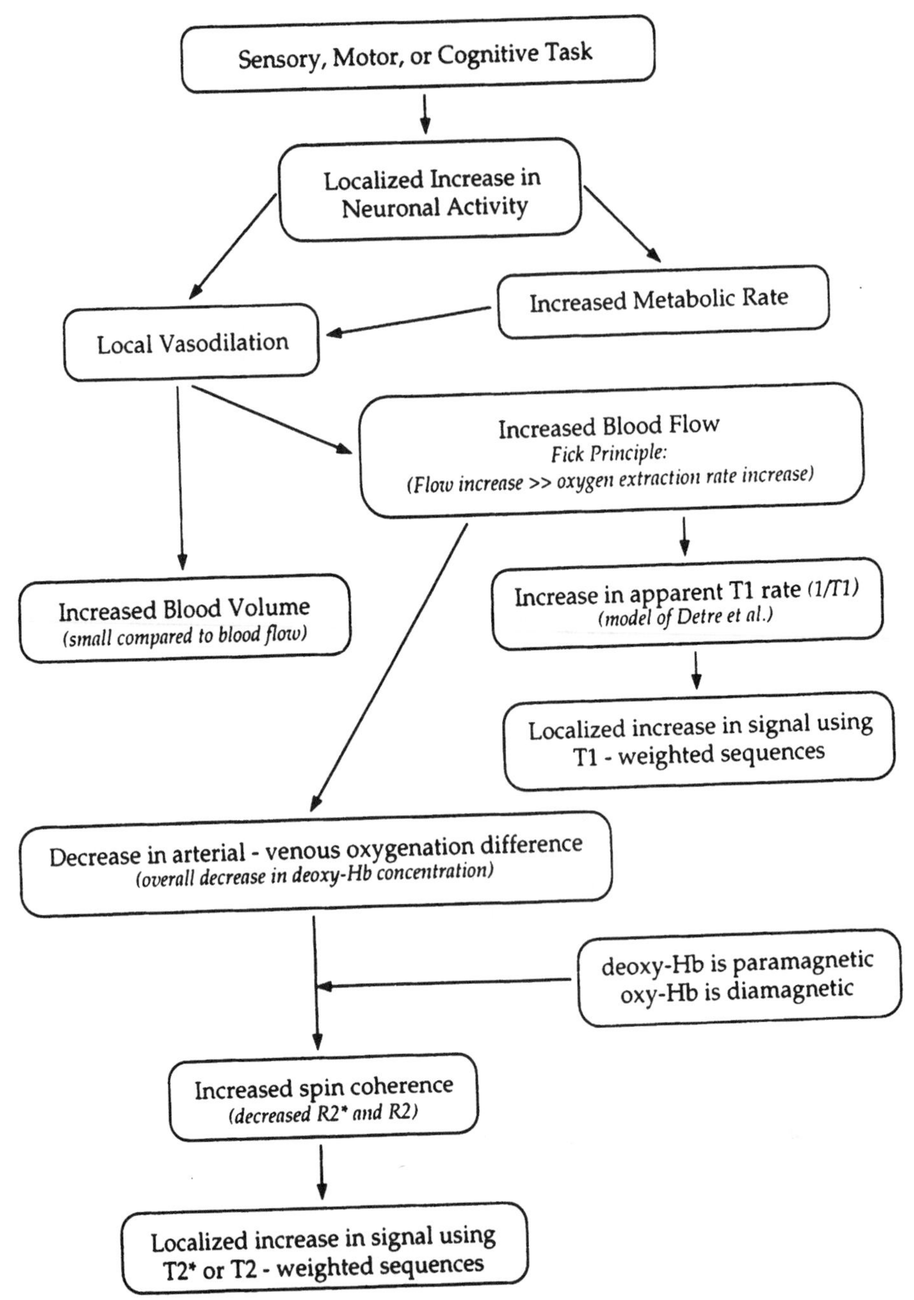

FIGURE 15.10. Flow chart summarizing the cascade of hemodynamic events that occur with brain activation, and their corresponding effects on the appropriately sensitized MRI signal. (Used with permission from Ref. [12].)

The biological origin of functional magnetic resonance imaging (fMRI) signals is not completely understood. One model for the generation of these signals is shown in Figure 15.10.

15.7. BRACHYTHERAPY

Brachytherapy denotes the use of sealed radioactive sources to treat cancerous tissue in the immediate vicinity of the sources. Radionuclides used for brachytherapy include those in Table 15.12. The characteristics of radionuclide sources used for various brachytherapy ap-

TABLE 15.12. Current brachytherapy radionuclides (with permission from Ref. [13])

Brachytherapy Source Materials						
Isotope	Max beta energy [MeV]	Effective or dominant photon energy [MeV]	Half-life	Specific exposure rate constant [R cm^2/mCi hr]	Half-value layer [mm Pb]	Theoretic max specific activity [Ci/g]
Ra-226	3.17/1.17	0.83 eff (0.5 mm Pt) 1.2 eff (1.0 mm Pt)	1622 yr	8.25 [R cm^2/mg hr] 7.71 [R cm^2/mg hr]	12	1 (Definition)
Co-60	0.31 (99%) 1.5 (0.1%)	1.33 & 1.17	5.26 yr	13	12	1,100
Cs-137	0.51 (93%) 1.18 (7%)	0.662 (85%)	30 yr	3.26	6	100
Ir-192	0.67 (46%) 0.24 (8%) 0.54 (41%) 0.39	0.38 eff dep. on jacket	73.8 d	4.66 or 4.60 dep. on jacket c. 4.8 naked	3	9,000
Am-241	0.054/0.022	0.06	432 yr	0.12	0.13	3
Rn-22	0.97	0.83 eff	3.83 d	8.35	12	150 000
Au-198	0.97	0.412	2.70 d	2.32	3	250 000
I-125	0.03 (90%)	0.035 (7%) 0.27 eff	60 d	1.45 normal 1.32 compen.	0.025	17 000
Pd-103	0.017/0.037	0.021 eff	17 d	1.48	0.008	75 000
Yb-169	0.298	0.093 eff	32 d	1.8	0.2	24 000
Ru-106	0.039/3.5	*	368 d	*		3,400
Re-186	1.07	0.137 (9%)	3.8 d	*		180 000
Sr-90, Y-90	0.6, 2.27	*	28 yr	*		140
Y-90	2.27	*	64.2 hr	*		540 000
I-131	0.61 (8%)	0.364	8 d	2.23	3	120 000
P-32	1.71	*	14.3 d	*		300 000
Sr-89	1.46	*	52.7	*		28 000
Cf-252	*	2(capture)	3.65 y			390

TABLE 15.13. General characteristics of brachytherapy sources (with permission from Ref. [13]).

Procedure	Nature	Characteristics of radionuclides
Interstitial implants	Permanent	Photon emitter, minimal electron contamination; low photon energy and/or short half-life; small size (high specific activity); biologically inert jacket.
	Temporary Low dose rate	Photon emitter, minimal electron contamination; small size (high specific activity).
	Temporary High dose rate	Photon emitter, minimal electron contamination; very high specific activity (small size); relatively long half-life.
Intracavity insertions	Temporary Low dose rate	Photon emitter, minimal electron contamination; relatively high photon energy.
	Temporary High dose rate	Photon emitter, minimal electron contamination; very high specific activity (small size); relatively high photon energy; relatively long half-life.
Intravascular treatments	Stents	Beta emitter, possibly with some photon component; Medium half-life; able to be incorporated permanently on or in a stent; biologically inert.
	Intracatheter High dose rate	Beta emitter, probably with some photon component; very high specific activity (small size); very flexible; relatively long half-life.
Surface applications	Skin	Photon emitter.
	Eye neoplasm	Photon emitter, low photon energy; small size (high specific activity).
	Eye corneal vasculature	Beta emitter, little photon contamination; high beta energy; high activity.

plications are given in Table 15.13. The ranges of radiation doses required to cure different types of cancer are given in Table 15.14.

A rapidly evolving application of brachytherapy is the intraluminal use of radioactive sources to suppress restenosis following angioplasty. Sources currently or potentially useful for intraluminal brachytherapy are listed in Table 15.15.

A major challenge in radiotherapy is to deliver curative doses to cancer while preserving the structure and function of normal tissue. This challenge is addressed by attention to dose thresholds (Table 15.16), above which treatment complications can be anticipated.

More precise treatments demand more exact definition of tissue volumes exposed to radiation. The current definitions of tissue volumes are given in Table 15.17.

TABLE 15.14. Approximate curative doses for different tumor types and sizes (with permission from Ref. [14]).

20–30 Gy
Seminoma Acute lymphocytic leukemia
30–40 Gy
Seminoma Wilm's tumor Neurobastoma
40–45 Gy
Hodgkin's disease Non-Hodgkin's lymphoma Skin cancer (basal and squamous)
50–60 Gy
Lymph nodes, metastatic (<1 cm) Squamous cell carcinomas (sub-clinical) Breast cancer (<1 cm) Medullobastoma Retinoblastoma Ewing's tumor
60–65 Gy
Larynx (<1 cm) Breast cancer, lumpectomy
70–75 Gy
Head and neck cancers (2–4 cm) Bladder cancers Cervix cancer Uterine fundal cancer Lymph nodes, metastatic (1–3 cm) Lung cancer (<3 cm)
80 Gy or more
Head and neck cancers (<4 cm) Breast cancer (>5 cm) Gliobastomas Osteogenic sarcomas Melanomas Thyroid cancer Lymph nodes, metastatic (>6 cm)

TABLE 15.15. Isotopes with demonstrated or potential application to intraluminal brachytherapy (with permission from Ref. [15]).

Element	Isotope	Emission	Energy range (keV)	Average energy (keV)	Half-life
Iridium	^{192}Ir	gamma	136–1,062	380	73.8 d
Iodine	^{125}I	x and gamma	27–35	28	59.6 d
Palladium	^{103}Pd	x and gamma	20–53	21	17.0 d
Phosphorus	^{32}P	beta−	1,710	690	14.3 d
Strontium	^{90}Sr/^{90}Y	beta−	2,270	970	28.9 d
Tungsten	^{188}W/^{188}Re	beta−	2,120	780	69.4 d
Vanadium	^{48}V	beta+	696	230	16.0 d

TABLE 15.16. Tolerance doses for fractionated doses to whole or partial organs (with permission from Ref. [14]).

Dose range/Target cells	Complication end point	$TD_{5/5} - TD_{50/5}$ (Gy)
2–10 Gy		
Lymphocytes and lymphoid	Lymphopenia	2–10
Testes spermatogonia	Sterility	1–2
Ovarian oocytes	Sterility	6–10
Diseased bone marrow	Severe leukopenia and thrombocytopenia	3–5
10–20 Gy		
Lens	Cataract	6–12
Bone marrow stem cells	Acute aplasia	15–20
20–30 Gy		
Kidney: renal glomeruli	Arterionephrosclerosis	23–28
Lung: Type II cells, vascular connective tissue stroma	Pneumonitis or fibrosis	20–30
30–40 Gy		
Liver: central veins	Hepatopathy	35–40
Bone marrow	Hypoplasia	25–35
40–50 Gy		
Heart (whole organ)	Pericarditis or pancarditis	43–50
Bone marrow microenvironments	Permanent aplasia	45–50
50–60 Gy		
Gastrointestinal	Infarction necrosis	50–55
Heart (partial organ)	Cardiomyopathy	55–65
Spinal cord	Myelopathy	50–60
60–70 Gy		
Brain	Encephalopathy	60–70
Mucosa	Ulcer	65–75
Rectum	Ulcer	65–75
Bladder	Ulcer	65–75
Mature bones	Fracture	65–75
Pancreas	Pancreatitis	>70

TABLE 15.17. Volumes of interest in treatment planning (with permission from Ref. [13]).

Gross tumor volume	Often defined as the palpable or imaged extent of the tumor.
Clinical target volume	This is the gross tumor volume with an added margin to account for subclinical disease.
Planning target volume	This is the clinical target volume with margin to account for error in positioning, and perhaps with adjustments to avoid critical structures.
Treated volume	Volume enclosed by the selected isodose surface to deliver the prescribed dose.
Irradiated volume	This is the volume that receives any significant amount of radiation throughout the course of treatment.

TABLE 15.18. TVL values for common shielding materials in megavoltage radiotherapy beams (with permission from Ref. [16]).

Nominal photon energy (MV)	TVL (cm)		
	Concrete	Steel	Lead
4	30.0	9.0	5.3
6	35.0	10.2	5.6
8	38.0	10.4	5.6
10	41.0	10.7	5.6
15	46.0	10.7	5.6
18	48.0	10.7	5.6
25	51.0	10.7	5.6

The tenth-value layer (TVL) is the thickness of a material required to reduce radiation intensity by a factor of 0.1 under conditions of good geometry. TVLs are shown in Table 15.18 for x-ray beams of various energies in concrete, steel, and lead.

15.8. NUCLEAR MEDICINE

Nuclear medicine is the practice of using radioactive pharmaceuticals for imaging procedures in medical diagnosis. Nuclear medicine is concerned primarily with biological function, and many different radioactive isotopes are used in conjunction with various pharmaceuticals to image different organ systems. Some of the more common nuclides used are shown in Table 15.19.

Nuclear medicine images reveal the distribution of a radioactive pharmaceutical in the body by detecting γ-rays emitted by the pharmaceutical. Attenuation of these γ-rays in various materials is described by the mass attenuation coefficients and HVLs shown in Table 15.20.

TABLE 15.19. Characteristics of common radionuclides used in nuclear medicine (with permission from Ref. [17]).

Nuclide	Half-life	Decay mode	Prominent emissions[a] energy (MeV), abundance (%)
^{11}C	20.4 m	β^+	(Annihilation γ) 0.511 (199.6%) (β^+) 0.960 (99.8%)
^{13}N	9.97 m	β^+	(Annihilation) 0.511 (199.6%) (β^+) 1.198 (99.8%)
^{15}O	2.04 m	β^+	(Annihilation) 0.511 (199.8%) (β^+) 1.731 (99.9%)
^{18}F	109.8 m	β^+	(Annihilation) 0.511 (194%) (β^+) 0.633 (97%)
^{32}P	14.26 d	β^-	(β^-) 1.710 (100%)
^{57}Co	271.8 d	EC	(γ) 0.122 (85.6%), 0.136 (19.7%)
^{67}Ga	3.26 d	EC	(γ) 0.093 (39.2%), 0.184 (21.2%), 0.300 (16.8%)
^{68}Ga	67.63 m	β^+, EC	(Annihilation γ) 0.511 (178.2%) (β^+) 1.899 (88.0%)
^{68}Ge	270.82 d	EC	Daughter radiation Ga-68 decay
^{89}Sr	50.53 d	β^-	(β^-) 1.492 (99.99%)
^{90}Y	64.10 h	β^-	(β^-) 2.280 (99.99%)
^{99}Mo	65.94 h	β^-	(γ) 0.140 (4.5%), 0.181 (5.99%) (γ) 0.739 (12.1%), 0.778 (4.3%) (β^-) 0.436 (16.4%), 1.214 (82.4%)
^{99m}Tc	6.01 h	IT	(γ) 0.1405 (89.1%)
^{111}In	2.805 d	EC	(γ) 0.171 (90.2%), 0.245 (94%) (x) 0.023 (67.8%), 0.026 (14.5%)
^{123}I	13.27 h	EC	(γ) 0.159 (83.3%) (x) 0.027 (70.6%), 0.031 (16%)
^{124}I	4.176 d	EC, β^+	(Annihilation γ) 0.511 (45.96%) (γ) 0.602 (62.9%), 0.722 (10.4%), 1.691 (10.9%) (β^+) 1.534 (11.79%), 2.137 (10.89%)
^{125}I	59.4 d	EC	(x) 0.027 (114%), 0.031 (25.8%) (γ) 0.0355 (6.7%)
^{127}Xe	36.4 d	EC	(γ) 0.172 (25.5%), 0.203 (68.3%), 0.375 (17.2%) (x) 0.028 (60.1%), 0.032 (13.7%)
^{131}I	8.02 d	β^-	(γ) 0.294 (6.1%), 0.364 (81.7%), 0.637 (7.2%) (β^-) 0.606 (89.9%), 0.096 (7.27%)
^{133}Xe	5.24 d	β^-	(γ) 0.081 (38.0%) (x) 0.030 (45.9%), 0.034 (10.6%) (β^-) 0.346 (99%)

(continued)

TABLE 15.19. Continued

Nuclide	Half-life	Decay mode	Prominent emissions[a] energy (MeV), abundance (%)
^{133}Ba	10.52 y	EC	(γ) 0.081 (34.1%), 0.276 (7.1%), 0.303 (18.3%) (γ) 0.356 (62.1%), 0.384 (8.9%) (x) 0.032 (44.0%), 0.036 (10.4%)
^{137}Cs	30.04 y	β^-	(γ) 0.662 (85.1%) (β^-) 0.514 (94.4%)
^{153}Gd	240.4 d	EC	(γ) 0.097 (29%), 0.103 (21.1%) (x) 0.041 (97.7%), 0.047 (24.6%)
^{153}Sm	46.5 h	β^-	(γ) 0.103 (29.8%) (β^-) 0.635 (32.2%), 0.705 (49.6%), 0.808 (17.5%)
^{201}Tl	72.9 h	EC	(x) 0.070 (73.7%), 0.080 (20.4%) (γ) 0.135 (2.6%), 0.167 (10.0%)

[a]x = x-rays; beta energies are maximum energies.

Administration of radioactive pharmaceuticals results in radiation doses to the patient and to the fetus in the case of pregnant patients. Fetal doses for selected nuclear medicine procedures are listed in Table 15.21.

15.9. REFERENCES

[1] R. Johnson, in *Biomedical Uses of Radiation*, edited by W. R. Hendee (Wiley–VCH, Weinheim, Germany, 1999).

[2] F. Wachsman and G. Drexler, *Graphs and Tables for Use in Radiotherapy* (Springer, Berlin, 1975).

[3] J. H. Hubbel, Radiat. Res. **70**, 58 (1977).

[4] D. J. Simpkin, B. R. Archer, and R. L. Dixon, in *Biomedical Uses of Radiation*, edited by W. R. Hendee (Wiley–VCH, Weinheim, Germany, 1999).

[5] W. R. Hendee, *Medical Radiation Physics*, 1st ed. (Mosby–Year Book, Chicago, 1970).

[6] C. W. Coffey II, J. L. Beach, D. J. Thompson, and M. Mendiano, Med. Phys. **7**, 716 (1980).

[7] S. C. Prasad, in *Biomedical Uses of Radiation*, edited by W. R. Hendee (Wiley–VCH, Weinheim, Germany, 1999).

[8] F. M. Edwards, in *Biomedical Uses of Radiation*, edited by W. R. Hendee (Wiley–VCH, Weinheim, Germany, 1999).

[9] W. R. Hendee, in *Health Effects of Exposure to Low-Level Ionizing Radiation*, edited by W. R. Hendee and F. M. Edwards (Institute of Physics Publishing, Bristol, U.K., 1996).

[10] C. Kimme-Smith, in *Biomedical Uses of Radiation*, edited by W. R. Hendee (Wiley–VCH, Weinheim, Germany, 1999).

[11] W. Riddle and H. Lee, in *Biomedical Uses of Radiation*, edited by W. R. Hendee (Wiley–VCH, Weinheim, Germany, 1999).

TABLE 15.20. Mass attenuation coefficients for various materials at energies used in nuclear medicine. Energies related to common nuclides ^{201}Tl (70-keV Hg x-rays), ^{99m}Tc (140 keV), ^{131}I (364 keV), and positron emitters (annihilation radiation 512 keV) (with permission from Ref. [18]).

Material	Density (g/cm^3)	Energy (keV)	Mass attenuation coefficient (cm^2/g)	Half-value layer (cm)
Water	1.0	70	0.193	3.6
		140	0.154	4.5
		364	0.110	6.3
		512	0.096	7.2
Bone	1.85	70	0.234	1.6
		140	0.154	2.4
		364	0.110	3.5
		512	0.092	4.1
Concrete	2.35	70	0.247	1.19
		140	0.149	1.98
		364	0.100	2.95
		512	0.0867	3.40
NaI	3.67	70	4.256	0.044
		140	0.722	0.262
		364	0.131	1.44
		512	0.093	2.03
Iron	7.8	70	0.821	0.108
		140	0.218	0.407
		364	0.099	0.900
		512	0.083	1.07
Copper	8.9	70	1.067	0.073
		140	0.250	0.311
		364	0.110	0.782
		512	0.083	0.936
Lead	11.3	70	3.395	0.018
		140	2.393	0.026
		364	0.278	0.221
		512	0.156	0.392
Tungsten	19.3	70	2.440	0.015
		140	1.885	0.019
		364	0.228	0.157
		512	0.134	0.268

TABLE 15.21. Fetal doses from selected nuclear medicine exams (with permission from Ref. [19]).

Exam	Activity MBq (mCi)	Early mGy (rad)	3 month mGy (rad)	6 month mGy (rad)	9 month mGy (rad)
^{99m}Tc MDP (Bone scan)	750 (20)	4.6 (0.46)	4.0 (0.40)	2.0 (0.20)	1.8 (0.18)
^{99m}Tc MIBI (Cardiac perfusion)	1100 (30)	17 (1.7)	13 (1.3)	9.2 (0.92)	5.9 (0.59)
^{99m}Tc RBC (Cardiac, GI bleed)	930 (25)	6.3 (0.63)	4.4 (0.44)	3.2 (0.32)	2.6 (0.26)
^{201}Tl Chloride (Cardiac perfusion)	150 (4)	15 (1.5)	8.7 (0.87)	7 (0.70)	4 (0.40)
Na^{131}I	370	26.4	24.3	85.6	100.0
^{18}F FDG	370 (10)	10 (1.0)	6.3 (0.63)	3.5 (0.35)	3.0 (0.30)

[12] R. B. Birn, K. M. Donahue, and P. A. Bandettini, in *Biomedical Uses of Radiation*, edited by W. R. Hendee (Wiley–VCH, Weinheim, Germany, 1999).

[13] B. Thomadsen and E. Hendee, in *Biomedical Uses of Radiation*, edited by W. R. Hendee (Wiley–VCH, Weinheim, Germany, 1999).

[14] P. Rubin and D. W. Siemann, in *Clinical Oncology: A Multidisciplinary Approach for Physicians and Students*, edited by R. Rubin, S. Mcdonald, and R. Qazi (W. B. Saunders Co., Philadelphia, 1993).

[15] G. Ibbott, in *Biomedical Uses of Radiation*, edited by W. R. Hendee (Wiley–VCH, Weinheim, Germany, 1999).

[16] J. Palta, in *Biomedical Uses of Radiation*, edited by W. R. Hendee (Wiley–VCH, Weinheim, Germany, 1999).

[17] R. R. Kinsey, National Nuclear Data Center, Brookhaven National Laboratory, Upton, NY, www.nndc.bnl.gov/nndc/nudat/.

[18] J. H. Hubbell, Int. J. Appl. Radiat. Isot. **33**, 1269–1290 (1982).

[19] J. R. Russel, M. G. Stabin, R. B. Sparks, and E. Watson, Health Phys. **73**, 756–759 (1997).

16

Molecular Spectroscopy and Structure

Peter F. Bernath
Departments of Chemistry and Physics, University of Waterloo

Contents

List of Tables

List of Figures

16.1. INTRODUCTION

Our understanding of the rotational-vibrational-electronic (rovibronic) spectra of molecules is based on the nonrelativistic Schrödinger equation, [1]

$$\hat{H}_0\psi = E\psi. \tag{16.1}$$

The Born-Oppenheimer approximation is used to separate electronic and nuclear motion, and then the nuclear motion is further assumed to be separable into vibrational and rotational motion, leading to the simple equations

$$E = E_{\text{el}} + E_{\text{vib}} + E_{\text{rot}} \tag{16.2}$$

and

$$\psi = \psi_{\text{el}}\psi_{\text{vib}}\psi_{\text{rot}}. \tag{16.3}$$

For molecules with net electronic spin and net electronic orbital angular momentum, additional terms such as spin-orbit coupling, need to be added to the Hamiltonian of Eq. (16.1).

The manifold of energy levels described by Eq. (16.2) are connected by transitions, as determined by the selection rules. More generally, [2] an absorption line between energy levels E_1 and E_0 is represented by Beer's law

$$I = I_0 e^{-\sigma(N_0 - N_1)\ell} = I_0 e^{-\alpha\ell}, \tag{16.4}$$

where I_0 is the initial radiation intensity, σ is the cross section (in m^2), $N_0 - N_1$ is the population density difference (m^{-3}), and ℓ is the path length (m). The intrinsic line strength of a transition is thus measured by a cross section, which is proportional to the square of a transition moment integral, i.e.,

$$\sigma \propto |\langle\psi_1|\, \hat{O}_p \,|\psi_0\rangle|^2 \propto A, \tag{16.5}$$

where $\hat{O}_p$ is a transition moment operator and A is the Einstein A factor for emission. Selection rules and line strengths are obtained by a detailed examination of Eq. (16.5).

16.2. ROTATIONAL SPECTROSCOPY

All gas phase molecules have quantized rotational energy levels, and pure rotational transitions are possible. A molecule can, in general, rotate about three geometric axes and can have three different moments of inertia relative to these axes. The moment of inertia about an axis is defined as

$$I = \sum_i m_i r_{i\perp}^2, \tag{16.6}$$

where m_i is the mass of the atom i and $r_{i\perp}$ is the perpendicular (shortest) distance between this atom and the axis. The internal axis system of a molecule is chosen to have its origin at the center of mass and is rotated so that the moment-of-inertia tensor is diagonal. [2] This is the principal-axis system for a rigid molecule. The three moments of inertia I_x, I_y, and I_z can be used to classify molecules into four different types of "tops":

1. Linear molecule (including diatomics), $I_x = I_y$; $I_z = 0$, e.g., CO, HCCH.
2. Spherical top, $I_x = I_y = I_z$, e.g., CH_4, SF_6.
3. Symmetric top, $I_x = I_y \neq I_z$, e.g., BF_3, CH_3Cl.
4. Asymmetric top, $I_x \neq I_y \neq I_z$, e.g., H_2O, CH_3OH.

The internal molecular axes x, y, and z are labeled according to a certain set of rules based on molecular symmetry. [3] An additional labeling scheme is also used that is based on the size of the moments of inertia. In particular, the axis labels A, B, and C are chosen to make the inequality $I_A \leq I_B \leq I_C$ true. Thus, molecular symmetry determines the x, y, and z labels, but it is the size of the moments of inertia that set the A, B, and C labels. In terms of the A, B, and C labels, it is conventional to classify molecules into five categories, as follows:

1. Linear molecules, $I_A = 0$, $I_B = I_C$.
2. Spherical tops, $I_A = I_B = I_C$.
3. Prolate symmetric tops, $I_A < I_B = I_C$, e.g., CH_3Cl.
4. Oblate symmetric tops, $I_A = I_B < I_C$, e.g., BF_3.
5. Asymmetric tops, $I_A < I_B < I_C$.

Underlying all of these considerations is the assumption that the concept of a molecular structure is useful. Floppy species such as Van der Waals molecules or molecules with internal rotors do not always have a well-defined molecular structure when zero-point motions are considered. Clearly, all molecules have a hypothetical equilibrium structure, but vibrational motion even in the zero-point level can destroy the concept of a molecular structure with well-defined bond lengths and bond angles. Fluxional molecules are best handled using concepts based on permutation-inversion group theory. [4]

16.2.1. Diatomics

For a rigid diatomic molecule in a ${}^1\Sigma^+$ electronic state (no net electron spin or orbital angular momentum), the rotational energy levels are given by

$$E_{\text{rot}} = BJ(J+1), \tag{16.7}$$

where B is the rotational constant and J, the rotational quantum number, has values $0, 1, 2, \ldots$. The units of (16.7) are determined by the units chosen for B, which are generally cm^{-1}, MHz, or (rarely) J (joules).

Various equations for B are

$$B(\text{joules}) = \frac{h^2}{8\pi^2 I}, \tag{16.8}$$

$$B(\text{MHz}) = \frac{h}{8\pi^2 I} \times 10^{-6} = \frac{505379.006}{I(\text{amu} \cdot \text{Å}^2)}, \tag{16.9}$$

$$B(\text{cm}^{-1}) = \frac{h}{8\pi^2 cI} \times 10^{-2} = \frac{16.85762908}{I(\text{amu} \cdot \text{Å}^2)}, \tag{16.10}$$

in which I, the moment of inertia, is defined by

$$I = \mu r^2 = \frac{m_A m_B}{m_A + m_B} r^2, \tag{16.11}$$

and the masses m_A and m_B are separated by a distance r. The reduced mass μ of the AB molecule

$$\mu = \frac{m_A m_B}{m_A + m_B} \tag{16.12}$$

is conventionally calculated using atomic (not nuclear) masses. [5] The use of a single symbol B for three separate physical quantities (energy, frequency, and wavenumber) is clearly confusing, but it is the spectroscopic custom. Thus, spectroscopists talk about "energy" levels, but locate them using cm^{-1} units.

A real molecule is not a rigid rotor because the bond between atoms A and B can stretch at the same time as the molecule rotates. As rotation increases, the centrifugal force stretches the bond, increasing r and decreasing the effective B-value. The bond length also depends, in an average sense, on the vibrational state v. The nonrigid rotor energy level equation for vibrational state v is

$$F_v(J) = B_v J(J+1) - D_v[J(J+1)]^2 + H_v[J(J+1)]^3 + L_v[J(J+1)]^4 + \cdots, \tag{16.13}$$

where D_v, H_v, and L_v are centrifugal distortion constants. The vibrational dependence of the rotation and distortion constants is parameterized by

$$B_v = B_e - \alpha_e(v + \tfrac{1}{2}) + \gamma_e(v + \tfrac{1}{2})^2 + \cdots, \tag{16.14}$$

$$D_v = D_e + \beta_e(v + \tfrac{1}{2}) + \cdots, \tag{16.15}$$

where α_e, β_e, and γ_e are vibration-rotation interaction constants and the B_e and D_e values refer to the extrapolated equilibrium values at the bottom of the potential energy curve. Much of the conventional notation for spectroscopic constants is based on Herzberg's three books, [6]–[8] the work of Mulliken [3] and recent updates. [5]

Each vibration state v has an effective internuclear separation r_v defined by the equations

$$B_v = \left\langle v \left| \frac{1}{r^2} \right| v \right\rangle \frac{h^2}{8\pi^2 \mu} \tag{16.16}$$

and

$$r_{\text{v}} = \sqrt{\frac{h^2}{8\pi^2 \mu B_{\text{v}}}}. \tag{16.17}$$

There is a useful relationship (due to Kratzer [6]) for estimating the centrifugal distortion constant:

$$D_e = \frac{4B_e^3}{\omega_e^2}, \tag{16.18}$$

where ω_e is the equilibrium vibrational constant, Eq. (16.60).

A typical potential energy curve is often approximated (for semiquantitative work) as that for a Morse oscillator, viz.,

$$U(r) = D_e[1 - \mathrm{e}^{-\beta(r-r_e)}]^2, \tag{16.19}$$

in which D_e is the equilibrium dissociation energy [not the equilibrium centrifugal distortion constant appearing in Eqs. (16.15) and (16.18)]. For the Morse oscillator, the main vibration-rotation interaction term α_e is given by the Pekeris relationship, [6]

$$\alpha_e = \frac{6\sqrt{(\omega_e x_e B_e^3)} - 6B_e^2}{\omega_e}, \tag{16.20}$$

with the vibrational constants ω_e and $\omega_e x_e$ given by Eq. (16.60).

The selection rules for a pure rotational transition are $\Delta J = \pm 1$, so that the frequencies for a $J+1 \leftarrow J$ transition are given by

$$\nu_{J+1\leftarrow J} = 2B_v(J+1) - (4D_v - 2H_v)(J+1)^3 + 6H_v(J+1)^5 + \cdots. \tag{16.21}$$

Thus, the pure rotational transitions are a series of lines separated by approximately $2B$. Excited vibrational levels create a similar series of "vibrational satellites" near the main transitions for the $v = 0$ vibrational level.

The cross section for an absorption transition $E_1 \leftarrow E_0$ is given by [2]

$$\sigma = \frac{2\pi^2 \nu M_{10}^2}{3\varepsilon_0 hc} g(\nu - \nu_{10}), \tag{16.22}$$

and the Einstein A coefficient for emission is

$$A_{1\to 0} = \frac{16\pi^3 \nu^3}{3\varepsilon_0 hc^3} M_{10}^2 g(\nu - \nu_{10}), \tag{16.23a}$$

$$A_{1\to 0} = 3.136 \times 10^{-7} (\tilde{\nu})^3 M_{10}^2, \tag{16.23b}$$

where ε_0 is the permittivity of the vacuum, M_{10} is the transition dipole moment, and $g(\nu - \nu_{10})$ is the normalized lineshape function. In the numerical expression for A (16.23b), $\tilde{\nu}$ is in cm^{-1} and M_{10} is in debye ($1\ \text{D} = 3.33564 \times 10^{-30}$ C m).

The most common lineshape functions (normalized) are Lorenztian (for pressure and natural lifetime broadening) and Gaussian (Doppler broadening), [2]

$$g_L(\nu - \nu_{10}) = \frac{\Delta\nu_{1/2}/(2\pi)}{(\Delta\nu_{1/2}/2)^2 + (\nu - \nu_{10})^2}, \tag{16.24}$$

$$g_D(\nu - \nu_{10}) = \frac{2}{\Delta\nu_D}\sqrt{\frac{\ln 2}{\pi}} e^{-4\ln 2[(\nu-\nu_0)/\Delta\nu_D]^2}, \tag{16.25}$$

where $\Delta\nu_{1/2}$ is the Lorentian full-width at half maximum and $\Delta\nu_D$ is the Doppler (Gaussian) full-width at half maximum. A convenient formula for the Doppler width is given by

$$\Delta\nu_D = 2\nu_{10}\sqrt{\frac{2kT\ln 2}{mc^2}} = 7.2 \times 10^{-7}\tilde{\nu}_{10}\sqrt{\frac{T}{M}}. \tag{16.26}$$

In the numerical formula (16.26), $\Delta\nu_D$ and $\tilde{\nu}_{10}$ are in cm^{-1}, and T in K and M is the total mass of the molecule in amu.

For pure rotational transitions of a diatomic molecule, the square of the transition dipole moment integral M_{10} is [6]

$$(M_{10})^2 = \mu^2 S_J = \mu^2(J + 1), \tag{16.27}$$

and the $2J + 1M$-degeneracy of each rotational level needs to be included so that the absorption coefficient α, Eq. (16.4), for the $J + 1 \leftarrow J$ transition becomes

$$\alpha = \frac{2\pi^2\nu}{3\varepsilon_0 hc}\left[\mu^2(J + 1)\right]\left(\frac{N_1}{2J + 3} - \frac{N_0}{2J + 1}\right) g(\nu - \nu_{10}). \tag{16.28}$$

In Eqs. (16.27) and (16.28) μ is the permanent dipole moment of the molecule and $S_J(= J + 1)$ is a Hönl-London factor (see below). Thus, for homonuclear molecules such as Cl_2, $\mu = 0$, and there are no allowed pure rotational transitions. For emission work for the transition $J + 1 \rightarrow J$, the expression for the Einstein A factor is

$$A = \frac{16\pi^3\nu^3}{3\varepsilon_0 hc^3}\left[\mu^2(J + 1)\right]\left(\frac{N_1}{2J + 3}\right) g(\nu - \nu_{10}). \tag{16.29}$$

The dipole moment of a diatomic molecule can be measured by the application of an electric field. The Stark effect partly lifts the M_J-rotational degeneracy, and each level splits into $J + 1$, $|M_J|$-components. The Stark effect adds the term [9, 10]

$$E^{(2)} = \frac{\mu^2 E^2}{2B}\left[\frac{J(J + 1) - 3M_J^2}{J(J + 1)(2J - 1)(2J + 3)}\right] \tag{16.30}$$

to the usual energy level expression. This is a second-order Stark effect because it depends on the square of the electric field E, and the energy level expression is derived using second-order perturbation theory. The measurement of Stark splittings in rotational transitions is one of the primary methods for measuring dipole moments.

We have ignored the possibility of a net electron spin or a net orbital angular momentum. If either one is present, all of the energy level expressions are modified and each rotational transition will have fine structure. For example, if a molecule has a single unpaired electron, $S = 1/2$ ($^2\Sigma^+$state), all of the energy levels and transitions will be doubled. [6]

The presence of nuclear spins in a molecule will also split the energy levels into components and hyperfine structure will appear in the rotational transitions. In general, a nuclear spin $\vec{I}$ will vector couple with the rotational angular moment $\vec{J}$, viz.,

$$\vec{F} = \vec{J} + \vec{I}, \tag{16.31}$$

to give a total angular moment $\vec{F}$. The hyperfine structure can split a line into a maximum of $2I + 1$ components ($J \geq I$) each labeled by an F-value. The study of the fine structure and hyperfine structure of rotational transitions is often a complicated, but well-understood, task. [9]–[11]

16.2.2. Linear Molecules

The rotational energy level expressions for diatomic molecules apply directly to linear molecules in $^1\Sigma^+$electronic states. The only change is that each subscript v, e.g., in B_v, is to be interpreted as a collection of vibrational quantum numbers, and there is the possibility of new effects for the excited vibrational states. Such an effect is ℓ-type doubling, [12] which adds a term

$$E_\ell = \pm \frac{q}{2} J(J+1) \tag{16.32}$$

to the energy level expression in the case of doubly degenerate bending vibrational levels for $\ell = 1$. The ℓ-type doubling constant q measures the splitting of the rotational line into two ℓ-doublet components. It is the presence of vibrational angular momentum $\vec{\ell}$ that is responsible for this effect.

The $3N$-5 vibrational modes in a linear molecule also modify the vibration-rotation interaction terms, and the expression for B_v becomes

$$B_\mathrm{v} = B_e - \sum \alpha_i \left(v_i + \frac{d_i}{2} \right), \tag{16.33}$$

where d_i is the degeneracy of the ith vibrational mode.

16.2.3. Symmetric tops

Symmetric top molecules have an additional rotational quantum number K that measures the component of $\vec{J}$ along the top (molecular symmetry) axis. Thus, the rotational angular momentum vector $\vec{J}$ has components $\hbar K$ along the top axis and $\hbar M_J$ along the laboratory z-axis. The rotational energy level expression for a rigid molecule is given by

$$E_{JK} = BJ(J+1) + (A - B)K_a^2 \quad \text{(prolate top)}, \tag{16.34}$$

$$E_{JK} = BJ(J+1) + (C - B)K_c^2 \quad \text{(oblate top)}, \tag{16.35}$$

with subscripts a and c added to K to distinguish the prolate and oblate cases. The rotational constants A, B, and C are defined by analogy with Eq. (16.8),

$$A = \frac{h^2}{8\pi^2 I_A}, \tag{16.36}$$

$$B = \frac{h^2}{8\pi^2 I_B}, \tag{16.37}$$

$$C = \frac{h^2}{8\pi^2 I_C}, \tag{16.38}$$

in energy units. The numerical formulae for B (Eqs. (16.9) and (16.10)) are also applicable for A and C. Each energy level defined by the quantum numbers J and K has a $(2J + 1)$-fold M_J-degeneracy and a 2-fold K-degeneracy ($K > 0$), in the absence of electric or magnetic fields. By custom, the quantum number K is positive, with the symbol k occasionally being used when a signed quantum number is needed (i.e., $K = |k|, k = -J, \ldots, 0, \ldots J$).

For a nonrigid symmetric top, the energy level expression becomes

$$F_v(J, K) = BJ(J+1) - D_J[J(J+1)]^2 + (A-B)K^2 - D_K K^4 - D_{JK}J(J+1)K^2 + \cdots \tag{16.39}$$

for the prolate case and C replaces A in the oblate case. The centrifugal distortion constants D_J, D_K, and D_{JK} and the rotational constants A, B, and C also depend on the vibrational state by analogy with the diatomic case.

The intensity of a pure rotational transition is proportional to the square of the permanent dipole moment, which (by symmetry) can only lie along the symmetry axis of the top. The selection rules are $\Delta K = 0$ and $\Delta J = \pm 1$, which result in rotational transitions spaced by approximately $2B$. For the nonrigid molecule, the rotational transition frequencies for the transition $J + 1, K \leftarrow J, K$ are given by

$$\nu_{J+1,K \leftarrow J,K} = 2B(J+1) - 4D_J(J+1)^3 - 2D_{JK}(J+1)K^2. \tag{16.40}$$

Thus, centrifugal distortion causes each line to split into $J + 1$-components for $K = 0, 1, \ldots J$. Because each set of constants applies to a particular vibrational level, vibrational satellites can also arise from population in an excited vibrational level.

The application of an electric field E results in a lifting of the M_J-degeneracy with an additional energy level term $E^{(1)}$, to first order

$$E^{(1)}_{JKM} = -E\frac{2\mu K M_J}{J(J+1)(J+2)}. \tag{16.41}$$

The 2-fold K-degeneracy for $K > 0$ results in a first-order Stark effect with energy splittings directly proportional to the product of the electric field strength E and the dipole moment μ for small fields. A good collection of dipole moments of molecules as determined by the Stark effect and other methods can be found in the *CRC Handbook of Chemistry and Physics*. [13]

16.2.4. Asymmetric tops

There is no general energy level formula available for an asymmetric top molecule. The rigid rotor Hamiltonian

$$\hat{H} = A\hat{J}_a^2 + B\hat{J}_b^2 + C\hat{J}_c^2 \tag{16.42}$$

commutes with $\hat{J}^2$ and $\hat{J}_z$ in the laboratory coordinates system so that J and M_J remain good quantum numbers. The Hamiltonian $\hat{H}$, however, does not commute with the components of $\vec{J}(\hat{J}_a, \hat{J}_b, \hat{J}_c)$ in the molecular coordinate system. This means that the 2-fold K-degeneracy of the symmetric top is lifted and each J splits into $2J + 1$ components. In the asymmetric top case, each of these levels is labeled by an index $\tau = J, \ldots, 0, \ldots - J$ (cf. k for a symmetric top) in order of decreasing energy. The energy levels of the asymmetric top are most easily derived using symmetric top basis functions, first deriving the matrix elements of $\hat{H}$ and then diagonalizing $\hat{H}$ to find the energy eigenvalues and eigenvectors. For each value of J, a $(2J+1) \times (2J+1)$ matrix results and the $2J+1$ eigenvalues are easily labeled by τ.

The τ labeling scheme, however, is not as popular as one based on a correlation diagram between the energy levels of a prolate symmetric top and an oblate symmetric top. The degree of asymmetry of a top is quantified by the asymmetry parameter κ with

$$\kappa = \frac{2B - A - C}{A - C}. \tag{16.43}$$

Values of κ range from -1 for a prolate top ($B = C$) and $+1$ for an oblate top ($A = B$), with values near 0 for highly asymmetric tops. The correlation diagram is displayed as Figure 16.1, with prolate tops (J, K_a) on the left and oblate tops (J, K_c) on the right. The

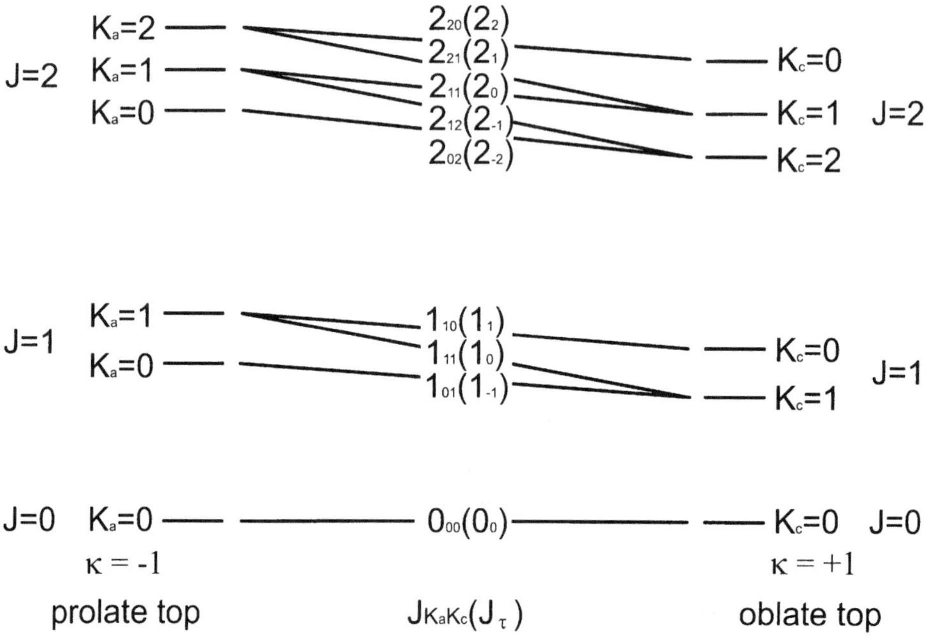

FIGURE 16.1. The correlation diagram and labeling of the rotational energy levels of an asymmetric top molecule.

parameter κ can be viewed as a continuous parameter with all asymmetric tops lying between these two extreme cases. The noncrossing rule is used to connect the energy levels, and the two limiting quantum numbers K_a and K_c serve to label each level. The relationship between the J_τ and $J_{K_a K_c}$ system of labels is $\tau = K_a - K_c$. Notice that $K_a + K_c = J$ or $J + 1$.

The energy levels for the rigid asymmetric top have exact analytical solutions for low values of J. A list of these energy levels for $J = 0, 1, 2$, and 3 are provided in Table 16.1. The energy levels for real molecules, however, require the addition of centrifugal distortion terms to the molecular Hamiltonian and are obtained by numerical solution of the resulting Hamiltonian matrices. [10, 11] Fortunately, many asymmetric top molecules are close to either the oblate or prolate limits.

The selection rules for asymmetric tops depend on the components (μ_a, μ_b, μ_c) of the permanent dipole moment vector $\vec{\mu}$ along the a, b, and c principal molecular axes. The selection rules can be divided into three general cases, as follows:

1. a-type transitions when $\mu_a \neq 0$, $\Delta K_a = 0(\pm 2, \pm 4, \ldots)$, $\Delta K_c = \pm 1(\pm 3, \pm 5, \ldots)$.
2. b-type transitions when $\mu_b \neq 0$, $\Delta K_a = \pm 1(\pm 3, \ldots)$, $\Delta K_c = \pm 1(\pm 3, \ldots)$.
3. c-type transitions when $\mu_c \neq 0$, $\Delta K_a = \pm 1(\pm 3, \pm 5, \ldots)$, $\Delta K_c = 0(\pm 2, \pm 4 \ldots)$.

The transitions in brackets are weaker than in the main transitions. A molecule of low symmetry can have $\mu_a \neq \mu_b \neq \mu_c \neq 0$ so that transitions of all three types can be found in its rotational spectrum. For very asymmetric tops, the pure rotational spectra have a very irregular appearance. We have ignored the complications of fine structure and hyperfine structure as well as any possible internal rotor structure.

TABLE 16.1. Rigid asymmetric rotor energy levels for $J = 0, 1, 2, 3$.

$J_{K_a K_c}$	J_τ	$F(J_\tau)$
0_{00}	0_0	0
1_{10}	1_1	$A + B$
1_{11}	1_0	$A + C$
1_{01}	1_{-1}	$B + C$
2_{20}	2_2	$2A + 2B + 2C + 2[(B - C)^2 + (A - C)(A - B)]^{1/2}$
2_{21}	2_1	$4A + B + C$
2_{11}	2_0	$A + 4B + C$
2_{12}	2_{-1}	$A + B + 4C$
2_{02}	2_{-2}	$2A + 2B + 2C - 2[(B - C)^2 + (A - C)(A - B)]^{1/2}$
3_{30}	3_3	$5A + 5B + 2C + 2[4(A - B)^2 + (A - C)(B - C)]^{1/2}$
3_{31}	3_2	$5A + 2B + 5C + 2[4(A - C)^2 - (A - B)(B - C)]^{1/2}$
3_{21}	3_1	$2A + 5B + 5C + 2[4(B - C)^2 + (A - B)(A - C)]^{1/2}$
3_{22}	3_0	$4A + 4B + 4C$
3_{12}	3_{-1}	$5A + 5B + 2C - 2[4(A - B)^2 + (A - C)(B - C)]^{1/2}$
3_{13}	3_{-2}	$5A + 2B + 5C - 2[4(A - C)^2 - (A - B)(B - C)]^{1/2}$
3_{03}	3_{-3}	$2A + 5B + 5C - 2[4(B - C)^2 + (A - B)(A - C)]^{1/2}$

Each vibrational level has a set of rotational constants whose vibrational dependence is parameterized by

$$A_v = A_e - \sum \alpha_i^A \left(v_i + \frac{d_i}{2} \right), \tag{16.44}$$

$$B_v = B_e - \sum \alpha_i^B \left(v_i + \frac{d_i}{2} \right), \tag{16.45}$$

$$C_v = C_e - \sum \alpha_i^C \left(v_i + \frac{d_i}{2} \right), \tag{16.46}$$

with d_i the degeneracy of the ith mode.

16.2.5. Spherical tops

The energy levels of a rigid spherical rotor are given by

$$F(J) = BJ(J+1). \tag{16.47}$$

Although this expression is identical to that for a rigid diatomic molecule, the degeneracy of each level is $(2J+1)^2$ rather than $2J+1$. There is a $(2J+1)$-fold M_J-degeneracy and a $(2J+1)$-fold K-degeneracy because the spherical top molecule is quantized, like a symmetric top, in both the laboratory and the molecular coordinate systems. For a nonrigid spherical top, the K-degeneracy can be partly lifted so that "cluster" splittings of the energy levels can be seen. [12] Because of its high degree of symmetry, a spherical top has no permanent dipole moment and, thus, no allowed pure rotational transitions.

16.3. VIBRATIONAL SPECTROSCOPY

A molecule must have a permanent dipole moment to have allowed pure rotational transitions. By contrast, a molecule needs to have its dipole moment change as it vibrates to have an allowed vibrational spectrum; i.e., vibrational spectra depend on dipole moment derivatives. This condition is much less restrictive, so that all molecules apart from homonuclear diatomics (which have both $\vec{\mu} = 0$ and $\frac{d\vec{\mu}}{dr} = 0$) have allowed vibrational spectra. For a polyatomic molecule, however, not all of the $3N$-6 vibrational modes (or $3N$-5 for a linear molecule) will necessarily be infrared active. For a molecule of sufficiently low symmetry, such as water, the three vibrations all appear in the infrared spectrum. The classification of molecules using a set of symmetry operations requires the application of group theory [2, 4, 14] and cannot be succinctly summarized. We will use the results of this symmetry classification as labels for vibrational and electronic states. [3] It can be shown that these labels, which are based on the irreducible representations of the appropriate molecular point group, can be applied to the individual rotational, vibrational, and electronic wavefunctions as well as to the overall product wavefunction,

$$\psi = \psi_{\text{el}} \psi_{\text{vib}} \psi_{\text{rot}}. \tag{16.48}$$

It is not customary to use these labels for rotational energy levels, and hence, we have not done so.

16.3.1. Diatomics

The potential energy curve for nuclear motion of a diatomic molecule can be approximated by a harmonic oscillator near the minimum, i.e., with the potential energy

$$U(r) = \frac{1}{2}k(r - r_e)^2, \tag{16.49}$$

with k as the force constant. The resulting classic vibrational frequency is

$$\nu = \frac{1}{2\pi}\sqrt{\frac{k}{\mu}}, \tag{16.50}$$

where μ is the reduced mass (not the dipole moment!). The quantum mechanical energy levels are, thus, given as

$$E = h\nu\left(v + \frac{1}{2}\right), \tag{16.51}$$

using energy units.

A more realistic potential energy curve is the three-parameter (D_e, r_e, and β) Morse oscillator of Eq. (16.19), that gives the vibrational energy levels

$$G(v) = \omega_e\left(v + \frac{1}{2}\right) - \omega_e x_e\left(v + \frac{1}{2}\right)^2 \tag{16.52}$$

now customarily in cm^{-1} units, with

$$\omega_e = \beta\left(\frac{D_e h \times 10^2}{2\pi^2 c\mu}\right)^{1/2}, \tag{16.53}$$

$$\omega_e x_e = \frac{h\beta^2 \times 10^2}{8\pi^2 \mu c}. \tag{16.54}$$

In this case, there is a single anharmonicity correction term $\omega_e x_e$ (which is always written as a single symbol). The vibration-rotation energy levels of the Morse oscillator (16.19) can be written as

$$\begin{aligned} E(v, J) = {} & \omega_e\left(v + \frac{1}{2}\right) - \omega_e x_e\left(v + \frac{1}{2}\right)^2 + B_e J(J+1) - D[J(J+1)]^2 \\ & - \alpha_e\left(v + \frac{1}{2}\right)J(J+1) + \cdots, \end{aligned} \tag{16.55}$$

in which B_e, D, and α_e are given by Eq. (16.14), (16.18), and (16.20), respectively. Note that although the vibrational energy expression (16.52) is exact for the Morse oscillator, the rotational terms appearing in Eq. (16.55) are not. [9] In Eqs. (16.53) and (16.54) all of the spectroscopic constants, including the Morse β parameter, are in cm^{-1} and the fundamental constants are in SI units. A Morse oscillator also has the dissociation energy $D_e = \omega_e^2/(4\omega_e x_e)$. In the above equations, the symbol D_e is customarily used for both the dissociation energy and the equilibrium centrifugal distortion constant, relying on the context to distinguish between the two. Here, in Eq. (16.55), the subscript e has been deleted for clarity.

Although a large number of potential energy functions have been proposed, the Dunham form

$$U(r) = a_0\xi^2(1 + a_1\xi + a_2\xi^2 + \cdots), \tag{16.56}$$

with

$$\xi = \frac{r - r_e}{r_e} \tag{16.57}$$

and

$$a_0 = \frac{kr_e^2}{2} = \frac{\omega_e^2}{4B_e}, \tag{16.58}$$

is the most widely used. The vibration-rotation energy levels of the Dunham potential are given by the double sum

$$E(v, J) = \sum Y_{jk}\left(v + \frac{1}{2}\right)^j [J(J+1)]^k. \tag{16.59}$$

The Dunham coefficients Y_{jk} can be related back to the Dunham potential parameters a_i [9] and to the customary spectroscopic constants as follows:

$$\begin{aligned}
&Y_{10} = \omega_e, && Y_{20} = -\omega_e x_e, && Y_{30} = \omega_e y_e, && Y_{40} = \omega_e z_e,\\
&Y_{01} = B_e, && Y_{11} = -\alpha_e, && Y_{21} = \gamma_e,\\
&Y_{02} = -D_e, && Y_{12} = -\beta_e,\\
&Y_{03} = H_e.
\end{aligned}$$

The customary vibrational energy level expression is

$$G(v) = \omega_e\left(v + \frac{1}{2}\right) - \omega_e x_e\left(v + \frac{1}{2}\right)^2 + \omega_e y_e\left(v + \frac{1}{2}\right)^3 + \omega_e z_e\left(v + \frac{1}{2}\right)^4 + \cdots. \tag{16.60}$$

The relationship between potential parameters and spectroscopic constants contains some correction terms, first derived by Dunham, and additional terms are needed to account for the breakdown of the Born-Oppenheimer approximation. [15] Notice that the use of Dunham Y_{jk} constants avoids the confusion created by constants with the same symbol, such as the equilibrium centrifugal distortion constant and the dissociation energy (both D_e), as well as the customary negative signs in front of $\omega_e x_e$, α_e, and D_e. The best collection of spectroscopic constants for diatomics remains the book by Huber and Herzberg. [16]

A harmonic oscillator has selection rules $\Delta v = \pm 1$, which leads to the fundamental vibrational band $v = 1 \leftrightarrow 0$ plus various hot bands, corresponding to $v = 2 \leftrightarrow 1, 3 \leftrightarrow 2, \ldots$. By definition, a hot band occurs between two excited vibrational levels. Vibrational bands can appear in absorption $v = 1 \leftarrow 0$ or emission $v = 1 \rightarrow 0$; it is customary to put the excited state quantum number first.

Real molecules are anharmonic oscillators because the potential energy function contains cubic and higher order terms (16.56), and because the dipole moment is not simply a linear function of the internuclear separation, but rather has the form

$$\mu(r) = \mu_e + \left.\frac{d\mu}{dr}\right|_{r_e} (r - r_e) + \frac{1}{2}\left.\frac{d^2\mu}{dr^2}\right|_{r_e} (r - r_e)^2 + \cdots. \tag{16.61}$$

A real diatomic molecule is thus both mechanically and electrically anharmonic. This anharmonicity allows overtone transitions to appear with $\Delta v = \pm 2, \pm 3, \ldots$. Each increase in Δv by one unit results in a decrease in intensity of an order of magnitude (or more). In terms of the vibrational constants, a fundamental band occurs at

$$\Delta G_{1/2} = \nu_{1-0} = \omega_e - 2\omega_e x_e + \cdots, \tag{16.62}$$

a hot band with $\Delta v = 1$ occurs at

$$\Delta G_{v+1/2} = \nu_{v+1 \leftarrow v} = \omega_e - 2\omega_e x_e (v + 1) + \cdots, \tag{16.63}$$

and the first overtone occurs at

$$\nu_{2 \leftarrow 0} = 2\omega_e - 6\omega_e x_e + \cdots, \tag{16.64}$$

with $\Delta G_{v+1/2}$ defined as

$$\Delta G_{v+1/2} = G_{v+1} - G_v. \tag{16.65}$$

A simple heteronuclear diatomic molecule such as HCl has a fundamental vibration-rotation transition that occurs in the infrared spectra region near 2900 cm^{-1}. The rotational selection rules are $\Delta J = \pm 1$. Transitions are labeled by their $\Delta J = J' - J'' = -3, -2, -1, 0, 1, 2, 3$ values as N, O, P, Q, R, S, and T, respectively. For electric dipole-allowed transitions, only $\Delta J = -1, 0, 1$, (P, Q, R) are possible. However, in the case of HCl, there is no net electron spin or orbital angular momentum so that $\Delta J = 0$ transitions are absent. Multiphoton transitions, Raman transitions, and electric quadrupole transitions have the possibility of occurring with $|\Delta J| > 1$.

It is customary to label the upper energy level with a single prime and the lower with a double prime, and to write the upper level first. The fundamental 1-0 vibrational band of HCl thus has energy levels

$$E(v' = 1, J') = \nu_{1\text{-}0} + B'J'(J' + 1), \tag{16.66}$$

$$E(v'' = 0, J'') = B''J''(J'' + 1), \tag{16.67}$$

with $\nu_{1\text{-}0} = \omega_e - 2\omega_e x_e + \cdots$ called the vibrational band origin.

The R-branch lines ($\Delta J = +1$) occur at

$$\nu_R(J+1 \leftarrow J) = \nu_{1\text{-}0} + 2B' + (3B' - B'')J + (B' - B'')J^2, \tag{16.68}$$

and the P-branch lines ($\Delta J = -1$) occur at

$$\nu_P(J-1 \leftarrow J) = \nu_{1\text{-}0} - (B' + B'')J + (B' - B'')J^2, \tag{16.69}$$

if the effects of centrifugal distortion are ignored. If the quantity $m = J + 1$ for the R-branch and $m = -J$ for the P-branch is defined, Eqs. (16.68) and (16.69) can be combined as

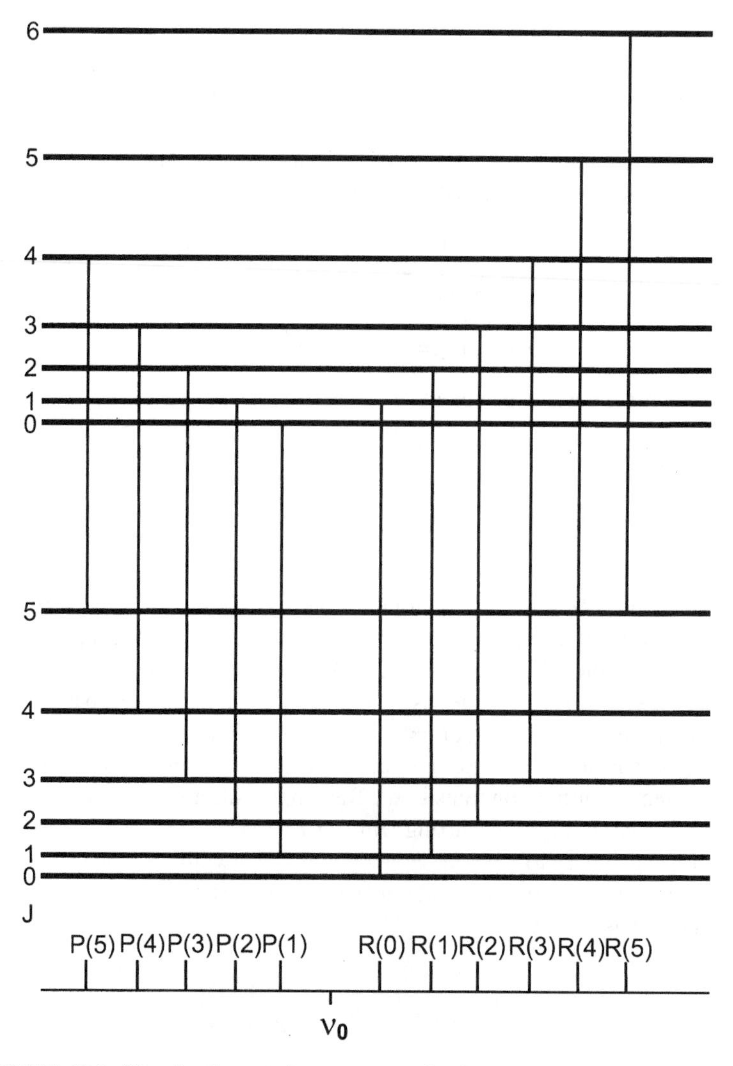

FIGURE 16.2. The vibration-rotation spectrum of a diatomic molecule, such as HCl.

$$\nu_{P \text{ and } R} = \nu_{1\text{-}0} + (B' + B'')m + (B' - B'')m^2. \tag{16.70}$$

Indeed, all infrared bands can be represented by a higher order polynomial in m when centrifugal distortion is included. If $B' \approx B''$, the R- and P-branch lines are spaced by about $2B$ with a "band gap" of $4B$ near the origin because of the missing $m = 0$ line (Figure 16.2).

16.3.2. Linear molecules

The vibrational motion of linear molecules (like other polyatomics) is approximated by a set of independent normal modes. Each normal mode is represented by a simple harmonic oscillator so that the vibrational energy is given by a sum over the $3N$-5 normal modes as

$$E(v_i) = \sum_{i=1}^{3N\text{-}5} h\nu_i \left(v_i + \frac{1}{2}\right), \tag{16.71}$$

where each normal mode i has a frequency ν_i and a vibrational quantum number v_i. Due to the degeneracy, the number of different vibrational frequencies can be less than $3N$-5, and hence, the sum is usually written in the form

$$E(v_i) = \sum h\nu_i \left(v_i + \frac{d_i}{2}\right), \tag{16.72}$$

where d_i is the degeneracy of the ith mode and the sum is over the distinct vibrational frequencies. For example, CO_2 has four normal modes, but the bending modes are doubly degenerate, so that there are only three fundamental frequencies corresponding to the symmetric stretch $\nu_1(\sigma_g)$ at 1388 cm^{-1}, the bending mode $\nu_2(\pi_u)$ at 667 cm^{-1} and the antisymmetric stretch $\nu_3(\sigma_u)$ at 2349 cm^{-1}. The infrared spectrum of CO_2, however, contains strong ν_2 and ν_3 bands because ν_1 has no oscillating dipole moment and is thus forbidden.

Degenerate vibrational modes have an additional complication because they have vibrational angular momentum $\vec{\ell}$ in addition to the rotational angular momentum, now called $\vec{R}$. Thus, the total angular momentum, $\vec{J}$, is the vector sum of rotational and vibrational contributions; i.e.,

$$\vec{J} = \vec{R} + \vec{\ell}\,. \tag{16.73}$$

The vibrational angular momentum quantum number ℓ can have values $\ell = v, v-2, \ldots 1$ or 0 for a degenerate bending mode with v quanta. Thus, $v = 0$ has $\ell = 0$, $v = 1$ has $\ell = 1$, $v = 2$ has $\ell = 2$ and 0, $v = 3$ has $\ell = 3$ and 1, and so on. Notice that all levels with $\ell > 0$ have a 2-fold degeneracy (associated with $\pm\ell$), even though only $|\,\ell\,|$ is used as a label.

The complete energy level expression for the vibrational energy levels of a real anharmonic molecule is

$$G(v_1 v_2 \ldots) = \sum \omega_r \left(v_r + \frac{d_r}{2}\right) + \sum_{r \le s} x_{rs} \left(v_r + \frac{d_r}{2}\right)\left(v_s + \frac{d_s}{2}\right) + \sum_{t \le t'} g_{tt'} \ell_t \ell_{t'}, \tag{16.74}$$

where d_r and d_s are the degeneracies of the rth and sth modes, and t applies to degenerate modes with vibrational angular momentum. For CO_2, this expression becomes

$$\begin{aligned} G(v_1 v_2 v_3) = {} & \omega_1\left(v_1 + \tfrac{1}{2}\right) + \omega_2(v_2 + 1) + \omega_3\left(v_3 + \tfrac{1}{2}\right) \\ & + x_{11}\left(v_1 + \tfrac{1}{2}\right)^2 + x_{22}(v_2 + 1)^2 + x_{33}\left(v_3 + \tfrac{1}{2}\right)^2 \\ & + x_{12}\left(v_1 + \tfrac{1}{2}\right)(v_2 + 1) + x_{13}\left(v_1 + \tfrac{1}{2}\right)\left(v_3 + \tfrac{1}{2}\right) \\ & + x_{23}(v_2 + 1)\left(v_3 + \tfrac{1}{2}\right) \\ & + g_2 \ell^2, \end{aligned} \tag{16.75}$$

in which the six x_{ij}'s and the g_2 term account for the anharmonic behavior and vibrational resonances are ignored. Indeed, most molecules display interactions between vibrational modes that shift some of the energy levels away from the values predicted by Eq. (16.75). These vibrational perturbations have been classified into various types such as Fermi resonance and Coriolis resonance interactions. [12] There is no convenient collection of vibrational frequencies of stable molecules other than the older work of Shimanouchi. [17] Jacox has compiled vibrational constants for transient polyatomics. [18]

Linear molecules have two basic types of vibrational motion: parallel to the linear axis (z-axis) and perpendicular to the linear axis (in the xy-plane). Those parallel to the axis involve bonding stretching motions, whereas those perpendicular involve bending motions. Stretching modes are thus called parallel bands and bending modes are called perpendicular bands because of the direction of the oscillating dipole moment. Parallel bands like the ν_3-mode of CO_2 have rotational selection rules $\Delta J = \pm 1$ and have spectra identical to those of a diatomic like HCl.

Perpendicular modes have rotational selection rules $\Delta J = 0, \pm 1$ and thus have P-, Q-, and R-branches. Ignoring the effect of the ℓ-type doubling term, Eq. (16.32), results in Eqs. (16.68) and (16.69) for P- and R-branches and

$$\nu_Q = \nu_{1\text{-}0} + (B' - B'')J(J + 1) \tag{16.76}$$

for the Q branch. The selection rule for vibrational angular momentum is $\Delta \ell = \pm 1$, and the ℓ-doubling of the excited state rotational levels does not appear directly in the spectrum because the two nearly degenerate levels have opposite total parity. Each J level of the ground state, thus, connects to only one of the two ℓ-doublets. A perpendicular band like ν_2 of CO_2, thus, has a strong Q branch piled up near the band origin in addition to the P- and R-branches.

The additional possible complications of fine structure and hyperfine structure have been ignored. Polyatomic molecules, however, have one additional feature that needs to be discussed: combination bands. For a collection of harmonic oscillators, the selection rule on v_i is $\Delta v_i = \pm 1$ and $\Delta v_j = 0$ for all $j \neq i$. Thus, only fundamental bands can appear in a cold spectrum. For a real anharmonic molecule, overtones similar to those in the diatomic case can appear. Combination modes are similar to overtones in that two or more vibrational quanta can be exchanged, but they can belong to different modes. For example, anharmonicity allows the $\nu_1 + \nu_3$- and the $\nu_1 + \nu_2$-modes of CO_2 to appear in the spectrum at 3716 cm^{-1} and 2076 cm^{-1}, respectively. Note that in the CO_2 case, the first overtone modes are all forbidden by symmetry although they can be located using Raman spectroscopy. [7]

16.3.3. Symmetric tops

The vibrational energy levels of a symmetric top are given, in general, by expressions (16.72) and (16.74) for linear polyatomics, although the number of modes is $3N$-6 rather than $3N$-5. Modes in symmetric top molecules can also be doubly degenerate and can have associated vibrational angular momentum.

In many ways, linear polyatomic molecules are special cases of symmetric top molecules. There are two types of vibrational modes in a symmetric top, parallel and perpendicular, depending on the orientation of the oscillating dipole moment relative to the top axis. The case of parallel bands is relatively simple and will be discussed first.

A typical prolate symmetric top molecule is CH_3F. CH_3F has six vibrational frequencies: $\nu_1(a_1)$ at 2965 cm^{-1}, the symmetric C-H stretch; $\nu_2(a_1)$ at 1475 cm^{-1}, the symmetric CH_3 umbrella bend; $\nu_3(a_1)$ at 1048 cm^{-1}, the C-F stretch; $\nu_4(e)$ at 2982 cm^{-1}, the antisymmetric C-H stretch; $\nu_5(e)$ at 1471 cm^{-1}, the antisymmetric bend; and $\nu_6(e)$ at 1196 cm^{-1}, the CH_3 rock. The ν_1-, ν_2-, and ν_3-modes are parallel modes, whereas ν_4, ν_5, and ν_6 are doubly-degenerate perpendicular modes. For the parallel modes, the selection rules are $\Delta K = 0$, $\Delta J = 0, \pm 1$ (except for $K = 0$ for which $\Delta J = \pm 1$). Because these parallel symmetric top selection rules are very similar to those for a perpendicular transition of a linear molecule, the bands will be similar. The bands, thus, have simple P-, Q-, R-branches, although for the symmetric top, each rotational line will show K-structure at high resolution. As in the pure rotational case, each rotational line will split into $J + 1$ K-components because $K = 0, 1, \ldots J$.

The perpendicular case is much more complicated and is difficult to summarize. The doubly degenerate excited vibrational states are strongly split by a large first-order Coriolis interaction. [12] The rotational energy levels are given either as

$$E = BJ(J+1) + (A-B)K_a^2 \mp 2A\zeta K_a \quad \text{(prolate top)} \tag{16.77}$$

or

$$E = BJ(J+1) + (C-B)K_c^2 \mp 2C\zeta K_c \quad \text{(oblate top)}, \tag{16.78}$$

where ζ is a Coriolis coupling constant. The vibrational angular momentum is now $\zeta\hbar$ units along the top axis and ζ need not be an integer (unlike the linear molecule case in which the projection is $\ell\hbar$, with ℓ integral). This large Coriolis splitting strongly shifts the K-structure associated with the perpendicular selection rules $\Delta J = 0, \pm 1$ and $\Delta K = \pm 1$. This results in widely shifted sub-bands, with each sub-band associated with a particular $K' \leftarrow K''$ transition and having a P-, Q-, and R-branch. The sub-band origins [7] are located at

$$\nu^{\text{sub}} = \nu_{1\text{-}0} + [A'(1-2\zeta) - B'] \pm 2[A'(1-\zeta) - B']K + [(A'-B') - (A''-B'')]K^2 \tag{16.79}$$

for $K+1 \leftarrow K$ (+ sign) and $K-1 \leftarrow K$ (− sign) for the prolate top. The K sub-bands are

TABLE 16.2. Hönl-London (symmetric top) rotational line strength factors.

S_{JK}	$\Delta K = +1$	$\Delta K = 0$	$\Delta K = -1$
$\Delta J = 1$	$\dfrac{(J''+2+K'')(J''+1+K'')}{4(J''+1)}$	$\dfrac{(J''+1+K'')(J''+1-K'')}{(J''+1)}$	$\dfrac{(J''+2-K'')(J''+1-K'')}{4(J''+1)}$
$\Delta J = 0$	$\dfrac{(J''+1+K'')(J''-K'')(2J''+1)}{4J''(J''+1)}$	$\dfrac{(2J''+1)(K'')^2}{J''(J''+1)}$	$\dfrac{(J''+1-K'')(J''+K'')(2J''+1)}{4J''(J''+1)}$
$\Delta J = -1$	$\dfrac{(J''-1-K'')(J''-K'')}{4J''}$	$\dfrac{(J''+K'')(J''-K'')}{J''}$	$\dfrac{(J''-1+K'')(J''+K'')}{4J''}$

thus approximately spaced by $2[A(1-\zeta)-B]$. The usual expressions (16.68), (16.69), and (16.76) for P-, Q-, and R-branch lines hold approximately for each sub-band. Depending on the magnitudes of A, B, and ζ these perpendicular bands can have well-separated sub-bands or a massively congested appearance with strongly overlapping sub-bands.

The absorption line intensities are given by the general expression Eq. (16.22), but the transition moment factor M_{10}^2 needs to be evaluated for the symmetric top. In particular,

$$M_{10}^2 = S_{JK}\mu_{10}^2, \tag{16.80}$$

where S_{JK} is a rotational line strength factor often called a Hönl-London factor and μ_{10} is to be interpreted as a vibrational transition dipole moment, rather than the permanent dipole moment, as in the pure rotational case. A collection of infrared band strengths of molecules can be found in the book edited by Rao and Weber. [19] The Hönl-London factors S_{JK} are provided in Table 16.2 for the nine different $J'K' \leftarrow J''K''$ cases. The absorption coefficient α for the $E_1 \leftarrow E_0$ transition is then given by

$$\alpha = \frac{2\pi^2\nu}{3\varepsilon_0 hc} S_{JK}\mu_{10}^2 \left(\frac{N_1}{d_1} - \frac{N_0}{d_0}\right) g(\nu - \nu_{10}), \tag{16.81}$$

where the degeneracy factors in the upper (d_1) and lower (d_0) states need to be included together with the line shape function $g(\nu - \nu_{10})$. If the upper state is unpopulated, then $N_1 \approx 0$. For a symmetric top, the degeneracy factors are $2J+1$ for $K=0$ and $2(2J+1)$ for $K > 0$. Note also that the effects of nuclear spin [4] are not included in Eq. (16.81). For emission, the expression for the Einstein A coefficient for emission from level 1 is

$$A = \frac{16\pi^3\nu^3}{3\varepsilon_0 hc^3} S_{JK}\mu_{10}^2 \left(\frac{N_1}{d_1}\right) g(\nu - \nu_{10}). \tag{16.82}$$

These Hönl-London factors and the intensity expressions are very useful in that they can be used for diatomics, linear molecules, and symmetric tops for pure rotational, vibrational, and electronic transitions. The quantum number K can be interpreted as the projection of any angular momentum along the symmetry axis. For example, for pure rotational transitions of diatomic and linear molecules in $^1\Sigma^+$ electronic states $K''=0$, $\Delta K''=0$, and $\Delta J = +1$, so that $S = J+1$, in agreement with Eq. (16.27). Similarly, pure rotational transitions of symmetric tops have $\Delta K = 0$, $\Delta J = +1$. For infrared transition of diatomic molecules in $^1\Sigma^+$ electronic states, $K''=0$, $\Delta K''=0$, and $\Delta J = \pm 1$ so that $S_J = J+1$ in the R-branch and $S_J = J$ in the P-branch. For linear polyatomics, the intensity expressions for parallel transitions are the same as for diatomics. For fundamental perpendicular transitions, however, $\ell = 1 = K''$ in the excited state and $\ell = 0 = K''$ in the ground state. Thus, the first column with $\Delta K = +1$ applies and $S_J = (J + 2)/4$, $(2J + 1)/4$, and $(J - 1)/4$ for the P-, Q-, and R-branches, respectively. Notice that in this case, the Q-branch lines are about twice as intense as the P- and R-branch lines.

16.3.4. Asymmetric tops

The vibrational energy levels of asymmetric tops are given by the general energy level expression (16.74). The symmetry of asymmetric tops is sufficiently low that degenerate vibrational levels and a resulting vibrational angular momentum does not occur. The energy level expression then simplifies to

$$G(v_i) = \sum_{i=1}^{3N\text{-}6} \omega_i \left(v_i + \frac{1}{2} \right) + \sum_{i \geq j}^{3N\text{-}6} \sum_{j=1}^{3N\text{-}6} x_{ij} \left(v_i + \frac{1}{2} \right) \left(v_j + \frac{1}{2} \right). \tag{16.83}$$

There are, thus, $3N$-6 distinct vibrational frequencies for asymmetric rotor molecules. These modes are allowed if they have an oscillating dipole moment.

The rotational energy levels and line strengths of the asymmetric rotor are not given by analytical formulae. Energy levels and intensities are, thus, computed numerically by diagonalizing the asymmetric rotor Hamiltonian using symmetric top basis functions. Line intensities are computed numerically based on the known Hönl-London factors that describe transition moment matrix elements between the symmetric top basis functions.

The bands of an asymmetric top are classified by whether they have oscillating dipole moments in the a, b, or c directions. Bands are a-type, b-type, or c-type if they have transition dipole moments μ_a, μ_b, or μ_c. These transition dipole moments are proportional to dipole moment derivatives in the a, b, and c directions for the particular motion described by the normal mode in question. For example, in H_2O, the symmetry axis is the b-axis so the symmetric stretching motion and the bending motion are both b-type bands. The antisymmetric stretching motion, in contrast, gives an oscillating dipole moment along the a-axis that lies in the plane of the molecule and is perpendicular to the symmetry axis. The rotational selection rules for a-, b-, and c-type transitions have already been given. In general, a molecule with sufficiently low symmetry can have a mode that has an oscillating dipole moment with components in a, b, and c directions, and thus, all three types of transitions will be found in this band. In general, asymmetric rotor bands have a complex appearance. However, if the molecule is close to a symmetric rotor (i.e., $|\,\kappa\,| \approx 1$), the bands will look like parallel or perpendicular (with $\zeta = 0$) bands of the corresponding symmetric top.

16.3.5. Spherical tops

Because of their high symmetry, spherical tops always have some degenerate vibrational modes and the full vibrational energy level expression (16.74) applies. Only triply degenerate fundamental vibrational modes are infrared active. For example, CH_4 has four vibrational frequencies: $\nu_1(a_1)$ at 2914 cm^{-1}, the symmetric C-H stretch; $\nu_2(e)$ at 1526 cm^{-1}, the doubly degenerate bend, $\nu_3(t_2)$ at 3020 cm^{-1}, the antisymmetric C-H stretch, and $\nu_4(t_2)$ at 1306 cm^{-1}, the triply degenerate bend. Only ν_3 and ν_4 appear strongly in the spectrum.

The triple vibrational degeneracy of the allowed modes causes complications in the excited state energy levels. [12] There is a strong first-order Coriolis effect that splits them into three components. The excited vibration-rotation energy levels are found at

$$E^+ = \nu_{1\text{-}0} + BJ(J+1) + 2B\zeta J, \tag{16.84}$$

$$E^0 = \nu_{1\text{-}0} + BJ(J+1) - 2B\zeta, \tag{16.85}$$

$$E^- = \nu_{1\text{-}0} + BJ(J+1) - 2B\zeta(J+1), \tag{16.86}$$

where $\nu_{1\text{-}0}$ is the band origin and there are $\zeta\hbar$ units of vibrational angular momentum present.

The rotational selection rule is $\Delta J = 0, \pm1$, but no obvious tripling is present in the spectrum. This is because only the E^+ levels are used for the P-branch, E^0 for the Q-branch, and E^- for the R-branch. The Coriolis effect is, thus, similar to that discussed for perpendicular bands of symmetric tops. At low resolution, spherical top bands have a *PQR* structure similar to a perpendicular band of a linear molecule or a parallel band of a symmetric top. At higher resolution, however, cluster splittings appear in the lines and they require a surprisingly sophisticated theory [20] to account for their magnitude.

16.3.6. Raman Spectroscopy

So far, no mention has been made of the Raman effect and other light scattering phenomena. When the nonresonant interaction of radiation with a molecule occurs, the scattered photon usually has the same energy as the incident photon (Rayleigh scattering). However, the scattering event can be inelastic and the molecule will, therefore, be left in a different vibrational, rotational, or electronic state. This phenomenon is the Raman effect, [7, 21, 22] and it has a typical scattering efficiency of 10^{-5}. Although the Raman effect is weak, the availability of high-power lasers has made observations routine. A Raman spectrum is obtained by measuring the energy lost (or gained) by the scattered photon. The Raman effect is particularly useful for measuring the vibrational frequencies of large molecules in condensed phases. Modern Raman spectroscopy is limited usually by the presence of trace amounts of fluorescent impurities in the sample.

The selection rules for the Raman effect are different from those that govern the normal dipolar interaction of radiation and matter. The transition operator, c.f. Eq. (16.5), involves matrix elements of the molecular polarizability tensor rather than the dipole pole moment (vector) operator. [22] This difference makes the selection rules for the Raman effect similar to those for two photon spectroscopy. [23] For example, the rotational selection rules for the vibration-rotation Raman spectrum of HCl are $\Delta J = 0, \pm2$ rather than the $\Delta J = \pm1$ dipole selection rules. For molecules like CO_2 that have a center of inversion symmetry, the infrared active modes are Raman inactive and vice versa. Raman spectroscopy, thus, has complementary selection rules to ordinary infrared spectroscopy and different intensities for the vibrational modes.

The vibrational frequencies of molecules are mainly deduced from infrared and Raman spectra as well as from vibronic transitions (Sec. 16.4). Vibrational frequencies vary

TABLE 16.3. Infrared group wavenumbers.

Group	ν/cm^{-1}	Group	ν/cm^{-1}
≡C–H	3300	–O–H	3600
=C(–H)	3020	>N–H	3350
≥C–H	2960	≥P=O	1295
–C ≡ C–	2050		
>C=C<	1650	>S=O	1310
≥C–C≤	900	≡C–H (bend)	700
–S–H	2500		
–N=N–	1600	=C(H)H (bend)	1100
>C=O	1700		
–C ≡ N	2100	–C(H)(H)H (bend)	1000
≥C–F	1100		
≥C–Cl	650	>C(H)H (bend)	1450
≥C–Br	560		
≥C–I	500	C≡C–C (bend)	300

from molecule to molecule, but certain regularities are obvious. In particular, certain types of chemical bonds and functional groups have characteristic stretching and bending frequencies. These characteristic group frequencies (Table 16.3 [2, 13]) are widely used for qualitative analysis in organic [24] and inorganic [25] chemical spectroscopy. The analysis of materials is the main practical application of vibrational spectroscopy.

16.4. ELECTRONIC SPECTRA

The electronic transitions of molecules show the greatest variety of all of the different types of spectra. This is largely because the electron spin and orbital angular momenta often change and the ground and excited states can have different geometries. In fact, transitions that change the symmetry of a molecule, such as linear to bent, are not uncommon. Changes in geometry, in addition to the necessity for considering effects such as spin-orbit and spin-rotation coupling, make the study of electronic spectra particularly complicated.

Electronic transitions have associated vibrational and rotational structures. A particular rovibronic transition occurs at a line position ν with

$$\nu = T_e + G'(v') + F'(J') - G''(v'') - F''(J''), \tag{16.87}$$

where $G(v)$ and $F(J)$ are the vibrational and rotational energy level expressions already discussed and T_e is the equilibrium transition energy between the states. This expression assumes, of course, that the electronic, vibrational, and rotational energies can be separated and that the states involved are not subject to some sort of fast dynamical process. If one or both of the states is, for example, predissociated or preionized, the rotational or even the vibrational structure may be intrinsically unresolvable. If a state participates in a fast dynamical process (time scale $\sim \Delta t$), the linewidth (ΔE or $\Delta \nu$) derived from Heisenberg's uncertainty principle

$$\Delta E \Delta t \geq \hbar \tag{16.88}$$

or

$$\Delta \nu \Delta t \geq \frac{1}{2\pi} \tag{16.89}$$

may be larger than the rotational or vibrational structure.

16.4.1. Diatomics

The various angular momenta in a diatomic molecule are illustrated in Figure 16.3. [2, 26] The total angular momentum (exclusive of nuclear spin) is always labeled as $\vec{J}$ and is the vector sum of nuclear rotational ($\vec{R}$), electron orbital ($\vec{L}$), and electron spin ($\vec{S}$) momenta, viz.

$$\vec{J} = \vec{R} + \vec{L} + \vec{S}. \tag{16.90}$$

The components of $\vec{L}$, $\vec{S}$, and $\vec{J}$ along the A-B-axis are called Λ, Σ, and Ω, respectively. Each state is identified by the term symbol $^{2S+1}\Lambda$. The degeneracy of the $^{2S+1}\Lambda$ term is $(2S+1)2$ because of the $2S+1$ different possible values of $\Sigma = S, S-1, \ldots -S$ and the

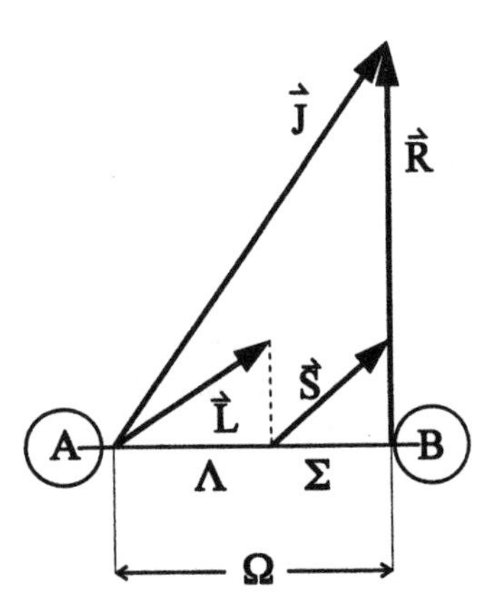

FIGURE 16.3. Angular momenta in a diatomic molecule.

2-fold $\pm|\Lambda|$ orbital degeneracy. For states with $\Lambda = 0$, i.e., ${}^{2S+1}\Sigma^{+}$ or ${}^{2S+1}\Sigma^{-}$ states, the degeneracy is $2S+1$. The superscripts $+$ or $-$ are added in this case to distinguish between electronic states that are symmetric ($+$) or antisymmetric ($-$) with respect to reflection in the symmetry plane containing the nuclei. For homonuclear molecules, an additional right subscript is added to the term symbol, ${}^{2S+1}\Lambda_g$ or ${}^{2S+1}\Lambda_u$. If the electronic wavefunction is symmetric with respect to inversion of the electrons through the center of symmetry, g is used, whereas u identifies the antisymmetric case. Notice that the g or u symmetry of an electronic state applies only to homonuclear molecules and is not to be confused with the total parity of a rovibronic state. Total parity is the symmetry associated with the inversion of all particles (electrons and nuclei) through the origin in the laboratory coordinate system. This is a symmetry operation for all molecules because the energy levels depend only on the relative positions of the particles, which are unchanged by this operation. [2]

Spin-orbit coupling is accounted for by considering the phenomenological spin-orbit Hamiltonian,

$$\hat{H}_{SO} = A\hat{L} \cdot \hat{S} \tag{16.91}$$

that causes an energy level splitting of

$$E_{SO} = A\Lambda\Sigma. \tag{16.92}$$

The effect of spin-orbit coupling is, thus, to lift the $(2S + 1)$-fold spin degeneracy for $\Lambda > 0$. Each spin-component of a ${}^{2S+1}\Lambda$ term is then labeled with Ω, which is written as a subscript, ${}^{2S+1}\Lambda_\Omega$. A 2-fold degeneracy remains to account for the $\pm|\Omega|$-possibilities. Note that, by custom, Λ and Ω are usually not signed but that Σ is signed to differentiate the $2S + 1$ spin components.

The selection rules for allowed one-photon electric dipole-allowed transitions are as follows:

1. $\Delta\Lambda = 0, \pm 1$.
2. $\Delta S = 0$ (light molecules only).
3. $\Delta\Omega = 0, \pm 1$.
4. Only $\Sigma^+ - \Sigma^+$, $\Sigma^- - \Sigma^-$, but $\Pi - \Sigma^+$ and $\Pi - \Sigma^-$ are both allowed.
5. Only $g \leftrightarrow u$ for homonuclear molecules.

Each electronic transition has associated vibrational structure and rotational structure. In the simplest approximation, the transition $v', J' \leftarrow v'', J''$ occurs at

$$\nu = T_e + \omega_e' \left(v' + \frac{1}{2}\right) + B'J'(J+1) - \omega_e'' \left(v'' + \frac{1}{2}\right) - B''J''(J''+1). \tag{16.93}$$

The line absorption or emission intensities are given by the usual formulae, (16.22) and (16.23). In the case of singlet-singlet electronic transitions (i.e., ${}^1\Lambda' -{}^1\Lambda''$), the transition dipole moment is given by

$$M_{10}^2 = R_e^2 S_J q_{v'v''}, \tag{16.94}$$

in which R_e^2 is the square of the electronic transition dipole moment, S_J is the Hönl-London factor (Table 16.2), and q is the Franck-Condon factor defined by

$$q_{v'v''} = \left| \int \psi^*_{v'} \psi_{v''} \, dr \right|^2 = |\langle v' \mid v'' \rangle|^2 \tag{16.95}$$

The Franck-Condon factor is the square of the overlap of the vibrational wavefunctions between the two electronic states. The Hönl-London factors are taken from Table 16.2, with Λ used for K. Thus, a ${}^1\Sigma^+ - {}^1\Sigma^+$ electronic transition has the same rotational line strength factors as an infrared transition of a diatomic and a ${}^1\Pi - {}^1\Sigma^+$ electronic transition has the same Hönl-London factors as a perpendicular vibrational transition of a linear polyatomic molecule. Notice that the electronic and rotational degeneracy is also required in the expression for the absorption coefficient or for the Einstein A factor.

The Franck-Condon factor q is a measure of how the electronic transition dipole moment is divided among the different vibrational bands. If the two electronic potential energy curves are similar in shape, only the diagonal ($\Delta v = 0$) vibrational bands are allowed because of the orthogonality of the vibrational wavefunctions, i.e., because

$$\left|\langle v' \mid v'' \rangle\right|^2 = \delta_{ij}. \tag{16.96}$$

This is generally not the case, however, and off-diagonal vibrational bands with $\Delta v \neq 0$ are usually found.

If the equilibrium bond length changes substantially, the rotational structure of each vibrational band in an electronic transition will appear very different from a typical infrared vibrational band. For most vibration-rotation transitions $B' \approx B''$, but for electronic transitions, B' is often very different from B''. This leads to the formation of band heads, in which the rotational lines in a branch pile up and turn around. [2, 6]

The expressions for line positions in P-, Q-, and R-branches (ignoring centrifugal distortion) are

$$\nu_{P,R} = \nu_0 + (B' + B'')m + (B' - B'')m^2 \tag{16.97}$$

and

$$\nu_Q = \nu_0 + (B' - B'')m(m+1), \tag{16.98}$$

with $m = J$ for the Q-branch, $m = J'' + 1$ for the R-branch and $m = -J''$ for the P-branch. These second-order polynomials in m are called Fortrat parabolas. [6] The band head forms at

$$m_H = -\frac{B' + B''}{2(B' - B'')}, \tag{16.99}$$

and the head (ν_H) to origin (ν_0) interval is given by

$$\nu_H - \nu_0 = -\frac{(B' + B'')^2}{4(B' - B'')}. \tag{16.100}$$

The appearance of nonsinglet bands is substantially more complicated and will not be discussed in detail. [2, 26] Basically, all of the rotational lines split into $2S+1$-components, but this splitting can be small [Hund's case (b)] or large [Hund's case (a)], depending on the size of the spin-orbit coupling parameter A. [2, 26] In addition, the 2-fold Λ (or Ω) degeneracy is lifted as the molecule begins to rotate, and the effects of Λ-doubling (or Ω-doubling) must also be considered. [26]

16.4.2. Polyatomics

The electronic spectroscopy of polyatomic molecules can be complex. In contrast to diatomic molecules, each electronic state has $3N$-6 (or $3N$-5) vibrational modes to be considered. The vibrational band structure of an electronic transition is then given by

$$\nu = T_e + G'(v_1', v_2', \ldots) - G''(v_1'', v_2'' \ldots). \tag{16.101}$$

A vibrational band is specified by providing the two sets of vibrational quantum numbers $(v_1', v_2' \ldots) - (v_1'', v_2'' \ldots)$ or, more compactly, $1^{v_1'}_{v_1''} 2^{v_2'}_{v_2''} \ldots$. The vibrational selection rule for allowed electronic transitions is $\Delta v_i = 0, \pm 1, \ldots$ for totally symmetric vibrations and $\Delta v_i = 0, \pm 2, \pm 4 \ldots$ for nonsymmetric vibrations. The intensities of each of these vibrational bands is approximately determined by the product of a set of Franck-Condon factors, one for each mode, so that

$$I \propto R_e^2 q_{v_1' v_1''} q_{v_2' v_2''} \cdots, \tag{16.102}$$

where R_e^2 is the magnitude squared of the electronic transition dipole moment. Simple formulae such as Eq. (16.102) are often not useful in a quantitative sense because of effects such as vibronic coupling, which prevent the separation of electronic and vibrational motion.

Each vibrational band has an associated rotational structure with energy levels described in simple cases by the formulae already discussed. The spectra of linear and symmetric tops can generally be classified as parallel and perpendicular, depending on the direction of the electronic transition dipole moments. For asymmetric tops, the rotational selection rules can be classified as a-type, b-type, and c-type, depending on the orientation of the electronic transition dipole moment. The electronic transition dipole moment can be evaluated using the electronic wavefunctions as

$$\vec{R}_e = \int (\psi_{\text{el}}')^* \vec{\mu} \psi_{\text{el}}'' \, d\tau. \tag{16.103}$$

where the integration is over the electronic coordinates. The possibility of geometry changes coupled with the large number of special effects, such as vibronic coupling, Jahn-Teller effect, and Renner-Teller effect, to say nothing of fine structure and hyperfine structure, make electronic spectroscopy of polatomics a fascinating and challenging area of study. [8]

16.5. STRUCTURE DETERMINATION

There is a direct relationship between the three moments of inertia and the molecular geometry, which may be expressed by the relations

$$I_A = \sum m_i(b_i^2 + c_i^2), \tag{16.104}$$

$$I_B = \sum m_i(a_i^2 + c_i^2), \tag{16.105}$$

$$I_C = \sum m_i(a_i^2 + b_i^2), \tag{16.106}$$

where the ith atom of mass m_i is located at (a_i, b_i, c_i) in the principal axis system. High-resolution spectroscopy is, thus, one of the most reliable methods for determining molecular geometry. [27] The main problem is that molecules have at most three moments of inertia, which is usually inadequate to determine the large number of bond angles and bond lengths. The solution is to record the spectra of isotopically substituted molecules and to assume that the geometry is invariant. In general, each isotopomer provides an additional three moments of inertia so that spectra of a sufficient number of isotopomers must be recorded to determine uniquely the unknown geometrical parameters. In this procedure, the center-of-mass equations

$$\sum m_i \vec{r}_i = 0, \tag{16.107}$$

with $\vec{r_i} = (a_i, b_i, c_i)$ for each isotopomer, are required as constraints.

Over the years, a number of techniques have been devised to determine molecular structure from moments of inertia. [11, 27] Each method provides a slightly different set of bond lengths and bond angles. The three most important types of structures are designated as r_0-, r_e-, and r_s-structures.

The "best" structure is considered to be the equilibrium or r_e-structure. Within the Born-Oppenheimer approximation, it is only the r_e-structure that does not change with isotopic substitution. For a diatomic molecule, it is easy to obtain an r_e-structure because, Eq. (16.14), only two B_v values are required to extrapolate to a B_e value, from which the r_e-value can then be calculated. For a polyatomic molecule, it is more difficult to determine A_e, B_e, and C_e because, in general, $3N$-6 α_e-values are required for each rotational constant; see Eqs. (16.44), (16.45), and (16.46). For larger molecules, this is a difficult task because a rotational analysis is required for each infrared fundamental (or equivalent information from a pure rotational or an electronic spectrum must be available).

If I_A^e, I_B^e, and I_C^e values are not available, structures may be calculated using the vibrationally averaged moments of inertia derived from A_0, B_0, and C_0 via Eqs. (16.36), (16.37), and (16.38). This is unfortunate because rigid-body relationships such as

$$I_C = I_A + I_B \tag{16.108}$$

for a planar molecular are most nearly true for equilibrium moments of inertia. In fact, Eq. (16.108) never holds exactly; it is used to define a moment of inertia defect via

$$\Delta = I_C - I_A - I_B \tag{16.109}$$

for planar molecules. Empirically, planar molecules should all have small positive Δ-values. Any deviations from the value of Δ expected empirically is taken as evidence for nonplanarity or for fluxional behavior.

Most structures of polyatomic molecules are r_0-structures. Even for diatomic molecules, r_0-values are reported if only a single B_0-value is available. For a bond between two heavy atoms, r_0 and r_e distances differ only slightly. For a hydrogen bond length, however, r_0- and r_e-values differ substantially. Deuterium bond lengths are shorter than corresponding H-bond lengths because of the significantly smaller zero point energy (by approximately $1/\sqrt{2}$) for a deuterium atom bonded to a heavy atom.

These problems with r_0-structures have led over the years to numerous schemes (some of them empirical) to estimate r_e-structures. The most important of these methods is based on isotopic substitution and results in r_s-structures (s for substitution). The basic ideas are due to Kraitchman and Costain. [11] By using the moments of inertia of a parent molecule and of the molecule with a single isotopic substitution in Kraitchman's equations, the distance to the substituted atom from the center of mass can be calculated. Thus, by repeated single isotopic substitution for each atom, a full substitution (r_s) structure is derived. The r_s-structure is a better approximation to the r_e-structure, and normally, the bond lengths obey the inequality $r_0 \geq r_s \geq r_e$. It is rare that a full substitution structure is determined

TABLE 16.4. Bond lengths in Å.

Single bonds

	H	C	N	O	S	F	Cl	Br	I
H	0.74	1.09	1.01	0.96	1.34	0.92	1.27	1.41	1.61
C		1.54	1.47	1.43	1.82	1.35	1.77	1.94	2.14
N			1.45	1.40	—	1.36	1.75	1.79	1.97
O				1.48	—	1.42	1.70	1.72	1.87
S					2.05	1.56	2.07	2.27	—
F						1.42	1.63	1.76	1.91
Cl							1.99	2.14	2.32
Br								2.28	2.47
I									2.67

Multiple bonds

Bond	Length/Å
C–C*	1.54
C = C	1.34
C ≡ C	1.20
C–N*	1.47
C = N	1.28
C ≡ N	1.16
C–O*	1.43
C = O	1.20
C ≡ O	1.13
N ≡ N	1.10
O = O	1.21

*Single bonds, repeated for comparison purposes.

because of the work involved in making and recording spectra of all possible singly substituted isotopomers. In addition, some elements have only a single stable isotope. Generally, a partial r_s-structure is determined using Kraitchman's equations for some of the atoms, and then the remaining geometrical parameters are derived using the moments-of-inertia equations, (16.104), (16.105), and (16.106), or the center-of-mass equation (16.107).

There are some convenient collections of molecular structures, including the *CRC Handbook of Chemistry and Physics*, [13] the paper of Harmony *et al.*, [28] and the Landolt-Bornstein series. [29] More recently, the MOGADOC database [30] has become available with structural data based mainly on electron diffraction and microwave spectroscopy. Additional spectroscopic data are available from various specialized databases, including HITRAN, [31] GEISA, [32] and the JPL catalog. [33]

Although bond lengths (and angles) like vibrational frequencies vary from molecule to molecule, some regularities can be discerned. For example, the bond length between a carbon and a hydrogen atom is about 1.09 Å in all molecules. Bond lengths are also inversely correlated with bond order, which is defined (approximately) as the number of electron pairs in the chemical bond holding two atoms together in a molecule. The concept of an average bond length is therefore useful, and a table of typical values [13, 34] is provided (Table 16.4). Actual molecules may have bond lengths that differ somewhat from the values reported in Table 16.4.

16.6. REFERENCES

[1] L. Cohen-Tannoudji, B. Diu, and F. Laloë, *Quantum Mechanics* (John Wiley and Sons, New York, 1977), Vols. 1 and 2.

[2] P. F. Bernath, *Spectra of Atoms and Molecules* (Oxford University Press, New York, 1995).

[3] R. S. Mulliken, J. Chem. Phys. **23**, 1997 (1955).

[4] P. R. Bunker and P. Jensen, *Molecular Symmetry and Spectroscopy*, 2nd ed. (NRC Press, Ottawa, 1998).

[5] I. Mills *et al.*, *Quantities, Units and Symbols in Physical Chemistry*, 2nd ed. (Blackwell, Oxford, 1993).

[6] G. Herzberg, *Spectra of Diatomic Molecules*, 2nd ed. (Van Nostrand Reinhold, New York, 1950).

[7] G. Herzberg, *Infrared and Raman Spectra of Polyatomic Molecules*, (Van Nostrand Reinhold, New York, 1945).

[8] G. Herzberg, *Electronic Spectra and Electronic Structure of Polyatomic Molecules* (Van Nostrand Reinhold, New York, 1967).

[9] C. H. Townes and A. L. Schawlow, *Microwave Spectroscopy* (Dover, New York, 1975).

[10] H.W. Kroto, *Molecular Rotation Spectra* (Dover, New York, 1992).

[11] W. Gordy and R. Cook, *Microwave Molecular Spectra*, 3rd ed. (Wiley, New York, 1984).

[12] D. Papousek and M. R. Aliev, *Molecular Vibrational-Rotational Spectra* (Elsevier, Amsterdam, 1982).

[13] D. R. Lide (Ed.), *CRC Handbook of Chemistry and Physics*, 82nd ed. (CRC Press, Boca Raton, FL, 2001).

[14] D. M. Bishop, *Group Theory and Chemistry* (Dover, New York, 1993).

[15] J. F. Ogilvie, *The Vibrational and Rotational Spectrometry of Diatomic Molecules* (Academic Press, San Diego, 1998).

[16] K. P. Huber and G. Herzberg, *Constants of Diatomic Molecules* (Van Nostrand Reinhold, New York, 1979); for recent references see http://diref.uwaterloo.ca/.

[17] T. Shimanouchi, J. Phys. Chem. Ref. Data. **9**, 1149 (1980) and references therein.

[18] M. Jacox, J. Phys. Chem. Ref. Data, Monograph 3 (1994); J. Phys. Chem. Ref. Data, **27**, 115 (1998); VEEL Database (NIST, Gaithersburg, MD); see http://webbook.nist.gov/chemistry/.

[19] M. A. H. Smith *et al.*, in *Spectroscopy of the Earth's Atmosphere and the Interstellar Medium*, edited by K. N. Rao and A. Weber (Academic Press, San Diego, 1992).

[20] J. P. Champion, M. Loëte, and G. Pierre, in *Spectroscopy of the Earth's Atmosphere and the Interstellar Medium*, edited by K. N. Rao and A. Weber (Academic Press, San Diego, 1992).

[21] N.B. Colthup, L. H. Daly, and S. E. Wilberley, *Introduction to Infrared and Raman Spectroscopy*, 3rd ed. (Academic Press, San Diego, 1990).

[22] D. A. Long, *Raman Spectroscopy* (McGraw-Hill, London, 1977).

[23] W. Demtröder, *Laser Spectroscopy*, 2nd ed. (Springer, Berlin, 1996).

[24] L. J. Bellamy, *The Infrared Spectra of Complex Molecules* (Chapman and Hall, London, 1975).

[25] K. Nakamoto, *Infrared and Raman Spectra of Inorganic and Coordination Compounds*, 4th ed. (Wiley, New York, 1986).

[26] H. Lefebvre-Brion and R.W. Field, *Perturbations in the Spectra of Diatomic Molecules* (Academic Press, Orlando, FL, 1986).

[27] A. Domenicano and I. Hargittai, *Accurate Molecular Structures* (Oxford University Press, New York, 1992).

[28] M. D. Harmony *et al.*, J. Phys. Chem. Ref. Data. **8**, 619 (1979).

[29] *Landolt-Bornstein Numerical Data and Functional Relationships in Science and Technology*, New Series, Group II, Vol. 15 (Springer, Berlin, 1987).

[30] J. Vögt, MOGADOC Database, Sekt. Spektren and Strukturdokumentation, University of Ulm, Germany; see http://www.uni-ulm.de/strudo/mogadoc/.

[31] L. S. Rothman *et al.*, J. Quant. Spectrosc. Radiat. Transfer. **60**, 665 (1998); see http://www.hitran.com/.

[32] N. Jacquinet-Husson *et al.*, J. Quant. Spectrosc. Radiat. Transfer. **59**, 511 (1998); see http://www.ara.polytechnique.fr/.

[33] H. M. Pickett *et al.*, J. Quant. Spectrosc. Radiat. Transfer. **60**, 891 (1998); see http://spec.jpl.nasa.gov/.

[34] S. R. Radel and M. H. Navidi, *Chemistry* (West, St. Paul, MN, 1990).

17

Nonlinear Physics and Complexity

Paul Manneville
Laboratoire d'Hydrodynamique CNRS & École Polytechnique, Palaiseau, France

Contents

List of Tables

List of Figures

On general grounds, nonlinearity essentially means that some effect (reaction) is not proportional to its cause (action), and that the *superposition principle* does not apply, so that the behavior of the system considered cannot be described in "simple" terms. Nonlinear processes are ubiquitous in nature, and the present use of the vocabulary of dynamical systems does not narrow the scope. On the contrary, most phenomena involving the *synergetic* coupling of state variables can be studied in this perspective, not only in physics and all its subfields, e.g., fluid dynamics (Chapter 13), where the study of instabilities, [1] pattern formation, [2] and turbulence [3, 4] has been a major source of progress, but also in chemistry, biology, [5] and fields outside of the natural sciences, and economics in particular. [6] See also Chapter 15.

Sec. 17.1 deals with problems depending on one independent variable interpreted as time. As a result of nonlinear *resonances*, the dynamics is no longer a mere superposition of simple motions but results in *complex behavior* (Sec. 17.2), with intrinsic randomness called *chaos* and scaling behavior associated to *fractals*. In Sec. 17.3, we extend this approach to systems involving several independent variables, i.e., space and time usually, with emphasis on pattern formation and related modeling issues. This chapter is an informal (mathematically sloppy) guide to concepts, tools, and formulas of sufficiently general usage. More detailed or rigorous approaches are to be found in textbooks, [7]–[12] monographs, [3], [13]–[16] review articles, [2, 17] or collections of original articles and reprints. [18]–[22]

17.1. DYNAMICAL SYSTEMS AND BIFURCATIONS

17.1.1. Preliminaries

Nonlinear dynamics deals with the general aspects of the evolution of systems supposed to be well described by a complete set of *state variables*. These variables $\{X_j; j = 1, 2, \ldots, d\}$ define points $\mathbf{X}$ of a d-dimensional manifold called the *phase space* $\mathbb{X}$, e.g., $\mathbb{X} \equiv \mathbb{R}^d$ (for a pictorial introduction, see [7]; note that in mechanics (Chapter 14), the phase space is defined as the product of coordinate space by momentum space).

Continuous-time dynamics can generally be expressed as a first-order differential system of the form

$$d\mathbf{X}/dt = \mathbf{F}(\mathbf{X}). \tag{17.1}$$

The initial-value problem, $\mathbf{X}(t_0) = \mathbf{X}_0$, is well posed (solution exists and is unique) under mild conditions; e.g., it is sufficient that the vector field $\mathbf{F}$ be $\mathcal{C}^1$ (continuous with continuous first derivatives). When the solution from any initial condition exists for all times, the integration of (17.1) defines a *flow*, i.e., an invertible map of $\mathbb{X}$ onto itself depending on the continuous parameter t. Systems like (17.1), for which $\mathbf{F}$ does not depend explicitly on time t, are called *autonomous*. Important examples of nonautonomous systems are *periodically forced* systems, for which $\mathbf{F}(\mathbf{X}, t+T) \equiv \mathbf{F}(\mathbf{X}, t)$ for all t.

Discrete-time dynamics is defined by a map (difference equation):

$$\mathbf{X}_{n+1} = \mathbf{G}(\mathbf{X}_n). \tag{17.2}$$

Reduction from continuous to discrete-time dynamics for periodically driven systems is achieved by means of *stroboscopic analysis* at the forcing period (Figure 17.1a). This reduction is a special case of *Poincaré sectioning* that defines the *first return map* as a map of a $(d-1)$-dimensional surface Σ in $\mathbb{X}$ onto itself by integration of (17.1) between successive crossings of Σ (Figure 17.1b). From now on, we only consider autonomous systems.

The *phase portrait* is a collective representation of *orbits* evidencing the global *qualitative* features of the dynamics. In phase space, permanent regimes are represented by *nonwandering* sets invariant under the dynamics. An *ω-limit set* (α-limit set) accounts for the asymptotic evolution as t or n go to $+\infty$ ($-\infty$). The *inset* (*outset*) of some limit set is the set of initial conditions that converge to it as $t \to +\infty$ ($-\infty$). An *attractor* is a stable ω-limit set (empty outset, inset containing an open neighborhood). Its *basin of attraction*, the closure of the set of points that converge to it, has finite volume.

The structure of a dynamical system usually depends on a set of *control parameters* $\underline{r}$, i.e., $\mathbf{F}(\mathbf{X}) \equiv \mathbf{F}(\mathbf{X}, \underline{r})$ or $\mathbf{G}(\mathbf{X}) \equiv \mathbf{G}(\mathbf{X}, \underline{r})$. Bifurcation theory is devoted to the study of how phase portraits change with $\underline{r}$. A dynamical system with $\underline{r} = \underline{r}_0$ is said to be *robust* or *structurally stable* if its qualitative behavior does not change in the neighborhood of $\underline{r}_0$. Otherwise, it is *structurally unstable*. The number of parameters that have to be varied to unfold such a situation, i.e., to recover structural stability, is called the *codimension* of the problem.

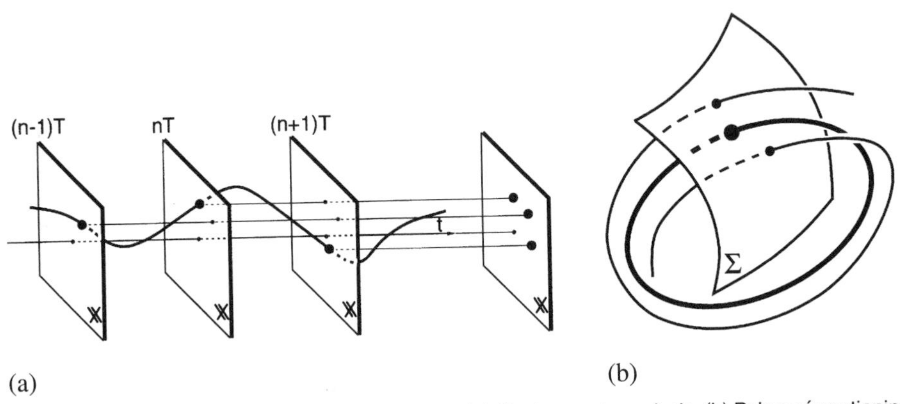

FIGURE 17.1. Reduction to discrete-time dynamics. (a) Stroboscopic analysis. (b) Poincaré sectioning. *Limit cycles* are periodic limit sets corresponding to closed loops in phase space and to fixed points on the Poincaré section.

17.1.2. Tangent dynamics and center-manifold reduction

Stability of fixed points

The *tangent* dynamics at a point $\mathbf{X}_0$ governs the evolution of infinitesimally close points $\mathbf{X} = \mathbf{X}_0 + \epsilon\mathbf{X}'$ with $\epsilon \to 0$:

$$\text{(a)} \quad d\mathbf{X}'/dt = D\mathbf{F} \cdot \mathbf{X}', \qquad \text{(b)} \quad \mathbf{X}'_{n+1} = D\mathbf{G} \cdot \mathbf{X}'_n, \tag{17.3}$$

where $D\mathbf{F} \equiv \partial\mathbf{F}/\partial\mathbf{X}$ ($D\mathbf{G} \equiv \partial\mathbf{G}/\partial\mathbf{X}$) is the Jacobian matrix of $\mathbf{F}$ ($\mathbf{G}$) computed at $\mathbf{X}_0$.

A continuous-time (discrete-time) system is locally contracting, volume-preserving, or expanding according to whether the trace of $D\mathbf{F}$ ($|\det(D\mathbf{G})|$) is < 0, $= 0$, or > 0 (< 1, $= 1$, or > 1). Systems that are everywhere volume preserving, for example, Hamiltonian systems (Chapter 14), are termed *conservative*, others are termed *dissipative*.

Stability properties of special nonevolving solutions $\mathbf{X}_*$ of (17.1) or (17.2), called *fixed points* and verifying $\mathbf{F}(\mathbf{X}_*) = 0$ or $\mathbf{X}_* = \mathbf{G}(\mathbf{X}_*)$, derive from the spectrum of the tangent operator evaluated at those points and denoted as $D\mathbf{F}_*$ or $D\mathbf{G}_*$ (Figure 17.2). Stable fixed points of time-continuous systems account for time-independent attractors.

The tangent space $\mathbb{X}'$ at a fixed point $\mathbf{X}_*$ can thus be decomposed into a direct sum of a stable subspace $\mathbb{X}'_\mathrm{s}$ ($\mathcal{R}e(s) < 0$ or $|\lambda| < 1$), a center subspace $\mathbb{X}'_\mathrm{c}$ ($\mathcal{R}e(s) = 0$ or $|\lambda| = 1$), and an unstable subspace $\mathbb{X}'_\mathrm{u}$ ($\mathcal{R}e(s) > 0$ or $|\lambda| > 1$). A fixed point, and more generally, a limit set, is said to be *hyperbolic* if its center subspace is empty.

Time-reversal implies $s \mapsto -s$ or $\lambda \mapsto 1/\lambda$, hence, exchanging the stable/unstable character of eigenmodes.

Fixed-point classification is illustrated in Figure 17.3 using a two-dimensional, real, linear system $d\mathbf{X}/dt = \mathbf{L} \cdot \mathbf{X}$, with two real or complex, distinct or degenerate, eigenvalues $s_m, m = 1, 2$. Extension to higher dimensional differential systems and to maps is straight-

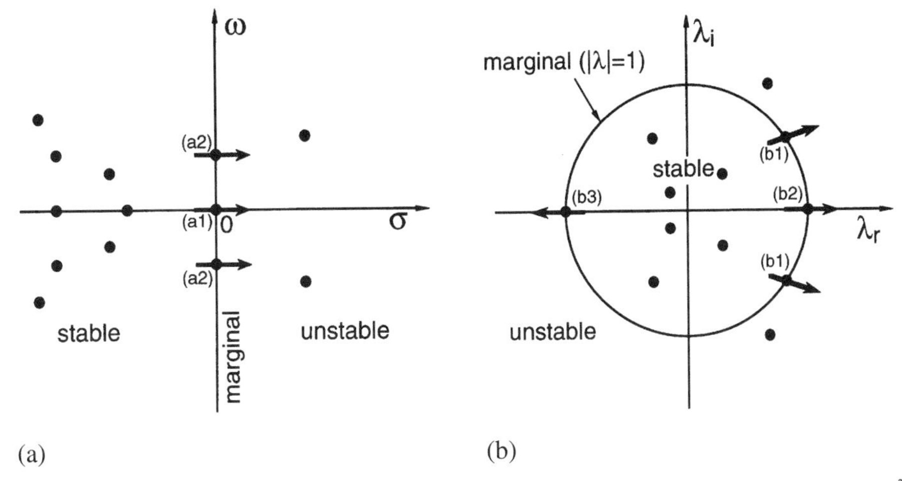

FIGURE 17.2. Linear stability spectra. (a) Continuous time: linear modes in the form $\mathbf{X}'(t) = \exp(st)\tilde{\mathbf{X}}'$, solutions of $s\tilde{\mathbf{X}}' = D\mathbf{F}_* \cdot \tilde{\mathbf{X}}'$ with eigenvalue $s = \sigma - i\omega$; (a1) stationary instability: $\omega = 0$; (a2) oscillatory instability: $\omega \neq 0$. (b) Discrete time: modes $\mathbf{X}'_n = \lambda^n\tilde{\mathbf{X}}'$, solutions of $\lambda\tilde{\mathbf{X}}' = D\mathbf{G}_* \cdot \tilde{\mathbf{X}}'$, with eigenvalue $\lambda = |\lambda|\exp(2\pi i\alpha)$; (b1) instability of a complex pair: $\lambda \notin \mathbb{R}$; (b2) synchronous instability: $\alpha = 0$, $\lambda = 1$; (b3) subharmonic instability: $\alpha = 1/2$, $\lambda = -1$.

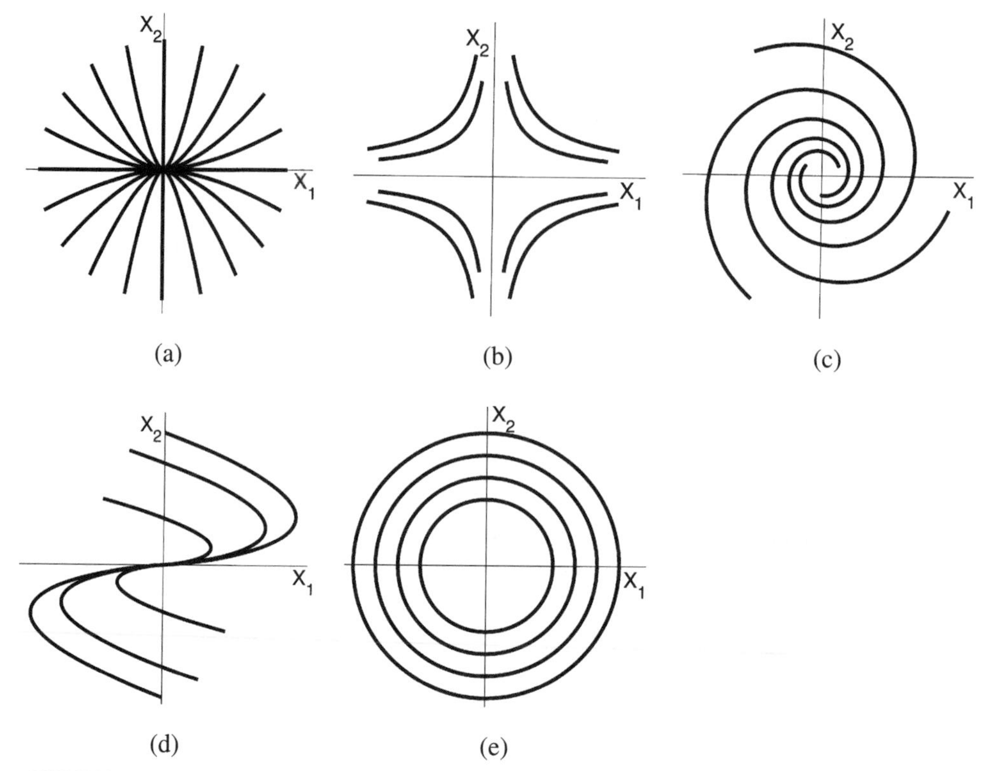

FIGURE 17.3. Fixed-point classification. (a) Node: $s_{1,2} \in \mathbb{R}$, $s_1 s_2 > 0$, stable when $s_m < 0$. (b) Saddle: $s_{1,2} \in \mathbb{R}$, $s_1 s_2 < 0$, always unstable. (c) Focus (or spiral point): $s_2 = s_1^* \in \mathbb{C}$, stable when $\mathcal{R}e(s_m) < 0$. (d) Improper node: $s_1 = s_2 \in \mathbb{R}$, no diagonal form. (e) Center (or elliptic point): $s_2 = s_1^* \in \mathbb{C}$, $\mathcal{R}e(s_m) = 0$.

forward. Nodes, saddles, and foci are hyperbolic [$\forall m$: $\mathcal{R}e(s_m) \neq 0$], whereas a center [$\mathcal{R}e(s_{1,2}) = 0$] is nonhyperbolic. As a matter of fact, the latter is marginal except when the physics enforces time-reversal invariance (Hamiltonian systems).

Center manifold and normal forms

The instability of a time-independent reference state associated to a fixed point $\mathbf{X}_*$ in phase space is the usual starting point of the cascade toward temporal complexity. Upon increasing a control parameter, some perturbation modes may become "dangerous" [$\mathcal{R}e(s) \approx 0$ or $|\lambda| \approx 1$], whereas the others remain strongly damped [$\mathcal{R}e(s) < 0$ and large or $|\lambda| \ll 1$]. The "slaving principle" asserts that an effective dynamics involving the sole dangerous (center) modes can be obtained by adiabatic elimination of stable modes. This can be put in asymptotically rigorous form by *center-manifold reduction*: [9] Assuming d_c center modes $\mathbf{X}_c$ and $d_s = d - d_c$ enslaved modes $\mathbf{X}_s$, one starts from a splitting of $d\mathbf{X}/dt = \mathbf{L}\cdot\mathbf{X} + \mathbf{N}(\mathbf{X})$ into

$$d\mathbf{X}_c/dt = \mathbf{L}_c \cdot \mathbf{X}_c + \mathbf{N}_c(\mathbf{X}_c, \mathbf{X}_s), \tag{17.4}$$

$$d\mathbf{X}_s/dt = \mathbf{L}_s \cdot \mathbf{X}_s + \mathbf{N}_s(\mathbf{X}_c, \mathbf{X}_s), \tag{17.5}$$

where $\mathbf{L}_{c,s}$ are the restrictions of $\mathbf{L}$ to subspaces $\mathbb{X}'_{c,s}$, and $\mathbf{N}_{c,s}(\mathbf{X}_c, \mathbf{X}_s)$ account for the nonlinear couplings. Enslaved modes are supposed to live on a slow manifold defined as

$$\mathbf{X}_s = \mathbf{H}(\mathbf{X}_c). \tag{17.6}$$

One finds the equation for $\mathbf{H}$ by inserting (17.6) into (17.5) and then substituting $d\mathbf{X}_c/dt$ from (17.4) to obtain

$$D\mathbf{H}(\mathbf{X}_c) \cdot [\mathbf{L}_c \cdot \mathbf{X}_c + \mathbf{N}_c(\mathbf{X}_c, \mathbf{H}(\mathbf{X}_c))] = \mathbf{L}_s \cdot \mathbf{H}(\mathbf{X}_c) + \mathbf{N}_s(\mathbf{X}_c, \mathbf{H}(\mathbf{X}_c)). \tag{17.7}$$

This equation is usually solved by representing $\mathbf{H}(\mathbf{X}_c)$ as a formal power series in $\mathbf{X}_c$. The *effective dynamics* for the $\mathbf{X}_c$ is finally given by

$$d\mathbf{X}_c/dt = \mathbf{L}_c \cdot \mathbf{X}_c + \mathbf{N}_c(\mathbf{X}_c, \mathbf{H}(\mathbf{X}_c)) = \mathbf{L}_c \cdot \mathbf{X}_c + \widetilde{\mathbf{N}}_c(\mathbf{X}_c). \tag{17.8}$$

The extension to more general situations leads to the theory of *inertial manifolds*. From now on, we tacitly assume restriction to center manifold and drop subscript "c." *Complexity* develops as additional dangerous modes are included and the dimension of the center manifold (17.6) is progressively increased.

Once the effective dynamics on the center manifold has been obtained, its expression can be simplified by means of polynomial changes of coordinates. Terms that cannot be eliminated through these changes are called *resonant*. The structure of the resulting *normal form* is system-independent and depends only on *resonance conditions* within the linear spectrum:

$$s_m = \sum\nolimits_{m'} n_{m'} s_{m'}, \quad n_{m'} \in \mathbb{N},\ n_{m'} \geq 0, \quad \sum\nolimits_{m'} n_{m'} = p, \quad p \geq 2, \tag{17.9}$$

where p is the order of the resonance. The behavior in the vicinity of the critical situation in parameter space is further unfolded by adding relevant small perturbations to the normal form. As an example, the unfolding of the dynamics at a double nondiagonalizable eigenvalue $s = 0$ (Takens-Bogdanov bifurcation [9]) reads as

$$dX_1/dt = X_2, \qquad dX_2/dt = \epsilon_1 X_1 + \epsilon_2 X_2 - X_1 X_2 \pm X_1^2. \tag{17.10}$$

17.1.3. Bifurcations

Gradient flows and catastrophes

The dynamics of $\mathbf{X}$ is said to derive from the potential $G(\mathbf{X})$ (*gradient flow*) when

$$d\mathbf{X}/dt = \nabla_{\mathbf{X}} G, \quad G \equiv G(\{X_i; i = 1, \ldots, d\}), \quad \text{i.e.,} \quad dX_i/dt = -\partial G/\partial X_i. \tag{17.11}$$

Reciprocally, the condition for (17.1) to derive from a potential is $\partial F_i/\partial X_j \equiv \partial F_j/\partial X_i$ for all i, j. The system then follows lines of steepest descent in the potential landscape ($dG/dt \leq 0$), so that the system eventually reaches one of the critical points of G, solutions of $\partial G/\partial X_i = 0$, $i = 1, \ldots, d$. The determination of normal forms of d-dimensional gradient flows depending on p control parameters is the subject of the *theory of catastrophes*; [12] see Table 17.1. The most frequent cases with $d = 1$ and degree $n = 2$, 3, and 5 are illustrated in Figure 17.4.

TABLE 17.1. The seven elementary catastrophes. $G = G_s + G_u$, where G_s is the singular part of G and G_u the perturbation.

d	p	$G_{\rm sing}$	$G_{\rm unfold}$	Name
1	1	X^3	r_0X	Fold
1	2	X^4	$r_0X + r_1X^2$	Cusp
1	3	X^5	$r_0X + r_1X^2 + r_2X^3$	Swallowtail
1	4	X^6	$r_0X + r_1X^2 + r_2X^3 + r_3X^4$	Butterfly
2	3	$X_1^3 + X_2^3$	$r_0X_1 + r_1X_2 + r_2X_1X_2$	Hyperbolic umbilic
2	3	$X_1^3 - 3X_1X_2^2$	$r_0X_1 + r_1X_2 + r_2(X_1^2 + X_2^2)$	Elliptic umbilic
2	4	$X_1^2X_2 + X_2^4$	$r_0X_1 + r_1X_2 + r_2X_1^2 + r_3X_2^2$	Parabolic umbilic

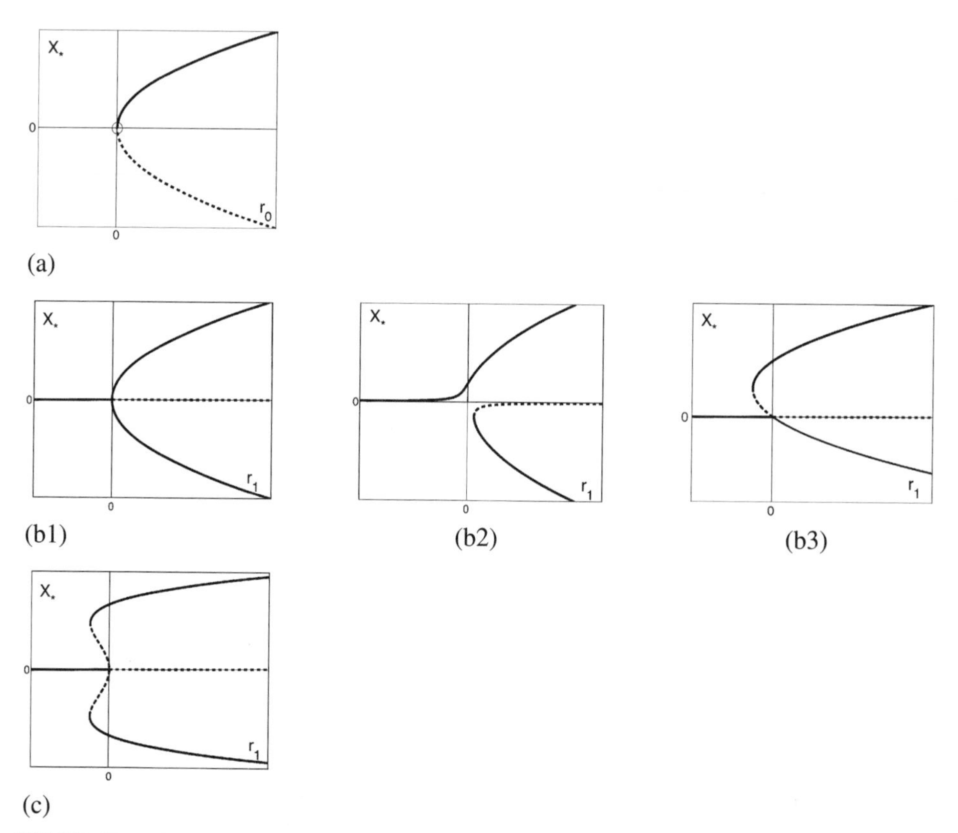

FIGURE 17.4. Elementary bifurcations of a single variable. Solid (dashed) lines indicate stable (unstable) branches of fixed points X_* as functions of control parameters. (a) Saddle-node bifurcation: $dX/dt = r_0 - X^2$. (b1) Fork bifurcation (supercritical): $dX/dt = r_1X - X^3$. (b2) Imperfect bifurcation: $dX/dt = r_0 + r_1X - X^3$ (r_0 given). (b3) Transcritical bifurcation (persistence of $X = 0$): $dX/dt = r_1X + r_2X^2 - X^3$ (r_2 given). (c) Symmetrical unfolding of $-X^5$: $dX/dt = r_1X + r_3X^3 - X^5$ with $r_3 > 0$ (local sub-critical fork, saddle-node bifurcation, hysteresis cycles on variation of r_1).

Development of time dependence

Relaxing the gradient flow condition allows for more complex than just time-independent asymptotic dynamics. In two dimensions, the behavior remains simple, i.e., not more complicated than periodic, as implied by the *Poincaré-Bendixson* theorem. [9] Time dependence usually sets in through a *Hopf* bifurcation [Figure 17.2, case (a2)]. The corresponding (continuous-time) normal form reads as

$$dZ/dt = (r - i\omega)Z - g|Z|^2 Z, \tag{17.12}$$

where $Z = X_1 + iX_2$.

The stability of a limit cycle in the full phase space is best studied *via* Poincaré sectioning (Figure 17.1b) and corresponding first-return maps. A bifurcation typically takes place when a pair of complex conjugate eigenvalues crosses the unit circle, with special cases for a crossing at ± 1 (Figure 17.2b). Elimination of enslaved modes with $|\lambda_m| < 1$ yields a return map in the form $Z_{n+1} = \lambda Z_n + \mathbf{N}_c(Z_n, Z_n^*)$, with $\lambda = \exp(2\pi i \alpha)$ at threshold for the remaining pair of complex (discrete time) center modes. The normal form is again obtained by performing nonlinear variable changes attempting to "kill" all nonresonant terms. The most dangerous trivial resonant term is $|Z|^2 Z$, present even when α is irrational. When α is rational, i.e., $\alpha = p/q$, $(Z^*)^{q-1}$ is the most relevant nontrivial resonant term. The resonance is said to be *weak* when this term is of a higher degree than is the trivial term, i.e., $q \geq 5$.

In the nonresonant case ($\alpha \notin \mathbb{Q}$), the system bifurcates toward a two-periodic regime with orbits winding on an invariant torus in phase space, whose intersection with Σ is a circle. The dynamics of $\theta = \arg(Z)$ along the circle is characterized by the winding number:

$$\hat{\alpha} = \lim_{n\to\infty} \frac{1}{n} \sum_{n'=0}^{n-1} (\theta_{n'+1} - \theta_{n'}). \tag{17.13}$$

The value of $\hat{\alpha}$ depends on α and the intensity of nonlinearity, itself a function of the distance to criticality. The regime is effectively quasi-periodic when $\hat{\alpha}$ is irrational; otherwise it is periodic, which corresponds to the *locking* phenomenon see Ref. [8, p. 93].

17.2. CHAOS AND FRACTALS

The transition from fixed points to limit cycles, and next to two-dimensional tori, are the first steps of the route to turbulence envisioned by Landau (in [22] p. 115) as a cascade of bifurcations introducing an infinite series of incommensurate oscillatory modes (n-periodicity, $n \to \infty$). This scenario was revised by Ruelle and Takens (in [22] p.120), who showed that a finite number of modes was sufficient to obtain chaos characterized by exponential decay of correlations and *sensitivity to initial conditions* (and infinitesimal perturbations). Resulting *strange attractors* display a fractal structure due to the combined effects of *stretching* and *folding* by the dynamics on the one hand, and of *squeezing* by dissipation on the other.

Chaos can be present already in one-dimensional noninvertible iterations (Figure 17.5 a–c). Invertible maps require $d \geq 2$ (Figure 17.5d). Chaos requires $d \geq 3$ to develop in differential systems. The Lorenz model (Figure 17.5e), is a classic example (see Lorenz, [21] p. 367). In that case, a special Poincaré map, called the *Lorenz map*, is obtained by plot-

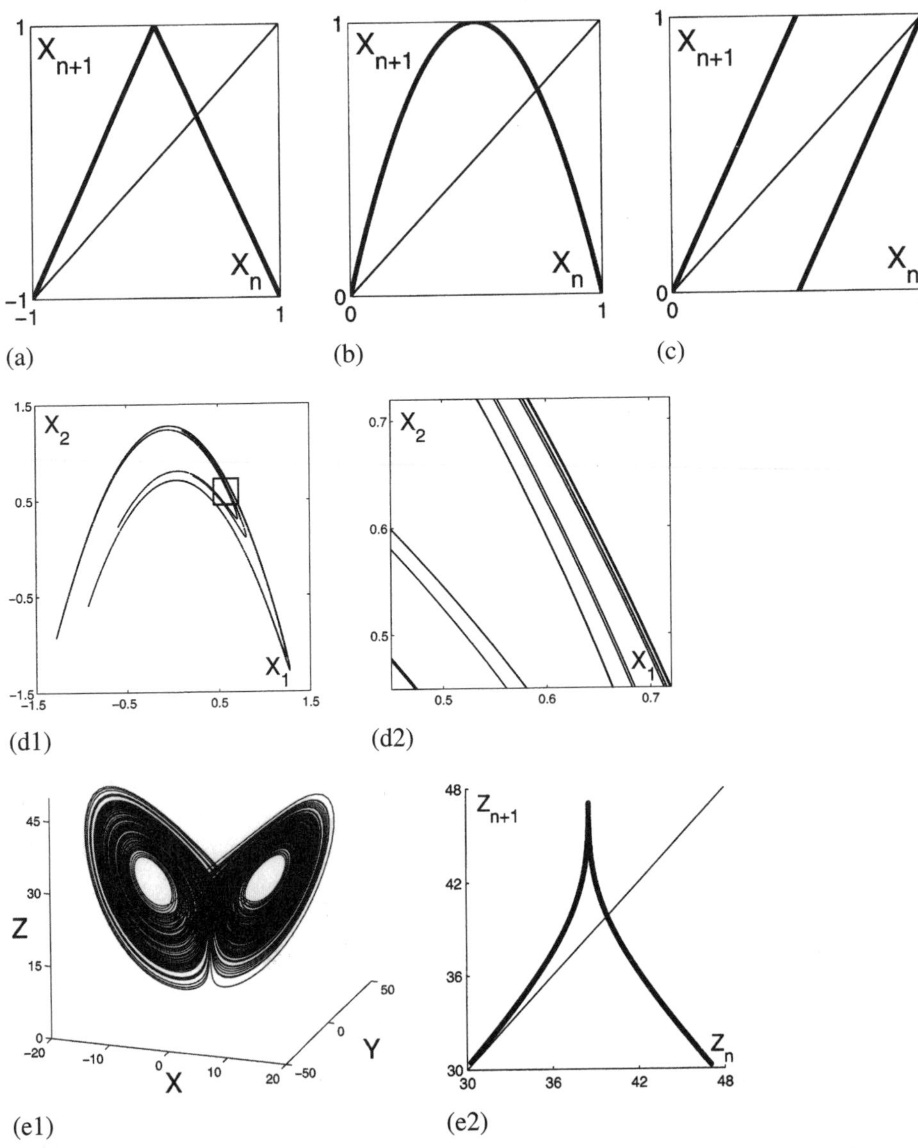

FIGURE 17.5. Chaos in maps and the Lorenz model. (a) Tent map $X \mapsto 1 - r|X|$, here for $r = 2$. (b) Logistic map $X \mapsto 4rX(1 - X)$, here for $r = 1$. (c) Dyadic map $X \mapsto 2X$ (mod 1). (d) Hénon's map $X_{1,n+1} = X_{2,n}$, $X_{2,n+1} = 1 - aX_{2,n}^2 + bX_{1,n}$ here for $a = 1.4$, $b = 0.3$, with zoom on boxed region (see Hénon, in [22, p. 235]. (e) Lorenz model $dX/dt = \sigma(Y - X)$, $dY/dt = rX - Y - XZ$, $dZ/dt = -bZ + XY$, here for $r = 28$, $\sigma = 10$, $b = 8/3$, and corresponding Lorenz map.

TABLE 17.2. Lyapunov signature of dynamical systems.[a]

Regime	Continuous time	Discrete time (Poincaré section)
Stationary (fixed point)	$- - - \cdots$	$- - \cdots$
Periodic (limit cycle)	$0 - - \cdots$	$- - \cdots$
Two-periodic (two-torus)	$0\,0 - \cdots$	$0 - \cdots$
Chaotic (strange attractor)	at least one $+$ e.g., $+\,0 - - \cdots$	at least one $+$ e.g., $+ - - \cdots$

[a] As soon as the attractor is not a fixed point, the Lyapunov spectrum has (at least) one zero exponent removed when taking the Poincaré section.

ting the successive maxima Z_n of $Z(t)$ against their predecessors $Z_{n+1} = f(Z_n)$ (Figure 17.5e2). It displays a tent-shaped map (Figure 17.5a) explaining the chaotic nature of the Lorenz attractor.

17.2.1. The nature of chaos

Divergence of trajectories and Lyapunov exponents

Long-term unpredictability is the direct consequence of the *sensitivity to initial conditions.* [17] Exponential divergence of neighboring trajectories is quantitatively measured by *Lyapunov exponents* that extend the concept of stability to aperiodic trajectories. For a one-dimensional map $X \mapsto f(X)$, the divergence rate away from a reference trajectory $\{X_n, n = 0, \ldots\}$ is given by

$$\lambda = \lim_{n\to\infty} \frac{1}{n} \sum_{n'=0}^{n-1} \ln\left(|\, f'(X_{n'})\,|\right), \tag{17.14}$$

where $f' \equiv df/dX$. It is readily shown that $\lambda = \ln(2) > 0$ for examples displayed in Figure 17.5a–c.

For a d-dimensional map $\mathbf{G}$, a whole *Lyapunov spectrum* $\{\lambda_m, m = 1, \ldots, d\}$ is obtained by considering the growth rate $\mu_m,\ m = 1, \ldots, d$, of volume elements of increasing dimension m: $\lambda_m = \mu_m - \mu_{m-1},\ \mu_1 = \lambda_1$ being the largest Lyapunov exponent derived from the evolution rate of the norm of $\delta\mathbf{X}_n = \prod_{n'=0}^{n-1} D\mathbf{G}(\mathbf{X}_{n'})\delta\mathbf{X}_0$.

Extension to differential dynamical systems is straightforward; see Table 17.2. The existence of at least one positive exponent is the simplest definition of *chaos.*

Symbolic dynamics and chaos

Consider the dyadic map $X \mapsto 2X$ of $[0, 1]$ (mod 1) (Figure 17.5c). Let X be given by its binary expansion, $X = \sum_{n\geq 1} \sigma_n 2^{-n}$, where σ_n is the nth bit. The first bit indicates whether X sits on $[0, 1/2[$ or $[1/2, 1[$. These two disjoint subintervals make a suitable *generating partition* of $[0, 1[$. Sensitivity to initial conditions arises from the unavoidable uncertainty on the initial bit sequence above some rank linked to experimental precision. The iteration then amounts to a *shift* of the bit sequence, which is an extreme form of chaos since the evolution comes to a random heads-and-tails drawing of bits (in the initial condition). This implies the existence of an infinite uncountable number of *aperiodic* trajectories besides the countable infinite number of periodic orbits corresponding to the coding for rational numbers in [0, 1[.

These properties are typical of chaotic attractors. The difficulty is to determine the appropriate generating partition permitting the reduction of the dynamics to a shift. In this perspective, dynamics can be transformed into a *language* in which *grammar rules* put constraints on possible finite sequences (words) of symbols (letters) in some finite set (alphabet). [15]

Statistics of chaos

Information about the long-term statistical properties of a given observable W are obtained by means of a *temporal average*

$$\langle W \rangle_{\mathrm{t}} = \lim_{n \to \infty} \frac{1}{n} \sum_{n'=0}^{n-1} W_{n'}. \tag{17.15}$$

Ergodicity [17] implies that one can replace this empirical average by an *ensemble average* over an invariant probability measure $\rho(\mathbf{X})d\mathbf{X}$ supported by the attractor; i.e.,

$$\langle W \rangle_{\mathrm{e}} = \int W(\mathbf{X})\rho(\mathbf{X})d\mathbf{X}. \tag{17.16}$$

The so-defined probability measure is identified as the *natural measure* (independent of initial conditions and robust against infinitesimal perturbations of the trajectories).

Operationally, $\rho(\mathbf{X})$ is obtained as

$$\rho(\mathbf{X}) = \lim_{n \to \infty} \frac{1}{n} \sum_{n'=0}^{n-1} \delta(\mathbf{X} - \mathbf{X}_{n'}), \tag{17.17}$$

where δ is the Dirac distribution. Thus, to a given region of space, the probability density $\rho(\mathbf{X})$ attributes a weight proportional to the fraction of time spent by a typical, arbitrarily long, trajectory in that region. Accordingly, it can (in principle) be determined by *box counting*, i.e., by measuring the relative occupation $\bar{\rho}_i(\epsilon) = \int_{B_i(\epsilon)} \rho(\mathbf{X})d\mathbf{X}$ of some element B_i of a partition of phase space in balls of diameter $\epsilon \to 0$.

17.2.2. Fractal properties and dimensions

Because of the instability of neighboring trajectories, a strange attractor is locally a continuous manifold with topological dimension d_{t} equal to the number of nonnegative Lyapunov exponents. However, the stretch-fold-squeeze nature of chaotic dynamics in dissipative systems implies the existence of a transverse *fractal structure* (see Figure 17.5d2).

On general grounds, [14] fractal sets of interest are (statistically) invariant under the combination of rotations, translations, and dilatations. Their properties are best introduced when a regular construction rule is given; so the *similarity dimension* is defined as $d_{\mathrm{sim}} = \ln(N_\varrho)/\ln(\varrho)$, where N_ϱ is the number of times the generating pattern is repeated on dilatation by a factor ϱ. This yields $d_{\mathrm{sim}} = \ln(2)/\ln(3)$ for the classic middle-third Cantor set. When the rule is unknown, one must rely on box counting. The *fractal dimension* or *capacity* is then defined as

$$d_{\mathrm{f}} = \lim_{\epsilon \to 0} \frac{\ln(\mathcal{N}(\epsilon))}{\ln(1/\epsilon)}, \tag{17.18}$$

where $\mathcal{N}(\epsilon)$ is the number of balls necessary to cover the set at resolution ϵ. A set is fractal as soon as $d_{\mathrm{f}} > d_{\mathrm{t}}$, which roughly expresses the fact that the set "occupies" the space more than suggested by d_{t}.

Complexity of a chaotic attractor is disclosed through the *multifractal* properties of its invariant probability density $\rho(\mathbf{X})$. They are measured by means of generalized "dimensions" d_q defined from its empirical estimate $\bar{\rho}_i(\epsilon)$ as

$$d_q = \frac{1}{q-1} \lim_{\epsilon \to 0} \frac{\ln\left(\sum_i \bar{\rho}_i^q\right)}{\ln(\epsilon)}, \qquad -\infty < q < +\infty. \tag{17.19}$$

The fractal dimension d_{f} of the attractor is recovered for $q = 0$. The limit $q \to 1$ yields the *information dimension* directly defined as $d_1 = d_{\mathrm{I}} = \lim_{\epsilon \to 0}\left[\sum_i \bar{\rho}_i \ln(\bar{\rho}_i)\right] / \ln(\epsilon)$. The correlation dimension corresponding to $q = 2$ is better obtained using the Grassberger-Procaccia method, which determines empirically the variation with phase-space distance R of the number of pairs of points closer than R on the attractor:

$$d_2 = \nu = \lim_{R \to 0} \frac{\ln[C(R)]}{\ln(R)}, \quad C(R) = \lim_{N \to \infty} \frac{1}{N^2} \sum_{n,n'} Y\left(R - \Delta(\mathbf{X}_n, \mathbf{X}_{n'})\right), \tag{17.20}$$

where Y is the Heaviside function [$Y(u) = 0$ when $u < 0$ and 1 otherwise] and Δ is a distance in phase space. On general grounds, one gets $0 < d_{\min} \le d_{q+1} \le d_q \le d_{\max} < d$. A homogeneous probability measure on a fractal is characterized by $d_q = d_{\mathrm{f}}$ for all q. The natural measure of a typical chaotic attractor is such that $d_{\min} < d_{\max}$, which indeed defines it as a multifractal.

The *multifractal formalism* relates dimensions d_q to the singularity spectrum $f(\alpha)$ defined as the fractal dimension of the set of points with pointwise dimension $d_{\mathrm{p}} = \alpha$. The *pointwise dimension* d_{p} at point $\mathbf{X}$ is defined as $d_{\mathrm{p}}(\mathbf{X}) = \lim_{\epsilon \to 0}\{\ln[\bar{\rho}\,(B(\mathbf{X}, \epsilon))] / \ln(\epsilon)\}$, where $\bar{\rho}\,(B(\mathbf{X}, \epsilon))$ is the mass of the attractor contained in a ball B of radius ϵ centered at that point.

Let $Z(q, \epsilon) = \lim_{\epsilon \to 0} \sum_i \bar{\rho}_i^q(\epsilon) \sim \epsilon^{\tau(q)}$, which defines $\tau(q)$ from the "partition function" $Z(q, \epsilon)$. Then, $\tau(q) \equiv (q-1)d_q$ and $f(\alpha)$ are related to each other by a Legendre transform:

$$\tau(q) = \min_\alpha\, [q\alpha - f(\alpha)], \qquad f(\alpha) = \min_q\, [\alpha q - \tau(q)]. \tag{17.21}$$

One has $d_0 = d_{\mathrm{f}} = \max_\alpha f(\alpha)$. The physical measure is peaked (smeared) on sets that contribute to $d_{+\infty} = \alpha_{\min}$ ($d_{-\infty} = \alpha_{\max}$). Finally, for $q = 1$, one has $\tau(1) = 0$ so that the $f(\alpha)$ curve is tangent to the line $f(\alpha) = \alpha$ (cf. Bohr and Tél in Vol. 2 of Ref. [18]).

From the spectrum of cumulated Lyapunov exponents $\mu_m = \sum_{m'=1}^{m} \lambda_{m'}$ (growth rate of an m-dimensional infinitesimal volume), one defines the *Lyapunov dimension* using the Kaplan-Yorke formula:

$$d_\mathrm{L} = \nu + \frac{\mu_\nu}{\mu_\nu - \mu_{\nu+1}} = \nu + \frac{1}{|\lambda_{\nu+1}|} \sum_{m=1}^{\nu} \lambda_m, \tag{17.22}$$

where ν is such that $\mu_\nu \geq 0$ and $\mu_{\nu+1} < 0$. For a two-dimensional chaotic attractor, this gives $d_\mathrm{L} = 1 + \lambda_1/|\lambda_2|$ (see Ref. [17] for details).

17.2.3. Routes to chaos

Several universal routes to chaos have been discovered. [13] The simplest scenarios rest on bifurcations of limit cycles and corresponding Poincaré maps (Figure 17.2b). The Ruelle-Takens scenario involves the first steps of a cascade of supercritical Hopf bifurcations [Figure 17.2, case (b1)]. The strongly resonant case $\lambda = -1$ [case (b3)] brings in chaos at the end of an infinite cascade of local subharmonic bifurcations with remarkable universal scaling properties. Other routes often imply more global ingredients.

Subharmonic cascade

Figure 17.6 displays the bifurcation diagram of the logistic map $X \mapsto 4rX(1-X)$, whose nontrivial fixed point $X_* = 1 - 1/4r$ bifurcates at $r_0 = 3/4$ with eigenvalue $\lambda = -1$ [strong resonance; Figure 17.2b, case (b3)]. A *direct cascade* of period doublings, $T = 2^k$, $k = 1, \ldots, \infty$, occurs at successive thresholds r_k up to $r_\infty = 0.89248\ldots$.

The cascade has *universal* scaling properties that can be obtained by *renormalization group* methods (see Feigenbaum, in [21] p. 49): The thresholds converge geometrically toward r_∞ with decrements

$$\delta_n = \frac{r_n - r_{n-1}}{r_{n+1} - r_n}, \qquad \lim_{n\to\infty} \delta_n = \delta = 4.6692016\ldots, \tag{17.23}$$

where δ is the first Feigenbaum constant. Maps corresponding to successive bifurcations can be locally mapped one onto the next asymptotically by rescaling. Close to r_∞, the

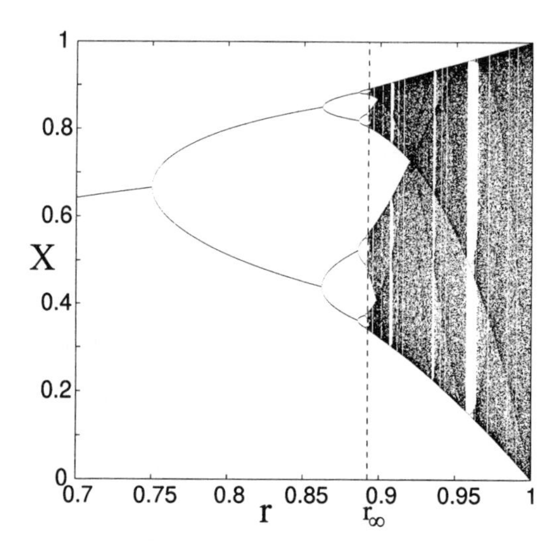

FIGURE 17.6. Subharmonic cascade of the logistic map.

rescaled maps tend locally to a universal function g, solution of $g(X) = -\alpha g\,(g(-X/\alpha))$ involving the second Feigenbaum constant $\alpha = 2.502907875\ldots$.

Beyond r_∞ up to $r = 1$, the direct cascade is mirrored by an *inverse cascade*, in which chaos can be observed outside windows of periodicity. Reentering periodicity sets in *via* intermittency (see below) and decays through a subharmonic cascade having the same global properties as the main cascade. A fractal distribution of periodicity windows on the interval $[r_\infty, 1]$ is visible in Figure 17.6.

Other scenarios

Contrasting with the local character of the period-doubling cascade, other scenarios generally involve processes depending on the global structure of phase space through the properties of distant limit sets.

In time series, the *intermittency* scenario is characterized by apparently random chaotic bursts interrupting nearly periodic intermissions. It develops when a limit cycle becomes unstable, whereas the global structure of the phase space ensures reinjection in the former basin of attraction of the limit cycle. Type I intermittency [13] is observed at the strong resonance $\lambda = 1$ [Figure 17.2b, case (b2)], whose unfolding produces the merging away of a pair of fixed points with opposite stability properties ($X \mapsto r + X + X^2$, saddle-node bifurcation, discrete-time version of Figure 17.4a). Type II (III) intermittency occurs when the Hopf (subharmonic) bifurcation [Figure 17.2b, case b1 (b3)] is subcritical.

Crises form another class of scenarios of frequent occurrence, in which the nature or size of an attractor is seen to change abruptly as a parameter is varied. [11] A *boundary crisis* takes place when a chaotic attractor collides with the unstable manifold of another limit set. At such a point, *transient chaos* is observed, which looks similar to chaos existing before the crisis, until the trajectory moves away. At an *interior crisis*, the attractor is suddenly enlarged by the merging of two or several components, with characteristic intermittent jumps from one of the formerly independent pieces to another (*crisis intermittency*).

17.2.4. Applied nonlinear dynamics

Empirical knowledge about a system is usually collected in the form of *time series* $\{W_n, n = 0, 1, \ldots\}$ of some observable W, regularly sampled in time at frequency $1/\tau$: $W_n = W(n\tau)$. The dynamics is then reconstructed by the *method of delays* [23] as an orbit in the d_e-dimensional space $\mathbb{V}$, where $\mathbf{V}_n = \{W_n, W_{n+1}, \ldots, W_{n+(d_e-1)}\}$. An effective strategy to choose the *embedding dimension* d_e is to consider a trial d'-dimensional reconstruction, i.e., $\{W_n, W_{n+1}, \ldots, W_{n+(d'-1)}\}$, and the $(d' + 1)$ reconstruction obtained by adding component $W_{n+d'}$; to further count the number of *false neighbors*, i.e., pairs of points that are neighbors in d' dimensions but no longer for $d' + 1$; and then increase d' up to the point when the number of false neighbors becomes negligible. [23]

Controlling nonchaotic systems usually requires large inputs to obtain large outputs. By contrast, low-power control is manageable in nonlinear systems operating in a potentially chaotic parameter range. Stabilization of unstable periodic orbits within a chaotic attractor can be achieved by using the Grebogi-Ott-Yorke algorithm. [24] The use of small parameter perturbations to control trajectories in the phase space of a chaotic system can further be exploited to target special orbits, with specific application to coding and communication when combined with symbolic dynamics properties (see Ref. [25] for extensive reviews of recent work).

17.3. SPACE-TIME DYNAMICAL SYSTEMS

The approach in terms of low-dimensional dynamical systems fails when the active modes become too numerous, e.g., in developed turbulence much beyond onset. [3] Here, we focus more specifically on composite systems distributed in (physical) space, [2, 16, 20] in order to establish a more immediate connection with statistical physics.

17.3.1. Classification of instabilities and the modeling issue

Infinitesimal perturbations $\mathbf{X}'$ to a stationary uniform reference state $\mathbf{X}_0$ in a translationally invariant domain are taken as $\mathbf{X}' = \exp(st + ikx)\,\hat{\mathbf{X}}'_k$, where $s = \sigma - i\omega$ is the (complex) growth rate and k is the (real) wave vector. In order that $\mathbf{X}'$ be a solution of the linearized problem $\partial_t \mathbf{X}' = \mathbf{L}(\partial_x, r) \cdot \mathbf{X}'$, s must be a root of the *dispersion relation* symbolically written here as $s = \sigma(k, r) - i\omega(k, r)$ (Figure 17.7a). The marginal stability curve corresponding to some branch of *normal modes* is obtained by solving for r the corresponding condition $\sigma(k, r) = 0$ (Figure 17.7b). The instability *threshold* r_c is defined as the minimum in k of the marginal stability curve of the most dangerous mode, reached at $k = k_c$ called the *critical wave vector*. The classification of instability modes follows from the values of k_c and $\omega_c = \omega(k_c, r_c)$, as given in Table 17.3. (From now on, subscript "c" indicates quantities evaluated at threshold.)

The specific space-time scales introduced by the instability mechanism suggest a new approach in which the system is analyzed as a collection of cells with sizes $\sim 2\pi/k_c$. A degree of freedom is attributed to each cell to account for local time behavior. However,

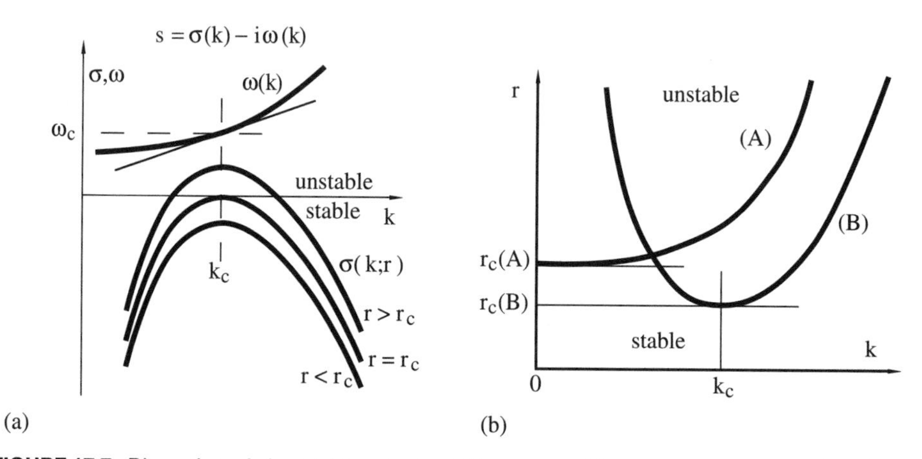

FIGURE 17.7. Dispersion relation and marginal stability curve. (a) Typical aspect of real and imaginary parts of the dispersion curve for a given normal-mode branch m [$\omega(k) \equiv 0$ for a stationary instability]. (b) Marginal stability curves: Curve (A) with $k_c = 0$ is for a homogeneous instability, curve (B) with $k_c \neq 0$ for a cellular instability ($\omega_c = 0$) or a dissipative wave ($\omega_c \neq 0$).

TABLE 17.3. Classification of instabilities.[a]

	$k_c = 0$	$k_c \neq 0$
$\omega_c = 0$	Stationary homogeneous	Cellular [Is]
$\omega_c \neq 0$	Oscillatory homogeneous [Io]	Dissipative wave [IIIo]

[a] Notations of [2] are given between brackets.

close to the linear threshold of a supercritical bifurcation, the coherence extends much beyond the size of individual cells, which legitimates a description in terms of modulations of an otherwise uniform bifurcated state. This points to a first, basically continuous, modeling strategy, with key words "envelope equations" and "phase dynamics" (Sec. 17.3.2). The complementary approach rests on a discrete description in terms of coupled cells (Sec. 17.3.3). Space discretization yields lattice models of subunits whose degrees of freedom are governed by dynamical systems of the type reviewed in Sec. 17.1. Further discretization of time yields coupled map lattices (CML). Cellular automata (CA) correspond to the ultimate modeling stage, where the local state variable can take on only a finite set of discrete values. This second strategy is natural when the original system appears as a spatially extended arrangement of physically distinct units. Otherwise, it is a good heuristic tool for spatially distributed systems with short-range space-time coherence, e.g., sufficiently far from a linear threshold or when the primary bifurcation is subcritical.

17.3.2. Continuous approach to space-time behavior

Spatial unfolding of classic bifurcations

In bounded media with overall size l, confinement effects have to be taken into account. The aspect ratio Γ is defined as the ratio of l to the critical wavelength $\lambda_c = 2\pi/k_c$. When $\Gamma \sim \mathcal{O}(1)$, the system is *confined*, the spatial structure of the unstable modes is frozen, and an effective discrete dynamical system is obtained for a small number of relevant mode amplitudes (full slaving, conventional center-manifold reduction, Sec. 17.1.2). By contrast, *extended systems* have $\Gamma \gg 1$, making the infinite-medium limit with fixed $k = k_c$ most relevant. Branching of such nonlinear solutions is governed by

$$dA/dt = L(k_c, r)A + N(A, k_c, r), \qquad (17.24)$$

where L accounts for the linear dynamics of the amplitude A of the critical mode. N describes the intrabranch couplings that result from the elimination of slaved modes belonging to the other noncritical branches. Slaving is here only partial, in the sense that modulations are allowed. In the neighborhood of the threshold (r_c, k_c, ω_c), the growth rate of modulations can be derived from the dispersion relation $s = \sigma - i\omega$, which expands as

$$\epsilon = \frac{r - r_c}{r_c}, \quad \delta k = k - k_c, \quad \tau_0\sigma = \epsilon - \xi_0^2\delta k^2,$$

$$\omega - \omega_c = \partial_k\omega_c\,\delta k + \tfrac{1}{2}\partial_{kk}\omega_c\,\delta k^2 + \cdots, \qquad (17.25)$$

where τ_0 is the natural time scale of the mode, ξ_0 is the coherence length, $V_g = \partial_k\omega_c$

is the group velocity, and $\frac{1}{2}\partial_{kk}\omega_c = \xi_0^2\alpha$ is the dispersion. Within the framework of a *multiscale expansion*, modulations are then associated to slow variables $(\bar{t}, \bar{x})$ through the correspondence $\partial_{\bar{t}} \mapsto \sigma$ and $\partial_{\bar{x}} \mapsto i\delta k$. The relation between $(\bar{t}, \bar{x})$ and (t, x) is given by a consistency condition as a power of the relative distance to threshold, i.e., usually $\bar{t} \propto \epsilon t$ and $\bar{x} \propto \epsilon^{1/2}x$.

- *Cellular instability*: In space dimension 1, one takes

$$\mathbf{X}(x,t) \propto \tfrac{1}{2}\left[A(t,x)\exp(ik_c x) + \text{c.c.}\right],$$

which yields A governed by

$$\tau_0\partial_t A = \epsilon A + \xi_0^2\partial_{xx}A - g|\,A\,|^2 A. \tag{17.26}$$

This equation derives from the potential $\mathcal{G}(A, A^*) = -\epsilon|\,A\,|^2 + \xi_0^2|\,\partial_x A\,|^2 + \frac{1}{2}g|\,A\,|^4$. Extension to space dimension 2 for a rotationally invariant instability is obtained on replacing ∂_x by $\partial_x + (1/2ik_c)\partial_{yy}$, yielding the *Newell-Whitehead-Segel* equation.

- *Homogeneous oscillatory instability*: The solution, taken as

$$\mathbf{X}(x,t) \propto \tfrac{1}{2}\left[A\exp(i\omega_c t) + \text{c.c.}\right],$$

is governed by the *complex Ginzburg-Landau* equation (for an introductory review, see van Saarloos, in [19] p. 19):

$$\tau_0\partial_t A = \epsilon A + \xi_0^2(1+i\alpha)\partial_{xx}A - (g_r + ig_i)|\,A\,|^2 A. \tag{17.27}$$

- *Dissipative waves* ($k_c \neq 0$, $\omega_c \neq 0$): In a parity-invariant medium, right (R) and left (L) waves propagating at group velocity $\pm V_g$ are generically governed by two coupled complex Ginzburg-Landau equations:

$$\begin{aligned}\tau_0(\partial_t \pm V_g\partial_x)A_{R,L} &= \epsilon A_{R,L} + \xi_0^2(1+i\alpha)\partial_{xx}A_{R,L}\\ &\quad - (g_r + ig_i)|\,A_{R,L}\,|^2 A_{R,L} - (g_r' + ig_i')|\,A_{L,R}\,|^2 A_{R,L}.\end{aligned} \tag{17.28}$$

Space-time behavior in the complex Ginzburg-Landau equation

The paradigm of systems displaying transition to spatiotemporal chaos is the cubic complex Ginzburg-Landau equation rewritten here in rescaled form:

$$\partial_t A = A + (1+i\alpha)\partial_{xx}A - (1+i\beta)|\,A\,|^2 A. \tag{17.29}$$

This equation admits uniform plane waves $A = A_q\exp(i\theta)$, called *phase winding* solutions, where A_q is the amplitude and $\theta = qx - \omega_q t$ is the phase of the wave, with $A_q = (1-q^2)^{1/2}$ and $\omega_q = \alpha q^2 + \beta(1-q^2)$. Space-time translational invariance implies that close to exact phase-winding solutions, the phase of the envelope is a slow variable governed by an effective nonlinear diffusion equation. For $q = 0$, it reads as

$$\partial_t\theta = D\partial_{xx}\theta - K\partial_{xxxx}\theta + g(\partial_x\theta)^2 + \text{h.o.t.}, \tag{17.30}$$

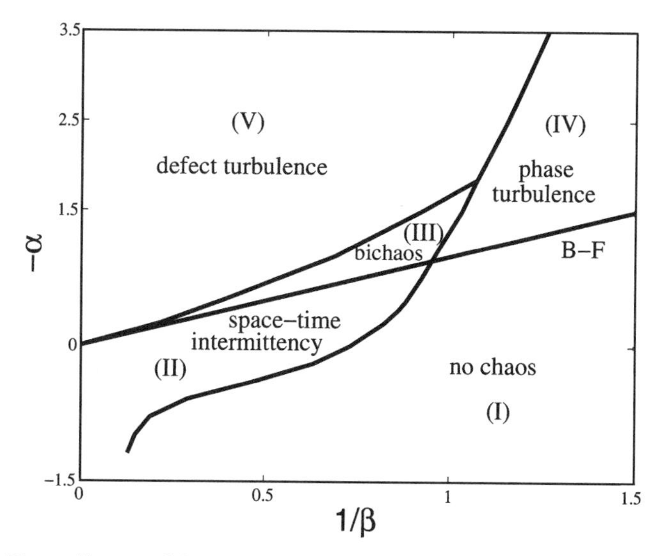

FIGURE 17.8. Phase diagram of the complex Ginzburg-Landau equation (17.29). After Chaté, in [19, p. 33].

with $D = 1+\alpha\beta$, $K = \frac{1}{2}\alpha^2(1+\beta^2)$, $g = \beta-\alpha$. $D < 0$ signals a *Benjamin-Feir instability*. Phase-winding solutions are all unstable when $1 + \alpha\beta < 0$ (Newell's criterion).

Figure 17.8 displays a diagram of the different possible steady-state regimes in the $(1/\beta, -\alpha)$ parameter plane. In region I, phase-winding solutions attract most initial conditions. Phase turbulence, characterized by weak modulations of $| A |$ enslaved to the phase perturbations, is present in region IV, close enough to the line corresponding to Newell's criterion (see [4] p. 67). Farther from this line, a "revolt" of $| A |$ ends in the formation of *defects* (phase singularities at zeroes of $| A |$). *Amplitude* turbulence is present in region V, where wild fluctuations of $| A |$ are observed. A weaker turbulent regime called *spatiotemporal intermittency* (STI, see [4] p. 111 for a general presentation) occupies region II. A similar regime called *bichaos* (STI + phase turbulence) takes place in region III, mixing behaviors in regions II and IV.

At lowest order, phase turbulence is governed by the *Kuramoto-Sivashinsky equation* (17.30), which, upon rescaling, reads as

$$\partial_t \phi + \partial_{xx}\phi + \partial_{xxxx}\phi + \tfrac{1}{2}(\partial_x \phi)^2 = 0, \tag{17.31}$$

and describes the propagation of interfaces and fronts. [26]

17.3.3. Discrete approach to space-time behavior

Coupled map lattices (CML)

CML are systems in which both time and space are discretized but the local state variables are continuous. They can be written in the general form $\mathbf{X}_{j,n+1} = \mathbf{G}_j(\mathbf{X}_{j',n};\ j' \in \mathcal{N}_j)$, where n is the iteration number (time), j and j' are nodes of the lattice (space), and $\mathcal{N}_j$

is a given neighborhood of site j. For homogeneous CMLs, definitions of the map and neighborhood do not depend on j. The coupling is *local* or *global* according to whether the neighborhood comprises a finite (and small) number of sites or the whole lattice. One-dimensional CMLs for a single local state variable X are frequently taken in the form

$$X_{j,n+1} = G(X_{j,n}) + g_{(+)}G(X_{j+1,n}) + g_{(0)}G(X_j, n) + g_{(-)}G(X_{j-1,n}), \qquad (17.32)$$

where $G(X)$ describes the evolution of the system in the absence of coupling (local reaction). The choice $g_{(+)} = g_{(-)} = g$, $g_{(0)} = -2g$ mimics diffusion, whereas the left-right symmetry can be broken to model advection in one direction ($g_{(+)} \neq g_{(-)}$). The form of the local reaction term G depends on the qualitative properties of the process under study. The case of the logistic map has been thoroughly studied. Pattern formation and the transition to space-time chaos follow roads similar to those observed in the complex Ginzburg-Landau equation, displaying frozen spatial chaos, random defect motion, STI, developed chaos, and so on, depending on the coupling g and the parameter of the logistic map. For a general review, see Ref. [20].

Cellular automata (CA), lattice gases, and neural nets

CA are systems of k-state discrete variables coupled on a (usually infinite) lattice and evolving according to a rule defined on some neighborhood $\mathcal{N}_j$ of each lattice node j. [16] Here, we consider Boolean ($k = 2$) automata on a one-dimensional lattice, $S_{j,n+1} = f(S_{j-r,n}; \ldots; S_{j+r,n})$, where r is the *range* of the coupling. Variables S_j are updated simultaneously. Rules are coded according to their action on the possible neighborhood configurations. The general coding scheme $\mathcal{R}_\mathrm{g}$ is explained in Table 17.4 using the example of rule $\mathcal{R}_\mathrm{g} = 150$. Extension to k-state automata is immediate. *Legal rules* are rules that preserve the null state and are symmetric under space reflection. The effect of a *totalistic rule* on some configuration C_j only depends on $s = \sum_{j'} S'_j$, $j' \in \mathcal{N}_j$. A simpler coding of totalistic rules is given by $\mathcal{R}_\mathrm{t} = \sum_{s=0}^{2r+1} 2^s f(s)$, $f(s)$ being the output of the application of the rule. There are 32 legal rules and 8 totalistic legal rules for Boolean CA with $r = 1$. Rule $\mathcal{R}_\mathrm{g} = 150$ is legal and totalistic, with $\mathcal{R}_\mathrm{t} = 1010_2 = 10_{10}$.

Evolution from an initial disordered configuration generates long-range correlations. A qualitative classification of the so-induced self-organization has been proposed by Wolfram according to the long-term behavior, as given in Table 17.5 and illustrated in Figure 17.9. Concepts defined in the study of chaos in dynamical systems, e.g., entropies and dimensions, can be translated to characterize space-time evolution of a given CA.

TABLE 17.4. Rule coding for one-dimensional range-*r* Boolean CA.[a]

$C_{j,n}$	111	110	101	100	011	010	001	000
$S_{j,n+1}$	1	0	0	1	0	1	1	0

[a] The 2^{2r+1} possible configurations $C_{j,n}$ at j are ordered according to their binary "value". The resulting state of site j is read as a 2^{2r+1}-digit binary number, further translated in decimal. Here for $r = 1$ and rule $S_{j,n+1} = S_{j-1,n} + S_{j,n} + S_{j+1,n}$ (mod 2), one has $\mathcal{R}_\mathrm{g} = 10010110_2 = 150_{10}$.

TABLE 17.5. Wolfram's classification of cellular automata.

Class	Asymptotic behavior	Difference pattern	Dynam. syst. analog
I	Homogeneous states	Dies out	Fixed point
II	Regular periodic structures	Persists but remain localized	Limit cycle
III	Chaotic aperiodic patterns	Expands at $\sim$ constant rate	Chaotic attractor
IV	Complex localized structures	Expands irregularly in time	

Lattice-gas CA are special CA describing the molecular dynamics of assemblies of particles with a finite number of states, moving and reacting on a lattice. Evolution rules are designed to account for microscopic properties of the medium to be modeled, i.e., free motion and collision-reaction rules. Macroscopic (partial differential) evolution equations, e.g., Navier-Stokes equations for a simple fluid, valid at large scale are recovered by coarse-grain averaging, which makes these systems popular models of complex spatially distributed media. [27]

Neural networks are special automata in which local units ("neurons") are coupled together and "fire" according to the rule

$$X_{j,n+1} = Y\left(\sum_{j} T_{jj'} X_{j',n} - \Xi_j\right), \tag{17.33}$$

where $Y(u)$ is the Heaviside function, $T_{jj'}$ is the synaptic weight matrix ($T_{jj} \equiv 0$, $T_{jj'} \equiv T_{j'j}$ for a symmetric network), and Ξ_j is the threshold of neuron j. In the symmetrical case, this is mapped on a spin problem through the variable change $S = 2X - 1$. The evolution rule then stipulates that the energy of a configuration

$$E = -\tfrac{1}{2}\sum_{j}\sum_{j'}\left(T_{jj'}S_{j'} - 2\Xi_j\right)S_j \tag{17.34}$$

decreases as time goes on. Asynchronous (or series) updating asymptotically yields fixed-

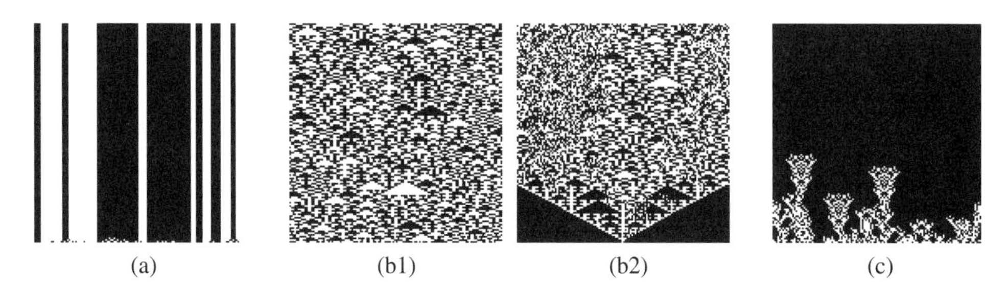

(a) (b1) (b2) (c)

FIGURE 17.9. Typical behavior of one-dimensional totalistic CA with $r = 2$ in Wolfram's classes II, III, and IV: (a) Class II, rule 56 (111000). (b) Class III, rule 38 (100110); the sensitivity to initial conditions for a chaotic rule is illustrated by the difference (b2) between the solution in (b1) and that issuing from an initial condition differing only by the state of the middle site. (c) Class IV, rule 20 (010100). Periodic boundary conditions, $L = 128$, time n running upward, 1 $\rightarrow$ 'white', 0 $\rightarrow$ 'black'.

point configurations. Synchronous (or parallel) updating leads to fixed-point or two-periodic configurations. Neural networks provide models of *associative memory*. Hebb's rule allows one to build the synaptic matrix of a network admitting a special set of M configurations to be retrieved upon asynchronous updating: $T_{jj'} = \sum_{m=1}^{M} S_j^m S_{j'}^m$. This simple-minded presentation points toward present-day studies of complexity in statistical physics of disordered media (e.g., glasses). [28]

17.4. REFERENCES

[1] *Hydrodynamic Instabilities and the Transition to Turbulence*, 2nd ed., edited by H.L. Swinney and J.P. Gollub (Springer-Verlag, New York, 1985).

[2] M. C. Cross and P.C. Hohenberg, Rev. Mod. Phys. **65**, 851–1112 (1993).

[3] U. Frisch, *Turbulence, the Legacy of A.N. Kolmogorov* (Cambridge University Press, Cambridge, UK, 1995).

[4] *Turbulence. A Tentative Dictionary*, edited by P. Tabeling and O. Cardoso (Plenum Press, New York, 1994).

[5] A. T. Winfree, *The Geometry of Biological Time* (Springer-Verlag, Berlin, 1990).

[6] *The Economy as an Evolving Complex System*, edited by P. W. Anderson, K. J. Arrow, and D. Pines (Addison-Wesley, Reading, MA, 1988).

[7] R. H. Abraham and C. D. Shaw, *Dynamics: The Geometry of Behavior*, 2nd ed. (Addison-Wesley, Reading, MA, 1992).

[8] G. L. Baker and J. P. Gollub, *Chaotic Dynamics, an Introduction*, 2nd ed. (Cambridge University Press, Cambridge, UK, 1996).

[9] J. Guckenheimer and Ph. Holmes, *Nonlinear Oscillations, Dynamical Systems, and Bifurcation of Vector Fields* (Springer-Verlag, New York, 1983).

[10] E. Ott, *Chaos in Dynamical Systems* (Cambridge University Press, Cambridge, UK, 1993).

[11] K. T. Alligood, T. D. Sauer, and J. A. Yorke, *Chaos, an Introduction to Dynamical Systems* (Springer-Verlag, Heidelberg, 1996).

[12] T. Poston and I. Stewart, *Catastrophe Theory and its Applications* (Pitman, London, 1978).

[13] H. G. Schuster, *Deterministic Chaos, an Introduction* (VCH, Weinheim, 1988).

[14] B. B. Mandelbrot, *The Fractal Geometry of Nature* (Freeman, San Francisco, 1982).

[15] R. Badii and A. Politi, *Complexity, Hierarchical Structures and Scaling in Physics* (Cambridge University Press, Cambridge, UK, 1997).

[16] *Theory and Application of Cellular Automata*, edited by S. Wolfram (World Scientific, Singapore, 1986).

[17] J. P. Eckmann and D. Ruelle, Rev. Mod. Phys. **57**, 617–656 (1985), addendum **57**, 1115 (1985).

[18] *Directions in Chaos*, edited by H. Bai-lin (World Scientific, Singapore), several volumes published starting 1987.

[19] *Spatio-Temporal Patterns in Nonequilibrium Complex Systems*, edited by P. E. Cladis and P. Palffy-Muhoray (Addison-Wesley, Reading, MA, 1995).

[20] *Theory and Application of Coupled Map Lattices*, edited by K. Kaneko (Wiley, Chichester, MA, 1993).

[21] *Universality in Chaos*, edited by P. Cvitanović (Adam Hilger, Bristol, 1989).

[22] *Chaos II*, edited by H. Bai-lin (World Scientific, Singapore, 1990).

[23] H. D. I. Abarbanel, *Analysis of Observed Chaotic Data* (Springer-Verlag, Heidelberg, 1996).

[24] T. Shinbrot, C. Grebogi, E. Ott, and J. A. Yorke, Nature **363**, 411–417 (1993).

[25] *Handbook of Chaos Control*, edited by H. G. Schuster (Wiley-VCH, Weinheim, 1999).

[26] *Solids Far from Equilibrium*, edited by C. Godrèche (Cambridge University Press, Cambridge, UK, 1992).

[27] D. H. Rothman and S. Zaleski, *Lattice-Gas Cellular Automata, Simple Models of Complex Hydrodynamics* (Cambridge University Press, Cambridge, UK, 1997).

[28] M. Mézard, G. Parisi, and M. A. Virasoro, *Spin Glass Theory and Beyond* (World Scientific, Singapore, 1987).

18

Nuclear Physics

Kenneth S. Krane
Oregon State University, Corvallis, Oregon

Contents

List of Tables

List of Figures

18.1. NUCLEAR PROPERTIES

Figure 18.1 illustrates the "nuclear landscape," the range of known stable and radioactive nuclei.

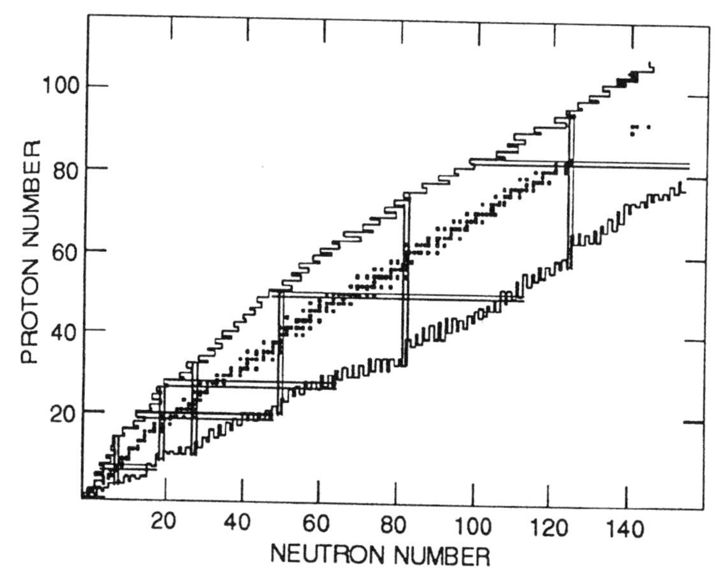

FIGURE 18.1. Chart of the nuclides shows the location of the proton and neutron drip lines. The filled squares show the stable isotopes. Vertical and horizontal bars indicate shell closures.

18.1.1. Size and shape of nuclei

The density of a spherical or near-spherical nucleus is often described by [1]

$$\rho(r) = \frac{\rho_0}{1 + \exp[(r - \bar{R})/a]}. \tag{18.1}$$

The central density ρ_0 remains nearly uniform throughout the range of nuclei at about 0.08 proton or neutron per fm^3. The mean radius or half radius $\bar{R}$ (which is here the radius where $\rho = \rho_0/2$) is often taken to be

$$\bar{R} = r_0 A^{1/3}, \tag{18.2}$$

where A is the mass number. Although there may be local variations in the shape that cause deviations from this value, the mean radius determined from Eq. (18.2) represents the average behavior across the range of nuclei. The radius parameter r_0 may take different values in the range 1.1–1.4 fm, depending on the type of measurement (e.g., Coulomb scattering versus neutron scattering); often, the value $r_0 = 1.2$ fm is chosen. The parameter a, sometimes called the diffuseness parameter, is related to the skin thickness t, the distance over which the density falls from $0.9\rho_0$ to $0.1\rho_0$, by

$$t = 4a \ln 3. \tag{18.3}$$

The parameters a and t show relatively little variation over the range of nuclei; typical values are $a = 0.5$–0.6 fm and $t = 2.2$–2.6 fm.

18.1.2. Mass and binding energy

For a nucleus of atomic number Z, mass number A, and atomic mass $M(Z, A)$, the binding energy B is [2]

$$B = [ZM(^1\mathrm{H}) + (A - Z)m_\mathrm{n} - M(Z, A)]c^2, \tag{18.4}$$

where $M(^1\mathrm{H})$ is the mass of a hydrogen atom and m_n is the mass of a neutron. This equation neglects a small correction due to the binding energy of the atomic electrons.

The binding energy per nucleon, B/A, varies only slightly in the range of 7–8 MeV for nuclei with $A > 12$. Figure 18.2 shows a plot of B/A versus A.

The binding energy can be approximately calculated from Weizsäcker's semiempirical formula:

$$B = a_\mathrm{v}A - a_\mathrm{s}A^{2/3} - a_\mathrm{c}Z(Z-1)A^{-1/3} - a_\mathrm{sym}(A - 2Z)^2/A + \delta, \tag{18.5}$$

where δ accounts for the pairing of like nucleons and has the value $+a_\mathrm{p}A^{-3/4}$ for Z and N both even, $-a_\mathrm{p}A^{-3/4}$ for Z and N both odd, and zero otherwise (A odd). The constants in this formula must be adjusted for the best agreement with data; typical values are $a_\mathrm{v} =$ 15.5 MeV, $a_\mathrm{s} = 16.8$ MeV, $a_\mathrm{c} = 0.72$ MeV, $a_\mathrm{sym} = 23$ MeV, and $a_\mathrm{p} = 34$ MeV.

18.1.3. Electromagnetic moments [3]

Magnetic dipole moment

The magnetic dipole operator is

$$\boldsymbol{\mu}_\mathrm{op} = \sum (g_l\mathbf{l} + g_s\mathbf{s}), \tag{18.6}$$

where the sum is carried out over all nucleons. The orbital g-factor g_l is $+1$ for protons and 0 for neutrons. The spin g-factor g_s is determined from the free proton and neutron magnetic moments:

$$\begin{aligned} \mu_\mathrm{p} &= +2.79284739 \quad \text{or} \quad g_{s\mathrm{p}} = +5.58569478, \\ \mu_\mathrm{n} &= -1.91304275 \quad \text{or} \quad g_{s\mathrm{n}} = -3.82608550. \end{aligned} \tag{18.7}$$

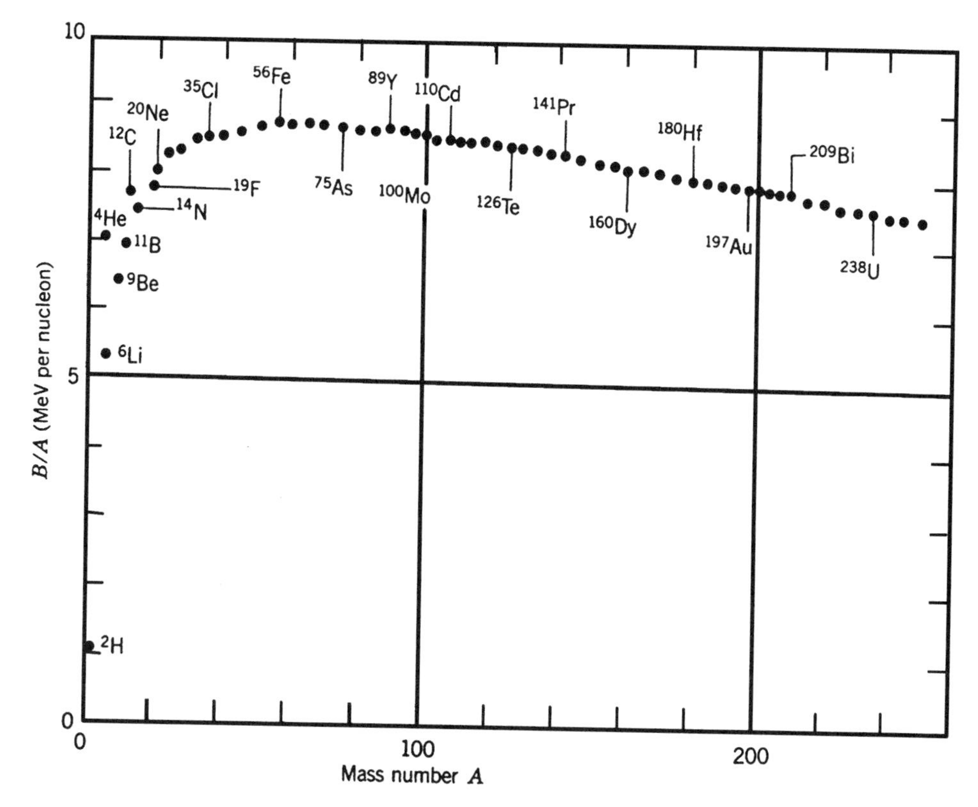

FIGURE 18.2. Selected values of the binding energy per nucleon (see Ref. [2]).

The nuclear magnetic dipole moment μ is the expectation value of $\boldsymbol{\mu}_{\rm op}$ in a state with maximum projection of angular momentum.

In the extreme single-particle model for a spherical odd-A nucleus, the magnetic dipole moment due to the valence nucleon is

$$\mu = j\left(g_l \pm \frac{g_s - g_l}{2l+1}\right)\mu_{\rm N} \quad \text{for} \quad j = l \pm 1/2. \tag{18.8}$$

The nuclear magneton $\mu_{\rm N}$ has the value 5.0508×10^{-27} J/T $= 3.1525 \times 10^{-8}$ eV/T.

Usually, the spin g-factor g_s is quenched in nuclei compared with its value for free nucleons. To account for this effect, g_s is often set equal to $0.6 g_s$(free).

The magnetic moment associated with collective motion of a state with angular momentum J is

$$\mu = g_R J, \tag{18.9}$$

where g_R, the rotational g-factor, is often very close to the value Z/A (≈ 0.4).

Nuclear magnetic dipole moments typically range in magnitude from 0 to 5 $\mu_{\rm N}$.

Electric quadrupole moment

The electric quadrupole operator is

$$eQ_{\text{op}} = \int \rho(r)(3z^2 - r^2)\, d^3r, \tag{18.10}$$

relative to a particular choice of the z axis, usually taken to be the cylindrical symmetry axis of the coordinate system fixed to the nucleus. Here, ρ refers to the density of electric charge in the nucleus. The expectation value of this quantity in the intrinsic (body-fixed) state is called the intrinsic quadrupole moment Q_0. The expectation value of Q_{op} in the laboratory system is usually called the spectroscopic quadrupole moment Q (or simply the quadrupole moment). For a nuclear state of laboratory angular momentum J having a projection K on the body-fixed axis, the relation between Q and Q_0 is

$$Q = Q_0 \frac{3K^2 - J(J+1)}{(J+1)(2J+3)} \tag{18.11}$$

or, for nuclear ground states with $J = K$,

$$Q = Q_0 \frac{J(2J-1)}{(J+1)(2J+3)}. \tag{18.12}$$

For a rotational ellipsoid of semiaxis c along the nuclear symmetry axis and semiaxis a perpendicular to the symmetry axis, the intrinsic quadrupole moment is

$$Q_0 = \tfrac{2}{5} Z(c^2 - a^2). \tag{18.13}$$

If the deviation from sphericity is not large, the nuclear surface can be represented in terms of the deformation parameter β as $R = \bar{R}(1+\beta Y_{20})$, in which case the intrinsic quadrupole moment is

$$Q_0 = \frac{3}{\sqrt{5\pi}} Z\bar{R}^2\beta \left(1 + \sqrt{\frac{5}{64\pi}}\,\beta\right), \tag{18.14}$$

neglecting higher order terms.

In a spherical nucleus with a valence proton, the quadrupole moment due to the odd particle in a shell-model state j is

$$Q = -\frac{2j-1}{2j+2}\langle r^2 \rangle. \tag{18.15}$$

For a uniformly charged sphere, $\langle r^2 \rangle = 0.6\bar{R}^2 A^{2/3}$.

Typical magnitudes of nuclear electric quadrupole moments are in the range 0 to 3 b (b = barn = 10^{-24} cm^2).

18.1.4. Isospin in nuclei

The charge independence of the nuclear force leads to a formalism in which nucleons are represented by the isospin, an abstract vector of length $t = 1/2$ defined such that protons have a projection $m_t = +1/2$ along an arbitrary axis (called the 3 axis) and neutrons have projection $m_t = -1/2$. The total 3 component of the isospin of a nucleus is thus $T_3 = (1/2)(Z - N)$, and the length of the corresponding vector T must be at least as great as $| T_3 |$.

The isospin T is usually a good quantum number in light nuclei, where the ground state generally has $T = | T_3 |$. The isospin can often be identified by the operation of selection rules; for example, nuclear reactions conserve T, E1 transitions have $\Delta T = 0, \pm 1$ with certain exceptions, and Fermi-type beta transitions (which do not change the nuclear spin or parity) are forbidden unless $\Delta T = 0$. In neighboring nuclei, multiplets appear having identical isospin and identical nucleon wave functions (except for the interchange of protons and neutrons); these are called isobaric analog states. For example, the spin-parity 0^+ $T = 1$ multiplet in the $A = 14$ system includes states with $T_3 = -1$ (ground state of ^{14}C), $T_3 = 0$ (2.31-MeV excited state of ^{14}N), and $T_3 = +1$ (ground state of ^{14}O).

18.2. RADIOACTIVE DECAY

18.2.1. Radioactive decay laws

The number of nuclei of a single radioactive species decays with time according to

$$N(t) = N_0 e^{-\lambda t}, \tag{18.16}$$

where N_0 is the original number of radioactive nuclei at $t = 0$ and λ is the decay constant (effectively, the decay probability per nucleus per second, assumed to be small and constant in time). The mean life τ and half-life $t_{1/2}$ are, respectively,

$$\tau = 1/\lambda \quad \text{and} \quad t_{1/2} = (\ln 2)/\lambda = 0.693/\lambda. \tag{18.17}$$

In practice, the measured quantity is usually the rate of decay or activity $A(t)$, rather than the number of radioactive nuclei. For a single decay, the activity also follows an exponential law:

$$A(t) \equiv \lambda N(t) = A_0 e^{-\lambda t}, \tag{18.18}$$

with the initial activity $A_0 = \lambda N_0$.

In a chain of decays A → B → C..., the activities of the daughters B, C, ... build up as A decays:

$$A_A(t) = A_{A0} e^{-\lambda_A t}, \tag{18.19}$$

$$A_B(t) = \frac{\lambda_B}{\lambda_B - \lambda_A} A_{A0}(e^{-\lambda_A t} - e^{-\lambda_B t}) + A_{B0} e^{-\lambda_B t}, \tag{18.20}$$

where $A_{A0}, A_{B0}, \ldots$ are the initial activities of the various species (often the initial activities of the species other than A are zero). The activities of subsequent members of the chain can be found by solving the rate equation $dN_i/dt = \lambda_{i-1} N_{i-1} - \lambda_i N_i$.

If radioactive nuclei of a given type are produced at a constant rate R by irradiation in a facility such as a reactor or a cyclotron, the activity of that particular nuclear species resulting after an irradiation time t is

$$A(t) = R(1 - e^{-\lambda t}). \tag{18.21}$$

18.2.2. Alpha decay [4]

The basic alpha decay process is ${}^{A}_{Z}X \to {}^{A-4}_{Z-2}X' + \alpha$, where $\alpha = {}^{4}_{2}He$. The Q-value for the decay is, using atomic masses,

$$Q = [M(X) - M(X') - M(^{4}He)]c^2. \tag{18.22}$$

For mass numbers $A \gg 4$, the kinetic energy of the alpha particle is

$$K_\alpha = Q(1 - 4/A). \tag{18.23}$$

The decay probability for alpha decay, which follows from a calculation based on tunneling through the Coulomb barrier, is approximately $\lambda_\alpha = (v/R)e^{-G}$, where v is the speed of the alpha particle inside the nucleus, R is the nuclear radius, and G is the Gamow factor

$$G = 2\sqrt{\frac{2m_\alpha}{\hbar^2 Q}}\,\frac{z(Z-z)e^2}{4\pi\epsilon_0}\left[\cos^{-1}\sqrt{Q/B} - \sqrt{(Q/B)(1-Q/B)}\right], \tag{18.24}$$

where $B = z(Z-z)e^2/4\pi\epsilon_0 R$ is the height of the Coulomb barrier for the alpha particles ($z = 2$). For most cases of interest, $Q \ll B$, in which case the Gamow factor is approximately

$$G \approx R\sqrt{2m_\alpha B/\hbar^2}\left[\pi\sqrt{B/Q} - 4\right]. \tag{18.25}$$

Small differences in the barrier thickness can lead to enormous differences in the decay probability; for example, in the thorium isotopes, a change in the decay energy by a factor of 2 (from 4 to 8 MeV) leads to a change in the decay probability by a factor of 10^{20}.

It is also possible for the nucleus to emit other heavy charged particles in a process similar to alpha decay. Near the proton drip line, for example, nuclei are unstable to the emission of protons [5]. In fact, the observation of such decays is a principal method for mapping the proton drip line. This decay process can be represented as ${}^{A}_{Z}X \to {}^{A-1}_{Z-1}X' + p$. Proton radioactivity has been observed in many nuclei with $A > 100$; the half-lives are typically in the range of μs to ms (although there are a few cases in the range of 1–10 s). Decay energies are typically of order 1 MeV. The Q-value for the decay can be written in a manner analogous to Eq. (18.22), and the decay probability can be calculated using Eq. (18.24) for the Gamow factor with $z = 1$ and m_α replaced by m_p.

Occasionally, it is energetically possible for the nucleus to emit clusters more massive than alphas in an analogous process [6]. As is the case with alpha decay, this decay mode is most commonly observed for nuclei with relatively small values of the binding energy per nucleon, that is, nuclei with $A > 200$. Emitted particles observed in this decay mode include ^{14}C, ^{24}Ne, ^{28}Mg, and ^{32}Si. Partial half-lives are typically in the range 10^{8}–10^{17} years; equivalently, the branching ratios for these decays are 10^{-9} to 10^{-12} relative to the competing alpha decay mode. This large reduction in the decay probability comes about from the Gamow factor [Eq. (18.24)] because the barrier height B increases more rapidly (in proportion to z) than does the Q-value.

18.2.3. Beta decay [7]

Three radioactive decay processes are classified as beta decays: negative beta decay (${}^{A}_{Z}X \to {}_{Z+1}^{\ \ A}X' + e^- + \bar{\nu}$), positive beta decay or positron decay (${}^{A}_{Z}X \to {}_{Z-1}^{\ \ A}X' + e^+ + \nu$), and electron capture (${}^{A}_{Z}X + e^- \to {}_{Z-1}^{\ \ A}X' + \nu$). The Q-values are (using atomic masses)

$$Q_{\beta^-} = [M(X) - M(X')]c^2, \tag{18.26}$$

$$Q_{\beta^+} = [M(X) - M(X') - 2m_e]c^2, \tag{18.27}$$

$$Q_{EC} = [M(X) - M(X')]c^2. \tag{18.28}$$

For allowed decays, in which the leptons (e and ν) carry no orbital angular momentum, the spectrum shape is (assuming massless neutrinos)

$$N(p)\,dp = \frac{g^2 |M_{fi}|^2}{2\pi^3 c^3 \hbar^7} F(p, Z') p^2 (Q - K)^2\,dp \tag{18.29}$$

where g is the weak-interaction coupling constant, M_{fi} is the nuclear matrix element of the transition, p is the electron momentum, K is the electron kinetic energy, and F is the Fermi function, which accounts for the distortion of the lepton wave function by the Coulomb field of the nucleus. For forbidden decays, this equation should also include a shape factor $S(p, q)$ for electron momentum p and neutrino momentum q. The reduced half-life or ft-value for the decay is

$$ft = \frac{(2\ln 2)\pi^3 \hbar^7}{g^2 m_e^5 c^4 |M_{fi}|^2}, \tag{18.30}$$

where t represents the half-life of the decay and the Fermi integral f is the integral over the spectrum of all energy-dependent factors. The ft-value gives a rough indication of the degree of forbiddenness of the decay, as indicated in Table 18.1.

It is in a few cases energetically possible for a nucleus to decay by double beta emission, for example, ${}^{A}_{Z}X \to {}_{Z+2}^{\ \ A}X' + 2e^- + 2\nu$ [8]. This type of decay has been observed in such

TABLE 18.1. Classifications of beta decays.

Type of decay	ΔJ	$\Delta\pi$	$\log ft$ range
Superallowed	0,1	no	3–4
Allowed	0,1	no	4–8
First forbidden	0,1,2	yes	6–9
Second forbidden	2,3	no	10–13
Third forbidden	3,4	yes	14–20
Fourth forbidden	4,5	no	> 20

nuclei as ^{76}Ge, ^{82}Se, ^{96}Zr, ^{100}Mo, ^{116}Cd, ^{128,130}Te, ^{150}Nd, and ^{238}U using geochemical, radiochemical, and direct counting techniques. Typical half-lives are in the range 10^{19}–10^{21} years. Searches have also been made for the neutrinoless version of this decay process, which is forbidden according to standard electroweak theory with Dirac spin-$\frac{1}{2}$ neutrinos (that is, ν is different from $\bar{\nu}$). If observed, neutrinoless double beta decay would suggest a violation of lepton-number conservation and the existence of neutrinos with mass. The present limit on the half-life for this decay process is at least 10^{22} years. A rough estimate for the half-life for double beta decay can be given in analogy with Eq. (18.30):

$$f^2 t = \frac{(4 \ln 2)\pi^6 \hbar^{13}}{g^4 m_e^9 c^6 | M_{fi} |^4} \tag{18.31}$$

which gives values of order 10^{17} years.

For nuclei far from stability, where the beta-decay Q-value is large, it is possible for the beta decay to lead to states in the daughter nuclei of such high excitation energies that they are unstable with respect to proton or neutron emission [9]. This process can occur only when Q_β exceeds the nucleon separation energy $S_{n,p} = [M(^{A-1}X') + m_{n,p} - M(^{A}X)]c^2$. These nucleons are generally emitted promptly from the excited states but follow the half-life of the original beta decay; hence, the process is called beta-delayed proton or neutron emission. Some states are even unstable with respect to the emission of two or more protons or neutrons (in fact, Rutherford observed beta-delayed alpha emission from ^{212}Bi in 1912). In light nuclei, beta-delayed proton emission can often help in identifying isobaric analog states populated in the beta decay, whereas beta-delayed single or multiple neutron emission is used to identify the neutron “halos” that may surround the cores of light nuclei.

18.2.4. Gamma decay [10]

When gamma-ray (photon) emission takes place between two states of energy difference $\Delta E = E_i - E_f$, the gamma energy is

$$E_\gamma = \Delta E - E_R \approx \Delta E - (\Delta E)^2/2Mc^2, \tag{18.32}$$

where E_R is the energy of nuclear recoil (which is usually a negligible correction).

The decay rate for emission of a multipole of type σ (electric E or magnetic M) and angular momentum L is

$$\lambda(\sigma L) = \frac{2(L+1)}{\epsilon_0 \hbar L(2J_i + 1)[(2L+1)!!]^2} \left(\frac{E_\gamma}{\hbar c}\right)^{2L+1} | M_{fi}(\sigma L) |^2, \tag{18.33}$$

where M_{fi} is the reduced nuclear matrix element corresponding to emission of the multipole σL. Often, this formula is written in terms of the reduced transition probability

$$B(\sigma L; J_i \to J_f) = (2J_i + 1)^{-1} | M_{fi}(\sigma L) |^2. \tag{18.34}$$

The angular momentum and parity selection rules for gamma emission from an initial state with angular momentum and parity J_i and π_i to a final state J_f and π_f are

$$| J_\mathrm{f} - J_\mathrm{i} | \le L \le J_\mathrm{f} + J_\mathrm{i} \text{ (no } L = 0),$$
$$\pi_\mathrm{i}\pi_\mathrm{f} = (-1)^L \text{ for E}L, \quad \text{or} \quad (-1)^{L+1} \text{ for M}L. \tag{18.35}$$

Often, several multipoles $\sigma L, \sigma' L', \ldots$ can contribute to a gamma transition; in which case, the total decay rate is $\lambda_\mathrm{T} = \lambda(\sigma L) + \lambda(\sigma' L') + \cdots$, and the relative contribution of a particular multipole to the intensity of the decay is $\lambda(\sigma L)/\lambda_\mathrm{T}$. In many applications, however, it is important to specify an amplitude ratio rather than an intensity ratio; the ratio of amplitudes L and L', called the mixing ratio δ, must have a carefully defined phase to enable comparison of theory and experiment:

$$\delta = i^{L'+\Lambda(\sigma')-L-\Lambda(\sigma)} \sqrt{\frac{L(L'+1)[(2L'+1)!!]^2}{L'(L+1)[(2L+1)!!]^2}} \left(\frac{E_\gamma}{\hbar c}\right)^{L'-L} \Delta, \tag{18.36}$$

where $\Lambda(\mathrm{E}) = 0$ and $\Lambda(\mathrm{M}) = 1$. Here, $\Delta = M_\mathrm{fi}(\sigma' L')/M_\mathrm{fi}(\sigma L)$, the ratio between the reduced matrix elements.

It is often convenient to express the transition probabilities of a given multipole order in terms of Weisskopf units, which are obtained by assuming that only a single valence proton is responsible for the transition and by making other simplifying assumptions about the spin-dependent factors in the transition probability. These units are as follows:

$$\lambda(\mathrm{E1}) = 1.0 \times 10^{14} A^{2/3} E_\gamma^3, \qquad \lambda(\mathrm{M1}) = 3.1 \times 10^{13} E_\gamma^3,$$
$$\lambda(\mathrm{E2}) = 7.3 \times 10^{7} A^{4/3} E_\gamma^5, \qquad \lambda(\mathrm{M2}) = 2.2 \times 10^{7} A^{2/3} E_\gamma^5,$$
$$\lambda(\mathrm{E3}) = 3.4 \times 10^{1} A^{2} E_\gamma^7 \qquad \lambda(\mathrm{M3}) = 1.0 \times 10^{1} A^{4/3} E_\gamma^7,$$
$$\lambda(\mathrm{E4}) = 1.1 \times 10^{-5} A^{8/3} E_\gamma^9, \qquad \lambda(\mathrm{M4}) = 3.3 \times 10^{-6} A^{2} E_\gamma^9. \tag{18.37}$$

Here λ is the transition probability in s^{-1} and E_γ is the transition energy in MeV.

18.2.5. Internal conversion [11]

In the internal-conversion process, an excited state de-excites through the ejection of an atomic electron. This process competes with gamma-ray emission. The kinetic energy of the emitted electron is the transition energy ΔE less the electron binding energy B:

$$K_\mathrm{e} = \Delta E - B. \tag{18.38}$$

The total decay probability λ_T must include contributions from both gamma emission and internal conversion:

$$\lambda_\mathrm{T} = \lambda_\gamma + \lambda_\mathrm{e} = \lambda_\gamma(1 + \alpha), \tag{18.39}$$

where $\alpha = \lambda_\mathrm{e}/\lambda_\gamma$ is the total internal-conversion coefficient, which can be expressed as the sum of the partial conversion coefficients corresponding to the atomic shell from which the electron is ejected:

$$\alpha = \alpha_\mathrm{K} + \alpha_\mathrm{L} + \alpha_\mathrm{M} + \cdots \tag{18.40}$$

Here, K, L, M, ... designate, respectively, the atomic shells with principal quantum number $n = 1, 2, 3, \ldots$. The conversion coefficients depend on the multipolarity and type of the transition, the atomic number, and the transition energy.

In contrast to gamma emission, the internal conversion process exists for electric monopole (E0) transitions. Because they sample the monopole distribution, E0 transitions are sensitive measures of changes in the nuclear shape.

18.2.6. Units for radioactivity

The activity A gives the rate at which a sample decays. The SI unit for activity is the becquerel (Bq): 1 becquerel = 1 decay/s. Another unit in common use is the curie (Ci), defined such that 1 Ci = 3.7×10^{10} decays/s = 37 GBq.

A measure of the effect of radiation is the exposure X, the charge produced in air by ionization. The SI unit is coulombs per kilogram, and the unit in more common use is the roentgen (R), where 1 R = 2.58×10^{-4} C/kg. The rate of exposure depends on the activity A of the source of radiation, its distance d from the air sample (often taken as 1 m), and the specific gamma-ray constant Γ, which depends on the energies of the emitted radiations:

$$\frac{\Delta X}{\Delta t} = \Gamma \frac{A}{d^2}. \tag{18.41}$$

Γ is typically of order 1 but can be much smaller (less than 0.1) for sources that emit only low-energy gamma rays.

For materials other than air, the significant factor is the absorbed dose, which measures the energy deposited in the material by the radiation per unit mass of the material. A common unit in use for absorbed dose is the rad (for radiation absorbed dose), which corresponds to an energy absorption of 100 ergs per gram of material. The SI unit for absorbed dose is the gray (Gy), corresponding to the absorption of one joule per kilogram of material. Thus, 1 Gy = 100 rad.

The effect of radiation on biological systems is given in terms of the dose equivalent, obtained by multiplying the absorbed dose by a dimensionless quality factor that accounts for the biological effect of different types of radiation. The quality factor, which varies with the type and energy of the radiation, ranges from 1 for low-energy gamma rays or x-rays to 5–10 for MeV protons or neutrons to about 20 for alpha particles. The dose equivalent is measured in rem (roentgen equivalent man) when the dose is in rads. The SI unit for dose equivalent is the sievert (Sv), where 1 Sv = 100 rem. See Chap. 15.

The various measures of radioactivity are summarized in Table 18.2.

TABLE 18.2. Different measures of the quantity of radioactivity.

Quantity	Measure of	Common unit	SI unit	Conversion
Activity	Decay rate	curie (Ci)	becquerel (Bq)	1 Bq = 27 pCi
Exposure	Ionization in air	roentgen (R)	coulomb/kilogram (C/kg)	1 C/kg = 3876 R
Absorbed dose	Energy absorption	rad	gray (Gy)	1 Gy = 100 rad
Dose equivalent	Biological effect	rem	sievert (Sv)	1 Sv = 100 rem

18.3. NUCLEAR MODELS [12]

18.3.1. The shell model [13]

The shell model assumes that the properties of a nucleus are due primarily to the valence nucleons, which are analyzed as if they move in a spherically symmetric mean field created by the remaining nucleons. One form of the potential energy of this interaction is

$$U(r) = U_0 f_{\mathrm{WS}}(r) + \frac{U_{\mathrm{SO}}}{\bar{R}^2}\,\frac{1}{r}\,\frac{df_{\mathrm{WS}}}{dr}\,\mathbf{l}\cdot\mathbf{s}. \tag{18.42}$$

Here, the radial form of the interaction, called the Woods-Saxon form, follows the nuclear charge or mass distribution of Eq. (18.1): $f_{\mathrm{WS}}(r) = \{1 + \exp[(r - \bar{R})/a]\}^{-1}$. The strength of the central interaction is taken to be $U_0 = -51 + 33(A - 2Z)/A$ MeV, and that of the spin-orbit term is $U_{\mathrm{SO}} = -0.44U_0$. Figure 18.3 shows the neutron states calculated from this potential energy.

Shell gaps occur at the so-called "magic numbers": 2, 8, 20, 28, 50, 82, 126.

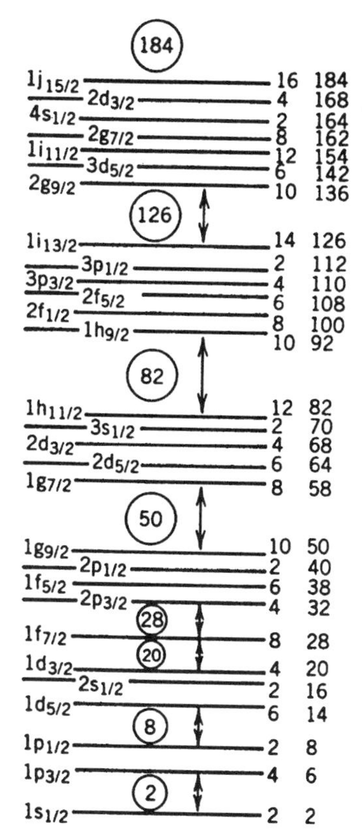

FIGURE 18.3. Representative levels in the nuclear shell model show the large energy gaps at the "magic numbers." The second column from the right shows the capacity of each level for neutrons or protons, and the rightmost column gives the cumulative capacity of all levels up to and including the indicated level (see Ref. [13]).

18.3.2. The deformed shell model [14]

For nonspherical nuclei, the central potential must take into account the deformed shape. Often, a harmonic-oscillator form is chosen for the interaction:

$$U_{\text{def}}(r) = \tfrac{1}{2}m(\omega_x^2 x^2 + \omega_y^2 y^2 + \omega_z^2 z^2). \tag{18.43}$$

For an axially symmetric shape with a deformation parameter $\delta = 2(c-a)(c+a)$, where c and a are the semiaxes, respectively, along and perpendicular to the symmetry axis, the oscillator frequencies are chosen to be

$$\omega_x = \omega_y = \omega_0(\delta)\left(1 + \tfrac{2}{3}\delta\right)^{1/2} \quad \text{and} \quad \omega_z = \omega_0(\delta)\left(1 - \tfrac{4}{3}\delta\right)^{1/2}. \tag{18.44}$$

Here, $\omega_0(\delta) = \bar{\omega}_0(1 + 2\delta^2/3)$, which ensures that the nuclear volume remains constant as the deformation is increased. The complete interaction includes not only $U_{\text{def}}(r)$, but also a spin-orbit term and another angular-momentum dependent term that gives a slight flattening of the oscillator interaction in the nuclear interior. Figure 18.4 shows a typical set of energy levels resulting from the calculation.

18.3.3. The collective model [15]

Nuclear vibrations

For vibrations about an equilibrium shape of mean radius $\bar{R}$, the surface is described by

$$R(\theta, \phi) = \bar{R}\left[1 + \sum_{\lambda\mu} \alpha_{\lambda\mu} Y_{\lambda\mu}^*(\theta, \phi)\right], \tag{18.45}$$

where $\lambda = 2, 3, 4\ldots$ gives the mode of vibration (referred to as quadrupole, octupole, hexadecapole, ...), $Y_{\lambda\mu}$ are spherical harmonics, and $\alpha_{\lambda\mu}$ are the time-dependent vibrational amplitudes. For any particular mode, the Hamiltonian is

$$H = \frac{B_\lambda}{2}\sum_\mu |\dot{\alpha}_{\lambda\mu}|^2 + \frac{C_\lambda}{2}\sum_\mu |\alpha_{\lambda\mu}|^2, \tag{18.46}$$

where B_λ is an inertial parameter and C_λ is a stiffness parameter, which in the case of an irrotational charged fluid are

$$B_\lambda = \frac{3Am\bar{R}^2}{4\pi\lambda} \quad \text{and} \quad C_\lambda = (\lambda - 1)(\lambda + 2)\bar{R}^2 a_{\text{s}} - \frac{3(\lambda - 1)Z^2e^2}{2\pi(2\lambda + 1)\bar{R}}. \tag{18.47}$$

Here, a_{s} is the surface energy parameter in the semiempirical binding energy formula, Eq. (18.5). The resulting energy spectrum is

$$E = \sum_\lambda \hbar\omega_\lambda \left(n_\lambda + \frac{1}{2}\right) \quad \text{with} \quad \omega_\lambda = \sqrt{C_\lambda/B_\lambda}, \tag{18.48}$$

where n_λ is the number of vibrational phonons of type λ. Each of these phonons carries

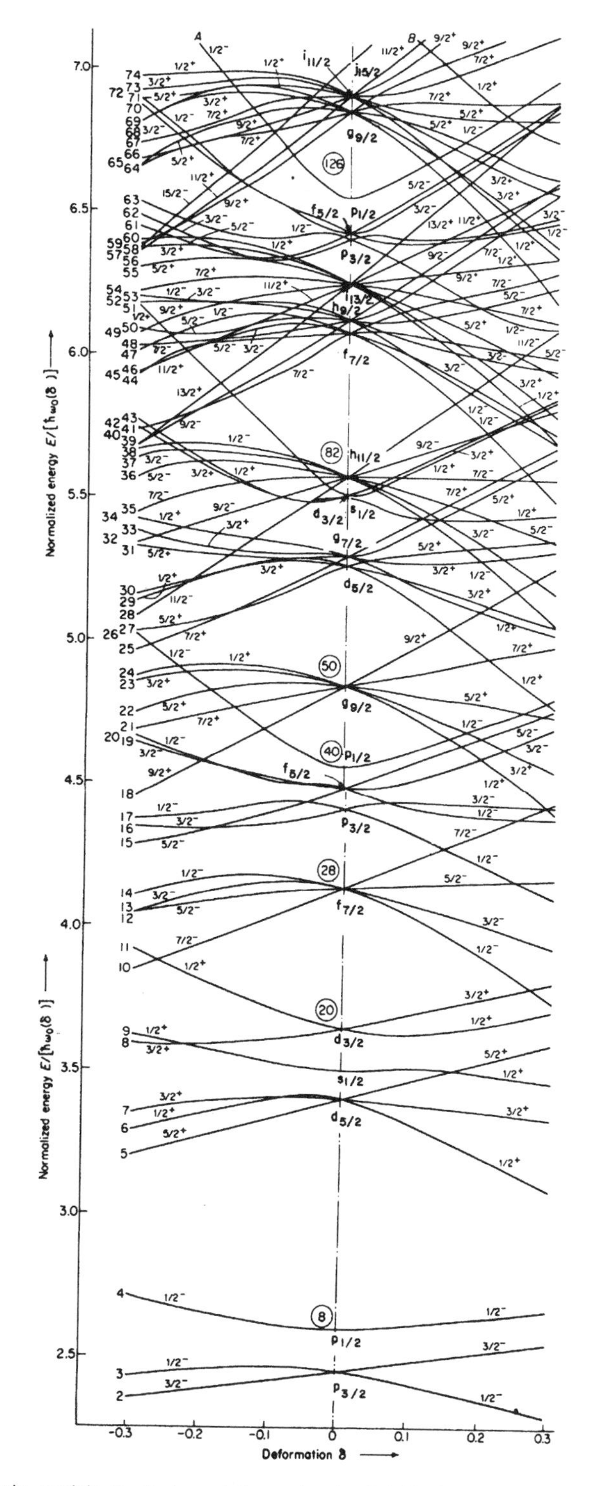

FIGURE 18.4. Single-particle levels in a deformed potential. Each shell-model state shown in Figure 18.3 splits into $j+1/2$ states whose energies depend on the deformation parameter δ (see Ref. [14]).

λ units of angular momentum, but the symmetrization of the total phonon state function suppresses certain values of the total angular momentum. For example, for quadrupole ($\lambda = 2$) vibrations of an even-even nucleus, the ground state ($n_2 = 0$) has spin-parity $J^\pi = 0^+$, the first excited state ($n_2 = 1$) has an energy of $\hbar\omega_2$ and $J^\pi = 2^+$, and the second excited state ($n_2 = 2$) is actually a triplet of states with $J^\pi = 0^+, 2^+, 4^+$ at an energy of $2\hbar\omega_2$. Electromagnetic matrix elements between vibrational states vanish unless $\Delta n_\lambda = \pm 1$. The reduced transition probability for a transition from a state with n_λ vibrational phonons is

$$B(E\lambda; n_\lambda \to n_\lambda - 1) = n_\lambda \left(\frac{3\bar{R}^\lambda Ze}{4\pi} \right)^2 \frac{\hbar}{2\omega_\lambda B_\lambda}. \tag{18.49}$$

Many nuclear reactions—for example, inelastic proton scattering and photonuclear processes—are observed to occur through a highly excited vibrational state known as a giant resonance. These resonances are populated too strongly to be associated with single-particle excitations and so must be collective in nature. Like other collective properties, the location and strength of these resonances on the average vary smoothly from one nucleus to another, especially in medium and heavy nuclei. The giant dipole resonance occurs at an energy of about $77A^{-1/3}$ MeV and has a width in the range of 4–8 MeV. This resonance corresponds to the protons and neutrons vibrating against one another. The total strength of the giant dipole resonance is given by the Thomas-Reiche-Kuhn sum rule:

$$\int_0^\infty \sigma(E)\, dE = \frac{2\pi^2\hbar^2\alpha}{m} \frac{NZ}{A} \approx 6.0 \frac{NZ}{A} \text{ MeV} \cdot \text{fm}^2, \tag{18.50}$$

where α is the fine structure constant ($\approx 1/137$) and m is the nucleon mass. Experimental values of the resonance strengths typically agree with this sum rule to within about $\pm 30\%$.

In the same energy region, there are other giant resonances, including the giant quadrupole resonance, which has two components: an isoscalar part, corresponding to protons and neutrons moving together and having an energy of approximately $65A^{-1/3}$ MeV, and an isovector part, corresponding to protons and neutrons moving oppositely and having an energy of about $130A^{-1/3}$ MeV.

Nuclear rotations

The surface of a nucleus having a deformed equilibrium shape can be described by

$$R(\theta, \phi) = \bar{R} \left[1 + \beta \sqrt{\frac{5}{16\pi}} [(\cos\gamma)(3\cos^2\theta - 1) + \sqrt{3}\, \sin\gamma \sin^2\theta \cos^2\phi] \right], \tag{18.51}$$

where β gives the deviation of the nucleus from sphericity and γ gives the deviation from axial symmetry. For $\gamma = 0°$, the nucleus is a prolate spheroid, whereas for $\gamma = 60°$, it is an oblate spheroid. For $0° < \gamma < 60°$, the shape is triaxial (lacking axial symmetry). Energy levels corresponding to the collective rotations and vibrations of a nucleus having this shape are described by

$$E_{n_\beta n_\gamma}(J) = \hbar\omega_\beta\left(n_\beta + \frac{1}{2}\right) + \hbar\omega_\gamma(2n_\gamma + 1) + \frac{\hbar^2}{2\Im}J(J+1). \tag{18.52}$$

Here, n_β and n_γ give, respectively, the number of excitations corresponding to the β and γ vibrations, and the last term represents the energy of a rigid rotor of rotational inertia $\Im$. To account for a lack of rigidity in the rotation, a correction term of the form $-B[J(J+1)]^2$ can be included in the energy.

The intrinsic (body-fixed) quadrupole moment of an axially symmetric ($\gamma = 0$) deformed nucleus was given in Eq. (18.14). The E2 transition probabilities within a rotational band are

$$B(E2;\, J_\mathrm{i} \to J_\mathrm{f}) = \frac{5}{16\pi}e^2 Q_0^2 \langle J_i K 2 0 \mid J_\mathrm{f} K\rangle^2, \tag{18.53}$$

which gives, in the case of a $K = 0$ band,

$$B(E2;\, J+2 \to J) = \frac{5}{16\pi}e^2 Q_0^2 \frac{3(J+1)(J+2)}{2(2J+3)(2J+5)} \tag{18.54}$$

or, for the $2 \to 0$ transition and using Eq. (18.14) to first order,

$$B(E2;\, 2 \to 0) = \frac{1}{5}\left(\frac{3Ze\bar{R}^2}{4\pi}\right)^2 \beta^2. \tag{18.55}$$

18.4. INTERACTION OF NUCLEAR RADIATION WITH MATTER [16]

18.4.1. Heavy charged particles

The stopping power, or energy loss by a particle per unit length of path in the material, is given by the Bethe-Bloch formula:

$$-\frac{dE}{dx} = \frac{z^2 Z e^4 n}{4\pi\epsilon_0^2 m_\mathrm{e} v^2}\left[\ln\frac{2m_\mathrm{e}v^2}{I(1-\beta^2)} - \beta^2 - \frac{C}{Z}\right] \tag{18.56}$$

for an incident particle of charge ze and velocity $v = \beta c$. Here, Z is the atomic number of the absorber, n is its number of atoms per unit volume, I is its mean ionization potential, and m_e is the electron mass. The term C/Z represents a shell correction term that is important only for low energies.

The range of the particle in the material can in principle be obtained from Eq. (18.56) by integrating the stopping power over all energies of the particle:

$$R = \int dx = \int_E^0 \frac{dE}{dE/dx}. \tag{18.57}$$

This integral cannot be performed because of uncertainties in Eq. (18.56) at low particle energies, and moreover, it ignores the effect of multiple scattering in altering the path. In practice, we therefore define an empirical mean range $\bar{R}$ as the thickness of material that reduces the intensity of a beam of particles by half. The distribution of ranges about the mean, called range straggling, can be represented as Gaussian:

$$N(R) \propto \exp[-(R - \bar{R})^2/\alpha_0^2], \tag{18.58}$$

where the range straggling parameter α_0 is approximately $0.010\bar{R}$–$0.015\bar{R}$.

The mean range (in cm) of protons with energies between a few MeV and 200 MeV in dry air at 1 atm and 15°C is approximately

$$\bar{R}_{\rm air} = 100[E(\text{MeV})]^{1.8}. \tag{18.59}$$

The relative mean ranges for a given type of particle in two different materials of densities ρ_1 and ρ_2 and atomic masses A_1 and A_2 can be approximated using the Bragg-Kleeman rule:

$$\bar{R}_1/\bar{R}_2 = \rho_2\sqrt{A_1}/\rho_1\sqrt{A_2}. \tag{18.60}$$

For different particles (of masses m_1 and m_2 and charges z_1 and z_2) having the same initial velocity v and traveling in the same material, the relative mean ranges are given by

$$\bar{R}_1/\bar{R}_2 = m_1 z_2^2/m_2 z_1^2. \tag{18.61}$$

18.4.2. Electrons

The total stopping power of electrons in matter can be written as the sum of two terms, one due to ionization by collision and the other due to radiation (bremsstrahlung):

$$-\frac{dE}{dx} = \left(-\frac{dE}{dx}\right)_{\rm c} + \left(-\frac{dE}{dx}\right)_{\rm r}. \tag{18.62}$$

The energy loss due to collisions is given by a Bethe-Bloch equation in a form similar to Eq. (18.56):

$$\left(-\frac{dE}{dx}\right)_{\rm c} = \frac{Ze^4 n}{8\pi\epsilon_0^2 m_{\rm e} v^2}\left[\ln\frac{m_{\rm e}v^2 K}{2I^2(1-\beta^2)} - \left(2\sqrt{1-\beta^2} - 1 + \beta^2\right)\ln 2 + 1 - \beta^2 + \frac{1}{8}\left(1 - \sqrt{1-\beta^2}\right)^2\right]. \tag{18.63}$$

Here, K is the kinetic energy of the electrons and all other symbols have the same meaning as in Eq. (18.56). The energy loss due to radiation is

$$\left(-\frac{dE}{dx}\right)_{\rm r} = \frac{\alpha n E Z(Z+1)e^4}{4\pi^2\epsilon_0^2(m_{\rm e}c^2)^2}\left[\ln\frac{2E}{m_{\rm e}c^2} - \frac{1}{3}\right], \tag{18.64}$$

with $E = K + mc^2$. Here, α is the fine-structure constant ($\approx 1/137$).

The radiative contribution to the energy loss becomes important only at high energies (typically above 10 MeV). The ratio between the stopping powers is approximately

$$\frac{(-dE/dx)_{\mathrm{r}}}{(-dE/dx)_{\mathrm{c}}} = \frac{EZ}{1600 m_{\mathrm{e}} c^2} \tag{18.65}$$

for E in MeV.

On account of multiple scattering, the range is generally not a useful quantity for electrons. Approximate formulas for the mean range (in g/cm^2) in aluminum are

$$\bar{R} = 0.412 K^{1.265 - 0.0954 \ln K} \qquad 0.01 < K < 3\ \mathrm{MeV}, \tag{18.66}$$

$$\bar{R} = 0.530 K - 0.106 \qquad 2.5\ \mathrm{MeV} < K < 20\ \mathrm{MeV}. \tag{18.67}$$

18.4.3. Electromagnetic radiation [17]

The intensity I of a beam of electromagnetic radiation after passing through a thickness x of a homogeneous material is given by

$$I(x) = I_0 \exp(-\mu x), \tag{18.68}$$

where I_0 is the initial beam intensity and μ is the total linear attenuation coefficient for all processes. This expression is often written in terms of the mass absorption coefficient $\mu_{\mathrm{m}} = \mu/\rho$, where ρ is the density of the material. The absorption coefficient can also be written in terms of the interaction cross section per atom σ as $\mu = n\sigma$, where n is the number of atoms per unit volume.

In the energy region of nuclear gamma rays, the interaction with matter is dominated by three processes: photoelectric, Compton, and pair production, such that the total interaction cross section is $\sigma = \sigma_{\mathrm{PE}} + \sigma_{\mathrm{C}} + \sigma_{\mathrm{PP}}$.

A nonrelativistic approximation for the cross section per atom for the photoelectric interaction in the K shell is

$$\sigma_{\mathrm{PE}} = \frac{32\pi\sqrt{2}\alpha^4 Z^5}{3} \left(\frac{m_{\mathrm{e}} c^2}{E_\gamma}\right)^{7/2} r_0^2, \tag{18.69}$$

where E_γ is the gamma-ray energy (here assumed to be small compared with $m_{\mathrm{e}}c^2$), α is the fine-structure constant, and r_0 is the classical radius of the electron (2.82 fm).

The Compton effect occurs when photons are scattered from free electrons. The energy E'_γ of a photon scattered at an angle θ_γ from the incident photon direction is

$$\frac{1}{E'_\gamma} = \frac{1}{E_\gamma} + \frac{1}{m_{\mathrm{e}} c^2}(1 - \cos\theta_\gamma) \tag{18.70}$$

or, in terms of the photon wavelength $\lambda = hc/E_\gamma$,

$$\lambda' - \lambda = \lambda_{\mathrm{C}}(1 - \cos\theta_\gamma), \tag{18.71}$$

where λ_{C} is the electron Compton wavelength (2.426 pm). The kinetic energy of the scattered electron is

$$K_{\mathrm{e}} = E_\gamma \frac{(E_\gamma/m_{\mathrm{e}}c^2)(1 - \cos\theta_\gamma)}{1 + (E_\gamma/m_{\mathrm{e}}c^2)(1 - \cos\theta_\gamma)}, \tag{18.72}$$

and the maximum energy that can be imparted to the scattered electron is

$$(K_{\rm e})_{\rm max} = \frac{E_\gamma}{1 + m_{\rm e}c^2/2E_\gamma}. \qquad (18.73)$$

The electron scattering angle is

$$\cot\theta_{\rm e} = (1 + E_\gamma/m_{\rm e}c^2)\tan(\theta_\gamma/2). \qquad (18.74)$$

The Compton-scattering differential cross section per electron for polarized incident radiation is

$$\left(\frac{d\sigma_{\rm C}}{d\Omega}\right)_{\rm pol} = \frac{r_0^2}{4}\left(\frac{E'_\gamma}{E_\gamma}\right)^2 \left(\frac{E_\gamma}{E'_\gamma} + \frac{E'_\gamma}{E_\gamma} + 4\cos^2\Theta - 2\right), \qquad (18.75)$$

where Θ is the angle between the directions of polarization of the incident and scattered radiation. For unpolarized incident radiation, the differential cross section is

$$\left(\frac{d\sigma}{d\Omega}\right)_{\rm unpol} = \frac{r_0^2}{2}\left(\frac{E'_\gamma}{E_\gamma}\right)^2 \left(\frac{E_\gamma}{E'_\gamma} + \frac{E'_\gamma}{E_\gamma} - \sin^2\theta_\gamma\right). \qquad (18.76)$$

The absorption cross section (per electron) for Compton scattering is

$$\sigma_{\rm C} = 2\pi r_0^2 \left[\frac{2(1+x)^2}{x^2(1+2x)} - \frac{1+3x}{(1+2x)^2} - \frac{(1+x)(2x^2-2x-1)}{x^2(1+2x)^2} - \frac{4x^2}{3(1+2x)^3} - \left(\frac{1+x}{x^3} - \frac{1}{2x} + \frac{1}{2x^3}\right)\ln(1+2x)\right], \qquad (18.77)$$

where $x = E_\gamma/m_{\rm e}c^2$.

The absorption cross section for pair production is (when $E_\gamma \gg m_{\rm e}c^2$)

$$\sigma_{\rm PP} = \alpha r_0^2 Z^2 \left(\frac{28}{9}\ln\frac{2E_\gamma}{m_{\rm e}c^2} - \frac{218}{27}\right), \qquad (18.78)$$

assuming that the two 511-keV photons from the positron annihilation are also totally absorbed.

18.4.4. Neutrons

The intensity of a monoenergetic beam of neutrons of initial intensity I_0 is diminished as it passes through a length x in matter according to

$$I(x) = I_0 e^{-\sigma_{\rm t} n x} = I_0 e^{-x/\lambda}, \qquad (18.79)$$

where n is the number of atoms per unit volume of the material and $\sigma_{\rm t}$ is the total cross section, which includes all possible processes: elastic scattering, inelastic scattering, capture, and so on. The quantity $\lambda = 1/\sigma_{\rm t} n$ is the mean free path for neutron interactions.

For elastic scattering from a nucleus of mass number A, the ratio between the final neutron energy E' and its initial energy E is

$$\frac{E'}{E} = \frac{A^2 + 1 + 2A\cos\theta}{(A+1)^2}, \tag{18.80}$$

where θ is the scattering angle. The average value of the logarithm of this ratio over all scattering angles gives the *lethargy* ξ:

$$\xi \equiv \left(\ln\frac{E'}{E}\right)_{\text{av}} = 1 + \frac{(A-1)^2}{2}\ln\frac{A-1}{A+1}. \tag{18.81}$$

The average value of $\ln E'$ is decreased by ξ after each collision.

18.5. NUCLEAR REACTIONS [18]

18.5.1. Nonrelativistic kinematics

Let particle 1 (mass M_1, laboratory kinetic energy K_1) be incident on particle 2 at rest, and let the reaction produce light particle 3 at laboratory angle θ_3 and heavy particle 4 at laboratory angle θ_4. The Q-value for the reaction is then

$$Q = (M_1 + M_2 - M_3 - M_4)c^2 = K_3 + K_4 - K_1. \tag{18.82}$$

The relationship between the energy of particle 3 and its direction is

$$K_3^{1/2} = \frac{(M_1 M_3 K_1)^{1/2}\cos\theta_3}{M_3 + M_4}\left[1 \pm \left(1 + \frac{1 + M_4/M_3}{\cos^2\theta_3}\left[\frac{M_4}{M_1}\left(1 + \frac{Q}{K_1}\right) - 1\right]\right)^{1/2}\right]. \tag{18.83}$$

Conversely, the Q-value can be found from the energy and direction of particle 3:

$$Q = K_3\left(1 + \frac{M_3}{M_4}\right) - K_1\left(1 - \frac{M_1}{M_4}\right) - \frac{2(M_1 M_3 K_1 K_3)^{1/2}}{M_4}\cos\theta_3. \tag{18.84}$$

If $Q < 0$, there is a threshold kinetic energy for particle 1, below which the reaction will not occur:

$$K_{1,\text{th}} = (-Q)\frac{M_3 + M_4}{M_3 + M_4 - M_1}. \tag{18.85}$$

The apparent double-valued behavior of Eq. (18.83) occurs only for $Q < 0$ and only for a range of incident energies between $K_{1,\text{th}}$ and $(-Q)M_4/(M_4 - M_1)$.

In the center-of-mass (CM) frame, which is indicated by primed variables, $\theta_3' = \pi - \theta_4'$. The energy of particle 3 in the CM frame is

$$K_3' = \frac{M_3}{M_4 + M_3}\left[Q + \left(1 - \frac{M_1}{M_3 + M_4}\right)K_1\right], \tag{18.86}$$

and the angle transformation is

$$\tan\theta_3 = \frac{\sin\theta_3'}{\cos\theta_3' + \gamma}, \tag{18.87}$$

where

$$\gamma = \left[\frac{M_1 M_3}{M_2 M_4} \frac{1}{1 + (Q/K_1)(1 + M_1/M_2)} \right]^{1/2}. \tag{18.88}$$

18.5.2. Cross sections

The differential (angular) cross section $d\sigma/d\Omega$, which can also be written as $\sigma(\theta, \phi)$ or just as $\sigma(\theta)$ for cases of axial symmetry, is given by the rate r_3 at which the emerging particles 3 appear in a given direction, normalized by the intensity I_1 of the beam of particle 1 and the number density N_2 of target nuclei per unit area:

$$\frac{d\sigma}{d\Omega} \equiv \sigma(\theta_3) = \frac{r_3(\theta_3)}{4\pi I_1 N_2}. \tag{18.89}$$

The relationship between the laboratory (unprimed) and CM (primed) cross sections is

$$\sigma(\theta_3') = \sigma(\theta_3) \frac{1 + \gamma \cos\theta_3'}{(1 + \gamma^2 + 2\gamma \cos\theta_3')^{3/2}}. \tag{18.90}$$

Coulomb scattering

The Coulomb (Rutherford) cross section is

$$\left(\frac{d\sigma}{d\Omega}\right)_{\text{Ruth}} = \left(\frac{zZe^2}{16\pi\epsilon_0 K_1'}\right)^2 \frac{1}{\sin^4(\theta_1'/2)} \tag{18.91}$$

for a nonrelativistic projectile of charge ze incident on a nucleus of charge Ze. Although this equation is written in the CM frame, it can be applied in the laboratory frame when $M_2 \gg M_1$ (that is, when the target is much more massive than is the projectile).

In the special case in which the two particles are identical ($z = Z$), a quantum-mechanical calculation gives the Mott scattering cross section:

$$\left(\frac{d\sigma}{d\Omega}\right)_{\text{Mott}} = \left(\frac{z^2 e^2}{8\pi\epsilon_0 K_1}\right)^2 \left(\frac{1}{\sin^4(\theta_1'/2)} + \frac{1}{\cos^4(\theta_1'/2)} + \frac{k \cos[(z^2\alpha/\beta) \ln \tan^2(\theta_1'/2)]}{[\sin^2(\theta_1'/2)][\cos^2(\theta_1'/2)]}\right), \tag{18.92}$$

where $k = -1$ for spin-$\frac{1}{2}$ particles or $+2$ for spin-0 particles, α is the fine-structure constant, and $\beta = v/c$.

Nuclear scattering

In nuclear scattering of a neutral spinless particle, the beam of incident particles can be represented as a plane wave $\exp(ikz)$ and the scattered wave as $f(\theta)\exp(ikr)/r$, where

$f(\theta)$ is the scattering amplitude. The differential cross section is then

$$\frac{d\sigma}{d\Omega} = |\, f(\theta)\,|^2. \tag{18.93}$$

For processes other than elastic scattering, the resulting reduction in the amplitude of the outgoing wave is accounted for by introducing the complex amplitude $\eta_l = \exp(2i\delta_l)$, where δ_l is the phase shift of the lth partial wave (representing orbital angular momentum quantum number l). The cross section can then be written as

$$\frac{d\sigma}{d\Omega} = \frac{1}{4k^2}\left|\sum_{l=0}^{\infty}(2l+1)i(1-\eta_l)P_l(\cos\theta)\right|^2, \tag{18.94}$$

where $k = 2\pi/\lambda$ is the wavenumber and P_l are Legendre polynomials. For elastic scattering, the total cross section (the differential cross section integrated over all angles) is

$$\sigma_{\mathrm{el}} = \int \frac{d\sigma}{d\Omega}\, d\Omega = \frac{4\pi}{k^2}\sum_{l=0}^{\infty}(2l+1)\sin^2\delta_l, \tag{18.95}$$

where the δ_l are strictly real. If other processes remove particles from the beam, $|\,\eta_l\,| < 1$ and the δ_l are complex. The reaction (or absorption) cross section is then

$$\sigma_{\mathrm{r}} = \frac{\pi}{k^2}\sum_{l=0}^{\infty}(2l+1)(1-|\,\eta_l\,|^2). \tag{18.96}$$

The total cross section for all processes is

$$\sigma_{\mathrm{t}} = \sigma_{\mathrm{el}} + \sigma_{\mathrm{r}} = \frac{2\pi}{k^2}\sum_{l=0}^{\infty}(2l+1)(1-\mathrm{Re}\,\eta_l). \tag{18.97}$$

Resonances occur where the total cross section is a maximum, that is, where $\eta_l \approx -1$ for a particular l. If the resonance occurs at energy E_{R} and has a width $\Gamma = 2/(\partial\delta_l/\partial E)$ evaluated at $E = E_{\mathrm{R}}$, then for elastic scattering the cross section near resonance has the Breit-Wigner shape:

$$\sigma_{\mathrm{el}} = \frac{\pi}{k^2}(2l+1)\frac{\Gamma^2}{(E-E_{\mathrm{R}})^2+\Gamma^2/4}. \tag{18.98}$$

Reaction or absorption cross sections near a resonance have a similar shape, with the factor Γ^2 in the numerator replaced by the product of the partial widths for the formation and decay of the resonant state.

18.6. COMPILATIONS OF NUCLEAR DATA

For a previous guide to nuclear compilations, see Ref. [19]. Summaries of experimental data on nuclear states can be found in the *Table of Isotopes* [20] or in the periodically updated compilations of the *Nuclear Data Sheets* [21]. The National Nuclear Data Center maintains a collection of nuclear structure references and evaluated data sets [22]. Analyses of a particular type of data set across a broad range of nuclei (often called a "horizontal compilation") are available in the *Atomic Data and Nuclear Data Tables* [23]. Review articles on various topics are published in the *Annual Reviews of Nuclear and Particle Science* [24]. Other collections of articles on theoretical and experimental aspects of nuclear physics can be found in *Nuclear Spectroscopy and Reactions* [25] and in *Advances in Nuclear Physics* [26].

Specific compilations of nuclear data can be found in Ref. [19] or in the following: atomic masses, [27]; nuclear moments, [28]; decay and other properties, [29]; internal conversion coefficients, [30] and references cited in [19]; photon absorption coefficients, [31]; deformation parameters, [32]; nuclear reaction analysis graphs and tables, [33]; neutron cross sections, [34].

18.7. REFERENCES

[1] R. C. Barrett and D. F. Jackson, *Nuclear Sizes and Structure* (Clarendon Press, Oxford, 1971).

[2] K. S. Krane, *Introductory Nuclear Physics* (Wiley, New York, 1987).

[3] H. Kopferman, *Nuclear Moments* (Academic Press, New York, 1958).

[4] J. O. Rasmussen, in *Alpha-, Beta- and Gamma-Ray Spectroscopy*, edited by K. Siegbahn (North-Holland, Amsterdam, 1965).

[5] P. J. Woods and C. A. Davids, Annu. Rev. Nucl. Part. Sci. **47**, 541 (1997).

[6] P. B. Price, Annu. Rev. Nucl. Part. Sci. **39**, 19 (1989).

[7] C. S. Wu and S. A. Moszkowski, *Beta Decay* (Wiley-Interscience, New York, 1966).

[8] M. Moe and P. Vogel, Annu. Rev. Nucl. Part. Sci. **44**, 247 (1994).

[9] J. Cerny and J. C. Hardy, Annu. Rev. Nucl. Sci. **27**, 333 (1977).

[10] K. Alder and R. M. Steffen, in *The Electromagnetic Interaction in Nuclear Spectroscopy*, edited by W. D. Hamilton (North-Holland, Amsterdam, 1975).

[11] H. C. Pauli, K. Alder, and R. M. Steffen, in *The Electromagnetic Interaction in Nuclear Spectroscopy*, edited by W. D. Hamilton (North-Holland, Amsterdam, 1975).

[12] J. M. Eisenberg and W. Greiner, *Nuclear Models* (North-Holland, Amsterdam, 1970).

[13] M. G. Mayer and J. H. D. Jensen, *Elementary Theory of Nuclear Shell Structure* (Wiley, New York, 1955).

[14] S. G. Nilsson, K. Dan. Vidensk. Selsk. Mat.-Fys. Medd. **29**, No. 16 (1955).

[15] A. Bohr and B. R. Mottelson, *Nuclear Structure* (Benjamin, Reading, MA, 1975).

[16] G. F. Knoll, *Radiation Detection and Measurement* (Wiley, New York, 1979).

[17] R. D. Evans, *The Atomic Nucleus* (McGraw-Hill, New York, 1955).

[18] G. R. Satchler, *Introduction to Nuclear Reactions* (Wiley, New York, 1980).

[19] F. Ajzenberg-Selove, in *Nuclear Spectroscopy and Reactions*, edited by J. Cerny (Academic Press, New York, 1974), part C, p. 551.

[20] R. B. Firestone, *Table of Isotopes*, 8th ed. (Wiley, New York, 1996).

[21] *Nuclear Data Sheets* (Academic Press, New York).

[22] http://www.nndc.bnl.gov.

[23] *Atomic Data and Nuclear Data Tables* (Academic Press, Orlando, FL).

[24] *Annual Reviews of Nuclear and Particle Science* (Annual Reviews Inc., Palo Alto, CA).

[25] *Nuclear Spectroscopy and Reactions*, edited by J. Cerny (Academic Press, New York, 1974).

[26] J. Negele and E. Vogt, editors, *Advances in Nuclear Physics* (Plenum Press, New York).

[27] G. Audi and A. H. Wapstra, Nucl. Phys. **A595**, 409 (1995).

[28] P. Raghavan, Atom. Data Nucl. Data Tables **42**, 189 (1989).

[29] G. Audi, O. Bersillon, J. Blachot, and A. H. Wapstra, Nucl. Phys. **A624**, 1 (1997).

[30] F. Rösel, H. M. Fries, K. Alder, and H. C. Pauli, Atom. Data Nucl. Data Tables **21**, 89 (1978).

[31] E. Storm and H. I. Israel, Nucl. Data Tables **A7**, 565 (1970).

[32] K. E. G. Löbner, M. Vetter, and V. Hönig, Nucl. Data Tables **A7**, 495 (1970).

[33] J. B. Marion and F. C. Young, *Nuclear Reaction Analysis* (North-Holland, Amsterdam, 1968).

[34] S. F. Mughabghab, M. Divadeenam, and N. E. Holder, editors, *Neutron Cross Sections* (Academic Press, New York, 1981).

19

Optics

Joseph Reader
National Institute of Standards and Technology, Gaithersburg, Maryland

Contents

List of Tables

List of Figures

19.1. REFLECTION AND REFRACTION

19.1.1. Reflection

When light is reflected from a plane surface (Figure 19.1) the angle of reflection θ_r is equal to the angle of incidence θ_i

$$\theta_r = \theta_i. \tag{19.1}$$

The incident ray, the normal, and the reflected ray lie in the *plane of incidence.*

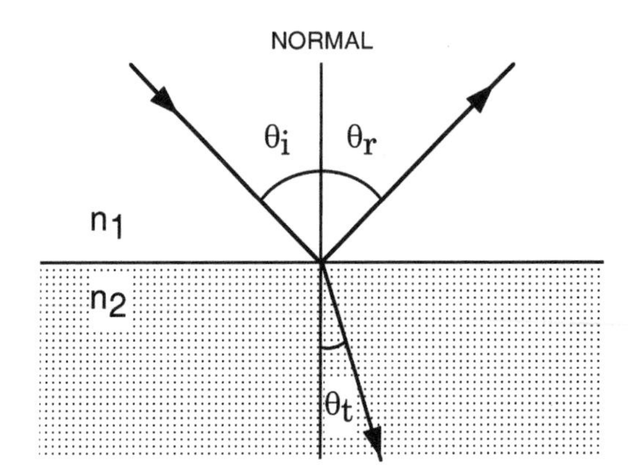

FIGURE 19.1. Reflection and refraction at a plane surface.

19.1.2. Index of refraction

The index of refraction n of a material is the ratio of the speed of light in vacuum c to the speed of light in that material v

$$n = c/v. \tag{19.2}$$

The value of the speed of light in vacuum is *defined as* [5]

$$c = 2.997\,924\,58 \times 10^8 \text{ m/s}. \tag{19.3}$$

Table 19.1 lists indices of refraction for a number of common materials.*

Complete treatments of most of the subjects discussed here can be found in the book by Jenkins and White [1] or the book by Born and Wolf. [2] Much of the above material was adapted from these books. Some material was also taken from the book by Thorne et al. [3] and the review article by Jacquinot. [4] The author would like to thank Craig Sansonetti for helpful discussions and assistance with the figures.

*Reference to commercial products in this table does not imply recommendation or endorsement by the National Institute of Standards and Technology.

TABLE 19.1. Index of refraction of common materials at various wavelengths (nm). n_H, $n_{G'}$, n_F, n_D, and n_C refer to the indices of refraction at the Fraunhofer *H*, *G'*, *F*, *D*, and *C* lines, respectively. Some values determined by polynomial fit to literature data.

Material	193	254	313	365	397 n_H	434 $n_{G'}$	486 n_F	589 n_D	633	656 n_C	706
Fused quartz[a]	1.560	1.505	1.485	1.475	1.471	1.467	1.463	1.458	1.457	1.456	1.455
Borosilicate crown-BK7[b]				1.536	1.531	1.527	1.522	1.517	1.515	1.514	1.513
Corning 7740-PYREX[c]					1.483	1.481	1.478	1.473	1.471	1.470	1.468
H_2O-Water (20°C)[d]					1.343	1.340	1.337	1.333	1.332	1.331	1.330
Polystyrene (20°C)[d]						1.616	1.605	1.591	1.587	1.586	1.584
Air[e] – $(n - 1) \times 10^4$	3.299	3.004	2.898	2.849	2.829	2.811	2.793	2.771	2.765	2.762	2.757

[a] W. S. Rodney and R. J. Spindler, *Index of Refraction of Fused-Quartz Glass for Ultraviolet, Visible, and Infrared Wavelengths*, J. Res. Nat. Bur. Std. (U.S.) **53**, 185 (1954).

[b] *Optical Glass* (Schott Optical Glass, Inc., Duryea, PA, 1982).

[c] *Material Properties of PYREX Glass Code 7740* (Corning Glass, New York, 1990).

[d] S. S. Ballard, K. A. McCarthy, and W. Brouwer, *Index of Refraction* in American of Institute of Physics Handbook, edited by D. E. Gray (McGraw-Hill, New York, 1957).

[e] Calculated from Eq. (19.5).

The *wavenumber* σ of a spectrum line is the number of waves per unit length *in vacuum*

$$\sigma = 1/\lambda_{\text{vac}}. \tag{19.4}$$

Wavenumbers are usually given in units of cm^{-1}; they are *always in vacuum.*

The index of refraction of air as a function of wavenumber $\sigma\,(\text{cm}^{-1})$ for wavelengths between 185 nm and 1700 nm at standard conditions, that is, dry air at 101325 Pa (760 torr), 15°C, 0.03% content of CO_2 by volume [(31 Pa (0.23 torr)], is [6]

$$(n-1)\times 10^8 = 8060.51 + \frac{2480900}{132.274 - 10^{-8}\sigma^2} + \frac{17455.7}{39.32957 - 10^{-8}\sigma^2}. \tag{19.5}$$

The index of refraction of standard air at 632.8165 nm is 1.00027652.

For temperatures t between 15°C and 30°C and pressures p between 93324 Pa (700 torr) and 106656 Pa (800 torr), the index of refraction is [7]

$$(n-1)_{tp} = (n-1)_s \left[\frac{0.00138823p}{1+0.003671t}\right], \tag{19.6}$$

where $(n-1)_s$ corresponds to the value for standard conditions, p is in torr, and t is in °C. More detailed formulas applying from 350 nm to 650 nm are given in Ref. [8].

The wavelength of a spectrum line in two different media is

$$n_1\lambda_1 = n_2\lambda_2. \tag{19.7}$$

The wavelength of a spectral line in air is

$$\lambda_{\text{air}} = \frac{\lambda_{\text{vac}}}{n}, \tag{19.8}$$

where n is the index of refraction of air at $\lambda_{\text{vac}} = 1/\sigma$. (Note that the quantity required to calculate n in Eq. (19.5) is the wavenumber σ. When starting from λ_{air}, an iterative procedure is required to obtain n.)

19.1.3. Refraction at a plane surface

When light is incident at a plane surface (Figure 19.1), the transmitted beam has an angle of refraction θ_t related to the angle of incidence θ_i according to

$$n_2 \sin\theta_t = n_1 \sin\theta_i. \tag{19.9}$$

This is Snell's law. The incident ray, the normal to the surface, and the transmitted ray lie in the *plane of incidence*.

19.1.4. Coefficients of reflectance and transmittance

If light is incident on a plane surface and I^i, I^r, and I^t are the intensities of the incident beam, the reflected beam, and the transmitted (or refracted) beam, respectively, the reflectance R_p and transmittance T_p for the component with electric vector parallel to the plane of incidence are

$$R_p = \frac{I_p^r}{I_p^i} = \frac{\tan^2(\theta_i - \theta_t)}{\tan^2(\theta_i + \theta_t)} \tag{19.10}$$

$$T_p = \frac{I_p^t}{I_p^i} = \frac{\sin 2\theta_i \sin 2\theta_t}{\sin^2(\theta_i + \theta_t)\cos^2(\theta_i - \theta_t)}. \tag{19.11}$$

The reflectance R_s and transmittance T_s for the component with electric vector perpendicular to the plane of incidence are

$$R_s = \frac{I_s^r}{I_s^i} = \frac{\sin^2(\theta_i - \theta_t)}{\sin^2(\theta_i + \theta_t)} \tag{19.12}$$

$$T_s = \frac{I_s^t}{I_s^i} = \frac{\sin 2\theta_i \sin 2\theta_t}{\sin^2(\theta_i + \theta_t)}. \tag{19.13}$$

The reflectances R_s, $(R_p + R_s)/2$, and R_p for fused quartz at 633 nm with index of refraction 1.457 are plotted in Figure 19.2.

19.1.5. Reflectance and transmittance at normal incidence

At normal incidence, there is no distinction between parallel and perpendicular polarization components. If light enters a medium with index of refraction n_2 from a medium with index of refraction n_1 at normal incidence and $n = n_2/n_1$

$$R = \left(\frac{n-1}{n+1}\right)^2 \tag{19.14}$$

$$T = \frac{4n}{(n+1)^2}. \tag{19.15}$$

For example, if light is incident from air on a flat piece of glass with index of refraction of 1.5, the reflectance is 0.04 and the transmittance is 0.96. When light exits from the glass into air, the reflectance will again be 0.04 and the transmittance will be 0.96. The total loss due to reflection in traversing the plate is 7.8%.

The quantity n_2/n_1 is known as the *relative index of refraction*. For objects in air, it is commonly taken to be the same as n, but for precision measurements, it must be kept in mind that the true index of refraction is *relative to vacuum*.

19.1.6. Brewster's angle of reflection

When a beam of light is incident on a plane surface at an angle such that the reflected and refracted beams are at 90° to each other ($\theta_r + \theta_t = \pi/2 = \theta_i + \theta_t$), the component of the beam with electric vector parallel to the plane of incidence has $R_p = 0$ and $T_p = 1$. That is, there is no loss of light due to reflection for this component. The angle at which this

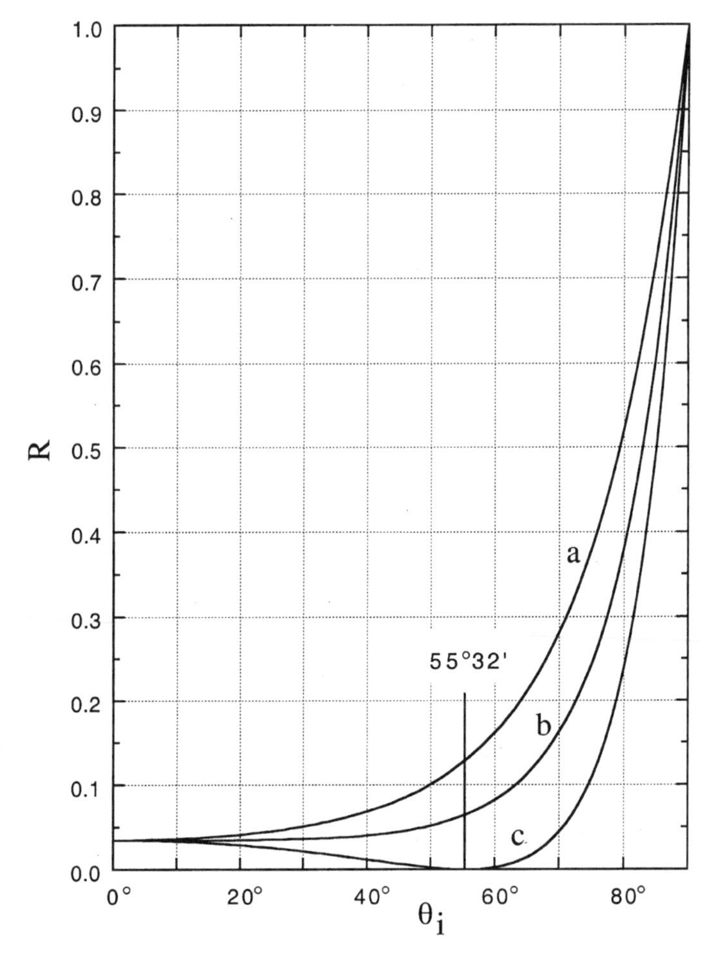

FIGURE 19.2. Intensity of reflected light as a function of angle of incidence for light polarized parallel to the plane of incidence and light polarized perpendicular to the plane of incidence for fused quartz at 633 nm, $n = 1.457$. a: R_s; b: $(R_s + R_p)/2$; c: R_p.

occurs is known as *Brewster's angle* and is given by

$$\theta_i = \arctan(n). \tag{19.16}$$

For fused silica at 633 nm, $n = 1.457$, and Brewster's angle is $55^\circ\ 32'$, as shown in Figure 19.2. Brewster angles are used in windows for laser tubes to minimize losses in the cavity due to reflection.

19.1.7. Total internal reflection

When light is incident on a plane surface going from a high-index material (n_2) to a low-index material (n_1), the light will undergo *total internal reflection* if the angle of incidence is equal to or greater than the *critical angle* θ_i^c. At this angle, all of the light is reflected back into the denser medium; no light is transmitted. The critical angle is

$$\theta_i^c = \arcsin(n_1/n_2) = \arcsin(1/n). \tag{19.17}$$

For glass, the critical angle varies from about 39° to 43°. Total internal reflection plays a role in many applications, such as in prisms for binoculars and in transmission of optical signals over long distances by optical fibers.

19.1.8. Beam displacement by a plane parallel plate

When a plane parallel beam of light traverses a plane parallel plate of thickness d, it emerges parallel to its original direction, but with a lateral displacement

$$\Delta = d \sin\theta_i \left(1 - \frac{n_i \cos\theta_i}{n_2 \cos\theta_t}\right), \tag{19.18}$$

where n_2 is the index of refraction of the plate and n_1 is the index of the surrounding medium.

If the original beam is divergent and strikes the plate with its central ray normal to the plate, for paraxial rays the transmitted beam will appear to originate at a point displaced toward the plate by a distance

$$\delta = d\left(1 - \frac{n_1}{n_2}\right). \tag{19.19}$$

This relation might be used when a filter is placed between a light source and a lens or between a lens and the slit of a spectrometer.

19.1.9. Prisms

When a plane parallel beam of light traverses a prism, it is deviated by an angle δ that depends on the index of refraction n, the prism angle α, and the angle of incidence of the light at the first surface. The amount of deviation can be determined by successive application of (19.9) to the two surfaces. The deviation is minimized if the beam passes through the prism symmetrically; that is, the angle of refraction at the first surface equals the angle of incidence at the second surface (Figure 19.3). For this case, the angle of *minimum deviation* $\delta_{\min}$ is

$$\sin \tfrac{1}{2}(\delta_{\min} + \alpha) = n \sin \tfrac{1}{2}\alpha. \tag{19.20}$$

This equation can be used to determine the index of refraction of a material that has been made into a prism. By measuring the prism angle α and the angle of minimum deviation for various wavelengths, the variation of index with wavelength $n(\lambda)$ can be determined for the material. The value of $dn/d\lambda$ at a particular wavelength is the *dispersion* of the material at that wavelength. The *angular dispersion* of the prism at minimum deviation is

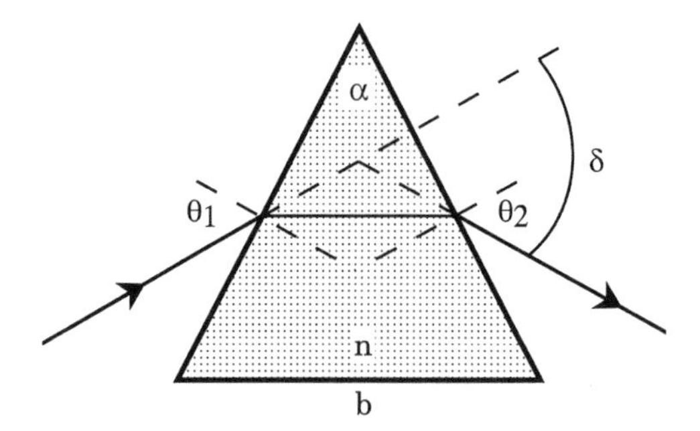

FIGURE 19.3. Passage of light through a prism at minimum deviation.

given by

$$\frac{d\theta_2}{d\lambda} = \frac{2\sin(\alpha/2)}{\cos\theta_1}\frac{dn}{d\lambda}. \tag{19.21}$$

The resolving power of a prism, or its ability to separate close wavelengths, is

$$\Re = \frac{\lambda}{\Delta\lambda} = b\frac{dn}{d\lambda}, \tag{19.22}$$

where $\Delta\lambda$ is the difference of two wavelengths just separated by the prism. That is, the resolving power at a particular wavelength depends only on the length of the base b and the dispersion of the material $dn/d\lambda$ at that wavelength, assuming that the light fills the entire prism. For a glass prism with a base length of 50 mm and a typical dispersion of 1.5×10^{-4}/nm, the resolving power is 7,500; at 500 nm it can resolve two lines separated by 0.15 nm.

19.1.10. Refraction at a spherical surface

If an object is placed in front of a spherical surface with radius of curvature r, as shown in Figure 19.4, an image will be formed in the medium according to

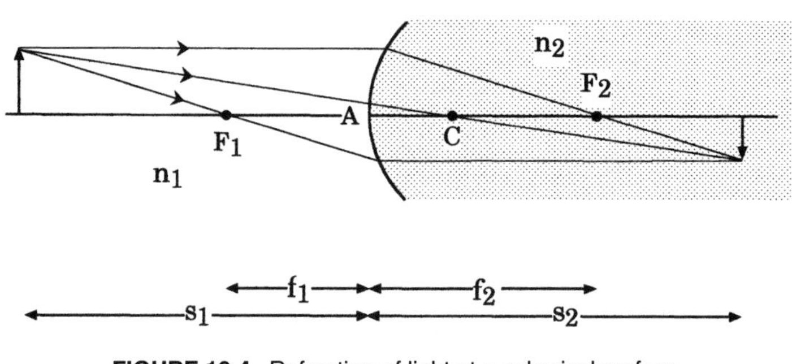

FIGURE 19.4. Refraction of light at a spherical surface.

$$\frac{n_1}{s_1} + \frac{n_2}{s_2} = \frac{n_2 - n_1}{r}. \qquad (19.23)$$

The distances are measured from the intersection of the surface with the axis A. The distance in front of the surface for which the refracted rays are formed parallel to the axis is the *primary focal length* f_1

$$\frac{n_1}{f_1} = \frac{n_2 - n_1}{r}. \qquad (19.24)$$

The distance in the medium at which rays entering the medium parallel to the axis are focused is the *secondary focal length* f_2

$$\frac{n_2}{f_2} = \frac{n_2 - n_1}{r}. \qquad (19.25)$$

Then,

$$\frac{n_1}{s_1} + \frac{n_2}{s_2} = \frac{n_1}{f_1} = \frac{n_2}{f_2}. \qquad (19.26)$$

The magnification is

$$m = -\frac{n_1 s_2}{n_2 s_1}. \qquad (19.27)$$

If the magnification is negative, the image is inverted.

19.2. ABSORPTION

19.2.1. Internal transmittance and total transmittance

If a material has an absorption coefficient α, its *internal transmittance* T_i is

$$T_i = \mathrm{e}^{-\alpha d}, \qquad (19.28)$$

where d is the distance that light travels in the material. For light incident on a flat plate, the *transparency or total transmittance* T (neglecting effects of multiple reflections) is

$$T = \frac{I_t}{I_i} = (1 - R)^2 \mathrm{e}^{-\alpha d}, \qquad (19.29)$$

where R is the coefficient of reflectance at each surface.

19.2.2. Optical density

If light with intensity I_i is incident on a transparent plate and I_t is the transmitted intensity, the plate is said to have an *optical density* D of

$$D = \log\left(\frac{1}{T}\right) = \log\left(\frac{I_i}{I_t}\right) \qquad (19.30)$$

or

$$T = \frac{I_t}{I_i} = 10^{-D}. \tag{19.31}$$

19.3. LENSES

19.3.1. Imaging by lenses

For *thin lenses* (Figures 19.5 and 19.6), the object distance s_1 and image distance s_2 are related according to

$$\frac{1}{s_1} + \frac{1}{s_2} = \frac{1}{f}, \tag{19.32}$$

where f is the focal length of the lens. The focal length is the distance from the lens at which parallel rays (object at infinity) are focused. This is the *Gaussian lens formula*. For diverging lenses (Figure 19.6), the focal length is considered to be negative.

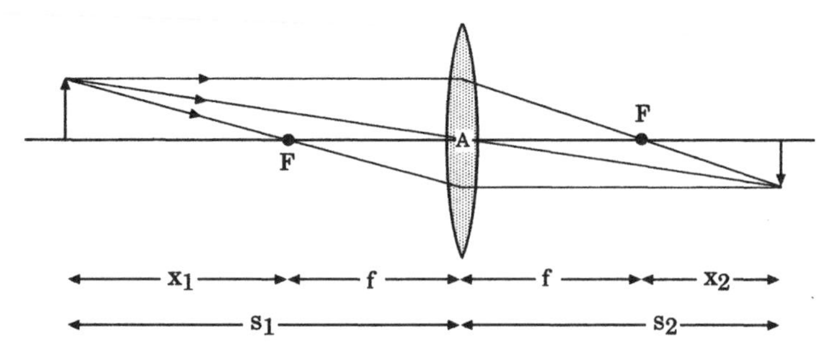

FIGURE 19.5. Formation of an image by a positive lens.

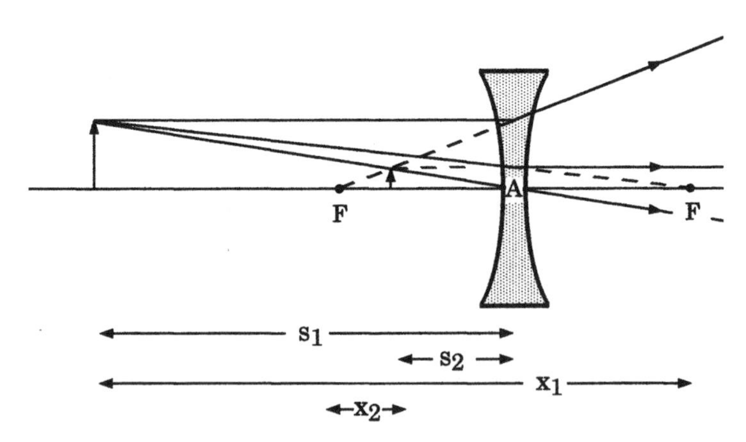

FIGURE 19.6. Formation of an image by a negative lens.

TABLE 19.2. Location of images for lenses (see Figure 19.7).

Object	Image	Real (R) Virtual (V)	Erect (E) Inverted (I)
Positive lenses			
L^∞	e	R	I
$L^\infty \leftrightarrow$ a	e $\leftrightarrow$ f	R	I
a	f	R	I
a $\leftrightarrow$ b	f $\leftrightarrow R^\infty$	R	I
b	R^∞	R	I
	L^∞	V	E
b $\leftrightarrow$ c	$L^\infty \leftrightarrow$ b	V	E
c	b	V	E
c $\leftrightarrow$ A	b $\leftrightarrow$ A	V	E
A $\leftrightarrow R^\infty$	A $\leftrightarrow$ e	R	E
Negative lenses			
L^∞	b	V	E
$L^\infty \leftrightarrow$ A	b $\leftrightarrow$ A	V	E
b	c	V	E
A $\leftrightarrow$ e	A $\leftrightarrow$ e	R	E
d	e	R	E
e	L^∞	V	I
	R^∞	R	E
e $\leftrightarrow R^\infty$	$L^\infty \leftrightarrow$ b	V	I

If x_1 and x_2 are the distances of the object and image from the focal point on each side of the lens,

$$x_1 x_2 = f^2. \tag{19.33}$$

This is the *Newtonian lens formula*. The lateral magnification m of the image is

$$m = -\frac{s_2}{s_1} = -\frac{f}{x_1} = -\frac{x_2}{f}. \tag{19.34}$$

The relationship between objects and images for lenses is given in Table 19.2 and Figure 19.7. When the distance between object and lens is less than the focal length, an

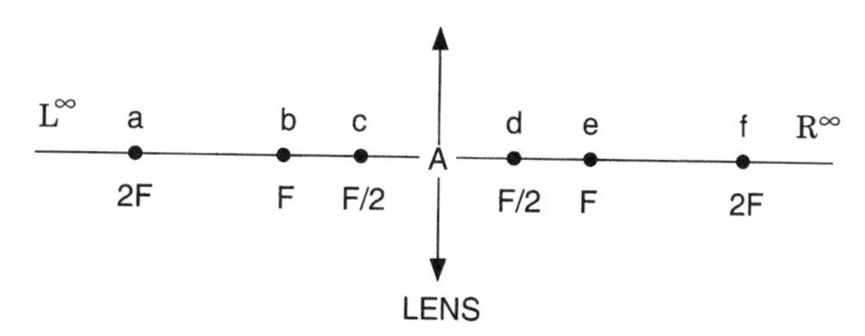

FIGURE 19.7. Object and image points for lenses (see Table 19.2).

enlarged virtual image is produced; this corresponds to the use of a positive lens as a *magnifying glass*.

19.3.2. Minimum focal distance

The minimum distance between an object and an image to produce a real image is

$$(s_1 + s_2)_{\min} = 4f. \tag{19.35}$$

This occurs for $s_1 = s_2 = 2f$.

19.3.3. Lens power and *F*-number

The power P of a lens in *diopters* is

$$P = \frac{1}{f}, \tag{19.36}$$

where the focal length f is in *meters*. Converging (positive) lenses have positive power; diverging (negative) lenses have negative power.

The light gathering power of a lens is its *F-number* or *focal ratio*. The F-number is the ratio of the focal length of the lens f to the lens diameter D

$$F\text{-number} = \frac{f}{D}. \tag{19.37}$$

If a lens has a diameter of 20 mm and focal length of 100 mm, it has an F-number of 5 and is called an $F/5$ lens. The F-number of a camera lens usually refers to its light gathering power at *maximum* aperture. Its other *F-stops* refer to F-numbers at smaller apertures that may be set by changing the opening of the diaphragm.

19.3.4. Lens maker's formula

If a thin lens has one surface (on the left) with radius of curvature r_1 and a second surface (on the right) with radius of curvature r_2, its focal length is given by

$$\frac{1}{f} = (n-1)\left(\frac{1}{r_1} - \frac{1}{r_2}\right). \tag{19.38}$$

Considering rays to travel from left to right, convex surfaces are taken to have positive radii; concave surfaces are taken to have negative radii. For a planoconvex or planoconcave lens, the radius of the flat surface is taken as infinite.

19.3.5. Thin lenses in combination

The effective focal length f_{eff} of a system of two thin lenses separated by a distance d is

$$\frac{1}{f_{\text{eff}}} = \frac{1}{f_1} + \frac{1}{f_2} - \frac{d}{f_1 f_2}. \tag{19.39}$$

19.3.6. Ray tracing for lenses

1. For both positive and negative lenses, a ray through the apex A is undeviated.
2. For a positive lens, a ray parallel to the axis from the left passes through the focal point on the right; a ray passing through the focal point on the left exits the lens on the right parallel to the axis.
3. For a negative lens, a ray parallel to the axis from the left exits the lens as if it had passed through the focal point on the left; a ray on the left directed to the focal point on the right exits the lens on the right parallel to the axis.

19.4. MIRRORS

19.4.1. Imaging by mirrors

The *focal length* of a spherical mirror of radius of curvature r is the distance from the mirror at which an object at infinity is focused, the *focal point*.

$$f = -\frac{r}{2}. \tag{19.40}$$

Objects and images are related as (Figures 19.8 and 19.9)

$$\frac{1}{s_1} + \frac{1}{s_2} = -\frac{2}{r} = \frac{1}{f}. \tag{19.41}$$

The *magnification* m is

$$m = -\frac{s_2}{s_1}. \tag{19.42}$$

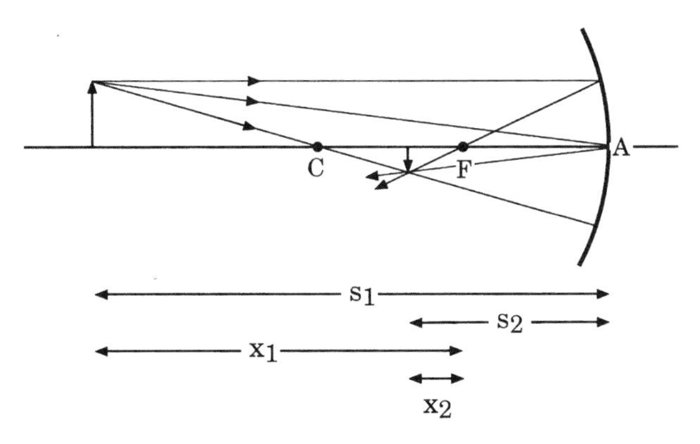

FIGURE 19.8. Formation of an image by a concave mirror.

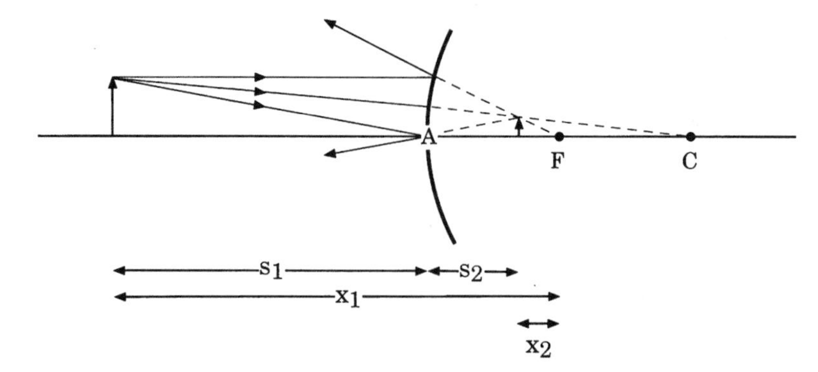

FIGURE 19.9. Formation of an image by a convex mirror.

1. The radius of concave mirrors is negative; the radius of convex mirrors is positive.
2. The focal length of concave mirrors is positive; the focal length of convex mirrors is negative.
3. s_1 is positive when the object lies to the left of the mirror and negative when it lies to the right (virtual object); s_2 is positive when the image lies to the left of the mirror and negative when it lies to the right (virtual image).

TABLE 19.3. Location of images for mirrors (see Figure 19.10).

Object	Image	Real (R) Virtual (V)	Erect (E) Inverted (I)
Concave Mirrors			
L^{∞}	b	R	I
$L^{\infty} \leftrightarrow$ a	b $\leftrightarrow$ a	R	I
a	a	R	I
a $\leftrightarrow$ b	a $\leftrightarrow L^{\infty}$	R	I
b	R^{∞}	V	E
	L^{∞}	R	I
b $\leftrightarrow$ A	A $\leftrightarrow R^{\infty}$	V	E
A $\leftrightarrow R^{\infty}$	A $\leftrightarrow$ b	R	E
Convex Mirrors			
L^{∞}	c	V	E
$L^{\infty} \leftrightarrow$ A	c $\leftrightarrow$ A	V	E
A $\leftrightarrow$ c	A $\leftrightarrow L^{\infty}$	R	E
c	L^{∞}	R	E
	R^{∞}	V	I
c $\leftrightarrow$ d	$R^{\infty} \leftrightarrow$ d	V	I
d	d	V	I
d $\leftrightarrow R^{\infty}$	d $\leftrightarrow$ c	V	I
R^{∞}	c	V	I

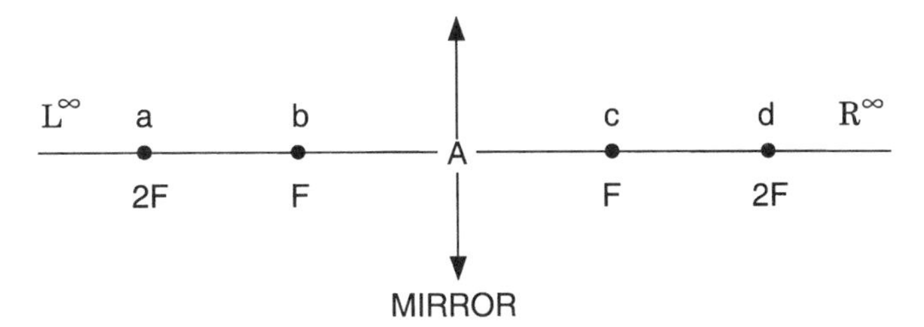

FIGURE 19.10. Object and image points for mirrors.

If object and image have distances x_1 and x_2 from the focal point

$$x_1 x_2 = f^2. \tag{19.43}$$

Distances x_1 and x_2 to the left of the focal point are positive; distances to the right of the focal point are negative. The relationship between objects and images for mirrors is given in Table 19.3 and Figure 19.10.

19.4.2. Ray tracing for mirrors

1. A ray directed at the apex A is reflected symmetrically about the axis.
2. A ray directed at the center of curvature is reflected back on itself.
3. For a concave mirror, a ray parallel to the axis from the left is reflected through the focal point. For a convex mirror, a ray parallel to the axis from the left is reflected as if it were coming from the focal point.
4. For a concave mirror, a ray passing through the focal point is reflected parallel to the axis. For a convex mirror, a ray directed at the focal point is reflected parallel to the axis. (These rays are not shown in Figures 19.8 and 19.9).

19.5. DIFFRACTION

19.5.1. Diffraction by a single slit

When a plane wave is incident on a slit of width a, the diffracted intensity has an angular distribution

$$I = I_0 \frac{\sin^2 \beta}{\beta^2}, \quad \text{where} \quad \beta = \frac{\pi a}{\lambda}(\sin\theta_i + \sin\theta_d), \tag{19.44}$$

where θ_i and θ_d are the angles of the incident and diffracted light, respectively, relative to the normal. I_0 is the intensity at the center of the pattern. A plot of the single slit diffraction pattern is given in Figure 19.11. Minima occur when $\beta = \pm m\pi (m = 1, 2, 3\ldots)$. Secondary maxima occur when $\tan\beta = \beta$. At normal incidence, $\theta_i = 0$, and for small

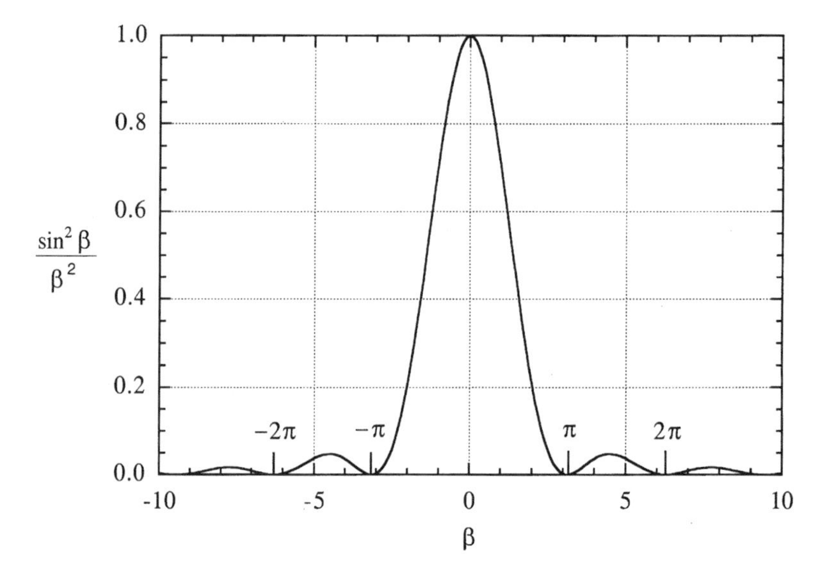

FIGURE 19.11. Diffraction pattern for a single slit.

angles, the first minimum occurs for the angle

$$\theta_d = \lambda/a. \tag{19.45}$$

19.5.2. Diffraction at a circular aperture

When light is incident normally at a circular aperture of diameter a, the angular distribution of the diffracted light is

$$I = I_0 \left[\frac{2J_1(\beta)}{\beta}\right]^2, \tag{19.46}$$

where J_1 is the Bessel function of order 1, and $\beta = (\pi a \sin\theta)/\lambda$, θ being the angular coordinate in the image space. I_0 is the intensity at the center of the pattern. The distribution function is plotted in Figure 19.12. The pattern consists of a bright central disk, *Airy's disk*, surrounded by a number of fainter rings. The first minimum in the distribution function occurs when $\beta = 1.22\pi$.

19.5.3. Resolving power of a telescope

According to the *Rayleigh criterion*, two diffraction patterns are considered to be resolved if the central maximum for one falls on the first minimum of the other. The angular resolution of a telescope is

$$\theta_{\min} = 1.22\frac{\lambda}{a}, \tag{19.47}$$

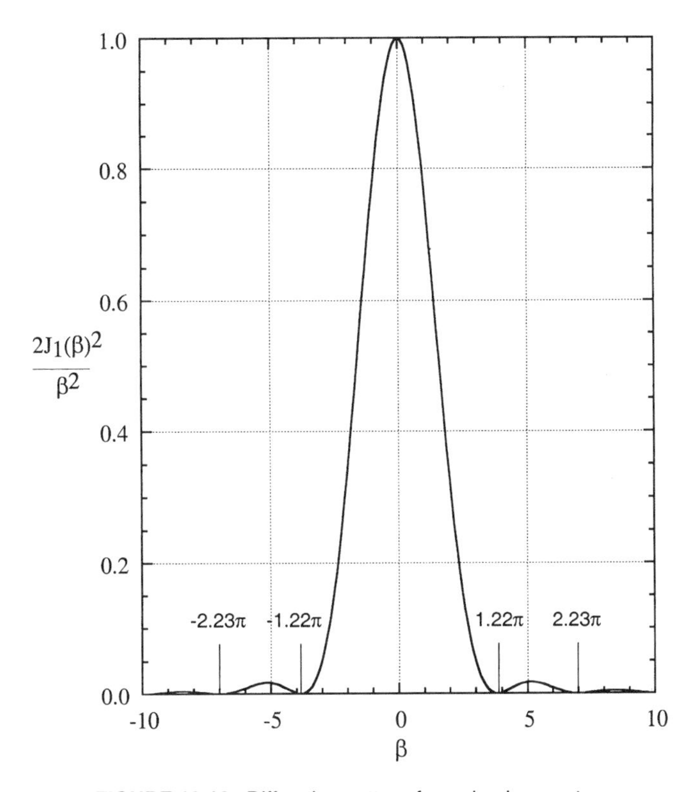

FIGURE 19.12. Diffraction pattern for a circular aperture.

where a is the *diameter* of the lens or mirror. When diffraction patterns of two objects of equal brightness are just resolved, the dip between the two maxima is 0.81 of the resultant height of the convoluted maxima.

19.5.4. Resolving power of a microscope

When two objects are illuminated by coherent light in a microscope, the minimum separation of the objects that allows the images to be distinguished is

$$d_{\text{min}} = \frac{\lambda}{2n \sin\theta}, \tag{19.48}$$

where n is the index of refraction of the medium surrounding the objects and θ is the half-angle subtended by the objective lens at the objects. The product $n \sin\theta$ is characteristic of a particular objective lens and is called the *numerical aperture*. For white light illumination and a numerical aperture of 1.6, $d_{\text{min}} \approx 0.0002$ mm.

19.6. INTERFERENCE

19.6.1. Double slit intensity distribution

If a planewave of light is incident at an angle θ_i on two parallel slits of width a and separation d, the diffracted intensity has a distribution of

$$I = I_0 \frac{\sin^2 \beta}{\beta^2} \cos^2 \gamma, \quad \text{where} \quad \beta = \frac{\pi a}{\lambda}(\sin\theta_i + \sin\theta_d)$$

$$\text{and} \quad \gamma = \frac{\pi d}{\lambda}(\sin\theta_i + \sin\theta_d). \tag{19.49}$$

This is the double slit pattern $\cos^2 \gamma$ overlaid with the single slit distribution. I_0 is the intensity at the center of the pattern. The double slit pattern has maxima when

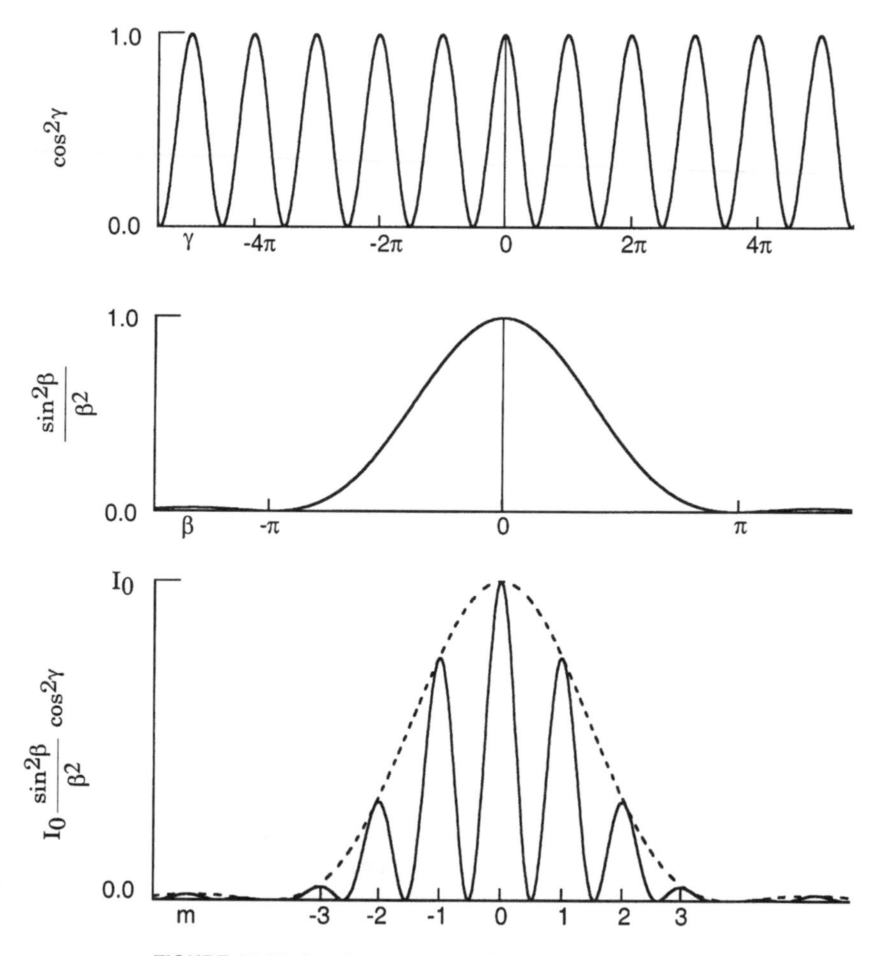

FIGURE 19.13. Interference pattern for a double slit; $d = 3a$.

$(\sin\theta_i + \sin\theta_d) = m\lambda/d(m = 0, 1, 2, \ldots)$. Minima in the double slit pattern occur for $(\sin\theta_i + \sin\theta_d) = (m + 1/2)\lambda/d$. The double slit interference pattern is shown in Figure 19.13.

19.6.2. Diffraction gratings

The intensity distribution for a diffraction grating (Figure 19.14) is an extension of the double slit pattern to a series of N slits (or grooves) of width a having equal separations d. The distribution is

$$I = I_0 \frac{\sin^2\beta}{\beta^2}\frac{\sin^2 N\gamma}{\sin^2\gamma}, \quad \text{where} \quad \beta = \frac{\pi a}{\lambda}(\sin\theta_i + \sin\theta_d)$$

$$\text{and} \quad \gamma = \frac{\pi d}{\lambda}(\sin\theta_i + \sin\theta_d). \tag{19.50}$$

θ_i and θ_d are the angles of incidence and diffraction, respectively. I_0 is the intensity at the center of the pattern. The angles are measured from the normal; if θ_d is on the opposite side of the normal from θ_i, θ_d is negative. The function $(\sin^2 N\gamma)/\sin^2\gamma$ has principal maxima for $\gamma = m\pi(m = 0, 1, 2, \ldots)$. A diffraction grating forms spectrum lines when

$$\sin\theta_i + \sin\theta_d = \frac{m\lambda}{d}, \tag{19.51}$$

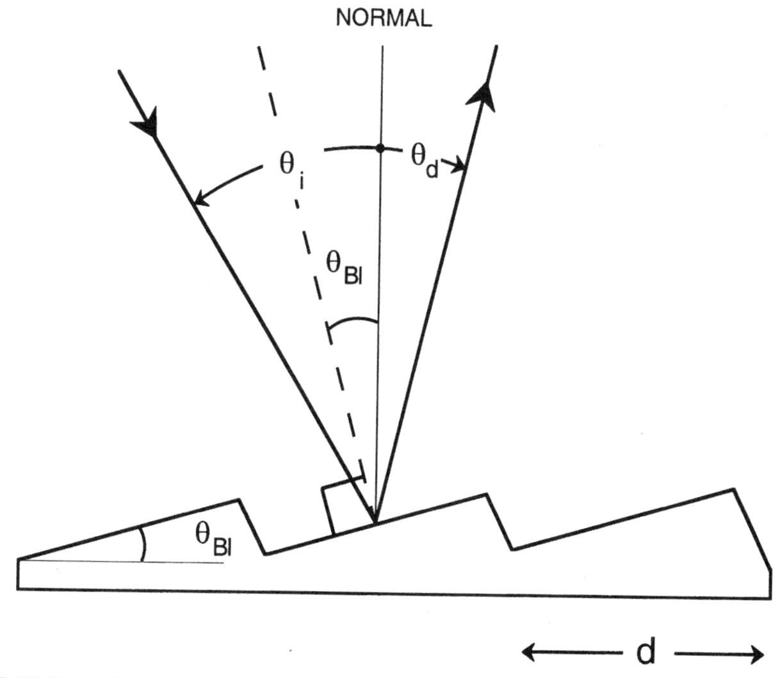

FIGURE 19.14. Incident and diffracted light at a diffraction grating with blaze angle θ_{BI}. d: grating spacing.

where m is the *order number* and d is the groove separation or the *grating spacing*. This is the *grating equation*. The image formed for $m = 0$ is the *direct image* or the *central image*. The prinicpal maxima are separated by secondary maxima whose intensities decrease with an increasing total number of grooves N; they are of negligible intensity for most gratings used in practice.

19.6.3. Dispersion of a diffraction grating

The variation of diffraction angle θ_d with wavelength for a fixed angle of incidence θ_i is

$$\frac{d\theta_d}{d\lambda} = \frac{m}{d\cos\theta_d}. \tag{19.52}$$

This is the *angular dispersion* of the grating. If the focal distance is f and the lateral coordinate of lines in the focal plane is x, the *linear dispersion* or *plate factor* is

$$\frac{d\lambda}{dx} = \frac{d\cos\theta_d}{fm}. \tag{19.53}$$

The linear dispersion is given in units of Å/mm or nm/mm.

19.6.4. Resolving power of a diffraction grating

Considering two spectral lines to be resolved if the central peak of the diffraction pattern of one line falls at the first minimum of the diffraction pattern of the other (Rayleigh criterion), the resolving power of a diffraction grating is

$$\Re = \frac{\lambda}{\Delta\lambda} = mN, \tag{19.54}$$

where $\Delta\lambda$ is the wavelength difference of two lines that can just be distinguished, m is the order number, and N is the total number of lines of the grating. For this separation, the dip between the lines is 0.81 of the intensity at the resultant peaks. In terms of the width of the grating W,

$$\Re = W(\sin\theta_i + \sin\theta_d)/\lambda. \tag{19.55}$$

That is, the *resolving power equals the number of wavelengths contained in the maximum path difference*, $W(\sin\theta_i + \sin\theta_d)$. This general principal is true for all instruments involving interference.

A typical concave grating may have a *groove density* of 1200 lines/mm. If it is 50 mm wide, it has a resolving power of 60,000 in the first order; at 500 nm, the resolving limit is 0.0083 nm. Echelle gratings have a low groove density; they achieve high resolution by working in high orders of diffraction. A typical echelle grating may have 300 lines/mm. If it is 50 mm wide, for $m = 20$, it has a resolving power of 300,000; at 500 nm the resolving limit is 0.0017 nm.

19.6.5. Free spectral range

The range of wavelengths that is free from disturbance by spectral lines in adjacent grating orders is the *free spectral range*

$$\Delta\lambda(\text{FSR}) = \lambda/m. \tag{19.56}$$

19.6.6. Optimum slit width

In a spectrometer, the width of entrance slit that produces maximum light throughput without loss of resolving power is approximately

$$\text{Optimum slit width} = \lambda f/W\ (\text{projected}) = \lambda F\text{-number}, \tag{19.57}$$

where W (projected) is the width of the grating as seen from the slit and f is the focal distance for the slit.

19.6.7. Grating blaze

If a reflection grating is ruled with its grooves as shown in Figure 19.14, it produces a strong diffracted intensity when light strikes it at angle θ_{Bl}, the *blaze angle*. The wavelength for which $\theta_i = \theta_d = \theta_{\text{Bl}}$ (retroreflectance at the blaze angle) is the *blaze wavelength*

$$\lambda_{\text{Bl}} = 2\,d \sin\theta_{\text{Bl}}. \tag{19.58}$$

If a grating is used at an angle of incidence α that differs from the blaze angle, the effective blaze wavelength λ'_{Bl} is

$$\lambda'_{\text{Bl}} = 2\,d \sin\theta_{\text{Bl}} \cos(\alpha - \theta_{\text{Bl}}) = \lambda_{\text{Bl}} \cos(\alpha - \theta_{\text{Bl}}). \tag{19.59}$$

19.6.8. Rowland circle

For a concave diffraction grating, when a slit parallel to the grating grooves is placed on the *Rowland circle*, the spectrum lines are in focus at points along the Rowland circle. The Rowland circle has a *diameter* equal to the *radius of curvature* of the grating; it is tangent to the grating at the midpoint of the grating.

19.6.9. Michelson interferometer

In a Michelson interferometer (Figure 19.15), the intensity at a point that subtends an angle θ with the axis is

$$I = I_0 \cos^2\left(\frac{2\pi\,d}{\lambda}\cos\theta\right), \tag{19.60}$$

where d is the path difference between the beams from the two mirrors. I_0 is the intensity at the center of the pattern. With the mirrors optically parallel, as in Figure 19.15, the interference fringes are circular and have maxima for

$$m\lambda = 2\,d\cos\theta. \tag{19.61}$$

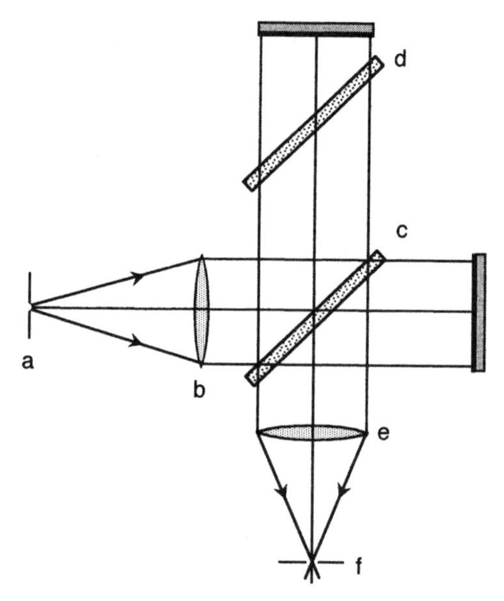

FIGURE 19.15. Michelson interferometer. a: entrance aperture; b: collimating lens; c: beam splitter; d; compensating plate; e: focussing lens; f: exit aperture.

At the center of the pattern, $\theta = 0$ and

$$m\lambda = 2d. \tag{19.62}$$

If a circular aperture is used to record the intensity at the center of the pattern as one mirror is moved with respect to the other as in a *Fourier transform spectrometer* (FTS), and the maximum effective displacement of the mirrors is D, the resolving power equals the number of wavelengths contained in the maximum path difference, $2D$. Thus, $\Re = 2D/\lambda$. In terms of wavenumbers, the resolving limit is

$$\Delta\sigma = \frac{1}{2D}. \tag{19.63}$$

For a 1 m maximum displacement of the two mirrors in a Fourier transform spectrometer, the resolving limit is 0.005 cm^{-1}, independent of wavelength. At 500 nm, $\Re = 4{,}000{,}000$ and the resolving limit is 0.0001 nm.

19.6.10. Fabry-Perot interferometer

In a Fabry-Perot interferometer (Figure 19.16), the intensity at a point that subtends an angle θ with the axis is

$$I = I_0 \frac{T^2}{(1-R)^2 + 4R\sin^2\delta} = I_0 \left[\frac{T}{(1-R)}\right]^2 \frac{1}{1 + (4R\sin^2\delta)/(1-R)^2}, \tag{19.64}$$

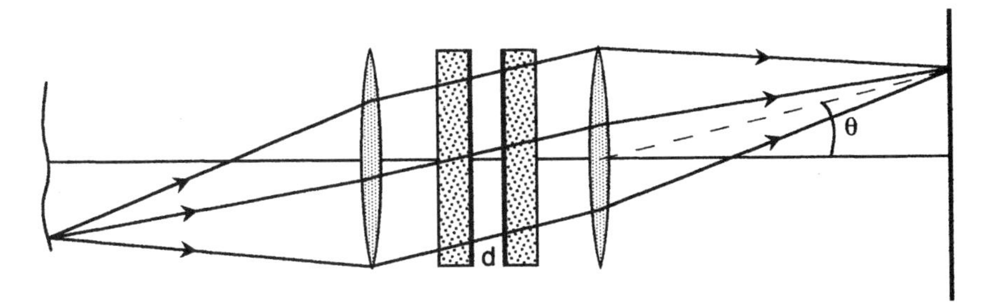

FIGURE 19.16. Fabry-Perot interferometer.

where T is the transmittance of the reflecting film, R is the reflectance of the reflecting film, and $\delta = (2\pi\, d/\lambda)\cos\theta$. I_0 is the incident intensity (or the intensity in the image plane when the interferometer is removed). If the medium between the plates has an index of refraction n (as in an interference filter), $\delta = (2\pi n\, d/\lambda_{\text{vac}})\cos\theta$. This is the Airy distribution. Maxima in the interference fringes occur when $\delta = m\pi\,(m = 0, 1, 2, \ldots)$ or

$$m\lambda = 2\,d\cos\theta. \tag{19.65}$$

In a pattern of circular Fabry-Perot fringes, the order numbers m increase toward the center of the pattern. For maxima, the intensity at the peak is

$$I_{\text{max}} = I_0\left(\frac{T}{1-R}\right)^2 = I_0\left[1 - \frac{A}{(1-R)}\right]^2, \tag{19.66}$$

where A is the absorptance of each reflecting film; $R + T + A = 1$. Minima occur for $\delta = (m + 1/2)\pi$. The intensity of minima is

$$I_{\text{min}} = I_0\left[\frac{T}{1+R}\right]^2. \tag{19.67}$$

The full-width at half maximum of each fringe in wave numbers is

$$\delta\sigma = \frac{1}{2d}\,\frac{(1-R)}{\pi\sqrt{R}}. \tag{19.68}$$

Two lines are said to be resolved if they are separated by the full-width at half maximum of the Airy distribution, $\delta\sigma$. The resolving power is

$$\Re = \frac{\sigma}{\delta\sigma} = m\frac{\pi\sqrt{R}}{(1-R)}. \tag{19.69}$$

When two spectral lines of equal intensity are separated by this amount, the dip between the two peaks is about 0.83 of the peak intensity. This is similar to the dip of 0.81 obtained when the Rayleigh criterion is applied to two circular diffraction patterns. By analogy with diffraction gratings, the number of interfering beams N_R is given by the *finesse*

$$\text{finesse} = N_R = \frac{\pi\sqrt{R}}{(1-R)}. \tag{19.70}$$

This is the effective number of interfering beams produced by the reflecting coatings. N_R is the *finesse of the coatings*. For a Fabry-Perot interferometer with $d = 25$ mm, at 500 nm, $m = 100{,}000$. If $R = 0.92$, $A = 0.01$, and $T = 0.07$, then $N_R = 38$, $\Re = mN_R = 3{,}800{,}000$, and $\delta\lambda = 0.00013$ nm.

In practice, the width of the spectrum lines may limit the effective resolving power that can be achieved. If the effective temperature of the light source in degrees Kelvin is T and the effective weight of the emitters in atomic mass units is M, the full width at half-maximum (FWHM) of the lines due to Doppler broadening is given by

$$\frac{\delta\sigma^D}{\sigma} = \frac{\delta\lambda^D}{\lambda} = 7.16 \times 10^{-7}\sqrt{\frac{T}{M}}. \tag{19.71}$$

For a discharge in iron $M = 56$, and if $T = 373$ K, then at $\lambda = 500$ nm $\delta\lambda^D = 0.00092$ nm. With this Doppler width, increasing the reflectivity of the films (at the expense of transmittance) would not reduce the widths of the observed lines, but would merely reduce the intensity of the peaks (Eq. 19.66).

Similarly, since interferometer plates are not perfectly flat, their deviation from flatness adds to the width of the fringes. If the plates deviate from perfect flatness by an average of δd, the plates produce a broadening of the fringes characterized by N_D, the *finesse of the plates*

$$N_D = \frac{\lambda}{2\,\delta d} = \frac{1}{2[\delta d/\lambda]}. \tag{19.72}$$

For interferometer plates having a flatness of one-seventy-sixth of a wave, $[\delta d/\lambda] = 1/76$, and the finesse of the plates is $N_D = 38$. For such plates, an increase of N_R over 38 would similarly result in needless loss of intensity at the peaks. The maximum useful value of R is about 0.92.

In the ring system of a Fabry-Perot interferometer, the diameters of the rings of a spectrum line depend on the wavelength λ and the order number m according to Eq. (19.65). If a second line with slightly different wavelength is also present, its rings will be displaced from those of the first line. The wavelength difference required to make a ring of the second line in order m coincide with the $m+1$ order of the first line is the FSR

$$\Delta\lambda(\text{FSR}) = \frac{\lambda^2}{2d}. \tag{19.73}$$

This is the range of wavelengths that can be spanned without overlapping from an adjacent order. In terms of wavenumbers,

$$\Delta\sigma(\text{FSR}) = \frac{1}{2d}. \tag{19.74}$$

19.7. SPECTRA

19.7.1. Important spectral lines

A list of selected important spectral lines is given in Table 19.4.

TABLE 19.4. Important spectral lines.

Spectrum	Wavelength[a,b] (nm)	Wavenumber (cm^{-1})	Classification[c,d]	
H I	121.5668	82259.30	$1s^2S_{1/2} - 2p\,^2P_{3/2}$	Ly α
	121.5674	82258.90	$1s\,^2S_{1/2} - 2p\,^2P_{1/2}$	Ly α
	434.047	23032.50	$n = 2 - n = 5$	Hγ
	486.133	20564.76	$n = 2 - n = 4$	Hβ
	656.279	15233.21	$n = 2 - n = 3$	Hα
He I	53.703	186209	$1s^2\,^1S_0 - 1s3p\,^1P_1$	
	58.433	171135	$1s^2\,^1S_0 - 1s2p\,^1P_1$	
	388.865	25708.59	$1s2s\,^1S_0 - 1s3p\,^3P_{2,1}$	
	447.148	22357.69	$1s2p\,^3P_{2,1} - 1s4d\,^3D_{3,2}$	
	447.168	22356.68	$1s2p\,^3P_0 - 1s4d\,^3D_1$	
	471.315	21211.32	$1s2p\,^3P_{2,1} - 1s4s\,^3S_1$	
	492.193	20311.56	$1s2p\,^1P_1 - 1s4d\,^1D_2$	
	501.568	19931.92	$1s2s\,^1S_0 - 1s3p\,^1P_1$	
	504.774	19805.33	$1s2p\,^1P_1 - 1s4s\,^1S_0$	
	587.562	17014.76	$1s2p\,^3P_{2,1} - 1s3d\,^3D_{3,2}$	
	587.597	17013.76	$1s2p\,^3P_0 - 1s3d\,^3D_1$	
	667.815	14970.07	$1s2p\,^1P_1 - 1s3d\,^1D_2$	
	706.519	14150.00	$1s2p\,^3P_{2,1} - 1s3s\,^3S_1$	
	706.571	14148.96	$1s2p\,^3P_0 - 1s3s\,^3S_1$	
	1083.205[b]	9231.859	$1s2s\,^3S_1 - 1s2p\,^3P_0$	
	1083.321[b]	9230.871	$1s2s\,^3S_1 - 1s2p\,^3P_1$	
	1083.330[b]	9230.795	$1s2s\,^3S_1 - 1s2p\,^3P_2$	
	2058.692[b]	4857.454	$1s2s\,^1S_0 - 1s2p\,^1P_1$	
Li I	323.266	30925.38	$2s\,^2S_{1/2} - 3p\,^2P_{1/2,3/2}$	
	670.776	14904.00	$2s\,^2S_{1/2} - 2p\,^2P_{3/2}$	
	670.791	14903.66	$2s\,^2S_{1/2} - 2p\,^2P_{1/2}$	
Ne I	62.682	159537	$2p^6\,^1S_0 - 2p^54s\,1/2[1/2]_1$	
	62.974	158798	$2p^6\,^1S_0 - 2p^54s\,3/2[3/2]_1$	
	73.590	135891	$2p^6\,^1S_0 - 2p^53s\,1/2[1/2]_1$	
	74.372	134461	$2p^6\,^1S_0 - 2p^53s\,3/2[3/2]_1$	
	632.8165	15798.002	$3p\,1/2[3/2]_2 - 5s1/2[1/2]_1$	

(continued)

TABLE 19.4. Continued

Spectrum	Wavelength[a,b] (nm)	Wavenumber (cm^{-1})	Classification[c,d]
Na I	330.237	30272.58	$3s\,^2S_{1/2} - 4p\,^2P_{3/2}$
	330.298	30266.99	$3s\,^2S_{1/2} - 4p\,^2P_{1/2}$
	588.995	16973.37	$3s\,^2S_{1/2} - 3p\,^2P_{3/2}$
	589.592	16956.17	$3s\,^2S_{1/2} - 3p\,^2P_{1/2}$
Ar I	86.680	115367	$3p^6\,^1S_0 - 3p^5 3d\,1/2[3/2]_1$
	86.975	114975	$3p^6\,^1S_0 - 3p^5 5s\,1/2[1/2]_1$
	87.606	114148	$3p^6\,^1S_0 - 3p^5 3d\,3/2[3/2]_1$
	87.995	113643	$3p^6\,^1S_0 - 3p^5 5s\,3/2[3/2]_1$
	89.431	111818	$3p^6\,^1S_0 - 3p^5 3d\,3/2[1/2]_1$
	104.822	95399.9	$3p^6\,^1S_0 - 3p^5 4s\,1/2[1/2]_1$
	106.666	93750.6	$3p^6\,^1S_0 - 3p^5 4s\,3/2[3/2]_1$
K I	404.414	24720.18	$4s\,^2S_{1/2} - 5p\,^2P_{3/2}$
	404.721	24701.43	$4s\,^2S_{1/2} - 5p\,^2P_{1/2}$
	766.490	13042.90	$4s\,^2S_{1/2} - 4p\,^2P_{3/2}$
	769.896	12985.20	$4s\,^2S_{1/2} - 4p\,^2P_{1/2}$
Cu I	324.754	30783.69	$4s\,^2S_{1/2} - 4p\,^2P_{3/2}$
	327.396	30535.30	$4s\,^2S_{1/2} - 4p\,^2P_{1/2}$
Kr I	94.654	105649	$4p^6\,^1S_0 - 4p^5 5d\,3/2[3/2]_1$
	95.106	105147	$4p^6\,^1S_0 - 4p^5 6s\,1/2[1/2]_1$
	95.340	104888	$4p^6\,^1S_0 - 4p^5 4d\,1/2[3/2]_1$
	96.337	103803	$4p^6\,^1S_0 - 4p^5 5d\,3/2[1/2]_1$
	100.106	99894.8	$4p^6\,^1S_0 - 4p^5 6s\,3/2[3/2]_1$
	100.355	99647.0	$4p^6\,^1S_0 - 4p^5 4d\,3/2[3/2]_1$
	103.002	97086.0	$4p^6\,^1S_0 - 4p^5 4d\,3/2[1/2]_1$
	116.487	85847.5	$4p^6\,^1S_0 - 4p^5 5s\,1/2[1/2]_1$
	123.584	80917.6	$4p^6\,^1S_0 - 4p^5 5s\,3/2[3/2]_1$
	557.029	17947.41	$5s\,3/2[3/2]_2 - 5p\,1/2[1/2]_1$
	587.091	17028.40	$5s\,3/2[3/2]_1 - 5p\,1/2[3/2]_2$
Rb I	420.180	23792.59	$5s\,^2S_{1/2} - 6p\,^2P_{3/2}$
	421.553	23715.08	$5s\,^2S_{1/2} - 6p\,^2P_{1/2}$
	780.027	12816.54	$5s\,^2S_{1/2} - 5p\,^2P_{3/2}$
	794.760	12578.95	$5s\,^2S_{1/2} - 5p\,^2P_{1/2}$
Ag I	328.068	30472.71	$5s\,^2S_{1/2} - 5p\,^2P_{3/2}$
	338.289	29552.05	$5s\,^2S_{1/2} - 5p\,^2P_{3/2}$

(continued)

TABLE 19.4. Continued

Spectrum	Wavelength[a,b] (nm)	Wavenumber (cm^{-1})	Classification[c,d]
Xe I	106.816	93618.8	$5p^6\,{}^1S_0 - 5p^5 5d\,1/2[3/2]_1$
	119.204	83890.5	$5p^6\,{}^1S_0 - 5p^5 5d\,3/2[3/2]_1$
	125.020	79987.2	$5p^6\,{}^1S_0 - 5p^5 5d\,3/2[1/2]_1$
	129.559	77185.6	$5p^6\,{}^1S_0 - 5p^5 6s\,1/2[1/2]_1$
	146.961	68045.7	$5p^6\,{}^1S_0 - 5p^5 6s\,3/2[3/2]_1$
Cs I	455.528	21946.40	$6s\,{}^2S_{1/2} - 7p\,{}^2P_{3/2}$
	459.317	21765.35	$6s\,{}^2S_{1/2} - 7p\,{}^2P_{1/2}$
	852.113	11732.31	$6s\,{}^2S_{1/2} - 6p\,{}^2P_{3/2}$
	894.347	11178.27	$6s\,{}^2S_{1/2} - 6p\,{}^2P_{1/2}$
Au I	242.795	41174.61	$6s\,{}^2S_{1/2} - 6p\,{}^2P_{3/2}$
	267.595	37358.99	$6s\,{}^2S_{1/2} - 6p\,{}^2P_{1/2}$
Hg I	184.950	54068.7	$6s^2\,{}^1S_0 - 6s6p\,{}^1P_1$
	253.652	39412.2	$6s^2\,{}^1S_0 - 6s6p\,{}^3P_1$
	296.728	33691.0	$6s6p\,{}^3P_0 - 6s6d\,{}^3D_1$
	312.567	31983.83	$6s6p\,{}^3P_1 - 6s6d\,{}^3D_2$
	313.156	31923.76	$6s6p\,{}^3P_1 - 6s6d\,{}^3D_1$
	313.184	31920.82	$6s6p\,{}^3P_1 - 6s6d\,{}^1D_2$
	365.016	27388.27	$6s6p\,{}^3P_2 - 6s6d\,{}^3D_3$
	365.484	27353.17	$6s6p\,{}^3P_2 - 6s6d\,{}^3D_2$
	366.328	27290.14	$6s6p\,{}^3P_2 - 6s6d\,{}^1D_2$
	404.656	24705.34	$6s6p\,{}^3P_0 - 6s7s\,{}^3S_1$
	407.783	24515.88	$6s6p\,{}^3P_1 - 6s7s\,{}^1S_0$
	435.833	22938.10	$6s6p\,{}^3P_1 - 6s7s\,{}^3S_1$
	491.607	20335.78	$6s6p\,{}^1P_1 - 6s8s\,{}^1S_0$
	546.074	18307.42	$6s6p\,{}^3P_2 - 6s7s\,{}^3S_1$
	576.960	17327.39	$6s6p\,{}^1P_1 - 6s6d\,{}^3D_2$
	578.966	17267.37	$6s6p\,{}^1P_1 - 6s6d\,{}^3D_1$
	579.066	17264.37	$6s6p\,{}^1P_1 - 6s6d\,{}^1D_2$
	1014.253[b]	9859.474	$6s6p\,{}^1P_1 - 6s7s\,{}^1S_0$
	1129.049[b]	8857.011	$6s7s\,{}^3S_1 - 6s7p\,{}^3P_2$
	2325.942[b]	4299.334	$6s7p\,{}^3P_1 - 6s8s\,{}^3S_1$

[a] J. Reader and C. H. Corliss, *Line Spectra of the Elements in CRC Handbook of Chemistry and Physics*, 82nd ed., edited by D. R. Lide (CRC Press, Boca Raton, 2001).

[b] Wavelengths between 200 nm and 1000 nm are in air; wavelengths shorter than 200 nm and longer than 1000 nm are in vacuum.

[c] C. E. Moore, *Atomic Energy Levels*, Natl. Bur. Std. (U.S.) Circ. 467 (U. S. Government Printing Office, Vol. I - 1949; Vol. II - 1952; Vol. III - 1958).

[d] Energy levels for excited states of rare gases are given in $J_1\ell$-coupling notation (pair coupling or J_1K-coupling). The j-value of the np^5 core J_1 is followed in brackets by K, the resultant of J_1 and the orbital angular momentum of the valence electron ℓ.

19.7.2. Common laser wavelengths

A list of wavelengths of commonly used lasers is given in Table 19.5.

TABLE 19.5. Wavelengths of commonly used lasers[a,b].

Laser	Wavelength[c]
He/Ne	543.5, 612.0, **632.8**, 1152.6, 1523.5, 3392.2
Ruby ($Cr:Al_2O_3$)	**694.3**
Cu vapor	**510.6**, **578.2**
He/Cd	325.0, **441.6**
Ar ion	351.4, 363.8, 454.5, 457.9, 465.8, 472.7, 476.5, **488.0**, 496.5, 501.7, **514.5**, 528.7
Kr ion	350.7, 406.7, 413.1, 415.4, 468.0, 476.2, 482.5, 520.8, **530.9**, 568.2, **647.1**, 676.4, 752.5, 799.3
XeF	**351**
N_2	**337**
XeCl	**308**
KrF	**248**
ArF	**193**
F_2	**157**
Nd:YAG ($Nd:Y_3Al_5O_{12}$)	**1.064**μm
Nd:glass (phosphate)	**1.054**μm
Nd:glass (silicate)	**1.06**μm
CO_2	**10.6**μm

[a] S. Svanberg, *Atomic and Molecular Spectroscopy* (Springer, Berlin, 1997).
[b] M. J. Weber, *Handbook of Laser Wavelengths* (CRC Press, Boca Raton, 1999).
[c] Wavelengths are in nm, unless otherwise noted. Principal lines are in bold type.

19.8. REFERENCES

[1] F. A. Jenkins and H. E. White, *Fundamentals of Optics* (McGraw-Hill, New York, 1950).

[2] M. Born and E. Wolf, *Principles of Optics* (Pergamon Press, New York, 1959).

[3] A. P. Thorne, U. Litzén, and S. Johansson, *Spectrophysics: Principles and Applications* (Springer, Berlin, 1999).

[4] P Jacquinot, *New Developments in Interference Spectroscopy*, in Reports Progress Phys. **23**, 267 (1960).

[5] R. Cohen and B. N. Taylor, Rev. Mod. Phys. **59**, 1121 (1987).

[6] E. R. Peck and K. Reeder, J. Opt. Soc. Am. **62**, 958 (1972).

[7] B. Edlén, Metrologia **2**, 71 (1966).

[8] K. P. Birch and M. J. Downs, Metrologia **31**, 315 (1994).

20

Particle Accelerators and Storage Rings

Kai Desler
DESY, Hamburg, Germany

Donald A. Edwards
Fermilab, Batavia, Illinois

Contents

List of Figures

20.1. INTRODUCTION

This article is a short overview of the physics of particle beams common to accelerators and storage rings. Sec. 20.2 gives a number of standard expressions relative to single-particle motion. The basic reference here is the article by Courant and Snyder [1]. Extended discussion and further references may be found in textbooks (e.g. Edwards and Syphers [2] and Wiedemann [3]).

When interaction between and among beam particles becomes important, the analytical approach generally is not sufficient by itself, and numerical simulation becomes necessary. Much of the material on multiparticle dynamics in Sec. 20.3 is thus introductory and illustrative in character. General references on these topics include the textbooks by Chao [4] and Reiser [5].

There is no account here of the technological foundation upon which the accelerator field rests. In a few pages, it is not possible to provide even a capsule survey of accelerator systems and components—magnets, accelerating structures, particle sources, beam diagnostics, and so on. The interested reader is referred to the *Handbook of Accelerator Physics and Engineering* [6] and to the references contained therein.

20.2. SINGLE-PARTICLE MOTION

An accelerator or storage ring will have a design trajectory; that is, a line in space that the "ideal" particle is to follow. Path length along the design trajectory is here denoted by s, which plays the role of the independent variable; the relationship between s and time is specified by the design. Particle positions are expressed relative to the ideal particle as shown in Figure 20.1. The design trajectory is assumed to lie in a plane, in order to simplify the discussion.

Any source provides particles with distributions in position, angle, and momentum. That particles initially close to the ideal particle remain close is a principal concern of beam dynamics. The need for transverse stability involves the shortest time scale, and so the equations of motion of Sec. 20.2.1 ignore other longer-term processes (e.g., synchrotron oscillations and resonances), which may subsequently be treated as perturbations.

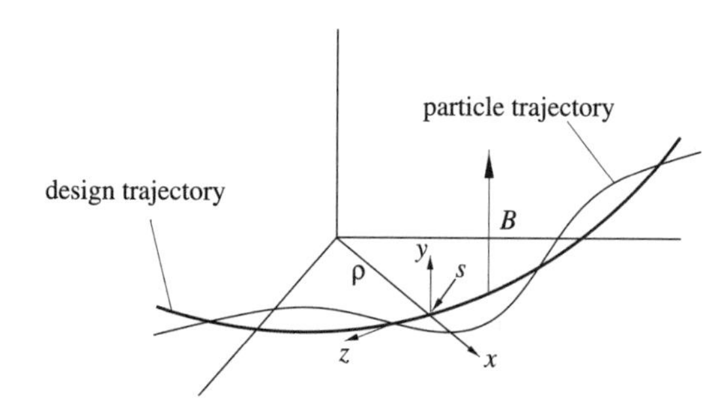

FIGURE 20.1. Trajectories of an ideal and a neighboring particle in an accelerator or storage ring.

20.2.1. Linear transverse motion

Present day high energy accelerators and colliders use alternating gradient focusing provided by quadrupole magnets. The stable excursions from the design trajectory are called *betatron* oscillations because they were first analyzed for that early type of accelerator.

Equations of motion

In the absence of coupling between the two transverse degrees of freedom, the equations of motion for transverse oscillations of particles of constant momentum and charge e are

$$\begin{aligned} x'' + K_x(s)x &= 0, & y'' + K_y(s)y &= 0, \\ x' &\equiv dx/ds, & y' &\equiv dy/ds, \\ K_x &\equiv B'/(B\rho) + 1/\rho(s)^2, & K_y &\equiv -B'/(B\rho), \\ B' &\equiv \partial B_y/\partial x = \partial B_x/\partial y, & (B\rho) &\equiv p/e, \\ 1/\rho(s) &= B_y(s)/(B\rho). \end{aligned} \tag{20.1}$$

As shown in Figure 20.1, the radius of curvature ρ lies in the x-s plane. The bending magnetic field B on the design trajectory is in the y direction. The quantity $B\rho$ is called the *magnetic rigidity*; if the momentum, p, is expressed in GeV/c, then $B\rho \approx (10/3)p$ T m.

The equations of motion are those appropriate to a simple harmonic oscillator, except that the "spring constants" are functions of the independent variable. These are Hill's equations in form, and the solutions may be presented in a variety of ways. Courant and Snyder [1] used three parameters, α, β, and γ, in their treatment, and their convention is followed here. There is a potential ambiguity in the use of "primes" to denote derivatives; in x' the derivative is with respect to s, while in B' the derivative is with respect to x. Again, these are conventions to which we adhere.

Solution in phase-amplitude form

$$x(s) = a\sqrt{\beta(s)}\cos[\psi(s) + \delta], \tag{20.2}$$

where a and δ are constants of integration.

The *amplitude function* β satisfies

$$2\beta\beta'' - \beta'^2 + 4\beta^2 K = 4. \tag{20.3}$$

The phase ψ advances according to $d\psi/ds = 1/\beta$. Thus, β plays the role of a position-dependent $\lambda/2\pi$ in the motion. Expressing a in terms of x, x' yields an invariant:

$$\begin{aligned} a^2 &= \gamma(s)x(s)^2 + 2\alpha(s)x(s)x'(s) + \beta(s)x'(s)^2 \\ &= \frac{1}{\beta}\left\{x(s)^2 + [\alpha(s)x(s) + \beta(s)x'(s)]^2\right\}, \end{aligned} \tag{20.4}$$

where

$$\alpha \equiv -\beta'/2, \qquad \gamma \equiv \frac{1+\alpha^2}{\beta}, \tag{20.5}$$

In x, x' (or y, y') coordinates, a particle moves on a surface of elliptical cross section; the orientation and aspect ratio changes with s, the area πa^2 remains constant.

The number of oscillations per turn in a circular accelerator is the *tune*, ν:

$$\nu \equiv \frac{1}{2\pi} \oint \frac{ds}{\beta}. \tag{20.6}$$

As in all oscillatory systems, the motion is susceptible to resonant perturbation, as is elaborated in Sec. 20.2.4. Variations in magnetic field gradient, through either error or intention, lead to a tune change

$$\Delta\nu = \pm\frac{1}{4\pi} \oint \beta \frac{\Delta B'}{(B\rho)}\, ds \tag{20.7}$$

where the $+$ sign is associated with the bend plane.

Propagation in matrix form

Transport of the state of a transversely displaced particle, represented in terms of x, x' or y, y', from one point in s to another can be expressed in matrix form:

$$\begin{pmatrix} x \\ x' \end{pmatrix}_{s_2} = M(s_1 \to s_2) \begin{pmatrix} x \\ x' \end{pmatrix}_{s_1}. \tag{20.8}$$

In terms of the Courant-Snyder parameters, $M(s_1 \to s_2)$ is represented by

$$\begin{pmatrix} \sqrt{\dfrac{\beta_2}{\beta_1}}\,(c + \alpha_1 s) & \sqrt{\beta_1\beta_2}\, s \\ -\dfrac{1+\alpha_1\alpha_2}{\sqrt{\beta_1\beta_2}}\, s + \dfrac{\alpha_1 - \alpha_2}{(\beta_1\beta_2)^{1/2}}\, c & \sqrt{\dfrac{\beta_1}{\beta_2}}\,(c - \alpha_2 s) \end{pmatrix}. \tag{20.9}$$

Here, $c \equiv \cos\,\Delta\psi$ and $s \equiv \sin\,\Delta\psi$. When $C \equiv s_2 - s_1$ is the distance between equivalent positions in the focusing system—e.g., circumference C in a synchrotron—M reduces to

$$M_C = I \cos\psi_C + J \sin\psi_C, \qquad J = \begin{pmatrix} \alpha & \beta \\ -\gamma & -\alpha \end{pmatrix}, \tag{20.10}$$

where ψ_C is the phase advance through the repetition length. Since $J^2 = -I$, this matrix can be represented as $e^{J\psi_C}$. The eigenvalue equation for the matrix yields the stability condition

$$|\,\mathrm{Tr}\, M\,| \le 2 \tag{20.11}$$

In terms of the matrix elements m_{ij} of $M(s_1 \to s_2)$, β, α, and γ transform according to

$$\begin{pmatrix} \beta \\ \alpha \\ \gamma \end{pmatrix}_2 = \mathcal{M} \begin{pmatrix} \beta \\ \alpha \\ \gamma \end{pmatrix}_1 \tag{20.12}$$

with $\mathcal{M}$ given by

$$\begin{pmatrix} m_{11}^2 & -2m_{11}m_{12} & m_{12}^2 \\ -m_{11}m_{12} & m_{11}m_{22}+m_{12}m_{21} & -m_{12}m_{22} \\ m_{21}^2 & -2m_{21}m_{22} & m_{22}^2 \end{pmatrix}, \tag{20.13}$$

and the phase advance from s_1 to s_2 is

$$\Delta\psi = \tan^{-1}\left(\frac{m_{12}}{\beta_1 m_{11} - \alpha_1 m_{12}}\right). \tag{20.14}$$

Momentum dispersion

In a synchrotron, particles with momenta other than that for the design trajectory follow a *displaced equilibrium orbit*. For momentum $p = p_0 + \Delta p$ the offset is expressed by the *momentum dispersion function*, D:

$$x(s) = D(p,s)\frac{\Delta p}{p_0}, \qquad D(p, s+C) = D(p,s), \tag{20.15}$$

and D is the periodic solution of the inhomogeneous Hill's equation

$$D'' + \left(K_x\frac{p_0}{p} - \frac{1}{\rho^2}\frac{\Delta p}{p}\right)D = \frac{1}{\rho}\frac{p_0}{p}. \tag{20.16}$$

Here, ρ is still the radius of curvature for the central momentum p_0. Transport of particles differing in momentum from that of the ideal particle is accommodated into the matrix representation through the use of a 3×3 matrix:

$$\begin{pmatrix} x_2 \\ x_2' \\ \dfrac{\Delta p}{p_0} \end{pmatrix} = \begin{pmatrix} m_{11} & m_{12} & m_{13} \\ m_{21} & m_{22} & m_{23} \\ 0 & 0 & 1 \end{pmatrix} \begin{pmatrix} x_1 \\ x_1' \\ \dfrac{\Delta p}{p_0} \end{pmatrix}, \tag{20.17}$$

where

$$\begin{aligned} m_{13} &= D_2 - m_{11}D_1 - m_{12}D_1', \\ m_{23} &= -m_{21}D_1 - m_{22}D_1' + D_2'. \end{aligned} \tag{20.18}$$

Propagation within beamline elements

Beamline elements usually have a near-constant value of K as defined in Eq. (20.1). Matrices for individual beamline elements are as follows:

$$\text{For } K > 0, \qquad \begin{pmatrix} c & \dfrac{1}{\sqrt{K}}s & \dfrac{B_0}{(B\rho)}\dfrac{1}{K}(1-c) \\ -\sqrt{K}\,s & c & \dfrac{B_0}{(B\rho)}\dfrac{1}{\sqrt{K}}s \\ 0 & 0 & 1 \end{pmatrix} \tag{20.19}$$

$$\text{for } K = 0 \qquad \begin{pmatrix} 1 & L & \frac{1}{2}\frac{B_0}{(B\rho)}L^2 \\ 0 & 1 & \frac{B_0}{(B\rho)}L \\ 0 & 0 & 1 \end{pmatrix};$$

$$\text{for } K < 0 \qquad \begin{pmatrix} u & \frac{1}{\sqrt{|K|}}v & \frac{B_0}{(B\rho)}\frac{1}{|K|}(u-1) \\ \sqrt{|K|}v & u & \frac{B_0}{(B\rho)}\frac{1}{\sqrt{|K|}}v \\ 0 & 0 & 1 \end{pmatrix};$$

where c, s, u, v are cos, sin, cosh, sinh, respectively, with the argument $\sqrt{|K|}\,L$. Here, L is the length of the element and B_0 is the value of the magnetic field on the design trajectory.

Differentiation of Eq. (20.3) with respect to s yields

$$\beta''' + 4K\beta' + 2\beta K' = 0. \tag{20.20}$$

Therefore, within a beamline element at constant K, the solution for β is one of the following three forms:

$$\begin{aligned} \beta &= a\cos 2\sqrt{K}\,s + b\sin 2\sqrt{K}\,s + c, \\ &= a + bs + cs^2, \\ &= a\cosh 2\sqrt{|K|}\,s + b\sinh 2\sqrt{|K|}\,s + c, \end{aligned} \tag{20.21}$$

depending on whether K is positive, zero, or negative, respectively.

Determination of Courant-Snyder parameters

By multiplication of the matrices for the individual elements of a repetition period of the focusing structure as set down in Eq. (20.19) and comparison of terms with Eq. (20.10), the Courant-Snyder parameters can be found. To illustrate, a standard repetitive structure or *cell* of a large synchrotron or linear accelerator is a series of equidistantly spaced focusing and defocusing quadrupoles, the so-called FODO cell. In the thin lens approximation, the matrices are

$$F = \begin{pmatrix} 1 & 0 \\ -1/f & 1 \end{pmatrix}, \quad D = \begin{pmatrix} 1 & 0 \\ 1/f & 1 \end{pmatrix}, \quad O = \begin{pmatrix} 1 & L \\ 0 & 1 \end{pmatrix}, \tag{20.22}$$

where f is the focal length of the quadrupole and L is the interlens spacing. Then

$$M = FODO = \begin{pmatrix} 1 + L/f & 2L + L^2/f \\ -L/f^2 & 1 - L/f - (L/f)^2 \end{pmatrix}, \tag{20.23}$$

If we call the phase advance through this cell μ, then comparison with Eq. (20.10) yields

$$\sin(\mu/2) = L/2f, \tag{20.24}$$

with the conclusion that the transverse oscillations through a sequence of these cells are stable provided the lens spacing is less than twice the focal length. By symmetry, the extrema in the amplitude functions occur at the lenses:

$$\beta_{\pm} = 2L\left(\frac{1 \pm \sin(\mu/2)}{\sin\mu}\right). \tag{20.25}$$

Transverse emittance and admittance

The *transverse emittance* ε is a parameter related to the area in x, x' (or y, y') phase space occupied by the particle distribution. Various definitions are in use, depending on the fraction of the distribution included. The *rms emittance* is defined in terms of the first and second moments of the distribution according to

$$\varepsilon = \sqrt{\langle x^2\rangle\langle x'^2\rangle - \langle xx'\rangle^2}\,. \tag{20.26}$$

For a Gaussian beam distribution with standard deviation σ, the phase space area containing a fraction F of the beam is

$$\varepsilon = -\frac{2\pi\sigma^2}{\beta}\ln(1-F). \tag{20.27}$$

In these terms, the definition in Eq. (20.26) contains 15% of the beam. Note that ε as defined in Eq. (20.26) is not the area of an ellipse with semi-axes $\langle x^2\rangle$ and $\langle x'^2\rangle$ (ignoring for the moment the correlation term); rather, it is that area divided by π. The presence or absence of π's in numbers quoted as emittances is a recurrent source of confusion. It is helpful to show emittances in units such as "π mm mrad" to allay the suspicion that the stated figure requires division by π.

The canonical conjugate to x is $p_x = \gamma m v_s x'$, and so a *normalized emittance* ε_N may be defined by

$$\varepsilon_N \equiv \left(\frac{v_s}{c}\gamma\right)\varepsilon. \tag{20.28}$$

The normalized emittance is an adiabatic invariant. With increase of energy, the amplitude of betatron oscillations will diminish as $p^{-1/2}$. In the long term—during beam storage, for example—there are many processes that may increase or decrease the normalized emittance. Emittance increase is discussed in Edwards and Syphers [2], and techniques for emittance reduction may be found in Sec. 20.3.3. The *admittance* or *acceptance*, denoted by A below, is the area of the largest ellipse that can propagate through the accelerator without loss. If the half-aperture available to the beam is $r(s)$, then

$$A = \left(\pi\frac{r(s)^2}{\beta(s)}\right)_{\min}, \tag{20.29}$$

which for constant r becomes $\pi r^2/\beta_{\max}$. To compare with Eq. (20.26), the π in the admittance relationship must be deleted.

The emittance and admittance should be *matched*. The α and β of the incoming beam ellipse must conform to that of the target focusing channel to achieve the most effective transfer; otherwise, the beam envelope will oscillate about the $\sqrt{\beta}$ pattern appropriate to the channel itself.

20.2.2. Longitudinal motion

Generally, in synchrotrons and linear accelerators, radio-frequency (RF) resonators or wave guides provide acceleration through electromagnetic fields that vary sinusoidally in time. Particles displaced in s or, equivalently, in RF phase from the ideal particle are limited in excursion in these variables by the principle of *phase stability* and describe *synchrotron oscillations*. The terminology differs somewhat between the two accelerator species. Synchrotron usage will be followed here; the underlying physics is the same.

Slip factor and transition

The longitudinal displacement is influenced both by the difference in speed and by the difference in orbit length from that of the ideal particle. If τ is the orbit period, then

$$\frac{\Delta\tau}{\tau} = \frac{\Delta C}{C} - \frac{\Delta v}{v} = \frac{\Delta C}{C} - \frac{1}{\gamma^2}\frac{\Delta p}{p}, \tag{20.30}$$

where C is the circumference. Orbits associated with differing momenta generally differ in path length. With the definition of *transition gamma* γ_t as a coefficient relating $\Delta C/C$ to $\Delta p/p$, Eq. (20.30) becomes

$$\frac{\Delta\tau}{\tau} = \eta\frac{\Delta p}{p}, \qquad n \equiv \frac{1}{\gamma_t^2} - \frac{1}{\gamma^2}, \tag{20.31}$$

where η is called the *slip factor*. When speed changes dominate, $\eta < 0$, and this regime is termed *below transition*. When path length changes dominate, $\eta > 0$, and this regime is termed *above transition*. The energy at which $\eta = 0$ is the *transition energy*, and the stability condition changes between below and above the transition energy.

Equations of motion for synchrotron oscillations

The difference equations relating the RF phase ϕ and the energy offset ΔE on successive orbits may be written

$$\begin{aligned} \phi_{n+1} &= \phi_n + \frac{2\pi h\eta}{\beta^2 E_s}\Delta E_{n+1}, \\ \Delta E_{n+1} &= \Delta E_n + eV(\sin\phi_n - \sin\phi_s). \end{aligned} \tag{20.32}$$

The subscript s denotes a synchronous quantity suitable to the ideal particle. The *harmonic number* h is the ratio of the orbit period to the RF cycle time. The maximum energy gain during an orbit is eV. These equations are nonlinear, and so one may expect both stable and unstable motion in longitudinal phase space. An iteration of the equations of motion is

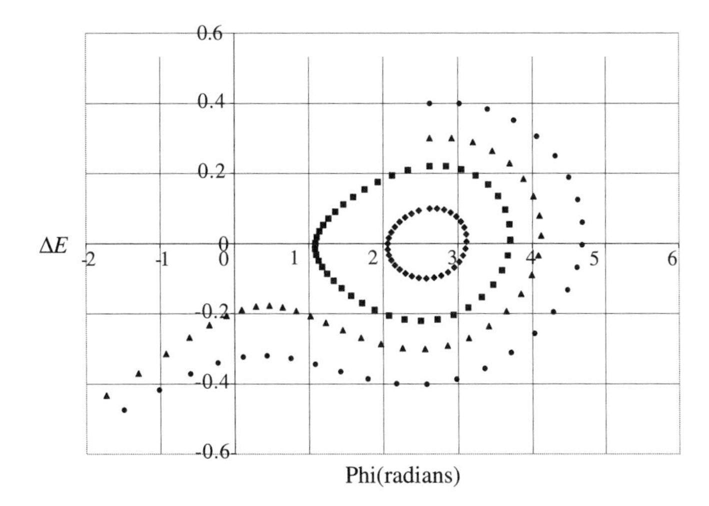

FIGURE 20.2. Iteration of Eqs. (20.32) demonstrates phase stability and instability. Initial conditions differ in ΔE (arbitrary units), but all are at the same initial phase $\phi_1 = \phi_S = (5/6)\pi$.

shown in Figure 20.2 for particles undergoing acceleration. Although the difference equations permit chaotic behavior, the parameters in accelerators are generally below the onset of chaos, and the differential equation approximation is reasonable:

$$\frac{d\phi}{dt} = \frac{2\pi h\eta}{\tau_0^2 E_s}(\tau\,\Delta E),$$

$$\frac{d}{dt}(\tau\,\Delta E) = eV(\sin\phi - \sin\phi_s), \tag{20.33}$$

where τ_0 is the transit time for $\beta = 1$. These combine to yield

$$\frac{d}{dt}\left(\frac{E_s}{\eta}\frac{d\phi}{dt}\right) = \frac{2\pi h}{\tau_0^2}eV(\sin\phi - \sin\phi_s). \tag{20.34}$$

Neglecting the time dependence of E_s and η, a first integral is

$$\frac{1}{2}\left(\frac{d\phi}{dt}\right)^2 + \frac{2\pi h\eta}{\tau_0^2}\frac{eV}{E}(\cos\phi + \phi\sin\phi_s) = \text{constant}. \tag{20.35}$$

The boundaries of phase stability lie between $\phi_1 = \pi - \phi_s$ and the solution ϕ_2 of $\cos\phi_2 + \phi_2\sin\phi_s = -\cos\phi_s + (\pi - \phi_s)\sin\phi_s$. The stable region in longitudinal phase space is termed a *bucket*.

For $\Delta\phi \equiv \phi - \phi_s \ll 1$, the equation of motion reduces to that of a simple harmonic oscillator. The number of synchrotron oscillations per turn is

$$\nu_s = \sqrt{-\frac{h\eta}{2\pi\beta_s^2 E_s}\,eV\cos\phi_s}. \tag{20.36}$$

This expression makes clear the relationship between transition and the synchronous phase; above transition, the slip factor is positive and so $\cos\phi_s$ must be negative.

For typical parameter sets, the synchrotron oscillation tune ν_s is two or more orders of magnitude smaller than the betatron oscillation tunes.

For slow variation of the factors in Eq. (20.36), the small oscillation amplitudes in energy and phase vary according to

$$\Delta\phi \propto k, \quad \Delta E \propto \frac{1}{\tau k}, \quad k \equiv \left(\frac{1}{E_s eV}\frac{-\eta}{\cos\phi_s}\right)^{\frac{1}{4}}. \tag{20.37}$$

As $\eta \to 0$ at the transition energy, adiabaticity fails; see Courant and Snyder [1].

Bucket area and longitudinal emittance

Contours in $\Delta E, \phi$ phase space follow from replacement of $d\phi/dt$ in Eq. (20.35) by ΔE using the first of the expressions in Eq. (20.33):

$$(\Delta E)^2 + \left(\frac{E_s eV}{\pi h \eta \beta_s^2}\right)[\cos\phi + \cos\phi_s + (\phi + \phi_s - \pi)\sin\phi_s] = 0. \tag{20.38}$$

The canonical conjugate to ΔE is

$$\Delta t \equiv \frac{\phi - \phi_s}{\omega_{\text{RF}}} = \frac{\phi - \phi_s}{2\pi h}\tau. \tag{20.39}$$

The longitudinal phase space area in ΔE, Δt coordinates, designated here by ε_L, is an adiabatic invariant, but shares with its transverse counterparts a susceptibility to dilution. In these units, the bucket area for $\phi_s = 0$ or $\phi_s = \pi$ (stationary bucket) is

$$\mathcal{A}_0 = \frac{16\tau}{2\pi h \beta_s}\left(\frac{E_s eV}{2\pi h \eta}\right)^{\frac{1}{2}}. \tag{20.40}$$

In this bucket, if $\widehat{\Delta\phi}$ is the maximum extent of the small oscillation, a beam bunch has a longitudinal emittance

$$\varepsilon_L = \pi \mathcal{A}_0 (\widehat{\Delta\phi})^2/16. \tag{20.41}$$

The bucket area and phase stability boundaries versus ϕ_s are shown in Figure 20.3.

20.2.3. Transverse coupling

The main sources of coupling between the transverse degrees of freedom are skew quadrupole terms arising from alignment errors of the quadrupole magnets that perform the alternating gradient focusing, and solenoid magnets associated with the experimental physics program. With these additions, Eqs. (20.1) become

$$x'' + K_x(s)x = S(s)y + R(s)y' + \tfrac{1}{2}R'(s)y,$$

$$y'' + K_y(s)y = S(s)x - R(s)y' - \tfrac{1}{2}R'(s)y,$$

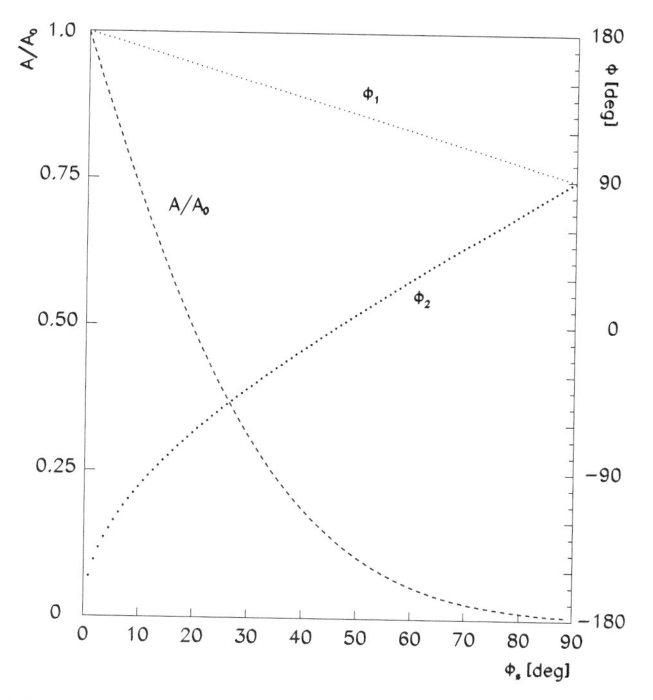

FIGURE 20.3. Ratio of bucket area to stationary bucket, and boundaries of phase stability, versus ϕ_S.

$$S = B'_{\text{skew}}/(B\rho), \quad R = B_s/(B\rho),$$
$$B' = \frac{\partial B_y}{\partial x}, \quad B'_{\text{skew}} = \frac{\partial B_x}{\partial x}. \tag{20.42}$$

The additional S and R terms arise from the coupling sources, skew quadrupole and solenoidal fields, respectively. Eqs. (20.42) are linear differential equations with periodic coeffcients and can be solved in a variety of ways (see Edwards and Teng [7], Ripken [8], and Wiedemann [3]). The matrices for transport through a skew quadrupole or a solenoid are

$$\begin{pmatrix} \frac{1}{2}(u+c) & \frac{1}{2\sqrt{K}}(v+s) & \frac{1}{2}(u-c) & \frac{1}{2\sqrt{K}}(v-s) \\ \frac{\sqrt{K}}{2}(v-s) & \frac{1}{2}(u+c) & \frac{\sqrt{K}}{2}(v+s) & \frac{1}{2}(u-c) \\ \frac{1}{2}(u-c) & \frac{1}{2\sqrt{K}}(v-s) & \frac{1}{2}(u+c) & \frac{1}{2\sqrt{K}}(v+s) \\ \frac{\sqrt{K}}{2}(v+s) & \frac{1}{2}(u-c) & \frac{\sqrt{K}}{2}(v-s) & \frac{1}{2}(u+c) \end{pmatrix} \tag{20.43}$$

and

$$\begin{pmatrix} c^2 & \frac{2}{K}sc & sc & \frac{2}{K}s^2 \\ -\frac{K}{2}sc & c^2 & -\frac{K}{2}s^2 & sc \\ -sc & -\frac{2}{K}s^2 & c^2 & \frac{2}{K}sc \\ \frac{K}{2}s^2 & -sc & -\frac{K}{2}sc & c^2 \end{pmatrix},$$

respectively. Here c, s, u, v are defined as in Eq. (20.19). For the skew quadrupole, $K = B'_{\text{skew}}/(B\rho)$ and for the solenoid, $K \equiv B_s/(B\rho)$. For elements of length L, the arguments of the circular and hyperbolic functions are $\sqrt{K}\,L$ for the skew quadrupole and $KL/2$ for the solenoid.

Half-integer resonances

If the unperturbed motion in x and y is given by

$$x = a\sqrt{\frac{\beta_x}{\beta_0}}\cos\psi_x, \quad y = b\sqrt{\frac{\beta_y}{\beta_0}}\cos\psi_y, \tag{20.44}$$

where β_0 is some convenient reference value of the amplitude function, and if the motion is perturbed by a distribution of coupling elements, then in lowest order, the first integrals of the motion are

$$a^2 + b^2 = \text{const.}, \qquad \nu_x - \nu_y = \text{integer}, \tag{20.45}$$

$$a^2 - b^2 = \text{const.}, \qquad \nu_x + \nu_y = \text{integer}. \tag{20.46}$$

For Eq. (20.45), the *difference resonance*, the sum of the squares of the amplitudes is constant and, therefore, the motion is bounded. Even though the motion is stable, operation at a difference resonance is usually avoided because of the difficult diagnosis and management of the coupled motion. But in Eq. (20.46), the *sum resonance*, the amplitudes may become arbitrarily large.

20.2.4. Nonlinear effects

The introduction of nonlinear fields into the particle path leads to equations of motion that are not soluble in closed form. Perturbative or numerical methods beyond the scope of this discussion must be used. Here, we limit ourselves to comment on two topics of significance under this heading.

Chromaticity and its control

Quadrupoles exhibit chromatic aberration; the K's in Eq. (20.1) vary inversely with momentum. The coefficient that expresses the variation of tune with particle momentum is the *chromaticity* ξ:

$$\Delta\nu = \xi(p)\frac{\Delta p}{p_0}. \tag{20.47}$$

The contribution to the chromaticity from the linear optics alone is

$$\xi = -\frac{1}{4\pi} \oint K\beta \, ds. \tag{20.48}$$

Chromaticity adjustment is carried out by use of sextupole magnets, in regions where $D \neq 0$. The sextupole magnet provides, in effect, a quadrupole term on the displaced equilibrium orbit. A single sextupole magnet of length l will contribute a change

$$\Delta\xi_x = \frac{1}{4\pi} D\beta \frac{B''l}{(B\rho)}, \qquad \Delta\xi_y = -\Delta\xi_x, \tag{20.49}$$

to the chromaticity. Chromaticity must be controlled for a variety of reasons: the tune spread arising from momentum spread in the beam may be undesirably large, avoidance of the head-tail instability requires slightly positive chromaticity, and so on. The price, however, is the introduction of intrinsically nonlinear fields into the accelerator, with consequent aberrations.

Resonances

Certain values of betatron tune must be avoided even if the equations of motion were linear. At an integer tune, any perturbation in the bending field of a synchrotron will result in an oscillation growing progressively from turn to turn. According to Eq. (20.46), half-integers (and the lines connecting them in the tune plane) are ruled out. With the introduction of nonlinear fields, the list of tunes at which motion may exhibit instability grows dramatically.

Transverse resonances

In the presence of a field perturbation $\Delta B(x, s)$, Eq. (20.1) for one degree of freedom becomes that of a driven harmonic oscillator through the *Floquet transformation*, $\zeta \equiv x/\sqrt{\beta}$ and $\phi \equiv \psi/\nu$:

$$\frac{d^2\zeta}{d\phi^2} + \nu^2\zeta = -\nu^2\beta^{3/2}\frac{\Delta B(\zeta, \phi)}{(B\rho)}. \tag{20.50}$$

Suppose that ΔB is due to a distribution of octopoles. Then,

$$\Delta B = \beta^{3/2}\frac{B'''(\phi)}{6}\zeta^3. \tag{20.51}$$

The factor $\beta^{3/2}B'''$ reflects the focusing structure. If it contains the nth harmonic in ϕ, then that *driving term* may combine with the 3ν-term in ζ^3 to produce a resonance at $4\nu = n$. Extension of this argument to other magnetic multipole terms in the right-hand side of Eq. (20.50) would suggest that resonances may exist for all tunes that are rational numbers.

In the case of two transverse degrees of freedom, the resonances are lines in the ν_x, ν_y plane of the form

$$M\nu_x + N\nu_y = P, \tag{20.52}$$

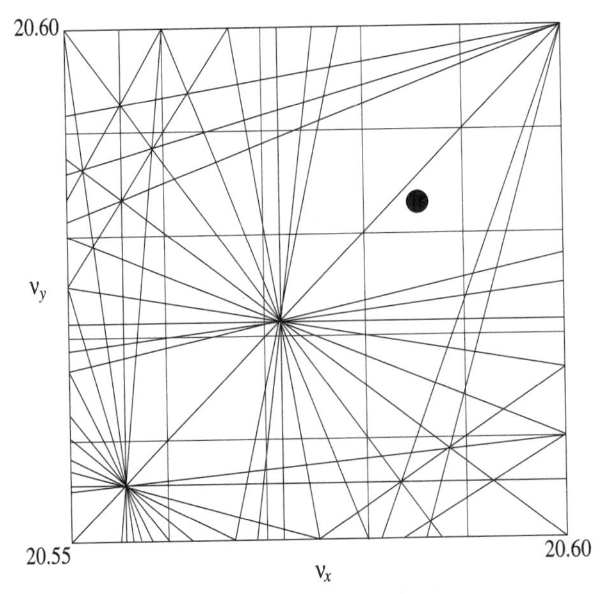

FIGURE 20.4. Resonance lines through 11th order in a 0.05 × 0.05 square of the tune plane. The Tevatron operates in the region shown as a circle in the upper right, between the fifth-order lines at 20.6 and the seventh-order cluster near $\nu_x = 20.57$ and slightly displaced from the $\nu_x = \nu_y$ coupling line.

where M, N, and P are integers of the same sign (one of the pair may be zero) for an unstable resonance. If M and N differ in sign, the motion, coupled, will be stable. The *order* of the resonance is $|M| + |N|$. In the example of the preceding paragraph, octopole fields led to a fourth-order resonance. In Figure 20.4 resonance lines of 11th order and below are drawn for a region in the tune plane, with the operating point of the Tevatron collider indicated. The perturbative argument above when carried to very high order would fill the tune diagram with resonance lines. In practice, achievable magnetic field quality, tune spread, tune variation with amplitude, and other processes not taken into account permit stable operation away from low-order resonances.

Unstable resonances may be turned to advantage in a process such as slow beam extraction. In Figure 20.5, a distribution of quadrupoles and octopoles is introduced into the otherwise stable focusing structure to establish a separatrix between an interior stable region and an unstable region in phase space. Extraction proceeds by slowly shrinking the phase-stable region onto the beam phase space. Near the fixed points, adiabaticity fails and particles stream away along the outgoing arms of the separatrix.

Synchro-betatron coupling

Synchro-betatron resonances may be excited when the tunes satisfy the relation

$$k\nu_x + l\nu_y + m\nu_s = n, \tag{20.53}$$

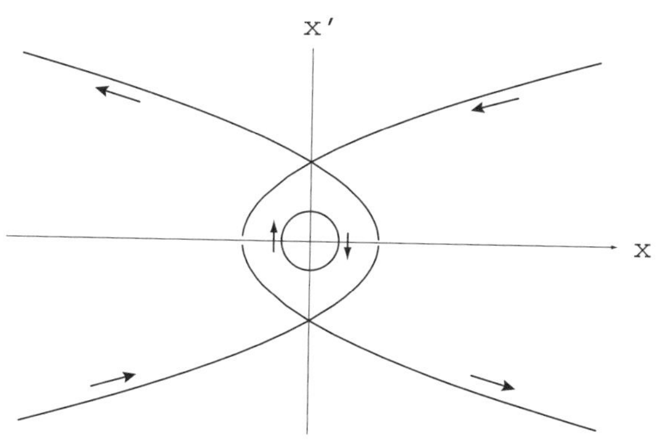

FIGURE 20.5. Phase space associated with resonant beam extraction. A distribution of quadrupoles and octopoles produces a separation into regions of stable and unstable motion. Gradual change in tune shrinks the stable region, and particles stream out along the outgoing extensions of the separatrix.

where k through n are integers, with $m \neq 0$. Coupling terms are provided by nonzero dispersion at RF cavities and by bunch-bunch collisions with a finite crossing angle, as examples (Piwinski [10]).

20.2.5. Synchrotron radiation

Radiation rate

A relativistic particle undergoing centripetal acceleration radiates at a rate given by the Larmor formula multiplied by the fourth power of the Lorentz factor:

$$P = \frac{1}{6\pi\varepsilon_0}\frac{e^2a^2}{c^3}\gamma^4 = \frac{1}{6\pi\varepsilon_0}\frac{e^2c}{\rho^2}\gamma^4 = \frac{1}{6\pi\varepsilon_0}\frac{e^4}{m^4c^5}B^2E^2. \tag{20.54}$$

In the second form, a has been replaced by the centripetal acceleration c^2/ρ for a particle of speed c. In the third form, $E = \gamma mc^2$ and B is the magnetic field producing the curvature of the particle path. In a synchrotron that has a constant radius of curvature in bending magnets, the radiation loss per turn for electrons is

$$U = 8.85 \times 10^{-5}\frac{E^4}{\rho} \quad \text{MeV per turn}, \tag{20.55}$$

where E is in GeV and ρ in km. Below, U will be reserved for the energy loss per turn in a synchrotron, and W will be used for radiated energy in general.

Bending magnet spectrum

$$\frac{d^2W}{dw\,d\Omega} = \frac{\alpha}{3\pi^2}\left(\frac{w\rho}{\hbar c}\right)^2 A^2 \times \left[K_{2/3}^2(\xi) + \frac{\theta^2}{A}K_{1/3}^2(\xi)\right],$$

$$\frac{dW}{d\Omega} = \frac{7}{16}\alpha\frac{\hbar c}{\rho}\frac{1}{A^{5/2}}\left[1+\frac{5}{7}\frac{\theta^2}{A}\right],$$

$$\frac{dW}{dw} = \alpha\sqrt{3}\,\gamma\frac{w}{w_c}\int_{w/w_c}^{\infty} K_{5/3}(v)\,dv, \tag{20.56}$$

where α is the fine-structure constant and

$$A \equiv \frac{1}{\gamma^2} + \theta^2,$$

$$\xi \equiv \frac{w\rho}{3\hbar c}A^{3/2},$$

$$w_c \equiv \frac{3}{2}\gamma^3\left(\frac{\hbar c}{\rho}\right),$$

and θ is the angle perpendicular to the bend plane. The quantity w_c is termed the *critical energy* of the spectrum. As in other relativistic radiation processes, the characteristic angle of synchrotron radiation is $1/\gamma$. The shape of the energy spectrum is shown in Figure 20.6, illustrating the falloff above the critical energy.

Undulator radiation

In order to achieve a near-line spectrum, the electrons are caused to make transverse oscillations perpendicular to the direction of motion in an *undulator magnet*. In a helical undulator, the transverse magnetic fields are described approximately by

$$B_x = B_0 \sin(2\pi s/\lambda_p),$$

$$B_y = B_0 \cos(2\pi s/\lambda_p), \tag{20.57}$$

with γ_p characterizing the periodicity of the magnetic field. A particle describes helical motion with transverse momentum at an angle $\theta = B_0\lambda_p/2\pi(B\rho)$ with respect to a straight path. The dimensionless undulator parameter K is defined as θ/γ. That is, for $K < 1$, the radiation is within the cone of the characteristic angle, and one expects radiation of

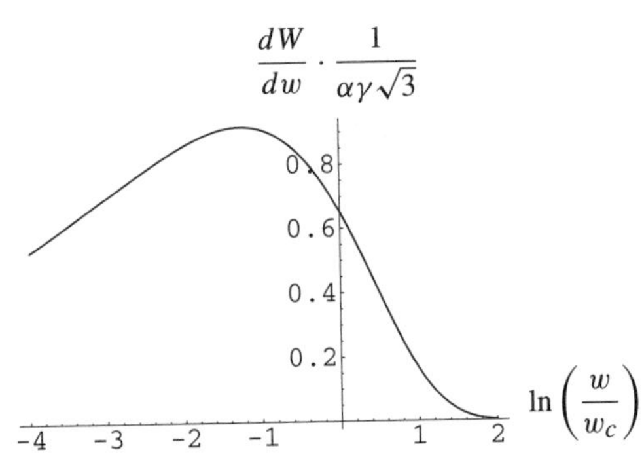

FIGURE 20.6. Energy spectrum of synchrotron radiation in the neighborhood of the critical energy w_c.

wavelength λ_p/γ to be produced in the rest frame of the electron. In the laboratory frame, the Doppler shift leads to a wavelength

$$\lambda = \frac{\lambda_p}{2\gamma_z^2}, \tag{20.58}$$

where γ_z is the Lorentz factor related to speed projected on the z axis. Expressed in terms of the Lorentz factor of the particle, the result is

$$\lambda = \frac{\lambda_p}{2\gamma^2}\left(1 + \frac{\theta^2}{\gamma^2}\right) = \frac{\lambda_p}{2\gamma^2}(1 + K^2). \tag{20.59}$$

In the planar, or flat, undulator one of the transverse field components is absent, and the resulting wavelength is

$$\lambda = \frac{\lambda_p}{2\gamma^2}\left(1 + \frac{K^2}{2}\right). \tag{20.60}$$

As K increases beyond 1, the transverse velocity becomes relativistic, and so the oscillation can no longer be described as simple harmonic motion; thus, harmonics appear in the radiation. For $K \gg 1$, the many harmonics merge into the continuous spectrum of the bend magnet. The device in this transition phase is called a *wiggler*.

The flux $\mathcal{F}$ is the number of photons per second in a fractional energy interval $\Delta w/w$. For a planar undulator of N periods traversed by a beam of current I, the flux in the central cone at the nth harmonic is

$$\begin{aligned}
\mathcal{F}^n &= \frac{\pi}{2}\alpha N \frac{\Delta w}{w}\frac{I}{e}(1 + K^2/2)\frac{F_n(K)}{n}, \\
F_n(K) &= \frac{K^2 n^2}{(1 + K^2/2)^2}\left[J_{\frac{n-1}{2}}(u) - J_{\frac{n+1}{2}}(u)\right]^2, \\
u &\equiv \frac{nK^2/4}{1 + K^2/2}.
\end{aligned} \tag{20.61}$$

The *brightness* (or *brilliance*) $\mathcal{B}$ is the phase-space density of the flux at the center of the cone:

$$\mathcal{B} = \frac{\mathcal{F}}{4\pi^2 \sigma_{Tx}\sigma_{Tx'}\sigma_{Ty}\sigma_{Ty'}}, \tag{20.62}$$

where

$$\begin{aligned}
\sigma_{Tx} &= \sqrt{\sigma_r^2 + \sigma_x^2}, \qquad & \sigma_{Tx'} &= \sqrt{\sigma_{r'}^2 + \sigma_{x'}^2}, \\
\sigma_{Ty} &= \sqrt{\sigma_r^2 + \sigma_y^2}, \qquad & \sigma_{Ty'} &= \sqrt{\sigma_{r'}^2 + \sigma_{y'}^2}.
\end{aligned} \tag{20.63}$$

The σ_T's represent the convolution between the diffractive limit for single particle emission (the σ_r's) and σ's related to the beam emittance:

$$\sigma_r = \frac{\sqrt{2\lambda L}}{4\pi}, \quad \sigma_{r'} = \sqrt{\frac{\lambda}{2L}}, \tag{20.64}$$

where λ is the wavelength at the harmonic of interest and L is the undulator length. The achievement of low beam emittance is critical to the production of high brightness. Figure 20.7 illustrates the peak brightness associated with various synchrotron radiation facilities, either existing or in preparation [10]. For the highest brightnesses in the figure, the undulator is a free-electron laser (FEL), wherein the brightness is enhanced as the laser action causes "micro bunches" of size comparable with a wavelength to form within a beam bunch, thereby permitting coherent radiation.

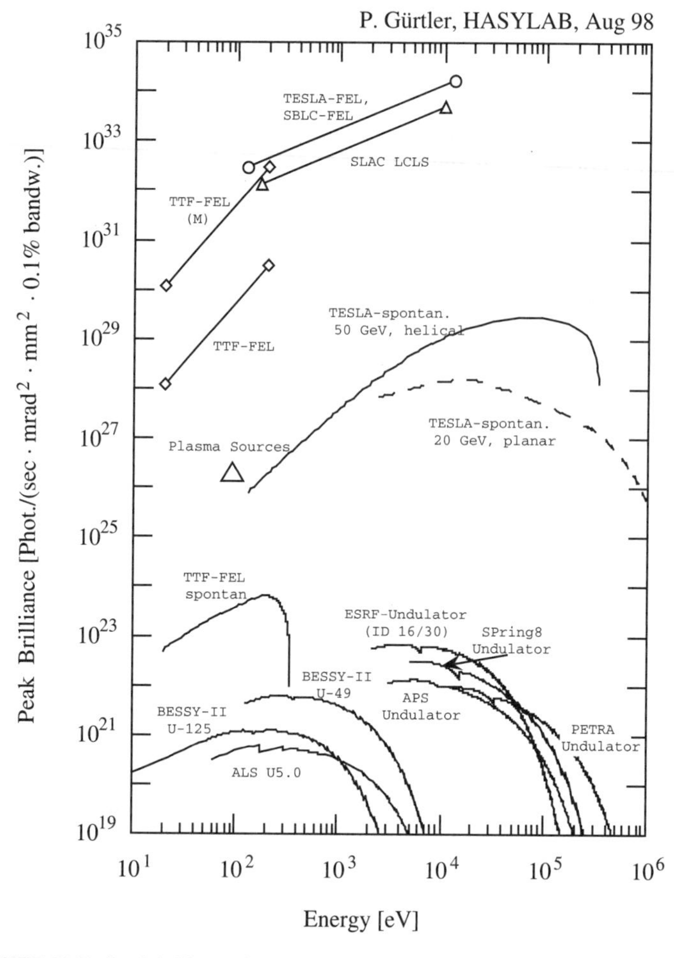

FIGURE 20.7. Peak brilliance for various sources. (Reproduced with permission.)

20.3. MULTIPARTICLE DYNAMICS

20.3.1. Space charge

Tune shifts

In the presence of space charge, the equation of motion in one transverse degree of freedom is

$$x'' + K(s)x = \frac{1}{\gamma m v^2} \times (\text{space charge force}). \tag{20.65}$$

For a round Gaussian beam having N particles of charge e per unit length with rms radius σ, the force is given by

$$F = \frac{e^2 N}{2\pi \varepsilon_0 \gamma^2 r}\left[1 - e^{-r^2/2\sigma^2}\right]. \tag{20.66}$$

The factor of γ^2 in the denominator reflects the offsetting effects of electric and magnetic fields.

For $r \ll \sigma$ in Eq. (20.66), the force is linear in displacement and the perturbation in tune is given by

$$\Delta\nu = \frac{N r_0 R}{2\varepsilon_N (v/c)\gamma^2}, \tag{20.67}$$

where R is the average radius of the accelerator, $\varepsilon_N = \sigma^2(\gamma v/c)/\beta$, and r_0 is the classical radius of the particle. Since particles at large amplitudes experience little change, this result expresses a tune range, often called an *incoherent* tune shift.

Charges and currents induced in the walls of the beam enclosure can produce tune shifts experienced by all of the particles in the beam, *coherent* tune shifts. For example, particles of a beam displaced a distance X in a circular conducting beam pipe of radius a will experience a force gradient

$$\frac{\partial eE}{\partial x} = \frac{e^2 N}{2\pi\varepsilon_0}\frac{X^2}{a^4}. \tag{20.68}$$

Typically, magnetic fields penetrate the chamber walls and the treatment of the magnetic images must include not only the external magnetic environment, but also the time constants for field penetration through conducting materials.

Intrabeam scattering

Multiple Coulomb scattering within a beam will lead to changes in the emittances during a store. This is a relatively slow process, and it is of importance in proton or heavy-ion storage rings. For a beam of nominal momentum p_0, the rms momentum spreads in the beam rest frame are $\sigma_{x'} p_0$, $\sigma_{y'} p_0$, and $\sigma_\delta p_0/\gamma$, with $\delta \equiv \Delta p/p$. The motion is nonrelativistic in the rest frame; for example, in the case of the Tevatron at 1 TeV, these momenta are 2 MeV/c, 2 MeV/c, and 0.14 MeV/c, respectively. Scattering would tend to transfer momentum from the transverse degrees of freedom to the longitudinal. However, since excitation of synchrotron motion is coupled to the bend plane through the dispersion function, for D large enough betatron amplitudes in the bend plane will also grow.

In the approximation of smooth focusing, an invariant is given by

$$-\eta\sigma_\delta^2 + \sigma_{x'}^2 + \sigma_{y'}^2 = \text{constant}, \tag{20.69}$$

where η is the slip factor defined in Eq. (20.31) (Piwinski [11]). Below transition energy, $\eta < 0$ and the sum is bounded; above transition, such is not the case. Expressions for the emittance growth rates are also given by Piwinski [11].

Envelope equations

An envelope equation describes the propagation of the beam size through the focusing structure. Equation (20.2) states that in the absence of space charge, $a\sqrt{\beta}$ traces out the maximum displacement of a particle. The equation satisfied by $w \equiv \sqrt{\beta}$ is

$$w^3(w'' + Kw) = 1. \tag{20.70}$$

As in Sec. 20.2.2, $\varepsilon \equiv a^2$ may be used as a parameter characterizing the phase space area that encloses the largest-amplitude particle. Multiplying Eq. (20.70) through by a and defining $aw = x_{\max}$ yields

$$x''_{\max} + Kx_{\max} = \frac{\varepsilon^2}{x^3_{\max}}. \tag{20.71}$$

Kapchinskij and Vladimirskij [12] extended the idea of an envelope equation to include space charge. Sacherer [13] showed that if the beam size were characterized as the rms displacement in each transverse coordinate, the resulting K-V style equations were essentially independent of the specific particle distribution function. With $\tilde{x} \equiv \sqrt{\langle x^2 \rangle}$, the result is

$$\tilde{x}'' + K(s)\tilde{x} - \frac{\varepsilon^2}{\tilde{x}^3} - \frac{\langle xF_s \rangle}{pv_s\tilde{x}} = 0, \tag{20.72}$$

where ε is defined in Eq. (20.26) and F_s is the x component of the self-force due to space charge.

For a ribbon beam having no y dependence and N particles per unit area projected onto the y-s plane, Eq. (20.72) can be written as

$$\tilde{x}'' + K(s)\tilde{x} - \frac{\varepsilon^2}{\tilde{x}^3} - \frac{e^2N}{2\varepsilon_0 m v_s^2 \gamma^3}\lambda = 0, \tag{20.73}$$

with the dimensionless parameter λ given by

$$\lambda \equiv \frac{2}{\tilde{x}} \int_{-\infty}^{\infty} xh(x)\,dx \int_0^x h(v)\,dv, \tag{20.74}$$

where h, normalized to unity, gives the shape of the distribution in x. For uniform, parabolic, Gaussian, and hollow ($h \propto x^2 \exp[-x^2/2]$) distributions, $\lambda = \sqrt{3}$ to within 2%.

In the case of an unbunched beam of elliptical cross section, the result corresponding to Eq. (20.73) is

$$\tilde{x}'' + K_x(s)\tilde{x} - \frac{\varepsilon_x^2}{\tilde{x}^3} - \frac{e^2 N}{4\pi\varepsilon_0 m v_s^2 \gamma^3}\frac{1}{\tilde{x}+\tilde{y}} = 0, \tag{20.75}$$

where N is the number of particles per unit length in s, with the equation in y obtained by interchange of x and y. The result assumes only that the particle distribution is of the form

$$n[R(x,y),t] = n\left[\frac{x^2}{a^2} + \frac{y^2}{b^2}, t\right]. \tag{20.76}$$

Finally, for a bunched beam of ellipsoidal symmetry, with distribution

$$n[R(x,y,s),t)] = n\left[\frac{x^2}{a^2} + \frac{y^2}{b^2} + \frac{s^2}{c^2}, t\right] \tag{20.77}$$

the envelope equation becomes

$$\tilde{x}'' + K_x(s)\tilde{x} - \frac{\varepsilon_x^2}{\tilde{x}^3} - \frac{e^2 N}{4\pi\varepsilon_0 m v_s^2 \gamma^3}\frac{1}{\tilde{x}^2} g_x\left(\frac{\tilde{y}}{\tilde{x}}, \frac{\tilde{s}}{\tilde{x}}\right)\lambda = 0, \tag{20.78}$$

where

$$g_x(u,v) = \frac{3}{2}\int_0^\infty \frac{ds}{(1+s)^{3/2}\sqrt{u^2+s}\,\sqrt{v^2+s}}. \tag{20.79}$$

The parameter λ depends on the distribution shape in the same spirit as in Eq. (20.73). Over the four distributions used as test cases, λ varies by no more than 5% from a value of $5^{3/2}$. Therefore, for many purposes, these equations can be used with confidence. In particular, computer programs for envelope propagation may include space charge kicks.

20.3.2. Collective instabilities

The charged particles in a beam set up charges and currents in the chambers through which they pass; these induced sources of electromagnetic fields in turn act back on the beam particles, and an instability may result. The literature on this diverse subject is extensive. Here we can only illustrate the flavor; the textbook by Chao [4] may be consulted for in-depth coverage.

Wake functions

A beam element may be described in terms of its azimuthal moments: a charge distributed over a thin ring of radius a according to $\cos m\theta$. That is, the charge density representing a multipole of order m is

$$\rho = \frac{Q_m}{\pi a^{m+1}(1+\delta_{0,m})}\,\delta(s-ct)\,\delta(r-a)\cos m\theta, \tag{20.80}$$

where Q_m is the multipole coeffcient. Note that the ultrarelativistic approximation is in use and

$$Q_m = \int \rho r^m \cos m\theta r \, dr \, d\theta \, ds. \tag{20.81}$$

Instability growth rate is slow compared with rapid fluctuations in the fields encountered by a spectator charge following the source of the disturbance; the result of averaging within the forces over the repetition length of the structure yields a remarkably simple result for the force on a spectator charge a distance Δs behind the charge distribution:

$$\begin{aligned} F_r &= -eQ_m m r^{m-1} \cos m\theta W_m(\Delta s), \\ F_\theta &= eQ_m m r^{m-1} \sin m\theta W_m(\Delta s), \\ F_z &= -eQ_m r^m \cos m\theta W'_m(\Delta s). \end{aligned} \tag{20.82}$$

Here, r and θ are the transverse coordinates of the spectator charge, and the W's are the *wake functions*. Often, W and W' are called the *transverse* and *longitudinal* wake functions, respectively. The solutions of Maxwell's equations that give the field distributions induced in the various structures through which a beam passes are contained within the wake functions and generally obtained by numerical methods.

In some simple geometries, analytical solutions are possible. The case of a resistive beam pipe of constant radius is one such solution. The wake function W_1 is given by

$$W_1 = -\frac{1}{\pi b^3}\sqrt{\frac{2\pi c}{\varepsilon_0 \sigma}}\frac{1}{\Delta s^{1/2}}, \tag{20.83}$$

where b is the radius of the beam pipe of conductivity σ. This result is used below in an example of a collective instability.

Macroparticle models

Insight can be gained into the physics of certain collective instabilities through a model in which a beam bunch consists of just two macroparticles. The transverse instability in linacs called *beam breakup* can be treated in an approximate fashion as follows. The first macroparticle, containing $N/2$ particles, is represented by an $m = 1$ charge distribution undergoing a betatron oscillation $x_1 = \hat{x}_1 \cos \omega_\beta t$. The force in the x direction on each constituent of the trailing macroparticle is $eQ_1 W_1(s)$, according to Eq. (20.82). In a high-energy linac, the synchrotron oscillation frequency is suffciently low that the longitudinal distance between particles is essentially fixed; so W_1 is a constant throughout the motion. With the use of Eq. (20.81),

$$Q_1 = \int \frac{Ne}{2}\delta(x - x_1)\,\delta(y)\,\delta(s) x \, dx \, dy \, ds = \frac{Ne}{2}x_1. \tag{20.84}$$

The equation of motion for the second macroparticle is then

$$\ddot{x}_2 + \omega_\beta^2 x_2 = \frac{-Ne^2 W_1}{2\gamma m}\hat{x}_1 \cos \omega_\beta t. \tag{20.85}$$

This is the equation of a harmonic oscillator driven on resonance, and so the amplitude of the trailing macroparticle will increase linearly with time. Inclusion of acceleration in the foregoing treatment will reduce the time dependence of the instability to logarithmic through the adiabatic damping of the betatron oscillations.

The instability may be avoided by moving off resonance. If ω_1 and ω_2 are the transverse oscillation frequencies of the head and the tail, respectively, the solution of the equation of motion for the trailing macroparticle is

$$x_2 = \hat{x}\cos\omega_2 t - \frac{Ne^2W_1}{2(\omega_2^2 - \omega_1^2)\gamma m}\hat{x}(\cos\omega_2 t - \cos\omega_1 t), \tag{20.86}$$

in which it is assumed that the two macroparticles have the same initial amplitude and phase. If the condition

$$\frac{Ne^2W_1}{2(\omega_2^2 - \omega_1^2)\gamma m} = -1 \tag{20.87}$$

is met, the tail just follows the head, and there is no instability. If the head and the tail of the bunch differ suffciently in energy, chromaticity provided by the transverse focusing may provide the needed tune split. This method is known as *BNS damping*, so-named after its inventors Balakin, Novokhatsky, and Smirnov [14].

20.3.3. Beam cooling

Synchrotron radiation cooling

Oscillations in all three degrees of freedom may exhibit adiabatic damping in a synchrotron; the focusing structure of storage rings is designed so that is the case. Quantum fluctuations in the radiation rate excite oscillations; therefore, an equilibrium emittance reflects a balance between damping and excitation.

The characteristic time for synchrotron radiation processes is the time during which the energy must be replenished by the acceleration system. If f_0 is the orbit frequency, this characteristic time is

$$\tau_0 = \frac{E}{f_0 U}, \tag{20.88}$$

where the radiated energy is given by Eq. (20.55).

Robinson's theorem [15] relates the damping time constants for the three degrees of freedom according to

$$\frac{1}{\tau_x} + \frac{1}{\tau_y} + \frac{1}{\tau_s} = \frac{2}{\tau_0}, \tag{20.89}$$

and the time constants are

$$\tau_x = \frac{2\tau_0}{2+\mathcal{D}}, \quad \tau_y = 2\tau_0, \quad \tau_s = \frac{2\tau_0}{1-\mathcal{D}}. \tag{20.90}$$

The function $\mathcal{D}$ is given by

$$\mathcal{D} \equiv \frac{\left\langle \frac{D}{\rho^2}\left(\frac{1}{\rho} + 2\frac{B'}{B}\right)\right\rangle}{\left\langle \frac{1}{\rho^2}\right\rangle}. \tag{20.91}$$

For a separated-function synchrotron, in which the focusing and bending are performed by separate elements, then $\mathcal{D} \approx 0$, and so for this case, $\tau_x \approx 2\tau_0$ and all three degrees of freedom damp.

The inclusion of the oscillation excitation due to the quantized character of the radiation leads to the following equilibrium values:

$$\varepsilon_x \equiv \frac{\sigma_x^2}{\beta_x} = \frac{55\sqrt{3}}{2^4 3^2}\left(\frac{\langle\mathcal{H}\rangle}{1-\mathcal{D}}\right)\frac{w_c}{E},$$

$$\varepsilon_y \equiv \frac{\sigma_y^2}{\beta_y} = 0,$$

$$\sigma_E/E = \left[\frac{55\sqrt{3}}{2^4 3^2}\left(\frac{1}{2+\mathcal{D}}\right)\frac{w_c}{E}\right]^{1/2}, \tag{20.92}$$

where

$$\mathcal{H} \equiv \gamma D^2 + 2\alpha D D' + \beta D'^2; \tag{20.93}$$

that is, $\mathcal{H}$ is the Courant-Snyder invariant of the dispersion function.

These idealized results imply that the vertical beam size is zero. In reality, some portion of the transverse emittance will be coupled into the vertical degree of freedom, as will some small part of the horizontal dispersion. Though the vertical beam size will not be zero, for a corrected lattice, it will be an order of magnitude or more smaller than the horizontal beam size. As a result, the words "ribbon beam" are often used to describe the transverse bunch cross section.

Stochastic cooling

The circumstance in which a beam contains a finite number of particles means that fluctuations in any of the degrees of freedom can be detected and compensated with a gentle "tap" back toward zero. The generic arrangement is that a signal picked up at one point on the ring cuts across on a chord so that an appropriate kick can be generated by the time the particles appear after their longer path.

Transverse (betatron) cooling illustrates the principle. The centroid position $\langle x\rangle$ of a sample of N_s particles is sensed at a pickup and conveyed across the ring to a kicker located an odd number of quarter wavelengths in betatron phase downstream of the pickup. A deflection is given to each particle of the sample such as to initiate a betatron oscillation of amplitude $g\langle x\rangle$, where g is a dimensionless gain.

If, between the kicker and its next appearance at the pickup, the sample rapidly exchanges particles with other samples, a centroid displacement will be regenerated. The process continues with emittance reduction of the sample proceeding according to

$$\frac{1}{\tau} \equiv -\frac{1}{\varepsilon}\frac{d\varepsilon}{dt} = \frac{2g - g^2}{N_s T} = \frac{2W}{N}(2g - g^2). \tag{20.94}$$

In the last form of the result, use has been made of the relationship among the sample size N_s, the total number of particles N uniformly distributed in the storage ring, the bandwidth W of the cooling system, and the revolution period T: $2TWN_s = N$.

To the above, add the effect of system noise with the assumption that the noise is equivalent to a position error x_n at the pickup. The parameter $U \equiv \langle x_n^2 \rangle / \langle x \rangle^2$ will express the influence of noise. Also, introduce the mixing parameter M, the number of revolutions for a particle of one sample to move to the next due to the variation of the revolution period with momentum. For "perfect mixing," $M = 1$. With these changes, the cooling rate becomes

$$\frac{1}{\tau} = \frac{2W}{N}[2g - g^2(M + U)]. \tag{20.95}$$

In longitudinal stochastic cooling, the pickup and subsequent signal processing detects the mean momentum of a sample and the kicker delivers an acceleration or deceleration as desired. This is the technique initially employed at the SPS at CERN and, subsequently, at the FNAL Tevatron to play the dual roles of longitudinal emittance reduction and antiproton accumulation. A fundamental reference is Möhl *et al.* [16].

Electron cooling

Cooling of a positive-ion beam in a storage ring can be accomplished by the overlap of the ion beam and an electron beam moving at the same speed over a portion of the ring. The "hot" ion gas is cooled by the "cold" electrons. In the particle frame, a "temperature" $T_i \equiv \langle p_i^2 \rangle / 2m$ may be defined for each degree of freedom. A thermionic cathode at 1000 K will produce an electron beam with $T_{x,y,s} \approx 0.1$ eV. Upon acceleration to 1 GeV, the transverse temperatures will remain unchanged while T_s drops to about 10^{-6} eV. In contrast, even a relatively low-emittance proton beam at 1 GeV with $\varepsilon_{x,y} \equiv \sigma\sigma' \approx 4 \times 10^{-6}$ m will have $T_{x,y} \approx 10$ eV.

For the theoretical analysis, see, for example, the article by Sørensen and Bonderup [17]. Here, we note only the scaling of the process with energy. For a proton with speed v_p greater than that of the electrons, both measured in the electron rest frame, the slowing-down time as observed in the laboratory is

$$\tau_L \equiv \frac{v_p}{dv_p/dt} = \frac{4\pi\varepsilon_0^2}{e^4}\frac{\gamma^2 mM}{FLn_L}v_p^3, \tag{20.96}$$

where M and m are the proton and electron masses, ε_0 is the permittivity of free space, n_L is the number density of the electrons as measured in the laboratory, and F is the fraction of the ring occupied by cooling, and the logarithmic factor L is given by

$$L = \ln\left(\frac{4\pi\varepsilon_0 m v_p^3}{e^2\omega_p}\right) - \frac{1}{2}. \tag{20.97}$$

In L, the plasma frequency $\omega_p = (ne^2/\varepsilon_0 m)^{1/2}$ is calculated for the electron density in the electron frame.

For a proton with transverse velocity v_x in the electron frame, Eq. (20.96) becomes

$$\tau_L = \frac{4\pi\varepsilon_0^2}{e^4}\frac{mM}{FLn_L}c^3\beta^3\gamma^5(x')^3, \tag{20.98}$$

from which it is clear that the steep dependence on γ favors low-energy cooling.

Ionization cooling

Ionization cooling has been proposed as a method for cooling large-emittance muon fluxes for a muon storage ring or collider. The principle closely resembles that of synchrotron radiation cooling, with absorbers replacing the photon emission process. An absorber reduces momentum in the direction of motion of the muon, while RF cavities provide momentum in the desired direction. The result is a net transverse damping. Longitudinal damping can be provided by a wedge-shaped absorber in a dispersive region, again in analogy with the corresponding synchrotron radiation process (see Neuffer [18]).

The short muon lifetime implies a considerable compression of the energy-loss-and-reacceleration process compared with that in an electron storage ring. With absorbers and accelerating cavities of suffciently short periodicity, the damping process can be described by differential equations. A characteristic time may be given by $\tau_0 \equiv E/cE'$, where E is the muon energy and E' is the energy loss due to ionization per unit length, appropriately reduced by the fraction of the path length occupied by the absorber. The muon speed is taken to be c.

The time constants for damping are given by

$$\tau_{\varepsilon_y} = \tau_0, \quad \tau_{\varepsilon_x} = \frac{\tau_0}{1 - \frac{D\delta'}{\delta_0}},$$
$$\tau_{\Delta E} = \frac{\tau_0}{\frac{d\ln E'}{d\ln E} + \frac{D\delta'}{\delta_0}}, \tag{20.99}$$

where the wedge absorbers are located at positions having dispersion function D and they vary in thickness with x according to $\delta = \delta_0 + \delta' x$. The time constants satisfy an invariant resembling Robinson's theorem:

$$\frac{1}{\tau_{\varepsilon_x}} + \frac{1}{\tau_{\varepsilon_y}} + \frac{1}{\tau_{\Delta E}} = 2 + \frac{d\ln E'}{d\ln E}. \tag{20.100}$$

20.3.4. Luminosity

The event rate R in a collider is proportional to the interaction cross section σ_{int}, and the factor of proportionality is the *luminosity*:

$$R = \mathcal{L}\sigma_{\text{int}}. \tag{20.101}$$

For two bunches containing n_1 and n_2 particles colliding with frequency f, the luminosity is

$$\mathcal{L} = f\frac{n_1 n_2}{4\pi\sigma_x\sigma_y} = \gamma f\frac{n_1 n_2}{4\pi\varepsilon_N(\beta_x^*\beta_y^*)^{1/2}}, \tag{20.102}$$

where σ_x and σ_y characterize Gaussian beam profiles in the two degrees of freedom transverse to the collision axis, and, in the second form, the β^*s are the amplitude functions at the interaction point. This version of Eq. (20.102) assumes that the two particle species have $v \approx c$, the same γ, and the same normalized emittance and that the beam cross sections are not altered in a collision.

In a circular collider, the luminosity is often limited by the beam-beam tune shift. One bunch presents a (nonlinear) lens to particles of the other. From Eq. (20.7), the resulting tune spread is $\Delta\nu = nr_0/(4\pi\varepsilon_N)$, where r_0 is the classical radius of the particle. The limiting value is in the range of 0.06–0.07 or so for electron-positron colliders, and over an order of magnitude less in the hadron-hadron case.

In linear collider designs, Eq. (20.102) is multiplied by the *disruption enhancement factor* H_D due to the pinch effect during the bunch-bunch passage. A gain of up to a factor of 2 has been achieved at the SLC. An important constraint is the energy spread due to *beamstrahlung*; the particles emit hard synchrotron radiation in the strong space charge field of the opposing bunch, resulting in a spread in the collision energy. This spread, δ_E, varies as $(\sigma_x + \sigma_y)^{-2}$. By choosing $\sigma_x \gg \sigma_y$, δ_E becomes independent of the vertical beam size and σ_y controls the luminosity.

20.4. REFERENCES

[1] E. D. Courant and H. S. Snyder, Ann. Phys. (New York) **3**, 1 (1958).

[2] D. A. Edwards and M. J. Syphers, *An Introduction to the Physics of High Energy Accelerators* (John Wiley and Sons, New York, 1993).

[3] H. Wiedemann, *Particle Accelerator Physics, I & II* (Springer-Verlag, Heidelberg, 1993 and 1995).

[4] A. W. Chao, *Physics of Collective Beam Instabilities in High Energy Accelerators* (John Wiley and Sons, New York, 1993).

[5] M. Reiser, *Theory and Design of Charged Particle Beams* (John Wiley and Sons, New York, 1994).

[6] A. W. Chao and M. Tigner (Eds.), *Handbook of Accelerator Physics and Engineering* (World Scientific Publishing Co., Singapore, 1999).

[7] D. A. Edwards and L. C. Teng, IEEE Trans. Nucl. Sci. **NS-20**, 885 (1973).

[8] G. Ripken, DESY Report R1-70/04, 1970.

[9] A. Piwinski, in *Synchro-betatron Resonances* (CERN Accelerator School) CERN-87-03, p. 187.

[10] P. Gürtler and J. Rossbach (private communication).

[11] A. Piwinski, in Proceedings of the 9th International Conference on High Energy Particle Acceleration, 1974, p. 405.

[12] I. M. Kapchinskij and V. V. Vladimirskij, Proceedings of the 2nd Int. Conf. on High Energy Acceleration and Inst., CERN, 1959, p. 274.

[13] F. Sacherer, IEEE Trans. Nucl. Sci. **NS-18**, 1105 (1971).

[14] V. Balakin, A. Novokhatsky, and V. Smirnov, Proc. 12th Int. Conf. High Energy Accel., Fermilab, 1983, p. 119.

[15] K. W. Robinson, Phys. Rev. **111**, 373 (1958).

[16] D. Möhl, G. Petrucci, L. Thorndahl, and S. van der Meer, Phys. Rep. **58**, 73 (1980).

[17] A. H. Sørensen and E. Bonderup, Nucl. Instrum. Methods **215**, 27 (1983).

[18] D. V. Neuffer, in Workshop on Beam Cooling and Applications, CERN 94-03, p. 49.

21

Plasma Physics

David L. Book
Naval Postgraduate School, Monterey, CA

Contents

List of Tables

List of Figures

21.1. FUNDAMENTAL PLASMA PARAMETERS

All quantities are in Gaussian cgs units except temperature (T, T_e, T_i), expressed in eV, and ion mass (m_i), expressed in units of the proton mass, $\mu = m_i/m_p$; Z is the charge state, K is wavenumber, γ is the adiabatic index, $\ln \Lambda$ is the Coulomb logarithm, k is Boltzmann's constant, $k = 1.38 \times 10^{-16}$ erg/K, and e is the elementary charge, $e = 4.803 \times 10^{-10}$ esu.

21.1.1. Frequencies

Electron gyrofrequency:

$$\omega_{ce} = eB/m_e c = 1.76 \times 10^7 B \text{ rad/s},$$

$$f_{ce} = \omega_{ce}/2\pi = 2.80 \times 10^6 B \text{ Hz}.$$

Ion gyrofrequency:

$$\omega_{ci} = ZeB/m_i c = 9.58 \times 10^3 Z\mu^{-1} B \text{ rad/s},$$

$$f_{ci} = \omega_{ci}/2\pi = 1.52 \times 10^3 Z\mu^{-1} B \text{ Hz}.$$

Electron plasma frequency:

$$\omega_{pe} = (4\pi n_e e^2/m_e)^{1/2} = 5.64 \times 10^4 n_e{}^{1/2} \text{ rad/s},$$

$$f_{pe} = \omega_{pe}/2\pi = 8.98 \times 10^3 n_e{}^{1/2} \text{ Hz}.$$

Ion plasma frequency:

$$\omega_{pi} = (4\pi n_i Z^2 e^2/m_i)^{1/2} = 1.32 \times 10^3\, Z(n_i/\mu)^{1/2} \text{rad/s},$$

$$f_{pi} = \omega_{pi}/2\pi = 2.10 \times 10^2 Z\mu^{-1/2} n_i{}^{1/2} \text{ Hz}.$$

Electron trapping rate:

$$\nu_{Te} = (eKE/m_e)^{1/2} = 7.26 \times 10^8 (KE)^{1/2} \text{ s}^{-1}.$$

Ion trapping rate:

$$\nu_{Ti} = (ZeKE/m_i)^{1/2} = 1.69 \times 10^7 (ZKE/\mu)^{1/2} \text{ s}^{-1}.$$

Electron collision rate:

$$\nu_e = 2.91 \times 10^{-6} n_e \ln\Lambda\, T_e{}^{-3/2} \text{ s}^{-1}$$

Ion collision rate:

$$\nu_i = 4.80 \times 10^{-8} Z^4 \mu^{-1/2} n_i \ln\Lambda\, T_i{}^{-3/2} \text{ s}^{-1}.$$

21.1.2. Lengths

Electron deBroglie length:

$$\lambda\!\!\!^{-} = \hbar/(m_e k T_e)^{1/2} = 2.76 \times 10^{-8} T_e{}^{-1/2} \text{ cm}.$$

Classical distance of minimum approach:

$$e^2/kT = 1.44 \times 10^{-7} T^{-1} \text{ cm}.$$

Electron gyroradius:

$$r_e = v_{Te}/\omega_{ce} = 2.38 T_e^{1/2} B^{-1} \text{ cm}.$$

Ion gyroradius:

$$r_i = v_{Ti}/\omega_{ci} = 1.02 \times 10^2 \mu^{1/2} Z^{-1} T_i^{1/2} B^{-1} \text{ cm}.$$

Plasma skin depth:

$$c/\omega_{pe} = 5.31 \times 10^5 n_e^{-1/2} \text{ cm}.$$

Debye length:

$$\lambda_D = (kT/4\pi n e^2)^{1/2} = 7.43 \times 10^2 T^{1/2} n^{-1/2} \text{ cm}.$$

21.1.3. Velocities

Electron thermal velocity:

$$v_{Te} = (kT_e/m_e)^{1/2} = 4.19 \times 10^7 T_e^{1/2} \text{ cm s}^{-1}.$$

Ion thermal velocity:

$$v_{Ti} = (kT_i/m_i)^{1/2} = 9.79 \times 10^5 (T_i/\mu)^{1/2} \text{ cm s}^{-1}.$$

Ion sound velocity:

$$C_s = (\gamma Z k T_e/m_i)^{1/2} = 9.79 \times 10^5 (\gamma Z T_e/\mu)^{1/2} \text{ cm s}^{-1}.$$

Alfvén velocity:

$$v_A = B/(4\pi n_i m_i)^{1/2} = 2.18 \times 10^{11} \mu^{-1/2} n_i^{1/2} B \text{ cm s}^{-1}.$$

21.1.4. Dimensionless

(Electron/proton mass ratio)$^{1/2}$:

$$(m_e/m_p)^{1/2} = 2.33 \times 10^{-2} = 1/42.9.$$

Number of particles in Debye sphere:

$$(4\pi/3) n {\lambda_D}^3 = 1.72 \times 10^9 n^{-1/2} T^{3/2}.$$

Alfvén velocity/speed of light:

$$v_A/c = 7.28\, \mu^{-1/2} {n_i}^{-1/2} B.$$

Electron plasma/gyrofrequency ratio:

$$\omega_{pe}/\omega_{ce} = 3.21 \times 10^{-3} n_e^{1/2} B^{-1}.$$

Ion plasma/gyrofrequency ratio:

$$\omega_{pi}/\omega_{ci} = 1.37 \times 10^{-1} \mu^{1/2} n_i^{1/2} B^{-1}.$$

Thermal/magnetic energy ratio:

$$\beta = 8\pi n k T/B^2 = 4.03 \times 10^{-11} n\, T\, B^{-2}.$$

Magnetic/ion rest energy ratio:

$$B^2/8\pi n_i m_i c^2 = 26.5\, \mu^{-1} n_i^{-1} B^2.$$

21.1.5. Miscellaneous

Bohm diffusion coefficient:

$$D_B = ckT/16eB = 6.25 \times 10^6 T B^{-1} \text{ cm}^2 \text{ s}^{-1}.$$

Transverse Spitzer resistivity:

$$\eta_\perp = 1.15 \times 10^{-14} Z \ln \Lambda\, T^{-3/2}\ \mathrm{s} = 1.03 \times 10^{-2} Z \ln \Lambda\, \mathrm{T}^{-3/2}\ \Omega\ \mathrm{cm}.$$

The anomalous collision rate due to low-frequency ion-sound turbulence is

$$\nu^* \approx \omega_{pe} \widetilde{W}/kT = 5.64 \times 10^4 n_e^{1/2} \widetilde{W}/kT\ \mathrm{s}^{-1},$$

where $\widetilde{W}$ is the total energy of waves with $\omega/K < v_{Ti}$.

Magnetic pressure is given by

$$P_{\mathrm{mag}} = B^2/8\pi = 3.98 \times 10^6 (B/B_0)^2\ \mathrm{dynes/cm^2} = 3.93 (B/B_0)^2\ \mathrm{atm},$$

where $B_0 = 10\ \mathrm{kG} = 1\ \mathrm{T}$.

Detonation energy of 1 kiloton of high explosive is

$$W_{\mathrm{kT}} = 10^{12}\ \mathrm{cal} = 4.2 \times 10^{19}\ \mathrm{erg}.$$

TABLE 21.1. Approximate magnitudes in some typical plasmas.

Plasma type	n cm^{-3}	T eV	ω_{pe} sec^{-1}	λ_D cm	$n\lambda_D^3$	ν_{ei} sec^{-1}
Interstellar gas	1	1	6×10^4	7×10^2	4×10^8	7×10^{-5}
Gaseous nebula	10^3	1	2×10^6	20	8×10^6	6×10^{-2}
Solar corona	10^9	10^2	2×10^9	0.2	8×10^6	60
Diffuse hot plasma	10^{12}	10^2	6×10^{10}	7×10^{-3}	4×10^5	40
Solar atmosphere, gas discharge	10^{14}	1	6×10^{11}	7×10^{-5}	40	2×10^9
Warm plasma	10^{14}	10	6×10^{11}	2×10^{-4}	8×10^2	10^7
Hot plasma	10^{14}	10^2	6×10^{11}	7×10^{-4}	4×10^4	4×10^6
Thermonuclear plasma	10^{15}	10^4	2×10^{12}	2×10^{-3}	8×10^6	5×10^4
Theta pinch	10^{16}	10^2	6×10^{12}	7×10^{-5}	4×10^3	3×10^8
Dense hot plasma	10^{18}	10^2	6×10^{13}	7×10^{-6}	4×10^2	2×10^{10}
Laser plasma	10^{20}	10^2	6×10^{14}	7×10^{-7}	40	2×10^{12}

Figure 21.1 gives comparable information in graphical form. [1]

21.2. PLASMA DISPERSION FUNCTION

21.2.1. Definition [2]

$$Z(\zeta) = \pi^{-1/2} \int_{-\infty}^{+\infty} dt \frac{\exp(-t^2)}{t - \zeta} = 2i \exp(-\zeta^2) \int_{-\infty}^{i\zeta} dt \exp(-t^2).$$

The first form is valid only for $\mathrm{Im}\,\zeta > 0$. Physically, $\zeta = x + iy$ is the ratio of wave phase velocity to thermal velocity.

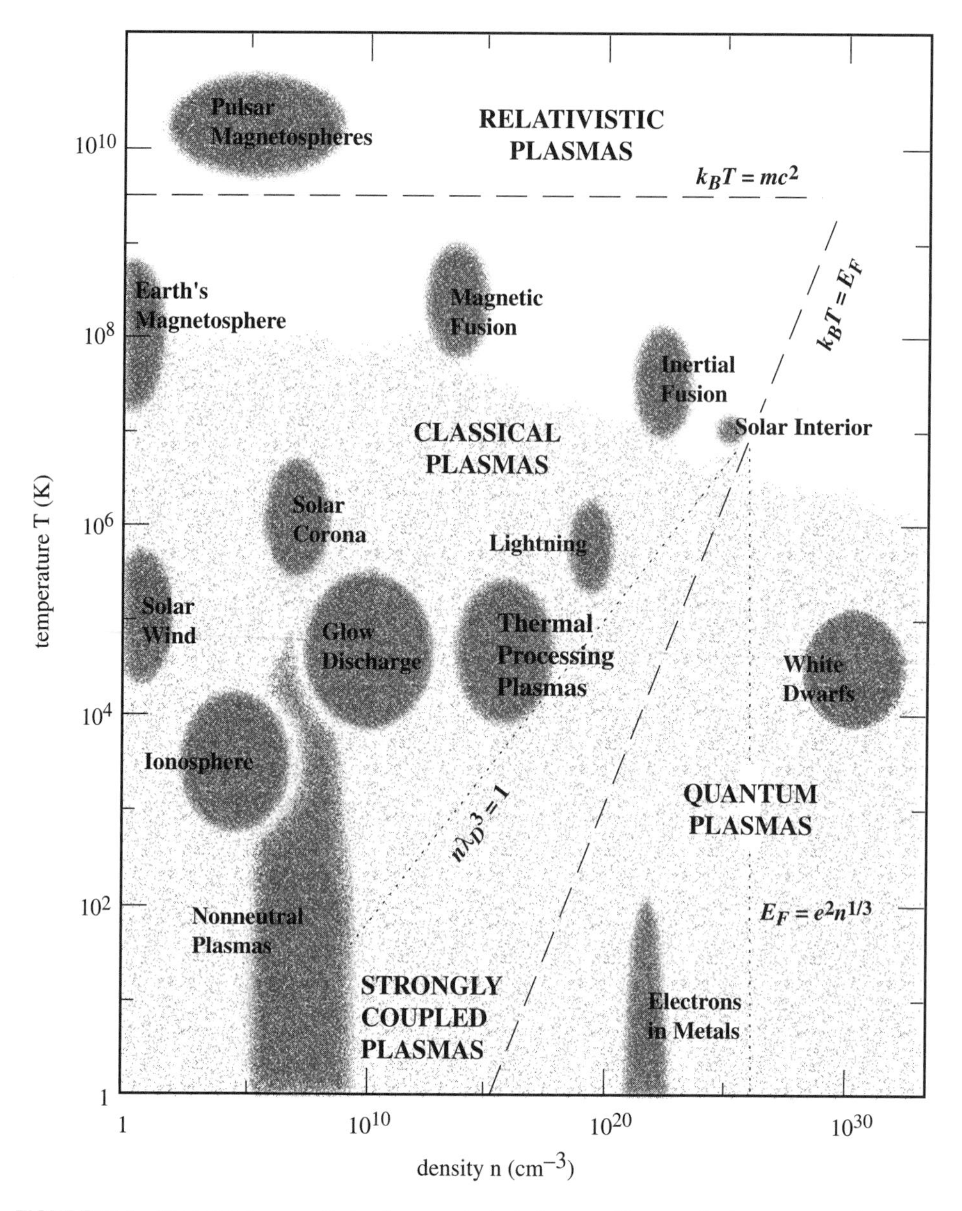

FIGURE 21.1. Approximate magnitudes in some typical plasmas. Plasmas that occur naturally or can be created in the laboratory are shown as a function of density (in particles per cubic centimeter) and temperature (in kelvin). The boundaries are approximate and indicate typical ranges of plasma parameters. Distinct plasma regimes are indicated.

Real argument ($y = 0$)

$$Z(x) = \exp\left(-x^2\right)\left[i\pi^{1/2} - 2\int_0^x dt \exp\left(t^2\right)\right].$$

Imaginary argument ($x = 0$)

$$Z(iy) = i\pi^{1/2}\exp\left(y^2\right)[1 - \mathrm{erf}(y)].$$

21.2.2. Differential equation

$$\frac{dZ}{d\zeta} = -2\left(1 + \zeta Z\right), Z(0) = i\pi^{1/2}; \quad \frac{d^2 Z}{d\zeta^2} + 2\zeta\frac{dZ}{d\zeta} + 2Z = 0.$$

21.2.3. Series expansions

Small argument

$$Z(\zeta) = i\pi^{1/2}\exp\left(-\zeta^2\right) - 2\zeta\left(1 - \tfrac{2}{3}\zeta^2 + \tfrac{4}{15}\zeta^4 - \tfrac{8}{105}\zeta^6 + \cdots\right).$$

Asymptotic series [3] $|\zeta| \gg 1$

$$Z(\zeta) = i\pi^{1/2}\sigma\exp\left(-\zeta^2\right) - \frac{1}{\zeta}\left(1 + \frac{1}{2\zeta^2} + \frac{3}{4\zeta^4} + \frac{15}{8\zeta^6} + \cdots\right),$$

where

$$\sigma = \begin{cases} 0; & y > |x| \\ 1; & |y| < |x|^{-1} \\ 2; & y < -|x| \end{cases}.$$

21.2.4. Symmetry properties

$$Z(\zeta^*) = -[Z(-\zeta)]^*,$$
$$Z(\zeta^*) = [Z(\zeta)]^* + 2i\pi^{1/2}\exp[-(\zeta^*)^2] \qquad (y > 0).$$

The asterisk denotes complex conjugation.

21.2.5. Two-pole approximations [4]

Valid for ζ in upper half-plane except when $x \gg 1$ and $y < \sqrt{\pi}\, x^2 e^{-x^2}$:

$$Z(\zeta) \approx \frac{0.50 + 0.81i}{a - \zeta} - \frac{0.50 - 0.81i}{a^* + \zeta}, \; a = 0.51 - 0.81i,$$

$$Z'(\zeta) \approx \frac{0.50 + 0.96i}{(b - \zeta)^2} + \frac{0.50 - 0.96i}{(b^* + \zeta)^2}, \; b = 0.48 - 0.91i.$$

21.3. COLLISIONS AND TRANSPORT

Temperatures are in eV, 1 K $= 8.62 \times 10^{-5}$ eV; Boltzmann's constant is $k = 1.60 \times 10^{-12}$ erg/eV; masses μ, μ' are in units of the proton mass; and $e_\alpha = Z_\alpha e$ is the charge of species α. All other units are cgs except where noted.

21.3.1. Relaxation rates

Rates are associated with four relaxation processes arising from the interaction of test particles (labeled α) streaming with velocity $\mathbf{v}_\alpha$ through a background of field particles (labeled β):

$$\text{slowing down} \qquad \frac{d\mathbf{v}_\alpha}{dt} = -\nu_s^{\alpha\backslash\beta}\mathbf{v}_\alpha,$$

$$\text{transverse diffusion} \qquad \frac{d}{dt}(\mathbf{v}_\alpha - \bar{\mathbf{v}}_\alpha)_\perp^2 = \nu_\perp^{\alpha\backslash\beta} v_\alpha^2,$$

$$\text{parallel diffusion} \qquad \frac{d}{dt}(\mathbf{v}_\alpha - \bar{\mathbf{v}}_\alpha)_\parallel^2 = \nu_\parallel^{\alpha\backslash\beta} v_\alpha^2,$$

$$\text{energy loss} \qquad \frac{d}{dt}{v_\alpha}^2 = -\nu_\epsilon^{\alpha\backslash\beta} v_\alpha^2,$$

where the averages are performed over an ensemble of test particles and a Maxwellian field particle distribution. The exact formulas may be written as [5]

$$\nu_s^{\alpha\backslash\beta} = (1 + m_\alpha/m_\beta)\psi(x^{\alpha\backslash\beta})\nu_0^{\alpha\backslash\beta}$$

$$\nu_\perp^{\alpha\backslash\beta} = 2\left[(1 - 1/2x^{\alpha\backslash\beta})\psi(x^{\alpha\backslash\beta}) + \psi'(x^{\alpha\backslash\beta})\right]\nu_0^{\alpha\backslash\beta}$$

$$\nu_\parallel^{\alpha\backslash\beta} = \left[\psi(x^{\alpha\backslash\beta})/x^{\alpha\backslash\beta}\right]\nu_0^{\alpha\backslash\beta}$$

$$\nu_\epsilon^{\alpha\backslash\beta} = 2\left[(m_\alpha/m_\beta)\psi(x^{\alpha\backslash\beta}) - \psi'(x^{\alpha\backslash\beta})\right]\nu_0^{\alpha\backslash\beta},$$

where

$$\nu_0^{\alpha\backslash\beta} = 4\pi e_\alpha^2 e_\beta^2 \lambda_{\alpha\beta} n_\beta/(m_\alpha^2 v_\alpha^3), \qquad x^{\alpha\backslash\beta} = m_\beta v_\alpha^2/2kT_\beta,$$

$$\psi(x) = \frac{2}{\sqrt{\pi}}\int_0^x dt\, t^{1/2} e^{-t}, \qquad \psi'(x) = \frac{d\psi}{dx} = 2\sqrt{\frac{x}{\pi}}\, e^{-x},$$

and $\lambda_{\alpha\beta} = \ln \Lambda_{\alpha\beta}$ is the Coulomb logarithm (see below). Limiting forms of ν_s, $\nu_\perp$, and $\nu_\parallel$ in units of s^{-1} are given in the following table. Test particle energy ϵ and field particle temperature T are both in eV; $\mu = m_i/m_p$, where m_p is the proton mass; Z is the ion charge state; and in ion–ion encounters field particle quantities are distinguished by a prime. The two expressions given below for each rate hold for very slow ($x^{\alpha\backslash\beta} \ll 1$) and very fast ($x^{\alpha\backslash\beta} \gg 1$) test particles, respectively.

Slow	Fast
Electron–electron	
$\nu_s^{e\backslash e}/n_e\lambda_{ee} \approx 5.8\times10^{-6}T^{-3/2}$	$\longrightarrow 7.7\times10^{-6}\epsilon^{-3/2}$,
$\nu_\perp^{e\backslash e}/n_e\lambda_{ee} \approx 5.8\times10^{-6}T^{-1/2}\epsilon^{-1}$	$\longrightarrow 7.7\times10^{-6}\epsilon^{-3/2}$,
$\nu_\parallel^{e\backslash e}/n_e\lambda_{ee} \approx 2.9\times10^{-6}T^{-1/2}\epsilon^{-1}$	$\longrightarrow 3.9\times10^{-6}T\epsilon^{-5/2}$.
Electron–ion	
$\nu_s^{e\backslash i}/n_iZ^2\lambda_{ei} \approx 0.23(\mu/T)^{3/2}$	$\longrightarrow 3.9\times10^{-6}\epsilon^{-3/2}$,
$\nu_\perp^{e\backslash i}/n_iZ^2\lambda_{ei} \approx 2.5\times10^{-4}(\mu/T)^{1/2}\epsilon^{-1}$	$\longrightarrow 7.7\times10^{-6}\epsilon^{-3/2}$,
$\nu_\parallel^{e\backslash i}/n_iZ^2\lambda_{ei} \approx 1.2\times10^{-4}(\mu/T)^{1/2}\epsilon^{-1}$	$\longrightarrow 2.1\times10^{-9}\mu^{-1}T\epsilon^{-5/2}$.
Ion–electron	
$\nu_s^{i\backslash e}/n_eZ^2\lambda_{ie} \approx 1.6\times10^{-9}\mu^{-1}T^{-3/2}$	$\longrightarrow 1.7\times10^{-4}\mu^{1/2}\epsilon^{-3/2}$,
$\nu_\perp^{i\backslash e}/n_eZ^2\lambda_{ie} \approx 3.2\times10^{-9}\mu^{-1}T^{-1/2}\epsilon^{-1}$	$\longrightarrow 1.8\times10^{-7}\mu^{-1/2}\epsilon^{-3/2}$,
$\nu_\parallel^{i\backslash e}/n_eZ^2\lambda_{ie} \approx 1.6\times10^{-9}\mu^{-1}T^{-1/2}\epsilon^{-1}$	$\longrightarrow 1.7\times10^{-4}\mu^{1/2}T\epsilon^{-5/2}$.
Ion–ion	
$\dfrac{\nu_s^{i\backslash i'}}{n_{i'}Z^2Z'^2\lambda_{ii'}} \approx 6.8\times10^{-8}\left(\mu'/\mu\right)^{1/2}(\mu'+\mu)^{-1/2}T^{-3/2}$	$\longrightarrow 9.0\times10^{-8}\left(\frac{1}{\mu}+\frac{1}{\mu'}\right)\frac{\mu^{1/2}}{\epsilon^{3/2}}$,
$\dfrac{\nu_\perp^{i\backslash i'}}{n_{i'}Z^2Z'^2\lambda_{ii'}} \approx 1.4\times10^{-7}\mu'^{1/2}\mu^{-1}T^{-1/2}\epsilon^{-1}$	$\longrightarrow 1.8\times10^{-7}\mu^{-1/2}\epsilon^{-3/2}$,
$\dfrac{\nu_\parallel^{i\backslash i'}}{n_{i'}Z^2Z'^2\lambda_{ii'}} \approx 6.8\times10^{-8}\mu'^{1/2}\mu^{-1}T^{-1/2}\epsilon^{-1}$	$\longrightarrow 9.0\times10^{-8}\mu^{1/2}\mu'^{-1}T\epsilon^{-5/2}$.

In the same limits, the energy transfer rate follows from the identity

$$\nu_\epsilon = 2\nu_s - \nu_\perp - \nu_\parallel,$$

except for the case of fast electrons or fast ions scattered by ions, where the leading terms cancel. Then, the appropriate forms are

$$\nu_\epsilon^{e\backslash i} \longrightarrow 4.2\times10^{-9}n_iZ^2\lambda_{ei}\left[\epsilon^{-3/2}\mu^{-1} - 8.9\times10^4(\mu/T)^{1/2}\epsilon^{-1}\exp(-1836\mu\epsilon/T)\right]\ \mathrm{s}^{-1}$$

and

$$\nu_\epsilon^{i\backslash i'} \longrightarrow 1.8\times10^{-7}n_{i'}Z^2Z'^2\lambda_{ii'}\left[\epsilon^{-3/2}\mu^{1/2}/\mu' - 1.1(\mu'/T)^{1/2}\epsilon^{-1}\exp(-\mu'\epsilon/T)\right]\ \mathrm{s}^{-1}.$$

In general, the energy transfer rate $\nu_\epsilon^{\alpha\backslash\beta}$ is positive for $\epsilon > \epsilon_\alpha^*$ and negative for $\epsilon < \epsilon_\alpha^*$, where $x^* = (m_\beta/m_\alpha)\epsilon_\alpha^*/T_\beta$ is the solution of $\psi'(x^*) = (m_\alpha/m_\beta)\psi(x^*)$. The ratio $\epsilon_\alpha^*/T_\beta$ is equal to 1.5, 0.98, 4.8×10^{-3}, 2.6×10^{-3}, 1.8×10^{-3}, and 1.4×10^{-3} for $\alpha\backslash\beta$ equal to

$i\backslash e$, $e\backslash e$ (or $i\backslash i$), $e\backslash p$, $e\backslash$D, $e\backslash$T (or $e\backslash$He^3), and $e\backslash$He^4, respectively. When both species are near Maxwellian, with $T_i \lesssim T_e$, there are just two characteristic collision rates. For $Z = 1$,

$$\nu_e = 2.9 \times 10^{-6} n\lambda T_e{}^{-3/2} \mathrm{s}^{-1},$$
$$\nu_i = 4.8 \times 10^{-8} n\lambda T_i{}^{-3/2} \mu^{-1/2} \mathrm{s}^{-1}.$$

21.3.2. Temperature isotropization

Isotropization is described by

$$\frac{dT_\perp}{dt} = -\frac{1}{2}\frac{dT_\parallel}{dt} = -\nu_T^\alpha (T_\perp - T_\parallel),$$

where, if $A \equiv T_\perp / T_\parallel - 1 > 0$,

$$\nu_T^\alpha = \frac{2\sqrt{\pi} e_\alpha{}^2 e_\beta{}^2 n_\alpha \lambda_{\alpha\beta}}{m_\alpha{}^{1/2} (kT_\parallel)^{3/2}} A^{-2} \left[-3 + (A+3) \frac{\tan^{-1}(A^{1/2})}{A^{1/2}} \right].$$

If $A < 0$, $\tan^{-1}(A^{1/2})/A^{1/2}$ is replaced by $\tanh^{-1}(-A)^{1/2}/(-A)^{1/2}$. For $T_\perp \approx T_\parallel \equiv T$,

$$\nu_T^e = 8.2 \times 10^{-7} n\lambda T^{-3/2}\ \mathrm{s}^{-1},$$
$$\nu_T^i = 1.9 \times 10^{-8} n\lambda Z^2 \mu^{-1/2} T^{-3/2}\ \mathrm{s}^{-1}.$$

21.3.3. Thermal equilibration

If the components of a plasma have different temperatures but no relative drift, equilibration is described by

$$\frac{dT_\alpha}{dt} = \sum_\beta \bar{\nu}_\epsilon^{\alpha\backslash\beta} (T_\beta - T_\alpha),$$

where

$$\bar{\nu}_\epsilon^{\alpha\backslash\beta} = 1.8 \times 10^{-19} \frac{(m_\alpha m_\beta)^{1/2} Z_\alpha{}^2 Z_\beta{}^2 n_\beta \lambda_{\alpha\beta}}{(m_\alpha T_\beta + m_\beta T_\alpha)^{3/2}}\ \mathrm{s}^{-1}.$$

For electrons and ions with $T_e \approx T_i \equiv T$, this implies

$$\bar{\nu}_\epsilon^{e\backslash i}/n_i = \bar{\nu}_\epsilon^{i\backslash e}/n_e = 3.2 \times 10^{-9} Z^2 \lambda / \mu T^{3/2}\ \mathrm{cm}^3\ \mathrm{s}^{-1}.$$

21.3.4. Coulomb logarithm

For test particles of mass m_α and charge $e_\alpha = Z_\alpha e$ scattering off field particles of mass m_β and charge $e_\beta = Z_\beta e$, the Coulomb logarithm is defined as $\lambda = \ln \Lambda \equiv \ln(r_{\max}/r_{\min})$. Here $r_{\min}$ is the larger of $e_\alpha e_\beta / m_{\alpha\beta}\bar{u}^2$ and $\hbar/2m_{\alpha\beta}\bar{u}$, averaged over both particle velocity distributions, where $m_{\alpha\beta} = m_\alpha m_\beta/(m_\alpha + m_\beta)$ and $\mathbf{u} = \mathbf{v}_\alpha - \mathbf{v}_\beta$; $r_{\max} = (4\pi \sum n_\gamma {e_\gamma}^2/kT_\gamma)^{-1/2}$, where the summation extends over all species γ for which $\bar{u}^2 < {v_{T\gamma}}^2 = kT_\gamma/m_\gamma$. If this inequality cannot be satisfied, or if either $\bar{u}{\omega_{c\alpha}}^{-1} < r_{\max}$ or $\bar{u}{\omega_{c\beta}}^{-1} < r_{\max}$, where ω_c is the cyclotron (gyro-) frequency, the theory breaks down. Typically, $\lambda \approx 10$–20. Corrections to the transport coefficients are $O(\lambda^{-1})$; hence the theory is good only to $\sim 10\%$, and it fails when $\lambda \sim 1$.

The following cases are of particular interest.

Thermal electron–electron collisions

$$\begin{aligned}\lambda_{ee} &= 23 - \ln({n_e}^{1/2}{T_e}^{-3/2}), && T_e \lesssim 10 \text{ eV};\\ &= 24 - \ln({n_e}^{1/2}{T_e}^{-1}), && T_e \gtrsim 10 \text{ eV}.\end{aligned}$$

Electron–ion collisions

$$\begin{aligned}\lambda_{ei} = \lambda_{ie} &= 23 - \ln\left({n_e}^{1/2} Z T_e^{-3/2}\right), && T_i m_e/m_i < T_e < 10Z^2 \text{ eV};\\ &= 24 - \ln\left({n_e}^{1/2} T_e^{-1}\right), && T_i m_e/m_i < 10Z^2\text{eV} < T_e\\ &= 30 - \ln\left({n_i}^{1/2} T_i^{-3/2} Z^2 \mu^{-1}\right), && T_e < T_i Z m_e/m_i.\end{aligned}$$

Mixed ion–ion collisions

$$\lambda_{ii'} = \lambda_{i'i} = 23 - \ln\left[\frac{ZZ'(\mu+\mu')}{\mu T_{i'} + \mu' T_i}\left(\frac{n_i Z^2}{T_i} + \frac{n_{i'} Z'^2}{T_{i'}}\right)^{1/2}\right].$$

Counterstreaming ions

Relative velocity $v_D = \beta_D c$, in the presence of warm electrons, $kT_i/m_i, kT_{i'}/m_{i'} < {v_D}^2 < kT_e/m_e$:

$$\lambda_{ii'} = \lambda_{i'i} = 35 - \ln\left[\frac{ZZ'(\mu+\mu')}{\mu\mu'{\beta_D}^2}\left(\frac{n_e}{T_e}\right)^{1/2}\right].$$

21.3.5. Fokker–Planck equation

$$\frac{Df^\alpha}{Dt} \equiv \frac{\partial f^\alpha}{\partial t} + \mathbf{v}\cdot\nabla f^\alpha + \frac{1}{m_\alpha}\mathbf{F}\cdot\nabla_{\mathbf{v}} f^\alpha = \left(\frac{\partial f^\alpha}{\partial t}\right)_{\text{coll}},$$

where $\mathbf{F}$ is an external force field. The general form of the collision integral is $(\partial f^\alpha/\partial t)_{\rm coll} = -\sum_\beta \nabla_{\mathbf{v}} \cdot \mathbf{J}^{\alpha\backslash\beta}$, with

$$\mathbf{J}^{\alpha\backslash\beta} = 2\pi\lambda_{\alpha\beta}\frac{{e_\alpha}^2{e_\beta}^2}{m_\alpha}\int d^3v'(u^2\mathbf{I} - \mathbf{uu})u^{-3} \cdot \left\{\frac{1}{m_\beta}f^\alpha(\mathbf{v})\nabla_{\mathbf{v}'}f^\beta(\mathbf{v}') - \frac{1}{m_\alpha}f^\beta(\mathbf{v}')\nabla_{\mathbf{v}}f^\alpha(\mathbf{v})\right\}$$

(Landau form), where $\mathbf{u} = \mathbf{v}' - \mathbf{v}$ and $\mathbf{I}$ is the unit dyad, or alternatively,

$$\mathbf{J}^{\alpha\backslash\beta} = 4\pi\lambda_{\alpha\beta}\frac{{e_\alpha}^2{e_\beta}^2}{{m_\alpha}^2}\left\{f^\alpha(\mathbf{v})\nabla_{\mathbf{v}}H(\mathbf{v}) - \tfrac{1}{2}\nabla_{\mathbf{v}}\cdot\left[f^\alpha(\mathbf{v})\nabla_{\mathbf{v}}\nabla_{\mathbf{v}}G(\mathbf{v})\right]\right\},$$

where the Rosenbluth potentials are

$$G(\mathbf{v}) = \int f^\beta(\mathbf{v}')u\,d^3v',$$

$$H(\mathbf{v}) = \left(1 + \frac{m_\alpha}{m_\beta}\right)\int f^\beta(\mathbf{v}')u^{-1}\,d^3v'.$$

If species α is a weak beam (number and energy density small compared with background) streaming through a Maxwellian plasma,

$$\mathbf{J}^{\alpha\backslash\beta} = -\frac{m_\alpha}{m_\alpha + m_\beta}\nu_s^{\alpha\backslash\beta}\mathbf{v}f^\alpha - \frac{1}{2}\nu_\parallel^{\alpha\backslash\beta}\mathbf{vv}\cdot\nabla_{\mathbf{v}}f^\alpha - \tfrac{1}{4}\nu_\perp^{\alpha\backslash\beta}\left(v^2\mathbf{I} - \mathbf{vv}\right)\cdot\nabla_{\mathbf{v}}f^\alpha.$$

21.3.6. B-G-K collision operator

For distribution functions with no large gradients in velocity space, the Fokker–Planck collision terms can be approximated according to

$$\frac{Df_e}{Dt} = \nu_{ee}(F_e - f_e) + \nu_{ei}(\bar{F}_e - f_e),$$

$$\frac{Df_i}{Dt} = \nu_{ie}(\bar{F}_i - f_i) + \nu_{ii}(F_i - f_i).$$

The respective slowing-down rates $\nu_s^{\alpha\backslash\beta}$ given in Sec. 21.3.1 can be used for $\nu_{\alpha\beta}$, assuming slow ions and fast electrons, with ϵ replaced by T_α. (For ν_{ee} and ν_{ii}, one can equally well use $\nu_\perp$, and the result is insensitive to whether the slow- or fast-test-particle limit is employed.) The Maxwellians F_α and $\bar{F}_\alpha$ are given by [6]

$$F_\alpha = n_\alpha\left(\frac{m_\alpha}{2\pi kT_\alpha}\right)^{3/2}\exp\left\{-\left[\frac{m_\alpha(\mathbf{v} - \mathbf{v}_\alpha)^2}{2kT_\alpha}\right]\right\},$$

$$\bar{F}_\alpha = n_\alpha\left(\frac{m_\alpha}{2\pi k\bar{T}_\alpha}\right)^{3/2}\exp\left\{-\left[\frac{m_\alpha(\mathbf{v} - \bar{\mathbf{v}}_\alpha)^2}{2k\bar{T}_\alpha}\right]\right\},$$

where n_α, $\mathbf{v}_\alpha$, and T_α are the number density, mean drift velocity, and effective temperature obtained by taking moments of f_α. Some latitude in the definition of $\bar{T}_\alpha$ and $\bar{\mathbf{v}}_\alpha$ is possible; one choice is $\bar{T}_e = T_i$, $\bar{T}_i = T_e$, $\bar{\mathbf{v}}_e = \mathbf{v}_i$, $\bar{\mathbf{v}}_i = \mathbf{v}_e$.

21.3.7. Transport coefficients

Transport equations for a multispecies plasma:

$$\frac{d^\alpha n_\alpha}{dt} + n_\alpha \nabla \cdot \mathbf{v}_\alpha = 0,$$

$$m_\alpha n_\alpha \frac{d^\alpha \mathbf{v}_\alpha}{dt} = -\nabla p_\alpha - \nabla \cdot \mathbf{P}_\alpha + Z_\alpha e n_\alpha \left[\mathbf{E} + \frac{1}{c}\mathbf{v}_\alpha \times \mathbf{B}\right] + \mathbf{R}_\alpha,$$

$$\frac{3}{2} n_\alpha \frac{d^\alpha k T_\alpha}{dt} + p_\alpha \nabla \cdot \mathbf{v}_\alpha = -\nabla \cdot \mathbf{q}_\alpha - \mathbf{P}_\alpha : \nabla \mathbf{v}_\alpha + Q_\alpha.$$

Here, $d^\alpha/dt \equiv \partial/\partial t + \mathbf{v}_\alpha \cdot \nabla$; $p_\alpha = n_\alpha k T_\alpha$, where k is Boltzmann's constant; $\mathbf{R}_\alpha = \sum_\beta \mathbf{R}_{\alpha\beta}$ and $Q_\alpha = \sum_\beta Q_{\alpha\beta}$, where $\mathbf{R}_{\alpha\beta}$ and $Q_{\alpha\beta}$ are, respectively, the momentum and energy gained by the αth species through collisions with the βth; $\mathbf{P}_\alpha$ is the stress tensor; and $\mathbf{q}_\alpha$ is the heat flow.

The transport coefficients in a simple two-component plasma (electrons and singly charged ions) are tabulated below. Here $\|$ and $\perp$ refer to the direction of the magnetic field $\mathbf{B} = \mathbf{b}B$; $\mathbf{u} = \mathbf{v}_e - \mathbf{v}_i$ is the relative streaming velocity; $n_e = n_i \equiv n$; $\mathbf{j} = -ne\mathbf{u}$ is the current; $\omega_{ce} = 1.76 \times 10^7 B\ \mathrm{sec}^{-1}$ and $\omega_{ci} = (m_e/m_i)\omega_{ce}$ are the electron and ion gyrofrequencies, respectively; and the basic collisional times are taken to be

$$\tau_e = \frac{3\sqrt{m_e}(kT_e)^{3/2}}{4\sqrt{2\pi}\, n\lambda e^4} = 3.44 \times 10^5 \frac{{T_e}^{3/2}}{n\lambda}\ \mathrm{s},$$

where λ is the Coulomb logarithm, and

$$\tau_i = \frac{3\sqrt{m_i}(kT_i)^{3/2}}{4\sqrt{\pi}\, n\lambda e^4} = 2.09 \times 10^7 \frac{{T_i}^{3/2}}{n\lambda}\mu^{1/2}\ \mathrm{s}.$$

In the limit of strong fields ($\omega_{c\alpha}\tau_\alpha \gg 1$, $\alpha = i, e$) the transport processes may be summarized as follows: [7]

Momentum transfer:

$$\mathbf{R}_{ei} = -\mathbf{R}_{ie} \equiv \mathbf{R} = \mathbf{R}_\mathbf{u} + \mathbf{R}_T.$$

Frictional force:

$$\mathbf{R}_\mathbf{u} = ne(\mathbf{j}_\|/\sigma_\| + \mathbf{j}_\perp/\sigma_\perp).$$

Electrical conductivities:

$$\sigma_\| = 1.96\sigma_\perp = 1.96\frac{ne^2\tau_e}{m_e}.$$

Thermal force:

$$\mathbf{R}_T = -0.71 n \nabla_{\parallel}(kT_e) - \frac{3n}{2\omega_{ce}\tau_e}\mathbf{b} \times \nabla_{\perp}(kT_e).$$

Ion heating:

$$Q_i = \frac{3m_e}{m_i}\frac{nk}{\tau_e}(T_e - T_i).$$

Electron heating:

$$Q_e = -Q_i - \mathbf{R} \cdot \mathbf{u}.$$

Ion heat flux:

$$\mathbf{q}_i = -\kappa_{\parallel}^i \nabla_{\parallel}(kT_i) - \kappa_{\perp}^i \nabla_{\perp}(kT_i) + \kappa_{\wedge}^i \mathbf{b} \times \nabla_{\perp}(kT_i).$$

Ion thermal conductivities:

$$\kappa_{\parallel}^i = 3.9\frac{nkT_i\tau_i}{m_i}, \quad \kappa_{\perp}^i = \frac{2nkT_i}{m_i\omega_{ci}^2\tau_i}, \quad \kappa_{\wedge}^i = \frac{5nkT_i}{2m_i\omega_{ci}}.$$

Electron heat flux:

$$\mathbf{q}_e = \mathbf{q}_{\mathbf{u}}^e + \mathbf{q}_T^e.$$

Frictional heat flux

$$\mathbf{q}_{\mathbf{u}}^e = 0.71 nkT_e \mathbf{u}_{\parallel} + \frac{3nkT_e}{2\omega_{ce}\tau_e}\mathbf{b} \times \mathbf{u}_{\perp}.$$

Thermal gradient heat flux:

$$\mathbf{q}_T^e = -\kappa_{\parallel}^e \nabla_{\parallel}(kT_e) - \kappa_{\perp}^e \nabla_{\perp}(kT_e) - \kappa_{\wedge}^e \mathbf{b} \times \nabla_{\perp}(kT_e).$$

Electron thermal conductivities:

$$\kappa_{\parallel}^e = 3.2\frac{nkT_e\tau_e}{m_e}, \quad \kappa_{\perp}^e = 4.7\frac{nkT_e}{m_e\omega_{ce}^2\tau_e}, \quad \kappa_{\wedge}^e = \frac{5nkT_e}{2m_e\omega_{ce}}.$$

Stress tensor (either species; the z-axis is defined parallel to $\mathbf{B}$):

$$P_{xx} = -\frac{\eta_0}{2}(W_{xx} + W_{yy}) - \frac{\eta_1}{2}(W_{xx} - W_{yy}) - \eta_3 W_{xy},$$

$$P_{yy} = -\frac{\eta_0}{2}(W_{xx} + W_{yy}) + \frac{\eta_1}{2}(W_{xx} - W_{yy}) + \eta_3 W_{xy},$$

$$P_{xy} = P_{yx} = -\eta_1 W_{xy} + \frac{\eta_3}{2}(W_{xx} - W_{yy}),$$

$$P_{xz} = P_{zx} = -\eta_2 W_{xz} - \eta_4 W_{yz},$$

$$P_{yz} = P_{zy} = -\eta_2 W_{yz} + \eta_4 W_{xz},$$

$$P_{zz} = -\eta_0 W_{zz}.$$

Ion viscosity:

$$\eta_0^i = 0.96nkT_i\tau_i, \quad \eta_1^i = \frac{3nkT_i}{10\omega_{ci}{}^2\tau_i}, \quad \eta_2^i = \frac{6nkT_i}{5\omega_{ci}{}^2\tau_i},$$

$$\eta_3^i = \frac{nkT_i}{2\omega_{ci}}, \quad \eta_4^i = \frac{nkT_i}{\omega_{ci}}.$$

Electron viscosity:

$$\eta_0^e = 0.73nkT_e\tau_e, \quad \eta_1^e = 0.51\frac{nkT_e}{\omega_{ce}^2\tau_e}, \quad \eta_2^e = 2.0\frac{nkT_e}{\omega_{ce}^2\tau_e},$$

$$\eta_3^e = -\frac{nkT_e}{2\omega_{ce}}, \quad \eta_4^e = -\frac{nkT_e}{\omega_{ce}}.$$

For both species, the rate-of-strain tensor is defined as

$$W_{jk} = \frac{\partial v_j}{\partial x_k} + \frac{\partial v_k}{\partial x_j} - \frac{2}{3}\delta_{jk}\nabla\cdot\mathbf{v}.$$

When $\mathbf{B} = 0$ the following simplifications occur:

$$\mathbf{R_u} = ne\mathbf{j}/\sigma_\parallel, \quad \mathbf{R}_T = -0.71n\nabla(kT_e), \quad \mathbf{q}_i = -\kappa_\parallel^i\nabla(kT_i),$$

$$\mathbf{q_u^e} = 0.71nkT_e\mathbf{u}; \quad \mathbf{q}_T^e = -\kappa_\parallel^e\nabla(kT_e); \quad P_{jk} = -\eta_0 W_{jk}.$$

For $\omega_{ce}\tau_e \gg 1 \gg \omega_{ci}\tau_i$ the electrons obey the high-field expressions and the ions obey the zero-field expressions.

Collisional transport theory is applicable when (1) macroscopic time rates of change satisfy $d/dt \ll 1/\tau$, where τ is the longest collisional time scale; and (in the absence of a magnetic field) (2) macroscopic length scales L satisfy $L \gg l$, where $l = \bar{v}\tau$ is the mean free path. In a strong field, $\omega_{ce}\tau \gg 1$, condition (2) is replaced by $L_\parallel \gg l$ and $L_\perp \gg \sqrt{lr_e}$ ($L_\perp \gg r_e$ in a uniform field), where $L_\parallel$ is a macroscopic scale parallel to the field $\mathbf{B}$ and $L_\perp$ is the smaller of $B/|\nabla_\perp B|$ and the transverse plasma dimension. In addition, the standard transport coefficients are valid only when (3) the Coulomb logarithm satisfies $\lambda \gg 1$; (4) the electron gyroradius satisfies $r_e \gg \lambda_D$, or $8\pi n_e m_e c^2 \gg B^2$; (5) relative drifts $\mathbf{u} = \mathbf{v}_\alpha - \mathbf{v}_\beta$ between two species are small compared with the thermal velocities, i.e., $u^2 \ll kT_\alpha/m_\alpha, kT_\beta/m_\beta$; and (6) anomalous transport processes owing to microinstabilities are negligible.

21.3.8. Weakly ionized plasmas

The collision frequency for the scattering of charged particles of species α by neutrals is

$$\nu_\alpha = n_0\sigma_s^{\alpha\backslash 0}(kT_\alpha/m_\alpha)^{1/2},$$

where n_0 is the neutral density and $\sigma_s^{\alpha\backslash 0}$ is the cross section; it is typically $\sim 5\times 10^{-15}$ cm^2 and weakly dependent on temperature.

For electrons with density n_e the rates for attachment, recombination, and collisions in air are, respectively,

$$\nu_a \approx 10^3 (n_0/10^{17})^2 \text{ s}^{-1},$$
$$\nu_r \approx 2 \times 10^{-7} n_e \text{ s}^{-1},$$
$$\nu_c \approx 1.7 \times 10^{-7} n_0 \text{ s}^{-1}.$$

When the system is small compared with a Debye length, $L \ll \lambda_D$, the charged-particle diffusion coefficients are

$$D_\alpha = kT_\alpha / m_\alpha \nu_\alpha,$$

In the opposite limit both species diffuse at the ambipolar rate

$$D_A = \frac{\mu_i D_e - \mu_e D_i}{\mu_i - \mu_e} = \frac{(T_i + T_e) D_i D_e}{T_i D_e + T_e D_i},$$

where $\mu_\alpha = e_\alpha / m_\alpha \nu_\alpha$ is the mobility. The conductivity σ_α satisfies $\sigma_\alpha = n_\alpha e_\alpha \mu_\alpha$.

In the presence of a magnetic field $\mathbf{B} = \mathbf{b}B$ the scalars μ and σ become tensors,

$$\mathbf{J}^\alpha = \boldsymbol{\sigma}^\alpha \cdot \mathbf{E} = \sigma_\parallel^\alpha \mathbf{E}_\parallel + \sigma_\perp^\alpha \mathbf{E}_\perp + \sigma_\wedge^\alpha \mathbf{E} \times \mathbf{b},$$

where $\sigma_\perp$ and $\sigma_\wedge$ are the Pedersen and Hall conductivities, respectively:

$$\sigma_\parallel^\alpha = n_\alpha {e_\alpha}^2 / m_\alpha \nu_\alpha,$$
$$\sigma_\perp^\alpha = \sigma_\parallel^\alpha {\nu_\alpha}^2 / ({\nu_\alpha}^2 + \omega_{c\alpha}^2),$$
$$\sigma_\wedge^\alpha = \sigma_\parallel^\alpha \nu_\alpha \omega_{c\alpha} / ({\nu_\alpha}^2 + \omega_{c\alpha}^2).$$

21.4. SOLAR AND IONOSPHERIC PHYSICS (see pp. 641–643 for Tables 21.3–21.4)

TABLE 21.2. General solar parameters. [8]

Parameter	Symbol	Numerical Value	Units
Total mass	$M_\odot$	1.99×10^{33}	g
Radius	$R_\odot$	6.96×10^{10}	cm
Surface gravity	$g_\odot$	2.74×10^4	cm s^{-2}
Escape speed	v_∞	6.18×10^7	cm s^{-1}
Upward mass flux in spicules	—	1.6×10^{-9}	g cm^{-2} s^{-1}
Vertically integrated atmospheric density	—	4.28	g cm^{-2}
Sunspot magnetic field strength	B_{max}	2500–3500	G
Surface effective temperature	T_0	5770	K
Radiant power	$\mathcal{L}_\odot$	3.83×10^{33}	erg s^{-1}
Radiant flux density	$\mathcal{F}$	6.28×10^{10}	erg cm^{-2} s^{-1}
Optical depth at 500 nm, measured from photosphere	τ_5	0.99	—
Astronomical unit (radius of earth's orbit)	AU	1.50×10^{13}	cm
Solar constant (intensity at 1 AU)	f	1.36×10^6	erg cm^{-2} s^{-1}

TABLE 21.3. Solar chromosphere and corona. [9]

Parameter (units)	Quiet sun	Coronal hole	Active region
Chromospheric radiation losses ($\text{erg cm}^{-2}\ \text{s}^{-1}$)			
Low chromosphere	2×10^6	2×10^6	$\gtrsim 10^7$
Middle chromosphere	2×10^6	2×10^6	10^7
Upper chromosphere	3×10^5	3×10^5	2×10^6
Total	4×10^6	4×10^6	$\gtrsim 2 \times 10^7$
Transition layer pressure (dyne/cm^2)	0.2	0.07	2
Coronal temperature (K) at 1.1 $R_\odot$	$(1.1\text{–}1.6) \times 10^6$	10^6	2.5×10^6
Coronal energy losses ($\text{erg cm}^{-2}\ \text{s}^{-1}$)			
Conduction	2×10^5	6×10^4	$10^5 - 10^7$
Radiation	10^5	10^4	5×10^6
Solar Wind	$\lesssim 5 \times 10^4$	7×10^5	$< 10^5$
Total	3×10^5	8×10^5	10^7
Solar wind mass loss ($\text{g cm}^{-2}\ \text{s}^{-1}$)	$\lesssim 2 \times 10^{-11}$	2×10^{-10}	$< 4 \times 10^{-11}$

The following table gives average nighttime values. Where two numbers are entered, the first refers to the lower and the second refers to the upper portion of the layer.

TABLE 21.4. Ionospheric parameters. [10]

Quantity (units)	E region	F region
Altitude (km)	90–160	160–500
Number density (m^{-3})	$1.5 \times 10^{10} - 3.0 \times 10^{10}$	$5 \times 10^{10} - 2 \times 10^{11}$
Height-integrated number density (m^{-2})	9×10^{14}	4.5×10^{15}
Ion–neutral collision frequency (s^{-1})	2000–100	0.5–0.05
Ion gyrofrequency/collision frequency κ_i	0.09–2.0	460–5000
Ion Pederson factor $\kappa_i/(1+\kappa_i{}^2)$	0.09–0.5	$2.2 \times 10^{-3} - 2 \times 10^{-4}$
Ion Hall factor $\kappa_i{}^2/(1+\kappa_i{}^2)$	$8 \times 10^{-4} - 0.8$	1.0
Electron–neutral collision frequency (s^{-1})	$1.5 \times 10^4 - 900$	80–10
Electron gyrofrequency/collision frequency κ_e	410–6900	$7.8 \times 10^4 - 6.2 \times 10^5$
Electron Pederson factor $\kappa_e/(1+\kappa_e{}^2)$	$2.7 \times 10^{-3} - 1.5 \times 10^{-4}$	$10^{-5} - 1.5 \times 10^{-6}$

(continued)

TABLE 21.4. Continued

Quantity (units)	E region	F region
Electron Hall factor $\kappa_e{}^2/(1+\kappa_e{}^2)$	1.0	1.0
Mean molecular weight	28–26	22–16
Ion gyrofrequency (s^{-1})	180–190	230–300
Neutral diffusion coefficient ($m^2\ s^{-1}$)	30–5000	10^5

The terrestrial magnetic field in the lower ionosphere at equatorial lattitudes is approximately $B_0 = 0.35 \times 10^{-4}$ tesla. The radius of the Earth is $R_E = 6371$ km.

21.5. THERMONUCLEAR FUSION [11]

21.5.1. Basic data and relationships

Natural isotope ratios

Hydrogen:

$$n_D/n_H = 1.5 \times 10^{-4}.$$

Helium:

$$n_{\mathrm{He}^3}/n_{\mathrm{He}^4} = 1.3 \times 10^{-6}.$$

Lithium:

$$n_{\mathrm{Li}^6}/n_{\mathrm{Li}^7} = 0.08.$$

Mass ratios:

$$m_e/m_D = 2.72 \times 10^{-4} = 1/3670,$$

$$(m_e/m_D)^{1/2} = 1.65 \times 10^{-2} = 1/60.6,$$

$$m_e/m_T = 1.82 \times 10^{-4} = 1/5496,$$

$$(m_e/m_T)^{1/2} = 1.35 \times 10^{-2} = 1/74.1.$$

Absorbed radiation dose

$$1 \text{ rad} = 10^2 \text{ erg/g}.$$

Radioactivity

1 curie (abbreviated Ci) $= 3.7 \times 10^{10}$ radioactive decays per second.

21.5.2. Fusion reactions [12]

Branching ratios are correct for energies near the cross-section peaks; a negative yield means the reaction is endothermic:

$$
\begin{aligned}
&(1a)\ \mathrm{D}+\mathrm{D} \xrightarrow[50\%]{} \mathrm{T}(1.01\ \mathrm{MeV}) + \mathrm{p}(3.02\ \mathrm{MeV}),\\
&(1b)\ \xrightarrow[50\%]{} \mathrm{He}^3(0.82\ \mathrm{MeV}) + \mathrm{n}(2.45\ \mathrm{MeV}),\\
&(2)\ \mathrm{D}+\mathrm{T} \longrightarrow \mathrm{He}^4(3.5\ \mathrm{MeV}) + \mathrm{n}(14.1\ \mathrm{MeV}),\\
&(3)\ \mathrm{D}+\mathrm{He}^3 \longrightarrow \mathrm{He}^4(3.6\ \mathrm{MeV}) + \mathrm{p}(14.7\ \mathrm{MeV}),\\
&(4)\ \mathrm{T}+\mathrm{T} \longrightarrow \mathrm{He}^4 + 2\mathrm{n} + 11.3\ \mathrm{MeV},\\
&(5a)\ \mathrm{He}^3+\mathrm{T} \xrightarrow[51\%]{} \mathrm{He}^4 + \mathrm{p} + \mathrm{n} + 12.1\ \mathrm{MeV},\\
&(5b)\ \xrightarrow[43\%]{} \mathrm{He}^4(4.8\ \mathrm{MeV}) + \mathrm{D}(9.5\ \mathrm{MeV}),\\
&(5c)\ \xrightarrow[6\%]{} \mathrm{He}^5(2.4\ \mathrm{MeV}) + \mathrm{p}(11.9\ \mathrm{MeV}),\\
&(6)\ \mathrm{p}+\mathrm{Li}^6 \longrightarrow \mathrm{He}^4(1.7\ \mathrm{MeV}) + \mathrm{He}^3(2.3\ \mathrm{MeV}),\\
&(7a)\ \mathrm{p}+\mathrm{Li}^7 \xrightarrow[20\%]{} 2\ \mathrm{He}^4 + 17.3\ \mathrm{MeV},\\
&(7b)\ \xrightarrow[80\%]{} \mathrm{Be}^7 + \mathrm{n} - 1.6\ \mathrm{MeV},\\
&(8)\ \mathrm{D}+\mathrm{Li}^6 \longrightarrow 2\ \mathrm{He}^4 + 22.4\ \mathrm{MeV},\\
&(9)\ \mathrm{p}+\mathrm{B}^{11} \longrightarrow 3\ \mathrm{He}^4 + 8.7\ \mathrm{MeV},\\
&(10)\ \mathrm{n}+\mathrm{Li}^6 \longrightarrow \mathrm{He}^4(2.1\ \mathrm{MeV}) + T(2.7\ \mathrm{MeV}).
\end{aligned}
$$

The total cross section in barns (1 barn $= 10^{-24}$ cm^2) as a function of E, the energy in keV of the incident particle [the first ion on the left side of Eqs. (1)–(5)], assuming the target ion at rest, can be fitted by [13]

$$\sigma_T(E) = \frac{A_5 + \left[(A_4 - A_3 E)^2 + 1\right]^{-1} A_2}{E\left[\exp(A_1 E^{-1/2}) - 1\right]},$$

where the Duane coefficients A_j for the principal fusion reactions are given in the accompanying table.

TABLE 21.5. Duane coefficients.

	D–D (1a)	D–D (1b)	D–T (2)	D–He^3 (3)	T–T (4)	T–He^3 (5a–c)
A_1	46.097	47.88	45.95	89.27	38.39	123.1
A_2	372	482	50200	25900	448	11250
A_3	4.36×10^{-4}	3.08×10^{-4}	1.368×10^{-2}	3.98×10^{-3}	1.02×10^{-3}	0
A_4	1.220	1.177	1.076	1.297	2.09	0
A_5	0	0	409	647	0	0

TABLE 21.6. Reaction rates $\overline{\sigma v}$ (in $cm^3\ s^{-1}$), averaged over Maxwellian distributions.

Temperature (keV)	D–D (1a + 1b)	D–T (2)	D–He^3 (3)	T–T (4)	T–He^3 (5a–c)
1.0	1.5×10^{-22}	5.5×10^{-21}	10^{-26}	3.3×10^{-22}	10^{-28}
2.0	5.4×10^{-21}	2.6×10^{-19}	1.4×10^{-23}	7.1×10^{-21}	10^{-25}
5.0	1.8×10^{-19}	1.3×10^{-17}	6.7×10^{-21}	1.4×10^{-19}	2.1×10^{-22}
10.0	1.2×10^{-18}	1.1×10^{-16}	2.3×10^{-19}	7.2×10^{-19}	1.2×10^{-20}
20.0	5.2×10^{-18}	4.2×10^{-16}	3.8×10^{-18}	2.5×10^{-18}	2.6×10^{-19}
50.0	2.1×10^{-17}	8.7×10^{-16}	5.4×10^{-17}	8.7×10^{-18}	5.3×10^{-18}
100.0	4.5×10^{-17}	8.5×10^{-16}	1.6×10^{-16}	1.9×10^{-17}	2.7×10^{-17}
200.0	8.8×10^{-17}	6.3×10^{-16}	2.4×10^{-16}	4.2×10^{-17}	9.2×10^{-17}
500.0	1.8×10^{-16}	3.7×10^{-16}	2.3×10^{-16}	8.4×10^{-17}	2.9×10^{-16}
1000.0	2.2×10^{-16}	2.7×10^{-16}	1.8×10^{-16}	8.0×10^{-17}	5.2×10^{-16}

For low energies ($T \lesssim 25$ keV), the data may be represented by

$$(\overline{\sigma v})_{DD} = 2.33 \times 10^{-14} T^{-2/3} \exp(-18.76 T^{-1/3})\ \text{cm}^3\ \text{s}^{-1},$$

$$(\overline{\sigma v})_{DT} = 3.68 \times 10^{-12} T^{-2/3} \exp(-19.94 T^{-1/3})\ \text{cm}^3\ \text{s}^{-1},$$

where T is measured in keV.

The power density released in the form of charged particles is

$$P_{DD} = 3.3 \times 10^{-13} n_D{}^2 (\overline{\sigma v})_{DD}\ \text{W cm}^{-3}\ \text{(including the subsequent D–T reaction)},$$

$$P_{DT} = 5.6 \times 10^{-13} n_D n_T (\overline{\sigma v})_{DT}\ \text{W cm}^{-3},$$

$$P_{DHe^3} = 2.9 \times 10^{-12} n_D n_{He^3} (\overline{\sigma v})_{DHe^3}\ \text{W cm}^{-3}.$$

21.6. ELECTRON AND ION BEAMS

Here $\gamma = (1 - \beta^2)^{-1/2}$ is the relativistic scaling factor; quantities in analytic formulas are expressed in SI or cgs units, as indicated; in numerical formulas I is in amperes (A), B is in gauss (G), electron linear density N is in cm^{-1}, temperature, voltage, and energy are in MeV, $\beta_z = v_z/c$, and k is Boltzmann's constant.

Relativistic electron gyroradius:

$$r_e = \frac{mc^2}{eB}(\gamma^2 - 1)^{1/2}\ \text{(cgs)} = 1.70 \times 10^3 (\gamma^2 - 1)^{1/2} B^{-1}\ \text{cm}.$$

Relativistic electron energy:

$$W = mc^2\gamma = 0.511\gamma\ \text{MeV}.$$

Bennett pinch condition:

$$I^2 = 2Nk(T_e + T_i)c^2 \text{ (cgs)} = 3.20 \times 10^{-4} N(T_e + T_i) \text{ A}^2.$$

Alfvén-Lawson limit:

$$I_A = (mc^3/e)\beta_z\gamma \text{ (cgs)} = (4\pi mc/\mu_0 e)\beta_z\gamma \text{ (SI)} = 1.70 \times 10^4 \beta_z\gamma \text{A}.$$

The ratio of net current to I_A is

$$\frac{I}{I_A} = \frac{\nu}{\gamma}.$$

Here $\nu = Nr_e$ is the Budker number, where $r_e = e^2/mc^2 = 2.82 \times 10^{-13}$ cm is the classical electron radius. Beam electron number density is

$$n_b = 2.08 \times 10^8 J\beta^{-1} \text{ cm}^{-3},$$

where J is the current density in A cm^{-2}. For a uniform beam of radius a (in cm),

$$n_b = 6.63 \times 10^7 Ia^{-2}\beta^{-1}\text{cm}^{-3}$$

and

$$\frac{2r_e}{a} = \frac{\nu}{\gamma}.$$

Child's law: nonrelativistic space-charge-limited current density between parallel plates with voltage drop V (in MV) and separation d (in cm) is

$$J = 2.34 \times 10^3 V^{3/2} d^{-2} \text{ A cm}^{-2}.$$

The saturated parapotential current (magnetically self-limited flow along equipotentials) in pinched diodes and transmission lines is [14]

$$I_p = 8.5 \times 10^3 G\gamma \ln\left[\gamma + (\gamma^2 - 1)^{1/2}\right] \text{ A},$$

where G is a geometrical factor depending on the diode structure:

$$G = \frac{w}{2\pi d}$$ for parallel plane cathode and anode of width w, separation d,

$$G = \left(\ln \frac{R_2}{R_1}\right)^{-1}$$ for cylinders of radii R_1 (inner) and R_2 (outer),

$$G = \frac{R_c}{d_0}$$ for conical cathode of radius R_c, maximum separation d_0 (at $r = R_c$) from plane anode.

For $\beta \to 0$ ($\gamma \to 1$), both I_A and I_p vanish.

The condition for a longitudinal magnetic field B_z to suppress filamentation in a beam of current density J (in A cm^{-2}) is

$$B_z > 47\beta_z(\gamma J)^{1/2}\ \mathrm{G}.$$

Voltage registered by Rogowski coil of minor cross-sectional area A (m^2), n turns, major radius a (m), inductance L (H), external resistance R (Ω), and capacitance C (F):

externally integrated $\quad V = (1/RC)(nA\mu_0 I/2\pi a),$

self-integrating $\quad V = (R/L)(nA\mu_0 I/2\pi a) = RI/n.$

X-ray production, target with average atomic number Z ($V \lesssim 5$MeV):

$$\eta \equiv \text{x-ray power/beam power} = 7 \times 10^{-4} ZV.$$

X-ray dose at 1 m generated by an e-beam depositing total charge Q coulombs while $V \geq 0.84 V_{\max}$ in material with charge state Z:

$$D = 150 V_{\max}^{2.8} Q Z^{1/2}\ \text{rads}.$$

In the following tables, subscripts e, i, d, b, and p stand for "electron," "ion," "drift," "beam," and "plasma," respectively. Thermal velocities are denoted by a bar. In addition, the following symbols are used:

m	electron mass	r_e, r_i	gyroradius
M	ion mass	β	plasma/magnetic energy density ratio
V	velocity		
T	temperature	V_A	Alfvén speed
n_e, n_i	number density	Ω_e, Ω_i	gyrofrequency
n	harmonic number	Ω_H	hybrid gyrofrequency, $\Omega_H{}^2 = \Omega_e\Omega_i$
$C_s = (T_e/M)^{1/2}$	ion sound speed		
ω_e, ω_i	plasma frequency	U	relative drift velocity of two ion species
λ_D	Debye length		

TABLE 21.7. Beam instabilities. [15]

Name	Conditions	Saturation mechanism	Parameters of most unstable mode: Growth rate	Frequency	Wavenumber	Group velocity
Electron-electron	$V_d > \overline{V}_{ej},\ j = 1, 2$	Electron trapping until $\overline{V}_{ej} \sim V_d$	$\frac{1}{2}\omega_e$	0	$0.9\frac{\omega_e}{V_d}$	0
Buneman	$V_d > (M/m)^{1/3}\overline{V}_i$, $V_d > \overline{V}_e$	Electron trapping until $\overline{V}_e \sim V_d$	$0.7\left(\frac{m}{M}\right)^{1/3}\omega_e$	$0.4\left(\frac{m}{M}\right)^{1/3}\omega_e$	$\frac{\omega_e}{V_d}$	$\frac{2}{3}V_d$
Beam-plasma	$V_b > (n_p/n_b)^{1/3}\overline{V}_b$	Trapping of beam electrons	$0.7\left(\frac{n_b}{n_p}\right)^{1/3}\omega_e$	$\omega_e - 0.4\left(\frac{n_b}{n_p}\right)^{1/3}\omega_e$	$\frac{\omega_e}{V_b}$	$\frac{2}{3}V_b$
Weak beam-plasma	$V_b < (n_p/n_b)^{1/3}\overline{V}_b$	Quasilinear or nonlinear (mode coupling)	$\frac{n_b}{2n_p}\left(\frac{V_b}{\overline{V}_b}\right)^2\omega_e$	ω_e	$\frac{\omega_e}{V_b}$	$\frac{3\overline{V}_e^{\,2}}{V_b}$
Beam-plasma (hot-electron)	$\overline{V}_e > V_b > \overline{V}_b$	Quasilinear or nonlinear	$\left(\frac{n_b}{n_p}\right)^{1/2}\frac{\overline{V}_e}{V_b}\omega_e$	$\frac{V_b}{\overline{V}_e}\omega_e$	λ_D^{-1}	V_b
Ion acoustic	$T_e \gg T_i,\ V_d \gg C_s$	Quasilinear, ion tail formation, nonlinear scattering, or resonance broadening.	$\left(\frac{m}{M}\right)^{1/2}\omega_i$	ω_i	λ_D^{-1}	C_s
Anisotropic temperature (hydro)	$T_{e\perp} > 2T_{e\parallel}$	Isotropization Ω_e	Ω_e	$\omega_e \cos\theta \sim \Omega_e$	r_e^{-1}	$\overline{V}_{e\perp}$
Ion cyclotron	$V_d > 20\overline{V}_i$ (for $T_e \approx T_i$)	Ion heating	$0.1\Omega_i$	$1.2\Omega_i$	r_i^{-1}	$\frac{1}{3}\overline{V}_i$
Beam-cyclotron (hydro)	$V_d > C_s$	Resonance broadening	$0.7\Omega_e$	$n\Omega_e$	$0.7\lambda_D^{-1}$	$\gtrsim V_d;\ \lesssim C_s$
Modified two-stream (hydro)	$V_d < (1+\beta)^{1/2}V_A$, $V_d > C_s$	Trapping	$\frac{1}{2}\Omega_H$	$0.9\Omega_H$	$1.7\frac{\Omega_H}{V_d}$	$\frac{1}{2}V_d$
Ion-ion (equal beams)	$U < 2(1+\beta)^{1/2}V_A$	Ion trapping	$0.4\Omega_H$	0	$1.2\frac{\Omega_H}{U}$	0
Ion-ion (equal beams)	$U < 2C_s$	Ion trapping	$0.4\omega_i$	0	$1.2\frac{\omega_i}{U}$	0

21.7. LASER–PLASMA INTERACTIONS

21.7.1. System parameters

Efficiencies and power levels are approximately state-of-the-art. [16]

Type	Wavelength (μm)	Efficiency	Power levels available (W)	
			Pulsed**	CW
CO_2	10.6	0.01–0.02 (pulsed)	$> 2 \times 10^{13}$	$> 10^5$
CO	5	0.4	$> 10^9$	> 100
Holmium	2.06	0.03†–0.1‡	$> 10^7$	30
Iodine	1.315	0.003	$> 10^{12}$	—
Nd-glass, YAG	1.06	0.001–0.06† > 0.1‡	$\sim 10^{15}$ (ten-beam system)	1–1000
*Color center	1–4	0.001	$> 10^6$	1
*Vibronic (Ti Sapphire)	0.7–0.9	$> 0.1\eta_p$	10^6	1–5
Ruby	0.6943	$< 10^{-3}$	10^{10}	1
He-Ne	0.6328	10^{-4}	—	0.001–0.05
*Argon ion	0.45–0.60	10^{-3}	5×10^4	1–20
*OPO	0.4–9.0	$> 0.1\,\eta_p$	10^6	1–5
N_2	0.3371	0.001–0.05	$10^5 - 10^6$	-
*Dye	0.3–1.1	10^{-3}	$> 10^6$	140
Kr-F	0.248	0.08	$> 10^{12}$	500
Xenon	0.175	0.02	$> 10^8$	—

*Tunable sources.

†Lamp-driven.

‡diode-driven.

**The power level for most pulsed systems can be raised by a factor of 10^2–10^3 by pulse compression ("chirping").

YAG: Yttrium–Aluminum Garnet. OPO: Optical Parametric Oscillator.

η_p: Pump laser efficiency.

21.7.2. Formulas

An electromagnetic wave with **k** $\parallel$ **B** has an index of refraction given by

$$n_\pm = [1 - \omega_{pe}^2/\omega(\omega \mp \omega_{ce})]^{1/2},$$

where $\pm$ refers to the helicity. The rate of change of polarization angle θ as a function of displacement s (Faraday rotation) is given by

$$d\theta/ds = (k/2)(n_- - n_+) = 2.36 \times 10^4 NBf^{-2}\ \mathrm{cm}^{-1},$$

where N (cm^{-3}) is the electron number density, B (G) is the field strength, and f (s^{-1}) is the wave frequency.

The quiver velocity of an electron in an electromagnetic field of angular frequency ω is

$$v_0 = eE_{\max}/m\omega = 25.6(I/\text{W cm}^{-2})^{1/2}(\lambda_0/\mu\text{m}) \text{ cm s}^{-1}$$

in terms of the laser flux $I = cE_{\max}^2/8\pi$. The ratio of quiver energy to thermal energy is

$$W_{\text{qu}}/W_{\text{th}} = m_e v_0^2/2kT = 1.81 \times 10^{-13}(\lambda_0/\mu\text{m})^2(I/\text{W/cm}^2)/(T/\text{eV}).$$

For example, if $I = 10^{15}$ W cm^{-2}, $\lambda_0 = 1\ \mu$m, and $T = 2$ keV, then $W_{\text{qu}}/W_{\text{th}} \approx 0.1$.

Pondermotive force:

$$\mathcal{F} = N\nabla\langle E^2\rangle/8\pi N_c,$$

where

$$N_c = 1.1 \times 10^{21}\lambda_0{}^{-2} \text{ cm}^{-3}.$$

For uniform illumination of a lens with f-number F, the diameter d at focus (85% of the energy) and the depth of focus l (distance to first zero in intensity) are given by

$$d \approx 2.44F\lambda\theta/\theta_{DL} \quad \text{and} \quad l \approx \pm 2F^2\lambda\theta/\theta_{DL}.$$

Here θ is the beam divergence containing 85% of energy and θ_{DL} is the diffraction-limited divergence:

$$\theta_{DL} = 2.44\lambda/b,$$

where b is the aperture. These formulas are modified for nonuniform (such as Gaussian) illumination of the lens or for pathological laser profiles.

21.8. ATOMIC PHYSICS AND RADIATION

Energies and temperatures are in eV; all other units are cgs except where noted. Z is the charge state ($Z = 0$ refers to a neutral atom). The subscript e labels electrons. N is number density and n is principal quantum number. Asterisk superscripts on level population densities denote local thermodynamic equilibrium (LTE) values. Thus, N_n^* is the LTE number density of atoms (or ions) in level n.

Characteristic atomic collision cross section:

$$\pi a_0{}^2 = 8.80 \times 10^{-17} \text{ cm}^2. \tag{21.1}$$

Binding energy of outer electron in level labeled by quantum numbers n, l:

$$E_\infty^Z(n, l) = -\frac{Z^2 E_\infty^H}{(n - \Delta_l)^2}, \tag{21.2}$$

where $E_\infty^H = 13.6$ eV is the hydrogen ionization energy and $\Delta_l = 0.75l^{-5}$, $l \gtrsim 5$, is the quantum defect.

21.8.1. Excitation and decay

Cross section (Bethe approximation) for electron excitation by dipole allowed transition $m \to n$ (Refs. [17] and [18]):

$$\sigma_{mn} = 2.36 \times 10^{-13} \frac{f_{nm} g(n, m)}{\epsilon \Delta E_{nm}} \text{ cm}^2, \tag{21.3}$$

where f_{nm} is the oscillator strength, $g(n, m)$ is the Gaunt factor, ϵ is the incident electron energy, and $\Delta E_{nm} = E_n - E_m$.

Electron excitation rate (Refs. [19] and [20]), averaged over Maxwellian velocity distribution, $X_{mn} = N_e \langle \sigma_{mn} v \rangle$,

$$X_{mn} = 1.6 \times 10^{-5} \frac{f_{nm} \langle g(n, m) \rangle N_e}{\Delta E_{nm} T_e^{1/2}} \exp\left(-\frac{\Delta E_{nm}}{T_e}\right) \text{ s}^{-1}, \tag{21.4}$$

where $\langle g(n, m) \rangle$ denotes the thermal averaged Gaunt factor (generally $\sim$1 for atoms, ~ 0.2 for ions).

Rate for electron collisional deexcitation:

$$Y_{nm} = (N_m{}^*/N_n{}^*) X_{mn}. \tag{21.5}$$

Here $N_m{}^*/N_n{}^* = (g_m/g_n) \exp(\Delta E_{nm}/T_e)$ is the Boltzmann relation for level population densities, where g_n is the statistical weight of level n.

Rate for spontaneous decay $n \to m$ (Einstein A coefficient): [19]

$$A_{nm} = 4.3 \times 10^7 (g_m/g_n) f_{mn} (\Delta E_{nm})^2 \text{ s}^{-1}. \tag{21.6}$$

Intensity emitted per unit volume from the transition $n \to m$ in an optically thin plasma:

$$I_{nm} = 1.6 \times 10^{-19} A_{nm} N_n \Delta E_{nm} \text{ W cm}^{-3}. \tag{21.7}$$

Condition for steady state in a corona model:

$$N_0 N_e \langle \sigma_{0n} v \rangle = N_n A_{n0}, \tag{21.8}$$

where the ground state is labeled by a zero subscript.

Hence for a transition $n \to m$ in ions, where $\langle g(n, 0) \rangle \approx 0.2$,

$$I_{nm} = 5.1 \times 10^{-25} \frac{f_{nm} g_0 N_e N_0}{g_m T_e^{1/2}} \left(\frac{\Delta E_{nm}}{\Delta E_{n0}}\right)^3 \exp\left(-\frac{\Delta E_{n0}}{T_e}\right) \text{ W cm}^{-3}. \tag{21.9}$$

21.8.2. Ionization and recombination

In a general time-dependent situation, the number density of the charge state Z satisfies

$$\frac{dN(Z)}{dt} = N_e \left[-S(Z)N(Z) - \alpha(Z)N(Z) + S(Z-1)N(Z-1) + \alpha(Z+1)N(Z+1)\right]. \tag{21.10}$$

Here $S(Z)$ is the ionization rate. The recombination rate $\alpha(Z)$ has the form $\alpha(Z) = \alpha_r(Z) + N_e\alpha_3(Z)$, where α_r and α_3 are the radiative and three-body recombination rates, respectively.

Classical ionization cross section [21] for any atomic shell j:

$$\sigma_i = 6 \times 10^{-14} b_j g_j(x)/U_j{}^2 \text{ cm}^2. \tag{21.11}$$

Here b_j is the number of shell electrons; U_j is the binding energy of the ejected electron; $x = \epsilon/U_j$, where ϵ is the incident electron energy; and g is a universal function with a minimum value $g_{\min} \approx 0.2$ at $x \approx 4$.

Ionization from the ion ground state, averaged over a Maxwellian electron distribution, for $0.02 \lesssim T_e/E_\infty^Z \lesssim 100$ (Ref. [20]):

$$S(Z) = 10^{-5} \frac{(T_e/E_\infty^Z)^{1/2}}{(E_\infty^Z)^{3/2}(6.0 + T_e/E_\infty^Z)} \exp\left(-\frac{E_\infty^Z}{T_e}\right) \text{cm}^3 \text{ s}^{-1}, \tag{21.12}$$

where E_∞^Z is the ionization energy.

Electron–ion radiative recombination rate $(e + N(Z) \to N(Z-1) + h\nu)$, $T_e/Z^2 \lesssim 400$ eV (Ref. [22]):

$$\alpha_r(Z) = 5.2 \times 10^{-14} Z \left(\frac{E_\infty^Z}{T_e}\right)^{1/2} \left[0.43 + \frac{1}{2}\ln(E_\infty^Z/T_e) + 0.469(E_\infty^Z/T_e)^{-1/3}\right] \text{cm}^3 \text{ s}^{-1}. \tag{21.13}$$

For 1 eV $< T_e/Z^2 <$ 15 eV this becomes approximately [20]

$$\alpha_r(Z) = 2.7 \times 10^{-13} Z^2 T_e{}^{-1/2} \text{ cm}^3 \text{ s}^{-1}. \tag{21.14}$$

Collisional (three-body) recombination rate for singly ionized plasma: [23]

$$\alpha_3 = 8.75 \times 10^{-27} T_e{}^{-4.5} \text{ cm}^6 \text{ s}^{-1}. \tag{21.15}$$

Photoionization cross section for ions in level n, l (short-wavelength limit):

$$\sigma_{\text{ph}}(n, l) = 1.64 \times 10^{-16} Z^5/n^3 K^{7+2l} \text{ cm}^2, \tag{21.16}$$

where K is the wavenumber in Rydbergs (1 Rydberg $= 1.0974 \times 10^5$ cm^{-1}).

21.8.3. Ionization equilibrium models

Saha equilibrium: [24]

$$\frac{N_e N_1^*(Z)}{N_n^*(Z-1)} = 6.0 \times 10^{21} \frac{g_1^Z T_e{}^{3/2}}{g_n^{Z-1}} \exp\left(-\frac{E_\infty^Z(n,l)}{T_e}\right) \text{ cm}^{-3}, \tag{21.17}$$

where g_n^Z is the statistical weight for level n of charge state Z and $E_\infty^Z(n, l)$ is the ionization energy of the neutral atom initially in level (n, l), given by Eq. (21.2).

In a steady state at high electron density,

$$\frac{N_e N^*(Z)}{N^*(Z-1)} = \frac{S(Z-1)}{\alpha_3}, \tag{21.18}$$

a function only of T.

Conditions for LTE: [24]

(i) Collisional and radiative excitation rates for a level n must satisfy

$$Y_{nm} \gtrsim 10 A_{nm}. \tag{21.19}$$

(ii) Electron density must satisfy

$$N_e \gtrsim 7 \times 10^{18} Z^7 n^{-17/2} (T/E_\infty^Z)^{1/2} \text{ cm}^{-3}. \tag{21.20}$$

Steady-state condition in corona model:

$$\frac{N(Z-1)}{N(Z)} = \frac{\alpha_r}{S(Z-1)}. \tag{21.21}$$

Corona model is applicable if [25]

$$10^{12} t_I{}^{-1} < N_e < 10^{16} T_e{}^{7/2} \text{ cm}^{-3}, \tag{21.22}$$

where t_I is the ionization time.

21.8.4. Radiation

N. B. Energies and temperatures are in eV; all other quantities are in cgs units except where noted. Z is the charge state ($Z = 0$ refers to a neutral atom). The subscript e labels electrons. N is number density; n is principal quantum number.

Average radiative decay rate of a state with principal quantum number n is

$$A_n = \sum_{m<n} A_{nm} = 1.6 \times 10^{10} Z^4 n^{-9/2} \text{ s}^{-1}. \tag{21.23}$$

Natural linewidth (ΔE in eV):

$$\Delta E \, \Delta t = h = 4.14 \times 10^{-15} \text{ eV s}, \tag{21.24}$$

where Δt is the lifetime of the line.

Doppler width:

$$\Delta\lambda/\lambda = 7.7 \times 10^{-5} (T/\mu)^{1/2}, \tag{21.25}$$

where μ is the mass of the emitting atom or ion scaled by the proton mass.

Optical depth for a Doppler-broadened line: [24]

$$\tau = 3.52 \times 10^{-13} f_{nm} \lambda (Mc^2/kT)^{1/2} NL = 5.4 \times 10^{-9} \lambda (\mu/T)^{1/2} NL, \tag{21.26}$$

where f_{nm} is the absorption oscillator strength, λ is the wavelength, and L is the physical depth of the plasma; M, N, and T are the mass, number density, and temperature of the absorber; μ is M divided by the proton mass. Optically thin means $\tau < 1$.

Resonance absorption cross section at center of line:

$$\sigma_{\lambda=\lambda_c} = 5.6 \times 10^{-13} \lambda^2 / \Delta\lambda \text{ cm}^2. \tag{21.27}$$

Wien displacement law (wavelength of maximum black-body emission):

$$\lambda_{\text{max}} = 2.50 \times 10^{-5} T^{-1} \text{ cm}. \tag{21.28}$$

Radiation from the surface of a black body at temperature T:

$$W = 1.03 \times 10^5 T^4 \text{ W cm}^{-2}. \tag{21.29}$$

Bremsstrahlung from hydrogen-like plasma: [11]

$$P_{\text{Br}} = 1.69 \times 10^{-32} N_e T_e{}^{1/2} \sum \left[Z^2 N(Z) \right] \text{ W cm}^{-3}, \tag{21.30}$$

where the sum is over all ionization states Z.

Bremsstrahlung optical depth: [26]

$$\tau = 5.0 \times 10^{-38} N_e N_i Z^2 \overline{g} L T^{-7/2}, \tag{21.31}$$

where $\overline{g} \approx 1.2$ is an average Gaunt factor and L is the physical path length.

Inverse bremsstrahlung absorption coefficient [27] for radiation of angular frequency ω:

$$\kappa = 3.1 \times 10^{-7} Z n_e{}^2 \ln \Lambda \, T^{-3/2} \omega^{-2} (1 - \omega_p^2/\omega^2)^{-1/2} \text{ cm}^{-1}; \tag{21.32}$$

here Λ is the electron thermal velocity divided by V, where V is the larger of ω and ω_p multiplied by the larger of Ze^2/kT and $\hbar/(mkT)^{1/2}$.

Recombination (free–bound) radiation:

$$P_r = 1.69 \times 10^{-32} N_e T_e{}^{1/2} \sum \left[Z^2 N(Z) \left(\frac{E_\infty^{Z-1}}{T_e} \right) \right] \text{ W cm}^{-3}. \tag{21.33}$$

Cyclotron radiation [11] in magnetic field $\mathbf{B}$:

$$P_c = 6.21 \times 10^{-28} B^2 N_e T_e \text{ W cm}^{-3}. \tag{21.34}$$

For $N_e k T_e = N_i k T_i = B^2/16\pi$ ($\beta = 1$, isothermal plasma), [11]

$$P_c = 5.00 \times 10^{-38} N_e^2 T_e^2 \text{ W cm}^{-3}. \tag{21.35}$$

Cyclotron radiation energy loss e-folding time for a single electron: [26]

$$t_c \approx \frac{9.0 \times 10^8 B^{-2}}{2.5 + \gamma} \text{ s}, \tag{21.36}$$

where γ is the kinetic plus rest energy divided by the rest energy mc^2.

Number of cyclotron harmonics [26] trapped in a medium of finite depth L:

$$m_{\rm tr} = (57\beta BL)^{1/6}, \tag{21.37}$$

where $\beta = 8\pi NkT/B^2$.

Line radiation is given by summing Eq. (21.9) over all species in the plasma.

21.9. REFERENCES

Most of this material is well known and for all practical purposes is in the "public domain." In the task of collecting and shaping it into the present form, I have been helped by colleagues and readers too numerous to list by name, all of whom I sincerely thank.

Several book-length compilations of data relevant to plasma physics are available. The following are particularly useful:

C. W. Allen, *Astrophysical Quantities*, 3rd ed. (Athlone Press, London, 1976).

A. Anders, *A Formulary for Plasma Physics* (Akademie–Verlag, Berlin, 1990).

K. R. Lang, *Astrophysical Formulae*, 2nd ed. (Springer, New York, 1980).

Additional material can also be found in D. L. Book, NRL Memorandum Report No. 3332, 1977.

[1] J. Sheffield, *Plasma Scattering of Electromagnetic Radiation* (Academic Press, New York, 1975), p. 6 (after J. W. Paul).

[2] The Z function is tabulated in B. D. Fried and S. D. Conte, *The Plasma Dispersion Function* (Academic Press, New York, 1961).

[3] T. H. Stix, *Waves in Plasmas* (American Institute of Physics, College Park, MD, 1992), pp. 207–210.

[4] B. D. Fried, C. L. Hedrick, and J. McCune, Phys. Fluids **11**, 249 (1968).

[5] B. A. Trubnikov, in *Reviews of Plasma Physics*, edited by M. A. Leontovich (Consultants Bureau, New York, 1965), Vol. 1, p. 105.

[6] J. M. Greene, Phys. Fluids **16**, 2022 (1973).

[7] S. I. Braginskii, in *Reviews of Plasma Physics*, edited by M. A. Leontovich (Consultants Bureau, New York, 1965), Vol. 1, p. 205.

[8] C. W. Allen, *Astrophysical Quantities*, 3rd ed. (Athlone Press, London, 1976), Ch. 9.

[9] G. L. Withbroe and R. W. Noyes, Ann. Rev. Astrophys. **15**, 363 (1977).

[10] K. H. Lloyd and G. Härendel, J. Geophys. Res. **78**, 7389 (1973).

[11] S. Glasstone and R. H. Lovberg, *Controlled Thermonuclear Reactions* (Van Nostrand, New York, 1960), Ch. 2.

[12] References to experimental measurements of branching ratios and cross sections are listed in F. K. McGowan *et al.*, Nucl. Data Tables **A6**, 353 (1969); **A8**, 199 (1970). The yields listed in the table are calculated directly from the mass defect.

[13] G. H. Miley, H. Towner, and N. Ivich, Report COO-2218-17, University of Illinois, Urbana, IL, 1974; B. H. Duane, Report BNWL-1685, Brookhaven National Laboratory, 1972.

[14] J. M. Creedon, J. Appl. Phys. **46**, 2946 (1975).

[15] A. B. Mikhailovskii, *Theory of Plasma Instabilities* (Consultants Bureau, New York, 1974), Vol. I. The table on p. 647 was compiled by K. Papadopoulos.

[16] Table prepared from data compiled by J. M. McMahon (private communication).

[17] M. J. Seaton, in *Atomic and Molecular Processes*, edited by D. R. Bates (New York, Academic Press, 1962), Ch. 11.

[18] H. Van Regemorter, Astrophys. J. **136**, 906 (1962).

[19] A. C. Kolb and R. W. P. McWhirter, Phys. Fluids **7**, 519 (1964).

[20] R. W. P. McWhirter, in *Plasma Diagnostic Techniques*, edited by R. H. Huddlestone and S. L. Leonard (Academic Press, New York, 1965).

[21] M. Gryzinski, Phys. Rev. **138A**, 336 (1965).

[22] M. J. Seaton, Mon. Not. Roy. Astron. Soc. **119**, 81 (1959).

[23] Ya. B. Zel'dovich and Yu. P. Raizer, *Physics of Shock Waves and High-Temperature Hydrodynamic Phenomena* (Academic Press, New York, 1966), Vol. I, p. 407.

[24] H. R. Griem, *Plasma Spectroscopy* (Academic Press, New York, 1966).

[25] T. F. Stratton, in *Plasma Diagnostic Techniques*, edited by R. H. Huddlestone and S. L. Leonard (Academic Press, New York, 1965).

[26] G. Bekefi, *Radiation Processes in Plasmas* (John Wiley and Sons, New York, 1966).

[27] T. W. Johnston and J. M. Dawson, Phys. Fluids **16**, 722 (1973).

[28] W. L. Wiese, M. W. Smith, and B. M. Glennon, *Atomic Transition Probabilities*, NSRDS-NBS 4 (U.S. Govt. Printing Office, Washington, DC, 1966), Vol. 1.

22

Polymer Physics

Stephen Z. D. Cheng
The University of Akron

Based on Ch. 13 in *Physicist's Desk Reference*, 2nd edition, prepared by Professor R. K. Eby.

Contents

List of Tables

List of Figures

22.1. INTRODUCTION

Polymer science as a coherent discipline of physics is barely 50 years old. Therefore, many of the concepts and experimental data are not in a final state. In this chapter, equations and data are presented, of which we believe to have wide acceptance. Space limitation precluded completeness and required us to omit a large amount of material that we wished to include. We believe, however, that the included material and references will prove useful to those working in polymer physics.

22.2. POLYMER MOLECULES

In the simplest cases, polymers are made by the chemical combination of small chemical units that are known as monomers (or repeating units). The polymer name is usually derived from that of the monomer. Thus, polyethylene has the structural formula

$$[-CH_2-CH_2-]_N,$$

where N represents the degree of polymerization or the number of times the unit is repeated in the molecule. Degrees of polymerization for polymers are often in the range 10^3–10^5. The range can even be exhibited within one sample. Therefore, polymers are often characterized by an average degree of polymerization or molecule mass. Other important structural variations, including sterochemical, geometrical, and compositional, are discussed in Refs. [1–3], in which structural formulas for many polymers are also given.

22.3. MOLECULAR-MASS AVERAGES

22.3.1. *k*th moment of a molecular-mass distribution *P*(*M*)

$$\langle M^k \rangle \equiv \int_0^\infty M^k P(M)\, dM.$$

22.3.2. Molecular-mass averages

1. Number average molecular mass:

$$M_n \equiv \langle M \rangle.$$

2. Weight average molecular mass:

$$M_w \equiv \langle M^2 \rangle / \langle M \rangle.$$

3. Z-average molecular mass:

$$M_Z \equiv \langle M^3 \rangle / \langle M^2 \rangle.$$

4. Viscosity average molecular mass:

$$M_v \equiv \left(\langle M^{1+\alpha} \rangle / \langle M^\alpha \rangle \right)^{1/\alpha}.$$

α is solvent dependent and usually lies between 0.5 and 0.8.

5. Polydispersity ratio:

$$R \equiv M_w / M_n \equiv \langle M^2 \rangle / \langle M \rangle^2, \; R - 1 \equiv \sigma^2 / \langle M \rangle^2,$$

where $\sigma^2 \equiv \langle M^2 \rangle - \langle M \rangle^2$.

Values of R are a measure of the width of the molecular-weight distribution.

22.4. SINGLE-CHAIN DIMENSIONS

A schematic representation of a homopolymer chain of $N + 1$ mass points is shown in Figure 22.1. For a detailed discussion, see Refs. [4] and [5].

End-to-end vector:

$$\mathbf{R} \equiv \sum_{i=1}^{N} \mathbf{r}_i.$$

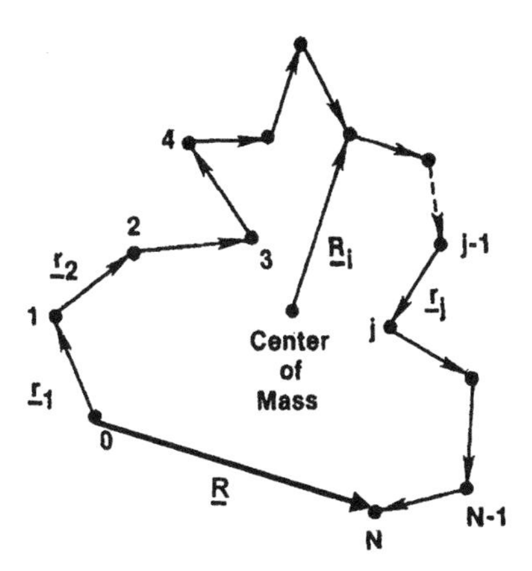

FIGURE 22.1. Schematic representation of homopolymer chain of $N + 1$ mass points.

Mean-square end-to-end distance:

$$\langle R^2 \rangle \equiv \sum_{i=1}^{N} \sum_{j=1}^{N} \langle \mathbf{r}_i \cdot \mathbf{r}_j \rangle.$$

Mean-square radius of gyration (homopolymer chain):

$$\begin{aligned}\langle S^2 \rangle &\equiv (N+1)^{-1} \sum_{i=0}^{N} \mathbf{R}_j^2 \\ &\equiv (N+1)^{-2} \sum_{i<j}^{N} \mathbf{R}_{ij}^2, \quad \mathbf{R}_{ij} \equiv \mathbf{R}_i - \mathbf{j} \\ &\equiv (N+1)^{-1} \sum_{i=1}^{N} \sum_{j=1}^{N} g_{ij} \langle \mathbf{r}_j \cdot \mathbf{R}_j \rangle,\end{aligned}$$

$$g_{ij} = \begin{cases} i[1 - j/(N+1)], & i \le j, \\ j[1 - i/(N+1)], & i > j. \end{cases}$$

Special values of $\langle S^2 \rangle$ valid for large N:

1. Linear, freely jointed chain: $\langle S^2 \rangle = \langle R^2 \rangle / 6 = Nb^2/6$, where b is the bond length.
2. Cyclic, freely jointed chain: $\langle S^2 \rangle = Nb^2/12$.
3. Freely jointed chain whose end-to-end distance is fixed at R: $\langle S^2 \rangle = (Nb^2 + R^2)/12$.
4. Linear, freely rotating chain with fixed bond angles of $\pi - \theta$.

$$\langle \mathbf{r}_i \cdot \mathbf{r}_{i+k} \rangle = b^2 \cos^k \theta, \ \langle S^2 \rangle = \frac{Nb^2}{6} \frac{1+\cos\theta}{1-\cos\theta}.$$

5. Rigid rod: $\langle S^2 \rangle = (Nb)^2/12$.

Characteristic ratio: [2]

$$C_\infty \equiv \lim_{N \to \infty} \langle R^2 \rangle / Nb^2.$$

Freely jointed chain: $C_\infty = 1$.

Freely rotating chain: $C_\infty = (1 + \cos\theta)/(1 - \cos\theta)$.

The characteristic ratio is a measure of chain backbone rigidity or stiffness.

22.5. θ SOLVENTS AND TEMPERATURES

A theta (θ) solvent is the generic name given to those polymer solvents in which upper critical temperatures are observed (typically, near room temperature). With increasing polymer molecular mass, the upper critical solution temperature (UCST) approaches a limiting value called the θ temperature.

A solvent is termed a *good solvent* if the temperature is well above the UCST. θ solvents are sometimes referred to as *poor solvents*. For large N, chain dimensions in solution vary as N^{γ}. In a θ solvent, $\gamma = 1$, whereas in a good solvent, $\gamma \approx 1.2$.

Selected θ solvents and temperatures are given in Table 22.1.

TABLE 22.1. Selected θ solvents and temperatures. (From Ref. [6].)

Polymer	Solvent	θ temp. (°C)
Polyethylene	n-pentane	85
	diphenyl	128
	diphenylmethane	142
	n-decanol	153
	dephenylether	165
Polyproylene		
Atactic	i-amyl acetate	34
	n-butyl acetate	58.5
	cyclohexanone	92
	diphenyl	129
Isotactic	i-amyl acetate	70
	l-cholronaphthalene	74
	diphenyl	125
Syndiotactic	i-amyl acetate	45
Poly(l-butene)		
Atactic	i-amyl acetate	23
	anisole	83
	diphenyl ether	141
Isotactic	anisle	89
	diphenyl ether	148
Polyisobutene	benzene	25
	ethylbenzene	−24
	toluene	−13
Polystyrene (atactic)	cyclohexane	35
	methylcyclohexane	60

22.6. MOLECULAR-WEIGHT CHARACTERIZATION

22.6.1. Solution viscosity

Specific viscosity:

$$\eta_{sp} \equiv (\eta - \eta_0)/\eta_0,$$

where η is the solution viscosity and η_0 is the solvent viscosity.

Limiting viscosity number or intrinsic viscosity:

$$[\eta] \equiv \lim_{c \to 0} \eta_{sp}/c,$$

where c is the polymer concentration.

Dependence on concentration:

$$\eta_{sp}/c = [\eta] + k[\eta]^2 c + \cdots,$$

where k is the Huggins constant; $0.3 < k < 0.4$ for most good solvents. Values of $[\eta]$ and k for polystyrene dissolved in toluene at 30°C are given in Table 22.2.

Mark-Houwink-Sakurada equation (semiempirical):

$$[\eta] = KM^{\alpha},$$

where K and α are constasts for a given polymer/solvent and are molecular-mass independent over a broad range of molecular mass ($\alpha = 0.5$ for a θ solvent). Tabulated values of K and α are available for many polymer/solvent pairs. [6]

22.6.2. Osmotic pressure π

Concentration, c, dependence:

$$\pi/cRT = 1/M_n + A_2 c + \cdots,$$

where A_2 is the second virial coefficient.

TABLE 22.2. Values of $[\eta]$ and k for polystyrene dissolved in toluene at 30°C. [From F. Danusso and G. Moraglio, J. Polym. Sci. 24, 161 (1957).]

M_n (g/mol)	$[\eta]$ (cm^3/g)	k
76 000	38.2	0.31
135 000	59.2	0.33
163 000	69.6	0.33
336 000	105.4	0.35
440 000	129.2	0.34
556 000	165.0	0.31
850 000	221.0	0.31

van't Hoff's law:

$$\lim_{c \to 0} \pi/cRT = 1/M_n.$$

22.6.3. Ultracentrifugation

An ultracentrifugation can be used to measure M_w and, under appropriate conditions, M_Z. [7] The method is based on the fact that heavier particles sediment more rapidly than do lighter particles.

22.6.4. Static light scattering [5,8–10]

Consider a polymer solution of concentration c scattering light of incident intensity I_0. If the incident light is unpolarized, the intensity $I(\theta)$, at a distance r from a unit volume of solution, of light scattered in a direction that makes an angle θ with the incident beam is

$$\frac{I(\theta)}{I_0} = \frac{K(1 + \cos^2\theta)c(1 - q^2\langle S^2\rangle/3 + \cdots)}{r^2(1/M_w + 2A_2c + \cdots)},$$

where $K \equiv 2\pi^2 n^2 (dn/dc)^2/N_A\lambda^4$, n is the refractive index of the solution, λ is the wavelength of incident light in vacuum, N_A is Avogadro's number, A_2 is the second virial coefficient, $\langle S^2\rangle$ is the mean-square radius of gyration, $q \equiv (4\pi/\lambda)\sin(\theta/2)$, and M_w is the weight average molecular mass.

Rayleigh ratio is defined as

$$R_\theta \equiv r^2 I(\theta)/I_0(1 + \cos^2\theta).$$

To determine molecular parameters, Rayleigh's ratio can be rewritten as

$$\frac{Kc}{R_\theta} = \left(\frac{1}{M_w} + 2A_2c + \cdots\right)\left(1 + \frac{q^2\langle S^2\rangle}{3} - \cdots\right),$$

where M_w, A_2, and $\langle S^2\rangle$ can be determined from experimental data through various graphical and regression procedures such as a Zimm plot. [6]

In practice, an absolute value of R_θ is not measured. Instead, a standard such as benzene is used to determine R_θ:

$$R_\theta = [I_\theta/I_B(90)](1 + \cos^2\theta)R_B,$$

where $I_B(90)$ is the scattering intensity and R_B is the Rayleigh ratio of benzene at a scattering angle of 90°.

22.6.5. Dynamic (quasielastic) light scattering [11–13]

This method can be used, under appropriate conditions, to determine the diffusion coefficient D of the center of mass of polymer molecules in dilute solution. As the concentration c approaches zero, the diffusion coefficient and molecular weight of a flexible monodisperse linear polymer are related by

$$\lim_{c \to 0} D(c) = AM^{-v},$$

where A is a constant and $v = 1/2$ for the θ condition. In general, v varies between 0.5 and 0.6.

22.7. CHARACTERIZATION BY SPECTROSCOPIC TECHNIQUES

22.7.1. Nuclear magnetic resonance

Data on solutions can be acquired such that relative intensities are proportional to the relative number of nuclei having a resonance frequency. Absolute concentrations are obtained by comparison against known standards. Resonances are influenced by the magnetic environment of nuclei (chemical shifts) providing information on chemical functionality and physical structure. Chemical shift ranges are available. [2,14–18] Proton-decoupled ^{13}C spectra usually provide the broadest information about geometrical, stereochemical, and compositional variations.

22.7.2. Vibrational spectroscopy

Vibrational spectroscopy (infrared and Raman) provides information on chemical composition, physical structure, and parameters of intermolecular and intramolecular potential energy functions. [16,19–21] Of particular importance is the use of vibrational frequencies and intensities to determine tacticity, degree of branching, crystallinity, and chain stem lengths in lamellar crystals.

22.8. CRYSTAL STRUCTURES

Most crystal structures of polymers are determined by wide angle x-ray diffraction experiments on uniaxially oriented polymer fibers or electron diffraction experiments on polymer single crystals. The crystal structures of a number of representative polymers are listed. Other extensive listings are available. [5,21–23] Note that a number of polymers exhibit polymorphism.

TABLE 22.3. Selected crystal structures of polymers.

Polymer	Crystal system space group	Unit cell axes (nm) and angles (deg)	Monomers units in unit cell	Chain conformation[a]	Density ρ_c (g/cm^3)
Polyethylene[b] $[CH_2–Ch_2–]_n$	orthorhomic *Pnam*	$a = 0.740$ $b = 0.493$ $c = 0.2534$	2	2* 1/1	1.000

(continued)

TABLE 22.3. Continued

Polymer	Crystal system space group	Unit cell axes (nm) and angles (deg)	Monomers units in unit cell	Chain conformation[a]	Density ρ_c (g/cm^3)
	monoclinic $C2/m$	$a = 0.809$ $b = 0.253$ (fiber) $c = 0.479$ $\beta = 107.9°$	2	2* 1/1	0.998
Isotactic Polypropylene[c] $[CH_2–CHCH_3–]_n$	monoclinic $C2/c$ or Cc	$a = 0.665$ $b = 2.096$ $c = 0.650$ $\beta = 99° 20'$	12	2* 3/1	0.936
	triclinic	$a = 0.654$ $b = 2.24$ $c = 0.650$ $\alpha = 89°$ $\beta = 99° 36$ $\gamma = 99°$	12	2* 3/1	0.954
Syndiotactic Polypropylene[d] $[CH_2–CHCH_3–]_n$	orthorhombic $C222_l$ (fiber)	$a = 1.450$ $b = 0.580$ $c = 0.740$	4	2* 4/1	0.90
	orthorhombic $Ibca$	$a = 1.450$ $b = 1.120$ $c = 0.740$	16	2* 4/1	0.90
	orthorhombic	$a = 0.522$ $b = 1.127$ $c = 0.506$	4	2* 2/1	0.945
Isotactic Polystyrene[c] $[CH_2–CHC_6CH_5–]_n$	trigonal $R\bar{3}$ or $R3c$	$a = 2.19$ $c = 0.665$	18	2* 3/1	1.126
Poly(vinylidene fluoride) $[CH_2–CF_2–]_n$					
α phase, form II[f]	orthorhombic $P2cm$	$a = 0.496$ $b = 0.964$ $c = 0.462$	4	$4(TGTG)$[g]	1.924
β phase, form I[h]	orthorhombic $Cm2m$	$a = 0.847$ $b = 0.490$ $c = 0.256$	2	2* 1/1	2.001
γ phase, form III[i]	orthorhombic $C2cm$	$a = 0.497$ $b = 0.966$ $c = 0.918$	8	$8(TTTGTTTTG')$	1.929
δ phase, form IV[j]	orthorhombic $P2_1cn$	$a = 0.496$ $b = 0.964$ $c = 0.462$	4	$4(TGTG')$	1.924

(continued)

TABLE 22.3. Continued

Polymer	Crystal system space group	Unit cell axes (nm) and angles (deg)	Monomers units in unit cell	Chain conformation[a]	Density ρ_c (g/cm^3)
Polytetrafluoroethylene (above 19°C)[k] $[CF_2–]_n$	trigonal	$a = 0.566$ $c = 1.950$	15	1* 15/7	2.302
trans 1,4-polybutadiene[l] $[CH_2–CH = CH–CH_2–]_n$	monoclinic $P2_1/a$	$a = 0.883$ $b = 0.911$ $c = 0.483$ $\beta = 114°$	4	4* 1/1	1.036
cis 1,4-polybutadiene[m] $[CH_2–CH = CH–CH_2–]_n$	monoclinic $C2/c$	$a = 0.460$ $b = 0.950$ $c = 0.860$ $\beta = 109°$	4	8* 1/1	1.01
Poly[1,2-bis(*p*-tolysulphonyloxymethylene)-1-butene-3-ynlene)][n] $[CH–C \equiv –CR =]_n$ ($R = –CH_2–O–SO_2–C_6H_4–CH_3$)	monoclinic $P2_1/c$	$a = 1.4493$ $b = 0.4910$ $c = 1.4936$ $\beta = 118.14°$	2	4* 1/1	1.483
Polyoxymethylene[o] $[CH_2–O–]_n$	trigonal $P3_1$ or $P3_2$	$a = 0.447$ $c = 17.39$	9	2* 9/5	1.49
Poly(ethylene oxide)[p] $[CH_2–CH_2–O–]_n$	monoclinic $P2_1/a$	$a = 0.805$ $b = 1.304$ $c = 1.948$ $\beta = 125.4°$	28	3* (7/2)	1.228
Poly(ethylene terephthalate)[q] $[(CH_2–)_2O–CO–C_6H_4–CO–)O–]_n$	triclinic $P\bar{1}$	$a = 0.456$ $b = 0.594$ $c = 1.075$ $\alpha = 98.5°$ $\beta = 118°$ $\gamma = 112°$	1	12* 1/1	1.455

[a]The notation $n^* p/q$ specifies the number (n) of skeletal atoms in the asymmetric unit of the chain and the number of such units (p) in q-turns of the helix in the crystallographic repeat.

[b]C. W. Bunn, Trans. Faraday Soc. **35**, 482 (1939) for the orthorhombic phase and S. Kavesh and J. J. Schultz, J. Polym. Sci., Part A, **8**, 243 (1970) for the monoclinic phase.

[c]G. Natta and P. Corradini, Nuovo Cimento Suppl. **15**, 40 (1960) for the monoclinic (α) phase and S.Bruckner and S. V. Meille, Nature **340**, 455 (1989) for the triclinic (γ) phase.

[d]G. Natta, I. Pasquon, P. Corradini, *et. al.*, Atti dell'Accademia Nazionale dei Lincei. Rendiconti **28**, 539 (1960), see also, P. Corradini, G. Natta, P. Ganis, and P. A.Temussi, J. Polym. Sci. **C16**, 2477 (1967). T. Seto, T. Hara, and K. Tanaki, Jpn. J. Appl. Phys. **7**, 31 (1968) for the isochiral orthorhombic phase. B. Lotz, A. J. Lovinger, and R. E. Cais, Macromolecules **21**, 2375 (1988). A. J. Lovinger, B. Lotz, and D. Davis, Polymer **31**, 2253 (1990). A. J. Lovinger, D. Davis, and B. Lotz, Macromolecules **24**, 552 (1991) for the anti-chiral orthorhombic phase.

[e]G. Natta and P. Corradini, Nuovo Cimento Suppl. **15**, 68 (1960).

(continued)

TABLE 22.3. Continued

Polymer	Crystal system space group	Unit cell axes (nm) and angles (deg)	Monomers units in unit cell	Chain conformation[a]	Density ρ_c (g/cm^3)
Poly(hexamethylene adipamide) (α form)[r] $[(CH_2-)_6NH-CO-(CH_2-)_4-NH-]_n$	triclinic $P\bar{1}$	$a = 0.49$ $b = 0.54$ $c = 1.72$ $\alpha = 48.5°$ $\beta = 77°$ $\gamma = 63.5°$	1	14* 1/1	1.24

[f] M. Bachmann and J. B. Lando, Macromolecules **14**, 40 (1981).
[g] T = *trans*, G = *gauche*.
[h] J. B. Lando, H. G. Olf, and A. Peterlin, J. Polym. Sci. Part A1 **4**, 941 (1966).
[i] S. Weinhold, M. H. Litt, and J. B. Lando, Macromolecules **13**, 1178 (1980).
[j] M. Bachmann, W. L. Gordon, S. Weinhold, and J. B. Lando, J. Appl. Phys. **51**, 5095 (1980).
[k] E. S. Clark and L. T. Muus, Z. Kryst, **117**, 119 (1962).
[l] S. Iwayanagi, I. Sakurai, T. Sakurai, and T. Seto, J. Macromol. Phys. B **2**, 163 (1968).
[m] G. Natta and P. Corradini, Nuovo Cimento Suppl. **15**, 111 (1960).
[n] D. Kobelt and E. F. Paulus, Acta Crystallogr. Sect. B **30**, 232 (1974).
[o] T. Uchida and H. Tadokoro, J. Polym. Sci. Part A2 **5**, 63 (1967).
[p] Y. Takahashi, and H. Tadorkoro, Macromolecules **6**, 672 (1973).
[q] R. de P. Daubeny, C. W. Bunn, and C. J. Brown, Proc. R. Soc. London Ser. A **226**, 531 (1954).
[r] C. W. Bunn and A. V. Garner, Proc. R. Soc. London Ser. A **189**, 39 (1947).

22.9. BOND LENGTHS AND ANGLES OF POLYMERS

Bond lengths and bond angles of polymers are not only important to determine the single chain dimensions, but also to calculate atomic positions in crystal structures. The selective data of bond lengths and angles of polymers are listed in Table 22.4.

TABLE 22.4. Bond lengths and angles for representative polymers.

Polymer	Bond or angle	Bond length or angle (nm or deg)	Polymer	Bond or angle	Bond length or angle (nm or deg)
Polyethylene[a]	C–C	0.153	Polypropylene[g]	C_1–C_2	0.154
	C–H	0.106	$C_1-C_2(-C_3)-C_1'$	C_2–C_3	0.154
	C–C–C	112.01		C_1–C_2–C_1	114
	H–C–H	109.29		C_1–C_2–C_3	110
	C–C–H	108.88			
Polytetrafluoroethylene[b]	C–C	0.155	Poly(ethylene oxide)[h]	C–C	0.154
	C–F	0.136		C–O	0.143
	C–C–C	113.85		C–H	0.109
	F–C–F	108		C–C–O	110
	C–C–F	108.7		C–O–C	112
				H–C–H	109.5

(continued)

TABLE 22.4. Continued

Polymer	Bond or angle	Bond length or angle (nm or deg)	Polymer	Bond or angle	Bond length or angle (nm or deg)
Polyamides[c]	C–N	0.147			
(C′ denotes amide carbon)	N–C′	0.132	Poly(ethylene terephthalate)[i]	C_1–C_2	0.149
	C′–O	0.124	O_6 C_4	C_1–C_3	0.134
	N–H	0.100	C_{5b}—C_{5a}—O_7—C_2—C_1	C_1–C_4	0.136
	N–C–C′	109.7	C_3	C_2–C_6	0.127
	N–C′–C	115.4		C_2–O_7	0.134
	C′–N–C	120.9		C_{5a}–O_7	0.144
	O–C′–C	121.0		C_{5a}–C_{5b}	0.149
	O–C′–N	123.6		C_2–C_1–C_3	125
	C′–N–H	123.0		C_2–C_1–C_4	118
	C–N–H	116.1		C_1–C_2–O_7	127
				C_1–C_2–O_7	110
Polyoxymethylene[d]	C–O	0.142		O_6–C_2–O_7	122
	C–H	0.109		C_2–O_7–C_{5a}	114
	C–O–C	112.4		O_7–C_{5a}–C_{5b}	104
	O–C–O	110.8		C_4–C_1–C_3	117
	H–C–H	109.5	Poly(vinylidene chloride)[j]	C–C	0.154
	H–C–O	108.5		C–F	0.134
		109.8		C–H	0.109
Poly[1,2-bis(*p*-	C–C	0.1428		F–C–F	109.5
tolysulphonyloxymethylene)-	C = C	0.1356		C–C–C	112.3
1-butene-3-ynylene)][e]	C ≡ C	0.1191	Polybutadiene[k]	C_1–C_2	0.153
C′					
C≡C–C=C	C–C = C	121.9	C_1–C_2–C_3 = C_4	C_2–C_3	0.154
	C′–C = C	120.3		C_3–C_4	0.115
	C– ≡ C	177.6		C_1–C_2–C_3	121
	C_2–C_3–C_4				142
Polystyrene[f]	C_1–C_2	0.154			
C_1–C_2–C_1'	C_2–C_3	0.154			
C_3	C_3–O_4	0.140			
C_4	C_1–C_2–C_1	116			
	C_3–C_2–C_1	108			
	C_3–C_2–C_1	111			

[a] J. D. Barnes and B. M Fanconi, J. Phys. Chem. Ref. Data **7**, 1309 (1978).
[b] M. J. Hannon, F. H. Boerio, and J. L. Koenig, J. Chem. Phys. **50**, 2829 (1969).
[c] S. Arnott, S. D. Dover, and A. Elliott, J. Mol. Biol. **30**, 201 (1967).
[d] Footnote *o* of Table 22.3.
[e] Footnote *n* of Table 22.3.
[f] Footnote *e* of Table 22.3.
[g] Footnote *c* of Table 22.3.
[h] Footnote *p* of Table 22.3.
[i] Footnote *q* of Table 22.3.
[j] Footnotes *h* and *i* of Table 22.3.
[k] Footnote *m* of Table 22.3.

22.10. MELTING AND CRYSTALLIZATION

22.10.1. Variation of melting point of thin crystals with thickness

For each crystalline polymer, the melting temperature can be very different depending on crystallization conditions and molecular characteristics. In particular, an equilibrium melting temperature of a crystalline polymer is very difficult to obtain since polymers do not crystallize in their equilibrium form as large crystals. On the contrary, the melting temperature is predominantly determined by the size of the crystals. In polymer crystals, lamellar-shaped crystals are most common morphological observations, and melting temperature and lamellar thickness can be expressed by

$$T_m = T_m^0[1 - 2\sigma_3/(\Delta h_f l)],$$

where T_m is the observed melting point, T_m^0 is the equilibrium melting point of infinitely thick crystal, σ_e is the lamellar large basal surface free energy, Δh_f is the heat of fusion, and l is the lamellar thickness.

22.10.2. Spherulitic growth rate controlled by secondary (surface) nucleation

$$G = G_0 \exp[-u^*/R(T - T_\infty) \exp(-nb\sigma\sigma_e/\Delta f kT),$$

where T is the temperature, T_∞ is the hypothetical temperature where viscous flow ceases ($\cong T_g - 30$ K), T_g is the glass transition temperature, l is the thickness of crystal (assumed to be small relative to the lateral dimensions) σ is the lateral surface energy, empirically approximated as $0.1\Delta h_f(ab)^{1/2}$, Δf is the free energy of fusion [$\simeq \Delta h_f(T_m^0 - T)/T_m^0$], G is the growth rate of spherulite, G_0 is the preexponential factor involving terms not strongly temperature dependent, u^* is the activation energy for transport of polymer in melt, R is the molar gas constant, b is the thickness of the crystallizing layer, d is the width of the crystalline molecular segment, and k is the Boltzmann constant. The value of n is a constant that represents the regimes of nucleation types in crystallization. The value of $n = 4$ represents either the mononucleation regime at high crystallization temperature (regime I) or rapid crystallization with very high density of secondary nucleation (regime III). When

TABLE 22.5. Typical values of crystallization parameters for crystalline polymers.

Polymer	T_m^0 (K)	T_g (K)	u^* (J/mol)	Δh_f (J/m^3)	σ_e (J/m^3)	b (nm)
Polyethylene	419.2	231	2.93×10^4	2.80×10^8	0.101	0.415
Polystyrene	515.2	363.5	6.53×10^3	9.11×10^7	0.035	0.55
Polyoxymethylene	459.2	213	6.28×10^3	1.86×10^8	0.061	0.386
Poly(ethylene oxide)	348.4	206	6.28×10^3	2.45×10^8	0.037	0.465
trans 1,4 polyisoprene	360.2	211	6.28×10^3	1.97×10^8	0.109	0.395

$n = 2$, the nucleation is comparable to or greater than subsequent growth (regime II). [24, 25] Other parameters have been illustrated in the previous equation. Although this crystal growth model only considers the surface-free energies as the nucleation barrier (entropic barrier), recent development in this area also takes entropic contributions to the nucleation barrier.

Specific values for the above parameters are subject to data selection and analysis. Typical values are given in Table 22.5. General reference is available. [24–28]

22.10.3. Avrami equation to describe overall kinetics of phase changes

$$\ln[1/(1-x)] = kt^n,$$

where x is the fraction transformed from liquid to crystal phase, t is the time, n is the exponent dependent on growth processes (see Table 22.6), and k is a constant dependent on nucleation rate, geometry of growing center, and growth rate. For spheres,

$$k = k_s = (\pi \rho_c \dot{N} G^3)/(3\rho_l).$$

For disks,

$$k - k_d = (\pi l_c \rho_c \dot{N} G^2)/(3\rho_l),$$

where ρ_c is the density of crystal phase, ρ_l is the density of liquid phase, $\dot{N}$ is the steady-state rate of nucleation, G is the linear growth rate of growing center, and l_c is the thickness of disk. General references are available. [28, 29]

TABLE 22.6. Values of *n* in Avrami equation for various types of nucleation and growth. (Steady-state nucleation rate in a completely crystallizable system is assumed. Impingement of growing centers has been accounted for.)

	Homogeneous nucleation	
Growth habit	Linear growth	Diffusion-controlled growth
Three dimensional	4	5/2
Two-dimensional	3	2
One-dimensional	2	3/2

22.11. LIQUID CRYSTALLINE TRANSITIONS

Liquid crystalline polymers are categorized to be main-chain, side-chain, and combined main-chain and side-chain types, depending on the locations of mesogenic groups. Furthermore, mesogenic groups can also be classified by geometric shapes, such as rod-

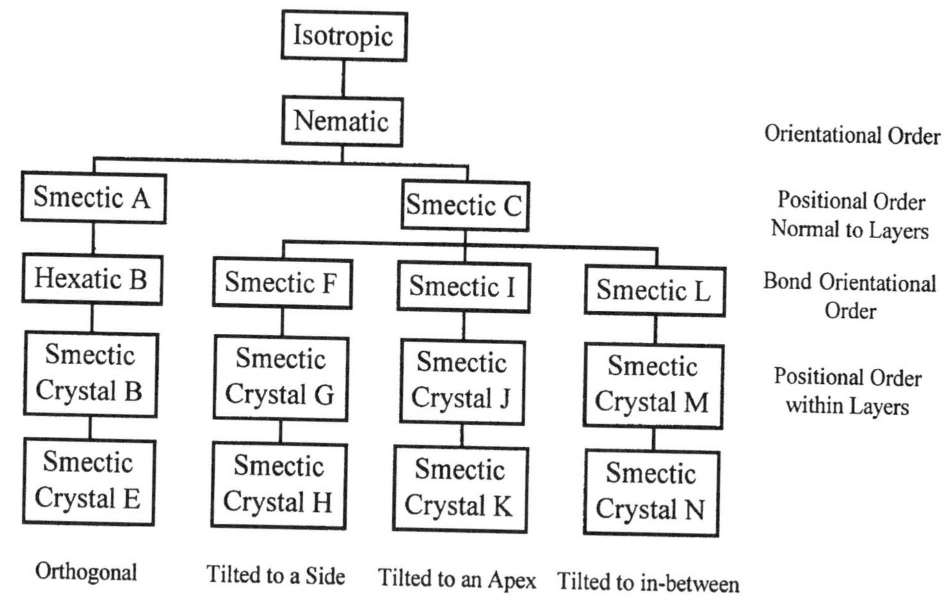

FIGURE 22.2. Phase relationships of liquid crystals.

like, ribbon-like, and disk-like. Liquid crystalline polymer phase behaviors follow the low molecular mass liquid crystalline materials. The order in liquid crystalline polymers can be distinguished to be molecular orientation, bond orientation, and positional orders. Almost all of the phases defined in the low molecular mass liquid crystals can find their analogues in polymers. The phase classification of the liquid crystalline phases in nonchiral liquid crystals can be summered in Figure 22.2. [30–32]

Most of the phase transformations observed experimentally related to the liquid crystalline polymers are thermodynamically first order. The transitions always take place near the equilibrium. For the nonracemic chiral crystalline polymers, the phase structures and transition behaviors are much richer than are the nonchiral ones.

22.12. HEAT CAPACITY AND THERMODYNAMIC FUNCTIONS

22.12.1. Heat capacity in solid and liquid states of semicrystalline polymers

The heat capacity of polymers with a mass fraction of crystalline component χ may be obtained from crystalline heat capacity $C_{p,c}$ and amorphous heat capacity $C_{p,a}$:

$$C_{p\chi} = \chi C_{p,c} + (1-\chi)C_{p,a},$$

$$\chi = \frac{\rho_c}{\rho}\frac{\rho-\rho_a}{\rho_c-\rho_a},$$

TABLE 22.7. Estimated heat capacities of crystalline and amorphous polyethylene. (From Ref. [33].)

Temp. (K)	Crystalline C_p ($\mathrm{J\,K^{-1}\,g^{-1}}$)	Amorphous C_p ($\mathrm{J\,K^{-1}\,g^{-1}}$)
1	0.000 008	0.000 028
5	0.000 98	0.003 7
10	0.007 6	0.024 3
25	0.087 8	0.138
50	0.319	0.436
100	0.678	0.694
150	0.900	1.043
200	1.103	1.407
		Glass transition
250	1.335	(2.01)
300	1.623	(2.20)
350	1.965	(2.36)
400	2.44	(2.51)
	Fusion	
450		2.67
500		2.82
550		2.97
600		3.13

where ρ is the sample density, ρ_c is the calculated density of the unit cell, and ρ_a is the density of amorphous state.

The estimated heat capacities of crystalline and amorphous polyethylene are given in Table 22.7.

22.12.2. General features of the heat capacity

Some general features applicable to glassy, C_{pg}, and crystalline heat capacities:

$$T < 1\ \mathrm{K},\ C_{pg} \propto T,$$

$$T \leq 10\ \mathrm{K},\ C_{pg}/T^3 \quad \text{reaches maximum,} \quad C_{p,g} \gg C_{p,c},$$

$$T \sim 25\ \mathrm{K},\ C_{p,g} - C_{p,c} \quad \text{reaches maximum,}$$

$$T \sim 70\ \mathrm{K} < T_g,\ C_{p,g} \sim C_{p,c},$$

$$T \sim T_g,\ C_{p,g} > C_{p,c}.$$

Thermal relaxations in glassy polymers begin to be observable at $\sim T_g - 50$ K. In many cases, C_p/T of solid polymer is in the vicinity of 4.2 mJ $\mathrm{K^{-2}\,g^{-1}}$ or C_p =

TABLE 22.8. Linear temperature dependence or proportionality of heat capacity.

Polymer	C_p/T (J K^{-2} g^{-1})	Range (K)	% deviation from linearity
Nylon 6	0.005 2	150–450	10
Phenolic resin (cured)	0.004 1	150–350	2
Polycarbonate	0.004 6	200–350	5
Polyethylene (crystalline)	0.005 6	160–375	5
Poly(methyl methacrylate)	0.004 6	200–350	5
Polystyrene	0.004 1	150–350	2
Poly(vinyl chloride)	$0.002\ 66 + 0.16/T$	80–340	1

$4.184T/1000$ J K^{-1} g^{-1} for a wide range of temperatures. The linear temperature dependence or proportionality of the heat capacity is given in Table 22.8.

22.12.3. Residual entropies at absolute zero for glass polymers and other properties

Glass polymers are expected to have residual entropies:

$$S_{0,g} = \int_0^{T_m} \frac{E_{p,c}}{T}\,dT + \Delta S_m - \int_{T_g}^{T_m} \frac{C_{p,l}}{T}\,dT - \int_0^{T_g} \frac{C_{p,g}}{T}\,dT,$$

where the subscripts l and m denote liquid and melting, respectively. Similarly, $H_{0,g} - H_{0,c} > 0$. When calculating Gibbs free energy as $G_T - H_{0,c}$, for glossy and semicrystalline polymers, contributions from $RS_{0,g}$ should be included. Typical values for $S_{0,g}$ are between $(R \ln 2)/2$ and $R \ln 2$ per chain atom. Some useful references are available. [33–35]

There also are attempts to calculate heat capacities in the solid states using vibration spectroscopy data and normal mode calculations. The vibration spectra can be practically constructed by both the caustic and optical vibration modes. Detailed procedures can be found in Refs. [34–36]. Calculations of the heat capacity in liquid states are more difficult since conformational and volume changes are necessary to be taken into account.

Knowing the heat capacity of polymers in the solid and liquid states, it is possible to calculate other thermodynamic functions, such as enthalpies, entropies, and free energies, based on well-known thermodynamic relationships.

22.13. GLASS TRANSITION

Amorphous polymer materials display phenomenologically a glass transition that appears to be a second-order transition in the Ehrenfest sense at a constant cooling and heating rate. [37] The apparent transition temperature T_g depends on the experimental technique and the time scale by which it is determined. High viscosities and long relaxation times characterize glasses (below T_g). The viscosity above T_g is given approximately by the universal WLF equation (Sec. 22.19.2). Table 22.9 lists the glass temperature for pure linear homopolymers of high molecular weight along with the three thermodynamic susceptibilities. [38] The glass temperature shows a strong variation as a function of molecular weight, amount and kind of diluent (plasticizer), pressure, composition of copolymers, composition of polymer blends, number of crosslinks in a rubber, and stretch ratio in a rubber. [39, 40] The state of a glass depends on the vitritication history. A glass formed isobarically at an increased pressure will have a greater density than one formed at atmospheric pressure and the same cooling rate. Data are available. [41–45] Recent publications give a good indication of the state of knowledge of polymer glasses. [46] Other polymers can also be found in Ref. [38]

TABLE 22.9. Thermal properties of glass transitions in selected polymers.

Polymers	T_g (K)	ΔC_p (J/K mol)
Polyethylene	237	10.5
Polytetrafluoroethylene	200	9.4
Polypropylene	270	19.2
Poly-l-butene	249	23.1
Poly(4-methy-l-pentene)	303	30.1
Poly-l,4-butadiene, *cis*	171	27.3
Poly-l,4 butadiene, *trans*	190	28.0
Poly(vinyl fluoride)	314	17.0
Poly(vinylidene fluoride)	212	21.2
Poly(vinyl chloride)	354	19.4
Poly(vinylidene chloride)	255	?
Polystyrene	373	30.8
Poly(α-methystyene	441	25.3
Polyoxymethylene	190	28.2
Polyoxyethylene	206	38.2
Poly(methyl acrylate)	279	42.3
Poly(methyl methacrylate	378	32.7
Poly(ethylene terephthalate	342	77.8
Poly(ethylene-2,6-naphthalene dicarboxylate)	390	81.6
Nylon 6	313	53.7
Nylon 6,6	323	115.5
Poly(thio-l,4-phenylene)	363	29.2
Poly-para-xylylene	286	37.6
Poly(oxy-2,6-dimethyl-l,4-phenylene)	482	31.9
Poly(aryl ether ether ketone)	419	78.1
Polycarbonate	424	48.8

22.14. THERMAL EXPANSION

The linear thermal expansion coefficient α

$$\alpha = \frac{\Delta L}{L\,\Delta T},$$

where ΔL is the change in length for a temperature change ΔT and L is the original length. The volume thermal expansion coefficient is thus 3α for isotropic materials. However, for polymers with anisotropic structures due to processing and other means, the linear thermal expansion coefficient along different directions may be different. For example, for thin films of relatively rigid macromolecules, in-plane orientation lads to a lower thermal expansion coefficient along the film surface and a higher one perpendicular to the film surface. The expansion coefficient depends on degree of crystallinity, thermal history, and mechanical history. [47] The value changes at the glass transition temperature. Values for polymers are in the range 0.5×10^{-5} to 30×10^{-5}/K, typically an order of magnitude larger than for other materials (fused silica: 0.54×10^{-6}/K; aluminum: 23×10^{-6}/K). Mismatched thermal expansion properties often contribute to the premature failure of objects manufactured from dissimilar materials. [47] Table 22.10 lists values for several common polymers. More extensive tabulations are available, both for pure polymers [47–49] and for engineering formulations. [49]

TABLE 22.10. Linear thermal expansion of polymers. (From Ref. [47].)

Polymer	α (K^{-1}; $\times 10^5$)
Polyethylene, high density	11–13
Low density	18–20
Polypropylene	5.8–10.2
Polytetrafluoroethylene	10
Poly(vinyl chloride), rigid	5–18
Flexible	7–25
Polystyrene	6–8
Poly(methyl methacrylate)	4.5
Nylon-6	5.9
Nylon-66	8
Nylon 6,10	9
Phenol-Formaldehyde	6.8
Epoxy resin (Shell 828-Z)	5.6

22.15. OPTICAL PROPERTIES OF POLYMERS

22.15.1. Orientation birefringence Δn in amorphous polymers [50]

$$\Delta n = \frac{2}{45}\pi \frac{(\bar{n}^2+2)^2}{\bar{n}} N(\lambda^2 - 1/\lambda)(b_1 - b_2),$$

where N is the number of network chains per unit volume, $\bar{n}$ is the average refractive index of the system, λ is the extension ratio, and b_1 and b_2 are the polarizabilities parallel (b_1) and perpendicular (b_2) to the axis of cylindrical statistical chain segments.

22.15.2. Stress optical coefficient C

$$C = \frac{\Delta n}{\sigma_a} = \frac{2}{45}\frac{\pi}{kT}\frac{(\bar{n}^2+2)^2}{\bar{n}}(\lambda^2 - 1/\lambda)(b_1 - b_2),$$

where σ_a is the true stress, k is the Boltzmann constant, and T is the absolute temperature.

22.15.3. Form birefringence Δn_f in two-phase systems [51]

For a two-phase system of optically *isotropic* rods (phase 1, volume fraction ϕ_1, refractive index n_1) parallel to one another and separated by an *isotropic* matrix material (phase 2, volume fraction ϕ_2, refractive index n_2), with the inter-rod distance between surfaces being small relative to the wavelength of light,

$$n_{\|}^2 = \phi_1 n_1^2 + \phi_2 n_2^2, \quad n_{\perp}^2 = \frac{n_2^2(\phi_1+1)n_1^2 + \phi_2 n_2^4}{(\phi_1+1)n_2^2 + \phi_2 n_1^2},$$

$$\Delta n_f = n_{\|} - n_{\perp},$$

where $n_{\|}$ and $n_{\perp}$ are the apparent refractive indices parallel and perpendicular to the rods.

22.15.4. Birefringence Δn of oriented crystalline polymers [50]

$$\Delta n = \Delta n_f + \phi_{\text{cr}} f_{\text{cr}} \Delta n_{\text{cr}}^0 + \phi_{\text{am}} f_{\text{am}} \Delta n_{\text{am}}^0,$$

where Δn_f if the form birefringence, ϕ_{cr} and ϕ_{am} are the volume fractions of crystalline and amorphous components [$\phi_{\text{cr}} = 1 - \phi_{\text{am}}$ and $\phi_{\text{cr}} = (\rho - \rho_a)/(\rho_c - \rho_a)$], ρ is the sample density, ρ_a is the density of polymer in the amorphous state, ρ_c is the calculated density based on unit cell structure of polymer, Δn_{cr}^0, and Δn_{am}^0 are the intrinsic birefringences of crystalline and amorphous components, and f_{cr}, and f_{am} are the orientation functions of crystalline and amorphous components. A general reference is available. [50]

22.15.5. Birefringence of spherulites Δn_{sph}

$$\Delta n_{\text{sph}} = n_{\|} - n_{\perp},$$

where $n_{\|}$ and $n_{\perp}$ are the refractive indices parallel and perpendicular to the spherulite radius. A general reference is available. [52] Also, a compendium of refractive indices of polymers is available. [6]

22.16. STRESS σ_{ij} AND DISPLACEMENT u_j AT CRACK TIPS

Three modes of crack-tip deformation for which a linear elastic stress field may be established are given in Ref. [53]. A coordinate system (r, α) is selected with the origin at the crack tip and with $r \ll a$, a is the crack length.

TABLE 22.11. Thermodynamic properties at the glass temperature[a]

Material/chemical formula	Expansion coefficient (10^{-4}/K) Liquid, α_e	Expansion coefficient (10^{-4}/K) Discontinuity, $\Delta\alpha$	Compressibility (10^{-5}/MPa) Liquid, β_e	Compressibility (10^{-5}/MPa) Discontinuity, $\Delta\beta$	Specific heat discontinuity ΔC_p (J/g K)	Glass temperature (K)
Polyethylene	5.1	3.2	50	20	0.60	140[b]
Polypropylene	6.8	4.4	38	9	0.48	244
Polyisobutylene	6.2	4.7	40	10	0.40	198
Poly(vinyl chloride)	5.7	3.7	44	20	0.30	350
Poly(vinyl acetate)	7.1	4.3	50	21	0.41	304
Poly(methylmethacrylate)	5.8	3.1	58	28	0.30	378
Polystyrene	5.1	2.9	61	29	0.34	362
Poly(α-methylstyrene	5.5	3.1	64	32	0.32	440
Polyisoprene	5.8	3.9	51	26	0.47	201
Polydimethylsiloxane	9.9	6.9	60	30	0.42	150
Poly(phenylene oxide)	5.3	3.2	50	20	0.24	480
Poly(ethylene terephthalate)	5.8	3.5	55	25	0.33	337
Polycarbonate[c]	6.0	3.44	51	18	0.23	424

[a] J. M. O'Reilley, J. Appl. Phys. **48**, 4047 (1977).
[b] Note that there is a diversity of opinion on the value of T_g for polyethylene.
[c] P. Zoller, J. Polym. Sci. Polym. Phys. Ed. **16**, 1261 (1978).

Model I:

$$\begin{Bmatrix} \sigma_{xx} \\ \sigma_{xy} \\ \sigma_{yy} \end{Bmatrix} = \frac{K_{\mathrm{I}}}{(2\pi r)^{1/2}} \cos\left(\frac{\alpha}{2}\right) \begin{Bmatrix} 1 - \sin\frac{\alpha}{2}\sin\frac{3\alpha}{2} \\ \sin\frac{\alpha}{2}\cos\frac{3\alpha}{2} \\ 1 + \sin\frac{\alpha}{2}\sin\frac{3\alpha}{2} \end{Bmatrix},$$

$$\begin{Bmatrix} U_x \\ U_y \end{Bmatrix} = \frac{K_{\text{I}}}{2G}\left(\frac{r}{2\pi}\right)^{1/2} \begin{Bmatrix} \cos\frac{\alpha}{2}\left(\kappa - 1 + 2\sin^2\frac{\alpha}{2}\right) \\ \sin\frac{\alpha}{2}\left(\kappa + 1 - 2\cos^2\frac{\alpha}{2}\right) \end{Bmatrix}.$$

Model II:

$$\begin{Bmatrix} \sigma_{xx} \\ \sigma_{xy} \\ \sigma_{yy} \end{Bmatrix} = \frac{K_{\text{II}}}{(2\pi r)}^{1/2} \begin{Bmatrix} -\sin\frac{\alpha}{2}\left(2 + \cos\frac{\alpha}{2}\right)\cos\frac{3\alpha}{2} \\ \cos\frac{\alpha}{2}\left(1 - \sin\frac{\alpha}{2}\sin\frac{3\alpha}{2}\right) \\ \sin\frac{\alpha}{2}\cos\frac{\alpha}{2}\cos\frac{3\alpha}{2} \end{Bmatrix},$$

$$\begin{Bmatrix} U_x \\ U_y \end{Bmatrix} = \frac{K_{\text{II}}}{2G}\left(\frac{r}{2\pi}\right)^{1/2} \begin{Bmatrix} \sin\frac{\alpha}{2}\left(\kappa + 1 + 2\cos^2\frac{\alpha}{2}\right) \\ -\cos\frac{\alpha}{2}\left(\kappa - 1 - 2\sin^2\frac{\alpha}{2}\right) \end{Bmatrix}.$$

Model III:

$$\begin{Bmatrix} \sigma_{xz} \\ \sigma_{yz} \end{Bmatrix} = \frac{2K_{\text{III}}}{(2\pi r)^{1/2}} \begin{Bmatrix} -\sin\frac{\alpha}{2} \\ \cos\frac{\alpha}{2} \end{Bmatrix},$$

$$u_z = \frac{K_{\text{III}}}{G}\left(\frac{r}{2\pi}\right)^{1/2}\sin\frac{\alpha}{2}.$$

K is the stress intensity factor [$= \sigma_0(\pi a)^{1/2} f(a/b)$, with $f(a/b)$ determined by geometry of specimen and mode, σ_0 is the applied stress, and b is the width of the specimen], G is the shear modulus, and

$$\kappa = (3 - \nu)/(1 + \nu) \qquad \text{for plane stress,}$$

$$\kappa = 3 - 4\nu \qquad \text{for plane stress}$$

where ν is Poisson's ratio.

22.17. INTERNAL FRICTION PEAKS IN SEMICRYSTALLINE POLYMERS

A frequently used measure of internal friction is the tangent of the phase angle δ between stress and strain under sinusoidal deformation. When measurements are made at constant frequency as a function of temperature, three peaks associated with various relaxation processes generally appear. The highest temperature peak is usually termed α, the next lower β, and so on, although this nomenclature is not always followed. Other peaks are sometimes observed, either at a lower temperature or as shoulders on these. The temperature of the peaks vary with the measure of mechanical loss used [tan δ, G, or J'' (see Chapter 13, Section 13.15 in *A Physicist's Desk Reference*, 2nd ed.)]. The temperature and magnitude

TABLE 22.12. Data on internal friction peaks for important polymers.

	α peak			β peak			γ peak		
Polymer	T (K)	tan δ	ν (Hz)	T (K)	tan δ	ν (Hz)	T (K)	tan δ	ν (Hz)
High-density polyethylenea	343	0.2	1.5	271	0.03	3.1	149	0.05	4.8
Low-density polyethylene[a]	339	0.25	1.8	262	0.03	6.8	147	0.05	13.7
Polyproylene[b]	323–378	0.11–0.15	0.25–0.38	277–285	0.06–0.11	0.6–1.0	213–223	0.016	0.8–1.3
Nylon 6, 6[c]	355	0.10	590	250	0.04	1050	165	0.041	1300
Poly(ethylene terephthalate)[d]	360	0.06[e]	1.0	213	0.03–0.5	1.0	(not observed)		
Poly(chlorotri-fluoroethylene)[f]	406	0.09	0.59	362–365	0.16–0.29	0.4–0.8	241	0.08	1.0–1.8

[a] A.A. Flocke, Kolloid Z. **180**, 118 (1962).
[b] E. Passaglia and G. M. Martin, J. Res. Nat. Bur. Stand. Sec. A **68**, 519 (1964).
[c] A. E. Woodward, J. A. Sauer, C. W. Deeley, and D. E. Kline, J. Colloid Sci. **12**, 363 (1957).
[d] K. H. Illers and H. Breuer, J. Colloid. Sci. **18**, 1 (1963).
[e] For high-crystallinity specimen; varies greatly with crystallinity.
[f] J. M. Crissman, and E. Passaglia, J. Polym. Sci. **14**, 237 (1966).

of the peaks are influenced by the frequency of measurement as well as by thermal treatment, moisture content, and other factors that influence the molecular mobility.

The activation energy for the γ process is generally of the order 60 kJ/mol. The β process is associated with the glass transition, and the frequency-temperature change of the maximum approximately follows the Williams-Landel-Ferry equation. [43] The activation energy of the α process varies from 160 to 330 kJ/mol depending on which portion of the process is investigated. Values as high as 600 kJ/mol have been reported. [54] Data for some important polymers are given in Table 22.12. A general reference is available. [55]

22.18. REPRESENTATIVE MECHANICAL PROPERTIES OF SOME COMMON STRUCTURAL POLYMERS

Table 22.13 gives mechanical properties often used for design purposes for some common structural polymers. The values are for room temperature and are obtained by various testing methods of the American Society for Testing and Materials. The values are strongly affected by temperature, time, the thermal history of the material, and other characteristics (such as molecular mass). For careful experiment work, the original literature should be consulted.

TABLE 22.13. Mechanical properties of common structural polymers.[a]

Polymer	Density (g/cm^3)	Yield strength (MN/m^2)	Young's modulus (GN/m^2)	Hardness (Rockwell)	Impact strength (kJ/m^2)
Polystyrene	1.04	35–70	3.3–3.5	M72	0.83[b]
Poly(methyl methacrylate)	1.17–1.29	84–120	2.5–3.5	M80–M102	1.33[b]
Polycarbonate	1.20	58	2.4	M70	4.8[b]
Polyethylene (low density)	0.91–0.925[c]	4.2–16[c]	0.1–0.25[c]	R10[c]	34.5[b]
Polypropylene	0.900–0.910	34–36	1.1–1.5	R80–R110	
6-6 nylon	1.13–1.15	60–90	2.7–3.3	R108–R118	5.1[b]
Polythylene (high density)	0.941–0.965[c]	22-38[c]	0.4–1.3[b]	D60–D70 (shore)	3.3[b]

[a]Data from *Handbook of Materials Science*, edited by Charles T. Lynch (Chemical Rubber, Cleveland, 1975).
[b]J. G. Williams, *Advances in Polymers* (Springer, Berlin, 1978).
[c]*Handbook of Plastics and Elastomers*, edited by Charles A. Harper (McGraw-Hill, New York, 1975).

22.19. RHEOLOGY

22.19.1. Introduction

The chapter on rheology (see Chapter 19 in *A Physicist's Desk Reference*, 2nd ed.) presents definitions and additional discussion of basic rheological terms, experiments, and material properties.

22.19.2. Linear viscoelasticity

(Zero-shear) viscosity [56]

The limiting value $\eta(0)$ of the viscosity function $\eta(\dot{\gamma})$ is known as the zero-shear viscosity η. For linear random coil polymers and their concentrated solutions, the empirical dependence on the absolute temperature T, the weight-average number of atoms (or groups) in the chain backbone Z_w and the volume fraction of polymer ϕ_2 is

$$\eta = F(X)\zeta,$$

where F is the structure factor and ζ is the friction factor per chain atom:

$$\begin{aligned} F(X) &= (N_A/6)X, \quad X < X_c, \\ &= (N_A/X_c^{2.4})X^{3.4}, \quad X > X_c, \\ X &= Z_w\phi_2(\langle S^2\rangle_0/M)\nu_2, \\ X_c &= 400 \times 10^{-17} \pm 10\%, \end{aligned}$$

$$\ln\zeta_0 + 1/\alpha(T - T_0), \quad T_g < T < T_g + 100,$$

where N_A is Avogadro's number, $\langle S^2\rangle_0$ is the unperturbed radius of gyration, M is the molecular weight, ν_2 is the specific volume of the polymer, and ζ_0, α, and T_0 are constants whose values for some polymers are listed in Table 22.14, together with other pertinent

TABLE 22.14. Parameters for zero-shear viscosity relations[a]

Polymer	T_g (K)	$10^4\alpha$ (K^{-1})	$\ln \zeta_0$	ν_2 (cm^3/g)	$10^{18}\langle S^2\rangle_0/M$ (cm^2 mol/g)	$10^{17}X_c$ (cm^5 mol/g^2)	Z_c
Polybutadiene	128	7.12	−10.90	1.11	12.6	360	330
Polydimethylsikoxane	30	7.12	−10.20	1.04	7.2	460	660
Polyethylene	0[c]	2.75[c]	−11.80[c]	1.307	17.0	350	270
Polyisobutylene	122	3.20	−11.74	1.123	8.7	420	540
Poly(methyl methacrylate)	308	3.88	−11.40	0.880	6.2	390	550
Polystyrene	313	5.57	−11.05	1.038	7.6	435	600
Poly(vinyl acetate)	248	4.58	−11.95	0.880	5.7	370	570

[a] From G. C. Berry and T. G. Fox, Adv. Polym. Sci. **5**, 261 (1968).
[b] 50% *cis*
[c] These values are uncertain, but represent the data over the limited temperature range for which they are available.

properties: ν_2, $\langle S^2\rangle_0/M$, and Z_c, the value of Z_w corresponding to X_c. The values of α and T_0 are listed for an undiluted polymer of high molecular mass ($Z > 80$).

Steady-state compliance J_e^0

For a linear viscoelastic fluid,

$$J_R(\infty) = J'(0) = \lim_{\omega\to 0} \frac{G'(\omega)/\omega^2}{\eta^2} = J_e^0.$$

For a nonlinear viscoelastic fluid in steady flow, $N_1(\dot\gamma)$ at low values of $\dot\gamma$ is given by

$$N_1(\dot\gamma) = 2\eta^2 J_e^0 J_e^0 \dot\gamma^2 + O(\dot\gamma^4).$$

For monodisperse linear random coil polymers and their concentrated solutions, [57]

$$J_e^0 = \frac{0.4M}{cRT}\left[1 + \left(\frac{cM}{\rho M_c'}\right)^2\right]^{-1/2},$$

where M is the molecular weight, R is the gas constant, T is the absolute temperature, c is the concentration of polymer, ρ is the density of the polymer, and M_c' is a characteristic molecular weight, values of which are included in Table 22.15.

For $M \gg M_c'$, J_e^0 is independent of molecular weight. Values for various undiluted monodisperse polymers are listed in Table 22.15. J_e^0 is highly sensitive to molecular-weight distribution, a dependence on $M_Z M_{Z+1}/M_w^2$ often being useful.

Rubber or entanglement plateau

For high-molecular-weight linear amorphous polymers, the compliance function have a plateau J_N^0. The modulus functions have a corresponding plateau $G_N^0(= 1/J_N^0)$. The value of J_N^0 is independent of molecular weight M for $M >\approx 2 \times 10^4$ and is associated with an entanglement network characterized by $M_e(= \rho RT J_N^0)$, the average molecular

TABLE 22.15. Characteristic viscoelastic values.[a]

	Rubber plateau				From steady-state compliance		
Polymer	°C	$\log J_N^0$ [a] (cm^2/dyn)	M_e (g/mol)	Z_e	°C	$\log J_e^0$ [b] (cm^2/dyn)	M_c' (g/mol)
Hevea rubber	25	−6.76	61 00	360	−30	−5.85	60 000
Polybutadiene ≈50% *cis*	25	−7.06	19 00	140	25	−6.60	13 800
Polydimethylsiloxane	25	−6.30	12 000	330	20	−6.00	61 000
Polyisobutylene	25	−6.46	7 600	270			
Poly(methyl methacrylate)	170	−6.94	4 700	94			
Polystyrene	140	−6.31	17 300	333	200	−5.76	130 000
					160	−5.85	
Poly(vinyl acetate)	60	−6.55	9 100	210	30	−5.90	86 000

[a] J. D. Ferry, *Viscoelastic Properties of Polymers* (Wiley, New York, 1980), 3rd edition, and W. W. Graessley, Adv. Polym. Sci. **16**, 1 (1974).
[b] J_e^0 for monodisperse polymers with $M > M_c'$.

weight between entanglement coupling point, or by Z_e, the corresponding number of chain atoms. The plateau occurs only if the molecular weight is considerably greater than M_e. Values of J_N^0, M_e, and Z_e, for several polymers are listed in Table 22.15.

Glass-rubber transition (or dispersion) zone

The loss functions have maxima in this zone, whereas other functions [$J(t)$, $G(t)$, $J'(\omega)$, and $G'(\omega)$] change by several orders of magnitude for amorphous polymers. There is little dependence on molecular weight. Table 12.1 of Ref. [43] lists values for some typical points in these functions.

Time-temperature shift factor a_{0T}

For many materials, the logarithmic plot of a viscoelastic function at the temperature T may be obtained from that at the temperature T_0 by shifting the curve along the $\log t$ or $\log \omega$ axis by the amount $\log a_{0T}$. Various expressions for a_{0T} are

Arrhenius: $\ln a_{0T} = \ln A + B/RT,$

Vogel or Fulcher: $\ln a_{0T} = \ln A + 1/\alpha(T - T_\infty),$

WLF (Williams-Landel-Ferry): $\log a_{0T} = -c_1^0(T - T_0)/(c_2^0 + T - T_0),$

where $T_\infty = T_0 - c_2^0$. The WLF equation and its equivalent, the Vogel equation, are valid for temperatures between T_g and $T_g + 100$ K. They have been interpreted in terms of the Doolittle free volume expression

$$\ln a_{0T} = B(f^{-1} - f_0^{-1}),$$

where B is an empirical constant usually assumed to be approximately unity, f and f_0 are the fractional free volumes at T and T_0, respectively, and f is assumed to be a linear function of T:

$$f = f_0 + \alpha_f(T - T_0),$$

where α_f is the thermal expansion coefficient for the free volume:

$$f_0 = B/2.303c_1^0, \quad \alpha_f = B/2.303c_1^0c_2^0.$$

See Table 22.16, which also lists f_g/B, the fractional free volume at T_g, which is 0.025 ± 0.005 for most systems. In the absence of other information, for a rough estimate of a_{0T}, T_0 may be chosen at T_g, $c_1^0 \approx 17$, and $c_2^0 \approx 52$ K, which are called "universal WLF parameters."

TABLE 22.16. Parameters characterizing temperature dependence of a_{0T}.[a]

Polymer	T_0 (K)	c_1^0	c_2^0 (K)	T_g (K)	f_g/B	α_f/B (10^4K^{-1})	T_∞ (K)
Butyl rubber[b]	298	9.03	201.6	205	0.026	2.4	96
Hevea rubber	248	8.86	101.6	200	0.026	4.8	146
	298	5.94	151.6				146
Polybutadiene[c]	298	3.64	186.5	172	0.039	6.4	112
Polydimethylsiloxane	303	1.90	222	150	0.071	10.3	81
Polyisobutylene	298	8.61	200.4	205	0.026	2.5	101
Poly(methyl methacrylate)	388	32.2	80	388	0.013	1.7	308
Polystyrene	373	13.7	50.0	373	0.032	6.3	323
Poly(vinyl acetate)	349	8.86	101.6	305	0.028	5.9	258
Styrene-butadiene copolymer[d]	298	4.57	113.6	210	0.021	8.2	184

[a] From J. D. Ferry, *Viscoelastic Properties of Polymers* (Wiley, New York, 1980), 3rd edition.
[b] Lightly vulcanized with sulfur.
[c] *cis:trans:* vinyl = 43:50:7.
[d] Styrene:butadiene = 23.5:76.5 random (by weight).

Rouse-Zimm theories for viscoelastic properties of dilute solutions of random coil macromolecules in θ solvents

The molecule (degree of polymerization P, molecular weight M) is represented as N-submolecules, where the end-to-end distance σ of the submolecule represents one statistical length.

$$\eta'(\omega) = \eta_s + \frac{cRT}{M}\sum_{p=1}^{N}\frac{\tau_p}{1+\omega^2\tau_p^2},$$

$$G'(\omega) = \frac{cRT}{M}\sum_{p=1}^{N}\frac{\omega^2\tau_p^2}{1+\omega^2\tau_p^2},$$

where c is the concentration of polymer in the solvent in g/ml and η_s is the viscosity of the solvent. In the Rouse (free draining) model,

$$\tau_p \cong \sigma^2 N^2 f_0 / 6\pi^2 p^2 kT,$$

where f_0 is the friction coefficient of the submolecule. In the Zimm theory, the perturbation of the velocity of the solvent is taken into account approximately by introducing a parameter h^* (a measure of the hydrodynamic interaction). Numerical values of τ_p must be obtained by computer calculation. Chapter 9 of Ref. [46] contains graphs of G' and G'' calculated from the Rouse and Zimm theories.

22.20. ELECTRICAL PROPERTIES

22.20.1. Dipole moments

Dipole moments of polymers in solution are given in Table 22.17.

TABLE 22.17. Dipole moments of polymers in solution, expressed as $(\bar{\mu}^2/N)^{1/2}$, the dipole moment per monomer unit. [$\phi = \bar{\mu}^2/N\bar{\mu}_0^2$, where μ_0 is the dipole moment of the isolated moment unit and N is the number of monomers per molecule. 1 D = 10^{-18} esu cm = 3.33564 x 10^{-30} cm. From *Polymer Handbook*, 2nd ed., edited by J. Brandrup and E. H. Immergut (Wiley, New York, 1975).] Reprinted by permission of John Wiley & Sons, Inc.

Polymer	Solvent	Temp. (°C)	$(\bar{\mu}^2/N)^{1/2}$ (D)	ϕ
Polystyrene	Carbon tetracholoride	25	0.26	0.56
Poly(*p*-chlorostyrene)	Benzene	30, 25	1.45	0.56
Poly(vinyl acetate)	Benzene	20	1.70	0.84
Poly(vinyl chloride)	Dioxane	20, 40	1.62	0.59
Poly(methyl acrylate)	Benzene	25	1.41–1.44	0.64–0.67
Poly(methyl methacrylate)				
Isotactic	Benzene	25–65	1.40–1.44	0.77–0.81
Atactic	Benzene	25–65	1.33–1.41	0.69–0.78
Syndiotactic	Benzene	25–65	1.34–1.41	0.70–0.78
cis 1,4-polyisoprene	Benzene	25	0.28	0.70
trans 1,4-polyisoprene	Benzene	25	0.31	0.82

22.20.2. Typical electrical properties

Table 22.18 gives some typical electrical property data. The values are to be considered estimates only. Dielectric properties are particularly sensitive to ionic impurities and may reflect effects of moisture content, thermal history, chemical variations, and details of the test method. For careful experimental work, the original literature should be consulted.

TABLE 22.18. Typical electric properties of selected polymers at room temperature. [From *Handbook of Plastics and Elastomers*, edited by C. A. Harper (McGraw-Hill, New York, 1975).]

Polymer	Volume resistivity (Ω cm)	Dielectric constant At 60 Hz	Dielectric strength, 3 mm thickness (kV/cm)	Dissipation factor at 60 Hz
Cellulose acetate	$10^{10} - 10^{12}$	3.2–7.5	120–240	0.01–0.1
Cellulose acetate butyrate	$10^{10} - 10^{12}$	3.2–6.4	100–160	0.01–0.04
Cellulose proprionate	$10^{12} - 10^{16}$	3.3–4.2	120–180	0.01–0.05
Polytetrafluoroethylene	$> 10^{18}$	2.1	160	< 0.0001
Nylon 6	$10^{14} - 10^{15}$	6.1	140	0.5
Nylon 6/6	$10^{14} - 10^{15}$	3.6–4.0	140	0.014
Nylon 6/10	$10^{14} - 10^{15}$	4.0–7.6	140	0.05
Polycarbonate	6×10^{15}	3.0	160	0.0001–0.0005
Polyethylene	$10^{15} - 10^{18}$	2.3	180–400	0.0001–0.006
Polyimide	$10^{16} - 10^{17}$	3.5	160	0.003
Polypropylene	$10^{15} - 10^{17}$	2.1–2.7	180–260	0.0007–0.005
Polystyrene	$10^{17} - 10^{21}$	2.5–2.65	200–280	0.0001–0.0005
Polystyrene, high impact	$10^{10} - 10^{17}$	2.5–3.5	200	0.003–0.005
Poly(vinyl chloride)				
Flexible	$10^{11} - 10^{15}$	5–9	120–400	0.08–0.15
Rigid	$10^{12} - 10^{16}$	3.4	170–400	0.01–0.02
Poly(vinylidene chloride), rigid	10^{15}	3.1	480–610	0.02
Poly(4-methylpentene-1)	$> 10^{16}$	2.12	280	0.001
Poly(acryl sulfone)	3×10^{16}	3.9	150	0.003
Poly(phenylene sulfide)	10^{16}	3.1	240	0.0004
Poly(phenylene oxide)	10^{17}	2.6	160–200	0.004
Polysulfone	5×10^{16}	2.8	160	0.007
Poly(ethersulfone)	$10^{17} - 10^{18}$	3.5	160	0.001

22.21. DIFFUSION AND PERMEATION

22.21.1. Diffusion into plane sheet

General solution

For Fickian-type diffusion, the amount of diffusant transferred M_t at time t between a sheet of thickness $2l$ and a stirred liquid of volume V_s is given by

$$\frac{M_1}{M_\infty} = 1 - \sum_{n=1}^{\infty} \frac{2\alpha(1-\alpha)}{1+\alpha+\alpha^2 q_n^2} e^{-q_n^2 \tau},$$

where M_∞ is the amount of diffusant transferred at equilibrium, $\tau = Dt/l^2$. D is the diffusion coefficient, K is the partition coefficient (ratio of diffusant concentration in the polymer to its concentration in the solvent), $\alpha = V_s/KV_p$, V_p is the volume of polymer, and q_n is the nonzero positive roots of $\tan q_n = -\alpha q_n$.

Approximation for $\tau < 0.1$

$$M_t/M_\infty = (1+\alpha)[1 - e^{\tau/\alpha^2}\mathrm{erf}(\tau^{1/2}/\alpha)].$$

Infinite bath for $\tau \le 0.1$

$$M_t/M_\infty = 2(\tau/\pi)^{1/2}.$$

22.21.2. Diffusion data

See Tables 22.19 and 22.20.

22.21.3. Gas transmission

The product of the diffusion constant and the solubility is the permeability coefficient P, useful for calculating the flux of gas through a unit area of polymer:

$$\dot{Q} = P(P_1 - P_0)/l.$$

$\dot{Q}$ is the rate of gas transmission, P_1, P_0 is the pressure on opposite sides of the membrane, and l is the membrane thickness. The diffusion coefficient is defined by Fick's law and the solubility coefficient by Henry's law. Equations for calculating transport under nonsteady-state conditions and for a variety of geometries are available. [58] Tabulations of permeability coefficients can be found. [6, 59]

22.22. NONLINEAR OPTICAL PROPERTIES

Polymers are an important class of materials for nonlinear optical processes. [60] Materials exhibit nonlinear optical effects when subjected to an intense oscillating electric field. Polarization generated in the material due to electric dipole interaction with the electric field can be expressed by a Taylor series expansion [61]

$$\begin{aligned}\vec{P} &= \overset{\rightrightarrows}{\chi}^{(1)} \cdot \vec{E} + \overset{\rightrightarrows}{\chi}^{(2)} : \vec{E}\,\vec{E} + \overset{\rightrightarrows}{\chi}^{(3)} \vdots\, \vec{E}\,\vec{E}\,\vec{E} \\ &= \overset{\rightrightarrows}{\chi}_{\text{eff}} \cdot \vec{E}\,. \end{aligned} \tag{22.1}$$

In the equation, $\overset{\rightrightarrows}{\chi}^{(1)}$ is the linear susceptibility that is generally adequate to describe the optical response to a weak field. It describes linear absorption, refraction, and scattering. The terms $\overset{\rightrightarrows}{\chi}^{(2)}$ and $\overset{\rightrightarrows}{\chi}^{(3)}$ are the second- and third-order nonlinear optical susceptibilities

TABLE 22.19. Typical values of diffusion coefficients and temperature dependence. [$D = A\exp(-E/RT)$, where R is the molar gas constant and T is the temperature. Diffusion coefficients are often strongly dependent on morphology and concentration. From J. H. Flynn, Polymer 13, 1325 (1982).]

Diffusing molecule	Molecular weight ($g\ mol^{-1}$)	Low-density polyethylene D(30°C) ($10^{-8}\ cm^2\ s^{-1}$)	Low-density polyethylene $-\log_{10} A$	Low-density polyethylene E ($kJ\ mol^{-1}$)	High-density polyethylene D(30°C) ($10^{-8}\ cm^2\ s^{-1}$)	High-density polyethylene $-\log_{10} A$	High-density polyethylene E ($kJ\ mol^{-1}$)	Polyisobutylene D(30°C) ($10^{-8}\ cm^2\ s^{-1}$)	Polyisobutylene $-\log_{10} A$	Polyisobutylene E ($kJ\ mol^{-1}$)
Methane	16.0	29	1.42	47	7.4	0.38	44			
Ethane	30.1	9.2	2.92	58	2.6	2.11	57			
Propane	44.1	4.7	2.26	56	0.72	1.66	57	0.30		
n-pentane	72.2	1.6	5.02	73	0.55	4.93	77	0.19	3.32	70
Benzene	78.1	2.0	3.28	66						
p-dioxane	88.1	0.84	4.90	75						
Bromoethane	95.0	8.3	1.84	52						
n-heptane	100.2	1.6	−1.09	45				0.20		
o-xylene	106.2	20	1.60	48	5.9	0.73	46			
p-xylene	106.2	56	−0.41	34	18	−0.25	44			
n-octane	114.2	0.66	2.42	62				0.19	4.37	76
n-decane	142.3	0.58	3.34	67						
Carbon tetrachloride	153.8	0.68	7.44	83						

TABLE 22.20. Self-diffusion of polyolefins. [$D = A\exp(-E/RT)$, where R is the molar gas constant and T is the temperature. Asterisks denote extrapolated values. From J. H. Flynn, Polymer 13, 1325 (1982).]

Polymer	Molecular weight ($\mathrm{g\ mol^{-1}}$)	D (30°C) ($10^{-8}\ \mathrm{cm^2\ s^{-1}}$)	D (140°C) ($10^{-8}\ \mathrm{cm^2\ s^{-1}}$)	$-\log_{10} A$	E ($\mathrm{kJ/mol^{-1}}$)
n-octadecane	254.5	310*	1800	2.67	175
n-dotriacontane	450.9	52*	630	2.22	24
Polyisobutylene	(~ 700)	5.9*	64	3.34	23
High-density polyethylene	(~ 4100)	1.3*	14	4.04	22
Low-density polyethylene	(~ 5800)	1.8*	10	4.91	16

that describe the nonlinear response. At optical frequencies, the refractive index η and dielectric constant $\varepsilon(\omega)$ are related by

$$\eta^2(\omega) = \varepsilon(\omega) = 1 + 4\pi\chi(\omega). \tag{22.2}$$

For a plane wave, the wave vector $k = \eta\omega/c$ and the phase velocity $\nu = c/\eta$. In a nonlinear medicum, $\overset{\rightrightarrows}{\chi}(\omega) = \overset{\rightrightarrows}{\chi}_{\mathrm{eff}}$ of Eq. 22.1 is dependent on E; therefore, η, k, and ν are all dependent on E. Two consequences of the second-order nonlinearity described by $\chi^{(2)}$ are second harmonic generation and Pockel's electrooptic effect. In second harmonic generation, an intense beam of frequency ω passing through a nonlinear medium of $\chi^{(2)} \neq 0$ generates an optical wave at frequency 2ω. The electrooptic effect describes the phase shift introduced by an applied low-frequency electric field. In other words, the refractive index of a material with nonvanishing $\chi^{(2)}$ can be modulated by the application of a DC or low-frequency AC field. Two consequences of the third-order optical nonlinearities represented by $\chi^{(3)}$ are third harmonic generation and intensity-dependent refractive index. Third harmonic generation describes the process in which an incident photon field of frequency ω generates through nonlinear polarization in the medium, a coherent optical field at 3ω. Through $\chi^{(3)}$ interaction, the refractive index of a nonlinear medium is given as $\eta = \eta_0 + \eta_2 I$, where η_2 describes the intensity dependence of refractive index; I is the intensity of the laser pulse.

Since χ^2 is a third-rank tensor, it vanishes in a centrosymmetric or isotropic medium. Therefore, second-order nonlinear optical processes can be observed only in noncentrosymmetric media, such as piezoelectric and ferroelectric systems. In contrast, there is no symmetric restriction on the third-order processes that can occur in all media.

At the microscopic level, the nonlinearity of the organic structures is described by the electric dipole interaction of the radiation field with the molecules. The resulting induced dipole moment is given as

$$\vec{\mu}_{\mathrm{ind}} = \overset{\rightrightarrows}{\alpha} \cdot \vec{E} + \overset{\rightrightarrows\rightarrow}{\beta} : \vec{E}\vec{E} + \overset{\rightrightarrows\rightrightarrows}{\gamma} \vdots \vec{E}\vec{E}\vec{E}. \tag{22.3}$$

In the equation, $\overset{\rightrightarrows}{\alpha}$ is the linear polarizability. The terms $\overset{\rightrightarrows\rightarrow}{\beta}$ and $\overset{\rightrightarrows\rightrightarrows}{\gamma}$, called first and second hyperpolarizabilities, describe the nonlinear optical interactions and are microscopic analogues of $\overset{\rightrightarrows\rightarrow}{\chi}^{(2)}$ and $\overset{\rightrightarrows\rightarrow}{\chi}^{(2)}$.

In the weak coupling limit, as in the case for most organic systems, each molecule can be treated as an independent source of nonlinear effects. Then, the macroscopic susceptibilities $\chi^{(3)}$ are derived from the microscopic nonlinearities β and γ by simple orientationally averaged site sums using appropriate local field correction factors that relate the applied field to the local field at the molecular site.

For second-order nonlinearity in organic systems, a system that consists of a π-electron structure containing an electron donor group on one side and an electron acceptor group on another side has been identified [62] as giving a large value of β. A typical example is

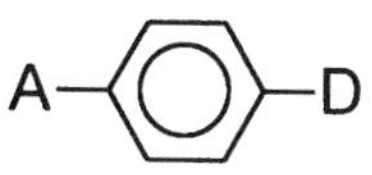

$$A-C_6H_4-D$$

in which A and D are, respectively, the electron acceptor and donor groups.

Second-order nonlinear processes have been observed in many types of organic structures and bulk systems. One does not need a polymeric structure. Small molecules can be used either in the crystalline form or in the form of Langmuir-Blodgett films. Although piezoelectric polymers, such as polyvinylidene fluoride, give rise to second-order effects, a polymeric medium has been used mainly as a passive host in which the nonlinearily active organic molecules are dispersed as a guest or attached to the polymeric structure by a flexible spacer (such as in a side-chain liquid crystalling polymers). [63] The second-order active nonlinear group is in the side-chain. The noncentrosymmetric arrangement of dipoles in the polymeric structure is created by electric field poling. The polymer is heated to its glass temperature, an electric field is applied to orient the dipoles, and then the polymer is cooled to freeze the oriented dipoles.

All macroscopic theoretical models predict the largest nonresonant, third-order optical nonlinearity from π-electrons. Therefore, conjugated polymeric structures that provide effective π-electron delocalization have become an important class of $\chi^{(3)}$ materials. In the nonresonant regime, the response of $\chi^{(3)}$ derived from the delocalized π-electrons is extremely fast (femtoseconds).

Some conjugated polymers, with their reported $\chi^{(3)}$-values are listed in Table 22.21, along with the method of measurement and the wavelength used.

TABLE 22.21. $\chi^{(3)}$-values for some polymers.

Common name	Structure	Measurement technique	λ (nm) or $h\nu$ (eV)	$\chi^{(3)}$ (esu)
p-toluene sulfonate (PTS) polydiacetylene[a]	R=—CH_2—O—SO_2—C_6H_4—CH_3	THG[h]	0.66eV	8.5×10^{-10} (parallel to chains)
poly-4-BCMU polydiacetylene[b]		DFWM[i]	2.07eV	4×10^{-10} (red form)
Polyacetylene[c]		THG	0.65eV	13×10^{-10}
			1.5eV	0.5×10^{-10}
poly(*p*-phenylene benzobiathiozole) (PBZT)[d,e]		DFWM	602nm	$\sim 10^{-11}$
poly(p-phenylene vinylene) (PPV)[f]		THG	0.67ev	7.8×10^{-12}
		DFWM	602nm	4×10^{-10} (parallel to draw direction)
Polythiophene[g]		DFWM	602nm	$\sim 5 \times 10^{-10}$

[a] C. Sauteret, J. P. Hermann, R. Frey, F. Pradire, J. Ducuing, R. Baughman, and R. R. Chance, Phys. Rev. Lett. **36**, 956 (1976).
[b] D. N. Rao, P. Chopra, S. K. Gheshal, J. Swiatkiewicz, and P. N. Prasad, J. Chem. Phys. **84**, 7049 (1986).
[c] F. Kajzar, S. Etemad, G. L. Baker, and J. Messier, Solid State Commun. **63**, 1113 (1987).
[d] D. N. Rao, J. Swiatkiewicz, P. Chopra, S. K. Gheshal, and P. N. Prasad, Appl. Phys. Lett. **48**, 1187 (1986),
[e] T. Kaino, K. I. Kobedera, S. Tomaru, T. Kurihara, S. Saito, T. Tsuitsui, and S. Tokito, Electron. Lett. **23**, 1095 (1987).
[f] B. P. Singh, P. N. Prasad, and F. E. Karasz, Polymer **29**, 1940 (1988).
[g] P. N. Prasad, M. K. Casstereus, J. Pfleger, and P. Logsdon, *Symposium on Multifunctional Materials, SPIE Proceedings* (1988), Vol. 878, p. 106.
[h] THG = Third harmonic generation.
[i] DFWM = Degenerate four wave mixing.

22.23. REFERENCES

[1] P. J. Flory, *Principle of Polymer Chemistry* (Cornell University Press, Ithaca, NY, 1971).

[2] F. A. Bovey and F. H. Winslow, editors, *Macromolecules: An Introduction to Polymer Science* (Academic Press, New York, 1979). See also "List of standard abbreviations for synthetic polymers and polymer materials," Pure Appl. Chem. **40**, 473 (1974).

[3] D. W. van Krevelen, *Properties of Polymers* (Elsevier, Amsterdam, 1997), 3rd edition.
[4] T. Birshtein, O. Ptitsyn, *Conformations of Macromolecules* (Wiley, New York, 1966).
[5] P. J. Flory, *Statistical Mechanics of Chain Molecules* (Wiley Interscience, New York, 1969).
[6] J. Brandrup, E. H. Immergut, and E. A. Grulke, editors, *Polymer Handbook* (Wiley, New York, 1999).
[7] C. H. Chervenka, *A Manual of Methods for Analytical Ultracentrifuge* (Beckman Instruments, Palo Alto, CA, 1973).
[8] M. G. Huglin, editor, *Light Scattering From Polymer Solutions* (Academic Press, New York, 1972).
[9] D. McIntyre and F. Gornick, editors, *Light Scattering From Dilute Polymer Solutions* (Gordon and Breach, New York, 1964).
[10] B. Chu, *Laser Scattering* (Academic Press, New York, 1974).
[11] B. Chu, *Laser Light Scattering* (Academic Press, New York, 1990), 2nd edition.
[12] B. Burne and R. Pecora, *Dynamic Light Scattering* (Wiley, New York, 1976).
[13] A. Z. Akcasu, M. Benmouna, and C. C. Han, Polymer **21**, 866 (1980).
[14] E. D. Becker, *High Resolution NMR* (Academic Press, New York, 1969).
[15] F. A. Bovey, *Nuclear Magnetic Resonance Spectroscopy* (Academic Press, New York, 1988), 2nd edition.
[16] J. L. Koenig, *Spectroscopy of Polymers* (American Chemical Society, Washington, DC, 1992).
[17] J. C. Randall, *Polymer Sequence Determination* (Academic Press, New York, 1977).
[18] J. C. Randall, in *Carbon-13 NMR in Polymer Science*, edited by W. M. Pasika, ACS Symp. Ser. **103** (1979), Chap. 14.
[19] H. W. Siesler and K. Holland-Moritz, *Practical Spectroscopy Vol. 4, Infrared and Raman Spectroscopy of Polymers* (Dekker, New York, 1980).
[20] P. C. Painter, M. M. Coleman, and J. L. Koenig, *The Theory of Vibrational Spectroscopy and Its Application to Polymeric Materials* (Wiley, New York, 1982).
[21] N. B. Colthup, L. H. Daly, and S. E. Wiberley, *Introduction to Infrared and Raman Spectroscopy* (Academic Press, New York, 1990), 3rd edition.
[22] H. Tadokoro, *Structure of Crystalline Polymers* (Wiley, New York, 1979).
[23] B. Wunderlich, *Macromolecular Physics* (Academic Press, New York, 1973), Vol. 1.
[24] J. D. Hoffman, G. T. Davis, and J. I. Lauritzen, Jr., *Treatise on Solid State Chemistry*, edited by N. B. Hannay (Plenum Press, New York, 1976), Vol. 3.
[25] J. D. Hoffman, Polymer **23**, 656 (1982), and **24**, 3 (1983).
[26] J. D. Hoffman and R. L. Miller, Macromolecules **21**, 3038 (1988).
[27] J. D. Hoffman and R. L. Miller, Polymer **38**, 3151 (1997).
[28] L. Mandelkern, *Crystallization of Polymers* (McGraw-Hill, New York, 1964).
[29] B. Wunderlich, *Macromolecular Physics* (Academic Press, New York, 1976), Vol. 2.
[30] G. W. Gray and J. W. Goodby, *Smectic Liquid Crystals: Textures and Structures* (Leonard Hill, London, 1984).
[31] P. S. Pershan, *Structure Of Liquid Crystal Phases* (World Scientific, Singapore, 1988).
[32] P. M. Chaikin and T. C. Lubersky, *Principles of Condensed Matter Physics* (Cambridge University Press, New York, 1995).
[33] B. Wunderlich and H. Bauer, Adv. Polymer Sci. **7**, 151 (1970); V. Gaur and B. Wunderlich, J. Phys. Chem. Ref. Data **10**, 19 (1981).

[34] B. Wunderlich, S. Z. D. Cheng, and K. Loufakis, in *Encyclopedia of Polymer Science and Engineering* (Wiley, New York, 1989), 2nd edition, Vol. 16, pp. 676–807.

[35] S. Z. D. Cheng, J. Appl. Polym. Sci. Appl Polym. Symp. **43**, 315 (1989).

[36] B. Wunderlich, Pure Appl. Chem. **67**, 1019 (1995).

[37] P. Ehrenfest, Leiden Comm. Suppl. **756** (1933).

[38] B. Wunderlich, *Thermal Analysis* (Academic Press, New York, 1991).

[39] E. A. DiMarzio, Annals N. Y. Acad. Sci. **371**, 1 (1981).

[40] P. R. Couchman, Macromolecules **11**, 1156 (178).

[41] J. E. McKinney and M. Goldstein, J. Res. Nat. Bur. Stand. **78A**, 331 (1974).

[42] J. E. McKinney and H. V. Belcher, J. Res. Nat. Bur. Stand. **67A**, 43 (1962).

[43] J. D. Ferry, *Viscoelastic Properties of Polymers* (Wiley, New York, 1980), 3rd edition.

[44] M. Goldstein, J. Phys. Chem. **77**, 667 (1973).

[45] J. E. McKinney and R. Simha, J. Res. Nat. Bur. Stand. **81A**, 283 (1977).

[46] *Conference on Structure and Mobility in Molecular and Atomic Glasses Proceedings*, Annals N. Y. Acad. Sci. **371**, 7 (1981).

[47] H. F. Mark, N. G. Gaylord, and N. M. Bikales, editors, *Encyclopedia of Polymer Science and Technology* (Wiley, New York, 1970), Vol. 13, p. 780.

[48] J. Brandrup and E. H. Immergut, editors, *Polymer Handbook*, (Wiley, New York, 1967).

[49] C. E. Harper, editor, *Handbook of Plastics and Elastomers* (McGraw-Hill, New York, 1975).

[50] R. S. Stein and G. L. Wilkes, *Structure and Properties of Oriented Polymers*, edited by I. M. Ward (Wiley, New York, 1975); and L. G. Treloar, *The Physics of Rubber Elasticity* (Oxford University Press, Oxford, 1958), 2nd edition.

[51] M. V. Folkes and A. Keller, Polymer **12**, 222 (1971).

[52] F. Khoury and E. Passaglia, *Treatise on Solid State Chemistry*, edited by N. B. Hannay (Plenum Press, New York, 1976), Vol. 3.

[53] James R. Rice, *Fracture, an Advanced Treatise*, edited by H. Liebowitz (Academic Press, New York, 1968), Vol. II.

[54] H. Kramer and K. E. Helf, Kolloid Z **180**, 115 (1962).

[55] N. G. McCrum, B. E. Read, and G. Williams, *Anelastic and Dielectric Effects in Polymeric Solids* (Wiley, New York, 1967).

[56] G. C. Berry and T. G. Fox, Adv. Polymer Sci. **5**, 261 (1968).

[57] W. W. Graessley, Adv. Polymer Sci. **16**, 1 (1974).

[58] J. Crank, *The Mathematics of Diffusion* (Clarendon Press, Oxford, 1976).

[59] S. T. Hwang, *et al.*, Sep. Sci. **9**, 461 (1974).

[60] *Nonlinear Optical and Electroactive Polymers*, edited by P. N. Prasad and D. R. Ulrich (Plenum Press, New York, 1988).

[61] Y. R. Shen, *The Principles of Nonlinear Optics* (Wiley, New York, 1988).

[62] P. N. Prasad and D. J. Williams, *Introduction to Nonlinear Optical Effects in Molecules and Polymers* (Wiley, New York, 1991).

[63] A. C. Griffin, A. M. Bhatti, and R. S. L. Hung, in *Nonlinear Optical and Electroactive Polymers*, edited by P. N. Prasad and D. A. Ulrich (Plenum Press, New York, 1988), p. 375.

23

Quantum Theory

M. P. Silverman
Trinity College, Hartford, Connecticut

R. L. Mallett
University of Connecticut, Storrs, Connecticut

Contents

List of Tables

List of Figures

PART I. QUANTUM MECHANICS

23.1. BASIC FORMALISMS

The principles of quantum theory were first expressed (1925–1926) in two superficially different mathematical formalisms: "matrix mechanics," a study of the algebra and eigenvalue spectra of dynamical operators, and "wave mechanics," a study of the solutions to differential equations of motion. The equivalence of the two formalisms was demonstrated by Schrödinger, and shortly afterward (1927) a general "transformation theory" for the construction of arbitrary representations of a quantum system was created by Dirac. Some two decades later, Feynman (1949) introduced the "space-time" or "path-integral" approach, whereby the dynamics of a quantum system follows from an integral equation containing the Lagrangian, and not a differential equation containing the Hamiltonian (as is characteristic of Dirac theory).

In general, the Dirac formalism is usually applied to the analysis of discrete quantum systems, whereas path integrals are most useful, if not essential, for treating problems in non-Abelian quantum field theory.

23.2. OPERATOR REPRESENTATIONS AND RELATIONSHIPS

23.2.1. Operator algebra

Quantum theory employs two distinct types of operators:

1. *Hermitian operators* are self-adjoint; i.e., $H = H^\dagger$, in which the adjoint is defined by the relation $H_{ij}^\dagger = H_{ji}^*$, with $H_{ij} = \langle i|\, H\, |j\rangle$. The eigenvalue spectrum of H is real-valued and corresponds to observable outcomes of measurements of dynamical variables.
2. *Unitary operators* are characterized by $U^{-1} = U^\dagger$; they transform quantum systems spatially, temporally, and between different representations. The general form of a unitary operator is

$$U(\lambda) = e^{iH\lambda}, \tag{23.1}$$

in which H is hermitian and λ is the transformation parameter. The exponential of operator A is defined by the infinite series

$$e^A = \sum_{n=0} \frac{A^n}{n!}. \tag{23.2}$$

The commutator of two operators, $[A, B] = AB - BA$, figures prominently in the time evolution, spatial transformations, and uncertainty relations of quantum systems. For arbitrary operators A, B, C,

$$[A, BC] = [A, B]C + B[A, C], \tag{23.3}$$

$$[A, [B, C]] + [C, [A, B]] + [B, [C, A]] = 0 \quad \text{[Jacobi identity]}, \tag{23.4}$$

$$e^A B e^{-A} = B + [A, B] + \tfrac{1}{2!}[A, [A, B]] + \tfrac{1}{3!}[A, [A, [A, B]]] + \ldots. \tag{23.5}$$

The relations

$$[A, B^n] = nB^{n-1}[A, B], \tag{23.6}$$

$$e^{\lambda A} e^{\lambda B} = e^{\lambda(A+B)+\frac{1}{2}\lambda^2[A,B]} \quad \text{[Baker-Campbell-Hausdorff (BCH) formula]} \tag{23.7}$$

are valid if A and B both commute with $[A, B]$. The BCH formula is one of a large class of relations that can be constructed from different parameterizations of Lie groups relevant to quantum theory. [1]

23.2.2. Coordinate and linear momentum operators

Representations

In a coordinate representation (with Cartesian coordinates), the (three-dimensional) coordinate operator $\mathbf{X}$ is diagonal with eigenvalue spectrum $\{\mathbf{x}\}$, and the linear momentum operator takes the form $\mathbf{P} = -i\hbar\nabla$. In a momentum representation, $\mathbf{P}$ is diagonal with eigenvalue spectrum $\{\mathbf{p}\}$ and the coordinate operator $\mathbf{X} = i\hbar\nabla_{\mathbf{P}}$ takes derivatives with respect to the components of $\mathbf{P}$. Irrespective of representation, the two operators satisfy the commutation relation

$$[\mathbf{X}, \mathbf{P}] = i\hbar\mathbf{1}, \tag{23.8}$$

where the unit operator

$$\mathbf{1} = \int |\,\mathbf{x}\rangle\langle\mathbf{x}\,|\,\mathbf{dx} = \int |\,\mathbf{p}\rangle\langle\mathbf{p}\,|\,\mathbf{dp} \tag{23.9}$$

can be expanded in a complete set of basis states. The transformation between coordinate and momentum representations is effected by the set of coefficients

$$\langle\mathbf{x} \mid \mathbf{p}\rangle = e^{i\,\mathbf{x}\cdot\mathbf{p}/\hbar}(2\pi\hbar)^{-3/2}. \tag{23.10}$$

What is familiarly called the "wave function" $\boldsymbol{\psi}(\mathbf{x})$ is the projection $\langle\mathbf{x} \mid \boldsymbol{\Psi}\rangle$ of the state vector $\boldsymbol{\Psi}$ onto a coordinate basis state. The momentum representation of the state vector

$$\varphi(\mathbf{p}) = \langle\mathbf{p} \mid \Psi\rangle = \int \langle\mathbf{p} \mid \mathbf{x}\rangle\langle\mathbf{x} \mid \Psi\rangle\, d\mathbf{x} = \frac{1}{(2\pi\hbar)^{3/2}} \int \psi(\mathbf{x}) e^{-i\,\mathbf{x}\cdot\mathbf{p}/\hbar}\, d\mathbf{x} \tag{23.11}$$

is a Fourier integral transformation.

The coordinate and momentum representations of an operator O are related by

$$\langle\mathbf{p}|\, O\, |\mathbf{p}'\rangle = \int \langle\mathbf{p}|\,\mathbf{x}\rangle\langle\mathbf{x} \mid O\, |\mathbf{x}'\rangle\langle\mathbf{x}' \mid \mathbf{p}'\rangle\, \mathbf{dx}\,\mathbf{dx}' \to \frac{1}{(2\pi\hbar)^3} \int O(\mathbf{x}) \mathbf{e}^{i(\mathbf{p}'-\mathbf{p})\cdot\mathbf{x}}\, \mathbf{dx}, \tag{23.12}$$

which reduces to a Fourier integral if the coordinate representation of O is diagonal:

$$\langle\mathbf{x}|\, O\, |\mathbf{x}'\rangle = O(\mathbf{x})\, \delta(\mathbf{x} - \mathbf{x}'). \tag{23.13}$$

[See Sec. 2.16 for the Dirac delta function.]

Translations

$\mathbf{P}$ is the generator of spatial translations, as expressed by the unitary transformation of a coordinate basis state by the displacement vector $\mathbf{a}$:

$$|\,\mathbf{x} + \mathbf{a}\rangle = e^{-i\,\mathbf{a}\cdot\mathbf{P}/\hbar}\, |\mathbf{x}\rangle. \tag{23.14}$$

The effect on the wave function is then

$$\langle\mathbf{x} + \mathbf{a}|\, \boldsymbol{\Psi}\rangle = \langle\mathbf{x} \mid \mathbf{e}^{i\,\mathbf{a}\cdot\mathbf{P}/\hbar}\, |\boldsymbol{\Psi}\rangle = e^{a\cdot\nabla}\psi(\mathbf{x}) = \boldsymbol{\psi}(\mathbf{x} + \mathbf{a}), \tag{23.15}$$

in which the third equality signifies a Taylor series expansion of the wave function about $\mathbf{a}$.

Canonical and kinetic momenta

For a quantum particle of mass m and electric charge q in the presence of an electromagnetic vector potential field $\mathbf{A}$, it is necessary to distinguish the kinetic linear momentum

$\mathbf{P}_k$, which corresponds to the classical expression of mass times velocity, from the canonical (or conjugate) linear momentum $\mathbf{P}_c = \mathbf{P}_k + (q/c)\mathbf{A}$ (where c is the speed of light), which takes the form $-i\hbar\nabla$ in a coordinate representation. In the Hamiltonian formulation of quantum theory, the kinetic energy K of a particle must be expressed in terms of $\mathbf{P}_c$ through the ("minimal coupling") substitution $\mathbf{P}_k = \mathbf{P}_c - (q/c)\mathbf{A}$. This substitution leads to diverse electromagnetic effects, such as the Zeeman effect, Stark effect, Kerr effect, Faraday effect, and others. [2]

23.2.3. Angular momentum

General properties

An axial vector operator $\mathbf{J}$ with components $\mathbf{J}_i (i = 1, 2, 3)$ that satisfy commutation relations

$$[\mathbf{J}_p, \mathbf{J}_q] = i\hbar \sum_r \varepsilon_{pqr} \mathbf{J}_r \tag{23.16}$$

is an angular momentum operator. The completely antisymmetric permutation symbol ε_{pqr} takes the value $+1\,(-1)$ for an even (odd) permutation of the indices $(1, 2, 3)$ and is 0 for any repetition of indices. [Thus, $\varepsilon_{123} = -\varepsilon_{213} = 1$; $\varepsilon_{223} = 0$.] Angular momentum eigenvectors $\mid jm\rangle$ are eigenvectors of the scalar operator $\mathbf{J}^2 = \mathbf{J}\cdot\mathbf{J} = \sum_i J_i^2$ and vector component $\mathbf{J}_3$:

$$\mathbf{J}^2 \mid jm\rangle = j(j+1)\hbar^2 \mid jm\rangle, \tag{23.17}$$

$$\mathbf{J}_3 \mid jm\rangle = m\hbar \mid jm\rangle, \tag{23.18}$$

where j is any positive integer $(0, 1, 2\ldots)$ or half-integer $(\frac{1}{2}, \frac{3}{2}, \frac{5}{2}\ldots)$, and m spans the range $-j \le m \le +j$ in unit steps. $\mathbf{J}^2$ commutes with all of the components of $\mathbf{J}$.

Linear combinations of vector components, $J_\pm = J_1 \pm iJ_2$, satisfy the commutation relations

$$[J_3, J_\pm] = \pm\hbar J_\pm, \tag{23.19}$$

$$[J_+, J_-] = 2\hbar J_3, \tag{23.20}$$

and constitute "ladder" operators,

$$J_\pm \mid jm\rangle = \sqrt{j(j+1) - m(m\pm 1)} \mid j\, m\pm 1\rangle, \tag{23.21}$$

between m-substates of a given j-manifold. Note that $J_+ \mid jj\rangle = J_- \mid j\,\bar{j}\rangle = 0$ (where the overbar signifies a negative quantum number). Use of ladder operators is often facilitated by the identity

$$\mathbf{J}^2 - J_3^2 \pm \hbar J_3 = J_\pm J_\mp. \tag{23.22}$$

Orbital angular momentum [3]

For a particle subject to a central force, the orbital angular momentum $\mathbf{L} = \mathbf{X}\times\mathbf{P}$ (where $\mathbf{P}$ is the canonical linear momentum) is a conserved quantity. Expressed in a coordinate representation with radial coordinate r, polar angle θ, and azimuthal angle φ, the components

$L_p = \sum_{q,r} \varepsilon_{pqr} X_q P_r$ are differential operators independent of r,

$$L_1 = -i\hbar \left(-\sin\varphi \frac{\partial}{\partial\theta} - \cot\theta \cos\varphi \frac{\partial}{\partial\varphi} \right), \tag{23.23}$$

$$L_2 = -i\hbar \left(\cos\varphi \frac{\partial}{\partial\theta} - \cot\theta \sin\varphi \frac{\partial}{\partial\varphi} \right), \tag{23.24}$$

$$L_3 = -i\hbar \frac{\partial}{\partial\varphi}, \tag{23.25}$$

and the scalar $\mathbf{L}^2$ is related to the Laplacian:

$$\mathbf{L}^2 = \frac{-\hbar^2 \nabla^2}{r^2} = -\hbar^2 \left[\frac{1}{\sin^2\theta} \frac{\partial^2}{\partial\varphi^2} + \frac{1}{\sin\theta} \frac{\partial}{\partial\theta} \left(\sin\theta \frac{\partial}{\partial\theta} \right) \right]. \tag{23.26}$$

The commutation between $\mathbf{L}$ and $\mathbf{X}$ takes the coordinate-independent form

$$[\mathbf{n} \cdot \mathbf{L}, \mathbf{X}] = -i\hbar(\mathbf{n} \times \mathbf{X}) \tag{23.27a}$$

(with $\mathbf{n}$ as an arbitrary unit vector) that reduces to

$$[L_p, X_q] = i\hbar \sum_r \varepsilon_{pqr} X_r \tag{23.27b}$$

in terms of Cartesian components. Expressions for the commutation of $\mathbf{L}$ with $\mathbf{P}$ are of identical form.

Solution of the eigenvalue problem for $\mathbf{L}^2$ leads to state vectors $|\, lm\rangle$ labeled by an integer orbital angular momentum quantum number ($l = 0, 1, 2. \ldots$). The wave functions obtained by projection of $|\, lm\rangle$ onto a coordinate basis (in spherical polar coordinates) are spherical harmonic functions

$$\psi_{lm}(\theta, \varphi) \equiv \langle \theta\varphi \mid lm \rangle = Y_l^m(\theta, \varphi) \tag{23.28}$$

defined by

$$Y_l^m(\theta, \varphi) = \sqrt{\frac{2l+1}{4\pi} \frac{(l-m)!}{(l+m)!}} (-1)^m e^{im\varphi} P_l^m(\cos\theta), \tag{23.29}$$

in which $P_l^m(\cos\theta)$ is an associated Legendre function. The properties of spherical harmonics are summarized in Sec. 2.20.5.

Spin angular momentum

Fundamental leptons (electron, muon, neutrino), quarks, and composite nuclear constituents (proton, neutron) all have an intrinsic spin-$\frac{1}{2}$ angular momentum represented by the operator $\mathbf{s} = \frac{1}{2}\hbar\boldsymbol{\sigma}$, where the components of $\boldsymbol{\sigma}$ are the Pauli spin matrices defined by the algebraic relation

$$\sigma_p \sigma_q = i \sum_r \varepsilon_{pqr} \sigma_r + \delta_{pq}. \tag{23.30}$$

(The Kronecker delta symbol δ_{pq} is unity if $p = q$ and vanishes otherwise.) A commonly employed representation of $\boldsymbol{\sigma}$ takes the form

$$\sigma_1 = \begin{pmatrix} 0 & 1 \\ 1 & 0 \end{pmatrix}, \quad \sigma_2 = \begin{pmatrix} 0 & -i \\ i & 0 \end{pmatrix}, \quad \sigma_3 = \begin{pmatrix} 1 & 0 \\ 0 & -1 \end{pmatrix}. \tag{23.31}$$

In this representation, the set of eigenvectors, or spinors, $|\,sm\rangle = \left\{|\,\frac{1}{2}\frac{1}{2}\rangle, |\frac{1}{2}\overline{\frac{1}{2}}\rangle\right\}$ are column vectors $\left\{\begin{pmatrix}1\\0\end{pmatrix}, \begin{pmatrix}0\\1\end{pmatrix}\right\}$.

If $\mathbf{A}$ and $\mathbf{B}$ are two vector operators that commute with $\boldsymbol{\sigma}$, the algebraic properties of Pauli matrices leads to the identity

$$(\boldsymbol{\sigma}\cdot\mathbf{A})(\boldsymbol{\sigma}\cdot\mathbf{A}) = \mathbf{A}\cdot\mathbf{B} + i\boldsymbol{\sigma}\cdot(\mathbf{A}\times\mathbf{B}). \tag{23.32}$$

Rotations

$\mathbf{J}$ is the generator of rotational transformations through the unitary operator $D = e^{-i\mathbf{n}\cdot\mathbf{J}\theta/\hbar}$, in which the unit vector $\mathbf{n}$ defines the axis of rotation and θ is the angular displacement. In the case of a spin-$\frac{1}{2}$ system, D reduces to the matrix

$$D = \mathbf{1}\cos\frac{\theta}{2} - i\mathbf{n}\cdot\boldsymbol{\sigma}\sin\frac{\theta}{2}. \tag{23.33}$$

For arbitrary j, a rotation of θ of about $\mathbf{n}$ can be accomplished by a suite of three rotations

$$D = e^{-i\alpha J_3/\hbar}e^{-i\beta J_2/\hbar}e^{-i\gamma J_3/\hbar}, \tag{23.34}$$

where (α, β, γ) are appropriate Euler angles, from which follow the rotation matrix elements

$$\langle jm|\,D\,|jn\rangle \equiv D^j_{mn}(\alpha\beta\gamma) = e^{-i(\alpha m+\gamma n)}\,d^j_{mn}(\beta), \tag{23.35}$$

with

$$d^j_{mn}(\beta) = \sum_q (-1)^q \frac{[(j+m)!(j-m)!(j+n)!(j-n)!]^{1/2}}{(j+m-q)!(j-n-q)!q!(q+n-m)!} \times \left(\cos\frac{\beta}{2}\right)^{2j+m-n-2q}\left(\sin\frac{\beta}{2}\right)^{2q+n-m}. \tag{23.36}$$

The above summation is taken over all values of q that lead to nonnegative factorials.

Addition of angular momentum [4]

The angular momentum states $\{|\,jm\rangle\}$ of a system comprising two subsystems of angular momenta j_1 and j_2 are constructed from the direct product of component states $\{|\,j_1m_1\rangle|\,j_2m_2\rangle\}$ by the series

$$|\,jm\rangle = \sum_{m_1,m_2} |\,j_1j_2m_1m_2\rangle\langle j_1j_2m_1m_2\mid jm\rangle, \tag{23.37}$$

in which the Clebsch-Gordan (CG) coefficient $\langle j_1 j_2 m_1 m_2 | \, jm\rangle$ vanishes unless $m = m_1 + m_2$ and j obeys the "triangular condition" $j_1 + j_2 \geq j \geq |\, j_1 - j_2 \,|$. A tabulation of selected CG coefficients is given in Sec. 2.20.7. The composite states are eigenstates of $\mathbf{J}^2$ and J_3, where $\mathbf{J} = \mathbf{J_1} + \mathbf{J_2}$.

Because of its high degree of symmetry, the Wigner 3-J symbol, a 3×2 array defined by

$$\begin{pmatrix} a & b & c \\ \alpha & \beta & \gamma \end{pmatrix} \sqrt{2c+1} = (-1)^{a-b-\gamma} \langle ab\alpha\beta \mid c \; -\gamma\rangle, \tag{23.38}$$

is often used in place of the CG coefficient. The 3-J symbol is invariant under an even (cyclic) permutation of its columns, incurs a phase factor $(-1)^{a+b+c}$ for an odd permutation of its columns, and incurs the same phase factor under a sign change of all substate quantum numbers, $(\alpha, \beta, \gamma) \to (-\alpha, -\beta, -\gamma)$. All three angular momenta (a, b, c) are displayed and manipulated on equal footing in a 3-J symbol.

3-J symbols can be calculated generally from the Wigner formula

$$\begin{aligned} \begin{pmatrix} a & b & c \\ \alpha & \beta & \gamma \end{pmatrix} &= (-1)^{a-b-\gamma} \delta_{\alpha+\beta,\gamma} \, \Delta(a\,b\,c) \\ &\quad \times [(a+\alpha)!(a-\alpha)!(b+\beta)!(b-\beta)!(c+\gamma)!(c-\gamma)!]^{1/2} \\ &\quad \times \sum_{\kappa} (-1)^{\kappa} [(a-\alpha-\kappa)!(c-b+\alpha+\kappa)!(b+\beta-\kappa)! \\ &\qquad (c-a-\beta+\kappa)!\kappa!(a+b-c-\kappa)!]^{-1} \end{aligned} \tag{23.39}$$

where the triangle function $\Delta(abc)$ is defined by

$$\Delta(abc) = \left[\frac{(a+b-c)!(a+c-b)!(b+c-a)!}{(a+b+c+1)!} \right]^{1/2}, \tag{23.40}$$

and the index κ of the summation runs over all integer values that do not lead to a negative factorial. 3-J symbols satisfy the orthogonality relations

$$\sum_{\alpha,\beta} (2c+1) \begin{pmatrix} a & b & c \\ \alpha & \beta & \gamma \end{pmatrix} \begin{pmatrix} a & b & c' \\ \alpha & \beta & \gamma' \end{pmatrix} = \delta_{cc'} \delta_{\gamma\gamma'}, \tag{23.41}$$

$$\sum_{c} (2c+1) \begin{pmatrix} a & b & c \\ \alpha & \beta & \gamma \end{pmatrix} \begin{pmatrix} a & b & c \\ \alpha' & \beta' & \gamma \end{pmatrix} = \delta_{\alpha\alpha'} \delta_{\beta\beta'}. \tag{23.42}$$

Tensor operators and the Wigner-Eckart theorem

An irreducible tensor operator T_{kq} of rank k has $2k+1$ components and transforms under a rotation according to $T'_{kq} = D T_{kq} D^{\dagger} = \sum_p T_{kp} D^k_{pq}(\alpha\beta\gamma)$. Under an infinitesimal transformation, the foregoing equation leads to the following commutation relations:

$$[J_3, T_{kq}] = q T_{kq}, \tag{23.43}$$

$$[J_{\pm}, T_{kq}] = \sqrt{(k \pm q + 1)(k \mp q)} \quad T_{k\,q\pm1}. \tag{23.44}$$

According to the Wigner-Eckart theorem, the matrix elements of a tensor operator factor into geometric and physical parts:

$$\langle \alpha jm|\, T_{kq}\, |\beta j'm'\rangle = (-1)^{2k}\langle \alpha j\|\, \mathbf{T}_{kq}\, \|\beta j'\rangle\langle jm \mid j'km'q\rangle \tag{23.45}$$

(in which α, β signify other quantum numbers defining the state vectors). Directional properties (dependent on substate quantum numbers) are contained in a CG coefficient; dynamical properties (independent of geometry) appear in a scalar matrix element $\langle \alpha j\|\, \mathbf{T}_{kq}\, \|\beta j'\rangle$ called the reduced matrix element.

One important application is the evaluation of orbital angular momentum matrix elements

$$\langle lm|\, Y_{LM}\, |l'm'\rangle = \langle lm|\, l'Lm'M\rangle\langle l\|\, \mathbf{Y}_L\, \|l'\rangle, \tag{23.46}$$

with

$$\langle l\|\, Y_L\, \|l'\rangle = \sqrt{\frac{(2l'+1)(2L+1)}{4\pi(2l+1)}}\langle l0 \mid Ll'00\rangle. \tag{23.47}$$

23.2.4. Hamiltonian

The Hamiltonian operator H is the generator of time translations. It is constructed as the sum of kinetic and potential energies of a system, expressed in terms of appropriately symmetrized canonically conjugate variables. For a nonrelativistic spinless charged particle in electromagnetic potentials $\mathbf{A}$ and φ, H takes the form

$$H = \frac{1}{2m}\left(\mathbf{P} - \frac{q}{c}\mathbf{A}(\mathbf{x}, t)\right)^2 + q\varphi(\mathbf{x}, t), \tag{23.48}$$

where $\mathbf{P}$ is the canonical linear momentum.

In a system for which total energy is conserved, the allowed energies are given by the eigenvalue spectrum of H. The Hamiltonian, however, is not necessarily the system energy operator (e.g., in the case of an open system with time-dependent interactions).

23.2.5. Commutation and uncertainty relations

General inequality

Whenever two dynamical variables, represented by hermitian operators A and B, do not commute, they cannot be measured simultaneously. Measured on a quantum state Ψ, the mean value and variance of A are

$$\langle A\rangle = \langle \Psi|\, A\, |\Psi\rangle, \tag{23.49}$$

$$(\Delta A)^2 = \langle \Psi \mid (A - \langle A\rangle)^2 \mid \Psi\rangle, \tag{23.50}$$

with corresponding expressions for B. If $[A, B] = iC$, the variances of A and B satisfy the inequality

$$(\Delta A)^2(\Delta B)^2 \geq \tfrac{1}{4}\langle C\rangle^2. \tag{23.51}$$

Coordinate-momentum uncertainty

The commutator of corresponding components of coordinate and canonical linear momentum $[X_j, P_j] = i\hbar(j = 1, 2, 3)$ is a constant and not an operator. Thus

$$\Delta X_j \, \Delta P_j \geq \hbar/2, \tag{23.52}$$

irrespective of the quantum state of the system.

Energy-time uncertainty [5]

The energy-time uncertainty relation connects the variance of the Hamiltonian H to a time interval characteristic of the rate of change of the system. Although H is a dynamical variable, time is just a parameter, and the correct uncertainty relation does not arise from commutation of H and t.

If A is a dynamical variable, the time variation of A in the Heisenberg representation leads to the relation

$$\Delta E \, \Delta t_A \geq \tfrac{1}{2}\hbar, \tag{23.53}$$

with time interval

$$\Delta t_A = \frac{\Delta A}{|\, d\langle A \rangle / dt \,|} \tag{23.54}$$

specific for A.

Amplitude-phase uncertainty

The quantum theory of oscillating systems like the harmonic oscillator or the electromagnetic field is conveniently expressed in terms of annihilation and creation operators $(\hat{a}, \hat{a}^\dagger)$ defined by the commutation relation

$$[\hat{a}, \hat{a}^\dagger] = 1. \tag{23.55a}$$

(For a multimode system, operators characterizing different modes commute.) The number of quanta in a mode (e.g., photons in the case of an electromagnetic field) is counted by the number operator $\hat{n} = \hat{a}^\dagger \hat{a}$ with commutation relations

$$[\hat{n}, \hat{a}^\dagger] = \hat{a}^\dagger, \quad [\hat{n}, \hat{a}] = -\hat{a}. \tag{23.55b}$$

Annihilation and creation operators can be written as products of amplitude and phase operators, $\hat{a} = \sqrt{\hat{n}+1}\,\hat{e}^{i\phi}$ and $\hat{a}^\dagger = \hat{e}^{-i\phi}\sqrt{\hat{n}+1}$. The two factors do not commute, signifying that an oscillator mode cannot have sharp values of amplitude and phase simultaneously. The number operator and hermitian phase operators defined by $\hat{c}(\phi) = \frac{1}{2}(\hat{e}^{i\,\phi} + \hat{e}^{-i\,\phi})$ and $\hat{s}(\phi) = \frac{1}{2i}(\hat{e}^{i\,\phi} - \hat{e}^{-i\,\phi})$ satisfy commutation relations [6]

$$[\hat{n}, \hat{c}(\phi)] = -i\,\hat{s}(\phi) \quad \text{and} \quad [\hat{n}, \hat{s}(\phi)] = i\,\hat{c}(\phi) \tag{23.55c}$$

from which follow uncertainty relations

$$\Delta n \, \Delta \cos\phi \geq \tfrac{1}{2} |\, \langle \hat{s}(\phi) \rangle \,|, \tag{23.56a}$$

$$\Delta n \, \Delta \sin\phi \geq \tfrac{1}{2} |\, \langle \hat{c}(\phi) \rangle \,|. \tag{23.56b}$$

23.3. QUANTUM DYNAMICS

23.3.1. Time-displacement operator

The evolution of a quantum system in time is determined by either local (i.e., differential) or global (i.e., integral) equations of motion. In the first category are the eponymous equations of Schrödinger, Heisenberg, Pauli, Dirac, and Klein-Gordon. The differential equations, distinguished by the choice of Hamiltonian H, derive from an infinitesimal time displacement of the state vector $\Psi(t+\varepsilon) = [1 - (i\varepsilon/\hbar)H(t)]\Psi(t)$ leading to the Schrödinger form of the equation of motion,

$$H(t)\,\Psi(t) = i\,\hbar\frac{d\Psi(t)}{dt}. \tag{23.57}$$

In the second category are various methods centered on the use of propagators and Green's functions in which the time development is derived from a finite unitary transformation

$$\Psi(t) = U(t, t_0)\Psi(t_0), \tag{23.58}$$

with time-development operator

$$\begin{aligned} U(t, t_0) &= \mathbf{T}\exp\left[-\frac{i}{\hbar}\int_{t_0}^{t} H(t')\,dt'\right] \qquad (23.59) \\ &= 1 + \sum_{1}^{\infty}\frac{1}{(i\hbar)}^{n}\int_{t_0}^{t} dt_1 \int_{t_0}^{t_1} dt_2 \ldots \int_{t_0}^{t_{n-1}} dt_n\,\mathbf{T}[H(t_1)\,H(t_2)\ldots H(t_n)], \end{aligned}$$

where $\mathbf{T}$ signifies the time-ordered product of operators

$$\mathbf{T}[A(t)\,B(t')] = \begin{cases} A(t)\,B(t'), & t \geq t', \\ B(t')A(t), & t' \geq t. \end{cases} \tag{23.60}$$

For time-independent H, one has simply $U(t, t_0) = e^{-i\,H(t-t_0)/\hbar}$.

23.3.2. Single-particle wave equations

Schrödinger equation

The equation of motion of a nonrelativistic, spinless, charged particle (mass m, charge q) with potential energy $V(\mathbf{X})$ and subject to an electromagnetic vector potential $\mathbf{A}$ is

$$\left[\frac{1}{2m}\left(\frac{\hbar}{i}\nabla - \frac{q}{c}\mathbf{A}\right)^2 + V(\mathbf{X})\right]\psi = i\,\hbar\frac{\partial\psi}{\partial t}. \tag{23.61}$$

If the potential energy arises from interaction with an electromagnetic scalar potential φ, then $V = q\varphi$.

Pauli equation

The equation of motion of a nonrelativistic spin-$1/2$ charged particle interacting with an applied magnetic field $\mathbf{B}$ (in addition to the preceding interactions) is

$$\left[\frac{1}{2m}\left(\frac{\hbar}{i}\nabla - \frac{q}{c}\mathbf{A}\right)^2 + V(\mathbf{X}) - \frac{g}{2}\frac{q\hbar}{2mc}\boldsymbol{\sigma}\cdot\mathbf{B}\right]\begin{pmatrix}\psi_\uparrow\\ \psi_\downarrow\end{pmatrix} = i\,\hbar\frac{\partial}{\partial t}\begin{pmatrix}\psi_\uparrow\\ \psi_\downarrow\end{pmatrix}, \qquad (23.62)$$

where the wave function is a two-component spinor representing spin-up and spin-down states. If the particle is an electron, the Bohr magneton is $q\hbar/2mc = -0.927 \times 10^{-20}$ erg/gauss, and the Landé g-factor is $g = 2$ (neglecting quantum electrodynamic corrections).

Relativistic Schrödinger equation

The equation of a relativistic spinless charged particle subject to electromagnetic scalar and vector potentials is

$$\left[\left(\frac{\hbar}{i}\nabla - \frac{q}{c}\mathbf{A}\right)^2 + m^2c^2\right]\psi = \frac{1}{c^2}\left[i\hbar\frac{\partial}{\partial t} - q\varphi\right]^2\psi \qquad (23.63a)$$

or, succinctly in relativistic notation with Einstein summation convention and $\mu = 0, 1, 2, 3$,

$$\left[\left(p^\mu - \frac{q}{c}\mathbf{A}^\mu\right)^2 - m^2c^2\right]\psi = 0. \qquad (23.63b)$$

$p^\mu = i\hbar\left(\frac{\partial}{c\partial t}, -\nabla\right)$ is the four-vector momentum, and $A^\mu = (\varphi, \mathbf{A})$ is the four-vector potential. [In the absence of electromagnetic potentials, Eq. (23.63a) or (23.63b) becomes the Klein-Gordon equation.] In the metric convention employed here, the scalar product $A^\mu B_\mu$ of two four-vectors $A^\mu = (A^0, \mathbf{A})$ and $B^\mu = (B^0, \mathbf{B})$ is $A^0B^0 - \mathbf{A}\cdot\mathbf{B}$. Equivalently, the covariant form of a contravariant four-vector B^μ is $B_\mu = (B^0, -\mathbf{B})$. Thus, $(p^\mu)^2 = p^\mu p_\mu = -\hbar^2((\partial^2/c^2\partial t^2) - \nabla^2)$.

Dirac equation [7]

The equation of motion of a relativistic spin-$1/2$ charged particle in electromagnetic potentials is

$$\left[c\boldsymbol{\alpha}\cdot\left(\frac{\hbar}{i}\nabla - \frac{q}{c}\mathbf{A}\right) + \beta mc^2\right]\begin{pmatrix}\psi_{+\uparrow}\\ \psi_{+\downarrow}\\ \psi_{-\uparrow}\\ \psi_{-\downarrow}\end{pmatrix} = \left(i\hbar\frac{\partial}{\partial t} - q\varphi\right)\begin{pmatrix}\psi_{+\uparrow}\\ \psi_{+\downarrow}\\ \psi_{-\uparrow}\\ \psi_{-\downarrow}\end{pmatrix}, \qquad (23.64a)$$

with 4×4 matrices $\boldsymbol{\alpha} = \begin{pmatrix}\mathbf{0} & \boldsymbol{\sigma}\\ \boldsymbol{\sigma} & \mathbf{0}\end{pmatrix}$, $\beta = \begin{pmatrix}\mathbf{1} & \mathbf{0}\\ \mathbf{0} & -\mathbf{1}\end{pmatrix}$. The wave function ψ is a four-component spinor representing spin-up and spin-down states of positive and negative energy.

In relativistic notation, the Dirac equation takes the compact form

$$\left[\gamma_\mu\left(p^\mu - \frac{q}{c}A^\mu\right) - mc\right]\psi = 0, \qquad (23.64b)$$

with matrices $\gamma^0 = \beta$ and $\boldsymbol{\gamma} = \begin{pmatrix} \mathbf{0} & \boldsymbol{\sigma} \\ -\boldsymbol{\sigma} & \mathbf{0} \end{pmatrix}$. Application of the charge conjugation operator $C = i\gamma^2\gamma^0$ to a negative energy solution generates a positive energy solution of the corresponding antiparticle with an opposite spin component.

Bilinear products of ψ and the adjoint spinor $\bar{\psi} = \psi^\dagger\gamma^0$ lead to quantities with the following distinct transformation properties:

$$\text{Scalar, } \bar{\psi}\psi; \tag{23.65a}$$

$$\text{Pseudoscalar, } \bar{\psi}\gamma^5\psi \text{ with } \gamma^5 = i\gamma^0\gamma^1\gamma^2\gamma^3 = \begin{pmatrix} \mathbf{0} & \mathbf{1} \\ \mathbf{1} & \mathbf{0} \end{pmatrix}; \tag{23.65b}$$

$$\text{Vector, } \bar{\psi}\gamma^\mu\psi, (\mu = 0, 1, 2, 3); \tag{23.65c}$$

$$\text{Pseudovector, } \bar{\psi}\gamma^\mu\gamma^5\psi, (\mu = 0, 1, 2, 3); \tag{23.65d}$$

$$\text{Antisymmetric tensor, } \bar{\psi}\sigma^{\mu\nu}\psi \text{ with } \sigma^{\mu\nu} = \tfrac{i}{2}(\gamma^\mu\gamma^\nu - \gamma^\nu\gamma^\mu)\ (\mu, \nu = 0, 1, 2, 3). \tag{23.65e}$$

23.3.3. Operator equations of motion

Heisenberg form

In the Heisenberg form (as contrasted with the Schrödinger form) of the equations of motion, the state vector is fixed and dynamical variables evolve in time according to the unitary transformation $O(t) = U(t, t_0)O(t_0)U(t, t_0)^\dagger$, which leads to

$$\frac{dO}{dt} = \frac{1}{i\hbar}[O, H] + \frac{\partial O}{\partial t}, \tag{23.66}$$

where H is the Hamiltonian. If an operator has no intrinsic time dependence, the last term is absent.

Ehrenfest's theorem [8]

Applied to the coordinate and linear momentum of a particle in a velocity-independent potential $V(\mathbf{X})$, the Heisenberg equation with Hamiltonian $H = \mathbf{P}^2/2m + V(\mathbf{X})$ leads to the operator equivalent of Newton's second law,

$$\frac{d\mathbf{X}}{dt} = \frac{1}{m}\mathbf{P}, \quad \frac{d\mathbf{P}}{dt} = -\nabla V. \tag{23.67}$$

The corresponding relation of expectation values, with definition $\langle d\mathbf{X}/dt\rangle = d\langle\mathbf{X}\rangle/dt$, is known as Ehrenfest's theorem.

For a spinless charged particle in an electromagnetic field (with the Hamiltonian of Sec. 23.2.4), the Heisenberg equation leads to the set of operator relations

$$\frac{d\mathbf{X}}{dt} = \frac{1}{m}\left(\mathbf{P}_c - \frac{q}{c}\mathbf{A}\right) = \frac{1}{m}\mathbf{P}_k, \tag{23.68}$$

$$\frac{d\mathbf{P}_k}{dt} = q\mathbf{E} + +\frac{q}{2c}\left(\frac{d\mathbf{X}}{dt} \times \mathbf{B} - \mathbf{B} \times \frac{d\mathbf{X}}{dt}\right), \tag{23.69}$$

which express the classical Lorentz force law with $\mathbf{E} = -\frac{1}{c}\frac{\partial \mathbf{A}}{\partial t} - \nabla\varphi, \mathbf{B} = \nabla \times \mathbf{A}$. Note that the velocity operator $d\mathbf{X}/dt$ contains the canonical linear momentum $\mathbf{P}_c$ and therefore does not commute with functions of $\mathbf{X}$.

Virial theorem

For a particle with Hamiltonian $H = \mathbf{P}^2/2m + V(\mathbf{X})$, the Heisenberg equation yields

$$\frac{d}{dt}(\mathbf{X} \cdot \mathbf{P}) = \frac{\mathbf{P}^2}{m} - \mathbf{X} \cdot \nabla V. \tag{23.70}$$

The expectation value of the left-hand side vanishes in a stationary state, resulting in the virial theorem

$$2\langle K \rangle = \langle \mathbf{X} \cdot \nabla V \rangle. \tag{23.71}$$

If V is a spherically symmetric potential proportional to r^n, then $2\langle K \rangle = n\langle V \rangle$.

Applied to a molecule in the Born-Oppenheimer approximation (in which nuclear coordinates $\mathbf{R}_k$ are vector parameters, not quantum operators), the virial theorem (together with the Hellman-Feynman theorem, Sec. 23.4.2) leads to

$$2\langle K_e \rangle + \langle V \rangle = -\sum_{k=1}^{N} \mathbf{R}_k \cdot \nabla_k E(\mathbf{R}_1, \mathbf{R}_2, \ldots, \mathbf{R}_N), \tag{23.72}$$

where K_e is the electron kinetic energy, V is the total potential energy of electrons and nuclei, and $\langle K_e \rangle + \langle V \rangle = E(\mathbf{R}_1, \mathbf{R}_2, \ldots, \mathbf{R}_N)$.

23.4. APPROXIMATE METHODS: STATIONARY STATES

23.4.1. Perturbation theory (bound states)

Rayleigh-Schrödinger theory

Nondegenerate stationary states The Hamiltonian takes the form $H = H_0 + \lambda V$, where $\{| n\rangle\}$ and $\{\varepsilon_n\}$ are the set of orthonormalized eigenvectors and spectrum of energy eigenvalues of H_0. The eigenvalues and eigenvectors of H take the form of series expansions in the perturbation parameter λ:

$$E_n = \varepsilon_n + \sum_{k=1} \lambda^k E_n^{(k)}, \tag{23.73}$$

$$| N\rangle = \sum_{k=0} \lambda^k \,| N^{(k)}\rangle = | n\rangle + \sum_{k=1} \lambda^k \sum_{m \neq n} |m\rangle\langle m \mid N^{(k)}\rangle, \tag{23.74}$$

where the state $| n\rangle$ is excluded from the last summation since $\langle n \mid m\rangle = 0$.

The terms of the series are determined recursively from the relations

$$E_n^{(k)} = \langle n| V |N^{(k-1)}\rangle, \tag{23.75}$$

$$\langle m \mid N^{(k)}\rangle = \frac{1}{\varepsilon_n - \varepsilon_m}(\langle m| V |N^{(k-1)}\rangle - E^{(1)}\langle m \mid N^{(k-1)}\rangle - \cdots - E^{(k-1)}\langle m \mid N^{(1)}\rangle). \tag{23.76}$$

To second order, the expressions yield eigenvalues

$$E_n^{(1)} = V_{nn}, \tag{23.77}$$

$$E_n^{(2)} = \sum_{m \neq n} \frac{|\, V_{nm}\,|^2}{\varepsilon_n - \varepsilon_m}, \tag{23.78}$$

and eigenvector components

$$\langle m \mid N^{(1)} \rangle = \frac{V_{mn}}{\varepsilon_n - \varepsilon_m}, \tag{23.79}$$

$$\langle m \mid N^{(2)} \rangle = \sum_{k \neq n} \frac{V_{mk} V_{kn}}{(\varepsilon_n - \varepsilon_k)(\varepsilon_n - \varepsilon_m)} - \frac{V_{mn} V_{nn}}{(\varepsilon_n - \varepsilon_m)^2}, \tag{23.80}$$

in which $V_{nm} \equiv \langle n|\, V\, |m\rangle$.

The states $|\, N\rangle$ are constructed with normalization $\langle n \mid N\rangle = 1$. Setting $|\, \bar{N}\rangle = Z^{1/2}|\, N\rangle$ with $Z = \partial E_n/\partial \varepsilon_n$ gives states with normalization $\langle \bar{N} \mid \bar{N}\rangle = 1$.

Degenerate stationary states The preceding perturbation formulas break down if there are any states $|\, m\rangle$ for which $\varepsilon_m = \varepsilon_n$. To remedy the problem, one constructs within the manifold of degenerate states $|\, n_i\rangle$ $(i = 1 \ldots g)$ linear combinations $|\, \nu_i\rangle$ such that $\langle \nu_i|\, V\, |\nu_j\rangle = 0$ for $i \neq j$. The nondegenerate perturbation formulas can then be used with the states $|\, \nu_i\rangle$ replacing $|\, n_i\rangle$.

The problem reduces to diagonalizing V within the group of degenerate states and leads to the solution $|\, \nu_i\rangle = \sum_{i=1}^{g} C_{ij}|\, n_j\rangle$, where C_{ij} is the jth component of the ith eigenvector of the $g \times g$ matrix with elements $\langle n_i|\, V\, |n_j\rangle$.

Brillouin-Wigner theory

Expanded in terms of the exact energy E_n, the state vector $|\, N\rangle$ is obtained recursively from the exact relation

$$|\, N\rangle = |\, n\rangle + \sum_{m \neq n} |\, m\rangle \frac{\lambda \langle m|\, V\, |N\rangle}{E_n - \varepsilon_m}, \tag{23.81a}$$

leading to the series

$$|\, N\rangle = |\, n\rangle + \lambda \sum_{m \neq n} |\, m\rangle \frac{V_{mn}}{E_n - \varepsilon_m} + \lambda^2 \sum_{j,m \neq n} |\, j\rangle \frac{V_{jm}}{E_n - \varepsilon_j} \frac{V_{mn}}{E_n - \varepsilon_m}$$

$$\lambda^3 \sum_{k,j,m \neq n} |\, k\rangle \frac{V_{kj}}{E_n - \varepsilon_k} \frac{V_{jm}}{E_n - \varepsilon_j} \frac{V_{mn}}{E_n - \varepsilon_m} + \cdots + . \tag{23.81b}$$

Substitution of the expression for $|\, N\rangle$ into the exact energy expression

$$E_n = \varepsilon_n + \langle n|\, V\, |N\rangle \tag{23.82}$$

leads to a nonlinear equation for E_n, which is often more accurate than the Rayleigh-Schrödinger expansion.

23.4.2. Variational Method

Rayleigh-Ritz variational principle

The variational method is a nonperturbative approach that can be tried when there is no closely related problem capable of exact solution or when the Schrödinger equation is nonseparable. If E_0 is the exact ground state eigenvalue of H, then for any wave function ψ,

$$\frac{\langle\psi|\, H\, |\psi\rangle}{\langle\psi \mid \psi\rangle} \geq E_0, \tag{23.83}$$

where the equality holds if ψ is the exact ground-state wave function. Operationally, the expectation value of H is minimized with respect to the parameter(s) of a normalized trial function to obtain the least upper bound on the ground-state energy. The best wave function is usually that which corresponds to the lowest energy. The procedure can also be applied to excited states by constructing a trial function that is orthogonal to the wave functions of lower energy eigenstates.

Hückel approximation

The state vector of a quantum system is represented as a linear superposition of basis states with adjustable coefficients, $|\,\Psi\rangle = \sum_{k=1}^{n} a_k |\,k\rangle$. The optimal set of parameters $\{a_k\}$ for the given basis is the solution to the eigenvalue problem

$$\sum_{k=1}^{n} a_k (H_{km} - E S_{km}) \quad (m = 1, 2, \ldots n), \tag{23.84}$$

with $H_{km} = \langle k|\, H\, |m\rangle$, $S_{km} = \langle k \mid m\rangle$; E is determined from the secular equation

$$\det(H_{km} - E S_{km}) = 0. \tag{23.85}$$

Hellman-Feynman theorem

If the hermitian operator $H(\lambda)$ depends on a parameter λ, and $\psi_k(\lambda)$ is a normalized eigenfunction with eigenvalue $E_k(\lambda)$, then

$$\frac{dE_k(\lambda)}{d\lambda} = \langle\psi_k(\lambda)|\, \frac{dH(\lambda)}{d\lambda}\, |\psi_k(\lambda)\rangle. \tag{23.86}$$

23.4.3. Wentzel-Kramers-Brillouin (WKB) theory

The WKB approximation can be applied to one-dimensional problems with slowly varying potential.

Connection formulas

Under the condition that particle momentum $p(x) = h/\lambda = \hbar k$ changes little over a de Broglie wavelength $\lambda(x)$, i.e., $\lambda(x)|\,\frac{dp/dx}{p(x)}\,| \ll 1$, the solution to the Schrödinger equation takes the form

$$\psi(x) = \frac{1}{\sqrt{k(x)}} \exp\left[\pm i \int^{x} k(x')\, dx'\right], \tag{23.87}$$

with $k(x) = \frac{1}{\hbar}\sqrt{2m[E - V(x)]}$ for $E > V(x)$, and $k(x) = \frac{-i}{\hbar}\sqrt{2m[V(x) - E]} = -i\kappa(x)$ for $E < V(x)$.

The connection formulas for establishing continuity of the wave function across a classical turning point, where $V(x) = E$, are as follows: [9]

(a) Turning point $x = a$ lies to the right of the classical region:

$$\frac{2}{\sqrt{k}}\cos\left(\int_x^a k\,dx - \frac{1}{4}\pi\right) \leftarrow\longleftrightarrow \frac{1}{\sqrt{k}}\exp\left(-\int_a^x \kappa\,dx\right), \tag{23.88a}$$

$$\frac{1}{\sqrt{k}}\sin\left(\int_x^a k\,dx - \frac{1}{4}\pi\right) \leftarrow\rightarrow\rightarrow \frac{-1}{\sqrt{\kappa}}\exp\left(\int_a^x \kappa\,dx\right). \tag{23.88b}$$

(b) Turning point $x = b$ lies to the left of the classical region:

$$\frac{1}{\sqrt{\kappa}}\exp\left(-\int_x^b \kappa\,dx\right) \leftarrow\rightarrow\rightarrow \frac{2}{\sqrt{k}}\cos\left(\int_b^x k\,dx - \frac{1}{4}\pi\right), \tag{23.88c}$$

$$\frac{-1}{\sqrt{\kappa}}\exp\left(\int_x^b \kappa\,dx\right) \leftarrow\longleftrightarrow \frac{1}{\sqrt{k}}\sin\left(\int_b^x k\,dx - \frac{1}{4}\pi\right). \tag{23.88d}$$

A particular connection formula should be used only in the direction of the double arrow, unless it is established that the other linearly independent solution in a given region is absent.

Phase integral

For bound-state problems (periodic motion with two turning points a and b), satisfaction of boundary conditions at infinity leads to the phase integral

$$\int_a^b k(x)\,dx = \left(n + \frac{1}{2}\right)\pi \tag{23.89a}$$

or, equivalently, to the contour integral over a closed curve in phase space

$$\oint p(x)\,dx = \left(n + \frac{1}{2}\right)h. \tag{23.89b}$$

Transmission through a barrier

The WKB transmission coefficient (i.e., relative flux) for particles incident from the right upon a potential barrier with classical turning points at $x = a$ and b $(b > a)$ is

$$T = \frac{4}{(2\xi + (1/2\xi))^2}, \tag{23.90}$$

where $\xi = \exp\left(\int_a^b \kappa(x)\,dx\right)$. No assumptions have been made regarding barrrier shape.

23.4.4. Scattering theory (stationary state)

Scattering amplitude

The outgoing wave function $\psi_{\mathbf{k}}^{(+)}(\mathbf{r})$ of a particle with mass m and initial momentum $\hbar\mathbf{k}$ scattered by a local potential $V(\mathbf{r})$ satisfies the exact integral equation (with $U = 2mV/\hbar^2$)

$$\psi_{\mathbf{k}}^{(+)}(\mathbf{r}) = \frac{1}{(2\pi)^{3/2}} e^{i\,\mathbf{k}\cdot\mathbf{r}} - \frac{1}{4\pi} \int \frac{\exp(ik|\,\mathbf{r}-\mathbf{r}'\,|)}{|\,\mathbf{r}-\mathbf{r}'\,|} U(\mathbf{r}')\psi_{\mathbf{k}}^{(+)}(\mathbf{r})\, d^3\mathbf{r}', \tag{23.91}$$

which may be solved by iterative substitution for $\psi_{\mathbf{k}}^{(+)}(\mathbf{r})$. The asymptotic form of the scattered wave function is

$$\psi_{\mathbf{k}}^{(+)}(\mathbf{r}) \frac{1}{(2\pi)^{3/2}} \left(e^{i\,\mathbf{k}\cdot\mathbf{r}} - \frac{e^{i\,kr}}{r} f_{\mathbf{k}}(\hat{\mathbf{r}}) \right), \tag{23.92}$$

in which

$$f_{\mathbf{k}}(\hat{\mathbf{r}}) = -\frac{(2\pi)^{3/2}}{4\pi} \int \exp(-ik\hat{\mathbf{r}}\cdot\mathbf{r}')\, U(\mathbf{r}')\psi_{\mathbf{k}}^{(+)}(\mathbf{r}')\, d^3\mathbf{r}' \tag{23.93}$$

is the scattering amplitude and $\hat{\mathbf{r}} = \mathbf{r}/r$ is the direction of scattering. The momentum of the outgoing particle is $\hbar\mathbf{k}'$ with $\mathbf{k}' = k\hat{\mathbf{r}}$. The differential scattering cross section is $d\sigma/d\Omega = |\, f_{\mathbf{k}}(\hat{\mathbf{r}})\,|^2$. (See also Sec. 6.8.2 on phase-shift analysis, Sec. 6.14.1 on Thomson and Rayleigh scattering, and Sec. 6.8.9 on Rutherford and Sec. 6.9.2 on Mott scattering.)

Born approximation

The first-order approximation obtained by substituting a plane wave function for $\psi_{\mathbf{k}}^{(+)}(\mathbf{r})$ leads to the Born amplitude

$$f_{\mathbf{k}\mathbf{k}'} = -\frac{m}{2\pi\hbar^2} \int e^{i\,(\mathbf{k}-\mathbf{k}')\cdot\mathbf{r}'}\, V(\mathbf{r}')\, d^3\mathbf{r}', \tag{23.94}$$

which is most reliable for weak scattering potentials or high incident energies. A particularly useful case is that of a spherically symmetric potential $V(r)$, for which the scattering amplitude

$$f_q(\theta) = -\frac{2m}{\hbar^2 q} \int_0^\infty r\, \sin(qr)\, V(r)\, dr, \quad q = 2k \sin\frac{1}{2}\theta, \tag{23.95}$$

depends on the angle θ between $\mathbf{k}$ and $\mathbf{k}'$. (See also Sec. 6.7.)

Optical theorem

As a general consequence of the conservation of probability (or particles), the total cross section σ and the forward scattering amplitude $f_{\mathbf{k}}(0)$ satisfy the equation

$$\sigma = \frac{4\pi}{k} \mathrm{Im}\{f_{\mathbf{k}}(0)\}. \tag{23.96}$$

23.5. TIME-DEPENDENT PERTURBATION THEORY

Time-dependent perturbation theory is applicable to a system with well-defined stationary states between which a weak external time-dependent interaction $H_I(t)$ drives transitions. The stationary states are eigenstates of a time-independent Hamiltonian H_0 with eigenvalue spectrum $\{\varepsilon_n\}$. The state vector, after the perturbation is turned on at time t_0, takes the form

$$| \Psi_t \rangle = \sum_k a_k(t) | k \rangle e^{-i\,\varepsilon_k t/\hbar}, \tag{23.97}$$

in which the initial state is $| \Psi_0 \rangle = | n \rangle$. It is assumed that the total transition probability to all states $k \neq n$ is sufficiently small that the initial state is not significantly depleted.

23.5.1. First-order transitions

The first-order amplitude for transition from state $| n \rangle$ to final state $| k \rangle$ is

$$a_k^{(1)}(t) = \frac{1}{i\hbar} \int_{t_0}^{t} \langle k | \, H_I(t') \, | n \rangle e^{i\,\omega_{kn} t'} \, dt', \tag{23.98}$$

with $\omega_{kn} = \varepsilon_k - \varepsilon_n / \hbar$, the Bohr angular frequency.

23.5.2. Second-order transitions

The second-order transition amplitude is

$$a_k^{(2)}(t) = \left(\frac{1}{i\hbar} \right)^2 \int_{t_0}^{t} dt_1 \int_{t_0}^{t_1} dt_2 \sum_m \langle k | \, H_I \, | m \rangle e^{i\,\omega_{km} t_1} \langle m | \, H_I \, | n \rangle e^{i\,\omega_{mn} t_2}, \tag{23.99}$$

where the sum is over a complete set of eigenstates of H_0.

23.5.3. Fermi Golden Rule

If $| k \rangle$ is one of a continuum of states about $| n \rangle$ with density $\rho(\varepsilon_k)$ and H_I is effectively constant after initiation at t_0, then the transition rate to$| k \rangle$ is given by the "Fermi Golden Rule":

$$\Gamma_{n \to k} = \frac{2\pi}{\hbar} \, | \, M_{kn} \, |^2 \delta(\varepsilon_k - \varepsilon_n) \tag{23.100a}$$

or, for transitions to all states within the group,

$$\Gamma_n = \frac{2\pi}{\hbar} \, [| \, M_{kn} \, |^2 \rho(\varepsilon_k)]_{\varepsilon_k = \varepsilon_n}, \tag{23.100b}$$

with matrix element (to second order)

$$M_{kn} = \langle k | \, H_I \, | n \rangle + \sum_{m \neq n} \frac{\langle k | \, H_I \, | m \rangle \langle m | \, H_I \, | n \rangle}{\varepsilon_n - \varepsilon_m} + \cdots \tag{23.101}$$

The golden rule is valid over the time interval $h/\Delta\varepsilon < t \ll h/\delta\varepsilon$, where $\Delta\varepsilon$ is the range of energies of the final states and $\delta\varepsilon$ is the energy spacing between states.

23.5.4. Density of states

The number of quantum states per unit energy E and volume is $\rho(E) = \eta p^2 dp\, d\Omega / h^3\, dE$, with $d\Omega$ as the differential element of solid angle and η as the number of spin (or polarization) degrees of freedom. For p independent of direction, the integral over $d\Omega$ leads to a factor of 4π. Two important cases are as follows:

Nonrelativistic free-particle states of mass m ($E = p^2/2m$; $\eta = 1$),

$$\rho(E) = \frac{4\pi\sqrt{2}\, m^{3/2} E^{1/2}}{h^3}. \tag{23.102}$$

Radiation field states ($E = pc$; $\eta = 2$),

$$\rho(E) = \frac{8\pi E^2}{c^3 h^3}. \tag{23.103}$$

23.5.5. Exponential decay

The probability that a system subject to perturbation H_I (constant after initiation) remains in initial state $|\, n\rangle$ is given by the amplitude

$$a_n(t) = e^{-\Gamma_n t/2} e^{-i E_n t/\hbar}, \tag{23.104}$$

where Γ_n is the total rate of transitions out of $|\, n\rangle$ and $E_n = \varepsilon_n + \Delta E_n$ is the energy of the state. To second order, the energy shift engendered by H_I is

$$\Delta E_n = \langle n|\, H_I \,|n\rangle + \mathbf{P} \sum_{k \neq n} \frac{|\, \langle n|\, H_I \,|k\rangle \,|^2}{\varepsilon_n - \varepsilon_k}, \tag{23.105}$$

where $\mathbf{P}$ signifies the Cauchy principal value.

23.6. RADIATION THEORY

23.6.1. Interaction Hamiltonian

The number of quanta N per unit area and time corresponding to a classical monochromatic plane-wave vector potential field $\mathbf{A} = A_0 \hat{\mathbf{e}}\, e^{i(\mathbf{k}\cdot\mathbf{r} - \omega t)}$ of polarization $\hat{\mathbf{e}}$ and angular frequency ω is given by $|\, \mathbf{A}\,|^2 = (hc/\omega)N$. The radiation intensity is $I = \hbar\omega N$.

The interaction Hamiltonian in the Lorentz gauge ($\nabla \cdot \mathbf{A} = 0$) for a one-electron atom is

$$H_I = (e/mc)\mathbf{A}\cdot\mathbf{P} + (e^2/2mc^2)\mathbf{A}^2, \tag{23.106}$$

with $e, m, \mathbf{P}$ as the electron charge, mass, and canonical linear momentum. The term quadratic in $\mathbf{A}$ does not contribute to single-quantum transitions.

23.6.2. Absorption and emission

Absorption and induced emission

To first order in $\mathbf{A}$, single-photon absorption at frequency $\omega = \omega_{kn}$, bringing the atom from initial state $|\,n\rangle$ to excited state $|\,k\rangle$, occurs at a rate of

$$\Gamma^{\text{abs}}_{n\to k} = \frac{(2\pi e)^2}{m^2 c\hbar\omega_{kn}} N(\omega_{kn})|\,\langle k\,|e^{i\,\mathbf{k}\cdot\mathbf{r}}\hat{\mathbf{e}}\cdot\mathbf{P}|\,n\rangle\,|^2. \tag{23.107}$$

The reverse process, single-photon induced (or stimulated) emission at frequency $\omega = -\omega_{nk}$, occurs at the numerically equal rate $\Gamma^{\text{ind}}_{k\to n} = \Gamma^{\text{abs}}_{n\to k}$.

Spontaneous emission

The transition probability per unit time for spontaneous decay of an atom from state $|\,k\rangle$ to state $|\,n\rangle$ with emission of a photon with wave vector $\mathbf{k}$ and polarization $\hat{\mathbf{e}}$ into solid angle $d\Omega$ is

$$d\,\Gamma^{\text{spon}}_{k\to n} = \frac{e^2\omega_{kn}}{m^2c^3\hbar}\,|\,\langle n\,|e^{-i\,\mathbf{k}\cdot\mathbf{r}}\hat{\mathbf{e}}\cdot\mathbf{P}|\,k\rangle\,|^2\,d\Omega. \tag{23.108}$$

23.6.3. Multipole transitions

Electric dipole

The electric dipole ($E1$) approximation $e^{i\,\mathbf{k}\cdot\mathbf{r}} \approx 1$ can be employed when $kr \ll 1$. The identity

$$\mathbf{P}_{kn} = \frac{m}{i\hbar}\langle k|\,[\mathbf{R}, H]\,|n\rangle = i\,m\omega_{kn}\mathbf{R}_{kn} \tag{23.109}$$

leads to transition rates (averaged over all light polarizations):

Induced:

$$\Gamma^{\text{ind}}_{k\to n} = \frac{4\pi^2e^2\omega_{kn}}{3\hbar c}N(\omega_{kn})\,|\,\mathbf{R}_{kn}\,|^2. \tag{23.110}$$

Spontaneous:

$$\Gamma^{\text{spon}}_{k\to n} = \frac{4e^2(\omega_{kn})^3}{3\hbar c^3}\,|\,\mathbf{R}_{kn}\,|^2. \tag{23.111}$$

See also Sec. 21.7.8.

Higher multipoles

If the $E1$ transition vanishes, further terms in the Taylor series or spherical Bessel series expansion of $e^{\pm i\,\mathbf{k}\cdot\mathbf{r}}$,

$$e^{i\,\mathbf{k}\cdot\mathbf{r}} = 1 + i\mathbf{k}\cdot\mathbf{r} + \frac{1}{2!}(i\mathbf{k}\cdot\mathbf{r})^2 + \cdots = \sum_{l=0}^{\infty} i^l(2l+1)j_l(kr)P_l(\cos\theta) \tag{23.112}$$

(with θ as the angle between **k** and **r**) give rise to magnetic dipole, electric quadrupole, and higher order processes. (See Sec. 7.8.4.) If the entire unapproximated integral vanishes, higher orders of perturbation theory lead to multiple-quantum processes. [10]

23.6.4. Sum rules [11]

The following closed-form expressions of sums over matrix elements facilitate the calculation of radiation intensity and other quantities.

Dipole moment

$$\sum_k |X_{kn}|^2 = X_{nn}^2. \tag{23.113}$$

Oscillator strength (Thomas-Reiche-Kuhn sum rule)

$$\sum_k f_{kn} \equiv \frac{2m}{\hbar} \sum_k \omega_{kn} |\mathbf{r}_{kn}|^2 = 3. \tag{23.114}$$

See, also, Sec. 21.8.1 and Sec. 7.8.2.

Linear momentum

$$\sum_k \omega_{kn}^2 |X_{kn}|^2 = \frac{1}{m^2} \sum_k |(P_x)_{kn}|^2. \tag{23.115}$$

If the state n is spherically symmetric, then

$$\sum_k \omega_{kn}^2 |\mathbf{r}_{kn}|^2 = \frac{2}{m}(E_n - V_{nn}). \tag{23.116}$$

Momentum times force

$$\sum_k \omega_{kn}^3 |X_{kn}|^2 = \frac{\hbar}{2m^2} \left(\frac{\partial^2 V}{\partial X^2}\right)_{nn}. \tag{23.117}$$

If the state n is spherically symmetric,

$$\sum_k \omega_{kn}^3 |\mathbf{r}_{kn}|^2 = \frac{2\pi\hbar e^2}{m^2} \rho(\mathbf{r})_{nn}, \tag{23.118}$$

where $e\rho(\mathbf{r})$ is the density of positive charge. For a bare nucleus, $\rho(\mathbf{r}) = Z\delta(\mathbf{r})$.

Force squared

$$\sum_k \omega_{kn}^4 |X_{kn}|^2 = \frac{1}{m^2} \sum_k \left|\left(\frac{\partial V}{\partial X}\right)_{kn}\right|^2. \tag{23.119}$$

If the state n is spherically symmetric,

$$\sum_k \omega_{kn}^4 | \mathbf{r}_{kn} |^2 = \frac{1}{m^2}[(\nabla V)^2]_{nn}. \tag{23.120}$$

23.7. ADDITIONAL LINKS TO QUANTUM SYSTEMS IN OTHER CHAPTERS

23.7.1. Coulomb potential

See Secs. 7.10.1 and 6.3.7 on the eigenvalues and wave functions of hydrogenlike atoms.

23.7.2. Quantum rotator

See Sec. 16.2 on rotational energies of diatomic and other rigid molecules.

23.7.3. Anharmonic oscillator

See Sec. 16.3 on vibrational energies of diatomic and polyatomic molecules.

PART II. QUANTUM FIELD THEORY

23.8. BRIEF HISTORY [12]

Quantum field theory (QFT) has its roots in quantum electrodynamics (QED), which established well-defined rules for predicting outcomes of all processes involving interacting photons and leptons. The excellent experimental agreement of QED motivated attempts to extend field-theoretic techniques to the weak and strong interactions. These efforts were successfully realized only with the development of gauge field theory, which provided a unified representation of elementary particle interactions as a coupling of gauge bosons and fermions. Quantum chromodynamics (QCD) could then be viewed as the interaction of (fermionic) quarks mediated by gauge bosonic gluons associated with local (i.e., space-time–dependent) color SU(3) symmetry. Correspondingly, the electroweak interaction was formulated as an exchange between fermions mediated by photons and massive intermediate vector bosons associated with local SU(2) $\times$ U(1) symmetry.

The following sections summarize QFT by means of the standard model. This is done by exhibiting the Feynman rules for calculating cross sections and transition rates in QED, QCD, and electroweak processes.

23.9. FEYNMAN RULES FOR GAUGE THEORIES [13]

23.9.1. The *S* matrix

The S matrix is defined as $S = U(\infty, -\infty)$, where U is the time-displacement operator of Sec. 23.3.1. For a transition between initial state i and final state f, the scattering amplitude takes the explicit form

$$S_{fi} = \delta_{fi} + (2\pi)^4 \delta^{(4)} \left(\sum_j p'_j - \sum_j p_j \right)$$
$$\times \prod_k (2VE_k)^{-1/2} \prod_{k'} (2VE'_{k'})^{-1/2} \prod_l (2m_l c^2)^{1/2} M_{fi}, \qquad (23.121)$$

with $p = (E/c, \mathbf{p})$ $[p' = (E'/c, \mathbf{p}')]$ as the four-momentum of an incoming (outgoing) particle, m_l as the mass of any incoming or outgoing lepton, V as the quantization volume, and M_{fi} as the Feynman amplitude. Cross sections and transition rates are then fully calculable once M_{fi} is known.

23.9.2. Cross sections

The cross section for the collision $1+2 \rightarrow 1'+2'+\cdots+N'$(where numbers label particles) takes the form

$$d\sigma = (2\pi)^4 \delta^{(4)} \left(\sum_{j=1'}^{N'} p'_j - \sum_{j=1}^{2} p_j \right) \prod_{k=1'}^{N'} \left(\frac{c\, d^3\mathbf{p}'_k}{(2\pi)^3 2E'_k} \right) \prod_l (2m_l c^2)$$
$$\times \frac{\hbar^2 S}{4E_1 E_2 v_{\text{rel}}} \mid M_{fi} \mid^2, \qquad (23.122)$$

with S as a statistical factor ($1/k!$ for each group of k identical particles in the final state). The relative velocity of the incoming particles is given by

$$v_{\text{rel}} = \frac{\sqrt{(p_1 \cdot p_2)^2 - (m_1 m_2 c^2)^2}}{E_1 E_2} = \begin{cases} \mid \mathbf{p}_1 \mid (E_1 + E_2)/E_1 E_2 & \text{c.m. frame,} \\ \mid \mathbf{p}_1 \mid / E_1 & \text{lab frame.} \end{cases} \qquad (23.123)$$

23.9.3. Decay rates

The decay rate for the process $1 \rightarrow 1' + 2' + \ldots + N'$ is determined from

$$d\Gamma = (2\pi)^4 \delta^{(4)} \left(\sum_{j=1'}^{N'} p'_j - p_1 \right) \prod_{k=1'}^{N'} \left(\frac{c\, d^3\mathbf{p}'_k}{(2\pi)^3 2E'_k} \right) \frac{S}{2\hbar m_1} \mid M_{fi} \mid^2, \qquad (23.124)$$

where the decaying particle is at rest. The particle lifetime is $\tau = \beta/\Gamma$ with branching ratio $\beta = \Gamma/\Sigma\Gamma$, where $\Sigma\Gamma$ is the sum of decay rates over all decay modes.

23.9.4. Diagrammatic construction of amplitudes

Table 23.1 summarizes the rules for the explicit construction of the Feynman amplitude M_{fi} in QED, QCD, and the standard electroweak model. Space-time indices are denoted by Greek letters $\alpha, \beta, \ldots \mu, \nu, \lambda$; gluon color indices are represented by Roman letters a, b, c with quark color indices i, j. (see Sec. 23.10). For the gluon propagator, the gauge

TABLE 23.1. Rules for constructing transition amplitudes.

Process	Diagram	Analytic factor
External Lines		
Incoming, outgoing scalar	 k ; k	1
Incoming, outgoing fermion	 p,s ; p,s	$u_s(p), \bar{u}_s(p)$
Incoming, outgoing antifermion	 p,s ; p,s	$v_s(p), \bar{v}_s(p)$
Incoming, outgoing		$\varepsilon_r^{\mu}(k), \varepsilon_r^{\mu}(k)^*$
Photon or	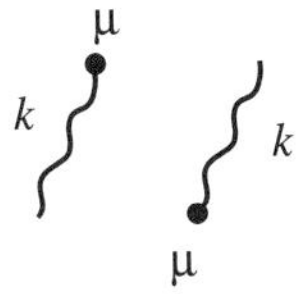 μ k ; k μ	
Vector boson	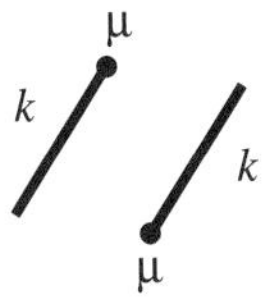 μ k ; k μ	
QED		
Photon	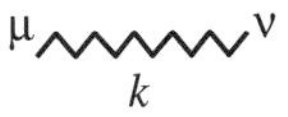 μ ν k	$\dfrac{-i\, g_{\mu\nu}}{k^2 + i\varepsilon}$
Lepton	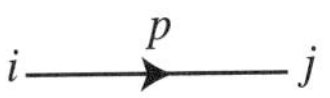 i p j	$\dfrac{i\ \not{p} + mc}{p^2 - (mc)^2 + i\varepsilon}$
Photon-lepton vertex	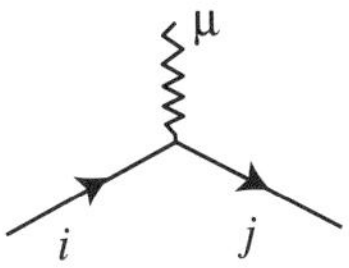 μ i j	$ig_e\gamma^{\mu}$
QCD		
Gluon	 a, μ k b, ν	$-i\left[g_{\mu\nu} - (1-\Lambda)\dfrac{p_\mu p_\nu}{p^2 + i\varepsilon}\right]\dfrac{\delta_{ab}}{p^2 + i\varepsilon}$
Quark	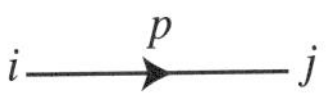 i p j	$\delta_{ab}\dfrac{i\ \not{p} + mc}{p^2 - (mc)^2 + i\varepsilon}$
Ghost[a]	 a k b	$\dfrac{i\delta_{ab}}{p^2 + i\varepsilon}$

(continued)

TABLE 23.1. Continued

Gluon-quark vertex	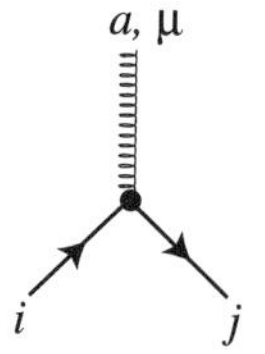 a, μ; i; j	$-\frac{1}{2}ig_s\gamma^\mu\lambda^a_{ij}$
Three-gluon vertex	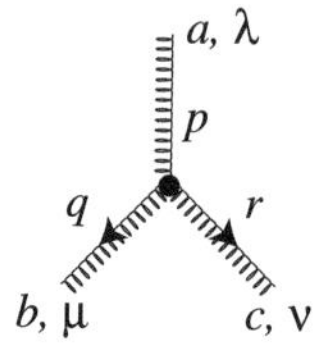 a, λ; p; q; r; b, μ; c, ν	$-g_s f_{abc}[(p-q)_\nu g_{\lambda\mu} + (q-r)_\lambda g_{\mu\nu} + (r-p)_\mu g_{\nu\lambda}]$
Four-gluon vertex	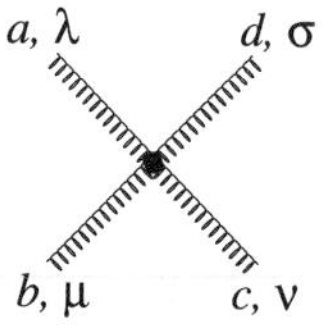 a, λ; d, σ; b, μ; c, ν	$-ig_s^2[f_{abc}f_{cde}(g_{\lambda\nu}g_{\mu\sigma} - g_{\lambda\sigma}g_{\mu\nu}) + f_{ace}f_{bde}(g_{\lambda\mu}g_{\nu\sigma} - g_{\lambda\sigma}g_{\mu\nu}) + f_{abe}f_{cde}(g_{\lambda\nu}g_{\mu\sigma} - g_{\lambda\mu}g_{\sigma\nu})]$
Gluon-ghost vertex	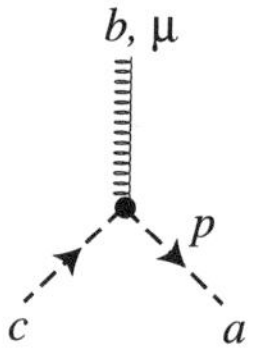 b, μ; p; c; a	$g_s f_{abc} p^\mu$
Standard Electroweak Model		
Massive vector boson	 α k β	$\dfrac{-i[g_{\alpha\beta} - \{k_\alpha k_\beta/(m_B c)^2\}]}{k^2 - (m_B c)^2 + i\varepsilon}$
Higgs scalar	 k	$\dfrac{i}{k^2 - (m_H c)^2 + i\varepsilon}$
Basic vertices:		
$e - e - Z$	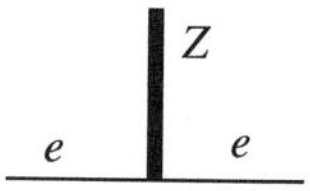 Z; e; e	$\dfrac{ig_W}{\sqrt{2}}\gamma^\mu\left([1 - 4\sin^2\theta_W] - \gamma^5\right)$
$\nu - \nu - Z$	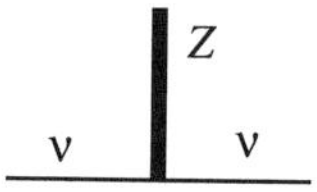 Z; ν; ν	$\dfrac{ig_W}{\sqrt{2}}\gamma^\mu\left(1 - \gamma^5\right)$

(continued)

TABLE 23.1. Continued

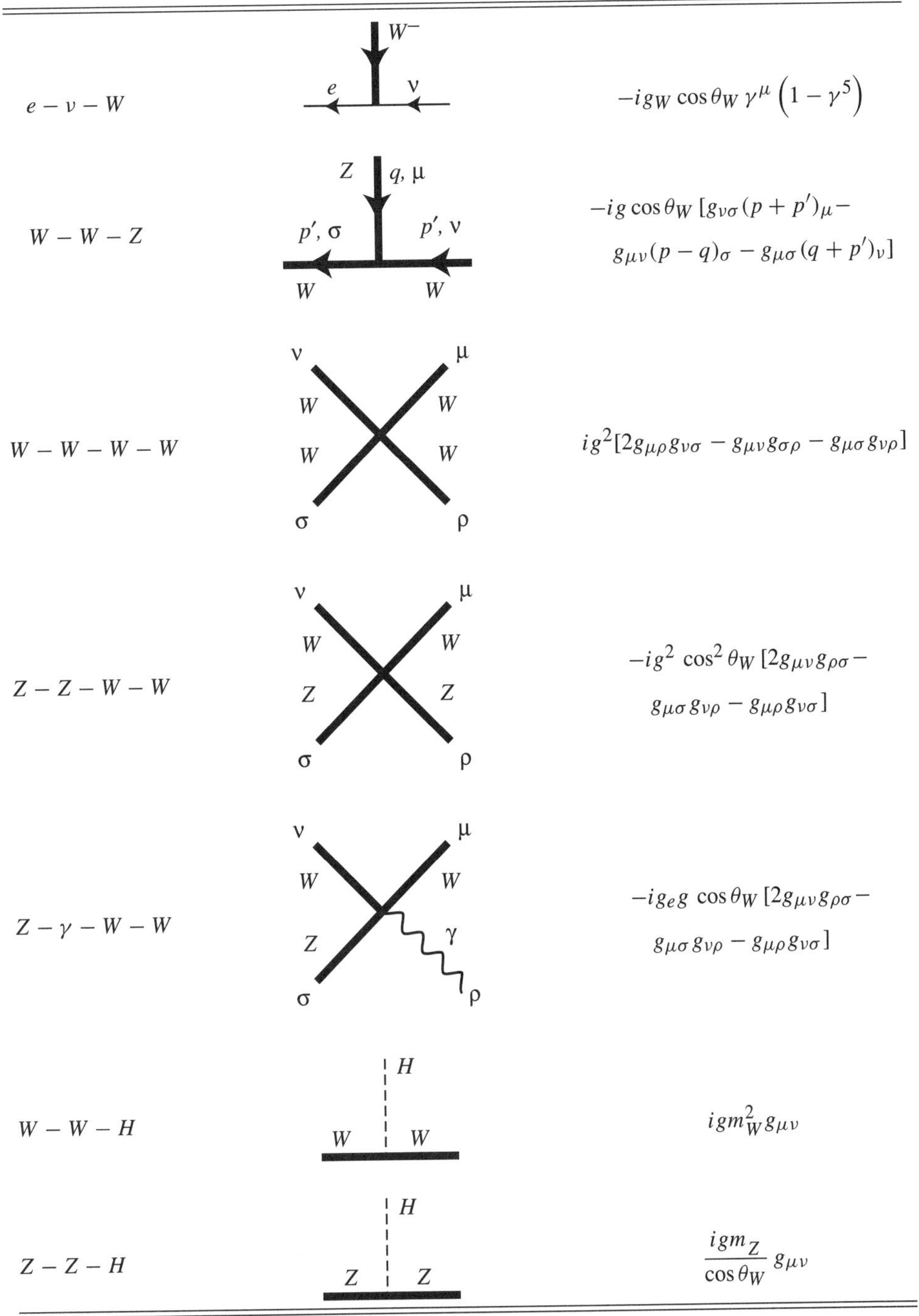

Vertex	Factor
$e-\nu-W$	$-ig_W\cos\theta_W\,\gamma^\mu\left(1-\gamma^5\right)$
$W-W-Z$	$-ig\cos\theta_W\,[g_{\nu\sigma}(p+p')_\mu-g_{\mu\nu}(p-q)_\sigma-g_{\mu\sigma}(q+p')_\nu]$
$W-W-W-W$	$ig^2[2g_{\mu\rho}g_{\nu\sigma}-g_{\mu\nu}g_{\sigma\rho}-g_{\mu\sigma}g_{\nu\rho}]$
$Z-Z-W-W$	$-ig^2\cos^2\theta_W\,[2g_{\mu\nu}g_{\rho\sigma}-g_{\mu\sigma}g_{\nu\rho}-g_{\mu\rho}g_{\nu\sigma}]$
$Z-\gamma-W-W$	$-ig_eg\cos\theta_W\,[2g_{\mu\nu}g_{\rho\sigma}-g_{\mu\sigma}g_{\nu\rho}-g_{\mu\rho}g_{\nu\sigma}]$
$W-W-H$	$igm_W^2g_{\mu\nu}$
$Z-Z-H$	$\dfrac{igm_Z}{\cos\theta_W}g_{\mu\nu}$

(continued)

TABLE 23.1. Continued

$W-W-H-H$	H H / W W	$i\frac{g^2}{2}g_{\mu\nu}$
$Z-Z-H-H$	H H / Z Z	$\frac{ig^2}{2\cos^2\theta_W}g_{\mu\nu}$
$e-e-H$	H / e e	$-i\frac{g}{2}\frac{m_e}{m_W}$
$H-H-H$	H / H H	$-i\frac{3g}{2}\frac{m_H^2}{m_W}$
$H-H-H-H$	H H / H H	$-i\frac{3g^2}{4}\frac{m_H^2}{m_W^2}$

[a]Faddeev-Popov ghosts in QCD are nonphysical particles introduced as part of a mathematical device for developing a perturbation series. In non-Abelian field theories, the ghost fields are necessary for gauge invariance and unitarity. Ghost fields do not couple to the gauge field in QED (an Abelian field theory) and, therefore, need not be introduced. Their role is nontrivial, however, in QCD.

parameter Λ equals 0 in the Landau gauge and 1 in the Feynman gauge. The prescription of adding the term $+i\varepsilon$ in all propagator denominators is for contour integration in the complex energy plane in which the limit $\varepsilon \to 0$ is taken in final results; this ensures that positive (negative) energy solutions propagate forward (backward) in time in accordance with causality.

23.9.5. Fermion spin sums

Dirac spinors for fermions $u_s(p)$ and antifermions $v_s(p)$ satisfy the completeness relations

$$\sum_{s=1,2} u_s(p)\bar{u}_s(p) = \frac{\not{p}+mc}{2mc} \qquad \sum_{s=1,2} v_s(p)\bar{v}_s(p) = \frac{\not{p}-mc}{2mc}, \tag{23.125}$$

with $\not{p} = \gamma^\mu p_\mu$; the following normalization applies here: $\bar{u}_s u_{s'} = \delta_{ss'}$, $\bar{v}_s v_{s'} = -\delta_{ss'}$.

23.9.6. Polarization sums

Photons

Photon polarization basis vectors satisfy the relation

$$\sum_{r=0}^{3} \eta_r \varepsilon_r^{\mu}(k)\varepsilon_r^{\nu}(k) = -g^{\mu\nu}, \tag{23.126}$$

with $\eta_0 = -1$, $\eta_1 = \eta_2 = \eta_3 = 1$, and (diagonal) metric $g^{\mu\nu} = (1, -1, -1, -1)$.

Massive boson polarization

For a vector boson of mass m_B ($B = W, Z$), the sum over polarization vectors yields

$$\sum_{r=0}^{3} \varepsilon_r^{\mu}(k)\varepsilon_r^{\nu}(k) = -g^{\mu\nu} + \frac{k^{\mu}k^{\nu}}{m_B^2}. \tag{23.127}$$

23.9.7. Contraction and trace relations

The following relations facilitate summing matrix elements over spin states. The Einstein summation convention is applied to repeated indices.

$$\gamma^{\mu}\gamma_{\mu} = 4 \quad (\mu = 0, 1, 2, 3), \tag{23.128a}$$

$$\gamma_{\lambda}\gamma^{\alpha}\gamma^{\lambda} = -2\gamma^{\alpha}, \tag{23.128b}$$

$$\gamma_{\lambda} \not{A} \gamma^{\lambda} = -2 \not{A}, \tag{23.128c}$$

$$\gamma_{\lambda} \not{A} \not{B} \gamma^{\lambda} = 4A \cdot B \quad (\text{where } A \cdot B = A^{\mu}B_{\mu}), \tag{23.128d}$$

$$\mathrm{Tr}\{\gamma^{\alpha}\gamma^{\beta}\} = 4g^{\alpha\beta} \quad (\text{where Tr signifies Trace}), \tag{23.128e}$$

$$\mathrm{Tr}\{\gamma^{\alpha}\gamma^{\beta}\gamma^{\gamma}\gamma^{\delta}\} = 4(g^{\alpha\beta}g^{\gamma\delta} - g^{\alpha\gamma}g^{\beta\delta} + g^{\alpha\delta}g^{\beta\gamma}), \tag{23.128f}$$

$$\mathrm{Tr}\{\not{A} \not{B}\} = 4A \cdot B, \tag{23.128g}$$

$$\mathrm{Tr}\{\not{A} \not{B} \not{C} \not{D}\} = 4\{(A \cdot B)(C \cdot D) - (A \cdot C)(B \cdot D) + (A \cdot D)(B \cdot C)\}, \tag{23.128h}$$

$$\mathrm{Tr}\{\gamma^5\} = \mathrm{Tr}\{\gamma^5\gamma^{\alpha}\} = \mathrm{Tr}\{\gamma^5\gamma^{\alpha}\gamma^{\beta}\} = \mathrm{Tr}\{\gamma^5\gamma^{\alpha}\gamma^{\beta}\gamma^{\gamma}\} = 0, \tag{23.128i}$$

$$\mathrm{Tr}\{\gamma^5\gamma^{\alpha}\gamma^{\beta}\gamma^{\gamma}\gamma^{\delta}\} = -4i\varepsilon^{\alpha\beta\gamma\delta} \tag{23.128j}$$

(where $\varepsilon^{\alpha\beta\gamma\delta}$ is the four-dimensional completely antisymetric tensor).

In general, the trace of the product of an odd number of gamma matrices is zero.

23.10. QUANTUM CHROMODYNAMICS

QCD describes the interactions between colored quarks mediated by spin-1 gluons. The color quantum number assumes three values ("red," "blue," "green"). Thus, a quark state is determined by the direct product of a Dirac spinor (giving momentum and spin) and a color vector. The strength of the chromodynamic force is set by a strong coupling constant g_s (which is the analog of g_e in QED; see Sec. 23.11.1).

The color gauge group SU(3) is generated by the Lie algebra:

$$\left[\frac{1}{2}\lambda_a, \frac{1}{2}\lambda_b\right] = i\, f_{abc}\frac{1}{2}\lambda_c, \quad a, b, c = 1, 2, \ldots 8, \tag{23.129}$$

with nonzero structure constants

$$\begin{aligned} f_{123} &= 1, & f_{246} &= \tfrac{1}{2}, & f_{367} &= -\tfrac{1}{2}, \\ f_{147} &= \tfrac{1}{2}, & f_{257} &= \tfrac{1}{2}, & f_{458} &= \tfrac{\sqrt{3}}{2}, \\ f_{156} &= -\tfrac{1}{2}, & f_{345} &= \tfrac{1}{2}, & f_{678} &= \tfrac{\sqrt{3}}{2}. \end{aligned} \tag{23.130}$$

23.11. STANDARD ELECTROWEAK MODEL

23.11.1. Coupling constants and fields

The standard electroweak model is based on spontaneously broken local SU(2) × U(1) symmetry. The local non-Abelian weak isospin group SU(2) has three gauge fields b^i_μ and a coupling constant g. The Abelian hypercharge group $U(1)$ has one gauge field a_μ and a coupling constant g'. Since the underlying symmetry is broken, the two neutral states can form linear superpositions corresponding to a massless particle (the photon) and a massive particle (the Z^0). The expression

$$\tan\theta_W = \frac{g'}{g} \tag{23.131}$$

defines the weak mixing angle θ_W. The standard QED coupling constant $g_e = \sqrt{4\pi\alpha_{\text{fs}}}$ (with $\alpha_{\text{fs}} = e^2/\hbar c$ as the fine structure constant) is related to g and g' by

$$g_e = g'\cos\theta_W = g\sin\theta_W. \tag{23.132a}$$

Derived coupling constants that appear in the literature include

$$g'' = \frac{g}{2\sqrt{2}\cos\theta_W}, \quad g_W = \frac{g_e}{\sin\theta_W}, \quad g_Z = \frac{g_e}{\sin\theta_W\cos\theta_W}. \tag{23.132b}$$

In terms of the four gauge fields, the physical charged and neutral boson fields are, respectively,

$$\begin{aligned} W^+_\mu &= \frac{1}{\sqrt{2}}(b^1_\mu - ib^2_\mu), & W^-_\mu &= \frac{1}{\sqrt{2}}(b^1_\mu + ib^2_\mu), \\ Z^0_\mu &= \frac{(-g'a_\mu + gb^3_\mu)}{\sqrt{g^2+g'^2}}, & A_\mu &= \frac{(ga_\mu + g'b^3_\mu)}{\sqrt{g^2+g'^2}}, \end{aligned} \tag{23.133}$$

with masses given by

$$m_W c^2 = \frac{g_e}{\sin\theta_W}\sqrt{\frac{\sqrt{2}}{8G_F}} = \frac{37.3\,Gev}{\sin\theta_W}, \qquad m_Z c^2 = \frac{m_W c^2}{\cos\theta_W}, \tag{23.134}$$

in which $G_F = \frac{\sqrt{2}}{8}(g/m_W c^2)^2 = 1.166\times 10^{-5}\,\text{GeV}^{-2}$ is the Fermi constant.

23.11.2. Example: Elastic neutrino-electron scattering

As an illustration of the rules for gauge theories, consider the leading contributions to $\nu_\mu e$ scattering involving the exchange of a Z^0 boson or Higgs particle (H), as shown in Figure 23.1, under the conditions of low momentum exchange $(\hbar k)^2 \ll (m_Z c)^2$, $(m_W c)^2$. The total Feynman amplitude is $M_{\nu e} = M^Z_{\nu e} + M^H_{\nu e}$, with

$$M^Z_{\nu e} = \frac{-iG_F}{2}[\bar{u}'_{\nu_\mu}\gamma^\alpha(1-\gamma_5)u_{\nu_\mu}][\bar{u}'_e\gamma^\alpha(g_V - g_A\gamma_5)u_e], \tag{23.135a}$$

$$M^H_{\nu e} = \frac{iG_F\sqrt{2}}{m_H^2}m_{\nu_\mu}m_e[\bar{u}'_{\nu_\mu}u_{\nu_\mu}][\bar{u}'_e u_e], \tag{23.135b}$$

and vector (V) and axial vector (A) coupling constants

$$g_V = 2\sin^2\theta_W - \frac{1}{2}, \quad g_A = -\frac{1}{2}. \tag{23.136}$$

Since $M^H_{\nu e}$ is of order $\frac{m_{\nu_\mu}m_e}{m_H^2}$ smaller than $M^Z_{\nu e}$, this amplitude can be neglected, leading to a cross section

$$\sigma(\nu_\mu e) = \frac{G_F^2\, s}{3\pi}(g_V^2 + g_V g_A + g_A^2), \tag{23.137}$$

with $s = 4\,(E_{\text{cm}}/c)^2 = 2m_e E_{\text{lab}} + m_e c$ as the square of the center of mass momentum; E_{cm} and E_{lab} are the neutrino energy in the center of mass frame and in the laboratory frame (where the target electron is at rest).

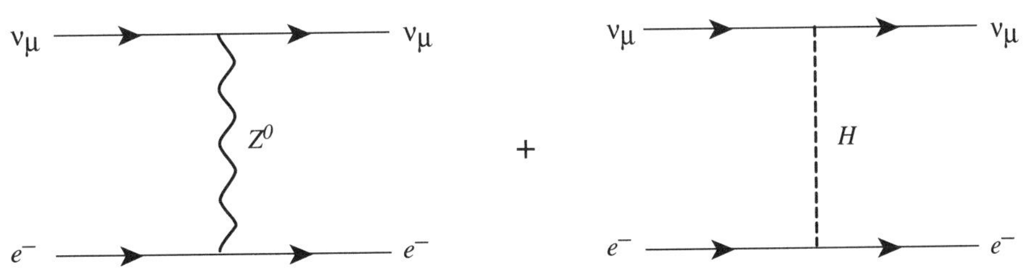

FIGURE 23.1. Example of weak neutral current interactions. Feynman diagrams for neutrino-electron scattering with exchange of either a Z^0 boson or Higgs particle.

23.12. REFERENCES

[1] R. Gilmore, *Lie Groups, Lie Algebras, and Some of Their Applications* (Wiley, New York, 1974) pp. 149–153, 461–463.

[2] M. P. Silverman, *Encyclopedia of Applied Physics* (Wiley, New York, 1998), Vol. 23, pp. 563–585.

[3] G. Baym, *Lectures on Quantum Mechanics* (W. A. Benjamin, New York, 1969), Ch. 6.

[4] D. M. Brink and G. R. Satchler, *Angular Momentum* (Clarendon, Oxford, 1968), Chs. 2 and 4.

[5] A. Messiah, *Quantum Mechanics* (Wiley, New York, 1965), Vol. 1, pp. 319–320.

[6] R. Loudon, *The Quantum Theory of Light*, 2nd ed. (Clarendon, Oxford, 1983), pp. 142–144.

[7] J. D. Bjorken and S. D. Drell, *Relativistic Quantum Mechanics* (McGraw-Hill, New York, 1964).

[8] L. I. Schiff, *Quantum Mechanics*, 3rd ed. (McGraw-Hill, New York, 1968), Sec. 24.

[9] E. Merzbacher, *Quantum Mechanics*, 2nd ed. (Wiley, New York, 1970), Ch. 7.

[10] M. P. Silverman, *Probing The Atom* (Princeton University Press, Princeton, NJ, 2000).

[11] H. A. Bethe and R. Jackiw, *Intermediate Quantum Mechanics*, 2nd ed. (W. A. Benjamin, Reading, MA, 1968), Ch. 11.

[12] S. S. Schweber, *QED and the Men Who Made It: Dyson, Feynman, Schwinger, and Tomonaga* (Princeton University Press, Princeton, NJ, 1994); L. Hoddeson *et al.*, *The Rise of the Standard Model* (Cambridge University Press, Cambridge, U.K., 1997).

[13] F. Mandl and G. Shaw, *Quantum Field Theory* (Wiley, New York, 1984); S. J. Chang, *Introduction to Quantum Field Theory* (World Scientific, Singapore, 1990); M. Guidry, *Gauge Field Theories* (Wiley, New York, 1991); E. Leader and E. Predazzi, *An Introduction to Gauge Theories and the New Physics* (Cambridge University Press, Cambridge, U.K., 1982); F. Gross, *Relativistic Quantum Mechanics and Field Theory* (Wiley, New York, 1993).

24

Solid State Physics

Costas M. Soukoulis
Ames Lab & Department of Physics and Astronomy, Iowa State University, Ames, Iowa

Eleftherios N. Economou
FORTH & University of Crete, Heraklion, Crete, Greece

Contents

List of Tables

List of Figures

Ames Laboratory is operated for the U.S. Department of Energy by Iowa State University under Contract W-7405-Eng-82. This work was also supported by EU grants. We would like to thank Dr. Maria Kafesaki for her assistance in preparing this article.

24.1. INTRODUCTION

Solid-state physics (SSP) or condensed matter physics (CMP) is concerned with the understanding of the properties of materials [1]–[3] starting from the microscopic structure of matter, i.e., electrons and nuclei, their motions, and their interactions with externally applied fields. The tools for this difficult task are electromagnetic theory (which gives information about the forces between the different particles), quantum mechanics (which allows us to predict the motion of the particles, once we know the forces), and statistical mechanics (which connects the microscopic motion with the macroscopic properties of solids). CMP provides the framework for the study of solids, liquids, polymers, biological materials, glasses, optoelectronic materials, and so on. In addition, SSP plays a central role in the development of high technology.

24.2. CLASSIFICATION OF SOLIDS ACCORDING TO THEIR BONDING CHARACTER

Simple solids can be classified according to five types of bonding: metallic, covalent, ionic, Van der Waals, and hydrogen bonding. The extent and form of the electron distribution distinguishes bonding mechanisms. There are enormous differences in physical and chemical properties of solids that have different types of bonding. A brief description of the five different types of solids is given.

24.2.1. Simple metals

In simple metals, the outer electrons (s and/or p), which are loosely bound, become free and travel throughout the solid. The ions form crystalline solids of fcc or bcc structures. The nearest-neighbor (nn) distance d between the ions is of the order 3–4 Å and is determined from the balance of two pressures. The repulsive one, which behaves as $1/d^5$, is due to the kinetic energy of the free electrons and is the result of the Heisenberg and Pauli principles. The attractive one, which behaves as $1/d^4$, is due to Coulomb interactions among electrons and ions.

24.2.2. Transition and rare-earth metals

In these metals, d electrons (for the transition metals) and/or f electrons (for rare-earth metals) also contribute to the bonding, in addition to outer s electrons, which are free to move all over the solid. The d or f electrons increase the bonding and therefore decrease the nn distance. These metals exhibit fcc, bcc, and hcp structures.

24.2.3. Covalent solids

In covalent solids, the outer s and p electrons detach from each ion and spread in the solid, but in an inhomogeneous way, preferentially along specific directions in which there are neighboring atoms. These solids crystallize in relatively open structures, such as the diamond, zincblende, and wurtzite structures.

24.2.4. Ionic solids

In ionic solids such as NaCl, an electron leaves one atom (i.e., Na) and becomes bound to another one (i.e., Cl); thus, the two charged ions (Na^+, Cl^-) have all of their shells full. There is a Coulomb attraction between the two oppositely charged ions, which increases as the ions approach each other. When the distance is very small, the overlap of the electrons of the nearest ions produces a strong repulsion. The balance between these two forces gives the equilibrium position, which is of the order 2–3 Å. The crystal structures of these solids result by alternating occupation of the sites of a simple cubic (or in some cases bcc) lattice by cations and anions.

24.2.5. Van der Waals solids

Inert elements and some molecular solids belong to this group. Since all of the atoms of the inert gases have all of their shells completely filled, it is not energetically favorable to detach electrons from the atoms. Thus, the atoms attract each other to form a crystal as a result of the instantaneously induced dipoles in neighbor atoms. This is a very weak force; thus, the bonding is weak, and the equilibrium nn distance is relatively large, of the order 3–4 Å. The crystal structure of the inert-gas crystals is mostly fcc. (The He atoms do not crystallize even at $T = 0$ in atmospheric pressure.)

24.2.6. Crystals with hydrogen bonding

Hydrogen forms a strong mixed covalent-ionic bond with one atom and a weaker "hydrogen bond" by polarizing a second nearby atom. This second bond is neither ionic (the high ionization potential of H does not allow it) nor covalent (because two electrons completely fill its 1s shell). Examples of solids that crystallize by employing the hydrogen bond include ice and large biomolecules, such as DNA.

24.3. APPROXIMATIONS

The core of SSP is based on the following four approximations:

1. *Adiabatic or Born-Oppenheimer Approximation.* For the study of the motion of the valence electrons in a solid, the ions are considered immobile.
2. *Independent-Electron Approximation.* Every electron moves independently of the other electrons.

3. *Harmonic Approximation.* The motion of the ions in a solid follows Hooke's law (i.e., $\mathbf{F}_\alpha = -\sum_\beta \mathbf{k}_{\alpha\beta} \cdot \mathbf{u}_\beta$; the force $\mathbf{F}_\alpha$ on the ion α depends linearly on the displacements $\mathbf{u}_\beta$ of the ions).
4. *Periodic Approximation.* Atoms in crystalline solids are arranged in a regular periodic array.

To further simplify the treatment of simple metals, the *jellium* (or free electron) *model* is often employed, according to which valence electrons are subject to no force.

24.4. ELECTRONS IN PERIODIC SOLIDS

24.4.1. Bloch's theorem; reciprocal lattice; Brillouin zone

The electronic wave functions $\psi(\mathbf{r})$ and the corresponding eigenenergies E are solutions of Schrödinger's equation, $-(\hbar^2/2m)\nabla^2\psi(\mathbf{r}) + V(\mathbf{r})\psi(\mathbf{r}) = E\psi(\mathbf{r})$. The potential $V(\mathbf{r})$ is assumed to be periodic; i.e., $V(\mathbf{r}) = V(\mathbf{r}+\mathbf{R}_n)$ for every lattice vector $\mathbf{R}_n = \sum_{i=1}^{3} n_i \mathbf{a}_i$, where $\mathbf{a}_1, \mathbf{a}_2, \mathbf{a}_3$ are the primitive noncoplanar lattice vectors and n_1, n_2, n_3 are any integers. Common cases of lattice structures are simple cubic (sc) [$\mathbf{a}_1 = \mathbf{i}a$ and cyclic permutation (cp) for $\mathbf{a}_2$ and $\mathbf{a}_3$, i.e., $\mathbf{a}_2 = \mathbf{j}a$ and $\mathbf{a}_3 = \mathbf{k}a$]; bcc [$\mathbf{a}_1 = a(-\mathbf{i} + \mathbf{j} + \mathbf{k})/2$ and c.p.]; fcc [$\mathbf{a}_1 = a(\mathbf{j} + \mathbf{k})/2$ and c.p.]; diamond or zincblende [two interpenetrated fcc lattices displaced by $a(\mathbf{i} + \mathbf{j} + \mathbf{k})/4$]; hcp [$\mathbf{a}_1 = \mathbf{i}a$, $\mathbf{a}_2 = \mathbf{i}(a/2) + \mathbf{j}(\sqrt{3}a/2)$, $\mathbf{a}_3 = \mathbf{k}c$]; NaCl [two fcc lattices displaced by $a(\mathbf{i}+\mathbf{j}+\mathbf{k})/2$]; and CsCl [two sc lattices displaced by $a(\mathbf{i}+\mathbf{j}+\mathbf{k})/2$]. For the lattice structures of selected solids, see Table 24.1.

The wavefunction $\psi(\mathbf{r})$, according to *Bloch's theorem*, resembles a plane wave:

$$\psi(\mathbf{r}) = w(\mathbf{r})\, e^{i\mathbf{k}\cdot\mathbf{r}}, \tag{24.1}$$

but with a modulated, periodic, complex, amplitude: $w(\mathbf{r}) = w(\mathbf{r} + \mathbf{R}_n)$. The wave vector $\mathbf{k}$ can be restricted without loss of generality to the so-called *first Brillouin zone* (1BZ) (see Figure 24.1), which is the smallest polyhedron symmetric around the point $\mathbf{k} = 0$ formed by planes bisecting the vectors $\mathbf{G}_m = \sum_{i=1}^{3} m_i \mathbf{b}_i$, where

$$\mathbf{b}_1 = 2\pi \frac{\mathbf{a}_2 \times \mathbf{a}_3}{(\mathbf{a}_1\mathbf{a}_2\mathbf{a}_3)} \tag{24.2}$$

(and $\mathbf{b}_2, \mathbf{b}_3$ are obtained by cyclic permutations of the indices 1, 2, 3), m_1, m_2, m_3 are any integers, and $(\mathbf{a}_1\mathbf{a}_2\mathbf{a}_3) = \mathbf{a}_1 \cdot (\mathbf{a}_2 \times \mathbf{a}_3) = V_o$; V_o is the volume of the primitive cell of the lattice $\{\mathbf{R}_n\}$. The vectors $\mathbf{G}_m$ form the so-called *reciprocal lattice* of the direct lattice $\{\mathbf{R}_n\}$. The reciprocal of the lattice $\{\mathbf{G}_m\}$ is the direct lattice $\{\mathbf{R}_n\}$. The planes bisecting the $\mathbf{G}_m$'s are known as *Bragg planes* because, to have diffracted beams, the tip of the wave vector $\mathbf{k}_i$ of the initial external beam must lie on a Bragg plane. This follows from the diffraction condition $\mathbf{k}_i - \mathbf{k}_f = \mathbf{G}_m$ (which is equivalent to the Bragg condition, $2d \sin\theta = n\lambda$), where $\mathbf{k}_f$ is the wave vector of a diffracted beam. By fixing $\mathbf{k}_i$ and measuring the direction of $\mathbf{k}_f$ (its magnitude equals $|\,\mathbf{k}_i\,|$), we determine the $\mathbf{G}_m$'s and, hence, the $\mathbf{R}_n$'s.

The restriction of $\mathbf{k}$ within the 1BZ is compensated by the existence of an infinite number of solutions for each $\mathbf{k}$. Thus, each wave function $\psi_{n,\mathbf{k}}(\mathbf{r})$, and the corresponding eigenenergy $E_n(\mathbf{k})$, are characterized by $\mathbf{k}$ and by the so-called band index n, where

TABLE 24.1. Crystal structures and lattice constants *a* and *c* for selected solids. *d* is the nearest-neighbor distance and n_i (in 10^{22} cm^{-3}) is the concentration of ions.

Solid	Structure	a (Å)	c (Å)	d (Å)	n_i
Na	bcc	4.225	—	3.66	2.652
Cu	fcc	3.610	—	2.56	8.45
Au	fcc	4.08	—	2.88	5.90
Mg	hcp	3.21	5.21	3.20	4.30
Ca	fcc	5.58	—	3.95	2.30
Nb	bcc	3.30	—	2.86	5.56
Fe	bcc	2.87	—	2.48	8.50
Co	hcp	2.51	4.07	2.50	9.00
Ni	fcc	3.52	—	2.49	9.14
Pd	fcc	3.89	—	2.75	6.80
C	diamond	3.567	—	1.54	17.6
Si	diamond	5.430	—	2.35	5.00
Ge	diamond	5.658	—	2.45	4.42
Pb	fcc	4.95	—	3.50	3.30
Eu	bcc	4.58	—	3.96	2.08
Gd	hcp	3.63	5.78	3.58	3.02
Nd	hexagonal	3.66	—	3.66	2.93
Pa	tetragonal	3.92	3.24	3.21	4.01
GaAs	zinc blende	5.65	—	2.45	4.44
AgI	zinc blende	6.47	—	2.80	2.95
MnO	sodium chloride	4.43	—	2.22	9.14
CsBr	cesium chloride	4.28	—	3.71	2.55
$YBa_2Cu_3O_7$	orthorhombic	3.86 ± 0.02	11.63	—	—
Tl2223[a]	b.c. tetragonal	3.822	36.26	—	—
$Nd_2Fe_{14}B$	tetragonal	8.792	12.172	—	—
$SmCo_5$	hexagonal	5.04	3.97	—	—

[a] Tl2223 = $Tl_{1.6}Ca_{1.8}Ba_2Cu_{3.1}O_{10.1}$.

$n = 1, 2, 3, \ldots$. The quantity $\hbar\mathbf{k}$ is called *crystal momentum*, because it bears some similarities with the ordinary momentum. As $\mathbf{k}$ takes all possible values within the 1BZ, each eigenenergy $E_n(\mathbf{k})$ covers a finite interval (from $E_{n,\min}$ to $E_{n,\max}$) called the nth *energy band*. Energy bands often overlap, forming composite bands. However, there are forbidden energy regions, called energy gaps or simply gaps, which do not belong to any band. Thus, the energy spectrum consists of alternating gaps and bands (usually composite bands). As we move up in energy, the tendency is for the bands to become wider and for the gaps to become narrower until they eventually disappear.

24.4.2. Density of states

A quantity of central importance in SSP is the so-called *density of states* (DOS), $\varrho(E)$, which gives the number of states per unit energy; i.e., $\varrho(E)\,dE$ is the number of differ-

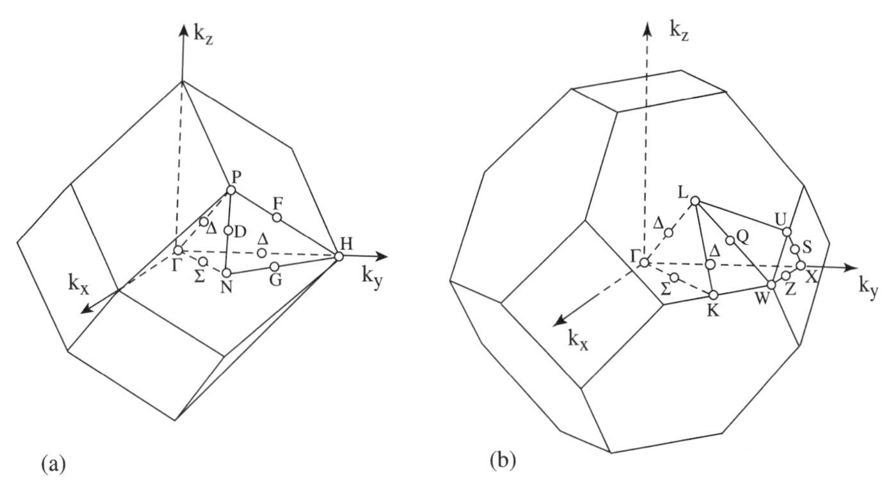

FIGURE 24.1. The first Brillouin zone for (a) an fcc and (b) a bcc lattice. Points and lines are labelled according to the standard notation.

ent states $\psi_{n,\mathbf{k}}$ with eigenenergy in the interval from E to $E + dE$. Obviously, $\varrho(E) = \sum_n \varrho_n(E)$, where $\varrho_n(E)$ is the partial DOS corresponding to the nth band. We define also the *number of states* $\mathcal{N}(E) = \int_{-\infty}^{E} dE' \varrho(E')$ and the partial number of states $\mathcal{N}_n(E) = \int_{-\infty}^{E} dE' \varrho_n(E')$ with eigenenergy less than E. The partial number of states can be obtained from the formula

$$\mathcal{N}_n(E) = V V_{n,\mathbf{k}}(E)/(2\pi)^3, \tag{24.3}$$

where V is the volume of the solid and $V_{n,\mathbf{k}}(E)$ is the volume of a region in the 1BZ such that for every $\mathbf{k}$ belonging to this region, $E_n(\mathbf{k}) \leq E$. From Eq. (24.3) and the fact that the volume V_{BZ} of the 1BZ equals $(2\pi)^3/V_0$, it follows that the number of states $\mathcal{N}_{BZ}$ for any given n within the whole 1BZ equals the total number N_c of the primitive cells in the crystalline solids, i.e., $\mathcal{N}_{BZ} = N_c$. Since each state can accommodate up to two electrons (one with spin up and the other with spin down), at $T = 0$ K, the lowest $N_e/2$ states will be completely filled with the N_e valence electrons present in the solid, whereas the remaining will be empty. If the highest occupied energy (at $T = 0$ K), known as the *Fermi energy* E_F, happens to be in the interior of a band (single or composite), the solid exhibits metallic properties. In contrast, if the highest occupied energy coincides with the bottom of a gap, the material exhibits (at $T = 0$ K) insulating properties. In this latter case, it is better to use for the Fermi energy E_F its general definition as the $T = 0$ K limit of the chemical potential, $\mu(T)$; then, it turns out that E_F lies exactly at the middle of the gap separating the highest occupied band (called the *valence band*) from the lowest empty band (known as the *conduction band*). (Either one or both of these bands can be composite bands.) The size E_g of this gap determines whether the solid is called an *insulator* (when $E_g \geq 3.5$ eV) or a *semiconductor* (when $E_g \leq 3.5$ eV) (see Table 24.2). If E_g is slightly negative, meaning a slight overlap of the valence and the conduction bands, the material is called *semimetal*.

TABLE 24.2. Electronic parameters of several elemental and compound semiconductors. i = indirect gap; d = direct gap; data are taken from Ref. [4].

Crystal	Eg (eV)	m^*_{e1}	m^*_{e2}	m^*_{lh}	m^*_{hh}	m^*_{soh}
C	5.48 (i)	0.36	1.4	0.36	1.08	0.15
Si	1.17 (i)	0.916	0.191	0.153	0.537	0.234
Ge	0.744 (i)	0.08	1.57	0.043	0.352	0.095
3*C*-SiC	2.42 (i)	0.247	0.677	—	—	—
C-BN	6.4 (i)	0.752	0.752	0.11–0.15	0.37–0.93	—
AlSb	1.686 (i)	0.26	1.8	0.09–0.12	0.3–0.88	—
GaAs	1.519 (d)	0.067	0.067	0.082	0.51	0.154
InSb	0.235 (d)	0.0136	0.0136	0.015	0.45	0.11
CdTe	1.607 (d)	0.090	0.090	0.12	0.35/0.72	—
γ-CuCl	3.395 (d)	0.415	0.415	~ 2	~ 20	—
InAs	0.418 (d)	0.023	0.023	0.026	0.41	0.08

The chemical potential μ is fixed by the relation $N_e = 2\int_{-\infty}^{\infty} dE\varrho(E)f(E)$, where $f(E) = [e^{\beta(E-\mu)} + 1]^{-1}$ is the Fermi-Dirac distribution, which at $T = 0$ K reduces to unity for $E < E_{\mathrm{F}}$ and to zero for $E > E_{\mathrm{F}}$. For $\beta(E - \mu) \gg 1$, $f(E)$ can be replaced by Boltzmann's distribution: $f(E) \simeq \exp[\beta(\mu - E)]$.

For metals, the concept of the *Fermi surface*, defined as the surface in $\mathbf{k}$ space separating the occupied states from the empty ones (at $T = 0$ K), is very important, because the response of each metal to external perturbations, especially in the presence of a static uniform magnetic field, depends on it. More often than not, the Fermi surface consists of more than one piece, each one associated with a band index n. The nth piece consists of the set of $\mathbf{k}$-points (if any) satisfying the equation $E_n(\mathbf{k}) = E_{\mathrm{F}}$.

For semiconductors and insulators, there is no Fermi surface, since there is no $\mathbf{k}$ (for any n) satisfying the equation $E_n(\mathbf{k}) = E_{\mathrm{F}}$.

24.4.3. Jellium model

In the jellium model (JM), where the potential $V(\mathbf{r})$ is a constant (taken as the zero of energy), the Bloch waves, Eq. (24.1), reduce to genuine plane waves with a constant amplitude equal to $1/\sqrt{V}$; the crystal momentum $\hbar\mathbf{k}$ becomes the ordinary momentum; the primitive lattice vectors can be taken as zero, so that the 1BZ extends over the whole $\mathbf{k}$ space with only one solution for each $\mathbf{k}$. The energy-momentum relation is the free-particle parabola, $E(k) = \hbar^2k^2/2m$. The quantity $V_{\mathbf{k}}(E)$ in Eq. (24.3) is the volume of a sphere of radius $k(E) = (2mE/\hbar^2)^{1/2}$. Thus, $\mathcal{N}(E) = (\sqrt{2}V/3\pi^2)(m/\hbar^2)^{3/2}E^{3/2}$ and

$$\rho(E) = \frac{V}{\pi^2\sqrt{2}}\left(\frac{m}{\hbar^2}\right)^{3/2} E^{1/2}, \quad E \geq 0. \tag{24.4}$$

The Fermi surface is a sphere of radius k_{F} ($\hbar k_{\mathrm{F}}$ is the so-called *Fermi momentum*, and $\hbar k_{\mathrm{F}}/m$ is the *Fermi velocity*). It follows from Eq. (24.3) that k_{F} satisfies the relation $N_e = 2[V/(2\pi)^3](4\pi k_{\mathrm{F}}^3/3)$ or $k_{\mathrm{F}} = (3\pi^2 n)^{1/3}$, where $n = N_e/V$ is the concentration of the valence electrons. The Fermi energy can be obtained from the relation $E_F = \hbar^2 k_F^2/2m$. Within the JM, the following relations are true: $\rho_F \equiv \rho(E_F) = 3N_e/4E_F$ and $K_t = 3N_e E_F/5$, where K_t is the total electronic kinetic energy. The total electrostatic energy can also be calculated with the concentration n the ionic radius r_c, and the valence as inputs. Thus, within the framework of the free-electron model, the total energy U_t (excluding the ionic zero-point motion), at $T = 0\ K$, is obtained: $U_t = A_1 n^{2/3} - A_2 n^{1/3}$. By minimizing the energy with respect to n, we obtain the equilibrium value of the concentration n, the mass density ρ_M, the bulk modulus B, and the quantity $c_o = (B/\rho_M)^{1/2}$, which is related to the longitudinal (c_l) and the transverse (c_t) sound velocities in an isotropic (polycrystalline or amorphous) solid:

$$c_l = c_o\left(1 + \frac{4}{3}x\right)^{1/2}, \quad c_t = c_0 x^{1/2}, \quad x \equiv \frac{\tilde{\mu}}{B}, \tag{24.5}$$

where $\tilde{\mu}$ is the shear elastic modulus known as the *shear Lamé coefficient*. Values of the ratio $x = \tilde{\mu}/B$ for most materials are in the range $0.1 \leq x \leq 0.7$; soft materials (e.g., Pb) have a smaller x, whereas in very hard materials (e.g., Be or diamond), x exceeds unity but is always less than 1.5. For monocrystalline cubic solids, three elastic constants (C_{11}, C_{12}, C_{14}) are needed $(B = (C_{11} + 2C_{12})/3)$. (For values of elastic constants, see Table 24.3.)

TABLE 24.3. Elastic constants and bulk modulus B (in 10^{11} N/m^2), and Debye temperature Θ_D (in K).

Crystal	C_{11}	C_{12}	C_{44}	B	Θ_D
Li	0.148	0.125	0.108	0.116	344
Na	0.070	0.061	0.045	0.068	158
Cu	1.68	1.21	0.75	1.37	345
Ag	1.24	0.93	0.46	1.00	225
Au	1.86	1.57	0.42	1.732	165
Al	1.07	0.61	0.28	0.722	426
Pb	0.46	0.39	0.144	0.430	96
C	10.76	1.25	5.77	4.43	2340
Si	1.66	0.64	0.80	0.988	645
Ge	1.24	0.413	0.683	0.772	374
V	2.29	1.19	0.43	1.619	380
Nb	2.47	1.35	0.287	1.702	275
Ni	2.45	1.40	1.25	1.86	440
Fe	2.34	1.36	1.18	1.683	460
LiCl	0.494	0.228	0.246	0.298	420
NaCl	0.487	0.124	0.126	0.240	297
KF	0.656	0.146	0.125	0.305	335
RbCl	0.361	0.062	0.047	0.156	194
InSb	0.670	0.365	0.302	0.467	202
InAs	0.833	0.453	0.396	0.580	249
GaAs	1.19	0.538	0.595	0.755	344

The free-electron model, essentially *ab initio*, and in any case without any adjustable parameters, obtains reasonably well fundamental quantities of the solid state, such as the density and the bulk modulus. The average deviation for the 70 metals of the periodic table is 25%. Furthermore, although this model fails to produce gaps, it is applicable even to semiconductors (for E near band edges) if the actual electronic mass is replaced by the *effective mass tensor* m^*, defined as follows:

$$\left(\frac{1}{m^*}\right)_{ij} \equiv \frac{1}{\hbar^2}\frac{\partial^2 E(\mathbf{k})}{\partial k_i\, \partial k_j}. \tag{24.6}$$

Fully occupied bands are inert in the sense that they do not contribute to electronic transport quantities, such as electrical and thermal conductivities. As a result, almost-fully occupied bands can be described more conveniently in terms of the few missing electrons, i.e., the *holes*, which are treated as particles of positive charge $|e|$, mass $m_h^* = -m_e^*$, wave vector $\mathbf{k}_h = -\mathbf{k}_e$, spin $\mathbf{s}_h = -\mathbf{s}_e$, eigenenergy $E_h(\mathbf{k}_h) = -E_e(\mathbf{k}_e)$, and average velocity $\mathbf{v}_h(\mathbf{k}_h) = \mathbf{v}_e(\mathbf{k}_e)$. In these formulae, the subscript e denotes the values of the corresponding quantities if an electron was occupying the empty state $\mathbf{k}_e$. The *average velocity* $\mathbf{v}$ is given by the group velocity:

$$\mathbf{v}_\mathbf{k} = \frac{1}{\hbar}\frac{\partial E(\mathbf{k})}{\partial \mathbf{k}}. \tag{24.7}$$

In Figure 24.2, we reproduce the bands and the DOS for three characteristic solids: (1) a transition metal with $2\rho_\mathrm{F} = 1.46/\mathrm{atom}\cdot\mathrm{eV}$, (2) a simple metal with almost free-like DOS and $2\rho_\mathrm{F} = 0.4/\mathrm{atom}\cdot\mathrm{eV}$, and (3) a semiconductor with $\rho_\mathrm{F} = 0$. For effective-mass values, see Table 24.2.

24.5. METHODS FOR BAND-STRUCTURE CALCULATIONS

24.5.1. General computational framework

To obtain theoretically the band structure, [5, 6] i.e., the $\psi_{n,\mathbf{k}}(\mathbf{r})$ and $E_n(\mathbf{k})$ for a crystalline solid, one should perform three interconnected tasks: (1) solve Schrödinger's equation in the presence of a potential $V(\mathbf{r})$, (2) determine $V(\mathbf{r})$ using the results of task (1) and the assumed ionic positions $\{\mathbf{R}_n\}$; and (3) minimize the total energy with respect to $\{\mathbf{R}_n\}$.

The potential $V(\mathbf{r})$ is the sum of three electrostatic terms: (1) the ionic, $V_I(\mathbf{r}; \{\mathbf{R}_n\})$, which depends on the ionic position; (2) the average electronic part, $V_H(\mathbf{r}) = e^2 \int d^3r' n(\mathbf{r}')/|\mathbf{r}-\mathbf{r}'|$; and (3) the so-called exchange and correlation term, $V_{xc}(\mathbf{r}; \{n(\mathbf{r})\})$, which is a complicated functional of the electronic concentration $n(\mathbf{r})$, where

$$n(\mathbf{r}) = 2\sum_{n,\mathbf{k}} |\psi_{n,\mathbf{k}}(\mathbf{r})|^2$$

and the summation is over the occupied states. Usually, the local-density approximation (LDA) is employed according to which V_{xc} is assumed to be the same functional of $n(\mathbf{r})$ as in the jellium model, where $n(\mathbf{r})$ is constant.

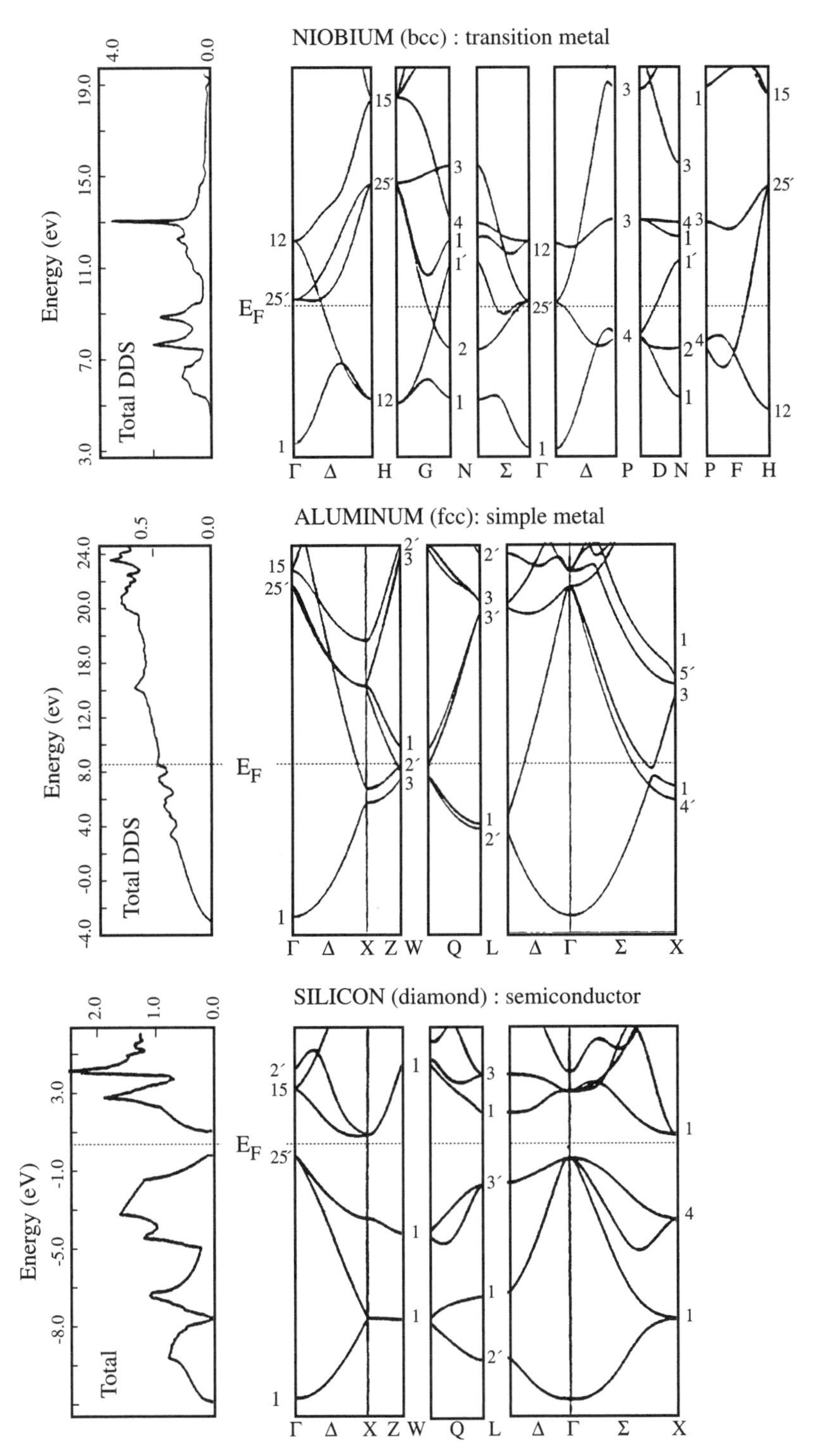

FIGURE 24.2. Band structure (E vs. $\mathbf{k}$) and the density of states (per atom and eV for both spins) of three characteristic solids. Taken from Ref. [5]. Reproduced with permission of Plenum Press, London, UK.

The solution of Schrödinger's equation is obtained by choosing a basis $\{\phi_j\}$, and transforming the differential equation to a matrix equation, $\sum_j H_{ij}c_j = Ec_i$, where $\psi = \sum_j c_j\phi_j$ and $H_{ij} =< \phi_i| H |\phi_j >$. Several choices for the basis are used. They are as follows.

24.5.2. Linear combination of atomic orbitals

The ϕ_i's are atomic or atomic-like orbitals. Then, the method is known as *linear combination of atomic orbitals* (LCAO) or *tight binding*. The exploitation of our knowledge regarding atomic levels and orbitals, the possibility of quick estimates, and the physical transparency of the results are among the many advantages of this method, which has been used extensively in chemistry and in SSP. Its main disadvantage is the difficulty in obtaining reliable *ab initio* values for the matrix elements H_{ij}. Thus, usually H_{ij} are obtained by fitting to accurate results obtained by other methods. [1, 5] A highly simplified but very fruitful version of the LCAO [6] identifies the diagonal matrix elements H_{ii} with atomic levels available from Hartree-Fock calculations. It determines the off-diagonal matrix elements H_{ij} between nn orbitals through expressions of the form $H_{ij} = \eta_{ij}\hbar^2/m\,d^2$, where d is the bond length and the η_{ij}'s depend on the type of orbitals (s, p) and their relative orientation with respect to the bond direction. The behavior of simple materials (involving no transition elements) can be qualitatively understood and semiquantitatively analyzed in terms of three parameters: (1) V_1, the difference (divided by 4) between the ϵ_p and ϵ_s levels of the same atom; (2) V_2, an off-diagonal matrix element between nn orbitals; and (3) V_3, half the difference between atomic levels of neighbor atoms. In terms of V_1, V_2, V_3, one can define a metallicity index α_m and a polarity index α_p:

$$\alpha_m = \frac{V_1}{\sqrt{{V_2}^2 + {V_3}^2}}, \qquad \alpha_p = \frac{V_3}{\sqrt{{V_2}^2 + {V_3}^2}}, \tag{24.8}$$

in terms of which tetravalent crystalline solids can be classified as shown in Figure 24.3.

24.5.3. Plane wave method

The ϕ_j's are *plane waves* (PW), $e^{i(\mathbf{k}+\mathbf{G})\cdot\mathbf{r}}/\sqrt{V}$. Because the ionic part of the potential $V(\mathbf{r})$ diverges as we approach each nucleus, the number of required PWs is very large. To avoid this difficulty, we replace the actual $V(\mathbf{r})$ by empirical or *ab initio pseudopotentials*. The latter are based on ionic pseudopotentials that produce the same valence atomic levels and the same valence orbitals (for $r \geq r_c$, but without nodes for $r \leq r_c$) as the actual ionic potential, while they remain finite for $r \leq r_c$. In terms of the Fourier component $\tilde{V}_{\mathbf{G}}$ of the pseudopotential, Schrödinger's equation becomes

$$\left(\frac{\hbar^2k^2}{2m} - E\right)c_{\mathbf{k}} + \sum_{\mathbf{G}} \tilde{V}_{\mathbf{G}}c_{\mathbf{k}-\mathbf{G}} = 0, \tag{24.9}$$

where the summation is over vectors of the reciprocal lattice.

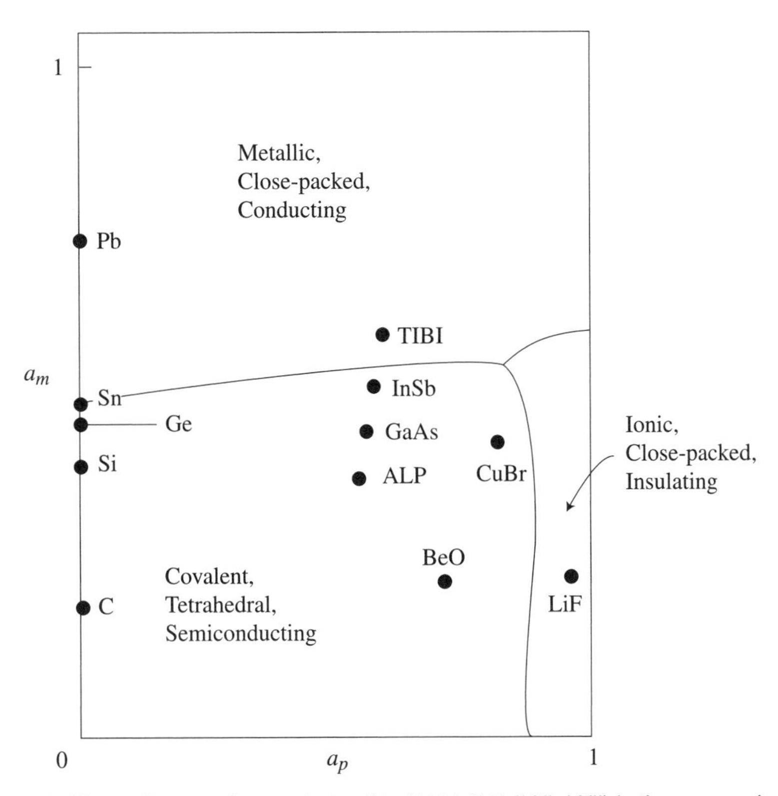

FIGURE 24.3. Phase diagram of tetravalent solids (IV-IV, III-V, II-VI, I-VII) in the a_p, a_m plane. From Ref. [6], with permission.

24.5.4. Other methods

The ϕ_j's are *augmented plane waves* (APW), meaning that each ϕ_j is a plane wave as in Sec. 24.5.3 in the region outside of the muffin-tin spheres, it is the actual solution of Schrödinger's equation inside each muffin-tin sphere, and it is continuous everywhere. The muffin-tin spheres are centered around each ion with a radius R_{MT} such that $r_c \leq R_{MT} \leq d/2$. This more complicated basis has no need for pseudopotentials. We mention also a fourth method, usually based on muffin-tin potentials, developed by Korringa, Kohn, and Rostoker (KKR).

24.6. IONIC VIBRATIONS

The ionic eigenfrequencies and eigenmodes are obtained by solving Newton's equation for the displacement $\mathbf{u}_{n\nu}$ of the atom (or ion) located at the site $\mathbf{r}_\nu$ of the $\mathbf{R}_n$ primitive cell:

$$-\omega^2 M_\nu \mathbf{u}_{n\nu} = \mathbf{F}_{n\nu}, \tag{24.10}$$

where, according to Hooke's law, the force $\mathbf{F}_{n\nu}$ is a linear function of the relative displacements $\mathbf{u}_{m\mu} - \mathbf{u}_{n\nu}$. Bloch's theorem, $\mathbf{u}_{n+l\nu} = \exp(i\mathbf{k} \cdot \mathbf{R}_l)\mathbf{u}_{n\nu}$, reduces the infinite set of equations (24.10) to a $3p \times 3p$ linear homogeneous system with $3p$ unknowns, the three components of the displacements of the p atoms within a primitive cell. Setting the determinant of this system equal to zero, we find for each value of $\mathbf{k}$ in the 1BZ $3p$ eigenmodes with the corresponding $3p$ eigenfrequencies. Three of those, known as *acoustic branches*,

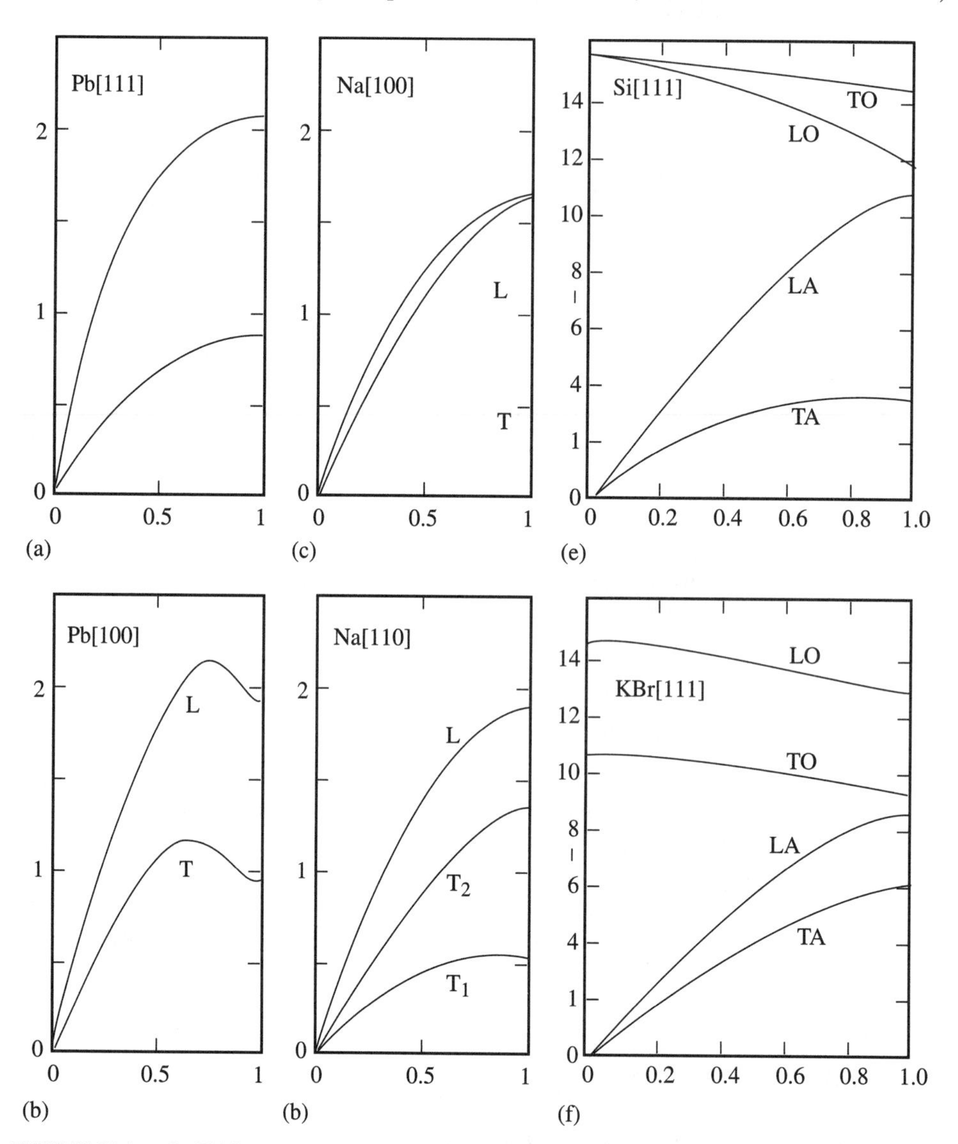

FIGURE 24.4. ν (in THz) vs. $k/k_{\max}$ for the lattice vibrations in three different materials (O stands for optical, A for acoustic, L for (predominantly or fully) longitudinal, and T for transverse). The transverse modes in Si and KBr in the 111 direction are doubly degenerate. Taken from Ref. [3]. Reproduced with permission of Academic Press, New York.

approach zero linearly with k: $\omega_l \approx c_l(\theta, \phi)k$, $\omega_{t_1} \approx c_{t_1}(\theta, \phi)k$, $\omega_{t_2} \approx c_{t_2}(\theta, \phi)k$, whereas the remaining $3(p-1)$ branches, called *optical branches*, approach nonzero values of ω as $k \to 0$. In Figure 24.4 the dispersion relations, ω vs. $\mathbf{k}$ (as $\mathbf{k}$ moves along the 100, or the 110, or the 111 direction), for four solids are shown.

The number of eigenmodes $\widetilde{\mathcal{N}}_l(\omega)$ belonging to the lth branch ($l = 1, 2, \ldots, 3p$) with frequency less than ω is given by Eq. (24.3). For low ω, only the three acoustic branches contribute; thus, the total number of modes with frequencies less than ω, $\widetilde{\mathcal{N}}(\omega)$, is given by

$$\widetilde{\mathcal{N}}(\omega) = \frac{V}{2\pi^2}\frac{\omega^3}{c^3}, \quad \frac{\omega}{c} \ll \frac{\pi}{a}, \tag{24.11}$$

where

$$\frac{1}{c^3} = \frac{1}{3}\sum_{s=1}^{3}\int \frac{d\Omega}{4\pi}\frac{1}{c_s^3(\theta, \phi)}. \tag{24.12}$$

Therefore, the density of modes $\mathcal{D}(\omega) \equiv d\widetilde{\mathcal{N}}(\omega)/d\omega$ equals $(3V/2\pi^2c^3)\omega^2$ for small ω. The maximum number of modes in each branch equals $\mathcal{N}_{BZ}$, which is the same as the number of primitive cells N_c. Thus, the total number of modes equals $3pN_c = 3N_i$, as expected, since $pN_c = N_i$ is the total number of atoms in the solid.

The *Debye approximation* assumes that Eq. (24.11) is valid all the way to a maximum cutoff frequency, known as the *Debye frequency* (ω_{D}), which is determined by the obvious requirement that $\widetilde{\mathcal{N}}(\omega_{\mathrm{D}}) = 3N_i$, from which it follows that $\omega_{\mathrm{D}} = c(6\pi^2N_i/V)^{1/3}$. The temperature $\Theta_{\mathrm{D}} \equiv \hbar\omega_{\mathrm{D}}/k_{\mathrm{B}}$ is called the *Debye temperature* (see Table 24.3). The Debye approximation reproduces correctly the spectrum at low frequencies (which guarantees accurate low-temperature thermodynamics) and preserves, by the definition of ω_{D}, the total number of modes (which makes certain that high-temperature thermodynamics is correctly reproduced). The energy of each eigenmode $\mathbf{k}, l$ is given by $\hbar\omega_l(\mathbf{k})(n_{kl} + 1/2)$, where the non-negative integer n_{kl} gives the corresponding number of the vibrational quanta called *phonons*.

24.7. THERMODYNAMIC QUANTITIES

Band-structure calculations give the total energy U_t (at $T = 0$ without the zero-point ionic energy) as a function of the lattice constant a for various plausible lattice structures:

$$U_t = K_t + \int V_I(\mathbf{r})n(\mathbf{r})d^3r + \frac{1}{2}\int V_H(\mathbf{r})n(\mathbf{r})d^3r + \int \epsilon_{xc}(\mathbf{r})n(\mathbf{r})d^3r + U_{II}, \tag{24.13}$$

where $K_t = 2\sum_\alpha \int \psi_\alpha^*(-\hbar^2/2m)\nabla^2\psi_a d^3r$ is the total electronic kinetic energy, U_{II} is the direct ion-ion electrostatic energy, and $\epsilon_{xc}(\mathbf{r})$ is connected to $V_{xc}(\mathbf{r})$ by the relation $\partial[n(\mathbf{r})\epsilon_{xc}(\mathbf{r})]/\partial n(\mathbf{r}) = V_{xc}(\mathbf{r})$.

Finding the minimum of U_t among the various lattices and with respect to the lattice constant, we obtain the equilibrium lattice structure and the lattice constant, the cohesive energy U_σ (Table 24.4), and the bulk modulus B; the latter is directly related to d^2U/da^2 at

TABLE 24.4. Cohesive energy U_σ, specific heat c_p, melting (T_m) and boiling (T_b) temperatures, and coefficient of linear expansion α_L (in 10^{-6} K^{-1}) of some elemental solids. Data are taken from Ref. [7].

Crystal	U_σ (eV/atom)	c_p (J/g K)	T_m (°C)	T_b (°C)	α_L
Ar	0.08	0.313	−189	−186	—
Na	1.11	1.228	98	883	71.0
Cu	3.49	0.385	1085	2563	16.5
Ag	2.95	0.235	962	2163	18.9
Au	3.81	0.129	1064	2857	14.2
Mg	1.51	1.023	650	1090	24.8
Ca	1.84	0.647	842	1494	22.3
Nb	7.57	0.265	2469	4744	7.30
Fe	4.28	0.449	1538	2862	11.8
Co	4.39	0.421	1495	2928	13.0
Ni	4.44	0.444	1455	2964	13.4
Pd	3.89	0.244	1555	2964	11.8
C	7.37	0.472	3827	4197	1.18
Si	4.63	0.709	1412	3267	4.68
Ge	3.85	0.320	937	2834	6.10
Pb	2.03	0.129	328	1750	28.9
Eu	1.86	0.182	822	1527	35.0
Gd	4.14	0.236	1313	3273	9.00
Nd	3.40	0.190	1021	3074	9.60

equilibrium. By imposing various periodic deformations and recalculating the total energy, one can obtain *ab initio* values for the "spring" constants required for the determination of the phononic dispersion relations $\omega_s = f_s(\mathbf{k})$.

Having the electronic and the phononic DOS, one can obtain the total energy at temperature T, $U_t(T) = U_t(0) + \Delta U_t(T)$.

Differentiating $\Delta U_t(T)$ with respect to T with V and N_e kept constant, we obtain the specific heat $C_V (\approx C_p)$:

$$C_V = k_B \int_0^{\omega_{\max}} d\omega \mathcal{D}(\omega) \frac{(\beta\hbar\omega)^2 e^{\beta\hbar\omega}}{(e^{\beta\hbar\omega} - 1)^2} + \frac{2\pi^2}{3} \rho_F (1 + \lambda) k_B^2 T. \qquad (24.14)$$

The first term in (24.14) is due to phonons and the second to electrons. The correction factor $1 + \lambda$, which is not present in (24.13), accounts for the electron-phonon interactions through the dimensionless strength parameter λ; $0.1 \lesssim \lambda \lesssim 1.5$ with a typical value of about 0.5 [see Table 24.4 for C_p values and Table 24.5 for $(2\pi^2/3)\rho_F(1 + \lambda)k_B^2$ values].

Note that for nonmetals, $\rho_F = 0$, and consequently, there is no significant electronic contribution to C_V. The phonon contribution to C_V reduces to $(12\pi^4/5)N_i k_B (T/\Theta_D)^3$ for $T \lesssim 0.1\Theta_D$, and to the classical limit $3N_i k_B$ for $T \gtrsim \Theta_D$.

The linear coefficient of thermal expansion a_L, defined as $(\partial V/\partial T)_P/3V$, can be expressed as $a_L = (\partial P/\partial T)_V/3B$. Thus, a_L can be calculated by differentiating $\Delta P(T)$ and dividing by the bulk modulus B. Here,

TABLE 24.5. The electronic specific heat constant, $\gamma = (2\pi^2/3)\rho_F(1+\lambda)k_B^2$ (in mJ mol^{-1} K^{-2}), the density of states at the Fermi energy ρ_F (in states/Ry atom), and the plasma frequency ω_p (in eV) of typical metals. Data are taken from Ref. [2] (for γ and ω_p) and Ref. [5] (for ρ_F and ω_p).

	γ	ρ_F	ω_p
Na	1.38	6.72	5.71
K	2.0	10.49	3.72
Al	1.35	5.46	15.3
Be	0.17	0.10	—
Cu	0.695	4.03	9.11
Ga	0.596	5.39	12.82
Fe	4.98	48.31	6.54
Ni	7.02	55.14	6.29
Nb	7.79	19.86	9.12
Mn	9.20	23.44	7.65

$$\Delta P = \hbar \int_0^{\omega_{max}} d\omega \left(\frac{\partial \tilde{\mathcal{N}}(\omega)}{\partial V} \right) \bar{n} + (2/V) \int_{-\infty}^{+\infty} dE \mathcal{N}(E)[f(E;T) - f(E;0)],$$

where $\bar{n} = (e^{\beta\hbar\omega} - 1)^{-1}$ is the average phonon number and $f(E;T)$ is the Fermi distribution. Within the JM and the Debye approximation, $\partial\tilde{\mathcal{N}}(\omega)/\partial V = 3\tilde{\mathcal{N}}(\omega)(4/3V) = \mathcal{D}(\omega)\omega(4/3V)$ and $\mathcal{N}(E) = (2/3)\rho(E)E$. Hence,

$$(\partial P/\partial T)_V = (4/3V)C_{V_p} + (2/3V)C_{V_e},$$

and

$$a_L \approx \frac{1}{3BV}\left(\frac{4}{3}C_{V_p} + \frac{2}{3}C_{V_e}\right) \approx \frac{4}{9BV}C_V, \tag{24.15}$$

where C_{V_p} (C_{V_e}) is the phononic (electronic) contribution to the specific heat [see Eq. (24.14)]. Equation (24.15) has the form of the so-called *Grüneisen relation*, according to which $a_L = \gamma_G C_V/3BV$, where γ_G is the Grüneisen parameter, which, in general, depends on the temperature.

24.8. LINEAR RESPONSE TO PERTURBATIONS

24.8.1. Dielectric function and conductivity

The dielectric function ϵ is related to the electrical conductivity σ as follows:

$$\epsilon = \epsilon_0 + \frac{i\sigma}{\omega} \quad \text{(SI)} \quad \text{or} \quad \epsilon = 1 + \frac{4\pi i\sigma}{\omega} \quad \text{(G-cgs)}. \tag{24.16}$$

For a uniform and isotropic system, the dielectric function is a scalar that depends on the frequency ω and the wavevector $\mathbf{k}$. In the regime $\omega/k \gg v_F$, the $\mathbf{k}$-dependence is weak and can be omitted. In this regime, one can easily calculate the average, field-induced, electronic velocity, $\mathbf{v}$: $m\dot{\mathbf{v}} = -m\mathbf{v}/\tau - e\mathbf{E}$. Since $\dot{\mathbf{v}} = -i\omega\mathbf{v}$, we obtain $\mathbf{v} = -ie\mathbf{E}/(m\omega + im/\tau)$. It follows that the electric current, $\mathbf{j} = -ne\mathbf{v}$, is equal to $ie^2n\mathbf{E}/m(\omega + i/\tau)$, and that the conductivity σ and the dielectric function $\epsilon(\omega)$ are given (in G-cgs) by

$$\sigma = \frac{ie^2n}{m(\omega + i/\tau)} \quad \text{and} \quad \epsilon(\omega) = 1 - \frac{\omega_p^2}{\omega^2 + i\omega/\tau}, \tag{24.17}$$

where $\omega_p = (4\pi e^2 n/m)^{1/2}$ is the so-called electronic plasma frequency and $1/\tau$ is the relaxation rate, i.e., a phenomenological parameter that describes the collision of the otherwise free electrons with phonons, impurities, and defects. For bound electrons, a restoring force, $-m\omega_j^2\mathbf{r}$, has to be added to the equation of motion, the net result of which is to replace ω^2 in the denominator of Eq. (24.17) by $\omega^2 - \omega_j^2$. Thus, if n_j is the concentration of those bound electrons, their contribution to $\epsilon(\omega)$ will be of the form $-\omega_{pj}^2/(\omega^2 - \omega_j^2 + i\omega/\tau_j)$, where $\omega_{pj}^2 = 4\pi e^2 n_j/m$. Actually, various terms of this type will be present, since the binding strength $-m\omega_j^2$ varies among the several groups of bound electrons. A quantum mechanical approach gives a similar expression, with $\hbar\omega_j$ being the energy difference between occupied and empty levels and ω_{pj}^2 being proportional to dipole transition probabilities between those levels.

Besides electrons, ions would also contribute to the ac electric current and the ac electrical conductivity. If in each primitive cell there is one ion of mass m_i and charge q with concentration n_i, the ionic contribution to the conductivity, by the same argument as before, would be equal to $iq^2n_i/m_i(\omega + i/\tau_i)$. In this case, the restoring force $-m_i\omega_{TA}^2\mathbf{u}_i$ goes to zero linearly with the wavenumber k: $\omega_{TA} \approx ck$, where c is constant. If there are two ions per primitive cell of opposite charges $q_1 = -q_2 = q'$, masses m_1, m_2, and concentrations $n_1 = n_2 = n_i'$, the contribution will be of the form $iq'^2n_i'/m_r(\omega + i/\tau_i')$, where $m_r^{-1} = m_1^{-1} + m_2^{-1}$. This contribution is due to out-of-phase motion of the two types of ions and involves a strong restoring force $-m_r\omega_{TO}^2(\mathbf{u}_1 - \mathbf{u}_2)$, even at $k = 0$. Collecting all of the terms together, we obtain that $\epsilon(\omega) = 1 + 4\pi\chi_e + 4\pi\chi_i$, where the electronic contribution

$$4\pi\chi_e = -\frac{\omega_p^2}{\omega^2 + i\omega/\tau} - \sum_j \frac{\omega_{pj}^2}{\omega^2 - \omega_j^2 + i\omega/\tau_j} \tag{24.18a}$$

is due to free and various groups of bound electrons, while the ionic contribution, $4\pi\chi_i$, is equal to

$$-\frac{\omega_{ip}^2}{\omega^2 - \omega_{TA}^2 + i\omega/\tau_i} \quad \text{or} \quad -\frac{\omega_{ip}'^2}{\omega^2 - \omega_{TO}^2 + i\omega/\tau_i'}, \tag{24.18b}$$

depending on whether we have one ion per primitive cell or two oppositely charged ions per primitive cell.

The ionic parameters are as follows: $\omega_{ip}^2 = 4\pi q^2 n_i/m_i$ and $\omega_{ip}'^2 = 4\pi q'^2 n_i'/m_r$. The ionic relaxation rate $1/\tau_i$ in metals equals $b\tau\omega^2$ for $kl \ll 1$ and $b'(m/m_i)^{1/2}\omega$ for $kl \gg 1$, where $l = v_F\tau$ is the electronic mean free path (to be examined below), k is the wave number of the ionic motion, and b, b' are numerical factors of the order of unity. Eq. (24.18a) is valid for $\omega/k \gg v_F$, whereas Eq. (24.18b) is valid for $\omega/k \gg v_i$, where v_i, the maximum ionic velocity, can be estimated by equating $m_i v_i^2$ to $\hbar\omega_D$, from which it follows that $v_i \approx (m/m_i)^{3/4} v_F$. For $\omega/k \ll v_F$, the first (free electron) term in (24.18a) should be replaced by $f(k)k_o^2/k^2$, where k_o is the screening parameter and $f(k)$ is a monotonically decreasing function of k such that $f(0) \to 1$ and $f(\infty) = 0$. The dielectric function $f(k)k_o^2/k^2$ transforms a static bare Coulomb field q/r to a screened field $(q/r)\exp(-k_o r)$ for $k_o r \gg 1$.

The knowledge of the dielectric function, $\epsilon = \epsilon_1 + i\epsilon_2$, permits us to

1. obtain the optical properties of solids, such as the velocity of light $\tilde{c} = c/\tilde{n}_1$ [where $\tilde{n}_1$ is the real part of the index of refraction, $\tilde{n}_1 + i\tilde{n}_2 \equiv \sqrt{\epsilon(\omega)}$], the absorption coefficient $\alpha = 2\tilde{n}_2\omega/c = \omega\epsilon_2/c\tilde{n}_1$, the reflection coefficient, etc.;
2. calculate the energy loss per unit length traveled by a fast charged particle following a straight trajectory in a solid;
3. calculate the differential inelastic scattering cross section $d\sigma/d\Omega d\epsilon_f$ of a charged particle by the solid: $d\sigma/d\Omega d\epsilon_f \propto -\mathrm{Im}\epsilon^{-1}(\omega, \mathbf{k})$, $\hbar\omega = \epsilon_i - \epsilon_f$, $\mathbf{k} = \mathbf{k}_i - \mathbf{k}_f$;
4. obtain an exact closed expression for the potential energy and for the total energy of the jellium model;
5. find the eigenfrequency(ies) of collective longitudinal charge density oscillations in a solid by setting $\epsilon(\omega, \mathbf{k}) = 0$;
6. find the eigenfrequency(ies) of collective transverse oscillations in a solid (within the electrostatic approximation) by setting $\epsilon(\omega, \mathbf{k}) = \infty$.

To prove Step (5), one shows that $\mathbf{D} = \mathbf{B} = \mathbf{H} = \nabla \times \mathbf{E} = 0$ and $\mathbf{E} \neq 0$ is a solution of Maxwell's equation when $\epsilon(\omega, \mathbf{k}) = 0$. To prove Step (6), one shows that Maxwell's equations can be satisfied when $\epsilon(\mathbf{k},\omega) = c^2k^2/\omega^2$. The electrostatic limit is obtained by setting $c = \infty$, i.e., $\epsilon(\omega, \mathbf{k}) = \infty$; the electrostatic approximation is valid if $\omega/k \ll c$ (or if $\omega/k \gg c$). When $\omega \approx ck$, the coupling of the collective mode's electrostatic field with the radiative field would be appreciable, resulting in a composite entity called a *polariton*, whose dispersion relation is given by the full equation $\epsilon(\mathbf{k}, \omega) = c^2k^2/\omega^2$. Combining Eq. (24.18b) with Step (6), we conclude that ω_{TA} and ω_{TO} are the frequencies of the transverse acoustical and transverse optical phonons, respectively.

For an ionic insulator or semiconductor, with one anion and one cation in the primitive cell, the free-electron term is absent and the smallest ω_j is no less than $E_g/\hbar$. Then, for $\hbar\omega \ll E_g$, Eq. (24.18b) gives $\epsilon(\omega) = \epsilon_\infty - \omega_{ip}'^2/(\omega^2 - \omega_{TO}^2)$, where $\epsilon_\infty = 1 + \sum_j \omega_{pj}^2/\omega_j^2$. The zero of $\epsilon(\omega)$, which, according to Step (5), coincides with the longitudinal optical phonon frequency ω_{LO} (for $k = 0$), is given by $\omega_{LO}^2 = \omega_{TO}^2 + \omega_{ip}'^2/\epsilon_\infty$; then, the dielectric function can be recast as $\epsilon(\omega) = \epsilon_\infty(\omega^2 - \omega_{LO}^2)/(\omega^2 - \omega_{TO}^2)$, an equation known for $\omega = 0$ as the Lyddane-Sachs-Teller relation, which can be tested using Table 24.6.

For a metal in the regime $v_i \ll \omega/k \ll v_F$ and $k \ll k_F$, Eqs. (24.18) give $\epsilon(\mathbf{k},\omega) \simeq \epsilon_\infty + (k_o^2/k^2) - \omega_{ip}^2/(\omega^2 - \omega_{TA}^2)$. Setting $\epsilon(\mathbf{k}, \omega) = 0$, we find the longitudinal acoustic

TABLE 24.6. Dielectric constants and lattice vibrational frequencies for typical semiconductors and insulators. Data are taken from Refs. [2] and [4].

Crystal	$\epsilon(0)$	$\epsilon(\infty)$	ω_{TO} $(10^{13} s^{-1})$	ω_{LO} $(10^{13} s^{-1})$
C	5.70	5.7	25.1	25.1
Si	11.8	11.8	9.8	9.8
Ge	16.0	16.0	5.7	5.7
3C-SiC	9.72	6.52	15.0	18.3
C-BN	7.10	4.5	19.9	24.6
AlSb	12.04	10.24	6.0	6.4
GaAs	12.40	10.60	5.1	5.4
InAs	15.15	12.25	4.1	4.5
InSb	16.8–18.0	15.68	3.5	3.7
CdTe	10.2	7.1	2.6	3.15
γ-CuCl	7.9	3.61	3.1	3.8
NaCl	5.9	2.25	3.1	5.0
CsCl	7.2	2.6	1.9	3.1
AgCl	11.2	3.92	2.0	3.7

phonon frequency ω_{LA}:

$$\omega_{LA}^2 \simeq (c_t^2 + \omega_{ip}^2/k_o^2)k^2. \tag{24.19}$$

Combining Eq. (24.19) with Eq. (24.5), we find that the screening parameter k_o is given by $\omega_{ip}/c_o(1 + x/3)^{1/2}$, which for the JM reduces to the Thomas-Fermi expression $k_o = k_{\mathrm{TF}} \equiv (4k_{\mathrm{F}}/\pi a_B)^{1/2}$.

For a metal at high frequencies, where the first term in Eq. (24.18a) dominates, the zero of $\epsilon(\omega)$ is equal to ω_p. (For typical values of ω_p, see Table 24.5.) This frequency corresponds to a collective longitudinal electronic oscillation, the quantum of which is known as a *plasmon*. A similar collective oscillation takes place in semiconductors, because the second term in (24.18a) for $\hbar\omega_p \gg E_g$ behaves as if the valence electrons were free.

24.8.2. Temperature dependence of the DC conductivity

The expression for the electronic DC conductivity is obtained by setting $\omega = 0$ in Eq. (24.17): $\sigma = e^2 n\tau/m$. For metals, it is convenient to rewrite it in terms of the electronic mean free path $l = \tau v_F$ and the area S_{F} of the Fermi surface (for JM, $S_{\mathrm{F}} = 4\pi k_{\mathrm{F}}^2$):

$$\sigma = \frac{1}{12\pi^3}\frac{e^2}{\hbar}S_{\mathrm{F}}l. \tag{24.20}$$

The mean free path l is obtained by the relation $l\sum_j n_{sj}\sigma_{sj} = 1$, where σ_{sj} is the total transport scattering cross section of an electron by a scattering center of type j and n_{sj} is the concentration of these scattering centers. Lattice vibrations, i.e., phonons, as well as impurities and defects provide such scattering centers. The phonon contribution to $1/l$ equals $n_p\sigma_p$. For temperatures $T \gtrsim \Theta_{\mathrm{D}}/3$, the concentration of phonons $n_p = \int_0^{\omega_{\max}} d\omega \mathcal{D}(\omega)(\bar{n}/V)$ is proportional to the temperature, whereas σ_p is practically tem-

TABLE 24.7. Electrical resistivity ρ (in $\mu\Omega$ cm) and thermal conductivity K (in W/cm K) at room temperature. Data are taken from Refs. [2] and [7].

Crystal	ρ	K	Crystal	ρ	K
Li	9.32	0.85	Pb	21.0	0.35
Na	4.75	1.41	Gd	134.0	0.11
Cu	1.70	4.01	C	Insulator	1.29
Ag	1.61	4.29	Si	semiconductor	1.48
Au	2.20	3.17	Ge	semiconductor	0.60
Nb	14.5	0.54	SiO_2	insulator	0.14
Fe	9.80	0.80	CaF_2	insulator	0.11
Pd	10.5	0.72	NaCl	insulator	0.064

perature independent. Thus, the phonon contribution to the resistivity, $\rho = 1/\sigma \sim l^{-1}$ for $T \gtrsim \Theta_D/3$, is proportional to T and for good metals dominates all other contributions (room-temperature values are given in Table 24.7).

At very low temperatures ($T \lesssim \Theta_D/10$), $n_p \sim T^3$ and $\sigma_p \sim T^2$. This last relation is due to the fact that at very low temperatures, only phonons with small wave vectors are involved. These phonons change the direction of incoming electrons by a small angle $\theta \sim T$. The change of electronic momentum in the forward direction, which is proportional to $1 - \cos\theta$, then equals $\theta^2 \sim T^2$. Thus, at very low temperatures, the impurity and defect scattering dominate; this scattering is usually temperature independent. A notable exception is scattering by independent local magnetic moments, which produce resonances in the cross section resulting in the logarithmic increase of ρ as T decreases in the range of very low temperatures.

For semiconductors or insulators, there is no Fermi surface and, according to Eq. (24.20), the electrical conductivity will be zero at $T = 0$. Whatever electronic conductivity will be observed is due to thermal (or photon) excitation of electrons to the conduction band and of holes to the valence band. Because the carrier (electron or hole) concentration is the dominant factor, we rewrite the expression $\sigma = e^2 n\tau/m$ as follows:

$$\sigma = e\mu_e(T)n_e(T) + e\mu_h(T)n_h(T), \tag{24.21}$$

where e is the proton charge, $\mu_i \equiv e\tau_i(T)/m_i^*$ ($i = e,\ h$) is the so-called electron or hole mobility (see Table 24.8), m_e^* is the electron effective mass in the conduction band, and m_h^* is the hole effective mass in the valence band.

TABLE 24.8. Carrier mobilities at room temperature, in $cm^2/V \cdot s$. Data are taken from Ref. [2].

Crystal	Electrons	Holes	Crystal	Electrons	Holes
C	1800	1200	InAs	30000	450
Si	1350	480	GaAs	8000	300
Ge	3600	1800	PbTe	2500	1000

The concentrations n_e, n_h are given by

$$n_e = A_e(k_B T)^{3/2} e^{\beta(\mu - E_g)}, \quad n_h = A_h(k_B T)^{3/2} e^{-\beta\mu}, \tag{24.22}$$

where $A_i = (1/\pi\sqrt{2\pi})(m_i^*/\hbar^2)^{3/2}$ $(i = e,\ h)$ and μ is the chemical potential. For intrinsic semiconductors, i.e., for those with no impurity levels in the gap, $n_e = n_h$, and $\mu = (E_g/2) + (3/4)k_B T \ln(m_h^*/m_e^*)$.

24.8.3. Thermal conductivity and thermoelectric power

In the simultaneous presence of an electric field **E** and a thermal gradient ∇T, the densities of electric **j** and thermal **q** currents are given by the following set of equations:

$$\widetilde{\mathbf{E}} = \rho\mathbf{j} + Q\nabla T \tag{24.23}$$

and

$$\mathbf{q} = \Pi\mathbf{j} - K\nabla T, \tag{24.24}$$

where $\widetilde{\mathbf{E}} \equiv \mathbf{E} + (1/e)\nabla\mu$, μ is the chemical potential, Q is the so-called thermoelectric power, Π is the Peltier coefficient related to Q by the general Kelvin formula $\Pi = TQ$, and K is the thermal conductivity.

The thermoelectric power in metals is given by the relation

$$Q = -(\pi^2 k_B^2 T/3e)\, d[\ln\sigma(E_F)]/dE_F,$$

whereas the simple kinetic theory gives $Q = -C_V/3eN_e$, a formula that is approximately applicable to semiconductors.

Both phonons and free electrons contribute to the thermal conductivity, typical values of which are given in Table 24.7. The simple kinetic theory gives $K = K_e + K_p$, where $K_\nu = (1/3)v_\nu l_\nu c_{V_\nu}$ $(\nu = e,\ p)$, $c_{V_\nu} = C_{V_\nu}/V$, $v_e \equiv v_F$, and v_p is an appropriate average of the sound velocities. For metals, the *Wiedeman-Franz law* holds according to which $K_e = L\sigma T$, where $L = (\pi^2/3)(k_B/e)^2 = 2.44 \times 10^{-8}$ W Ω/K^2 is the Lorenz number. The phonon contribution to K at low temperatures behaves as T^3, since $C_{V_p} \sim T^3$ and v_p and l_p are temperature independent; at high temperatures ($T \gtrsim \Theta_D$), v_p and C_{V_p} are temperature independent, but l_p is proportional to $T^{-\nu}$ with $1 \leq \nu \leq 2$ because at these temperatures, phonon-phonon scattering dominates.

24.8.4. Hall effect and magnetoresistance

The presence of a static magnetic field B (in the z direction) modifies the electrical resistivity. The field-dependent off-diagonal term ρ_{yx} is called Hall resistivity and is given by RB, where R, the Hall coefficient, at high fields equals $-1/nec$; n is the total electron minus the total hole concentration. The simple formula $R = -1/nec$ fails if there is at least one band in which neither all occupied nor all empty orbits are closed (or if $n = 0$, in which case, $\rho_{yx} \sim B^2$).

At metal-oxide semiconductor structures or at GaAs/AlGaAs interfaces, the Hall resistance $-R_{yx} = -\rho_{yx}/d$ increases with the field B following a stair-like curve around the straight line $-(R/d)B$, where d is the microscopic width of the interface. The horizontal segments of this curve are given by the expression $(p/q)(h/e^2)$, where q is a positive integer and p is either one (in which case, the phenomenon is known as the integral quantum Hall effect, IQHE) or a small integer (larger than unity), in which case, we have the fractional quantum Hall effect (FQHE).

The magnetoresistance is defined by the relation $[\rho_{ii}(B) - \rho_{ii}(0)]/\rho_{ii}(0)$ $(i = x, y, z)$. For $i = x, y$, we have the transverse magnetoresistance $\delta\rho_\perp/\rho(0)$, whereas for $i = z$, we have the longitudinal one, $\delta\rho_{||}/\rho(0)$. For low fields ($\omega_c\tau \ll 1$, where $\omega_c = eB/m^*c$), the magnetoresistance $\delta\rho$ is usually proportional to B^2 and positive, whereas for high fields ($\omega_c\tau \gg 1$), it approaches a field independent value if $R = -1/nec$. If this last formula fails, the high-field magnetoresistance is proportional to B^2.

24.8.5. Cyclotron resonance, ESR, and NMR

The simple free-electron expression for the frequency of the classical orbit perpendicular to the field, $\omega_c = eB/mc$, is valid if the mass m is replaced by a cyclotron effective mass $m_c^* \equiv (\hbar^2/2\pi)\partial A(E, k_z)/\partial E$, where A is the area enclosed by the electron or hole orbit in $\mathbf{k}$ space. [E and k_z are the electron (or hole) energy and the z component of $\mathbf{k}$, which are conserved in the presence of $\mathbf{B} = (0, 0, B)$.]

The precession of a local (electronic or ionic) magnetic moment, $\mathbf{m} = -g\mu_B\mathbf{J}$, around the field $\mathbf{B}$ creates another characteristic frequency, $\omega_o = \gamma B$, where γ, the gyromagnetic ratio, is given by $\gamma = g\mu_B/\hbar$; $\mu_B = (e\hbar/2mc) = 5.7884 \times 10^{-9}$ eV G^{-1} (or $e\hbar/2m = 9.2740\times10^{-24}$ J T^{-1} in the SI system) is the so-called Bohr magneton, $\mathbf{J} = \mathbf{L}+\mathbf{S}$ is the total angular momentum, and $g = 1+[J(J+1)+S(S+1)-L(L+1)]/2J(J+1)$ is the Lande g-factor. The same is true for a nuclear magnetic moment, for which the characteristic frequency $\omega_I = \gamma_I B$, where $\gamma_I = g_I\mu_N/\hbar$, $\mu_N = e\hbar/2m_pc$, and g_I for the free proton is $2 \times 2.7928 = 5.5857$. The response of the material to an external electromagnetic field of frequency ω in the presence of a static magnetic field $\mathbf{B}$ exhibits a resonance if $\omega = \omega_c$ (cyclotron resonance), or $\omega = \omega_o$ [electron spin resonance (ESR), or electron paramagnetic resonance (EPR)] or $\omega = \omega_I$ [nuclear magnetic resonance (NMR)]. For ESR, $\omega_o/2\pi = \nu_o$(GHz)$= 28.0B$(Tesla), whereas for proton NMR ν_o(MHz)$= 42.58B$(Tesla). In the NMR case, the s electrons produce an extra effective field at the nucleus, thus shifting the NMR frequency by a factor $1 + K$, where K, the so-called Knight shift, is given by $K = (8\pi/3)(|\psi(0)|^2/n)\chi_p$; $\psi(0)$ is the value of the s wave electronic function at the nucleus.

24.9. DISORDERED SYSTEMS

24.9.1. Localization and metal-insulator transition

Disordered materials lack the periodicity of crystals. However, they also possess some universal features, [8] such as the experimentally observed exponential behavior of the tails in the DOS (Urbach tails). The role of the disorder to second order in perturbation theory is to shift the band edge by an amount proportional to W^2, the variance of the disorder. In addition to this rigid shift of the DOS, both the valence and the conduction bands develop tails toward the gap. In most cases, as a result of many independent sources of disorder, one expects the distribution of the disorder to be Gaussian, which gives that the DOS $\rho(E) \sim \exp(-E/E_0)$, with $E_0 \sim W^2$. Experiments yield that E_0 is of the order of 50 meV for most amorphous semiconductors.

Disorder influences the electronic wave functions as well. If the disorder is weak, the wave function will look like a plane wave on a short length scale. The distance over which the phase of the wave function deviates appreciably from that of the plane wave is called the mean free path. As the disorder increases, the eigenstates start to exhibit amplitude fluctuations at scales considerably larger than atomic size and some of them become localized, i.e., their amplitudes decay exponentially away from a center. Thus, regions of localized states appear, delineated from regions of propagating states by critical energies E_c, called mobility edges. As the disorder increases, the Fermi energy E_F may cross a mobility edge. In this case, a metal-insulator transition (MIT) will take place at $T = 0$. Ioffe and Regel have argued that at the mobility edge, $kl \simeq 1$. The approximate scaling theory of localization, [9] as well as numerical results, [10, 11] show that the amplitude of the wave function develops strong fluctuations as one approaches the mobility edge. At the mobility edge, the amplitude is fractal at all length scales [12] with a fractal dimensionality $D(2) = 1.70 \pm 0.30$ for a random tight-binding model. The fluctuations can be characterized by a length ξ, which has a power-law divergence as the localization transition is approached, $\xi \sim (E - E_c)^{-\nu}$. There is a similar divergence for the localization length l_c, in the localized site of the MIT. The length ξ determines the dc conductivity σ, through the relation $\sigma = (e^2/\hbar)(\alpha/\xi + \beta/L)$, where L is the linear dimension of the system and α and β are numerical constants. So the divergence of ξ causes critical behavior in σ of the form $\sigma \sim (E - E_c)^{\nu}$. The scaling theory of localization predicts a value of $\nu = 1$, in contrast to numerical work, which suggests $\nu \simeq 1.5$, and to experiments on uncompensated Si:P, which give $\nu \simeq 1/2$. It is possible that this last discrepancy might be due to Coulomb interactions. The scaling theory of localization also predicts that all states are localized in disordered one-dimensional (1D) and two-dimensional (2D) systems.

An MIT can also take place in systems, called Mott-Hubbard insulators, [13] where the electron-electron (el-el) interactions are very strong. The most pronounced effects of el-el interactions appear in narrow bands, as in the case of disorder. Thus, a satisfactory theory must take into account simultaneously the effects of randomness and el-el interaction, more conveniently described by the Hubbard model $H = \sum_i \epsilon_i | i\rangle\langle i | + \sum_{i,j} t_{ij} | i\rangle\langle j | + U \sum_i n_{i\uparrow} n_{i\downarrow}$, where U is the on-site Coulomb repulsion and $n_{i\uparrow}$ and $n_{i\downarrow}$ are the number operators at site i with spin up and spin down, respectively. For one electron per atom (half-filled case), the Fermi level lies in the middle of the gap, provided that U exceeds a critical value U_c of the order of half the bandwidth; thus, el-el interactions induce an MIT for a half-filled band. It must be pointed out that even without the U term, an MIT appears when the disorder W exceeds the critical value W_c. The Hubbard model for a less than half-filled band has been studied extensively over the last ten years because of its connection to strongly correlated systems and, possibly, to high-T_c superconductors.

24.9.2. Metal-insulator transition in 2D disordered systems

The scaling theory of localization predicts that all states are localized in a disordered 2D system in the absence of magnetic fields or spin-orbit interactions. Experiments done back in the 1980s generally confirmed the predictions of the scaling theory. However, in a series of recent experiments, Kravchenko and colleagues [14] reported evidence for an MIT in a 2D electron gas in zero magnetic field. The new ingredient in these recent experiments [14] on MOSFETs is the much higher electron mobilities than those in the 1980s experiments. The experimental reports for the 2D MIT challenge the accepted picture and seem to suggest that el-el interactions play a more important role in 2D systems than previously realized.

24.10. MAGNETISM

Electrons and nuclei in solids produce magnetic fields, because they are moving charges and have intrinsic magnetic dipole moments. The magnetic fields produced by the electrons are much larger than are those produced by the nuclei. When an external magnetic field **H** is applied, a macroscopic field is induced. The magnetic susceptibility is defined as $\chi = (\partial M/\partial H)$, usually in the limit of $H \to 0$, where M is the magnetization. The magnetic permeability $\mu = B/H$ is related to χ as $\mu = \mu_0(1 + \chi)$ (SI) or $\mu = 1 + 4\pi\chi$ (G-cgs). Very briefly, we discuss the magnetic classification of materials.

Diamagnetic materials have a negative magnetic susceptibility χ. These materials have atoms or ions with complete shells, and their diamagnetic behavior is due to the fact that the external magnetic field distorts the orbital motion of the electrons. The susceptibility per mole is given (in G-cgs) by

$$\chi^{\text{mol}} = -0.79\, Z_i \times 10^{-6}\langle (r/a_{\text{B}})^2\rangle \text{ cm}^3/\text{mole}. \tag{24.25}$$

The quantity $\langle (r/a_{\text{B}})^2\rangle$ is of the order of unity, $a_{\text{B}} = 0.529$ Å is the Bohr radius, and Z_i is the total number of electrons in the ion (or atom).

Paramagnetic materials have a positive χ. These materials have ions with incomplete shells, thus, their paramagnetic behavior. The magnetization of these materials is given by

$$M = n_i g\mu_{\text{B}} J B_J(g\mu_{\text{B}} JH/k_{\text{B}}T), \tag{24.26}$$

where the Brillouin function $B_J(x)$ is defined by

$$B_J(x) = \frac{2J+1}{2J}\coth\frac{2J+1}{2J}x - \frac{1}{2J}\coth\frac{x}{2J}. \tag{24.27}$$

When $g\mu_{\text{B}}H \ll k_{\text{B}}T$, the small-$x$ expansion gives the susceptibility

$$\chi^{\text{mol}} = \frac{N_{\text{A}}}{3}\frac{p^2\mu_{\text{B}}^2}{k_{\text{B}}T} = \frac{0.125p^2}{T}\frac{\text{cm}^3}{\text{mol}}, \quad \text{G-cgs}, \tag{24.28}$$

where p, the *effective Bohr magneton number*, is given by $p = g[J(J+1)]^{1/2}$, N_{A} is the Avogadro number, and T is the temperature in K. The expression $\chi = C/T$ is the *Curie law*, and $C = N_{\text{A}}\mu_{\text{B}}^2 p^2/3k_{\text{B}}$ is the Curie constant. Once C is obtained from experimental data, it can be used to estimate p. For rare-earth ions, there is an excellent agreement between theory and experiment. However, for transition-metal ions with $3d$ electrons, agreement with experiment is possible if $\mathbf{J} = \mathbf{S}$, i.e., $\mathbf{L} = 0$, called *quenching of the orbital angular momentum*. This is due to the strong crystal-field effects that are present in the transition metals.

At high magnetic fields B, very low temperatures, and very pure monocrystalline specimens of metals, the susceptibility χ exhibits a periodic behavior as a function of $1/B$ with a period, $\Delta(1/B)$, given by $2\pi e/\hbar c A$, where A is the maximum or minimum area of the cross section of the Fermi surface perpendicular to the magnetic field. This so-called *de Haas-Van Alphen effect* is a very useful tool for determining experimentally the shape and the size of the Fermi surface.

In metals, conduction electrons make two contributions to the susceptibility. The first, called *Pauli paramagnetism*, arises from the fact that each conduction electron carries a spin magnetic moment, which tends to align with the field; it is given (in G-cgs) by

$$\chi_{\mathrm{P}} = \frac{2\mu_{\mathrm{B}}^2 \rho_{\mathrm{F}}}{V} = \frac{2.59}{(r_s/a_{\mathrm{B}})} \times 10^{-6}, \tag{24.29}$$

where $r_s = (3/4\pi n)^{1/3}$ with n as the number of conduction electrons per unit volume. This expression shows that χ_{P} is much smaller than the paramagnetic contribution of the ions.

The second, χ_{L}, known as the *Landau diamagnetism*, is due to the orbital motion of the electrons in the presence of $\mathbf{B}$ and for free electrons satisfies the relation $\chi_{\mathrm{L}} = -\frac{1}{3}\chi_{\mathrm{P}}$. In Table 24.9, values of χ for some solids are given.

Ferromagnetic materials have a spontaneous magnetization below a transition temperature T_c, where the atomic dipole moments tend to be aligned in some favorable direction of the crystal. Only some transition and rare-earth elements are ferromagnetic, and therefore, the phenomenon is due to the unfilled $3d$ and $4f$ shells in these substances. Above T_c, the susceptibility follows the Curie-Weiss law $\chi = C/(T - T_c)$, whereas very close to T_c, a variety of critical exponents can be defined and measured. For example, the magnetization $M(T) \sim (T - T_c)^{\beta}$, $\chi(T) \sim (T - T_c)^{-\gamma}$, and the magnetic specific heat $C(T) \sim (T - T_c)^{-\alpha}$.

TABLE 24.9. Magnetic susceptibilities (in 10^{-6} cm^3/mol) of typical metals and semiconductors at room temperature.

Crystal	χ	Crystal	χ	Crystal	χ
Li	14.2	Mo	89	C	−5.9
Na	16.0	Cr	180	Si	−3.9
K	20.8	Nb	195	Sn	−37.0
Al	16.5	V	255	Ge	−76.9
Cu	−5.46	Mn	529	SiC	−12.8
Au	−28.0	MnO	4850	La_2O_3	−78.0
NaCl	−30.3	Sm	1860	La	95.9
Pd	567.4	Nd	5930	Gd	185000

TABLE 24.10. Transition temperatures (T_c or T_N) for ferromagnetic and antiferromagnetic crystals. Data are taken from Ref. [2] and recent publications.

Crystal	$T_c(K)$	Crystal	$T_N(K)$
Fe	1043	Cr	308
Co	1388	Mn	100
Ni	627	MnO	116
Gd	293	FeO	198
CrO_2	386	$LaMnO_3$	140
$SmCo_5$	993	La_2CuO_4	340
$Nd_2Fe_{14}B$	588	$YBa_2Cu_3O_6$	500

Antiferromagnetic materials have atomic dipole moments ordered antiparallel (with zero net magnetization) at temperatures below the ordering temperature T_N (Néel temperature). Above T_N, $\chi = C/(T + T_N)$. Values for T_c and T_N are given in Table 24.10.

In *ferrimagnetic materials*, the magnitudes of the oppositely directed atomic magnetic moments are not the same, and so the magnetization does not vanish as in the antiferromagnetic case.

24.11. SUPERCONDUCTIVITY

Superconductivity, the disappearance of the electrical resistivity of a material and the expulsion of the magnetic field **B** from its interior below a critical temperature T_c, is a very fascinating phenomenon. Once a current is established in a closed superconducting ring, it goes forever. Superconductors exhibit quantum and wave properties at a macroscopic scale, and their explanation is one of the triumphs of modern physics. Superconductivity has a lot of important applications, mainly in superconducting magnets, medical resonance imaging (MRI) devices, superconducting quantum interference devices (SQUIDS), and so on.

The mechanism responsible for the phenomenon of superconductivity is a dynamical, phonon-mediated, el-el attraction that at a certain frequency interval overcomes the Coulomb repulsion and creates bound electronic pairs in the presence of the Fermi sea.

The BCS theory gives the following expression for T_c:

$$T_c = 1.13\Theta_D \exp(-1/\lambda), \qquad (24.30)$$

where Θ_D is the Debye temperature, $\lambda = \rho_F V$ is the same quantity as in Eq. (24.14), and V is the strength of the attractive el-el interaction. BCS found that the superconducting energy gap at $T = 0$, $2\Delta_0$, is equal to $3.53k_B T_c$, which is obeyed for weakly coupled conventional superconductors. BCS also calculated the T dependence of the energy gap 2Δ. For the ratio Δ/Δ_0, they obtained $\Delta/\Delta_0 = \tanh(T_c\Delta/T\Delta_0)$, and for $T \simeq T_c$, $\Delta/\Delta_0 = 1.76(1 - T/T_c)^{1/2}$. BCS also calculated the T dependence of the critical magnetic field $H_c(T)$ (Table 24.11), i.e., the field needed to destroy the superconducting state, $H_c(T) = H_c(0)[1 - 1.06(T/T_c)^2]$ for $T \ll T_c$ and $H_c(T) = 1.74H_c(0)[1 - (T/T_c)]$ for

TABLE 24.11. Transition temperature T_c and upper critical field $H_{C2}(0)$ of typical supeconductors. Data are taken from Ref. [2] and recent publications.

Crystal	T_c (K)	H_{c2} (T)	Crystal	T_c (K)	H_{c2} (T)
Hg	4.153	0.04	$La_{2-x}Sr_xCuO_4$	38	≥ 80
Nb	9.50	0.20	$YBa_2Cu_3O_7$	92	≥ 150
Pb	7.193	0.08	$Bi_2Sr_2Ca_2Cu_3O_{10}$	110	≥ 250
NbN	16.0	13.2	$TlBa_2Ca_2Cu_3O_9$	110	≥ 100
V_3Si	17.1	23.0	$Tl_2Ba_2Ca_2Cu_3O_{10}$	125	≥ 150
Nb_3Ge	23.2	38.0	$HgBa_2Ca_2Cu_3O_8$	133	≥ 150

$T \simeq T_c$. Heat capacity measurements have shown that the electronic part of the superconducting state $C_{es} \propto \exp[-\Delta/k_B T]$ for $T \ll T_c$, and BCS give $(C_{es} - C_{en})/C_{en} = 1.43$ at $T = T_c$. This is in good agreement with experiments in conventional superconductors. The ratio of ultrasonic attenuation coefficients α in the superconducting and normal states is given by $(2\alpha_n/\alpha_s) = \exp(\Delta/k_B T) + 1$.

The flux through a superconducting ring is quantized in units $\phi_0 = hc/2e = 2.07 \times 10^{-7}$ G cm^2, called the flux quantum. This is a manifestation of quantum phenomena on a macroscopic scale and beautiful, direct experimental evidence of the idea of Cooper pairs.

Tunneling experiments in a metal-insulator-superconductor junction give direct information of the superconducting band gaps. In the so-called Josephson junction with two superconductors separated by a thin (1 nm) insulating barrier with an applied voltage V_a, the current density is given by $J = J_0 \sin[\delta(0) - \omega_a t]$, where $\omega_a = 2eV_a/\hbar$. Thus, Cooper pair tunneling produces a dc current with zero voltage and an ac current with dc voltage.

If we connect two Josephson junctions in parallel, one can observe interference effects, similar to those observed in Young's double-slit experiments. The phase difference $\delta_A - \delta_B$ between the two junctions A and B equals $2\pi(\Phi/\phi_0) = (2e/\hbar c)\Phi$, where Φ is the externally applied flux. So

$$\begin{aligned} J_{\text{tot}} = J_A + J_B &= J_0[\sin\delta_A + \sin\delta_B] \\ &= 2J_0 \sin\delta_0 \cos(e\Phi/\hbar c), \end{aligned} \tag{24.31}$$

where $\delta_0 = (\delta_A + \delta_B)/2$. Thus, the total current depends periodically on the external flux or external magnetic field. Such devices are called SQUIDS and can be used to detect extremely weak magnetic fields, even those produced by brain activity.

The discovery of high-temperature superconductivity in 1986 in $La_{2-x}Ba_xCuO_4$ at 35 K and in 1987 in $YBa_2Cu_3O_7$ at 93 K, above the temperature of liquid nitrogen (77 K), can be considered one of the most exciting developments in modern physics. These high-T_c materials are ceramics, which are too brittle to be made into wires for transmission lines or electromagnets. However, they can be deposited as thin films for small-scale applications, such as electronic components. Although these ceramics show great promise, much developmental work remains to be done before we will see commercial applications. Much work is also needed to explain the mechanism of the occurrence of superconductivity in these materials, although some progress has been made regarding their normal-state properties.

24.12. ELEMENTARY EXCITATIONS

Besides phonons, plasmons, and polaritons, there are also other composite elementary excitations in solids.

24.12.1. Excitons

When an electron jumps from the valence band to the conduction band, it leaves behind a hole. The electron and the hole have opposite charge and so attract each other and possibly form a bound state, referred to as the *exciton*. One can argue that excitons are like hydrogen atoms with the hole playing the role of the proton. The binding energy and the radius of the exciton are

$$E_{ex} = \frac{m^*}{m}\frac{e^2}{2a_B}\frac{1}{\epsilon^2}, \qquad a_{ex} = \frac{\hbar^2\epsilon}{m^*e^2} = \epsilon\frac{m}{m^*}a_B, \tag{24.32}$$

where ϵ is the dielectric constant and m^* is the reduced effective mass ($1/m^* = 1/m_e + 1/m_h$). In Table 24.12 some characteristic values of E_{ex} are given. For expressions (24.32) to be valid, we must have $a_{ex} \gg a_B$, which is obeyed when either ϵ is large or m^* is relatively small. This type of weakly bound exciton is known as a *Mott-Wannier exciton.* For typical Mott-Wannier excitons, $\epsilon \simeq 10$ and $m/m^* \simeq 10$; therefore, $a_{ex} \simeq 50$ Å and $E_{ex} \simeq 10$ meV. The energy of the photon involved in exciton absorption in a semiconductor is $h\nu = E_g - E_{ex}$, where E_g is the width of the gap. The exciton spectrum consists of a sharp line, just below the bottom of the conduction band. If $a_{ex} \approx a_B$, we have strongly bound excitons, which are called *Frenkel excitons.*

24.12.2. Polarons and bipolarons

An electron plus its induced lattice polarization is called a *polaron.* When an electron (or hole) is added in an ionic crystal, the charge of the electron will deform locally the surrounding medium, by attracting the positive (or negative) ions. The effective mass of the polaron is larger than that of the charge carrier (i.e., electron or hole). If the polaron coupling constant α is relatively strong ($\alpha \geq 6$), we are in the regime of the small polaron, which is strongly localized, and its transport takes place by *variable range hopping.* If $1 \leq \alpha \leq 6$, we are in the regime of a large polaron, where the band structure picture is still valid, but the mobility of the polarons is small. Values of α are given in Table 24.12.

TABLE 24.12. Binding energy of excitons in meV, and polaron coupling constant α (in parentheses). Data are taken from Ref. [2] and from Ref. [15].

Si	14.7	KI	480 (2.5)	AgBr	20 (1.69)
Ge	4.1	KCl	400 (3.97)	AgCl	30 (1.84)
GaAs	4.2 (0.06)	KBr	400 (3.52)	BaO	56 (—)
GdTe	15.0 (0.31)	RbCl	440 (3.81)	$SrTiO_3$	— (4.5)
InSb	0.4 (0.02)	LiF	1000 (—)		

When two electrons (or two holes) interact with each other simultaneously through the Coulomb force and via the electron-phonon-electron interaction, either two independent polarons can occur or a bound state of the two polarons, the so-called *bipolaron*. Bipolarons have been recently considered as candidates in playing a role to explain high-T_c superconductivity.

24.12.3. Spin waves

The excitations in magnetically ordered systems are called *spin waves*, and their quanta are referred to as *magnons*. A formalism similar to that used for phonons can be set up, and for low energies, it can be shown that the spin-wave or magnon energy is equal to $\omega_q \simeq Dq^2$ for ferromagnets and $\omega_q \simeq Dq$ for antiferromagnets. D is called the *spin wave stiffness* and, similar to phonons, can be obtained from inelastic neutron-scattering measurements. Knowledge of these elementary excitations allows us to calculate the temperature dependence of the magnetic contribution to the properties (such as specific heat, resistivity, etc.) of magnetic systems.

24.13. ARTIFICIAL SOLID STRUCTURES AND PHOTONIC CRYSTALS

It is well known that new materials and devices are needed for electronic applications. [16] There is also a trend in the miniaturization of these devices. Therefore, there is a need for artificial structures that can alter the physical properties (e.g., the mobility) of the electrons or holes. Various kinds of 2D (superlattice), 1D (quantum wires), and *zero*-dimension (quantum dots or quantum wells) structures can be fabricated with very small dimensions (of the order of 100 Å or less). These artificial structures are produced by molecular beam epitaxy (MBE). They have led to interesting new physics and to new classes of electronic devices, and they have contributed to the success of the semiconductor industry.

For example, a quantum well with width L of the order of 50 Å or less has been fabricated and the individual energy levels have been measured by optical absorption. It was found experimentally that the energy levels vary as L^{-2}, which indicates that quantum effects are observed. Also, exciton effects have been measured in these systems, with very pronounced peaks.

Resonant tunneling experiments have been performed in double-barrier structures, where the current versus voltage is measured. If the energy of the incident electron coincides with one of the energy levels of the quantum well, the electron tunnels across the barrier without attenuation. The differential conductance, defined as dI/dV, has also been measured in quantum-well structures. A negative differential conductance is observed, which has potential device applications. The splitting of the electronic and phonon band into subbands by a periodic superlattice of a large period gives rise to new $k \simeq 0$ phonons, which have been observed by Raman or infrared techniques.

The success of the semiconductor industry has motivated the researchers to try to fabricate new devices with higher speeds. This can be accomplished by using photons rather than electrons. In semiconductor technology, one tries to tailor the properties either by changing their bandgaps or by doping. Recently, new materials, called *photonic crys-*

tals, [17, 18] have been fabricated that prevent certain bands of electromagnetic waves from propagating through the structure. The photonic crystals are of interest to the opto-electronic industry, since they can efficiently manipulate photons and therefore can be used in a variety of applications. Photonic crystals do not occur naturally; thus, one must fabricate them by constructing periodic dielectric structures. Because crystals must be built at the same scale as the photon wavelength, the difficulty of producing photonic crystals tends to increase with higher frequency. As a result, to date, most successful fabrications of 3D photonic crystals are in the microwave and far-infrared regimes.

24.14. REFERENCES

[1] N. W. Ashcroft and N. D. Mermin, *Solid State Physics* (Holt, Rinehart and Winston, New York, 1976).

[2] C. Kittel, *Introduction to Solid State Physics*, 6th ed. (Wiley, New York, 1986).

[3] G. Burns, *Solid State Physics*, (Academic Press, New York, 1985).

[4] O. Madelung (Ed.), *Semiconductors—Basic Data* (Springer, Berlin, 1996).

[5] D. A. Papaconstantopoulos, *Handbook of the Band Structure of Elemental Solids* (Plenum Press, London, 1986).

[6] W. A. Harrison, *Electronic Structure and the Properties of Solids* (Freeman, San Francisco, 1980); Also in (Dover, New York, 1989).

[7] D. R. Lide (Ed.), *Handbook of Chemistry and Physics*, 73rd ed. (CRC Press, Boca Raton, FL, 1992).

[8] C. M. Soukoulis and E. N. Economou, in *Encyclopedia of Applied Physics* (VCH, New York, 1993), Vol. 5, p. 549.

[9] P. A. Lee and T. R. Ramakrishnan, *Rev. Mod. Phys.* **57**, 287 (1985).

[10] For a recent review, see B. Kramer and A. Mackinnon, *Rep. Prog. Phys.* **56**, 1469 (1993).

[11] E. N. Economou, *Green's Functions in Quantum Physics* (Springer-Verlag, Heidelberg, 1983).

[12] C. M. Soukoulis and E. N. Economou, *Phys. Rev. Lett.* **52**, 565 (1984).

[13] N. F. Mott, *Metal-Insulator Transitions* (Taylor and Francis, London, 1974).

[14] S. V. Kravchenko, *et al.*, *Phys. Rev. Lett.* **77**, 4938 (1996); D. Simonian, S. V. Kravchenko, M. P. Sarachik, and V. M. Pudalov, *Phys. Rev. Lett.* **79**, 2304 (1997).

[15] J. T. Devreese, in *Encyclopedia of Applied Physics* (VCH, New York, 1996), Vol. 14, p. 383.

[16] C. Weisbusch and B. Vinter, *Quantum Semiconductor Devices* (Academic Press, New York, 1991).

[17] See the articles in *Photonic Band Gap Materials*, edited by C. M. Soukoulis (NATO Advanced Studies Institute Series E, Materials, Kluwer Academic Publishers, Dordrecht, 1996), Vol. 315. Also the articles in *Photonic Crystals and Light Localization in the 21st Century*, edited by C. M. Soukoulis (NATO Science Series C, Kluwer Academic Publishers, Dordrecht, 2001), Vol. 563.

[18] J. D. Joannopoulos, R. D. Meade, and J. N. Winn, *Photonic Crystals* (University Press, Princeton, NJ, 1995).

25

Surfaces and Films

Roland Resch
Infineon Technologies, Villach, Austria

Bruce E. Koel
University of Southern California, Los Angeles, California

Contents

List of Figures

25.1. INTRODUCTION

Surfaces control a vast array of chemical and physical processes in nature. All chemical reactions between two phases occur at the interfacial boundary between the phases—the only place that each atom or molecule is not surrounded by a relatively homogeneous environment. This situation has prompted identifying surfaces or interfaces as the "seat of communication between two phases."

At an interface, changes occur in bonding, coordination number, and electrical potential. The inhomogeneous environment at a surface imparts unique chemical and physical properties to this region; i.e., chemical and physical processes at the surface are often different from those in the bulk phase. For example, the geometric and electronic structure of crystalline materials can be strongly altered at the surface, and phase diagrams for bulk compound formation and miscibility are not applicable at the surface. For interfaces, one has to add surface tension as an intensive thermodynamic variable and surface area as a new extensive variable.

Possible surfaces or interfaces are gas-liquid, gas-solid, liquid-liquid, liquid-solid, and solid-solid. Surface science is the study of surfaces, with principal applications in materials science (e.g., design, fabrication, and processing of electronic materials for integrated circuits) and heterogeneous catalysis (the heart of the huge petrochemical industry and the basis for catalytic converters used in automobile exhaust gas control). Colloids are a special

This work was supported by the National Science Foundation, Division of Chemical Sciences. We thank Dr. Matthias Batzill for a critical reading of the manuscript and Dr. Santanu Banerjee for helpful comments.

class of matter that are intimately related to surfaces and that have important applications in suspensions, blood, detergents, gels, emulsions, foams, and so on, but it is beyond our scope to treat this topic here.

Processes that occur at vacuum/gas-solid interfaces are the best understood. This situation exists primarily because spectroscopic probes of solid surfaces are still much better developed than are probes that are compatible with liquids. In addition, crystalline solids have a well-defined structure and composition that facilitates fundamental microscopic understanding.

25.2. SURFACE ANALYSIS: PROBING SURFACES AND FILMS [1]–[10]

Many analytical techniques exist for determining the geometric and electronic structure and composition of solid surfaces. These have been used to study gas adsorption, atomic motions at surfaces, and nucleation and growth of thin films. It is testimony to the ingenuity of physicists that such a vast array of tools exists to probe surfaces and films (although the alphabet soup can be bewildering). We shall only be able to discuss a few of the most used methods here.

Difficulties in surface analysis include both absolute sensitivity and background discrimination. Solids have atomic densities ρ of about 10^{22} cm^{-3}, whereas surfaces have densities of about $\rho^{2/3}$ or 10^{15} cm^{-2}. Thus, one has to be able to detect 10^{12} to 10^{15} species in a background of 10^7-to-1. With the main exception of recent developments in microscopy, these techniques commonly involve low-energy (1–1000 eV) charged particles (electrons or ions) as a source of excitation or as an analyzed product. This is because inelastic scattering mean free paths for electrons in solids are 4–10Å for energies of 1–1000 eV and even less for ions of these energies. Surface sensitivity comes inherently from resolving features that do not arise from inelastic scattering events.

25.2.1. Electron spectroscopy [11]–[13]

In electron spectroscopy, the kinetic energy of electrons emitted from the surface is analyzed. A simple one-electron picture often suffices for photoexcitation by either x rays (x-ray photoelectron spectroscopy or XPS) or UV light (UV photoelectron spectroscopy or UPS). An incoming photon of energy $h\nu$ is absorbed by the target atom, photoionizing the atom and ejecting an electron of binding energy E_b (defined relative to the Fermi energy E_{F}) with kinetic energy E_k according to

$$E_k = h\nu - E_b - \varphi, \tag{25.1}$$

where φ is the work function of the solid. UV light (typically, He(I) at 22.40 eV and He(II) at 40.89 eV) is used to maximize sensitivity to the valence levels of surface atoms and adsorbed molecules to probe electronic structure and bonding. Information on the occupied density of states (DOS) is obtained to the extent that the wave function in the final state can be approximated by a plane wave. Angle-resolved UPS (ARUPS), especially when used with tunable light from a synchrotron, is a powerful probe of metal band structure and adsorbate molecular orbital symmetry.

X-ray irradiation (typically, Al K_α at 1486 eV and Mg K_α at 1254 eV) is used to ionize deep, atomic-like core levels. Because core-level binding energies are characteristic of each element, and photoabsorption cross sections for these levels are independent of the chemical environment, XPS is a powerful, quantitative probe for elemental analysis. Furthermore, changes in valence electronic state (i.e., the oxidation state) cause chemical shifts in the core-level binding energies, and so chemical changes such as oxidation and reduction can be studied. The measured electron binding energy E_b is the difference in energy between initial-state E_i and final-state energies E_f after ejection of the photoelectron:

$$E_b = E_i - E_f. \tag{25.2}$$

The binding energy for an atom in a solid can be expressed as

$$E_b = E_{\text{atom}} + \Delta E_{\text{chem}} + \Delta E_{\text{Mad}} + \Delta E_r, \tag{25.3}$$

where E_{atom} corresponds to the binding energy in the free atom, ΔE_{chem} accounts for changes in the local chemical environment, ΔE_{Mad} is a contribution from the Madelung potential (for ionic crystals), and ΔE_r is due to several relaxation or screening (dynamical) effects. It is important to remember that in assigning oxidation states to binding energy shifts, one makes a sudden approximation that $\Delta E_b = \Delta E_i$. Changes in screening of the final state core hole from interatomic polarization and charge transfer and intraatomic relaxation (valence excitation/ ionization) causes shifts and satellite peaks.

In Auger electron spectroscopy (AES), the kinetic energies of emitted Auger electrons are analyzed. The Auger process is a two-electron, nonradiative decay of atoms raised into an excited ionic state by the creation of a core hole via photoionization by incident x-rays or ionization by a high-energy (1–5 keV) electron beam. This initial state relaxes as this hole is subsequently filled with an electron from a higher level and the excess energy is carried away by the kinetic energy of the ejected Auger electron. The kinetic energy of the Auger electron $E(KL_1L_2)$ is given by

$$E(KL_1L_2) = E_K - E_{L1} - E_{L2} - \Delta, \tag{25.4}$$

where E_K is the binding energy of the initially ionized electron and E_{L1} and E_{L2} are the binding energies of electrons in the L_1 and L_2 energy levels involved in the Auger process. A hole-hole final-state interaction term Δ ranges from 0 to 20 eV. Calculations of Auger energies often use an equivalent-cores approximation that, for an element of atomic number n, uses E_{L1} and E_{L2} for the $n + 1$ element to approximate the effect of the core-level ion initial-state. AES is used widely for elemental analysis because E_K, and often E_{L1} and E_{L2}, are well-defined core levels unique to each atom. Chemical shifts occur in AES, but these are not correlated simply with oxidation state. Because AES uses an intense focused incident electron beam, it is often more sensitive than XPS and can be readily used in a scanning mode in scanning Auger microscopy (SAM) to give surface elemental analysis with a lateral resolution down to 10 nm.

25.2.2. Ion spectroscopy [14]–[17]

Ion scattering spectroscopy (ISS) is characterized by the incident ion energy. In low-energy ion scattering (LEIS), monoenergetic ions in the 0.2–5 keV range impinge on the sample surface and the backscattered incident ions are energy-analyzed at a known scattering angle. In medium-energy ion scattering (MEIS), ions in the 10–200 keV range are used. Often, inert-gas ions (e.g., ^{4}He and ^{20}Ne) are used. Alkali ions (e.g., ^{7}Li and ^{23}Na) have strong trajectory-dependent neutralization cross sections and give relatively intense angular-dependent signals in alkali ion scattering (ALISS) studies of surface atomistic structure. Mean free paths of ions in solids at low energies are extremely short, such that only the topmost layer composition is probed in LEIS (in contrast to XPS and AES). Because collision times are very short (10^{-15} to 10^{-16} s), the interactions can be approximated as elastic binary collisions between the incident ion and a single surface atom (i.e., with an effective mass equal to the atomic mass). The basic equation in ISS for the scattered ion energy E_1 upon collision of an incident ion of mass M_1 and energy E_o with a target surface atom of mass M_2 is

$$\frac{E_1}{E_0} = \frac{1}{(1 + M_2/M_1)^2} \left(\cos\theta \pm \sqrt{(M_2/M_1)^2 - \sin^2\theta} \right)^2, \tag{25.5}$$

where θ is the scattering angle. For those cases in which the target atom is heavier than the incident ion ($M_2/M_1 > 1$), only the plus sign applies and each target mass can be identified by a single peak in the spectrum. If $M_2/M_1 < 1$, both signs apply and each target mass gives rise to two peaks at different energies (detection of forward-scattered, recoil ions in elastic recoil spectrometry (ERS) has been particularly important for analysis of surface hydrogen). Equation (25.4) is used to identify the mass of target atoms from the energies of peaks in the scattered ion distributions. LEIS provides the highest surface sensitivity, but the physics of the scattering process is complex. MEIS gives relatively high surface sensitivity and accurate structural information about the top several layers of a crystalline material.

High-energy ion scattering (HEIS), known as Rutherford backscattering spectrometry (RBS), is one of the few techniques that can probe buried interfaces (such as in multilayer films for mirror coatings and semiconductor-based devices). RBS is less readily available because an ion accelerator is needed to produce the 1–4 MeV He^+ incident ion beam and nuclear particle detectors are required for the collection of the backscattered He^+ ions. Surface sensitivity is compromised, but RBS is an excellent quantitative tool because the absolute scattering cross sections are known. Equation (25.5) describes RBS too, providing information on the identity and concentration of target atoms. The depth and distribution of target atoms within the first micron of the surface can be ascertained because high-energy ions lose energy when traveling through solids primarily through inelastic collisions with electrons. This energy loss ΔE of an ion of energy E_0 is usually expressed by

$$\Delta E = \varepsilon N \Delta x, \tag{25.6}$$

where ε is the stopping cross section (in units of eV $\cdot$ cm^2), N is the atomic density, and Δx is the distance traveled by the ion in the solid (twice the depth of the target atom). For example, 2-MeV $^{4}He^+$ ions in Al have $\Delta E/\Delta x = 27$ eV/Å. RBS is generally regarded as nondestructive.

In secondary-ion mass spectrometry (SIMS), incident primary ions (with energies of 0.1–5 keV) hit the sample surface, losing energy and momentum through elastic and inelastic interactions with target atoms. These target atoms are displaced, and they induce the movement of other atoms (recoils). The comparatively low energy of the late recoils is sufficient to eject atoms and molecular clusters into the gas phase. The focus in SIMS is on the number and identity of the sputtered target atoms (ions) rather than on the scattered incident ions. Although the greatest portion of the sputtered flux has zero charge, a small fraction (10^{-1} to 10^{-6}) exists as positive and negative ions depending on the composition of the surface. Incident beams for SIMS are made typically using electron-impact (EI) ion sources (e.g., for Xe^+ and Ar^+) and liquid-metal ion guns (e.g., for Ga^+ and In^+).

The advantages of SIMS include high sensitivity (10^{-6} monolayer), enormous dynamic range, detection of all elements, including H, the ability to perform isotope analysis (key for studies of oxidation using O^{16} and O^{18} passivated films), and scanning or imaging mode operation. Disadvantages include large differences (10^3) in signal intensities for different chemical states that makes quantitative analysis difficult and the use of inherently more destructive probing ions compared with photons or electrons. Careful calibration with ion-implanted elemental standards and low total beam fluxes (below 5×10^{12} atoms/cm^2 in so-called "static-SIMS") work around these problems.

25.2.3. Electron diffraction [18]–[22]

Surface crystallography can be carried out by using low-energy electron diffraction (LEED). Electrons with energy E have a de Broglie wavelength given by

$$\lambda = \sqrt{\frac{150.4 \text{ eV}}{E}} \text{ Å}. \tag{25.7}$$

Low-energy electrons (20–500 eV) have $\lambda = 0.5$–3 Å. This is of the order of atomic distances at surfaces. Consequently, low-energy electrons are diffracted at crystalline surfaces and they interfere with each other. The condition for constructive interference is given by

$$n\lambda = a \sin\theta, \tag{25.8}$$

where n is the spectrum order, a is the surface periodicity, and θ is the angle of diffraction. Elastic backscattering from an ordered array of atoms produces a diffraction pattern on a fluorescent screen that reveals the size and orientation of the surface unit mesh. Data taken on the diffraction spot intensity as a function of the incident electron energy, commonly called current versus voltage (I-V) curves, for several diffraction spots can provide information on the basis of the unit cell and surface defects.

Use of low-energy electrons gives LEED excellent surface sensitivity, but severely complicates analysis of the diffraction pattern because of strong multiple scattering effects. It cannot be assumed that each scattered electron has interacted with only one scattering center, as is done in the kinematic approximation for x-ray crystallography. I-V curves of unreconstructed fcc and bcc surfaces show strong "Bragg peaks" from scattering of electrons by the first few layers of the crystal lattice at energies compatible with the Bragg condition of $n\lambda = 2d \sin\theta$, where d is the interlayer spacing. However, usually many more "multiple-scattering" peaks are present. Thus, only small unit cells and relatively few atoms per unit cell (<50) can be handled.

25.2.4. Field emission [23]–[25]

Very high electric fields $\mathbf{E}$ of the order of $100\ \mathrm{V/nm}$ can cause emission of electrons from a cold sample. This is called field emission, in contrast to thermionic emission. Such fields can be reached by using a potential V of the order of 10 keV applied at a sharp metal tip with radius $r \sim 10^{-5}$ cm. The field at the tip is given by

$$\mathbf{E} = \frac{V}{kr}\frac{\mathbf{r}}{r}, \tag{25.9}$$

where k is a constant ~ 5. Field-emitted electrons are accelerated radially from the tip onto a hemispherical fluorescent screen. The image thus formed in field emission microscopy (FEM) is an electron emission map of the tip, magnified by an amount given by cx/r, where c is a compression factor ~ 0.6 and x is the tip-screen distance. Linear magnifications of 10^6 are achievable, but with the resolution limited to ~ 20 Å. Regions of different crystal faces can be identified because they emit with different intensities due to differences in work function.

In field ion microscopy (FIM), the tip is positively charged and it strips electrons from adsorbed gas to generate ions that are then accelerated onto a fluorescent screen. Atomic resolution in FIM is possible by operating at cryogenic temperatures.

25.2.5. Electron microscopy [26]–[32]

High-resolution images of surface topography, with excellent depth of field, can be produced using a highly focused, scanning (primary) electron beam. Primary electrons with an energy of 0.5–50 keV that enter a solid can be transmitted or generate low-energy secondary electrons, backscattered electrons, Auger electrons, and x-rays. The intensity of the secondary electrons is largely governed by the surface topography of the sample. An image of the sample surface can be constructed by measuring the secondary electron intensity as a function of the position of the scanning primary beam in scanning electron microscopy (SEM). High spatial resolution is possible because the primary electron beam can be focused to a small spot of 1–50 nm in size. High sensitivity to topographic features on the outermost surface (< 5 nm) is achieved when using a primary electron beam with energy of < 1 keV. The analysis of characteristic x-rays emitted from the sample under electron bombardment, called energy-dispersive x-ray spectroscopy (EDS), gives quantitative elemental information with a sampling depth of 20 nm–5 μm. High-resolution transmission electron microscopy (TEM) can achieve atomic resolution by using high energies (~ 300 keV) in combination with very thin samples and ultrahigh vacuum (UHV).

25.2.6. Scanning probe microscopy [33]–[36]

A general scheme for scanning-probe microscopy (SPM) involves a sharp probe (e.g., tip, metal wire, optical fiber, pipette) that is raster-scanned across a sample surface by means of piezoelectric transducers to generate a signal that is recorded by the probe at each raster point.

In scanning tunneling microscopy (STM), a bias voltage is applied between a sharp metal tip and a conducting sample. Bringing the tip and sample surface to within a separation of only a few Ångstrøms permits overlap of electron wave functions and a quantum mechanical tunneling current I across the gap before mechanical point contact between tip and sample is reached. This is described by

$$I = \int_0^{eV} \rho_s(r, E)\rho_t(r, E - eV)T(r, E, eV)\, dE, \tag{25.10}$$

where $\rho_s(r, E)$ and $\rho_t(r, E)$ are the electronic DOS of the sample and the tip at location r and energy E, respectively, V is the applied bias voltage, and T is the tunneling transmission probability given by

$$T(E, eV) = \exp\left(-\frac{2z\sqrt{2m_e}}{\hbar}\sqrt{\frac{\varphi_s + \varphi_t}{2} + \frac{eV}{2} - E}\right), \tag{25.11}$$

where φ_s and φ_t are the work functions of the sample and tip, respectively, and z is the tip-sample separation. If one assumes $\varphi_s = \varphi_t$, then in the limit of low values of V,

$$\begin{aligned} T(E, eV) &= \exp\left(-\frac{2z\sqrt{2m_e}}{\hbar}\sqrt{\varphi}\right) \\ &= \exp\left(-1.02\sqrt{\varphi} \cdot z\right), \end{aligned} \tag{25.12}$$

where φ has units of eV and z is given in Å. In constant-current STM, the tip follows the complex contour line composed of the surface density of states and transmission probability. If φ is 4 eV, an increase in z of 1 Å decreases I by a little more than one order of magnitude. Thus, the electron tunneling current can be used to probe physical properties locally at the sample surface as well as to control the separation between the tip and sample surface.

Contrary to most other techniques, each atom is sensed locally and data are not generated by an average over ensembles of many atoms, as shown in Figure 25.1. The electrons involved in STM have energies of a few electron volts, often smaller than chemical bond energies, allowing nondestructive atomic resolution imaging. STM can be operated in air and liquids, as well as in vacuum, because no free electrons are involved in the STM experiment.

The common operating principle of scanning force microscopes (SFM) is to sense forces or force gradients between a sharp tip and a sample surface. Such forces can be measured by mounting the tip on a thin cantilever beam, which acts as a spring, and sensing the deflection of this beam due to interaction of the tip and sample. The force is proportional to the deflection and depends on the spring constant of the cantilever (Hooke's law). This detection scheme can be considered a quasi-static operational mode (sometimes referred to as "dc mode"). Another option is to drive the cantilever at a certain oscillation frequency (sometimes referred to as "dynamic mode" or "ac mode"). As interaction forces or force gradients are encountered, the oscillation will be damped. Changes of the resonant frequency or the vibration amplitude are sensitive measures of the forces acting on the tip. An advantage of SFM over STM is the ability to study nonconductive samples.

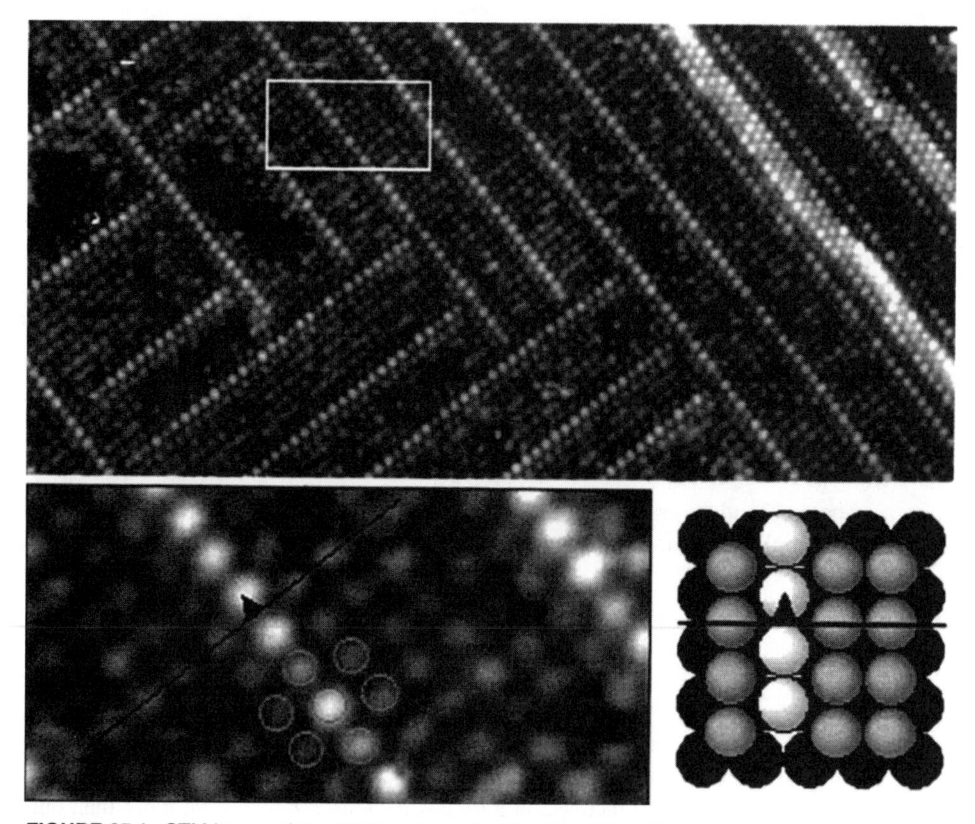

FIGURE 25.1. STM image of the (100) surface of a $Pt_{40}Ni_{60}(100)$ alloy shows the pseudohexagonal reconstruction of a Pt(100) surface [W. Hebenstreit, G. Ritz, M. Schmid, A. Biedermann, and P. Varga, Surf. Sci. 388 (1997) 150]. Used with permission of P. Varga.

25.3. STRUCTURE AND COMPOSITION OF SURFACES [37]–[46]

25.3.1. Thermodynamics of one-component surfaces

Equilibrium thermodynamics gives a macroscopic approach to describe the morphology and other properties of surfaces. In a homogeneous bulk phase, the force in all directions is equal, but this is not true when describing the interfacial region between two phases. The force $F_{\perp}$ perpendicular to an interfacial plane of area b^2 differs from the forces across any plane parallel to the interface and may be written as

$$F_{\perp} = pbt - \gamma b, \tag{25.13}$$

where γ defines the interfacial or surface tension, p is the pressure, and t is the interface thickness. Considering that the volume V^s of the interface region changes in relation to a change in the surface area, the total work can be written as

$$-pA\,dt - (pt - \gamma)\,dA = -p(A\,dt + t\,dA) + \gamma dA = -p\,dV^s + \gamma\,dA. \tag{25.14}$$

Then the Gibbs free energy for a surface or an interface can be expressed as

$$G^s = \sum_i \mu_i n_i^s = F^s + pV - \gamma A \tag{25.15}$$

and the variation of the surface free energy as

$$dG^s = S^s\, dT\, V^s\, dp - A\, d\gamma + \sum \mu_i\, dn_i^s. \tag{25.16}$$

The surface tension γ can be regarded as an excess free energy per unit area. It defines the reversible work of formation of a unit area of surface or interface at constant system parameters (V, T, μ, and number of components involved).

Surface properties of crystalline solids depend on the orientation of the surface. For example, breaking bonds in generating additional surface area depends on the crystal direction. Thus, γ depends on the particular surface orientation.

25.3.2. Surface morphology, defects, and dislocations

Surfaces of crystalline solids exhibit structural imperfections (defects) due to the emergence of bulk dislocations and defect structures at the surface and a variety of other point defects due to adatoms and vacancies. A schematic depiction of several defects is shown in Figure 25.2. In addition, cleavage of a semiconductor or mineral crystal or cutting of a metal crystal produces a surface that contains steps and kinks (intersecting steps) in addition to the flat terraces of high-density (low free energy) planes. The density of these structures can be controlled to some extent by preparing vicinal surfaces. Defect structures at a surface play an important role, because local variations affect the physical and chemical properties of surfaces.

25.3.3. Surface lattices and superstructures

All symmetry properties of surfaces are two-dimensional (2D). To describe surfaces and surface crystallography, one has to consider 2D point groups and 2D Bravais nets or lattices. Point-group operations are rotation axes perpendicular to the surface and mirror planes normal to the surface. Ten different point-group symmetries have to be considered. Their operation on a 2D lattice reveals five Bravais lattices. A different periodicity can be

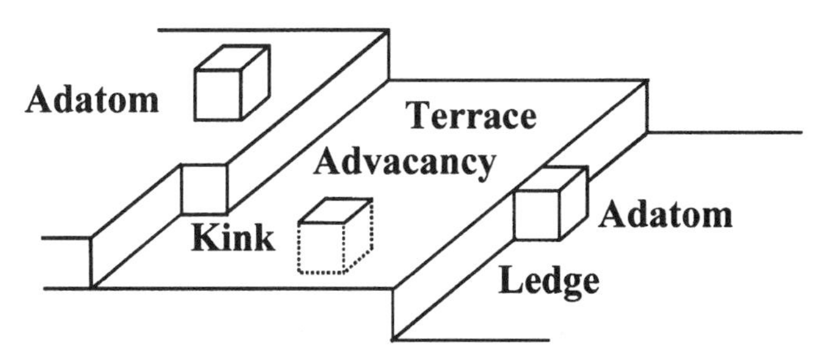

FIGURE 25.2. Several defects on surfaces.

present in the outermost layer(s) compared with that expected from the ideal termination of the bulk lattice of the crystalline solid because of surface reconstruction or formation of an adsorbed layer. A surface lattice is then superimposed on the substrate lattice. This surface net of the outermost layer can be described by unit-cell vectors as

$$\mathbf{b}_1 = m_{11}\mathbf{a}_1 + m_{12}\mathbf{a}_2,$$
$$\mathbf{b}_2 = m_{21}\mathbf{a}_1 + m_{22}\mathbf{a}_2, \tag{25.17}$$

where $\mathbf{a}_1$, $\mathbf{a}_2$, and $\mathbf{b}_1$, $\mathbf{b}_2$, are the two unit-mesh vectors obtained by the bulk projection and the topmost layer, respectively. A matrix $\mathbf{M}$ can be defined by

$$\begin{pmatrix} \mathbf{b}_1 \\ \mathbf{b}_2 \end{pmatrix} = \mathbf{M} \begin{pmatrix} \mathbf{a}_1 \\ \mathbf{a}_2 \end{pmatrix}. \tag{25.18}$$

The determinant of $\mathbf{M}$ is given by

$$\det \mathbf{M} = |\,\mathbf{b}_1 \times \mathbf{b}_2\,| \,/\, |\,\mathbf{a}_1 \times \mathbf{a}_2\,|\,. \tag{25.19}$$

It can be used as follows to distinguish three cases for a surface unit mesh that is different from the bulk mesh:

- Simple superlattice, $\det \mathbf{M}$ is an integer
- Coincidence lattice, $\det \mathbf{M}$ is a rational number
- Incommensurate lattice, $\det \mathbf{M}$ is an irrational number

In Wood's notation, the superlattice is labeled p or $c(n \times m)R\Phi°$. The designation of p or c refers to primitive or centered unit cells, and n and m are scale factors that relate the surface unit cell to that of the substrate.

Diffraction

The diffraction pattern is given by the reciprocal lattice. The reciprocal translation vector can be written as

$$G_{hk} = h\mathbf{a}_1^* + k\mathbf{a}_2^*, \tag{25.20}$$

where the integer numbers h and k are the Miller indices. Translational vectors $\mathbf{a}_1^*$ and $\mathbf{a}_2^*$ are related to the real-space lattice vectors $\mathbf{a}_1$ and $\mathbf{a}_2$ by

$$\mathbf{a}_1^* = \frac{2\pi(\mathbf{a}_2 \times \mathbf{n})}{\mathbf{a}_1 \cdot (\mathbf{a}_2 \times \mathbf{n})}, \quad \mathbf{a}_2^* = \frac{2\pi(\mathbf{n} \times \mathbf{a}_1)}{\mathbf{a}_1 \cdot (\mathbf{a}_2 \times \mathbf{n})}, \tag{25.21}$$

where $\mathbf{n}$ is the unit vector normal to the surface.

$$\mathbf{a}_i^* \cdot \mathbf{a}_j = 2\pi\,\delta_{ij}$$

and

$$|\,\mathbf{a}_i^*\,| = 2\pi[|a_i|\sin\theta(\mathbf{a}_i\mathbf{a}_j)]^{-1} \quad \text{with} \quad i, j = 1, 2. \tag{25.22}$$

25.3.4. Atomistic structure: relaxation and reconstruction

The equilibrium geometric structure of atoms forming the topmost layer of a solid is often considerably changed compared with that in the bulk. Two characteristic rearrangements occur at a surface, as shown in Figure 25.3. Relaxation denotes a contraction (or expansion) of the separation between the topmost layer(s) normal to the surface. Reconstruction refers to a rearrangement of atoms within the surface layer(s) so that the net of the topmost layer(s) has different dimensions from the bulk 2D net. Reconstruction can include large-scale atomic displacements, such as missing rows of atoms or exposure of several layers of the solid, such as occurs in the Si(111) − (7 × 7) reconstruction.

25.3.5. Surfaces of compounds and metal alloys

Surfaces in ambient environments are always covered with impurities from adsorbed layers or segregation (accumulation) of bulk impurities at the surface. In addition, even after careful cleaning under vacuum conditions, the surface composition of multicomponent compounds and metal alloy systems is often found different from that of the bulk. These changes are primarily driven by minimizing the surface free energy. Assuming ideal solution (simplest) behavior, one can write for a two-component system

$$\frac{X_1^s}{X_2^s} = \frac{X_1^b}{X_2^b} \exp\left[\frac{(\gamma_1 - \gamma_2)a}{RT}\right]. \tag{25.23}$$

The atom fractions of the two components are given by X_1^s and X_2^s at the surface and by X_1^b and X_2^b in the bulk. The surface tensions of the pure components 1 and 2 are given by γ_1 and γ_2, and a is the area/mole. The component with the lower surface tension has a higher concentration at the surface.

There is a good correlation for metals between the molar surface tension $\gamma_m = \gamma a$ and the heat of sublimation ΔH_{subl}, with the result that $\gamma_m \simeq 0.16\,\Delta H_{\text{subl}}$. Thus, the metal constituent with the lower heat of sublimation tends to have enhanced concentration at the surface.

Quantitative prediction of surface enrichment at alloy surfaces requires more rigorous considerations, including the role of the heat of mixing of the two components and the reduction of strain in the lattice due to size differences of the two components.

25.4. ELECTRONIC STRUCTURE AT SURFACES

The loss of three-dimensional periodicity at an interface, along with any reconstruction of the atomic layers, causes formation of electronic surface states that are localized in space only at the surface. These states can have energies within the band gap of the bulk-solid band structure. The presence of electrons or holes in these surface states causes a distortion of the bulk band structure of nonmetals in the surface region, known as band bending.

Processes that remove electrons from a solid, such as thermionic emission, field emission, formation of contact potentials, and chemical bonding to adsorbates, are influenced

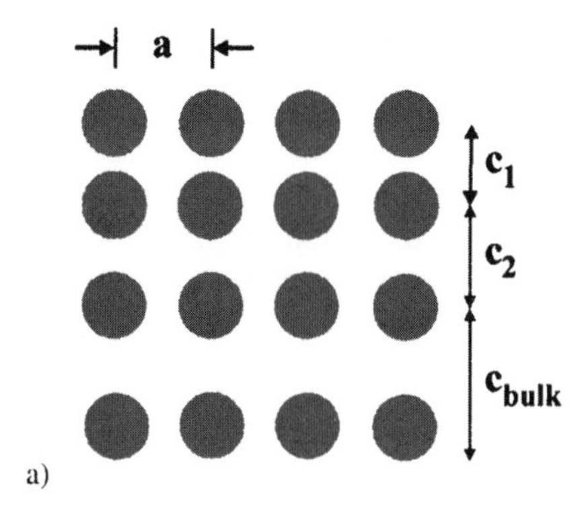

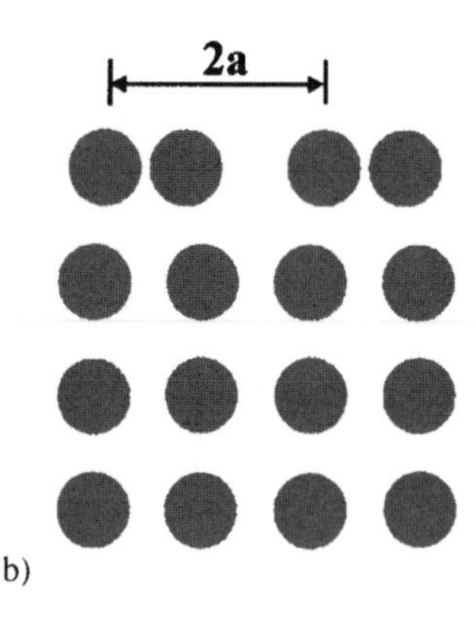

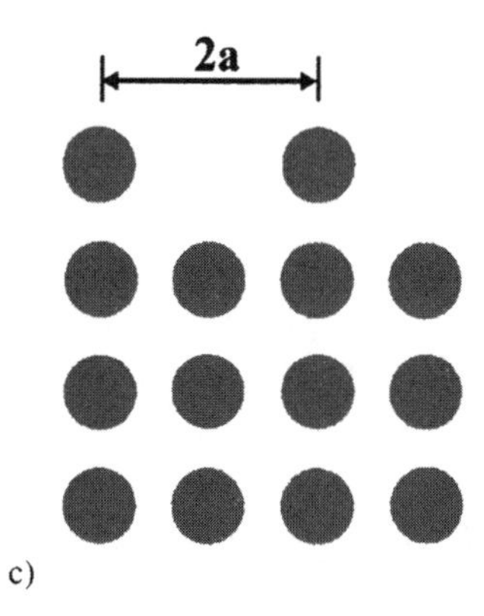

FIGURE 25.3. Characteristic rearrangements of surface atoms in a cubic lattice. (a) relaxation, (b) reconstruction, and (c) missing row reconstruction.

by the work function φ of a solid. This parameter can be defined as the difference in energy between an electron at E_F for metals, or the weakest bound electron for semiconductors and insulators, and an electron at rest in the vacuum just outside the surface. The photoelectric effect is a clear manifestation of the work required to remove such an electron from a solid, and the energy needed $E = h\nu \geq e\varphi$. Several contributions give rise to the value of φ. This includes a chemical part (a few eV) from the attractive potential for the electron by the ion cores in the bulk, and a surface contribution. The latter arises partly from the electrostatic image potential V

$$V = \frac{1}{2}\left(\frac{e^2}{2d}\right) \tag{25.24}$$

of an electron at distance d above a perfect conductor and partly (a few tenths of an eV) from the surface dipole layer formed by the asymmetrical charge distribution produced by the electron wave function tails extending past the surface ion cores and spilling over into the vacuum. In addition, adsorption of gases or the presence of impurity atoms at the surface can form a dipole layer that can strongly change (by several eV) the work function of a clean surface.

Measurements of the change in work function $\Delta\varphi$ that occurs upon formation of an adsorbed layer can be used to obtain qualitative information about charge transfer between the adsorbate and substrate. The change in surface potential is given by $\Delta\chi = 4\pi n\mu$, where n is the dipole concentration and μ is the dipole moment considering that the adsorbed layer acts as a parallel-plate capacitor. By convention, $-\Delta\varphi = \Delta\chi$, and converting μ to units of Debye (D) leads to

$$-\Delta\varphi = 3.77 \times 10^{-15} n\,\mu(\mathrm{D}) \quad [\mathrm{eV}]. \tag{25.25}$$

The dipole moment is positive for excess positive charge outward from the surface and this causes a decrease in the work function. This simple formula uses an "effective" dipole moment that relates to the perpendicular component of the surface dipole, treats the adsorbed layer as a single phase, and does not consider depolarization within the dipole layer that occurs at high concentrations.

25.5. THE GAS-SOLID INTERFACE

25.5.1. Solid-gas interactions

The interaction of a molecule and a surface is usually discussed in terms of a one-dimensional potential function that describes the electronic ground state of the adsorbed complex. The Lennard-Jones (LJ or 6,12) potential is a pairwise potential for two gaseous molecules that is reasonably realistic and commonly used. Integrating this function over all atoms in the solid yields the potential

$$E(z) = 4\pi\varepsilon N\sigma^3\left[\frac{1}{45}\left(\frac{\sigma}{z}\right)^9 - \left(\frac{\sigma}{z}\right)^3\right], \tag{25.26}$$

where ε is the well depth, N is the number of atoms per unit volume in the solid, σ is the hard sphere diameter, and z is the perpendicular distance between the surface and the center of the molecule. The interaction at long range is attractive, and this leads to exothermic adsorption, i.e., the localization of molecules at the interface. Because the solid is more polarizable than is another molecule, the interaction with the surface is usually stronger than that for condensation of the gas.

Two types of adsorption can be distinguished. Physisorption occurs via long-range, nonlocal weak electrostatic or van der Waals forces with energies typically lower than 30–40 kJ/mol. Chemisorption involves chemical bonding interactions (covalent, ionic, or dative) with energies usually greater than 40 kJ/mol. In contrast to physisorption, chemisorption is chemically specific and depends strongly on the surface and the adsorbate identity and structure. Adsorbates occupy specific locations (adsorption sites) at the surface.

Potential energy diagrams for chemisorption systems contain several energy curves: physisorption, molecular chemisorption, and dissociative chemisorption. Each curve is characterized by its values for ε and σ, and it correlates to different gas phase species best describing the electronic ground state of the adsorbed complex. Adiabatic adsorption paths are normally considered in which the incoming adsorbate follows the lowest energy curve at each position above the surface. Intersection of these curves form energetic barriers to surface processes.

The four elementary processes in the interaction of gaseous species with solid surfaces are adsorption, diffusion, reaction, and desorption.

25.5.2. Energy accommodation and adsorption

The accommodation coefficient α describes the change in energy of a gas by interaction with a surface:

$$\alpha = (T_i - T_f)/(T_i - T_s), \tag{25.27}$$

where T_i and T_f denote the temperatures of the molecule in the gas phase before and after the collision, respectively, and T_s is the temperature of the surface.

In an initial step for any surface reaction, gas molecules impinge on the surface with an incoming flux Z at pressure P given by kinetic theory to be

$$Z = \frac{P}{\sqrt{2\pi mkT}} = 3.52 \times 10^{22} P \text{ (torr) molec cm}^{-2} s^{-1}, \tag{25.28}$$

where m is the mass, k is the Boltzmann constant, and T is the gas phase temperature. Molecules impinging on the surface are either captured or reflected. The sticking coefficient S is defined as the ratio of the number of molecules that stick to the number of molecules that hit, or in terms of rates

$$S = \frac{R_a}{Z} = \frac{k_a P}{Z'P} = \frac{k_a}{Z'}, \tag{25.29}$$

where R_a is the adsorption rate, k_a is the adsorption rate constant, and Z' is Z/P. S is nominally pressure independent, but is a function of the temperature and the structure of the substrate. Sticking leads to accumulation of the adsorbate at the interface in an adsorbed layer or film. Concentration in this layer is defined as coverage $\theta = n/n_s$, where n is the number of adsorbates/cm^2 and n_s is the total number of adsorption sites/cm^2, i.e., the monolayer coverage. In cases in which the incoming gaseous species have hyperthermal energies (0.5–5 eV), collision-induced dissociation (CID) of the colliding molecule may occur as the incident molecule leaves the surface and the fragments may appear in the gas phase.

Adsorption isotherms

Adsorption isotherms are measurements of θ versus the gas pressure P at a constant surface temperature T_s. Adsorption isobars are measurements of θ versus T_s at constant P. Adsorption isosteres are measurements of P as a function of T_s for constant values of θ.

The coverage θ can be simply calculated as $\theta = Z\tau$, assuming $\alpha = 1$, where τ is the mean residence time of the adsorbate on a surface and is given by

$$\tau = \tau_o \exp\left(\frac{E_d}{RT_s}\right), \tag{25.30}$$

where E_d is the activation energy for desorption (equal to the adsorption energy for nonactivated adsorption) and $\tau_o \sim 10^{-13}$ sec. The resulting expression for θ leads to the simplest isotherm,

$$\theta = kP, \tag{25.31}$$

where $k = Z'\tau$ and $Z' = Z/P$. This expression is a limiting case at low pressures for weakly interacting, physisorbed systems, e.g., Ar/SiO_2. A more realistic isotherm is required for adsorption that terminates on completion of one monolayer, which is the case for all chemisorption. The Langmuir isotherm can be derived by making the additional assumptions that the surface is homogeneous, all adsorption sites are identical, the adsorption energy does not depend on θ, and gas species hitting an adsorbate (an occupied site) reflect instead of stick, i.e., $S = S_o(1 - \theta)$ This leads to

$$\theta = \frac{bP}{1 + bP} \quad \text{or} \quad 1/n = 1/(kn_0P) + 1/n_0, \tag{25.32}$$

where P is the partial pressure of the adsorbing gas and b is a constant. At low pressures (and high temperatures), one has $\theta = bP$, and at high pressures (and low temperatures), $\theta = 1$. Langmuirian behavior is observed in cases of physisorption, but the assumption that the adsorption energy does not depend on θ is much too restrictive for most chemisorption systems. Other isotherms work better for these systems. For example, the Tempkin isotherm is derived by assuming that the adsorption energy decreases linearly with θ.

The Brunauer-Emmett-Teller (BET) isotherm is used extensively to measure the surface areas of porous solids. Derivation of the BET isotherm leads to a two-parameter equation that treats the important case of multilayer adsorption:

$$\frac{P}{n(P_o - P)} = \frac{1}{n_m c} + \frac{c-1}{n_m c}\left(\frac{P}{P_o}\right), \tag{25.33}$$

where n is the number of moles of gas adsorbed, P_o is the saturation vapor pressure of the adsorbing gas, n_m is the number of moles of gas adsorbed in the monolayer, and c is a constant.

A practical (useful) form of this equation is

$$\theta = \frac{n}{n_m} = \frac{cx}{(1-x)[1+(c-1)x]}, \qquad x = \frac{P}{P_o}. \tag{25.34}$$

25.5.3. Desorption

An expression for the desorption rate r_d can be written in the form

$$r_d = k_d \theta^n, \tag{25.35}$$

where n is the kinetic order ($n = 0$ for sublimation, $n = 1$ for nonreactive desorption, and $n = 2$ for recombinative desorption), with the desorption rate constant k_d expressed by an Arrhenius expression containing the desorption activation energy E_d

$$k_d = \nu^n \exp^{-E_d/RT}. \tag{25.36}$$

The mean-stay time τ of an atom or molecule on the surface depends on E_d because $\tau = 1/k_d$. The value of E_d is often strongly coverage dependent.

Temperature-programmed desorption (TPD) or thermal desorption spectroscopy (TDS) is a widely used technique in surface science for measuring E_d (which can often be related to adsorption energies and barriers for surface reactions). In TPD, the substrate is heated linearly with time and the desorbing flux of previously adsorbed species is simultaneously monitored by means of a mass spectrometer. The identity of desorbing gases, desorption rates, and values for E_d, ν^n, and n can be measured. Simple expressions have been derived by Redhead to relate the temperature of the maximum desorption rate T_m to the value of E_d, e.g., for $n = 1$,

$$E_d/(RT_m^2) = (\nu/\beta)\exp(-E_d/RT_m), \tag{25.37}$$

where β is the heating rate.

25.5.4. Surface diffusion

Diffusion is a dominant mechanism for material transport at a surface and can be the rate-determining step in a variety of processes. Reconstructive phase transitions, annealing of surface roughness, segregation, and crystal growth are processes that all involve material (atom, molecule, etc.) transport in which diffusion plays an essential role.

The diffusion coefficient $D(B)$ for moving particles B on a surface A is defined by Fick's first law,

$$J(B) = -D(B)\partial[B]/\partial x, \tag{25.38}$$

where $J(B)$ is the diffusion flux (in one dimension) due to a coverage gradient (i.e., density gradient in two dimensions) $\partial[B]/\partial x$. A more general expression would substitute the chemical potential for the concentration gradient. Fick's second law,

$$\partial[B]/\partial t = \partial(D\partial[B]/\partial x)/\partial x, \tag{25.39}$$

expresses the time rate of change of surface concentration in terms of D.

Diffusion of surface atoms or molecules at moderate temperatures is an activated process, and experimental results are often expressed in the form

$$D(B) = D_0 \exp(-E_m/kT), \tag{25.40}$$

where D_0 is a pre-exponential factor. The energetic barrier to diffusion E_m is usually a small fraction (0.1–0.5) of the sublimation energy, and so diffusion occurs at much lower temperatures than does desorption or sublimation. Many surface species are mobile at room temperature or above.

Two different microscopic mechanisms for diffusion can be distinguished: a hop-site mechanism and a place-exchange mechanism. For hopping between sites separated by distance a, the mean intersite-jump frequency Γ can be expressed according to transition state theory by

$$\Gamma = \frac{kT}{h} \exp\left(\frac{\Delta S_m^*}{k}\right) \exp\left(\frac{-E_m}{kT}\right), \tag{25.41}$$

with ΔS_m^* as the activation entropy and E_m as the intersite migration activation energy. The residence time at a specific site τ is given by $1/\Gamma$. The root mean square displacement $\langle x^2 \rangle^{1/2}$ for a random walk depends on the time t according to

$$\langle x^2 \rangle^{1/2} = (\Gamma t)^{1/2} a/\alpha, \tag{25.42}$$

where $\alpha = 1$ or 2, depending on the dimensions considered. The Einstein relation

$$\langle x^2 \rangle = 2\,Dt \tag{25.43}$$

can be used to show that

$$D = a^2 \Gamma / 2\alpha. \tag{25.44}$$

25.5.5. Chemical reactions at solid surfaces

Two cases of chemisorption can be distinguished: nondissociative chemisorption, in which a molecule adsorbs on the surface and remains intact, and dissociative chemisorption. In a chemical reaction, bonds in the incoming molecule are broken and bonds are formed to the surface to produce surface-bound atoms.

Surface reactions can occur without any change of the surface,

$$\mathrm{CO}(g) + \mathrm{H_2O}(g) \rightarrow \mathrm{CO_2}(g) + \mathrm{H_2}(g), \tag{25.45}$$

or by direct involvement of the surface in the reaction and compound formation,

$$2\mathrm{Ni}(s) + \mathrm{CO}(g) \rightarrow \mathrm{NiC}(s) + \mathrm{NiO}(s). \tag{25.46}$$

Catalysis

Catalysis is acceleration of the rate of a chemical reaction without net consumption of the catalyst. The catalyst does not alter the overall reaction thermodynamics, but it lowers rate-limiting barriers and speeds reactions by stabilization of transition states and reaction intermediates. Homogeneous catalysis describes a liquid phase process in which the catalyst, reactant, and product are present in a single phase. Heterogeneous catalysis involves two or more phases, and it is desirable to have a high interfacial area. Metal catalysts are often supported on high surface-area oxides as small particles to achieve a high dispersion, i.e., surface/volume ratio. Catalysts are characterized by their activity (amount of product formed per unit time) and selectivity (amount of desired product formed compared with the amount of reactant consumed). Promoters are additives that enhance performance, and poisons are those that decrease performance.

In general, three major types of catalysts can be distinguished:

- *Electronic*—transition-metals (Fe, Pt, Ni) or semiconductors (NiO, ZnO) that have free or easily excited electrons for oxidation/reduction involving electron transfer
- *Acidic*—insulators (Al_2O_3, SiO_2) and ionic solids with acidic functional groups
- *Structural*—zeolites and shape-selective catalysts

Oxidation and corrosion

In the oxidation of metals, oxygen-containing molecules first have to chemisorb dissociatively on the surface. The adatoms then might migrate across the surface or diffuse into the crystal, leading to a rearrangement of surface atoms and successive nucleation and growth of oxide islands. Oxidation reactions also involve halogens, chalcogenides, and carbon-containing molecules.

Extensive oxidation processes, which consume large amounts of the metal surface, are called corrosion. Dissolution of the solid or flaking of the oxidation products often leads to a loss of material from the solid.

25.6. SOLID-LIQUID AND LIQUID-LIQUID INTERFACES

25.6.1. Electrochemical processes and the double layer

Electrochemical phenomena determine the behavior of solid surfaces in solution and control reactions at this interface. The distribution of an electrolyte A^+B^- is given by

$$n^- = n_0 e^{ze\Psi/kT} \quad \text{and} \quad n^+ = n_0 e^{-ze\Psi/kT}, \tag{25.47}$$

where Ψ is the electrical potential in solution and n_0 is the concentration at the surface and n^i is the concentration in solution for an ion of charge i. A positively charged surface repels positive ions from the surface and attracts negatively charged ions. This leads to an excess of negative ions at the surface. The net value of the charge density ρ on the surface is given by

$$\rho = -2n_0 ze \sinh(ze\Psi/kT). \tag{25.48}$$

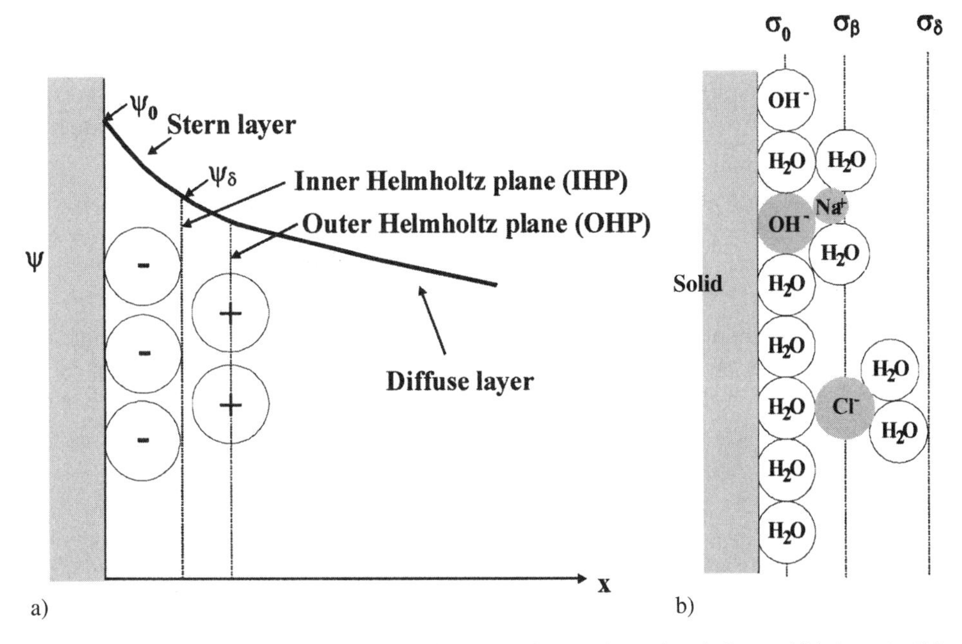

FIGURE 25.4. (a) The electric double layer (Stern layer) on surfaces in solution and (b) the potential-determining ions at an oxide interface.

Integration over all space results in the Poisson-Boltzmann equation

$$\nabla^2 \Psi = -4\pi \frac{\rho}{\varepsilon} = \frac{8\pi n_o z e}{\varepsilon} \sinh\left(\frac{ze\Psi}{kT}\right). \tag{25.49}$$

Several solutions of this equation to mathematically describe the resulting diffuse double layer have been studied. One approach, as shown in Figure 25.4, is to divide the solution into a bulk part and a region close to the surface (Stern layer). The charge density σ_S of occupied sites in this layer is then

$$\sigma_S = \frac{\varepsilon'}{4\delta}(\Psi_o - \Psi_\delta), \tag{25.50}$$

where Ψ_o is the potential at the surface, Ψ_δ is the potential with a thickness δ, and ε' is a local dielectric constant. If surface adsorption is only considered to be that of balancing the double layer charge, then the free energy G^S of a diffuse double layer can be written as

$$dG^S (= d\gamma) = \sigma_S \, d\Psi_o. \tag{25.51}$$

Electrode currents

The net current J through a metal electrode in the vicinity of ions in solution is given by the anodic minus cathodic current and can be described by the Butler-Volmer equation

$$J = J_0 \left[\exp\left(\tfrac{1}{2}\eta q/kT\right) - \exp\left(-\tfrac{1}{2}\eta q/kT\right) \right], \tag{25.52}$$

where J_0 is the current passing through the electrode in equilibrium, η is the voltage necessary to maintain zero net current, and $q = ze$. The Butler-Volmer equation for a semiconductor electrode can be written as

$$J = J_0[1 - \exp(-\eta q/kT)]. \tag{25.53}$$

25.6.2. Solid-liquid interactions

A liquid drop placed on a solid surface will assume a shape that depends only on the surface free energies of the materials involved and is defined by an angle of contact θ. This phenomenon can be described by consideration of the mechanical equilibrium between the interfacial tensions or forces which yields Young's equation,

$$\gamma_{LV} \cos\theta = \gamma_{SV} - \gamma_{SL}, \tag{25.54}$$

where γ_{SL}, γ_{SV}, and γ_{LV} are the solid-liquid, solid-vapor, and liquid-vapor surface tension, respectively. "Wetting" of the surface occurs if the contact angle is zero. Contact-angle hysteresis refers to the general observation of differences in the contact angle for a liquid moving on a surface due to the roughness of the surface.

Adsorption of Electrolytes

The adsorption of electrolytes at charged surfaces can be understood by rewriting Eq. (25.49) to give the Stern equation

$$\sigma_S/(\sigma_0 - \sigma_S) = N_s \exp[(ze\Psi_\delta + \Phi)/kT], \tag{25.55}$$

where $\sigma_S/(\sigma_0 - \sigma_S)$ is the ratio of occupied to unoccupied sites, N_s is the mole fraction of the solute, Ψ_δ is the potential at the border between the compact and diffuse layer, and Φ is an additional chemical adsorption potential. This can be rewritten in a form similar to the Langmuir isotherm:

$$\theta/(1 - \theta) = C_2 \exp[(ze\Psi_\delta + \Phi)/kT]. \tag{25.56}$$

Adsorption of nonelectrolytes

The adsorption process from solution can be written as

$$N_2 + N_1^s = N_2^s + N_1, \tag{25.57}$$

where N_1 and N_2 are the mole fractions of solvent and solute in solution, respectively, and N_1^s and N_2^s are the mole fractions of adsorbed solvent and adsorbed solute, respectively. An adsorption isotherm can be derived as

$$N_2^s = \theta = ba_2/(1 + ba_2), \tag{25.58}$$

where a_1 and a_2 are the solute and solvent activities (concentrations) in solution, respectively, $b = K/a_1$, K is the equilibrium constant for this process, and θ is the fraction of the surface occupied by the solute. Adsorption on a heterogeneous surface is often better described by a Freundlich isotherm.

25.6.3. Solid-liquid reactions

Corrosion

Many surfaces naturally form passive films (e.g., nanometer-thick oxide films) that greatly reduce the rate of corrosion. However, localized breakdown can result in accelerated dissolution of the underlying metal. An attack initiated on an open surface is called pitting corrosion, whereas an attack on an occluded site is called crevice corrosion. Characteristic electrochemical potentials exist; stable pits form and grow at potentials higher than the pitting potential and the repassivation potential. Furthermore, corrosion will only occur in the presence of an aggressive anionic species. The temperature and the composition of the surface are also critical factors. The breakdown of a passive film and successive pit formation can be described by three main mechanisms: passive film penetration, film breaking, and adsorption.

Electrochemical theory describes metal corrosion as a combination of anodic oxidation (metal dissolution) and cathodic reduction (oxygen ionization or hydrogen ion discharge). An electron is transferred from the highest occupied molecular orbital (HOMO) of an electron-donor species to the lowest unoccupied molecular orbital (LUMO) of another species acting as an electron-acceptor. The corrosion process also involves acid-base reactions, i.e., interaction between an occupied donor orbital (Lewis base) and an unoccupied acceptor orbital of a reaction partner (Lewis acid). The two general mechanisms below for metal dissolution emphasize the important role of adsorption of water molecules and reactive anions on the metal M surface:

Hydroxide mechanism:

$$\mathrm{M + H_2O \rightarrow MOH_{(a)} + H^+ + e^-}$$

$$\mathrm{MOH_{(a)} \rightarrow MOH^+ + e^-}$$

$$\mathrm{MOH^+ \rightarrow M^{2+} + OH^-}$$

Chloride mechanism:

$$\mathrm{M + Cl^- \rightarrow MCl_{(a)} + e^-}$$

$$\mathrm{MCl_{(a)} \rightarrow MCl^+ + e^-}$$

$$\mathrm{MCl^+ \rightarrow M^{2+} + Cl^-}$$

Dissolution

Crystallization and dissolution are two opposing processes and can be described by

$$z\mathrm{A}^{y+}(aq) + y\mathrm{B}^{z-}(aq) \longleftrightarrow \mathrm{A}_z\mathrm{B}_y \text{ (lattice)}. \tag{25.59}$$

Crystal growth or dissolution depends on the Gibbs free energy of the corresponding reaction $\Delta G = -RT \ln S$, where S is the saturation ratio given by

$$S = (I_p/K_a)^{1/\nu}, \tag{25.60}$$

in which I_p is the ionic activity product, K_a is the solubility product, and ν is the total number of ions. If $S < 1$, then an undersaturated solution exists and dissolution takes place.

High stress areas due to defects offer favorable sites for dissolution when a solid (e.g., mineral) surface is in contact with a solution. Stress energy is released on dissolution. Specific chemical factors need to be considered in addition to the theory of the electric double layer. Fundamental chemical interaction of solutes with natural surfaces occurs through the formation of coordinative bonds. Thus, the basic concept of the surface coordination model is that surface functional groups form on natural hydrous surfaces. The type of species on the surface determines its reactivity.

25.7. FILM FORMATION AND STRUCTURE

25.7.1. Nucleation and growth modes [47]–[51]

Considering heterogeneous nucleation on a planar substrate, the change in total free energy, ΔG, accompanying the formation of a solid aggregate of mean dimension r is given by

$$\Delta G = a_3 r^3 \Delta G_{\mathrm{v}} + a_1 r^2 \gamma_{\mathrm{vf}} + a_2 r^2 \gamma_{\mathrm{fs}} - a_2 r^2 \gamma_{\mathrm{sv}} \tag{25.61}$$

where ΔG_{v} is the chemical free energy change per unit volume of the condensate, and γ_{vf}, γ_{fs} and γ_{sv} are the interfacial tensions of the vapor-film, film-surface, and surface-vapor interfaces, respectively. This is similar to that for homogeneous nucleation, but now several γ values are involved and one has to assume a geometrical shape for the aggregate. For a spherical cap-shaped nucleus with contact angle θ, the geometric constants in Eq. (25.61) are $a_1 = 2\pi(1 - \cos\theta)$, $a_2 = \pi \sin^2\theta$, and $a_3 = \pi(2 - 3\cos\theta + \cos^3\theta)/3$. The critical nucleus size r^*, i.e., the value of r for which $d\Delta G/dr = 0$, is given by

$$r^* = -2(a_1\gamma_{\mathrm{vf}} + a_2\gamma_{\mathrm{fs}} - a_2\gamma_{\mathrm{sv}})/3a_3\Delta G_{\mathrm{v}}. \tag{25.62}$$

The nucleation rate N' (number of nuclei of a critical size r^* that form on the substrate per unit time) is given by $N' = N^* A^* \omega$, where N^* is the equilibrium concentration of stable, critical nuclei of surface area A^* and ω is the surface-diffusion related impingement rate of atoms onto these nuclei. A convenient thermodynamic expression for N' can be written as

$$N' = 2\pi r^* a_0 \sin\theta \left[P/(2\pi mkT)^{1/2}\right] \times n_{\mathrm{s}} \exp\left[(E_{\mathrm{d}} - E_{\mathrm{m}} - \Delta G^*)/kT\right] \tag{25.63}$$

where a_0 is an atomic dimension, n_{s} is the total nucleation site density, E_{d} and E_{m} are activation energies for adatom desorption and diffusion, respectively, and ΔG^* is ΔG evaluated at $r = r^*$, i.e., the activation barrier to forming these critical nuclei. Atomistic theory is straightforward when the critical nucleus is very small, and using a kinetic approach, and neglecting coalescence processes, the nucleation rate can be written for mobile monomers as

$$\frac{dN_1}{dt} = R - \frac{N_1}{\tau} - k_1 N_1^2 - N_1 \sum_{i=2}^{\infty} k_i N_i \tag{25.64}$$

where R' is the deposition rate, N_1 is the monomer density, τ is the surface-residence time, and k_i is the monomer-capture rate constant by stable islands N_i. These rate constants contain capture numbers σ (cross sections) and adatom diffusion coefficients D. In the initial stages of growth, the density of stable nuclei increases with time to a maximum level and then decreases. This decrease in cluster density is mainly caused by coalescence processes. In general, coalescence can be characterized by the following features:

- Decrease of the total projected area of nuclei on the substrate surface
- Increase in the height of surviving clusters
- Rounding of nuclei with well-defined crystallographic facets
- Orientation of the composite that is defined by the orientation of the larger island of the two that merge
- Frequently liquid-like appearance of islands merging and undergoing shape-changes in droplet motion fashion
- Process described as cluster-mobility coalescence prior to impact and union

Several mass-transport mechanisms have been proposed to account for these coalescence phenomena. These are ripening, sintering, and cluster migration. The growing (ripening) of larger islands at the expense of smaller islands is called Ostwald ripening. This process is driven by minimization of the surface free energy of the island structure. Diffusion of individual atoms proceeds from the smaller to the larger islands until the former disappears entirely. The free energy per atom μ_i is given by

$$\mu_i = dG_\gamma/dn_i = 8\pi r_i \gamma\, dr_i/(4\pi r_i^2\, dr_i/\Omega) = (2\Omega\gamma)/r_i, \tag{25.65}$$

where Ω is the atomic volume. From this, the Gibbs-Thomson equation can be derived:

$$kT \ln(a_i/a_\infty) = 2\Omega\gamma/r_i, \tag{25.66}$$

where a_i is the activity (effective concentration) of an isolated adatom on the substrate and a_∞ is the activity of an adatom in equilibrium with a planar island ($r_i = \infty$).

Sintering is a coalescence mechanism involving islands in contact. A neck between the particles forms immediately and then thickens successively as atoms are transported into the region. Again, the driving force is the reduction of the total surface energy of the system. Effectively, a concentration gradient between the neck and the islands develops, because atoms on the convex island surfaces have a greater activity than do atoms situated in the concave neck. Variations in an island's curvature also give rise to the local concentration differences. Mass transport likely involves self-diffusion through the bulk of the islands according to

$$x^5/r^2 = 28\, D_L \gamma \Omega t/kT, \tag{25.67}$$

or via the surface of the islands following

$$x^7/r^3 = 28\, D_S \gamma \Omega^{4/3} t/kT, \tag{25.68}$$

where x is the neck radius and D_L and D_S are the diffusion coefficients for the bulk lattice and surface, respectively.

The coalescence mechanisms discussed above fail to explain enhanced coalescence in the presence of an applied electric field in the substrate plane, e.g., in electromigration.

Other driving forces are operative. Another mechanism occurs as a result of collisions between individual clusters during random migration across the surface. Cluster migration is thermally activated with a barrier E_c related to the energy of self-diffusion. Smaller clusters move more rapidly than do larger ones. An effective diffusion coefficient for clusters of radius r can be written as

$$D(r) = \left(B(T)/r^y\right) \exp(-E_c/kT), \tag{25.69}$$

where $B(T)$ is a T-dependent constant and y ranges from 1 to 3.

Interfacial energies are primarily responsible for generating different morphologies in film growth. Young's equation, reproduced from Eq. (25.54),

$$\gamma_{sv} = \gamma_{fs} + \gamma_{vf} \cos\theta \tag{25.70}$$

allows distinguishing between the three basic modes that describe film growth: three-dimensional (3D) island (Volmer-Weber), layer-by-layer (Frank-van der Merwe), and layer-plus-island (Stranski-Krastanov). For three-dimensional island growth, $\theta > 0$ and Eq. (25.70) is

$$\gamma_{sv} < \gamma_{fs} + \gamma_{vf}. \tag{25.71}$$

For layer-by-layer growth, $\theta \leq 0$, and

$$\gamma_{sv} \geq \gamma_{fs} + \gamma_{vf}. \tag{25.72}$$

To explain layer-plus-island growth, additional energies must be considered. For example, strain relaxation in the film is known to cause island formation on top of a wetting layer.

The growth mode depends on the supersaturation

$$S = p/p_e = R/R_e, \tag{25.73}$$

where p is the pressure of the vapor beam, p_e is the equilibrium vapor pressure of the deposit at the substrate temperature, and R and R_e are the respective arrival rates (molecules $\text{cm}^{-2}\,\text{sec}^{-1}$). The mode can change with the degree of supersaturation and Eqns. (25.71) and (25.72) must be replaced by

$$\gamma_{sv} < \gamma_{fs} + \gamma_{vf} - \Delta\mu/\text{const.}, \tag{25.74}$$

$$\gamma_{sv} \geq \gamma_{fs} + \gamma_{vf} - \Delta\mu/\text{const.}, \tag{25.75}$$

where $\Delta\mu = kT \ln S$ is the variation of the free energy at the transition from the vapor to the solid phase.

It is important to recognize that film growth can often be carried out under conditions far from thermodynamic equilibrium, and growth may be governed mainly by the kinetics of surface processes occurring when impinging material reacts with the outermost layers of the substrate.

25.7.2. Structure and properties of thin films

Thin films can grow with amorphous, polycrystalline, or single-crystalline structures depending on the nature of the substrate, substrate temperature, deposition rate, and film material. Process parameters that raise the mobility of adatoms on the surface cause an increase in the grain size of the film. This may arise, for instance, from an increase in the substrate temperature. Decreasing the temperature correspondingly decreases the grain size. Defects or impurities on the surface can remarkably reduce the mobility of adatoms. Films grown on single-crystal substrates often exhibit a relationship between the orientation of the substrate and that of the film. This phenomenon is called epitaxy or epitaxial growth. Under favorable conditions, the entire film can be deposited as a single crystal. Columnar grain structure appears in polycrystalline thin films when the mobility of deposited atoms is limited. This leads to a network of high-density columns surrounded by low-density regions (grain boundaries). Amorphous or glassy materials have a structure that exhibits only short-range order. In general, amorphous films are produced at very high deposition rates and low substrate temperatures. Large doses of ion-implanted dopants can also amorphize surfaces and films.

25.7.3. Film growth from the gaseous phase [52]–[61]

Fabrication of microelectronics devices for production of integrated circuits entails a variety of processing steps for producing multilayer, patterned structures. Many of these steps involve the growth of thin films from the gaseous phase. In chemical vapor deposition (CVD), chemical reactions between gaseous reactants (precursors) takes place on the surface to grow a film on a substrate maintained at a particular temperature. Three fundamental processes are involved: transportation of the reactants to the surface, supplying heat to the reaction site, and removal of surplus precursors and reaction byproducts. CVD reactor designs can be classified into atmospheric pressure and low-pressure categories. Usually, the flow of the gaseous reactants is viscous, and the chemicals reach the substrate surface by diffusion through the gas phase. Thin films deposited by CVD include GaN (by using $GaCl_3 + N_2$), Si_3N_4 (by using $SiH_2Cl_2 + NH_3$), W (by using $WF_6 + Si$), and $C_{diamond}$ (by using $C_6H_6 + Ar$).

Metalorganic chemical vapor deposition (MOCVD)

Organometallic compounds are used as precursors in MOCVD, as shown in Figure 25.5, especially in the preparation of compound semiconductor heterostructures. MOCVD techniques have the advantage of being able to grow films over large areas and offer precise control of epitaxial deposition. The growth of thin films by MOCVD entails the transport of gas-phase organometallic precursors, like $Ga(CH_3)_3$, to a heated substrate on which the precursors are pyrolysed and the film is deposited. The underlying chemistry is complex, involving both gas-phase and surface reactions. In atmospheric or low-pressure MOCVD, precursors are partially dissociated in a stream of H_2 prior to hitting the surface and diffuse to the surface through a stagnant boundary layer above the substrate. Those atoms (elements) required for growth migrate into appropriate lattice sites following additional

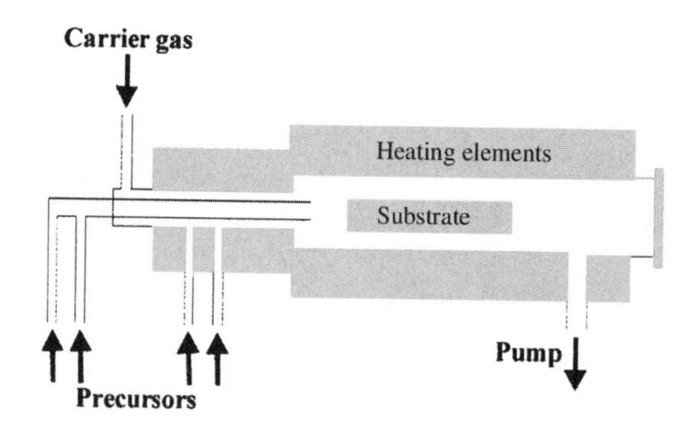

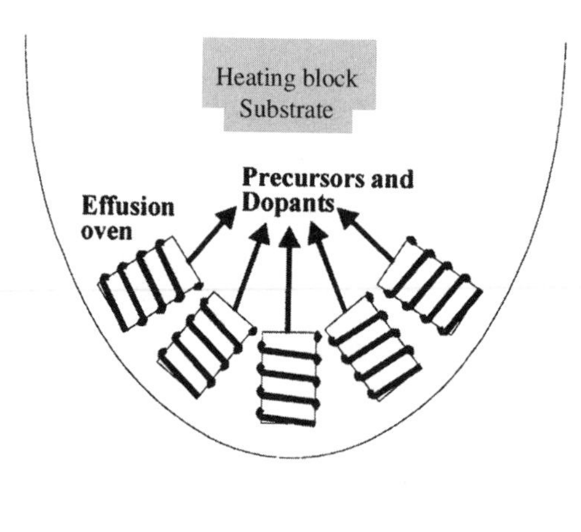

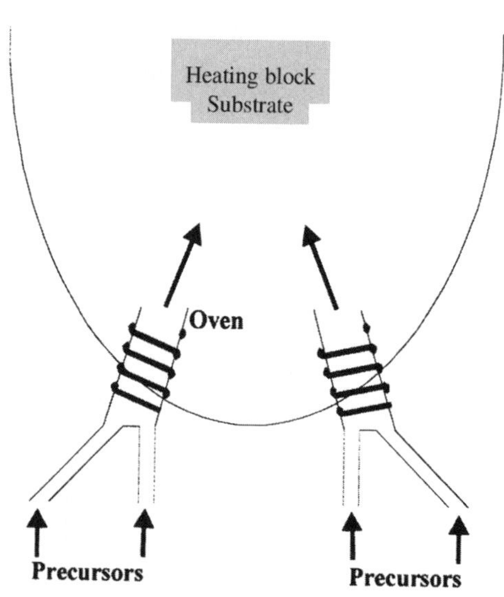

FIGURE 25.5. Instrumental setup in MOCVD or ALE (top), MBE, and CBE (bottom).

decomposition reactions at the surface. Diffusion of the precursors through the boundary layer limits the growth rate.

Molecular beam epitaxy (MBE)

MBE is a highly controlled, UHV, evaporation technique for growing thin epitaxial structures made of semiconductors, metals, or insulators. Reactants are generated inside the growth chamber by evaporation of solid elemental sources. At low ($<10^{-4}$ mbar) ambient pressures, mean free paths between molecular collisions are larger than the source-inlet to substrate distance and gas transport is "collision free." Atomic (e.g., Al, Ga, In) and molecular (As_4) beams are used. The "beam" nature of the process allows growth of semiconductor heterostructures in a layer-by-layer fashion. The growth mechanism in MBE is very different from that in CVD. In the growth of III-V semiconductors like GaAs, growth must usually occur in the presence of excess flux of the group-V component because of their lower surface lifetimes.

Chemical beam epitaxy (CBE)

CBE combines the beam nature of MBE and the use of all-vapor sources as in MOCVD. For III-V growth, a beam of Group III-alkyl molecules at a pressure $\approx 10^{-4}$ mbar impinge onto a heated substrate. There is no boundary layer in front of the substrate nor molecular collisions in the beam. After a molecule strikes the surface, it can either dissociate or desorb (intact or partially dissociated), depending on the substrate temperature and arrival rate of the precursor. The advantages of CBE over MBE, i.e., precise, stable and reproducible flows of the precursors, are important in growth of lattice-matched materials, such as $As_{y-1}P_y$. CBE also shows a stronger dependence of the growth rate on the substrate temperature than does MBE or MOCVD. This makes it easier to obtain desirable films. Furthermore, surface-catalyzed reactions that govern CBE growth make the process highly sensitive to the nature of the substrate surface. Consequently, if part of the surface is masked, epitaxial growth will only occur on exposed regions of the substrate.

Atomic layer epitaxy (ALE)

The basic idea in ALE is that sequential, self-limiting surface reactions cause stepwise growth of monolayers. Generally, it can be described as a molecular precursor approach that is most appropriate for binary compounds. Binary CVD reaction can easily be separated into two half-reactions. Using the growth of ZnS film as an example, the reaction

$$ZnCl_2(g) + H_2S(g) \rightarrow ZnS(s) + 2HCl(g) \tag{25.76}$$

can be divided into two half-reactions

$$ZnCl^*(s) + H_2S(g) \rightarrow ZnSH^*(s) + HCl(s) \tag{25.77}$$

and

$$SH^*(s) + ZnCl_2(g) \rightarrow SZnCl^*(s) + HCl(g), \tag{25.78}$$

where the * designates surface-bound species. During the first half-reaction, the surface is exposed to H_2S. This leads to the deposition of S and H-termination of the surface. During the subsequent half-reaction, Zn is deposited and the surface is reconverted into a Cl-terminated surface. Usually, the growth is carried out in a flow reactor at high pressure and a purging step is implemented after every precursor cycle. The continued repetition of these cycles leads to the deposition of ZnS to the desired film thickness. It is important to choose conditions such that only the desired reaction occurs between the solid surface and the reactive species, and that no physisorption in additional layers takes place. The amount of deposited material is dependent on the coverage of surface functional groups and the ALE process is "self-limiting."

25.7.4. Film growth from the liquid phase

Liquid-phase growth of films is increasingly important because of its advantage in depositing thin films under ambient conditions without the use of expensive growth chambers or reactors.

Self-assembled monolayers (SAMs) [62]–[66]

SAMs are molecular assemblies that form spontaneously by the immersion of an appropriate substrate into a solution of an active surfactant dissolved in an organic solvent. The most extensively studied SAMs are alkanethiols (RSH) deposited on Au, Ag, and Cu surfaces. Many other SAMs form, including organosilicon polymers on hydroxylated (OH) surfaces, alkyl sulfides (RSR′; R, R′ = hydrocarbon groups) and disulfides (RSSR′) on Au, alcohols (ROH) and amines (RNH_2) on Pt, and carboxylic acids (RCO_2H) on Ag and Al_2O_3 (alumina). In general, self-assembling surfactant molecules have three parts. The first is a head group that provides a connection to a specific site on the substrate through chemisorption via covalent bonds (e.g., Au-S bonding) or ionic bonds (e.g., $R\text{-}CO_2^-Ag^+$ bonding). These active functional groups cause the precursors to occupy as many binding sites on the substrate as possible until adsorbate-adsorbate repulsive interactions stop adsorption upon formation of a high-density monolayer. Formation of ordered, closely packed, crystalline assemblies involves pushing already-deposited molecules together and requires some adsorbate mobility. Small van der Waals interactions ($\leq$ 10 kcal/mol) and sometimes long-range electrostatic interactions of the second part of the molecule, often a long hydrocarbon chain, influence packing in the monolayer. The third component is the functional group at the outermost surface of the SAM, e.g., a methyl (CH_3) group for alkyl chains. These are not so involved in the formation of the SAM, but of course determine completely the subsequent chemistry that can occur at the surface of the SAM.

Successive ionic layer adsorption and reaction (SILAR) [62]–[64]

SILAR or Liquid-Phase Atomic Layer Epitaxy (LPALE) is used to deposit inorganic thin films. It is based on heterogeneous reactions between adsorbed ions and solvated ions at the solid-liquid interface and can be seen as a liquid-phase analog to ALE. The substrate is treated separately with each precursor solution so that the individual steps of adsorption and reaction can take place separately. Film thickness is controlled by the number of SILAR cycles. An advantage of this method over gas-phase methods is that the whole process can be performed under ambient conditions of temperature and pressure. The SILAR method has been mainly used to grow ZnS, CdS, ZnS:Mn, PbS, and PbSe films.

25.7.5. Film growth at the gas-liquid interface [70]–[71]

Organic molecules with a "head" that is polar and soluble in water (hydrophilic) and a "tail" that is nonpolar and not soluble in water (hydrophobic) can attain a high degree of orientation at gas-liquid interfaces. Surface films are formed with a thickness of one monolayer, i.e., a Langmuir layer or a Langmuir-Blodgett (LB) film. This minimizes the surface energy. A classic monolayer-forming molecule is n-octadecanoic acid ($C_{17}H_{35}COOH$ or stearic acid). The polar COOH group is hydrophilic, and the hydrocarbon chain is hydrophobic. The orienting forces are modulated by the hydrophobic chain length and the charge of the polar group. Most monolayers are applied to a subphase by dissolving them in a suitable organic solvent and bringing the solution in contact with water. The evaporation of the organic solvent compresses the organic molecules and forms a floating 2D film in which the molecules are uniformly oriented. Vertical movement of a solid substrate through the monolayer/air interface deposits the film on the substrate. The substrate is often hydrophilic, e.g., glass, and the first monolayer transfers like a carpet with the polar group binding to the substrate upon moving the substrate vertically upward. Raising the substrate through a monolayer composed of a single type of molecule and then subsequently lowering it through a monolayer composed of a different molecule can produce multilayers with a rich, but controlled, structure, as shown in Figure 25.6.

LB layers can also be deposited by compressing the LB film at the water-air interface and then placing the substrate on this film. LB films can be formed from aromatic or heterocyclic molecules, porphyrins, phthalocyanines, fullerenes, polymers, and biological compounds (e.g., phospholipids, pigments, peptides, and proteins).

25.8. MECHANICAL PROPERTIES OF SURFACES AND THIN FILMS [72]–[74]

Interfaces and thin films affect the mechanical properties of materials. Deformation phenomena like elasticity, flow and creep (continuous deformations at constant load), adhesion, e.g., peeling away an adherent film from a substrate, and friction and lubrication phenomena during sliding of two solid objects over each other with and without lubricating layers are of particular importance.

25.8.1. Friction

The resistance to motion that a solid object experiences while moving across a surface is called friction. The frictional force F_R always acts in a direction opposite to that of the relative velocity of the surfaces and is proportional to the normal force $F_R = \mu F_N$, with μ as the friction coefficient. Exceptions occur with very hard (diamond) or very soft (Teflon) materials. Then, a relation such as

$$F_R = cF_N^x, \tag{25.79}$$

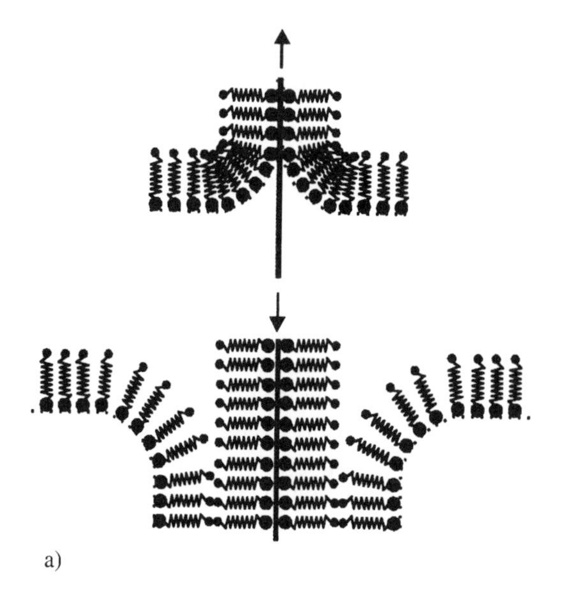

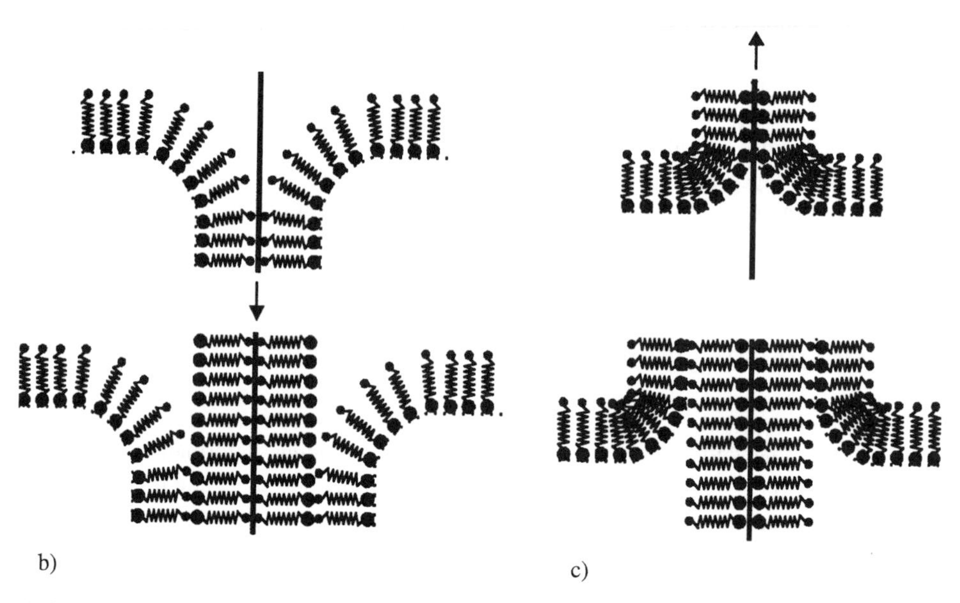

FIGURE 25.6. Deposition scheme of a Langmuir Blodgett film: (a) Y-type, (b) X-type, and (c) Z-type.

with c as a constant and $x = 2/3$ to 1 can be observed. In general, F_R is independent of the contact area (except on very soft materials) and the sliding velocity. If one surface is soft, deformation occurs until the pressure has fallen to a characteristic yield pressure p_m. The contact area A is then given by

$$A = F_N/p_m. \tag{25.80}$$

In a typical friction measurement, the force to move a block across a surface is measured. This force comprises two terms. One is a shear force F_s that is required to shear the junctions at points of contact, which is given by

$$F_s = As_m, \tag{25.81}$$

with s_m as the shear strength per unit area. The second is a plowing force F_p, which can be obtained by using a thin slider to minimize the shear contribution. F_p is proportional to the cross-sectional area of the slider A^*

$$F_p = kA^*. \tag{25.82}$$

25.8.2. Lubrication

Lubrication is used to reduce the friction between moving solids in contact. Two situations can be distinguished: fluid lubrication (no contact between the solids) and boundary lubrication (contact between the solids). Fluid lubrication can be divided into three forms: hydrostatic lubrication in which a lubricating film is forced between the solids by pumping, hydrodynamic lubrication in which the film arises from the sliding, and elastohydrodynamic lubrication, in which the pressure of the fluid deforms the surface of one solid. A soft, solid film can also be introduced between two solids, and this is referred to as solid film lubrication.

The friction behavior will be influenced directly since shear takes place and the friction coefficient of the film μ_l can be expressed as

$$\mu_l = s_l / p_m, \tag{25.83}$$

where s_l is the shear strength of the lubricant film.

25.8.3. Wear

Four main forms of wear can be described: adhesive wear when two smooth bodies slide over each other and fragments that are formed from one surface adhere to the second, abrasive wear in which a hard surface slides on a softer surface and ploughs grooves into the surface, corrosive wear in which sliding takes place in a corrosive environment and wears away a passive film or speeds the corrosion process, and surface fatigue wear when repeated loading and unloading cycles form surface cracks.

25.8.4. Adhesion

Adhesion describes the phenomena that a normal tensile force is required to separate two solid surfaces that have been brought into contact. The work necessary to separate two phases A and B with a common interface is known as the work of adhesion W_{AB} between the two phases and is given (most simply) by

$$W_{AB} = \gamma_A + \gamma_B - \gamma_{AB}. \tag{25.84}$$

Analogous expressions for the energy of adhesion can be given, although these equations ignore important electrostatic contributions that can occur through charging of the surfaces upon separation. These energies are large for strong chemical bonding at the interface, and they can exceed the cohesive energies of the pure phases; i.e., adhesive joints can fail by formation of an interface within the bulk of one of the two phases.

25.9. REFERENCES

[1] L. C. Feldman and J. W. Mayer, *Fundamentals of Surface and Thin Film Analysis* (North-Holland, New York, 1986).

[2] J. C. Rivière and S. Myhra, editors, *Handbook of Surface and Interface Analysis: Methods for Problem-Solving* (Marcel Dekker, New York, 1998).

[3] A. A. R. Elshabini-Riad and F. D. Barlow III, *Thin Film Technology Handbook* (McGraw-Hill, New York, 1998).

[4] J. C. Vickerman, editor, *Surface Analysis: The Principal Techniques* (Wiley, New York, 1997).

[5] R. Rosei, editor, *Chemical, Structural, and Electronic Analysis of Heterogeneous Surfaces on Nanometer Scale* (Kluwer Academic, Dordrecht, 1997).

[6] D. J. O'Connor, B. A. Sexton, and R. St. C. Smart, editors, *Surface Analysis Methods in Materials Science* (Springer, Berlin, 1992).

[7] D. Briggs and M. P. Seah, editors, *Practical Surface Analysis* (Wiley, Chichester, 1990).

[8] L. A. Casper and C. J. Powell, editors, *Industrial Applications of Surface Analysis* (American Chemical Society, Washington, D.C., 1982).

[9] A. W. Czanderna, editor, *Methods of Surface Analysis* (Elsevier Scientific, Amsterdam, 1975).

[10] K. L. Mittal and K.-W. Lee, editors, *Polymer Surfaces and Interfaces: Characterization, Modification, and Application* (VSP, Utrecht, The Netherlands, 1997).

[11] Z. L. Wang, *Reflection Electron Microscopy and Spectroscopy for Surface Analysis* (Cambridge University Press, Cambridge, England, 1996).

[12] T. L. Barr, *Modern ESCA: The Principles and Practice of X-Ray Photoelectron Spectroscopy* (CRC Press, Boca Raton, FL, 1994).

[13] G. C. Smith, *Surface Analysis by Electron Spectroscopy: Measurement and Interpretation* (Plenum Press, New York, 1994).

[14] D. Briggs, *Surface Analysis of Polymers by XPS and Static SIMS* (Cambridge University Press, Cambridge, England, 1998).

[15] A. W. Czanderna and David M. Hercules, *Ion Spectroscopies for Surface Analysis* (Plenum Press, New York, 1991).

[16] J. P. Maier, editor, *Ion and Cluster Ion Spectroscopy and Structure* (Elsevier, Amsterdam, 1989).

[17] L. Fiermans, J. Vennik, and W. Dekeyser, editors, *Electron and Ion Spectroscopy of Solids* (Plenum Press, New York, 1978).

[18] Z. L. Wang, *Elastic and Inelastic Scattering in Electron Diffraction and Imaging* (Plenum Press, New York, 1995).

[19] M. A. Van Hove, W. H. Weinberg, and C.-M. Chan, *Low-Energy Electron Diffraction: Experiment, Theory, and Surface Structure Determination* (Springer-Verlag, Berlin, 1986).

[20] L. J. Clarke, *Surface Crystallography: An Introduction to Low Energy Electron Diffraction* (Wiley, New York, 1985).

[21] J. M. Cowley, *Electron Diffraction Techniques* (Oxford University Press, Oxford, 1992).

[22] P. M. Marcus and F. Jona, editors, *Determination of Surface Structure by LEED* (Plenum Press, New York, 1984).

[23] R. Gomer, *Field Emission and Field Ionization* (American Institute of Physics, New York, 1993).

[24] M. D. Morris, editor, *Microscopic and Spectroscopic Imaging of the Chemical State* (M. Dekker, New York, 1993).

[25] T. T. Tsong, *Atom-Probe Field Ion Microscopy: Field Ion Emission and Surfaces and Interfaces at Atomic Resolution* (Cambridge University Press, Cambridge, England, 1990).

[26] L. Reimer, *Scanning Electron Microscopy: Physics of Image Formation and Microanalysis* (Springer, Berlin, 1998).

[27] D. Shindo and K. Hiraga, *High-Resolution Electron Microscopy for Materials Science* (Springer, Tokyo, 1998).

[28] I. M. Watt, *The Principles and Practice of Electron Microscopy* (Cambridge University Press, Cambridge, England, 1997).

[29] L. Reimer, *Transmission Electron Microscopy: Physics of Image Formation and Microanalysis* (Springer, Berlin, 1997).

[30] R.E. Lee, *Scanning Electron Microscopy and X-Ray Microanalysis* (Prentice-Hall, Englewood Cliffs, NJ, 1993).

[31] M. J. Dykstra, *Biological Electron Microscopy: Theory, Techniques, and Troubleshooting* (Plenum Press, New York, 1992).

[32] L. E. Murr, *Electron and Ion Microscopy and Microanalysis: Principles and Applications* (M. Dekker, New York, 1991).

[33] R. Wiesendanger, *Scanning Probe Microscopy and Spectroscopy: Methods and Applications* (Cambridge University Press, Cambridge, England, 1994).

[34] R. Wiesendanger, editor, *Scanning Probe Microscopy: Analytical Methods* (Springer-Verlag, Berlin, 1998).

[35] D. Sarid, *Scanning Force Microscopy: With Applications to Electric, Magnetic, and Atomic Forces* (Oxford University Press, New York, 1991).

[36] C. Bai, *Scanning Tunneling Microscopy and Its Application* (Scientific and Technical Publishers, Shanghai, 1995).

[37] H. Lüth, *Surfaces and Interfaces of Solid Materials* (Springer, Berlin, 1995).

[38] A. W. Adamson and A. P. Gast, *Physical Chemistry of Surfaces* (Wiley, New York, 1997), 6th edition.

[39] S. R. Morrison, *The Chemical Physics of Surfaces* (Plenum Press, New York, 1990), 2nd edition.

[40] R. J. MacDonald, E. C. Taglauer, and K. R. Wandelt, editors, *Surface Science: Principles and Current Applications* (Springer, Berlin, 1996).

[41] J. B. Hudson, *Surface Science: An Introduction* (Wiley, New York, 1998).

[42] K. S. Birdi, editor, *Handbook of Surface and Colloid Chemistry* (CRC Press, Boca Raton, FL, 1997).

[43] A. Kiejna and K. F. Wojciechowski, *Metal Surface Electron Physics* (Elsevier, Oxford, 1996).

[44] M. Prutton, *Introduction to Surface Physics* (Clarendon Press, Oxford; University Press, New York, 1994).

[45] M.-C. Desjonquères and D. Spanjaard, *Concepts in Surface Physics* (Springer-Verlag, Berlin, 1993).

[46] S. R. Morrison, *Electrochemistry at Semiconductor and Oxidized Metal Electrodes* (Plenum Press, New York, 1980).

[47] J. M. Venables, Surface Sci. **299/300**, 798 (1994).

[48] M. Ohring, *The Materials Science of Thin Films* (Academic Press, New York, 1992).

[49] D. T. J. Hurle, editor, *Handbook of Crystal Growth* (North-Holland, Amsterdam, 1993).

[50] J. L. Vossen and W. Kern, editors, *Thin Film Processes* (Academic Press, New York, 1978–1991).

[51] R. W. Vook, International Metals Rev. **27**, 209 (1982).

[52] A. A. R. Elshabini-Riad and F. D. Barlow III, *Thin Film Technology Handbook* (McGraw-Hill, New York, 1998).

[53] J. S. Ford, G. J. Davies, and W. T. Tsang, editors, *Chemical Beam Epitaxy and Related Techniques* (Wiley, New York, 1997).

[54] W. T. Tsang, editor, *Semiconductors and Semimetals* (Academic Press, Orlando, FL, 1985), Vol. 22A.

[55] M. Razeghi, *The MOCVD Challenge* (Adam Hilger, Bristol, 1989).

[56] E. Veuhoff, J. Crystal Growth **188**, 231 (1998).

[57] D. J. Eaglesham, J. Appl. Phys. **77**, 3597 (1995).

[58] O. Ambacher, J. Phys. D: Appl. Phys. **31**, 2653 (1998).

[59] M. Pessa, R. Makela, and T. Suntola, Appl. Phys. Lett. **38**, 131 (1981).

[60] S. M. George, A. W. Ott, and J. W. Klaus, J. Phys. Chem. **100**, 13121 (1996).

[61] C. L. Goodman and M. Pessa, J. Appl. Phys. **60**, R65 (1986).

[62] A. Ulman, *An Introduction to Ultrathin Organic Films* (Academic Press, New York, 1991).

[63] C. W. Frank, editor, *Organic Thin Films: Structure and Applications* (American Chemical Society, Washington, DC, 1998).

[64] A. Ulman and L. E. Fitzpatrick, editors, *Characterization of Organic Thin Films* (Butterworth-Heinemann, Boston, 1995).

[65] A. Ulman, editor, *Organic Thin Films and Surfaces: Directions for the Nineties* (Academic Press, San Diego, CA, 1995).

[66] R. H. Tredgold, *Order in Thin Organic Film* (Cambridge University Press, New York, 1994).

[67] Y. F. Nicolau, Appl Surf Sci, **22/23**, 1061 (1985).

[68] Y. F. Nicolau and J. C. Menard, J. Cryst. Growth. **92**, 128 (1988).

[69] S. Lindroos, T. Kanniainen, and M. Leskelä, Appl. Surf. Sci. **75**, 70 (1994).

[70] M. C. Petty, *Langmuir-Blodgett Films: An Introduction* (Cambridge University Press, New York, 1996).

[71] G. Roberts, editor, *Langmuir-Blodgett Films* (Plenum Press, New York, 1990).

[72] E. Rabinowicz, *Friction and Wear of Materials* (Wiley, New York, 1995).

[73] J. N. Israelachvili, *Intermolecular and Surface Forces* (Academic Press, London, 1991).

[74] B. Bhushan, editor, *Handbook of Micro/Nano Tribology* (CRC Press, Boca Raton, FL, 1999).

26

Thermodynamics and Thermophysics

J. P. Martin Trusler
Imperial College of Science, Technology and Medicine
William A. Wakeham
University of Southampton

Contents

List of Tables

26.1. INTRODUCTION

Thermodynamics and thermophysics constitute a very broad and largely mature area of scientific knowledge. In this chapter, we review classical thermodynamics, statistical thermodynamics, transport processes, and kinetic theory, with emphasis on the fundamentals. Given the constraint on space, we give only a few examples of specific applications to fluid systems and none at all for solids.

26.2. CLASSICAL THERMODYNAMICS

Thermodynamics is a formal discipline with its own precisely defined concepts and terminology in terms of which it is possible to formulate the laws of thermodynamics. We begin by defining these elements. A thermodynamic *system* is a part of real space, which we choose to consider; everything else constitutes the *surroundings* of that system. An *open* system is able to exchange matter with its surroundings; a *closed* system is not. An *adiabatic enclosure* is one whose contents are thermally isolated from the surroundings, and an *isolated system* is closed, adiabatic, and not subject to external work. The *state* of a system is defined by the values of its macroscopic measurable properties.[1] A *state variable* is a quantity, the change of which between any two equilibrium states depends only on the initial and final states and not on the path followed between them. A system is said to be in *equilibrium* when its state is no longer changing and it is subject to neither external work nor an exchange of heat with its surroundings. This may be contrasted with a *steady state* in which the macroscopic measurable properties of the system are constant, but external work or an exchange of heat with the surroundings is nevertheless taking place. An *extensive* property is one for which the value for the system as a whole is the sum of the values for all parts into which the system might be subdivided, whereas an *intensive* property is one that may have the same value for the system as a whole as for any subdivision. Any part of the system throughout which the intensive properties are constant is called a *phase*. A thermodynamic system consisting of a single phase is called *homogeneous*; all others are *heterogeneous*.

26.2.1. The laws of thermodynamics

The laws of thermodynamics were postulated on the basis of empirical evidence obtained primarily as a result of efforts to gain a quantitative understanding of the performance of heat engines. These laws govern the relation between various equilibrium states of a thermodynamic system and, in particular, describe quantities of heat and work associated with a change from one equilibrium state to another. The *zeroth*, *first*, and *second* laws of thermodynamics can be regarded as general principles, universally obeyed by macroscopic systems in states of thermodynamic equilibrium. The *third* law of thermodynamics has a slightly different status owing to the existence of a number of apparent exceptions, but

[1]The number of macroscopic variables that are independent is determined for a system in equilibrium by the phase rule; see Sec. 26.2.4.

nevertheless, it embodies a physical principle not established by the zeroth, first, or second laws.

The zeroth law of thermodynamics defines an intensive quantity called temperature that determines whether two subsystems are in thermal equilibrium. We know that when two initially isolated systems A and B are brought into thermal contact, in general, the properties of both will change until a condition of equilibrium is attained. The two subsystems A and B are then said to be in thermal equilibrium and there is a relation, which did not exist before, between the state ψ_{A} of A and the state ψ_{B} of B, which may be written as

$$F(\psi_{\mathrm{A}}, \psi_{\mathrm{B}}) = 0. \tag{26.1}$$

The zeroth law states that if an isolated system is divided into subsystems A, B, and C by thermally conducting walls and B has reached thermal equilibrium with both A and C, then A and C must also be in thermal equilibrium with each other. It follows that relations of the form of (26.1) exist between each of the three possible pairs formed from A, B, and C and hence that the function $F(\psi_{\mathrm{A}}, \psi_{\mathrm{B}})$ must be of the form

$$F(\psi_{\mathrm{A}}, \psi_{\mathrm{B}}) = \theta_{\mathrm{A}} - \theta_{\mathrm{B}}, \tag{26.2}$$

where θ is an intensive property that depends on the state of the subsystem. Furthermore, the zeroth law implies that the quantity θ assumes the same value for all subsystems that are in thermal equilibrium with each other.

The empirical temperature θ introduced here is a quantity that might be defined in a number of ways consistent with the zeroth law. A special case is that of the "perfect gas temperature" Θ obtained from measurements with a gas thermometer. In the gas thermometer, one measures for a fixed amount of substance n the pressure p as a function of volume V at a constant empirical temperature θ. It is found experimentally that the limiting value of (pV/n) as p tends to zero depends only on θ and is entirely independent of the nature of the thermometric gas. Accordingly, the limiting value,

$$\Theta = \lim_{p \to 0} (pV/n), \tag{26.3}$$

is a quantitative measure of the temperature that is always positive and unique. It turns out to be particularly convenient to define a thermodynamic temperature T as a separate physical quantity with $T \propto \Theta$. In the SI system, the unit of thermodynamic temperature is the kelvin, which is defined such that $T = 273.16$ K at the temperature of the triple point of water.[2]

The first law of thermodynamics is a generalization of the principle of the conservation of energy to encompass the flow of energy associated with the existence of a difference in temperature. It states that, for a closed system, any change ΔU in the internal energy U obeys the relation

$$\Delta U = W + Q, \tag{26.4}$$

where W is work done on the system by the surroundings and Q is the *heat* flowing into the system from the surroundings. Thus, although work is a flow of energy into the system

[2]That is the unique temperature at which pure solid, liquid, and gaseous water are in mutual thermodynamic equilibrium.

caused by the action of some generalized external driving "force", heat is a flow of energy into the system caused by a difference in temperature between the system and its surroundings. For an adiabatically enclosed system, $Q = 0$, whereas, for an isolated system, $Q = 0$ and $W = 0$. The internal energy is a state variable, but heat and work are not; thus, when equation Eq. (26.4) is written for an infinitesimal change, it is customary to denote the infinitesimal quantities of heat and work by $đQ$ and $đW$ to remind us that the quantities of heat and work necessary to advance from one state to another are path dependent:

$$dU = đW + đQ. \tag{26.5}$$

An infinitesimal change in the state of the system involving heat $đQ$ and work $đW$ is said to be *reversible* if the provision of heat $-đQ$ and work $-đW$ returns the system to exactly its initial state.

The second law is a powerful universal principle that may be stated in a number of equivalent ways. It is closely connected with the thermodynamic property called *entropy*, the concept of reversibility, the criterion for thermodynamic equilibrium, and the availability of energy to do work. The second law may be properly stated by means of a fundamental equation and a fundamental inequality. [1] The former asserts that the equation for an infinitesimal change in the state of single phase is

$$dU = T\,dS - pdV + \sum_i \mu_i dn_i, \tag{26.6}$$

where S is the entropy, n_i is the amount of the ith chemical substance, and μ_i is the chemical potential of that substance. The entropy, which is completely defined by Eq. (26.6), is thus (like V and the amounts of substance) an extensive state variable, whereas T, p and the chemical potentials are intensive properties. The term containing the chemical potential of the components is necessary to describe the change in energy accompanying a change in composition as a result of either an exchange of material across the boundaries of the phase or a change in the extent of one or more chemical reactions occurring within the phase. In a system containing two or more phases, Eq. (26.6) may be applied separately for each phase to obtain dU for the entire system. If forms of work other than the pressure-volume term are involved, the fundamental equation must be generalized by inclusion of the corresponding terms on the right-hand side.

The second law further states that the equilibrium state of an isolated system is one of maximum entropy. In other words, if anything is happening in an isolated system, the entropy is increasing, and if nothing is happening, the system is at equilibrium and the entropy is at a maximum. Accordingly, the fundamental inequality is

$$(\partial S/\partial t)_{U,V,N} \geq 0, \tag{26.7}$$

where t denotes time and N denotes the total material content of the system regardless of the number of phases present or the state of physical or chemical aggregation.

When interest lies with the thermodynamics of matter, formulae (26.6) and (26.7) constitute a particularly convenient way of stating the second law. However, an alternative statement of the second law, seemingly more relevant to heat engines, is often preferred: *It is impossible to construct a machine operating in a cyclical manner whose sole effect is to absorb heat from its surroundings and convert it into work.* It can be shown rigorously, albeit at some length, that this statement of the second law is entirely equivalent to the fundamental equation (26.6), generalized if necessary to include additional work terms, and the fundamental inequality (26.7). [2]

The third law of thermodynamics is usually taken to be a statement of the unattainability of the absolute zero of temperature: *It is impossible to reduce the thermodynamic temperature of a system to zero in any finite number of finite steps.* Stated in this form, the third law is never violated. A closely related theorem due to Nernst states that the entropy change ΔS associated with an isothermal process tends to zero as $T \to 0$, and this assertion has also sometimes been taken as a statement of the third law. Nernst's heat theorem in this original form is usually, but not always, obeyed, and in order to elevate it to the status of a general law, it is necessary to exclude certain exceptional processes involving metastable, rather than truly stable, states. The most restrictive modern statements limit the applicability of the theorem to processes that involve only perfectly ordered solids in the limit $T \to 0$. This form of Nernst's heat theorem is always obeyed, although it fails to encompass certain other processes for which ΔS also tends to zero as $T \to 0$. The theorem suggests but does not require that $S(T \to 0) = 0$ for all perfectly ordered solids, and this assignment is adopted in the tabulation of so-called "conventional" entropies. Apart from the difficulty of knowing in advance whether a process involves only truly stable states as $T \to 0$, the problem for the third law lies in the fact that although the unattainability of zero temperature is implied by Nernst's heat theorem, the two are not entirely equivalent. [3]

26.2.2. Consequences of the first and second laws

We consider first a phase of constant composition, for which Eq. (26.6) reduces to

$$dU = T\,dS - p\,dV. \tag{26.8}$$

When the phase is compressed adiabatically by means of a piston, it follows from the first law that the change in internal energy is $dU = -p_e\,dV$, where p_e is the *external* pressure acting through the piston on the system. In order that the compression be reversible, it is necessary that $p_e = p$ throughout the process, where p is the equilibrium pressure *of* the system; only then is the increase in stored energy available to do useful work equal to the energy expended in the initial compression. When this condition is satisfied, it follows from Eq. (26.8) that $dU = -p\,dV$ and $dS = 0$, which demonstrates that an adiabatic and reversible process is necessarily isentropic. Returning to the state just after the initial compression and maintaining the adiabatic condition, we have an isolated system. According to the fundamental inequality (26.7), any process occurring within the isolated system must lead to an increase in entropy, to a failure of the compression process to be reversible, and to a situation in which the stored energy available to do useful work is less than the energy expended in the initial compression.

If the change of state is reversible but not adiabatic, we still have $đW = -p\,dV$ and, consequently, $T\,dS = đQ$ for such a process. Thus, one can determine experimentally the entropy difference ΔS between any two states of a phase of constant composition by evaluating the integral

$$\Delta S = \int_{\text{rev}} đQ/T \tag{26.9}$$

along any reversible path connecting those states. The independence of the value of this integral to the course of the reversible path is an experimentally verifiable consequence of the second law, which defines S as a state variable.

The fundamental inequality (26.7) may be used to demonstrate rigorously that the conditions for thermal and hydrostatic equilibrium within an isolated system containing no internal partitions are those of uniform temperature and uniform pressure. Thus, the quantity T, which appears in the fundamental equation, is a temperature by the criterion of the zeroth law; one can further verify that T is proportional to the perfect-gas temperature defined by Eq. (26.3). The fundamental inequality also establishes that the condition for the diffusive equilibrium of component i between two phases A and B in thermal equilibrium is $\mu_i^{(A)} = \mu_i^{(B)}$, and this relation is fundamental to the solution of problems involving phase equilibrium. Finally, inequality (26.7) establishes the condition for equilibrium of the general chemical reaction $\sum_i \nu_i B_i = 0$ within an isolated single-phase system, where B_i denotes a chemical substance having stoichiometric number ν_i. That condition is $\sum_i \nu_i \mu_i = 0$ for each such reaction.

One of the most important applications of the first and second laws is the calculation of the efficiency of a heat engine. We consider a cyclical process carried out on a fixed mass of the working fluid and note first that $\Delta U = 0$ for a complete cycle. Thus, the work done by the system on its surroundings is given by

$$-W = |\, Q_1 \,| - |\, Q_2 \,|, \qquad (26.10)$$

where Q_1 is the sum of all positive quantities of heat and Q_2 is the sum of all negative quantities of heat, associated with individual steps in the cycle.[3] We may then define the efficiency η of the engine as the ratio of work done to heat supplied:

$$\eta = -W/|\, Q_1 \,| = (|\, Q_1 \,| - |\, Q_2 \,|)/|\, Q_1 \,|. \qquad (26.11)$$

The second law clearly states that this efficiency must be less than unity, and thus, every cyclical heat engine must have both a source from which it absorbs heat and a sink to which it rejects heat. Carnot's cycle is an idealized process, involving a working fluid of fixed composition, in which heat is absorbed only at temperature T_1 in a reversible isothermal process and rejected only at temperature T_2 in another reversible isothermal process. The other steps in the cycle are an isentropic expansion and an isentropic compression. For the reversible isothermal steps, we see from Eq. (26.6) that the corresponding entropy change is Q_i/T_i. Thus, since $\Delta S = 0$ for a complete cycle, $|\, Q_1 \,|/T_1 = |\, Q_2 \,|/T_2$, and hence,

$$\eta = (T_1 - T_2)/T_1. \qquad (26.12)$$

Carnot's principle, which may be proved starting from the second law, states that no cycle operating between heat reservoirs at T_1 and T_2 can exceed the efficiency given by Eq. (26.12), and all reversible heat engines operating between heat reservoirs at T_1 and T_2 have the same efficiency.

[3] We stick rigorously to the sign convention for heat so that heat supplied to the system is positive, whereas heat rejected by the system is negative.

26.2.3. Dependence of the thermodynamic properties on temperature, pressure, and composition

In the fundamental equation (26.6), S, V and the amounts of substance (n_i) are the independent variables. However, this is not always the most convenient choice and auxiliary state functions are therefore defined in terms of which the fundamental equation may be rewritten with the independent variables of choice. Thus, the enthalpy H of a phase is defined as $U + pV$, the Helmholtz energy A of a phase is defined as $U - TS$, and the Gibbs energy G of a phase is defined as $U + pV - TS$. These quantities lead to three alternative forms of the fundamental equation:

$$dH = T\,dS + V dp + \sum_i \mu_i\, dn_i, \tag{26.13a}$$

$$dA = -S\,dT - p\,dV + \sum_i \mu_i\, dn_i, \tag{26.13b}$$

$$dG = -S\,dT + V\,dp + \sum_i \mu_i\, dn_i; \tag{26.13c}$$

and to three alternative forms of the fundamental inequality:

$$(\partial H/\partial t)_{S,p,N} \leq 0, \tag{26.14a}$$

$$(\partial A/\partial t)_{T,V,N} \leq 0, \tag{26.14b}$$

$$(\partial G/\partial t)_{T,p,N} \leq 0. \tag{26.14c}$$

It should be remembered that Eqs. (26.13a–c) apply to each individual phase, whereas inequalities (26.14a–c) apply to the entire system. Enthalpy is the most useful state variable for problems involving steady-state flow or isobaric calorimetry, whereas the Helmholtz energy is most useful in connection with equations of state in which T and V are independent variables, and the Gibbs function is most useful in connection with equations of state in which T and p are independent variables. We see from Eq. (26.14b) that equilibrium in an isothermal-isochoric system corresponds to a state of minimum Helmholtz energy, whereas from Eq. (26.14c), it follows that equilibrium in an isothermal-isobaric system corresponds to a state of minimum Gibbs energy.

It is essential to establish how thermodynamic quantities are measured, for only measurable properties are physically significant. Changes in energy and enthalpy accompanying a changes in temperature, pressure, or composition may be determined by a combination of calorimetry and pVT measurements. [4] Equation (26.9) provides a general route to the experimental determination of a change in entropy accompanying a change in temperature or pressure with composition fixed. In order to explore changes in entropy consequent upon changes in composition at constant temperature and pressure, it is best to consider first the Gibbs function and the associated chemical potentials. It follows from the fundamental equation (26.13c) that the chemical potential μ_i of component i in a mixture is identical with the partial molar Gibbs energy for that component:

$$\mu_i = (\partial G/\partial n_i)_{T,p,n_{j \neq i}}. \tag{26.15}$$

Differences in the chemical potential of a component in a mixture arising from changes in composition upon either mixing or a chemical reaction under conditions of constant temperature and pressure may be measured experimentally. The techniques are various and in some cases convoluted; the specialist literature should be consulted for a detailed description of the experimental methods. [1] Since T and p are intensive variables, Eq. (26.13c) may be integrated to give

$$G = \sum_i \mu_i n_i . \tag{26.16}$$

Consequently, changes in Gibbs energy accompanying either mixing or a chemical reaction at constant temperature and pressure are also experimentally accessible. The corresponding changes in enthalpy may be measured directly in a calorimeter. Finally, since $G = H - TS$, it follows that changes in entropy upon either mixing or a chemical reaction may be determined from experimental measurements.

Numerous additional equations may be obtained interrelating state variables and various derived quantities. Some of these relations are summarized in Table 26.1. The Maxwell equations, which follow from the equality of the cross derivative $\{\partial(\partial G/\partial T)_p/\partial p\}_T$ and $\{\partial(\partial G/\partial p)_T/\partial T\}_p$, are particularly useful.

None of the four state variables U, H, A, and G has a naturally defined zero in classical thermodynamics, and it is only differences in these quantities that are of physical significance. Similarly, only differences in entropy and in chemical potential may be measured. The problem of constructing a database of thermodynamic properties may be simplified by defining for each component a standard state that depends only on the temperature in question. One then tabulates for pure substances quantities such as $\{H(T, p) - H^{\ominus}(T_{\mathrm{ref}})\}$ and $\{S(T, p) - S^{\ominus}(T_{\mathrm{ref}})\}$, where superscript $^{\ominus}$ denotes a property evaluated in the standard state and T_{ref} is a specified reference temperature, such as $T = 298.15$ K. For components that are gases at 298.15 K, the standard state is taken to be that of the hypothetical perfect gas at the temperature in question and at the standard pressure $p^{\ominus}$, whereas, for components that are liquids or solids at 298.15 K, the standard state is that of the actual pure substance at T and $p^{\ominus}$. The standard pressure is usually, but not always, taken to be $p^{\ominus} = 10^5$ Pa. Solutions are generally treated differently from other mixtures with separately defined standard states for solutes and for the solvent which we will not consider further here.

It is often useful to separate a thermodynamic property X, for example, into a perfect-gas part X^{pg} and a residual part X^{res}. [5] The perfect gas contribution is defined as the value of the property for the hypothetical perfect gas at the specified temperature, pressure, and composition. All observable thermodynamic properties of the perfect gas may be evaluated from exact thermodynamic relations plus knowledge of the isobaric perfect-gas heat capacity as a function of temperature and the equation of state of the perfect gas. Residual properties may be obtained from the identity

$$X^{\mathrm{res}} = \int_{\infty}^{V_{\mathrm{m}}} [(\partial X/\partial V_{\mathrm{m}})_T - (\partial X^{\mathrm{pg}}/\partial V_{\mathrm{m}})_T]\, dV_{\mathrm{m}} + \int_{RT/p}^{V_{\mathrm{m}}} (\partial X^{\mathrm{pg}}/\partial V_{\mathrm{m}})_T\, dV_{\mathrm{m}}, \tag{26.17}$$

where V_{m} is the molar volume. The partial derivatives that appear here can be obtained from the equations of state of the real fluid and of the perfect gas. A typical equation of state gives the pressure as a function of temperature, molar volume, and composition, and

TABLE 26.1. Thermodynamic properties of a phase.

Maxwell Relations	*State Functions*
$(\partial S/\partial p)_T = -(\partial V/\partial T)_p$	$H = U + pV$ Enthalpy
$(\partial S/\partial V)_T = (\partial p/\partial T)_V$	$A = U - TS$ Helmholtz free energy
	$G = U + pV - TS$ Gibbs free energy
Gibbs-Helmholtz Equation	*Gibbs-Duhem Equation for a Single Phase*
$H = G - T(\partial G/\partial T)_p$	$0 = S\,dT - V\,dp + \sum_i n_i\,d\mu_i$
Heat Capacities	*Expansivity and Compressibility*
$C_x = T(\partial S/\partial T)_x$ Heat capacity at constant x	$\alpha = V^{-1}(\partial V/\partial T)_p$ Isobaric expansivity
$C_p = (\partial H/\partial T)_p$ Isobaric heat capacity	$\kappa_T = -V^{-1}(\partial V/\partial p)_T$ Isothermal compressibility
$C_V = (\partial U/\partial T)_V$ Isochoric heat capacity	$\kappa_S = -V^{-1}(\partial V/\partial p)_s$ Isentropic compressibility
$C_p - C_V = T\alpha^2 V/\kappa_T$	$\kappa_T - \kappa_S = T\alpha^2 V/C_p$
Joule-Thomson Expansion	*Partial Molar Quantities*
$\mu_{\mathrm{JT}} = (\partial T/\partial p)_H = -\{V - T(\partial V/\partial T)_p\}/C_p$	$X_i = (\partial X/\partial n_i)_{T,p,n_{j\neq i}}$
$\phi_{\mathrm{JT}} = (\partial H/\partial p)_T = V - T(\partial V/\partial T)_p$	where X is an extensive variable
Perfect Gas Mixture	*Ideal Liquid Mixture*[a]
$pV = (\sum_i n_i)RT$	$\mu_i^{\mathrm{id}} = \mu_i^{\ominus} + RT\ln x_i$
$\mu_i^{\mathrm{pg}} = \mu_i^{\ominus} + RT\ln(x_i p/p^{\ominus})$	$\Delta_{\mathrm{mix}} H_m^{\mathrm{id}} = 0$
Fugacity[b]	*Activity Coefficients*
$f_i = (x_i p)\exp\{(\mu_i - \mu_i^{\mathrm{pg}})/RT\}$	$\gamma_i = \exp\{(\mu_i - \mu_i^{\mathrm{id}})/RT\}$
$RT\ln(f_i/x_i p) = \int_0^p (V_i - RT/p)\,dp$	$\gamma_i = f_i/(x_i f_i^{\ominus})$
$= \int_V^{\infty}\{(\partial p/\partial n_i)_{T,V,n_{j\neq i}} - RT/V\}\,dV - RT\ln Z$	

[a] $\Delta_{\mathrm{mix}} X$ denotes the change in X upon mixing at constant T and p.
[b] Z denotes the compression (or compressibility) factor: $Z = pV/(RT\sum_i n_i)$.

it may be necessary to use a Maxwell relation to obtain the partial derivative required in Eq. (26.17). The observable thermodynamic properties of a fluid are thus determined fully once the equation of state and the isobaric perfect-gas heat capacity are known.

The chemical potential of a component in a mixture is fundamental to the solution of problems involving phase equilibria or chemical equilibria. Since we have no absolute scale of chemical potential it is convenient to define two additional quantities in terms of departures from hypothetical states of ideality. The first is the fugacity f_i of a component which is related (see Table 26.1) to the difference $\mu_i - \mu_i^{\mathrm{pg}}$, where μ_i^{pg} is the chemical potential of component i in the hypothetical perfect gas at the same temperature, pressure and composition. The fugacity has the units of pressure and reduces to the partial pressure $x_i p$ in a perfect-gas mixture. The second quantity is an activity coefficient γ_i for component i in a liquid mixture. This quantity is related (see Table 26.1) to the difference $\mu_i - \mu_i^{\mathrm{id}}$ where μ_i^{id} is the chemical potential of a component in an ideal liquid mixture. Numerous empirical and semi-empirical methods have been devised for the prediction of the thermodynamic properties in multi-component mixtures (especially f_i and γ_i). [5]

26.2.4. Phase equilibria

The criteria for equilibrium between two phases (A) and (B) in the absence of chemical reactions are as follows:

$$\left.\begin{aligned} T^{(A)} &= T^{(B)} \\ p^{(A)} &= p^{(B)} \\ \mu_i^{(A)} &= \mu_i^{(B)} \quad i = 1, 2, \cdots N_{\mathrm{C}} \end{aligned}\right\} \tag{26.18}$$

Here, N_{C} is the number of components present, and the condition requiring equality of chemical potentials also implies equality of the corresponding fugacities. The number of independent variables F (again in the absence of chemical reactions) is given by the phase rule:

$$F = N_{\mathrm{C}} + 2 - N_{\mathrm{p}}, \tag{26.19}$$

where N_{P} is the number of phases present. Thus, for a single component we have two degrees of freedom in single-phase regions of the phase diagram, one degree of freedom when two phases are present, and no degrees of freedom at a triple point. For mixtures, phase equilibrium problems can be solved with the aid of an equation of state from which the fugacity of a component in a mixture of specified temperature, pressure, and composition may be calculated. The number of independent variables that must be specified is equal to the number of degrees of freedom F obtained from the phase rule; the remaining (dependent) variables are then adjusted to satisfy the criteria for phase equilibrium.

Often, we do not have an equation of state that can be applied to a liquid mixture with sufficient accuracy. In that case, vapor-liquid equilibrium problems are typically solved with the aid of an activity-coefficient model for the liquid phase and an equation of state for the vapor phase. Numerous empirical activity-coefficient models exist containing parameters that are either fitted to vapor-liquid equilibrium data for binary mixtures or calculated from structural information on the molecules involved. [5, 6] An important simplification arises when the liquid (L) and vapor (V) phases are both taken to be ideal. In that case, $f_i^{(\mathrm{V})} = y_i p$, $f_i^{\ominus} = p_i^{\mathrm{sat}}$, $\gamma_i = 1$, and, hence, $f_i^{(L)} = x_i p_i^{\mathrm{sat}}$, where x_i and y_i are the mole fractions of i in the liquid and vapor phases, and p_i^{sat} is the saturated vapor pressure of component i at the temperature in question. The equality of fugacities then reduces to Raoult's law:

$$y_i p = x_i p_i^{\mathrm{sat}}. \tag{26.20}$$

26.2.5. Chemical equilibria

We now consider again a general chemical reaction $\sum_i \nu_i B_i = 0$. The condition for equilibrium of this reaction is

$$\sum_i \nu_i \mu_i^{\text{eq}}(T, p) = 0 \tag{26.21}$$

where the superscript "eq" denotes the equilibrium composition. In terms of chemical potential differences, Eq. (26.21) is

$$\sum_i \nu_i \{\mu_i^{\text{eq}}(T, p) - \mu_i^{\ominus}(T)\} = -\sum_i \nu_i \mu_i^{\ominus}(T) = RT \ln K^{\ominus}(T), \tag{26.22}$$

and this relation defines the standard equilibrium constant $K^{\ominus}(T)$, which is closely related to the composition of the system at equilibrium. Making use of the relation between fugacity f_i and chemical potential μ_i (see Table 26.1), one can show that

$$K^{\ominus} = \prod_i (f_i/p^{\ominus})^{\nu_i}. \tag{26.23}$$

The standard equilibrium constant may be obtained from measurements of the equilibrium composition of the system combined with either a thermodynamic model or sufficient additional experimental data to relate the mole fractions to the corresponding fugacities at the temperature and pressure in question. In the case of a reaction among perfect gases, where $f_i = x_i p$, this calculation is particularly simple.

Equation (26.22) is also used to define a standard molar Gibbs energy change $\Delta G_{\text{m}}^{\ominus} = \sum_i \nu_i \mu_i^{\ominus}$, and there is also a corresponding standard molar enthalpy change $\Delta H_{\text{m}}^{\ominus} = \sum_i \nu_i H_i^{\ominus}$ and a standard molar entropy change $\Delta S_{\text{m}}^{\ominus} = \sum_i \nu_i S_i^{\ominus}$ for the reaction. In view of the relation between G, H, and S, we have

$$\Delta G_{\text{m}}^{\ominus} = \Delta H_{\text{m}}^{\ominus} - T\,\Delta S_{\text{m}}^{\ominus} = RT \ln K^{\ominus}. \tag{26.24}$$

A special case of interest is the reaction in which a particular compound is formed as a pure substance (in a specified phase) from the corresponding elements in their standard states. The associated standard molar Gibbs energy, enthalpy, and entropy for that reaction are known as the standard molar Gibbs energy, enthalpy, and entropy of formation, and these quantities are usually contained in a primary thermodynamic table or database.

26.3. STATISTICAL THERMODYNAMICS

The objective of statistical thermodynamics is to calculate certain long-time averages of microscopic properties that correspond to observable macroscopic properties of the system. It is recognized that the possible microscopic states of a system are stationary quantum states, each characterized by an energy E_j, where j denotes the full set of quantum numbers. Although the solutions of the underlying mechanical problems are not usually known exactly, we have good quantum-mechanical models for systems of noninteracting molecules, and for systems of interacting molecules, semiclassical results are available.

Accordingly, we proceed on the assumption that the allowed energies are known with useful accuracy. Here, we shall discuss the basis of the ensemble method and review some of the important results for both noninteracting and interacting molecules.

26.3.1. Postulates of statistical thermodynamics

The concept of an ensemble of systems is central to the development of the statistical thermodynamic formalism. An ensemble is a hypothetical assembly of a number $\mathcal{N}$ of systems, each of which is a replica on a macroscopic scale of the actual system of interest. Thus, every system in the ensemble shares common values of the chosen independent thermodynamic variables, and the type of ensemble is classified according to the choice of these variables. Here, we shall consider only the canonical ensemble, which is representative of a closed isothermal system with specified values of N_i, V, and $T(i = 1, 2, \ldots N_c)$, where N_i denotes the number of molecules of type i in a system of N_c-components. Other examples include the microcanonical ensemble, representative of an isolated systems with fixed E, V, and N_i, and the grand-canonical ensemble, which is representative of an open isothermal system with fixed μ_i, V, and T. [7]

The ensemble is a useful device only in connection with two basic postulates, the first of which states: *As the number of systems $\mathcal{N}$ in the ensemble becomes very large, the ensemble average of a mechanical variable approaches the long-time average of that variable in the real system.* By this postulate, we replace the long-time average of a mechanical variable in one system by an instantaneous average over all systems in the ensemble. Thus, for example, the thermodynamic internal energy is given by

$$U = \operatorname*{Lim}_{\mathcal{N}\to\infty} \sum_j \mathcal{P}_j E_j, \tag{26.25}$$

where $\mathcal{P}_j$ is the probability that a system selected from the ensemble will be in the quantum stated labeled j with energy E_j. The second postulate states that: *The only dynamical variable upon which the quantum states of the entire canonical depend is the total ensemble energy $\mathcal{E}$.* One can define a *distribution* by specifying for every possible quantum state of an individual system the number of systems in the ensemble that exist in that state when the total system energy is $\mathcal{E}$. Each distribution is associated with a multiplicity Ω equal to the number of ways in which it can be realized, and according to the second postulate, the probability of observing a particular distribution is simply proportional to that multiplicity. One can show that, as $\mathcal{N} \to \infty$, the distribution with the greatest value of Ω comes to dominate all others so that the distribution of system states approaches that distribution with the greatest multiplicity. [7] Furthermore, in that case, the probability that a system selected at random from the ensemble will be in the quantum state labeled j turns out to be

$$\mathcal{P}_j(E_j) = \exp(-\beta E_j)/Q. \tag{26.26}$$

Here, β is a positive intensive quantity and Q is the canonical ensemble partition function that is given by

$$Q = \sum_j \exp(-\beta E_j). \tag{26.27}$$

It turns out that although Q has little physical meaning, all of the thermodynamic properties of the system may be expressed in terms of this quantity. In particular, the thermodynamic internal energy for a system of constant composition obtained from Eqs. (26.25) and (26.26) may be written as

$$U = -(\partial \ln Q/\partial \beta)_{N,V}, \tag{26.28}$$

where N is the total number of molecules in the system. A change in internal energy, again subject to the condition of constant composition, can occur only if either the probability function $\mathcal{P}_j$ or the energy levels E_j change:

$$dU = \sum_j E_j\, d\mathcal{P}_j + \sum_j \mathcal{P}_j dE_j. \tag{26.29}$$

Since, in the absence of external fields, the energy levels depend only on the system volume, a comparison with Eq. (26.8) shows that

$$S = -k_B \sum_j \mathcal{P}_j \ln \mathcal{P}_j, \tag{26.30}$$

$$p = -\sum_j \mathcal{P}_j (\partial E_j/\partial V)_N, \tag{26.31}$$

and that $\beta = 1/k_B T$, where k_B is Boltzmann's constant. Like U, the entropy S may be expressed in terms of the canonical partition function Q and one can then show that the Helmholtz free energy A is given by the simple formula

$$A = -k_B T \ln Q. \tag{26.32}$$

Since, according to Eq. (26.13b), A is the characteristic state function corresponding to the independent variables N_i, T, and $V (i = 1, 2, \ldots N_c)$, all of the thermodynamic properties of the system may be obtained from appropriate manipulations of Eq. (26.32).

The entire canonical ensemble is an isolated "supersystem" containing $\mathcal{N}N$-molecules in volume $\mathcal{N}V$ with fixed total energy $\mathcal{E}$. One can show, starting from the relations above, that the entryopy of this supersystem is just $k_B \ln \Omega$, thereby proving Boltzmann's famous equation for the entropy of a system containing N molecules in volume V with fixed energy E:

$$S(N, V, E) = k_B \ln \Omega(N, V, E). \tag{26.33}$$

The second postulate tells us that all allowed states of an isolated system are equally probable and leads to Eq. (26.33) for the entropy of such a system. Thus, if an isolated system begins in an equilibrium state with multiplicity Ω and, by the removal of some constraint, additional quantum states become available such that the multiplicity increases to Ω', the system will advance spontaneously to a new equilibrium state and the accompanying change in entropy will be $\Delta S = k_B \ln(\Omega'/\Omega) > 0$. That this entropy change is greater than zero is the statistical analogue of the second law of thermodynamics. Statistical thermodynamics does not actually forbid a spontaneous process in the isolated system associated with a negative entropy change. However, a macroscopically significant decrease corresponds to Ω' being many orders of magnitude less than Ω and is therefore associated with a vanishingly small probability.

26.3.2. The perfect gas

For the case of noninteracting molecules, the canonical partition function factorizes into a product of N one-molecule functions because each molecule becomes an independent subsystem. Furthermore, if the molecules are indistinguishable, and the number of possible states of the ensemble is much larger than the number of molecules, the canonical partition function simplifies to [7]

$$Q = q^N / N!. \tag{26.34}$$

Here, q is a molecular partition function given by the sum over the quantum states j (each having energy ε_j) of one molecule:

$$q = \sum_j \exp(-\varepsilon_j / kT). \tag{26.35}$$

The problem is further simplified by writing the total energy of a molecule as a sum of translational, rotational, vibrational, and electronic terms, each of which is usually treated as independent of the others. This assumption of independent modes results in a factorization of the molecular partition function into a product of terms:

$$q = q_\mathrm{t} q_\mathrm{r} q_\mathrm{v} q_\mathrm{e} q_0 \exp(-\varepsilon_0 / k_B T). \tag{26.36}$$

Here, subscript t, r, v, and e denote translational, rotational, vibrational, and electronic factors, ε_0 is the energy, and g_0 is the degeneracy of the lowest available molecular energy level. Each of the factors in q is obtained from a summation of the kind indicated in Eq. (26.35) carried out over the available quantum states of the specified kind (t, r, v, or e) only, and quantum-mechanical models may be adopted to obtain analytic approximations for each term. Since the term $g_0 \exp(-\varepsilon_0 / k_B T)$ accounts for the ground state of the molecule, energy levels are measured relative to the lowest allowed energy of the specified kind. The translational energy levels may be determined easily for a particle of mass m contained within a (usually cubic) box of volume V. These energy levels are very close together when the box is of macroscopic dimensions, and the summation over translational states may be replaced by an integral with negligible uncertainty. Molecular rotations are often treated according to the rigid rotor model, and except for light molecules at low temperatures, an analytic approximation (obtained by integration) is satisfactory. In cases such as H_2, HD, and D_2, where the moment of inertia is small and the rotational energy levels are not very close together compared with $k_B T$, a numerical summation is employed. According to the harmonic oscillator model, molecular vibrations may be separated into s_v independent harmonic modes, where $s_\mathrm{v} = 3s - 5$ for a linear molecule or $3s - 6$ for a nonlinear molecule containing s atoms, and the summation over states required to obtain the corresponding factors of q_v may be evaluated exactly. Finally, in the case of electronic terms, it is usually necessary to consider only a few low-lying energy levels, and the summation over states is then evaluated explicitly. With a few exceptions, excited electronic states are much higher in energy than is the electronic ground state, and at "ordinary" temperatures, there is no significant electronic excitation so that $q_e = 1$.

The results for the rigid-rotor harmonic-oscillator model without electronic excitation are given in Table 26.2. The derived thermodynamic properties are often accurate for small molecules. Results of even higher accuracy may be obtained, again for small molecules, by relaxing the assumptions of harmonic vibrations and rigid rotations and resorting to a numerical treatment of the coupled rotation-vibration problem with values of the molecular constants obtained from spectroscopy. For larger molecules, the treatment of all internal deformations as harmonic or nearly harmonic oscillations is unsatisfactory, and it is necessary to recognize, for example, internal rotations of one part of the molecule relative to another and to treat these modes by a more appropriate quantum-mechanical model [8].

Boltzmann's distribution law in the form of Eq. (26.34) incorporates quantum mechanics, through the discrete nature of the energy levels, but it ignores quantum statistical restrictions on the states that may be adopted by an assembly of indistinguishable molecules. These restrictions are unimportant when the number of quantum states consistent with the total energy is much greater than is the number of molecules. However, at low enough temperatures, the number of available states is no longer much larger than is the number of molecules, and the restrictions imposed by quantum statistics must be considered. We then recognize that only accessible quantum states should be included in the calculation of Q and that, according to quantum statistics, the accessible states are either all symmetrical or

TABLE 26.2. Properties of the perfect gas in the rigid-rotator, harmonic-oscillator approximation.

Translational Terms	*Vibrational Terms*
$q_t = (2\pi m k_B T/h^2)^{3/2} V$	$q_v = \prod_i (1 - e^{-u_i}) \quad i = 1, 2, 3 \cdots s_v$
$A_t = -Nk_B T \ln[(2\pi m k_B T/h^2)^{3/2}(Ve/N)]$	$A_v = Nk_B T \sum_i \ln(1 - e^{-u_i})$
$U_t = \frac{3}{2} Nk_B T$	$U_v = Nk_B T \sum_i \{u_i/(e^{u_i} - 1)\}$
$S_t = Nk_B \ln[(2\pi m k_B T/h^2)^{3/2}(Ve^{5/2}/N)]$	$S_v = Nk_B \sum_i [\{u_i/(e^{u_i} - 1)\} - \ln(1 - e^{-u_i})]$
$C_{Vt} = \frac{3}{2} Nk_B$	$C_{V_v} = Nk_B \sum_i u_i^2 e^{u_i}/(e^{u_i} - 1)^2$
$P = Nk_B T/V$	$u_i = (h\upsilon_i/k_B T)$, υ_i = vibrational frequency
Rotational Terms (linear molecule)	*Rotational Terms* (non-linear molecule)
$q_r = (T/\sigma\theta_r)$	$q_r = (\pi^{1/2}/\sigma)(T^3/\theta_A\theta_B\theta_C)^{1/2}$
$A_r = Nk_B T \ln(T/\sigma\theta_r)$	$A_r = -Nk_B T \ln[(\pi^{1/2}/\sigma)(T^3/\theta_A\theta_B\theta_C)^{1/2}]$
$U_r = Nk_B T$	$U_r = \frac{3}{2} Nk_B T$
$S_r = Nk_B \ln(eT/\sigma\theta_r)$	$S_r = Nk_B \ln[(\pi^{1/2}/\sigma)(e^3 T^3/\theta_A\theta_B\theta_C)^{1/2}]$
$C_{Vr} = Nk_B$	$C_{Vr} = \frac{3}{2} Nk_B$
$\theta_r = (\hbar^2/2I)$, I = moment of inertia	$\theta_i = (\hbar^2/2I_i)$, I_i = principal moment of inertia

all antisymmetrical with respect to an exchange of the coordinates of any two identical particles. For particles of integer spin (e.g., ^{4}He), only the symmetrical states are accessible, whereas, for particles with half-integer spin (e.g., ^{3}He), only the antisymmetrical states are accessible. The former results in Bose-Einstein statistics and the latter in Fermi-Dirac statistics; both reduce to Boltzmann statistics at sufficiently high temperatures.

26.3.3. Real gases

For real fluids, it is necessary to incorporate the intermolecular potential energy, and this formally couples the states of the molecules so that we revert to having an N-body problem for which there is no general solution. However, at least for small rigid molecules, it remains true that the internal quantum states of the molecules are essentially independent so that the partition function of the system may be written as

$$Q = Q_{\mathrm{cm}} q_{\mathrm{int}}^N, \tag{26.37}$$

where Q_{cm} is a factor (dependent on N, V, and T) that accounts for the motion and interactions of the centers of mass and q_{int} is a factor (dependent only on T) that accounts for the internal states of one molecule. The latter may be evaluated by the methods of the last section, and being independent of N and V, it has no effect on the pressure. A semiclassical treatment of the center-of-mass term results in the following:

$$Q_{\mathrm{cm}} = \frac{(2\pi m k_B T/h^2)^{3N/2}}{N!} \int \underset{V}{\cdots} \int \exp(-\mathcal{U}/k_B T)\, d\mathbf{r}_1 \Lambda\, d\mathbf{r}_N, \tag{26.38}$$

where $\mathcal{U}$ is the total intermolecular potential energy of the system and $\mathbf{r}_i$ is the position vector of molecule i. [5]

The configuration integral that appears in Eq. (26.38) and the associated contribution to the thermodynamic properties of the fluid may be estimated for real gases and liquids either by molecular simulation [9] or from the principle or corresponding states. [10] It reduces to V^N in the limit $(N/V) \to 0$, but otherwise an exact evaluation is impossible. However, for gases, an expansion of the configuration in powers of (N/V) is found to be useful and results in the following expression for the pressure:

$$P = (nRT/V) \sum_{i=1}^{\infty} B_i (n/V)^{i-1}, \tag{26.39}$$

where $n = N/N_A$ and N_A is Avagadro's constant. Equation (26.39) is the virial equation of state and the coefficients B_i, called virial coefficients, are a function of temperature and (in a mixture) composition; each is related to interactions in a cluster of i-molecules. Clearly, B_1 is unity. The second virial coefficient of a gas composed of spherical molecules is given by

$$B_2 = -2\pi N_A \int_V f_{12} r_{12}^2\, dr_{12}, \tag{26.40}$$

where $f_{ij} = \exp(-u_{ij}/k_B T) - 1$ is the Mayer function and u_{ij} is the pair potential energy between molecules i and j when they are separated by distance r_{ij}. For nonspherical molecules, u_{ij} depends upon both orientation and separation, and the formulae for the virial coefficients involve integrals over the relative orientations of the molecules in the appropriate clusters. General expressions for B_i are known in the approximation that the potential energy of a cluster of molecules is the simple sum of the pair interaction energies. [11] When this is not the case, additional terms arise and these have been worked out for the third and fourth virial coefficients. Equation (26.37) is based on a semiclassical treatment and, for light molecules at low temperatures, quantum effects may be important. Corrections may be introduced to account adequately for these effects in most cases. However, for hydrogen and helium at low temperatures, a fully quantum-mechanical treatment is required.

The lower order virial coefficients may be evaluated fairly easily by numerical integration. Analytical results are known for some model potential-energy functions, the most well studied of which is the hard-sphere gas. The virial coefficients of the hard-sphere gas are given in Table 26.3. In this case, the virial series converges for all accessible densities and the calculated pressure agrees with the results of molecular simulation up to the density at which the latter shows solidification occurring. [12] For potential models with an attractive branch, a liquid phase is possible, but the virial series does not converge at liquid densities. In practice, the series is usually truncated after the second, third, or fourth term, and the results are then useful only at densities well below the critical.

TABLE 26.3. Virial coefficients of hard spheres of diameter *d* [21].

$B_2 = 2\pi N_A d^3/3 = b_0$	$B_5 = (0.110252 \pm 0.000001)b_0^4$	$B_8 = (0.00432 \pm 0.00010)b_0^7$
$B_3 = (5/8)b_0^2$	$B_6 = (0.038808 \pm 0.000055)b_0^5$	$B_9 \approx 0.00142 b_0^8$
$B_4 = 0.2869495 b_0^3$	$B_7 = (0.013071 \pm 0.000070)b_0^6$	$B_{10} \approx 0.00047 b_0^9$

26.4. TRANSPORT PROPERTIES

26.4.1. Fluxes and gradients

Phenomenologically, fluxes of mass, momentum, or energy in a material not in equilibrium arise from gradients of concentration, flow velocity, or temperature, respectively. Empirically, it is found that for sufficiently small perturbations from equilibrium, the relationship between the fluxes **J** and gradients (driving forces) **X** is a linear one. For an arbitrary mixture of pure components [13] and in the absence of external influences, such as magnetic or electric fields,

$$\mathbf{J} = \begin{vmatrix} L_{11} & L_{12} & \cdots & L_{1n} \\ L_{21} & L_{22} & \cdots & L_{2n} \\ \vdots & \vdots & & \\ L_{n1} & L_{n2} & \cdots & L_{nn} \end{vmatrix} \begin{pmatrix} \mathbf{X}_1 \\ \mathbf{X}_2 \\ \vdots \\ \mathbf{X}_n \end{pmatrix}, \tag{26.41}$$

where the L_{ij} are scalar quantities similar to a conductance, reciprocal resistance, or affinity. The diagonal coefficients L_{ii} represent the direct effects, and the off-diagonal coefficients $L_{ij} (i \neq j)$ represent the coupled effects.

Considering first the transport of heat and mass, suitable choices of driving forces corresponding to the fluxes of heat (q) and matter and species (j) are

$$\mathbf{X}_\mathrm{q} = \nabla(1/T) \tag{26.42}$$

and

$$\mathbf{X}_j = \nabla(-\mu_j/T), \tag{26.43}$$

where T is the thermodynamic temperature and μ_j is the chemical potential of species j in the mixture. With these choices, the expressions for the heat $\mathbf{J}_\mathrm{q}$ and material fluxes of species i $\mathbf{J}_i$ then become

$$\mathbf{J}_\mathrm{q} = L_\mathrm{qq}\nabla(1/T) + \sum_j L_{\mathrm{q}j}\nabla(-\mu_j/T) \tag{26.44}$$

and

$$\mathbf{J}_i = L_{i\mathrm{q}}\nabla(1/T) + \sum_j L_{ij}\nabla(-\mu_j/T) \tag{26.45}$$

The generalized transport coefficients $L_{\mathrm{q}j}$, L_{ij} are then such that they obey reciprocity relationships first derived by Onsager [13]

$$L_{ij} = L_{ji} \quad \text{and} \quad L_{i\mathrm{q}} = L_{\mathrm{q}i}. \tag{26.46}$$

Equations (26.44) and (26.45) demonstrate the nature of direct and indirect transport processes. For example, we see from Eq. (26.44) that a flux of heat can arise from a gradient in the chemical potential of the species in a mixture (a coupled effect known as the Dufour effect) as well as from a temperature gradient (a direct effect called thermal conduction). In the same way, Eq. (26.45) shows that transfer of material of one species can arise from a gradient in the chemical potential of that species or others (diffusion) or from a gradient of temperature (thermal diffusion or the Soret effect). The four transport processes thus identified may take place alone or in combination and are common to all nonequilibrium states of matter.

In practice, the generalized transport coefficients defined above are rarely employed directly because they are not associated with gradients in the characteristics of a material or its state which are most easily measured. In the particular case of a binary mixture consisting of species 1 and 2 with molar masses M_1 and M_2, it is particularly simple and helpful to relate the generalized transport coefficients to quantities more frequently measured. For example, the thermal conductivity of a system is defined by the phenomenological Fourier equation

$$\mathbf{J}_\mathrm{q} = -\lambda\nabla T, \tag{26.47}$$

from which it follows that

$$\lambda = -\frac{L_\mathrm{qq}}{T^2}. \tag{26.48}$$

For the diffusive transport processes, a similar identification with measurable quantities is possible, but there are a number of additional possibilities and complications. First, gradients of chemical potential are generally replaced by a more convenient measure of composition. However, there are a number of different composition measures in common use, including molar concentration (usually employed for the liquid state) and mole fraction (usually employed for the gaseous state). Furthermore, for diffusive processes, there are several different inertial reference frames that can be employed to describe the process of motion of material. Obviously, the most easily realized in practice is a reference frame that is fixed with respect to a cell in which the diffusive process takes place. However, this reference frame is not the most convenient theoretically because, for example, in some systems, mixing is accompanied by a change of volume, which means that a volume-fixed reference frame does not coincide with the cell-fixed frame. For these reasons, there is a wide variety of definitions of the transport coefficients that are associated with diffusive processes, and here we provide just some of them.

One transport process is associated with the flux of linear momentum within a fluid that is proportional to the gradient of a macroscopic flow velocity. Naturally, this transport process is confined to those materials that are in the fluid state. There is no coupling between the momentum transport process and other transport processes, such as diffusion and heat conduction. Furthermore, the process of momentum transport is the transport mechanism that most frequently reveals a nonlinear relationship within the phenomenological law.

26.4.2. Definitions of transport coefficients

The thermal conductivity of a system is defined as the coefficient of proportionality λ in the linear Fourier Law of Heat Conduction relating heat flux $\mathbf{J}_\mathrm{q}$ to the temperature gradient

$$\mathbf{J}_\mathrm{q} = -\lambda \nabla T. \tag{26.49}$$

The viscosity η of a fluid is defined as the coefficient of proportionality in the linear Law of Newton, which relates the deviatoric shear stress tensor in a fluid $\underline{\underline{\tau}}$, equivalent to a momentum flux, to the gradient of linear velocity

$$\underline{\underline{\tau}} = -\eta \nabla \mathbf{u}. \tag{26.50}$$

It should be noted that the tensorial nature of the momentum flux is consistent with its failure to couple to the vectorial fluxes of heat and material.

In the case of diffusive fluxes, it is convenient to consider only a binary mixture here for a particular reference frame and to refer the reader elsewhere for other definitions. [14, 15] The process of interdiffusion of two components in which the diffusive flow is measured relative to a suitable reference frame R is described in terms of a molar flux by the equation:

$$-(\mathbf{J}_i)_R = (D_{i\mathrm{m}})_R \nabla C_i, \tag{26.51}$$

where C_i is the molar concentration of species i. For a system in which there is no change of volume upon mixing, then,

$$D_{1\mathrm{m}} = D_{2\mathrm{m}} = D_{12} \tag{26.52}$$

and D_{12} is the interdiffusion coefficient (or mutual diffusion coefficient). Diffusion measurements are sometimes performed so that an isotopic labeled species is used to initiate a diffusion process in an otherwise uniform mixture. The diffusional process is then characterized by a so-called tracer diffusion coefficient for that species.

For coupled, indirect phenomena in liquids, it is common to work with a reference frame in which one of the components is stationary (solvent-fixed or Hittorf reference frame). In that case, the diffusion flux of component 1 (the solute in a two-component system) is written as

$$\mathbf{J}_1^{\mathrm{H}} = -D_1^{\mathrm{H}} \nabla C_1 - C_1 D_1^{\mathrm{T}} \nabla T. \tag{26.53}$$

Here, the appropriate diffusion coefficient D_1^{H} is given in terms of the generalized coefficients as

$$D_1^{\mathrm{H}} = \frac{L_{11}}{T} \left(\frac{\partial \mu_1}{\partial C_1} \right)_{T,P}, \tag{26.54}$$

and the thermal diffusion coefficient is

$$D_1^{\mathrm{T}} = \frac{L_{1\mathrm{q}}}{T^2 C_1}. \tag{26.55}$$

The Soret coefficient σ is more usually employed in liquids than is D_1^{T}, and it is defined as

$$\sigma = D_1^{\mathrm{T}} / D_1^{\mathrm{H}}. \tag{26.56}$$

The Dufour coefficient D^{D} can be related to the generalized coefficients defined earlier by the equation for a binary mixture:

$$D^{\mathrm{D}} = L_{\mathrm{q}i} \frac{x_1 M_1 + x_1 M_2}{\rho x_1 x_2 T^2}, \tag{26.57}$$

where x denotes mole fraction and ρ is the density of the mixture. For gases in which there is no volume change on mixing, it is more common to use the thermal diffusion factor in place of the Soret coefficient. It is defined in terms of the thermal diffusion coefficient and the interdiffusion coefficient by the relationship

$$\alpha_{\mathrm{T}} = \frac{D^{\mathrm{T}} T}{D_{12}}. \tag{26.58}$$

This relationship is given a direct interpretation because it follows that

$$\nabla x_1 = -x_1 x_2 \alpha_{\mathrm{T}} \nabla \ln T, \tag{26.59}$$

which shows that a gradient of temperature generates a mole fraction gradient in the gas.

26.4.3. Measurement of transport properties

All of the transport coefficients described above depend on the thermodynamic state of the system, characterized by the pressure, temperature, and composition. They have been the subject of an experimental investigation for over 100 years. However, the process of measurement is far from straightforward since it is usually necessary simultaneously to impose a gradient of a state variable, such as temperature, composition, or pressure, and to maintain the material sufficiently close to an equilibrium state so that the thermodynamic state of the measurement is well defined and that the transport process remains linear. Furthermore, in the case of measurements on fluids, it is necessary to ensure that no convective motion of the fluid, driven by buoyancy forces, interferes with the transport process.

Naturally, the minimum level of gradient that it is possible to impose is set by the limits of detection of the primary effect of the gradient. These limits have slowly been reduced following technological advances so that the precision of the measurements has steadily improved. However, the accuracy of measurements has been limited rather more by the adequacy or otherwise of the working equations used to describe the circumstances of a particular experiment. Only in the last 30 years have accurate measurements become routine as a result of the development of techniques of measurement for which accurate theoretical descriptions have been possible. Many of these make use, not of the steady-state equations (26.49), (26.50), and (26.51) that define the transport coefficients, but, rather, of time-dependent versions that express the temporal rate of change of a measurable quantity in terms of the transport property. The transient hot-wire technique for the measurement of the thermal conductivity of fluids and the oscillating-body viscometers represent examples of this class of method. A complete description of modern methods for the measurement of the transport properties of fluids can be found in the text prepared by the Wakeham *et al.* [16] Measurement techniques for solids have been treated by Maglic and Cezairliyan. [17]

TABLE 26.4. Transport properties of ethane along the saturation line [18].

$\frac{T}{K}$	$\frac{P}{\text{MPa}}$	$\frac{\rho_g}{\text{mol L}^{-1}}$	$\frac{\eta_g}{\mu\text{Pa s}}$	$\frac{\lambda_g}{\text{mW m}^{-1}\text{K}^{-1}}$	$\frac{\rho_l}{\text{mol}^{-1}\text{L}^{-1}}$	$\frac{\eta_l}{\mu\text{Pa s}}$	$\frac{\lambda_l}{\text{mW m}^{-1}\text{K}^{-1}}$
100	1.1(−5)[a]	1.3(−5)			21.323	876.02	239.76
120	3.55(−4)	3.56(−4)			20.597	486.51	229.69
140	3.83(−3)	3.30(−3)			19.852	322.16	212.15
160	0.02145	0.01631			19.081	232.45	191.55
180	0.07872	0.05412			18.276	175.74	170.05
200	0.2174	0.1387	6.020		17.423	136.73	150.71
220	0.4293	0.3001	6.684		16.498	108.24	132.00
230	0.7005	0.4221	7.055	14.49	15.999	96.62	123.21
240	0.9671	0.5810	7.472	16.00	15.467	86.27	114.78
250	1.3012	0.7867	7.958	17.70	14.892	76.91	106.73
260	1.7120	1.0534	8.544	19.70	14.261	68.29	99.07
270	2.2097	1.4038	9.278	22.19	13.551	60.17	91.82
280	2.8058	1.8787	10.25	25.59	12.723	52.28	85.14
290	3.5144	2.5703	11.64	31.07	11.683	44.19	79.39
295	3.9169	3.0723	12.67	35.78	11.010	39.77	76.97
300	4.3560	3.8129	14.23	44.68	10.101	34.63	75.12
302	4.5432	4.2616	15.23	52.00	9.5867	32.09	75.30
304	4.7377	4.9547	16.91	71.97	8.8412	28.85	79.81

[a] $(-n)$ means 10^{-n}.

The transport properties vary greatly with both the state and the composition of the material. Extensive tabulations of the transport properties of materials are available often in a computerized form. An extensive summary of those available in several countries and those that have international approval is contained in Chapter 17 of Ref. [18]. As an example, Table 26.4 shows the behavior of the viscosity and thermal conductivity of ethane along the saturation line in both vapor and liquid phases. In the gas phase, the properties are relatively weak functions of temperature and density (pressure). This remains true of the thermal conductivity in the liquid phase, but the viscosity of the liquid changes by a factor of 30 in just 200 K along the saturation line.

The values of diffusion coefficient, thermal diffusion factors, and Soret coefficients vary so widely that no brief tabulation of them is useful. The reader is therefore referred to Chapters 9 and 10 of Ref. [16] for a suitable bibliography of data compilations.

26.5. KINETIC THEORY

26.5.1. The Boltzmann equation

The theoretical description of the transport properties is the task of a branch of nonequilibrium statistical mechanics that is known as the kinetic theory. The nature of this theory varies dramatically depending on the density and the state of aggregation of the material being investigated. At one extreme, the kinetic theory of "dilute" gases assumes that the mean-free path of the molecules that comprise the gas is long compared with molecular diameter and yet small compared with the dimensions of the container. This means that the density of the gas must be low enough to ensure that only binary molecular encounters exist but high enough that these collisions are very much more frequent than are collisions with walls. At the other extreme, in a solid, molecules are generally very closely packed in a regular array and molecular motion comprises occasional jumps from one "cell" to another. In each cell, molecules are closely interacting near the minimum of their potential energy of interaction at all times, and the effects of walls are irrelevant. At intermediate densities, for example, in the liquid state, the behavior is also intermediate between these two extremes since the mean free path of molecules is comparable with their dimensions and yet the molecules are not constrained to oscillate about equilibrium positions. In addition, near the gas-liquid critical point, thermodynamic fluctuations contribute to the transport properties in a singular fashion so that both the viscosity and thermal conductivity of a pure fluid become infinite at this unique state. [18]

There is no single theory that is able to describe the transport properties in such a range of thermodynamic states. There is therefore a range of theoretical descriptions that are of variable rigor as one passes from the dilute state of a monatomic gas to a liquid and then to a crystalline solid. The foundations of the theory of the two extremes are sound, but for the liquid state, there is essentially no rigorous theory as yet, although for the critical region, the singularities are rather well described by the means adopted from the treatment of other critical behavior. A convenient summary of the state of the theories of transport properties for dense gases, liquids, and the critical point can be found in Chapters 5 to 13 of Ref. [18]. This same text describes methods of estimating the transport properties of fluids in the absence of direct measurements. Here we confine ourselves to a treatment of the dilute gas state in which the theory is essentially exact.

In the semiclassical description of transport in a dilute gas state, the kinetic theory description makes use of a distribution function $f_i(\mathbf{c}, E_i, \mathbf{r}, t)$ that is such that $f\, d\mathbf{c}\, d\mathbf{r}$ is the number of molecules with internal energy E_i (characterized by a set of quantum numbers $\{i\}$) with a translational velocity between $\mathbf{c}$ and $\mathbf{c} + d\mathbf{c}$ and position between $\mathbf{r}$ and $\mathbf{r} + d\mathbf{r}$. This distribution function must obey an integrodifferential equation first derived by Boltzmann for monatomic gases ($E_i = 0$ for all i), which expresses how the distribution function evolves by virtue of molecular motions and binary collisions toward the equilibrium distribution function for molecular energies described earlier.

The nonequilibrium distribution function is written as

$$f_i = f_i^0(1 + \varphi_i), \tag{26.60}$$

where the local equilibrium distribution function $f_i^{(0)}$ is

$$f_i^{(0)} = \frac{N}{Q}\left(\frac{m}{2\pi k_B T}\right)^{3/2} \exp(-\mathcal{C}^2 - \varepsilon_i), \tag{26.61}$$

where $Q = \sum_i \exp(-\varepsilon_i)$ is the internal state partition function, $\varepsilon_i = E_i/k_B T$ is the reduced energy of the state i, and $\mathcal{C}$ is the reduced peculiar velocity

$$\mathcal{C} = \left(\frac{m}{2k_B T}\right)^{1/2} (\mathbf{c} - \mathbf{c}_0), \tag{26.62}$$

and $\mathbf{c}_0$ is the average velocity.

For molecules with internal energy, the Boltzmann equation is replaced by the Wang Chang-Uhlenbeck equation, which reads

$$\mathcal{D} f_i^{(0)} + f_i^{(0)} \mathcal{D}(\varphi_i) = -f_i^{(0)} n \mathcal{R}(\varphi_i + \varphi_j), \tag{26.63}$$

where $\mathcal{D}$ is the operator $(\partial/\partial t) + \mathbf{c} \cdot \nabla$ and $\mathcal{R}$ is the collision operator defined by

$$n f_i^{(0)} \mathcal{R}(Y) = \sum_{jkl} \iiint f_i^{(0)} f_j^{(0)} [Y - Y'] g \sigma_{ij}^{kl}(g, \chi, \psi) \sin\chi\, d\chi\, d\psi\, d\mathbf{c}_j \tag{26.64}$$

and is related to a binary collision in which two colliding molecules initially in states i and j approach each other with the relative velocity $\mathbf{g}$. After collision, the relative velocity has changed to $\mathbf{g}'$ and is rotated through the polar angle χ and the azimuthal angle ψ while the internal states of the molecules have changed to k and l. Furthermore, $\sigma_{ij}^{kl}(g, \chi\psi)$ is the differential scattering cross section for the collision process.

26.5.2. The transport properties

The Wang Chang-Uhlenbeck or Boltzmann equations can be solved by means of the assumption that the perturbation of the local equilibrium solution depends on gradients of the thermodynamic state variables and the flow velocity; so

$$\varphi_i = \frac{1}{n}\mathbf{A} \cdot \nabla \ln T - \frac{1}{n}\mathbf{B} \cdot \nabla \mathbf{c}_0 \tag{26.65}$$

for a pure fluid. The solution methodology [19] is to expand the coefficients **A** and **B** in terms of an infinite series of polynomials, which leads to an infinite set of linear equations to be solved to evaluate the transport properties of the dilute gas.

It is found, for example, that the thermal conductivity of the fluid is related just to A by the equation

$$\lambda = \frac{2k_B^2 T}{3m}[\mathbf{A} \cdot \mathcal{R}(\mathbf{A})], \tag{26.66}$$

where the notation $[\mathbf{A} \cdot \mathcal{R}(\mathbf{A})]$ denotes a so-called bracket-integral. [19] This equation makes it clear that the thermal conductivity of a gas is intimately connected with the details of the binary encounters between the molecules of the gas through the collision operator $\mathcal{R}$. Explicit evaluation of Eq. (26.66) and of other transport properties is possible by means of some algebra to produce exact results for the transport properties of a gas.

26.5.3. Monatomic gases

The greatest amount of work has been performed for monatomic gases, in which the absence of internal energy in the molecules leads to the greatest simplicity of calculation from first principles.

The viscosity of a monatomic gas is given to a first-order approximation ($[\eta_1]$) (within 1%) by the expression

$$[\eta]_1 = \frac{k_B T}{\langle c \rangle \mathcal{S}(2000)}, \tag{26.67}$$

where $\mathcal{S}(2000)$ is a so-called effective cross section, whereas the thermal conductivity is given by

$$[\lambda]_1 = \frac{15k_B^2 T}{4m\langle c \rangle \mathcal{S}(2000)} \tag{26.68}$$

so that

$$[\lambda]_1 = \frac{15k_B[\eta]_1}{4m} \tag{26.69}$$

and the self-diffusion coefficient by

$$[D]_1 = \frac{6A^*[\eta]_1}{4Nm}, \tag{26.70}$$

where

$$A^* = \frac{5}{6}\frac{\mathcal{S}(2000)}{\mathcal{S}(1000)} \tag{26.71}$$

and $\mathcal{S}(1000)$ is another effective cross section.

All of the effective cross sections can be evaluated from integrations with respect to all possible collision parameters over the differential cross section σ_{ij}^{kl} introduced earlier. The differential cross section must be evaluated by a detailed calculation of the dynamics of binary collision under the influence of a particular intermolecular pair potential, such as those for the rare gases. [20] Standard programs exist for the evaluation of the effective cross sections for a series of intermolecular pair potential models. [19] Normally, for convenience, the two effective cross sections of interest here are tabulated in the form of reduced collision integrals $\Omega^{(l,n)^*}$. These quantities are related to the requisite effective cross sections by the equations

$$\mathcal{S}(2000) = \frac{4}{5}\sigma^2\Omega^{(2,2)^*}(T^*) \tag{26.72}$$

and

$$\mathcal{S}(1000) = \frac{2}{3}\sigma^2\Omega^{(1,1)^*}(T^*). \tag{26.73}$$

Here, σ is the separation of two atoms for which their intermolecular potential energy is zero and

$$T^* = k_B T/\varepsilon, \tag{26.74}$$

where ε is the depth of the same intermolecular potential energy minimum.

The application of the more general kinetic theory to polyatomic gases is very much more complicated since, in addition to the nonspherically-symmetric intermolecular pair potential, which complicates the calculation process considerably, the unique set of energy levels for each molecular species means generalized calculations are not possible. A summary of the latest developments is outlined in Ref. [18]

26.5.4. Dense fluids

For moderately dense gases, Enskog [18] attempted to adapt the dilute gas theory for hard-sphere molecules by considering only the fact that the finite size of the molecules increased the collision frequency and allowed both momentum and energy to be transferred over the distance of a molecular diameter on collision without the corresponding transport of a molecule. This original step has been modified subsequently to allow for the fact that real molecules are not rigid spheres and to correct the theory even for hard spheres since it neglects any correlations of molecular velocities between collisions, which must arise at high densities. All of the modifications rely on approximation or empiricism to some extent and are therefore inadequate for an accurate description of the transport properties in any fundamental manner. The application of the same ideas to liquids has had some empirical successes [18] but lacks a fundamental basis.

26.6. REFERENCES

[1] M. L McGlashan, *Chemical Thermodynamics* (Academic Press, London, 1979).

[2] K. E. Bett, J. S. Rowlinson, and G. Saville, *Thermodynamics for Chemical Engineers* (Athlone Press, London, 1975).

[3] E. A. Guggenheim, *Thermodynamics*, 5th ed. (North Holland, Amsterdam, 1967).

[4] B. Le Neindre and B. Vodar (Eds.), *Experimental Thermodynamics* (Butterworths, London, 1975), Vol. II.

[5] M. J. Assaiel, J. P. M. Trusler, and T. F. Tsolakis, *Thermophysical Properties* (Imperial College Press, London, 1996).

[6] A. Fredenslund, J. Gmehling, and P. Rasmussen, *Vapor-Liquid-Equilibria Using UNIFAC* (Elsevier, Amsterdam, 1977).

[7] T. L. Hill, *An Introductin to Statistical Thermodynamics* (Dover, New York, 1986).

[8] G. J. Janz, *Thermodynamics Properties of Organic Compounds* (Academic Press, New York, 1967).

[9] K. E. Gubbins and N. Quirke, *Molecular Simulation and Industrial Applications* (Gordon and Breach, Amsterdam, 1996).

[10] J. S. Rowlinson and I. D. Watson, Chem. Eng. Sci. **24**, 1565–1574 (1969).

[11] E. A. Mason and T. H. Spurling, *The Virial Equation of State* (Pergamon Press, Oxford, 1969).

[12] J. P. Hansen and I. R. McDonald, *Theory of Simple Liquids* (Academic Press, New York, 1976).

[13] J. R. de Groot and P. Mazur, *Non-Equilibrium Thermodynamics* (Interscience, New York, 1962).

[14] R. B. Bird, W. E. Stewart, and E. N. Lightfoot, *Transport Phenomena* (Wiley, New York, 1960).

[15] H. J. V. Tyrrell and K. R. Harris, *Diffusion in Liquids* (Butterworths, London, 1984).

[16] W. A. Wakeham, A. Nagashima, and J. V. Sengers (Eds.), *Experimental Thermodynamics* (Blackwell Scientific, Oxford, 1991), Vol. III.

[17] K. D. Maglic and A. Cezairliyan (Eds.), *Compendium on Thermophysical Properties Measurement Methods* (Plenum, New York, 1991), Vol. 2.

[18] J. Millat, J. H. Dymond, and C. A. Nieto de Castro (Eds.), *Transport Properties of Fluids: Their Correlation, Prediction and Estimation* (Cambridge University Press, Cambridge, U. K., 1996).

[19] G. C. Maitland, M. Rigby, E. B. Smith, and W. A. Wakeham, *Intermolecular Forces* (Oxford University Press, Oxford, U.K., 1981).

[20] R. A. Aziz, in *Inert Gases, Potentials, Dynamics and Energy Transfer in Doped Crystals*, edited by M. Klein (Springer Verlag, Berlin, 1984), pp. 5–86.

[21] E. J. Janse van Rensburg, *Journal of Physics* A, Vol. 26, pp. 4805–4818 (1993).

27

Practical Laboratory Data

David R. Lide
Gaithersburg, Maryland

Contents

27.1. INTRODUCTION

This chapter contains tables of data that are useful in designing experiments, calibrating instruments, constructing equipment, and interpreting experimental measurements. Other useful sources of such data are:

1. D. R. Lide (Ed.), *CRC Handbook of Chemistry and Physics*, 82nd ed. (CRC Press, Boca Raton, FL, 2001).
2. D. E. Gray (Ed.), *American Institute of Physics Handbook* (McGraw-Hill, New York, 1972).
3. G. W. C. Kaye and T. H. Laby, *Tables of Physical and Chemical Constants*, 16th ed. (Longman, London, 1995).

4. *J. Phys. Chem. Ref. Data* (published jointly by National Institute of Standards and Technology and American Institute of Physics).
5. Websites of the National Institute of Standards and Technology: <physics.nist.gov> and <webbook.nist.gov>.

27.2. TABLE: PERIODIC TABLE

Key to Chart:

- Atomic Number → 50
- Symbol → Sn
- 1995 Atomic Weight → 118.710
- Oxidation States → +2 +4
- Electron Configuration → -18-18-4

New Notation	1	2	3	4	5	6	7	8	9	10	11	12	13	14	15	16	17	18	
Previous IUPAC Form	Group IA	IIA	IIIA	IVA	VA	VIA	VIIA		VIIIA		IB	IIB	IIIB	IVB	VB	VIB	VIIB	VIIIA	
CAS Version			IIIB	IVB	VB	VIB	VIIB		VIII		IB	IIB	IIIA	IVA	VA	VIA	VIIA		Shell
	1 H +1 -1 1.00794 1																	2 He 0 4.002602 2	K
	3 Li +1 6.941 2-1	4 Be +2 9.012182 2-2											5 B +3 10.811 2-3	6 C +2 +4 -4 12.0107 2-4	7 N +1 +2 +3 +4 +5 -1 -2 -3 14.00674 2-5	8 O -2 15.9994 2-6	9 F -1 18.9984032 2-7	10 Ne 0 20.1797 2-8	K-L
	11 Na +1 22.989770 2-8-1	12 Mg +2 24.3050 2-8-2											13 Al +3 26.981538 2-8-3	14 Si +2 +4 -4 28.0855 2-8-4	15 P +3 +5 -3 30.973761 2-8-5	16 S +4 +6 -2 32.066 2-8-6	17 Cl +1 +5 +7 -1 35.4527 2-8-7	18 Ar 0 39.948 2-8-8	K-L-M
	19 K +1 39.0983 -8-8-1	20 Ca +2 40.078 -8-8-2	21 Sc +3 44.955910 -8-9-2	22 Ti +2 +3 +4 47.867 -8-10-2	23 V +2 +3 +4 +5 50.9415 -8-11-2	24 Cr +2 +3 +6 51.9961 -8-13-1	25 Mn +2 +3 +4 +7 54.938049 -8-13-2	26 Fe +2 +3 55.845 -8-13-2	27 Co +2 +3 58.933200 -8-15-2	28 Ni +2 +3 58.6934 -8-16-2	29 Cu +1 +2 63.546 -8-18-1	30 Zn +2 65.39 -8-18-2	31 Ga +3 69.723 -8-18-3	32 Ge +2 +4 72.61 -8-18-4	33 As +3 +5 -3 74.92160 -8-18-5	34 Se +4 +6 -2 78.96 -8-18-6	35 Br +1 +5 -1 79.904 -8-18-7	36 Kr 0 83.80 -8-18-8	-L-M-N
	37 Rb +1 85.4678 -18-8-1	38 Sr +2 87.62 -18-8-2	39 Y +3 88.90585 -18-9-2	40 Zr +4 91.224 -18-10-2	41 Nb +3 +5 92.90638 -18-12-1	42 Mo +6 95.94 -18-13-1	43 Tc +4 +6 +7 (98) -18-13-2	44 Ru +3 101.07 -18-15-1	45 Rh +3 102.90550 -18-16-1	46 Pd +2 +3 106.42 -18-18-0	47 Ag +1 107.8682 -18-18-1	48 Cd +2 112.411 -18-18-2	49 In +3 114.818 -18-18-3	50 Sn +2 +4 118.710 -18-18-4	51 Sb +3 +5 -3 121.760 -18-18-5	52 Te +4 +6 -2 127.60 -18-18-6	53 I +1 +5 +7 -1 126.90447 -18-18-7	54 Xe 0 131.29 -18-18-8	-M-N-O
	55 Cs +1 132.90545 -18-8-1	56 Ba +2 137.327 -18-8-2	57* La +3 138.9055 -18-9-2	72 Hf +4 178.49 -32-10-2	73 Ta +5 180.9479 -32-11-2	74 W +6 183.84 -32-12-2	75 Re +4 +6 +7 186.207 -32-13-2	76 Os +3 +4 190.23 -32-14-2	77 Ir +3 +4 192.217 -32-15-2	78 Pt +2 +4 195.078 -32-17-1	79 Au +1 +3 196.96655 -32-18-1	80 Hg +1 +2 200.59 -32-18-2	81 Tl +1 +3 204.3833 -32-18-3	82 Pb +2 +4 207.2 -32-18-4	83 Bi +3 +5 208.98038 -32-18-5	84 Po +2 +4 (209) -32-18-6	85 At (210) -32-18-7	86 Rn 0 (222) -32-18-8	-N-O-P
	87 Fr +1 (223) -18-8-1	88 Ra +2 (226) -18-8-2	89** Ac +3 (227) -18-9-2	104 Rf +4 (261) -32-10-2	105 Db (262) -32-11-2	106 Sg (266) -32-12-2	107 Bh (264) -32-13-2	108 Hs (269) -32-14-2	109 Mt (268) -32-15-2	110 Uun (271) -32-16-2	111 Uuu (272)	112 Uub							-O-P-Q

															Shell
* Lanthanides	58 Ce +3 +4 140.116 -19-9-2	59 Pr +3 140.90765 -21-8-2	60 Nd +3 144.24 -22-8-2	61 Pm +3 (145) -23-8-2	62 Sm +2 +3 150.36 -24-8-2	63 Eu +2 +3 151.964 -25-8-2	64 Gd +3 157 .25 -25-9-2	65 Tb +3 158.92534 -27-8-2	66 Dy +3 162.50 -28-8-2	67 Ho +3 164.93032 -29-8-2	68 Er +3 167.26 -30-8-2	69 Tm +3 168.93421 -31-8-2	70 Yb +2 +3 173.04 -32-8-2	71 Lu +3 174.967 -32-9-2	-N-O-P
** Actinides	90 Th +4 232.0381 -18-10-2	91 Pa +5 +4 231.03588 -20-9-2	92 U +3 +4 +5 +6 238.0289 -21-9-2	93 Np +3 +4 +5 +6 (237) -22-9-2	94 Pu +3 +4 +5 +6 (244) -24-8-2	95 Am +3 +4 +5 +6 (243) -25-8-2	96 Cm +3 (247) -25-9-2	97 Bk +3 +4 (247) -27-8-2	98 Cf +3 (251) -28-8-2	99 Es +3 (252) -29-8-2	100 Fm +3 (257) -30-8-2	101 Md +2 +3 (258) -31-8-2	102 No +2 +3 (259) -32-8-2	103 Lr +3 (262) -32-9-2	-O-P-Q

27.3. TABLE: PHYSICAL CONSTANTS OF ELEMENTS AND COMPOUNDS

Formula	Name	$t_m/°C$	$t_b/°C$	$t_c/°C$	p_c/MPa	$\rho/g\ cm^{-3}$
Elements						
Ac	Actinium	1051	3198			10
Ag	Silver	961.78	2162			10.5
Al	Aluminum	660.32	2519			2.70
Am	Americium	1176	2011			12
Ar	Argon	−189.36 tp	−185.85	−122.28	4.9	1.633 g/L
As	Arsenic	817 tp	603 sp			5.75
At	Astatine	302				
Au	Gold	1064.18	2856			19.3
B	Boron	2075	4000			2.34
Ba	Barium	727	1897			3.62
Be	Beryllium	1287	2471			1.85
Bi	Bismuth	271.4	1564			9.79
Bk	Berkelium	1050				14.78
Br_2	Bromine	−7.2	58.8	315	10.34	3.1028
C	Carbon	4489 tp	3825 sp			2.2
Ca	Calcium	842	1484			1.54
Cd	Cadmium	321.07	767			8.69
Ce	Cerium	798	3443			6.770
Cf	Californium	900				15.1
Cl_2	Chlorine	−101.5	−34.04	143.8	7.99	2.898 g/L
Cm	Curium	1345	≈ 3100			13.51
Co	Cobalt	1495	2927			8.86
Cr	Chromium	1907	2671			7.15
Cs	Cesium	28.5	671	1665	9.4	1.93
Cu	Copper	1084.62	2562			8.96
Dy	Dysprosium	1412	2567			8.55
Er	Erbium	1529	2868			9.07
Es	Einsteinium	860				
Eu	Europium	822	1529			5.24
F_2	Fluorine	−219.67 tp	−188.12	−129.02	5.17	1.553 g/L
Fe	Iron	1538	2861			7.87
Fm	Fermium	1527				
Fr	Francium	27				
Ga	Gallium	29.76	2204			5.91
Gd	Gadolinium	1313	3273			7.90
Ge	Germanium	938.25	2833			5.3234
H_2	Hydrogen	−259.34	−252.87	−240.18	1.29	0.082 g/L
He	Helium		−268.93	−267.96	0.23	0.164 g/L
Hf	Hafnium	2233	4603			13.3
Hg	Mercury	−38.83	356.73	1477	172	13.5336
Ho	Holmium	1474	2700			8.80
I_2	Iodine	113.7	184.4	546		4.933

(continued)

Formula	Name	$t_m/°C$	$t_b/°C$	$t_c/°C$	p_c/MPa	$\rho/g\ cm^{-3}$
In	Indium	156.6	2072			7.31
Ir	Iridium	2446	4428			22.5
K	Potassium	63.5	759	≈ 1950	≈ 16	0.89
Kr	Krypton	−157.38 tp	−153.22	−63.74	5.5	3.425 g/L
La	Lanthanum	918	3464			6.15
Li	Lithium	180.5	1342	≈ 2950	≈ 67	0.534
Lr	Lawrencium	1627				
Lu	Lutetium	1663	3402			9.84
Md	Mendelevium	827				
Mg	Magnesium	650	1090			1.74
Mn	Manganese	1246	2061			7.3
Mo	Molybdenum	2623	4639			10.2
N_2	Nitrogen	−210	−195.79	−146.94	3.39	1.145 g/L
Na	Sodium	97.8	883	≈ 2300	≈ 35	0.97
Nb	Niobium	2477	4744			8.57
Nd	Neodymium	1021	3074			7.01
Ne	Neon	−248.61 tp	−246.08	−228.7	2.76	0.825 g/L
Ni	Nickel	1455	2913			8.90
No	Nobelium	827				
Np	Neptunium	644				20.2
O_2	Oxygen	−218.79	−182.95	−118.56	5.04	1.308 g/L
Os	Osmium	3033	5012			22.59
P	Phosphorus	44.15	280.5			1.823
Pa	Protactinium	1572				15.4
Pb	Lead	327.46	1749			11.3
Pd	Palladium	1554.9	2963			12.0
Pm	Promethium	1042	3000			7.26
Po	Polonium	254	962			9.20
Pr	Praseodymium	931	3520			6.77
Pt	Platinum	1768.4	3825			21.5
Pu	Plutonium	640	3228			19.7
Ra	Radium	700				5
Rb	Rubidium	39.3	688	≈ 1820	≈ 16	1.53
Re	Rhenium	3186	5596			20.8
Rh	Rhodium	1964	3695			12.4
Rn	Radon	−71	−61.7	104	6.28	9.074 g/L
Ru	Ruthenium	2334	4150			12.1
S	Sulfur	115.21	444.6			2.07
Sb	Antimony	630.63	1587			6.68
Sc	Scandium	1541	2836			2.99
Se	Selenium	221	685			4.39
Si	Silicon	1414	3265			2.3290
Sm	Samarium	1074	1794			7.52
Sn	Tin	231.93	2602			7.265

(continued)

Formula	Name	$t_m/°C$	$t_b/°C$	$t_c/°C$	p_c/MPa	$\rho/g\ cm^{-3}$
Sr	Strontium	777	1382			2.64
Ta	Tantalum	3017	5458			16.4
Tb	Terbium	1356	3230			8.23
Tc	Technetium	2157	4265			11
Te	Tellurium	449.51	988			6.24
Th	Thorium	1750	4788			11.7
Ti	Titanium	1668	3287			4.506
Tl	Thallium	304	1473			11.8
Tm	Thulium	1545	1950			9.32
U	Uranium	1135	4131			19.1
V	Vanadium	1910	3407			6.0
W	Tungsten	3422	5555			19.3
Xe	Xenon	−111.79 tp	−108.12	16.62	5.84	5.366 g/L
Y	Yttrium	1522	3345			4.47
Yb	Ytterbium	819	1196			6.90
Zn	Zinc	419.53	907			7.14
Zr	Zirconium	1855	4409			6.52
Inorganic Compounds						
BCl_3	Boron trichloride	−107	12.65	182	3.87	4.789 g/L
BF_3	Boron trifluoride	−126.8	−101	−12.3	4.98	2.771 g/L
B_2H_6	Diborane	−165.5	−92.4	16.7	4.05	1.131 g/L
CO	Carbon monoxide	−205.02	−191.5	−140.24	3.5	1.145 g/L
COS	Carbon oxysulfide	−138.8	−50	102	5.88	2.456 g/L
CO_2	Carbon dioxide	−56.56 tp	−78.4 sp	30.98	7.38	1.799 g/L
CS_2	Carbon disulfide	−112.1	46	279	7.9	1.2555
GeF_4	Germanium tetrafluoride	−15 tp	−36.5 sp			6.074 g/L
HBr	Hydrogen bromide	−86.8	−66.38	90.1	8.55	3.307 g/L
HCN	Hydrogen cyanide	−13.29	26	183.6	5.39	0.684
HCl	Hydrogen chloride	−114.17	−85	51.6	8.31	1.490 g/L
$HClO_4$	Perchloric acid	−112	dec ≈ 90			1.77
HF	Hydrogen fluoride	−83.35	20	188	6.48	0.818 g/L
HI	Hydrogen iodide	−50.76	−35.55	150.9	8.31	5.228 g/L
HNO_3	Nitric acid	−41.6	83			1.55
HN_3	Hydrazoic acid	−80	35.7			
H_2O	Water	0.00	100.0	373.99	22.06	0.9970
H_2O_2	Hydrogen peroxide	−0.43	150.2	≈ 455	≈ 22	1.44
H_2S	Hydrogen sulfide	−85.5	−59.55	100.1	8.94	1.393 g/L
H_2SO_4	Sulfuric acid	10.31	337			1.8
H_2Se	Hydrogen selenide	−65.73	−41.25	138	8.92	3.310 g/L
NH_3	Ammonia	−77.73	−33.33	132.4	11.35	0.696 g/L
NO	Nitric oxide	−163.6	−151.74	−93	6.48	1.226 g/L
NO_2	Nitrogen dioxide		see N_2O_4			1.880 g/L
N_2H_4	Hydrazine	1.4	113.55	380	14.7	1.0036
N_2O	Nitrous oxide	−90.8	−88.48	36.42	7.26	1.799 g/L

(continued)

Formula	Name	t_m/°C	t_b/°C	t_c/°C	p_c/MPa	ρ/g cm^{-3}
N_2O_3	Nitrogen trioxide	−101.1	≈ 3 dec			1.4 (2°C)
N_2O_4	Nitrogen tetroxide	−9.3	21.15	158	10.1	1.45 (20°C)
N_2O_5	Nitrogen pentoxide		33 sp			2.0
O_3	Ozone	−193	−111.35	−12.0	5.57	1.962 g/L
PCl_3	Phosphorus trichloride	−112	75.95	290		1.574
PCl_5	Phosphorus pentachloride	167 tp	160 sp	373		2.1
PF_3	Phosphorus trifluoride	−151.5	−101.5	−1.9	4.33	3.596 g/L
PF_5	Phosphorus pentafluoride	−93.8	−84.6			5.149 g/L
PH_3	Phosphine	−133	−87.75	51.4	6.54	1.390 g/L
SF_6	Sulfur hexafluoride	−50.7 tp	−63.8 sp	45.54	3.77	5.970 g/L
SO_2	Sulfur dioxide	−75.5	−10.05	157.7	7.88	2.619 g/L
SO_2Cl_2	Sulfuryl chloride	−51	69.4			1.680
SO_3	Sulfur trioxide	16.8	45	217.9	8.2	1.92
$SiCl_4$	Tetrachlorosilane	−68.74	57.65	235.0	3.59	1.5
SiF_4	Tetrafluorosilane	−90.2	−86	−14.1	3.72	4.254 g/L
SiH_4	Silane	−185	−111.9			1.313 g/L
UF_6	Uranium hexafluoride	64.0 tp	56.5 sp	232.7	4.66	5.09
Organic compounds						
CCl_4	Carbon tetrachloride	−22.62	76.8	283.5	4.52	1.5940
$CHBr_3$	Bromoform	8.69	149.1			2.8788
$CHCl_3$	Chloroform	−63.41	61.17	263.3	5.47	1.4788
CH_2Br_2	Dibromomethane	−52.5	97			2.4969
CH_2Cl_2	Dichloromethane	−97.2	40	237	6.1	1.3266
CH_2O	Formaldehyde	−92	−19.1			0.815 (−20°C)
CH_2O_2	Formic acid	8.3	101	315		1.220
CH_3Br	Methyl bromide	−93.68	3.5			1.6755
CH_3Cl	Methyl chloride	−97.7	−24.09	143.1	6.68	0.911*
CH_3I	Methyl iodide	−66.4	42.43	255		2.2789
CH_3NO_2	Nitromethane	−28.38	101.19	315	5.87	1.1371
CH_4	Methane	−182.47	−161.4	−82.59	4.6	0.4228 (−162°C)
CH_4O	Methanol	−97.53	64.6	239.4	8.08	0.7914
CH_5N	Methylamine	−93.5	−6.32	157.6	7.61	0.656*
C_2Cl_4	Tetrachloroethylene	−22.3	121.3	347.1		1.6230
C_2HCl_3	Trichloroethylene	−84.7	87.21	271.1	5.02	1.4642
C_2H_3Br	Vinyl bromide	−139.54	15.8			1.4933
C_2H_3Cl	Vinyl chloride	−153.84	−13.3			0.9106
C_2H_3N	Acetonitrile	−43.82	81.65	272.4	4.85	0.7857
C_2H_4	Ethylene	−169.15	−103.7	9.19	5.04	0.5678 (−104°C)
$C_2H_4Cl_2$	1,2-Dichloroethane	−35.7	83.5	288	5.4	1.2454
C_2H_4O	Ethylene oxide	−112.5	10.6	196	7.19	0.8821 (10°C)
$C_2H_4O_2$	Acetic acid	16.64	117.9	319.56	5.79	1.0446
$C_2H_4O_2$	Methyl formate	−99	31.7	214.1	6	0.9713
C_2H_5Br	Ethyl bromide	−118.6	38.5	230.8	6.23	1.4604
C_2H_5Cl	Ethyl chloride	−138.4	12.3	187.3	5.3	0.8902*

(continued)

Formula	Name	$t_m/°C$	$t_b/°C$	$t_c/°C$	p_c/MPa	$\rho/g\ cm^{-3}$
C_2H_5NO	*N*-Methylformamide	−3.8	199.51			1.011
C_2H_6	Ethane	−182.79	−88.6	32.17	4.87	0.5446 (−89°C)
C_2H_6O	Ethanol	−114.14	78.29	240.9	6.14	0.7893
C_2H_6O	Dimethyl ether	−141.5	−24.8	126.9	5.37	
C_2H_6OS	Dimethyl sulfoxide	17.89	189			1.1010
$C_2H_6O_2$	Ethylene glycol	−12.69	197.3	445		1.1135
C_2H_7N	Dimethylamine	−92.18	6.88	164.07	5.34	0.6804 (0°C)
C_2H_7N	Ethylamine	−80.5	16.5	183	5.62	0.677*
C_2H_7NO	Ethanolamine	10.5	171			1.0180
C_3H_6	Propylene	−185.24	−47.69	91.8	4.6	0.505*
C_3H_6	Cyclopropane	−127.58	−32.81	124.9	5.54	0.617*
C_3H_6O	Acetone	−94.7	56.05	235.0	4.700	0.7845
$C_3H_6O_2$	Methyl acetate	−98.25	56.87	233.4	4.75	0.9342
C_3H_7NO	*N*, *N*-Dimethylformamide	−60.48	153	376.5		0.9445
C_3H_8	Propane	−187.63	−42.1	96.68	4.25	0.493*
C_3H_8O	1-Propanol	−124.39	97.2	263.7	5.17	0.7997
$C_3H_8O_2$	1,2-Propylene glycol	−60	187.6			1.0361
$C_3H_8O_3$	Glycerol	18.1	290			1.2613
C_3H_9N	Trimethylamine	−117.1	2.87	159.64	4.09	0.627*
C_4H_4O	Furan	−85.61	31.5	217.1	5.5	0.9514
C_4H_5N	Pyrrole	−23.39	129.79	366.6	6.34	0.9698
C_4H_8O	Methyl ethyl ketone	−86.64	79.59	263.63	4.21	0.7999
C_4H_8O	Tetrahydrofuran	−108.44	65	267.0	5.19	0.8833
$C_4H_8O_2$	1,4-Dioxane	11.85	101.5	314	5.21	1.0337
$C_4H_8O_2$	Ethyl acetate	−83.8	77.11	250.2	3.88	0.9003
C_4H_{10}	Butane	−138.3	−0.5	151.97	3.8	0.573*
C_4H_{10}	Isobutane	−159.4	−11.73	134.7	3.64	0.5510*
$C_4H_{10}O$	1-Butanol	−88.6	117.73	289.9	4.41	0.8095
$C_4H_{10}O$	Diethyl ether	−116.2	34.5	193.59	3.64	0.7138
$C_4H_{10}O_3$	Diethylene glycol	−10.4	245.8			1.1197
$C_4H_{11}N$	Diethylamine	−49.8	55.5	226.84	3.76	0.7056
$C_4H_{11}NO_2$	Diethanolamine	28	268.8			1.0966
$C_4H_{12}Si$	Tetramethylsilane	−99.06	26.6	175.49	2.82	0.648
C_5H_5N	Pyridine	−41.70	115.23	346.9	5.67	0.9819
$C_5H_6O_2$	Furfuryl alcohol	−14.6	171			1.1296
C_5H_{10}	Cyclopentane	−93.4	49.3	238.6	4.51	0.7457
C_5H_{12}	Pentane	−129.67	36.06	196.6	3.37	0.6262
$C_5H_{12}O$	1-Pentanol	−77.6	137.98	315.0	3.9	0.8144
C_6H_5Cl	Chlorobenzene	−45.31	131.72	359.3	4.52	1.1058
$C_6H_5NO_2$	Nitrobenzene	5.7	210.8			1.2037
C_6H_6	Benzene	5.49	80.09	288.9	4.9	0.8765
C_6H_6O	Phenol	40.89	181.87	421.1	6.13	1.0545 (45°C)
C_6H_7N	Aniline	−6.02	184.17	426	4.89	1.0217
$C_6H_{10}O$	Cyclohexanone	−27.9	155.43	379.9	4.0	0.9478

(continued)

Formula	Name	t_m/°C	t_b/°C	t_c/°C	p_c/MPa	ρ/g cm^{-3}
C_6H_{12}	Cyclohexane	6.59	80.73	280.7	4.08	0.7739
$C_6H_{12}O$	Cyclohexanol	25.93	160.84	374.0	4.4	0.9624
$C_6H_{12}O$	Methyl isobutyl ketone	−84	116.5	298	3.27	0.7965
C_6H_{14}	Hexane	−95.35	68.73	234.5	3.03	0.6606
$C_6H_{14}O$	Diisopropyl ether	−85.4	68.51	227.17	2.83	0.7192
$C_6H_{14}O$	1-Hexanol	−47.4	157.6	337.2	3.42	0.8136
$C_6H_{15}N$	Triethylamine	−114.7	89	262.5	3.03	0.7275
C_7H_6O	Benzaldehyde	−26	179	422	4.65	1.0415 (10°C)
C_7H_8	Toluene	−94.95	110.63	318.6	4.11	0.8668
C_7H_{16}	Heptane	−90.55	98.4	267.1	2.74	0.6795
C_8H_{10}	*o*-Xylene	−25.2	144.5	357.2	3.73	0.8802 (10°C)
C_8H_{10}	*m*-Xylene	−47.8	139.12	343.9	3.54	0.8642
C_8H_{10}	*p*-Xylene	13.25	138.37	343.1	3.51	0.8566
C_8H_{18}	Octane	−56.82	125.67	295.6	2.49	0.6986
$C_8H_{18}O$	1-Octanol	−14.8	195.16	379.4	2.78	0.8262
$C_{10}H_8$	Naphthalene	80.26	217.9	475.3	4.05	1.0253

t_m is melting point; tp indicates a solid-liquid-gas triple point; t_b is normal boiling point; sp indicates a sublimation point, where the vapor pressure of the solid reaches 101.325 kPa; t_c is liquid-gas critical point temperature; p_c is the critical pressure; ρ is density at ambient temperature (20–25°C) in g/cm^3 unless otherwise indicated. Values of density followed by g/L are gas densities at 25°C and 101.325 kPa. An * indicates a liquid at its saturated vapor pressure (greater than 1 atm).

27.4. TABLE: THERMAL AND ELECTRICAL PROPERTIES OF METALS

Metal	$\alpha/10^{-6}\ K^{-1}$	$c_p/J\ g^{-1}\ K^{-1}$	$\lambda/W\ cm^{-1}\ K^{-1}$	$\rho/10^{-8}\Omega\ m$
Aluminum	23.1	0.897	2.37	2.71
Antimony	11.0	0.207	0.24	39*
Barium	20.6	0.204	0.18	34.0
Beryllium	11.3	1.825	2.00	3.70
Bismuth	13.4	0.122	0.08	107*
Cadmium	30.8	0.232	0.97	6.8*
Calcium	22.3	0.647	2.00	3.42
Cerium	5.2	0.192	0.11	82.8
Cesium		0.242	0.36	20.8
Chromium	4.9	0.449	0.94	12.6
Cobalt	13.0	0.421	1.00	5.6*
Copper	16.5	0.385	4.01	1.71
Dysprosium	9.9	0.173	0.11	92.6
Erbium	12.2	0.168	0.15	86.0
Europium	35.0	0.182	0.14	90.0
Gadolinium	9.4	0.236	0.11	131
Gallium		0.371	0.41	13.6*
Gold	14.2	0.129	3.17	2.26
Hafnium	5.9	0.144	0.23	33.7
Holmium	11.2	0.165	0.16	81.4
Indium	32.1	0.233	0.82	8.0*
Iridium	6.4	0.131	1.47	4.7*
Iron	11.8	0.449	0.8	9.87
Lanthanum	12.1	0.195	0.13	61.5
Lead	28.9	0.129	0.35	21.1
Lithium	46	3.582	0.85	9.47
Lutetium	9.9	0.154	0.16	58.2
Magnesium	24.8	1.023	1.56	4.48
Manganese	21.7	0.479	0.08	144
Mercury	60.4	0.139	0.08	96.1
Molybdenum	4.8	0.251	1.38	5.47
Neodymium	9.6	0.190	0.17	64.3
Nickel	13.4	0.444	0.91	7.12
Niobium	7.3	0.265	0.54	15.2*
Osmium	5.1	0.130	0.88	8.1*
Palladium	11.8	0.246	0.72	10.73
Platinum	8.8	0.133	0.72	10.7
Plutonium	46.7		0.07	
Potassium	83.3	0.757	1.02	7.39
Praseodymium	6.7	0.193	0.13	70.0
Promethium	11		0.15	75
Rhenium	6.2	0.137	0.48	17.2*

(continued)

Metal	$\alpha/10^{-6}\ \mathrm{K}^{-1}$	$c_p/\mathrm{J\ g^{-1}\ K^{-1}}$	$\lambda/\mathrm{W\ cm^{-1}\ K^{-1}}$	$\rho/10^{-8}\Omega\ \mathrm{m}$
Rhodium	8.2	0.243	1.50	4.3*
Rubidium		0.363	0.58	13.1
Ruthenium	6.4	0.238	1.17	7.1*
Samarium	12.7	0.197	0.13	94.0
Scandium	10.2	0.568	0.16	56.2
Silver	18.9	0.235	4.29	1.62
Sodium	71	1.228	1.41	4.88
Strontium	22.5	0.301	0.35	13.4
Tantalum	6.3	0.140	0.58	13.4
Technetium			0.51	
Terbium	10.3	0.182	0.11	115
Thallium	29.9	0.129	0.46	15*
Thorium	11.0	0.113	0.540	14.7*
Thulium	13.3	0.160	0.17	67.6
Tin	22.0	0.228	0.67	11.5*
Titanium	8.6	0.523	0.22	39*
Tungsten	4.5	0.132	1.74	5.39
Uranium	13.9	0.116	0.28	28*
Vanadium	8.4	0.489	0.31	20.1
Ytterbium	26.3	0.155	0.39	25.0
Yttrium	10.6	0.298	0.17	59.6
Zinc	30.2	0.388	1.16	6.01
Zirconium	5.7	0.278	0.227	42.9

α is linear thermal expansion coefficient at 25°C; c_p is specific heat capacity at 25°C; λ is thermal conductivity at 25°C; ρ is electrical resistivity; values with * are at 0°C, otherwise at 25°C.

27.5. TABLE: DIELECTRIC CONSTANT (RELATIVE PERMITTIVITY) OF LIQUIDS

		T/K	ε
Inorganic compounds			
Ar	Argon	140	1.32
Br_2	Bromine	297.9	3.15
CO_2	Carbon dioxide	295	1.45
CS_2	Carbon disulfide	293.2	2.63
Cl_2	Chlorine	208	2.15
F_2	Fluorine	53.48	1.49
HCl	Hydrogen chloride	158.9	14.3
HF	Hydrogen fluoride	273.2	83.6
H_2	Hydrogen	13.52	1.28
H_2O	Water	293.2	80.10
He	Helium	2.06	1.06
I_2	Iodine	391.25	11.08
Kr	Krypton	119.8	1.66
NH_3	Ammonia	293.2	16.61
N_2	Nitrogen	63.15	1.47
N_2O_4	Nitrogen tetroxide	293.2	2.44
Ne	Neon	26.11	1.19
O_2	Oxygen	54.48	1.57
O_3	Ozone	90.2	4.75
P	Phosphorus	307.2	4.10
S	Sulfur	407.2	3.50
SF_6	Sulfur hexafluoride	223.2	1.81
SO_2	Sulfur dioxide	293.2	14.3
Se	Selenium	510.65	5.44
$SiCl_4$	Tetrachlorosilane	273.2	2.25
Xe	Xenon	161.35	1.88
XeF_6	Xenon hexafluoride	328.2	4.10
Organic compounds			
CCl_4	Carbon tetrachloride	293.2	2.24
$CHCl_3$	Chloroform	293.2	4.81
CH_3NO_2	Nitromethane	293.2	37.27
CH_4	Methane	91	1.68
CH_4O	Methanol	293.2	33.0
C_2H_4	Ethylene	270	1.48
$C_2H_4O_2$	Acetic acid	293.2	6.20
C_2H_5Cl	Chloroethane	293.2	9.45
C_2H_6	Ethane	95	1.94
C_2H_6O	Ethanol	293.2	25.3
$C_2H_6O_2$	Ethylene glycol	293.2	41.4
C_3H_6O	Acetone	293.2	21.01
C_3H_8	Propane	293.19	1.67

(continued)

		T/K	ε
C_3H_8O	Isopropyl alcohol	293.2	20.18
$C_3H_8O_3$	Glycerol	293.2	46.53
C_4H_{10}	Butane	295	1.77
$C_4H_{10}O$	Diethyl ether	293.2	4.27
C_5H_5N	Pyridine	293.2	13.26
C_6H_{12}	Cyclohexane	293.2	2.02
$C_6H_{14}O$	1-Hexanol	293.2	13.03
C_6H_5Cl	Chlorobenzene	293.2	5.69
C_6H_6	Benzene	293.2	2.28
C_6H_6O	Phenol	303.2	12.4
C_7H_8	Toluene	296.35	2.38

See D. R. Lide, *CRC Handbook of Chemistry and Physics*, 82nd ed. (CRC Press, Boca Raton, FL, 2001) for a more extensive table. Data refer to atmospheric pressure except when the temperature is above the normal boiling point, in which case the vapor pressure at that temperature is understood.

27.6. TABLE: VISCOSITY OF LIQUIDS AND GASES

Liquids–Viscosity in mPa s		−25°C	0°C	25°C	50°C	75°C	100°C
H_2O	Water		1.793	0.890	0.547	0.378	0.282
Hg	Mercury			1.526	1.402	1.312	1.245
CCl_4	Carbon tetrachloride		1.321	0.908	0.656	0.494	
$CHCl_3$	Chloroform	0.988	0.706	0.537	0.427		
CH_4O	Methanol	1.258	0.793	0.544			
CS_2	Carbon disulfide		0.429	0.352			
C_2HCl_3	Trichloroethylene		0.703	0.545	0.444	0.376	
$C_2H_4O_2$	Acetic acid			1.056	0.786	0.599	0.464
C_2H_5Cl	Chloroethane	0.416	0.319				
C_2H_6O	Ethanol	3.26	1.786	1.074	0.694	0.476	
$C_2H_6O_2$	Ethylene glycol			16.06	6.55	3.34	1.975
C_3H_6O	Acetone	0.540	0.395	0.306	0.247		
C_3H_8O	2-Propanol		4.62	2.04	1.028	0.576	
$C_3H_8O_3$	Glycerol			934	152	39.8	14.76
C_4H_8O	Tetrahydrofuran	0.849	0.605	0.456	0.359		
$C_4H_8O_2$	1,4-Dioxane			1.177	0.787	0.569	
$C_5H_{12}O$	1-Pentanol	25.4	8.51	3.62	1.820	1.035	0.646
C_6H_5Cl	Chlorobenzene	1.703	1.058	0.753	0.575	0.456	0.369
$C_6H_5NO_2$	Nitrobenzene		3.04	1.863	1.262	0.918	0.704
C_6H_6	Benzene			0.604	0.436	0.335	
C_6H_6O	Phenol				3.44	1.784	1.099
C_6H_{12}	Cyclohexane			0.894	0.615	0.447	
C_6H_{14}	Hexane		0.405	0.300	0.240		
C_7H_8	Toluene	1.165	0.778	0.560	0.424	0.333	0.270
C_8H_{18}	Octane		0.700	0.508	0.385	0.302	0.243
$C_8H_{18}O$	1-Octanol			7.29	3.23	1.681	0.991

(continued)

TABLE 27.6. (Continued)

Gases–Viscosity in μPa s		100 K	200 K	300 K	400 K	500 K	600 K
(Data valid at pressures up to one atmosphere)							
	Air	7.1	13.3	18.6	23.1	27.1	30.8
Ar	Argon	8.0	15.9	22.9	28.8	34.2	39.0
H_2	Hydrogen	4.2	6.8	9.0	10.9	12.7	14.4
D_2	Deuterium	5.9	9.6	12.6	15.4	17.9	20.3
H_2O	Water			10.0	13.3	17.3	21.4
He	Helium	9.7	15.3	20.0	24.4	28.4	32.3
Kr	Krypton	8.8	17.1	25.6	33.1	39.8	45.9
N_2	Nitrogen		12.9	17.9	22.2	26.1	29.6
Ne	Neon	14.4	24.3	32.1	38.9	45.0	50.8
O_2	Oxygen	7.5	14.6	20.8	26.1	30.8	35.1
SF_6	Sulfur hexafluoride			15.3	19.8	23.9	27.7
Xe	Xenon	8.3	15.4	23.2	30.7	37.6	44.0
CO	Carbon monoxide	6.7	12.9	17.8	22.1	25.8	29.1
CO_2	Carbon dioxide		10.0	15.0	19.7	24.0	28.0
CH_4	Methane		7.7	11.2	14.3	17.0	19.4
C_2H_6	Ethane		6.4	9.5	12.3	14.9	17.3
C_6H_{14}	Hexane				8.6	10.8	12.8

More extensive tables may be found in D. R. Lide (Ed.), *CRC Handbook of Chemistry and Physics*, 82nd ed. (CRC Press, Boca Raton, FL, 2001).

27.7. TABLE: VAPOR PRESSURE OF THE ELEMENTS AND SELECTED COMPOUNDS

		Temperature in °C for the indicated pressure					
		1 Pa	10 Pa	100 Pa	1 kPa	10 kPa	100 kPa
Elements							
Ag	Silver	1010	1140	1302	1509	1782	2160
Al	Aluminum	1209	1359	1544	1781	2091	2517
Ar	Argon		−226.4 s	−220.3 s	−212.4 s	−201.7 s	−186.0
As	Arsenic	280 s	323 s	373 s	433 s	508 s	601 s
At	Astatine	88 s	119 s	156 s	202 s	258 s	334
Au	Gold	1373	1541	1748	2008	2347	2805
B	Boron	2075	2289	2549	2868	3272	3799
Ba	Barium	638 s	765	912	1115	1413	1897
Be	Beryllium	1189 s	1335	1518	1750	2054	2469
Bi	Bismuth	668	768	892	1052	1265	1562
Br_2	Bromine	−87.7 s	−71.8 s	−52.7 s	−29.3 s	2.5	58.4
C	Carbon (graphite)		2566 s	2775 s	3016 s	3299 s	3635 s
Ca	Calcium	591 s	683 s	798 s	954	1170	1482
Cd	Cadmium	257 s	310 s	381	472	594	767
Ce	Cerium	1719	1921	2169	2481	2886	3432
Cl_2	Chlorine	−145 s	−133.7 s	−120.2 s	−103.6 s	−76.1	−34.2

(continued)

		Temperature in °C for the indicated pressure					
		1 Pa	10 Pa	100 Pa	1 kPa	10 kPa	100 kPa
Co	Cobalt	1517	1687	1892	2150	2482	2925
Cr	Chromium	1383 s	1534 s	1718 s	1950	2257	2669
Cs	Cesium	144.5	195.6	260.9	350.0	477.1	667.0
Cu	Copper	1236	1388	1577	1816	2131	2563
Dy	Dysprosium	1105 s	1250 s	1431	1681	2031	2558
Er	Erbium	1231 s	1390 s	1612	1890	2279	2859
Eu	Europium	590 s	684 s	799 s	961	1179	1523
F_2	Fluorine	−235 s	−229.5 s	−222.9 s	−214.8	−204.3	−188.3
Fe	Iron	1455 s	1617	1818	2073	2406	2859
Fr	Francium	131	181	246	335	465	673
Ga	Gallium	1037	1175	1347	1565	1852	2245
Gd	Gadolinium	1563	1755	1994	2300	2703	3262
Ge	Germanium	1371	1541	1750	2014	2360	2831
H_2	Hydrogen					−258.6	−252.8
He	Helium					−270.6	−268.9
Hf	Hafnium	2416	2681	3004	3406	3921	4603
Hg	Mercury	42.0	76.6	120.0	175.6	250.3	355.9
Ho	Holmium	1159 s	1311 s	1502	1767	2137	2691
I_2	Iodine	−12.8 s	9.3 s	35.9 s	68.7 s	108 s	184.0
In	Indium	923	1052	1212	1417	1689	2067
Ir	Iridium	2440 s	2684	2979	3341	3796	4386
K	Potassium	200.2	256.5	328	424	559	756.2
Kr	Krypton	−214.0 s	−208.0 s	−199.4 s	−188.9 s	−174.6 s	−153.6
La	Lanthanum	1732	1935	2185	2499	2905	3453
Li	Lithium	524.3	612.3	722.1	871.2	1064.3	1337.1
Lu	Lutetium	1633 s	1829.8	2072.8	2380	2799	3390
Mg	Magnesium	428 s	500 s	588 s	698	859	1088
Mn	Manganese	955 s	1074 s	1220 s	1418	1682	2060
Mo	Molybdenum	2469 s	2721	3039	3434	3939	4606
N_2	Nitrogen	−236 s	−232 s	−226.8 s	−220.2 s	−211.1 s	−195.9
Na	Sodium	280.6	344.2	424.3	529	673	880.2
Nb	Niobium	2669	2934	3251	3637	4120	4740
Nd	Neodymium	1322.3	1501.2	1725.3	2023	2442	3063
Ne	Neon	−261 s	−260 s	−258 s	−255 s	−252 s	−246.1
Ni	Nickel	1510	1677	1881	2137	2468	2911
O_2	Oxygen				−211.9	−200.5	−183.1
Os	Osmium	2887 s	3150	3478	3875	4365	4983
P	Phosphorus (white)	6 s	34 s	69	115	180	276
P	Phosphorus (red)	182 s	216 s	256 s	303 s	362 s	431 s
Pb	Lead	705	815	956	1139	1387	1754
Pd	Palladium	1448 s	1624	1844	2122	2480	2961
Po	Polonium				573	730.2	963.3
Pr	Praseodymium	1497.7	1699.4	1954	2298	2781	3506
Pt	Platinum	2057	2277	2542	2870	3283	3821

(continued)

		Temperature in °C for the indicated pressure					
		1 Pa	10 Pa	100 Pa	1 kPa	10 kPa	100 kPa
Pu	Plutonium	1483	1680	1925	2238	2653	3226
Ra	Radium	546 s	633 s	764	936	1173	1526
Rb	Rubidium	160.4	212.5	278.9	368	496.1	685.3
Re	Rhenium	3030 s	3341	3736	4227	4854	5681
Rh	Rhodium	2015	2223	2476	2790	3132	3724
Rn	Radon	−163 s	−152 s	−139 s	−121.4 s	−97.6 s	−62.3
Ru	Ruthenium	2315 s	2538	2814	3151	3572	4115
S	Sulfur	102 s	135	176	235	318	444
Sb	Antimony	534 s	603 s	738	946	1218	1585
Sc	Scandium	1372 s	1531 s	1733	1993	2340	2828
Se	Selenium	227	279	344	431	540	685
Si	Silicon	1635	1829	2066	2363	2748	3264
Sm	Samarium	728 s	833 s	967 s	1148	1402	1788
Sn	Tin	1224	1384	1582	1834	2165	2620
Sr	Strontium	523 s	609 s	717 s	866	1072	1373
Ta	Tantalum	3024	3324	3684	4122	4666	5361
Tb	Terbium	1516.1	1706.1	1928	2232	2640	3218
Tc	Technetium	2454	2725	3051	3453	3961	4621
Te	Tellurium			502	615	768.8	992.4
Th	Thorium	2360	2634	2975	3410	3986	4782
Ti	Titanium	1709	1898	2130	2419	2791	3285
Tl	Thallium	609	704	824	979	1188	1485
Tm	Thulium	844 s	962 s	1108 s	1297 s	1548	1944
U	Uranium	2052	2291	2586	2961	3454	4129
V	Vanadium	1828 s	2016	2250	2541	2914	3406
W	Tungsten	3204 s	3500	3864	4306	4854	5550
Xe	Xenon	−190 s	−181 s	−170 s	−155.8 s	−136.6 s	−108.4
Y	Yttrium	1610.1	1802.3	2047	2354	2763	3334
Yb	Ytterbium	463 s	540 s	637 s	774 s	993	1192
Zn	Zinc	337 s	397 s	477	579	717	912
Zr	Zirconium	2366	2618	2924	3302	3780	4405
Inorganic compounds							
H_2O	Water & ice	−60.7 s	−42.2 s	−20.3 s	7.0	45.8	99.6
NH_3	Ammonia	−139 s	−127 s	−112 s	−94.5 s	−71.3	−33.6
SiH_4	Silane			−181	−165.4	−143.7	−111.8
O_3	Ozone	−189	−182	−172	−158	−139.7	−111.5
CO	Carbon monoxide			−223 s	−216.5 s	−207.2 s	−191.7
CO_2	Carbon dioxide	−159.1 s	−148.9 s	−136.7 s	−121.6 s	−103.1 s	−78.6 s
N_2O	Nitrous oxide	−167 s	−157 s	−145.4 s	−131.1 s	−112.9 s	−88.7
SO_2	Sulfur dioxide			−98 s	−80 s	−52.2	−10.3
SF_6	Sulfur hexafluoride	−158 s	−147 s	−133.6 s	−116.6 s	−94.4 s	−64.1 s
H_2S	Hydrogen sulfide		−149 s	−136 s	−118.9 s	−95.9 s	−60.5
H_2SO_4	Sulfuric acid	72	103	140	187	248	330

(continued)

		Temperature in °C for the indicated pressure					
		1 Pa	10 Pa	100 Pa	1 kPa	10 kPa	100 kPa
HCN	Hydrogen cyanide			−77 s	−52.6 s	−22.7 s	25.4
HF	Hydrogen fluoride				−71.1	−33.7	19.2
HCl	Hydrogen chloride				−138.2 s	−118.0	−85.2
HBr	Hydrogen bromide		−153.3 s	−140.4 s	−123.8 s	−101.5 s	−67.0
HI	Hydrogen iodide	−146 s	−135.2 s	−120.8 s	−101.9 s	−75.9 s	−35.9
LiF	Lithium fluoride	801 s	896	1024	1188	1395	1672
NaF	Sodium fluoride		920 s	1058	1218	1426	1702
KF	Potassium fluoride			869	1017	1216	1499
RbF	Rubidium fluoride			910	1001	1145	1409
CsF	Cesium fluoride				825	999	1249
$LiCl$	Lithium chloride		649	761	905	1101	1381
$NaCl$	Sodium chloride	653 s	733 s	835	987	1182	1461
KCl	Potassium chloride	625 s	704 s	804	945	1137	1411
$RbCl$	Rubidium chloride			777	916	1105	1379
$CsCl$	Cesium chloride			730	864	1043	1297
Al_2O_3	Aluminum oxide			2122	2351	2629	2975
SiO_2	Silicon dioxide	1966	2149	2368			
SrO	Strontium oxide	1789 s	1903 s	2047 s	2235 s	2488 s	
Organic compounds							
CCl_4	Carbon tetrachloride	−79.4 s	−70.8 s	−53.5 s	−24.4 s	15.8	76.2
CF_4	Carbon tetrafluoride	−199.9 s	−193 s	−183.9 s	−171.6	−153.9	−128.3
CH_4O	Methanol	−87	−69	−47.5	−20.4	15.2	64.2
CH_5N	Methylamine				−76.7	−48.1	−6.6
$C_2H_4O_2$	Acetic acid	−42.8 s	−26.7 s	−8 s	14.2 s	55.9	117.5
C_2H_6O	Ethanol	−73	−56	−34	−7	29.2	78.0
$C_2H_6O_2$	Ethylene glycol	2	24	51.1	86.1	132.5	196.9
C_3H_6O	Acetone	−95	−81.8	−62.8	−35.6	1.3	55.7
C_3H_8	Propane	−156.9	−145.6	−130.9	−111.4	−83.8	−42.3
C_3H_8O	Isopropyl alcohol	−65	−49	−28	−1.3	33.6	82.0
$C_3H_8O_3$	Glycerol	96	113	136	168	213.4	287
C_4H_{10}	Butane	−134.3	−121.0	−103.9	−81.1	−49.1	−0.8
$C_4H_{10}O$	Diethyl ether	−111	−96	−77	−52.6	−17.8	34.1
C_6H_6	Benzene			−40 s	−15.1 s	20.0	79.7
C_6H_{12}	Cyclohexane	−85.6 s	−68.9 s	−47.6 s	−19.8 s	19.3	80.4
C_7H_8	Toluene	−78.1	−57.1	−31.3	1.5	45.2	110.1
$C_{10}H_8$	Naphthalene	3.2 s	24.1 s	49.3 s	80.7	135.6	217.5

An "s" following an entry indicates the substance is solid at that temperature.

27.8. TABLE: VAPOR PRESSURE OF CRYOGENIC FLUIDS

Helium									
1.0 K	1.5 K	2.0 K	2.5 K	3.0 K	3.5 K	4.0 K	4.5 K	5.0 K	5.1 K
0.0156	0.0472	3.130	10.23	24.05	47.05	81.62	130.3	196.0	211.6

Hydrogen									
14 K	16 K	18 K	20 K	22 K	24 K	26 K	28 K	30 K	32 K
7.90	21.53	48.08	93.26	163.2	264.2	402.9	586.1	821.4	1119.0

Neon									
25 K	26 K	27 K	28 K	29 K	30 K	32 K	34 K	36 K	38 K
51.3	71.8	98.5	132.1	173.5	223.8	355.2	535.2	772.8	1078

Other Substances									
	50 K	60 K	70 K	80 K	90 K	100 K	125 K	150 K	175 K
Ar	0.1 s	0.8 s	7.7 s	40.7 s	134	324	1584	4736	
CH_4			0.3	2.1	10.6	34.5	269	1041	2779
CF_4					0.1	0.8	19.3	141	537
C_2H_6							0.7	9.7	59.0
CO	0.1 s	2.6 s	21.0	83.7	239	545	2400		
F_2		1.5	12.3	55.3	172	420	2108		
F_2O				0.2	1.2	5.3	76.7	444	1541
HCl							0.3	5.8	45.1
Kr				0.4 s	2.7 s	12.1 s	150	655	1877
NF_3					0.2	0.9	21.1	150	581
NO					0.4 s	3.8 s	151	1231	5089
N_2	0.4 s	6.3 s	38.6	137	361	779	3209		
O_2		0.73	6.26	30.1	99.3	254	1351	4219	
O_3						0.1	3.8	43.7	222
Rn								1.5	12.5
SF_6								0.4 s	7.1 s
SiF_4							0.1 s	3.8 s	74.2 s
SiH_4						0.2	6.1	50.3	210
Xe						0.1 s	2.7 s	34.2 s	173

All pressures are given in kilopascals (kPa). For conversion, 1 kPa = 0.01 bar = 7.50062 torr. An "s" following a pressure entry indicates the substance is solid at the indicated temperature.

27.9. TABLE: AQUEOUS SOLUBILITY OF SOLIDS AND LIQUIDS

Inorganic compounds	0°C	20°C	25°C	50°C	100°C
$CaCl_2$	36.70	42.13	44.83	56.05	59.94
CaF_2	0.0013		0.0016		
$Ca(NO_3)_2$	50.1	56.7	59.0	77.8	78.5
$CaSO_4$	0.174	0.202	0.205	0.207	0.163
CsBr			55.2		
CsCl	61.83	64.96	65.64	68.60	72.96
CsI	30.9	43.2	45.9	57.3	69.2
$CuCl_2$	40.8	42.6	43.1	46.0	52.7
$Cu(NO_3)_2$	45.2	56.3	59.2	63.1	71.0
$CuSO_4$	12.4	16.7	18.0	25.4	43.5
$HgCl_2$	4.24	6.17	6.81	12.02	36.62
Hg_2SO_4	0.038	0.048	0.051	0.065	0.093
KBr	35.0	39.4	40.4	44.8	50.8
KCl	21.74	25.39	26.22	30.04	36.05
KF	30.90	47.3	50.41		
KI	56.0	59.0	59.7	62.8	67.4
$KMnO_4$	2.74	5.96	7.06	14.42	
KNO_3	12.0	24.2	27.7	45.7	70.8
K_2CO_3	51.3	52.3	52.7	54.9	61.0
K_2SO_4	7.11	9.95	10.7	14.2	19.3
LiBr	58.4	62.7	64.4	68.3	72.8
LiCl	40.45	45.29	45.81	48.47	56.34
LiF	0.120	0.131	0.134		
LiI	59.4	61.7	62.3	65.8	82.6
$LiNO_3$	34.8	42.7	50.5	62.2	69.7
Li_2CO_3	1.54	1.33	1.28	1.07	0.72
Li_2SO_4	26.3	25.6	25.5	24.8	23.6
$MgCl_2$	33.96	35.58	35.90	37.34	42.15
MgF_2			0.013		
$MgSO_4$	18.2	25.1	26.3	33.4	33.3
NH_4Br	37.5	42.7	43.9	49.4	57.4
NH_4Cl	22.92	27.27	28.34	33.50	43.24
NH_4F	41.7	44.7	45.5	49.3	
NH_4I	60.7	63.4	64.0	66.8	71.1
$(NH_4)_2SO_4$	41.3	42.9	43.3	45.6	50.5
NaBr	44.4	47.7	48.6	53.7	54.9
NaCl	26.28	26.41	26.45	26.84	28.05
NaF	3.52	3.89	3.97	4.34	4.82
$NaHCO_3$	6.48	8.73	9.32	12.40	19.10
NaI	61.2	63.9	64.8	69.8	75.1
$NaNO_3$	42.2	46.6	47.7	53.2	63.8

(continued)

Inorganic compounds	0°C	20°C	25°C	50°C	100°C
Na_2CO_3	6.44	17.9	23.5	32.2	30.9
Na_2SO_4		16.13	21.94	31.55	29.67
$PbCl_2$	0.66	0.98	1.07	1.64	3.42
$Pb(NO_3)_2$	28.46	35.67	37.38	45.17	56.75
RbBr	47.4	52.6	53.8	58.8	65.9
RbCl	43.58	47.53	48.42	52.34	58.15
RbF		75			
RbI	55.8	61.1	62.3	67.2	73.8
$RbNO_3$	16.4	34.6	39.4	60.8	81.2
Rb_2SO_4	27.3	32.5	33.7	38.7	44.9
$ZnCl_2$		79.0	80.3	82.4	86.0
ZnF_2			1.53		
$ZnSO_4$	29.1	35.0	36.6	43.0	37.6

Organic compounds		0°C	20°C	25°C	50°C
CCl_4	Carbon tetrachloride			0.065	
$CHCl_3$	Chloroform			0.80	
CS_2	Carbon disulfide		0.21		
$C_4H_{10}O$	1-Butanol	10.4		7.4	6.4
$C_4H_{10}O$	Diethyl ether			6.04	
$C_5H_{12}O$	1-Pentanol	3.1		2.20	1.8
C_6H_5Cl	Chlorobenzene			0.0495	0.0882
C_6H_6	Benzene			0.177	0.208
C_6H_6O	Phenol			8.66	
C_6H_7N	Aniline			3.38	
C_6H_{12}	Cyclohexane			0.0058	
C_6H_{14}	Hexane			0.0011	
$C_6H_{14}O$	1-Hexanol	0.79		0.60	0.51
C_7H_8	Toluene			0.053	
$C_{10}H_8$	Naphthalene			0.0031	0.0082

All values are given in mass percent, i.e., 100 times mass of solute (as specified by the formula) divided by total mass of the saturated solution. In many cases the solid phase in equilibrium with the solution is a hydrate; the particular hydrate may be different at different temperatures.

27.10. TABLE: SOLUBILITY OF GASES IN WATER

	Solubility in mole fraction		
	15°C	25°C	35°C
H_2	1.510×10^{-5}	1.411×10^{-5}	1.350×10^{-5}
D_2	1.595	1.460	
N_2	1.386	1.183	1.047
O_2	2.756	2.293	1.982
He	0.712	0.700	0.699
Ne	0.870	0.815	0.783
Ar	3.025	2.519	2.169
Kr	5.696	4.512	3.725
Xe	10.519	7.890	6.212
Rn	22.99	16.71	12.88
N_2O	59.48	43.67	33.48
CO	2.095	1.774	1.562
CO_2	82.5	62.2	48.7
CH_4	3.122	2.552	2.180

Values are for a gas pressure of 101.325 kPa (1 atm) in equilibrium with the solution. For pressures below 1 atm the solubility is proportional to pressure to a good approximation.

27.11. TABLE: PROPERTIES OF MISCELLANEOUS SOLID MATERIALS

	$\rho/\mathrm{g\ cm^{-3}}$	$\lambda/\mathrm{W\ m^{-1}\ K^{-1}}$	Hardness
Agate	2.5–2.7		6.5
Amber	1.06–1.11		
Asbestos	2.0–2.8	0.09	
Asphalt	1.1–1.5	0.06	
Basalt	2.4–3.1	1	
Beeswax	0.96–0.97		
Beryl	2.69–2.70		7.8
Bone	1.7–2.0		
Brasses	8.44–8.75	120	
Brick	1.4–2.2	0.04	
Bronzes	8.74–8.89	110	
Carborundum	3.16		9.3
Cardboard	0.69		
Celluloid	1.4	0.02	
Cement, set	2.7–3.0	0.5	

(continued)

	ρ/g cm^{-3}	λ/W m^{-1} K^{-1}	Hardness
Chalk	1.9–2.8	0.9	
Clay	1.8–2.6		
Coal, anthracite	1.4–1.8		
bituminous	1.2–1.5	0.26	
Coke	1.0–1.7		
Cork	0.22–0.26	0.06	
Corundum	3.9–4.0		9
Diamond (type I)	3.51	900	10
Dolomite	2.84		3.5
Emery	4		
Feldspar	2.55–2.75		6
Flint	2.63		7
Garnet	3.15–4.3		8.5
Glass, common	2.4–2.8	0.90–1.10	
lead	3–4	0.50–0.60	
Pyrex	2.23	1.10	
Granite	2.64–2.76	2.2	
Graphite	2.30–2.72		0.5
Gypsum	2.31–2.33		2
Iron, cast	7.0–7.4	47	
Ivory	1.83–1.92		
Kaolin	2.6		
Lime, slaked	1.3–1.4		
Limestone	2.68–2.76	1	
Marble	2.6–2.84		3.5
Mica	2.6–3.2	0.6–0.7	2.8
Opal	2.2		5
Paper	0.7–1.15		
Paraffin	0.87–0.91		
Polyamides (Nylons)	1.15–1.25	0.30	
Polyethylene	0.92–0.97		
Poly(methyl methacrylate)	1.19		
Polypropylene	0.91–0.94		
Polystyrene	1.06–1.12		
Polytetrafluoroethylene	2.28–2.30	0.26	
Polyvinylacetate	1.19		
Polyvinylchloride	1.39–1.42		
Porcelain	2.3–2.5	1	
Quartz	2.65	7–12	7
Rock salt	2.18	6.4	2
Rubber, hard	1.19	0.16	
soft	1.1	0.15	
pure gum	0.91–0.93	0.13	
neoprene	1.23–1.25	0.19	
Sandstone	2.14–2.36	1.3	

(continued)

	ρ/g cm^{-3}	λ/W m^{-1} K^{-1}	Hardness
Silica, fused	2.21	1.4	6.5
Silicon carbide	3.16		
Slate	2.6–3.3	1.5–2.5	
Solder	8.7–9.4		
Steel, stainless	7.8	14	
Talc	2.7–2.8		1
Tar	1.02		
Tungsten carbide	14.0–15.0		9
Wax, sealing	1.8		
Wood (seasoned)			
ash	0.65–0.85		
balsa	0.11–0.14	0.04	
bamboo	0.31–0.40		
birch	0.51–0.77		
cherry	0.70–0.90		
ebony	1.11–1.33		
elm	0.54–0.60		
hickory	0.60–0.93		
mahogany	0.66–0.85		
maple	0.62–0.75		
oak	0.60–0.90	0.16	
pine, white	0.35–0.50	0.11–0.26	
poplar	0.35–0.50		
teak, Indian	0.66–0.98		
walnut	0.64–0.70	0.14	
Wood's metal	9.7		

ρ is density at room temperature; λ is thermal conductivity at room temperature; hardness is on the Mohs scale. Data taken from D. R. Lide (Ed.), *Handbook of Chemistry and Physics*, 82nd ed. (CRC Press, Boca Raton, FL, 2001).

27.12. TABLE: DENSITIES KNOWN TO HIGH ACCURACY

	0°C	20°C	25°C	40°C	50°C	100°C
Water	0.9998426	0.9982063	0.9970480	0.9922204	0.98803	0.95840
Mercury	13.59508	13.54585	13.53399	13.49690	13.47251	13.35142
Cyclohexane		0.778583	0.773896	0.759624	0.749960	
Silicon		2.32907				

Densities given in g/cm^3. See D. R. Lide (Ed.), *Handbook of Chemistry and Physics*, 82nd ed. (CRC Press, Boca Raton, FL, 2001) for values at other temperatures.

27.13. FIXED POINTS ON THE INTERNATIONAL TEMPERATURE SCALE OF 1990

The ITS-90 is based on the vapor pressure of helium at temperatures up to 5 K. Beyond that, a helium gas thermometer and platinum resistance thermometer are used to interpolate between the following fixed points:

H_2 triple point	13.8033 K	In freezing point	429.7485
Ne triple point	24.5561	Sn freezing point	505.078
O_2 triple point	54.3584	Zn freezing point	692.677
Ar triple point	83.8058	Al freezing point	933.473
Hg triple point	234.3156	Ag freezing point	1234.93
H_2O triple point	273.16	Au freezing point	1337.33
Ga melting point	302.9146	Cu freezing point	1357.77

Above 1234.93 K, the ITS-90 is defined in terms of the Planck radiation law, using the freezing point temperature of either Ag, Au, or Cu as the reference temperature. Details are given in *Metrologia*, **27**, 3 (1990); **27**, 107 (1990). For conversion from previous scales to ITS-90, see G. W. Burns, *et al.*, in *Temperature: Its Measurement and Control in Science and Industry*, Vol. 6, edited by J. F. Schooley (AIP Press, New York, 1993). See also D. R. Lide (Ed.), *CRC Handbook of Chemistry and Physics*, 82nd ed. pp. 1–21 to 1–24 (2001). (CRC Press, Boca Raton, FL, 2001).

27.14. TABLE: PROPERTIES OF LIQUID HELIUM

T/K	p/kPa	ρ/g cm^{-3}	C_s/J mol^{-1}K^{-1}	$\Delta_{vap}H$/J mol^{-1}	ε	σ/mN m^{-1}	$10^3\alpha$/K^{-1}	η/μPa s	λ/W m^{-1} K^{-1}
0.0		0.1451397	0	59.83	1.057255		0.000		
0.5		0.1451377	0.010	70.24	1.057254	0.3530	0.107		
1.0	0.01558	0.1451183	0.415	80.33	1.057246	0.3471	0.309	3.873	
1.5	0.4715	0.1451646	4.468	89.35	1.057265	0.3322	−2.36	1.346	
2.0	3.130	0.1456217	21.28	93.07	1.057449	0.3021	−12.2	1.468	
2.5	10.23	0.1448402	9.083	92.50	1.057135	0.2623	39.4	3.259	0.01497
3.0	24.05	0.1412269	9.944	94.11	1.055683	0.2161	61.5	3.517	0.01717
3.5	47.05	0.1360736	12.37	92.84	1.053615	0.1626	88.7	3.509	0.01868
4.0	81.62	0.1289745	15.96	87.00	1.050770	0.1095	129	3.319	0.01965
4.5	130.3	0.1188552	21.8	75.86	1.046725	0.0609	211		
5.0	196.0		44.7	47.67		0.0157			

All values refer to liquid helium (^4He) at saturated vapor pressure; temperatures are on the ITS-90 scale. Several properties show a singularity at the lambda point (2.1768 K). Data taken from R. J. Donnelly and C. F. Barenghi, *J. Phys. Chem. Reference Data* **27**, 1217 (1998).

p is vapor pressure; σ is surface tension; ρ is density; α is coefficient of linear expansion; C_s is heat capacity; η is viscosity; $\Delta_{vap}H$ is enthalpy of vaporization; λ is thermal conductivity; ε is relative permittivity (dielectric constant).

27.15. TABLE: PROPERTIES OF WATER AND ICE

	Temperature in °C						
	−20	−10	0 (ice)	0 (liq)	10	20	30
ρ/g cm^{-3}	0.9203	0.9187	0.9167	0.9998	0.9997	0.9982	0.9957
$\alpha/10^{-3}$ °C^{-1}	0.149	0.155	0.159			0.206	0.302
$\kappa/10^{-4}$ MPa^{-1}	1.27	1.28	1.30			4.591	4.475
c_p/J g^{-1} °C^{-1}	1.96	2.03	2.11	4.218	4.192	4.182	4.178
$\Delta_{vap}H$/kJ g^{-1}			2.838	2.501	2.477	2.454	2.430
$\Delta_{fus}H$/kJ g^{-1}			0.3336				
ε	97.5	94.4	91.6	87.90	83.96	80.20	76.60
η/mPa s				1.793	1.307	1.002	0.798
λ/W °C^{-1} m^{-1}	2.4	2.3	2.14	0.561	0.580	0.598	0.615
p/kPa	0.103	0.260	0.61	0.61	1.23	2.34	4.25
σ/mN m^{-1}				75.64	74.23	72.75	71.20

ρ is density; ε is dielectric constant; α is volume expansion coefficient; η is viscosity; κ is adiabatic compressibility; λ is thermal conductivity; c_p is specific heat capacity; p is vapor pressure; $\Delta_{vap}H$ is enthalpy of vaporization; σ is surface tension; $\Delta_{fus}H$ is enthalpy of fusion.

27.16. TABLE: VAPOR PRESSURE OF WATER ON THE ITS-90 TEMPERATURE SCALE

t/°C	0.01	5	10	15	20	25	30	35	40	45	50
p/kPa	0.612	0.873	1.228	1.706	2.339	3.170	4.247	5.629	7.385	9.595	12.352

t/°C	55	60	65	70	75	80	85	90	95	99.974	100
p/kPa	15.762	19.946	25.042	31.201	38.595	47.414	57.867	70.182	84.608	101.325	101.42

Index

Note: Most of the terms in this index are concepts, theories, properties, laws, equations, and general subject terms. With the exception of ubiquitous substances such as hydrogen, helium, water, carbon dioxide, etc., the index does not list individual elements, chemical compounds or other materials. Information on properties of substances and materials can be located through the appropriate property name. Chapter titles are in boldface.

B

C

F

G

J

K

M

N

O

P

Q

S

T

U

V

W

X

Y

Z